Table of Integrals

W9-AWN-232

ELEMENTARY FORMS

1. $\int u\,dv = uv - \int v\,du$

2. $\int u^n\,du = \dfrac{1}{n+1}u^{n+1} + C$ if $n \neq -1$

3. $\int \dfrac{du}{u} = \ln|u| + C$

4. $\int e^u\,du = e^u + C$

5. $\int a^u\,du = \dfrac{a^u}{\ln a} + C$

6. $\int \sin u\,du = -\cos u + C$

7. $\int \cos u\,du = \sin u + C$

8. $\int \sec^2 u\,du = \tan u + C$

9. $\int \csc^2 u\,du = -\cot u + C$

10. $\int \sec u \tan u\,du = \sec u + C$

11. $\int \csc u \cot u\,du = -\csc u + C$

12. $\int \tan u\,du = \ln|\sec u| + C$

13. $\int \cot u\,du = \ln|\sin u| + C$

14. $\int \sec u\,du = \ln|\sec u + \tan u| + C$

15. $\int \csc u\,du = \ln|\csc u - \cot u| + C$

16. $\int \dfrac{du}{\sqrt{a^2 - u^2}} = \sin^{-1}\dfrac{u}{a} + C$

17. $\int \dfrac{du}{a^2 + u^2} = \dfrac{1}{a}\tan^{-1}\dfrac{u}{a} + C$

18. $\int \dfrac{du}{a^2 - u^2} = \dfrac{1}{2a}\ln\left|\dfrac{u + a}{u - a}\right| + C$

TRIGONOMETRIC FORMS

19. $\int \sin^2 u\,du = \dfrac{1}{2}u - \dfrac{1}{4}\sin 2u + C$

20. $\int \cos^2 u\,du = \dfrac{1}{2}u + \dfrac{1}{4}\sin 2u + C$

21. $\int \tan^2 u\,du = \tan u - u + C$

22. $\int \cot^2 u\,du = -\cot u - u + C$

23. $\int \sin^3 u\,du = -\dfrac{1}{3}(2 + \sin^2 u)\cos u + C$

24. $\int \cos^3 u\,du = \dfrac{1}{3}(2 + \cos^2 u)\sin u + C$

25. $\int \tan^3 u\,du = \dfrac{1}{2}\tan^2 u + \ln|\cos u| + C$

26. $\int \cot^3 u\,du = -\dfrac{1}{2}\cot^2 u - \ln|\sin u| + C$

27. $\int \sec^3 u\,du = \dfrac{1}{2}\sec u \tan u + \dfrac{1}{2}\ln|\sec u + \tan u| + C$

28. $\int \csc^3 u\,du = -\dfrac{1}{2}\csc u \cot u + \dfrac{1}{2}\ln|\csc u - \cot u| + C$

29. $\int \sin au \sin bu\,du = \dfrac{\sin(a-b)u}{2(a-b)} - \dfrac{\sin(a+b)u}{2(a+b)} + C$ if $a^2 \neq b^2$

(Continued on Rear Endpaper)

DIFFERENTIAL EQUATIONS AND BOUNDARY VALUE PROBLEMS

Computing and Modeling

DIFFERENTIAL EQUATIONS AND BOUNDARY VALUE PROBLEMS

Computing and Modeling

Third Edition

C. Henry Edwards

David E. Penney

The University of Georgia

PEARSON
Prentice Hall

PEARSON EDUCATION, INC.
Upper Saddle River, NJ 07458

Library of Congress Cataloging-in-Publication Data

Edwards, C. H. (Charles Henry).
 Differential equations and boundary value problems : computing and modeling / C. Henry
Edwards, David E. Penney.—3rd ed.
 p. cm.
 Includes bibliographical references and index.
 ISBN 0-13-065245-8
 1. Differential equations. 2. Boundary value problems. I. Penney, David E. II. Title.
QA371.E28 2004
515'.35—dc21
 2003051262

Acquisitions Editor: *George Lobell*
Editorial/Production Supervision: *Bayani Mendoza de Leon*
Editor-in-Chief: *Sally Yagan*
Assistant Vice President of Production and Manufacturing: *David W. Riccardi*
Senior Managing Editor: *Linda Mihatov Behrens*
Executive Managing Editor: *Kathleen Schiaparelli*
Manufacturing Buyer: *Michael Bell*
Manufacturing Manager: *Trudy Pisciotti*
Marketing Manager: *Halee Dinsey*
Marketing Assistant: *Rachel Beckman*
Director of Creative Services: *Paul Belfanti*
Art Editor: *Tom Benfatti*
Creative Director: *Carole Anson*
Art Director/Interior Designer/Cover Designer: *John Christiana*
Editorial Assistant: *Jennifer Brady*

TRADEMARK INFORMATION

MATLAB is a registered trademark of The MathWorks, Inc.
For MATLAB product information, please contact:
The MathWorks, Inc.
3 Apple Hill Drive
Natick, MA, 01760-2098 USA
Tel: 508-647-7000
Fax: 508-647-7101
E-mail: info@mathworks.com
Web: www.mathworks.com/

Maple is a registered trademark of Waterloo Maple, Inc.

Mathematica is a registered trademark of Wolfram Research, Inc.

Printed in the United States of America
10 9 8 7 6 5 4 3 2 1

ISBN 0-13-065245-8

Pearson Education LTD., *London*
Pearson Education Australia Pty, Limited, *Sydney*
Pearson Education Singapore, Pte. Ltd
Pearson Education North Asia Ltd., *Hong Kong*
Pearson Education Canada, Ltd., *Toronto*
Pearson Education de Mexico, S.A. de C.V.
Pearson Education Japan, *Tokyo*
Pearson Education Malaysia, Pte. Ltd.

CONTENTS

v

APPLICATION MODULES

The modules listed here follow the indicated sections in the text. Most provide computing projects that illustrate the content of the corresponding text sections. *Maple*, *Mathematica*, and MATLAB versions of these investigations are included in the Applications Manual that accompanies this text.

PREFACE

Many introductory differential equations courses in the recent past have emphasized the formal solution of standard types of differential equations using a (seeming) grab-bag of systematic solution techniques. Many students have concentrated on learning to match memorized methods with memorized equations. The evolution of the present text is based on experience teaching a course with a greater emphasis on conceptual ideas and the use of applications and computing projects to involve students in more intense and sustained problem-solving experiences.

The availability of technical computing environments like *Maple*, *Mathematica*, and MATLAB is reshaping the role and applications of differential equations in science and engineering and has shaped our approach in this text. New technology motivates a shift in emphasis from traditional manual methods to both qualitative and computer-based methods that

- render accessible a wider range of more realistic applications;
- permit the use of both numerical computation and graphical visualization to develop greater conceptual understanding; and
- encourage empirical investigations that involve deeper thought and analysis than standard textbook problems.

Major Features

The following features of this text are intended to support a contemporary differential equations course that augments traditional core skills with conceptual perspectives that students will need for the effective use of differential equations in their subsequent work and study:

- Coverage of seldom-used topics has been trimmed and new topics added to place a greater emphasis on core techniques as well as qualitative aspects of the subject associated with direction fields, solution curves, phase plane portraits, and dynamical systems. We combine symbolic, graphic, and numeric solution methods wherever it seems advantageous. A fresh computational flavor should be evident in figures, examples, problems, and applications throughout the text. About 15% of the examples in the text are new or newly revised for this edition.

- The organization of the book places an increased emphasis on linear systems of differential equations, which are covered in Chapters 4 and 5 (together with the necessary linear algebra), followed by a substantial treatment in Chapter 6 of nonlinear systems and phenomena (including chaos in dynamical systems).

- This book begins and ends with discussions and examples of the mathematical modeling of real-world phenomena. Students learn through mathematical

modeling and empirical investigation to balance the questions of what equation to formulate, how to solve it, and whether a solution will yield useful information.

- The first course in differential equations should also be a window on the world of mathematics. While it is neither feasible nor desirable to include proofs of the fundamental existence and uniqueness theorems along the way in an elementary course, students need to see precise and clear-cut statements of these theorems and to understand their role in the subject. We include appropriate existence and uniqueness proofs in the Appendix and occasionally refer to them in the main body of the text.

- While our approach reflects the widespread use of new computer methods for the solution of differential equations, certain elementary analytical methods of solution (as in Chapters 1 and 3) are important for students to learn. Effective and reliable use of numerical methods often requires preliminary analysis using standard elementary techniques; the construction of a realistic numerical model often is based on the study of a simpler analytical model. We therefore continue to stress the mastery of traditional solution techniques (especially through the inclusion of extensive problem sets).

Computing Features

The following features highlight the flavor of computing technology that distinguishes much of our exposition.

- Almost 700 *computer-generated figures*—over half of them new for this edition and most constructed using Mathematica or MATLAB—show students vivid pictures of direction fields, solution curves, and phase plane portraits that bring symbolic solutions of differential equations to life. For instance, the cover graphic shows an eigenfunction of the three-dimensional wave equation that illustrates surface waves on a spherical planet and was constructed using associated Legendre functions (see Section 10.5).

- About 45 *application modules* follow key sections throughout the text. Most of these applications outline "technology neutral" investigations illustrating the use of technical computing systems and seek to actively engage students in the application of new technology.

- A fresh *numerical emphasis* that is afforded by the early introduction of numerical solution techniques in Chapter 2 (on mathematical models and numerical methods). Here and in Chapter 4, where numerical techniques for systems are treated, a concrete and tangible flavor is achieved by the inclusion of numerical algorithms presented in parallel fashion for systems ranging from graphing calculators to MATLAB.

- A *conceptual perspective* shaped by the availability of computational aids, which permits a leaner and more streamlined coverage of certain traditional manual topics (like exact equations and variation of parameters) in Chapters 1, 3, and 5.

Modeling Features

Mathematical modeling is a goal and constant motivation for the study of differential equations. To sample the range of applications in this text, take a look at the following questions:

- What explains the commonly observed time lag between indoor and outdoor daily temperature oscillations? (Section 1.5)

- What makes the difference between doomsday and extinction in alligator populations? (Section 2.1)

- How do a unicycle and a two-axle car react differently to road bumps? (Sections 3.7 and 5.3)

- How can you predict the time of next perihelion passage of a newly observed comet? (Section 4.3)

- Why might an earthquake demolish one building and leave standing the one next door? (Section 5.3)

- What determines whether two species will live harmoniously together, or whether competition will result in the extinction of one of them and the survival of the other? (Section 6.3)

- Why and when does nonlinearity lead to chaos in biological and mechanical systems? (Section 6.5)

- If a mass on a spring is periodically struck with a hammer, how does the behavior of the mass depend on the frequency of the hammer blows? (Section 7.6)

- Why are flagpoles hollow instead of solid? (Section 8.6)

- What explains the difference in the sounds of a guitar, a xylophone, and drum? (Sections 9.6, 10.2, and 10.4)

Organization and Content

We have reshaped the usual approach and sequence of topics to accommodate new technology and new perspectives. For instance,

- After a precis of first-order equations in Chapter 1 (though with the coverage of certain traditional symbolic methods streamlined a bit), Chapter 2 offers an early introduction to mathematical modeling, stability and qualitative properties of differential equations, and numerical methods—a combination of topics that frequently are dispersed later in an introductory course.

- Chapters 4 and 5 provide a flexible treatment of linear systems. Motivated by current trends in science and engineering education and practice, Chapter 4 offers an early, intuitive introduction to first-order systems, models, and numerical approximation techniques. Chapter 5 begins with a self-contained treatment of the linear algebra that is needed and then presents the eigenvalue approach to linear systems. It includes a wide range of applications (ranging from railway cars to earthquakes) of all the various cases of the eigenvalue method. Section 5.5 includes a fairly extensive treatment of matrix exponentials, which are exploited in Section 5.6 on nonhomogeneous linear systems.

- Chapter 6 on nonlinear systems and phenomena ranges from phase plane anal-

ysis to ecological and mechanical systems to a concluding section on chaos and bifurcation in dynamical systems. Section 6.5 presents an elementary introduction to such contemporary topics as period doubling in biological and mechanical systems, the pitchfork diagram, and the Lorenz strange attractor (all illustrated with vivid computer graphics).

- Laplace transform methods (Chapter 7) and power series methods (Chapter 8) follow the material on linear and nonlinear systems but can be covered at any earlier point (after Chapter 3) the instructor desires.

- Chapters 9 and 10 treat the applications of Fourier series, separation of variables, and Sturm-Liouville theory to partial differential equations and boundary value problems. After the introduction of Fourier series, the three classical equations—the wave and heat equations and Laplace's equation—are discussed in the last three sections of Chapter 9. The Sturm-Liouville methods of Chapter 10 are developed sufficiently to include some rather significant and realistic applications.

This book includes enough material appropriately arranged for different courses varying in length from one quarter to two semesters. The briefer version, *Differential Equations: Computing and Modeling*, ends with Chapter 7 on Laplace transform methods (and thus omits the material on power series methods, Fourier series, separation of variables and partial differential equations).

Problems, Applications, and Solutions Manuals

Almost 20% of the text's over 1900 problems are new for this edition or are newly revised to include graphic or qualitative content. Accordingly, the answer section now includes almost 300 new computer-generated figures illustrating those which students are expected to construct.

The answer section for this revision has been expanded considerably to increase its value as a learning aid. It now includes the answers to most odd-numbered problems plus a good many even-numbered ones. The 625-page *Instructor's Solutions Manual* (0-13-047578-5) accompanying this book provides worked-out solutions for most of the problems in the book, and the 375-page *Student Solutions Manual* (0-13-047579-3) contains solutions for most of the odd-numbered problems.

The approximately 45 application modules in the text contain additional problem and project material designed largely to engage students in the exploration and application of computational technology. These investigations are expanded considerably in the 325-page *Applications Manual* (0-13-047577-7) that accompanies the text and supplements it with additional and sometimes more challenging investigations. Each section in this manual has parallel subsections "Using Maple," "Using Mathematica," and "Using MATLAB" that detail the applicable methods and techniques of each system and will afford student users an opportunity to compare the merits and styles of different computational systems.

Technology Manuals and Website

The author-written solutions and applications manuals described previously, as well as the additional technology manuals listed next, are available shrink-wrapped free with the textbook upon order using the indicated ISBN numbers:

- Text with *Student Solutions Manual* (0-13-114492-8)
- Text with *Applications Manual* (0-13-114491-X)
- Text with David Calvis, *Mathematica for Differential Equations: Projects, Insights, Syntax, and Animations* (0-13-114489-8)
- Text with Selwyn Hollis, *A Mathematica Companion for Differential Equations* (0-13-178327-0)
- Text with Robert Gilbert & George Hsiao, *Maple Projects for Differential Equations* (0-13-178326-2)
- Text with John Polking & David Arnold, *Ordinary Differential Equations Using* MATLAB, 2nd edition (0-13-075668-7)

Notebooks and worksheets supporting these manuals—plus additional software including a package of Maple worksheets keyed to this text by John Maloney—are available for downloading at the website **www.prenhall.com/edwards**. Many of the figures in this text were computer generated using Polking's MATLAB programs **dfield** and **pplane** that are linked at the site. Another MATLAB-based ODE package that has impressive graphical capabilities and is referenced in the text is **Iode** (see **www.math.uiuc.edu/iode**).

Acknowledgments

In preparing this revision we profited greatly from the advice and assistance of the following very perceptive reviewers:

David Calvis, *Baldwin-Wallace College*
Mila Čenkl, *Northeastern University*
Christopher French, *University of Illinois at Urbana-Champaign*
Moses Glasner, *Penn State University*
Richard Laugesen, *University of Illinois at Urbana-Champaign*
Juan Lopez, *Arizona State University*
James Moseley, *West Virginia University*
Peter Mucha, *Georgia Institute of Technology*
Arthur Wasserman, *University of Michigan*

We thank Bayani DeLeon for his usual efficient supervision of the process of book production. We are especially grateful to our editor, George Lobell, for his enthusiastic encouragement and advice that has shaped many aspects of this book in its successive editions. And it is a pleasure to credit Dennis Kletzing and his extraordinary TEXpertise for the attractive presentation of both the text and the art in this book.

C. H. E.
hedwards@math.uga.edu
Athens, Georgia, U.S.A.

D. E. P.
dpenney@math.uga.edu
Athens, Georgia, U.S.A.

CHAPTER 1

First-Order Differential Equations

1.1 Differential Equations and Mathematical Models

The laws of the universe are written in the language of mathematics. Algebra is sufficient to solve many static problems, but the most interesting natural phenomena involve change and are described by equations that relate changing quantities.

Because the derivative $dx/dt = f'(t)$ of the function f is the rate at which the quantity $x = f(t)$ is changing with respect to the independent variable t, it is natural that equations involving derivatives are frequently used to describe the changing universe. An equation relating an unknown function and one or more of its derivatives is called a **differential equation**.

Example 1 The differential equation

$$\frac{dx}{dt} = x^2 + t^2$$

involves both the unknown function $x(t)$ and its first derivative $x'(t) = dx/dt$. The differential equation

$$\frac{d^2y}{dx^2} + 3\frac{dy}{dx} + 7y = 0$$

involves the unknown function y of the independent variable x and the first two derivatives y' and y'' of y. ∎

The study of differential equations has three principal goals:

1. To discover the differential equation that describes a specified physical situation.
2. To find—either exactly or approximately—the appropriate solution of that equation.
3. To interpret the solution that is found.

In algebra, we typically seek the unknown *numbers* that satisfy an equation such as $x^3 + 7x^2 - 11x + 41 = 0$. By contrast, in solving a differential equation, we are challenged to find the unknown *functions* $y = y(x)$ for which an identity such as $y'(x) = 2xy(x)$—that is, the differential equation

$$\frac{dy}{dx} = 2xy$$

—holds on some interval of real numbers. Ordinarily, we will want to find *all* solutions of the differential equation, if possible.

Example 2 If C is a constant and

$$y(x) = Ce^{x^2}, \qquad (1)$$

then

$$\frac{dy}{dx} = C\left(2xe^{x^2}\right) = (2x)\left(Ce^{x^2}\right) = 2xy.$$

Thus every function $y(x)$ of the form in Eq. (1) *satisfies*—and thus is a solution of—the differential equation

$$\frac{dy}{dx} = 2xy \qquad (2)$$

for all x. In particular, Eq. (1) defines an *infinite* family of different solutions of this differential equation, one for each choice of the arbitrary constant C. By the method of separation of variables (Section 1.4) it can be shown that every solution of the differential equation in (2) is of the form in Eq. (1). ■

Differential Equations and Mathematical Models

The following three examples illustrate the process of translating scientific laws and principles into differential equations. In each of these examples the independent variable is time t, but we will see numerous examples in which some quantity other than time is the independent variable.

Example 3 Newton's law of cooling may be stated in this way: The *time rate of change* (the rate of change with respect to time t) of the temperature $T(t)$ of a body is proportional to the difference between T and the temperature A of the surrounding medium (Fig. 1.1.1). That is,

$$\frac{dT}{dt} = -k(T - A), \qquad (3)$$

where k is a positive constant. Observe that if $T > A$, then $dT/dt < 0$, so the temperature is a decreasing function of t and the body is cooling. But if $T < A$, then $dT/dt > 0$, so that T is increasing.

Thus the physical law is translated into a differential equation. If we are given the values of k and A, we should be able to find an explicit formula for $T(t)$, and then—with the aid of this formula—we can predict the future temperature of the body. ■

Temperature A

Temperature T

FIGURE 1.1.1. Newton's law of cooling, Eq. (3), describes the cooling of a hot rock in water.

Example 4 Torricelli's law implies that the *time rate of change* of the volume V of water in a draining tank (Fig. 1.1.2) is proportional to the square root of the depth y of water in the tank:

$$\frac{dV}{dt} = -k\sqrt{y}, \qquad (4)$$

where k is a constant. If the tank is a cylinder with vertical sides and cross-sectional area A, then $V = Ay$, so $dV/dt = A \cdot (dy/dt)$. In this case Eq. (4) takes the form

$$\frac{dy}{dt} = -h\sqrt{y}, \qquad (5)$$

where $h = k/A$ is a constant. ∎

Example 5

The *time rate of change* of a population $P(t)$ with constant birth and death rates is, in many simple cases, proportional to the size of the population. That is,

$$\frac{dP}{dt} = kP, \qquad (6)$$

where k is the constant of proportionality. ∎

Let us discuss Example 5 further. Note first that each function of the form

$$P(t) = Ce^{kt} \qquad (7)$$

is a solution of the differential equation

$$\frac{dP}{dt} - kP$$

FIGURE 1.1.2. Torricelli's law of draining, Eq. (4), describes the draining of a water tank.

in (6). We verify this assertion as follows:

$$P'(t) = Cke^{kt} = k\left(Ce^{kt}\right) = kP(t)$$

for all real numbers t. Because substitution of each function of the form given in (7) into Eq. (6) produces an identity, all such functions are solutions of Eq. (6).

Thus, even if the value of the constant k is known, the differential equation $dP/dt = kP$ has *infinitely many* different solutions of the form $P(t) = Ce^{kt}$, one for each choice of the "arbitrary" constant C. This is typical of differential equations. It is also fortunate, because it may allow us to use additional information to select from among all these solutions a particular one that fits the situation under study.

Example 6

Suppose that $P(t) = Ce^{kt}$ is the population of a colony of bacteria at time t, that the population at time $t = 0$ (hours, h) was 1000, and that the population doubled after 1 h. This additional information about $P(t)$ yields the following equations:

$$1000 = P(0) = Ce^0 = C,$$
$$2000 = P(1) = Ce^k.$$

It follows that $C = 1000$ and that $e^k = 2$, so $k = \ln 2 \approx 0.693147$. With this value of k the differential equation in (6) is

$$\frac{dP}{dt} = (\ln 2)P \approx (0.693147)P.$$

Substitution of $k = \ln 2$ and $C = 1000$ in Eq. (7) yields the particular solution

$$P(t) = 1000e^{(\ln 2)t} = 1000(e^{\ln 2})^t = 1000 \cdot 2^t \qquad \text{(because } e^{\ln 2} = 2)$$

that satisfies the given conditions. We can use this particular solution to predict future populations of the bacteria colony. For instance, the predicted number of bacteria in the population after one and a half hours (when $t = 1.5$) is

$$P(1.5) = 1000 \cdot 2^{3/2} \approx 2828.$$ ∎

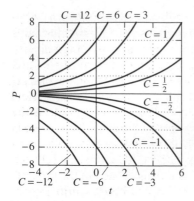

FIGURE 1.1.3. Graphs of $P(t) = Ce^{kt}$ with $k = \ln 2$.

The condition $P(0) = 1000$ in Example 6 is called an **initial condition** because we frequently write differential equations for which $t = 0$ is the "starting time." Figure 1.1.3 shows several different graphs of the form $P(t) = Ce^{kt}$ with $k = \ln 2$. The graphs of all the infinitely many solutions of $dP/dt = kP$ in fact fill the entire two-dimensional plane, and no two intersect. Moreover, the selection of any one point P_0 on the P-axis amounts to a determination of $P(0)$. Because exactly one solution passes through each such point, we see in this case that an initial condition $P(0) = P_0$ determines a unique solution agreeing with the given data.

Mathematical Models

Our brief discussion of population growth in Examples 5 and 6 illustrates the crucial process of *mathematical modeling* (Fig. 1.1.4), which involves the following:

1. The formulation of a real-world problem in mathematical terms; that is, the construction of a mathematical model.
2. The analysis or solution of the resulting mathematical problem.
3. The interpretation of the mathematical results in the context of the original real-world situation; for example, answering the question originally posed.

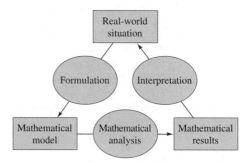

FIGURE 1.1.4. The process of mathematical modeling.

In the population example, the real-world problem is that of determining the population at some future time. A **mathematical model** consists of a list of variables (P and t) that describe the given situation, together with one or more equations relating these variables ($dP/dt = kP$, $P(0) = P_0$) that are known or are assumed to hold. The mathematical analysis consists of solving these equations (here, for P as a function of t). Finally, we apply these mathematical results to attempt to answer the original real-world question.

Nevertheless, it is quite possible that no one solution of the differential equation fits *all* the known information. In such a case we must suspect that the differential equation may not adequately describe the real world. For instance, the solutions of Eq. (6) are of the form $P(t) = Ce^{kt}$, where C is a positive constant, but for *no* choice of the constants k and C does $P(t)$ accurately describe the actual growth of the human population of the world over the past few centuries. We must therefore write a perhaps more complicated differential equation, one that takes into account the effects of population pressure on the birth rate, the declining food supply, and other factors. This should not be regarded as a failure of the model in Example 5, but as an insight into what additional factors must be considered in studying the growth of populations. Indeed, Eq. (6) is quite accurate under certain circumstances—for example, the growth of a bacterial population under conditions of unlimited food and space.

But in our population example we ignored the effects of such factors as varying birth and death rates. This made the mathematical analysis quite simple, perhaps unrealistically so. A satisfactory mathematical model is subject to two contradictory requirements: It must be sufficiently detailed to represent the real-world situation with relative accuracy, yet it must be sufficiently simple to make the mathematical analysis practical. If the model is so detailed that it fully represents the physical situation, then the mathematical analysis may be too difficult to carry out. If the model is too simple, the results may be so inaccurate as to be useless. Thus there is an inevitable tradeoff between what is physically realistic and what is mathematically possible. The construction of a model that adequately bridges this gap between realism and feasibility is therefore the most crucial and delicate step in the process. Ways must be found to simplify the model mathematically without sacrificing essential features of the real-world situation.

Mathematical models are discussed throughout this book. The remainder of this introductory section is devoted to simple examples and to standard terminology used in discussing differential equations and their solutions.

Examples and Terminology

Example 7 If C is a constant and $y(x) = 1/(C - x)$, then

$$\frac{dy}{dx} = \frac{1}{(C - x)^2} = y^2$$

if $x \neq C$. Thus

$$y(x) = \frac{1}{C - x} \tag{8}$$

defines a solution of the differential equation

$$\frac{dy}{dx} = y^2 \tag{9}$$

on any interval of real numbers not containing the point $x = C$. Actually, Eq. (8) defines a *one-parameter family* of solutions of $dy/dx = y^2$, one for each value of the arbitrary constant or "parameter" C. With $C = 1$ we get the particular solution

$$y(x) = \frac{1}{1 - x}$$

that satisfies the initial condition $y(0) = 1$. As indicated in Fig. 1.1.5, this solution is continuous on the interval $(-\infty, 1)$ but has a vertical asymptote at $x = 1$. ∎

Example 8 Verify that the function $y(x) = 2x^{1/2} - x^{1/2} \ln x$ satisfies the differential equation

$$4x^2 y'' + y = 0 \tag{10}$$

for all $x > 0$.

Solution First we compute the derivatives

$$y'(x) = -\tfrac{1}{2}x^{-1/2} \ln x \quad \text{and} \quad y''(x) = \tfrac{1}{4}x^{-3/2} \ln x - \tfrac{1}{2}x^{-3/2}.$$

Then substitution into Eq. (10) yields

$$4x^2 y'' + y = 4x^2 \left(\tfrac{1}{4}x^{-3/2} \ln x - \tfrac{1}{2}x^{-3/2}\right) + 2x^{1/2} - x^{1/2} \ln x = 0$$

if x is positive, so the differential equation is satisfied for all $x > 0$. ∎

The fact that we can write a differential equation is not enough to guarantee that it has a solution. For example, it is clear that the differential equation

$$(y')^2 + y^2 = -1 \tag{11}$$

has *no* (real-valued) solution, because the sum of nonnegative numbers cannot be negative. For a variation on this theme, note that the equation

$$(y')^2 + y^2 = 0 \tag{12}$$

obviously has only the (real-valued) solution $y(x) \equiv 0$. In our previous examples any differential equation having at least one solution indeed had infinitely many.

The **order** of a differential equation is the order of the highest derivative that appears in it. The differential equation of Example 8 is of second order, those in Examples 2 through 7 are first-order equations, and

$$y^{(4)} + x^2 y^{(3)} + x^5 y = \sin x$$

is a fourth-order equation. The most general form of an **n th-order** differential equation with independent variable x and unknown function or dependent variable $y = y(x)$ is

$$F\left(x, y, y', y'', \ldots, y^{(n)}\right) = 0, \tag{13}$$

where F is a specific real-valued function of $n + 2$ variables.

Our use of the word *solution* has been until now somewhat informal. To be precise, we say that the continuous function $u = u(x)$ is a **solution** of the differential equation in (13) **on the interval** I provided that the derivatives $u', u'', \ldots, u^{(n)}$ exist on I and

$$F\left(x, u, u', u'', \ldots, u^{(n)}\right) = 0$$

for all x in I. For the sake of brevity, we may say that $u = u(x)$ **satisfies** the differential equation in (13) on I.

Remark: Recall from elementary calculus that a differentiable function on an open interval is necessarily continuous there. This is why only a continuous function can qualify as a (differentiable) solution of a differential equation on an interval. ∎

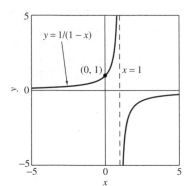

FIGURE 1.1.5. The solution of $y' = y^2$ defined by $y(x) = 1/(1 - x)$.

Example 7
Continued

Figure 1.1.5 shows the two "connected" branches of the graph $y = 1/(1 - x)$. The left-hand branch is the graph of a (continuous) solution of the differential equation $y' = y^2$ that is defined on the interval $(-\infty, 1)$. The right-hand branch is the graph of a *different* solution of the differential equation that is defined (and continuous) on the different interval $(1, \infty)$. So the single formula $y(x) = 1/(1 - x)$ actually defines two different solutions (with different domains of definition) of the same differential equation $y' = y^2$. ∎

Example 9

If A and B are constants and

$$y(x) = A \cos 3x + B \sin 3x, \tag{14}$$

then two successive differentiations yield

$$y'(x) = -3A \sin 3x + 3B \cos 3x,$$
$$y''(x) = -9A \cos 3x - 9B \sin 3x = -9y(x)$$

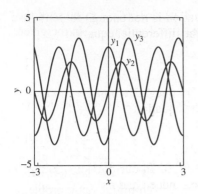

FIGURE 1.1.6. The three solutions $y_1(x) = 3\cos 3x$, $y_2(x) = 2\sin 3x$, and $y_3(x) = -3\cos 3x + 2\sin 3x$ of the differential equation $y'' + 9y = 0$.

for all x. Consequently, Eq. (14) defines what it is natural to call a *two-parameter family* of solutions of the second-order differential equation

$$y'' + 9y = 0 \qquad (15)$$

on the whole real number line. Figure 1.1.6 shows the graphs of several such solutions. ∎

Although the differential equations in (11) and (12) are exceptions to the general rule, we will see that an nth-order differential equation ordinarily has an n-parameter family of solutions—one involving n different arbitrary constants or parameters.

In both Eqs. (11) and (12), the appearance of y' as an implicitly defined function causes complications. For this reason, we will ordinarily assume that any differential equation under study can be solved explicitly for the highest derivative that appears; that is, that the equation can be written in the so-called *normal form*

$$y^{(n)} = G\left(x, y, y', y'', \ldots, y^{(n-1)}\right), \qquad (16)$$

where G is a real-valued function of $n + 1$ variables. In addition, we will always seek only real-valued solutions unless we warn the reader otherwise.

All the differential equations we have mentioned so far are **ordinary** differential equations, meaning that the unknown function (dependent variable) depends on only a *single* independent variable. If the dependent variable is a function of two or more independent variables, then partial derivatives are likely to be involved; if they are, the equation is called a **partial** differential equation. For example, the temperature $u = u(x, t)$ of a long thin uniform rod at the point x at time t satisfies (under appropriate simple conditions) the partial differential equation

$$\frac{\partial u}{\partial t} = k\frac{\partial^2 u}{\partial x^2},$$

where k is a constant (called the *thermal diffusivity* of the rod). In Chapters 1 through 8 we will be concerned only with *ordinary* differential equations and will refer to them simply as differential equations.

In this chapter we concentrate on *first-order* differential equations of the form

$$\frac{dy}{dx} = f(x, y). \qquad (17)$$

We also will sample the wide range of applications of such equations. A typical mathematical model of an applied situation will be an **initial value problem,** consisting of a differential equation of the form in (17) together with an **initial condition** $y(x_0) = y_0$. Note that we call $y(x_0) = y_0$ an initial condition whether or not $x_0 = 0$. To **solve** the initial value problem

$$\frac{dy}{dx} = f(x, y), \quad y(x_0) = y_0 \qquad (18)$$

means to find a differentiable function $y = y(x)$ that satisfies both conditions in Eq. (18) on some interval containing x_0.

Example 10

Given the solution $y(x) = 1/(C - x)$ of the differential equation $dy/dx = y^2$ discussed in Example 7, solve the initial value problem

$$\frac{dy}{dx} = y^2, \quad y(1) = 2.$$

Solution We need only find a value of C so that the solution $y(x) = 1/(C - x)$ satisfies the initial condition $y(1) = 2$. Substitution of the values $x = 1$ and $y = 2$ in the given solution yields

$$2 = y(1) = \frac{1}{C - 1},$$

so $2C - 2 = 1$, and hence $C = \frac{3}{2}$. With this value of C we obtain the desired solution

$$y(x) = \frac{1}{\frac{3}{2} - x} = \frac{2}{3 - 2x}.$$

Figure 1.1.7 shows the two branches of the graph $y = 2/(3 - 2x)$. The left-hand branch is the graph on $(-\infty, \frac{3}{2})$ of the solution of the given initial value problem $y' = y^2$, $y(1) = 2$. The right-hand branch passes through the point $(2, -2)$ and is therefore the graph on $(\frac{3}{2}, \infty)$ of the solution of the different initial value problem $y' = y^2$, $y(2) = -2$. ∎

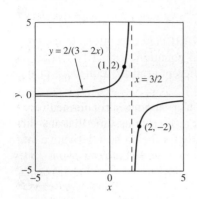

FIGURE 1.1.7. The solutions of $y' = y^2$ defined by $y(x) = 2/(3 - 2x)$.

The central question of greatest immediate interest to us is this: If we are given a differential equation known to have a solution satisfying a given initial condition, how do we actually *find* or *compute* that solution? And, once found, what can we do with it? We will see that a relatively few simple techniques—separation of variables (Section 1.4), solution of linear equations (Section 1.5), elementary substitution methods (Section 1.6)—are enough to enable us to solve a variety of first-order equations having impressive applications.

1.1 Problems

In Problems 1 through 12, verify by substitution that each given function is a solution of the given differential equation. Throughout these problems, primes denote derivatives with respect to x.

1. $y' = 3x^2$; $y = x^3 + 7$
2. $y' + 2y = 0$; $y = 3e^{-2x}$
3. $y'' + 4y = 0$; $y_1 = \cos 2x$, $y_2 = \sin 2x$
4. $y'' = 9y$; $y_1 = e^{3x}$, $y_2 = e^{-3x}$
5. $y' = y + 2e^{-x}$; $y = e^x - e^{-x}$
6. $y'' + 4y' + 4y = 0$; $y_1 = e^{-2x}$, $y_2 = xe^{-2x}$
7. $y'' - 2y' + 2y = 0$; $y_1 = e^x \cos x$, $y_2 = e^x \sin x$
8. $y'' + y = 3\cos 2x$, $y_1 = \cos x - \cos 2x$, $y_2 = \sin x - \cos 2x$
9. $y' + 2xy^2 = 0$; $y = \dfrac{1}{1 + x^2}$
10. $x^2 y'' + xy' - y = \ln x$; $y_1 = x - \ln x$, $y_2 = \dfrac{1}{x} - \ln x$
11. $x^2 y'' + 5xy' + 4y = 0$; $y_1 = \dfrac{1}{x^2}$, $y_2 = \dfrac{\ln x}{x^2}$
12. $x^2 y'' - xy' + 2y = 0$; $y_1 = x \cos(\ln x)$, $y_2 = x \sin(\ln x)$

In Problems 13 through 16, substitute $y = e^{rx}$ into the given differential equation to determine all values of the constant r for which $y = e^{rx}$ is a solution of the equation.

13. $3y' = 2y$
14. $4y'' = y$
15. $y'' + y' - 2y = 0$
16. $3y'' + 3y' - 4y = 0$

In Problems 17 through 26, first verify that $y(x)$ satisfies the given differential equation. Then determine a value of the constant C so that $y(x)$ satisfies the given initial condition. Use a computer or graphing calculator (if desired) to sketch several typical solutions of the given differential equation, and highlight the one that satisfies the given initial condition.

17. $y' + y = 0$; $y(x) = Ce^{-x}$, $y(0) = 2$
18. $y' = 2y$; $y(x) = Ce^{2x}$, $y(0) = 3$
19. $y' = y + 1$; $y(x) = Ce^x - 1$, $y(0) = 5$
20. $y' = x - y$; $y(x) = Ce^{-x} + x - 1$, $y(0) = 10$
21. $y' + 3x^2 y = 0$; $y(x) = Ce^{-x^3}$, $y(0) = 7$
22. $e^y y' = 1$; $y(x) = \ln(x + C)$, $y(0) = 0$
23. $x\dfrac{dy}{dx} + 3y = 2x^5$; $y(x) = \frac{1}{4}x^5 + Cx^{-3}$, $y(2) = 1$
24. $xy' - 3y = x^3$; $y(x) = x^3(C + \ln x)$, $y(1) = 17$
25. $y' = 3x^2(y^2 + 1)$; $y(x) = \tan(x^3 + C)$, $y(0) = 1$
26. $y' + y \tan x = \cos x$; $y(x) = (x + C) \cos x$, $y(\pi) = 0$

In Problems 27 through 31, a function $y = g(x)$ is described by some geometric property of its graph. Write a differential equation of the form $dy/dx = f(x, y)$ having the function g as its solution (or as one of its solutions).

27. The slope of the graph of g at the point (x, y) is the sum of x and y.
28. The line tangent to the graph of g at the point (x, y) intersects the x-axis at the point $(x/2, 0)$.

29. Every straight line normal to the graph of g passes through the point $(0, 1)$.

30. The graph of g is normal to every curve of the form $y = x^2 + k$ (k is a constant) where they meet.

31. The line tangent to the graph of g at (x, y) passes through the point $(-y, x)$.

In Problems 32 through 36, write—in the manner of Eqs. (3) through (6) of this section—a differential equation that is a mathematical model of the situation described.

32. The time rate of change of a population P is proportional to the square root of P.

33. The time rate of change of the velocity v of a coasting motorboat is proportional to the square of v.

34. The acceleration dv/dt of a Lamborghini is proportional to the difference between 250 km/h and the velocity of the car.

35. In a city having a fixed population of P persons, the time rate of change of the number N of those persons who have heard a certain rumor is proportional to the number of those who have not yet heard the rumor.

36. In a city with a fixed population of P persons, the time rate of change of the number N of those persons infected with a certain contagious disease is proportional to the product of the number who have the disease and the number who do not.

In Problems 37 through 42, determine by inspection at least one solution of the given differential equation. That is, use your knowledge of derivatives to make an intelligent guess. Then test your hypothesis.

37. $y'' = 0$ **38.** $y' = y$

39. $xy' + y = 3x^2$ **40.** $(y')^2 + y^2 = 1$

41. $y' + y = e^x$ **42.** $y'' + y = 0$

43. (a) If k is a constant, show that a general (1-parameter) solution of the differential equation

$$\frac{dx}{dt} = kx^2$$

is given by $x(t) = 1/(C - kt)$ where C is an arbitrary constant.

(b) Determine by inspection a solution of the initial value problem $x' = kx^2$, $x(0) = 0$.

44. (a) Continuing Problem 43, assume that k is positive, and then sketch graphs of solutions of $x' = kx^2$ with several typical positive values of $x(0)$.

(b) How would these solutions differ if the constant k were negative?

45. Suppose a population P of rodents satisfies the differential equation $dP/dt = kP^2$. Initially, there are $P(0) = 2$ rodents, and their number is increasing at the rate of $dP/dt = 1$ rodent per month when there are $P = 10$ rodents. How long will it take for this population to grow to a hundred rodents? To a thousand? What's happening here?

46. Suppose the velocity v of a motorboat coasting in water satisfies the differential equation $dv/dt = kv^2$. The initial speed of the motorboat is $v(0) = 10$ meters per second (m/s), and v is decreasing at the rate of 1 m/s^2 when $v = 5$ m/s. How long does it take for the velocity of the boat to decrease to 1 m/s? To $\frac{1}{10}$ m/s? When does the boat come to a stop?

47. In Example 7 we saw that $y(x) = 1/(C - x)$ defines a one-parameter family of solutions of the differential equation $dy/dx = y^2$. **(a)** Determine a value of C so that $y(10) = 10$. **(b)** Is there a value of C such that $y(0) = 0$? Can you nevertheless find by inspection a solution of $dy/dx = y^2$ such that $y(0) = 0$? **(c)** Figure 1.1.8 shows typical graphs of solutions of the form $y(x) = 1/(C - x)$. Does it appear that these solution curves fill the entire xy-plane? Can you conclude that, given any point (a, b) in the plane, the differential equation $dy/dx = y^2$ has exactly one solution $y(x)$ satisfying the condition $y(a) = b$?

48. (a) Show that $y(x) = Cx^4$ defines a one-parameter family of differentiable solutions of the differential equation $xy' = 4y$ (Fig. 1.1.9). **(b)** Show that

$$y(x) = \begin{cases} -x^4 & \text{if } x < 0, \\ x^4 & \text{if } x \geq 0 \end{cases}$$

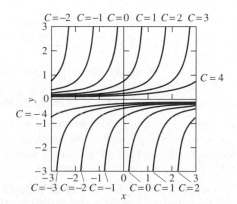

FIGURE 1.1.8. Graphs of solutions of the equation $dy/dx = y^2$.

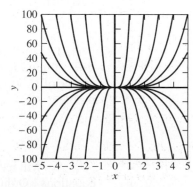

FIGURE 1.1.9. The graph $y = Cx^4$ for various values of C.

defines a differentiable solution of $xy' = 4y$ for all x, but is not of the form $y(x) = Cx^4$. **(c)** Given any two real numbers a and b, explain why—in contrast to the situation in part (c) of Problem 47—there exist infinitely many differentiable solutions of $xy' = 4y$ that all satisfy the condition $y(a) = b$.

1.2 Integrals As General And Particular Solutions

The first-order equation $dy/dx = f(x, y)$ takes an especially simple form if the function f is independent of the dependent variable y:

$$\frac{dy}{dx} = f(x). \tag{1}$$

In this special case we need only integrate both sides of Eq. (1) to obtain

$$y(x) = \int f(x)\,dx + C. \tag{2}$$

This is a **general solution** of Eq. (1), meaning that it involves an arbitrary constant C, and for every choice of C it is a solution of the differential equation in (1). If $G(x)$ is a particular antiderivative of f—that is, if $G'(x) \equiv f(x)$—then

$$y(x) = G(x) + C. \tag{3}$$

The graphs of any two such solutions $y_1(x) = G(x) + C_1$ and $y_2(x) = G(x) + C_2$ on the same interval I are "parallel" in the sense illustrated by Figs. 1.2.1 and 1.2.2. There we see that the constant C is geometrically the vertical distance between the two curves $y(x) = G(x)$ and $y(x) = G(x) + C$.

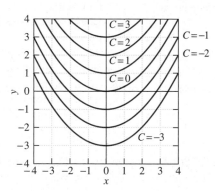

FIGURE 1.2.1. Graphs of $y = \frac{1}{4}x^2 + C$ for various values of C.

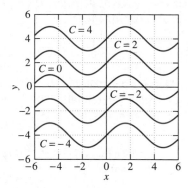

FIGURE 1.2.2. Graphs of $y = \sin x + C$ for various values of C.

To satisfy an initial condition $y(x_0) = y_0$, we need only substitute $x = x_0$ and $y = y_0$ into Eq. (3) to obtain $y_0 = G(x_0) + C$, so that $C = y_0 - G(x_0)$. With this choice of C, we obtain the **particular solution** of Eq. (1) satisfying the initial value problem

$$\frac{dy}{dx} = f(x), \quad y(x_0) = y_0.$$

We will see that this is the typical pattern for solutions of first-order differential equations. Ordinarily, we will first find a *general solution* involving an arbitrary constant C. We can then attempt to obtain, by appropriate choice of C, a *particular solution* satisfying a given initial condition $y(x_0) = y_0$.

Remark: As the term is used in the previous paragraph, a *general solution* of a first-order differential equation is simply a one-parameter family of solutions. A natural question is whether a given general solution contains *every* particular solution of the differential equation. When this is known to be true, we call it **the** general solution of the differential equation. For example, because any two antiderivatives of the same function $f(x)$ can differ only by a constant, it follows that every solution of Eq. (1) is of the form in (2). Thus Eq. (2) serves to define **the** general solution of (1). ∎

| **Example 1** | Solve the initial value problem |

$$\frac{dy}{dx} = 2x + 3, \quad y(1) = 2.$$

Solution Integration of both sides of the differential equation as in Eq. (2) immediately yields the general solution

$$y(x) = \int (2x + 3)\, dx = x^2 + 3x + C.$$

Figure 1.2.3 shows the graph $y = x^2 + 3x + C$ for various values of C. The particular solution we seek corresponds to the curve that passes through the point $(1, 2)$, thereby satisfying the initial condition

$$y(1) = (1)^2 + 3 \cdot (1) + C = 2.$$

It follows that $C = -2$, so the desired particular solution is

$$y(x) = x^2 + 3x - 2.$$ ∎

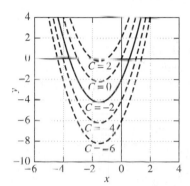

FIGURE 1.2.3. Solution curves for the differential equation in Example 1.

Second-order Equations. The observation that the special first-order equation $dy/dx = f(x)$ is readily solvable (provided that an antiderivative of f can be found) extends to second-order differential equations of the special form

$$\frac{d^2 y}{dx^2} = g(x), \tag{4}$$

in which the function g on the right-hand side involves neither the dependent variable y nor its derivative dy/dx. We simply integrate once to obtain

$$\frac{dy}{dx} = \int y''(x)\, dx = \int g(x)\, dx = G(x) + C_1,$$

where G is an antiderivative of g and C_1 is an arbitrary constant. Then another integration yields

$$y(x) = \int y'(x)\, dx = \int [G(x) + C_1]\, dx = \int G(x)\, dx + C_1 x + C_2.$$

where C_2 is a second arbitrary constant. In effect, the second-order differential equation in (4) is one that can be solved by solving successively the *first-order* equations

$$\frac{dv}{dx} = g(x) \quad \text{and} \quad \frac{dy}{dx} = v(x).$$

Velocity and Acceleration

Direct integration is sufficient to allow us to solve a number of important problems concerning the motion of a particle (or *mass point*) in terms of the forces acting on it. The motion of a particle along a straight line (the x-axis) is described by its **position function**

$$x = f(t) \tag{5}$$

giving its x-coordinate at time t. The **velocity** of the particle is defined to be

➤ $$v(t) = f'(t); \quad \text{that is,} \quad v = \frac{dx}{dt}. \tag{6}$$

Its **acceleration** $a(t)$ is $a(t) = v'(t) = x''(t)$; in Leibniz notation,

➤ $$a = \frac{dv}{dt} = \frac{d^2x}{dt^2}. \tag{7}$$

Newton's *second law of motion* says that if a force $F(t)$ acts on the particle and is directed along its line of motion, then

$$ma(t) = F(t); \quad \text{that is,} \quad F = ma, \tag{8}$$

where m is the mass of the particle. If the force F is known, then the equation $x''(t) = F(t)/m$ can be integrated twice to find the position function $x(t)$ in terms of two constants of integration. These two arbitrary constants are frequently determined by the **initial position** $x_0 = x(0)$ and the **initial velocity** $v_0 = v(0)$ of the particle.

Constant acceleration. For instance, suppose that the force F, and therefore the acceleration $a = F/m$, are *constant*. Then we begin with the equation

$$\frac{dv}{dt} = a \quad (a \text{ is a constant}) \tag{9}$$

and integrate both sides to obtain

$$v(t) = \int a \, dt = at + C_1.$$

We know that $v = v_0$ when $t = 0$, and substitution of this information into the preceding equation yields the fact that $C_1 = v_0$. So

$$v(t) = \frac{dx}{dt} = at + v_0. \tag{10}$$

A second integration gives

$$x(t) = \int v(t) \, dt = \int (at + v_0) \, dt = \tfrac{1}{2}at^2 + v_0 t + C_2,$$

and the substitution $t = 0$, $x = x_0$ gives $C_2 = x_0$. Therefore,

$$x(t) = \tfrac{1}{2}at^2 + v_0 t + x_0. \tag{11}$$

Thus, with Eq. (10) we can find the velocity, and with Eq. (11) the position, of the particle at any time t in terms of its *constant* acceleration a, its initial velocity v_0, and its initial position x_0.

Example 2

A lunar lander is falling freely toward the surface of the moon at a speed of 450 meters per second (m/s). Its retrorockets, when fired, provide a constant deceleration of 2.5 meters per second per second (m/s^2) (the gravitational acceleration produced by the moon is assumed to be included in the given deceleration). At what height above the lunar surface should the retrorockets be activated to ensure a "soft touchdown" ($v = 0$ at impact)?

Solution

We denote by $x(t)$ the height of the lunar lander above the surface, as indicated in Fig. 1.2.4. We let $t = 0$ denote the time at which the retrorockets should be fired. Then $v_0 = -450$ (m/s, negative because the height $x(t)$ is decreasing), and $a = +2.5$ because an upward thrust increases the velocity v (although it decreases the *speed* $|v|$). Then Eqs. (10) and (11) become

$$v(t) = 2.5t - 450 \tag{12}$$

and

$$x(t) = 1.25t^2 - 450t + x_0, \tag{13}$$

where x_0 is the height of the lander above the lunar surface at the time $t = 0$ when the retrorockets should be activated.

From Eq. (12) we see that $v = 0$ (soft touchdown) occurs when $t = 450/2.5 = 180$ s (that is, 3 minutes); then substitution of $t = 180$, $x = 0$ into Eq. (13) yields

$$x_0 = 0 - (1.25)(180)^2 + 450(180) = 40{,}500$$

meters, that is, $x_0 = 40.5$ km $\approx 25\frac{1}{6}$ miles. Thus the retrorockets should be activated when the lunar lander is 40.5 kilometers above the surface of the moon, and it will touch down softly on the lunar surface after 3 minutes of decelerating descent. ∎

FIGURE 1.2.4. The lunar lander of Example 2.

Physical Units

Numerical work requires units for the measurement of physical quantities such as distance and time. We sometimes use ad hoc units—such as distance in miles or kilometers and time in hours—in special situations (such as in a problem involving an auto trip). However, the foot-pound-second (fps) and meter-kilogram-second (mks) unit systems are used more generally in scientific and engineering problems. In fact, fps units are commonly used only in the United States (and a few other countries), while mks units constitute the standard international system of scientific units.

	fps units	**mks units**
Force	pound (lb)	newton (N)
Mass	slug	kilogram (kg)
Distance	foot (ft)	meter (m)
Time	second (s)	second (s)
g	32 ft/s^2	9.8 m/s^2

The last line of this table gives values for the gravitational acceleration g at the surface of the earth. Although these approximate values will suffice for most examples and problems, more precise values are 9.7805 m/s^2 and 32.088 ft/s^2 (at sea level at the equator).

Both systems are compatible with Newton's second law $F = ma$. Thus 1 N is (by definition) the force required to impart an acceleration of 1 m/s^2 to a mass of 1 kg. Similarly, 1 slug is (by definition) the mass that experiences an acceleration of 1 ft/s^2 under a force of 1 lb. (We will use mks units in all problems requiring mass units, and thus will rarely need slugs to measure mass.)

Inches and centimeters (as well as miles and kilometers) also are commonly used in describing distances. For conversions between fps and mks units it helps to remember that

$$1 \text{ in.} = 2.54 \text{ cm (exactly)} \quad \text{and} \quad 1 \text{ lb} \approx 4.448 \text{ N.}$$

For instance,

$$1 \text{ ft} = 12 \text{ in.} \times 2.54 \frac{\text{cm}}{\text{in.}} = 30.48 \text{ cm,}$$

and it follows that

$$1 \text{ mi} = 5280 \text{ ft} \times 30.48 \frac{\text{cm}}{\text{ft}} = 160934.4 \text{ cm} \approx 1.609 \text{ km.}$$

Thus a posted U.S. speed limit of 50 mi/h means that—in international terms—the legal speed limit is about $50 \times 1.609 \approx 80.45$ km/h.

Vertical Motion with Gravitational Acceleration

The **weight** W of a body is the force exerted on the body by gravity. Substitution of $a = g$ and $F = W$ in Newton's second law $F = ma$ gives

$$W = mg \tag{14}$$

for the weight W of the mass m at the surface of the earth (where $g \approx 32$ ft/s$^2 \approx 9.8$ m/s^2). For instance, a mass of $m = 20$ kg has a weight of $W = (20 \text{ kg})(9.8 \text{ m/s}^2) = 196$ N. Similarly, a mass m weighing 100 pounds has mks weight

$$W = (100 \text{ lb})(4.448 \text{ N/lb}) = 444.8 \text{ N,}$$

so its mass is

$$m = \frac{W}{g} = \frac{444.8 \text{ N}}{9.8 \text{ m/s}^2} \approx 45.4 \text{ kg.}$$

To discuss vertical motion it is natural to choose the y-axis as the coordinate system for position, frequently with $y = 0$ corresponding to "ground level." If we choose the *upward* direction as the positive direction, then the effect of gravity on a vertically moving body is to decrease its height and also to decrease its velocity $v = dy/dt$. Consequently, if we ignore air resistance, then the acceleration $a = dv/dt$ of the body is given by

$$\frac{dv}{dt} = -g. \tag{15}$$

This acceleration equation provides a starting point in many problems involving vertical motion. Successive integrations (as in Eqs. (10) and (11)) yield the velocity and height formulas

$$v(t) = -gt + v_0 \tag{16}$$

and

$$y(t) = -\tfrac{1}{2}gt^2 + v_0 t + y_0. \tag{17}$$

Here, y_0 denotes the initial ($t = 0$) height of the body and v_0 its initial velocity.

Example 3

(a) Suppose that a ball is thrown straight upward from the ground ($y_0 = 0$) with initial velocity $v_0 = 96$ (ft/s, so we use $g = 32$ ft/s^2 in fps units). Then it reaches its maximum height when its velocity (Eq. (16)) is zero,

$$v(t) = -32t + 96 = 0,$$

and thus when $t = 3$ s. Hence the maximum height that the ball attains is

$$y(3) = -\tfrac{1}{2} \cdot 32 \cdot 3^2 + 96 \cdot 3 + 0 = 144 \text{ (ft)}$$

(with the aid of Eq. (17)).

(b) If an arrow is shot straight upward from the ground with initial velocity $v_0 = 49$ (m/s, so we use $g = 9.8$ m/s^2 in mks units), then it returns to the ground when

$$y(t) = -\tfrac{1}{2} \cdot (9.8)t^2 + 49t = (4.9)t(-t + 10) = 0,$$

and thus after 10 s in the air. ■

A Swimmer's Problem

Figure 1.2.5 shows a northward-flowing river of width $w = 2a$. The lines $x = \pm a$ represent the banks of the river and the y-axis its center. Suppose that the velocity v_R at which the water flows increases as one approaches the center of the river, and indeed is given in terms of distance x from the center by

$$v_R = v_0 \left(1 - \frac{x^2}{a^2}\right). \tag{18}$$

FIGURE 1.2.5. A swimmer's problem (Example 4).

You can use Eq. (18) to verify that the water does flow the fastest at the center, where $v_R = v_0$, and that $v_R = 0$ at each riverbank.

Suppose that a swimmer starts at the point $(-a, 0)$ on the west bank and swims due east (relative to the water) with constant speed v_S. As indicated in Fig. 1.2.5, his velocity vector (relative to the ground) has horizontal component v_S and vertical component v_R. Hence the swimmer's direction angle α is given by

$$\tan \alpha = \frac{v_R}{v_S}.$$

Because $\tan \alpha = dy/dx$, substitution using (18) gives the differential equation

$$\frac{dy}{dx} = \frac{v_0}{v_S} \left(1 - \frac{x^2}{a^2}\right) \tag{19}$$

for the swimmer's trajectory $y = y(x)$ as he crosses the river.

Example 4

Suppose that the river is 1 mile wide and that its midstream velocity is $v_0 = 9$ mi/h. If the swimmer's velocity is $v_S = 3$ mi/h, then Eq. (19) takes the form

$$\frac{dy}{dx} = 3(1 - 4x^2).$$

Integration yields

$$y(x) = \int (3 - 12x^2)\, dx = 3x - 4x^3 + C$$

for the swimmer's trajectory. The initial condition $y\left(-\frac{1}{2}\right) = 0$ yields $C = 1$, so

$$y(x) = 3x - 4x^3 + 1.$$

Then

$$y\left(\tfrac{1}{2}\right) = 3\left(\tfrac{1}{2}\right) - 4\left(\tfrac{1}{2}\right)^3 + 1 = 2,$$

so the swimmer drifts 2 miles downstream while he swims 1 mile across the river. ∎

1.2 Problems

In Problems 1 through 10, find a function $y = f(x)$ satisfying the given differential equation and the prescribed initial condition.

1. $\dfrac{dy}{dx} = 2x + 1;\ y(0) = 3$

2. $\dfrac{dy}{dx} = (x - 2)^2;\ y(2) = 1$

3. $\dfrac{dy}{dx} = \sqrt{x};\ y(4) = 0$

4. $\dfrac{dy}{dx} = \dfrac{1}{x^2};\ y(1) = 5$

5. $\dfrac{dy}{dx} = \dfrac{1}{\sqrt{x + 2}};\ y(2) = -1$

6. $\dfrac{dy}{dx} = x\sqrt{x^2 + 9};\ y(-4) = 0$

7. $\dfrac{dy}{dx} = \dfrac{10}{x^2 + 1};\ y(0) = 0$

8. $\dfrac{dy}{dx} = \cos 2x;\ y(0) = 1$

9. $\dfrac{dy}{dx} = \dfrac{1}{\sqrt{1 - x^2}};\ y(0) = 0$

10. $\dfrac{dy}{dx} = xe^{-x};\ y(0) = 1$

In Problems 11 through 18, find the position function $x(t)$ of a moving particle with the given acceleration $a(t)$, initial position $x_0 = x(0)$, and initial velocity $v_0 = v(0)$.

11. $a(t) = 50,\ v_0 = 10,\ x_0 = 20$
12. $a(t) = -20,\ v_0 = -15,\ x_0 = 5$
13. $a(t) = 3t,\ v_0 = 5,\ x_0 = 0$
14. $a(t) = 2t + 1,\ v_0 = -7,\ x_0 = 4$
15. $a(t) = 4(t + 3)^2,\ v_0 = -1,\ x_0 = 1$
16. $a(t) = \dfrac{1}{\sqrt{t + 4}},\ v_0 = -1,\ x_0 = 1$
17. $a(t) = \dfrac{1}{(t + 1)^3},\ v_0 = 0,\ x_0 = 0$
18. $a(t) = 50 \sin 5t,\ v_0 = -10,\ x_0 = 8$

In Problems 19 through 22, a particle starts at the origin and travels along the x-axis with the velocity function $v(t)$ whose graph is shown in Figs. 1.2.6 through 1.2.9. Sketch the graph of the resulting position function $x(t)$ for $0 \leq t \leq 10$.

19.

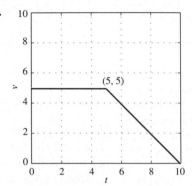

FIGURE 1.2.6. Graph of the velocity function $v(t)$ of Problem 19.

20.

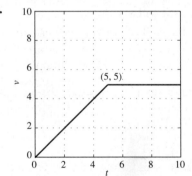

FIGURE 1.2.7. Graph of the velocity function $v(t)$ of Problem 20.

21.

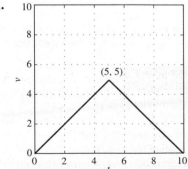

FIGURE 1.2.8. Graph of the velocity function $v(t)$ of Problem 21.

22.

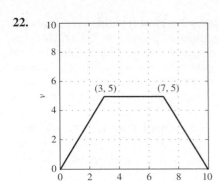

FIGURE 1.2.9. Graph of the velocity function $v(t)$ of Problem 22.

23. What is the maximum height attained by the arrow of part (b) of Example 3?

24. A ball is dropped from the top of a building 400 ft high. How long does it take to reach the ground? With what speed does the ball strike the ground?

25. The brakes of a car are applied when it is moving at 100 km/h and provide a constant deceleration of 10 meters per second per second (m/s^2). How far does the car travel before coming to a stop?

26. A projectile is fired straight upward with an initial velocity of 100 m/s from the top of a building 20 m high and falls to the ground at the base of the building. Find (a) its maximum height above the ground; (b) when it passes the top of the building; (c) its total time in the air.

27. A ball is thrown straight downward from the top of a tall building. The initial speed of the ball is 10 m/s. It strikes the ground with a speed of 60 m/s. How tall is the building?

28. A baseball is thrown straight downward with an initial speed of 40 ft/s from the top of the Washington Monument (555 ft high). How long does it take to reach the ground, and with what speed does the baseball strike the ground?

29. A diesel car gradually speeds up so that for the first 10 s its acceleration is given by

$$\frac{dv}{dt} = (0.12)t^2 + (0.6)t \quad (\text{ft/s}^2).$$

If the car starts from rest ($x_0 = 0$, $v_0 = 0$), find the distance it has traveled at the end of the first 10 s and its velocity at that time.

30. A car traveling at 60 mi/h (88 ft/s) skids 176 ft after its brakes are suddenly applied. Under the assumption that the braking system provides constant deceleration, what is that deceleration? For how long does the skid continue?

31. The skid marks made by an automobile indicated that its brakes were fully applied for a distance of 75 m before it came to a stop. The car in question is known to have a constant deceleration of 20 m/s^2 under these conditions. How fast—in km/h—was the car traveling when the brakes were first applied?

32. Suppose that a car skids 15 m if it is moving at 50 km/h when the brakes are applied. Assuming that the car has the same constant deceleration, how far will it skid if it is moving at 100 km/h when the brakes are applied?

33. On the planet Gzyx, a ball dropped from a height of 20 ft hits the ground in 2 s. If a ball is dropped from the top of a 200-ft-tall building on Gzyx, how long will it take to hit the ground? With what speed will it hit?

34. A person can throw a ball straight upward from the surface of the earth to a maximum height of 144 ft. How high could this person throw the ball on the planet Gzyx of Problem 29?

35. A stone is dropped from rest at an initial height h above the surface of the earth. Show that the speed with which it strikes the ground is $v = \sqrt{2gh}$.

36. If a woman has enough "spring" in her legs to jump vertically to a height of 2.25 ft on the earth, how high could she jump on the moon, where the surface gravitational acceleration is (approximately) 5.3 ft/s^2?

37. At noon a car starts from rest at point A and proceeds at constant acceleration along a straight road toward point B. If the car reaches B at 12:50 P.M. with a velocity of 60 mi/h, what is the distance from A to B?

38. At noon a car starts from rest at point A and proceeds with constant acceleration along a straight road toward point C, 35 miles away. If the constantly accelerated car arrives at C with a velocity of 60 mi/h, at what time does it arrive at C?

39. If $a = 0.5$ mi and $v_0 = 9$ mi/h as in Example 4, what must the swimmer's speed v_S be in order that he drifts only 1 mile downstream as he crosses the river?

40. Suppose that $a = 0.5$ mi, $v_0 = 9$ mi/h, and $v_S = 3$ mi/h as in Example 4, but that the velocity of the river is given by the fourth-degree function

$$v_R = v_0 \left(1 - \frac{x^4}{a^4} \right)$$

rather than the quadratic function in Eq. (18). Now find how far downstream the swimmer drifts as he crosses the river.

41. A bomb is dropped from a helicopter hovering at an altitude of 800 feet above the ground. From the ground directly beneath the helicopter, a projectile is fired straight upward toward the bomb, exactly 2 seconds after the bomb is released. With what initial velocity should the projectile be fired, in order to hit the bomb at an altitude of exactly 400 feet?

42. A spacecraft is in free fall toward the surface of the moon at a speed of 1000 mph (mi/h). Its retrorockets, when fired provide a constant deceleration of 20,000 mi/h^2. At what height above the lunar surface should the astronauts fire the retrorockets to insure a soft touchdown? (As in Example 2, ignore the moon's gravitational field.)

43. Arthur Clarke's *The Wind from the Sun* (1963) describes Diana, a spacecraft propelled by the solar wind. Its aluminized sail provides it with a constant acceleration of $0.001g = 0.0098$ m/s^2. Suppose this spacecraft starts

from rest at time $t = 0$ and simultaneously fires a projectile (straight ahead in the same direction) that travels at one-tenth of the speed $c = 3 \times 10^8$ m/s of light. How long will it take the spacecraft to catch up with the projectile, and how far will it have traveled by then?

44. A driver involved in an accident claims he was going only 25 mph. When police tested his car, they found that when its brakes were applied at 25 mph, the car skidded only 45 feet before coming to a stop. But the driver's skid marks at the accident scene measured 210 feet. Assuming the same (constant) deceleration, determine the speed he was actually traveling just prior to the accident.

1.3 Slope Fields and Solution Curves

Consider a differential equation of the form

$$\frac{dy}{dx} = f(x, y) \qquad (1)$$

where the right-hand function $f(x, y)$ involves both the independent variable x and the dependent variable y. We might think of integrating both sides in (1) with respect to x, and hence write $y(x) = \int f(x, y(x))\,dx + C$. However, this approach does not lead to a solution of the differential equation, because the indicated integral involves the *unknown* function $y(x)$ itself, and therefore cannot be evaluated explicitly. Actually, there exists *no* straightforward procedure by which a general differential equation can be solved explicitly. Indeed, the solutions of such a simple-looking differential equation as $y' = x^2 + y^2$ cannot be expressed in terms of the ordinary elementary functions studied in calculus textbooks. Nevertheless, the graphical and numerical methods of this and later sections can be used to construct *approximate* solutions of differential equations that suffice for many practical purposes.

Slope Fields and Graphical Solutions

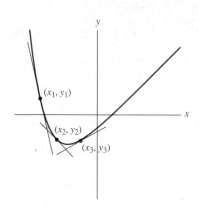

FIGURE 1.3.1. A solution curve for the differential equation $y' = x - y$ together with tangent lines having

- slope $m_1 = x_1 - y_1$ at the point (x_1, y_1);
- slope $m_2 = x_2 - y_2$ at the point (x_2, y_2); and
- slope $m_3 = x_3 - y_3$ at the point (x_3, y_3).

There is a simple geometric way to think about solutions of a given differential equation $y' = f(x, y)$. At each point (x, y) of the xy-plane, the value of $f(x, y)$ determines a slope $m = f(x, y)$. A solution of the differential equation is simply a differentiable function whose graph $y = y(x)$ has this "correct slope" at each point $(x, y(x))$ through which it passes—that is, $y'(x) = f(x, y(x))$. Thus a **solution curve** of the differential equation $y' = f(x, y)$—the graph of a solution of the equation—is simply a curve in the xy-plane whose tangent line at each point (x, y) has slope $m = f(x, y)$. For instance, Fig. 1.3.1 shows a solution curve of the differential equation $y' = x - y$ together with its tangent lines at three typical points.

This geometric viewpoint suggests a *graphical method* for constructing *approximate* solutions of the differential equation $y' = f(x, y)$. Through each of a representative collection of points (x, y) in the plane we draw a short line segment having the proper slope $m = f(x, y)$. All these line segments constitute a **slope field** (or a **direction field**) for the equation $y' = f(x, y)$.

| **Example 1** | Figures 1.3.2 (a)–(d) show slope fields and solution curves for the differential equation |

$$\frac{dy}{dx} = ky \qquad (2)$$

with the values $k = 2, 0.5, -1$, and -3 of the parameter k in Eq. (2). Note that each slope field yields important qualitative information about the set of all solutions of the differential equation. For instance, Figs. 1.3.2(a) and (b) suggest that each solution $y(x)$ approaches $\pm\infty$ as $x \to +\infty$ if $k > 0$, whereas Figs. 1.3.2(c) and

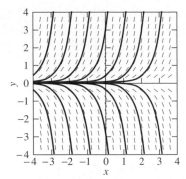

FIGURE 1.3.2(a) Slope field and solution curves for $y' = 2y$.

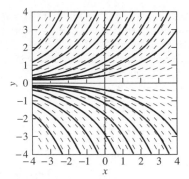

FIGURE 1.3.2(b) Slope field and solution curves for $y' = (0.5)y$.

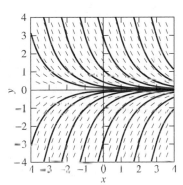

FIGURE 1.3.2(c) Slope field and solution curves for $y' = -y$.

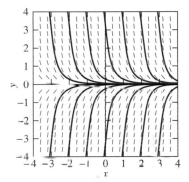

FIGURE 1.3.2(d) Slope field and solution curves for $y' = -3y$.

(d) suggest that $y(x) \to 0$ as $x \to +\infty$ if $k < 0$. Moreover, although the sign of k determines the *direction* of increase or decrease of $y(x)$, its absolute value $|k|$ appears to determine the rate of change of $y(x)$. All this is apparent from slope fields like those in Fig. 1.3.2 even without knowing that the general solution of Eq. (2) is given explicitly by $y(x) = Ce^{kx}$. ∎

A slope field suggests visually the general shapes of solution curves of the differential equation. Through each point a solution curve should proceed in such a direction that its tangent line is nearly parallel to the nearby line segments of the slope field. Starting at any initial point (a, b), we can attempt to sketch freehand an approximate solution curve that threads its way through the slope field, following the visible line segments as closely as possible.

Example 2 Construct a slope field for the differential equation $y' = x - y$ and use it to sketch an approximate solution curve that passes through the point $(-4, 4)$.

Solution Solution Figure 1.3.3 shows a table of slopes for the given equation. The numerical slope $m = x - y$ appears at the intersection of the horizontal x-row and the vertical y-column of the table. If you inspect the pattern of upper-left to lower-right diagonals in this table, you can see that it was easily and quickly constructed. (Of course, a more complicated function $f(x, y)$ on the right-hand side of the differential equation would necessitate more complicated calculations.) Figure 1.3.4 shows

$x \backslash y$	-4	-3	-2	-1	0	1	2	3	4
-4	0	-1	-2	-3	-4	-5	-6	-7	-8
-3	1	0	-1	-2	-3	-4	-5	-6	-7
-2	2	1	0	-1	-2	-3	-4	-5	-6
-1	3	2	1	0	-1	-2	-3	-4	-5
0	4	3	2	1	0	-1	-2	-3	-4
1	5	4	3	2	1	0	-1	-2	-3
2	6	5	4	3	2	1	0	-1	-2
3	7	6	5	4	3	2	1	0	-1
4	8	7	6	5	4	3	2	1	0

FIGURE 1.3.3. Values of the slope $y' = x - y$ for $-4 \leqq x, \; y \leqq 4$.

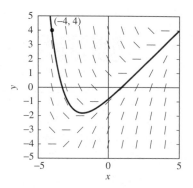

FIGURE 1.3.4. Slope field for $y' = x - y$ corresponding to the table of slopes in Fig. 1.3.3.

FIGURE 1.3.5. The solution curve through $(-4, 4)$.

the corresponding slope field, and Fig. 1.3.5 shows an approximate solution curve sketched through the point $(-4, 4)$ so as to follow as this slope field as closely as possible. At each point it appears to proceed in the direction indicated by the nearby line segments of the slope field. ∎

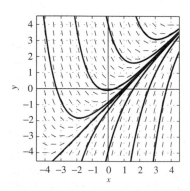

FIGURE 1.3.6. Slope field and typical solution curves for $y' = x - y$.

Although a spreadsheet program (for instance) readily constructs a table of slopes as in Fig. 1.3.3, it can be quite tedious to plot by hand a sufficient number of slope segments as in Fig. 1.3.4. However, most computer algebra systems include commands for quick and ready construction of slope fields with as many line segments as desired; such commands are illustrated in the application material for this section. The more line segments are constructed, the more accurately solution curves can be visualized and sketched. Figure 1.3.6 shows a "finer" slope field for the differential equation $y' = x - y$ of Example 2, together with typical solution curves treading through this slope field.

If you look closely at Fig. 1.3.6, you may spot a solution curve that appears to be a straight line! Indeed, you can verify that the linear function $y = x - 1$ is a solution of the equation $y' = x - y$, and it appears likely that the other solution curves approach this straight line as an asymptote as $x \to +\infty$. This inference illustrates the fact that a slope field can suggest tangible information about solutions that is not at all evident from the differential equation itself. Can you, by tracing the appropriate solution curve in this figure, infer that $y(3) \approx 2$ for the solution $y(x)$ of the initial value problem $y' = x - y, \; y(-4) = 4$?

Applications of Slope fields

The next two examples illustrate the use of slope fields to glean useful information in physical situations that are modeled by differential equations. Example 3 is based on the fact that a baseball moving through the air at a moderate speed v (less than about 300 ft/s) encounters air resistance that is approximately proportional to v. If the baseball is thrown straight downward from the top of a tall building or from a hovering helicopter, then it experiences both the downward acceleration of gravity and an upward acceleration of air resistance. If the y-axis is directed *downward*, then the ball's velocity $v = dy/dt$ and its gravitational acceleration $g = 32$ ft/s^2 are both positive, while its acceleration due to air resistance is negative. Hence its total acceleration is of the form

$$\frac{dv}{dt} = g - kv. \tag{3}$$

A typical value of the air resistance proportionality constant might be $k = 0.16$.

Example 3

FIGURE 1.3.7. Slope field and typical solution curves for $v' = 32 - 0.16v$.

Suppose you throw a baseball straight downward from a helicopter hovering at an altitude of 3000 feet. You wonder whether someone standing on the ground below could conceivably catch it. In order to estimate the speed with which the ball will land, you can use your laptop's computer algebra system to construct a slope field for the differential equation

$$\frac{dv}{dt} = 32 - 0.16v. \tag{4}$$

The result is shown in Fig. 1.3.7, together with a number of solution curves corresponding to different values of the initial velocity $v(0)$ with which you might throw the baseball downward. Note that all these solution curves appear to approach the horizontal line $v = 200$ as an asymptote. This implies that—however you throw it—the baseball should approach the *limiting velocity* $v = 200$ ft/s instead of accelerating indefinitely (as it would in the absence of any air resistance). The handy fact that 60 mi/h = 88 ft/s yields

$$v = 200\frac{\text{ft}}{\text{s}} \times \frac{60\text{ mi/h}}{88\text{ ft/s}} \approx 136.36\,\frac{\text{mi}}{\text{h}}.$$

Perhaps a catcher accustomed to 100 mi/h fastballs would have some chance of fielding this speeding ball. ∎

Comment If the ball's initial velocity is $v(0) = 200$, then Eq. (4) gives $v'(0) = 32 - (0.16)(200) = 0$ so the ball experiences *no* initial acceleration. Its velocity therefore remains unchanged, and hence $v(t) \equiv 200$ is a constant "equilibrium solution" of the differential equation. If the initial velocity is greater than 200, then the initial acceleration given by Eq. (4) is negative, so the ball slows down as it falls. But if the initial velocity is less than 200, then the initial acceleration given by (4) is positive, so the ball speeds up as it falls. It therefore seems quite reasonable that, because of air resistance, the baseball will approach a limiting velocity of 200 ft/s—whatever initial velocity it starts with. You might like to verify that—in the absence of air resistance—this ball would hit the ground at over 300 mi/h. ∎

In Section 2.1 we will discuss in detail the logistic differential equation

$$\frac{dP}{dt} = kP(M - P) \tag{5}$$

that often is used to model a population $P(t)$ that inhabits an environment with *carrying capacity M*. This means that M is the maximum population that this environment can sustain on a long-term basis (in terms of the maximum available food, for instance).

Example 4 If we take $k = 0.0004$ and $M = 150$, then the logistic equation in (5) takes the form

$$\frac{dP}{dt} = 0.0004P(150 - P) = 0.06P - 0.0004P^2. \tag{6}$$

The positive term $0.06P$ on the right in (6) corresponds to natural growth at a 6% annual rate (with time t measured in years). The negative term $-0.0004P^2$ represents the inhibition of growth due to limited resources in the environment.

Figure 1.3.8 shows a slope field for Eq. (6), together with a number of solution curves corresponding to possible different values of the initial population $P(0)$. Note that all these solution curves appear to approach the horizontal line $P = 150$ as an asymptote. This implies that—whatever the initial population—the population $P(t)$ approaches the *limiting population $P = 150$* as $t \to \infty$. ∎

FIGURE 1.3.8. Slope field and typical solution curves for $P' = 0.06P - 0.0004P^2$.

Comment If the initial population is $P(0) = 150$, then Eq. (6) gives

$$P'(0) = 0.0004(150)(150 - 150) = 0,$$

so the population experiences *no* initial (instantaneous) change. It therefore remains unchanged, and hence $P(t) \equiv 150$ is a constant "equilibrium solution" of the differential equation. If the initial population is greater than 150, then the initial rate of change given by (6) is negative, so the population immediately begins to decrease. But if the initial population is less than 150, then the initial rate of change given by (6) is positive, so the population immediately begins to increase. It therefore seems quite reasonable to conclude that the population will approach a limiting value of 150—whatever the (positive) initial population. ∎

Existence and Uniqueness of Solutions

Before one spends much time attempting to solve a given differential equation, it is wise to know that solutions actually *exist*. We may also want to know whether there is only one solution of the equation satisfying a given initial condition—that is, whether its solutions are *unique*.

Example 5 **(a)** [Failure of existence] The initial value problem

$$y' = \frac{1}{x}, \quad y(0) = 0 \tag{7}$$

has *no* solution, because no solution $y(x) = \int (1/x)\,dx = \ln|x| + C$ of the differential equation is defined at $x = 0$.
(b) [Failure of uniqueness] On the other hand, you can readily verify that the initial value problem

$$y' = 2\sqrt{y}, \quad y(0) = 0 \tag{8}$$

has the *two* different solutions $y_1(x) = x^2$ and $y_2(x) \equiv 0$ (see Problem 27). ∎

Example 5 illustrates the fact that, before we can speak of "the" solution of an initial value problem, we need to know that it has *one and only one* solution. Questions of existence and uniqueness of solutions also bear on the process of mathematical modeling. Suppose that we are studying a physical system whose behavior is completely determined by certain initial conditions, but that our proposed mathematical model involves a differential equation *not* having a unique solution satisfying those conditions. This raises an immediate question as to whether the mathematical model adequately represents the physical system.

The theorem stated below implies that the initial value problem $y' = f(x, y)$, $y(a) = b$ has one and only one solution defined near the point $x = a$ on the x-axis, provided that both the function f and its partial derivative $\partial f/\partial y$ are continuous near the point (a, b) in the xy-plane. Methods of proving existence and uniqueness theorems are discussed in the Appendix.

THEOREM 1: Existence and Uniqueness of Solutions

Suppose that both the function $f(x, y)$ and its partial derivative $D_y f(x, y)$ are continuous on some rectangle R in the xy-plane that contains the point (a, b) in its interior. Then, for some open interval I containing the point a, the initial value problem

$$\frac{dy}{dx} = f(x, y), \quad y(a) = b \tag{9}$$

has one and only one solution that is defined on the interval I. ▲

Remark 1: In the case of the differential equation $dy/dx = -y$ of Example 1 and Fig. 1.3.2(c), both the function $f(x, y) = -y$ and the partial derivative $\partial f/\partial y = -1$ are continuous everywhere, so Theorem 1 implies the existence of a unique solution for any initial data (a, b). Although the theorem ensures existence only on some open interval containing $x = a$, each solution $y(x) = Ce^{-x}$ actually is defined for all x.

Remark 2: In the case of the differential equation $dy/dx = -2\sqrt{y}$ of Example 5(b) and Eq. (8), the function $f(x, y) = -2\sqrt{y}$ is continuous wherever $y > 0$, but the partial derivative $\partial f/\partial y = 1/\sqrt{y}$ is discontinuous when $y = 0$, and hence at the point $(0, 0)$. This is why it is possible for there to exist two different solutions $y_1(x) = x^2$ and $y_2(x) \equiv 0$, each of which satisfies the initial condition $y(0) = 0$.

Remark 3: In Example 7 of Section 1.1 we examined the especially simple differential equation $dy/dx = y^2$. Here we have $f(x, y) = y^2$ and $\partial f/\partial y = 2y$. Both of these functions are continuous everywhere in the xy-plane, and in particular on the rectangle $-2 < x < 2, 0 < y < 2$. Because the point $(0, 1)$ lies in the interior of this rectangle, Theorem 1 guarantees a unique solution—necessarily a continuous function—of the initial value problem

$$\frac{dy}{dx} = y^2, \quad y(0) = 1 \tag{10}$$

on some open x-interval containing $a = 0$. Indeed this is the solution

$$y(x) = \frac{1}{1 - x}$$

that we discussed in Example 7. But $y(x) = 1/(1 - x)$ is discontinuous at $x = 1$, so our unique continuous solution does not exist on the entire interval $-2 < x < 2$. Thus the solution interval I of Theorem 1 may not be as wide as the rectangle R where f and $\partial f / \partial y$ are continuous. Geometrically, the reason is that the solution curve provided by the theorem may leave the rectangle—wherein solutions of the differential equation are guaranteed to exist—before it reaches the one or both ends of the interval (see Fig. 1.3.9). ∎

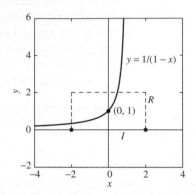

FIGURE 1.3.9. The solution curve through the initial point $(0, 1)$ leaves the rectangle R before it reaches the right end of the interval I.

The following example shows that, if the function $f(x, y)$ and/or its partial derivative $\partial f / \partial y$ fail to satisfy the continuity hypothesis of Theorem 1, then the initial value problem in (9) may have *either* no solution *or* many—even infinitely many—solutions.

Example 6 Consider the first-order differential equation

$$x \frac{dy}{dx} = 2y. \tag{11}$$

Applying Theorem 1 with $f(x, y) = 2y/x$ and $\partial f / \partial y = 2/x$, we conclude that Eq. (11) must have a unique solution near any point in the xy-plane where $x \neq 0$. Indeed, we see immediately by substitution in (11) that

$$y(x) = Cx^2 \tag{12}$$

satisfies Eq. (11) for any value of the constant C and for all values of the variable x. In particular, the initial value problem

$$x \frac{dy}{dx} = 2y, \quad y(0) = 0 \tag{13}$$

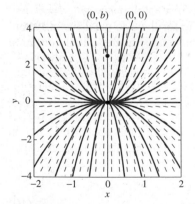

FIGURE 1.3.10. There are infinitely many solution curves through the point $(0, 0)$, but no solution curves through the point $(0, b)$ if $b \neq 0$.

has infinitely many different solutions, whose solution curves are the parabolas $y = Cx^2$ illustrated in Fig. 1.3.10. (In case $C = 0$ the "parabola" is actually the x-axis $y = 0$.)

Observe that all these parabolas pass through the origin $(0, 0)$, but none of them passes through any other point on the y-axis. It follows that the initial value problem in (13) has infinitely many solutions, but the initial value problem

$$x\frac{dy}{dx} = 2y, \quad y(0) = b \tag{14}$$

has no solution if $b \neq 0$.

Finally, note that through any point off the y-axis there passes only one of the parabolas $y = Cx^2$. Hence, if $a \neq 0$ then the initial value problem

$$x\frac{dy}{dx} = 2y, \quad y(a) = b \tag{15}$$

has a unique solution on any interval that contains the point $x = a$ but not the origin $x = 0$ In summary, the initial value problem in (15) has

- a unique solution near (a, b) if $a \neq 0$;
- no solution if $a = 0$ but $b \neq 0$;
- infinitely many solutions if $a = b = 0$. ∎

Still more can be said about the initial value problem in (15). Consider a typical initial point off the y axis—for instance the point $(-1, 1)$ indicated in Fig. 1.3.11. Then for any value of the constant C the function defined by

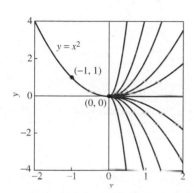

FIGURE 1.3.11. There are infinitely many solution curves through the point $(1, -1)$.

$$y(x) = \begin{cases} x^2 & \text{if } x \leq 0, \\ Cx^2 & \text{if } x > 0 \end{cases} \tag{16}$$

is continuous and satisfies the initial value problem

$$x\frac{dy}{dx} = 2y, \quad y(-1) = 1. \tag{17}$$

For a particular value of C, the solution curve defined by (16) consists of the left half of the parabola $y = x^2$ and the right half of the parabola $y = Cx^2$. Thus the unique solution curve near $(-1, 1)$ branches at the origin into the infinitely many solution curves illustrated in Fig. 1.3.11.

We therefore see that Theorem 1 (if its hypotheses are satisfied) guarantees uniqueness of the solution near the initial point (a, b), but a solution curve through (a, b) may eventually branch elsewhere so that uniqueness is lost. Thus a solution may exist on a larger interval than one on which the solution is unique. For instance, the solution $y(x) = x^2$ of the initial value problem in (17) exists on the whole x-axis, but this solution is unique only on the negative x-axis $-\infty < x < 0$.

1.3 Problems

In Problems 1 through 10, we have provided the slope field of the indicated differential equation, together with one or more solution curves. Sketch likely solution curves through the additional points marked in each slope field.

1. $\dfrac{dy}{dx} = -y - \sin x$

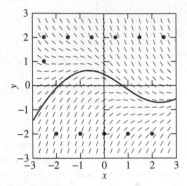

FIGURE 1.3.12.

2. $\dfrac{dy}{dx} = x + y$

FIGURE 1.3.13.

3. $\dfrac{dy}{dx} = y - \sin x$

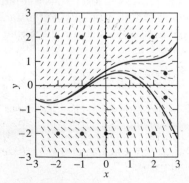

FIGURE 1.3.14.

4. $\dfrac{dy}{dx} = x - y$

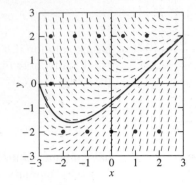

FIGURE 1.3.15.

5. $\dfrac{dy}{dx} = y - x + 1$

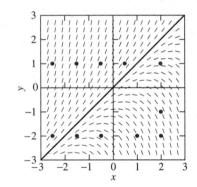

FIGURE 1.3.16.

6. $\dfrac{dy}{dx} = x - y + 1$

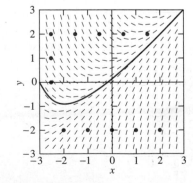

FIGURE 1.3.17.

7. $\dfrac{dy}{dx} = \sin x + \sin y$

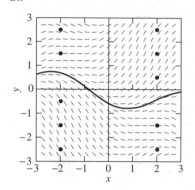

FIGURE 1.3.18.

8. $\dfrac{dy}{dx} = x^2 - y$

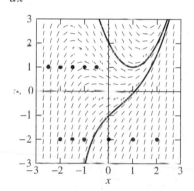

FIGURE 1.3.19.

9. $\dfrac{dy}{dx} = x^2 - y - 2$

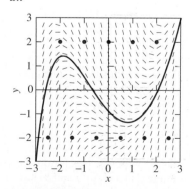

FIGURE 1.3.20.

10. $\dfrac{dy}{dx} = -x^2 + \sin y$

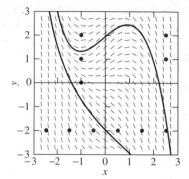

FIGURE 1.3.21.

In Problems 11 through 20, determine whether Theorem 1 does or does not guarantee existence of a solution of the given initial value problem. If existence is guaranteed, determine whether Theorem 1 does or does not guarantee uniqueness of that solution.

11. $\dfrac{dy}{dx} = 2x^2y^2; \quad y(1) = -1$

12. $\dfrac{dy}{dx} = x \ln y; \quad y(1) = 1$

13. $\dfrac{dy}{dx} = \sqrt[3]{y}; \quad y(0) = 1$

14. $\dfrac{dy}{dx} = \sqrt[3]{y}; \quad y(0) = 0$

15. $\dfrac{dy}{dx} = \sqrt{x - y}; \quad y(2) = 2$

16. $\dfrac{dy}{dx} = \sqrt{x - y}; \quad y(2) = 1$

17. $y\dfrac{dy}{dx} = x - 1; \quad y(0) = 1$

18. $y\dfrac{dy}{dx} = x - 1; \quad y(1) = 0$

19. $\dfrac{dy}{dx} = \ln(1 + y^2); \quad y(0) = 0$

20. $\dfrac{dy}{dx} = x^2 - y^2; \quad y(0) = 1$

In Problems 21 and 22, first use the method of Example 2 to construct a slope field for the given differential equation. Then sketch the solution curve corresponding to the given initial condition. Finally, use this solution curve to estimate the desired value of the solution $y(x)$.

21. $y' = x + y, \quad y(0) = 0; \quad y(-4) =?$
22. $y' = y - x, \quad y(4) = 0; \quad y(-4) =?$

Problems 23 and 24 are like Problems 21 and 22, but now use a computer algebra system to plot and print out a slope field for the given differential equation. If you wish (and know how) you can check your manually sketched solution curve by plotting it with the computer.

23. $y' = x^2 + y^2 - 1$, $y(0) = 0$; $y(2) =$?

24. $y' = x + \dfrac{1}{2}y^2$, $y(-2) = 0$; $y(2) =$?

25. You bail out of the helicopter of Example 3 and pull the ripcord of your parachute. Now $k = 1.6$ in Eq. (3), so your downward velocity satisfies the initial value problem

$$\frac{dv}{dt} = 32 - 1.6v, \quad v(0) = 0.$$

In order to investigate your chances of survival, construct a slope field for this differential equation and sketch the appropriate solution curve. What will your limiting velocity be? Will a strategically located haystack do any good? How long will it take you to reach 95% of your limiting velocity?

26. Suppose the deer population $P(t)$ in a small forest satisfies the logistic equation

$$\frac{dP}{dt} = 0.0225P - 0.0003P^2.$$

Construct a slope field and appropriate solution curve to answer the following questions: If there are 25 deer at time $t = 0$ and t is measured in months, how long will it take the number of deer to double? What will be the limiting deer population?

The next seven problems illustrate the fact that, if the hypotheses of Theorem 1 are not satisfied, then the initial value problem $y' = f(x, y)$, $y(a) = b$ may have either no solutions, finitely many solutions, or infinitely many solutions.

27. (a) Verify that if c is a constant, then the function defined piecewise by

$$y(x) = \begin{cases} 0 & \text{for } x \leq c, \\ (x - c)^2 & \text{for } x > c \end{cases}$$

satisfies the differential equation $y' = 2\sqrt{y}$ for all x (including the point $x = c$). Construct a figure illustrating the fact that the initial value problem $y' = 2\sqrt{y}$, $y(0) = 0$ has infinitely many different solutions. **(b)** For what values of b does the initial value problem $y' = 2\sqrt{y}$, $y(0) = b$ have (i) no solution, (ii) a unique solution that is defined for all x?

28. Verify that if k is a constant, then the function $y(x) \equiv kx$ satisfies the differential equation $xy' = y$ for all x. Construct a slope field and several of these straight line solution curves. Then determine (in terms of a and b) how many different solutions the initial value problem $xy' = y$, $y(a) = b$ has—one, none, or infinitely many.

29. Verify that if c is a constant, then the function defined piecewise by

$$y(x) = \begin{cases} 0 & \text{for } x \leq c, \\ (x - c)^3 & \text{for } x > c \end{cases}$$

satisfies the differential equation $y' = 3y^{2/3}$ for all x. Can you also use the "left half" of the cubic $y = (x - c)^3$ in piecing together a solution curve of the differential equation? (See Fig. 1.3.22.) Sketch a variety of such solution curves. Is there a point (a, b) of the xy-plane such that the initial value problem $y' = 3y^{2/3}$, $y(a) = b$ has either no solution or a unique solution that is defined for all x? Reconcile your answer with Theorem 1.

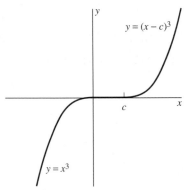

FIGURE 1.3.22. A suggestion for Problem 29.

30. Verify that if c is a constant, then the function defined piecewise by

$$y(x) = \begin{cases} +1 & \text{if } x \leq c, \\ \cos(x - c) & \text{if } c < x < c + \pi, \\ -1 & \text{if } x \geq c \end{cases}$$

satisfies the differential equation $y' = -\sqrt{1 - y^2}$ for all x. (Perhaps a preliminary sketch with $c = 0$ will be helpful.) Sketch a variety of such solution curves. Then determine (in terms of a and b) how many different solutions the initial value problem $y' = -\sqrt{1 - y^2}$, $y(a) = b$ has.

31. Carry out an investigation similar to that in Problem 30, except with the differential equation $y' = +\sqrt{1 - y^2}$. Does it suffice simply to replace $\cos(x - c)$ with $\sin(x - c)$ in piecing together a solution that is defined for all x?

32. Verify that if $c > 0$, then the function defined piecewise by

$$y(x) = \begin{cases} 0 & \text{if } x^2 \leq c, \\ (x^2 - c)^2 & \text{if } x^2 > c \end{cases}$$

satisfies the differential equation $y' = 4x\sqrt{y}$ for all x. Sketch a variety of such solution curves for different values of c. Then determine (in terms of a and b) how many different solutions the initial value problem $y' = 4x\sqrt{y}$, $y(a) = b$ has.

33. If $c \neq 0$ verify that the function defined by $y(x) = x/(cx - 1)$ (with graph illustrated in Fig. 1.3.23) satisfies the differential equation $x^2 y' + y^2 = 0$ if $x \neq 1/c$. Sketch a variety of such solution curves for different values of c. Also, note the constant-valued function $y(x) \equiv 0$ that does not result from any choice of the constant c. Finally, determine (in terms of a and b) how many different solutions the initial value problem $x^2 y' + y^2 = 0$, $y(a) = b$ has.

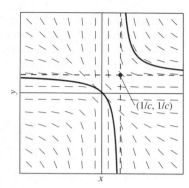

FIGURE 1.3.23. Slope field for $x^2 y' + y^2 = 0$ and graph of a solution $y(x) = x/(cx - 1)$.

1.3 Application: Computer-Generated Slope Fields and Solution Curves

Widely available computer algebra systems and technical computing environments include facilities to automate the construction of slope fields and solution curves, as do some graphing calculators (see Fig. 1.3.24).

FIGURE 1.3.24. Slope field and solution curves for the differential equation

$$\frac{dy}{dx} = \sin(x - y)$$

with initial points $(0, b)$, $b = -3, -1, -2, 0, 2, 4$ and window $-5 \le x, y \le 5$ on a TI-89 graphing calculator.

The applications manual accompanying this textbook includes discussion of *Maple*©, *Mathematica*©, and MATLAB©resources for the investigation of differential equations. For instance, the *Maple* command

```
with(DEtools):
DEplot(diff(y(x),x)=sin(x-y(x)), y(x), x=-5..5, y=-5..5);
```

and the *Mathematica* command

```
<< Graphics\PlotField.m
PlotVectorField[{1, Sin[x-y]}, {x, -5, 5}, {y, -5, 5}]
```

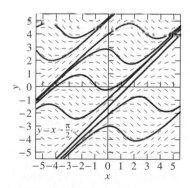

FIGURE 1.3.25. Computer-generated slope field and solution curves for the differential equation $y' = \sin(x - y)$.

produce slope fields similar to the one shown in Fig. 1.3.25. Figure 1.3.25 itself was generated with the MATLAB program `dfield` (John Polking and David Arnold, *Ordinary Differential Equations Using* MATLAB (2nd edition), Prentice

Hall, 1999) that is freely available for educational use (**math.rice.edu/~dfield**). When a differential equation is entered in the **dfield** setup menu (Fig. 1.3.26), you can (with mouse button clicks) plot both a slope field and the solution curve (or curves) through any desired point (or points). Another freely available and user-friendly MATLAB-based ODE package with impressive graphical capabilities is **Iode** (**www.math.uiuc.edu/iode**).

FIGURE 1.3.26. MATLAB **dfield** setup to construct slope field and solution curves for $y' = \sin(x - y)$.

Use a graphing calculator or computer system in the following investigations. You might warm up by generating the slope fields and some solution curves for Problems 1 through 10 in this section.

INVESTIGATION A: Plot a slope field and typical solution curves for the differential equation $dy/dx = \sin(x - y)$, but with a larger window than that of Fig. 1.3.25. With $-10 \leqq x \leqq 10$, $-10 \leqq y \leqq 10$, for instance, a number of apparent straight line solution curves should be visible.

(a) Substitute $y = ax + b$ in the differential equation to determine what the coefficients a and b must be in order to get a solution.

(b) A computer algebra system gives the general solution

$$y(x) = x - 2 \tan^{-1}\left(\frac{x - 2 - C}{x - C}\right).$$

Can you determine a value of the arbitrary constant C that yields the linear solution $y = x - (\pi/2)$ corresponding to the initial condition $y(\pi/2) = 0$?

INVESTIGATION B: For your own personal investigation, let n be the *smallest* digit in your student ID number that is greater than 1, and consider the differential equation

$$\frac{dy}{dx} = \frac{1}{n}\cos(x - ny).$$

(a) First investigate (as in part (a) of Investigation A) the possibility of straight line solutions.

(b) Then generate a slope field for this differential equation, with the viewing window chosen so that you can picture some of these straight lines, plus a sufficient number of nonlinear solution curves that you can formulate a conjecture about what happens to $y(x)$ as $x \to +\infty$. State your inference as plainly as you can. Given the initial value $y(0) = y_0$, try to predict (perhaps in terms of y_0) how $y(x)$ behaves as $x \to +\infty$.

(c) A computer algebra system gives the general solution

$$y(x) = \frac{1}{n}\left[x + 2\tan^{-1}\left(\frac{1}{x - C}\right)\right].$$

Can you make a connection between this symbolic solution and your graphically generated solution curves (straight lines or otherwise)?

1.4 Separable Equations and Applications

The first-order differential equation

$$\frac{dy}{dx} = H(x, y) \tag{1}$$

is called **separable** provided that $H(x, y)$ can be written as the product of a function of x and a function of y:

➤
$$\frac{dy}{dx} = g(x)h(y) = \frac{g(x)}{f(y)},$$

where $h(y) = 1/f(y)$. In this case the variables x and y can be *separated*—isolated on opposite sides of an equation—by writing informally the equation

$$f(y)\,dy = g(x)\,dx,$$

which we understand to be concise notation for the differential equation

$$f(y)\frac{dy}{dx} = g(x). \tag{2}$$

It is easy to solve this special type of differential equation simply by integrating both sides with respect to x:

$$\int f(y(x))\frac{dy}{dx}\,dx = \int g(x)\,dx + C;$$

equivalently,

$$\int f(y)\,dy = \int g(x)\,dx + C. \tag{3}$$

All that is required is that the antiderivatives

$$F(y) = \int f(y)\,dy \quad \text{and} \quad G(x) = \int g(x)\,dx$$

can be found. To see that Eqs. (2) and (3) are equivalent, note the following consequence of the chain rule:

$$D_x[F(y(x))] = F'(y(x))y'(x) = f(y)\frac{dy}{dx} = g(x) = D_x[G(x)],$$

which in turn is equivalent to

$$F(y(x)) = G(x) + C, \tag{4}$$

because two functions have the same derivative on an interval if and only if they differ by a constant on that interval.

Example 1 Solve the initial value problem

$$\frac{dy}{dx} = -6xy, \quad y(0) = 7.$$

Solution Informally, we divide both sides of the differential equation by y and multiply each side by dx to get

$$\frac{dy}{y} = -6x\,dx.$$

Hence

$$\int \frac{dy}{y} = \int (-6x)\,dx;$$
$$\ln|y| = -3x^2 + C.$$

We see from the initial condition $y(0) = 7$ that $y(x)$ is positive near $x = 0$, so we may delete the absolute value symbols:

$$\ln y = -3x^2 + C,$$

and hence

$$y(x) = e^{-3x^2+C} = e^{-3x^2}e^C = Ae^{-3x^2},$$

where $A = e^C$. The condition $y(0) = 7$ yields $A = 7$, so the desired solution is

$$y(x) = 7e^{-3x^2}.$$

This is the upper emphasized solution curve shown in Fig. 1.4.1. ∎

FIGURE 1.4.1. Slope field and solution curves for $y' = -6xy$ in Example 1.

Remark: Suppose, instead, that the initial condition in Example 1 had been $y(0) = -4$. Then it would follow that $y(x)$ is *negative* near $x = 0$. We should therefore replace $|y|$ with $-y$ in the integrated equation $\ln|y| = -3x^2 + C$ to obtain

$$\ln(-y) = -3x^2 + C.$$

The initial condition then yields $C = \ln 4$, so $\ln(-y) = -3x^2 + \ln 4$, and hence

$$y(x) = -4e^{-3x^2}.$$

This is the lower emphasized solution curve in Fig. 1.4.1. ∎

Example 2

Solve the differential equation

$$\frac{dy}{dx} = \frac{4 - 2x}{3y^2 - 5}.$$ (5)

Solution

When we separate the variables and integrate both sides, we get

$$\int (3y^2 - 5)\,dy = \int (4 - 2x)\,dx;$$

$$y^3 - 5y = 4x - x^2 + C.$$ (6)

This equation is not readily solved for y as an explicit function of x. ∎

As Example 2 illustrates, it may or may not be possible or practical to solve Eq. (4) explicitly for y in terms of x. If not, then we call (4) an *implicit solution* of the differential equation in (2). Thus Eq. (6) gives an implicit solution of the differential equation in (5). Although it is not convenient to solve Eq. (6) explicitly in terms of x, we see that each solution curve $y = y(x)$ lies on a contour (or level) curve where the function

$$H(x, y) = x^2 - 4x + y^3 - 5y$$

is constant. Figure 1.4.2 shows several of these contour curves.

FIGURE 1.4.2. Slope field and solution curves for $y' = (4 - 2x)/(3y^2 - 5)$ in Example 2.

Example 3

To solve the initial value problem

$$\frac{dy}{dx} = \frac{4 - 2x}{3y^2 - 5}, \quad y(1) = 3,$$ (7)

we substitute $x = 1$ and $y = 3$ in Eq. (6) and get $C = 9$. Thus the desired particular solution $y(x)$ is defined implicitly by the equation

$$y^3 - 5y = 4x - x^2 + 9.$$ (8)

The corresponding solution curve $y = y(x)$ lies on the upper contour curve in Fig. 1.4.2—the one passing through $(1, 3)$. Because the graph of a differentiable solution cannot have a vertical tangent line anywhere, it appears from the figure that this particular solution is defined on the interval $(-1, 5)$ but not on the interval $(-3, 7)$. ∎

FIGURE 1.4.3. Graph of $f(y) = y^3 - 5y - 9$.

Remark 1: When a specific value of x is substituted in Eq. (8), we can attempt to solve numerically for y. For instance, $x = 4$ yields the equation

$$f(y) = y^3 - 5y - 9 = 0.$$

Figure 1.4.3 shows the graph of f. With a graphing calculator we can solve for the single real root $y \approx 2.8552$. This yields the value $y(4) \approx 2.8552$ of the particular solution in Example 3.

Remark 2: If the initial condition in (7) is replaced with the condition $y(1) = 0$, then the resulting particular solution of the differential equation in (5) lies on the lower "half" of the oval contour curve in Fig. 1.4.2. It appears that this particular solution through $(1, 0)$ is defined on the interval $(0, 4)$ but not on the

interval $(-1, 5)$. On the other hand, with the initial condition $y(1) = -2$ we get the lower contour curve in Fig. 1.4.2. This particular solution is defined for all x. Thus the initial condition can determine whether a particular solution is defined on the whole real line or only on some bounded interval. With a computer algebra system one can readily calculate a table of values of the y-solutions of Eq. (8) for x-values at desired increments from $x = -1$ to $x = 5$ (for instance). Such a table of values serves effectively as a "numerical solution" of the initial value problem in (7). ■

Implicit Solutions and Singular Solutions

In general, the equation $K(x, y) = 0$ is called an **implicit solution** of a differential equation if it is satisfied (on some interval) by some solution $y = y(x)$ of the differential equation. But note that a particular solution $y = y(x)$ of $K(x, y) = 0$ may or may not satisfy a given initial condition. For example, differentiation of $x^2 + y^2 = 4$ yields

$$x + y\frac{dy}{dx} = 0,$$

so $x^2 + y^2 = 4$ is an implicit solution of the differential equation $x + yy' = 0$. But only the first of the two explicit solutions

$$y(x) = +\sqrt{4 - x^2} \quad \text{and} \quad y(x) = -\sqrt{4 - x^2}$$

satisfies the initial condition $y(0) = 2$ (Fig. 1.4.4).

The argument preceding Example 1 shows that every particular solution of Eq. (2) satisfies Eq. (4) for some choice of C; *this* is why it is appropriate to call (4) the **general solution** of Eq. (2).

Warning: Suppose, however, that we begin with the differential equation

$$\frac{dy}{dx} = g(x)h(y) \tag{9}$$

and divide by $h(y)$ to obtain the separated equation

$$\frac{1}{h(y)}\frac{dy}{dx} = g(x). \tag{10}$$

If y_0 is a root of the equation $h(y) = 0$—that is, if $h(y_0) = 0$—then the constant function $y(x) \equiv y_0$ is clearly a solution of Eq. (9), but may *not* be contained in the general solution of Eq. (10). Thus solutions of a differential equation may be lost upon division by a vanishing factor. (Indeed, false solutions may be gained upon multiplication by a vanishing factor. This phenomenon is similar to the introduction of extraneous roots in solving algebraic equations.)

In Section 1.5 we shall see that every particular solution of a *linear* first-order differential equation is contained in its general solution. By contrast, it is common for a nonlinear first-order differential equation to have both a general solution involving an arbitrary constant C and one or several particular solutions that cannot be obtained by selecting a value for C. These exceptional solutions are frequently called **singular solutions**. In Problem 30 we ask you to show that the general solution of the differential equation $(y')^2 = 4y$ yields the family of parabolas $y = (x - C)^2$ illustrated in Fig. 1.4.5, and to observe that the constant-valued function $y(x) \equiv 0$ is a singular solution that cannot be obtained from the general solution by any choice of the arbitrary constant C.

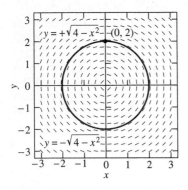

FIGURE 1.4.4. Slope field and solution curves for $y' = -x/y$.

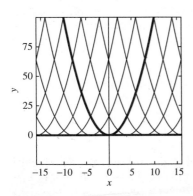

FIGURE 1.4.5. The general solution curves $y = (x - C)^2$ and the singular solution curve $y = 0$ of the differential equation $(y')^2 = 4y$.

Example 4 Find all solutions of the differential equation

$$\frac{dy}{dx} = 6x(y-1)^{2/3}.$$

Solution Separation of variables gives

$$\int \frac{1}{3(y-1)^{2/3}}\, dy = \int 2x\, dx;$$

$$(y-1)^{1/3} = x^2 + C;$$

$$y(x) = 1 + (x^2 + C)^3.$$

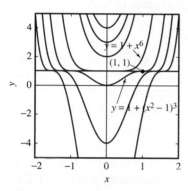

FIGURE 1.4.6. General and singular solution curves for $y' = 6x(y-1)^{2/3}$.

Positive values of the arbitrary constant C give the solution curves in Fig. 1.4.6 that lie above the line $y = 1$, whereas negative values yield those that dip below it. The value $C = 0$ gives the solution $y(x) = 1 + x^6$, but *no* value of C gives the singular solution $y(x) \equiv 1$ that was lost when the variables were separated. Note that the two different solutions $y(x) \equiv 1$ and $y(x) = 1 + (x^2 - 1)^3$ both satisfy the initial condition $y(1) = 1$. Indeed, the whole singular solution curve $y = 1$ consists of points where the solution is not unique and where the function $f(x, y) = 6x(y-1)^{2/3}$ is not differentiable. ∎

Natural Growth and Decay

The differential equation

$$\frac{dx}{dt} = kx \quad (k \text{ a constant}) \tag{11}$$

serves as a mathematical model for a remarkably wide range of natural phenomena— any involving a quantity whose time rate of change is proportional to its current size. Here are some examples.

POPULATION GROWTH: Suppose that $P(t)$ is the number of individuals in a population (of humans, or insects, or bacteria) having *constant* birth and death rates β and δ (in births or deaths per individual per unit of time). Then, during a short time interval Δt, approximately $\beta P(t)\Delta t$ births and $\delta P(t)\Delta t$ deaths occur, so the change in $P(t)$ is given approximately by

$$\Delta P \approx (\beta - \delta)P(t)\Delta t,$$

and therefore

$$\frac{dP}{dt} = \lim_{\Delta t \to 0} \frac{\Delta P}{\Delta t} = kP, \tag{12}$$

where $k = \beta - \delta$.

COMPOUND INTEREST: Let $A(t)$ be the number of dollars in a savings account at time t (in years), and suppose that the interest is *compounded continuously* at an annual interest rate r. (Note that 10% annual interest means that $r = 0.10$.) Continuous compounding means that during a short time interval Δt, the amount of interest added to the account is approximately $\Delta A = r A(t)\Delta t$, so that

$$\frac{dA}{dt} = \lim_{\Delta t \to 0} \frac{\Delta A}{\Delta t} = rA. \tag{13}$$

RADIOACTIVE DECAY: Consider a sample of material that contains $N(t)$ atoms of a certain radioactive isotope at time t. It has been observed that a constant fraction of those radioactive atoms will spontaneously decay (into atoms of another element or into another isotope of the same element) during each unit of time. Consequently, the sample behaves exactly like a population with a constant death rate and no births. To write a model for $N(t)$, we use Eq. (12) with N in place of P, with $k > 0$ in place of δ, and with $\beta = 0$. We thus get the differential equation

$$\frac{dN}{dt} = -kN. \tag{14}$$

The value of k depends on the particular radioactive isotope.

The key to the method of *radiocarbon dating* is that a constant proportion of the carbon atoms in any living creature is made up of the radioactive isotope ^{14}C of carbon. This proportion remains constant because the fraction of ^{14}C in the atmosphere remains almost constant, and living matter is continuously taking up carbon from the air or is consuming other living matter containing the same constant ratio of ^{14}C atoms to ordinary ^{12}C atoms. This same ratio permeates all life, because organic processes seem to make no distinction between the two isotopes.

The ratio of ^{14}C to normal carbon remains constant in the atmosphere because, although ^{14}C is radioactive and slowly decays, the amount is continuously replenished through the conversion of ^{14}N (ordinary nitrogen) to ^{14}C by cosmic rays bombarding the upper atmosphere. Over the long history of the planet, this decay and replenishment process has come into nearly steady state.

Of course, when a living organism dies, it ceases its metabolism of carbon and the process of radioactive decay begins to deplete its ^{14}C content. There is no replenishment of this ^{14}C, and consequently the ratio of ^{14}C to normal carbon begins to drop. By measuring this ratio, the amount of time elapsed since the death of the organism can be estimated. For such purposes it is necessary to measure the **decay constant** k. For ^{14}C, it is known that $k \approx 0.0001216$ if t is measured in years.

(Matters are not as simple as we have made them appear. In applying the technique of radiocarbon dating, extreme care must be taken to avoid contaminating the sample with organic matter or even with ordinary fresh air. In addition, the cosmic ray levels apparently have not been constant, so the ratio of ^{14}C in the atmosphere has varied over the past centuries. By using independent methods of dating samples, researchers in this area have compiled tables of correction factors to enhance the accuracy of this process.)

DRUG ELIMINATION: In many cases the amount $A(t)$ of a certain drug in the bloodstream, measured by the excess over the natural level of the drug, will decline at a rate proportional to the current excess amount. That is,

$$\frac{dA}{dt} = -\lambda A, \tag{15}$$

where $\lambda > 0$. The parameter λ is called the **elimination constant** of the drug.

The Natural Growth Equation

The prototype differential equation $dx/dt = kx$ with $x(t) > 0$ and k a constant (either negative or positive) is readily solved by separating the variables and inte-

grating:

$$\int \frac{1}{x}\,dx = \int k\,dt;$$

$$\ln x = kt + C.$$

Then we solve for x:

$$e^{\ln x} = e^{kt+C}; \quad x = x(t) = e^C e^{kt} = Ae^{kt}.$$

Because C is a constant, so is $A = e^C$. It is also clear that $A = x(0) = x_0$, so the particular solution of Eq. (11) with the initial condition $x(0) = x_0$ is simply

➤
$$x(t) = x_0 e^{kt}. \tag{16}$$

Because of the presence of the natural exponential function in its solution, the differential equation

➤
$$\frac{dx}{dt} = kx \tag{17}$$

is often called the **exponential** or **natural growth equation.** Figure 1.4.7 shows a typical graph of $x(t)$ in the case $k > 0$; the case $k < 0$ is illustrated in Fig. 1.4.8.

FIGURE 1.4.7. Natural growth.

FIGURE 1.4.8. Natural decay.

Example 5

According to data listed at **www.census.gov**, the world's total population reached 6 billion persons in mid-1999, and was then increasing at the rate of about 212 thousand persons each day. Assuming that natural population growth at this rate continues, we want to answer these questions:

(a) What is the annual growth rate k?

(b) What will be the world population at the middle of the 21st century?

(c) How long will it take the world population to increase tenfold—thereby reaching the 60 billion that some demographers believe to be the maximum for which the planet can provide adequate food supplies?

Solution

(a) We measure the world population $P(t)$ in billions and measure time in years. We take $t = 0$ to correspond to (mid) 1999, so $P_0 = 6$. The fact that P is increasing by 212,000, or 0.000212 billion, persons per day at time $t = 0$ means that

$$P'(0) = (0.000212)(365.25) \approx 0.07743$$

billion per year. From the natural growth equation $P' = kP$ with $t = 0$ we now obtain

$$k = \frac{P'(0)}{P(0)} \approx \frac{0.07743}{6} \approx 0.0129.$$

Thus the world population was growing at the rate of about 1.29% annually in 1999. This value of k gives the world population function

$$P(t) = 6e^{0.0129t}.$$

(b) With $t = 51$ we obtain the prediction

$$P(51) = 6e^{(0.0129)(51)} \approx 11.58 \text{ (billion)}$$

for the world population in mid-2050 (so the population will almost have doubled in the just over a half-century since 1999).

(c) The world population should reach 60 billion when

$$60 = 6e^{0.0129t}; \quad \text{that is, when } t = \frac{\ln 10}{0.0129} \approx 178;$$

and thus in the year 2177. ■

Note: Actually, the rate of growth of the world population is expected to slow somewhat during the next half-century, and the best current prediction for the 2050 population is "only" 9.1 billion. A simple mathematical model cannot be expected to mirror precisely the complexity of the real world.

The decay constant of a radioactive isotope is often specified in terms of another empirical constant, the *half-life* of the isotope, because this parameter is more convenient. The **half-life** τ of a radioactive isotope is the time required for *half* of it to decay. To find the relationship between k and τ, we set $t = \tau$ and $N = \frac{1}{2}N_0$ in the equation $N(t) = N_0 e^{kt}$, so that $\frac{1}{2}N_0 = N_0 e^{k\tau}$. When we solve for τ, we find that

$$\tau = \frac{\ln 2}{k}. \tag{18}$$

For example, the half-life of ^{14}C is $\tau \approx (\ln 2)/(0.0001216)$, approximately 5700 years.

Example 6 A specimen of charcoal found at Stonehenge turns out to contain 63% as much ^{14}C as a sample of present-day charcoal of equal mass. What is the age of the sample?

Solution We take $t = 0$ as the time of the death of the tree from which the Stonehenge charcoal was made and N_0 as the number of ^{14}C atoms that the Stonehenge sample contained then. We are given that $N = (0.63)N_0$ now, so we solve the equation $(0.63)N_0 = N_0 e^{-kt}$ with the value $k = 0.0001216$. Thus we find that

$$t = -\frac{\ln(0.63)}{0.0001216} \approx 3800 \text{ (years)}.$$

Thus the sample is about 3800 years old. If it has any connection with the builders of Stonehenge, our computations suggest that this observatory, monument, or temple—whichever it may be—dates from 1800 B.C. or earlier. ■

Cooling and Heating

According to Newton's law of cooling (Eq. (3) of Section 1.1), the time rate of change of the temperature $T(t)$ of a body immersed in a medium of constant temperature A is proportional to the difference $A - T$. That is,

$$\frac{dT}{dt} = k(A - T), \tag{19}$$

where k is a positive constant. This is an instance of the linear first-order differential equation with constant coefficients:

$$\frac{dx}{dt} = ax + b. \tag{20}$$

It includes the exponential equation as a special case ($b = 0$) and is also easy to solve by separation of variables.

Example 7 A 4-lb roast, initially at $50°$F, is placed in a $375°$F oven at 5:00 P.M. After 75 minutes it is found that the temperature $T(t)$ of the roast is $125°$F. When will the roast be $150°$F (medium rare)?

Solution We take time t in minutes, with $t = 0$ corresponding to 5:00 P.M. We also assume (somewhat unrealistically) that at any instant the temperature $T(t)$ of the roast is uniform throughout. We have $T(t) < A = 375$, $T(0) = 50$, and $T(75) = 125$. Hence

$$\frac{dT}{dt} = k(375 - T);$$

$$\int \frac{1}{375 - T} \, dT = \int k \, dt;$$

$$-\ln(375 - T) = kt + C,$$

$$375 - T = Be^{-kt}.$$

Now $T(0) = 50$ implies that $B = 325$, so $T(t) = 375 - 325e^{-kt}$. We also know that $T = 125$ when $t = 75$. Substitution of these values in the preceding equation yields

$$k = -\tfrac{1}{75} \ln \left(\tfrac{250}{325} \right) \approx 0.0035.$$

Hence we finally solve the equation

$$150 = 375 - 325e^{(-0.0035)t}$$

for $t = -[\ln(225/325)]/(0.0035) \approx 105$ (min), the total cooking time required. Because the roast was placed in the oven at 5:00 P.M., it should be removed at about 6:45 P.M. ∎

Torricelli's Law

Suppose that a water tank has a hole with area a at its bottom, from which water is leaking. Denote by $y(t)$ the depth of water in the tank at time t, and by $V(t)$ the volume of water in the tank then. It is plausible—and true, under ideal conditions— that the velocity of water exiting through the hole is

$$v = \sqrt{2gy}, \tag{21}$$

which is the velocity a drop of water would acquire in falling freely from the surface of the water to the hole (see Problem 35 of Section 1.2). One can derive this formula beginning with the assumption that the sum of the kinetic and potential energy of the system remains constant. Under real conditions, taking into account the constriction of a water jet from an orifice, $v = c\sqrt{2gy}$, where c is an empirical constant between 0 and 1 (usually about 0.6 for a small continuous stream of water). For simplicity we take $c = 1$ in the following discussion.

As a consequence of Eq. (21), we have

➤
$$\frac{dV}{dt} = -av = -a\sqrt{2gy}; \tag{22a}$$

equivalently,

➤
$$\frac{dV}{dt} = -k\sqrt{y} \quad \text{where} \quad k = a\sqrt{2g}. \tag{22b}$$

This is a statement of Torricelli's law for a draining tank. If $A(y)$ denotes the horizontal cross-sectional area of the tank at height y above the hole, the method of volume by cross sections gives

$$V = \int_0^y A(y)\,dy,$$

so the fundamental theorem of calculus implies that $dV/dy = A(y)$ and therefore that

$$\frac{dV}{dt} = \frac{dV}{dy} \cdot \frac{dy}{dt} = A(y)\frac{dy}{dt}. \tag{23}$$

From Eqs. (22) and (23) we finally obtain

$$A(y)\frac{dy}{dt} = -a\sqrt{2gy} = -k\sqrt{y}, \tag{24}$$

an alternative form of Torricelli's law. ∎

Example 8 A hemispherical bowl has top radius 4 ft and at time $t = 0$ is full of water. At that moment a circular hole with diameter 1 in. is opened in the bottom of the tank. How long will it take for all the water to drain from the tank?

Solution From the right triangle in Fig. 1.4.9, we see that

$$A(y) = \pi r^2 = \pi\left[16 - (4 - y)^2\right] = \pi(8y - y^2).$$

With $g = 32$ ft/s^2, Eq. (24) becomes

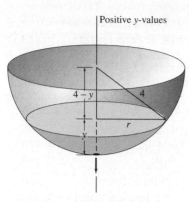

Positive y-values

$4-y$

4

r

y

FIGURE 1.4.9. Draining a hemispherical tank.

$$\pi(8y - y^2)\frac{dy}{dt} = -\pi \left(\tfrac{1}{24}\right)^2 \sqrt{2 \cdot 32y};$$

$$\int (8y^{1/2} - y^{3/2}) \, dy = -\int \tfrac{1}{72} \, dt;$$

$$\tfrac{16}{3}y^{3/2} - \tfrac{2}{5}y^{5/2} = -\tfrac{1}{72}t + C.$$

Now $y(0) = 4$, so

$$C = \tfrac{16}{3} \cdot 4^{3/2} - \tfrac{2}{5} \cdot 4^{5/2} = \tfrac{448}{15}.$$

The tank is empty when $y = 0$, thus when

$$t = 72 \cdot \tfrac{448}{15} \approx 2150 \text{ (s)};$$

that is, about 35 min 50 s. So it takes slightly less than 36 min for the tank to drain. ∎

1.4 Problems

Find general solutions (implicit if necessary, explicit if convenient) of the differential equations in Problems 1 through 18. Primes denote derivatives with respect to x.

1. $\dfrac{dy}{dx} + 2xy = 0$

2. $\dfrac{dy}{dx} + 2xy^2 = 0$

3. $\dfrac{dy}{dx} = y \sin x$

4. $(1 + x)\dfrac{dy}{dx} = 4y$

5. $2\sqrt{x}\,\dfrac{dy}{dx} = \sqrt{1 - y^2}$

6. $\dfrac{dy}{dx} = 3\sqrt{xy}$

7. $\dfrac{dy}{dx} = (64xy)^{1/3}$

8. $\dfrac{dy}{dx} = 2x \sec y$

9. $(1 - x^2)\dfrac{dy}{dx} = 2y$

10. $(1 + x)^2 \dfrac{dy}{dx} = (1 + y)^2$

11. $y' = xy^3$

12. $yy' = x(y^2 + 1)$

13. $y^3 \dfrac{dy}{dx} = (y^4 + 1) \cos x$

14. $\dfrac{dy}{dx} = \dfrac{1 + \sqrt{x}}{1 + \sqrt{y}}$

15. $\dfrac{dy}{dx} = \dfrac{(x - 1)y^5}{x^2(2y^3 - y)}$

16. $(x^2 + 1)(\tan y)y' = x$

17. $y' = 1 + x + y + xy$ (*Suggestion*: Factor the right-hand side.)

18. $x^2y' = 1 - x^2 + y^2 - x^2y^2$

Find explicit particular solutions of the initial value problems in Problems 19 through 28.

19. $\dfrac{dy}{dx} = ye^x, \quad y(0) = 2e$

20. $\dfrac{dy}{dx} = 3x^2(y^2 + 1), \quad y(0) = 1$

21. $2y\dfrac{dy}{dx} = \dfrac{x}{\sqrt{x^2 - 16}}, \quad y(5) = 2$

22. $\dfrac{dy}{dx} = 4x^3y - y, \quad y(1) = -3$

23. $\dfrac{dy}{dx} + 1 = 2y, \quad y(1) = 1$

24. $(\tan x)\dfrac{dy}{dx} = y, \quad y\left(\tfrac{1}{2}\pi\right) = \tfrac{1}{2}\pi$

25. $x\dfrac{dy}{dx} - y = 2x^2y, \quad y(1) = 1$

26. $\dfrac{dy}{dx} = 2xy^2 + 3x^2y^2, \quad y(1) = -1$

27. $\dfrac{dy}{dx} = 6e^{2x - y}, \quad y(0) = 0$

28. $2\sqrt{x}\dfrac{dy}{dx} = \cos^2 y, \quad y(4) = \pi/4$

29. **(a)** Find a general solution of the differential equation $dy/dx = y^2$. **(b)** Find a singular solution that is not included in the general solution. **(c)** Inspect a sketch of typical solution curves to determine the points (a, b) for which the initial value problem $y' = y^2$, $y(a) = b$ has a unique solution.

30. Solve the differential equation $(dy/dx)^2 = 4y$ to verify the general solution curves and singular solution curve that are illustrated in Fig. 1.4.5. Then determine the points (a, b) in the plane for which the initial value problem $(y')^2 = 4y$, $y(a) = b$ has **(a)** no solution, **(b)** infinitely many solutions that are defined for all x, **(c)** on some neighborhood of the point $x = a$, only finitely many solutions.

31. Discuss the difference between the differential equations $(dy/dx)^2 = 4y$ and $dy/dx = 2\sqrt{y}$. Do they have the same solution curves? Why or why not? Determine the points (a, b) in the plane for which the initial value problem $y' = 2\sqrt{y}$, $y(a) = b$ has **(a)** no solution, **(b)** a unique solution, **(c)** infinitely many solutions.

32. Find a general solution and any singular solutions of the differential equation $dy/dx = y\sqrt{y^2 - 1}$. Determine the

points (a, b) in the plane for which the initial value problem $y' = y\sqrt{y^2 - 1}$, $y(a) = b$ has **(a)** no solution, **(c)** a unique solution, **(b)** infinitely many solutions.

33. (Population growth) A certain city had a population of 25000 in 1960 and a population of 30000 in 1970. Assume that its population will continue to grow exponentially at a constant rate. What population can its city planners expect in the year 2000?

34. (Population growth) In a certain culture of bacteria, the number of bacteria increased sixfold in 10 h. How long did it take for the population to double?

35. (Radiocarbon dating) Carbon extracted from an ancient skull contained only one-sixth as much ^{14}C as carbon extracted from present-day bone. How old is the skull?

36. (Radiocarbon dating) Carbon taken from a purported relic of the time of Christ contained 4.6×10^{10} atoms of ^{14}C per gram. Carbon extracted from a present-day specimen of the same substance contained 5.0×10^{10} atoms of ^{14}C per gram. Compute the approximate age of the relic. What is your opinion as to its authenticity?

37. (Continuously compounded interest) Upon the birth of their first child, a couple deposited $5000 in an account that pays 8% interest compounded continuously. The interest payments are allowed to accumulate. How much will the account contain on the child's eighteenth birthday?

38. (Continuously compounded interest) Suppose that you discover in your attic an overdue library book on which your grandfather owed a fine of 30 cents 100 years ago. If an overdue fine grows exponentially at a 5% annual rate compounded continuously, how much would you have to pay if you returned the book today?

39. (Drug elimination) Suppose that sodium pentobarbitol is used to anesthetize a dog. The dog is anesthetized when its bloodstream contains at least 45 milligrams (mg) of sodium pentobarbitol per kilogram of the dog's body weight. Suppose also that sodium pentobarbitol is eliminated exponentially from the dog's bloodstream, with a half-life of 5 h. What single dose should be administered in order to anesthetize a 50-kg dog for 1 h?

40. The half-life of radioactive cobalt is 5.27 years. Suppose that a nuclear accident has left the level of cobalt radiation in a certain region at 100 times the level acceptable for human habitation. How long will it be until the region is again habitable? (Ignore the probable presence of other radioactive isotopes.)

41. Suppose that a mineral body formed in an ancient cataclysm—perhaps the formation of the earth itself—originally contained the uranium isotope ^{238}U (which has a half-life of 4.51×10^9 years) but no lead, the end product of the radioactive decay of ^{238}U. If today the ratio of ^{238}U atoms to lead atoms in the mineral body is 0.9, when did the cataclysm occur?

42. A certain moon rock was found to contain equal numbers of potassium and argon atoms. Assume that all the argon is the result of radioactive decay of potassium (its half-life

is about 1.28×10^9 years) and that one of every nine potassium atom disintegrations yields an argon atom. What is the age of the rock, measured from the time it contained only potassium?

43. A pitcher of buttermilk initially at $25°C$ is to be cooled by setting it on the front porch, where the temperature if $0°C$. Suppose that the temperature of the buttermilk has dropped to $15°C$ after 20 min. When will it be at $5°C$?

44. When sugar is dissolved in water, the amount A that remains undissolved after t minutes satisfies the differential equation $dA/dt = -kA$ ($k > 0$). If 25% of the sugar dissolves after 1 min, how long does it take for half of the sugar to dissolve?

45. The intensity I of light at a depth of x meters below the surface of a lake satisfies the differential equation $dI/dx = (-1.4)I$. **(a)** At what depth is the intensity half the intensity I_0 at the surface (where $x = 0$)? **(b)** What is the intensity at a depth of 10 m (as a fraction of I_0)? **(c)** At what depth will the intensity be 1% of that at the surface?

46. The barometric pressure p (in inches of mercury) at an altitude x miles above sea level satisfies the initial value problem $dp/dx = (-0.2)p$, $p(0) = 29.92$. **(a)** Calculate the barometric pressure at 10000 ft and again at 30000 ft. **(b)** Without prior conditioning, few people can survive when the pressure drops to less than 15 in. of mercury. How high is that?

47. A certain piece of dubious information about phenylethylamine in the drinking water began to spread one day in a city with a population of 100,000. Within a week, 10,000 people had heard this rumor. Assume that the rate of increase of the number who have heard the rumor is proportional to the number who have not yet heard it. How long will it be until half the population of the city has heard the rumor?

48. According to one cosmological theory, there were equal amounts of the two uranium isotopes ^{235}U and ^{238}U at the creation of the universe in the "big bang." At present there are 137.7 atoms of ^{238}U for each atom of ^{235}U. Using the half-lives 4.51×10^9 years for ^{238}U and 7.10×10^8 years for ^{235}U, calculate the age of the universe.

49. A cake is removed from an oven at $210°F$ and left to cool at room temperature, which is $70°F$. After 30 min the temperature of the cake is $140°F$. When will it be $100°F$?

50. The amount $A(t)$ of atmospheric pollutants in a certain mountain valley grows naturally and is tripling every 7.5 years.

 (a) If the initial amount is 10 pu (pollutant units), write a formula for $A(t)$ giving the amount (in pu) present after t years.

 (b) What will be the amount (in pu) of pollutants present in the valley atmosphere after 5 years?

 (c) If it will be dangerous to stay in the valley when the amount of pollutants reaches 100 pu, how long will this take?

51. An accident at a nuclear power plant has left the surrounding area polluted with radioactive material that decays naturally. The initial amount of radioactive material present is 15 su (safe units), and 5 months later it is still 10 su.

(a) Write a formula giving the amount $A(t)$ of radioactive material (in su) remaining after t months.

(b) What amount of radioactive material will remain after 8 months?

(c) How long—total number of months or fraction thereof—will it be until $A = 1$ su, so it is safe for people to return to the area?

52. There are now about 3300 different human "language families" in the whole world. Assume that all these are derived from a single original language, and that a language family develops into 1.5 language families every 6 thousand years. About how long ago was the single original human language spoken?

53. Thousands of years ago ancestors of the Native Americans crossed the Bering Strait from Asia and entered the western hemisphere. Since then, they have fanned out across North and South America. The single language that the original Native Americans spoke has since split into many Indian "language families." Assume (as in Problem 52) that the number of these language families has been multiplied by 1.5 every 6000 years. There are now 150 Native American language families in the western hemisphere. About when did the ancestors of today's Native Americans arrive?

54. A tank is shaped like a vertical cylinder; it initially contains water to a depth of 9 ft, and a bottom plug is removed at time $t = 0$ (hours). After 1 h the depth of the water has dropped to 4 ft. How long does it take for all the water to drain from the tank?

55. Suppose that the tank of Problem 48 has a radius of 3 ft and that its bottom hole is circular with radius 1 in. How long will it take the water (initially 9 ft deep) to drain completely?

56. At time $t = 0$ the bottom plug (at the vertex) of a full conical water tank 16 ft high is removed. After 1 h the water in the tank is 9 ft deep. When will the tank be empty?

57. Suppose that a cylindrical tank initially containing V_0 gallons of water drains (through a bottom hole) in T minutes. Use Torricelli's law to show that the volume of water in the tank after $t \leq T$ minutes is $V = V_0 [1 - (t/T)]^2$.

58. A water tank has the shape obtained by revolving the curve $y = x^{4/3}$ around the y-axis. A plug at the bottom is removed at 12 noon, when the depth of water in the tank is 12 ft. At 1 P.M. the depth of the water is 6 ft. When will the tank be empty?

59. A water tank has the shape obtained by revolving the parabola $x^2 = by$ around the y-axis. The water depth is 4 ft at 12 noon, when a circular plug in the bottom of the tank is removed. At 1 P.M. the depth of the water is 1 ft. (a) Find the depth $y(t)$ of water remaining after t hours. (b) When will the tank be empty? (c) If the initial radius

of the top surface of the water is 2 ft, what is the radius of the circular hole in the bottom?

60. A cylindrical tank with length 5 ft and radius 3 ft is situated with its axis horizontal. If a circular bottom hole with a radius of 1 in. is opened and the tank is initially half full of xylene, how long will it take for the liquid to drain completely?

61. A spherical tank of radius 4 ft is full of gasoline when a circular bottom hole with radius 1 in. is opened. How long will be required for all the gasoline to drain from the tank?

62. Suppose that an initially full hemispherical water tank of radius 1 m has its flat side as its bottom. It has a bottom hole of radius 1 cm. If this bottom hole is opened at 1 P.M., when will the tank be empty?

63. Consider the initially full hemispherical water tank of Example 8, except that the radius r of its circular bottom hole is now unknown. At 1 P.M. the bottom hole is opened and at 1:30 P.M. the depth of water in the tank is 2 ft. (a) Use Torricelli's law in the form $dV/dt = -(0.6)\pi r^2 \sqrt{2gy}$ (taking constriction into account) to determine when the tank will be empty. (b) What is the radius of the bottom hole?

64. (The *clepsydra*, or water clock) A 12-h water clock is to be designed with the dimensions shown in Fig. 1.4.10, shaped like the surface obtained by revolving the curve $y = f(x)$ around the y-axis. What should be this curve, and what should be the radius of the circular bottom hole, in order that the water level will fall at the *constant* rate of 4 inches per hour (in./h)?

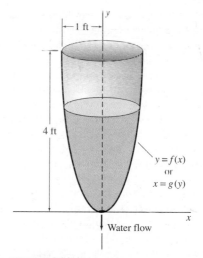

FIGURE 1.4.10. The clepsydra.

65. Just before midday the body of an apparent homicide victim is found in a room that is kept at a constant temperature of 70°F. At 12 noon the temperature of the body is 80°F and at 1 P.M. it is 75°F. Assume that the temperature of the body at the time of death was 98.6°F and that it has cooled in accord with Newton's law. What was the time of death?

66. Early one morning it began to snow at a constant rate. At 7 A.M. a snowplow set off to clear a road. By 8 A.M. it had traveled 2 miles, but it took two more hours (until 10 A.M.) for the snowplow to go an additional 2 miles. (a) Let $t = 0$ when it began to snow and let x denote the distance traveled by the snowplow at time t. Assuming that the snowplow clears snow from the road at a constant rate (in cubic feet per hour, say), show that

$$k\frac{dx}{dt} = \frac{1}{t}$$

where k is a constant. (b) What time did it start snowing? (*Answer*: 6 A.M.)

67. A snowplow sets off at 7 A.M. as in Problem 60. Suppose now that by 8 A.M. it had traveled 4 miles and that by 9 A.M. it had moved an additional 3 miles. What time did it start snowing? This is a more difficult snowplow problem because now a transcendental equation must be solved numerically to find the value of k. (*Answer*: 4:27 A.M.)

FIGURE 1.4.11. A bead sliding down
a wire—the brachistochrone problem.

68. Figure 1.4.11 shows a bead sliding down a frictionless wire from point P to point Q. The *brachistochrone problem* asks what shape the wire should be in order to minimize the bead's time of descent from P to Q. In June of 1696, John Bernoulli proposed this problem as a public challenge, with a 6-month deadline (later extended to Easter 1697 at George Leibniz's request). Isaac Newton, then retired from academic life and serving as Warden of the Mint in London, received Bernoulli's challenge on January 29, 1697. The very next day he communicated his own solution—the curve of minimal descent time is an arc of an inverted cycloid—to the Royal Society of London. For a modern derivation of this result, suppose the bead starts from rest at the origin P and let $y = y(x)$ be the equation of the desired curve in a coordinate system with the y-axis pointing downward. Then a mechanical analogue of Snell's law in optics implies that

$$\frac{\sin \alpha}{v} = \text{constant}, \tag{i}$$

where α denotes the angle of deflection (from the vertical) of the tangent line to the curve—so $\cot \alpha = y'(x)$ (why?)—and $v = \sqrt{2gy}$ is the bead's velocity when it has descended a distance y vertically (from KE $= \frac{1}{2}mv^2 = mgy = -$PE).

(a) First derive from Eq. (i) the differential equation

$$\frac{dy}{dx} = \sqrt{\frac{2a - y}{y}} \tag{ii}$$

where a is an appropriate positive constant.

(b) Substitute $y = 2a \sin^2 t$, $dy = 4a \sin t \cos t\, dt$ in (ii) to derive the solution

$$x = a(2t - \sin 2t), \quad y = a(1 - \cos 2t) \tag{iii}$$

for which $t = y = 0$ when $x = 0$. Finally, the substitution of $\theta = 2a$ in (iii) yields the standard parametric equations $x = a(\theta - \sin\theta)$, $y = a(1 - \cos\theta)$ of the cycloid that is generated by a point on the rim of a circular wheel of radius a as it rolls along the x-axis. [See Example 5 in Section 10.4 of Edwards and Penney, *Calculus*, 6th edition (Upper Saddle River, NJ: Prentice Hall, 2002).]

69. Suppose a uniform flexible cable is suspended between two points $(\pm L, H)$ at equal heights located symmetrically on either side of the x-axis (Fig. 1.4.12). Principles of physics can be used to show that the shape $y = y(x)$ of the hanging cable satisfies the differential equation

$$a\frac{d^2 y}{dx^2} = \sqrt{1 + \left(\frac{dy}{dx}\right)^2},$$

where the constant $a = T/\rho$ is the ratio of the cable's tension T at its lowest point $x = 0$ (where $y'(0) = 0$) and its (constant) linear density ρ. If we substitute $v = dy/dx$, $dv/dx = d^2y/dx^2$ in this second-order differential equation, we get the first-order equation

$$a\frac{dv}{dx} = \sqrt{1 + v^2}.$$

Solve this differential equation for $y'(x) = v(x) = \sinh(x/a)$. Then integrate to get the shape function

$$y(x) = a\cosh\left(\frac{x}{a}\right) + C$$

of the hanging cable. This curve is called a *catenary*, from the Latin word for *chain*.

FIGURE 1.4.12. The catenary.

1.4 **Application: The Logistic Equation**

As in Eq. (3) of this section, the solution of a separable differential equation reduces to the evaluation of two indefinite integrals. It is tempting to use a symbolic algebra system for this purpose. We illustrate this approach using the *logistic differential equation*

$$\frac{dx}{dt} = ax - bx^2 \tag{1}$$

that models a population $x(t)$ with births (per unit time) proportional to x and deaths proportional to x^2. Here we concentrate on the solution of Eq. (1) and defer discussion of population applications to Section 2.1.

If $a = 0.01$ and $b = 0.0001$, for instance, Eq. (1) is

$$\frac{dx}{dt} = (0.01)x - (0.0001)x^2 = \frac{x}{10000}(100 - x). \tag{2}$$

Separation of variables leads to

$$\int \frac{1}{x(100 - x)}\, dx = \int \frac{1}{10000}\, dt = \frac{t}{10000} + C. \tag{3}$$

FIGURE 1.4.13. TI-89 screen showing the integral in Eq. (3).

We can evaluate the integral on the left by using the *Maple* command

 int(1/(x*(100 - x)), x);

the *Mathematica* command

 Integrate[1/(x*(100 - x)), x]

or the MATLAB command

 syms x; int(1/(x*(100 - x)))

Any computer algebra system gives a result of the form

$$\frac{1}{100}\ln x - \frac{1}{100}\ln(x - 100) = \frac{t}{10000} + C \tag{4}$$

equivalent to the graphing calculator result shown in Fig. 1.4.13.

You can now apply the initial condition $x(0) = x_0$, combine logarithms, and finally exponentiate to solve Eq. (4) for the particular solution

$$x(t) = \frac{100 x_0 e^{t/100}}{100 - x_0 + x_0 e^{t/100}} \tag{5}$$

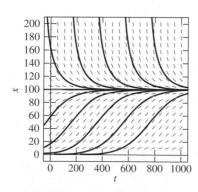

FIGURE 1.4.14. Slope field and solution curves for $x' = (0.01)x - (0.0001)x^2$.

of Eq. (2). The slope field and solution curves shown in Fig. 1.4.14 suggest that, whatever is the initial value x_0, the solution $x(t)$ approaches 100 as $t \to +\infty$. Can you use Eq. (5) to verify this conjecture?

INVESTIGATION: For your own personal logistic equation, take $a = m/n$ and $b = 1/n$ in Eq. (1), with m and n being the *largest* two distinct digits (in either order) in your student ID number.

(a) First generate a slope field for your differential equation and include a sufficient number of solution curves that you can see what happens to the population as $t \to +\infty$. State your inference plainly.

(b) Next use a computer algebra system to solve the differential equation symbolically; then use the symbolic solution to find the limit of $x(t)$ as $t \to +\infty$. Was your graphically-based inference correct?

(c) Finally, state and solve a numerical problem using the symbolic solution. For instance, how long does it take x to grow from a selected initial value x_0 to a given target value x_1?

1.5 Linear First-Order Equations

In Section 1.4 we saw how to solve a separable differential equation by integrating *after* multiplying both sides by an appropriate factor. For instance, to solve the equation

$$\frac{dy}{dx} = 2xy \quad (y > 0), \tag{1}$$

we multiply both sides by the factor $1/y$ to get

$$\frac{1}{y} \cdot \frac{dy}{dx} = 2x; \quad \text{that is,} \quad D_x(\ln y) = D_x\left(x^2\right). \tag{2}$$

Because each side of the equation in (2) is recognizable as a *derivative* (with respect to the independent variable x), all that remains are two simple integrations, which yield $\ln y = x^2 + C$. For this reason, the function $\rho(y) = 1/y$ is called an *integrating factor* for the original equation in (1). An **integrating factor** for a differential equation is a function $\rho(x, y)$ such that the multiplication of each side of the differential equation by $\rho(x, y)$ yields an equation in which each side is recognizable as a derivative.

With the aid of the appropriate integrating factor, there is a standard technique for solving the **linear first-order equation**

$$\blacktriangleright \qquad \frac{dy}{dx} + P(x)y = Q(x) \tag{3}$$

on an interval on which the coefficient functions $P(x)$ and $Q(x)$ are continuous. We multiply each side in Eq. (3) by the integrating factor

$$\blacktriangleright \qquad \rho(x) = e^{\int P(x)\,dx}. \tag{4}$$

The result is

$$e^{\int P(x)\,dx}\frac{dy}{dx} + P(x)e^{\int P(x)\,dx}y = Q(x)e^{\int P(x)\,dx}. \tag{5}$$

Because

$$D_x\left[\int P(x)\,dx\right] = P(x),$$

the left-hand side is the derivative of the *product* $y(x) \cdot e^{\int P(x)\,dx}$, so Eq. (5) is equivalent to

$$D_x\left[y(x) \cdot e^{\int P(x)\,dx}\right] = Q(x)e^{\int P(x)\,dx}.$$

Integration of both sides of this equation gives

$$y(x)e^{\int P(x)\,dx} = \int \left(Q(x)e^{\int P(x)\,dx}\right)dx + C.$$

Finally, solving for y, we obtain the general solution of the linear first-order equation in (3):

$$y(x) = e^{-\int P(x)\,dx}\left[\int \left(Q(x)e^{\int P(x)\,dx}\right)dx + C\right]. \tag{6}$$

This formula should **not** be memorized. In a specific problem it generally is simpler to use the *method* by which we developed the formula. That is, in order to solve an equation that can be written in the form in Eq. (3) with the coefficient functions $P(x)$ and $Q(x)$ displayed explicitly, you should attempt to carry out the following steps.

METHOD: SOLUTION OF FIRST-ORDER EQUATIONS

1. Begin by calculating the integrating factor $\rho(x) = e^{\int P(x)\,dx}$.
2. Then multiply both sides of the differential equation by $\rho(x)$.
3. Next, recognize the left-hand side of the resulting equation as the derivative of a product:
$$D_x\left[\rho(x)y(x)\right] = \rho(x)Q(x).$$
4. Finally, integrate this equation,
$$\rho(x)y(x) = \int \rho(x)Q(x)\,dx + C,$$

then solve for y to obtain the general solution of the original differential equation.

Remark 1: Given an initial condition $y(x_0) = y_0$, you can (as usual) substitute $x = x_0$ and $y = y_0$ into the general solution and solve for the value of C yielding the particular solution that satisfies this initial condition.

Remark 2: You need not supply explicitly a constant of integration when you find the integrating factor $\rho(x)$. For if we replace

$$\int P(x)\,dx \quad \text{with} \quad \int P(x)\,dx + K$$

in Eq. (4), the result is

$$\rho(x) = e^{K+\int P(x)\,dx} = e^K e^{\int P(x)\,dx}.$$

But the constant factor e^K does not affect materially the result of multiplying both sides of the differential equation in (3) by $\rho(x)$, so we might as well take $K = 0$. You may therefore choose for $\int P(x)\,dx$ *any* convenient antiderivative of $P(x)$, without bothering to add a constant of integration. ■

Example 1 Solve the initial value problem

$$\frac{dy}{dx} - y = \tfrac{11}{8}e^{-x/3}, \quad y(0) = -1.$$

Solution Here we have $P(x) \equiv -1$ and $Q(x) = \tfrac{11}{8}e^{-x/3}$, so the integrating factor is

$$\rho(x) = e^{\int(-1)\,dx} = e^{-x}.$$

Multiplication of both sides of the given equation by e^{-x} yields

$$e^{-x}\frac{dy}{dx} - e^{-x}y = \tfrac{11}{8}e^{-4x/3}, \tag{7}$$

which we recognize as

$$\frac{d}{dx}\left(e^{-x}y\right) = \tfrac{11}{8}e^{-4x/3}.$$

Hence integration with respect to x gives

$$e^{-x}y = \int \tfrac{11}{8}e^{-4x/3}\,dx = -\tfrac{33}{32}e^{-4x/3} + C,$$

and multiplication by e^x gives the general solution

$$y(x) = Ce^x - \tfrac{33}{32}e^{-x/3}. \tag{8}$$

Substitution of $x = 0$ and $y = -1$ now gives $C = \tfrac{1}{32}$, so the desired particular solution is

$$y(x) = \tfrac{1}{32}e^x - \tfrac{33}{32}e^{-x/3} = \tfrac{1}{32}\left(e^x - 33e^{-x/3}\right). \qquad\blacksquare$$

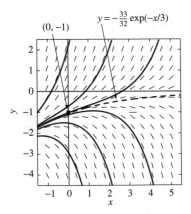

FIGURE 1.5.1. Slope field and solution curves for $y' = y + \tfrac{11}{8}e^{-x/3}$.

Remark: Figure 1.5.1 shows a slope field and typical solution curves for Eq. (7), including the one passing through the point $(0, -1)$. Note that some solutions grow rapidly in the positive direction as x increases, while others grow rapidly in the negative direction. The behavior of a given solution curve is determined by its initial condition $y(0) = y_0$. The two types of behavior are separated by the particular solution $y(x) = -\tfrac{33}{32}e^{-x/3}$ for which $C = 0$ in Eq. (8), so $y_0 = -\tfrac{33}{32}$ for the solution curve that is dashed in Fig. 1.5.1. If $y_0 > -\tfrac{33}{32}$, then $C > 0$ in Eq. (8), so the term e^x eventually dominates the behavior of $y(x)$, and hence $y(x) \to +\infty$ as $x \to +\infty$. But if $y_0 < -\tfrac{33}{32}$, then $C < 0$, so both terms in $y(x)$ are negative and therefore $y(x) \to -\infty$ as $x \to +\infty$. Thus the initial condition $y_0 = -\tfrac{33}{32}$ is *critical* in the sense that solutions that start above $-\tfrac{33}{32}$ on the y-axis grow in the positive direction, while solutions that start lower than $-\tfrac{33}{32}$ grow in the negative direction as $x \to +\infty$. The interpretation of a mathematical model often hinges on finding such a critical condition that separates one kind of behavior of a solution from a different kind of behavior. \blacksquare

Example 2 Find a general solution of

$$(x^2 + 1)\frac{dy}{dx} + 3xy = 6x. \tag{9}$$

Solution After division of both sides of the equation by $x^2 + 1$, we recognize the result

$$\frac{dy}{dx} + \frac{3x}{x^2+1}y = \frac{6x}{x^2+1}$$

as a first-order linear equation with $P(x) = 3x/(x^2+1)$ and $Q(x) = 6x/(x^2+1)$. Multiplication by

$$\rho(x) = \exp\left(\int \frac{3x}{x^2+1}\,dx\right) = \exp\left(\tfrac{3}{2}\ln(x^2+1)\right) = (x^2+1)^{3/2}$$

yields

$$(x^2+1)^{3/2}\frac{dy}{dx} + 3x(x^2+1)^{1/2}y = 6x(x^2+1)^{1/2},$$

and thus

$$D_x\left[(x^2+1)^{3/2}y\right] = 6x(x^2+1)^{1/2}.$$

Integration then yields

$$(x^2+1)^{3/2}y = \int 6x(x^2+1)^{1/2}\,dx = 2(x^2+1)^{3/2} + C.$$

Multiplication of both sides by $(x^2+1)^{-3/2}$ gives the general solution

$$y(x) = 2 + C(x^2+1)^{-3/2}. \tag{10}$$

∎

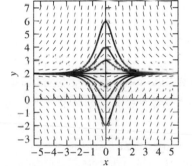

FIGURE 1.5.2. Slope field and solution curves for the differential equation in Eq. (9).

Remark: Figure 1.5.2 shows a slope field and typical solution curves for Eq. (9). Note that, as $x \to +\infty$, all other solution curves approach the constant solution curve $y(x) \equiv 2$ that corresponds to $C = 0$ in Eq. (10). This constant solution can be described as an *equilibrium solution* of the differential equation, because $y(0) = 2$ implies that $y(x) = 2$ for all x (and thus the value of the solution remains forever where it starts). ∎

A Closer Look at the Method

The preceding derivation of the solution in Eq. (6) of the linear first-order equation $y' + Py = Q$ bears closer examination. Suppose that the coefficient functions $P(x)$ and $Q(x)$ are continuous on the (possibly unbounded) open interval I. Then the antiderivatives

$$\int P(x)\,dx \quad \text{and} \quad \int \left(Q(x)e^{\int P(x)\,dx}\right)dx$$

exist on I. Our derivation of Eq. (6) shows that if $y = y(x)$ is a solution of Eq. (3) on I, *then* $y(x)$ is given by the formula in Eq. (6) for some choice of the constant C. Conversely, you may verify by direct substitution (Problem 31) that the function $y(x)$ given in Eq. (6) satisfies Eq. (3). Finally, given a point x_0 of I and any number y_0, there is—as previously noted—a unique value of C such that $y(x_0) = y_0$. Consequently, we have proved the following existence-uniqueness theorem.

> **THEOREM 1:** The Linear First-Order Equation

If the functions $P(x)$ and $Q(x)$ are continuous on the open interval I containing the point x_0, then the initial value problem

$$\frac{dy}{dx} + P(x)y = Q(x), \quad y(x_0) = y_0 \tag{11}$$

has a unique solution $y(x)$ on I, given by the formula in Eq. (6) with an appropriate value of C. ▲

Remark 1: Theorem 1 gives a solution on the *entire* interval I for a *linear* differential equation, in contrast with Theorem 1 of Section 1.3, which guarantees only a solution on a possibly smaller interval.

Remark 2: Theorem 1 tells us that every solution of Eq. (3) is included in the general solution given in Eq. (6). Thus a *linear* first-order differential equation has *no* singular solutions.

Remark 3: The appropriate value of the constant C in Eq. (6)—as needed to solve the initial value problem in Eq. (11)—can be selected "automatically" by writing

$$\rho(x) = \exp\left(\int_{x_0}^{x} P(t)\,dt\right),$$

$$y(x) = \frac{1}{\rho(x)}\left[y_0 + \int_{x_0}^{x} \rho(t)Q(t)\,dt\right]. \tag{12}$$

The indicated limits x_0 and x effect a choice of indefinite integrals in Eq. (6) that guarantees in advance that $\rho(x_0) = 1$ and that $y(x_0) = y_0$ (as you can verify directly by substituting $x = x_0$ in Eq. (12)). ∎

Example 3 Solve the initial value problem

$$x^2\frac{dy}{dx} + xy = \sin x, \quad y(1) = y_0. \tag{13}$$

Solution Division by x^2 gives the linear first-order equation

$$\frac{dy}{dx} + \frac{1}{x}y = \frac{\sin x}{x^2}$$

with $P(x) = 1/x$ and $Q(x) = (\sin x)/x^2$. With $x_0 = 1$ the integrating factor in (12) is

$$\rho(x) = \exp\left(\int_1^x \frac{1}{t}\,dt\right) = \exp(\ln x) = x,$$

so the desired particular solution is given by

$$y(x) = \frac{1}{x}\left[y_0 + \int_1^x \frac{\sin t}{t}\,dt\right]. \tag{14}$$

In accord with Theorem 1, this solution is defined on the whole positive x-axis. ∎

Comment: In general, an integral such as the one in Eq. (14) would (for given x) need to be approximated numerically—using Simpson's rule, for instance—to find the value $y(x)$ of the solution at x. In this case, however, we have the sine integral function

$$\text{Si}(x) = \int_0^x \frac{\sin t}{t}\, dt,$$

which appears with sufficient frequency in applications that its values have been tabulated. A good set of tables of special functions is Abramowitz and Stegun, *Handbook of Mathematical Functions* (New York: Dover, 1965). Then the particular solution in Eq. (14) reduces to

$$y(x) = \frac{1}{x}\left[y_0 + \int_0^x \frac{\sin t}{t}\, dt - \int_0^1 \frac{\sin t}{t}\, dt \right] = \frac{1}{x}\left[y_0 + \text{Si}(x) - \text{Si}(1) \right]. \quad (15)$$

FIGURE 1.5.3. Typical solution curves defined by Eq. (15).

The sine integral function is available in most scientific computing systems and can be used to plot typical solution curves defined by Eq. (15). Figure 1.5.3 shows a selection of solution curves with initial values $y(1) = y_0$ ranging from $y_0 = -3$ to $y_0 = 3$. It appears that on each solution curve, $y(x) \to 0$ as $x \to +\infty$, and this is in fact true because the sine integral function is bounded. ∎

In the sequel we will see that it is the exception—rather than the rule—when a solution of a differential equation can be expressed in terms of elementary functions. We will study various devices for obtaining good approximations to the values of the nonelementary functions we encounter. In Chapter 2 we will discuss numerical integration of differential equations in some detail.

Mixture Problems

As a first application of linear first-order equations, we consider a tank containing a solution—a mixture of solute and solvent—such as salt dissolved in water. There is both inflow and outflow, and we want to compute the *amount* $x(t)$ of solute in the tank at time t, given the amount $x(0) = x_0$ at time $t = 0$. Suppose that solution with a concentration of c_i grams of solute per liter of solution flows into the tank at the constant rate of r_i liters per second, and that the solution in the tank—kept thoroughly mixed by stirring—flows out at the constant rate of r_o liters per second.

To set up a differential equation for $x(t)$, we estimate the change Δx in x during the brief time interval $[t, t + \Delta t]$. The amount of solute that flows into the tank during Δt seconds is $r_i c_i \Delta t$ grams. To check this, note how the cancellation of dimensions checks our computations:

Input: r_i L/s, c_i g/L

Amount $x(t)$
Volume $V(t)$
Concentration $c_o(t) = \frac{x}{V}$

Output:
r_o L/s,
c_o g/L

FIGURE 1.5.4. The single-tank mixture problem.

$$\left(r_i\, \frac{\text{liters}}{\text{second}} \right) \left(c_i\, \frac{\text{grams}}{\text{liter}} \right) (\Delta t\ \text{seconds})$$

yields a quantity measured in grams.

The amount of solute that flows out of the tank during the same time interval depends on the concentration $c_o(t)$ of solute in the solution at time t. But as noted in Fig. 1.5.4, $c_o(t) = x(t)/V(t)$, where $V(t)$ denotes the volume (not constant unless $r_i = r_o$) of solution in the tank at time t. Then

$$\Delta x = \{\text{grams input}\} - \{\text{grams output}\} \approx r_i c_i \Delta t - r_o c_o \Delta t.$$

We now divide by Δt:

$$\frac{\Delta x}{\Delta t} \approx r_i c_i - r_o c_o.$$

Finally, we take the limit as $\Delta t \to 0$; if all the functions involved are continuous and $x(t)$ is differentiable, then the error in this approximation also approaches zero, and we obtain the differential equation

$$\blacktriangleright \qquad \frac{dx}{dt} = r_i c_i - r_o c_o, \qquad (16)$$

in which r_i, c_i, and r_o are constants, but c_o denotes the variable concentration

$$c_o(t) = \frac{x(t)}{V(t)} \qquad (17)$$

of solute in the tank at time t. Thus the amount $x(t)$ of solute in the tank satisfies the differential equation

$$\frac{dx}{dt} = r_i c_i - \frac{r_o}{V} x. \qquad (18)$$

If $V_0 = V(0)$, then $V(t) = V_0 + (r_i - r_o)t$, so Eq. (18) is a linear first-order differential equation for the amount $x(t)$ of solute in the tank at time t.

Important: Equation (18) need not be committed to memory. It is the process we used to obtain that equation—examination of the behavior of the system over a short time interval $[t, t + \Delta t]$—that you should strive to understand, because it is a very useful tool for obtaining all sorts of differential equations.

Remark: It was convenient for us to use g/L mass/volume units in deriving Eq. (18). But any other consistent system of units can be used to measure amounts of solute and volumes of solution. In the following example we measure both in cubic kilometers. ∎

Example 4 Assume that Lake Erie has a volume of 480 km^3 and that its rate of inflow (from Lake Huron) and outflow (to Lake Ontario) are both 350 km^3 per year. Suppose that at the time $t = 0$ (years), the pollutant concentration of Lake Erie—caused by past industrial pollution that has now been ordered to cease—is five times that of Lake Huron. If the outflow henceforth is perfectly mixed lake water, how long will it take to reduce the pollution concentration in Lake Erie to twice that of Lake Huron?

Solution Here we have

$$V = 480 \quad (km^3),$$
$$r_i = r_o = r = 350 \,(km^3/yr),$$
$$c_i = c \text{ (the pollutant concentration of Lake Huron), and}$$
$$x_0 = x(0) = 5cV,$$

and the question is this: When is $x(t) = 2cV$? With this notation, Eq. (18) is the separable equation

$$\frac{dx}{dt} = rc - \frac{r}{V} x, \qquad (19)$$

which we rewrite in the linear first-order form

$$\frac{dx}{dt} + px = q \qquad (20)$$

with constant coefficients $p = r/V$, $q = rc$, and integrating factor $\rho = e^{pt}$. You can either solve this equation directly or apply the formula in (12). The latter gives

$$x(t) = e^{-pt} \left[x_0 + \int_0^t q e^{pt} \, dt \right] = e^{-pt} \left[x_0 + \frac{q}{p} \left(e^{pt} - 1 \right) \right]$$

$$= e^{-rt/V} \left[5cV + \frac{rc}{r/V} \left(e^{rt/V} - 1 \right) \right];$$

$$x(t) = cV + 4cV e^{-rt/V}. \tag{21}$$

To find when $x(t) = 2cV$, we therefore need only solve the equation

$$cV + 4cV e^{-rt/V} = 2cV \quad \text{for} \quad t = \frac{V}{r} \ln 4 = \frac{480}{350} \ln 4 \approx 1.901 \text{ (years).} \qquad \blacksquare$$

Example 5 A 120-gallon (gal) tank initially contains 90 lb of salt dissolved in 90 gal of water. Brine containing 2 lb/gal of salt flows into the tank at the rate of 4 gal/min, and the well-stirred mixture flows out of the tank at the rate of 3 gal/min. How much salt does the tank contain when it is full?

Solution The interesting feature of this example is that, due to the differing rates of inflow and outflow, the volume of brine in the tank increases steadily with $V(t) = 90 + t$ gallons. The change Δx in the amount x of salt in the tank from time t to time $t + \Delta t$ (minutes) is given by

$$\Delta x \approx (4)(2)\Delta t - 3 \left(\frac{x}{90 + t} \right) \Delta t,$$

so our differential equation is

$$\frac{dx}{dt} + \frac{3}{90 + t} x = 8.$$

An integrating factor is

$$\rho(x) = \exp \left(\int \frac{3}{90 + t} \, dt \right) = e^{3 \ln(90+t)} = (90 + t)^3,$$

which gives

$$D_t \left[(90 + t)^3 x \right] = 8(90 + t)^3;$$
$$(90 + t)^3 x = 2(90 + t)^4 + C.$$

Substitution of $x(0) = 90$ gives $C = -(90)^4$, so the amount of salt in the tank at time t is

$$x(t) = 2(90 + t) - \frac{90^4}{(90 + t)^3}.$$

The tank is full after 30 min, and when $t = 30$, we have

$$x(30) = 2(90 + 30) - \frac{90^4}{120^3} \approx 202 \text{ (lb)}$$

of salt in the tank. \blacksquare

1.5 Problems

Find general solutions of the differential equations in Problems 1 through 25. If an initial condition is given, find the corresponding particular solution. Throughout, primes denote derivatives with respect to x.

1. $y' + y = 2$, $y(0) = 0$
2. $y' - 2y = 3e^{2x}$, $y(0) = 0$
3. $y' + 3y = 2xe^{-3x}$
4. $y' - 2xy = e^{x^2}$
5. $xy' + 2y = 3x$, $y(1) = 5$
6. $xy' + 5y = 7x^2$, $y(2) = 5$
7. $2xy' + y = 10\sqrt{x}$
8. $3xy' + y = 12x$
9. $xy' - y = x$, $y(1) = 7$
10. $2xy' - 3y = 9x^3$
11. $xy' + y = 3xy$, $y(1) = 0$
12. $xy' + 3y = 2x^5$, $y(2) = 1$
13. $y' + y = e^x$, $y(0) = 1$
14. $xy' - 3y = x^3$, $y(1) = 10$
15. $y' + 2xy = x$, $y(0) = -2$
16. $y' = (1 - y)\cos x$, $y(\pi) = 2$
17. $(1 + x)y' + y = \cos x$, $y(0) = 1$
18. $xy' = 2y + x^3 \cos x$
19. $y' + y \cot x = \cos x$
20. $y' = 1 + x + y + xy$, $y(0) = 0$
21. $xy' = 3y + x^4 \cos x$, $y(2\pi) = 0$
22. $y' = 2xy + 3x^2 \exp(x^2)$, $y(0) = 5$
23. $xy' + (2x - 3)y = 4x^4$
24. $(x^2 + 4)y' + 3xy = x$, $y(0) = 1$
25. $(x^2 + 1)\dfrac{dy}{dx} + 3x^3y = 6x \exp\left(-\tfrac{3}{2}x^2\right)$, $y(0) = 1$

Solve the differential equations in Problems 26 through 28 by regarding y as the independent variable rather than x.

26. $(1 - 4xy^2)\dfrac{dy}{dx} = y^3$ 27. $(x + ye^y)\dfrac{dy}{dx} = 1$

28. $(1 + 2xy)\dfrac{dy}{dx} = 1 + y^2$

29. Express the general solution of $dy/dx = 1 + 2xy$ in terms of the **error function**
$$\text{erf}(x) = \frac{2}{\sqrt{\pi}} \int_0^x e^{-t^2}\, dt.$$

30. Express the solution of the initial value problem
$$2x\frac{dy}{dx} = y + 2x \cos x, \quad y(1) = 0$$
as an integral as in Example 3 of this section.

31. **(a)** Show that
$$y_c(x) = Ce^{-\int P(x)\,dx}$$
is a general solution of $dy/dx + P(x)y = 0$. **(b)** Show that
$$y_p(x) = e^{-\int P(x)\,dx}\left[\int \left(Q(x)e^{\int P(x)\,dx}\right)dx\right]$$

is a particular solution of $dy/dx + P(x)y = Q(x)$. **(c)** Suppose that $y_c(x)$ is any general solution of $dy/dx + P(x)y = 0$ and that $y_p(x)$ is any particular solution of $dy/dx + P(x)y = Q(x)$. Show that $y(x) = y_c(x) + y_p(x)$ is a general solution of $dy/dx + P(x)y = Q(x)$.

32. **(a)** Find constants A and B such that $y_p(x) = A \sin x + B \cos x$ is a solution of $dy/dx + y = 2 \sin x$. **(b)** Use the result of part (a) and the method of Problem 31 to find the general solution of $dy/dx + y = 2 \sin x$. **(c)** Solve the initial value problem $dy/dx + y = 2 \sin x$, $y(0) = 1$.

33. A tank contains 1000 liters (L) of a solution consisting of 100 kg of salt dissolved in water. Pure water is pumped into the tank at the rate of 5 L/s, and the mixture—kept uniform by stirring— is pumped out at the same rate. How long will it be until only 10 kg of salt remains in the tank?

34. Consider a reservoir with a volume of 8 billion cubic feet (ft^3) and an initial pollutant concentration of 0.25%. There is a daily inflow of 500 million ft^3 of water with a pollutant concentration of 0.05% and an equal daily outflow of the well-mixed water in the reservoir. How long will it take to reduce the pollutant concentration in the reservoir to 0.10%?

35. Rework Example 4 for the case of Lake Ontario. The only differences are that this lake has a volume of 1640 km^3 and an inflow-outflow rate of 410 km^3/year.

36. A tank initially contains 60 gal of pure water. Brine containing 1 lb of salt per gallon enters the tank at 2 gal/min, and the (perfectly mixed) solution leaves the tank at 3 gal/min; thus the tank is empty after exactly 1 h. **(a)** Find the amount of salt in the tank after t minutes. **(b)** What is the maximum amount of salt ever in the tank?

37. A 400-gal tank initially contains 100 gal of brine containing 50 lb of salt. Brine containing 1 lb of salt per gallon enters the tank at the rate of 5 gal/s, and the well-mixed brine in the tank flows out at the rate of 3 gal/s. How much salt will the tank contain when it is full of brine?

38. Consider the *cascade* of two tanks shown in Fig. 1.5.5, with $V_1 = 100$ (gal) and $V_2 = 200$ (gal) the volumes of brine in the two tanks. Each tank also initially contains 50 lb of salt. The three flow rates indicated in the figure are each 5 gal/min, with pure water flowing into tank 1. **(a)** Find the amount $x(t)$ of salt in tank 1 at time t. **(b)** Suppose that $y(t)$ is the amount of salt in tank 2 at time t. Show first that
$$\frac{dy}{dt} = \frac{5x}{100} - \frac{5y}{200},$$
and then solve for $y(t)$, using the function $x(t)$ found in part (a). **(c)** Finally, find the maximum amount of salt ever in tank 2.

FIGURE 1.5.5. A cascade of two tanks.

39. Suppose that in the cascade shown in Fig. 1.5.5, tank 1 initially contains 100 gal of pure ethanol and tank 2 initially contains 100 gal of pure water. Pure water flows into tank 1 at 10 gal/min, and the other two flow rates are also 10 gal/min. **(a)** Find the amounts $x(t)$ and $y(t)$ of ethanol in the two tanks at time $t \geq 0$. **(b)** Find the maximum amount of ethanol ever in tank 2.

40. A multiple cascade is shown in Fig. 1.5.6. At time $t = 0$, tank 0 contains 1 gal of ethanol and 1 gal of water; all the remaining tanks contain 2 gal of pure water each. Pure water is pumped into tank 0 at 1 gal/min, and the varying mixture in each tank is pumped into the one below it at the same rate. Assume, as usual, that the mixtures are kept perfectly uniform by stirring. Let $x_n(t)$ denote the amount of ethanol in tank n at time t.

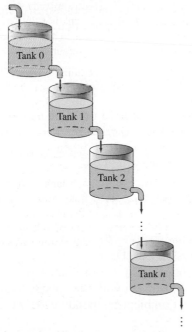

FIGURE 1.5.6. A multiple cascade.

(a) Show that $x_0(t) = e^{-t/2}$. **(b)** Show by induction on n

that

$$x_n(t) = \frac{t^n e^{-t/2}}{n! \, 2^n} \quad \text{for } n > 0.$$

(c) Show that the maximum value of $x_n(t)$ for $n > 0$ is $M_n = n^n e^{-n}/n!$. **(d)** Conclude from **Stirling's approximation** $n! \approx n^n e^{-n}\sqrt{2\pi n}$ that $M_n \approx (2\pi n)^{-1/2}$.

41. A 30-year-old woman accepts an engineering position with a starting salary of \$30,000 per year. Her salary $S(t)$ increases exponentially, with $S(t) = 30e^{t/20}$ thousand dollars after t years. Meanwhile, 12% of her salary is deposited continuously in a retirement account, which accumulates interest at a continuous annual rate of 6%. **(a)** Estimate ΔA in terms of Δt to derive the differential equation satisfied by the amount $A(t)$ in her retirement account after t years. **(b)** Compute $A(40)$, the amount available for her retirement at age 70.

42. Suppose that a falling hailstone with density $\delta = 1$ starts from rest with negligible radius $r = 0$. Thereafter its radius is $r = kt$ (k is a constant) as it grows by accretion during its fall. Set up and solve the initial value problem

$$\frac{d}{dt}(mv) = mg, \quad v(0) = 0,$$

where m is the variable mass of the hailstone, $v = dy/dt$ is its velocity, and the positive y-axis points downward. Then show that $dv/dt = g/4$. Thus the hailstone falls as though it were under *one-fourth* the influence of gravity.

43. Figure 1.5.7 shows a slope field and typical solution curves for the equation $y' = x - y$. **(a)** Show that every solution curve approaches the straight line $y = x - 1$ as $x \to +\infty$. **(b)** For each of the five values $y_1 = 3.998$, 3.999, 4.000, 4.001, and 4.002, determine the initial value y_0 (accurate to four decimal places) such that $y(5) = y_1$ for the solution satisfying the initial condition $y(-5) = y_0$.

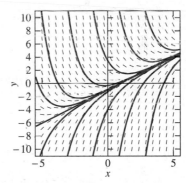

FIGURE 1.5.7. Slope field and solution curves for $y' = x - y$.

44. Figure 1.5.8 shows a slope field and typical solution curves for the equation $y' = x + y$. **(a)** Show that every solution curve approaches the straight line $y = -x - 1$ as $x \to -\infty$. **(b)** For each of the five values $y_1 = -10$, -5, 0, 5, and 10, determine the initial value y_0 (accurate to five decimal places) such that $y(5) = y_1$ for the solution satisfying the initial condition $y(-5) = y_0$.

FIGURE 1.5.8. Slope field and solution curves for $y' = x + y$.

Problems 45 and 46 deal with a shallow reservoir that has a one square kilometer water surface and an average water depth of 2 meters. Initially it is filled with fresh water, but at time $t = 0$ water contaminated with a liquid pollutant begins flowing into the reservoir at the rate of 200 thousand cubic meters per month. The well-mixed water in the reservoir flows out at the same rate. Your first task is to find the amount $x(t)$ of pollutant (in millions of liters) in the reservoir after t months.

45. The incoming water has a pollutant concentration of $c(t) = 10$ liters per cubic meter (L/m^3). Verify that the graph of $x(t)$ resembles the steadily rising curve in Fig. 1.5.9, which approaches asymptotically the graph of the equilibrium solution $x(t) \equiv 20$ that corresponds to the reservoir's long-term pollutant content. How long does it take the pollutant concentration in the reservoir to reach 5 L/m^3?

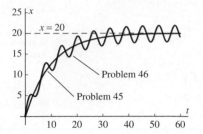

FIGURE 1.5.9. Graphs of solutions in Problems 45 and 46.

46. The incoming water has pollutant concentration $c(t) = 10(1 + \cos t)$ L/m^3 that varies between 0 and 20, with an average concentration of 10 L/m^3 and a period of oscillation of slightly over $6\frac{1}{4}$ months. Does it seem predictable that the lake's pollutant content should ultimately oscillate periodically about an average level of 20 million liters? Verify that the graph of $x(t)$ does, indeed, resemble the oscillatory curve shown in Fig. 1.5.9. How long does it take the pollutant concentration in the reservoir to reach 5 L/m^3?

1.5 Application: Indoor Temperature Oscillations

For an interesting applied problem that involves the solution of a linear differential equation, consider indoor temperature oscillations that are driven by outdoor temperature oscillations of the form

$$A(t) = a_0 + a_1 \cos \omega t + b_1 \sin \omega t. \tag{1}$$

If $\omega = \pi/12$, then these oscillations have a period of 24 hours (so that the cycle of outdoor temperatures repeats itself daily). For instance, for a typical July day in Athens, GA with a minimum temperature of 70°F when $t = 4$ (4 A.M.) and a maximum of 90°F when $t = 16$ (4 P.M.), we would take

$$A(t) = 80 - 10 \cos \omega(t - 4) = 80 - 5 \cos \omega t - 5\sqrt{3} \sin \omega t. \tag{2}$$

We derived Eq. (2) by using the identity $\cos(\alpha - \beta) = \cos \alpha \cos \beta + \sin \alpha \sin \beta$ to get $a_0 = 80$, $a_1 = -5$, and $b_1 = -5\sqrt{3}$ in Eq. (1).

If we write Newton's law of cooling (Eq. (3) of Section 1.1) for the corresponding indoor temperature $u(t)$ at time t, but with the outside temperature $A(t)$ given by Eq. (1) instead of a constant ambient temperature A, we get the linear first-order differential equation

$$\frac{du}{dt} = -k(u - A(t));$$

that is,

$$\frac{du}{dt} + ku = k(a_0 + a_1 \cos \omega t + b_1 \sin \omega t) \tag{3}$$

with coefficient functions $P(t) \equiv k$ and $Q(t) = kA(t)$. Typical values of the proportionality constant k range from 0.2 to 0.5 (although k might be greater than 0.5 for a poorly insulated building with open windows, or less than 0.2 for a well-insulated building with tightly sealed windows).

SCENARIO: Suppose that our air conditioner fails at time $t_0 = 0$ one midnight, and we cannot afford to have it repaired until payday at the end of the month. We therefore want to investigate the resulting indoor temperatures that we must endure for the next several days.

Begin your investigation by solving Eq. (3) with the initial condition $u(0) = u_0$ (the indoor temperature at the time of the failure of the air conditioner). You may want to use the integral formulas in 49 and 50 of the endpapers, or possibly a computer algebra system. You should get the solution

$$u(t) = a_0 + c_0 e^{-kt} + c_1 \cos \omega t + d_1 \sin \omega t, \tag{4}$$

where

$$c_0 = u_0 - a_0 - \frac{k^2 a_1 - k\omega b_1}{k^2 + \omega^2},$$

$$c_1 = \frac{k^2 a_1 - k\omega b_1}{k^2 + \omega^2}, \qquad d_1 = \frac{k\omega a_1 + k^2 b_1}{k^2 + \omega^2}$$

with $\omega = \pi/12$.

With $a_0 = 80$, $a_1 = -5$, $b_1 = -5\sqrt{3}$ (as in Eq. (2)), $\omega = \pi/12$, and $k = 0.2$ (for instance), this solution reduces (approximately) to

$$u(t) = 80 + e^{-t/5}(u_0 - 82.3351) + (2.3351) \cos \frac{\pi t}{12} - (5.6036) \sin \frac{\pi t}{12}. \tag{5}$$

Observe first that the damped exponential term in Eq. (5) approaches zero as $t \to +\infty$, leaving the long-term "steady periodic" solution

$$u_{\text{sp}}(t) = 80 + (2.3351) \cos \frac{\pi t}{12} - (5.6036) \sin \frac{\pi t}{12}. \tag{6}$$

Consequently, the long-term indoor temperatures oscillate every 24 hours around the same average temperature $80°F$ as the average outdoor temperature.

Figure 1.5.10 shows a number of solution curves corresponding to possible initial temperatures u_0 ranging from $65°F$ to $95°F$. Observe that—whatever the initial temperature—the indoor temperature "settles down" within about 18 hours to a periodic daily oscillation. But the amplitude of temperature variation is less indoors than outdoors. Indeed, using the trigonometric identity mentioned earlier, Eq. (6) can be rewritten (verify this!) as

$$u(t) = 80 - (6.0707) \cos \left(\frac{\pi t}{12} - 1.9656 \right)$$

$$= 80 - (6.0707) \cos \frac{\pi}{12}(t - 7.5082). \tag{7}$$

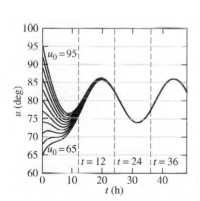

FIGURE 1.5.10. Solution curves given by Eq. (5) with $u_0 = 65, 68, 71, \ldots, 92, 95$.

FIGURE 1.5.11. Comparison of indoor and outdoor temperature oscillations.

Do you see that this implies that the indoor temperature varies between a minimum of about $74°$F and a maximum of about $86°$F?

Finally, comparison of Eqs. (2) and (7) indicates that the indoor temperature lags behind the outdoor temperature by about $7.5082 - 4 \approx 3.5$ hours, as illustrated in Fig. 1.5.11. Thus the temperature inside the house continues to rise until about 7:30 P.M. each evening, so the hottest part of the day inside is early evening rather than late afternoon (as outside).

For a personal problem to investigate, carry out a similar analysis using average July daily maximum/minimum figures for your own locale and a value of k appropriate to your own home. You might also consider a winter day instead of a summer day. (What is the winter-summer difference for the indoor temperature problem?)

1.6 Substitution Methods and Exact Equations

The first-order differential equations we have solved in the previous sections have all been either separable or linear. But many applications involve differential equations that are neither separable nor linear. In this section we illustrate (mainly with examples) substitution methods that sometimes can be used to transform a given differential equation into one that we already know how to solve.

For instance, the differential equation

$$\frac{dy}{dx} = f(x, y),\tag{1}$$

with dependent variable y and independent variable x, may contain a conspicuous combination

$$v = \alpha(x, y)\tag{2}$$

of x and y that suggests itself as a new independent variable v. Thus the differential equation

$$\frac{dy}{dx} = (x + y + 3)^2$$

practically demands the substitution $v = x + y + 3$ of the form in Eq. (2).

If the substitution relation in Eq. (2) can be solved for

$$y = \beta(x, v),\tag{3}$$

then application of the chain rule—regarding v as an (unknown) function of x—yields

$$\frac{dy}{dx} = \frac{\partial \beta}{\partial x}\frac{dx}{dx} + \frac{\partial \beta}{\partial v}\frac{dv}{dx} = \beta_x + \beta_v\frac{dv}{dx},\tag{4}$$

where the partial derivatives $\partial \beta / \partial x = \beta_x(x, v)$ and $\partial \beta / \partial v = \beta_v(x, v)$ are *known* functions of x and v. If we substitute the right-hand side in (4) for dy/dx in Eq. (1) and then solve for dv/dx, the result is a *new* differential equation of the form

$$\frac{dv}{dx} = g(x, v)\tag{5}$$

with *new* dependent variable v. If this new equation is either separable or linear, then we can apply the methods of preceding sections to solve it.

If $v = v(x)$ is a solution of Eq. (5), then $y = \beta(x, v(x))$ will be a solution of the original Eq. (1). The trick is to select a substitution such that the transformed Eq. (5) is one we can solve. Even when possible, this is not always easy; it may require a fair amount of ingenuity or trial and error.

Example 1 Solve the differential equation

$$\frac{dy}{dx} = (x + y + 3)^2.$$

Solution As indicated earlier, let's try the substitution

$$v = x + y + 3; \quad \text{that is,} \quad y = v - x - 3.$$

Then

$$\frac{dy}{dx} = \frac{dv}{dx} - 1,$$

so the transformed equation is

$$\frac{dv}{dx} = 1 + v^2.$$

This is a separable equation, and we have no difficulty in obtaining its solution

$$x = \int \frac{dv}{1 + v^2} = \tan^{-1} v + C.$$

So $v = \tan(x - C)$. Because $v = x + y + 3$, the general solution of the original equation $dy/dx = (x + y + 3)^2$ is $x + y + 3 = \tan(x - C)$; that is,

$$y(x) = \tan(x - C) - x - 3. \qquad \blacksquare$$

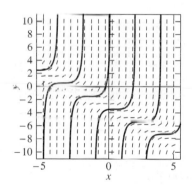

FIGURE 1.6.1. Slope field and solution curves for $y' = (x + y + 3)^2$.

Remark: Figure 1.6.1 shows a slope field and typical solution curves for the differential equation of Example 1. We see that, although the function $f(x, y) = (x+y+3)^2$ is continuously differentiable for all x and y, each solution is continuous only on a bounded interval. In particular, because the tangent function is continuous on the open interval $(-\pi/2, \pi/2)$, the particular solution with arbitrary constant value C is continuous on the interval where $-\pi/2 < x - C < \pi/2$; that is, $C - \pi/2 < x < C + \pi/2$. $\qquad \blacksquare$

Example 1 illustrates the fact that any differential equation of the form

$$\frac{dy}{dx} = F(ax + by + c) \qquad (6)$$

can be transformed into a separable equation by use of the substitution $v = ax + by + c$ (see Problem 31). The paragraphs that follow deal with other classes of first-order equations for which there are standard substitutions that are known to succeed.

Homogeneous Equations

A **homogeneous** first-order differential equation is one that can be written in the form

➤
$$\frac{dy}{dx} = F\left(\frac{y}{x}\right). \tag{7}$$

If we make the substitutions

$$v = \frac{y}{x}, \quad y = vx, \quad \frac{dy}{dx} = v + x\frac{dv}{dx}, \tag{8}$$

then Eq. (7) is transformed into the *separable* equation

$$x\frac{dv}{dx} = F(v) - v.$$

Thus every homogeneous first-order differential equation can be reduced to an integration problem by means of the substitutions in (8).

Example 2 Solve the differential equation

$$2xy\frac{dy}{dx} = 4x^2 + 3y^2.$$

Solution This equation is neither separable nor linear, but we recognize it as a homogeneous equation by writing it in the form

$$\frac{dy}{dx} = \frac{4x^2 + 3y^2}{2xy} = 2\left(\frac{x}{y}\right) + \frac{3}{2}\left(\frac{y}{x}\right).$$

The substitutions in (8) then take the form

$$y = vx, \quad \frac{dy}{dx} = v + x\frac{dv}{dx}, \quad v = \frac{y}{x}, \quad \text{and} \quad \frac{1}{v} = \frac{x}{y}.$$

These yield

$$v + x\frac{dv}{dx} = \frac{2}{v} + \frac{3}{2}v,$$

and hence

$$x\frac{dv}{dx} = \frac{2}{v} + \frac{v}{2} = \frac{v^2 + 4}{2v};$$

$$\int \frac{2v}{v^2 + 4}\,dv = \int \frac{1}{x}\,dx;$$

$$\ln(v^2 + 4) = \ln|x| + \ln C.$$

We apply the exponential function to both sides of the last equation to obtain

$$v^2 + 4 = C|x|;$$

$$\frac{y^2}{x^2} + 4 = C|x|;$$

$$y^2 + 4x^2 = kx^3.$$

Note that the left-hand side of this equation is necessarily nonnegative. It follows that $k > 0$ in the case of solutions that are defined for $x > 0$, while $k < 0$ for solutions where $x < 0$. Indeed, the family of solution curves illustrated in Fig. 1.6.2 exhibits symmetry about both coordinate axes. Actually, there are positive-valued and negative-valued solutions of the forms $y(x) = \pm\sqrt{kx^3 - 4x^2}$ that are defined for $x > 4/k$ if the constant k is positive, for $x < 4/k$ if k is negative. ∎

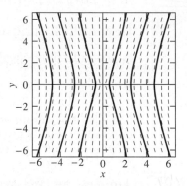

FIGURE 1.6.2. Slope field and solution curves for $2xyy' = 4x^2 + 3y^2$.

Example 3 Solve the initial value problem

$$x\frac{dy}{dx} = y + \sqrt{x^2 - y^2}, \quad y(x_0) = 0$$

where $x_0 > 0$.

Solution We divide both sides by x and find that

$$\frac{dy}{dx} = \frac{y}{x} + \sqrt{1 - \left(\frac{y}{x}\right)^2},$$

so we make the substitutions in (8); we get

$$v + x\frac{dv}{dx} = v + \sqrt{1 - v^2};$$

$$\int \frac{1}{\sqrt{1 - v^2}}\,dv = \int \frac{1}{x}\,dx;$$

$$\sin^{-1} v = \ln x + C.$$

We need not write $\ln|x|$ because $x > 0$ near $x = x_0 > 0$. Now note that $v(x_0) = y(x_0)/x_0 = 0$, so $C = \sin^{-1} 0 - \ln x_0 = -\ln x_0$. Hence

$$v = \frac{y}{x} = \sin(\ln x - \ln x_0) = \sin\left(\ln\frac{x}{x_0}\right),$$

and therefore

$$y(x) = x\sin\left(\ln\frac{x}{x_0}\right)$$

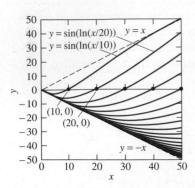

FIGURE 1.6.3. Solution curves for $xy' = y + \sqrt{x^2 - y^2}$.

is the desired particular solution. Because the range of $\sin^{-1} v = \ln(x/x_0)$ is the interval $-\pi/2 \leqq \ln(x/x_0) \leqq \pi/2$, it follows that this particular solution is defined on the interval $x_0 e^{-\pi/2} \leqq x \leqq x_0 e^{\pi/2}$. Figure 1.6.3 shows some typical solution curves. Because of the radical in the differential equation, these solution curves are confined to the indicated triangular region $x \geqq |y|$. You can check that the boundary lines $y = x$ and $y = -x$ (for $x > 0$) are singular solution curves that consist of points of tangency with the solution curves found earlier. ∎

Bernoulli Equations

A first-order differential equation of the form

$$\frac{dy}{dx} + P(x)y = Q(x)y^n \tag{9}$$

is called a **Bernoulli equation.** If either $n = 0$ or $n = 1$, then Eq. (9) is linear. Otherwise, as we ask you to show in Problem 56, the substitution

$$v = y^{1-n} \tag{10}$$

transforms Eq. (9) into the linear equation

$$\frac{dv}{dx} + (1 - n)P(x)v = (1 - n)Q(x).$$

Rather than memorizing the form of this transformed equation, it is more efficient to make the substitution in Eq. (10) explicitly, as in the following examples.

Example 4

If we rewrite the homogeneous equation $2xyy' = 4x^2 + 3y^2$ of Example 2 in the form

$$\frac{dy}{dx} - \frac{3}{2x}y = \frac{2x}{y},$$

we see that it is also a Bernoulli equation with $P(x) = -3/(2x)$, $Q(x) = 2x$, $n = -1$, and $1 - n = 2$. Hence we substitute

$$v = y^2, \quad y = v^{1/2}, \quad \text{and} \quad \frac{dy}{dx} = \frac{dy}{dv}\frac{dv}{dx} = \frac{1}{2}v^{-1/2}\frac{dv}{dx}.$$

This gives

$$\frac{1}{2}v^{-1/2}\frac{dv}{dx} - \frac{3}{2x}v^{1/2} = 2xv^{-1/2}.$$

Then multiplication by $2v^{1/2}$ produces the linear equation

$$\frac{dv}{dx} - \frac{3}{x}v = 4x$$

with integrating factor $\rho = e^{\int(-3/x)\,dx} = x^{-3}$. So we obtain

$$D_x(x^{-3}v) = \frac{4}{x^2};$$

$$x^{-3}v = -\frac{4}{x} + C;$$

$$x^{-3}y^2 = -\frac{4}{x} + C;$$

$$y^2 = -4x^2 + Cx^3. \qquad ∎$$

Example 5 The equation

$$x\frac{dy}{dx} + 6y = 3xy^{4/3}$$

is neither separable nor linear nor homogeneous, but it is a Bernoulli equation with $n = \frac{4}{3}$, $1 - n = -\frac{1}{3}$. The substitutions

$$v = y^{-1/3}, \quad y = v^{-3}, \quad \text{and} \quad \frac{dy}{dx} = \frac{dy}{dv}\frac{dv}{dx} = -3v^{-4}\frac{dv}{dx}$$

transform it into

$$-3xv^{-4}\frac{dv}{dx} + 6v^{-3} = 3xv^{-4}.$$

Division by $-3xv^{-4}$ yields the linear equation

$$\frac{dv}{dx} - \frac{2}{x}v = -1$$

with integrating factor $\rho = e^{\int(-2/x)\,dx} = x^{-2}$. This gives

$$D_x(x^{-2}v) = -\frac{1}{x^2}; \quad x^{-2}v = \frac{1}{x} + C; \quad v = x + Cx^2;$$

and finally,

$$y(x) = \frac{1}{(x + Cx^2)^3}.$$ ∎

Example 6 The equation

$$2xe^{2y}\frac{dy}{dx} = 3x^4 + e^{2y} \tag{11}$$

is neither separable, nor linear, nor homogeneous, nor is it a Bernoulli equation. But we observe that y appears only in the combinations e^{2y} and $D_x(e^{2y}) = 2e^{2y}y'$. This prompts the substitution

$$v = e^{2y}, \quad \frac{dv}{dx} = 2e^{2y}\frac{dy}{dx}$$

that transforms Eq. (11) into the linear equation $xv'(x) = 3x^4 + v(x)$; that is,

$$\frac{dv}{dx} - \frac{1}{x}v = 3x^3.$$

After multiplying by the integrating factor $\rho = 1/x$, we find that

$$\frac{1}{x}v = \int 3x^2\,dx = x^3 + C, \quad \text{so} \quad e^{2y} = v = x^4 + Cx,$$

and hence

$$y(x) = \tfrac{1}{2}\ln\left|x^4 + Cx\right|.$$ ∎

Flight Trajectories

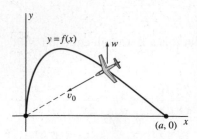

FIGURE 1.6.4. The airplane headed for the origin.

Suppose that an airplane departs from the point $(a, 0)$ located due east of its intended destination—an airport located at the origin $(0, 0)$. The plane travels with constant speed v_0 relative to the wind, which is blowing due north with constant speed w. As indicated in Fig. 1.6.4, we assume that the plane's pilot maintains its heading directly toward the origin.

Figure 1.6.5 helps us derive the plane's velocity components relative to the ground. They are

$$\frac{dx}{dt} = -v_0 \cos \theta = -\frac{v_0 x}{\sqrt{x^2 + y^2}},$$

$$\frac{dy}{dt} = -v_0 \sin \theta + w = -\frac{v_0 y}{\sqrt{x^2 + y^2}} + w.$$

Hence the trajectory $y = f(x)$ of the plane satisfies the differential equation

$$\frac{dy}{dx} = \frac{dy/dt}{dx/dt} = \frac{1}{v_0 x} \left(v_0 y - w\sqrt{x^2 + y^2} \right). \tag{12}$$

FIGURE 1.6.5. The components of the velocity vector of the airplane.

If we set

$$k = \frac{w}{v_0}, \tag{13}$$

the ratio of the windspeed to the plane's airspeed, then Eq. (12) takes the homogeneous form

$$\frac{dy}{dx} = \frac{y}{x} - k \left[1 + \left(\frac{y}{x} \right)^2 \right]^{1/2}. \tag{14}$$

The substitution $y = xv$, $y' = v + xv'$ then leads routinely to

$$\int \frac{dv}{\sqrt{1 + v^2}} = -\int \frac{k}{x} dx. \tag{15}$$

By trigonometric substitution, or by consulting a table for the integral on the left, we find that

$$\ln \left(v + \sqrt{1 + v^2} \right) = -k \ln x + C, \tag{16}$$

and the initial condition $v(a) = y(a)/a = 0$ yields

$$C = k \ln a. \tag{17}$$

According to Problem 68, the result of substituting (17) in Eq. (16) and then solving for v is

$$v = \frac{1}{2} \left[\left(\frac{x}{a} \right)^{-k} - \left(\frac{x}{a} \right)^{k} \right]. \tag{18}$$

Because $y = xv$, we finally obtain

$$y(x) = \frac{a}{2} \left[\left(\frac{x}{a} \right)^{1-k} - \left(\frac{x}{a} \right)^{1+k} \right] \tag{19}$$

for the equation of the plane's trajectory.

Note that only in the case $k < 1$ (that is, $w < v_0$) does the curve in Eq. (19) pass through the origin, so that the plane reaches its destination. If $w = v_0$ (so that $k = 1$), then Eq. (19) takes the form $y(x) = \frac{1}{2}a(1 - x^2/a^2)$, so the plane's trajectory approaches the point $(0, a/2)$ rather than $(0, 0)$. The situation is even worse if $w > v_0$ (so $k > 1$)—in this case it follows from Eq. (19) that $y \to +\infty$ as $x \to 0$.

Example 7 If $a = 200$ mi, $v_0 = 500$ mi/h, and $w = 100$ mi/h, then $k = w/v_0 = \frac{1}{5}$, so the plane will succeed in reaching the airport at $(0, 0)$. With these values, Eq. (19) yields

$$y(x) = 100 \left[\left(\frac{x}{200} \right)^{4/5} - \left(\frac{x}{200} \right)^{6/5} \right]. \tag{20}$$

Now suppose that we want to find the maximum amount by which the plane is blown off course during its trip. That is, what is the maximum value of $y(x)$ for $0 \le x \le 200$?

Solution Differentiation of the function in Eq. (20) yields

$$\frac{dy}{dx} = \frac{1}{2} \left[\frac{4}{5} \left(\frac{x}{200} \right)^{-1/5} - \frac{6}{5} \left(\frac{x}{200} \right)^{1/5} \right],$$

and we readily solve the equation $y'(x) = 0$ to obtain $(x/200)^{2/5} = \frac{2}{3}$. Hence

$$y_{\max} = 100 \left[\left(\frac{2}{3} \right)^2 - \left(\frac{2}{3} \right)^3 \right] = \frac{400}{27} \approx 14.81.$$

Thus the plane is blown almost 15 mi north at one point during its westward journey. (The graph of the function in Eq. (20) is the one used to construct Fig. 1.6.4. The vertical scale there is exaggerated by a factor of 4.) ∎

Exact Differential Equations

We have seen that a general solution $y(x)$ of a first-order differential equation is often defined implicitly by an equation of the form

$$F(x, y(x)) = C, \tag{21}$$

where C is a constant. On the other hand, given the identity in (21), we can recover the original differential equation by differentiating each side with respect to x. Provided that Eq. (21) implicitly defines y as a differentiable function of x, this gives the original differential equation in the form

$$\frac{\partial F}{\partial x} + \frac{\partial F}{\partial y} \frac{dy}{dx} = 0;$$

that is,

$$M(x, y) + N(x, y) \frac{dy}{dx} = 0, \tag{22}$$

where $M(x, y) = F_x(x, y)$ and $N(x, y) = F_y(x, y)$.

It is sometimes convenient to rewrite Eq. (22) in the more symmetric form

$$M(x, y)\, dx + N(x, y)\, dy = 0, \tag{23}$$

called its **differential** form. The general first-order differential equation $y' = f(x, y)$ can be written in this form with $M = f(x, y)$ and $N \equiv -1$. The preceding discussion shows that, *if* there exists a function $F(x, y)$ such that

$$\frac{\partial F}{\partial x} = M \quad \text{and} \quad \frac{\partial F}{\partial y} = N,$$

then the equation

$$F(x, y) = C$$

implicitly defines a general solution of Eq. (23). In this case, Eq. (23) is called an **exact differential equation**—the differential

$$dF = F_x\, dx + F_y\, dy$$

of $F(x, y)$ is exactly $M\,dx + N\,dy$.

Natural questions are these: How can we determine whether the differential equation in (23) is exact? And if it is exact, how can we find the function F such that $F_x = M$ and $F_y = N$? To answer the first question, let us recall that if the mixed second-order partial derivatives F_{xy} and F_{yx} are continuous on an open set in the xy-plane, then they are equal: $F_{xy} = F_{yx}$. If Eq. (23) is exact and M and N have continuous partial derivatives, it then follows that

$$\frac{\partial M}{\partial y} = F_{xy} = F_{yx} = \frac{\partial N}{\partial x}.$$

Thus the equation

$$\frac{\partial M}{\partial y} = \frac{\partial N}{\partial x} \tag{24}$$

is a *necessary condition* that the differential equation $M\,dx + N\,dy = 0$ be exact. That is, if $M_y \neq N_x$, then the differential equation in question is not exact, so we need not attempt to find a function $F(x, y)$ such that $F_x = M$ and $F_y = N$—there is no such function.

| Example 8 | The differential equation |

$$y^3\, dx + 3xy^2\, dy = 0 \tag{25}$$

is exact because we can immediately see that the function $F(x, y) = xy^3$ has the property that $F_x = y^3$ and $F_y = 3xy^2$. Thus a general solution of Eq. (25) is

$$xy^3 = C;$$

if you prefer, $y(x) = kx^{-1/3}$. ∎

But suppose that we divide each term of the differential equation in Example 8 by y^2 to obtain

$$y\,dx + 3x\,dy = 0. \tag{26}$$

This equation is not exact because, with $M = y$ and $N = 3x$, we have

$$\frac{\partial M}{\partial y} = 1 \neq 3 = \frac{\partial N}{\partial x}.$$

Hence the necessary condition in Eq. (24) is not satisfied.

We are confronted with a curious situation here. The differential equations in (25) and (26) are essentially equivalent, and they have exactly the same solutions, yet one is exact and the other is not. In brief, whether a given differential equation is exact or not is related to the precise form $M\,dx + N\,dy = 0$ in which it is written. Theorem 1 tells us that (subject to differentiability conditions usually satisfied in practice) the necessary condition in (24) is also a *sufficient* condition for exactness. In other words, if $M_y = N_x$, then the differential equation $M\,dx + N\,dy = 0$ is exact.

THEOREM 1: Criterion for Exactness

Suppose that the functions $M(x, y)$ and $N(x, y)$ are continuous and have continuous first-order partial derivatives in the open rectangle $R: a < x < b, c < y < d$. Then the differential equation

$$M(x, y)\,dx + N(x, y)\,dy = 0 \tag{23}$$

is exact in R if and only if

$$\frac{\partial M}{\partial y} = \frac{\partial N}{\partial x} \tag{24}$$

at each point of R. That is, there exists a function $F(x, y)$ defined on R with $\partial F/\partial x = M$ and $\partial F/\partial y = N$ if and only if Eq. (24) holds on R.

Proof: We have seen already that it is necessary for Eq. (24) to hold if Eq. (23) is to be exact. To prove the converse, we must show that if Eq. (24) holds, then we can construct a function $F(x, y)$ such that $\partial F/\partial x = M$ and $\partial F/\partial y = N$. Note first that, for *any* function $g(y)$, the function

$$F(x, y) = \int M(x, y)\,dx + g(y) \tag{27}$$

satisfies the condition $\partial F/\partial x = M$. (In Eq. (27), the notation $\int M(x, y)\,dx$ denotes an antiderivative of $M(x, y)$ with respect to x.) We plan to choose $g(y)$ so that

$$N = \frac{\partial F}{\partial y} = \left(\frac{\partial}{\partial y} \int M(x, y)\,dx \right) + g'(y)$$

as well; that is, so that

$$g'(y) = N - \frac{\partial}{\partial y} \int M(x, y)\,dx. \tag{28}$$

To see that there *is* such a function of y, it suffices to show that the right-hand side in Eq. (28) is a function of y alone. We can then find $g(y)$ by integrating with respect to y. Because the right-hand side in Eq. (28) is defined on a rectangle, and hence on an interval as a function of x, it suffices to show that its derivative with respect to x is identically zero. But

$$\frac{\partial}{\partial x}\left(N - \frac{\partial}{\partial y}\int M(x, y)\, dx\right) = \frac{\partial N}{\partial x} - \frac{\partial}{\partial x}\frac{\partial}{\partial y}\int M(x, y)\, dx$$

$$= \frac{\partial N}{\partial x} - \frac{\partial}{\partial y}\frac{\partial}{\partial x}\int M(x, y)\, dx$$

$$= \frac{\partial N}{\partial x} - \frac{\partial M}{\partial y} = 0$$

by hypothesis. So we can, indeed, find the desired function $g(y)$ by integrating Eq. (28). We substitute this result in Eq. (27) to obtain

$$F(x, y) = \int M(x, y)\, dx + \int \left(N(x, y) - \frac{\partial}{\partial y}\int M(x, y)\, dx\right) dy \qquad (29)$$

as the desired function with $F_x = M$ and $F_y = N$. ▲

Instead of memorizing Eq. (29), it is usually better to solve an exact equation $M\, dx + N\, dy = 0$ by carrying out the process indicated by Eqs. (27) and (28). First we integrate $M(x, y)$ with respect to x and write

$$F(x, y) = \int M(x, y)\, dx + g(y),$$

thinking of the function $g(y)$ as an "arbitrary constant of integration" as far as the variable x is concerned. Then we determine $g(y)$ by imposing the condition that $\partial F/\partial y = N(x, y)$. This yields a general solution in the implicit form $F(x, y) = C$.

Example 9 Solve the differential equation

$$(6xy - y^3)\, dx + (4y + 3x^2 - 3xy^2)\, dy = 0. \qquad (30)$$

Solution Let $M(x, y) = 6xy - y^3$ and $N(x, y) = 4y + 3x^2 - 3xy^2$. The given equation is exact because

$$\frac{\partial M}{\partial y} = 6x - 3y^2 = \frac{\partial N}{\partial x}.$$

Integrating $\partial F/\partial x = M(x, y)$ with respect to x, we get

$$F(x, y) = \int (6xy - y^3)\, dx = 3x^2 y - xy^3 + g(y).$$

Then we differentiate with respect to y and set $\partial F/\partial y = N(x, y)$. This yields

$$\frac{\partial F}{\partial y} = 3x^2 - 3xy^2 + g'(y) = 4y + 3x^2 - 3xy^2,$$

and it follows that $g'(y) = 4y$. Hence $g(y) = 2y^2 + C_1$, and thus

$$F(x, y) = 3x^2y - xy^3 + 2y^2 + C_1.$$

Therefore, a general solution of the differential equation is defined implicitly by the equation

$$3x^2y - xy^3 + 2y^2 = C \tag{31}$$

(we have absorbed the constant C_1 into the constant C). ∎

Remark: Figure 1.6.6 shows a rather complicated structure of solution curves for the differential equation of Example 9. The solution satisfying a given initial condition $y(x_0) = y_0$ is defined implicitly by Eq. (31), with C determined by substituting $x = x_0$ and $y = y_0$ in the equation. For instance, the particular solution satisfying $y(0) = 1$ is defined implicitly by the equation $3x^2y - xy^3 + 2y^2 = 2$. The other two special points in the figure—at $(0, 0)$ and near $(0.75, 2.12)$—are ones where both coefficient functions in Eq. (30) vanish, so the theorem of Section 1.3 does not guarantee a unique solution. ∎

FIGURE 1.6.6. Slope field and solution curves for the exact equation in Example 9.

Reducible Second-Order Equations

A *second-order differential equation* involves the second derivative of the unknown function $y(x)$, and thus has the general form

$$F(x, y, y', y'') = 0. \tag{32}$$

If *either* the dependent variable y *or* the independent variable x is missing from a second-order equation, then it is easily reduced by a simple substitution to a first-order equation that may be solvable by the methods of this chapter.

Dependent variable y missing. If y is missing, then Eq. (32) takes the form

➤ $$F(x, y', y'') = 0. \tag{33}$$

Then the substitution

➤ $$p = y' = \frac{dy}{dx}, \quad y'' = \frac{dp}{dx} \tag{34}$$

results in the *first-order* differential equation

$$F(x, p, p') = 0.$$

If we can solve this equation for a general solution $p(x, C_1)$ involving an arbitrary constant C_1, then we need only write

$$y(x) = \int y'(x)\,dx = \int p(x, C_1)\,dx + C_2$$

to get a solution of Eq. (33) that involves two arbitrary constants C_1 and C_2 (as is to be expected in the case of a second-order differential equation).

xample 10 Solve the equation $xy'' + 2y' = 6x$ in which the dependent variable y is missing.

Solution The substitution defined in (34) gives the first-order equation

$$x\frac{dp}{dx} + 2p = 6x; \quad \text{that is, } \frac{dp}{dx} + \frac{2}{x}p = 6.$$

Observing that the equation on the right here is linear, we multiply by its integrating factor $\rho = \exp\left(\int (2/x)\, dx\right) = e^{2\ln x} = x^2$ and get

$$D_x(x^2 p) = 6x^2,$$
$$x^2 p = 2x^3 + C_1,$$
$$p = \frac{dy}{dx} = 2x + \frac{C_1}{x^2}.$$

A final integration with respect to x yields the general solution

$$y(x) = x^2 + \frac{C_1}{x} + C_2$$

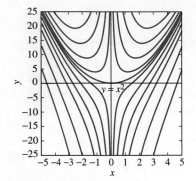

FIGURE 1.6.7. Solution curves of the form $y = x^2 + \dfrac{C_1}{x}$ for $C_1 = 0, \pm 3, \pm 10, \pm 20, \pm 35, \pm 60, \pm 100$.

of the second-order equation $xy'' + 2y' = 6x$. Solution curves with $C_1 = 0$ are simply vertical translates of the parabola $y = x^2$. Fig. 1.6.7 shows typical solution curves with $C_2 = 0$. Solution curves with C_1 and C_2 both nonzero are vertical translates of those shown in Fig. 1.6.7. ∎

Independent variable x missing. If x is missing, then Eq. (32) takes the form

$$\blacktriangleright \qquad\qquad F(y, y', y'') = 0. \qquad\qquad (35)$$

Then the substitution

$$\blacktriangleright \qquad p = y' = \frac{dy}{dx}, \quad y'' = \frac{dp}{dx} = \frac{dp}{dy}\frac{dy}{dx} = p\frac{dp}{dy} \qquad (36)$$

results in the *first-order* differential equation

$$F\left(y, p, p\frac{dp}{dy}\right) = 0$$

for p as a function of y. If we can solve this equation for a general solution $p(y, C_1)$ involving an arbitrary constant C_1, then (assuming that $y' \neq 0$) we need only write

$$x(y) = \int \frac{dx}{dy}\, dx = \int \frac{1}{dy/dx}\, dx = \int \frac{1}{p}\, dy = \int \frac{dy}{p(y, C_1)} + C_2.$$

If the final integral $P = \int (1/p)\, dy$ can be evaluated, the result is an implicit solution $x(y) = P(y, C_1) + C_2$ of our second-order differential equation.

Example 11 Solve the equation $yy'' = (y')^2$ in which the independent variable x is missing.

Solution We assume temporarily that y and y' are both nonnegative, and then point out at the end that this restriction is unnecessary. The substitution defined in (36) gives the first-order equation

$$yp\frac{dp}{dy} = p^2.$$

Then separation of variables gives

$$\int \frac{dp}{p} = \int \frac{dy}{y},$$
$$\ln p = \ln y + C \quad \text{(because } y > 0 \text{ and } p = y' > 0\text{)},$$
$$p = C_1 y$$

where $C_1 = e^C$. Hence

$$\frac{dx}{dy} = \frac{1}{p} = \frac{1}{C_1 y},$$
$$C_1 x = \int \frac{dy}{y} = \ln y + C_1.$$

The resulting general solution of the second-order equation $yy'' = (y')^2$ is

$$y(x) = \exp(C_1 x - C_2) = Ae^{Bx},$$

where $A = e^{-C_2}$ and $B = C_1$. Despite our temporary assumptions, which imply that the constants A and B are both positive, we readily verify that $y(x) = Ae^{Bx}$ satisfies $yy'' = (y')^2$ for *all* real values of A and B. With $B = 0$ and different values of A, we get all horizontal lines in the plane as solution curves. The upper half of Fig. 1.1.8 shows the solution curves obtained with $A = 1$ (for instance) and different positive values of B. With $A = -1$ these solution curves are reflected in the x-axis. And with negative values of B they are reflected in the y-axis. In particular, we see that we get solutions of $yy'' = (y')^2$ allowing both positive and negative possibilities for both y and y'. ∎

FIGURE 1.6.8. The solution curves $y = Ae^{Bx}$ with $B = 0$ and $A = 0, \pm 1$ are the horizontal lines $y = 0, \pm 1$. The exponential curves with $B > 0$ and $A = \pm 1$ are in color, those with $B < 0$ and $A = \pm 1$ are black.

1.6 Problems

Find general solutions of the differential equations in Problems 1 through 30. Primes denote derivatives with respect to x throughout.

1. $(x + y)y' = x - y$
2. $2xyy' = x^2 + 2y^2$
3. $xy' = y + 2\sqrt{xy}$
4. $(x - y)y' = x + y$
5. $x(x + y)y' = y(x - y)$
6. $(x + 2y)y' = y$
7. $xy^2 y' = x^3 + y^3$
8. $x^2 y' = xy + x^2 e^{y/x}$
9. $x^2 y' = xy + y^2$
10. $xyy' = x^2 + 3y^2$
11. $(x^2 - y^2)y' = 2xy$
12. $xyy' = y^2 + x\sqrt{4x^2 + y^2}$
13. $xy' = y + \sqrt{x^2 + y^2}$
14. $yy' + x = \sqrt{x^2 + y^2}$
15. $x(x + y)y' + y(3x + y) = 0$

16. $y' = \sqrt{x + y + 1}$
17. $y' = (4x + y)^2$
18. $(x + y)y' = 1$
19. $x^2 y' + 2xy = 5y^3$
20. $y^2 y' + 2xy^3 = 6x$
21. $y' = y + y^3$
22. $x^2 y' + 2xy = 5y^4$
23. $xy' + 6y = 3xy^{4/3}$
24. $2xy' + y^3 e^{-2x} = 2xy$
25. $y^2(xy' + y)(1 + x^4)^{1/2} = x$
26. $3y^2 y' + y^3 = e^{-x}$
27. $3xy^2 y' = 3x^4 + y^3$
28. $xe^y y' = 2(e^y + x^3 e^{2x})$
29. $(2x \sin y \cos y)y' = 4x^2 + \sin^2 y$
30. $(x + e^y)y' = xe^{-y} - 1$

In Problems 31 through 42, verify that the given differential equation is exact; then solve it.

31. $(2x + 3y)\,dx + (3x + 2y)\,dy = 0$

32. $(4x - y)\,dx + (6y - x)\,dy = 0$

33. $(3x^2 + 2y^2)\,dx + (4xy + 6y^2)\,dy = 0$

34. $(2xy^2 + 3x^2)\,dx + (2x^2y + 4y^3)\,dy = 0$

35. $\left(x^3 + \dfrac{y}{x}\right)dx + (y^2 + \ln x)\,dy = 0$

36. $(1 + ye^{xy})\,dx + (2y + xe^{xy})\,dy = 0$

37. $(\cos x + \ln y)\,dx + \left(\dfrac{x}{y} + e^y\right)dy = 0$

38. $(x + \tan^{-1} y)\,dx + \dfrac{x + y}{1 + y^2}\,dy = 0$

39. $(3x^2y^3 + y^4)\,dx + (3x^3y^2 + y^4 + 4xy^3)\,dy = 0$

40. $(e^x \sin y + \tan y)\,dx + (e^x \cos y + x \sec^2 y)\,dy = 0$

41. $\left(\dfrac{2x}{y} - \dfrac{3y^2}{x^4}\right)dx + \left(\dfrac{2y}{x^3} - \dfrac{x^2}{y^2} + \dfrac{1}{\sqrt{y}}\right)dy = 0$

42. $\dfrac{2x^{5/2} - 3y^{5/3}}{2x^{5/2}y^{2/3}}\,dx + \dfrac{3y^{5/3} - 2x^{5/2}}{3x^{3/2}y^{5/3}}\,dy = 0$

Find a general solution of each reducible second-order differential equation in Problems 43–54. Assume x, y and/or y' positive where helpful (as in Example 11).

43. $xy'' = y'$

44. $yy'' + (y')^2 = 0$

45. $y'' + 4y = 0$

46. $xy'' + y' = 4x$

47. $y'' = (y')^2$

48. $x^2y'' + 3xy' = 2$

49. $yy'' + (y')^2 = yy'$

50. $y'' = (x + y')^2$

51. $y'' = 2y(y')^3$

52. $y^3y'' = 1$

53. $y'' = 2yy'$

54. $yy'' = 3(y')^2$

55. Show that the substitution $v = ax + by + c$ transforms the differential equation $dy/dx = F(ax + by + c)$ into a separable equation.

56. Suppose that $n \neq 0$ and $n \neq 1$. Show that the substitution $v = y^{1-n}$ transforms the Bernoulli equation $dy/dx + P(x)y = Q(x)y^n$ into the linear equation

$$\frac{dv}{dx} + (1 - n)P(x)v(x) = (1 - n)Q(x).$$

57. Show that the substitution $v = \ln y$ transforms the differential equation $dy/dx + P(x)y = Q(x)(y \ln y)$ into the linear equation $dv/dx + P(x) = Q(x)v(x)$.

58. Use the idea in Problem 57 to solve the equation

$$x\frac{dy}{dx} - 4x^2y + 2y \ln y = 0.$$

59. Solve the differential equation

$$\frac{dy}{dx} = \frac{x - y - 1}{x + y + 3}$$

by finding h and k so that the substitutions $x = u + h$, $y = v + k$ transform it into the homogeneous equation

$$\frac{dv}{du} = \frac{u - v}{u + v}.$$

60. Use the method in Problem 59 to solve the differential equation

$$\frac{dy}{dx} = \frac{2y - x + 7}{4x - 3y - 18}.$$

61. Make an appropriate substitution to find a solution of the equation $dy/dx = \sin(x - y)$. Does this general solution contain the linear solution $y(x) = x - \pi/2$ that is readily verified by substitution in the differential equation?

62. Show that the solution curves of the differential equation

$$\frac{dy}{dx} = -\frac{y(2x^3 - y^3)}{x(2y^3 - x^3)}$$

are of the form $x^3 + y^3 = 3Cxy$.

63. The equation $dy/dx = A(x)y^2 + B(x)y + C(x)$ is called a **Riccati equation.** Suppose that one particular solution $y_1(x)$ of this equation is known. Show that the substitution

$$y = y_1 + \frac{1}{v}$$

transforms the Riccati equation into the *linear* equation

$$\frac{dv}{dx} + (B + 2Ay_1)v = -A.$$

Use the method of Problem 63 to solve the equations in Problems 64 and 65, given that $y_1(x) = x$ is a solution of each.

64. $\dfrac{dy}{dx} + y^2 = 1 + x^2$

65. $\dfrac{dy}{dx} + 2xy = 1 + x^2 + y^2$

66. An equation of the form

$$y = xy' + g(y') \tag{37}$$

is called a **Clairaut equation.** Show that the one-parameter family of straight lines described by

$$y(x) = Cx + g(C) \tag{38}$$

is a general solution of Eq. (37).

67. Consider the Clairaut equation

$$y = xy' - \tfrac{1}{4}(y')^2$$

for which $g(y') = -\tfrac{1}{4}(y')^2$ in Eq. (37). Show that the line

$$y = Cx - \tfrac{1}{4}C^2$$

is tangent to the parabola $y = x^2$ at the point $\left(\tfrac{1}{2}C, \tfrac{1}{4}C^2\right)$. Explain why this implies that $y = x^2$ is a singular solution of the given Clairaut equation. This singular solution and the one-parameter family of straight line solutions are illustrated in Fig. 1.6.9.

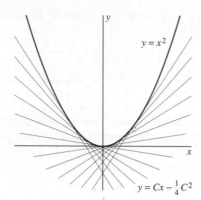

FIGURE 1.6.9. Solutions of the Clairaut equation of Problem 67. The "typical" straight line with equation $y = Cx - \frac{1}{4}C^2$ is tangent to the parabola at the point $(\frac{1}{2}C, \frac{1}{4}C^2)$.

68. Derive Eq. (18) in this section from Eqs. (16) and (17).

69. In the situation of Example 7, suppose that $a = 100$ mi, $v_0 = 400$ mi/h, and $w = 40$ mi/h. Now how far northward does the wind blow the airplane?

70. As in the text discussion, suppose that an airplane maintains a heading toward an airport at the origin. If $v_0 = 500$ mi/h and $w = 50$ mi/h (with the wind blowing due north), and the plane begins at the point $(200, 150)$, show that its trajectory is described by

$$y + \sqrt{x^2 + y^2} = 2(200x^9)^{1/10}.$$

71. A river 100 ft wide is flowing north at w feet per second. A dog starts at $(100, 0)$ and swims at $v_0 = 4$ ft/s, always heading toward a tree at $(0, 0)$ on the west bank directly across from the dog's starting point. **(a)** If $w = 2$ ft/s, show that the dog reaches the tree. **(b)** If $w = 4$ ft/s, show that the dog reaches instead the point on the west bank 50 ft north of the tree. **(c)** If $w = 6$ ft/s, show that the dog never reaches the west bank.

72. In the calculus of plane curves, one learns that the *curvature* κ of the curve $y = y(x)$ at the point (x, y) is given by

$$\kappa = \frac{|y''(x)|}{[1 + y'(x)^2]^{3/2}},$$

and that the curvature of a circle of radius r is $\kappa = 1/r$. [See Example 3 in Section 12.6 of Edwards and Penney, *Calculus*, 6th edition (Upper Saddle River, NJ: Prentice Hall, 2002).] Conversely, substitute $\rho = y'$ to derive a general solution of the second-order differential equation

$$ry'' = [1 + (y')^2]^{3/2}$$

(with r constant) in the form

$$(x - a)^2 + (y - b)^2 = r^2.$$

Thus a circle of radius r (or a part thereof) is the *only* plane curve with constant curvature $1/r$.

1.6 Application: Computer Algebra Solutions

Computer algebra systems typically include commands for the "automatic" solution of differential equations. But two different such systems often give different results whose equivalence is not clear, and a single system may give the solution in an overly complicated form. Consequently, computer algebra solutions of differential equations often require considerable "processing" or simplification by a human user in order to yield concrete and applicable information. Here we illustrate these issues using the interesting differential equation

$$\frac{dy}{dx} = \sin(x - y) \tag{1}$$

that appeared in the Section 1.3 Application. The *Maple* command

```
dsolve( D(y)(x) = sin(x - y(x)), y(x));
```

yields the simple and attractive result

$$y(x) = x - 2\tan^{-1}\left(\frac{x - 2 - C1}{x - C1}\right) \tag{2}$$

that was cited there. But the supposedly equivalent *Mathematica* command

```
DSolve[ y'[x] == Sin[x - y[x]], y[x], x]
```

returns the more complicated result

$$y(x) = x \pm 2\cos^{-1}\left(\pm\frac{\sqrt{4 + 4x + x^2 + 4C1 + 2xC1 + C1^2}}{\sqrt{4 + 4x + 2x^2 + 4C1 + 4xC1 + 2C1^2}}\right). \quad (3)$$

This apparent disparity is not unusual, and with another differential equation *Mathematica* might well produce the simpler form of the solution. As an alternative to attempting to reconcile the results in Eqs. (2) and (3), a common tactic is to simplify the differential equation before submitting it to a computer algebra system.

EXERCISE 1: Show that the plausible substitution $v = x - y$ in Eq. (1) yields the separable equation

$$\frac{dv}{dx} = 1 - \sin v. \quad (4)$$

Now the *Mathematica* command

```
Integrate[1/(1 - Sin[v]), v] // TrigReduce
```

yields

$$\int \frac{1}{1 - \sin v}\, dv = \frac{\cos v}{1 - \sin v} \quad (5)$$

(*Mathematica* omits the constant of integration unless you specifically ask for it.)

EXERCISE 2: Apply the trigonometric identities $\sin 2A = 2\sin A \cos A$ and $\cos 2A = \cos^2 A - \sin^2 A$ to deduce from Eq. (5) that

$$\int \frac{1}{1 - \sin v}\, dv = \frac{1 + \tan(v/2)}{1 - \tan(v/2)} + C. \quad (6)$$

EXERCISE 3: Deduce from (6) that Eq. (4) has the general solution

$$v(x) = 2\tan^{-1}\left(\frac{x - 1 + C}{x + 1 + C}\right),$$

and hence that Eq. (1) has the general solution

$$y(x) = x - 2\tan^{-1}\left(\frac{x - 1 + C}{x + 1 + C}\right). \quad (7)$$

EXERCISE 4: Finally, reconcile the forms in Eq. (2) and Eq. (7). What is the relation between the constants C and $C1$?

FIGURE 1.6.10. Implicit solution of $y' = \sin(x - y)$ generated by a TI-89 graphing calculator.

EXERCISE 5: Show that the integral in Eq. (5) yields immediately the graphing calculator implicit solution shown in Fig. 1.6.10.

INVESTIGATION: For your own personal differential equation, let p and q be two distinct nonzero digits in your student ID number, and consider the differential equation

$$\frac{dy}{dx} = \frac{1}{p}\cos(x - qy). \tag{8}$$

(a) Find a symbolic general solution using a computer algebra system and/or some combination of the techniques listed in this project.

(b) Determine the symbolic particular solution corresponding to several typical initial conditions of the form $y(x_0) = y_0$.

(c) Determine the possible values of a and b such that the straight line $y = ax + b$ is a solution curve of Eq. (8).

(d) Plot a direction field and some typical solution curves. Can you make a connection between the symbolic solution and your (linear and nonlinear) solution curves?

Chapter 1: Summary

In this chapter we have discussed applications of and solution methods for several important types of first-order differential equations, including those that are separable (Section 1.4), linear (Section 1.5), or exact (Section 1.6). In Section 1.6 we also discussed substitution techniques that can sometimes be used to transform a given first-order differential equation into one that is either separable, linear, or exact.

Lest it appear that these methods constitute a "grab bag" of special and unrelated techniques, it is important to note that they are all versions of a single idea. Given a differential equation

$$f(x, y, y') = 0, \tag{1}$$

we attempt to write it in the form

$$\frac{d}{dx}[G(x, y)] = 0. \tag{2}$$

It is precisely to obtain the form in Eq. (2) that we multiply the terms in Eq. (1) by an appropriate integrating factor (even if all we are doing is separating the variables). But once we have found a function $G(x, y)$ such that Eqs. (1) and (2) are equivalent, a general solution is defined implicitly by means of the equation

$$G(x, y) = C \tag{3}$$

that one obtains by integrating Eq. (2).

Given a specific first-order differential equation to be solved, we can attack it by means of the following steps:

- Is it *separable*? If so, separate the variables and integrate (Section 1.4).
- Is it *linear*? That is, can it be written in the form

$$\frac{dy}{dx} + P(x)y = Q(x)?$$

If so, multiply by the integrating factor $\rho = \exp\left(\int P\,dx\right)$ of Section 1.5.

- Is it *exact*? That is, when the equation is written in the form $Mdx + Ndy = 0$, is $\partial M/\partial y = \partial N/\partial x$ (Section 1.6)?
- If the equation as it stands is not separable, linear, or exact, is there a plausible substitution that will make it so? For instance, is it homogeneous (Section 1.6)?

Many first-order differential equations succumb to the line of attack outlined here. Nevertheless, many more do not. Because of the wide availability of computers, numerical techniques are commonly used to *approximate* the solutions of differential equations that cannot be solved readily or explicitly by the methods of this chapter. Indeed, most of the solution curves shown in figures in this chapter were plotted using numerical approximations rather than exact solutions. Several numerical methods for the appropriate solution of differential equations will be discussed in Chapter 2.

Chapter 1: Review Problems

Find general solutions of the differential equations in Problems 1 through 30. Primes denote derivatives with respect to x.

1. $x^3 + 3y - xy' = 0$
2. $xy^2 + 3y^2 - x^2 y' = 0$
3. $xy + y^2 - x^2 y' = 0$
4. $2xy^3 + e^x + (3x^2 y^2 + \sin y)y' = 0$
5. $3y + x^4 y' = 2xy$ 6. $2xy^2 + x^2 y' = y^2$
7. $2x^2 y + x^3 y' = 1$ 8. $2xy + x^2 y' = y^2$
9. $xy' + 2y = 6x^2 \sqrt{y}$
10. $y' = 1 + x^2 + y^2 + x^2 y^2$
11. $x^2 y' = xy + 3y^2$
12. $6xy^3 + 2y^4 + (9x^2 y^2 + 8xy^3)y' = 0$
13. $4xy^2 + y' = 5x^4 y^2$ 14. $x^3 y' = x^2 y - y^3$
15. $y' + 3y = 3x^2 e^{-3x}$ 16. $y' = x^2 - 2xy + y^2$
17. $e^x + ye^{xy} + (e^y + xe^{yx})y' = 0$
18. $2x^2 y - x^3 y' = y^3$
19. $3x^5 y^2 + x^3 y' = 2y^2$
20. $xy' + 3y = 3x^{-3/2}$
21. $(x^2 - 1)y' + (x - 1)y = 1$
22. $xy' = 6y + 12x^4 y^{2/3}$
23. $e^y + y\cos x + (xe^y + \sin x)y' = 0$

24. $9x^2 y^2 + x^{3/2} y' = y^2$
25. $2y + (x + 1)y' = 3x + 3$
26. $9x^{1/2} y^{4/3} - 12x^{1/5} y^{3/2} + (8x^{3/2} y^{1/3} - 15x^{6/5} y^{1/2})y' = 0$
27. $3y + x^3 y^4 + 3xy' = 0$
28. $y + xy' = 2e^{2x}$
29. $(2x + 1)y' + y = (2x + 1)^{3/2}$
30. $y' = \sqrt{x + y}$

Each of the differential equations in Problems 31 through 36 is of two different types considered in this chapter—separable, linear, homogeneous, Bernoulli, exact, etc. Hence, derive general solutions for each of these equations in two different ways; then reconcile your results.

31. $\dfrac{dy}{dx} = 3(y + 7)x^2$ 32. $\dfrac{dy}{dx} = xy^3 - xy$

33. $\dfrac{dy}{dx} = -\dfrac{3x^2 + 2y^2}{4xy}$ 34. $\dfrac{dy}{dx} = \dfrac{x + 3y}{y - 3x}$

35. $\dfrac{dy}{dx} = \dfrac{2xy + 2x}{x^2 + 1}$ 36. $\dfrac{dy}{dx} = \dfrac{\sqrt{y} - y}{\tan x}$

C H A P T E R

Mathematical Models and Numerical Methods

2

2.1 Population Models

In Section 1.4 we introduced the exponential differential equation $dP/dt = kP$, with solution $P(t) = P_0 e^{kt}$, as a mathematical model for natural population growth that occurs as a result of constant birth and death rates. Here we present a more general population model that accommodates birth and death rates that are not necessarily constant. As before, however, our population function $P(t)$ will be a *continuous* approximation to the actual population, which of course grows by integral increments.

Suppose that the population changes only by the occurrence of births and deaths—there is no immigration or emigration from outside the country or environment under consideration. It is customary to track the growth or decline of a population in terms of its *birth rate* and *death rate* functions defined as follows:

- $\beta(t)$ is the number of births per unit of population per unit of time at time t;
- $\delta(t)$ is the number of deaths per unit of population per unit of time at time t.

Then the numbers of births and deaths that occur during the time interval $[t, t + \Delta t]$ is given (approximately) by

$$\text{births:} \quad \beta(t) \cdot P(t) \cdot \Delta t, \qquad \text{deaths:} \quad \delta(t) \cdot P(t) \cdot \Delta t.$$

Hence the change ΔP in the population during the time interval $[t, t + \Delta t]$ of length Δt is

$$\Delta P = \{\text{births}\} - \{\text{deaths}\} \approx \beta(t) \cdot P(t) \cdot \Delta t - \delta(t) \cdot P(t) \cdot \Delta t,$$

so

$$\frac{\Delta P}{\Delta t} \approx [\beta(t) - \delta(t)] \, P(t).$$

77

The error in this approximation should approach zero as $\Delta t \to 0$, so—taking the limit—we get the differential equation

➤
$$\frac{dP}{dt} = (\beta - \delta)P, \qquad (1)$$

in which we write $\beta = \beta(t)$, $\delta = \delta(t)$, and $P = P(t)$ for brevity. Equation (1) is the **general population equation.** If β and δ are constant, Eq. (1) reduces to the natural growth equation with $k = \beta - \delta$. But it also includes the possibility that β and δ are variable functions of t. The birth and death rates need not be known in advance; they may well depend on the unknown function $P(t)$.

Example 1	Suppose that an alligator population numbers 100 initially, and that its death rate is $\delta = 0$ (so none of the alligators is dying). If the birth rate is $\beta = (0.0005)P$—and thus increases as the population does—then Eq. (1) gives the initial value problem

$$\frac{dP}{dt} = (0.0005)P^2, \quad P(0) = 100$$

(with t in years). Then upon separating the variables we get

$$\int \frac{1}{P^2}\,dP = \int (0.0005)\,dt;$$

$$-\frac{1}{P} = (0.0005)t + C.$$

Substitution of $t = 0$, $P = 100$ gives $C = -1/100$, and then we readily solve for

$$P(t) = \frac{2000}{20 - t}.$$

For instance, $P(10) = 2000/10 = 200$, so after 10 years the alligator population has doubled. But we see that $P \to +\infty$ as $t \to 20$, so a real "population explosion" occurs in 20 years. Indeed, the direction field and solution curves shown in Fig. 2.1.1 indicate that a population explosion always occurs, whatever the size of the (positive) initial population $P(0) = P_0$. In particular, it appears that the population always becomes unbounded in a *finite* period of time. ∎

FIGURE 2.1.1. Slope field and solution curves for the equation $dP/dt = (0.0005)P^2$ in Example 1.

Bounded Populations and the Logistic Equation

In situations as diverse as the human population of a nation and a fruit fly population in a closed container, it is often observed that the birth rate decreases as the population itself increases. The reasons may range from increased scientific or cultural sophistication to a limited food supply. Suppose, for example, that the birth rate β is a *linear* decreasing function of the population size P, so that $\beta = \beta_0 - \beta_1 P$, where β_0 and β_1 are positive constants. If the death rate $\delta = \delta_0$ remains constant, then Eq. (1) takes the form

$$\frac{dP}{dt} = (\beta_0 - \beta_1 P - \delta_0)P;$$

that is,

$$\frac{dP}{dt} = aP - bP^2, \tag{2}$$

where $a = \beta_0 - \delta_0$ and $b = \beta_1$.

If the coefficients a and b are both positive, then Eq. (2) is called the **logistic equation.** For the purpose of relating the behavior of the population $P(t)$ to the values of the parameters in the equation, it is useful to rewrite the logistic equation in the form

$$\frac{dP}{dt} = kP(M - P), \tag{3}$$

where $k = b$ and $M = a/b$ are constants.

Example 2

In Example 4 of Section 1.3 we explored graphically a population that is modeled by the logistic equation

$$\frac{dP}{dt} = 0.0004P(150 - P) = 0.06P - 0.0004P^2. \tag{4}$$

To solve this differential equation symbolically, we separate the variables and integrate. We get

$$\int \frac{dP}{P(150 - P)} = \int 0.0004 \, dt,$$

$$\frac{1}{150} \int \left(\frac{1}{P} + \frac{1}{150 - P} \right) dP = \int 0.0004 \, dt, \qquad \text{[partial fractions]}$$

$$\ln|P| - \ln|150 - P| = 0.06t + C,$$

$$\frac{P}{150 - P} = \pm e^C e^{0.06t} = Be^{0.06t} \quad \text{[where } B = \pm e^C\text{]}.$$

If we substitute $t = 0$ and $P = P_0 \neq 150$ into this last equation we find that $B = P_0/(150 - P_0)$. Hence

$$\frac{P}{150 - P} = \frac{P_0 e^{0.06t}}{150 - P_0}.$$

Finally, this equation is easy to solve for the population

$$P(t) = \frac{150P_0}{P_0 + (150 - P_0)e^{-0.06t}} \tag{5}$$

FIGURE 2.1.2. Typical solution curves for the logistic equation $P' = 0.06P - 0.0004P^2$.

at time t in terms of the initial population $P_0 = P(0)$. Figure 2.1.2 shows a number of solution curves corresponding to different values of the initial population ranging from $P_0 = 20$ to $P_0 = 300$. Note that all these solution curves appear to approach the horizontal line $P = 150$ as an asymptote. Indeed, you should be able to see directly from Eq. (5) that $\lim_{t \to \infty} P(t) = 150$, whatever the initial value $P_0 > 0$. ∎

Limiting Populations and Carrying Capacity

The finite limiting population noted in Example 2 is characteristic of logistic populations. In Problem 32 we ask you to use the method of solution of Example 2 to show that the solution of the logistic initial value problem

$$\frac{dP}{dt} = kP(M - P), \quad P(0) = P_0 \tag{6}$$

is

$$P(t) = \frac{MP_0}{P_0 + (M - P_0)e^{-kMt}}. \tag{7}$$

Actual animal populations are positive valued. If $P_0 = M$, then (7) reduces to the unchanging (constant-valued) "equilibrium population" $P(t) \equiv M$. Otherwise, the behavior of a logistic population depends on whether $0 < P_0 < M$ or $P_0 > M$. If $0 < P_0 < M$, then we see from (6) and (7) that $P' > 0$ and

$$P(t) = \frac{MP_0}{P_0 + (M - P_0)e^{-kMt}} = \frac{MP_0}{P_0 + \{\text{pos. number}\}} < \frac{MP_0}{P_0} = M.$$

However, if $P_0 > M$, then we see from (6) and (7) that $P' < 0$ and

$$P(t) = \frac{MP_0}{P_0 + (M - P_0)e^{-kMt}} = \frac{MP_0}{P_0 + \{\text{neg. number}\}} > \frac{MP_0}{P_0} = M.$$

In either case, the "positive number" or "negative number" in the denominator has absolute value less than P_0 and—because of the exponential factor—approaches 0 as $t \to +\infty$. It follows that

$$\lim_{t \to +\infty} P(t) = \frac{MP_0}{P_0 + 0} = M. \tag{8}$$

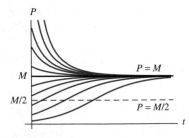

FIGURE 2.1.3. Typical solution curves for the logistic equation $P' = kP(M - P)$. Each solution curve that starts below the line $P = M/2$ has an inflection point on this line. (See Problem 34.)

Thus a population that satisfies the logistic equation does *not* grow without bound like a naturally growing population modeled by the exponential equation $P' = kP$. Instead, it approaches the finite **limiting population** M as $t \to +\infty$. As illustrated by the typical logistic solution curves in Fig. 2.1.3, the population $P(t)$ steadily increases and approaches M from below if $0 < P_0 < M$, but steadily decreases and approaches M from above if $P_0 > M$. Sometimes M is called the **carrying capacity** of the environment, considering it to be the maximum population that the environment can support on a long-term basis.

| Example 3 | Suppose that in 1885 the population of a certain country was 50 million and was growing at the rate of 750,000 people per year at that time. Suppose also that in 1940 its population was 100 million and was then growing at the rate of 1 million per year. Assume that this population satisfies the logistic equation. Determine both the limiting population M and the predicted population for the year 2000. |

| Solution | We substitute the two given pairs of data in Eq. (3) and find that |

$$0.75 = 50k(M - 50), \quad 1.00 = 100k(M - 100).$$

We solve simultaneously for $M = 200$ and $k = 0.0001$. Thus the limiting population of the country in question is 200 million. With these values of M and k, and with $t = 0$ corresponding to the year 1940 (in which $P_0 = 100$), we find that— according to Eq. (7)—the population in the year 2000 will be

$$P(60) = \frac{100 \cdot 200}{100 + (200 - 100)e^{-(0.0001)(200)(60)}},$$

about 153.7 million people. ∎

Historical Note

The logistic equation was introduced (around 1840) by the Belgian mathematician and demographer P. F. Verhulst as a possible model for human population growth. In the next two examples we compare natural growth and logistic model fits to the 19th-century U.S. population census data, then compare projections for the 20th century.

| Example 4 | The U.S. population in 1800 was 5.308 million and in 1900 was 76.212 million. If we take $P_0 = 5.308$ (with $t = 0$ in 1800) in the natural growth model $P(t) = P_0e^{rt}$ and substitute $t = 100$, $P = 76.212$, we find that |

$$76.212 = 5.308e^{100r}, \quad \text{so} \quad r = \frac{1}{100} \ln \frac{76.212}{5.308} \approx 0.026643.$$

Thus our natural growth model for the U.S. population during the 19th century is

$$P(t) = (5.308)e^{(0.026643)t} \tag{9}$$

(with t in years and P in millions). Because $e^{0.026643} \approx 1.02700$, the average population growth between 1800 and 1900 was about 2.7% per year. ∎

| Example 5 | The U.S. population in 1850 was 23.192 million. If we take $P_0 = 5.308$ and substitute the data pairs $t = 50$, $P = 23.192$ (for 1850) and $t = 100$, $P = 76.212$ (for 1900) in the logistic model formula in Eq. (7), we get the two equations |

$$\frac{(5.308)M}{5.308 + (M - 5.308)e^{-50kM}} = 23.192,$$

$$\frac{(5.308)M}{5.308 + (M - 5.308)e^{-100kM}} = 76.212 \tag{10}$$

in the two unknowns k and M. Nonlinear systems like this ordinarily are solved numerically using an appropriate computer system. But with the right algebraic trick (Problem 36 in this section) the equations in (10) can be solved manually for $k = 0.000167716$, $M = 188.121$. Substitution of these values in Eq. (7) yields the logistic model

$$P(t) = \frac{998.546}{5.308 + (182.813)e^{-(0.031551)t}}. \tag{11}$$

The table in Fig. 2.1.4 compares the actual 1800–1990 U.S. census population figures with those predicted by the exponential growth model in (9) and the logistic model in (11). Both agree well with the 19th-century figures. But the exponential model diverges appreciably from the census data in the early decades of the 20th century, whereas the logistic model remains accurate until 1940. By the end of the 20th century the exponential model vastly overestimates the actual U.S. population—predicting over a billion in the year 2000—whereas the logistic model somewhat underestimates it.

Year	Actual U.S. Pop.	Exponential Model	Exponential Error	Logistic Model	Logistic Error
1800	5.308	5.308	0.000	5.308	0.000
1810	7.240	6.929	0.311	7.202	0.038
1820	9.638	9.044	0.594	9.735	−0.097
1830	12.861	11.805	1.056	13.095	−0.234
1840	17.064	15.409	1.655	17.501	−0.437
1850	23.192	20.113	3.079	23.192	0.000
1860	31.443	26.253	5.190	30.405	1.038
1870	38.558	34.268	4.290	39.326	−0.768
1880	50.189	44.730	5.459	50.034	0.155
1890	62.980	58.387	4.593	62.435	0.545
1900	76.212	76.212	0.000	76.213	−0.001
1910	92.228	99.479	−7.251	90.834	1.394
1920	106.022	129.849	−23.827	105.612	0.410
1930	123.203	169.492	−46.289	119.834	3.369
1940	132.165	221.237	−89.072	132.886	−0.721
1950	151.326	288.780	−137.454	144.354	6.972
1960	179.323	376.943	−197.620	154.052	25.271
1970	203.302	492.023	−288.721	161.990	41.312
1980	226.542	642.236	−415.694	168.316	58.226
1990	248.710	838.308	−589.598	173.252	76.458
2000	281.422	1094.240	−812.818	177.038	104.384

FIGURE 2.1.4. Comparison of exponential growth and logistic models with U.S. census populations (in millions).

In order to measure the extent to which a given model fits actual data, it is customary to define the **average error** (in the model) as *the square root of the average of the squares of the individual errors* (the latter appearing in the fourth and sixth columns of the table in Fig. 2.1.4). Using only the 1800–1900 data, this definition gives 3.162 for the average error in the exponential model, while the average error

in the logistic model is only 0.452. Consequently, even in 1900 we might well have anticipated that the logistic model would predict the U.S. population growth during the 20th century more accurately than the exponential model. ∎

The moral of Examples 4 and 5 is simply that one should not expect too much of models that are based on severely limited information (such as just a pair of data points). Much of the science of *statistics* is devoted to the analysis of large "data sets" to formulate useful (and perhaps reliable) mathematical models.

More Applications of the Logistic Equation

We next describe some situations that illustrate the varied circumstances in which the logistic equation is a satisfactory mathematical model.

1. *Limited environment situation.* A certain environment can support a population of at most M individuals. It is then reasonable to expect the growth rate $\beta - \delta$ (the combined birth and death rates) to be proportional to $M - P$, because we may think of $M - P$ as the potential for further expansion. Then $\beta - \delta = k(M - P)$, so that

$$\frac{dP}{dt} = (\beta - \delta)P = kP(M - P).$$

 The classic example of a limited environment situation is a fruit fly population in a closed container.

2. *Competition situation.* If the birth rate β is constant but the death rate δ is proportional to P, so that $\delta = \alpha P$, then

$$\frac{dP}{dt} = (\beta - \alpha P)P = kP(M - P).$$

 This might be a reasonable working hypothesis in a study of a cannibalistic population, in which all deaths result from chance encounters between individuals. Of course, competition between individuals is not usually so deadly, nor its effects so immediate and decisive.

3. *Joint proportion situation.* Let $P(t)$ denote the number of individuals in a constant-size susceptible population M who are infected with a certain contagious and incurable disease. The disease is spread by chance encounters. Then $P'(t)$ should be proportional to the product of the number P of individuals having the disease and the number $M - P$ of those not having it, and therefore $dP/dt = kP(M - P)$. Again we discover that the mathematical model is the logistic equation. The mathematical description of the spread of a rumor in a population of M individuals is identical.

Example 6

Suppose that at time $t = 0$, 10 thousand people in a city with population $M = 100$ thousand people have heard a certain rumor. After 1 week the number $P(t)$ of those who have heard it has increased to $P(1) = 20$ thousand. Assuming that $P(t)$ satisfies a logistic equation, when will 80% of the city's population have heard the rumor?

Solution

Substituting $P_0 = 10$ and $M = 100$ (thousand) in Eq. (7), we get

$$P(t) = \frac{1000}{10 + 90e^{-100kt}}. \tag{12}$$

Then substitution of $t = 1$, $P = 20$ gives the equation

$$20 = \frac{1000}{10 + 90e^{-100k}}$$

that is readily solved for

$$e^{-100k} = \tfrac{4}{9}, \quad \text{so} \quad k = \tfrac{1}{100} \ln \tfrac{9}{4} \approx 0.008109.$$

With $P(t) = 80$, Eq. (12) takes the form

$$80 = \frac{1000}{10 + 90e^{-100kt}},$$

which we solve for $e^{-100kt} = \tfrac{1}{36}$. It follows that 80% of the population has heard the rumor when

$$t = \frac{\ln 36}{100k} = \frac{\ln 36}{\ln \tfrac{9}{4}} \approx 4.42,$$

thus after about 4 weeks and 3 days. ∎

Doomsday versus Extinction

Consider a population $P(t)$ of unsophisticated animals in which females rely solely on chance encounters to meet males for reproductive purposes. It is reasonable to expect such encounters to occur at a rate that is proportional to the product of the number $P/2$ of males and the number $P/2$ of females, hence at a rate proportional to P^2. We therefore assume that births occur at the rate kP^2 (per unit time, with k constant). The birth rate (births/time/population) is then given by $\beta = kP$. If the death rate δ is constant, then the general population equation in (1) yields the differential equation

➤
$$\frac{dP}{dt} = kP^2 - \delta P = kP(P - M) \tag{13}$$

(where $M = \delta/k > 0$) as a mathematical model of the population.

Note that the right-hand side in Eq. (13) is the *negative* of the right-hand side in the logistic equation in (3). We will see that the constant M is now a **threshold population**, with the way the population behaves in the future depending critically on whether the initial population P_0 is less than or greater than M.

Example 7 Consider an animal population $P(t)$ that is modeled by the equation

$$\frac{dP}{dt} = 0.0004P(P - 150) = 0.0004P^2 - 0.06P. \tag{14}$$

We want to find $P(t)$ if (a) $P(0) = 200$; (b) $P(0) = 100$.

Solution To solve the equation in (14), we separate the variables and integrate. We get

$$\int \frac{dP}{P(P-150)} = \int 0.0004\,dt,$$

$$-\frac{1}{150}\int\left(\frac{1}{P}-\frac{1}{P-150}\right)dP = \int 0.0004\,dt, \quad \text{[partial fractions]}$$

$$\ln|P| - \ln|P-150| = -0.06t + C,$$

$$\frac{P}{P-150} = \pm e^{C}e^{-0.06t} = Be^{-0.06t} \qquad \text{[where } B = \pm e^{C}\text{].} \qquad (15)$$

(a) Substitution of $t = 0$ and $P = 200$ into (15) gives $B = 4$. With this value of B we solve Eq. (15) for

$$P(t) = \frac{600e^{-0.06t}}{4e^{-0.06t}-1}. \qquad (16)$$

Note that, as t increases and approaches $T = \ln(4)/0.06 \approx 23.105$, the positive denominator on the right in (16) decreases and approaches 0. Consequently $P(t) \to +\infty$ as $t \to T^{-}$. This is a *doomsday* situation—a real population *explosion*.
(b) Substitution of $t = 0$ and $P = 100$ into (15) gives $B = -2$. With this value of B we solve Eq. (15) for

$$P(t) = \frac{300e^{-0.06t}}{2e^{-0.06t}+1} = \frac{300}{2+e^{0.06t}}. \qquad (17)$$

Note that, as t increases without bound, the positive denominator on the right in (16) approaches $+\infty$. Consequently, $P(t) \to 0$ as $t \to +\infty$. This is an (eventual) *extinction* situation. ∎

Thus the population in Example 7 either explodes or is an endangered species threatened with extinction, depending on whether or not its initial size exceeds the threshold population $M = 150$. An approximation to this phenomenon is sometimes observed with animal populations, such as the alligator population in certain areas of the southern United States.

Figure 2.1.5 shows typical solution curves that illustrate the two possibilities for a population $P(t)$ satisfying Eq. (13). If $P_0 = M$ (exactly!), then the population remains constant. However, this equilibrium situation is very unstable. If P_0 exceeds M (even slightly), then $P(t)$ rapidly increases without bound, whereas if the initial (positive) population is less than M (however slightly), then it decreases (more gradually) toward zero as $t \to +\infty$. See Problem 33.

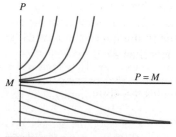

FIGURE 2.1.5. Typical solution curves for the explosion-extinction equation $P' = kP(P - M)$.

2.1 Problems

Separate variables and use partial fractions to solve the initial value problems in Problems 1–8. Use either the exact solution or a computer-generated slope field to sketch the graphs of several solutions of the given differential equation, and highlight the indicated particular solution.

1. $\dfrac{dx}{dt} = x - x^2$, $x(0) = 2$ **2.** $\dfrac{dx}{dt} = 10x - x^2$, $x(0) = 1$

3. $\dfrac{dx}{dt} = 1 - x^2$, $x(0) = 3$ **4.** $\dfrac{dx}{dt} = 9 - 4x^2$, $x(0) = 0$

5. $\dfrac{dx}{dt} = 3x(5 - x)$, $x(0) = 8$

6. $\dfrac{dx}{dt} = 3x(x - 5)$, $x(0) = 2$

7. $\dfrac{dx}{dt} = 4x(7 - x)$, $x(0) = 11$

8. $\dfrac{dx}{dt} = 7x(x - 13)$, $x(0) = 17$

9. The time rate of change of a rabbit population P is proportional to the square root of P. At time $t = 0$ (months) the population numbers 100 rabbits and is increasing at the rate of 20 rabbits per month. How many rabbits will there be one year later?

10. Suppose that the fish population $P(t)$ in a lake is attacked by a disease at time $t = 0$, with the result that the fish cease to reproduce (so that the birth rate is $\beta = 0$) and the death rate δ (deaths per week per fish) is thereafter proportional to $1/\sqrt{P}$. If there were initially 900 fish in the lake and 441 were left after 6 weeks, how long did it take all the fish in the lake to die?

11. Suppose that when a certain lake is stocked with fish, the birth and death rates β and δ are both inversely proportional to \sqrt{P}. **(a)** Show that

$$P(t) = \left(\tfrac{1}{2}kt + \sqrt{P_0} \right)^2,$$

where k is a constant. **(b)** If $P_0 = 100$ and after 6 months there are 169 fish in the lake, how many will there be after 1 year?

12. The time rate of change of an alligator population P in a swamp is proportional to the square of P. The swamp contained a dozen alligators in 1988, two dozen in 1998. When will there be four dozen alligators in the swamp? What happens thereafter?

13. Consider a prolific breed of rabbits whose birth and death rates, β and δ, are each proportional to the rabbit population $P = P(t)$, with $\beta > \delta$. **(a)** Show that

$$P(t) = \frac{P_0}{1 - kP_0t}, \quad k \text{ constant.}$$

Note that $P(t) \to +\infty$ as $t \to 1/(kP_0)$. This is doomsday. **(b)** Suppose that $P_0 = 6$ and that there are nine rabbits after ten months. When does doomsday occur?

14. Repeat part (a) of Problem 13 in the case $\beta < \delta$. What now happens to the rabbit population in the long run?

15. Consider a population $P(t)$ satisfying the logistic equation $dP/dt = aP - bP^2$, where $B = aP$ is the time rate at which births occur and $D = bP^2$ is the rate at which deaths occur. If the initial population is $P(0) = P_0$, and B_0 births per month and D_0 deaths per month are occurring at time $t = 0$, show that the limiting population is $M = B_0 P_0 / D_0$.

16. Consider a rabbit population $P(t)$ satisfying the logistic equation as in Problem 15. If the initial population is 120 rabbits and there are 8 births per month and 6 deaths per month occurring at time $t = 0$, how many months does it take for $P(t)$ to reach 95% of the limiting population M?

17. Consider a rabbit population $P(t)$ satisfying the logistic equation as in Problem 15. If the initial population is 240 rabbits and there are 9 births per month and 12 deaths per month occurring at time $t = 0$, how many months does it take for $P(t)$ to reach 105% of the limiting population M?

18. Consider a population $P(t)$ satisfying the extinction-explosion equation $dP/dt = aP^2 - bP$, where $B = aP^2$ is the time rate at which births occur and $D = bP$ is the rate at which deaths occur. If the initial population is $P(0) = P_0$ and B_0 births per month and D_0 deaths per month are occurring at time $t = 0$, show that the threshold population is $M = D_0 P_0 / B_0$.

19. Consider an alligator population $P(t)$ satisfying the extinction/explosion equation as in Problem 18. If the initial population is 100 alligators and there are 10 births per month and 9 deaths per months occurring at time $t = 0$, how many months does it take for $P(t)$ to reach 10 times the threshold population M?

20. Consider an alligator population $P(t)$ satisfying the extinction/explosion equation as in Problem 18. If the initial population is 110 alligators and there are 11 births per month and 12 deaths per month occurring at time $t = 0$, how many months does it take for $P(t)$ to reach 10% of the threshold population M?

21. Suppose that the population $P(t)$ of a country satisfies the differential equation $dP/dt = kP(200 - P)$ with k constant. Its population in 1940 was 100 million and was then growing at the rate of 1 million per year. Predict this country's population for the year 2000.

22. Suppose that at time $t = 0$, half of a "logistic" population of 100,000 persons have heard a certain rumor, and that the number of those who have heard it is then increasing at the rate of 1000 persons per day. How long will it take for this rumor to spread to 80% of the population? (*Suggestion*: Find the value of k by substituting $P(0)$ and $P'(0)$ in the logistic equation, Eq. (3).)

23. As the salt KNO_3 dissolves in methanol, the number $x(t)$ of grams of the salt in a solution after t seconds satisfies the differential equation $dx/dt = 0.8x - 0.004x^2$.

(a) What is the maximum amount of the salt that will ever dissolve in the methanol?

(b) If $x = 50$ when $t = 0$, how long will it take for an additional 50 g of salt to dissolve?

24. Suppose that a community contains 15,000 people who are susceptible to Michaud's syndrome, a contagious disease. At time $t = 0$ the number $N(t)$ of people who have developed Michaud's syndrome is 5000 and is increasing at the rate of 500 per day. Assume that $N'(t)$ is proportional to the product of the numbers of those who have caught the disease and of those who have not. How long will it take for another 5000 people to develop Michaud's syndrome?

25. The data in the table in Fig. 2.1.6 are given for a certain population $P(t)$ that satisfies the logistic equation in (3). **(a)** What is the limiting population M? (*Suggestion*: Use the approximation

$$P'(t) \approx \frac{P(t+h) - P(t-h)}{2h}$$

with $h = 1$ to estimate the values of $P'(t)$ when $P = 25.00$ and when $P = 47.54$. Then substitute these values in the logistic equation and solve for k and M.) **(b)** Use the values of k and M found in part (a) to determine when $P = 75$. (*Suggestion*: Take $t = 0$ to correspond to the year 1925.)

Year	P (millions)
1924	24.63
1925	25.00
1926	25.38
⋮	⋮
1974	47.04
1975	47.54
1976	48.04

FIGURE 2.1.6. Population data for Problem 25.

26. A population $P(t)$ of small rodents has birth rate $\beta = (0.001)P$ (births per month per rodent) and *constant* death rate δ. If $P(0) = 100$ and $P'(0) = 8$, how long (in months) will it take this population to double to 200 rodents? (*Suggestion*: First find the value of δ.)

27. Consider an animal population $P(t)$ with constant death rate $\delta = 0.01$ (deaths per animal per month) and with birth rate β proportional to P. Suppose that $P(0) = 200$ and $P'(0) = 2$. **(a)** When is $P = 1000$? **(b)** When does doomsday occur?

28. Suppose that the number $x(t)$ (with t in months) of alligators in a swamp satisfies the differential equation $dP/dt = 0.0001x^2 - 0.01x$.

(a) If initially there are 25 alligators in the swamp, solve this differential equation to determine what happens to the alligator population in the long run.

(b) Repeat part (a), except with 150 alligators initially.

29. During the period from 1790 to 1930, the U.S. population $P(t)$ (t in years) grew from 3.9 million to 123.2 million. Throughout this period, $P(t)$ remained close to the solution of the initial value problem

$$\frac{dP}{dt} = 0.03135P - 0.0001489P^2, \quad P(0) = 3.9.$$

(a) What 1930 population does this logistic equation predict?

(b) What limiting population does it predict?

(c) Has this logistic equation continued since 1930 to accurately model the U.S. population?

[This problem is based on a computation by Verhulst, who in 1845 used the 1790–1840 U.S. population data to predict accurately the U.S. population through the year 1930 (long after his own death, of course).]

30. A tumor may be regarded as a population of multiplying cells. It is found empirically that the "birth rate" of the cells in a tumor decreases exponentially with time, so that $\beta(t) = \beta_0 e^{-\alpha t}$ (where α and β_0 are positive constants), and hence

$$\frac{dP}{dt} = \beta_0 e^{-\alpha t} P, \quad P(0) = P_0.$$

Solve this initial value problem for

$$P(t) = P_0 \exp\left(\frac{\beta_0}{\alpha}(1 - e^{-\alpha t})\right).$$

Observe that $P(t)$ approaches the finite limiting population $P_0 \exp(\beta_0/\alpha)$ as $t \to +\infty$.

31. For the tumor of Problem 30, suppose that at time $t = 0$ there are $P_0 = 10^6$ cells and that $P(t)$ is then increasing at the rate of 3×10^5 cells per month. After 6 months the tumor has doubled (in size and in number of cells). Solve numerically for α, and then find the limiting population of the tumor.

32. Derive the solution

$$P(t) = \frac{MP_0}{P_0 + (M - P_0)e^{-kMt}}$$

of the logistic initial value problem $P' = kP(M - P)$, $P(0) = P_0$. Make it clear how your derivation depends on whether $0 < P_0 < M$ or $P_0 > M$.

33. (a) Derive the solution

$$P(t) = \frac{MP_0}{P_0 + (M - P_0)e^{kMt}}$$

of the extinction-explosion initial value problem $P' = kP(P - M)$, $P(0) = P_0$.

(b) How does the behavior of $P(t)$ as t increases depend on whether $0 < P_0 < M$ or $P_0 > M$?

34. If $P(t)$ satisfies the logistic equation in (3), use the chain rule to show that

$$P''(t) = 2k^2 P(P - \tfrac{1}{2}M)(P - M).$$

Conclude that $P'' > 0$ if $0 < P < \tfrac{1}{2}M$; $P'' = 0$ if $P = \tfrac{1}{2}M$; $P'' < 0$ if $\tfrac{1}{2}M < P < M$; and $P'' > 0$ if $P > M$. In particular, it follows that any solution curve that crosses the line $P = \tfrac{1}{2}M$ has an inflection point where it crosses that line, and therefore resembles one of the lower S-shaped curves in Fig. 2.1.3.

35. Consider two population functions $P_1(t)$ and $P_2(t)$, both of which satisfy the logistic equation with the same limiting population M, but with different values k_1 and k_2 of the constant k in Eq. (3). Assume that $k_1 < k_2$. Which population approaches M the most rapidly? You can reason *geometrically* by examining slope fields (especially if appropriate software is available), *symbolically* by analyzing the solution given in Eq. (7), or *numerically* by substituting successive values of t.

36. To solve the two equations in (10) for the values of k and M, begin by solving the first equation for the quantity $x = e^{-50kM}$ and the second equation for $x^2 = e^{-100kM}$. Upon equating the two resulting expressions for x^2 in terms of M, you get an equation that is readily solved for M. With M now known, either of the original equations is readily solved for k. This technique can be used to "fit" the logistic equation to any three population values P_0, P_1, and P_2 corresponding to *equally spaced* times $t_0 = 0$, t_1, and $t_2 = 2t_1$.

37. Use the method of Problem 36 to fit the logistic equation to the actual U.S. population data (Fig. 2.1.4) for the years 1850, 1900, and 1950. Solve the resulting logistic equation and compare the predicted and actual populations for the years 1990 and 2000.

38. Fit the logistic equation to the actual U.S. population data (Fig. 2.1.4) for the years 1900, 1930, and 1960. Solve the resulting logistic equation, then compare the predicted and actual populations for the years 1980, 1990, and 2000.

39. Birth and death rates of animal populations typically are not constant; instead, they vary periodically with the passage of seasons. Find $P(t)$ if the population P satisfies the differential equation

$$\frac{dP}{dt} = (k + b\cos 2\pi t)P,$$

where t is in years and k and b are positive constants. Thus the growth rate function $r(t) = k + b\cos 2\pi t$ varies periodically about its mean value k. Construct a graph that contrasts the growth of this population with one that has the same initial value P_0 but satisfies the natural growth equation $P' = kP$ (same constant k). How would the two populations compare after the passage of many years?

2.1 Application: Logistic Modeling of Population Data

These investigations deal with the problem of fitting a logistic model to given population data. Thus we want to determine the numerical constants a and b so that the solution $P(t)$ of the initial value problem

$$\frac{dP}{dt} = aP + bP^2, \quad P(0) = P_0 \tag{1}$$

approximates the given values P_0, P_1, ..., P_n of the population at the times $t_0 = 0$, t_1, \ldots, t_n. If we rewrite Eq. (1) (the logistic equation with $kM = a$ and $k = -b$) in the form

$$\frac{1}{P}\frac{dP}{dt} = a + bP, \tag{2}$$

then we see that the points

$$\left(P(t_i), \frac{P'(t_i)}{P(t_i)}\right), \quad i = 0, 1, 2, \ldots, n,$$

should all lie on the straight line with y-intercept a and slope b (as determined by the function of P on the right-hand side in Eq. (2)).

This observation provides a way to find a and b. If we can determine the approximate values of the derivatives P'_1, P'_2, ... corresponding to the given population data, then we can proceed with the following agenda:

- First plot the points $(P_1, P_1'/P_1)$, $(P_2, P_2'/P_2)$, ... on a sheet of graph paper with horizontal P-axis.
- Then use a ruler to draw a straight line that appears to approximate these points well.
- Finally, measure this straight line's y-intercept a and slope b.

But where are we to find the needed values of the derivative $P'(t)$ of the (as yet) unknown function P? It is easiest to use the approximation

$$P_i' = \frac{P_{i+1} - P_{i-1}}{t_{i+1} - t_{i-1}} \tag{3}$$

suggested by Fig. 2.1.7. For instance, if we take $i = 0$ corresponding to the year 1790, then the U.S. population data in Fig. 2.1.8 give

$$P_1' = \frac{P_2 - P_0}{t_2 - t_0} = \frac{7.240 - 3.929}{20} \approx 0.166$$

for the slope at (t_1, P_1) corresponding to the year 1800.

FIGURE 2.1.7. The symmetric difference approximation
$\dfrac{P_{i+1} - P_{i-1}}{t_{i+1} - t_{i-1}}$ to the derivative $P'(t_i)$.

INVESTIGATION A: Use Eq. (3) to verify the slope figures shown in the final column of the table in Fig. 2.1.8, then plot the points $(P_1, P_1'/P_1)$, ..., $(P_{11}, P_{11}'/P_{11})$ indicated by the asterisks in Fig. 2.1.9. If an appropriate graphing calculator, spreadsheet, or computer program is available, use it to find the straight line $y = a + bP$ as in (2) that best fits these points. If not, draw your own straight line approximating these points, and then measure its intercept a and slope b as accurately as you can. Next, solve the logistic equation in (1) with these numerical parameters, taking $t = 0$ corresponding to the year 1800. Finally, compare the predicted 20th-century U.S. population figures with the actual data listed in Fig. 2.1.4.

INVESTIGATION B: Repeat Investigation A, but take $t = 0$ in 1900 and use only 20th-century population data. Do you get a better approximation for the U.S. population during the final decades of the 20th century?

INVESTIGATION C: Model similarly the world population data shown in Fig. 2.1.10. The Population Division of the United Nations predicts a world population of 8.177 billion in the year 2025. What do you predict?

Year	i	t_i	Population P_i	Slope P_i'
1790	0	−10	3.929	
1800	1	0	5.308	0.166
1810	2	10	7.240	0.217
1820	3	20	9.638	0.281
1830	4	30	12.861	0.371
1840	5	40	17.064	0.517
1850	6	50	23.192	0.719
1860	7	60	31.443	0.768
1870	8	70	38.558	0.937
1880	9	80	50.189	1.221
1890	10	90	62.980	1.301
1900	11	100	76.212	1.462
1910	12	110	92.228	

FIGURE 2.1.8. U.S. population data (in millions) and approximate slopes.

FIGURE 2.1.9. Points and approximating straight line for U.S. population data from 1800 to 1900.

Year	World Population (billions)
1960	3.049
1965	3.358
1970	3.721
1975	4.103
1980	4.473
1985	4.882
1990	5.249
1995	5.679
2000	6.127

FIGURE 2.1.10. World population data.

2.2 Equilibrium Solutions and Stability

In previous sections we have often used explicit solutions of differential equations to answer specific numerical questions. But even when a given differential equation is difficult or impossible to solve explicitly, it often is possible to extract *qualitative* information about general properties of its solutions. For example, we may be able to establish that every solution $x(t)$ grows without bound as $t \to +\infty$, or approaches a finite limit, or is a periodic function of t. In this section we introduce—mainly by consideration of simple differential equations that *can* be solved explicitly—some of the more important qualitative questions that can sometimes be answered for equations that are difficult or impossible to solve.

Example 1

Let $x(t)$ denote the temperature of a body with initial temperature $x(0) = x_0$. At time $t = 0$ this body is immersed in a medium with constant temperature A. Assuming Newton's law of cooling,

$$\frac{dx}{dt} = -k(x - A) \qquad (k > 0 \quad \text{constant}), \tag{1}$$

we readily solve (by separation of variables) for the explicit solution

$$x(t) = A + (x_0 - A)e^{-kt}.$$

It follows immediately that

$$\lim_{t \to \infty} x(t) = A, \tag{2}$$

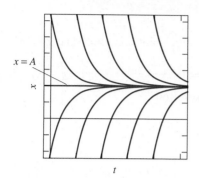

FIGURE 2.2.1. Typical solution curves for the equation of Newton's law of cooling, $dx/dt = -k(x - A)$.

so the temperature of the body approaches that of the surrounding medium (as is evident to one's intuition). Note that the constant function $x(t) \equiv A$ is a solution of Eq. (1); it corresponds to the temperature of the body when it is in thermal equilibrium with the surrounding medium. In Fig. 2.2.1 the limit in (2) means that every other solution curve approaches the equilibrium solution curve $x = A$ asymptotically as $t \to +\infty$. ∎

Remark: The behavior of solutions of Eq. (1) is summarized briefly by the **phase diagram** in Fig. 2.2.2. The right-hand side $f(x) = -k(x - A) = k(A - x)$ is positive if $x < A$, negative if $x > A$. This observation corresponds to the fact that solutions starting above the line $x = A$ and those starting below it both approach the limiting solution $x(t) \equiv A$ as t increases (as indicated by the arrows). ∎

FIGURE 2.2.2. Phase diagram for the equation $dx/dt = f(x) = k(A - x)$.

In Section 2.1 we introduced the general population equation

$$\frac{dx}{dt} = (\beta - \delta)x, \tag{3}$$

where β and δ are the birth and death rates, respectively, in births or deaths per individual per unit of time. The question of whether a population $x(t)$ is bounded or unbounded as $t \to +\infty$ is of evident interest. In many situations—like the logistic and explosion/extinction populations of Section 2.1—the birth and death rates are known functions of x. Then Eq. (3) takes the form

$$\frac{dx}{dt} = f(x). \tag{4}$$

This is an **autonomous** first-order differential equation—one in which the independent variable t does not appear explicitly. As in Example 1, the solutions of the equation $f(x) = 0$ play an important role and are called **critical points** of the autonomous differential equation $dx/dt = f(x)$.

If $x = c$ is a critical point of Eq. (4), then the differential equation has the constant solution $x(t) \equiv c$. A constant solution of a differential equation is sometimes called an **equilibrium solution** (one may think of a population that remains constant because it is in "equilibrium" with its environment). Thus the critical point $x = c$, a number, corresponds to the equilibrium solution $x(t) \equiv c$, a constant-valued function.

Example 2 illustrates the fact that the qualitative behavior (as t increases) of the solutions of an autonomous first-order equation can be described in terms of its critical points.

Example 2 Consider the logistic differential equation

$$\frac{dx}{dt} = kx(M - x) \tag{5}$$

(with $k > 0$ and $M > 0$). It has two critical points—the solutions $x = 0$ and $x = M$ of the equation

$$f(x) = kx(M - x) = 0.$$

In Section 2.1 we discussed the logistic-equation solution

$$x(t) = \frac{Mx_0}{x_0 + (M - x_0)e^{-kMt}} \tag{6}$$

satisfying the initial condition $x(0) = x_0$. Note that the initial values $x_0 = 0$ and $x_0 = M$ yield the equilibrium solutions $x(t) \equiv 0$ and $x(t) \equiv M$ of Eq. (5).

We observed in Section 2.1 that if $x_0 > 0$, then $x(t) \to M$ as $t \to +\infty$. But if $x_0 < 0$, then the denominator in Eq. (6) initially is positive, but vanishes when

$$t = t_1 = \frac{1}{kM} \ln \frac{M - x_0}{-x_0} > 0.$$

Because the numerator in (6) is negative in this case, it follows that

$$\lim_{t \to t_1^-} x(t) = -\infty \quad \text{if} \quad x_0 < 0.$$

It follows that the solution curves of the logistic equation in (5) look as illustrated in Fig. 2.2.3. ∎

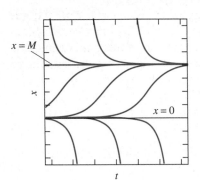

FIGURE 2.2.3. Typical solution curves for the logistic equation $dx/dt = kx(M - x)$.

Stability of Critical Points

Figure 2.2.3 illustrates the concept of *stability*. A critical point $x = c$ of an autonomous first-order equation is said to be *stable* provided that, if the initial value x_0 is sufficiently close to c, then $x(t)$ remains close to c for all $t > 0$. More precisely, the critical point c is **stable** if, for each $\epsilon > 0$, there exists $\delta > 0$ such that

$$|x_0 - c| < \delta \quad \text{implies that} \quad |x(t) - c| < \epsilon \tag{7}$$

for all $t > 0$. The critical point $x = c$ is **unstable** if it is not stable.

Figure 2.2.4 shows a "wider view" of the solution curves of a logistic equation with $k = 1$ and $M = 4$. Note that the strip $3.5 < x < 4.5$ enclosing the stable equilibrium curve $x = 4$ acts like a *funnel*—solution curves (moving from left to right) enter this strip and thereafter remain within it. By contrast, the strip $-0.5 < x < 0.5$ enclosing the unstable solution curve $x = 0$ acts like a *spout*—solution curves leave this strip and thereafter remain outside it. Thus the critical point $x = M$ is stable, whereas the critical point $x = 0$ is unstable.

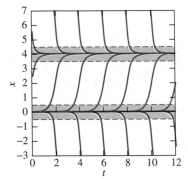

FIGURE 2.2.4. Solution curves, funnel, and spout for $dx/dt = 4x - x^2$.

Remark 1: We can summarize the behavior of solutions of the logistic equation in (5)—in terms of theirinitial values—by means of the phase diagram shown in Fig. 2.2.5. It indicates that $x(t) \to M$ as $t \to +\infty$ if either $x_0 > M$ or $0 < x_0 < M$, whereas $x(t) \to -\infty$ as t increases if $x_0 < 0$. The fact that M is a stable critical point would be important, for instance, if we wished to conduct an experiment with a population of M bacteria. It is impossible to count precisely M bacteria for M large, but any initially positive population will approach M as t increases.

$x' < 0 \qquad x' > 0 \qquad x' < 0$

$x = 0 \qquad\qquad x = M$
Unstable Stable

FIGURE 2.2.5. Phase diagram for the logistic equation $dx/dt = f(x) = kx(M - x)$.

Remark 2: Related to the stability of the limiting solution $M = a/b$ of the logistic equation

$$\frac{dx}{dt} = ax - bx^2 \tag{8}$$

is the "predictability" of M for an actual population. The coefficients a and b are unlikely to be known precisely for an actual population. But if they are replaced with close approximations a^\star and b^\star—derived perhaps from empirical measurements—then the approximate limiting population $M^\star = a^\star/b^\star$ will be close to the actual limiting population $M = a/b$. We may therefore say that the value M of the limiting population predicted by a logistic equation not only is a stable critical point of the differential equation; this value also is "stable" with respect to small perturbations of the constant coefficients in the equation. (Note that one of these two statements involves changes in the initial value x_0; the other involves changes in the coefficients a and b). ∎

Example 3 Consider now the explosion/extinction equation

$$\frac{dx}{dt} = kx(x - M) \tag{9}$$

of Eq. (10) in Section 2.1. Like the logistic equation, it has the two critical points $x = 0$ and $x = M$ corresponding to the equilibrium solutions $x(t) \equiv 0$ and $x(t) \equiv M$. According to Problem 33 in Section 2.1, its solution with $x(0) = x_0$ is given by

$$x(t) = \frac{Mx_0}{x_0 + (M - x_0)e^{kMt}} \tag{10}$$

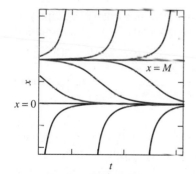

FIGURE 2.2.6. Typical solution curves for the explosion/extinction equation $dx/dt = kx(x - M)$.

(with only a single difference in sign from the logistic solution in (6)). If $x_0 < M$, then (because the coefficient of the exponential in the denominator is positive) it follows immediately from Eq. (10) that $x(t) \to 0$ as $t \to +\infty$. But if $x_0 > M$, then the denominator in (10) initially is positive, but vanishes when

$$t = t_1 = \frac{1}{kM} \ln \frac{x_0}{x_0 - M} > 0.$$

Because the numerator in (10) is positive in this case, it follows that

$$\lim_{t \to t_1^-} x(t) = +\infty \quad \text{if} \quad x_0 > M.$$

Therefore, the solution curves of the explosion/extinction equation in (9) look as illustrated in Fig. 2.2.6. A narrow band along the equilibrium curve $x = 0$ (as in Fig. 2.2.4) would serve as a funnel, while a band along the solution curve $x = M$ would serve as a spout for solutions. The behavior of the solutions of Eq. (9) is summarized by the phase diagram in Fig. 2.2.7, where we see that the critical point $x = 0$ is stable and the critical point $x = M$ is unstable. ∎

FIGURE 2.2.7. Phase diagram for the explosion/extinction equation $dx/dt = f(x) = kx(x - M)$.

Harvesting a Logistic Population

The autonomous differential equation

$$\frac{dx}{dt} = ax - bx^2 - h \tag{11}$$

(with a, b, and h all positive) may be considered to describe a logistic population *with harvesting*. For instance, we might think of the population of fish in a lake from which h fish per year are removed by fishing.

Example 4 Let us rewrite Eq. (11) in the form

$$\frac{dx}{dt} = kx(M - x) - h, \tag{12}$$

which exhibits the limiting population M in the case $h = 0$ of no harvesting. Assuming hereafter that $h > 0$, we can solve the quadratic equation $-kx^2 + kMx - h = 0$ for the two critical points

$$H, N = \frac{kM \pm \sqrt{(kM)^2 - 4hk}}{2k} = \frac{1}{2}\left(M \pm \sqrt{M^2 - 4h/k}\right), \tag{13}$$

assuming that the harvesting rate h is sufficiently small that $4h < kM^2$, so both roots H and N are real with $0 < H < N < M$. Then we can rewrite Eq. (12) in the form

$$\frac{dx}{dt} = k(N - x)(x - H). \tag{14}$$

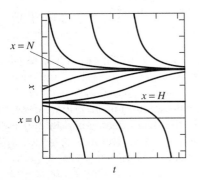

For instance, the number of critical points of the equation may change abruptly as the value of a parameter is changed. In Problem 24 we ask you to solve this equation for the solution

$$x(t) = \frac{N(x_0 - H) - H(x_0 - N)e^{-k(N-H)t}}{(x_0 - H) - (x_0 - N)e^{-k(N-H)t}} \tag{15}$$

in terms of the initial value $x(0) = x_0$.

Note that the exponent $-k(N - H)t$ is negative for $t > 0$. If $x_0 > N$ then each of the coefficients within parentheses in Eq. (15) is positive; it follows that

$$\text{If} \quad x_0 > N \quad \text{then} \quad x(t) \to N \quad \text{as} \quad t \to +\infty. \tag{16}$$

FIGURE 2.2.8. Typical solution curves for the logistic harvesting equation $dx/dt = k(N - x)(x - H)$.

In Problem 25 we ask you to deduce also from Eq. (15) that

$$\text{If} \quad H < x_0 < N \quad \text{then} \quad x(t) \to N \quad \text{as} \quad t \to +\infty, \quad \text{whereas} \tag{17}$$

$$\text{if} \quad x_0 < H \quad \text{then} \quad x(t) \to -\infty \quad \text{as} \quad t \to t_1 \tag{18}$$

FIGURE 2.2.9. Phase diagram for the logistic harvesting equation $dx/dt = f(x) = k(N - x)(x - H)$.

for a positive value t_1 that depends on x_0. It follows that the solution curves of Eq. (12)—still assuming that $4h < kM^2$—look as illustrated in Fig. 2.2.8. (Can you visualize a funnel along the line $x = N$ and a spout along the line $x = H$?) Thus the constant solution $x(t) \equiv N$ is an equilibrium *limiting solution,* whereas $x(t) \equiv H$ is a *threshold solution* that separates different behaviors—the population approaches N if $x_0 > H$, while it becomes extinct because of harvesting if $x_0 < H$. Finally, the stable critical point $x = N$ and the unstable critical point $x = H$ are illustrated in the phase diagram in Fig. 2.2.9. ∎

Example 5

For a concrete application of our stability conclusions in Example 4, suppose that $k = 1$ and $M = 4$ for a logistic population $x(t)$ of fish in a lake, measured in hundreds after t years. Without any fishing at all, the lake would eventually contain nearly 400 fish, whatever the initial population. Now suppose that $h = 3$, so that 300 fish are "harvested" annually (at a constant rate throughout the year). Equation (12) is then $dx/dt = x(4 - x) - 3$, and the quadratic equation

$$-x^2 + 4x - 3 = (3 - x)(x - 1) = 0$$

has solutions $H = 1$ and $N = 3$. Thus the threshold population is 100 fish and the (new) limiting population is 300 fish. In short, if the lake is stocked initially with more than 100 fish, then as t increases the fish population will approach a limiting value of 300 fish. But if the lake is stocked initially with fewer than 100 fish, then the lake will be "fished out" and the fish will disappear entirely within a finite period of time. ■

Bifurcation and Dependence on Parameters

A biological or physical system that is modeled by a differential equation may depend crucially on the numerical values of certain coefficients or parameters that appear in the equation. For instance, the number of critical points of the equation may change abruptly as the value of a parameter is changed.

Example 6

The differential equation

$$\frac{dx}{dt} = x(4 - x) - h \tag{19}$$

(with x in hundreds) models the harvesting of a logistic population with $k = 1$ and limiting population $M = 4$ (hundred). In Example 5 we considered the case of harvesting level $h = 3$, and found that the new limiting population is $N = 3$ hundred and the threshold population is $H = 1$ hundred. Typical solution curves, including the equilibrium solutions $x(t) \equiv 3$ and $x(t) \equiv 1$, then look like those pictured in Fig. 2.2.8.

Now let's investigate the dependence of this picture upon the harvesting level h. According to Eq. (13) with $k = 1$ and $M = 4$, the limiting and threshold populations N and H are given by

$$H, N = \frac{1}{2}\left(4 \pm \sqrt{16 - 4h}\right) = 2 \pm \sqrt{4 - h}. \tag{20}$$

If $h < 4$—we can consider negative values of h to describe stocking rather than harvesting the fish—then there are distinct equilibrium solutions $x(t) \equiv N$ and $x(t) \equiv H$ with $N > H$ as in Fig. 2.2.8.

But if $h = 4$, then Eq. (20) gives $N = H = 2$, so the differential equation has only the single equilibrium solution $x(t) \equiv 2$. In this case the solution curves of the equation look like those illustrated in Fig. 2.2.10. If the initial number x_0 (in hundreds) of fish exceeds 2, then the population approaches a limiting population of 2 (hundred fish). However, any initial population $x_0 < 2$ (hundred) results in extinction with the fish dying out as a consequence of the harvesting of 4 hundred fish annually.

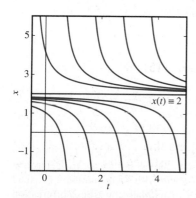

FIGURE 2.2.10. Solution curves of the equation $x' = x(4 - x) - h$ with critical harvesting $h = 4$.

FIGURE 2.2.11. Solution curves of the equation $x' = x(4 - x) - h$ with excessive harvesting $h = 5$.

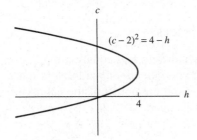

FIGURE 2.2.12. The parabola $(c - 2)^2 = 4 - h$ is the bifurcation diagram for the differential equation $x' = x(4 - x) - h$.

If, finally, $h > 4$, then the quadratic equation corresponding to (20) has no real solutions and the differential equation in (19) has no equilibrium solutions. The solution curves then look like those illustrated in Fig. 2.2.11, and (whatever the initial number of fish) the population dies out as a result of the excessive harvesting. ∎

If we imagine turning a dial to gradually increase the value of the parameter h in Eq. (10), then the picture of the solution curves changes from one like Fig. 2.2.8 with $h < 4$, to Fig. 2.2.10 with $h = 4$, to one like Fig. 2.2.11 with $h > 4$. Thus the differential equation has

- two critical points if $h < 4$;
- one critical point if $h = 4$;
- no critical point if $h > 4$.

The value $h = 4$—for which the qualitative nature of the solutions changes as h increases—is called a **bifurcation point** for the differential equation containing the parameter h. A common way to visualize the corresponding "bifurcation" in the solutions is to plot the **bifurcation diagram** consisting of all points (h, c), where c is a critical point of the equation $x' = x(4 - x) + h$. For instance, if we rewrite Eq. (20) as

$$c = 2 \pm \sqrt{4 - h},$$
$$(c - 2)^2 = 4 - h,$$

where either $c = N$ or $c = H$, then we get the equation of the parabola that is shown in Fig. 2.2.12. This parabola is then the bifurcation diagram for our differential equation that models a logistic fish population with harvesting at the level specified by the parameter h.

2.2 Problems

In Problems 1 through 12 first solve the equation $f(x) = 0$ to find the critical points of the given autonomous differential equation $dx/dt = f(x)$. Then analyze the sign of $f(x)$ to determine whether each critical point is stable or unstable, and construct the corresponding phase diagram for the differential equation. Next, solve the differential equation explicitly for $x(t)$ in terms of t. Finally, use either the exact solution or a computer-generated slope field to sketch typical solution curves for the given differential equation, and verify visually the stability of each critical point.

1. $\dfrac{dx}{dt} = x - 4$

2. $\dfrac{dx}{dt} = 3 - x$

3. $\dfrac{dx}{dt} = x^2 - 4x$

4. $\dfrac{dx}{dt} = 3x - x^2$

5. $\dfrac{dx}{dt} = x^2 - 4$

6. $\dfrac{dx}{dt} = 9 - x^2$

7. $\dfrac{dx}{dt} = (x - 2)^2$

8. $\dfrac{dx}{dt} = -(3 - x)^2$

9. $\dfrac{dx}{dt} = x^2 - 5x + 4$

10. $\dfrac{dx}{dt} = 7x - x^2 - 10$

11. $\dfrac{dx}{dt} = (x - 1)^3$

12. $\dfrac{dx}{dt} = (2 - x)^3$

In Problems 13 through 18, use a computer system or graphing calculator to plot a slope field and/or enough solution curves to indicate the stability or instability of each critical point of the given differential equation.

13. $\dfrac{dx}{dt} = (x + 2)(x - 2)^2$

14. $\dfrac{dx}{dt} = x(x^2 - 4)$

15. $\dfrac{dx}{dt} = (x^2 - 4)^2$

16. $\dfrac{dx}{dt} = (x^2 - 4)^3$

17. $\dfrac{dx}{dt} = x^2(x^2 - 4)$

18. $\dfrac{dx}{dt} = x^3(x^2 - 4)$

19. The differential equation $dx/dt = \frac{1}{10}x(10 - x) - h$ models a logistic population with harvesting at rate h. Determine (as in Example 6) the dependence of the number of critical points on the parameter h, and then construct a bifurcation diagram like Fig. 2.2.12.

20. The differential equation $dx/dt = \frac{1}{100}x(x - 5) + s$ models a population with stocking at rate s. Determine the dependence of the number of critical points c on the parameter s,

and then construct the corresponding bifurcation diagram in the sc-plane.

21. Consider the differential equation $dx/dt = kx - x^3$. **(a)** If $k \leq 0$, show that the only critical value $c = 0$ of x is stable. **(b)** If $k > 0$, show that the critical point $c = 0$ is now unstable, but that the critical points $c = \pm\sqrt{k}$ are stable. Thus the qualitative nature of the solutions changes at $k = 0$ as the parameter k increases, and so $k = 0$ is a bifurcation point for the differential equation with parameter k. The plot of all points of the form (k, c) where c is a critical point of the equation $x' = kx - x^3$ is the "pitchfork diagram" shown in Fig. 2.2.13.

22. Consider the differential equation $dx/dt = x + kx^3$ containing the parameter k. Analyze (as in Problem 21) the dependence of the number and nature of the critical points on the value of k, and construct the corresponding bifurcation diagram.

23. Suppose that the logistic equation $dx/dt = kx(M - x)$ models a population $x(t)$ of fish in a lake after t months during which no fishing occurs. Now suppose that, because of fishing, fish are removed from the lake at the rate of hx fish per month (with h a positive constant). Thus fish are "harvested" at a rate proportional to the existing fish population, rather than at the constant rate of Example 4. **(a)** If $0 < h < kM$, show that the population is still logistic. What is the new limiting population? **(b)** If $h \geq kM$, show that $x(t) \to 0$ are $t \to +\infty$, so the lake is eventually fished out.

24. Separate variables in the logistic harvesting equation $dx/dt = k(N - x)(x - H)$ and then use partial fractions to derive the solution given in Eq. (15).

25. Use the alternative forms

$$x(t) = \frac{N(x_0 - H) + H(N - x_0)e^{-k(N-H)t}}{(x_0 - H) + (N - x_0)e^{-k(N-H)t}}$$

$$= \frac{H(N - x_0)e^{-k(N-H)t} - N(H - x_0)}{(N - x_0)e^{-k(N-H)t} - (H - x_0)}$$

of the solution in (15) to establish the conclusions stated in (17) and (18).

Example 4 dealt with the case $4h > kM^2$ in the equation $dx/dt = kx(M - x) - h$ that describes constant-rate harvesting of a logistic population. Problems 26 and 27 deal with the other cases.

26. If $4h = kM^2$, show that typical solution curves look as illustrated in Fig. 2.2.14. Thus if $x_0 \geq M/2$, then $x(t) \to M/2$ as $t \to +\infty$. But if $x_0 < M/2$, then $x(t) = 0$ after a finite period of time, so the lake is fished out. The critical point $x = M/2$ might be called *semistable*, because it looks stable from one side, unstable from the other.

27. If $4h > kM^2$, show that $x(t) = 0$ after a finite period of time, so the lake is fished out (whatever the initial population). *Suggestion:* Complete the square to rewrite the differential equation in the form $dx/dt = -k[(x - a)^2 + b^2]$. Then solve explicitly by separation of variables. The results of this and the previous problem (together with Example 4) show that $h = \frac{1}{4}kM^2$ is a critical harvesting rate for a logistic population. At any lesser harvesting rate the population approaches a limiting population N that is less than M (why?), whereas at any greater harvesting rate the population reaches extinction.

28. This problem deals with the differential equation $dx/dt = kx(x - M) - h$ that models the harvesting of an unsophisticated population (such as alligators). Show that this equation can be rewritten in the form $dx/dt = k(x - H)(x - K)$, where

$$H = \frac{1}{2}\left(M + \sqrt{M^2 + 4h/k}\right) > 0,$$

$$K = \frac{1}{2}\left(M - \sqrt{M^2 + 4h/k}\right) < 0.$$

Show that typical solution curves look as illustrated in Fig. 2.2.15.

29. Consider the two differential equations

$$\frac{dx}{dt} = (x - a)(x - b)(x - c) \qquad (21)$$

and

$$\frac{dx}{dt} = (a - x)(b - x)(c - x), \qquad (22)$$

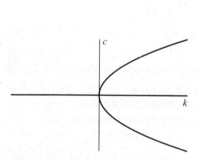

FIGURE 2.2.13. Bifurcation diagram for $dx/dt = kx - x^3$.

FIGURE 2.2.14. Solution curves for harvesting a logistic population with $4h = kM^2$.

FIGURE 2.2.15. Solution curves for harvesting a population of alligators.

each having the critical points a, b, and c; suppose that $a < b < c$. For one of these equations, only the critical point b is stable; for the other equation, b is the only unstable critical point. Construct phase diagrams for the two equations to determine which is which. Without at-

tempting to solve either equation explicitly, make rough sketches of typical solution curves for each. You should see two funnels and a spout in one case, two spouts and a funnel in the other.

2.3 Acceleration-Velocity Models

In Section 1.2 we discussed vertical motion of a mass m near the surface of the earth under the influence of constant gravitational acceleration. If we neglect any effects of air resistance, then Newton's second law ($F = ma$) implies that the velocity v of the mass m satisfies the equation

$$m\frac{dv}{dt} = F_G, \tag{1}$$

where $F_G = -mg$ is the (downward-directed) force of gravity, where the gravitational acceleration is $g \approx 9.8$ m/s² (in mks units; $g \approx 32$ ft/s² in fps units).

Example 1 Suppose that a crossbow bolt is shot straight upward from the ground ($y_0 = 0$) with initial velocity $v_0 = 49$ (m/s). Then Eq. (1) with $g = 9.8$ gives

$$\frac{dv}{dt} = -9.8, \quad \text{so} \quad v(t) = -(9.8)t + v_0 = -(9.8)t + 49.$$

Hence the bolt's height function $y(t)$ is given by

$$y(t) = \int [-(9.8)t + 49]\,dt = -(4.9)t^2 + 49t + y_0 = -(4.9)t^2 + 49t.$$

The bolt reaches its maximum height when $v = -(9.8)t + 49 = 0$, hence when $t = 5$ (s). Thus its maximum height is

$$y_{max} = y(5) = -(4.9)(5^2) + (49)(5) = 122.5 \text{ (m)}.$$

The bolt returns to the ground when $y = -(4.9)t(t - 10) = 0$, and thus after 10 seconds aloft. ∎

Now we want to take account of air resistance in a problem like Example 1. The force F_R exerted by air resistance on the moving mass m must be added in Eq. (1), so now

$$m\frac{dv}{dt} = F_G + F_R. \tag{2}$$

Newton showed in his *Principia Mathematica* that certain simple physical assumptions imply that F_R is proportional to the *square* of the velocity: $F_R = kv^2$. But empirical investigations indicate that the actual dependence of air resistance on velocity can be quite complicated. For many purposes it suffices to assume that

$$F_R = kv^p,$$

where $1 \leqq p \leqq 2$ and the value of k depends on the size and shape of the body, as well as the density and viscosity of the air. Generally speaking, $p = 1$ for relatively

low speeds and $p = 2$ for high speeds, whereas $1 < p < 2$ for intermediate speeds. But how slow "low speed" and how fast "high speed" are depend on the same factors that determine the value of the coefficient k.

Thus air resistance is a complicated physical phenomenon. But the simplifying assumption that F_R is exactly of the form given here, with either $p = 1$ or $p = 2$, yields a tractable mathematical model that exhibits the most important qualitative features of motion with resistance.

Resistance Proportional to Velocity

Let us first consider the vertical motion of a body with mass m near the surface of the earth, subject to two forces: a downward gravitational force F_G and a force F_R of air resistance that is proportional to velocity (so that $p = 1$) and of course directed opposite the direction of motion of the body. If we set up a coordinate system with the positive y-direction upward and with $y = 0$ at ground level, then $F_G = -mg$ and

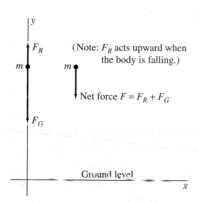

(Note: F_R acts upward when the body is falling.)

Net force $F = F_R + F_G$

FIGURE 2.3.1. Vertical motion with air resistance.

$$F_R = -kv, \tag{3}$$

where k is a positive constant and $v = dy/dt$ is the velocity of the body. Note that the minus sign in Eq. (3) makes F_R positive (an upward force) if the body is falling (v is negative) and makes F_R negative (a downward force) if the body is rising (v is positive). As indicated in Fig. 2.3.1, the net force acting on the body is then

$$F = F_R + F_G = -kv - mg,$$

and Newton's law of motion $F = m(dv/dt)$ yields the equation

$$m\frac{dv}{dt} = -kv - mg.$$

Thus

$$\frac{dv}{dt} = -\rho v - g, \tag{4}$$

where $\rho = k/m > 0$. You should verify for yourself that if the positive y-axis were directed downward, then Eq. (4) would take the form $dv/dt = -\rho v + g$.

Equation (4) is a separable first-order differential equation, and its solution is

$$v(t) = \left(v_0 + \frac{g}{\rho}\right) e^{-\rho t} - \frac{g}{\rho}. \tag{5}$$

Here, $v_0 = v(0)$ is the initial velocity of the body. Note that

$$v_\tau = \lim_{t \to \infty} v(t) = -\frac{g}{\rho}. \tag{6}$$

Thus the speed of a body falling with air resistance does *not* increase indefinitely; instead, it approaches a *finite* limiting speed, or **terminal speed,**

$$|v_\tau| = \frac{g}{\rho} = \frac{mg}{k}. \tag{7}$$

This fact is what makes a parachute a practical invention; it even helps explain the occasional survival of people who fall without parachutes from high-flying airplanes.

We now rewrite Eq. (5) in the form

$$\frac{dy}{dt} = (v_0 - v_\tau)e^{-\rho t} + v_\tau. \tag{8}$$

Integration gives

$$y(t) = -\frac{1}{\rho}(v_0 - v_\tau)e^{-\rho t} + v_\tau t + C.$$

We substitute 0 for t and let $y_0 = y(0)$ denote the initial height of the body. Thus we find that $C = y_0 + (v_0 - v_\tau)/\rho$, and so

$$y(t) = y_0 + v_\tau t + \frac{1}{\rho}(v_0 - v_\tau)(1 - e^{-\rho t}). \tag{9}$$

Equations (8) and (9) give the velocity v and height y of a body moving vertically under the influence of gravity and air resistance. The formulas depend on the initial height y_0 of the body, its initial velocity v_0, and the *drag coefficient* ρ, the constant such that the acceleration due to air resistance is $a_R = -\rho v$. The two equations also involve the terminal velocity v_τ defined in Eq. (6).

For a person descending with the aid of a parachute, a typical value of ρ is 1.5, which corresponds to a terminal speed of $|v_\tau| \approx 21.3$ ft/s, or about 14.5 mi/h. With an unbuttoned overcoat flapping in the wind in place of a parachute, an unlucky skydiver might increase ρ to perhaps as much as 0.5, which gives a terminal speed of $|v_\tau| \approx 65$ ft/s, about 44 mi/h. See Problems 10 and 11 for some parachute-jump computations.

Example 2 We again consider a bolt shot straight upward with initial velocity $v_0 = 49$ m/s from a crossbow at ground level. But now we take air resistance into account, with $\rho = 0.04$ in Eq. (4). We ask how the resulting maximum height and time aloft compare with the values found in Example 1.

Solution We substitute $y_0 = 0$, $v_0 = 49$, and $v_\tau = -g/\rho = -245$ in Eqs. (5) and (9), and obtain

$$v(t) = 294e^{-t/25} - 245,$$
$$y(t) = 7350 - 245t - 7350e^{-t/25}.$$

To find the time required for the bolt to reach its maximum height (when $v = 0$), we solve the equation
$$v(t) = 294e^{-t/25} - 245 = 0$$

for $t_m = 25\ln(294/245) \approx 4.558$ (s). Its maximum height is then $y_{max} = v(t_m) \approx 108.280$ meters (as opposed to 122.5 meters without air resistance). To find when the bolt strikes the ground, we must solve the equation

$$y(t) = 7350 - 245t - 7350e^{-t/25} = 0.$$

Using Newton's method, we can begin with the initial guess $t_0 = 10$ and carry out the iteration $t_{n+1} = t_n - y(t_n)/y'(t_n)$ to generate successive approximations to the root. Or we can simply use the **Solve** command on a calculator or computer. We

find that the bolt is in the air for $t_f \approx 9.411$ seconds (as opposed to 10 seconds without air resistance). It hits the ground with a reduced speed of $|v(t_f)| \approx 43.227$ m/s (as opposed to its initial velocity of 49 m/s).

Thus the effect of air resistance is to decrease the bolt's maximum height, the total time spent aloft, and its final impact speed. Note also that the bolt now spends more time in descent ($t_f - t_m \approx 4.853$ s) than in ascent ($t_m \approx 4.558$ s). ■

Resistance Proportional to Square of Velocity

Now we assume that the force of air resistance is proportional to the *square* of the velocity:

➤ $$F_R = \pm k v^2, \tag{10}$$

with $k > 0$. The choice of signs here depends on the direction of motion, which the force of resistance always opposes. Taking the positive y-direction as upward, $F_R < 0$ for upward motion (when $v > 0$) while $F_R > 0$ for downward motion (when $v < 0$). Thus the sign of F_R is always opposite that of v, so we can rewrite Eq. (10) as

$$F_R = -kv|v|. \tag{10'}$$

Then Newton's second law gives

$$m\frac{dv}{dt} = F_G + F_R = -mg - kv|v|;$$

that is,

$$\frac{dv}{dt} = -g - \rho v|v|, \tag{11}$$

where $\rho = k/m > 0$. We must discuss the cases of upward and downward motion separately.

UPWARD MOTION: Suppose that a projectile is launched straight upward from the initial position y_0 with initial velocity $v_0 > 0$. Then Eq. (11) with $v > 0$ gives the differential equation

$$\frac{dv}{dt} = -g - \rho v^2 = -g\left(1 + \frac{\rho}{g}v^2\right). \tag{12}$$

In Problem 13 we ask you to make the substitution $u = v\sqrt{\rho/g}$ and apply the familiar integral

$$\int \frac{1}{1+u^2}\,du = \tan^{-1} u + C$$

to derive the projectile's velocity function

$$v(t) = \sqrt{\frac{g}{\rho}}\tan\left(C_1 - t\sqrt{\rho g}\right) \quad \text{with} \quad C_1 = \tan^{-1}\left(v_0\sqrt{\frac{\rho}{g}}\right). \tag{13}$$

Because $\int \tan u\,du = -\ln|\cos u| + C$, a second integration (see Problem 14) yields the position function

$$y(t) = y_0 + \frac{1}{\rho}\ln\left|\frac{\cos\left(C_1 - t\sqrt{\rho g}\right)}{\cos C_1}\right|. \tag{14}$$

DOWNWARD MOTION: Suppose that a projectile is launched (or dropped) straight downward from the initial position y_0 with initial velocity $v_0 \leqq 0$. Then Eq. (11) with $v < 0$ gives the differential equation

$$\frac{dv}{dt} = -g + \rho v^2 = -g \left(1 - \frac{\rho}{g} v^2 \right).$$ (15)

In Problem 15 we ask you to make the substitution $u = v\sqrt{\rho/g}$ and apply the integral

$$\int \frac{1}{1 - u^2} \, du = \tanh^{-1} u + C$$

to derive the projectile's velocity function

$$v(t) = \sqrt{\frac{g}{\rho}} \tanh \left(C_2 - t\sqrt{\rho g} \right) \quad \text{with} \quad C_2 = \tanh^{-1} \left(v_0 \sqrt{\frac{\rho}{g}} \right).$$ (16)

Because $\int \tanh u \, du = \ln |\cosh u| + C$, another integration (Problem 16) yields the position function

$$y(t) = y_0 - \frac{1}{\rho} \ln \left| \frac{\cosh \left(C_2 - t\sqrt{\rho g} \right)}{\cosh C_2} \right|.$$ (17)

(Note the analogy between Eqs. (16) and (17) and Eqs. (13) and (14) for upward motion.)

If $v_0 = 0$, then $C_2 = 0$, so $v(t) = -\sqrt{g/\rho} \tanh \left(t\sqrt{\rho g} \right)$. Because

$$\lim_{x \to \infty} \tanh x = \lim_{x \to \infty} \frac{\sinh x}{\cosh x} = \lim_{x \to \infty} \frac{\frac{1}{2}(e^x - e^{-x})}{\frac{1}{2}(e^x + e^{-x})} = 1,$$

it follows that in the case of downward motion the body approaches the terminal speed

$$|v_\tau| = \sqrt{\frac{g}{\rho}}$$ (18)

(as compared with $|v_\tau| = g/\rho$ in the case of downward motion with linear resistance described by Eq. (4)).

Example 3 We consider once more a bolt shot straight upward with initial velocity $v_0 = 49$ m/s from a crossbow at ground level, as in Example 2. But now we assume air resistance proportional to the square of the velocity, with $\rho = 0.0011$ in Eqs. (12) and (15). In Problems 17 and 18 we ask you to verify the entries in the last line of the following table.

Air Resistance	Maximum Height (ft)	Time Aloft (s)	Ascent Time (s)	Descent Time (s)	Impact Speed (ft/s)
0.0	122.5	10	5	5	49
$(0.04)v$	108.28	9.41	4.56	4.85	43.23
$(0.0011)v^2$	108.47	9.41	4.61	4.80	43.49

Comparison of the last two lines of data here indicates little difference—for the motion of our crossbow bolt—between linear air resistance and air resistance proportional to the square of the velocity. And in Fig. 2.3.2, where the corresponding

FIGURE 2.3.2. The height functions in Example 1 (without air resistance), Example 2 (with linear air resistance), and Example 3 (with air resistance proportional to the square of the velocity) are all plotted. The graphs of the latter two are visually indistinguishable.

height functions are graphed, the difference is hardly visible. However, the difference between linear and nonlinear resistance can be significant in more complex situations—such as, for instance, the atmospheric reentry and descent of a space vehicle. ∎

Variable Gravitational Acceleration

Unless a projectile in vertical motion remains in the immediate vicinity of the earth's surface, the gravitational acceleration acting on it is not constant. According to Newton's law of gravitation, the gravitational force of attraction between two point masses M and m located at a distance r apart is given by

$$F = \frac{GMm}{r^2},\tag{19}$$

where G is a certain empirical constant ($G \approx 6.6726 \times 10^{-11}$ N·(m/kg)2 in mks units). The formula is also valid if either or both of the two masses are homogeneous spheres; in this case, the distance r is measured between the centers of the spheres.

The following example is similar to Example 2 in Section 1.2, but now we take account of lunar gravity.

Example 4 A lunar lander is free-falling towards the moon, and at an altitude of 53 kilometers above the lunar surface its downward velocity is measured at 1477 km/h. Its retrorockets, when fired in free space, provide a deceleration of $T = 4$ m/s^2. At what height above the lunar surface should the retrorockets be activated to ensure a "soft touchdown" ($v = 0$ at impact)?

Solution Let $r(t)$ denote the lander's distance from the center of the moon at time t (Fig. 2.3.3). When we combine the (constant) thrust acceleration T and the (negative) lunar acceleration $F/m = GM/r^2$ of Eq. (19), we get the (acceleration) differential equation

$$\frac{d^2r}{dt^2} = T - \frac{GM}{r^2},\tag{20}$$

where $M = 7.35 \times 10^{22}$ (kg) is the mass of the moon, which has a radius of $R = 1.74 \times 10^6$ meters (or 1740 km, a little over a quarter of the earth's radius).

Noting that this second-order differential equation does not involve the independent variable t, we substitute

$$v = \frac{dr}{dt}, \quad \frac{d^2r}{dt^2} = \frac{dv}{dt} = \frac{dv}{dr} \cdot \frac{dr}{dt} = v\frac{dv}{dr}$$

(as in Eq. (36) of Section 1.6) and obtain the first-order equation

$$v\frac{dv}{dr} = T - \frac{GM}{r^2}$$

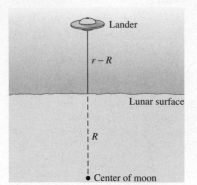

FIGURE 2.3.3. The lunar lander descending to the surface of the moon.

with the new independent variable r. Integration with respect to r now yields the equation

$$\frac{1}{2}v^2 = Tr + \frac{GM}{r} + C \qquad (21)$$

that we can apply both before ignition ($T = 0$) and after ignition ($T = 4$).

Before ignition: Substitution of $T = 0$ in (21) gives the equation

$$\frac{1}{2}v^2 = \frac{GM}{r} + C_1 \qquad (21a)$$

where the constant is given by $C_1 = v_0^2/2 - GM/r_0$ with

$$v_0 = -1477\frac{\text{km}}{\text{h}} \times 1000\frac{\text{m}}{\text{km}} \times \frac{1 \text{ h}}{3600 \text{ s}} = -\frac{14770}{36}\frac{\text{m}}{\text{s}}$$

and $r_0 = (1.74 \times 10^6) + 53000 = 1.793 \times 10^6$ m (from the initial velocity-position measurement).

After ignition: Substitution of $T = 4$ and $v = 0$, $r = R$ (at touchdown) into (21) gives

$$\frac{1}{2}v^2 = 4r + \frac{GM}{r} + C_2 \qquad (21b)$$

where the constant $C_2 = -4R - GM/R$ is obtained by substituting the values $v = 0$, $r = R$ at touchdown.

At the instant of ignition the lunar lander's position and velocity satisfy both (21a) and (21b). Therefore we can find its desired height h above the lunar surface at ignition by equating the right-hand sides in (21a) and (21b). This gives $r = \frac{1}{4}(C_1 - C_2) = 1.78187 \times 10^6$ and finally $h = r - R = 41870$ meters (that is, 41.87 kilometers—just over 26 miles). Moreover, substitution of this value of r in (21a) gives the velocity $v = -450$ m/s at the instant of ignition. ∎

Escape Velocity

In his novel *From the Earth to the Moon* (1865), Jules Verne raised the question of the initial velocity necesary for a projectile fired from the surface of the earth to reach the moon. Similarly, we can ask what initial velocity v_0 is necessary for the projectile to escape from the earth altogether. This will be so if its velocity $v = dr/dt$ remains *positive* for all $t > 0$, so it continues forever to move away from the earth. With $r(t)$ denoting the projectile's distance from the earth's center at time t (Fig. 2.3.4), we have the equation

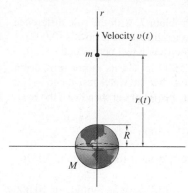

$$\frac{dv}{dt} = \frac{d^2r}{dt^2} = -\frac{GM}{r^2}, \tag{22}$$

FIGURE 2.3.4. A mass m at a great distance from the earth.

similar to Eq. (20), but with $T = 0$ (no thrust) and with $M = 5.975 \times 10^{24}$ (kg) denoting the mass of the earth, which has an equatorial radius of $R = 6.378 \times 10^6$ (m). Substitution of the chain rule expression $dv/dt = v(dv/dr)$ as in Example 4 gives

$$v\frac{dv}{dr} = -\frac{GM}{r^2}.$$

Then integration of both sides with respect to r yields

$$\frac{1}{2}v^2 = \frac{GM}{r} + C.$$

Now $v = v_0$ and $r = R$ when $t = 0$, so $C = \frac{1}{2}v_0^2 - GM/R$, and hence solution for v^2 gives

$$v^2 = v_0^2 + 2GM\left(\frac{1}{r} - \frac{1}{R}\right). \tag{23}$$

This implicit solution of Eq. (22) determines the projectile's velocity v as a function of its distance r from the earth's center. In particular,

$$v^2 > v_0^2 - \frac{2GM}{R},$$

so v will remain positive provided that $v_0^2 \geq 2GM/R$. Therefore, the **escape velocity** from the earth is given by

$$v_0 = \sqrt{\frac{2GM}{R}}. \tag{24}$$

In Problem 27 we ask you to show that, if the projectile's initial velocity exceeds $\sqrt{2GM/R}$, then $r(t) \to \infty$ as $t \to \infty$, so it does, indeed, "escape" from the earth. With the given values of G and the earth's mass M and radius R, this gives $v_0 \approx 11180$ (m/s) (about 36680 ft/s, about 6.95 mi/s, about 25000 mi/h).

Remark: Equation (24) gives the escape velocity for any other (spherical) planetary body when we use *its* mass and radius. For instance, when we use the mass M and radius R for the moon given in Example 4, we find that escape velocity from the lunar surface is $v_0 \approx 2375$ m/s. This is just over one-fifth of the escape velocity from the earth's surface, a fact that greatly facilitates the return trip ("From the Moon to the Earth"). ∎

2.3 Problems

1. The acceleration of a Maserati is proportional to the difference between 250 km/h and the velocity of this sports car. If this machine can accelerate from rest to 100 km/h in 10 s, how long will it take for the car to accelerate from rest to 200 km/h?

2. Suppose that a body moves through a resisting medium with resistance proportional to its velocity v, so that $dv/dt = -kv$. **(a)** Show that its velocity and position at time t are given by

$$v(t) = v_0 e^{-kt}$$

and

$$x(t) = x_0 + \left(\frac{v_0}{k}\right)(1 - e^{-kt}).$$

(b) Conclude that the body travels only a finite distance, and find that distance.

3. Suppose that a motorboat is moving at 40 ft/s when its motor suddenly quits, and that 10 s later the boat has slowed to 20 ft/s. Assume, as in Problem 2, that the resistance it encounters while coasting is proportional to its velocity. How far will the boat coast in all?

4. Consider a body that moves horizontally through a medium whose resistance is proportional to the *square* of the velocity v, so that $dv/dt = -kv^2$. Show that

$$v(t) = \frac{v_0}{1 + v_0 kt}$$

and that

$$x(t) = x_0 + \frac{1}{k}\ln(1 + v_0 kt).$$

Note that, in contrast with the result of Problem 2, $x(t) \rightarrow +\infty$ as $t \rightarrow +\infty$.

5. Assuming resistance proportional to the square of the velocity (as in Problem 4), how far does the motorboat of Problem 3 coast in the first minute after its motor quits?

6. Assume that a body moving with velocity v encounters resistance of the form $dv/dt = -kv^{3/2}$. Show that

$$v(t) = \frac{4v_0}{\left(kt\sqrt{v_0} + 2\right)^2}$$

and that

$$x(t) = x_0 + \frac{2}{k}\sqrt{v_0}\left(1 - \frac{2}{kt\sqrt{v_0} + 2}\right).$$

Conclude that under a $\frac{3}{2}$-power resistance a body coasts only a finite distance before coming to a stop.

7. Suppose that a car starts from rest, its engine providing an acceleration of 10 ft/s², while air resistance provides 0.1 ft/s² of deceleration for each foot per second of the car's velocity. **(a)** Find the car's maximum possible (limiting) velocity. **(b)** Find how long it takes the car to attain 90% of its limiting velocity, and how far it travels while doing so.

8. Rework both parts of Problem 7, with the sole difference that the deceleration due to air resistance now is $(0.001)v^2$ ft/s² when the car's velocity is v feet per second.

9. A motorboat weighs 32,000 lb and its motor provides a thrust of 5000 lb. Assume that the water resistance is 100 pounds for each foot per second of the speed v of the boat. Then

$$1000\frac{dv}{dt} = 5000 - 100v.$$

If the boat starts from rest, what is the maximum velocity that it can attain?

10. A woman bails out of an airplane at an altitude of 10,000 ft, falls freely for 20 s, then opens her parachute. How long will it take her to reach the ground? Assume linear air resistance ρv ft/s², taking $\rho = 0.15$ without the parachute and $\rho = 1.5$ with the parachute. (*Suggestion:* First determine her height and velocity when the parachute opens.)

11. According to a newspaper account, a paratrooper survived a training jump from 1200 ft when his parachute failed to open but provided some resistance by flapping unopened in the wind. Allegedly he hit the ground at 100 mi/h after falling for 8 s. Test the accuracy of this account. (*Suggestion:* Find ρ in Eq. (4) by assuming a terminal velocity of 100 mi/h. Then calculate the time required to fall 1200 ft.)

12. It is proposed to dispose of nuclear wastes—in drums with weight $W = 640$ lb and volume 8 ft³—by dropping them into the ocean ($v_0 = 0$). The force equation for a drum falling through water is

$$m\frac{dv}{dt} = -W + B + F_R,$$

where the buoyant force B is equal to the weight (at 62.5 lb/ft³) of the volume of water displaced by the drum (Archimedes' principle) and F_R is the force of water resistance, found empirically to be 1 lb for each foot per second of the velocity of a drum. If the drums are likely to burst upon an impact of more than 75 ft/s, what is the maximum depth to which they can be dropped in the ocean without likelihood of bursting?

13. Separate variables in Eq. (12) and substitute $u = v\sqrt{\rho/g}$ to obtain the upward-motion velocity function given in Eq. (13) with initial condition $v(0) = v_0$.

14. Integrate the velocity function in Eq. (13) to obtain the upward-motion position function given in Eq. (14) with initial condition $y(0) = y_0$.

15. Separate variables in Eq. (15) and substitute $u = v\sqrt{\rho/g}$ to obtain the downward-motion velocity function given in Eq. (16) with initial condition $v(0) = v_0$.

16. Integrate the velocity function in Eq. (16) to obtain the downward-motion position function given in Eq. (17) with initial condition $y(0) = y_0$.

17. Consider the crossbow bolt of Example 3, shot straight upward from the ground ($y = 0$) at time $t = 0$ with initial velocity $v_0 = 49$ m/s. Take $g = 9.8$ m/s^2 and $\rho = 0.0011$ in Eq. (12). Then use Eqs. (13) and (14) to show that the bolt reaches its maximum height of about 108.47 m in about 4.61 s.

18. Continuing Problem 17, suppose that the bolt is now dropped ($v_0 = 0$) from a height of $y_0 = 108.47$ m. Then use Eqs. (16) and (17) to show that it hits the ground about 4.80 s later with an impact speed of about 43.49 m/s.

19. A motorboat starts from rest (initial velocity $v(0) = v_0 = 0$). Its motor provides a constant acceleration of 4 ft/s^2, but water resistance causes a deceleration of $v^2/400$ ft/s^2. Find v when $t = 10$ s, and also find the *limiting velocity* as $t \to +\infty$ (that is, the maximum possible speed of the boat).

20. An arrow is shot straight upward from the ground with an initial velocity of 160 ft/s. It experiences both the deceleration of gravity and deceleration $v^2/800$ due to air resistance. How high in the air does it go?

21. If a ball is projected upward from the ground with initial velocity v_0, deduce from Eq. (14) that the maximum height it attains is

$$y_{\max} = \frac{1}{2\rho} \ln\left(1 + \frac{\rho v_0^2}{g}\right).$$

22. Suppose that $\rho = 0.075$ (in fps units, with $g = 32$ ft/s^2) in Eq. (15) for a paratrooper falling with parachute open. If he jumps from an altitude of 10000 ft and opens his parachute immediately, what will be his terminal speed? How long will it take him to reach the ground?

23. Suppose that the paratrooper of Problem 22 falls freely for 30 s with $\rho = 0.00075$ before opening his parachute. How long will it now take him to reach the ground?

24. The mass of the sun is 329,320 times that of the earth and its radius is 109 times the radius of the earth. **(a)** To what radius (in meters) would the earth have to be compressed in order for it to become a *black hole*—the escape velocity from its surface equal to the velocity $c = 3 \times 10^8$ m/s of light? **(b)** Repeat part (a) with the sun in place of the earth.

25. **(a)** Show that if a projectile is launched straight upward from the surface of the earth with initial velocity v_0 less than escape velocity $\sqrt{2GM/R}$, then the maximum distance from the center of the earth attained by the projectile is

$$r_{\max} = \frac{2GMR}{2GM - Rv_0^2},$$

where M and R are the mass and radius of the earth, respectively. **(b)** With what initial velocity v_0 must such a projectile be launched to yield a maximum altitude of 100 kilometers above the surface of the earth? **(c)** Find the maximum distance from the center of the earth, expressed in terms of earth radii, attained by a projectile launched from the surface of the earth with 90% of escape velocity.

26. Suppose that you are stranded—your rocket engine has failed—on an asteroid of diameter 3 miles, with density equal to that of the earth with radius 3960 miles. If you

have enough spring in your legs to jump 4 feet straight up on earth while wearing your space suit, can you blast off from this asteroid using leg power alone?

27. **(a)** Suppose a projectile is launched vertically from the surface $r = R$ of the earth with initial velocity $v_0 = \sqrt{2GM/R}$ so $v_0^2 = k^2/R$ where $k^2 = 2GM$. Then solve the differential equation $dr/dt = k/\sqrt{r}$ (from Eq. (23) in this section) explicitly to deduce that $r(t) \to \infty$ as $t \to \infty$.

(b) If the projectile is launched vertically with initial velocity $v_0 > \sqrt{2GM/R}$, deduce that

$$\frac{dr}{dt} = \sqrt{\frac{k^2}{r} + \alpha} > \frac{k}{\sqrt{r}}.$$

Why does it again follow that $r(t) \to \infty$ as $t \to \infty$?

28. **(a)** Suppose that a body is dropped ($v_0 = 0$) from a distance $r_0 > R$ from the earth's center, so its acceleration is $dv/dt = -GM/r^2$. Ignoring air resistance, show that it reaches the height $r < r_0$ at time

$$t = \sqrt{\frac{r_0}{2GM}} \left(\sqrt{rr_0 - r^2} + r_0 \cos^{-1}\sqrt{\frac{r}{r_0}}\right).$$

(*Suggestion:* Substitute $r = r_0 \cos^2\theta$ to evaluate $\int \sqrt{r/(r_0 - r)}\, dr$.) **(b)** If a body is dropped from a height of 1000 km above the earth's surface and air resistance is neglected, how long does it take to fall and with what speed will it strike the earth's surface?

29. Suppose that a projectile is fired straight upward from the surface of the earth with initial velocity $v_0 < \sqrt{2GM/R}$. Then its height $y(t)$ above the surface satisfies the initial value problem

$$\frac{d^2y}{dt^2} = -\frac{GM}{(y+R)^2}; \quad y(0) = 0, \quad y'(0) = v_0.$$

Substitute $dv/dt = v(dv/dy)$ and then integrate to obtain

$$v^2 = v_0^2 - \frac{2GMy}{R(R+y)}$$

for the velocity v of the projectile at height y. What maximum altitude does it reach if its initial velocity is 1 km/s?

30. In Jules Verne's original problem, the projectile launched from the surface of the earth is attracted by both the earth and the moon, so its distance $r(t)$ from the center of the earth satisfies the initial value problem

$$\frac{d^2r}{dt^2} = -\frac{GM_e}{r^2} + \frac{GM_m}{(S-r)^2}; \quad r(0) = R, \quad r'(0) = v_0$$

where M_e and M_m denote the masses of the earth and the moon, respectively; R is the radius of the earth and $S = 384{,}400$ km is the distance between the centers of the earth and the moon. To reach the moon, the projectile must only just pass the point between the moon and earth where its net acceleration vanishes. Thereafter it is "under the control" of the moon, and falls from there to the lunar surface. Find the *minimal* launch velocity v_0 that suffices for the projectile to make it "From the Earth to the Moon."

2.3 Application: Rocket Propulsion

FIGURE 2.3.5. An ascending rocket.

Suppose that the rocket of Fig.2.3.5 blasts off straight upward from the surface of the earth at time $t = 0$. We want to calculate its height y and velocity $v = dy/dt$ at time t. The rocket is propelled by exhaust gases that exit (rearward) with constant speed c (relative to the rocket). Because of the combustion of its fuel, the mass $m = m(t)$ of the rocket is variable.

To derive the equation of motion of the rocket, we use Newton's second law in the form

$$\frac{dP}{dt} = F \tag{1}$$

where P is momentum (the product of mass and velocity) and F denotes net external force (gravity, air resistance, etc.). If m is constant, then Eq. (1) takes the standard form $F = m(dv/dt) = ma$.

But here m is not constant. Suppose m changes to $m + \Delta m$ and v to $v + \Delta v$ during the short time interval from t to $t + \Delta t$. Then the change in the momentum *of the rocket itself* is

$$\Delta P \approx (m + \Delta m)(v + \Delta v) - mv = m\,\Delta v + v\,\Delta m + \Delta m\,\Delta v.$$

But the system also includes the exhaust gases expelled during this time interval, with mass $-\Delta m$ and approximate velocity $v - c$. Hence the total change in momentum during the time interval Δt is

$$\Delta P \approx (m\,\Delta v + v\,\Delta m + \Delta m\,\Delta v) + (-\Delta m)(v - c)$$
$$= m\,\Delta v + c\,\Delta m + \Delta m\,\Delta v.$$

If we divide by Δt and take the limit as $\Delta t \to 0$ and then substitute the resulting expression for dP/dt in (1), we get the **rocket propulsion equation**

$$m\frac{dv}{dt} + c\frac{dm}{dt} = F. \tag{2}$$

If $F = F_G + F_R$, where $F_G = -mg$ is a constant force of gravity and $F_R = -kv$ is a force of air resistance proportional to velocity, then Eq. (2) finally gives

$$m\frac{dv}{dt} + c\frac{dm}{dt} = -mg - kv. \tag{3}$$

Constant Thrust

Now suppose that the rocket fuel is consumed at the constant "burn rate" β during the time interval $[0, t_1]$, during which time the mass of the rocket decreases from m_0 to m_1. Thus

$$m(0) = m_0, \qquad m(t_1) = m_1,$$
$$m(t) = m_0 - \beta t, \qquad \frac{dm}{dt} = -\beta \quad \text{for } t \leq t_1, \tag{4}$$

with burnout occurring at time $t = t_1$.

PROBLEM 1 Substitute the expressions in (4) into Eq. (3) to obtain the differential equation

$$(m - \beta t)\frac{dv}{dt} + kv = \beta c - (m_0 - \beta t)g. \tag{5}$$

Solve this linear equation for

$$v(t) = v_0 M^{k/\beta} + \frac{\beta c}{k}(1 - M^{k/\beta}) + \frac{gm_0}{\beta - k}(1 - M^{k/\beta}), \tag{6}$$

where $v_0 = v(0)$ and

$$M = \frac{m(t)}{m_0} = \frac{m_0 - mt}{m_0}$$

denotes the rocket's **fractional mass** at time t.

No Resistance

PROBLEM 2 For the case of no air resistance, set $k = 0$ in Eq. (5) and integrate to obtain

$$v(t) = v_0 - gt + c \ln \frac{m_0}{m_0 - \beta t}. \tag{7}$$

Because $m_0 - \beta t_1 = m_1$, it follows that the velocity of the rocket at burnout ($t = t_1$) is

$$v_1 = v(t_1) = v_0 - gt_1 + c \ln \frac{m_0}{m_1}. \tag{8}$$

PROBLEM 3 Start with Eq. (7) and integrate to obtain

$$y(t) = (v_0 + c)t - \frac{1}{2}gt^2 - \frac{c}{\beta}(m_0 - \beta t) \ln \frac{m_0}{m_0 - \beta t}. \tag{9}$$

It follows that the rocket's altitude at burnout is

$$y_1 = y(t_1) = (v_0 + c)t_1 - \frac{1}{2}gt_1^2 - \frac{cm_1}{\beta} \ln \frac{m_0}{m_1}. \tag{10}$$

PROBLEM 4 The V-2 rocket that was used to attack London in World War II had an initial mass of 12,850 kg, of which 68.5% was fuel. This fuel burned uniformly for 70 seconds with an exhaust velocity of 2 km/s. Assume it encounters air resistance of 1.45 N per m/s of velocity. Then find the velocity and altitude of the V-2 at burnout under the assumption that it is launched vertically upward from rest on the ground.

Free Space

Suppose finally that the rocket is accelerating in free space, where there is neither gravity nor resistance, so $g = k = 0$. With $g = 0$ in Eq. (8) we see that, as the mass of the rocket decreases from m_0 to m_1, its increase in velocity is

$$\Delta v = v_1 - v_0 = c \ln \frac{m_0}{m_1}. \tag{11}$$

Note that Δv depends only on the exhaust gas speed c and the initial-to-final mass ratio m_0/m_1, but does not depend on the burn rate β. For example, if the rocket blasts off from rest ($v_0 = 0$) and $c = 5$ km/s and $m_0/m_1 = 20$, then its velocity at burnout is $v_1 = 5 \ln 20 \approx 15$ km/s. Thus if a rocket initially consists predominantly of fuel, then it can attain velocities significantly greater than the (relative) velocity of its exhaust gases.

2.4 Numerical Approximation: Euler's Method

It is the exception rather than the rule when a differential equation of the general form

$$\frac{dy}{dx} = f(x, y)$$

can be solved exactly and explicitly by elementary methods like those discussed in Chapter 1. For example, consider the simple equation

$$\frac{dy}{dx} = e^{-x^2}. \tag{1}$$

A solution of Eq. (1) is simply an antiderivative of e^{-x^2}. But it is known that every antiderivative of $f(x) = e^{-x^2}$ is a **nonelementary** function—one that cannot be expressed as a finite combination of the familiar functions of elementary calculus. Hence no particular solution of Eq. (1) is finitely expressible in terms of elementary functions. Any attempt to use the symbolic techniques of Chapter 1 to find a simple explicit formula for a solution of (1) is therefore doomed to failure.

As a possible alternative, a computer plotter can be programmed to draw a solution curve that starts at the initial point (x_0, y_0) and attempts to thread its way through the slope field of a given differential equation $y' = f(x, y)$. The procedure the plotter carries out can be described as follows.

- The plotter pen starts at the initial point (x_0, y_0) and moves a tiny distance along the slope segment though (x_0, y_0). This takes it to the point (x_1, y_1).
- At (x_1, y_1) the pen changes direction, and now moves a tiny distance along the slope segment through this new starting point (x_1, y_1). This takes it to the next starting point (x_2, y_2).
- At (x_2, y_2) the pen changes direction again, and now moves a tiny distance along the slope segment through (x_2, y_2). This takes it to the next starting point (x_3, y_3).

Fig. 2.4.1 illustrates the result of continuing in this fashion—by a sequence of discrete straight-line steps from one starting point to the next. In this figure we see a polygonal curve consisting of line segments that connect the successive points (x_0, y_0), (x_1, y_1), (x_2, y_2), (x_3, y_3), However, suppose that each "tiny distance" the pen travels along a slope segment—before the midcourse correction that sends it along a fresh new slope segment—is so very small that the naked eye cannot distinguish the individual line segments constituting the polygonal curve. Then the resulting polygonal curve looks like a smooth, continuously turning solution curve of the differential equation. Indeed, this is (in essence) how most of the solution curves shown in the figures of Chapter 1 were computer generated.

Euler did not have a computer plotter (in the 18th century), and his idea was to do all this numerically rather than graphically. In order to approximate the solution of the initial value problem

$$\frac{dy}{dx} = f(x, y), \quad y(x_0) = y_0, \tag{2}$$

we first choose a fixed (horizontal) **step size** h to use in making each step from one point to the next. Suppose we've started at the initial point (x_0, y_0) and after

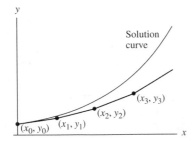

FIGURE 2.4.1. The first few steps in approximating a solution curve.

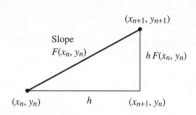

FIGURE 2.4.2. The step from (x_n, y_n) to (x_{n+1}, y_{n+1}).

n steps have reached the point (x_n, y_n). Then the step from (x_n, y_n) to the next point (x_{n+1}, y_{n+1}) is illustrated in Fig. 2.4.2. The slope of the direction segment through (x_n, y_n) is $m = f(x_n, y_n)$. Hence a horizontal change of h from x_n to x_{n+1} corresponds to a vertical change of $m \cdot h = h \cdot f(x_n, y_n)$ from y_n to y_{n+1}. Therefore the coordinates of the new point (x_{n+1}, y_{n+1}) are given in terms of the old coordinates by

$$x_{n+1} = x_n + h, \quad y_{n+1} = y_n + h \cdot f(x_n, y_n).$$

Given the initial value problem in (2), **Euler's method** with step size h consists of starting with the initial point (x_0, y_0) and applying the formulas

$$
\begin{aligned}
x_1 &= x_0 + h & y_1 &= y_0 + h \cdot f(x_0, y_0) \\
x_2 &= x_1 + h & y_2 &= y_1 + h \cdot f(x_1, y_1) \\
x_3 &= x_2 + h & y_3 &= y_2 + h \cdot f(x_2, y_2) \\
&\;\;\vdots & &\;\;\vdots
\end{aligned}
$$

to calculate successive points (x_1, y_1), (x_2, y_2), (x_3, y_3), ... on an approximate solution curve.

However, we ordinarily do not sketch the corresponding polygonal approximation. Instead, the numerical result of applying Euler's method is the sequence of *approximations*

$$y_1, y_2, y_3, \ldots, y_n, \ldots$$

to the *true values*

$$y(x_1), y(x_2), y(x_3), \ldots, y(x_n), \ldots$$

at the points $x_1, x_2, x_3, \ldots, x_n, \ldots$ of the *exact* (though unknown) solution $y(x)$ of the initial value problem. These results typically are presented in the form of a table of approximate values of the desired solution.

ALGORITHM: The Euler Method

Given the initial value problem

$$\blacktriangleright \qquad \frac{dy}{dx} = f(x, y), \quad y(x_0) = y_0, \tag{2}$$

Euler's method with step size h consists of applying the iterative formula

$$\blacktriangleright \qquad y_{n+1} = y_n + h \cdot f(x_n, y_n) \quad (n \geq 0) \tag{3}$$

to calculate successive approximations y_1, y_2, y_3, ... to the [true] values $y(x_1)$, $y(x_2)$, $y(x_3)$, ... of the [exact] solution $y = y(x)$ at the points x_1, x_2, x_3, \ldots, respectively. ▲

The iterative formula in (3) tells us how to make the typical step from y_n to y_{n+1} and is the heart of Euler's method. Although the most important applications of Euler's method are to nonlinear equations, we first illustrate the method with a simple initial value problem whose exact solution is available for purpose of comparison of approximate and actual solutions.

Example 1 Apply Euler's method to approximate the solution of the initial value problem

$$\frac{dy}{dx} = x + \frac{1}{5}y, \quad y(0) = -3, \tag{4}$$

(a) first with step size $h = 1$ on the interval $[0, 5]$,
(b) then with step size $h = 0.2$ on the interval $[0, 1]$.

Solution **(a)** With $x_0 = 0$, $y_0 = -3$, $f(x, y) = x + \frac{1}{5}y$, and $h = 1$ the iterative formula in (3) yields the approximate values

$$y_1 = y_0 + h \cdot [x_0 + \tfrac{1}{5}y_0] = (-3) + (1)[0 + \tfrac{1}{5}(-3)] = -3.6,$$

$$y_2 = y_1 + h \cdot [x_1 + \tfrac{1}{5}y_1] = (-3.6) + (1)[1 + \tfrac{1}{5}(-3.6)] = -3.32,$$

$$y_3 = y_2 + h \cdot [x_2 + \tfrac{1}{5}y_2] = (-3.32) + (1)[2 + \tfrac{1}{5}(-3.32)] = -1.984,$$

$$y_4 = y_3 + h \cdot [x_3 + \tfrac{1}{5}y_3] = (-1.984) + (1)[3 + \tfrac{1}{5}(-1.984)] = 0.6192, \quad \text{and}$$

$$y_5 = y_4 + h \cdot [x_4 + \tfrac{1}{5}y_4] = (0.6912) + (1)[4 + \tfrac{1}{5}(0.6912)] \approx 4.7430$$

at the points $x_1 = 1$, $x_2 = 2$, $x_3 = 3$, $x_4 = 4$, and $x_5 = 5$. Note how the result of each calculation feeds into the next one. The resulting table of approximate values is

x	0	1	2	3	4	5
Approx. y	-3	-3.6	-3.32	-1.984	0.6912	4.7430

Figure 2.4.3 shows the graph of this approximation, together with the graphs of the Euler approximations obtained with step sizes $h = 0.2$ and 0.05, as well as the graph of the exact solution

$$y(x) = 22e^{x/5} - 5x - 25$$

that is readily found using the linear-equation technique of Section 1.5. We see that decreasing the step size increases the accuracy, but with any single approximation, the accuracy decreases with distance from the initial point.

FIGURE 2.4.3. Graphs of Euler approximations with step sizes $h = 1$, $h = 0.2$, and $h = 0.05$.

(b) Starting afresh with $x_0 = 0$, $y_0 = -3$, $f(x, y) = x + \frac{1}{5}y$, and $h = 0.2$, we get the approximate values

$$y_1 = y_0 + h \cdot [x_0 + \tfrac{1}{5}y_0] = (-3) + (0.2)[0 + \tfrac{1}{5}(-3)] = -3.12,$$

$$y_2 = y_1 + h \cdot [x_1 + \tfrac{1}{5}y_1] = (-3.12) + (0.2)[0.2 + \tfrac{1}{5}(-3.12)] \approx -3.205,$$

$$y_3 = y_2 + h \cdot [x_2 + \tfrac{1}{5}y_2] \approx (-3.205) + (0.2)[0.4 + \tfrac{1}{5}(-3.205)] \approx -3.253,$$

$$y_4 = y_3 + h \cdot [x_3 + \tfrac{1}{5}y_3] \approx (-3.253) + (0.2)[0.6 + \tfrac{1}{5}(-3.253)] \approx -3.263,$$

$$y_5 = y_4 + h \cdot [x_4 + \tfrac{1}{5}y_4] \approx (-3.263) + (0.2)[0.8 + \tfrac{1}{5}(-3.263)] \approx -3.234$$

at the points $x_1 = 0.2$, $x_2 = 0.4$, $x_3 = 0.6$, $x_4 = 0.8$, and $x_5 = 1$. The resulting table of approximate values is

x	0	0.2	0.4	0.6	0.8	1
Approx. y	-3	-3.12	-3.205	-3.253	-3.263	-3.234

■

High accuracy with Euler's method usually requires a very small step size hence and a larger number of steps than can reasonably be carried out by hand. The application material for this section contains calculator and computer programs for automating Euler's method. One of these programs was used to calculate the table entries shown in Fig. 2.4.4. We see that 500 Euler steps (with step size $h = 0.002$) from $x = 0$ to $x = 1$ yield values that are accurate to within 0.001.

x	Approx y with $h = 0.2$	Approx y with $h = 0.02$	Approx y with $h = 0.002$	Actual value of y
0	-3.000	-3.000	-3.000	-3.000
0.2	-3.120	-3.104	-3.102	-3.102
0.4	-3.205	-3.172	-3.168	-3.168
0.6	-3.253	-3.201	-3.196	-3.195
0.8	-3.263	-3.191	-3.184	-3.183
1	-3.234	-3.140	-3.130	-3.129

FIGURE 2.4.4. Euler approximations with step sizes $h = 0.2$, $h = 0.02$, and $h = 0.002$.

Example 2

Suppose the baseball of Example 3 in Section 1.3 is simply dropped (instead of being thrown downward) from the helicopter. Then it's velocity $v(t)$ after t seconds satisfies the initial value problem

$$\frac{dv}{dt} = 32 - 0.16v, \quad v(0) = 0. \tag{5}$$

We use Euler's method with $h = 1$ to track the ball's increasing velocity at 1-second intervals for the first 10 seconds of fall. With $t_0 = 0$, $v_0 = 0$, $F(t, v) = 32 - 0.16v$, and $h = 1$ the iterative formula in (3) yields the approximate values

$$v_1 = v_0 + h \cdot [32 - 0.16v_0] = (0) + (1)[32 - 0.16(0)] = 32,$$

$$v_2 = v_1 + h \cdot [32 - 0.16v_1] = (32) + (1)[32 - 0.16(32)] = 58.88,$$

$$v_3 = v_2 + h \cdot [32 - 0.16v_2] = (58.88) + (1)[32 - 0.16(58.88)] \approx 81.46,$$

$$v_4 = v_3 + h \cdot [32 - 0.16v_3] = (81.46) + (1)[32 - 0.16(81.46)] \approx 100.43, \quad \text{and}$$

$$v_5 = v_4 + h \cdot [32 - 0.16v_4] = (100.43) + (1)[32 - 0.16(100.43)] \approx 116.36.$$

Continuing in this fashion, we complete the $h = 1$ column of v-values shown in the table of Fig. 2.4.5—where we have rounded off velocity entries to the nearest foot per second. The values corresponding to $h = 0.1$ were calculated using a computer, and we see that they are accurate to within about 1 ft/s. Note also that after 10 seconds the falling ball has attained about 80% of its limiting velocity of 200 ft/s. ∎

t	Approx v with $h = 1$	Approx v with $h = 0.1$	Actual value of v
1	32	30	30
2	59	55	55
3	81	77	76
4	100	95	95
5	116	111	110
6	130	124	123
7	141	135	135
8	150	145	144
9	158	153	153
10	165	160	160

FIGURE 2.4.5. Euler approximations in Example 2 with step sizes $h = 1$ and $h = 0.1$.

Local and Cumulative Errors

There are several sources of error in Euler's method that may make the approximation y_n to $y(x_n)$ unreliable for large values of n, those for which x_n is not sufficiently close to x_0. The error in the linear approximation formula

$$y(x_{n+1}) \approx y_n + h \cdot f(x_n, y_n) = y_{n+1} \tag{6}$$

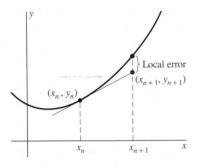

FIGURE 2.4.6. The local error in Euler's method.

is the amount by which the tangent line at (x_n, y_n) departs from the solution curve through (x_n, y_n), as illustrated in Fig. 2.4.6. This error, introduced at each step in the process, is called the **local error** in Euler's method.

The local error indicated in Fig. 2.4.6 *would be* the total error in y_{n+1} *if* the starting point y_n in (6) were an exact value, rather than merely an approximation to the actual value $y(x_n)$. But y_n itself suffers from the accumulated effects of all the local errors introduced at the previous steps. Thus the tangent line in Fig. 2.4.6 is tangent to the "wrong" solution curve—the one through (x_n, y_n) rather than the actual solution curve through the initial point (x_0, y_0). Figure 2.4.7 illustrates this **cumulative error** in Euler's method; it is the amount by which the polygonal stepwise path from (x_0, y_0) departs from the actual solution curve through (x_0, y_0).

The usual way of attempting to reduce the cumulative error in Euler's method is to decrease the step size h. The table in Fig. 2.4.8 shows the results obtained in approximating the exact solution $y(x) = 2e^x - x - 1$ of the initial value problem

$$\frac{dy}{dx} = x + y, \quad y(0) = 1,$$

using the successively smaller step sizes $h = 0.1$, $h = 0.02$, $h = 0.005$, and $h = 0.001$. We show computed values only at intervals of $\Delta x = 0.1$. For instance,

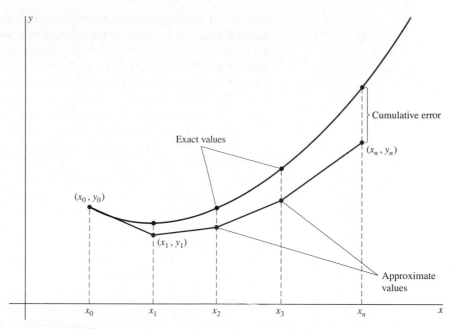

FIGURE 2.4.7. The cumulative error in Euler's method.

x	y with $h = 0.1$	y with $h = 0.02$	y with $h = 0.005$	y with $h = 0.001$	Actual y
0.1	1.1000	1.1082	1.1098	1.1102	1.1103
0.2	1.2200	1.2380	1.2416	1.2426	1.2428
0.3	1.3620	1.3917	1.3977	1.3993	1.3997
0.4	1.5282	1.5719	1.5807	1.5831	1.5836
0.5	1.7210	1.7812	1.7933	1.7966	1.7974
0.6	1.9461	2.0227	2.0388	2.0431	2.0442
0.7	2.1974	2.2998	2.3205	2.3261	2.3275
0.8	2.4872	2.6161	2.6422	2.6493	2.6511
0.9	2.8159	2.9757	3.0082	3.0170	3.0192
1.0	3.1875	3.3832	3.4230	3.4238	3.4266

FIGURE 2.4.8. Approximating the solution of $dy/dx = x + y$, $y(0) = 1$ with successively smaller step sizes

with $h = 0.001$, the computation required 1000 Euler steps, but the value y_n is shown only when n is a multiple of 100, so that x_n is an integral multiple of 0.1.

By scanning the column in Fig. 2.4.8 we observe that, for each fixed step size h, the error $y_{\text{actual}} - y_{\text{approx}}$ increases as x gets farther from the starting point $x_0 = 0$. But by scanning the rows of the table we see that for each fixed x, the error decreases as the step size h is reduced. The percentage errors at the final point $x = 1$ range from 7.25% with $h = 0.1$ down to only 0.08% with $h = 0.001$. Thus the smaller the step size, the more slowly does the error grow with increasing distance from the starting point.

The column of data for $h = 0.1$ in Fig. 2.4.8 requires only 10 steps, so Euler's method can be carried out with a hand-held calculator. But 50 steps are required to reach $x = 1$ with $h = 0.02$, 200 steps with $h = 0.005$, and 1000 steps with

$h = 0.001$. A computer is almost always used to implement Euler's method when more than 10 or 20 steps are required. Once an appropriate computer program has been written, one step size is—in principle—just as convenient as another; after all, the computer hardly cares how many steps it is asked to carry out.

Why, then, do we not simply choose an exceedingly small step size (such as $h = 10^{-12}$), with the expectation that very great accuracy will result? There are two reasons for not doing so. The first is obvious: the time required for the computation. For example, the data in Fig. 2.4.8 were obtained using a hand-held calculator that carried out nine Euler steps per second. Thus it required slightly over one second to approximate $y(1)$ with $h = 0.1$ and about 1 min 50 s with $h = 0.001$. But with $h = 10^{-12}$ it would require over 3000 years!

The second reason is more subtle. In addition to the local and cumulative errors discussed previously, the computer itself will contribute **roundoff error** at each stage because only finitely many significant digits can be used in each calculation. An Euler's method computation with $h = 0.0001$ will introduce roundoff errors 1000 times as often as one with $h = 0.1$. Hence with certain differential equations, $h = 0.1$ might actually produce more accurate results than those obtained with $h = 0.0001$, because the cumulative effect of roundoff error in the latter case might exceed combined cumulative and roundoff error in the case $h = 0.1$.

The "best" choice of h is difficult to determine in practice as well as in theory. It depends on the nature of the function $f(x, y)$ in the initial value problem in (2), on the exact code in which the program is written, and on the specific computer used. With a step size that is too large, the approximations inherent in Euler's method may not be sufficiently accurate, whereas if h is too small, then roundoff errors may accumulate to an unacceptable degree or the program may require too much time to be practical. The subject of *error propagation* in numerical algorithms is treated in numerical analysis courses and textbooks.

The computations in Fig. 2.4.8 illustrate the common strategy of applying a numerical algorithm, such as Euler's method, several times in succession, beginning with a selected number n of subintervals for the first application, then doubling n for each succeeding application of the method. Visual comparison of successive results often can provide an "intuitive feel" for their accuracy. In the next two examples we present graphically the results of successive applications of Euler's method.

Example 3 The exact solution of the logistic initial value problem

$$\frac{dy}{dx} = \tfrac{1}{3}y(8 - y), \quad y(0) = 1$$

is $y(x) = 8/(1 + 7e^{-8x/3})$. Figure 2.4.9 shows both the exact solution curve and approximate solution curves obtained by applying Euler's method on the interval $0 \le x \le 5$ with $n = 5, n = 10$, and $n = 20$ subintervals. Each of these "curves" actually consists of line segments joining successive points (x_n, y_n) and (x_{n+1}, y_{n+1}). The Euler approximation with 5 subintervals is poor, and the approximation with 10 subintervals also overshoots the limiting value $y = 8$ of the solution before leveling off, but with 20 subintervals we obtain fairly good qualitative agreement with the actual behavior of the solution. ■

Example 4 The exact solution of the initial value problem

$$\frac{dy}{dx} = y \cos x, \quad y(0) = 1$$

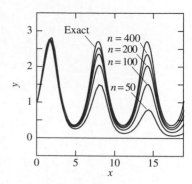

FIGURE 2.4.9. Approximating a logistic solution using Euler's method with $n = 5, n = 10$, and $n = 20$ subintervals.

FIGURE 2.4.10. Approximating the exact solution $y = e^{\sin x}$ using Euler's method with 50, 100, 200, and 400 subintervals.

is the periodic function $y(x) = e^{\sin x}$. Figure 2.4.10 shows both the exact solution curve and approximate solution curves obtained by applying Euler's method on the interval $0 \leq x \leq 6\pi$ with $n = 50, n = 100, n = 200$, and $n = 400$ subintervals. Even with this many subintervals, Euler's method evidently has considerable difficulty keeping up with the oscillations in the actual solution. Consequently, the more accurate methods discussed in succeeding sections are needed for serious numerical investigations. ■

A Word of Caution

The data shown in Fig. 2.4.8 indicate that Euler's method works well in approximating the solution of $dy/dx = x + y$, $y(0) = 1$ on the interval $[0, 1]$. That is, for each fixed x it appears that the approximate values approach the actual value of $y(x)$ as the step size h is decreased. For instance, the approximate values in the rows corresponding to $x = 0.3$ and $x = 0.5$ suggest that $y(0.3) \approx 1.40$ and $y(0.5) \approx 1.80$, in accord with the actual values shown in the final column of the table.

Example 5, in contrast, shows that some initial value problems are not so well behaved.

Example 5 Use Euler's method to approximate the solution of the initial value problem

$$\frac{dy}{dx} = x^2 + y^2, \quad y(0) = 1 \tag{7}$$

on the interval $[0, 1]$.

Solution Here $f(x, y) = x^2 + y^2$, so the iterative formula of Euler's method is

$$y_{n+1} = y_n + h \cdot (x_n^2 + y_n^2). \tag{8}$$

With step size $h = 0.1$ we obtain

$$y_1 = 1 + (0.1) \cdot [(0)^2 + (1)^2] = 1.1,$$
$$y_2 = 1.1 + (0.1) \cdot [(0.1)^2 + (1.1)^2] = 1.222,$$
$$y_3 = 1.222 + (0.1) \cdot [(0.2)^2 + (1.222)^2] \approx 1.3753,$$

and so forth. Rounded to four decimal places, the first ten values obtained in this manner are

$$y_1 = 1.1000 \qquad y_6 = 2.1995$$
$$y_2 = 1.2220 \qquad y_7 = 2.7193$$
$$y_3 = 1.3753 \qquad y_8 = 3.5078$$
$$y_4 = 1.5735 \qquad y_9 = 4.8023$$
$$y_5 = 1.8371 \qquad y_{10} = 7.1895$$

But instead of naively accepting these results as accurate approximations, we decided to use a computer to repeat the computations with smaller values of h. The table in Fig. 2.4.11 shows the results obtained with step sizes $h = 0.1$, $h = 0.02$, and $h = 0.005$. Observe that now the "stability" of the procedure in Example 1 is missing. Indeed, it seems obvious that something is going wrong near $x = 1$.

Figure 2.4.12 provides a graphical clue to the difficulty. It shows a slope field for $dy/dx = x^2 + y^2$, together with a solution curve through $(0, 1)$ plotted using one of the more accurate approximation methods of the following two sections. It appears that this solution curve may have a vertical asymptote near $x = 0.97$. Indeed, an exact solution using Bessel functions (see Problem 16 in Section 8.6) can be used to show that $y(x) \rightarrow +\infty$ as $x \rightarrow 0.969811$ (approximately). Although Euler's method gives values (albeit spurious ones) at $x = 1$, the actual solution does not exist on the entire interval $[0, 1]$. Moreover, Euler's method is unable to "keep up" with the rapid changes in $y(x)$ that occur as x approaches the infinite discontinuity near 0.969811. ■

x	y with $h = 0.1$	y with $h = 0.02$	y with $h = 0.005$
0.1	1.1000	1.1088	1.1108
0.2	1.2220	1.2458	1.2512
0.3	1.3753	1.4243	1.4357
0.4	1.5735	1.6658	1.6882
0.5	1.8371	2.0074	2.0512
0.6	2.1995	2.5201	2.6104
0.7	2.7193	3.3612	3.5706
0.8	3.5078	4.9601	5.5763
0.9	4.8023	9.0000	12.2061
1.0	7.1895	30.9167	1502.2090

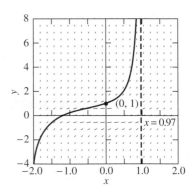

FIGURE 2.4.11. Attempting to approximate the solution of $dy/dx = x^2 + y^2$, $y(0) = 1$.

FIGURE 2.4.12. Solution of $dy/dx = x^2 + y^2$, $y(0) = 1$.

The moral of Example 5 is that there are pitfalls in the numerical solution of certain initial value problems. Certainly it's pointless to attempt to approximate a solution on an interval where it doesn't even exist (or where it is not unique, in which case there's no general way to predict which way the numerical approximations will branch at a point of nonuniqueness). One should never accept as accurate the results of applying Euler's method with a single fixed step size h. A second "run" with smaller step size ($h/2$, say, or $h/5$, or $h/10$) may give seemingly consistent results, thereby suggesting their accuracy, or it may—as in Example 5—reveal the presence

of some hidden difficulty in the problem. Many problems simply require the more accurate and powerful methods that are discussed in the final two sections of this chapter.

2.4 Problems

In Problems 1 through 10, an initial value problem and its exact solution $y(x)$ are given. Apply Euler's method twice to approximate to this solution on the interval $[0, \frac{1}{2}]$, first with step size $h = 0.25$, then with step size $h = 0.1$. Compare the 3-decimal-place values of the two approximations at $x = \frac{1}{2}$ with the value $y(\frac{1}{2})$ of the actual solution.

1. $y' = -y$, $y(0) = 2$; $y(x) = 2e^{-x}$
2. $y' = 2y$, $y(0) = \frac{1}{2}$; $y(x) = \frac{1}{2}e^{2x}$
3. $y' = y + 1$, $y(0) = 1$; $y(x) = 2e^x - 1$
4. $y' = x - y$, $y(0) = 1$; $y(x) = 2e^{-x} + x - 1$
5. $y' = y - x - 1$, $y(0) = 1$; $y(x) = 2 + x - e^x$
6. $y' = -2xy$, $y(0) = 2$; $y(x) = 2e^{-x^2}$
7. $y' = -3x^2 y$, $y(0) = 3$; $y(x) = 3e^{-x^3}$
8. $y' = e^{-y}$, $y(0) = 0$; $y(x) = \ln(x + 1)$
9. $y' = \frac{1}{4}(1 + y^2)$, $y(0) = 1$; $y(x) = \tan \frac{1}{4}(x + \pi)$
10. $y' = 2xy^2$, $y(0) = 1$; $y(x) = \dfrac{1}{1 - x^2}$

Note: *The application following this problem set lists illustrative calculator/computer programs that can be used in the remaining problems.*

A programmable calculator or a computer will be useful for Problems 11 through 16. In each problem find the exact solution of the given initial value problem. Then apply Euler's method twice to approximate (to four decimal places) this solution on the given interval, first with step size $h = 0.01$, then with step size $h = 0.005$. Make a table showing the approximate values and the actual value, together with the percentage error in the more accurate approximation, for x an integral multiple of 0.2. Throughout, primes denote derivatives with respect to x.

11. $y' = y - 2$, $y(0) = 1$; $0 \le x \le 1$
12. $y' = \frac{1}{2}(y - 1)^2$, $y(0) = 2$; $0 \le x \le 1$
13. $yy' = 2x^3$, $y(1) = 3$; $1 \le x \le 2$
14. $xy' = y^2$, $y(1) = 1$; $1 \le x \le 2$
15. $xy' = 3x - 2y$, $y(2) = 3$; $2 \le x \le 3$
16. $y^2 y' = 2x^5$, $y(2) = 3$; $2 \le x \le 3$

A computer with a printer is required for Problems 17 through 24. In these initial value problems, use Euler's method with step sizes $h = 0.1$, 0.02, 0.004, and 0.0008 to approximate to four decimal places the values of the solution at ten equally spaced points of the given interval. Print the results in tabular

form with appropriate headings to make it easy to gauge the effect of varying the step size h. Throughout, primes denote derivatives with respect to x.

17. $y' = x^2 + y^2$, $y(0) = 0$; $0 \le x \le 1$
18. $y' = x^2 - y^2$, $y(0) = 1$; $0 \le x \le 2$
19. $y' = x + \sqrt{y}$, $y(0) = 1$; $0 \le x \le 2$
20. $y' = x + \sqrt[3]{y}$, $y(0) = -1$; $0 \le x \le 2$
21. $y' = \ln y$, $y(1) = 2$; $1 \le x \le 2$
22. $y' = x^{2/3} + y^{2/3}$, $y(0) = 1$; $0 \le x \le 2$
23. $y' = \sin x + \cos y$, $y(0) = 0$; $0 \le x \le 1$
24. $y' = \dfrac{x}{1 + y^2}$, $y(-1) = 1$; $-1 \le x \le 1$

25. You bail out of the helicopter of Example 2 and immediately pull the ripcord of your parachute. Now $k = 1.6$ in Eq. (5), so your downward velocity satisfies the initial value problem

$$\frac{dv}{dt} = 32 - 1.6v, \quad v(0) = 0$$

(with t in seconds and v in ft/sec). Use Euler's method with a programmable calculator or computer to approximate the solution for $0 \le t \le 2$, first with step size $h = 0.01$ and then with $h = 0.005$, rounding off approximate v-values to 1 decimal place. What percentage of the limiting velocity 20 ft/sec has been attained after 1 second? After 2 seconds?

26. Suppose the deer population $P(t)$ in a small forest initially numbers 25 and satisfies the logistic equation

$$\frac{dP}{dt} = 0.0225P - 0.0003P^2$$

(with t in months). Use Euler's method with a programmable calculator or computer to approximate the solution for 10 years, first with step size $h = 1$ and then with $h = 0.5$, rounding off approximate P-values to integral numbers of deer. What percentage of the limiting population of 75 deer has been attained after 5 years? After 10 years?

Use Euler's method with a computer system to find the desired solution values in Problems 27 and 28. Start with step size $h = 0.1$, and then use successively smaller step sizes until successive approximate solution values at $x = 2$ agree rounded off to 2 decimal places.

27. $y' = x^2 + y^2 - 1$, $y(0) = 0$; $y(2) = ?$

28. $y' = x + \frac{1}{2}y^2$, $y(-2) = 0$; $y(2) = ?$

29. Consider the initial value problem

$$7x\frac{dy}{dx} + y = 0, \quad y(-1) = 1.$$

(a) Solve this problem for the exact solution

$$y(x) = -\frac{1}{x^{1/7}},$$

which has an infinite discontinuity at $x = 0$. **(b)** Apply Euler's method with step size $h = 0.15$ to approximate this solution on the interval $-1 \leq x \leq 0.5$. Note that, from these data alone, you might not suspect any difficulty near $x = 0$. The reason is that the numerical approximation "jumps across the discontinuity" to another solution of $7xy' + y = 0$ for $x > 0$. **(c)** Finally, apply Euler's method with step sizes $h = 0.03$ and $h = 0.006$, but still printing results only at the original points $x = -1.00$,

-0.85, -0.70, ..., 1.20, 1.35. and 1.50. Would you now suspect a discontinuity in the exact solution?

30. Apply Euler's method with successively smaller step sizes on the interval $[0, 2]$ to verify empirically that the solution of the initial value problem

$$\frac{dy}{dx} = x^2 + y^2, \quad y(0) = 0$$

has a vertical asymptote near $x = 2.003147$. (Contrast this with Example 2, in which $y(0) = 1$.)

31. The general solution of the equation

$$\frac{dy}{dx} = (1 + y^2)\cos x$$

is $y(x) = \tan(C + \sin x)$. With the initial condition $y(0) = 0$ the solution $y(x) = \tan(\sin x)$ is well behaved. But with $y(0) = 1$ the solution $y(x) = \tan\left(\frac{1}{4}\pi + \sin x\right)$ has a vertical asymptote at $x = \sin^{-1}(\pi/4) \approx 0.90334$. Use Euler's method to verify this fact empirically.

2.4 Application: Implementing Euler's Method

Construction of a calculator or computer program to implement a numerical algorithm can sharpen one's understanding of the algorithm. Figure 2.4.13 lists TI-85 and BASIC programs implementing Euler's method to approximate the solution of the initial value problem

$$\frac{dy}{dx} = x + y, \quad y(0) = 1$$

considered in this section. The comments provided in the final column should make these programs intelligible even if you have little familiarity with the BASIC and TI calculator programming languages.

TI-85	BASIC	Comment
PROGRAM:EULER	Program EULER	Program title
:10→N	N = 10	Number of steps
:0→X	X = 0	Initial x
:1→Y	Y = 1	Initial y
:1→X1	X1 = 1	Final x
:(X1-X)/N→H	H = (X1-X)/N	Step size
:For(I,1,N)	FOR I=1 TO N	Begin loop
:X+Y→F	F = X + Y	Function value
:Y+H*F→Y	Y = Y + H*F	Euler iteration
:X+H→X	X = X + H	New x
:Disp X,Y	PRINT X,Y	Display results
:End	NEXT I	End loop

FIGURE 2.4.13. TI-85 and BASIC Euler's method programs.

To increase the number of steps (and thereby decrease the step size) you need only change the value of **N** specified in the first line of the program. To apply Euler's method to a different equation $dy/dx = f(x, y)$, you need only change the single line that calculates the function value **F**.

Any other procedural programming language (such as FORTRAN or Pascal) would follow the pattern illustrated by the parallel lines of TI-85 and BASIC code in Fig. 2.4.13. Some of the modern functional programming languages mirror standard mathematical notation even more closely. Figure 2.4.14 shows a MATLAB implementation of Euler's method. The **euler** function takes as input the initial value **x**, the initial value **y**, the final value **x1** of x, and the desired number **n** of subintervals. For instance, the MATLAB command

```
[X, Y] = euler(0, 1, 1, 10)
```

then generates the x_n- and y_n-data shown in the first two columns of the table of Fig. 2.4.8.

```
function yp = f(x,y)
yp = x + y;              % yp = y'

function [X,Y] = euler(x,y,x1,n)
h = (x1 - x)/n;          % step size
X = x;                   % initial x
Y = y;                   % initial y
for i = 1:n              % begin loop
    y = y + h*f(x,y);    % Euler iteration
    x = x + h;           % new x
    X = [X;x];           % update x-column
    Y = [Y;y];           % update y-column
end                      % end loop
```

FIGURE 2.4.14. MATLAB implementation of Euler's method.

You should begin this project by implementing Euler's method with your own calculator or computer system. Test your program by first applying it to the initial value problem in Example 1, then to some of the problems for this section.

Famous Numbers Investigation

The following problems describe the numbers $e \approx 2.71828$, $\ln 2 \approx 0.69315$, and $\pi \approx 3.14159$ as specific values of solutions of certain initial value problems. In each case, apply Euler's method with $n = 50, 100, 200, \ldots$ subintervals (doubling n each time). How many subintervals are needed to obtain—twice in succession—the correct value of the target number rounded to three decimal places?

1. The number $e = y(1)$, where $y(x)$ is the solution of the initial value problem $dy/dx = y$, $y(0) = 1$.

2. The number $\ln 2 = y(2)$, where $y(x)$ is the solution of the initial value problem $dy/dx = 1/x$, $y(1) = 0$.

3. The number $\pi = y(1)$, where $y(x)$ is the solution of the initial value problem $dy/dx = 4/(1 + x^2)$, $y(0) = 0$.

Also explain in each problem what the point is—why the indicated famous number is the expected numerical result.

2.5 A Closer Look at the Euler Method

The Euler method as presented in Section 2.4 is not often used in practice, mainly because more accurate methods are available. But Euler's method has the advantage of simplicity, and a careful study of this method yields insights into the workings of more accurate methods, because many of the latter are extensions or refinements of the Euler method.

To compare two different methods of numerical approximation, we need some way to measure the accuracy of each. Theorem 1 tells what degree of accuracy we can expect when we use Euler's method.

THEOREM 1: The Error in the Euler Method

Suppose that the initial value problem

$$\frac{dy}{dx} = f(x, y), \quad y(x_0) = y_0 \tag{1}$$

has a unique solution $y(x)$ on the closed interval $[a, b]$ with $a = x_0$, and assume that $y(x)$ has a continuous second derivative on $[a, b]$. (This would follow from the assumption that f, f_x, and f_y are all continuous for $a \leq x \leq b$ and $c \leq y \leq d$, where $c \leq y(x) \leq d$ for all x in $[a, b]$.) Then there exists a constant C such that the following is true: If the approximations $y_1, y_2, y_3, \ldots, y_k$ to the actual values $y(x_1)$, $y(x_2), y(x_3), \ldots, y(x_k)$ at points of $[a, b]$ are computed using Euler's method with step size $h > 0$, then

$$|y_n - y(x_n)| \leq Ch \tag{2}$$

for each $n = 1, 2, 3, \ldots, k$. ▲

Remark: The **error**

$$y_{\text{actual}} - y_{\text{approx}} = y(x_n) - y_n$$

in (2) denotes the [cumulative] error in Euler's method after n steps in the approximation, *exclusive* of roundoff error (as though we were using a perfect machine that made no roundoff errors). The theorem can be summarized by saying that *the error in Euler's method is of order* h; that is, the error is bounded by a [predetermined] constant C multiplied by the step size h. It follows, for instance, that (on a given closed interval) halving the step size cuts the maximum error in half; similarly, with step size $h/10$ we get 10 times the accuracy (that is, $1/10$ the maximum error) as with step size h. Consequently, we can—in principle—get any degree of accuracy we want by choosing h sufficiently small. ■

We will omit the proof of this theorem, but one can be found in Chapter 7 of G. Birkhoff and G.-C. Rota, *Ordinary Differential Equations,* 4th ed. (New York: John Wiley, 1989). The constant C deserves some comment. Because C tends to increase as the maximum value of $|y''(x)|$ on $[a, b]$ increases, it follows that C must depend in a fairly complicated way on y, and actual computation of a value of C such that the inequality in (2) holds is usually impractical. In practice, the following type of procedure is commonly employed.

1. Apply Euler's method to the initial value problem in (1) with a reasonable value of h.
2. Repeat with $h/2$, $h/4$, and so forth, at each stage halving the step size for the next application of Euler's method.
3. Continue until the results obtained at one stage agree—to an appropriate number of significant digits—with those obtained at the previous stage. Then the approximate values obtained at this stage are considered likely to be accurate to the indicated number of significant digits.

Example 1 Carry out this procedure with the initial value problem

$$\frac{dy}{dx} = -\frac{2xy}{1+x^2}, \quad y(0) = 1 \tag{3}$$

to approximate accurately the value $y(1)$ of the solution at $x = 1$.

Solution Using an Euler method program, perhaps one of those listed in Figs. 2.4.7 and 2.4.8, we begin with a step size $h = 0.04$ requiring $n = 25$ steps to reach $x = 1$. The table in Fig. 2.5.1 shows the approximate values of $y(1)$ obtained with successively smaller values of h. The data suggest that the true value of $y(1)$ is exactly 0.5. Indeed, the exact solution of the initial value problem in (3) is $y(x) = 1/(1+x^2)$, so the true value of $y(1)$ is exactly $\frac{1}{2}$. ∎

| h | Approximate $y(1)$ | Actual $y(1)$ | $|\text{Error}|/h$ |
|---|---|---|---|
| 0.04 | 0.50451 | 0.50000 | 0.11 |
| 0.02 | 0.50220 | 0.50000 | 0.11 |
| 0.01 | 0.50109 | 0.50000 | 0.11 |
| 0.005 | 0.50054 | 0.50000 | 0.11 |
| 0.0025 | 0.50027 | 0.50000 | 0.11 |
| 0.00125 | 0.50013 | 0.50000 | 0.10 |
| 0.000625 | 0.50007 | 0.50000 | 0.11 |
| 0.0003125 | 0.50003 | 0.50000 | 0.10 |

FIGURE 2.5.1. Table of values in Example 1.

The final column of the table in Fig. 2.5.1 displays the ratio of the magnitude of the error to h; that is, $|y_{\text{actual}} - y_{\text{approx}}|/h$. Observe how the data in this column substantiate Theorem 1—in this computation, the error bound in (2) appears to hold with a value of C slightly larger than 0.1.

An Improvement in Euler's Method

As Fig. 2.5.2 shows, Euler's method is rather unsymmetrical. It uses the predicted slope $k = f(x_n, y_n)$ of the graph of the solution at the left-hand endpoint of the interval $[x_n, x_n + h]$ as if it were the actual slope of the solution over that entire interval. We now turn our attention to a way in which increased accuracy can easily be obtained; it is known as the *improved Euler method*.

Given the initial value problem

$$\frac{dy}{dx} = f(x, y), \quad y(x_0) = y_0, \tag{4}$$

FIGURE 2.5.2. True and predicted values in Euler's method.

suppose that after carrying out n steps with step size h we have computed the approximation y_n to the actual value $y(x_n)$ of the solution at $x_n = x_0 + nh$. We can use the Euler method to obtain a first estimate—which we now call u_{n+1} rather than y_{n+1}—of the value of the solution at $x_{n+1} = x_n + h$. Thus

➤
$$u_{n+1} = y_n + h \cdot f(x_n, y_n) = y_n + h \cdot k_1.$$

Now that $u_{n+1} \approx y(x_{n+1})$ has been computed, we can take

➤
$$k_2 = f(x_{n+1}, u_{n+1})$$

as a second estimate of the slope of the solution curve $y = y(x)$ at $x = x_{n+1}$.

Of course, the approximate slope $k_1 = f(x_n, y_n)$ at $x = x_n$ has already been calculated. Why not *average* these two slopes to obtain a more accurate estimate of the average slope of the solution curve over the entire subinterval $[x_n, x_{n+1}]$? This idea is the essence of the *improved* Euler method. Figure 2.5.3 shows the geometry behind this method.

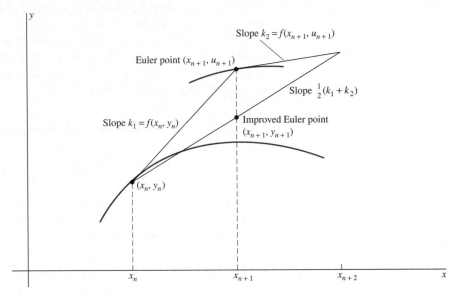

FIGURE 2.5.3. The improved Euler method: Average the slopes of the tangent lines at (x_n, y_n) and (x_{n+1}, u_{n+1}).

ALGORITHM: The Improved Euler Method

Given the initial value problem

$$\frac{dy}{dx} = f(x, y), \quad y(x_0) = y_0,$$

the **improved Euler method with step size** h consists in applying the iterative formulas

$$
\begin{aligned}
k_1 &= f(x_n, y_n), \\
u_{n+1} &= y_n + h \cdot k_1, \\
k_2 &= f(x_{n+1}, u_{n+1}), \\
y_{n+1} &= y_n + h \cdot \tfrac{1}{2}(k_1 + k_2)
\end{aligned}
\tag{5}
$$

to compute successive approximations y_1, y_2, y_3, ... to the [true] values $y(x_1)$, $y(x_2)$, $y(x_3)$, ... of the [exact] solution $y = y(x)$ at the points x_1, x_2, x_3, ..., respectively. ▲

Remark: The final formula in (5) takes the "Euler form"

$$y_{n+1} = y_n + h \cdot k$$

if we write

$$k = \frac{k_1 + k_2}{2}$$

for the approximate *average slope* on the interval $[x_n, x_{n+1}]$. ■

The improved Euler method is one of a class of numerical techniques known as **predictor-corrector** methods. First a predictor u_{n+1} of the next y-value is computed; then it is used to correct itself. Thus the **improved Euler method** with step size h consists of using the **predictor**

➤ $$u_{n+1} = y_n + h \cdot f(x_n, y_n) \qquad (6)$$

and the **corrector**

➤ $$y_{n+1} = y_n + h \cdot \tfrac{1}{2} \left[f(x_n, y_n) + f(x_{n+1}, u_{n+1}) \right] \qquad (7)$$

iteratively to compute successive approximations y_1, y_2, y_2, ... to the values $y(x_1)$, $y(x_2)$, $y(x_3)$, ... of the actual solution of the initial value problem in (4).

Remark: Each improved Euler step requires two evaluations of the function $f(x, y)$, as compared with the single function evaluation required for an ordinary Euler step. We naturally wonder whether this doubled computational labor is worth the trouble.

Answer: Under the assumption that the exact solution $y = y(x)$ of the initial value problem in (4) has a continuous third derivative, it can be proved—see Chapter 7 of Birkhoff and Rota—that the error in the improved Euler method is of order h^2. This means that on a given bounded interval $[a, b]$, each approximate value y_n satisfies the inequality

➤ $$|y(x_n) - y_n| \leq Ch^2, \qquad (8)$$

where the constant C does not depend on h. Because h^2 is much smaller than h if h itself is small, this means that the improved Euler method is more accurate than Euler's method itself. This advantage is offset by the fact that about twice as many computations are required. But the factor h^2 in (8) means that halving the step size results in 1/4 the maximum error, and with step size $h/10$ we get 100 times the accuracy (that is, 1/100 the maximum error) as with step size h. ■

Example 2 Figure 2.4.8 shows results of applying Euler's method to the initial value problem

$$\frac{dy}{dx} = x + y, \quad y(0) = 1 \qquad (9)$$

with exact solution $y(x) = 2e^x - x - 1$. With $f(x, y) = x + y$ in Eqs. (6) and (7), the predictor-corrector formulas for the improved Euler method are

$$u_{n+1} = y_n + h \cdot (x_n + y_n),$$
$$y_{n+1} = y_n + h \cdot \tfrac{1}{2}\left[(x_n + y_n) + (x_{n+1} + u_{n+1})\right].$$

With step size $h = 0.1$ we calculate

$$u_1 = 1 + (0.1) \cdot (0 + 1) = 1.1,$$
$$y_1 = 1 + (0.05) \cdot [(0 + 1) + (0.1 + 1.1)] = 1.11,$$
$$u_2 = 1.11 + (0.1) \cdot (0.1 + 1.11) = 1.231,$$
$$y_2 = 1.11 + (0.05) \cdot [(0.1 + 1.11) + (0.2 + 1.231)] = 1.24205,$$

and so forth. The table in Fig. 2.5.4 compares the results obtained using the improved Euler method with those obtained previously using the "unimproved" Euler method. When the same step size $h = 0.1$ is used, the error in the Euler approximation to $y(1)$ is 7.25%, but the error in the improved Euler approximation is only 0.24%.

	Euler method, $h = 0.1$	Euler method, $h = 0.005$	Improved Euler, $h = 0.1$	
x	Values of y	Values of y	Values of y	Actual y
0.1	1.1000	1.1098	1.1100	1.1103
0.2	1.2200	1.2416	1.2421	1.2428
0.3	1.3620	1.3977	1.3985	1.3997
0.4	1.5282	1.5807	1.5818	1.5836
0.5	1.7210	1.7933	1.7949	1.7974
0.6	1.9431	2.0388	2.0409	2.0442
0.7	2.1974	2.3205	2.3231	2.3275
0.8	2.4872	2.6422	2.6456	2.6511
0.9	2.8159	3.0082	3.0124	3.0192
1.0	3.1875	3.4230	3.4282	3.4366

FIGURE 2.5.4. Euler and improved Euler approximations to the solution of $dy/dx = x + y$, $y(0) = 1$.

	Improved Euler, Approximate	Actual
x	y	y
0.0	1.00000	1.00000
0.1	1.11034	1.11034
0.2	1.24280	1.24281
0.3	1.39971	1.39972
0.4	1.58364	1.58365
0.5	1.79744	1.79744
0.6	2.04423	2.04424
0.7	2.32749	2.32751
0.8	2.65107	2.65108
0.9	3.01919	3.01921
1.0	3.43654	3.43656

FIGURE 2.5.5. Improved Euler approximation to the solution of Eq. (9) with step size $h = 0.005$.

Indeed, the improved Euler method with $h = 0.1$ is more accurate (in this example) than the original Euler method with $h = 0.005$. The latter requires 200 evaluations of the function $f(x, y)$, but the former requires only 20 such evaluations, so in this case the improved Euler method yields greater accuracy with only about one-tenth the work.

Figure 2.5.5 shows the results obtained when the improved Euler method is applied to the initial value problem in (9) using step size $h = 0.005$. Accuracy of five significant figures is apparent in the table. This suggests that, in contrast with the original Euler method, the improved Euler method is sufficiently accurate for certain practical applications—such as plotting solution curves. ∎

An improved Euler program (similar to the ones listed in the project material for this section) was used to compute approximations to the exact value $y(1) = 0.5$

of the solution $y(x) = 1/(1 + x^2)$ of the initial value problem

$$\frac{dy}{dx} = -\frac{2xy}{1 + x^2}, \quad y(0) = 1 \tag{3}$$

of Example 1. The results obtained by successively halving the step size appear in the table in Fig. 2.5.6. Note that the final column of this table impressively corroborates the form of the error bound in (8), and that each halving of the step size reduces the error by a factor of almost exactly 4, as should happen if the error is proportional to h^2.

	Improved Euler Approximation				
h	to $y(1)$	Error	$	Error	/h^2$
0.04	0.500195903	−0.000195903	0.12		
0.02	0.500049494	−0.000049494	0.12		
0.01	0.500012437	−0.000012437	0.12		
0.005	0.500003117	−0.000003117	0.12		
0.0025	0.500000780	−0.000000780	0.12		
0.00125	0.500000195	−0.000000195	0.12		
0.000625	0.500000049	−0.000000049	0.12		
0.0003125	0.500000012	−0.000000012	0.12		

FIGURE 2.5.6. Improved Euler approximation to $y(1)$ for $dy/dx = -2xy/(1 + x^2)$, $y(0) = 1$.

In the following two examples we exhibit graphical results obtained by employing this strategy of successively halving the step size, and thus doubling the number of subintervals of a fixed interval on which we are approximating a solution.

Example 3 In Example 3 of Section 2.4 we applied Euler's method to the logistic initial value problem

$$\frac{dy}{dx} = \tfrac{1}{3}y(8 - y), \quad y(0) = 1.$$

Figure 2.4.9 shows an obvious difference between the exact solution $y(x) = 8/(1 + 7e^{-8x/3})$ and the Euler approximation on $0 \leq x \leq 5$ using $n = 20$ subintervals. Figure 2.5.7 shows approximate solution curves plotted using the improved Euler's method.

The approximation with five subintervals is still bad—perhaps worse! It appears to level off considerably short of the actual limiting population $M = 8$. You should carry out at least the first two improved Euler steps manually to see for yourself how it happens that, after increasing appropriately during the first step, the approximate solution *decreases* in the second step rather than continuing to increase (as it should). In the project for this section we ask you to show empirically that the improved Euler approximate solution with step size $h = 1$ levels off at $y \approx 4.3542$.

In contrast, the approximate solution curve with $n = 20$ subintervals tracks the exact solution curve rather closely, and with $n = 40$ subintervals the exact and approximate solution curves are indistinguishable in Fig. 2.5.7. The table in Fig. 2.5.8 indicates that the improved Euler approximation with $n = 200$ subintervals is accurate rounded to three decimal places (that is, four significant digits) on the interval

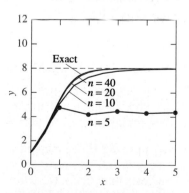

FIGURE 2.5.7. Approximating a logistic solution using the improved Euler method with $n = 5$, $n = 10$, $n = 20$, and $n = 40$ subintervals.

x	Actual $y(x)$	Improved Euler with $n = 200$
0	1.0000	1.0000
1	5.3822	5.3809
2	7.7385	7.7379
3	7.9813	7.9812
4	7.9987	7.9987
5	7.9999	7.9999

FIGURE 2.5.8. Using the improved Euler method to approximate the actual solution of the initial value problem in Example 3.

$0 \leq x \leq 5$. Because discrepancies in the fourth significant digit are not visually apparent at the resolution of an ordinary computer screen, the improved Euler method (using several hundred subintervals) is considered adequate for many graphical purposes. ■

Example 4 In Example 4 of Section 2.4 we applied Euler's method to the initial value problem

$$\frac{dy}{dx} = y \cos x, \quad y(0) = 1.$$

Figure 2.4.10 shows obvious visual differences between the periodic exact solution $y(x) = e^{\sin x}$ and the Euler approximations on $0 \leq x \leq 6\pi$ with as many as $n = 400$ subintervals.

Figure 2.5.9 shows the exact solution curve and approximate solution curves plotted using the improved Euler method with $n = 50$, $n = 100$, and $n = 200$ subintervals. The approximation obtained with $n = 200$ is indistinguishable from the exact solution curve, and the approximation with $n = 100$ is only barely distinguishable from it. ■

Although Figs. 2.5.7 and 2.5.9 indicate that the improved Euler method can provide accuracy that suffices for many graphical purposes, it does not provide the higher-precision numerical accuracy that sometimes is needed for more careful investigations. For instance, consider again the initial value problem

$$\frac{dy}{dx} = -\frac{2xy}{1 + x^2}, \quad y(0) = 1$$

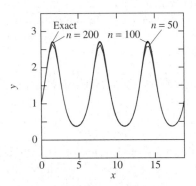

FIGURE 2.5.9. Approximating the exact solution $y = e^{\sin x}$ using the improved Euler method with $n = 50$, 100, and 200 subintervals.

of Example 1. The final column of the table in Fig. 2.5.6 suggests that, if the improved Euler method is used on the interval $0 \leq x \leq 1$ with n subintervals and step size $h = 1/n$, then the resulting error E in the final approximation $y_n \approx y(1)$ is given by

$$E = |y(1) - y_n| \approx (0.12)h^2 = \frac{0.12}{n^2}.$$

If so, then 12-place accuracy (for instance) in the value $y(1)$ would require that $(0.12)n^{-2} < 5 \times 10^{-13}$, which means that $n \geq 489{,}898$. Thus, roughly half a million steps of length $h \approx 0.000002$ would be required. Aside from the possible impracticality of this many steps (using available computational resources), the roundoff error resulting from so many successive steps might well overwhelm the

cumulative error predicted by theory (which assumes exact computations in each separate step). Consequently, still more accurate methods than the improved Euler method are needed for such high-precision computations. Such a method is presented in Section 2.6.

2.5 Problems

A hand-held calculator will suffice for Problems 1 through 10, where an initial value problem and its exact solution are given. Apply the improved Euler method to approximate this solution on the interval [0, 0.5] with step size $h = 0.1$. Construct a table showing 4-decimal place values of the approximate solution and actual solution at the points $x = 0.1, 0.2, 0.3, 0.4, 0.5$.

1. $y' = -y, y(0) = 2; y(x) = 2e^{-x}$
2. $y' = 2y, y(0) = \frac{1}{2}; y(x) = \frac{1}{2}e^{2x}$
3. $y' = y + 1, y(0) = 1; y(x) = 2e^x - 1$
4. $y' = x - y, y(0) = 1; y(x) = 2e^{-x} + x - 1$
5. $y' = y - x - 1, y(0) = 1; y(x) = 2 + x - e^x$
6. $y' = -2xy, y(0) = 2; y(x) = 2e^{-x^2}$
7. $y' = -3x^2 y, y(0) = 3; y(x) = 3e^{-x^3}$
8. $y' = e^{-y}, y(0) = 0; y(x) = \ln(x + 1)$
9. $y' = \frac{1}{4}(1 + y^2), y(0) = 1; y(x) = \tan \frac{1}{4}(x + \pi)$
10. $y' = 2xy^2, y(0) = 1, y(x) = \dfrac{1}{1 - x^2}$

Note: *The application following this problem set lists illustrative calculator/computer programs that can be used in Problems 11 through 24.*

A programmable calculator or a computer will be useful for Problems 11 through 16. In each problem find the exact solution of the given initial value problem. Then apply the improved Euler method twice to approximate (to five decimal places) this solution on the given interval, first with step size $h = 0.01$, then with step size $h = 0.005$. Make a table showing the approximate values and the actual value, together with the percentage error in the more accurate approximations, for x an integral multiple of 0.2. Throughout, primes denote derivatives with respect to x.

11. $y' = y - 2, y(0) = 1; 0 \le x \le 1$
12. $y' = \frac{1}{2}(y - 1)^2, y(0) = 2; 0 \le x \le 1$
13. $yy' = 2x^3, y(1) = 3; 1 \le x \le 2$
14. $xy' = y^2, y(1) = 1; 1 \le x \le 2$
15. $xy' = 3x - 2y, y(2) = 3; 2 \le x \le 3$
16. $y^2 y' = 2x^5, y(2) = 3; 2 \le x \le 3$

A computer with a printer is required for Problems 17 through 24. In these initial value problems, use the improved Euler method with step sizes $h = 0.1, 0.02, 0.004,$ and 0.0008 to approximate to five decimal places the values of the solution at ten equally spaced points of the given interval. Print the results in tabular form with appropriate headings to make it easy to gauge the effect of varying the step size h. Throughout, primes denote derivatives with respect to x.

17. $y' = x^2 + y^2, y(0) = 0; 0 \le x \le 1$
18. $y' = x^2 - y^2, y(0) = 1; 0 \le x \le 2$
19. $y' = x + \sqrt{y}, y(0) = 1; 0 \le x \le 2$
20. $y' = x + \sqrt[3]{y}, y(0) = -1; 0 \le x \le 2$
21. $y' = \ln y, y(1) = 2; 1 \le x \le 2$
22. $y' = x^{2/3} + y^{2/3}, y(0) = 1; 0 \le x \le 2$
23. $y' = \sin x + \cos y, y(0) = 0; 0 \le x \le 1$
24. $y' = \dfrac{x}{1 + y^2}, y(-1) = 1; -1 \le x \le 1$

25. As in Problem 25 of Section 2.4, you bail out a helicopter and immediately open your parachute, so your downward velocity satisfies the initial value problem

$$\frac{dv}{dt} = 32 - 1.6v, \quad v(0) = 0$$

(with t in seconds and v in ft/s). Use the improved Euler method with a programmable calculator or computer to approximate the solution for $0 \le t \le 2$, first with step size $h = 0.01$ and then with $h = 0.005$, rounding off approximate v-values to 3 decimal places. What percentage of the limiting velocity 20 ft/s has been attained after 1 second? After 2 seconds?

26. As in Problem 26 of Section 2.4, suppose the deer population $P(t)$ in a small forest initially numbers 25 and satisfies the logistic equation

$$\frac{dP}{dt} = 0.0225P - 0.0003P^2$$

(with t in months). Use the improved Euler method with a programmable calculator or computer to approximate the solution for 10 years, first with step size $h = 1$ and then with $h = 0.5$, rounding off approximate P-values to 3 decimal places. What percentage of the limiting population of 75 deer has been attained after 5 years? After 10 years?

Use the improved Euler method with a computer system to find the desired solution values in Problems 27 and 28. Start with step size $h = 0.1$, and then use successively smaller step sizes until successive approximate solution values at $x = 2$ agree rounded off to 4 decimal places.

27. $y' = x^2 + y^2 - 1, y(0) = 0; y(2) = ?$
28. $y' = x + \frac{1}{2}y^2, y(-2) = 0; y(2) = ?$

29. Consider the crossbow bolt of Example 2 in Section 2.3, shot straight upward from the ground with an initial velocity of 49 m/s. Because of linear air resistance, its velocity function $v(t)$ satisfies the initial value problem

$$\frac{dv}{dt} = -(0.04)v - 9.8, \quad v(0) = 49$$

with exact solution $v(t) = 294e^{-t/25} - 245$. Use a calculator or computer implementation of the improved Euler method to approximate $v(t)$ for $0 \le t \le 10$ using both $n = 50$ and $n = 100$ subintervals. Display the results at intervals of 1 second. Do the two approximations—each rounded to two decimal places—agree both with each other and with the exact solution? If the exact solution were unavailable, explain how you could use the improved Euler method to approximate closely **(a)** the bolt's time of ascent to its apex (given in Section 2.3 as 4.56 s) and **(b)** its impact velocity after 9.41 s in the air.

30. Consider now the crossbow bolt of Example 3 in Section 2.3. It still is shot straight upward from the ground with an initial velocity of 49 m/s, but because of air resistance proportional to the square of its velocity, its velocity function $v(t)$ satisfies the initial value problem

$$\frac{dv}{dt} = -(0.0011)v|v| - 9.8, \quad v(0) = 49.$$

The symbolic solution discussed in Section 2.3 required separate investigations of the bolt's ascent and its descent, with $v(t)$ given by a tangent function during ascent and by a hyperbolic tangent function during descent. But the improved Euler method requires no such distinction. Use a calculator or computer implementation of the improved Euler method to approximate $v(t)$ for $0 \le t \le 10$ using both $n = 100$ and $n = 200$ subintervals. Display the results at intervals of 1 second. Do the two approximations—each rounded to two decimal places—agree with each other? If an exact solution were unavailable, explain how you could use the improved Euler method to approximate closely **(a)** the bolt's time of ascent to its apex (given in Section 2.3 as 4.61 s) and **(b)** its impact velocity after 9.41 s in the air.

2.5 Application: Improved Euler Implementation

Figure 2.5.10 lists TI-85 and BASIC programs implementing the improved Euler method to approximate the solution of the initial value problem

$$\frac{dy}{dx} = x + y, \quad y(0) = 1$$

TI-85	BASIC	Comment
PROGRAM:IMPEULER	Program IMPEULER	Program title
:F=X+Y	DEF FN F(X,Y) = X + Y	Define function f
:10→N	N = 10	No. of steps
:0→X	X = 0	Initial x
:1→Y	Y = 1	Initial y
:1→X1	X1 = 1	Final x
:(X1-X)/N→H	H = (X1-X)/N	Step size
:For(I,1,N)	FOR I=1 TO N	Begin loop
:Y→Y0	Y0 = Y	Save previous y
:F→K1	K1 = FNF(X,Y)	First slope
:Y0+H*K1→Y	Y = Y0 + H*K1	Predictor
:X+H→X	X = X + H	New x
:F→K2	K2 = FNF(X,Y)	Second slope
:(K1+K2)/2→K	K = (K1 + K2)/2	Average slope
:Y0+H*K→Y	Y = Y0 + H*K	Corrector
:Disp X,Y	PRINT X,Y	Display results
:End	NEXT I	End loop

FIGURE 2.5.10. TI-85 and BASIC improved Euler programs.

considered in Example 2 of this section. The comments provided in the final column should make these programs intelligible even if you have little familiarity with the BASIC and TI programming languages.

To apply the improved Euler method to a differential equation $dy/dx = f(x, y)$, one need only change the initial line of the program, in which the function f is defined. To increase the number of steps (and thereby decrease the step size) one need only change the value of **N** specified in the second line of the program.

Figure 2.5.11 exhibits one MATLAB implementation of the improved Euler method. The **impeuler** function takes as input the initial value **x**, the initial value **y**, the final value **x1** of x, and the desired number **n** of subintervals. As output it produces the resulting column vectors **X** and **Y** of x- and y-values. For instance, the MATLAB command

```
[X, Y] = impeuler(0, 1, 1, 10)
```

then generates the first and fourth columns of data shown in the table of Fig. 2.5.4.

```
function yp = f(x,y)
yp = x + y;                          % yp = y'

function [X,Y] = impeuler(x,y,x1,n)
h = (x1 - x)/n;                      % step size
X = x;                               % initial x
Y = y;                               % initial y
for i = 1:n;                         % begin loop
    k1 = f(x,y);                     % first slope
    k2 = f(x+h,y+h*k1);              % second slope
    k = (k1 + k2)/2;;                % average slope
    x = x + h;                       % new x
    y = y + h*k;                     % new y
    X = [X;x];                       % update x-column
    Y = [Y;y];                       % update y-column
    end                              % end loop
```

FIGURE 2.5.11. MATLAB implementation of improved Euler method.

You should begin this project by implementing the improved Euler method with your own calculator or computer system. Test your program by first applying it to the initial value problem of Example 1, then to some of the problems for this section.

Famous Numbers Revisited

The following problems describe the numbers $e \approx 2.7182818$, $\ln 2 \approx 0.6931472$, and $\pi \approx 3.1415927$ as specific values of certain initial value problems. In each case, apply the improved Euler method with $n = 10, 20, 40, \ldots$ subintervals (doubling n each time). How many subintervals are needed to obtain—twice in succession—the correct value of the target number rounded to five decimal places?

1. The number $e = y(1)$, where $y(x)$ is the solution of the initial value problem $dy/dx = y$, $y(0) = 1$.
2. The number $\ln 2 = y(2)$, where $y(x)$ is the solution of the initial value problem $dy/dx = 1/x$, $y(1) = 0$.

3. The number $\pi = y(1)$, where $y(x)$ is the solution of the initial value problem $dy/dx = 4/(1 + x^2)$, $y(0) = 0$.

Logistic Population Investigation

Apply your improved Euler program to the initial value problem $dy/dx = \frac{1}{3}y(8 - y)$, $y(0) = 1$ of Example 3. In particular, verify (as claimed) that the approximate solution with step size $h = 1$ levels off at $y \approx 4.3542$ rather than at the limiting value $y = 8$ of the exact solution. Perhaps a table of values for $0 \leqq x \leqq 100$ will make this apparent.

For your own logistic population to investigate, consider the initial value problem

$$\frac{dy}{dx} = \frac{1}{n}y(m - y), \quad y(0) = 1,$$

where m and n are (for instance) the largest and smallest nonzero digits in your student ID number. Does the improved Euler approximation with step size $h = 1$ level off at the "correct" limiting value of the exact solution? If not, find a smaller value of h so that it does.

Periodic Harvesting and Restocking

The differential equation

$$\frac{dy}{dt} = ky(M - y) - h \sin\left(\frac{2\pi t}{P}\right)$$

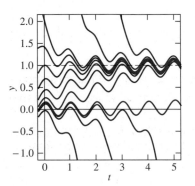

FIGURE 2.5.12. Solution curves of $dy/dt = y(1 - y) - \sin 2\pi t$.

models a logistic population that is periodically harvested and restocked with period P and maximal harvesting/restocking rate h. A numerical approximation program was used to plot the typical solution curves for the case $k = M = h = P = 1$ that are shown in Fig. 2.5.12. This figure suggests—although it does not prove—the existence of a threshold initial population such that

- Beginning with an initial population above this threshold, the population oscillates (perhaps with period P?) around the (unharvested) stable limiting population $y(t) \equiv M$, whereas
- The population dies out if it begins with an initial population below this threshold.

Use an appropriate plotting utility to investigate your own logistic population with periodic harvesting and restocking (selecting typical values of the parameters k, M, h, and P). Do the observations indicated here appear to hold for your population?

2.6 The Runge-Kutta Method

We now discuss a method for approximating the solution $y = y(x)$ of the initial value problem

$$\frac{dy}{dx} = f(x, y), \quad y(x_0) = y_0 \tag{1}$$

that is considerably more accurate than the improved Euler method and is more widely used in practice than any of the numerical methods discussed in Sections 2.4

and 2.5. It is called the *Runge-Kutta method*, after the German mathematicians who developed it, Carl Runge (1856–1927) and Wilhelm Kutta (1867–1944).

With the usual notation, suppose that we have computed the approximations $y_1, y_2, y_3, \ldots, y_n$ to the actual values $y(x_1), y(x_2), y(x_3), \ldots, y(x_n)$ and now want to compute $y_{n+1} \approx y(x_{n+1})$. Then

$$y(x_{n+1}) - y(x_n) = \int_{x_n}^{x_{n+1}} y'(x)\,dx = \int_{x_n}^{x_n+h} y'(x)\,dx \tag{2}$$

by the fundamental theorem of calculus. Next, Simpson's rule for numerical integration yields

$$y(x_{n+1}) - y(x_n) \approx \frac{h}{6}\left[y'(x_n) + 4y'\left(x_n + \frac{h}{2}\right) + y'(x_{n+1})\right]. \tag{3}$$

Hence we want to define y_{n+1} so that

$$y_{n+1} \approx y_n + \frac{h}{6}\left[y'(x_n) + 2y'\left(x_n + \frac{h}{2}\right) + 2y'\left(x_n + \frac{h}{2}\right) + y'(x_{n+1})\right]; \tag{4}$$

we have split $4y'\left(x_n + \frac{1}{2}h\right)$ into a sum of two terms because we intend to approximate the slope $y'\left(x_n + \frac{1}{2}h\right)$ at the midpoint $x_n + \frac{1}{2}h$ of the interval $[x_n, x_{n+1}]$ in two different ways.

On the right-hand side in (4), we replace the [true] slope values $y'(x_n)$, $y'\left(x_n + \frac{1}{2}h\right)$, $y'\left(x_n + \frac{1}{2}h\right)$, and $y'(x_{n+1})$, respectively, with the following estimates.

➤
$$k_1 = f(x_n, y_n). \tag{5a}$$

- This is the Euler method slope at x_n.

➤
$$k_2 = f\left(x_n + \tfrac{1}{2}h, y_n + \tfrac{1}{2}hk_1\right) \tag{5b}$$

- This is an estimate of the slope at the midpoint of the interval $[x_n, x_{n+1}]$ using the Euler method to predict the ordinate there.

➤
$$k_3 = f\left(x_n + \tfrac{1}{2}h, y_n + \tfrac{1}{2}hk_2\right) \tag{5c}$$

- This is an improved Euler value for the slope at the midpoint.

➤
$$k_4 = f(x_{n+1}, y_n + hk_3) \tag{5d}$$

- This is the Euler method slope at x_{n+1}, using the improved slope k_3 at the midpoint to step to x_{n+1}.

When these substitutions are made in (4), the result is the iterative formula

➤
$$y_{n+1} = y_n + \frac{h}{6}(k_1 + 2k_2 + 2k_3 + k_4). \tag{6}$$

The use of this formula to compute the approximations y_1, y_2, y_3, \ldots successively constitutes the **Runge-Kutta method.** Note that Eq. (6) takes the "Euler form"

$$y_{n+1} = y_n + h \cdot k$$

if we write

$$k = \frac{1}{6}(k_1 + 2k_2 + 2k_3 + k_4) \tag{7}$$

for the approximate *average slope* on the interval $[x_n, x_{n+1}]$.

The Runge-Kutta method is a *fourth-order* method—it can be proved that the cumulative error on a bounded interval $[a, b]$ with $a = x_0$ is of order h^4. (Thus the iteration in (6) is sometimes called the *fourth-order* Runge-Kutta method because it is possible to develop Runge-Kutta methods of other orders.) That is,

$$\blacktriangleright \qquad\qquad |y(x_n) - y_n| \leqq Ch^4, \tag{8}$$

where the constant C depends on the function $f(x, y)$ and the interval $[a, b]$, but does not depend on the step size h. The following example illustrates this high accuracy in comparison with the lower-order accuracy of our previous numerical methods.

| Example 1 | We first apply the Runge-Kutta method to the illustrative initial value problem

$$\frac{dy}{dx} = x + y, \quad y(0) = 1 \tag{9}$$

that we considered in Fig. 2.4.8 of Section 2.4 and again in Example 2 of Section 2.5. The exact solution of this problem is $y(x) = 2e^x - x - 1$. To make a point we use $h = 0.5$, a larger step size than in any previous example, so only two steps are required to go from $x = 0$ to $x = 1$.

In the first step we use the formulas in (5) and (6) to calculate

$$k_1 = 0 + 1 = 1,$$
$$k_2 = (0 + 0.25) + (1 + (0.25) \cdot (1)) = 1.5,$$
$$k_3 = (0 + 0.25) + (1 + (0.25) \cdot (1.5)) = 1.625,$$
$$k_4 = (0.5) + (1 + (0.5) \cdot (1.625)) = 2.3125,$$

and then

$$y_1 = 1 + \frac{0.5}{6}[1 + 2 \cdot (1.5) + 2 \cdot (1.625) + 2.3125] \approx 1.7969.$$

Similarly, the second step yields $y_2 \approx 3.4347$.

Figure 2.6.1 presents these results together with the results (from Fig. 2.5.4) of applying the improved Euler method with step size $h = 0.1$. We see that even with the larger step size, the Runge-Kutta method gives (for this problem) four to five times the accuracy (in terms of relative percentage errors) of the improved Euler method. ∎

	Improved Euler		Runge-Kutta		
x	y with $h = 0.1$	Percent Error	y with $h = 0.5$	Percent Error	Actual y
0.0	1.0000	0.00%	1.0000	0.00%	1.0000
0.5	1.7949	0.14%	1.7969	0.03%	1.7974
1.0	3.4282	0.24%	3.4347	0.05%	3.4366

FIGURE 2.6.1. Runge-Kutta and improved Euler results for the initial value problem $dy/dx = x + y$, $y(0) = 1$.

It is customary to measure the computational labor involved in solving $dy/dx = f(x, y)$ numerically by counting the number of evaluations of the function $f(x, y)$ that are required. In Example 1, the Runge-Kutta method required eight evaluations of $f(x, y) = x + y$ (four at each step), whereas the improved Euler method required 20 such evaluations (two for each of 10 steps). Thus the Runge-Kutta method gave over four times the accuracy with only 40% of the labor.

Computer programs implementing the Runge-Kutta method are listed in the project material for this section. Figure 2.6.2 shows the results obtained by applying the improved Euler and Runge-Kutta methods to the problem $dy/dx = x + y$, $y(0) = 1$ with the same step size $h = 0.1$. The relative error in the improved Euler value at $x = 1$ is about 0.24%, but for the Runge-Kutta value it is 0.00012%. In this comparison the Runge-Kutta method is about 2000 times as accurate, but requires only twice as many function evaluations, as the improved Euler method.

x	Improved Euler y	Runge-Kutta y	Actual y
0.1	1.1100	1.110342	1.110342
0.2	1.2421	1.242805	1.242806
0.3	1.3985	1.399717	1.399718
0.4	1.5818	1.583648	1.583649
0.5	1.7949	1.797441	1.797443
0.6	2.0409	2.044236	2.044238
0.7	2.3231	2.327503	2.327505
0.8	2.6456	2.651079	2.651082
0.9	3.0124	3.019203	3.019206
1.0	3.4282	3.436559	3.436564

FIGURE 2.6.2. Runge-Kutta and improved Euler results for the initial value problem $dy/dx = x + y$, $y(0) = 1$, with the same step size $h = 0.1$.

The error bound

$$|y(x_n) - y_n| \leq Ch^4 \tag{8}$$

for the Runge-Kutta method results in a rapid decrease in the magnitude of errors when the step size h is reduced (except for the possibility that very small step sizes may result in unacceptable roundoff errors). It follows from the inequality in (8) that (on a fixed bounded interval) halving the step size decreases the absolute error by a factor of $\left(\frac{1}{2}\right)^4 = \frac{1}{16}$. Consequently, the common practice of successively halving the step size until the computed results "stabilize" is particularly effective with the Runge-Kutta method.

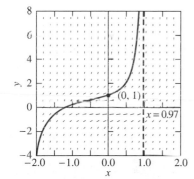

FIGURE 2.6.3. Solutions of $dy/dx = x^2 + y^2$, $y(0) = 1$.

Example 2

In Example 5 of Section 2.4 we saw that Euler's method is not adequate to approximate the solution $y(x)$ of the initial value problem

$$\frac{dy}{dx} = x^2 + y^2, \quad y(0) = 1 \tag{10}$$

as x approaches the infinite discontinuity near $x = 0.969811$ (see Fig. 2.6.3). Now we apply the Runge-Kutta method to this initial value problem.

Figure 2.6.4 shows Runge-Kutta results on the interval $[0.0, 0.9]$, computed with step sizes $h = 0.1$, $h = 0.05$, and $h = 0.025$. There is still some difficulty near $x = 0.9$, but it seems safe to conclude from these data that $y(0.5) \approx 2.0670$.

x	y with $h = 0.1$	y with $h = 0.05$	y with $h = 0.025$
0.1	1.1115	1.1115	1.1115
0.3	1.4397	1.4397	1.4397
0.5	2.0670	2.0670	2.0670
0.7	3.6522	3.6529	3.6529
0.9	14.0218	14.2712	14.3021

FIGURE 2.6.4. Approximating the solution of the initial value problem in Eq. (10).

We therefore begin anew and apply the Runge-Kutta method to the initial value problem

$$\frac{dy}{dx} = x^2 + y^2, \quad y(0.5) = 2.0670. \tag{11}$$

Figure 2.6.5 shows results on the interval $[0.5, 0.9]$, obtained with step sizes $h = 0.01$, $h = 0.005$, and $h = 0.0025$. We now conclude that $y(0.9) \approx 14.3049$.

x	y with $h = 0.01$	y with $h = 0.005$	y with $h = 0.0025$
0.5	2.0670	2.0670	2.0670
0.6	2.6440	2.6440	2.6440
0.7	3.6529	3.6529	3.6529
0.8	5.8486	5.8486	5.8486
0.9	14.3048	14.3049	14.3049

FIGURE 2.6.5. Approximating the solution of the initial value problem in Eq. (11).

Finally, Fig. 2.6.6 shows results on the interval $[0.90, 0.95]$ for the initial value problem

$$\frac{dy}{dx} = x^2 + y^2, \quad y(0.9) = 14.3049, \tag{12}$$

obtained using step sizes $h = 0.002$, $h = 0.001$, and $h = 0.0005$. Our final approximate result is $y(0.95) \approx 50.4723$. The actual value of the solution at $x = 0.95$ is $y(0.95) \approx 50.471867$. Our slight overestimate results mainly from the fact that the four-place initial value in (12) is (in effect) the result of rounding *up* the actual value $y(0.9) \approx 14.304864$; such errors are magnified considerably as we approach the vertical asymptote. ∎

x	y with $h = 0.002$	y with $h = 0.001$	y with $h = 0.0005$
0.90	14.3049	14.3049	14.3049
0.91	16.7024	16.7024	16.7024
0.92	20.0617	20.0617	20.0617
0.93	25.1073	25.1073	25.1073
0.94	33.5363	33.5363	33.5363
0.95	50.4722	50.4723	50.4723

FIGURE 2.6.6. Approximating the solution of the initial value problem in Eq. (12).

Example 3 A skydiver with a mass of 60 kg jumps from a helicopter hovering at an initial altitude of 5 kilometers. Assume that she falls vertically with initial velocity zero and experiences an upward force F_R of air resistance given in terms of her velocity v (in meters per second) by

$$F_R = (0.0096)(100v + 10v^2 + v^3)$$

(in newtons, and with the coordinate axis directed downward so that $v > 0$ during her descent to the ground). If she does not open her parachute, what will be her terminal velocity? How fast will she be falling after 5 s have elapsed? After 10 s? After 20 s?

Solution Newton's law $F = ma$ gives

$$m\frac{dv}{dt} = mg - F_R;$$

that is,

$$60\frac{dv}{dt} = (60)(9.8) - (0.0096)(100v + 10v^2 + v^3) \tag{13}$$

because $m = 60$ and $g = 9.8$. Thus the velocity function $v(t)$ satisfies the initial value problem

$$\frac{dv}{dt} = f(v), \quad v(0) = 0, \tag{14}$$

where

$$f(v) = 9.8 - (0.00016)(100v + 10v^2 + v^3). \tag{15}$$

The skydiver reaches her terminal velocity when the forces of gravity and air resistance balance, so $f(v) = 0$. We can therefore calculate her terminal velocity immediately by solving the equation

$$f(v) = 9.8 - (0.00016)(100v + 10v^2 + v^3) = 0. \tag{16}$$

Figure 2.6.7 shows the graph of the function $f(v)$ and exhibits the single real solution $v \approx 35.5780$ (found graphically or by using a calculator or computer `Solve` procedure). Thus the skydiver's terminal speed is approximately 35.578 m/s, about 128 km/h (almost 80 mi/h).

Figure 2.6.8 shows the results of Runge-Kutta approximations to the solution of the initial value problem in (14); the step sizes $h = 0.2$ and $h = 0.1$ yield the same results (to three decimal places). Observe that the terminal velocity is effectively attained in only 15 s. But the skydiver's velocity is 91.85% of her terminal velocity after only 5 s, and 99.78% after 10 s. ■

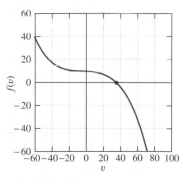

FIGURE 2.6.7. Graph of $f(v) = 9.8 - (0.00016)(100v + 10v^2 + v^3)$.

The final example of this section contains a *warning*: For certain types of initial value problems, the numerical methods we have discussed are not nearly so successful as in the previous examples.

t (s)	v (m/s)	t (s)	v (m/s)
0	0	11	35.541
1	9.636	12	35.560
2	18.386	13	35.569
3	25.299	14	35.574
4	29.949	15	35.576
5	32.678	16	35.577
6	34.137	17	35.578
7	34.875	18	35.578
8	35.239	19	35.578
9	35.415	20	35.578
10	35.500		

FIGURE 2.6.8. The skydiver's velocity data.

Example 4 Consider the seemingly innocuous initial value problem

$$\frac{dy}{dx} = 5y - 6e^{-x}, \quad y(0) = 1 \tag{17}$$

whose exact solution is $y(x) = e^{-x}$. The table in Fig. 2.6.9 shows the results obtained by applying the Runge-Kutta method on the interval $[0, 4]$ with step sizes $h = 0.2$, $h = 0.1$, and $h = 0.05$. Obviously these attempts are spectacularly unsuccessful. Although $y(x) = e^{-x} \to 0$ as $x \to +\infty$, it appears that our numerical approximations are headed toward $-\infty$ rather than zero.

x	Runge-Kutta y with $h = 0.2$	Runge-Kutta y with $h = 0.1$	Runge-Kutta y with $h = 0.05$	Actual y
0.4	0.66880	0.67020	0.67031	0.67032
0.8	0.43713	0.44833	0.44926	0.44933
1.2	0.21099	0.29376	0.30067	0.30199
1.6	−0.46019	0.14697	0.19802	0.20190
2.0	−4.72142	−0.27026	0.10668	0.13534
2.4	−35.53415	−2.90419	−0.12102	0.09072
2.8	−261.25023	−22.05352	−1.50367	0.06081
3.2	−1,916.69395	−163.25077	−11.51868	0.04076
3.6	−14059.35494	−1205.71249	−85.38156	0.02732
4.0	−103,126.5270	−8903.12866	−631.03934	0.01832

FIGURE 2.6.9. Runge-Kutta attempts to solve numerically the initial value problem in Eq. (17).

The explanation lies in the fact that the general solution of the equation $dy/dx = 5y - 6e^{-x}$ is

$$y(x) = e^{-x} + Ce^{5x}. \tag{18}$$

The particular solution of (17) satisfying the initial condition $y(0) = 1$ is obtained with $C = 0$. But any departure, however small, from the exact solution

$y(x) = e^{-x}$—even if due only to roundoff error—introduces [in effect] a nonzero value of C in Eq. (18). And as indicated in Fig. 2.6.10, all solution curves of the form in (18) with $C \neq 0$ diverge rapidly away from the one with $C = 0$, even if their initial values are close to 1. ∎

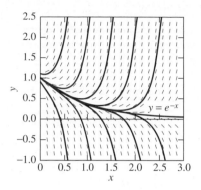

FIGURE 2.6.10. Direction field and solution curves for $dy/dx = 5y - 6e^{-x}$.

Difficulties of the sort illustrated by Example 4 sometimes are unavoidable, but one can at least hope to recognize such a problem when it appears. Approximate values whose order of magnitude varies with changing step size are a common indicator of such instability. These difficulties are discussed in numerical analysis textbooks and are the subject of current research in the field.

2.6 Problems

A hand-held calculator will suffice for Problems 1 through 10, where an initial value problem and its exact solution are given. Apply the Runge-Kutta method to approximate this solution on the interval $[0, 0.5]$ with step size $h = 0.25$. Construct a table showing 5-decimal place values of the approximate solution and actual solution at the points $x = 0.25$ and 0.5.

1. $y' = -y$, $y(0) = 2$; $y(x) = 2e^{-x}$
2. $y' = 2y$, $y(0) = \frac{1}{2}$; $y(x) = \frac{1}{2}e^{2x}$
3. $y' = y + 1$, $y(0) = 1$; $y(x) = 2e^x - 1$
4. $y' = x - y$, $y(0) = 1$; $y(x) = 2e^{-x} + x - 1$
5. $y' = y - x - 1$, $y(0) = 1$; $y(x) = 2 + x - e^x$
6. $y' = -2xy$, $y(0) = 2$; $y(x) = 2e^{-x^2}$
7. $y' = -3x^2 y$, $y(0) = 3$; $y(x) = 3e^{-x^3}$
8. $y' = e^{-y}$, $y(0) = 0$; $y(x) = \ln(x + 1)$
9. $y' = \frac{1}{4}(1 + y^2)$, $y(0) = 1$; $y(x) = \tan \frac{1}{4}(x + \pi)$
10. $y' = 2xy^2$, $y(0) = 1$; $y(x) = \dfrac{1}{1 - x^2}$

Note: *The application following this problem set lists illustrative calculator/computer programs that can be used in the remaining problems.*

A programmable calculator or a computer will be useful for Problems 11 through 16. In each problem find the exact solution of the given initial value problem. Then apply the Runge-Kutta method twice to approximate (to five decimal places) this

solution on the given interval, first with step size $h = 0.2$, then with step size $h = 0.1$. Make a table showing the approximate values and the actual value, together with the percentage error in the more accurate approximation, for x an integral multiple of 0.2. Throughout, primes denote derivatives with respect to x.

11. $y' = y - 2$, $y(0) = 1$; $0 \leq x \leq 1$
12. $y' = \frac{1}{2}(y - 1)^2$, $y(0) = 2$; $0 \leq x \leq 1$
13. $yy' = 2x^3$, $y(1) = 3$; $1 \leq x \leq 2$
14. $xy' = y^2$, $y(1) = 1$; $1 \leq x \leq 2$
15. $xy' = 3x - 2y$, $y(2) = 3$; $2 \leq x \leq 3$
16. $y^2 y' = 2x^5$, $y(2) = 3$; $2 \leq x \leq 3$

A computer with a printer is required for Problems 17 through 24. In these initial value problems, use the Runge-Kutta method with step sizes $h = 0.2, 0.1, 0.05,$ and 0.025 to approximate to six decimal places the values of the solution at five equally spaced points of the given interval. Print the results in tabular form with appropriate headings to make it easy to gauge the effect of varying the step size h. Throughout, primes denote derivatives with respect to x.

17. $y' = x^2 + y^2$, $y(0) = 0$; $0 \leq x \leq 1$
18. $y' = x^2 - y^2$, $y(0) = 1$; $0 \leq x \leq 2$
19. $y' = x + \sqrt{y}$, $y(0) = 1$; $0 \leq x \leq 2$

20. $y' = x + \sqrt[3]{y}$, $y(0) = -1$; $0 \leq x \leq 2$

21. $y' = \ln y$, $y(1) = 2$; $1 \leq x \leq 2$

22. $y' = x^{2/3} + y^{2/3}$, $y(0) = 1$; $0 \leq x \leq 2$

23. $y' = \sin x + \cos y$, $y(0) = 0$; $0 \leq x \leq 1$

24. $y' = \dfrac{x}{1 + y^2}$, $y(-1) = 1$; $-1 \leq x \leq 1$

25. As in Problem 25 of Section 2.5, you bail out a helicopter and immediately open your parachute, so your downward velocity satisfies the initial value problem

$$\frac{dv}{dt} = 32 - 1.6v, \quad v(0) = 0$$

(with t in seconds and v in ft/s). Use the Runge-Kutta method with a programmable calculator or computer to approximate the solution for $0 \leq t \leq 2$, first with step size $h = 0.1$ and then with $h = 0.05$, rounding off approximate v-values to 3 decimal places. What percentage of the limiting velocity 20 ft/s has been attained after 1 second? After 2 seconds?

26. As in Problem 26 of Section 2.5, suppose the deer population $P(t)$ in a small forest initially numbers 25 and satisfies the logistic equation

$$\frac{dP}{dt} = 0.0225P - 0.0003P^2$$

(with t in months). Use the Runge-Kutta method with a programmable calculator or computer to approximate the solution for 10 years, first with step size $h = 6$ and then with $h = 3$, rounding off approximate P-values to 4 decimal places. What percentage of the limiting population of 75 deer has been attained after 5 years? After 10 years?

Use the Runge-Kutta method with a computer system to find the desired solution values in Problems 27 and 28. Start with step size $h = 1$, and then use successively smaller step sizes until successive approximate solution values at $x = 2$ agree rounded off to 5 decimal places.

27. $y' = x^2 + y^2 - 1$, $y(0) = 0$; $y(2) = ?$

28. $y' = x + \frac{1}{2}y^2$, $y(-2) = 0$; $y(2) = ?$

Velocity-Acceleration Problems

In Problems 25 and 26, the linear acceleration $a = dv/dt$ of a moving particle is given by a formula $dv/dt = f(t, v)$, where the velocity $v = dy/dt$ is the derivative of the function $y = y(t)$ giving the position of the particle at time t. Suppose that the velocity $v(t)$ is approximated using the Runge-Kutta method to solve numerically the initial value problem

$$\frac{dv}{dt} = f(t, v), \quad v(0) = v_0. \tag{19}$$

That is, starting with $t_0 = 0$ and v_0, the formulas in Eqs. (5) and (6) are applied—with t and v in place of x and y—to calculate the successive approximate velocity values v_1, v_2, v_3, \ldots, v_m at the successive times t_1, t_2, t_3, \ldots, t_m (with $t_{n+1} = t_n + h$). Now suppose that we also want to approximate

the distance $y(t)$ *traveled by the particle. We can do this by beginning with the initial position $y(0) = y_0$ and calculating*

$$y_{n+1} = y_n + v_n h + \tfrac{1}{2}a_n h^2 \tag{20}$$

$(n = 1, 2, 3, \ldots)$, *where $a_n = f(t_n, v_n) \approx v'(t_n)$ is the particle's approximate acceleration at time t_n. The formula in (20) would give the correct increment (from y_n to y_{n+1}) if the acceleration a_n remained constant during the time interval $[t_n, t_{n+1}]$.*

Thus, once a table of approximate velocities has been calculated, Eq. (20) provides a simple way to calculate a table of corresponding successive positions. This process is illustrated in the project for this section, by beginning with the velocity data in Fig. 2.6.8 (Example 3) and proceeding to follow the skydiver's position during her descent to the ground.

29. Consider again the crossbow bolt of Example 2 in Section 2.3, shot straight upward from the ground with an initial velocity of 49 m/s. Because of linear air resistance, its velocity function $v = dy/dt$ satisfies the initial value problem

$$\frac{dv}{dt} = -(0.04)v - 9.8, \quad v(0) = 49$$

with exact solution $v(t) = 294e^{-t/25} - 245$. **(a)** Use a calculator or computer implementation of the Runge-Kutta method to approximate $v(t)$ for $0 \leq t \leq 10$ using both $n = 100$ and $n = 200$ subintervals. Display the results at intervals of 1 second. Do the two approximations—each rounded to four decimal places—agree both with each other and with the exact solution? **(b)** Now use the velocity data from part (a) to approximate $y(t)$ for $0 \leq t \leq 10$ using $n = 200$ subintervals. Display the results at intervals of 1 second. Do these approximate position values—each rounded to two decimal places—agree with the exact solution

$$y(t) = 7350\left(1 - e^{-t/25}\right) - 245t?$$

(c) If the exact solution were unavailable, explain how you could use the Runge-Kutta method to approximate closely the bolt's times of ascent and descent and the maximum height it attains.

30. Now consider again the crossbow bolt of Example 3 in Section 2.3. It still is shot straight upward from the ground with an initial velocity of 49 m/s, but because of air resistance proportional to the square of its velocity, its velocity function $v(t)$ satisfies the initial value problem

$$\frac{dv}{dt} = -(0.0011)v|v| - 9.8, \quad v(0) = 49.$$

Beginning with this initial value problem, repeat parts (a) through (c) of Problem 25 (except that you may need $n = 200$ subintervals to get four-place accuracy in part (a) and $n = 400$ subintervals for two-place accuracy in part (b)). According to the results of Problems 17 and 18 in Section 2.3, the bolt's velocity and position functions

during ascent and descent are given by the following formulas.

Ascent:
$$v(t) = (94.388)\tan(0.478837 - [0.103827]t),$$
$$y(t) = 108.465$$
$$+ (909.091)\ln(\cos(0.478837 - [0.103827]t));$$

Descent:
$$v(t) = -(94.388)\tanh(0.103827[t - 4.6119]),$$
$$y(t) = 108.465$$
$$- (909.091)\ln(\cosh(0.103827[t - 4.6119])).$$

2.6 Application: Runge-Kutta Implementation

Figure 2.6.11 lists TI-85 and BASIC programs implementing the Runge-Kutta method to approximate the solution of the initial value problem

$$\frac{dy}{dx} = x + y, \quad y(0) = 1$$

considered in Example 1 of this section. The comments provided in the final column should make these programs intelligible even if you have little familiarity with the BASIC and TI programming languages.

TI-85	BASIC	Comment
PROGRAM:RK	Program RK	Program title
:F=X+Y	DEF FN F(X,Y) = X + Y	Define function f
:10→N	N = 10	No. of steps
:0→X	X = 0	Initial x
:1→Y	Y = 1	Initial y
:1→X1	X1 = 1	Final x
:(X1-X)/N→H	H = (X1-X)/N	Step size
:For(I,1,N)	FOR I=1 TO N	Begin loop
:X→X0	X0 = X	Save previous x
:Y→Y0	Y0 = Y	Save previous y
:F→K1	K1 = FNF(X,Y)	First slope
:X0+H/2→X	X = X0 + H/2	Midpoint
:Y0+H*K1/2→Y	Y = Y0 + H*K1/2	Midpt predictor
:F→K2	K2 = FNF(X,Y)	Second slope
:Y0+H*K2/2→Y	Y = Y0 + H*K2/2	Midpt predictor
:F→K3	K3 = FNF(X,Y)	Third slope
:X0+H→X	X = X0 + H	New x
:Y0+H*K3→Y	Y = Y0 + H*K3	Endpt predictor
:F→K4	K4 = FNF(X,Y)	Fourth slope
:(K1+2*K2+2*K3 +K4)/6→K	K = (K1+2*K2+2*K3 +K4)/6	Average slope
:Y0+H*K→Y	Y = Y0 + K*K	Corrector
:Disp X,Y	PRINT X,Y	Display results
:End	NEXT I	End loop

FIGURE 2.6.11. TI-85 and BASIC Runge-Kutta programs.

To apply the Runge-Kutta method to a different equation $dy/dx = f(x, y)$, one need only change the initial line of the program, in which the function f is

defined. To increase the number of steps (and thereby decrease the step size), one need only change the value of N specified in the second line of the program.

Figure 2.6.12 exhibits a MATLAB implementation of the Runge-Kutta method. Suppose that the function f describing the differential equation $y' = f(x, y)$ has been defined. Then the rk function takes as input the initial value x, the initial value y, the final value x1 of x, and the desired number n of subintervals. As output it produces the resulting column vectors X and Y of x- and y-values. For instance, the MATLAB command

```
[X, Y] = rk(0, 1, 1, 10)
```

then generates the first and third columns of data shown in the table in Fig. 2.6.2.

```
function yp = f(x,y)
yp = x + y;                              % yp = y'

function [X,Y] = rk(x,y,x1,n)
h = (x1 - x)/n;                          % step size
X = x;                                   % initial x
Y = y;                                   % initial y
for i = 1:n                              % begin loop
    k1 = f(x,y);                         % first slope
    k2 = f(x+h/2,y+h*k1/2);              % second slope
    k3 = f(x+h/2,y+h*k2/2);              % third slope
    k4 = f(x+h,y+h*k3);                  % fourth slope
    k  = (k1+2*k2+2*k3+k4)/6;            % average slope
    x  = x + h;                          % new x
    y  = y + h*k;                        % new y
    X = [X;x];                           % update x-column
    Y = [Y;y];                           % update y-column
end                                      % end loop
```

FIGURE 2.6.12. MATLAB implementation of the Runge-Kutta method.

You should begin this project by implementing the Runge-Kutta method with your own calculator or computer system. Test your program by applying it first to the initial value problem in Example 1, then to some of the problems for this section.

Famous Numbers Revisited, One Last Time

The following problems describe the numbers

$$e \approx 2.71828182846, \quad \ln 2 \approx 0.69314718056, \quad \text{and} \quad \pi \approx 3.14159265359$$

as specific values of certain initial value problems. In each case, apply the Runge-Kutta method with $n = 10, 20, 40, \ldots$ subintervals (doubling n each time). How many subintervals are needed to obtain—twice in succession—the correct value of the target number rounded to nine decimal places?

1. The number $e = y(1)$, where $y(x)$ is the solution of the initial value problem $dy/dx = y$, $y(0) = 1$.
2. The number $\ln 2 = y(2)$, where $y(x)$ is the solution of the initial value problem $dy/dx = 1/x$, $y(1) = 0$.

3. The number $\pi = y(1)$, where $y(x)$ is the solution of the initial value problem $dy/dx = 4/(1 + x^2)$, $y(0) = 0$.

The Skydiver's Descent

The following MATLAB function describes the skydiver's acceleration function in Example 3.

```
function    vp = f(t,v)
vp = 9.8 - 0.00016*(100*v + 10*v^2 + v^3);
```

Then the commands

```
k = 200                      % 200 subintervals
[t,v] = rk(0, 20, 0, k);     % Runge-Kutta approximation
[t(1:10:k+1); v(1:10:k+1)]   % Display every 10th entry
```

produce the table of approximate velocities shown in Fig. 2.6.8. Finally, the commands

t (s)	v (m/s)	y (m)
0	0	0
2	18.386	18.984
4	29.949	68.825
6	34.137	133.763
8	35.239	203.392
10	35.500	274.192
12	35.560	345.266
14	35.574	416.403
16	35.577	487.555
18	35.578	558.710
20	35.578	629.866

```
y = zeros(k+1,1):              % initialize y
h = 0.1;                       % step size
for n = 1:k                    % for n = 1 to k
  a = f(t(n),v(n)):            % acceleration
  y(n+1) = y(n) + v(n)*h + 0.5*a*h^2;   % Equation (20)
end                            % end loop
[t(1:20:k+1),v(1:20:k+1),y(1:20:k+1)]   % each 20th entry
```

FIGURE 2.6.13. The skydiver's velocity and position data.

carry out the position function calculations described in Eq. (20) in the instructions for Problems 25 and 26. The results of these calculations are shown in the table in Fig. 2.6.13. It appears that the skydiver falls 629.866 m during her first 20 s of descent, and then free falls the remaining 4370.134 meters to the ground at her terminal speed of 35.578 m/s. Hence her total time of descent is $20 + (4370.134/35.578) \approx 142.833$ s, or about 2 min 23 s.

For an individual problem to solve after implementing these methods using an available computer system, analyze your own skydive (perhaps from a different height), using your own mass m and a plausible air resistance force of the form $F_R = av + bv^2 + cv^3$.

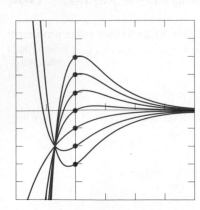

C H A P T E R 3

Linear Equations of Higher Order

3.1 Introduction: Second-Order Linear Equations

In Chapters 1 and 2 we investigated first-order differential equations. We now turn to equations of higher order $n \geq 2$, beginning in this chapter with equations that are linear. The general theory of linear differential equations parallels the second-order case ($n = 2$), which we outline in this initial section.

Recall that a second-order differential equation in the (unknown) function $y(x)$ is one of the form

$$G(x, y, y', y'') = 0. \tag{1}$$

This differential equation is said to be **linear** provided that G is linear in the dependent variable y and its derivatives y' and y''. Thus a linear second-order equation takes (or can be written in) the form

➤ $$A(x)y'' + B(x)y' + C(x)y = F(x). \tag{2}$$

Unless otherwise noted, we will always assume that the (known) coefficient functions $A(x)$, $B(x)$, $C(x)$, and $F(x)$ are continuous on some open interval I (perhaps unbounded) on which we wish to solve this differential equation, but we do *not* require that they be linear functions. Thus the differential equation

$$e^x y'' + (\cos x)y' + (1 + \sqrt{x})y = \tan^{-1} x$$

is linear because the dependent variable y and its derivatives y' and y'' appear linearly. By contrast, the equations

$$y'' = yy' \quad \text{and} \quad y'' + 3(y')^2 + 4y^3 = 0$$

are not linear because products and powers of y or its derivatives appear.

If the function $F(x)$ on the right-hand side of Eq. (2) vanishes identically on I, then we call Eq. (2) a **homogeneous** linear equation; otherwise, it is **nonhomogeneous**. For example, the second-order equation

$$x^2 y'' + 2xy' + 3y = \cos x$$

is nonhomogeneous; its *associated* homogeneous equation is

$$x^2 y'' + 2xy' + 3y = 0.$$

In general, the homogeneous linear equation **associated** with Eq. (2) is

$$\blacktriangleright \qquad A(x)y'' + B(x)y' + C(x)y = 0. \qquad (3)$$

In case the differential equation in (2) models a physical system, the nonhomogeneous term $F(x)$ frequently corresponds to some *external* influence on the system.

A Typical Application

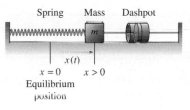

Spring Mass Dashpot

$x(t)$

$x = 0$ $x > 0$

Equilibrium position

FIGURE 3.1.1. A mass-spring-dashpot system.

m

$F_S \longleftarrow \quad \longleftarrow F_R$

$x, v > 0$

FIGURE 3.1.2. Directions of the forces acting on m.

Linear differential equations frequently appear as mathematical models of mechanical systems and electrical circuits. For example, suppose that a mass m is attached both to a spring that exerts on it a force F_S and to a dashpot (shock absorber) that exerts a force F_R on the mass (Fig. 3.1.1). Assume that the restoring force F_S of the spring is proportional to the displacement x (positive to the right, negative to the left) of the mass from equilibrium, and that the dashpot force F_R is proportional to the velocity $v = dx/dt$ of the mass. With the aid of Fig. 3.1.2 we also get the appropriate directions of action of these two forces:

$$F_S = -kx \quad \text{and} \quad F_R = -cv \quad (k, c > 0).$$

The minus signs are correct—F_S is negative when x is positive, F_R is negative when v is positive. Newton's law $F = ma$ now gives

$$mx'' = F_S + F_R; \qquad (4)$$

that is,

$$\blacktriangleright \qquad m\frac{d^2 x}{dt^2} + c\frac{dx}{dt} + kx = 0. \qquad (5)$$

Thus we have a differential equation satisfied by the position function $x(t)$ of the mass m. This homogeneous second-order linear equation governs the *free vibrations* of the mass; we will return to this problem in detail in Section 3.4.

If, in addition to F_S and F_R, the mass m is acted on by an external force $F(t)$—which must then be added to the right-hand side in Eq. (4)—the resulting equation is

$$\blacktriangleright \qquad m\frac{d^2 x}{dt^2} + c\frac{dx}{dt} + kx = F(t). \qquad (6)$$

This nonhomogeneous linear differential equation governs the *forced vibrations* of the mass under the influence of the external force $F(t)$.

Homogeneous Second-Order Linear Equations

Consider the general second-order linear equation

$$A(x)y'' + B(x)y' + C(x)y = F(x), \tag{7}$$

where the coefficient functions A, B, C, and F are continuous on the open interval I. Here we assume in addition that $A(x) \neq 0$ at each point of I, so we can divide each term in Eq. (7) by $A(x)$ and write it in the form

$$\blacktriangleright \qquad y'' + p(x)y' + q(x)y = f(x). \tag{8}$$

We will discuss first the associated homogeneous equation

$$\blacktriangleright \qquad y'' + p(x)y' + q(x)y = 0. \tag{9}$$

A particularly useful property of this *homogeneous* linear equation is the fact that the sum of any two solutions of Eq. (9) is again a solution, as is any constant multiple of a solution. This is the central idea of the following theorem.

THEOREM 1: Principle of Superposition for Homogeneous Equations

Let y_1 and y_2 be two solutions of the homogeneous linear equation in (9) on the interval I. If c_1 and c_2 are constants, then the linear combination

$$y = c_1 y_1 + c_2 y_2 \tag{10}$$

is also a solution of Eq. (9) on I.

Proof: The conclusion follows almost immediately from the linearity of the operation of differentiation, which gives

$$y' = c_1 y_1' + c_2 y_2' \quad \text{and} \quad y'' = c_1 y_1'' + c_2 y_2''.$$

Then

$$
\begin{aligned}
y'' + py' + qy &= (c_1 y_1 + c_2 y_2)'' + p(c_1 y_1 + c_2 y_2)' + q(c_1 y_1 + c_2 y_2) \\
&= (c_1 y_1'' + c_2 y_2'') + p(c_1 y_1' + c_2 y_2') + q(c_1 y_1 + c_2 y_2) \\
&= c_1(y_1'' + py_1' + qy_1) + c_2(y_2'' + py_2' + qy_2) \\
&= c_1 \cdot 0 + c_2 \cdot 0 = 0
\end{aligned}
$$

because y_1 and y_2 are solutions. Thus $y = c_1 y_1 + c_2 y_2$ is also a solution. ▲

Example 1 We can see by inspection that

$$y_1(x) = \cos x \quad \text{and} \quad y_2(x) = \sin x$$

are two solutions of the equation

$$y'' + y = 0.$$

Theorem 1 tells us that any linear combination of these solutions, such as

$$y(x) = 3y_1(x) - 2y_2(x) = 3\cos x - 2\sin x,$$

is also a solution. We will see later that, conversely, *every* solution of $y'' + y = 0$ is a linear combination of these two particular solutions y_1 and y_2. Thus a general solution of $y'' + y = 0$ is given by

$$y(x) = c_1 \cos x + c_2 \sin x.$$

It is important to understand that this single formula for the general solution encompasses a "twofold infinity" of particular solutions, because the two coefficients c_1 and c_2 can be selected independently. Figures 3.1.3 through 3.1.5 illustrate some of the possibilities, with either c_1 or c_2 set equal to zero, or with both nonzero. ■

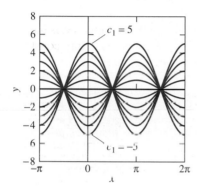

FIGURE 3.1.3. Solutions $y(x) = c_1 \cos x$ of $y'' + y = 0$.

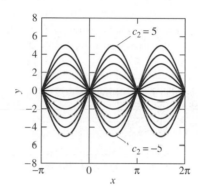

FIGURE 3.1.4. Solutions $y(x) = c_2 \sin x$ of $y'' + y = 0$.

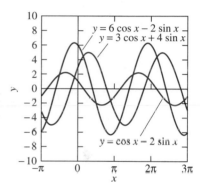

FIGURE 3.1.5. Solutions of $y'' + y = 0$ with c_1 and c_2 both nonzero.

Earlier in this section we gave the linear equation $mx'' + cx' + kx = F(t)$ as a mathematical model of the motion of the mass shown in Fig. 3.1.1. Physical considerations suggest that the motion of the mass should be determined by its initial position and initial velocity. Hence, given any preassigned values of $x(0)$ and $x'(0)$, Eq. (6) ought to have a *unique* solution satisfying these initial conditions. More generally, in order to be a "good" mathematical model of a deterministic physical situation, a differential equation must have unique solutions satisfying any appropriate initial conditions. The following existence and uniqueness theorem (proved in the Appendix) gives us this assurance for the general second-order equation.

THEOREM 2: Existence and Uniqueness for Linear Equations

Suppose that the functions p, q, and f are continuous on the open interval I containing the point a. Then, given any two numbers b_0 and b_1, the equation

$$y'' + p(x)y' + q(x)y = f(x) \tag{8}$$

has a unique (that is, one and only one) solution on the entire interval I that satisfies the initial conditions

$$y(a) = b_0, \quad y'(a) = b_1. \tag{11}$$

▲

Remark 1: Equation (8) and the conditions in (11) constitute a second-order linear **initial value problem.** Theorem 2 tells us that any such initial value problem has a unique solution on the *whole* interval I where the coefficient functions in (8) are continuous. Recall from Section 1.3 that a *nonlinear* differential equation generally has a unique solution on only a smaller interval.

Remark 2: Whereas a *first-order* differential equation $dy/dx = F(x, y)$ generally admits only a single solution curve $y = y(x)$ passing through a given initial point (a, b), Theorem 2 implies that the *second-order* equation in (8) has infinitely many solution curves passing through the point (a, b_0)—namely, one for each (real number) value of the initial slope $y'(a) = b_1$. That is, instead of there being only one line through (a, b_0) tangent to a solution curve, *every* nonvertical straight line through (a, b_0) is tangent to some solution curve of Eq. (8). Figure 3.1.6 shows a number of solution curves of the equation $y'' + 3y' + 2y = 0$ all having the same initial value $y(0) = 1$, while Fig. 3.1.7 shows a number of solution curves all having the same initial slope $y'(0) = 1$. The application at the end of this section suggests how to construct such families of solution curves for a given homogeneous second-order linear differential equation. ∎

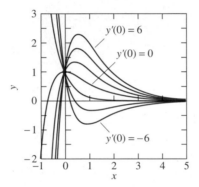

FIGURE 3.1.6. Solutions of $y'' + 3y' + 2y = 0$ with the same initial value $y(0) = 1$ but different initial slopes.

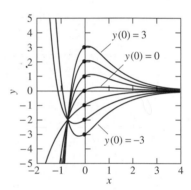

FIGURE 3.1.7. Solutions of $y'' + 3y' + 2y = 0$ with the same initial slope $y'(0) = 1$ but different initial values.

Example 1

Continued

We saw in the first part of Example 1 that $y(x) = 3\cos x - 2\sin x$ is a solution (on the entire real line) of $y'' + y = 0$. It has the initial values $y(0) = 3$, $y'(0) = -2$. Theorem 2 tells us that this is the *only* solution with these initial values. More generally, the solution

$$y(x) = b_0 \cos x + b_1 \sin x$$

satisfies the *arbitrary* initial conditions $y(0) = b_0$, $y'(0) = b_1$; this illustrates the *existence* of such a solution, also as guaranteed by Theorem 2. ∎

Example 1 suggests how, given a *homogeneous* second-order linear equation, we might actually find the solution $y(x)$ whose existence is assured by Theorem 2. First, we find two "essentially different" solutions y_1 and y_2; second, we attempt to impose on the general solution

$$y = c_1 y_1 + c_2 y_2 \tag{12}$$

the initial conditions $y(a) = b_0$, $y'(a) = b_1$. That is, we attempt to solve the simultaneous equations

$$c_1 y_1(a) + c_2 y_2(a) = b_0,$$
$$c_1 y_1'(a) + c_2 y_2'(a) = b_1$$

(13)

for the coefficients c_1 and c_2.

| **Example 2** | Verify that the functions |

$$y_1(x) = e^x \quad \text{and} \quad y_2(x) = xe^x$$

are solutions of the differential equation

$$y'' - 2y' + y = 0,$$

and then find a solution satisfying the initial conditions $y(0) = 3$, $y'(0) = 1$.

Solution The verification is routine; we omit it. We impose the given initial conditions on the general solution

$$y(x) = c_1 e^x + c_2 x e^x,$$

for which

$$y'(x) = (c_1 + c_2)e^x + c_2 x e^x,$$

to obtain the simultaneous equations

$$y(0) = c_1 \qquad = 3,$$
$$y'(0) = c_1 + c_2 = 1.$$

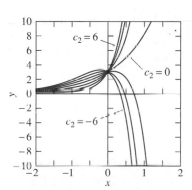

FIGURE 3.1.8. Different solutions $y(x) = 3e^x + c_2 x e^x$ of $y'' - 2y' + y = 0$ with the same initial value $y(0) = 3$.

The resulting solution is $c_1 = 3$, $c_2 = -2$. Hence the solution of the original initial value problem is

$$y(x) = 3e^x - 2xe^x.$$

Figure 3.1.8 shows several additional solutions of $y'' - 2y' + y = 0$, all having the same initial value $y(0) = 3$. ∎

In order for the procedure of Example 2 to succeed, the two solutions y_1 and y_2 must have the elusive property that the equations in (13) can always be solved for c_1 and c_2, no matter what the initial conditions b_0 and b_1 might be. The following definition tells precisely how different the two functions y_1 and y_2 must be.

Definition: Linear Independence of Two Functions

Two functions defined on an open interval I are said to be **linearly independent** on I provided that neither is a constant multiple of the other.

Two functions are said to be **linearly dependent** on an open interval provided that they are not linearly independent there; that is, one of them is a constant multiple of the other. We can always determine whether two given functions f and g are linearly dependent on an interval I by noting at a glance whether either of the two quotients f/g or g/f is a constant-valued function on I.

Example 3 Thus it is clear that the following pairs of functions are linearly independent on the entire real line:

$$\sin x \quad \text{and} \quad \cos x;$$
$$e^x \quad \text{and} \quad e^{-2x};$$
$$e^x \quad \text{and} \quad xe^x;$$
$$x+1 \quad \text{and} \quad x^2;$$
$$x \quad \text{and} \quad |x|.$$

That is, neither $\sin x / \cos x = \tan x$ nor $\cos x / \sin x = \cot x$ is a constant-valued function; neither $e^x / e^{-2x} = e^{3x}$ nor e^{-2x} / e^x is a constant-valued function; and so forth. But the identically zero function $f(x) \equiv 0$ and any other function g are linearly dependent on every interval, because $0 \cdot g(x) = 0 = f(x)$. Also, the functions

$$f(x) = \sin 2x \quad \text{and} \quad g(x) = \sin x \cos x$$

are linearly dependent on any interval because $f(x) = 2g(x)$ is the familiar trigonometric identity $\sin 2x = 2 \sin x \cos x$. ∎

General Solutions

But does the homogeneous equation $y'' + py' + qy = 0$ always have two linearly independent solutions? Theorem 2 says yes! We need only choose y_1 and y_2 so that

$$y_1(a) = 1, \ y_1'(a) = 0 \quad \text{and} \quad y_2(a) = 0, \ y_2'(a) = 1.$$

It is then impossible that either $y_1 = ky_2$ or $y_2 = ky_1$ because $k \cdot 0 \neq 1$ for any constant k. Theorem 2 tells us that two such linearly independent solutions *exist*; actually finding them is a crucial matter that we will discuss briefly at the end of this section, and in greater detail beginning in Section 3.3.

We want to show, finally, that given *any* two linearly independent solutions y_1 and y_2 of the homogeneous equation

$$y''(x) + p(x)y'(x) + q(x)y(x) = 0, \tag{9}$$

every solution y of Eq. (9) can be expressed as a linear combination

$$y = c_1 y_1 + c_2 y_2 \tag{12}$$

of y_1 and y_2. This means that the function in (12) is a *general solution* of Eq. (9)—it provides all possible solutions of the differential equation.

As suggested by the equations in (13), the determination of the constants c_1 and c_2 in (12) depends on a certain 2×2 determinant of values of y_1, y_2, and their derivatives. Given two functions f and g, the **Wronskian** of f and g is the determinant

$$W = \begin{vmatrix} f & g \\ f' & g' \end{vmatrix} = fg' - f'g.$$

We write either $W(f, g)$ or $W(x)$, depending on whether we wish to emphasize the two functions or the point x at which the Wronskian is to be evaluated. For example,

$$W(\cos x, \sin x) = \begin{vmatrix} \cos x & \sin x \\ -\sin x & \cos x \end{vmatrix} = \cos^2 x + \sin^2 x = 1$$

and

$$W(e^x, xe^x) = \begin{vmatrix} e^x & xe^x \\ e^x & e^x + xe^x \end{vmatrix} = e^{2x}.$$

These are examples of linearly *independent* pairs of solutions of differential equations (see Examples 1 and 2). Note that in both cases the Wronskian is everywhere *nonzero*.

On the other hand, if the functions f and g are linearly dependent, with $f = kg$ (for example), then

$$W(f, g) = \begin{vmatrix} kg & g \\ kg' & g' \end{vmatrix} = kgg' - kg'g \equiv 0.$$

Thus the Wronskian of two linearly *dependent* functions is identically zero. In Section 3.2 we will prove that, if the two functions y_1 and y_2 are solutions of a homogeneous second-order linear equation, then the strong converse stated in part (b) of Theorem 3 holds.

THEOREM 3: Wronskians of Solutions

Suppose that y_1 and y_2 are two solutions of the homogeneous second-order linear equation (Eq. (9))

$$y'' + p(x)y' + q(x)y = 0$$

on an open interval I on which p and q are continuous.

(a) If y_1 and y_2 are linearly dependent, then $W(y_1, y_2) \equiv 0$ on I.

(b) If y_1 and y_2 are linearly independent, then $W(y_1, y_2) \neq 0$ at each point of I. ▲

Thus, given two solutions of Eq. (9), there are just two possibilities: The Wronskian W is identically zero if the solutions are linearly dependent; the Wronskian is never zero if the solutions are linearly independent. The latter fact is what we need to show that $y = c_1y_1 + c_2y_2$ is the general solution of Eq. (9) if y_1 and y_2 are linearly independent solutions.

THEOREM 4: General Solutions of Homogeneous Equations

Let y_1 and y_2 be two linearly independent solutions of the homogeneous equation (Eq. (9))

$$y'' + p(x)y' + q(x)y = 0$$

with p and q continuous on the open interval I. If Y is any solution whatsoever of Eq. (9) on I, then there exist numbers c_1 and c_2 such that

$$Y(x) = c_1y_1(x) + c_2y_2(x)$$

for all x in I. ▲

In essence, Theorem 4 tells us that when we have found *two* linearly independent solutions of the second-order homogeneous equation in (9), then we have found *all* of its solutions. We therefore call the linear combination $Y = c_1y_1 + c_2y_2$ a *general solution* of the differential equation.

Proof of Theorem 4: Choose a point a of I, and consider the simultaneous equations

$$c_1 y_1(a) + c_2 y_2(a) = Y(a),$$
$$c_1 y_1'(a) + c_2 y_2'(a) = Y'(a).$$
(14)

The determinant of the coefficients in this system of linear equations in the unknowns c_1 and c_2 is simply the Wronskian $W(y_1, y_2)$ evaluated at $x = a$. By Theorem 3, this determinant is nonzero, so by elementary algebra it follows that the equations in (14) can be solved for c_1 and c_2. With these values of c_1 and c_2, we define the solution

$$G(x) = c_1 y_1(x) + c_2 y_2(x)$$

of Eq. (9); then

$$G(a) = c_1 y_1(a) + c_2 y_2(a) = Y(a)$$

and

$$G'(a) = c_1 y_1'(a) + c_2 y_2'(a) = Y'(a).$$

Thus the two solutions Y and G have the same initial values at a, as do Y' and G'. By the uniqueness of a solution determined by such initial values (Theorem 2), it follows that Y and G agree on I. Thus we see that

$$Y(x) \equiv G(x) = c_1 y_1(x) + c_2 y_2(x),$$

as desired. ▲

Example 4 If $y_1(x) = e^{2x}$ and $y_2(x) = e^{-2x}$, then

$$y_1'' = (2)(2)e^{2x} = 4e^{2x} = 4y_1 \quad \text{and} \quad y_2'' = (-2)(-2)e^{-2x} = 4e^{-2x} = 4y_2.$$

Therefore, y_1 and y_2 are linearly independent solutions of

$$y'' - 4y = 0.$$
(15)

But $y_3(x) = \cosh 2x$ and $y_4(x) = \sinh 2x$ are also solutions of Eq. (15), because

$$\frac{d^2}{dx^2}(\cosh 2x) = \frac{d}{dx}(2\sinh 2x) = 4\cosh 2x$$

and, similarly, $(\sinh 2x)'' = 4\sinh 2x$. It therefore follows from Theorem 4 that the functions $\cosh 2x$ and $\sinh 2x$ can be expressed as linear combinations of $y_1(x) = e^{2x}$ and $y_2(x) = e^{-2x}$. Of course, this is no surprise, because

$$\cosh 2x = \tfrac{1}{2}e^{2x} + \tfrac{1}{2}e^{-2x} \quad \text{and} \quad \sinh 2x = \tfrac{1}{2}e^{2x} - \tfrac{1}{2}e^{-2x}$$

by the definitions of the hyperbolic cosine and hyperbolic sine. ■

Remark: Because e^{2x}, e^{-2x} and $\cosh x$, $\sinh x$ are two different pairs of linearly independent solutions of the equation $y'' - 4y = 0$ in (15), Theorem 4 implies that every particular solution $Y(x)$ of this equation can be written both in the form

$$Y(x) = c_1 e^{2x} + c_2 e^{-2x}$$

and in the form

$$Y(x) = a \cosh x + b \sinh x.$$

Thus these two different linear combinations (with arbitrary constant coefficients) provide two different descriptions of the set of all solutions of the same differential equation $y'' - 4y = 0$. Hence each of these two linear combinations is a general solution of the equation. Indeed, this is why it is accurate to refer to a specific such linear combination as "*a* general solution" rather than as "*the* general solution." ■

Linear Second-Order Equations with Constant Coefficients

As an illustration of the general theory introduced in this section, we discuss the homogeneous second-order linear differential equation

$$ay'' + by' + cy = 0 \tag{16}$$

with *constant* coefficients a, b, and c. We first look for a *single* solution of Eq. (16) and begin with the observation that

$$(e^{rx})' = re^{rx} \quad \text{and} \quad (e^{rx})'' = r^2 e^{rx}, \tag{17}$$

so any derivative of e^{rx} is a constant multiple of e^{rx}. Hence, if we substituted $y = e^{rx}$ in Eq. (16), then each term would be a constant multiple of e^{rx}, with the constant coefficients dependent on r and the coefficients a, b, and c. This suggests that we try to find a value of r so that these multiples of e^{rx} will have sum zero. If we succeed, then $y = e^{rx}$ will be a solution of Eq. (16).

For example, if we substitute $y = e^{rx}$ in the equation

$$y'' - 5y' + 6y = 0,$$

we obtain

$$r^2 e^{rx} - 5r e^{rx} + 6e^{rx} = 0.$$

Thus

$$(r^2 - 5r + 6)e^{rx} = 0; \quad (r - 2)(r - 3)e^{rx} = 0.$$

Hence $y = e^{rx}$ will be a solution if either $r = 2$ or $r = 3$. So, in searching for a single solution, we actually have found two solutions: $y_1(x) = e^{2x}$ and $y_2(x) = e^{3x}$.

To carry out this procedure in the general case, we substitute $y = e^{rx}$ in Eq. (16). With the aid of the equations in (17), we find the result to be

$$ar^2 e^{rx} + br e^{rx} + c e^{rx} = 0.$$

Because e^{rx} is never zero, we conclude that $y(x) = e^{rx}$ will satisfy the differential equation in (16) precisely when r is a root of the *algebraic* equation

$$ar^2 + br + c = 0. \tag{18}$$

This quadratic equation is called the **characteristic equation** of the homogeneous linear differential equation

$$ay'' + by' + cy = 0. \tag{16}$$

If Eq. (18) has two *distinct* (unequal) roots r_1 and r_2, then the corresponding solutions $y_1(x) = e^{r_1 x}$ and $y_2(x) = e^{r_2 x}$ of (16) are linearly independent. (Why?) This gives the following result.

THEOREM 5: Distinct Real Roots

If the roots r_1 and r_2 of the characteristic equation in (18) are real and distinct, then

$$y(x) = c_1 e^{r_1 x} + c_2 e^{r_2 x} \tag{19}$$

is a general solution of Eq. (16). ▲

Example 5 Find the general solution of

$$2y'' - 7y' + 3y = 0.$$

Solution We can solve the characteristic equation

$$2r^2 - 7r + 3 = 0$$

by factoring:

$$(2r - 1)(r - 3) = 0.$$

The roots $r_1 = \frac{1}{2}$ and $r_2 = 3$ are real and distinct, so Theorem 5 yields the general solution

$$y(x) = c_1 e^{x/2} + c_2 e^{3x}.$$ ∎

Example 6 The differential equation $y'' + 2y' = 0$ has characteristic equation

$$r^2 + 2r = r(r + 2) = 0$$

with distinct real roots $r_1 = 0$ and $r_2 = -2$. Because $e^{0 \cdot x} \equiv 1$, we get the general solution

$$y(x) = c_1 + c_2 e^{-2x}.$$

Figure 3.1.9 shows several different solution curves with $c_1 = 1$, all appearing to approach the solution curve $y(x) \equiv 1$ (with $c_2 = 0$) as $x \to +\infty$. ∎

Remark: Note that Theorem 5 changes a problem involving a differential equation into one involving only the solution of an *algebraic* equation. ∎

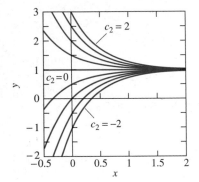

FIGURE 3.1.9. Solutions $y(x) = 1 + c_2 e^{-2x}$ of $y'' + 2y' = 0$ with different values of c_2.

If the characteristic equation in (18) has equal roots $r_1 = r_2$, we get (at first) only the single solution $y_1(x) = e^{r_1 x}$ of Eq. (16). The problem in this case is to produce the "missing" second solution of the differential equation.

A double root $r = r_1$ will occur precisely when the characteristic equation is a constant multiple of the equation

$$(r - r_1)^2 = r^2 - 2r_1 r + r_1^2 = 0.$$

Any differential equation with this characteristic equation is equivalent to

$$y'' - 2r_1 y' + r_1^2 y = 0. \qquad (20)$$

But it is easy to verify by direct substitution that $y = xe^{r_1 x}$ is a second solution of Eq. (20). It is clear (but you should verify) that

$$y_1(x) = e^{r_1 x} \quad \text{and} \quad y_2(x) = xe^{r_1 x}$$

are linearly independent functions, so the general solution of the differential equation in (20) is

$$y(x) = c_1 e^{r_1 x} + c_2 x e^{r_1 x}.$$

THEOREM 6: Repeated Roots

If the characteristic equation in (18) has equal (necessarily real) roots $r_1 = r_2$, then

➤
$$y(x) = (c_1 + c_2 x)e^{r_1 x} \tag{21}$$

is a general solution of Eq. (16). ▲

Example 7 To solve the initial value problem

$$y'' + 2y' + y = 0;$$
$$y(0) = 5, \quad y'(0) = -3,$$

we note first that the characteristic equation

$$r^2 + 2r + 1 = (r + 1)^2 = 0$$

has equal roots $r_1 = r_2 = -1$. Hence the general solution provided by Theorem 6 is

$$y(x) = c_1 e^{-x} + c_2 x e^{-x}.$$

Differentiation yields

$$y'(x) = -c_1 e^{-x} + c_2 e^{-x} - c_2 x e^{-x},$$

so the initial conditions yield the equations

$$y(0) = c_1 = 5,$$
$$y'(0) = -c_1 + c_2 = -3,$$

which imply that $c_1 = 5$ and $c_2 = 2$. Thus the desired particular solution of the initial value problem is

$$y(x) = 5e^{-x} + 2xe^{-x}.$$

This particular solution, together with several others of the form $y(x) = c_1 e^{-x} + 2xe^{-x}$, is illustrated in Fig. 3.1.10. ■

The characteristic equation in (18) may have either real or complex roots. The case of complex roots will be discussed in Section 3.3.

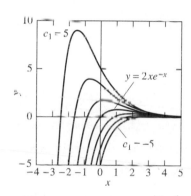

FIGURE 3.1.10. Solutions $y(x) = c_1 e^{-x} + 2xe^{-x}$ of $y'' + 2y' + y = 0$ with different values of c_1.

3.1 Problems

In Problems 1 through 16, a homogeneous second-order linear differential equation, two functions y_1 and y_2, and a pair of initial conditions are given. First verify that y_1 and y_2 are solutions of the differential equation. Then find a particular solution of the form $y = c_1 y_1 + c_2 y_2$ that satisfies the given initial conditions. Primes denote derivatives with respect to x.

1. $y'' - y = 0$; $y_1 = e^x$, $y_2 = e^{-x}$; $y(0) = 0$, $y'(0) = 5$
2. $y'' - 9y = 0$; $y_1 = e^{3x}$, $y_2 = e^{-3x}$; $y(0) = -1$, $y'(0) = 15$
3. $y'' + 4y = 0$; $y_1 = \cos 2x$, $y_2 = \sin 2x$; $y(0) = 3$, $y'(0) = 8$
4. $y'' + 25y = 0$; $y_1 = \cos 5x$, $y_2 = \sin 5x$; $y(0) = 10$, $y'(0) = -10$
5. $y'' - 3y' + 2y = 0$; $y_1 = e^x$, $y_2 = e^{2x}$; $y(0) = 1$, $y'(0) = 0$
6. $y'' + y' - 6y = 0$; $y_1 = e^{2x}$, $y_2 = e^{-3x}$; $y(0) = 7$, $y'(0) = -1$
7. $y'' + y' = 0$; $y_1 = 1$, $y_2 = e^{-x}$; $y(0) = -2$, $y'(0) = 8$
8. $y'' - 3y' = 0$; $y_1 = 1$, $y_2 = e^{3x}$; $y(0) = 4$, $y'(0) = -2$
9. $y'' + 2y' + y = 0$; $y_1 = e^{-x}$, $y_2 = xe^{-x}$; $y(0) = 2$, $y'(0) = -1$
10. $y'' - 10y' + 25y = 0$; $y_1 = e^{5x}$, $y_2 = xe^{5x}$; $y(0) = 3$, $y'(0) = 13$

11. $y'' - 2y' + 2y = 0$; $y_1 = e^x \cos x$, $y_2 = e^x \sin x$; $y(0) = 0$, $y'(0) = 5$

12. $y'' + 6y' + 13y = 0$; $y_1 = e^{-3x} \cos 2x$, $y_2 = e^{-3x} \sin 2x$; $y(0) = 2$, $y'(0) = 0$

13. $x^2 y'' - 2xy' + 2y = 0$; $y_1 = x$, $y_2 = x^2$; $y(1) = 3$, $y'(1) = 1$

14. $x^2 y'' + 2xy' - 6y = 0$; $y_1 = x^2$, $y_2 = x^{-3}$; $y(2) = 10$, $y'(2) = 15$

15. $x^2 y'' - xy' + y = 0$; $y_1 = x$, $y_2 = x \ln x$; $y(1) = 7$, $y'(1) = 2$

16. $x^2 y'' + xy' + y = 0$; $y_1 = \cos(\ln x)$, $y_2 = \sin(\ln x)$; $y(1) = 2$, $y'(1) = 3$

The following three problems illustrate the fact that the superposition principle does not generally hold for nonlinear equations.

17. Show that $y = 1/x$ is a solution of $y' + y^2 = 0$, but that if $c \neq 0$ and $c \neq 1$, then $y = c/x$ is not a solution.

18. Show that $y = x^3$ is a solution of $yy'' = 6x^4$, but that if $c^2 \neq 1$, then $y = cx^3$ is not a solution.

19. Show that $y_1 \equiv 1$ and $y_2 = \sqrt{x}$ are solutions of $yy'' + (y')^2 = 0$, but that their sum $y = y_1 + y_2$ is not a solution.

Determine whether the pairs of functions in Problems 20 through 26 are linearly independent or linearly dependent on the real line.

20. $f(x) = \pi$, $g(x) = \cos^2 x + \sin^2 x$

21. $f(x) = x^3$, $g(x) = x^2 |x|$

22. $f(x) = 1 + x$, $g(x) = 1 + |x|$

23. $f(x) = xe^x$, $g(x) = |x|e^x$

24. $f(x) = \sin^2 x$, $g(x) = 1 - \cos 2x$

25. $f(x) = e^x \sin x$, $g(x) = e^x \cos x$

26. $f(x) = 2\cos x + 3\sin x$, $g(x) = 3\cos x - 2\sin x$

27. Let y_p be a particular solution of the nonhomogeneous equation $y'' + py' + qy = f(x)$ and let y_c be a solution of its associated homogeneous equation. Show that $y = y_c + y_p$ is a solution of the given nonhomogeneous equation.

28. With $y_p = 1$ and $y_c = c_1 \cos x + c_2 \sin x$ in the notation of Problem 27, find a solution of $y'' + y = 1$ satisfying the initial conditions $y(0) = -1 = y'(0)$.

29. Show that $y_1 = x^2$ and $y_2 = x^3$ are two different solutions of $x^2 y'' - 4xy' + 6y = 0$, both satisfying the initial conditions $y(0) = 0 = y'(0)$. Explain why these facts do not contradict Theorem 2 (with respect to the guaranteed uniqueness).

30. (a) Show that $y_1 = x^3$ and $y_2 = |x^3|$ are linearly independent solutions on the real line of the equation $x^2 y'' - 3xy' + 3y = 0$. (b) Verify that $W(y_1, y_2)$ is identically zero. Why do these facts not contradict Theorem 3?

31. Show that $y_1 = \sin x^2$ and $y_2 = \cos x^2$ are linearly independent functions, but that their Wronskian vanishes at $x = 0$. Why does this imply that there is *no* differential equation of the form $y'' + p(x)y' + q(x)y = 0$, with both p and q continuous everywhere, having both y_1 and y_2 as solutions?

32. Let y_1 and y_2 be two solutions of $A(x)y'' + B(x)y' + C(x)y = 0$ on an open interval I where A, B, and C are continuous and $A(x)$ is never zero. (a) Let $W = W(y_1, y_2)$. Show that

$$A(x) \frac{dW}{dx} = (y_1)(Ay_2'') - (y_2)(Ay_1'').$$

Then substitute for Ay_2'' and Ay_1'' from the original differential equation to show that

$$A(x) \frac{dW}{dx} = -B(x)W(x).$$

(b) Solve this first-order equation to deduce **Abel's formula**

$$W(x) = K \exp\left(-\int \frac{B(x)}{A(x)} \, dx\right),$$

where K is a constant. (c) Why does Abel's formula imply that the Wronskian $W(y_1, y_2)$ is either zero everywhere or nonzero everywhere (as stated in Theorem 3)?

Apply Theorems 5 and 6 to find general solutions of the differential equations given in Problems 33 through 42. Primes denote derivatives with respect to x.

33. $y'' - 3y' + 2y = 0$

34. $y'' + 2y' - 15y = 0$

35. $y'' + 5y' = 0$

36. $2y'' + 3y' = 0$

37. $2y'' - y' - y = 0$

38. $4y'' + 8y' + 3y = 0$

39. $4y'' + 4y' + y = 0$

40. $9y'' - 12y' + 4y = 0$

41. $6y'' - 7y' - 20y = 0$

42. $35y'' - y' - 12y = 0$

Each of Problems 43 through 48 gives a general solution $y(x)$ of a homogeneous second-order differential equation $ay'' + by' + cy = 0$ with constant coefficients. Find this equation.

43. $y(x) = c_1 + c_2 e^{-10x}$

44. $y(x) = c_1 e^{10x} + c_2 e^{-10x}$

45. $y(x) = c_1 e^{-10x} + c_2 x e^{-10x}$

46. $y(x) = c_1 e^{10x} + c_2 e^{100x}$

47. $y(x) = c_1 + c_2 x$

48. $y(x) = e^x \left(c_1 e^{x\sqrt{2}} + c_2 e^{-x\sqrt{2}}\right)$

Problems 49 and 50 deal with the solution curves of $y'' + 3y' + 2y = 0$ shown in Figs. 3.1.6 and 3.1.7.

49. Find the highest point on the solution curve with $y(0) = 1$ and $y'(0) = 6$ in Fig. 3.1.6.

50. Find the third-quadrant point of intersection of the solution curves shown in Fig. 3.1.7.

51. A second-order **Euler equation** is one of the form

$$ax^2 y'' + bxy' + cy = 0 \tag{22}$$

where a, b, c are constants. (a) Show that if $x > 0$ then the substitution $v = \ln x$ transforms Eq. (22) into the constant-coefficient linear equation

$$a\frac{d^2 y}{dv^2} + (b - a)\frac{dy}{dv} + cy = 0 \tag{23}$$

with independent variable v. (b) If the roots r_1 and r_2 of the characteristic equation of Eq. (23) are real and distinct, conclude that a general solution of the Euler equation in (22) is $y(x) = c_1 x^{r_1} + c_2 x^{r_2}$.

Make the substitution $v = \ln x$ of Problem 51 to find general solutions (for $x > 0$) of the Euler equations in Problems 52–56.

52. $x^2 y'' + xy' - y = 0$ **53.** $x^2 y'' + 2xy' - 12y = 0$
54. $4x^2 y'' + 8xy' - 3y = 0$ **55.** $x^2 y'' + xy' = 0$
56. $x^2 y'' - 3xy' + 4y = 0$

3.1 Application: Plotting Second-Order Solution Families

This application deals with the plotting by computer of families of solutions such as those illustrated in Figs. 3.1.6 and 3.1.7. Show first that the general solution of the differential equation

$$y'' + 3y' + 2y = 0 \tag{1}$$

is

$$y(x) = c_1 e^{-x} + c_2 e^{-2x}. \tag{2}$$

For Fig. 3.1.6, show that the particular solution of Eq. (1) satisfying the initial conditions $y(0) = 1$, $y'(0) = b$ is given by

$$y(x) = (b + 2)e^{-x} - (b + 1)e^{-2x}. \tag{3}$$

The MATLAB loop

```
x = - 1 : 0.02 : 5     % x-vector from x = - 1 to x = 5
for b = -6 : 2 : 6     % for b = -6 to 6 with db = 2 do
   y = (b + 2)*exp(-x) - (b + 1)*exp(-2*x);
   plot(x,y)
end
```

was used to generate Fig. 3.1.6.

For Fig. 3.1.7, show that the particular solution of Eq. (1) satisfying the initial conditions $y(0) = a$, $y'(0) = 1$ is given by

$$y(x) = (2a + 1)e^{-x} - (a + 1)e^{-2x}. \tag{4}$$

The MATLAB loop

```
x = -2 : 0.02 : 4      % x-vector from x = -2 to x = 4
for a = -3 : 1 : 3     % for a = -3 to 3 with da = 1 do
   y = (2*a + 1)*exp(-x) - (a + 1)*exp(-2*x);
   plot(x,y)
end
```

was used to generate Fig. 3.1.7.

Computer algebra systems such as *Maple* and *Mathematica*, as well as graphing calculators, have commands to carry out for-loops such as the two shown here. Begin by reproducing Figs. 3.1.6 and 3.1.7. Then, for each of the following differential equations, construct both a family of solution curves satisfying $y(0) = 1$ and a family of solution curves satisfying the initial condition $y'(0) = 1$.

1. $y'' - y = 0$

2. $y'' - 3y' + 2y = 0$

3. $2y'' + 3y' + y = 0$

4. $y'' + y = 0$ (see Example 1)

5. $y'' + 2y' + 2y = 0$, which has general solution $y(x) = e^{-x}(c_1 \cos x + c_2 \sin x)$.

3.2 General Solutions of Linear Equations

We now show that our discussion in Section 3.1 of second-order linear equations generalizes in a very natural way to the general *n*th-order **linear** differential equation of the form

$$P_0(x)y^{(n)} + P_1(x)y^{(n-1)} + \cdots + P_{n-1}(x)y' + P_n(x)y = F(x). \qquad (1)$$

Unless otherwise noted, we will always assume that the coefficient functions $P_i(x)$ and $F(x)$ are continuous on some open interval I (perhaps unbounded) where we wish to solve the equation. Under the additional assumption that $P_0(x) \neq 0$ at each point of I, we can divide each term in Eq. (1) by $P_0(x)$ to obtain an equation with leading coefficient 1, of the form

$$\blacktriangleright \qquad y^{(n)} + p_1(x)y^{(n-1)} + \cdots + p_{n-1}(x)y' + p_n(x)y = f(x). \qquad (2)$$

The **homogeneous** linear equation **associated with** Eq. (2) is

$$\blacktriangleright \qquad y^{(n)} + p_1(x)y^{(n-1)} + \cdots + p_{n-1}(x)y' + p_n(x)y = 0. \qquad (3)$$

Just as in the second-order case, a *homogeneous* *n*th-order linear differential equation has the valuable property that any superposition, or *linear combination,* of solutions of the equation is again a solution. The proof of the following theorem is essentially the same—a routine verification—as that of Theorem 1 of Section 3.1.

THEOREM 1: Principle of Superposition for Homogeneous Equations

Let y_1, y_2, \ldots, y_n be n solutions of the homogeneous linear equation in (3) on the interval I. If c_1, c_2, \ldots, c_n are constants, then the linear combination

$$\blacktriangleright \qquad y = c_1 y_1 + c_2 y_2 + \cdots + c_n y_n \qquad (4)$$

is also a solution of Eq. (3) on I. ▲

Example 1

It is easy to verify that the three functions

$$y_1(x) = e^{-3x}, \quad y_2(x) = \cos 2x, \quad \text{and} \quad y_3(x) = \sin 2x$$

are all solutions of the homogeneous third-order equation

$$y^{(3)} + 3y'' + 4y' + 12y = 0$$

on the entire real line. Theorem 1 tells us that any linear combination of these solutions, such as

$$y(x) = -3y_1(x) + 3y_2(x) - 2y_3(x) = -3e^{-3x} + 3\cos 2x - 2\sin 2x,$$

is also a solution on the entire real line. We will see that, conversely, every solution of the differential equation of this example is a linear combination of the three particular solutions y_1, y_2, and y_3. Thus a general solution is given by

$$y(x) = c_1 e^{-3x} + c_2 \cos 2x + c_3 \sin 2x. \qquad \blacksquare$$

Existence and Uniqueness of Solutions

We saw in Section 3.1 that a particular solution of a *second-order* linear differential equation is determined by *two* initial conditions. Similarly, a particular solution of an nth-order linear differential equation is determined by n initial conditions. The following theorem, proved in the Appendix, is the natural generalization of Theorem 2 of Section 3.1.

THEOREM 2: Existence and Uniqueness for Linear Equations

Suppose that the functions p_1, p_2, \ldots, p_n, and f are continuous on the open interval I containing the point a. Then, given n numbers $b_0, b_1, \ldots, b_{n-1}$, the nth-order linear equation (Eq. (2))

$$y^{(n)} + p_1(x)y^{(n-1)} + \cdots + p_{n-1}(x)y' + p_n(x)y = f(x)$$

has a unique (that is, one and only one) solution on the entire interval I that satisfies the n initial conditions

$$y(a) = b_0, \quad y'(a) = b_1, \quad \ldots, \quad y^{(n-1)}(a) = b_{n-1}. \tag{5}$$

▲

Equation (2) and the conditions in (5) constitute an nth-order **initial value problem.** Theorem 2 tells us that any such initial value problem has a unique solution on the *whole* interval I where the coefficient functions in (2) are continuous. It tells us nothing, however, about how to find this solution. In Section 3.3 we will see how to construct explicit solutions of initial value problems in the *constant* coefficient case that occurs often in applications.

Example 1

Continued

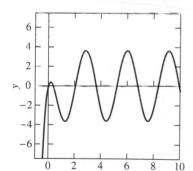

FIGURE 3.2.1. The particular solution $y(x) = -3e^{-3x} + 3\cos 2x - 2\sin 2x$.

We saw earlier that

$$y(x) = -3e^{-3x} + 3\cos 2x - 2\sin 2x$$

is a solution of

$$y^{(3)} + 3y'' + 4y' + 12y = 0$$

on the entire real line. This particular solution has initial values $y(0) = 0$, $y'(0) = 5$, and $y''(0) = -39$, and Theorem 2 implies that there is no other solution with these same initial values. Note that its graph (in Fig. 3.2.1) looks periodic on the right. Indeed, because of the negative exponent, we see that $y(x) \approx 3\cos 2x - 2\sin 2x$ for large positive x. ■

Remark: Because its general solution involves the three arbitrary constants c_1, c_2, and c_3, the third-order equation in Example 1 has a "threefold infinity" of solutions, including three families of especially simple solutions:

- $y(x) = c_1 e^{-3x}$ (obtained from the general solution with $c_2 = c_3 = 0$),
- $y(x) = c_2 \cos 2x$ (with $c_1 = c_3 = 0$), and
- $y(x) = c_3 \sin 2x$ (with $c_1 = c_2 = 0$).

Alternatively, Theorem 2 suggests a threefold infinity of particular solutions corresponding to independent choices of the three initial values $y(0) = b_0$, $y'(0) = b_1$, and $y''(0) = b_2$. Figures 3.2.2 through 3.2.4 illustrate three corresponding families of solutions—for each of which, two of these three initial values are zero. ■

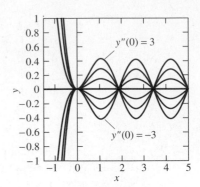

FIGURE 3.2.2. Solutions of $y^{(3)} + 3y'' + 4y' + 12y = 0$ with $y'(0) = y''(0) = 0$ but with different values for $y(0)$.

FIGURE 3.2.3. Solutions of $y^{(3)} + 3y'' + 4y' + 12y = 0$ with $y(0) = y''(0) = 0$ but with different values for $y'(0)$.

FIGURE 3.2.4. Solutions of $y^{(3)} + 3y'' + 4y' + 12y = 0$ with $y(0) = y'(0) = 0$ but with different values for $y''(0)$.

Note that Theorem 2 implies that the *trivial* solution $y(x) \equiv 0$ is the only solution of the *homogeneous* equation

$$y^{(n)} + p_1(x)y^{(n-1)} + \cdots + p_{n-1}(x)y' + p_n(x)y = 0 \tag{3}$$

that satisfies the *trivial* initial conditions

$$y(a) = y'(a) = \cdots = y^{(n-1)}(a) = 0.$$

Example 2 It is easy to verify that

$$y_1(x) = x^2 \quad \text{and} \quad y_2(x) = x^3$$

are two different solutions of

$$x^2 y'' - 4xy' + 6y = 0,$$

and that both satisfy the initial conditions $y(0) = y'(0) = 0$. Why does this not contradict the uniqueness part of Theorem 2? It is because the leading coefficient in this differential equation vanishes at $x = 0$, so this equation cannot be written in the form of Eq. (3) with coefficient functions *continuous* on an open interval containing the point $x = 0$. ∎

Linearly Independent Solutions

On the basis of our knowledge of general solutions of second-order linear equations, we anticipate that a general solution of the *homogeneous* nth-order linear equation

$$\blacktriangleright \qquad y^{(n)} + p_1(x)y^{(n-1)} + \cdots + p_{n-1}(x)y' + p_n(x)y = 0 \tag{3}$$

will be a linear combination

$$\blacktriangleright \qquad y = c_1 y_1 + c_2 y_2 + \cdots + c_n y_n, \tag{4}$$

where y_1, y_2, \ldots, y_n are particular solutions of Eq. (3). But these n particular solutions must be "sufficiently independent" that we can always choose the coefficients

c_1', c_2, \ldots, c_n in (4) to satisfy arbitrary initial conditions of the form in (5). The question is this: What should be meant by *independence* of three or more functions?

Recall that *two* functions f_1 and f_2 are linearly *dependent* if one is a constant multiple of the other; that is, if either $f_1 = kf_2$ or $f_2 = kf_1$ for some constant k. If we write these equations as

$$(1)f_1 + (-k)f_2 = 0 \quad \text{or} \quad (k)f_1 + (-1)f_2 = 0,$$

we see that the linear dependence of f_1 and f_2 implies that there exist two constants c_1 and c_2 *not both zero* such that

$$c_1 f_1 + c_2 f_2 = 0. \tag{6}$$

Conversely, if c_1 and c_2 are not both zero, then Eq. (6) certainly implies that f_1 and f_2 are linearly dependent.

In analogy with Eq. (6), we say that n functions f_1, f_2, \ldots, f_n are *linearly dependent* provided that some *nontrivial* linear combination

$$c_1 f_1 + c_2 f_2 + \cdots + c_n f_n$$

of them vanishes identically; *nontrivial* means that *not all* of the coefficients c_1, c_2, \ldots, c_n are zero (although *some* of them may be zero).

Definition: Linear Dependence of Functions

The n functions f_1, f_2, \ldots, f_n are said to be **linearly dependent** on the interval I provided that there exist constants c_1, c_2, \ldots, c_n not all zero such that

$$c_1 f_1 + c_2 f_2 + \cdots + c_n f_n = 0 \tag{7}$$

on I; that is,

$$c_1 f_1(x) + c_2 f_2(x) + \cdots + c_n f_n(x) = 0$$

for all x in I.

If not all the coefficients in Eq. (7) are zero, then clearly we can solve for at least one of the functions as a linear combination of the others, and conversely. Thus the functions f_1, f_2, \ldots, f_n are linearly dependent if and only if at least one of them is a linear combination of the others.

Example 3 The functions

$$f_1(x) = \sin 2x, \quad f_2(x) = \sin x \cos x, \quad \text{and} \quad f_3(x) = e^x$$

are linearly dependent on the real line because

$$(1)f_1 + (-2)f_2 + (0)f_3 = 0$$

(by the familiar trigonometric identity $\sin 2x = 2 \sin x \cos x$). ∎

The n functions f_1, f_2, \ldots, f_n are called **linearly independent** on the interval I provided that they are not linearly dependent there. Equivalently, they are linearly independent on I provided that the identity

$$c_1 f_1 + c_2 f_2 + \cdots + c_n f_n = 0 \tag{7}$$

holds on I only in the trivial case

$$c_1 = c_2 = \cdots = c_n = 0;$$

that is, *no* nontrivial linear combination of these functions vanishes on I. Put yet another way, the functions f_1, f_2, \ldots, f_n are linearly independent if no one of them is a linear combination of the others. (Why?)

Sometimes one can show that n given functions are linearly dependent by finding, as in Example 3, nontrivial values of the coefficients so that Eq. (7) holds. But in order to show that n given functions are linearly independent, we must prove that nontrivial values of the coefficients *cannot* be found, and this is seldom easy to do in any direct or obvious manner.

Fortunately, in the case of n solutions of a homogeneous nth-order linear equation, there is a tool that makes the determination of their linear dependence or independence a routine matter in many examples. This tool is the Wronskian determinant, which we introduced (for the case $n = 2$) in Section 3.1. Suppose that the n functions f_1, f_2, \ldots, f_n are each $n - 1$ times differentiable. Then their **Wronskian** is the $n \times n$ determinant

$$\blacktriangleright \qquad W = \begin{vmatrix} f_1 & f_2 & \cdots & f_n \\ f_1' & f_2' & \cdots & f_n' \\ \vdots & \vdots & & \vdots \\ f_1^{(n-1)} & f_2^{(n-1)} & \cdots & f_n^{(n-1)} \end{vmatrix}. \tag{8}$$

We write $W(f_1, f_2, \ldots, f_n)$ or $W(x)$, depending on whether we wish to emphasize the functions or the point x at which their Wronskian is to be evaluated. The Wronskian is named after the Polish mathematician J. M. H. Wronski (1778–1853).

We saw in Section 3.1 that the Wronskian of two linearly dependent functions vanishes identically. More generally, *the Wronskian of n linearly dependent functions f_1, f_2, \ldots, f_n is identically zero*. To prove this, assume that Eq. (7) holds on the interval I for some choice of the constants c_1, c_2, \ldots, c_n not all zero. We then differentiate this equation $n - 1$ times in succession, obtaining the n equations

$$\begin{aligned} c_1 f_1(x) + c_2 f_2(x) & + \cdots + & c_n f_n(x) = 0, \\ c_1 f_1'(x) + c_2 f_2'(x) & + \cdots + & c_n f_n'(x) = 0, \\ & & \vdots \\ c_1 f_1^{(n-1)}(x) + c_2 f_2^{(n-1)}(x) & + \cdots + & c_n f_n^{(n-1)}(x) = 0, \end{aligned} \tag{9}$$

which hold for all x in I. We recall from linear algebra that a system of n linear *homogeneous* equations in n unknowns has a nontrivial solution if and only if the determinant of coefficients vanishes. In Eq. (9) the unknowns are the constants c_1, c_2, \ldots, c_n and the determinant of coefficients is simply the Wronskian

$W(f_1, f_2, \ldots, f_n)$ evaluated at the typical point x of I. Because we know that the c_i are not all zero, it follows that $W(x) \equiv 0$, as we wanted to prove.

Therefore, to show that the functions f_1, f_2, \ldots, f_n are *linearly independent* on the interval I, it suffices to show that their Wronskian is nonzero at just one point of I.

Example 4

Show that the functions $y_1(x) = e^{-3x}$, $y_2(x) = \cos 2x$, and $y_3(x) = \sin 2x$ (of Example 1) are linearly independent.

Solution

Their Wronskian is

$$W = \begin{vmatrix} e^{-3x} & \cos 2x & \sin 2x \\ -3e^{-3x} & -2\sin 2x & 2\cos 2x \\ 9e^{-3x} & -4\cos 2x & -4\sin 2x \end{vmatrix}$$

$$= e^{-3x} \begin{vmatrix} -2\sin 2x & 2\cos 2x \\ -4\cos 2x & -4\sin 2x \end{vmatrix} + 3e^{-3x} \begin{vmatrix} \cos 2x & \sin 2x \\ -4\cos 2x & -4\sin 2x \end{vmatrix}$$

$$+ 9e^{-3x} \begin{vmatrix} \cos 2x & \sin 2x \\ -2\sin 2x & 2\cos 2x \end{vmatrix} = 26e^{-3x} \neq 0.$$

Because $W \neq 0$ everywhere, it follows that y_1, y_2, and y_3 are linearly independent on any open interval (including the entire real line). ∎

Example 5

Show first that the three solutions

$$y_1(x) = x, \quad y_2(x) = x\ln x, \quad \text{and} \quad y_3(x) = x^2$$

of the third-order equation

$$x^3 y^{(3)} - x^2 y'' + 2xy' - 2y = 0 \tag{10}$$

are linearly independent on the open interval $x > 0$. Then find a particular solution of Eq. (10) that satisfies the initial conditions

$$y(1) = 3, \quad y'(1) = 2, \quad y''(1) = 1. \tag{11}$$

Solution

Note that for $x > 0$, we could divide each term in (10) by x^3 to obtain a homogeneous linear equation of the standard form in (3). When we compute the Wronskian of the three given solutions, we find that

$$W = \begin{vmatrix} x & x\ln x & x^2 \\ 1 & 1+\ln x & 2x \\ 0 & \dfrac{1}{x} & 2 \end{vmatrix} = x.$$

Thus $W(x) \neq 0$ for $x > 0$, so y_1, y_2, and y_3 are linearly independent on the interval $x > 0$. To find the desired particular solution, we impose the initial conditions in (11) on

$$y(x) = c_1 x + c_2 x \ln x \quad + c_3 x^2,$$

$$y'(x) = c_1 \quad + c_2(1 + \ln x) + 2c_3 x,$$

$$y''(x) = 0 \quad + \frac{c_2}{x} \quad + 2c_3.$$

This yields the simultaneous equations

$$y(1) \ = c_1 \qquad + c_3 \ = 3,$$

$$y'(1) = c_1 + c_2 + 2c_3 = 2,$$

$$y''(1) = \qquad c_2 + 2c_3 = 1;$$

we solve to find $c_1 = 1$, $c_2 = -3$, and $c_3 = 2$. Thus the particular solution in question is

$$y(x) = x - 3x \ln x + 2x^2 \qquad \blacksquare$$

Provided that $W(y_1, y_2, \ldots, y_n) \neq 0$, it turns out (Theorem 4) that we can always find values of the coefficients in the linear combination

$$y = c_1 y_1 + c_2 y_2 + \cdots + c_n y_n$$

that satisfy any given initial conditions of the form in (5). Theorem 3 provides the necessary nonvanishing of W in the case of linearly independent solutions.

THEOREM 3: Wronskians of Solutions

Suppose that y_1, y_2, \ldots, y_n are n solutions of the homogeneous nth-order linear equation

$$y^{(n)} + p_1(x)y^{(n-1)} + \cdots + p_{n-1}(x)y' + p_n(x)y = 0 \qquad (3)$$

on an open interval I where each p_i is continuous. Let

$$W = W(y_1, y_2, \ldots, y_n).$$

(a) If y_1, y_2, \ldots, y_n are linearly dependent, then $W \equiv 0$ on I.

(b) If y_1, y_2, \ldots, y_n are linearly independent, then $W \neq 0$ at each point of I.

Thus there are just two possibilities: Either $W = 0$ everywhere on I, or $W \neq 0$ everywhere on I.

Proof: We have already proven part (a). To prove part (b), it is sufficient to assume that $W(a) = 0$ at some point of I, and show this implies that the solutions y_1, y_2, \ldots, y_n are linearly dependent. But $W(a)$ is simply the determinant of coefficients of the system of n homogeneous linear equations

$$c_1 y_1(a) + \quad c_2 y_2(a) + \cdots + \quad c_n y_n(a) = 0,$$

$$c_1 y_1'(a) + \quad c_2 y_2'(a) + \cdots + \quad c_n y_n'(a) = 0,$$

$$\vdots$$

$$c_1 y_1^{(n-1)}(a) + c_2 y_2^{(n-1)}(a) + \cdots + c_n y_n^{(n-1)}(a) = 0$$

(12)

in the n unknowns c_1, c_2, \ldots, c_n. Because $W(a) = 0$, the basic fact from linear algebra quoted just after (9) implies that the equations in (12) have a nontrivial solution. That is, the numbers c_1, c_2, \ldots, c_n are not all zero.

We now use these values to define the particular solution

$$Y(x) = c_1 y_1(x) + c_2 y_2(x) + \cdots + c_n y_n(x) \tag{13}$$

of Eq. (3). The equations in (12) then imply that Y satisfies the trivial initial conditions

$$Y(a) = Y'(a) = \cdots = Y^{(n-1)}(a) = 0.$$

Theorem 2 (uniqueness) therefore implies that $Y(x) \equiv 0$ on I. In view of (13) and the fact that c_1, c_2, \ldots, c_n are not all zero, this is the desired conclusion that the solutions y_1, y_2, \ldots, y_n are linearly dependent. This completes the proof of Theorem 3. ▲

General Solutions

We can now show that every solution of a *homogeneous* nth-order equation is a linear combination of n given linearly independent solutions. Using the fact from Theorem 3 that the Wronskian of n linearly independent solutions is nonzero, the proof of the following theorem is essentially the same as the proof of Theorem 4 of Section 3.1 (the case $n = 2$).

THEOREM 4: General Solutions of Homogeneous Equations

Let y_1, y_2, \ldots, y_n be n linearly independent solutions of the homogeneous equation

$$y^{(n)} + p_1(x) y^{(n-1)} + \cdots + p_{n-1}(x) y' + p_n(x) y = 0 \tag{3}$$

on an open interval I where the p_i are continuous. If Y is any solution whatsoever of Eq. (3), then there exist numbers c_1, c_2, \ldots, c_n such that

$$Y(x) = c_1 y_1(x) + c_2 y_2(x) + \cdots + c_n y_n(x)$$

for all x in I. ▲

Thus *every* solution of a homogeneous nth-order linear differential equation is a linear combination

$$\blacktriangleright \qquad y = c_1 y_1 + c_2 y_2 + \cdots + c_n y_n$$

of any n given linearly independent solutions. On this basis we call such a linear combination a **general solution** of the differential equation.

Example 6 According to Example 4, the particular solutions $y_1(x) = e^{-3x}$, $y_2(x) = \cos 2x$, and $y_3(x) = \sin 2x$ of the linear differential equation $y^{(3)} + 3y'' + 4y' + 12y = 0$ are linearly independent. Now Theorem 2 says that—given b_0, b_1, and b_2—there exists a particular solution $y(x)$ satisfying the initial conditions $y(0) = b_0$, $y'(0) = b_1$, and $y''(0) = b_2$. Hence Theorem 4 implies that *this* particular solution is a linear combination of y_1, y_2, and y_3. That is, there exist coefficients c_1, c_2, and c_3 such that

$$y(x) = c_1 e^{-3x} + c_2 \cos 2x + c_3 \sin 2x.$$

Upon successive differentiation and substitution of $x = 0$, we discover that to find these coefficients, we need only solve the three linear equations

$$\begin{aligned} c_1 + c_2 \quad\quad &= b_0, \\ -3c_1 \quad\quad + 2c_3 &= b_1, \\ 9c_1 - 4c_2 \quad\quad &= b_2. \end{aligned}$$

(See the application for this section.) ■

Nonhomogeneous Equations

We now consider the *nonhomogeneous* nth-order linear differential equation

➤ $$Ly = y^{(n)} + p_1(x)y^{(n-1)} + \cdots + p_{n-1}(x)y' + p_n(x)y = f(x) \qquad (2)$$

with associated homogeneous equation

➤ $$Ly = y^{(n)} + p_1(x)y^{(n-1)} + \cdots + p_{n-1}(x)y' + p_n(x)y = 0. \qquad (3)$$

Here we introduce the symbol L to represent an **operator**; given an n times differentiable function y, L *operates* on y (as suggested in Fig. 3.2.5) to produce the linear combination

$$Ly = y^{(n)} + p_1(x)y^{(n-1)} + \cdots + p_{n-1}(x)y' + p_n(x)y \qquad (16)$$

FIGURE 3.2.5. The idea of L "operating" on the function y.

of y and its first n derivatives. The principle of superposition (Theorem 1) means simply that the operator L is linear; that is,

➤ $$L(c_1 y_1 + c_2 y_2) = c_1 L y_1 + c_2 L y_2 \qquad (17)$$

if c_1 and c_2 are constants.

Suppose that a single particular solution y_p of the nonhomogeneous equation in (2) is known and that Y is any other solution of Eq. (2). Then Eq. (15) implies that

$$L(Y - y_p) = LY - Ly_p = f - f = 0.$$

Thus $y_c = Y - y_p$ is a solution of the associated homogeneous equation in (3). Then

➤ $$Y = y_c + y_p, \qquad (18)$$

and it follows from Theorem 4 that

➤ $$y_c = c_1 y_1 + c_2 y_2 + \cdots + c_n y_n, \qquad (19)$$

where y_1, y_2, \ldots, y_n are linearly independent solutions of the associated *homogeneous* equation. We call y_c a **complementary function** of the nonhomogeneous equation and have thus proved that a *general solution* of the nonhomogeneous equation in (2) is the sum of its complementary function y_c and a single particular solution y_p of Eq. (2).

> **THEOREM 5:** Solutions of Nonhomogeneous Equations
>
> Let y_p be a particular solution of the nonhomogeneous equation in (2) on an open interval I where the functions p_i and f are continuous. Let y_1, y_2, \ldots, y_n be linearly independent solutions of the associated homogeneous equation in (3). If Y is any solution whatsoever of Eq. (2) on I, then there exist numbers c_1, c_2, \ldots, c_n such that
>
> $$Y(x) = c_1 y_1(x) + c_2 y_2(x) + \cdots + c_n y_n(x) + y_p(x) \qquad (20)$$
>
> for all x in I. ▲

Example 7 It is evident that $y_p = 3x$ is a particular solution of the equation

$$y'' + 4y = 12x, \qquad (21)$$

and that $y_c(x) = c_1 \cos 2x + c_2 \sin 2x$ is its complementary solution. Find a solution of Eq. (19) that satisfies the initial conditions $y(0) = 5$, $y'(0) = 7$.

Solution The general solution of Eq. (19) is

$$y(x) = c_1 \cos 2x + c_2 \sin 2x + 3x.$$

Now

$$y'(x) = -2c_1 \sin 2x + 2c_2 \cos 2x + 3.$$

Hence the initial conditions give

$$y(0) = c_1 = 5,$$
$$y'(0) = 2c_2 + 3 = 7.$$

We find that $c_1 = 5$ and $c_2 = 2$. Thus the desired solution is

$$y(x) = 5 \cos 2x + 2 \sin 2x + 3x. \qquad ■$$

3.2 Problems

In Problems 1 through 6, show directly that the given functions are linearly dependent on the real line. That is, find a non-trivial linear combination of the given functions that vanishes identically.

1. $f(x) = 2x$, $g(x) = 3x^2$, $h(x) = 5x - 8x^2$
2. $f(x) = 5$, $g(x) = 2 - 3x^2$, $h(x) = 10 + 15x^2$
3. $f(x) = 0$, $g(x) = \sin x$, $h(x) = e^x$
4. $f(x) = 17$, $g(x) = 2 \sin^2 x$, $h(x) = 3 \cos^2 x$
5. $f(x) = 17$, $g(x) = \cos^2 x$, $h(x) = \cos 2x$
6. $f(x) = e^x$, $g(x) = \cosh x$, $h(x) = \sinh x$

In Problems 7 through 12, use the Wronskian to prove that the given functions are linearly independent on the indicated interval.

7. $f(x) = 1$, $g(x) = x$, $h(x) = x^2$; the real line
8. $f(x) = e^x$, $g(x) = e^{2x}$, $h(x) = e^{3x}$; the real line
9. $f(x) = e^x$, $g(x) = \cos x$, $h(x) = \sin x$; the real line
10. $f(x) = e^x$, $g(x) = x^{-2}$, $h(x) = x^{-2} \ln x$; $x > 0$
11. $f(x) = x$, $g(x) = xe^x$, $h(x) = x^2 e^x$; the real line
12. $f(x) = x$, $g(x) = \cos(\ln x)$, $h(x) = \sin(\ln x)$; $x > 0$

In Problems 13 through 20, a third-order homogeneous linear equation and three linearly independent solutions are given. Find a particular solution satisfying the given initial conditions.

13. $y^{(3)} + 2y'' - y' - 2y = 0$; $y(0) = 1$, $y'(0) = 2$, $y''(0) = 0$; $y_1 = e^x$, $y_2 = e^{-x}$, $y_3 = e^{-2x}$

14. $y^{(3)} - 6y'' + 11y' - 6y = 0$; $y(0) = 0$, $y'(0) = 0$, $y''(0) = 3$; $y_1 = e^x$, $y_2 = e^{2x}$, $y_3 = e^{3x}$.

15. $y^{(3)} - 3y'' + 3y' - y = 0$; $y(0) = 2$, $y'(0) = 0$, $y''(0) = 0$; $y_1 = e^x$, $y_2 = xe^x$, $y_3 = x^2e^x$

16. $y^{(3)} - 5y'' + 8y' - 4y = 0$; $y(0) = 1$, $y'(0) = 4$, $y''(0) = 0$; $y_1 = e^x$, $y_2 = e^{2x}$, $y_3 = xe^{2x}$

17. $y^{(3)} + 9y' = 0$; $y(0) = 3$, $y'(0) = -1$, $y''(0) = 2$; $y_1 = 1$, $y_2 = \cos 3x$, $y_3 = \sin 3x$

18. $y^{(3)} - 3y'' + 4y' - 2y = 0$; $y(0) = 1$, $y'(0) = 0$, $y''(0) = 0$; $y_1 = e^x$, $y_2 = e^x \cos x$, $y_3 = e^x \sin x$.

19. $x^3 y^{(3)} - 3x^2 y'' + 6xy' - 6y = 0$; $y(1) = 6$, $y'(1) = 14$, $y''(1) = 22$; $y_1 = x$, $y_2 = x^2$, $y_3 = x^3$

20. $x^3 y^{(3)} + 6x^2 y'' + 4xy' - 4y = 0$; $y(1) = 1$, $y'(1) = 5$, $y''(1) = -11$; $y_1 = x$, $y_2 = x^{-2}$, $y_3 = x^{-2} \ln x$

In Problems 21 through 24, a nonhomogeneous differential equation, a complementary solution y_c, and a particular solution y_p are given. Find a solution satisfying the given initial conditions.

21. $y'' + y = 3x$; $y(0) = 2$, $y'(0) = -2$; $y_c = c_1 \cos x + c_2 \sin x$; $y_p = 3x$

22. $y'' - 4y = 12$; $y(0) = 0$, $y'(0) = 10$; $y_c = c_1 e^{2x} + c_2 e^{-2x}$; $y_p = -3$

23. $y'' - 2y' - 3y = 6$; $y(0) = 3$, $y'(0) = 11$; $y_c = c_1 e^{-x} + c_2 e^{3x}$; $y_p = -2$

24. $y'' - 2y' + 2y = 2x$; $y(0) = 4$, $y'(0) = 8$; $y_c = c_1 e^x \cos x + c_2 e^x \sin x$; $y_p = x + 1$

25. Let $Ly = y'' + py' + qy$. Suppose that y_1 and y_2 are two functions such that

$$Ly_1 = f(x) \quad \text{and} \quad Ly_2 = g(x).$$

Show that their sum $y = y_1 + y_2$ satisfies the nonhomogeneous equation $Ly = f(x) + g(x)$.

26. (a) Find by inspection particular solutions of the two nonhomogeneous equations

$$y'' + 2y = 4 \quad \text{and} \quad y'' + 2y = 6x.$$

(b) Use the method of Problem 25 to find a particular solution of the differential equation $y'' + 2y = 6x + 4$.

27. Prove directly that the functions

$$f_1(x) \equiv 1, \quad f_2(x) = x, \quad \text{and} \quad f_3(x) = x^2$$

are linearly independent on the whole real line. (*Suggestion*: Assume that $c_1 + c_2 x + c_3 x^2 = 0$. Differentiate this equation twice. You now have three equations that must be satisfied by every real number x, including $x = 0$. Conclude that $c_1 = c_2 = c_3 = 0$.)

28. Generalize the method of Problem 27 to prove directly that the functions

$$f_0(x) \equiv 1, \quad f_1(x) = x, \quad f_2(x) = x^2, \ldots, \quad f_n(x) = x^n$$

are linearly independent on the real line.

29. Use the result of Problem 28 and the definition of linear independence to prove directly that, for any constant r, the functions

$$f_0(x) = e^{rx}, \quad f_1(x) = xe^{rx}, \ldots, \quad f_n(x) = x^n e^{rx}$$

are linearly independent on the whole real line.

30. Verify that $y_1 = x$ and $y_2 = x^2$ are linearly independent solutions on the entire real line of the equation

$$x^2 y'' - 2xy' + 2y = 0,$$

but that $W(x, x^2)$ vanishes at $x = 0$. Why do these observations not contradict part (b) of Theorem 3?

31. This problem indicates why we can impose *only n* initial conditions on a solution of an nth-order linear differential equation. **(a)** Given the equation

$$y'' + py' + qy = 0,$$

explain why the value of $y''(a)$ is determined by the values of $y(a)$ and $y'(a)$. **(b)** Prove that the equation

$$y'' - 2y' - 5y = 0$$

has a solution satisfying the conditions

$$y(0) = 1, \quad y'(0) = 0, \quad \text{and} \quad y''(0) = C$$

if and only if $C = 5$.

32. Prove that an nth-order homogeneous linear differential equation satisfying the hypotheses of Theorem 2 has n linearly independent solutions y_1, y_2, \ldots, y_n. (*Suggestion*: Let y_i be the unique solution such that

$$y_i^{(i-1)}(a) = 1 \quad \text{and} \quad y_i^{(k)}(a) = 0 \quad \text{if } k \neq i - 1.)$$

33. Suppose that the three numbers r_1, r_2, and r_3 are distinct. Show that the three functions $\exp(r_1 x)$, $\exp(r_2 x)$, and $\exp(r_3 x)$ are linearly independent by showing that their Wronskian

$$W = \exp([r_1 + r_2 + r_3]x) \cdot \begin{vmatrix} 1 & 1 & 1 \\ r_1 & r_2 & r_3 \\ r_1^2 & r_2^2 & r_3^2 \end{vmatrix}$$

is nonzero for all x.

34. Assume as known that the **Vandermonde determinant**

$$V = \begin{vmatrix} 1 & 1 & \cdots & 1 \\ r_1 & r_2 & \cdots & r_n \\ r_1^2 & r_2^2 & \cdots & r_n^2 \\ \vdots & \vdots & & \vdots \\ r_1^{n-1} & r_2^{n-1} & \cdots & r_n^{n-1} \end{vmatrix}$$

is nonzero if the numbers r_1, r_2, \ldots, r_n are distinct. Prove by the method of Problem 33 that the functions

$$f_i(x) = \exp(r_i x), \quad 1 \leq i \leq n$$

are linearly independent.

35. According to Problem 32 of Section 3.1, the Wronskian $W(y_1, y_2)$ of two solutions of the second-order equation

$$y'' + p_1(x)y' + p_2(x)y = 0$$

is given by Abel's's formula

$$W(x) = K \exp\left(-\int p_1(x)\,dx\right)$$

for some constant K. It can be shown that the Wronskian of n solutions y_1, y_2, \ldots, y_n of the nth-order equation

$$y^{(n)} + p_1(x)y^{(n-1)} + \cdots + p_{n-1}(x)y' + p_n(x)y = 0$$

satisfies the same identity. Prove this for the case $n = 3$ as follows: **(a)** The derivative of a determinant of functions is the sum of the determinants obtained by separately differentiating the rows of the original determinant. Conclude that

$$W' = \begin{vmatrix} y_1 & y_2 & y_3 \\ y_1' & y_2' & y_3' \\ y_1^{(3)} & y_2^{(3)} & y_3^{(3)} \end{vmatrix}.$$

(b) Substitute for $y_1^{(3)}$, $y_2^{(3)}$, and $y_3^{(3)}$ from the equation

$$y^{(3)} + p_1 y'' + p_2 y' + p_3 y = 0,$$

and then show that $W' = -p_1 W$. Integration now gives Abel's formula.

36. Suppose that one solution $y_1(x)$ of the homogeneous second-order linear differential equation

$$y'' + p(x)y' + q(x)y = 0 \qquad (22)$$

is known (on an interval I where p and q are continuous functions). The method of **reduction of order** consists of substituting $y_2(x) = v(x)y_1(x)$ in (20) and attempting to determine the function $v(x)$ so that $y_2(x)$ is a second linearly independent solution of (20). After substituting $y = v(x)y_1(x)$ in Eq. (20), use the fact that $y_1(x)$ is a solution to deduce that

$$y_1 v'' + (2y_1' + py_1)v' = 0. \qquad (23)$$

If $y_1(x)$ is known, then (21) is a separable equation that is readily solved for the derivative $v'(x)$ of $v(x)$. Integration of $v'(x)$ then gives the desired (nonconstant) function $v(x)$.

37. Before applying Eq. (21) with a given homogeneous second-order linear differential equation and a known solution $y_1(x)$, the equation must first be written in the form of (20) with leading coefficient 1 in order to correctly determine the coefficient function $p(x)$. Frequently it is more convenient to simply substitute $y = v(x)y_1(x)$ in the given differential equation and then proceed directly to find $v(x)$. Thus, starting with the readily verified solution $y_1(x) = x^3$ of the equation

$$x^2 y'' - 5xy' + 9y = 0 \qquad (x > 0),$$

substitute $y = vx^3$ and deduce that $xv'' + v' = 0$. Thence solve for $v(x) = C \ln x$, and thereby obtain (with $C = 1$) the second solution $y_2(x) = x^3 \ln x$.

In each of Problems 38 through 45, a differential equation and one solution y_1 are given. Use the method of reduction of order as in Problem 37 to find a second linearly independent solution y_2.

38. $x^2 y'' + xy' - 9y = 0$ $(x > 0)$; $y_1(x) = x^3$

39. $4y'' - 4y' + y = 0$; $y_1(x) = e^{x/2}$

40. $x^2 y'' - x(x+2)y' + (x+2)y = 0$ $(x > 0)$; $y_1(x) = x$

41. $(x+1)y'' - (x+2)y' + y = 0$ $(x > -1)$; $y_1(x) = e^x$

42. $(1-x^2)y'' + 2xy' - 2y = 0$ $(-1 < x < 1)$; $y_1(x) = x$

43. First note that $y_1(x) = x$ is one solution of Legendre's equation of order 1,

$$(1 - x^2)y'' - 2xy' + 2y = 0.$$

Then use the method of reduction of order to derive the second solution

$$y_2(x) = 1 - \frac{x}{2} \ln \frac{1+x}{1-x} \qquad \text{(for } -1 < x < 1\text{)}.$$

44. First verify by substitution that $y_1(x) = x^{-1/2} \cos x$ is one solution (for $x > 0$) of Bessel's equation of order $\frac{1}{2}$,

$$x^2 y'' + xy' + (x^2 - \tfrac{1}{4})y = 0.$$

Then derive by reduction of order the second solution $y_2(x) = x^{-1/2} \sin x$.

3.2 Application: Plotting Third-Order Solution Families

This application deals with the plotting by computer of families of solutions such as those illustrated in Figs. 3.2.2 through 3.2.4. We know from Example 6 that the general solution of

$$y^{(3)} + 3y'' + 4y' + 12y = 0 \qquad (1)$$

is

$$y(x) = c_1 e^{-3x} + c_2 \cos 2x + c_3 \sin 2x. \qquad (2)$$

For Fig. 3.2.2, use the method of Example 6 to show that the particular solution of Eq. (1) satisfying the initial conditions $y(0) = a$, $y'(0) = 0$, and $y''(0) = 0$ is given by

$$y(x) = \frac{a}{13}\left(4e^{-3x} + 9\cos 2x + 6\sin 2x\right). \tag{3}$$

The MATLAB loop

```
x = -1.5 : 0.02 : 5      % x-vector from x = -1.5 to x = 5
for a = -3 : 1 : 3       % for a = -3 to 3 with da = 1 do
    c1 = 4*a/13;
    c2 = 9*a/13;
    c3 = 6*a/13;
    y = c1*exp(-3*x) + c2*cos(2*x) + c3*sin(2*x);
    plot(x,y)
end
```

was used to generate Fig. 3.2.2.

For Fig. 3.2.3, show that the particular solution of Eq. (1) satisfying the initial conditions $y(0) = 0$, $y'(0) = b$, and $y''(0) = 0$ is given by

$$y(x) = \frac{b}{2}\sin 2x, \tag{4}$$

and alter the preceding for-loop accordingly.

For Fig. 3.2.3, show that the particular solution of Eq. (1) satisfying the initial conditions $y(0) = 0$, $y'(0) = 0$, and $y''(0) = c$ is given by

$$y(x) = \frac{c}{26}\left(2e^{-3x} - 2\cos 2x + 3\sin 2x\right). \tag{5}$$

Computer algebra systems such as *Maple* and *Mathematica,* as well as graphing calculators, have commands to carry out for-loops such as the one shown here. Begin by reproducing Figs. 3.2.2 through 3.2.4. Then plot similar families of solution curves for the differential equations in Problems 13 through 20.

3.3 Homogeneous Equations with Constant Coefficients

In Section 3.2 we saw that a general solution of an nth-order homogeneous linear equation is a linear combination of n linearly independent particular solutions, but we said little about how actually to find even a single solution. The solution of a linear differential equation with *variable* coefficients ordinarily requires numerical methods (Chapter 2) or infinite series methods (Chapter 8). But we can now show how to find, explicitly and in a rather straightforward way, n linearly independent solutions of a given nth-order linear equation if it has *constant* coefficients. The general such equation may be written in the form

$$\blacktriangleright \qquad a_n y^{(n)} + a_{n-1}y^{(n-1)} + \cdots + a_2 y'' + a_1 y' + a_0 y = 0, \tag{1}$$

where the coefficients $a_0, a_1, a_2, \ldots, a_n$ are real constants with $a_n \neq 0$.

The Characteristic Equation

We first look for a *single* solution of Eq. (1), and begin with the observation that

$$\frac{d^k}{dx^k}(e^{rx}) = r^k e^{rx},\tag{2}$$

so any derivative of e^{rx} is a constant multiple of e^{rx}. Hence, if we substituted $y = e^{rx}$ in Eq. (1), each term would be a constant multiple of e^{rx}, with the constant coefficients depending on r and the coefficients a_i. This suggests that we try to find r so that all these multiples of e^{rx} will have sum zero, in which case $y = e^{rx}$ will be a solution of Eq. (1).

For example, in Section 3.1 we substituted $y = e^{rx}$ in the second-order equation

$$ay'' + by' + cy = 0$$

to derive the characteristic equation

$$ar^2 + br + c = 0$$

that r must satisfy.

To carry out this technique in the general case, we substitute $y = e^{rx}$ in Eq. (1), and with the aid of Eq. (2) we find the result to be

$$a_n r^n e^{rx} + a_{n-1} r^{n-1} e^{rx} + \cdots + a_2 r^2 e^{rx} + a_1 r e^{rx} + a_0 e^{rx} = 0;$$

that is,

$$e^{rx}\left(a_n r^n + a_{n-1} r^{n-1} + \cdots + a_2 r^2 + a_1 r + a_0\right) = 0.$$

Because e^{rx} is never zero, we see that $y = e^{rx}$ will be a solution of Eq. (1) precisely when r is a root of the equation

$$a_n r^n + a_{n-1} r^{n-1} + \cdots + a_2 r^2 + a_1 r + a_0 = 0.\tag{3}$$

This equation is called the **characteristic equation** or **auxiliary equation** of the differential equation in (1). Our problem, then, is reduced to the solution of this purely algebraic equation.

According to the fundamental theorem of algebra, every nth-degree polynomial— such as the one in Eq. (3)—has n zeros, though not necessarily distinct and not necessarily real. Finding the exact values of these zeros may be difficult or even impossible; the quadratic formula is sufficient for second-degree equations, but for equations of higher degree we may need either to spot a fortuitous factorization or to apply a numerical technique such as Newton's method (or use a calculator/computer `solve` command).

Distinct Real Roots

Whatever the method we use, let us suppose that we have solved the characteristic equation. Then we can always write a general solution of the differential equation. The situation is slightly more complicated in the case of repeated roots or complex roots of Eq. (3), so let us first examine the simplest case—in which the characteristic equation has n distinct (no two equal) *real* roots r_1, r_2, \ldots, r_n. Then the functions

$$e^{r_1 x}, e^{r_2 x} \ldots, e^{r_n x}$$

are all solutions of Eq. (1), and (by Problem 34 of Section 3.2) these n solutions are linearly independent on the entire real line. In summary, we have proved Theorem 1.

THEOREM 1: Distinct Real Roots

If the roots r_1, r_2, \ldots, r_n of the characteristic equation in (3) are real and distinct, then

$$\blacktriangleright \qquad y(x) = c_1 e^{r_1 x} + c_2 e^{r_2 x} + \cdots + c_n e^{r_n x} \qquad (4)$$

is a general solution of Eq. (1). $\qquad \blacktriangle$

Example 1 Solve the initial value problem

$$y^{(3)} + 3y'' - 10y' = 0;$$
$$y(0) = 7, \quad y'(0) = 0, \quad y''(0) = 70.$$

Solution The characteristic equation of the given differential equation is

$$r^3 + 3r^2 - 10r = 0.$$

We solve by factoring:

$$r(r^2 + 3r - 10) = r(r+5)(r-2) = 0,$$

and so the characteristic equation has the three distinct real roots $r = 0$, $r = -5$, and $r = 2$. Because $e^0 = 1$, Theorem 1 gives the general solution

$$y(x) = c_1 + c_2 e^{-5x} + c_3 e^{2x}.$$

Then the given initial conditions yield the linear equations

$$
\begin{aligned}
y(0) = c_1 + {} & c_2 + {} & c_3 &= 7, \\
y'(0) = {} & -5c_2 + {} & 2c_3 &= 0, \\
y''(0) = {} & 25c_2 + {} & 4c_3 &= 70
\end{aligned}
$$

in the coefficients c_1, c_2, and c_3. The last two equations give $y''(0) - 2y'(0) = 35c_2 = 70$, so $c_2 = 2$. Then the second equation gives $c_3 = 5$, and finally the first equation gives $c_1 = 0$. Thus the desired particular solution is

$$y(x) = 2e^{-5x} + 5e^{2x}. \qquad \blacksquare$$

Polynomial Operators

If the roots of the characteristic equation in (3) are *not* distinct—there are repeated roots—then we cannot produce n linearly independent solutions of Eq. (1) by the method of Theorem 1. For example, if the roots are 1, 2, 2, and 2, we obtain only the *two* functions e^x and e^{2x}. The problem, then, is to produce the missing linearly independent solutions. For this purpose it is convenient to adopt the operator notation introduced near the conclusion of Section 3.2. Equation (1) corresponds to the operator equation $Ly = 0$, where L is the operator

$$L = a_n \frac{d^n}{dx^n} + a_{n-1} \frac{d^{n-1}}{dx^{n-1}} + \cdots + a_2 \frac{d^2}{dx^2} + a_1 \frac{d}{dx} + a_0. \tag{5}$$

We also denote by $D = d/dx$ the operation of differentiation with respect to x, so that

$$Dy = y', \quad D^2 y = y'', \quad D^3 y = y^{(3)},$$

and so on. In terms of D, the operator L in (5) may be written

$$L = a_n D^n + a_{n-1} D^{n-1} + \cdots + a_2 D^2 + a_1 D + a_0, \tag{6}$$

and we will find it useful to think of the right-hand side in Eq. (6) as a (formal) nth-degree polynomial in the "variable" D; it is a **polynomial operator.**

A first-degree polynomial operator has the form $D - a$ where a is a real number. It operates on the function $y = y(x)$ to produce

$$(D - a)y = Dy - ay = y' - ay.$$

The important fact about such operators is that any two of them *commute*:

$$(D - a)(D - b)y = (D - b)(D - a)y \tag{7}$$

for any twice differentiable function $y = y(x)$. The proof of the formula in (7) is the following computation:

$$\begin{aligned}
(D - a)(D - b)y &= (D - a)(y' - by) \\
&= D(y' - by) - a(y' - by) \\
&= y'' - (b + a)y' + aby = y'' - (a + b)y' + bay \\
&= D(y' - ay) - b(y' - ay) \\
&= (D - b)(y' - ay) = (D - b)(D - a)y.
\end{aligned}$$

Repeated Roots

Let us now consider the possibility that the characteristic equation

$$a_n r^n + a_{n-1} r^{n-1} + \cdots + a_2 r^2 + a_1 r + a_0 = 0 \tag{3}$$

has *repeated* roots. For example, suppose that Eq. (3) has only two distinct roots, r_0 of multiplicity 1 and r_1 of multiplicity $k > 1$. Then (after dividing by a_n) Eq. (3) can be rewritten in the form

$$(r - r_1)^k (r - r_0) = (r - r_0)(r - r_1)^k = 0. \tag{8}$$

Similarly, the corresponding operator L in (6) can be written as

$$L = (D - r_1)^k (D - r_0) = (D - r_0)(D - r_1)^k, \qquad (9)$$

the order of the factors making no difference because of the formula in (7).

Two solutions of the differential equation $Ly = 0$ are certainly $y_0 = e^{r_0 x}$ and $y_1 = e^{r_1 x}$. This is, however, not sufficient; we need $k + 1$ linearly independent solutions in order to construct a general solution, because the equation is of order $k + 1$. To find the missing $k - 1$ solutions, we note that

$$Ly = (D - r_0)[(D - r_1)^k y] = 0.$$

Consequently, *every* solution of the kth-order equation

$$(D - r_1)^k y = 0 \qquad (10)$$

will also be a solution of the original equation $Ly = 0$. Hence our problem is reduced to that of finding a general solution of the differential equation in (10).

The fact that $y_1 = e^{r_1 x}$ is one solution of Eq. (10) suggests that we try the substitution

$$y(x) = u(x)y_1(x) = u(x)e^{r_1 x}, \qquad (11)$$

where $u(x)$ is a function yet to be determined. Observe that

$$(D - r_1)\left[ue^{r_1 x}\right] = (Du)e^{r_1 x} + u(r_1 e^{r_1 x}) - r_1(ue^{r_1 x}) = (Du)e^{r_1 x}. \qquad (12)$$

Upon k applications of this fact, it follows that

$$(D - r_1)^k \left[ue^{r_1 x}\right] = (D^k u)e^{r_1 x} \qquad (13)$$

for any sufficiently differentiable function $u(x)$. Hence $y = ue^{r_1 x}$ will be a solution of Eq. (10) if and only if $D^k u = u^{(k)} = 0$. But this is so if and only if

$$u(x) = c_1 + c_2 x + c_3 x^2 + \cdots + c_k x^{k-1},$$

a polynomial of degree at most $k - 1$. Hence our desired solution of Eq. (10) is

$$y(x) = ue^{r_1 x} = (c_1 + c_2 x + c_3 x^2 + \cdots + c_k x^{k-1})e^{r_1 x}.$$

In particular, we see here the additional solutions $xe^{r_1 x}, x^2 e^{r_1 x}, \ldots, x^{k-1}e^{r_1 x}$ of the original differential equation $Ly = 0$.

The preceding analysis can be carried out with the operator $D - r_1$ replaced with an arbitrary polynomial operator. When this is done, the result is a proof of the following theorem.

THEOREM 2: Repeated Roots

If the characteristic equation in (3) has a repeated root r of multiplicity k, then the part of a general solution of the differential equation in (1) corresponding to r is of the form

$$(c_1 + c_2 x + c_3 x^2 + \cdots + c_k x^{k-1})e^{rx}. \qquad (14)$$

We may observe that according to Problem 29 of Section 3.2, the k functions e^{rx}, xe^{rx}, x^2e^{rx}, ..., and $x^{k-1}e^{rx}$ involved in (14) are linearly independent on the real line. Thus a root of multiplicity k corresponds to k linearly independent solutions of the differential equation.

Example 2 Find a general solution of the fifth-order differential equation

$$9y^{(5)} - 6y^{(4)} + y^{(3)} = 0.$$

Solution The characteristic equation is

$$9r^5 - 6r^4 + r^3 = r^3(9r^2 - 6r + 1) = r^3(3r-1)^2 = 0.$$

It has the triple root $r = 0$ and the double root $r = \frac{1}{3}$. The triple root $r = 0$ contributes

$$c_1e^{0\cdot x} + c_2xe^{0\cdot x} + c_3x^2e^{0\cdot x} = c_1 + c_2x + c_3x^2$$

to the solution, while the double root $r = \frac{1}{3}$ contributes $c_4e^{x/3} + c_5xe^{x/3}$. Hence a general solution of the given differential equation is

$$y(x) = c_1 + c_2x + c_3x^2 + c_4e^{x/3} + c_5xe^{x/3}. \qquad \blacksquare$$

Complex-Valued Functions and Euler's Formula

Because we have assumed that the coefficients of the differential equation and its characteristic equation are real, any complex (nonreal) roots will occur in complex conjugate pairs $a + bi$ where a and b are real and $i = \sqrt{-1}$. This raises the question as to what might be meant by an exponential such as $\exp[(a+bi)x]$.

To answer this question, we recall from elementary calculus the Taylor series for the exponential function

$$e^t = \sum_{n=0}^{\infty} \frac{t^n}{n!} = 1 + t + \frac{t^2}{2!} + \frac{t^3}{3!} + \frac{t^4}{4!} + \cdots.$$

If we substitute $t = ix$ in this series and recall that $i^2 = -1$, $i^3 = -i$, $i^4 = 1$, etc., we get

$$e^{ix} = \sum_{n=0}^{\infty} \frac{(ix)^n}{n!}$$

$$= 1 + ix - \frac{x^2}{2!} - \frac{ix^3}{3!} + \frac{x^4}{4!} + \frac{ix^5}{5!} - \cdots$$

$$= \left(1 - \frac{x^2}{2!} + \frac{x^4}{4!} - \cdots\right) + i\left(x - \frac{x^3}{3!} + \frac{x^5}{5!} - \cdots\right).$$

Because the two real series in the last line are the Taylor series for $\cos x$ and $\sin x$, respectively, this implies that

$$e^{ix} = \cos x + i \sin x. \tag{15}$$

This result is known as **Euler's formula**. Because of it, we *define* the exponential function e^z, for $z = x + iy$ an arbitrary complex number, to be

$$e^z = e^{x+iy} = e^x e^{iy} = e^x (\cos y + i \sin y). \tag{16}$$

Thus it appears that complex roots of the characteristic equation will lead to complex-valued solutions of the differential equation. A **complex-valued function** F of the real variable x associates with each real number x (in its domain of definition) the complex number

$$F(x) = f(x) + ig(x). \tag{17}$$

The real-valued functions f and g are called the **real** and **imaginary** parts, respectively, of F. If they are differentiable, we define the **derivative** F' of F by

$$F'(x) = f'(x) + ig'(x). \tag{18}$$

Thus we simply differentiate the real and imaginary parts of F separately.

We say that the complex-valued function $F(x)$ satisfies the homogeneous linear differential equation $L[F(x)] = 0$ provided that its real and imaginary parts in (17) separately satisfy this equation—so $L[F(x)] = L[f(x)] + iL[g(x)] = 0$.

The particular complex-valued functions of interest here are of the form $F(x) = e^{rx}$, where $r = a \pm bi$. We note from Euler's formula that

$$e^{(a+bi)x} = e^{ax} e^{ibx} = e^{ax} (\cos bx + i \sin bx) \tag{19a}$$

and

$$e^{(a-bi)x} = e^{ax} e^{-ibx} = e^{ax} (\cos bx - i \sin bx). \tag{19b}$$

The most important property of e^{rx} is that

$$D_x(e^{rx}) = re^{rx}, \tag{20}$$

if r is a complex number. The proof of this assertion is a straightforward computation based on the definitions and formulas given earlier:

$$\begin{aligned}
D_x(e^{rx}) &= D_x(e^{ax} \cos bx) + i D_x(e^{ax} \sin bx) \\
&= \left(ae^{ax} \cos bx - be^{ax} \sin bx\right) + i \left(ae^{ax} \sin bx + be^{ax} \cos bx\right) \\
&= (a + bi)(e^{ax} \cos bx + ie^{ax} \sin bx) = re^{rx}.
\end{aligned}$$

Complex Roots

It follows from Eq. (20) that when r is complex (just as when r is real), e^{rx} will be a solution of the differential equation in (1) if and only if r is a root of its characteristic equation. If the complex conjugate pair of roots $r_1 = a + bi$ and $r_2 = a - bi$ are simple (nonrepeated), then the corresponding part of a general solution of Eq. (1) is

$$\begin{aligned}
y(x) &= C_1 e^{r_1 x} + C_2 e^{r_2 x} = C_1 e^{(a+bi)x} + C_2 e^{(a-bi)x} \\
&= C_1 e^{ax} (\cos bx + i \sin bx) + C_2 e^{ax} (\cos bx - i \sin bx) \\
y(x) &= (C_1 + C_2) e^{ax} \cos bx + i(C_1 - C_2) e^{ax} \sin bx,
\end{aligned}$$

where the arbitrary constants C_1 and C_2 can be complex. For instance, the choice $C_1 = C_2 = \frac{1}{2}$ gives the real-valued solution $y_1(x) = e^{ax} \cos bx$, while the choice $C_1 = -\frac{1}{2}i$, $C_2 = \frac{1}{2}i$ gives the independent real-valued solution $y_2(x) = e^{ax} \sin bx$. This yields the following result.

THEOREM 3: Complex Roots

If the characteristic equation in (3) has an unrepeated pair of complex conjugate roots $a \pm bi$ (with $b \neq 0$), then the corresponding part of a general solution of Eq. (1) has the form

➤
$$e^{ax}(c_1 \cos bx + c_2 \sin bx). \tag{21}$$

▲

Example 3

The characteristic equation of

$$y'' + b^2 y = 0 \quad (b > 0)$$

is $r^2 + b^2 = 0$, with roots $r = \pm bi$. So Theorem 3 (with $a = 0$) gives the general solution

$$y(x) = c_1 \cos bx + c_2 \sin bx.$$

∎

Example 4

Find the particular solution of

$$y'' - 4y' + 5y = 0$$

for which $y(0) = 1$ and $y'(0) = 5$.

Solution The characteristic equation is

$$r^2 - 4r + 5 = (r - 2)^2 + 1 = 0,$$

with roots $2 + i$ and $2 - i$. Hence a general solution is

$$y(x) = e^{2x}(c_1 \cos x + c_2 \sin x).$$

Then
$$y'(x) = 2e^{2x}(c_1 \cos x + c_2 \sin x) + e^{2x}(-c_1 \sin x + c_2 \cos x),$$
so the initial conditions give

$$y(0) = c_1 = 1 \quad \text{and} \quad y'(0) = 2c_1 + c_2 = 5.$$

It follows that $c_2 = 3$, and so the desired particular solution is

$$y(x) = e^{2x}(\cos x + 3 \sin x).$$

∎

Example 5 Find a general solution of $y^{(4)} + 4y = 0$.

Solution The characteristic equation is

$$r^4 + 4 = (r^2 + 2i)(r^2 - 2i) = 0,$$

and its four roots are $\pm\sqrt{\pm 2i}$. Now $i = e^{i\pi/2}$ and $-i = e^{3i\pi/2}$, so

$$\sqrt{i} = \left(e^{i\pi/2}\right)^{1/2} = e^{i\pi/4} = \frac{1+i}{\sqrt{2}}$$

and

$$\sqrt{-i} = \left(e^{3i\pi/2}\right)^{1/2} = e^{3i\pi/4} = \frac{-1+i}{\sqrt{2}}.$$

Thus the four (distinct) roots of the characteristic equation are $r = \pm(1 \pm i)$. These two pairs of complex conjugate roots, $1 \pm i$ and $-1 \pm i$, give a general solution

$$y(x) = e^x(c_1 \cos x + c_2 \sin x) + e^{-x}(c_3 \cos x + c_4 \sin x)$$

of the differential equation $y^{(4)} + 4y = 0$. ∎

In Example 5 we employed the polar form

$$x + iy = re^{i\theta} \tag{22}$$

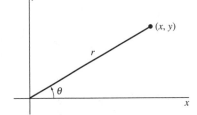

FIGURE 3.3.1. Modulus and argument of the complex number $x + iy$.

of a complex number. The relations between the real and imaginary parts x and y and the **modulus** r and **argument** θ are indicated in Fig. 3.3.1. One consequence of Eq. (22) is that the nonzero complex number $x + iy$ has the two square roots

$$\pm(x + iy)^{1/2} = \pm(re^{i\theta})^{1/2} = \pm r^{1/2}e^{i\theta/2}. \tag{23}$$

Repeated Complex Roots

Theorem 2 holds for repeated complex roots. If the conjugate pair $a \pm bi$ has multiplicity k, then the corresponding part of the general solution has the form

$$(A_1 + A_2x + \cdots + A_kx^{k-1})e^{(a+bi)x} + (B_1 + B_2x + \cdots + B_kx^{k-1})e^{(a-bi)x}$$

$$= \sum_{p=0}^{k-1} x^p e^{ax}(c_i \cos bx + d_i \sin bx). \tag{24}$$

It can be shown that the $2k$ functions

$$x^p e^{ax} \cos bx, \quad x^p e^{ax} \sin bx, \quad 0 \le p \le k - 1$$

that appear in Eq. (24) are linearly independent.

Example 6 Find a general solution of $(D^2 + 6D + 13)^2 y = 0$.

Solution By completing the square, we see that the characteristic equation

$$(r^2 + 6r + 13)^2 = [(r + 3)^2 + 4]^2 = 0$$

has as its roots the conjugate pair $-3 \pm 2i$ of multiplicity $k = 2$. Hence Eq. (24) gives the general solution

$$y(x) = e^{-3x}(c_1 \cos 2x + d_1 \sin 2x) + xe^{-3x}(c_2 \cos 2x + d_2 \sin 2x). \qquad \blacksquare$$

In applications we are seldom presented in advance with a factorization as convenient as the one in Example 6. Often the most difficult part of solving a homogeneous linear equation is finding the roots of its characteristic equation. Example 7 illustrates an approach that may succeed when a root of the characteristic equation can be found by inspection. The project material for this section illustrates other possibilities.

Example 7 The characteristic equation of the differential equation

$$y^{(3)} + y' - 10y = 0$$

is the cubic equation

$$r^3 + r - 10 = 0.$$

By a standard theorem of elementary algebra, the only possible rational roots are the factors ± 1, ± 2, ± 5, and ± 10 of the constant term 10. By trial and error (if not by inspection) we discover the root 2. The factor theorem of elementary algebra implies that $r - 2$ is a factor of $r^3 + r - 10$, and division of the former into the latter produces as quotient the quadratic polynomial

$$r^2 + 2r + 5 = (r + 1)^2 + 4.$$

The roots of this quotient are the complex conjugates $-1 \pm 2i$. The three roots we have found now yield the general solution

$$y(x) = c_1 e^{2x} + e^{-x}(c_2 \cos 2x + c_3 \sin 2x). \qquad \blacksquare$$

Example 8 The roots of the characteristic equation of a certain differential equation are 3, -5, 0, 0, 0, 0, -5, $2 \pm 3i$, and $2 \pm 3i$. Write a general solution of this homogeneous differential equation.

Solution The solution can be read directly from the list of roots. It is

$$y(x) = c_1 + c_2 x + c_3 x^2 + c_4 x^3 + c_5 e^{3x} + c_6 e^{-5x} + c_7 x e^{-5x}$$
$$+ e^{2x}(c_8 \cos 3x + c_9 \sin 3x) + x e^{2x}(c_{10} \cos 3x + c_{11} \sin 3x). \qquad \blacksquare$$

3.3 Problems

Find the general solutions of the differential equations in Problems 1 through 20.

1. $y'' - 4y = 0$
2. $2y'' - 3y' = 0$
3. $y'' + 3y' - 10y = 0$
4. $2y'' - 7y' + 3y = 0$
5. $y'' + 6y' + 9y = 0$
6. $y'' + 5y' + 5y = 0$
7. $4y'' - 12y' + 9y = 0$
8. $y'' - 6y' + 13y = 0$
9. $y'' + 8y' + 25y = 0$
10. $5y^{(4)} + 3y^{(3)} = 0$
11. $y^{(4)} - 8y^{(3)} + 16y'' = 0$
12. $y^{(4)} - 3y^{(3)} + 3y'' - y' = 0$
13. $9y^{(3)} + 12y'' + 4y' = 0$
14. $y^{(4)} + 3y'' - 4y = 0$
15. $y^{(4)} - 8y'' + 16y = 0$
16. $y^{(4)} + 18y'' + 81y = 0$
17. $6y^{(4)} + 11y'' + 4y = 0$
18. $y^{(4)} = 16y$
19. $y^{(3)} + y'' - y' - y = 0$
20. $y^{(4)} + 2y^{(3)} + 3y'' + 2y' + y = 0$ (*Suggestion*: Expand $(r^2 + r + 1)^2$.)

Solve the initial value problems given in Problems 21 through 26.

21. $y'' - 4y' + 3y = 0$; $y(0) = 7$, $y'(0) = 11$
22. $9y'' + 6y' + 4y = 0$; $y(0) = 3$, $y'(0) = 4$
23. $y'' - 6y' + 25y = 0$; $y(0) = 3$, $y'(0) = 1$
24. $2y^{(3)} - 3y'' - 2y' = 0$; $y(0) = 1$, $y'(0) = -1$, $y''(0) = 3$
25. $3y^{(3)} + 2y'' = 0$; $y(0) = -1$, $y'(0) = 0$, $y''(0) = 1$
26. $y^{(3)} + 10y'' + 25y' = 0$; $y(0) = 3$, $y'(0) = 4$, $y''(0) = 5$

Find general solutions of the equations in Problems 27 through 32. First find a small integral root of the characteristic equation by inspection; then factor by division.

27. $y^{(3)} + 3y'' - 4y = 0$
28. $2y^{(3)} - y'' - 5y' - 2y = 0$
29. $y^{(3)} + 27y = 0$
30. $y^{(4)} - y^{(3)} + y'' - 3y' - 6y = 0$
31. $y^{(3)} + 3y'' + 4y' - 8y = 0$
32. $y^{(4)} + y^{(3)} - 3y'' - 5y' - 2y = 0$

In Problems 33 through 36, one solution of the differential equation is given. Find the general solution.

33. $y^{(3)} + 3y'' - 54y = 0$; $y = e^{3x}$
34. $3y^{(3)} - 2y'' + 12y' - 8y = 0$; $y = e^{2x/3}$
35. $6y^{(4)} + 5y^{(3)} + 25y'' + 20y' + 4y = 0$; $y = \cos 2x$
36. $9y^{(3)} + 11y'' + 4y' - 14y = 0$; $y = e^{-x} \sin x$
37. Find a function $y(x)$ such that $y^{(4)}(x) = y^{(3)}(x)$ for all x and $y(0) = 18$, $y'(0) = 12$, $y''(0) = 13$, and $y^{(3)}(0) = 7$.
38. Solve the initial value problem

$$y^{(3)} - 5y'' + 100y' - 500y = 0;$$
$$y(0) = 0, \quad y'(0) = 10, \quad y''(0) = 250$$

given that $y_1(x) = e^{5x}$ is one particular solution of the differential equation.

In Problems 39 through 42, find a linear homogeneous constant-coefficient equation with the given general solution.

39. $y(x) = (A + Bx + Cx^2)e^{2x}$
40. $y(x) = Ae^{2x} + B \cos 2x + C \sin 2x$
41. $y(x) = A \cos 2x + B \sin 2x + C \cosh 2x + D \sinh 2x$
42. $y(x) = (A + Bx + Cx^2) \cos 2x + (D + Ex + Fx^2) \sin 2x$

Problems 43 through 47 pertain to the solution of differential equations with complex coefficients.

43. (a) Use Euler's formula to show that every complex number can be written in the form $re^{i\theta}$, where $r \geq 0$ and $-\pi < \theta \leq \pi$. (b) Express the numbers 4, -2, $3i$, $1 + i$, and $-1 + i\sqrt{3}$ in the form $re^{i\theta}$. (c) The two square roots of $re^{i\theta}$ are $\pm e^{i\theta/2}\sqrt{r}$. Find the square roots of the numbers $2 - 2i\sqrt{3}$ and $-2 + 2i\sqrt{3}$.
44. Use the quadratic formula to solve the following equations. Note in each case that the roots are not complex conjugates.

(a) $x^2 + ix + 2 = 0$ (b) $x^2 - 2ix + 3 = 0$

45. Find a general solution of $y'' - 2iy' + 3y = 0$.
46. Find a general solution of $y'' - iy' + 6y = 0$
47. Find a general solution of $y'' = (-2 + 2i\sqrt{3})y$.
48. Solve the initial value problem

$$y^{(3)} = y; \quad y(0) = 1, \quad y'(0) = y''(0) = 0.$$

(*Suggestion*: Impose the given initial conditions on the general solution

$$y(x) = Ae^x + Be^{\alpha x} + Ce^{\beta x},$$

where α and β are the complex conjugate roots of $r^3 - 1 = 0$, to discover that

$$y(x) = \frac{1}{3}\left(e^x + 2e^{-x/2}\cos\frac{x\sqrt{3}}{2}\right)$$

is a solution.)

49. Solve the initial value problem

$$y^{(4)} = y^{(3)} + y'' + y' + 2y;$$
$$y(0) = y'(0) = y''(0) = 0, \ y^{(3)}(0) = 30.$$

50. The differential equation

$$y'' + (\text{sgn}\,x)y = 0 \tag{25}$$

has the discontinuous coefficient function

$$\text{sgn}\,x = \begin{cases} +1 & \text{if } x > 0, \\ -1 & \text{if } x < 0. \end{cases}$$

Show that Eq. (25) nevertheless has two linearly independent solutions $y_1(x)$ and $y_2(x)$ defined for all x such that

- Each satisfies Eq. (25) at each point $x \neq 0$;
- Each has a continuous derivative at $x = 0$;
- $y_1(0) = y_2'(0) = 1$ and $y_2(0) = y_1'(0) = 0$.

(*Suggestion*: Each $y_i(x)$ will be defined by one formula for $x < 0$ and by another for $x \geqq 0$.) The graphs of these two solutions are shown in Fig. 3.3.2.

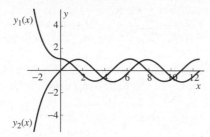

FIGURE 3.3.2. Graphs of $y_1(x)$ and $y_2(x)$ in Problem 50.

51. According to Problem 51 in Section 3.1, the substitution $v = \ln x$ $(x > 0)$ transforms the second-order Euler equation $ax^2 y'' + bxy' + cy = 0$ to a constant-coefficient homogeneous linear equation. Show similarly that this same substitution transforms the third-order Euler equation

$$ax^3 y''' + bx^2 y'' + cxy' + dy = 0$$

(where a, b, c, d are constants) into the constant-coefficient equation

$$a\frac{d^3 y}{dv^3} + (b - 3a)\frac{d^2 y}{dv^2} + (c - b + 2a)\frac{dy}{dv} + dy = 0.$$

Make the substitution $v = \ln x$ of Problem 51 to find general solutions (for $x > 0$) of the Euler equations in Problems 52 through 58.

52. $x^2 y'' + xy' + 9y = 0$
53. $x^2 y'' + 7xy' + 25y = 0$
54. $x^3 y''' + 6x^2 y'' + 4xy' = 0$
55. $x^3 y''' - x^2 y'' + xy' = 0$
56. $x^3 y''' + 3x^2 y'' + xy' = 0$
57. $x^3 y''' - 3x^2 y'' + xy' = 0$
58. $x^3 y''' + 6x^2 y'' + 7xy' + y = 0$

3.3 Application: Approximate Solutions of Linear Equations

To meet the needs of applications such as those of this section, polynomial-solving utilities are now a common feature of calculator and computer systems, and can be used to solve a characteristic equation numerically even when no simple factorization is evident or even possible. For instance, suppose that we want to solve the homogeneous linear differential equation

$$y^{(3)} - 3y'' + y = 0 \tag{1}$$

with characteristic equation

$$r^3 - 3r^2 + 1 = 0. \tag{2}$$

A typical graphing calculator has a `solve` command that can be used to find the approximate roots of a polynomial equation. As indicated in Figs. 3.3.3 and 3.3.4, we find that the roots of Eq. (2) are given by $r \approx -0.5321$, 0.6527, and 2.8794. Some analogous computer algebra system commands are

```
solve(R^3-3R²+1,
R,-1)
              -.5321
solve(R^3-3R²+1,
R,1)
               .6527
solve(R^3-3R²+1,
R,3)
```

FIGURE 3.3.3. Solving the equation $r^3 - 3r^2 + 1 = 0$ with a TI-83 calculator that requires an estimate of each root.

```
F1-  F2-  F3-  F4-  F5   F6-
Tools Algebra Calc Other PrgmIO Clean Up

■ solve(r³ - 3·r² + 1 = 0, r)
   r = 2.8794 or r = .6527 or ▶
solve(r^3-3r^2+1=0,r)
MAIN       RAD AUTO      DE     1/30
```

FIGURE 3.3.4. Solving the characteristic equation with a TI-89 calculator having a more sophisticated `solve` facility.

```
fsolve(r^3 - 3*r^2 + 1 = 0, r);        (Maple)
NSolve[r^3 - 3*r^2 + 1 == 0, r]        (Mathematica)
roots([1   -3   0   1])                (MATLAB)
```

(In the MATLAB command, one enters the polynomial's vector `[1 -3 0 1]` of coefficients, listed in descending order.) However we find these approximate roots, it follows that a general solution of the differential equation in (1) is given (approximately) by

$$y(x) = c_1 e^{-(0.5321)x} + c_2 e^{(0.6527)x} + c_3 e^{(2.8794)x}. \tag{3}$$

Use calculator or computer methods like those indicated here to find general solutions (in approximate numerical form) of the following differential equations.

1. $y^{(3)} - 3y' + y = 0$
2. $y^{(3)} + 3y'' - 3y = 0$
3. $y^{(3)} + y' + y = 0$
4. $y^{(3)} + 3y' + 5y = 0$
5. $y^{(4)} + 2y^{(3)} - 3y = 0$
6. $y^{(4)} + 3y' - 4y = 0$

3.4 Mechanical Vibrations

FIGURE 3.4.1. A mass-spring-dashpot system.

The motion of a mass attached to a spring serves as a relatively simple example of the vibrations that occur in more complex mechanical systems. For many such systems, the analysis of these vibrations is a problem in the solution of linear differential equations with constant coefficients.

We consider a body of mass m attached to one end of an ordinary spring that resists compression as well as stretching; the other end of the spring is attached to a fixed wall, as shown in Fig. 3.4.1. Assume that the body rests on a frictionless horizontal plane, so that it can move only back and forth as the spring compresses and stretches. Denote by x the distance of the body from its **equilibrium position**—its position when the spring is unstretched. We take $x > 0$ when the spring is stretched, and thus $x < 0$ when it is compressed.

According to Hooke's law, the restorative force F_S that the spring exerts on the mass is proportional to the distance x that the spring has been stretched or compressed. Because this is the same as the displacement x of the mass m from its equilibrium position, it follows that

$$F_S = -kx. \tag{1}$$

The positive constant of proportionality k is called the **spring constant.** Note that F_S and x have opposite signs: $F_S < 0$ when $x > 0$, $F_S > 0$ when $x < 0$.

Figure 3.4.1 shows the mass attached to a dashpot—a device, like a shock absorber, that provides a force directed opposite to the instantaneous direction of motion of the mass m. We assume the dashpot is so designed that this force F_R is proportional to the velocity $v = dx/dt$ of the mass; that is,

$$F_R = -cv = -c\frac{dx}{dt}. \tag{2}$$

The positive constant c is the **damping constant** of the dashpot. More generally, we may regard Eq. (2) as specifying frictional forces in our system (including air resistance to the motion of m).

If, in addition to the forces F_S and F_R, the mass is subjected to a given **external force** $F_E = F(t)$, then the total force acting on the mass is $F = F_S + F_R + F_E$. Using Newton's law

$$F = ma = m\frac{d^2x}{dt^2} = mx'',$$

we obtain the second-order linear differential equation

$$\quad mx'' + cx' + kx = F(t) \qquad (3)$$

that governs the motion of the mass.

If there is no dashpot (and we ignore all frictional forces), then we set $c = 0$ in Eq. (3) and call the motion **undamped**; it is **damped** motion if $c > 0$. If there is no external force we replace $F(t)$ with 0 in Eq. (3). We refer to the motion as **free** in this case and **forced** in the case $F(t) \neq 0$. Thus the homogeneous equation

$$\quad mx'' + cx' + kx = 0 \qquad (4)$$

describes free motion of a mass on a spring with dashpot but with no external forces applied. We will defer discussion of forced motion until Section 3.6.

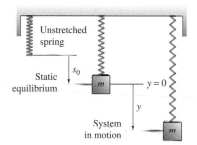

For an alternative example, we might attach the mass to the lower end of a spring that is suspended vertically from a fixed support, as in Fig. 3.4.2. In this case the weight $W = mg$ of the mass would stretch the spring a distance s_0 determined by Eq. (1) with $F_S = -W$ and $x = s_0$. That is, $mg = ks_0$, so that $s_0 = mg/k$. This gives the **static** equilibrium position of the mass. If y denotes the displacement of the mass in motion, measured downward from its static equilibrium position, then we ask you to show in Problem 9 that y satisfies Eq. (3); specifically, that

$$my'' + cy' + ky = F(t) \qquad (5)$$

FIGURE 3.4.2. A mass suspended vertically from a spring.

if we include damping and external forces.

The Simple Pendulum

The importance of the differential equation that appears in Eqs. (3) and (5) stems from the fact that it describes the motion of many other simple mechanical systems. For example, a **simple pendulum** consists of a mass m swinging back and forth on the end of a string (or better, a *massless rod*) of length L, as shown in Fig. 3.4.3. We may specify the position of the mass at time t by giving the counterclockwise angle $\theta = \theta(t)$ that the string or rod makes with the vertical at time t. To analyze the motion of the mass m, we will apply the law of the *conservation of mechanical energy*, according to which the sum of the kinetic energy and the potential energy of m remains constant.

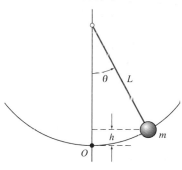

The distance along the circular arc from 0 to m is $s = L\theta$, so the velocity of the mass is $v = ds/dt = L(d\theta/dt)$, and therefore its kinetic energy is

FIGURE 3.4.3. The simple pendulum.

$$T = \frac{1}{2}mv^2 = \frac{1}{2}m\left(\frac{ds}{dt}\right)^2 = \frac{1}{2}mL^2\left(\frac{d\theta}{dt}\right)^2.$$

We next choose as reference point the lowest point O reached by the mass (see Fig. 3.4.3). Then its potential energy V is the product of its weight mg and its vertical height $h = L(1 - \cos\theta)$ above O, so

$$V = mgL(1 - \cos\theta).$$

The fact that the sum of T and V is a constant C therefore gives

$$\frac{1}{2}mL^2\left(\frac{d\theta}{dt}\right)^2 + mgL(1 - \cos\theta) = C.$$

We differentiate both sides of this identity with respect to t to obtain

$$mL^2\left(\frac{d\theta}{dt}\right)\left(\frac{d^2\theta}{dt^2}\right) + mgL(\sin\theta)\frac{d\theta}{dt} = 0,$$

so

$$\frac{d^2\theta}{dt^2} + \frac{g}{L}\sin\theta = 0 \tag{6}$$

after removal of the common factor $mL^2(d\theta/dt)$.

Now recall that $\sin\theta \approx \theta$ when θ is small; in fact, $\sin\theta$ and θ agree to two decimal places when $|\theta|$ is at most $\pi/12$ (that is, $15°$). In a typical pendulum clock, for example, θ would never exceed $15°$. It therefore seems reasonable to simplify our mathematical model of the simple pendulum by replacing $\sin\theta$ with θ in Eq. (6). If we also insert a term $c\theta'$ to account for the frictional resistance of the surrounding medium, the result is an equation in the form of Eq. (4):

$$\theta'' + c\theta' + k\theta = 0, \tag{7}$$

where $k = g/L$. Note that this equation is independent of the mass m on the end of the rod. We might, however, expect the effects of the discrepancy between θ and $\sin\theta$ to accumulate over a period of time, so that Eq. (7) will probably not describe accurately the actual motion of the pendulum over a long period of time.

In the remainder of this section, we first analyze free undamped motion and then free damped motion.

Free Undamped Motion

If we have only a mass on a spring, with neither damping nor external force, then Eq. (3) takes the simpler form

$$mx'' + kx = 0. \tag{8}$$

It is convenient to define

$$\omega_0 = \sqrt{\frac{k}{m}}, \tag{9}$$

and rewrite Eq. (8) as

$$x'' + \omega_0^2 x = 0. \tag{8'}$$

The general solution of Eq. (8′) is

$$x(t) = A\cos\omega_0 t + B\sin\omega_0 t. \tag{10}$$

To analyze the motion described by this solution, we choose constants C and α so that

$$C = \sqrt{A^2 + B^2}, \quad \cos\alpha = \frac{A}{C}, \quad \text{and} \quad \sin\alpha = \frac{B}{C}, \quad (11)$$

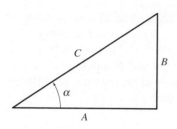

FIGURE 3.4.4. The angle α.

as indicated in Fig. 3.4.4. Note that, although $\tan\alpha = B/A$, the angle α is *not* given by the principal branch of the inverse tangent function (which gives values only in the interval $-\pi/2 < x < \pi/2$). Instead, α is the angle between 0 and 2π whose cosine and sine have the signs given in (11), where either A or B or both may be negative. Thus

$$\alpha = \begin{cases} \tan^{-1}(B/A) & \text{if } A > 0, B > 0 \text{ (first quadrant)}, \\ \pi + \tan^{-1}(B/A) & \text{if } A < 0 \text{ (second or third quadrant)}, \\ 2\pi + \tan^{-1}(B/A) & \text{if } A > 0, B < 0 \text{ (fourth quadrant)}, \end{cases}$$

where $\tan^{-1}(B/A)$ is the angle in $(-\pi/2, \pi/2)$ given by a calculator or computer. In any event, from (10) and (11) we get

$$x(t) = C\left(\frac{A}{C}\cos\omega_0 t + \frac{B}{C}\sin\omega_0 t\right) = C(\cos\alpha\cos\omega_0 t + \sin\alpha\sin\omega_0 t).$$

With the aid of the cosine addition formula, we find that

$$x(t) = C\cos(\omega_0 t - \alpha). \quad (12)$$

Thus the mass oscillates to and fro about its equilibrium position with

1. **Amplitude** C,
2. **Circular frequency** ω_0, and
3. **Phase angle** α.

Such motion is called **simple harmonic motion**.

If time t is measured in seconds, the circular frequency ω_0 has dimensions of radians per second (rad/s). The **period** of the motion is the time required for the system to complete one full oscillation, so is given by

$$T = \frac{2\pi}{\omega_0} \quad (13)$$

seconds; its **frequency** is

$$\nu = \frac{1}{T} = \frac{\omega_0}{2\pi} \quad (14)$$

in hertz (Hz), which measures the number of complete cycles per second. Note that frequency is measured in cycles per second, whereas circular frequency has the dimensions of radians per second.

A typical graph of a simple harmonic position function

$$x(t) = C\cos(\omega_0 t - \alpha) = C\cos\left(\omega_0\left(t - \frac{\alpha}{\omega_0}\right)\right) = C\cos(\omega_0(t - \delta))$$

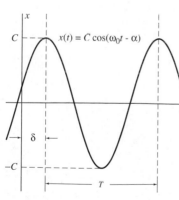

FIGURE 3.4.5. Simple harmonic motion.

is shown in Fig. 3.4.5, where the geometric significance of the amplitude C, the period T, and the **time lag**

$$\delta = \frac{\alpha}{\omega_0}$$

are indicated.

If the initial position $x(0) = x_0$ and initial velocity $x'(0) = v_0$ of the mass are given, we first determine the values of the coefficients A and B in Eq. (10), then find the amplitude C and phase angle α by carrying out the transformation of $x(t)$ to the form in Eq. (12), as indicated previously.

Example 1 A body with mass $m = \frac{1}{2}$ kilogram (kg) is attached to the end of a spring that is stretched 2 meters (m) by a force of 100 newtons (N). It is set in motion with initial position $x_0 = 1$ (m) and initial velocity $v_0 = -5$ (m/s). (Note that these initial conditions indicate that the body is displaced to the right and is moving to the left at time $t = 0$.) Find the position function of the body as well as the amplitude, frequency, period of oscillation, and time lag of its motion.

Solution The spring constant is $k = (100 \text{ N})/(2 \text{ m}) = 50$ (N/m), so Eq. (8) yields $\frac{1}{2}x'' + 50x = 0$; that is,

$$x'' + 100x = 0.$$

Consequently, the circular frequency of the resulting simple harmonic motion of the body will be $\omega_0 = \sqrt{100} = 10$ (rad/s). Hence it will oscillate with period

$$T = \frac{2\pi}{\omega_0} = \frac{2\pi}{10} \approx 0.6283 \text{ s}$$

and with frequency

$$\nu = \frac{1}{T} = \frac{\omega_0}{2\pi} = \frac{10}{2\pi} \approx 1.5915 \text{ Hz.}$$

We now impose the initial conditions $x(0) = 1$ and $x'(0) = -5$ on the position function

$$x(t) = A \cos 10t + B \sin 10t \quad \text{with} \quad x'(t) = -10A \sin 10t + 10B \cos 10t.$$

It follows readily that $A = 1$ and $B = -\frac{1}{2}$, so the position function of the body is

$$x(t) = \cos 10t - \frac{1}{2} \sin 10t.$$

Hence its amplitude of motion is

$$C = \sqrt{(1)^2 + (-\tfrac{1}{2})^2} = \tfrac{1}{2}\sqrt{5} \text{ m.}$$

To find the time lag, we write

$$x(t) = \frac{\sqrt{5}}{2} \left(\frac{2}{\sqrt{5}} \cos 10t - \frac{1}{\sqrt{5}} \sin 10t \right) = \frac{\sqrt{5}}{2} \cos(10t - \alpha),$$

where the phase angle α satisfies

$$\cos \alpha = \frac{2}{\sqrt{5}} > 0 \quad \text{and} \quad \sin \alpha = -\frac{1}{\sqrt{5}} < 0.$$

Hence α is the fourth-quadrant angle

$$\alpha = 2\pi + \tan^{-1}\left(\frac{-1/\sqrt{5}}{2/\sqrt{5}}\right) = 2\pi - \tan^{-1}(\tfrac{1}{2}) \approx 5.8195,$$

and the time lag of the motion is

$$\delta = \frac{\alpha}{\omega_0} \approx 0.5820 \text{ s.}$$

With the amplitude and approximate phase angle shown explicitly, the position function of the body takes the form

$$x(t) \approx \tfrac{1}{2}\sqrt{5}\cos(10t - 5.8195),$$

and its graph is shown in Fig. 3.4.6. ∎

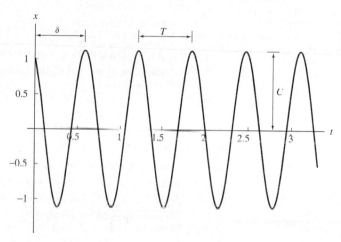

FIGURE 3.4.6. Graph of the position function $x(t) = C\cos(\omega_0 t - \alpha)$ in Example 1, with amplitude $C \approx 1.118$, period $T \approx 0.628$, and time lag $\delta \approx 0.582$.

Free Damped Motion

With damping but no external force, the differential equation we have been studying takes the form $mx'' + cx' + kx = 0$; alternatively,

$$x'' + 2px' + \omega_0^2 x = 0, \tag{15}$$

where $\omega_0 = \sqrt{k/m}$ is the corresponding *undamped* circular frequency and

$$p = \frac{c}{2m} > 0. \tag{16}$$

The characteristic equation $r^2 + 2pr + \omega_0^2 = 0$ of Eq. (15) has roots

$$r_1, \ r_2 = -p \pm (p^2 - \omega_0^2)^{1/2} \tag{17}$$

that depend on the sign of

$$p^2 - \omega_0^2 = \frac{c^2}{4m^2} - \frac{k}{m} = \frac{c^2 - 4km}{4m^2}. \tag{18}$$

The **critical damping** c_{cr} is given by $c_{cr} = \sqrt{4km}$, and we distinguish three cases, according as $c > c_{cr}$, $c = c_{cr}$, or $c < c_{cr}$.

OVERDAMPED CASE: $c > c_{cr}(c^2 > 4km)$. Because c is relatively large in this case, we are dealing with a strong resistance in comparison with a relatively weak spring or a small mass. Then (17) gives distinct real roots r_1 and r_2, both of which are negative. The position function has the form

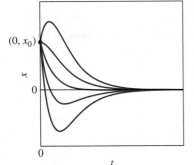

$$x(t) = c_1 e^{r_1 t} + c_2 e^{r_2 t}. \tag{19}$$

FIGURE 3.4.7. Overdamped motion: $x(t) = c_1 e^{r_1 t} + c_2 e^{r_2 t}$ with $r_1 < 0$ and $r_2 < 0$. Solution curves are graphed with the same initial position x_0 and different initial velocities.

It is easy to see that $x(t) \to 0$ as $t \to +\infty$ and that the body settles to its equilibrium position without any oscillations (Problem 29). Figure 3.4.7 shows some typical graphs of the position function for the overdamped case; we chose x_0 a fixed positive number and illustrated the effects of changing the initial velocity v_0. In every case the would-be oscillations are damped out.

CRITICALLY DAMPED CASE: $c = c_{cr}(c^2 = 4km)$. In this case, (17) gives equal roots $r_1 = r_2 = -p$ of the characteristic equation, so the general solution is

$$x(t) = e^{-pt}(c_1 + c_2 t). \tag{20}$$

Because $e^{-pt} > 0$ and $c_1 + c_2 t$ has at most one positive zero, the body passes through its equilibrium position at most once, and it is clear that $x(t) \to 0$ as $t \to +\infty$. Some graphs of the motion in the critically damped case appear in Fig. 3.4.8, and they resemble those of the overdamped case (Fig. 3.4.7). In the critically damped case, the resistance of the dashpot is just large enough to damp out any oscillations, but even a slight reduction in resistance will bring us to the remaining case, the one that shows the most dramatic behavior.

FIGURE 3.4.8. Critically damped motion: $x(t) = (c_1 + c_2 t)e^{-pt}$ with $p > 0$. Solution curves are graphed with the same initial position x_0 and different initial velocities.

UNDERDAMPED CASE: $c < c_{cr}(c^2 < 4km)$. The characteristic equation now has two complex conjugate roots $-p \pm i\sqrt{\omega_0^2 - p^2}$, and the general solution is

$$x(t) = e^{-pt}(A \cos \omega_1 t + B \sin \omega_1 t), \tag{21}$$

where

$$\omega_1 = \sqrt{\omega_0^2 - p^2} = \frac{\sqrt{4km - c^2}}{2m}. \tag{22}$$

Using the cosine addition formula as in the derivation of Eq. (12), we may rewrite Eq. (20) as

$$x(t) = C e^{-pt}\left(\frac{A}{C}\cos \omega_1 t + \frac{B}{C}\sin \omega_1 t\right),$$

so

$$x(t) = C e^{-pt}\cos(\omega_1 t - \alpha) \tag{23}$$

FIGURE 3.4.9. Underdamped oscillations:
$x(t) = Ce^{-pt}\cos(\omega_1 t - \alpha).$

where

$$C = \sqrt{A^2 + B^2}, \quad \cos\alpha = \frac{A}{C}, \quad \text{and} \quad \sin\alpha = \frac{B}{C}.$$

The solution in (22) represents exponentially damped oscillations of the body around its equilibrium position. The graph of $x(t)$ lies between the "amplitude envelope" curves $x = -Ce^{-pt}$ and $x = Ce^{-pt}$ and touches them when $\omega_1 t - \alpha$ is an integral multiple of π. The motion is not actually periodic, but it is nevertheless useful to call ω_1 its **circular frequency** (more properly, its **pseudofrequency**), $T_1 = 2\pi/\omega_1$ its **pseudoperiod** of oscillation, and Ce^{-pt} its **time-varying amplitude.** Most of these quantities are shown in the typical graph of underdamped motion shown in Fig. 3.4.9. Note from Eq. (21) that in this case ω_1 is less than the undamped circular frequency ω_0, so T_1 is larger than the period T of oscillation of the same mass without damping on the same spring. Thus the action of the dashpot has at least two effects:

1. It exponentially damps the oscillations, in accord with the time-varying amplitude.
2. It slows the motion; that is, the dashpot decreases the frequency of the motion.

As the following example illustrates, damping typically also delays the motion further—that is, increases the time lag—as compared with undamped motion with the same initial conditions.

Example 2 The mass and spring of Example 1 are now attached also to a dashpot that provides 1 N of resistance for each meter per second of velocity. The mass is set in motion with the same initial position $x(0) = 1$ and initial velocity $x'(0) = -5$ as in Example 1. Now find the position function of the mass, its new frequency and pseudoperiod of motion, its new time lag, and the times of its first four passages through the initial position $x = 0$.

Solution Rather than memorizing the various formulas given in the preceding discussion, it is better practice in a particular case to set up the differential equation and then solve it directly. Recall that $m = \frac{1}{2}$ and $k = 50$; we are now given $c = 1$ in mks units. Hence Eq. (4) is $\frac{1}{2}x'' + x' + 50x = 0$; that is,

$$x'' + 2x' + 100x = 0.$$

The characteristic equation $r^2 + 2r + 100 = (r+1)^2 + 99 = 0$ has roots r_1, $r_2 = -1 \pm \sqrt{99}\,i$, so the general solution is

$$x(t) = e^{-t}(A\cos\sqrt{99}\,t + B\sin\sqrt{99}\,t). \tag{24}$$

Consequently, the new circular (pseudo)frequency is $\omega_1 = \sqrt{99} \approx 9.9499$ (as compared with $\omega_0 = 10$ in Example 1). The new (pseudo)period and frequency are

$$T_1 = \frac{2\pi}{\omega_1} = \frac{2\pi}{\sqrt{99}} \approx 0.6315 \text{ s}$$

and

$$\nu_1 = \frac{1}{T_1} = \frac{\omega_1}{2\pi} = \frac{\sqrt{99}}{2\pi} \approx 1.5836 \text{ Hz}$$

(as compared with $T \approx 0.6283 < T_1$ and $\nu \approx 1.5915 > \nu_1$ in Example 1).

We now impose the initial conditions $x(0) = 1$ and $x'(0) = -5$ on the position function in (23) and the resulting velocity function

$$x'(t) = -e^{-t}(A \cos \sqrt{99}\,t + B \sin \sqrt{99}\,t) + \sqrt{99}\,e^{-t}(-A \sin \sqrt{99}\,t + B \cos \sqrt{99}\,t).$$

It follows that

$$x(0) = A = 1 \quad \text{and} \quad x'(0) = -A + B\sqrt{99} = -5,$$

whence we find that $A = 1$ and $B = -4/\sqrt{99}$. Thus the new position function of the body is

$$x(t) = e^{-t}\left(\cos \sqrt{99}\,t - \frac{4}{\sqrt{99}} \sin \sqrt{99}\,t\right).$$

Hence its time-varying amplitude of motion is

$$C_1 e^{-t} = \sqrt{(1)^2 + \left(-\frac{4}{\sqrt{99}}\right)^2}\, e^{-t} = \sqrt{\frac{115}{99}}\, e^{-t}.$$

We therefore write

$$x(t) = \frac{\sqrt{115}}{\sqrt{99}} e^{-t}\left(\frac{\sqrt{99}}{\sqrt{115}} \cos \sqrt{99}\,t - \frac{4}{\sqrt{115}} \sin \sqrt{99}\,t\right)$$

$$= \sqrt{\frac{115}{99}} e^{-t} \cos(\sqrt{99}\,t - \alpha_1),$$

where the phase angle α_1 satisfies

$$\cos \alpha_1 = \frac{\sqrt{99}}{\sqrt{115}} > 0 \quad \text{and} \quad \sin \alpha_1 = -\frac{4}{\sqrt{115}} < 0.$$

Hence α_1 is the fourth-quadrant angle

$$\alpha_1 = 2\pi + \tan^{-1}\left(\frac{-4/\sqrt{115}}{\sqrt{99}/\sqrt{115}}\right) = 2\pi - \tan^{-1}\left(\frac{4}{\sqrt{99}}\right) \approx 5.9009,$$

and the time lag of the motion is

$$\delta_1 = \frac{\alpha_1}{\omega_1} \approx 0.5931 \text{ s}$$

(as compared with $\delta \approx 0.5820 < \delta_1$ in Example 1). With the time-varying amplitude and approximate phase angle shown explicitly, the position function of the mass takes the form

$$x(t) \approx \sqrt{\frac{115}{99}} e^{-t} \cos(\sqrt{99}\,t - 5.9009), \tag{25}$$

and its graph is the damped exponential that is shown in Fig. 3.4.10 (in comparison with the undamped oscillations of Example 1.

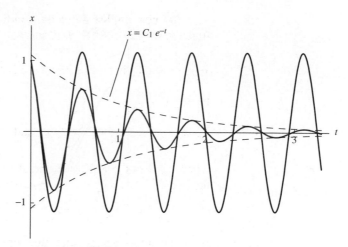

FIGURE 3.4.10. Graphs of the position function $x(t) = C_1 e^{-t} \cos(\omega_1 t - \alpha_1)$ of Example 2 (damped oscillations), the position function $x(t) = C \cos(\omega_0 t - \alpha)$ of Example 1 (undamped oscillations), and the envelope curves $x(t) = \pm C_1 e^{-t}$.

From (24) we see that the mass passes through its equilibrium position $x = 0$ when $\cos(\omega_1 t - \alpha_1) = 0$, and thus when

$$\omega_1 t - \alpha_1 = -\frac{3\pi}{2}, -\frac{\pi}{2}, \frac{\pi}{2}, \frac{3\pi}{2}, \ldots, \quad \text{that is, when}$$

$$t = \delta_1 - \frac{3\pi}{2\omega_1}, \delta_1 - \frac{\pi}{2\omega_1}, \delta_1 + \frac{\pi}{2\omega_1}, \delta_1 + \frac{3\pi}{2\omega_1}, \ldots.$$

We see similarly that the undamped mass of Example 1 passes through equilibrium when

$$t = \delta_0 - \frac{3\pi}{2\omega_0}, \delta_0 - \frac{\pi}{2\omega_0}, \delta_0 + \frac{\pi}{2\omega_0}, \delta_0 + \frac{3\pi}{2\omega_0}, \ldots.$$

The following table compares the first four values t_1, t_2, t_3, t_4 we calculate for the undamped and damped cases, respectively.

n	1	2	3	4
t_n (undamped)	0.1107	0.4249	0.7390	1.0532
t_n (damped)	0.1195	0.4352	0.7509	1.0667

Accordingly, in Fig. 3.4.11 (where only the first three equilibrium passages are shown) we see the damped oscillations lagging slightly behind the undamped ones. ∎

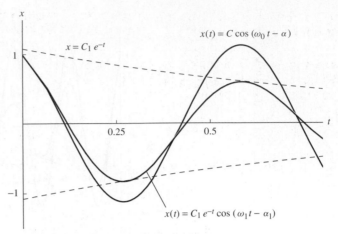

FIGURE 3.4.11. Graphs on the interval $0 \leq t \leq 0.8$ illustrating the additional delay associated with damping.

3.4 Problems

1. Determine the period and frequency of the simple harmonic motion of a 4-kg mass on the end of a spring with spring constant 16 N/m.

2. Determine the period and frequency of the simple harmonic motion of a body of mass 0.75 kg on the end of a spring with spring constant 48 N/m.

3. A mass of 3 kg is attached to the end of a spring that is stretched 20 cm by a force of 15 N. It is set in motion with initial position $x_0 = 0$ and initial velocity $v_0 = -10$ m/s. Find the amplitude, period, and frequency of the resulting motion.

4. A body with mass 250 g is attached to the end of a spring that is stretched 25 cm by a force of 9 N. At time $t = 0$ the body is pulled 1 m to the right, stretching the spring, and set in motion with an initial velocity of 5 m/s to the left. **(a)** Find $x(t)$ in the form $C \cos(\omega_0 t - \alpha)$. **(b)** Find the amplitude and period of motion of the body.

In Problems 5 through 8, assume that the differential equation of a simple pendulum of length L is $L\theta'' + g\theta = 0$, where $g = GM/R^2$ is the gravitational acceleration at the location of the pendulum (at distance R from the center of the earth; M denotes the mass of the earth).

5. Two pendulums are of lengths L_1 and L_2 and—when located at the respective distances R_1 and R_2 from the center of the earth—have periods p_1 and p_2. Show that

$$\frac{p_1}{p_2} = \frac{R_1 \sqrt{L_1}}{R_2 \sqrt{L_2}}.$$

6. A certain pendulum keeps perfect time in Paris, where the radius of the earth is $R = 3956$ (mi). But this clock loses 2 min 40 s per day at a location on the equator. Use the result of Problem 5 to find the amount of the equatorial bulge of the earth.

7. A pendulum of length 100.10 in., located at a point at sea level where the radius of the earth is $R = 3960$ (mi), has the same period as does a pendulum of length 100.00 in. atop a nearby mountain. Use the result of Problem 5 to find the height of the mountain.

8. Most grandfather clocks have pendulums with adjustable lengths. One such clock loses 10 min per day when the length of its pendulum is 30 in. With what length pendulum will this clock keep perfect time?

9. Derive Eq. (5) describing the motion of a mass attached to the bottom of a vertically suspended spring. (*Suggestion:* First denote by $x(t)$ the displacement of the mass below the unstretched position of the spring; set up the differential equation for x. Then substitute $y = x - x_0$ in this differential equation.)

10. Consider a floating cylindrical buoy with radius r, height h, and uniform density $\rho \leq 0.5$ (recall that the density of water is 1 g/cm³). The buoy is initially suspended at rest with its bottom at the top surface of the water and is released at time $t = 0$. Thereafter it is acted on by two forces: a downward gravitational force equal to its weight $mg = \rho \pi r^2 hg$ and an upward force of buoyancy equal to the weight $\pi r^2 xg$ of water displaced, where $x = x(t)$ is the depth of the bottom of the buoy beneath the surface at time t (Fig. 3.4.12). Conclude that the buoy undergoes simple harmonic motion around its equilibrium position $x_e = \rho h$ with period $p = 2\pi \sqrt{\rho h/g}$. Compute p and the amplitude of the motion if $\rho = 0.5$ g/cm³, $h = 200$ cm, and $g = 980$ cm/s².

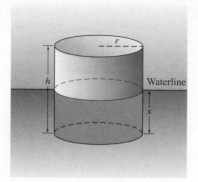

FIGURE 3.4.12. The buoy of Problem 10.

11. A cylindrical buoy weighing 100 lb (thus of mass $m = 3.125$ slugs in ft-lb-s (fps) units) floats in water with its axis vertical (as in Problem 10). When depressed slightly and released, it oscillates up and down four times every 10 s. Assume that friction is negligible. Find the radius of the buoy.

12. Assume that the earth is a solid sphere of uniform density, with mass M and radius $R = 3960$ (mi). For a particle of mass m within the earth at distance r from the center of the earth, the gravitational force attracting m toward the center is $F_r = -GM_r m/r^2$, where M_r is the mass of the part of the earth within a sphere of radius r. **(a)** Show that $F_r = -GMmr/R^3$. **(b)** Now suppose that a small hole is drilled straight through the center of the earth, thus connecting two antipodal points on its surface. Let a particle of mass m be dropped at time $t = 0$ into this hole with initial speed zero, and let $r(t)$ be its distance from the center of the earth at time t (Fig. 3.4.13). Conclude from Newton's second law and part (a) that $r''(t) = -k^2 r(t)$, where $k^2 = GM/R^3 = g/R$.

FIGURE 3.4.13. A mass m falling down a hole through the center of the earth (Problem 12).

(c) Take $g = 32.2$ ft/s^2, and conclude from part (b) that the particle undergoes simple harmonic motion back and forth between the ends of the hole, with a period of about 84 min. **(d)** Look up (or derive) the period of a satellite that just skims the surface of the earth; compare with the result in part (c). How do you explain the coincidence? Or *is* it a coincidence? **(e)** With what speed (in miles

per hour) does the particle pass through the center of the earth? **(f)** Look up (or derive) the orbital velocity of a satellite that just skims the surface of the earth; compare with the result in part (e). How do you explain the coincidence? Or *is* it a coincidence?

13. Suppose that the mass in a mass-spring-dashpot system with $m = 10$, $c = 9$, and $k = 2$ is set in motion with $x(0) = 0$ and $x'(0) = 5$. **(a)** Find the position function $x(t)$ and show that its graph looks as indicated in Fig. 3.4.14. **(b)** Find how far the mass moves to the right before starting back toward the origin.

FIGURE 3.4.14. The position function $x(t)$ of Problem 13.

14. Suppose that the mass in a mass-spring-dashpot system with $m = 25$, $c = 10$, and $k = 226$ is set in motion with $x(0) = 20$ and $x'(0) = 41$. **(a)** Find the position function $x(t)$ and show that its graph looks as indicated in Fig. 3.4.15. **(b)** Find the pseudoperiod of the oscillations and the equations of the "envelope curves" that are dashed in the figure.

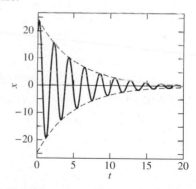

FIGURE 3.4.15. The position function $x(t)$ of Problem 14.

The remaining problems in this section deal with free damped motion. In Problems 15 through 21, a mass m is attached to both a spring (with given spring constant k) and a dashpot (with given damping constant c). The mass is set in motion with initial position x_0 and initial velocity v_0. Find the position function $x(t)$ and determine whether the motion is overdamped, critically damped, or underdamped. If it is underdamped, write the position function in the form $x(t) = C_1 e^{-pt} \cos(\omega_1 t - \alpha_1)$. Also, find the undamped position

function $u(t) = C_0 \cos(\omega_0 t - \alpha_0)$ that would result if the mass on the spring were set in motion with the same initial position and velocity, but with the dashpot disconnected (so $c = 0$). Finally, construct a figure that illustrates the effect of damping by comparing the graphs of $x(t)$ and $u(t)$.

15. $m = \frac{1}{2}, c = 3, k = 4; x_0 = 2, v_0 = 0$
16. $m = 3, c = 30, k = 63; x_0 = 2, v_0 = 2$
17. $m = 1, c = 8, k = 16; x_0 = 5, v_0 = -10$
18. $m = 2, c = 12, k = 50; x_0 = 0, v_0 = -8$
19. $m = 4, c = 20, k = 169; x_0 = 4, v_0 = 16$
20. $m = 2, c = 16, k = 40; x_0 = 5, v_0 = 4$
21. $m = 1, c = 10, k = 125; x_0 = 6, v_0 = 50$
22. A 12-lb weight (mass $m = 0.375$ slugs in fps units) is attached both to a vertically suspended spring that it stretches 6 in. and to a dashpot that provides 3 lb of resistance for every foot per second of velocity. **(a)** If the weight is pulled down 1 ft below its static equilibrium position and then released from rest at time $t = 0$, find its position function $x(t)$. **(b)** Find the frequency, time-varying amplitude, and phase angle of the motion.
23. This problem deals with a highly simplified model of a car of weight 3200 lb (mass $m = 100$ slugs in fps units). Assume that the suspension system acts like a single spring and its shock absorbers like a single dashpot, so that its vertical vibrations satisfy Eq. (4) with appropriate values of the coefficients. **(a)** Find the stiffness coefficient k of the spring if the car undergoes free vibrations at 80 cycles per minute (cycles/min) when its shock absorbers are disconnected. **(b)** With the shock absorbers connected the car is set into vibration by driving it over a bump, and the resulting damped vibrations have a frequency of 78 cycles/min. After how long will the time-varying amplitude be 1% of its initial value?

Problems 24 through 34 deal with a mass-spring-dashpot system having position function $x(t)$ satisfying Eq. (4). We write $x_0 = x(0)$ and $v_0 = x'(0)$ and recall that $p = c/(2m)$, $\omega_0^2 = k/m$, and $\omega_1^2 = \omega_0^2 - p^2$. The system is critically damped, overdamped, or underdamped, as specified in each problem.

24. (Critically damped) Show in this case that
$$x(t) = (x_0 + v_0 t + px_0 t)e^{-pt}.$$

25. (Critically damped) Deduce from Problem 24 that the mass passes through $x = 0$ at some instant $t > 0$ if and only if x_0 and $v_0 + px_0$ have opposite signs.
26. (Critically damped) Deduce from Problem 24 that $x(t)$ has a local maximum or minimum at some instant $t > 0$ if and only if v_0 and $v_0 + px_0$ have the same sign.
27. (Overdamped) Show in this case that
$$x(t) = \frac{1}{2\gamma}\left[(v_0 - r_2 x_0)e^{r_1 t} - (v_0 - r_1 x_0)e^{r_2 t}\right],$$
where $r_1, r_2 = -p \pm \sqrt{p^2 - \omega_0^2}$ and $\gamma = (r_1 - r_2)/2 > 0$.
28. (Overdamped) If $x_0 = 0$, deduce from Problem 27 that
$$x(t) = \frac{v_0}{\gamma}e^{-pt}\sinh \gamma t.$$

29. (Overdamped) Prove that in this case the mass can pass through its equilibrium position $x = 0$ at most once.
30. (Underdamped) Show that in this case
$$x(t) = e^{-pt}\left(x_0 \cos \omega_1 t + \frac{v_0 + px_0}{\omega_1}\sin \omega_1 t\right).$$

31. (Underdamped) If the damping constant c is small in comparison with $\sqrt{8mk}$, apply the binomial series to show that
$$\omega_1 \approx \omega_0\left(1 - \frac{c^2}{8mk}\right).$$

32. (Underdamped) Show that the local maximum and minima of
$$x(t) = Ce^{-pt}\cos(\omega_1 t - \alpha)$$
occur where
$$\tan(\omega_1 t - \alpha) = -\frac{p}{\omega_1}.$$
Conclude that $t_2 - t_1 = 2\pi/\omega_1$ if two consecutive maxima occur at times t_1 and t_2.
33. (Underdamped) Let x_1 and x_2 be two consecutive local maximum values of $x(t)$. Deduce from the result of Problem 32 that
$$\ln \frac{x_1}{x_2} = \frac{2\pi p}{\omega_1}.$$
The constant $\Delta = 2\pi p/\omega_1$ is called the **logarithmic decrement** of the oscillation. Note also that $c = m\omega_1 \Delta/\pi$ because $p = c/(2m)$.

Note: The result of Problem 33 provides an accurate method for measuring the viscosity of a fluid, which is an important parameter in fluid dynamics, but which is not easy to measure directly. According to Stokes's drag law, a spherical body of radius a moving at a (relatively slow) speed through a fluid of viscosity μ experiences a resistive force $F_R = 6\pi \mu a v$. Thus if a spherical mass on a spring is immersed in the fluid and set in motion, this drag resistance damps its oscillations with damping constant $c = 6\pi a\mu$. The frequency ω_1 and logarithmic decrement Δ of the oscillations can be measured by direct observation. The final formula in Problem 33 then gives c and hence the viscosity of the fluid.

34. (Underdamped) A body weighing 100 lb (mass $m = 3.125$ slugs in fps units) is oscillating attached to a spring and a dashpot. Its first two maximum displacements of 6.73 in. and 1.46 in. are observed to occur at times 0.34 s and 1.17 s, respectively. Compute the damping constant (in pound-seconds per foot) and spring constant (in pounds per foot).

Differential Equations and Determinism
Given a mass m, a dashpot constant c, and a spring constant k, Theorem 2 of Section 3.1 implies that the equation
$$mx'' + cx' + kx = 0 \qquad (26)$$

has a unique *solution for* $t \geqq 0$ *satisfying given initial conditions* $x(0) = x_0$, $x'(0) = v_0$. *Thus the future motion of an ideal mass-spring-dashpot system is completely determined by the differential equation and the initial conditions. Of course in a real physical system it is impossible to measure the parameters* m, c, *and* k *precisely. Problems 35 through 38 explore the resulting uncertainty in predicting the future behavior of a physical system.*

35. Suppose that $m = 1$, $c = 2$, and $k = 1$ in Eq. (25). Show that the solution with $x(0) = 0$ and $x'(0) = 1$ is

$$x_1(t) = te^{-t}.$$

36. Suppose that $m = 1$ and $c = 2$ but $k = 1 - 10^{-2n}$. Show that the solution of Eq. (25) with $x(0) = 0$ and $x'(0) = 1$ is

$$x_2(t) = 10^n e^{-t} \sinh 10^{-n}t.$$

37. Suppose that $m = 1$ and $c = 2$ but that $k = 1 + 10^{-2n}$. Show that the solution of Eq. (25) with $x(0) = 0$ and $x'(0) = 1$ is

$$x_3(t) = 10^n e^{-t} \sin 10^{-n}t.$$

38. Whereas the graphs of $x_1(t)$ and $x_2(t)$ resemble those shown in Figs. 3.4.7 and 3.4.8, the graph of $x_3(t)$ exhibits damped oscillations like those illustrated in Fig. 3.4.9, but with a very long pseudoperiod. Nevertheless, show that for each fixed $t > 0$ it is true that

$$\lim_{n \to \infty} x_2(t) = \lim_{n \to \infty} x_3(t) = x_1(t).$$

Conclude that *on a given finite time interval* the three solutions are in "practical" agreement if n is sufficiently large.

3.5 Nonhomogeneous Equations and Undetermined Coefficients

We learned in Section 3.3 how to solve homogeneous linear equations with constant coefficients, but we saw in Section 3.4 that an external force in a simple mechanical system contributes a nonhomogeneous term to its differential equation. The general nonhomogeneous nth-order linear equation with constant coefficients has the form

$$a_n y^{(n)} + a_{n-1} y^{(n-1)} + \cdots + a_1 y' + a_0 y = f(x). \tag{1}$$

By Theorem 5 of Section 3.2, a general solution of Eq. (1) has the form

$$y = y_c + y_p \tag{2}$$

where the complementary function $y_c(x)$ is a general solution of the associated homogeneous equation

$$a_n y^{(n)} + a_{n-1} y^{(n-1)} + \cdots + a_1 y' + a_0 y = 0, \tag{3}$$

and $y_p(x)$ is a particular solution of Eq. (1). Thus our remaining task is to find y_p.

The **method of undetermined coefficients** is a straightforward way of doing this when the given function $f(x)$ in Eq. (1) is sufficiently simple that we can make an intelligent guess as to the general form of y_p. For example, suppose that $f(x)$ is a polynomial of degree m. Then, because the derivatives of a polynomial are themselves polynomials of lower degree, it is reasonable to suspect a particular solution

$$y_p(x) = A_m x^m + A_{m-1} x^{m-1} + \cdots + A_1 x + A_0$$

that is also a polynomial of degree m, but with as yet undetermined coefficients. We may, therefore, substitute this expression for y_p into Eq. (1), and then—by equating coefficients of like powers of x on the two sides of the resulting equation—attempt to determine the coefficients A_0, A_1, \ldots, A_m so that y_p will, indeed, be a particular solution of Eq. (1).

Similarly, suppose that

$$f(x) = a \cos kx + b \sin kx.$$

Then it is reasonable to expect a particular solution of the same form:

$$y_p(x) = A \cos kx + B \sin kx,$$

a linear combination with undetermined coefficients A and B. The reason is that any derivative of such a linear combination of $\cos kx$ and $\sin kx$ has the same form. We may therefore substitute this form of y_p in Eq. (1), and then—by equating coefficients of $\cos kx$ and $\sin kx$ on both sides of the resulting equation—attempt to determine the coefficients A and B so that y_p will, indeed, be a particular solution.

It turns out that this approach does succeed whenever all the derivatives of $f(x)$ have the same form as $f(x)$ itself. Before describing the method in full generality, we illustrate it with several preliminary examples.

Example 1 Find a particular solution of $y'' + 3y' + 4y = 3x + 2$.

Solution Here $f(x) = 3x + 2$ is a polynomial of degree 1, so our guess is that

$$y_p(x) = Ax + B.$$

Then $y_p' = A$ and $y_p'' = 0$, so y_p will satisfy the differential equation provided that

$$(0) + 3(A) + 4(Ax + B) = 3x + 2.$$

This is so if and only if $4A = 3$ and $3A + 4B = 2$. These two equations yield $A = \frac{3}{4}$ and $B = -\frac{1}{16}$, so we have found the particular solution

$$y_p(x) = \frac{3}{4}x - \frac{1}{16}. \qquad \blacksquare$$

Example 2 Find a particular solution of $y'' - 4y = 2e^{3x}$.

Solution Any derivative of e^{3x} is a constant multiple of e^{3x}, so it is reasonable to try

$$y_p(x) = Ae^{3x}.$$

Then $y_p'' = 9Ae^{3x}$, so the given differential equation will be satisfied provided that

$$9Ae^{3x} - 4(Ae^{3x}) = 2e^{3x};$$

that is, $5A = 2$, so that $A = \frac{2}{5}$. Thus our particular solution is $y_p(x) = \frac{2}{5}e^{3x}$. \blacksquare

Example 3 Find a particular solution of $3y'' + y' - 2y = 2 \cos x$.

Solution A first guess might be $y_p(x) = A \cos x$, but the presence of y' on the left-hand side signals that we probably need a term involving $\sin x$ as well. So we try

$$
\begin{aligned}
y_p(x) &= A \cos x + B \sin x; \\
y_p'(x) &= -A \sin x + B \cos x, \\
y_p''(x) &= -A \cos x - B \sin x.
\end{aligned}
$$

Then substitution of y_p and its derivatives into the given differential equation yields

$$3(-A\cos x - B\sin x) + (-A\sin x + B\cos x) - 2(A\cos x + B\sin x) = 2\cos x.$$

We equate the coefficient of $\cos x$ on the left-hand side with that of $\cos x$ on the right and do the same with $\sin x$ to get the equations

$$-5A + B = 2,$$
$$-A - 5B = 0,$$

with solution $A = -\frac{5}{13}$, $B = \frac{1}{13}$. Hence a particular solution is

$$y_p(x) = -\tfrac{5}{13}\cos x + \tfrac{1}{13}\sin x. \qquad \blacksquare$$

The following example, which superficially resembles Example 2, indicates that the method of undetermined coefficients is not always quite so simple as we have made it appear.

Example 4 Find a particular solution of $y'' - 4y = 2e^{2x}$.

Solution If we try $y_p(x) = Ae^{2x}$, we find that

$$y_p'' - 4y_p = 4Ae^{2x} - 4Ae^{2x} = 0 \neq 2e^{2x}.$$

Thus, no matter how A is chosen, Ae^{2x} cannot satisfy the given nonhomogeneous equation. In fact, the preceding computation shows that Ae^{2x} satisfies instead the associated *homogeneous* equation. Therefore, we should begin with a trial function $y_p(x)$ whose derivative involves both e^{2x} *and something else* that can cancel upon substitution into the differential equation to leave the e^{2x} term that we need. A reasonable guess is

$$y_p(x) = Axe^{2x},$$

for which

$$y_p'(x) = Ae^{2x} + 2Axe^{2x} \quad \text{and} \quad y_p''(x) = 4Ae^{2x} + 4Axe^{2x}.$$

Substitution into the original differential equation yields

$$(4Ae^{2x} + 4Axe^{2x}) - 4(Axe^{2x}) = 2e^{2x}.$$

The terms involving xe^{2x} obligingly cancel, leaving only $4Ae^{2x} = 2e^{2x}$, so that $A = \frac{1}{2}$. Consequently, a particular solution is

$$y_p(x) = \tfrac{1}{2}xe^{2x}. \qquad \blacksquare$$

The General Approach

Our initial difficulty in Example 4 resulted from the fact that $f(x) = 2e^{2x}$ satisfies the associated homogeneous equation. Rule 1, given shortly, tells what to do when we do not have this difficulty, and Rule 2 tells what to do when we do have it.

The method of undetermined coefficients applies whenever the function $f(x)$ in Eq. (1) is a linear combination of (finite) products of functions of the following three types:

1. A polynomial in x;
2. An exponential function e^{rx}; (4)
3. $\cos kx$ or $\sin kx$.

Any such function, for example

$$f(x) = (3 - 4x^2)e^{5x} - 4x^3 \cos 10x,$$

has the crucial property that only *finitely* many linearly independent functions appear as terms (summands) in $f(x)$ and its derivatives of all orders. In Rules 1 and 2 we assume that $Ly = f(x)$ is a nonhomogeneous linear equation with constant coefficients and that $f(x)$ is a function of this kind.

RULE 1: Method of Undetermined Coefficients

Suppose that no term appearing either in $f(x)$ or in any of its derivatives satisfies the associated homogeneous equation $Ly = 0$. Then take as a trial solution for y_p a linear combination of all linearly independent such terms and their derivatives. Then determine the coefficients by substitution of this trial solution into the nonhomogeneous equation $Ly = f(x)$. ▲

Note that this rule is not a theorem requiring proof; it is merely a procedure to be followed in searching for a particular solution y_p. If we succeed in finding y_p, then nothing more need be said. (It can be proved, however, that this procedure will always succeed under the conditions specified here.)

In practice we check the supposition made in Rule 1 by first using the characteristic equation to find the complementary function y_c, and then write a list of all the terms appearing in $f(x)$ and its successive derivatives. If none of the terms in this list duplicates a term in y_c, then we proceed with Rule 1.

Example 5 Find a particular solution of

$$y'' + 4y = 3x^3.$$ (5)

Solution The (familiar) complementary solution of Eq. (5) is

$$y_c(x) = c_1 \cos 2x + c_2 \sin 2x.$$

The function $f(x) = 3x^3$ and its derivatives are constant multiples of the linearly independent functions x^3, x^2, x, and 1. Because none of these appears in y_c, we try

$$y_p = Ax^3 + Bx^2 + Cx + D,$$
$$y_p' = 3Ax^2 + 2Bx + C,$$
$$y_p'' = 6Ax + 2B.$$

Substitution in Eq. (5) gives

$$y_p'' + 4y_p = (6Ax + 2B) + 4(Ax^3 + Bx^2 + Cx + D)$$
$$= 4Ax^3 + 4Bx^2 + (6A + 4C)x + (2B + D) = 3x^3.$$

We equate coefficients of like powers of x in the last equation to get

$$4A = 3, \qquad 4B = 0,$$
$$6A + 4C = 0, \qquad 2B + D = 0$$

with solution $A = \frac{3}{4}$, $B = 0$, $C = -\frac{9}{8}$, and $D = 0$. Hence a particular solution of Eq. (5) is

$$y_p(x) = \tfrac{3}{4}x^3 - \tfrac{9}{8}x. \qquad\blacksquare$$

Example 6 Solve the initial value problem

$$y'' - 3y' + 2y = 3e^{-x} - 10\cos 3x;$$
$$y(0) = 1, \quad y'(0) = 2. \tag{6}$$

Solution The characteristic equation $r^2 - 3r + 2 = 0$ has roots $r = 1$ and $r = 2$, so the complementary function is

$$y_c(x) = c_1 e^x + c_2 e^{2x}.$$

The terms involved in $f(x) = 3e^{-x} - 10\cos 3x$ and its derivatives are e^{-x}, $\cos 3x$, and $\sin 3x$. Because none of these appears in y_c, we try

$$y_p = Ae^{-x} + B\cos 3x + C\sin 3x,$$
$$y_p' = -Ae^{-x} - 3B\sin 3x + 3C\cos 3x,$$
$$y_p'' = Ae^{-x} - 9B\cos 3x - 9C\sin 3x.$$

After we substitute these expressions into the differential equation in (6) and collect coefficients, we get

$$y_p'' - 3y_p' + 2y_p = 6Ae^{-x} + (-7B - 9C)\cos 3x + (9B - 7C)\sin 3x$$
$$= 3e^{-x} - 10\cos 3x.$$

We equate the coefficients of the terms involving e^{-x}, those involving $\cos 3x$, and those involving $\sin 3x$. The result is the system

$$6A = 3,$$
$$-7B - 9C = -10,$$
$$9B - 7C = 0$$

with solution $A = \frac{1}{2}$, $B = \frac{7}{13}$, and $C = \frac{9}{13}$. This gives the particular solution

$$y_p(x) = \tfrac{1}{2}e^{-x} + \tfrac{7}{13}\cos 3x + \tfrac{9}{13}\sin 3x,$$

which, however, does not have the required initial values in (6).

To satisfy those initial conditions, we begin with the *general* solution

$$y(x) = y_c(x) + y_p(x)$$
$$= c_1 e^x + c_2 e^{2x} + \tfrac{1}{2} e^{-x} + \tfrac{7}{13} \cos 3x + \tfrac{9}{13} \sin 3x,$$

with derivative

$$y'(x) = c_1 e^x + 2c_2 e^{2x} - \tfrac{1}{2} e^{-x} - \tfrac{21}{13} \sin 3x + \tfrac{27}{13} \cos 3x.$$

The initial conditions in (6) lead to the equations

$$y(0) = c_1 + c_2 + \tfrac{1}{2} + \tfrac{7}{13} = 1,$$
$$y'(0) = c_1 + 2c_2 - \tfrac{1}{2} + \tfrac{27}{13} = 2$$

with solution $c_1 = -\tfrac{1}{2}$, $c_2 = \tfrac{6}{13}$. The desired particular solution is therefore

$$y(x) = -\tfrac{1}{2} e^x + \tfrac{6}{13} e^{2x} + \tfrac{1}{2} e^{-x} + \tfrac{7}{13} \cos 3x + \tfrac{9}{13} \sin 3x. \qquad \blacksquare$$

Example 7 Find the general form of a particular solution of

$$y^{(3)} + 9y' = x \sin x + x^2 e^{2x}. \qquad (7)$$

Solution The characteristic equation $r^3 + 9r = 0$ has roots $r = 0$, $r = -3i$, and $r = 3i$. So the complementary function is

$$y_c(x) = c_1 + c_2 \cos 3x + c_3 \sin 3x.$$

The derivatives of the right-hand side in Eq. (7) involve the terms

$$\cos x, \quad \sin x, \quad x \cos x, \quad x \sin x,$$
$$e^{2x}, \quad x e^{2x}, \quad \text{and} \quad x^2 e^{2x}.$$

Because there is no duplication with the terms of the complementary function, the trial solution takes the form

$$y_p(x) = A \cos x + B \sin x + Cx \cos x + Dx \sin x + E e^{2x} + Fx e^{2x} + Gx^2 e^{2x}.$$

Upon substituting y_p in Eq. (7) and equating coefficients of like terms, we get seven equations determining the seven coefficients A, B, C, D, E, F, and G. \blacksquare

The Case of Duplication

Now we turn our attention to the situation in which Rule 1 does not apply: Some of the terms involved in $f(x)$ and its derivatives satisfy the associated homogeneous equation. For instance, suppose that we want to find a particular solution of the differential equation

$$(D - r)^3 y = (2x - 3)e^{rx}. \tag{8}$$

Proceeding as in Rule 1, out first guess would be

$$y_p(x) = Ae^{rx} + Bxe^{rx}. \tag{9}$$

This form of $y_p(x)$ will not be adequate because the complementary function of Eq. (8) is

$$y_c(x) = c_1 e^{rx} + c_2 x e^{rx} + c_3 x^2 e^{rx}, \tag{10}$$

so substitution of (9) in the left-hand side of (8) would yield zero rather than $(2x - 3)e^{rx}$.

To see how to amend our first guess, we observe that

$$(D - r)^2 [(2x - 3)e^{rx}] = [D^2(2x - 3)]e^{rx} = 0$$

by Eq. (13) of Section 3.3. If $y(x)$ is *any* solution of Eq. (8) and we apply the operator $(D - r)^2$ to both sides, we see that $y(x)$ is also a solution of the equation $(D - r)^5 y = 0$. The general solution of this *homogeneous* equation can be written as

$$y(x) = \underbrace{c_1 e^{rx} + c_2 x e^{rx} + c_3 x^2 e^{rx}}_{y_c} + \underbrace{Ax^3 e^{rx} + Bx^4 e^{rx}}_{y_p}.$$

Thus *every* solution of our original equation in (8) is the sum of a complementary function and a *particular solution* of the form

$$y_p(x) = Ax^3 e^{rx} + Bx^4 e^{rx}. \tag{11}$$

Note that the right-hand side in Eq. (11) can be obtained by multiplying each term of our first guess in (9) by the least positive integral power of x (in this case, x^3) that suffices to eliminate duplication between the terms of the resulting trial solution $y_p(x)$ and the complementary function $y_c(x)$ given in (10). This procedure succeeds in the general case.

To simplify the general statement of Rule 2, we observe that to find a particular solution of the nonhomogeneous linear differential equation

$$Ly = f_1(x) + f_2(x), \tag{12}$$

it suffices to find *separately* particular solutions $Y_1(x)$ and $Y_2(x)$ of the two equations

$$Ly = f_1(x) \quad \text{and} \quad Ly = f_2(x), \tag{13}$$

respectively. For linearity then gives

$$L[Y_1 + Y_2] = LY_1 + LY_2 = f_1(x) + f_2(x),$$

and therefore $y_p = Y_1 + Y_2$ is a particular solution of Eq. (12). (This is a type of "superposition principle" for nonhomogeneous linear equations.)

Now our problem is to find a particular solution of the equation $Ly = f(x)$, where $f(x)$ is a linear combination of products of the elementary functions listed in (4). Thus $f(x)$ can be written as a sum of terms each of the form

$$P_m(x)e^{rx} \cos kx \quad \text{or} \quad P_m(x)e^{rx} \sin kx, \tag{14}$$

where $P_m(x)$ is a polynomial in x of degree m. Note that any derivative of such a term is of the same form but with *both* sines and cosines appearing. The procedure by which we arrived earlier at the particular solution in (11) of Eq. (8) can be generalized to show that the following procedure is always successful.

RULE 2: Method of Undetermined Coefficients

If the function $f(x)$ is of either form in (14), take as the trial solution

$$\begin{aligned} y_p(x) = x^s[&(A_0 + A_1 x + A_2 x^2 + \cdots + A_m x^m)e^{rx} \cos kx \\ &+ (B_0 + B_1 x + B_2 x^2 + \cdots + B_m x^m)e^{rx} \sin kx], \end{aligned} \tag{15}$$

where s is the smallest nonnegative integer such that such that no term in y_p duplicates a term in the complementary function y_c. Then determine the coefficients in Eq. (15) by substituting y_p into the nonhomogeneous equation. ▲

In practice we seldom need to deal with a function $f(x)$ exhibiting the full generality in (14). The table in Fig. 3.5.1 lists the form of y_p in various common cases, corresponding to the possibilities $m = 0$, $r = 0$, and $k = 0$.

On the other hand, it is common to have

$$f(x) = f_1(x) + f_2(x),$$

where $f_1(x)$ and $f_2(x)$ are different functions of the sort listed in the table in Fig. 3.5.1. In this event we take as y_p the sum of the trial solutions for $f_1(x)$ and $f_2(x)$, choosing s *separately* for each part to eliminate duplication with the complementary function. This procedure is illustrated in Examples 8 through 10.

$f(x)$	y_p
$P_m = b_0 + b_1 x + b_2 x^2 + \cdots + b_m x^m$	$x^s(A_0 + A_1 x + A_2 x^2 + \cdots + A_m x^m)$
$a \cos kx + b \sin kx$	$x^s(A \cos kx + B \sin kx)$
$e^{rx}(a \cos kx + b \sin kx)$	$x^s e^{rx}(A \cos kx + B \sin kx)$
$P_m(x)e^{rx}$	$x^s(A_0 + A_1 x + A_2 x^2 + \cdots + A_m x^m)e^{rx}$
$P_m(x)(a \cos kx + b \sin kx)$	$x^s[(A_0 + A_1 x + \cdots + A_m x^m) \cos kx$
	$+ (B_0 + B_1 x + \cdots + B_m x^m) \sin kx]$

FIGURE 3.5.1. Substitutions in the method of undetermined coefficients.

Example 8 Find a particular solution of

$$y^{(3)} + y'' = 3e^x + 4x^2. \tag{16}$$

Solution The characteristic equation $r^3 + r^2 = 0$ has roots $r_1 = r_2 = 0$ and $r_3 = -1$, so the complementary function is

$$y_c(x) = c_1 + c_2 x + c_3 e^{-x}.$$

As a first step toward our particular solution, we form the sum

$$(Ae^x) + (B + Cx + Dx^2).$$

The part Ae^x corresponding to $3e^x$ does not duplicate any part of the complementary function, but the part $B + Cx + Dx^2$ must be multiplied by x^2 to eliminate duplication. Hence we take

$$y_p = Ae^x + Bx^2 + Cx^3 + Dx^4,$$
$$y_p' = Ae^x + 2Bx + 3Cx^2 + 4Dx^3,$$
$$y_p'' = Ae^x + 2B + 6Cx + 12Dx^2, \quad \text{and}$$
$$y_p^{(3)} = Ae^x + 6C + 24Dx.$$

Substitution of these derivatives in Eq. (16) yields

$$2Ae^x + (2B + 6C) + (6C + 24D)x + 12Dx^2 = 3e^x + 4x^2.$$

The system of equations

$$2A = 3, \qquad 2B + 6C = 0,$$
$$6C + 24D = 0, \qquad 12D = 4$$

has the solution $A = \frac{3}{2}$, $B = 4$, $C = -\frac{4}{3}$, and $D = \frac{1}{3}$. Hence the desired particular solution is

$$y_p(x) = \frac{3}{2}e^x + 4x^2 - \frac{4}{3}x^3 + \frac{1}{3}x^4. \qquad \blacksquare$$

Example 9 Determine the appropriate form for a particular solution of

$$y'' + 6y' + 13y = e^{-3x}\cos 2x.$$

Solution The characteristic equation $r^2 + 6r + 13 = 0$ has roots $-3 \pm 2i$, so the complementary function is
$$y_c(x) = e^{-3x}(c_1\cos 2x + c_2\sin 2x).$$

This is the same form as a first attempt $e^{-3x}(A\cos 2x + B\sin 2x)$ at a particular solution, so we must multiply by x to eliminate duplication. Hence we would take

$$y_p(x) = e^{-3x}(Ax\cos 2x + Bx\sin 2x). \qquad \blacksquare$$

Example 10 Determine the appropriate form for a particular solution of the fifth-order equation

$$(D-2)^3(D^2+9)y = x^2 e^{2x} + x\sin 3x.$$

Solution The characteristic equation $(r - 2)^3(r^2 + 9) = 0$ has roots $r = 2, 2, 2, 3i,$ and $-3i$, so the complementary function is

$$y_c(x) = c_1 e^{2x} + c_2 x e^{2x} + c_3 x^2 e^{2x} + c_4 \cos 3x + c_5 \sin 3x.$$

As a first step toward the form of a particular solution, we examine the sum

$$[(A + Bx + Cx^2)e^{2x}] + [(D + Ex)\cos 3x + (F + Gx)\sin 3x].$$

To eliminate duplication with terms of $y_c(x)$, the first part—corresponding to $x^2 e^{2x}$—must be multiplied by x^3, and the second part—corresponding to $x \sin 3x$—must be multiplied by x. Hence we would take

$$y_p(x) = (Ax^3 + Bx^4 + Cx^5)e^{2x} + (Dx + Ex^2)\cos 3x + (Fx + Gx^2)\sin 3x. \quad \blacksquare$$

Variation of Parameters

Finally, let us point out the kind of situation in which the method of undetermined coefficients cannot be used. Consider, for example, the equation

$$y'' + y = \tan x, \tag{17}$$

which at first glance may appear similar to those considered in the preceding examples. Not so; the function $f(x) = \tan x$ has *infinitely many* linearly independent derivatives

$$\sec^2 x, \quad 2\sec^2 x \tan x, \quad 4\sec^2 x \tan^2 x + 2\sec^4 x, \quad \dots .$$

Therefore, we do not have available a *finite* linear combination to use as a trial solution.

We discuss here the method of **variation of parameters**, which—in principle (that is, if the integrals that appear can be evaluated)—can always be used to find a particular solution of the nonhomogeneous linear differential equation

$$y^{(n)} + p_{n-1}(x)y^{(n-1)} + \cdots + p_1(x)y' + p_0(x)y = f(x), \tag{18}$$

provided that we already know the general solution

$$y_c = c_1 y_1 + c_2 y_2 + \cdots + c_n y_n \tag{19}$$

of the associated homogeneous equation

$$y^{(n)} + p_{n-1}(x)y^{(n-1)} + \cdots + p_1(x)y' + p_0(x)y = 0. \tag{20}$$

Here, in brief, is the basic idea of the method of variation of parameters. Suppose that we replace the constants, or *parameters,* c_1, c_2, \dots, c_n in the complementary function in Eq. (19) with *variables*: functions u_1, u_2, \dots, u_n of x. We ask whether it is possible to choose these functions in such a way that the combination

$$y_p(x) = u_1(x)y_1(x) + u_2(x)y_2(x) + \cdots + u_n(x)y_n(x) \tag{21}$$

is a particular solution of the nonhomogeneous equation in (18). It turns out that this *is* always possible.

The method is essentially the same for all orders $n \geq 2$, but we will describe it in detail only for the case $n = 2$. So we begin with the second-order nonhomogeneous equation

$$\blacktriangleright \qquad L[y] = y'' + P(x)y' + Q(x)y = f(x) \qquad (22)$$

with complementary function

$$y_c(x) = c_1 y_1(x) + c_2 y_2(x) \qquad (23)$$

on some open interval I where the functions P and Q are continuous. We want to find functions u_1 and u_2 such that

$$\blacktriangleright \qquad y_p(x) = u_1(x)y_1(x) + u_2(x)y_2(x) \qquad (24)$$

is a particular solution of Eq. (22).

One condition on the two functions u_1 and u_2 is that $L[y_p] = f(x)$. Because two conditions are required to determine two functions, we are free to impose an additional condition of our choice. We will do so in a way that simplifies the computations as much as possible. But first, to impose the condition $L[y_p] = f(x)$, we must compute the derivatives y_p' and y_p''. The product rule gives

$$y_p' = (u_1 y_1' + u_2 y_2') + (u_1' y_1 + u_2' y_2).$$

To avoid the appearance of the second derivatives u_1'' and u_2'', the additional condition that we now impose is that the second sum here must vanish:

$$u_1' y_1 + u_2' y_2 = 0. \qquad (25)$$

Then

$$y_p' = u_1 y_1' + u_2 y_2', \qquad (26)$$

and the product rule gives

$$y_p'' = (u_1 y_1'' + u_2 y_2'') + (u_1' y_1' + u_2' y_2'). \qquad (27)$$

But both y_1 and y_2 satisfy the homogeneous equation

$$y'' + Py' + Qy = 0$$

associated with the nonhomogeneous equation in (22), so

$$y_i'' = -Py_i' - Qy_i \qquad (28)$$

for $i = 1, 2$. It therefore follows from Eq. (27) that

$$y_p'' = (u_1' y_1' + u_2' y_2') - P \cdot (u_1 y_1' + u_2 y_2') - Q \cdot (u_1 y_1 + u_2 y_2).$$

In view of Eqs. (24) and (26), this means that

$$y_p'' = (u_1' y_1' + u_2' y_2') - Py_p' - Qy_p;$$

hence

$$L[y_p] = u_1' y_1' + u_2' y_2'. \tag{29}$$

The requirement that y_p satisfy the nonhomogeneous equation in (22)—that is, that $L[y_p] = f(x)$—therefore implies that

$$u_1' y_1' + u_2' y_2' = f(x). \tag{30}$$

Finally, Eqs. (25) and (30) determine the functions u_1 and u_2 that we need. Collecting these equations, we obtain a system

$$\begin{aligned} u_1' y_1 + u_2' y_2 &= 0, \\ u_1' y_1' + u_2' y_2' &= f(x) \end{aligned} \tag{31}$$

of two linear equations in the two *derivatives* u_1' and u_2'. Note that the determinant of coefficients in (31) is simply the Wronskian $W(y_1, y_2)$. Once we have solved the equations in (31) for the derivatives u_1' and u_2', we integrate each to obtain the functions u_1 and u_2 such that

$$y_p = u_1 y_1 + u_2 y_2 \tag{32}$$

is the desired particular solution of Eq. (22). In Problem 63 we ask you to carry out this process explicitly and thereby verify the formula for $y_p(x)$ in the following theorem.

THEOREM 1: Variation of Parameters

If the nonhomogeneous equation $y'' + P(x)y' + Q(x)y = f(x)$ has complementary function $y_c(x) = c_1 y_1(x) + c_2 y_2(x)$, then a particular solution is given by

$$y_p(x) = -y_1(x) \int \frac{y_2(x) f(x)}{W(x)} \, dx + y_2(x) \int \frac{y_1(x) f(x)}{W(x)} \, dx, \tag{33}$$

where $W = W(y_1, y_2)$ is the Wronskian of the two independent solutions y_1 and y_2 of the associated homogeneous equation. ▲

Example 11 Find a particular solution of the equation $y'' + y = \tan x$.

Solution The complementary function is $y_c(x) = c_1 \cos x + c_2 \sin x$, and we could simply substitute directly in Eq. (33). But it is more instructive to set up the equations in (31) and solve for u_1' and u_2', so we begin with

$$\begin{aligned} y_1 &= \cos x, & y_2 &= \sin x, \\ y_1' &= -\sin x, & y_2' &= \cos x. \end{aligned}$$

Hence the equations in (31) are

$$\begin{aligned} (u_1')(\cos x) + (u_2')(\sin x) &= 0, \\ (u_1')(-\sin x) + (u_2')(\cos x) &= \tan x. \end{aligned}$$

We easily solve these equations for

$$u_1' = -\sin x \tan x = -\frac{\sin^2 x}{\cos x} = \cos x - \sec x,$$

$$u_2' = \cos x \tan x = \sin x.$$

Hence we take

$$u_1 = \int (\cos x - \sec x)\, dx = \sin x - \ln|\sec x + \tan x|$$

and

$$u_2 = \int \sin x\, dx = -\cos x.$$

(Do you see why we choose the constants of integration to be zero?) Thus our particular solution is

$$y_p(x) = u_1(x)y_1(x) + u_2(x)y_2(x)$$
$$= (\sin x - \ln|\sec x + \tan x|)\cos x + (-\cos x)(\sin x);$$

that is,

$$y_p(x) = -(\cos x)\ln|\sec x + \tan x|. \qquad \blacksquare$$

3.5 Problems

In Problems 1 through 20, find a particular solution y_p of the given equation. In all these problems, primes denote derivatives with respect to x.

1. $y'' + 16y = e^{3x}$
2. $y'' - y' - 2y = 3x + 4$
3. $y'' - y' - 6y = 2\sin 3x$
4. $4y'' + 4y' + y = 3xe^x$
5. $y'' + y' + y = \sin^2 x$
6. $2y'' + 4y' + 7y = x^2$
7. $y'' - 4y = \sinh x$
8. $y'' - 4y = \cosh 2x$
9. $y'' + 2y' - 3y = 1 + xe^x$
10. $y'' + 9y = 2\cos 3x + 3\sin 3x$
11. $y^{(3)} + 4y' = 3x - 1$
12. $y^{(3)} + y' = 2 - \sin x$
13. $y'' + 2y' + 5y = e^x \sin x$
14. $y^{(4)} - 2y'' + y = xe^x$
15. $y^{(5)} + 5y^{(4)} - y = 17$
16. $y'' + 9y = 2x^2 e^{3x} + 5$
17. $y'' + y = \sin x + x\cos x$
18. $y^{(4)} - 5y'' + 4y = e^x - xe^{2x}$
19. $y^{(5)} + 2y^{(3)} + 2y'' = 3x^2 - 1$
20. $y^{(3)} - y = e^x + 7$

In Problems 21 through 30, set up the appropriate form of a particular solution y_p, but do not determine the values of the coefficients.

21. $y'' - 2y' + 2y = e^x \sin x$
22. $y^{(5)} - y^{(3)} = e^x + 2x^2 - 5$
23. $y'' + 4y = 3x\cos 2x$
24. $y^{(3)} - y'' - 12y' = x - 2xe^{-3x}$
25. $y'' + 3y' + 2y = x(e^{-x} - e^{-2x})$

26. $y'' - 6y' + 13y = xe^{3x}\sin 2x$
27. $y^{(4)} + 5y'' + 4y = \sin x + \cos 2x$
28. $y^{(4)} + 9y'' = (x^2 + 1)\sin 3x$
29. $(D - 1)^3(D^2 - 4)y = xe^x + e^{2x} + e^{-2x}$
30. $y^{(4)} - 2y'' + y = x^2 \cos x$

Solve the initial value problems in Problems 31 through 40.

31. $y'' + 4y = 2x;\ y(0) = 1,\ y'(0) = 2$
32. $y'' + 3y' + 2y = e^x;\ y(0) = 0,\ y'(0) = 3$
33. $y'' + 9y = \sin 2x;\ y(0) = 1,\ y'(0) = 0$
34. $y'' + y = \cos x;\ y(0) = 1,\ y'(0) = -1$
35. $y'' - 2y' + 2y = x + 1;\ y(0) = 3,\ y'(0) = 0$
36. $y^{(4)} - 4y'' = x^2;\ y(0) = y'(0) = 1,\ y''(0) = y^{(3)}(0) = -1$
37. $y^{(3)} - 2y'' + y' = 1 + xe^x;\ y(0) = y'(0) = 0,\ y''(0) = 1$
38. $y'' + 2y' + 2y = \sin 3x;\ y(0) = 2,\ y'(0) = 0$
39. $y^{(3)} + y'' = x + e^{-x};\ y(0) = 1,\ y'(0) = 0,\ y''(0) = 1$
40. $y^{(4)} - y = 5;\ y(0) = y'(0) = y''(0) = y^{(3)}(0) = 0$
41. Find a particular solution of the equation

$$y^{(4)} - y^{(3)} - y'' - y' - 2y = 8x^5.$$

42. Find the solution of the initial value problem consisting of the differential equation of Problem 41 and the initial conditions

$$y(0) = y'(0) = y''(0) = y^{(3)}(0) = 0.$$

43. (a) Write

$$\cos 3x + i \sin 3x = e^{3ix} = (\cos x + i \sin x)^3$$

by Euler's formula, expand, and equate real and imaginary parts to derive the identities

$$\cos^3 x = \tfrac{3}{4} \cos x + \tfrac{1}{4} \cos 3x,$$
$$\sin^3 x = \tfrac{3}{4} \sin x - \tfrac{1}{4} \sin 3x.$$

(b) Use the result of part (a) to find a general solution of

$$y'' + 4y = \cos^3 x.$$

Use trigonometric identities to find general solutions of the equations in Problems 44 through 46.

44. $y'' + y' + y = \sin x \sin 3x$
45. $y'' + 9y = \sin^4 x$
46. $y'' + y = x \cos^3 x$

In Problems 47 through 56, use the method of variation of parameters to find a particular solution of the given differential equation.

47. $y'' + 3y' + 2y = 4e^x$ **48.** $y'' - 2y' - 8y = 3e^{-2x}$
49. $y'' - 4y' + 4y = 2e^{2x}$ **50.** $y'' - 4y = \sinh 2x$
51. $y'' + 4y = \cos 3x$ **52.** $y'' + 9y = \sin 3x$
53. $y'' + 9y = 2 \sec 3x$ **54.** $y'' + y = \csc^2 x$
55. $y'' + 4y = \sin^2 x$ **56.** $y'' - 4y = xe^x$

57. You can verify by substitution that $y_c = c_1 x + c_2 x^{-1}$ is a complementary function for the nonhomogeneous second-order equation

$$x^2 y'' + xy' - y = 72x^5.$$

But before applying the method of variation of parameters, you must first divide this equation by its leading coefficient x^2 to rewrite it in the standard form

$$y'' + \frac{1}{x} y' - \frac{1}{x^2} y = 72x^3.$$

Thus $f(x) = 72x^3$ in Eq. (22). Now proceed to solve the equations in (31) and thereby derive the particular solution $y_p = 3x^5$.

In Problems 58 through 62, a nonhomogeneous second-order linear equation and a complementary function y_c are given. Apply the method of Problem 57 to find a particular solution of the equation.

58. $x^2 y'' - 4xy' + 6y = x^3$; $y_c = c_1 x^2 + c_2 x^3$
59. $x^2 y'' - 3xy' + 4y = x^4$; $y_c = x^2(c_1 + c_2 \ln x)$
60. $4x^2 y'' - 4xy' + 3y = 8x^{4/3}$; $y_c = c_1 x + c_2 x^{3/4}$
61. $x^2 y'' + xy' + y = \ln x$; $y_c = c_1 \cos(\ln x) + c_2 \sin(\ln x)$
62. $(x^2 - 1)y'' - 2xy' + 2y = x^2 - 1$; $y_c = c_1 x + c_2(1 + x^2)$
63. Carry out the solution process indicated in the text to derive the variation of parameters formula in (33) from Eqs. (31) and (32).
64. Apply the variation of parameters formula in (33) to find the particular solution $y_p(x) = -x \cos x$ of the nonhomogeneous equation $y'' + y = 2 \sin x$.

3.5 Application: Automated Variation of Parameters

The variation of parameters formula in (33) is especially amenable to implementation in a computer algebra system when the indicated integrals would be too tedious or inconvenient for manual evaluation. For example, suppose that we want to find a particular solution of the nonhomogeneous equation

$$y'' + y = \tan x$$

of Example 11, with complementary function $y_c(x) = c_1 \cos x + c_2 \sin x$. Then the *Maple* commands

```
y1 := cos(x):
y2 := sin(x):
f := tan(x):
W := y1*diff(y2,x) - y2*diff(y1,x):
W := simplify(W):
yp := -y1*int(y2*f/W,x) + y2*int(y1*f/W,x):
simplify(yp);
```

implement (33) and produce the result

$$y_p(x) = -(\cos x) \ln \left(\frac{1 + \sin x}{\cos x} \right)$$

equivalent to the result $y_p(x) = -(\cos x)\ln(\sec x + \tan x)$ found in Example 11. The analogous *Mathematica* commands

```
y1 = Cos[x];
y2 = Sin[x];
f = Tan[x];
W = y1*D[y2,x] - y2*D[y1,x]   //   Simplify
yp = -y1*Integrate[y2*f/W,x] + y2*Integrate[y1*f/W,x];
Simplify[yp]
```

produce the result

$$y_p(x) = -(\cos x)\ln\left(\frac{\cos(x/2) + \sin(x/2)}{\cos(x/2) - \sin(x/2)}\right),$$

which (by the usual difference-of-squares technique) also is equivalent to the result found in Example 11.

To solve similarly a second-order linear equation $y'' + P(x)y' + Q(x)y = f(x)$ whose complementary function $y_c(x) = c_1 y_1(x) + c_2 y_2(x)$ is known, we need only insert the corresponding definitions of $y_1(x)$, $y_2(x)$, and $f(x)$ in the initial lines shown here. Find in this way the indicated particular solution $y_p(x)$ of the nonhomogeneous equations in Problems 1 through 6.

1. $y'' + y = 2\sin x$ $y_p(x) = -x\cos x$
2. $y'' + y = 4x\sin x$ $y_p(x) = x\sin x - x^2\cos x$
3. $y'' + y = 12x^2\sin x$ $y_p(x) = 3x^2\sin x + (3x - 2x^3)\cos x$
4. $y'' - 2y' + 2y = 2e^x\sin x$ $y_p(x) = -xe^x\cos x$
5. $y'' - 2y' + 2y = 4xe^x\sin x$ $y_p(x) = e^x(x\sin x - x^2\cos x)$
6. $y'' - 2y' + 2y = 12x^2 e^x\sin x$ $y_p(x) = e^x\left[3x^2\sin x + (3x - 2x^3)\cos x\right]$

3.6 Forced Oscillations and Resonance

In Section 3.4 we derived the differential equation

$$mx'' + cx' + kx = F(t) \tag{1}$$

that governs the one-dimensional motion of a mass m that is attached to a spring (with constant k) and a dashpot (with constant c) and is also acted on by an external force $F(t)$. Machines with rotating components commonly involve mass-spring systems (or their equivalents) in which the external force is simple harmonic:

$$F(t) = F_0\cos\omega t \quad \text{or} \quad F(t) = F_0\sin\omega t, \tag{2}$$

where the constant F_0 is the amplitude of the periodic force and ω is its circular frequency.

For an example of how a rotating machine component can provide a simple harmonic force, consider the cart with a rotating vertical flywheel shown in Fig. 3.6.1. The cart has mass $m - m_0$, not including the flywheel of mass m_0. The centroid of the flywheel is off center at a distance a from its center, and its angular speed is ω radians per second. The cart is attached to a spring (with constant k) as shown. Assume that the centroid of the cart itself is directly beneath the center

FIGURE 3.6.1. The cart-with-flywheel system.

of the flywheel, and denote by $x(t)$ its displacement from its equilibrium position (where the spring is unstretched). Figure 3.6.1 helps us to see that the displacement \bar{x} of the centroid of the combined cart plus flywheel is given by

$$\bar{x} = \frac{(m - m_0)x + m_0(x + a\cos\omega t)}{m} = x + \frac{m_0 a}{m}\cos\omega t.$$

Let us ignore friction and apply Newton's second law $m\bar{x}'' = -kx$, because the force exerted by the spring is $-kx$. We substitute for \bar{x} in the last equation to obtain

$$mx'' - m_0 a\omega^2 \cos\omega t = -kx;$$

that is,

$$mx'' + kx = m_0 a\omega^2 \cos\omega t. \tag{3}$$

Thus the cart with its rotating flywheel acts like a mass on a spring under the influence of a simple harmonic external force with amplitude $F_0 = m_0 a\omega^2$. Such a system is a reasonable model of a front-loading washing machine with the clothes being washed loaded off center. This illustrates the practical importance of analyzing solutions of Eq. (1) with external forces as in (2).

Undamped Forced Oscillations

To study undamped oscillations under the influence of the external force $F(t) = F_0 \cos\omega t$, we set $c = 0$ in Eq. (1), and thereby begin with the equation

➤
$$mx'' + kx = F_0 \cos\omega t \tag{4}$$

whose complementary function is $x_c = c_1 \cos\omega_0 t + c_2 \sin\omega_0 t$. Here,

➤
$$\omega_0 = \sqrt{\frac{k}{m}}$$

is the (circular) **natural frequency** of the mass-spring system. Let us assume initially that the external and natural frequencies are *unequal*: $\omega \neq \omega_0$. We substitute $x_p = A\cos\omega t$ in Eq. (4) to find a particular solution. (No sine term is needed in x_p because there is no term involving x' on the left-hand side in Eq. (4).) This gives

$$-m\omega^2 A\cos\omega t + kA\cos\omega t = F_0 \cos\omega t,$$

so

$$A = \frac{F_0}{k - m\omega^2} = \frac{F_0/m}{\omega_0^2 - \omega^2}, \tag{5}$$

and thus

$$x_p(t) = \frac{F_0/m}{\omega_0^2 - \omega^2}\cos\omega t. \tag{6}$$

Therefore, the general solution $x = x_c + x_p$ is given by

$$x(t) = c_1 \cos\omega_0 t + c_2 \sin\omega_0 t + \frac{F_0/m}{\omega_0^2 - \omega^2}\cos\omega t, \tag{7}$$

where the constants c_1 and c_2 are determined by the initial values $x(0)$ and $x'(0)$. Equivalently, as in Eq. (12) of Section 3.4, we can rewrite Eq. (7) as

$$x(t) = C \cos(\omega_0 t - \alpha) + \frac{F_0/m}{\omega_0^2 - \omega^2} \cos \omega t, \tag{8}$$

so we see that the resulting motion is a superposition of two oscillations, one with natural circular frequency ω_0, the other with the frequency ω of the external force.

Example 1

Suppose that $m = 1$, $k = 9$, $F_0 = 80$, and $\omega = 5$, so the differential equation in (4) is

$$x'' + 9x = 80 \cos 5t.$$

Find $x(t)$ if $x(0) = x'(0) = 0$.

Solution

Here the natural frequency $\omega_0 = 3$ and the frequency $\omega = 5$ of the external force are unequal, as in the preceding discussion. First we substitute $x_p = A \cos 5t$ in the differential equation and find that $-25A + 9A = 80$, so that $A = -5$. Thus a particular solution is

$$x_p(t) = -5 \cos 5t.$$

The complementary function is $x_c = c_1 \cos 3t + c_2 \sin 3t$, so the general solution of the given nonhomogeneous equation is

$$x(t) = c_1 \cos 3t + c_2 \sin 3t - 5 \cos 5t,$$

with derivative

$$x'(t) = -3c_1 \sin 3t + 3c_2 \cos 3t + 25 \sin 5t.$$

The initial conditions $x(0) = 0$ and $x'(0) = 0$ now yield $c_1 = 5$ and $c_2 = 0$, so the desired particular solution is

$$x(t) = 5 \cos 3t - 5 \cos 5t.$$

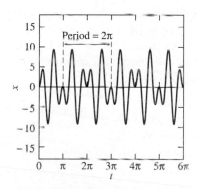

FIGURE 3.6.2. The response $x(t) = 5 \cos 3t - 5 \cos 5t$ in Example 1.

As indicated in Fig. 3.6.2, the period of $x(t)$ is the least common multiple 2π of the periods $2\pi/3$ and $2\pi/5$ of the two cosine terms. ∎

Beats

If we impose the initial conditions $x(0) = x'(0) = 0$ on the solution in (7), we find that

$$c_1 = -\frac{F_0}{m(\omega_0^2 - \omega^2)} \quad \text{and} \quad c_2 = 0,$$

so the particular solution is

$$x(t) = \frac{F_0}{m(\omega_0^2 - \omega^2)}(\cos \omega t - \cos \omega_0 t). \tag{9}$$

The trigonometric identity $2 \sin A \sin B = \cos(A - B) - \cos(A + B)$, applied with $A = \frac{1}{2}(\omega_0 + \omega)t$ and $B = \frac{1}{2}(\omega_0 - \omega)t$, enables us to rewrite Eq. (9) in the form

$$x(t) = \frac{2F_0}{m(\omega_0^2 - \omega^2)} \sin \tfrac{1}{2}(\omega_0 - \omega)t \sin \tfrac{1}{2}(\omega_0 + \omega)t. \tag{10}$$

Suppose now that $\omega \approx \omega_0$, so that $\omega_0 + \omega$ is very large in comparison with $|\omega_0 - \omega|$. Then $\sin \frac{1}{2}(\omega_0 + \omega)t$ is a *rapidly* varying function, whereas $\sin \frac{1}{2}(\omega_0 - \omega)t$ is a *slowly* varying function. We may therefore interpret Eq. (10) as a rapid oscillation with circular frequency $\frac{1}{2}(\omega_0 + \omega)$,

$$x(t) = A(t) \sin \tfrac{1}{2}(\omega_0 + \omega)t,$$

but with a slowly varying amplitude

$$A(t) = \frac{2F_0}{m(\omega_0^2 - \omega^2)} \sin \tfrac{1}{2}(\omega_0 - \omega)t.$$

Example 2 With $m = 0.1$, $F_0 = 50$, $\omega_0 = 55$, and $\omega = 45$, Eq. (10) gives

$$x(t) = \sin 5t \sin 50t.$$

Figure 3.6.3 shows the corresponding oscillation of frequency $\frac{1}{2}(\omega_0 + \omega) = 50$ that is "modulated" by the amplitude function $A(t) = \sin 5t$ of frequency $\frac{1}{2}(\omega_0 - \omega) = 5$. ∎

A rapid oscillation with a (comparatively) slowly varying periodic amplitude exhibits the phenomenon of *beats*. For example, if two horns not exactly attuned to one another simultaneously play their middle C, one at $\omega_0/(2\pi) = 258$ Hz and the other at $\omega/(2\pi) = 254$ Hz, then one hears a beat—an audible variation in the *amplitude* of the combined sound—with a frequency of

$$\frac{(\omega_0 - \omega)/2}{2\pi} = \frac{258 - 254}{2} = 2 \text{ (Hz)}.$$

FIGURE 3.6.3. The phenomenon of beats.

Resonance

Looking at Eq. (6), we see that the amplitude A of x_p is large when the natural and external frequencies ω_0 and ω are approximately equal. It is sometimes useful to rewrite Eq. (5) in the form

$$A = \frac{F_0}{k - m\omega^2} = \frac{F_0/k}{1 - (\omega/\omega_0)^2} = \pm\frac{\rho F_0}{k}, \tag{11}$$

where F_0/k is the **static displacement** of a spring with constant k due to a *constant* force F_0, and the **amplification factor** ρ is defined to be

$$\rho = \frac{1}{|1 - (\omega/\omega_0)^2|}. \tag{12}$$

It is clear that $\rho \to +\infty$ as $\omega \to \omega_0$. This is the phenomenon of **resonance**—the increase without bound (as $\omega \to \omega_0$) in the amplitude of oscillations of an undamped system with natural frequency ω_0 in response to an external force with frequency $\omega \approx \omega_0$.

We have been assuming that $\omega \neq \omega_0$. What sort of catastrophe should one expect if ω and ω_0 are precisely equal? Then Eq. (4), upon division of each term by m, becomes

$$x'' + \omega_0^2 x = \frac{F_0}{m} \cos \omega_0 t. \tag{13}$$

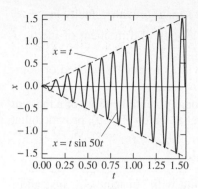

FIGURE 3.6.4. The phenomenon of resonance.

Because $\cos \omega_0 t$ is a term of the complementary function, the method of undetermined coefficients calls for us to try

$$x_p(t) = t(A \cos \omega_0 t + B \sin \omega_0 t).$$

We substitute this in Eq. (13), and thereby find that $A = 0$ and $B = F_0/(2m\omega_0)$. Hence the particular solution is

$$x_p(t) = \frac{F_0}{2m\omega_0} t \sin \omega_0 t. \tag{14}$$

The graph of $x_p(t)$ in Fig. 3.6.4 (in which $m = 1$, $F_0 = 100$, and $\omega_0 = 50$) shows vividly how the amplitude of the oscillation theoretically would increase without bound in this case of *pure resonance,* $\omega = \omega_0$. We may interpret this phenomenon as reinforcement of the natural vibrations of the system by externally impressed vibrations at the same frequency.

Example 3 Suppose that $m = 5$ kg and that $k = 500$ N/m in the cart with the flywheel of Fig. 3.6.1. Then the natural frequency is $\omega_0 = \sqrt{k/m} = 10$ rad/s; that is, $10/(2\pi) \approx 1.59$ Hz. We would therefore expect oscillations of very large amplitude to occur if the flywheel revolves at about $(1.59)(60) \approx 95$ revolutions per minute (rpm). ∎

In practice, a mechanical system with very little damping can be destroyed by resonance vibrations. A spectacular example can occur when a column of soldiers marches in step over a bridge. Any complicated structure such as a bridge has many natural frequencies of vibration. If the frequency of the soldiers' cadence is approximately equal to one of the natural frequencies of the structure, then—just as in our simple example of a mass on a spring—resonance will occur. Indeed, the resulting resonance vibrations can be of such large amplitude that the bridge will collapse. This has actually happened—for example, the collapse of Broughton Bridge near Manchester, England, in 1831—and it is the reason for the now-standard practice of breaking cadence when crossing a bridge. Resonance may have been involved in the 1981 Kansas City disaster in which a hotel balcony (called a *skywalk*) collapsed with dancers on it. The collapse of a building in an earthquake is sometimes due to resonance vibrations caused by the ground oscillating at one of the natural frequencies of the structure; this happened to many buildings in the Mexico City earthquake of September 19, 1985. On occasion an airplane has crashed because of resonant wing oscillations caused by vibrations of the engines. It is reported that for some of the first commercial jet aircraft, the natural frequency of the vertical vibrations of the airplane during turbulence was almost exactly that of the mass-spring system consisting of the pilot's head (mass) and spine (spring). Resonance occurred, causing pilots to have difficulty in reading the instruments. Large modern commercial jets have different natural frequencies, so that this resonance problem no longer occurs.

Modeling Mechanical Systems

The avoidance of destructive resonance vibrations is an ever-present consideration in the design of mechanical structures and systems of all types. Often the most important step in determining the natural frequency of vibration of a system is the formulation of its differential equation. In addition to Newton's law $F = ma$, the principle of conservation of energy is sometimes useful for this purpose (as in the derivation of the pendulum equation in Section 3.4). The following kinetic and potential energy formulas are often useful.

1. *Kinetic energy*: $T = \frac{1}{2}mv^2$ for translation of a mass m with velocity v;

2. *Kinetic energy*: $T = \frac{1}{2}I\omega^2$ for rotation of a body of a moment of inertia I with angular velocity ω;

3. *Potential energy*: $V = \frac{1}{2}kx^2$ for a spring with constant k stretched or compressed a distance x;

4. *Potential energy*: $V = mgh$ for the gravitational potential energy of a mass m at height h above the reference level (the level at which $V = 0$), provided that g may be regarded as essentially constant.

Example 4 Find the natural frequency of a mass m on a spring with constant k if, instead of sliding without friction, it is a uniform disk of radius a that rolls without slipping, as shown in Fig 3.6.5.

Solution With the preceding notation, the principle of conservation of energy gives

$$\tfrac{1}{2}mv^2 + \tfrac{1}{2}I\omega^2 + \tfrac{1}{2}kx^2 = E$$

where E is a constant (the total mechanical energy of the system). We note that $v = a\omega$ and recall that $I = ma^2/2$ for a uniform circular disk. Then we may simplify the last equation to

$$\tfrac{3}{4}mv^2 + \tfrac{1}{2}kx^2 = E.$$

Differentiation ($v = x'$, $v' = x''$) now gives

$$\tfrac{3}{2}mx'x'' + kxx' = 0.$$

FIGURE 3.6.5. The rolling disk.

We divide each term by $\frac{3}{2}mx'$ to obtain

$$x'' + \frac{2k}{3m}x = 0.$$

Thus the natural circular frequency is $\omega_0 = \sqrt{2k/(3m)}$, which is $\sqrt{2/3} \approx 0.8165$ times the frequency in the previous situation of sliding without friction. ∎

Example 5 Suppose that a car oscillates vertically as if it were a mass $m = 800$ kg on a single spring (with constant $k = 7 \times 10^4$ N/m), attached to a single dashpot (with constant $c = 3000$ N·s/m). Suppose that this car with the dashpot *disconnected* is driven along a washboard road surface with an amplitude of 5 cm and a wavelength of $L = 10$ m (Fig. 3.6.6). At what car speed will resonance vibrations occur?

Solution We think of the car as a unicycle, as pictured in Fig. 3.6.7. Let $x(t)$ denote the upward displacement of the mass m from its equilibrium position; we ignore the force of gravity, because it merely displaces the equilibrium position as in Problem 9 of Section 3.4. We write the equation of the road surface as

$$y = a\cos\frac{2\pi s}{L} \qquad (a = 0.05 \text{ m}, L = 10 \text{ m}). \tag{15}$$

FIGURE 3.6.6. The washboard road surface of Example 5.

FIGURE 3.6.7. The "unicycle model" of a car.

When the car is in motion, the spring is stretched by the amount $x - y$, so Newton's second law, $F = ma$, gives

$$mx'' = -k(x - y);$$

that is,

$$mx'' + kx = ky \tag{16}$$

If the velocity of the car is v, then $s = vt$ in Eq. (15), so Eq. (16) takes the form

$$mx'' + kx = ka \cos \frac{2\pi vt}{L}. \tag{16'}$$

This is the differential equation that governs the vertical oscillations of the car. In comparing it with Eq. (4), we see that we have forced oscillations with circular frequency $\omega = 2\pi v/L$. Resonance will occur when $\omega = \omega_0 = \sqrt{k/m}$. We use our numerical data to find the speed of the car at resonance:

$$v = \frac{L}{2\pi} \sqrt{\frac{k}{m}} = \frac{10}{2\pi} \sqrt{\frac{7 \times 10^4}{800}} \approx 14.89 \text{ (m/s)};$$

that is, about 33.3 mi/h (using the conversion factor of 2.237 mi/h per m/s). ∎

Damped Forced Oscillations

In real physical systems there is always some damping, from frictional effects if nothing else. The complementary function x_c of the equation

➤ $$mx'' + cx' + kx = F_0 \cos \omega t \tag{17}$$

is given by Eq. (18), (19), or (20) of Section 3.4, depending on whether $c > c_{cr} = \sqrt{4km}$, $c = c_{cr}$, or $c < c_{cr}$. The specific form is not important here. What is important is that, in any case, these formulas show that $x_c(t) \to 0$ as $t \to +\infty$. Thus x_c is a **transient solution** of Eq. (17)—one that dies out with the passage of time, leaving only the particular solution x_p.

The method of undetermined coefficients indicates that we should substitute

$$x(t) = A \cos \omega t + B \sin \omega t$$

in Eq. (17). When we do, collect terms, and equate coefficients of $\cos \omega t$ and $\sin \omega t$, we obtain the two equations

$$(k - m\omega^2)A + c\omega B = F_0, \qquad -c\omega A + (k - m\omega^2)B = 0 \tag{18}$$

that we solve without difficulty for

$$A = \frac{(k - m\omega^2)F_0}{(k - m\omega^2)^2 + (c\omega)^2}, \qquad B = \frac{c\omega F_0}{(k - m\omega^2)^2 + (c\omega)^2}. \tag{19}$$

If we write

$$A \cos \omega t + B \sin \omega t = C(\cos \omega t \cos \alpha + \sin \omega t \sin \alpha) = C \cos(\omega t - \alpha)$$

as usual, we see that the resulting steady periodic oscillation

$$x_p(t) = C \cos(\omega t - \alpha) \tag{20}$$

has amplitude

$$C = \sqrt{A^2 + B^2} = \frac{F_0}{\sqrt{(k - m\omega^2)^2 + (c\omega)^2}}. \tag{21}$$

Now (19) implies that $\sin \alpha = B/C > 0$, so it follows that the phase angle α lies in the first or second quadrant. Thus

$$\tan \alpha = \frac{B}{A} = \frac{c\omega}{k - m\omega^2} \quad \text{with} \quad 0 < \alpha < \pi, \tag{22}$$

so

$$\alpha = \begin{cases} \tan^{-1} \dfrac{c\omega}{k - m\omega^2} & \text{if } k > m\omega^2, \\[2ex] \pi + \tan^{-1} \dfrac{c\omega}{k - m\omega^2} & \text{if } k < m\omega^2 \end{cases}$$

(whereas $\alpha = \pi/2$ if $k = m\omega^2$).

Note that if $c > 0$, then the "forced amplitude"—defined as a function $C(\omega)$ by (21)—always remains finite, in contrast with the case of resonance in the undamped case when the forcing frequency ω equals the critical frequency $\omega_0 = \sqrt{k/m}$. But the forced amplitude may attain a maximum for some value of ω, in which case we speak of *practical resonance*. To see if and when practical resonance occurs, we need only graph C as a function of ω and look for a global maximum. It can be shown (Problem 27) that C is a steadily decreasing function of ω if $c \geqq \sqrt{2km}$. But if $c < \sqrt{2km}$, then the amplitude of C attains a maximum value—and so practical resonance occurs—at some value of ω less than ω_0, and then approaches zero as $\omega \to +\infty$. It follows that an underdamped system typically will undergo forced oscillations whose amplitude is

- Large if ω is close to the critical resonance frequency;
- Close to F_0/k if ω is very small;
- Very small if ω is very large.

Example 6

Find the transient motion and steady periodic oscillations of a damped mass-and-spring system with $m = 1$, $c = 2$, and $k = 26$ under the influence of an external force $F(t) = 82 \cos 4t$ with $x(0) = 6$ and $x'(0) = 0$. Also investigate the possibility of practical resonance for this system.

Solution

The resulting motion $x(t) = x_{tr}(t) + x_{sp}(t)$ of the mass satisfies the initial value problem

$$x'' + 2x' + 26x = 82 \cos 4t; \quad x(0) = 6, \, x'(0) = 0. \tag{23}$$

Instead of applying the general formulas derived earlier in this section, it is better in a concrete problem to work it directly. The roots of the characteristic equation

$$r^2 + 2r + 26 = (r + 1)^2 + 25 = 0$$

are $r = -1 \pm 5i$, so the complementary function is

$$x_c(t) = e^{-t}(c_1 \cos 5t + c_2 \sin 5t).$$

When we substitute the trial solution

$$x(t) = A \cos 4t + B \sin 4t$$

in the given equation, collect like terms, and equate coefficients of $\cos 4t$ and $\sin 4t$, we get the equations

$$10A + 8B = 82,$$
$$-8A + 10B = 0$$

with solution $A = 5$, $B = 4$. Hence the general solution of the equation in (23) is

$$x(t) = e^{-t}(c_1 \cos 5t + c_2 \sin 5t) + 5 \cos 4t + 4 \sin 4t.$$

At this point we impose the initial conditions $x(0) = 6$, $x'(0) = 0$ and find that $c_1 = 1$ and $c_2 = -3$. Therefore, the transient motion and the steady periodic oscillation of the mass are given by

$$x_{tr}(t) = e^{-t}(\cos 5t - 3 \sin 5t)$$

and

$$x_{sp}(t) = 5 \cos 4t + 4 \sin 4t = \sqrt{41}\left(\frac{5}{\sqrt{41}} \cos 4t + \frac{4}{\sqrt{41}} \sin 4t\right)$$
$$= \sqrt{41} \cos(4t - \alpha)$$

where $\alpha = \tan^{-1}\left(\frac{4}{5}\right) \approx 0.6747$.

Figure 3.6.8 shows graphs of the solution $x(t) = x_{tr}(t) + x_{sp}(t)$ of the initial value problem

$$x'' + 2x' + 26x = 82 \cos 4t, \quad x(0) = x_0, \, x'(0) = 0 \tag{24}$$

for the different values $x_0 = -20, -10, 0, 10$, and 20 of the initial position. Here we see clearly what it means for the transient solution $x_{tr}(t)$ to "die out with the passage of time," leaving only the steady periodic motion $x_{sp}(t)$. Indeed, because

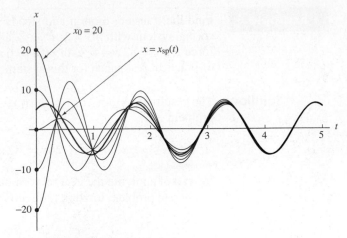

FIGURE 3.6.8. Solutions of the initial value problem in (24) with $x_0 = -20, -10, 0, 10,$ and 20.

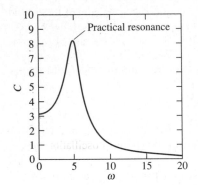

FIGURE 3.6.9. Plot of amplitude C versus external frequency ω.

$x_{\mathrm{tr}}(t) \to 0$ exponentially, within a very few cycles the full solution $x(t)$ and the steady periodic solution $x_{\mathrm{sp}}(t)$ are virtually indistinguishable.

To investigate the possibility of practical resonance in the given system, we substitute the values $m = 1$, $c = 2$, and $k = 26$ in (21) and find that the forced amplitude at frequency ω is

$$C(\omega) = \frac{82}{\sqrt{676 - 48\omega^2 + \omega^4}}.$$

The graph of $C(\omega)$ is shown in Fig. 3.6.9. The maximum amplitude occurs when

$$C'(\omega) = \frac{-41(4\omega^3 - 96\omega)}{(676 - 48\omega^2 + \omega^4)^{3/2}} = \frac{-164\omega(\omega^2 - 24)}{(676 - 48\omega^2 + \omega^4)^{3/2}} = 0.$$

Thus practical resonance occurs when the external frequency is $\omega = \sqrt{24}$ (a bit less than the mass-and-spring's undamped critical frequency of $\omega_0 = \sqrt{k/m} = \sqrt{26}$). ∎

3.6 Problems

In Problems 1 through 6, express the solution of the given initial value problem as a sum of two oscillations as in Eq. (8). Throughout, primes denote derivatives with respect to time t. In Problems 1–4, graph the solution function $x(t)$ in such a way that you can identify and label (as in Fig. 3.6.2) its period.

1. $x'' + 9x = 10\cos 2t$; $x(0) = x'(0) = 0$
2. $x'' + 4x = 5\sin 3t$; $x(0) = x'(0) = 0$
3. $x'' + 100x = 225\cos 5t + 300\sin 5t$; $x(0) = 375$, $x'(0) = 0$
4. $x'' + 25x = 90\cos 4t$; $x(0) = 0, x'(0) = 90$
5. $mx'' + kx = F_0 \cos\omega t$ with $\omega \neq \omega_0$; $x(0) = x_0, x'(0) = 0$
6. $mx'' + kx = F_0 \cos\omega t$ with $\omega = \omega_0$; $x(0) = 0, x'(0) = v_0$

In each of Problems 7 through 10, find the steady periodic solution $x_{\mathrm{sp}}(t) = C\cos(\omega t - \alpha)$ of the given equation $mx'' + cx' + kx = F(t)$ with periodic forcing function $F(t)$ of frequency ω. Then graph $x_{\mathrm{sp}}(t)$ together with (for comparison) the adjusted forcing function $F_1(t) = F(t)/m\omega$.

7. $x'' + 4x' + 4x = 10\cos 3t$
8. $x'' + 3x' + 5x = -4\cos 5t$
9. $2x'' + 2x' + x = 3\sin 10t$
10. $x'' + 3x' + 3x = 8\cos 10t + 6\sin 10t$

In each of Problems 11 through 14, find and plot both the steady periodic solution $x_{\mathrm{sp}}(t) = C\cos(\omega t - \alpha)$ of the given differential equation and the transient solution $x_{\mathrm{tr}}(t)$ that satisfies the given initial conditions.

11. $x'' + 4x' + 5x = 10 \cos 3t$; $x(0) = x'(0) = 0$

12. $x'' + 6x' + 13x = 10 \sin 5t$; $x(0) = x'(0) = 0$

13. $x'' + 2x' + 26x = 600 \cos 10t$; $x(0) = 10, x'(0) = 0$

14. $x'' + 8x' + 25x = 200 \cos t + 520 \sin t$; $x(0) = -30$, $x'(0) = -10$

Each of Problems 15 through 18 gives the parameters for a forced mass-spring-dashpot system with equation $mx'' + cx' + kx = F_0 \cos \omega t$. Investigate the possibility of practical resonance of this system. In particular, find the amplitude $C(\omega)$ of steady periodic forced oscillations with frequency ω. Sketch the graph of $C(\omega)$ and find the practical resonance frequency ω (if any).

15. $m = 1, c = 2, k = 2, F_0 = 2$

16. $m = 1, c = 4, k = 5, F_0 = 10$

17. $m = 1, c = 6, k = 45, F_0 = 50$

18. $m = 1, c = 10, k = 650, F_0 = 100$

19. A mass weighing 100 lb (mass $m = 3.125$ slugs in fps units) is attached to the end of a spring that is stretched 1 in. by a force of 100 lb. A force $F_0 \cos \omega t$ acts on the mass. At what frequency (in hertz) will resonance oscillations occur? Neglect damping.

20. A front-loading washing machine is mounted on a thick rubber pad that acts like a spring; the weight $W = mg$ (with $g = 9.8$ m/s^2) of the machine depresses the pad exactly 0.5 cm. When its rotor spins at ω radians per second, the rotor exerts a vertical force $F_0 \cos \omega t$ newtons on the machine. At what speed (in revolutions per minute) will resonance vibrations occur? Neglect friction.

21. Figure 3.6.10 shows a mass m on the end of a pendulum (of length L) also attached to a horizontal spring (with constant k). Assume small oscillations of m so that the spring remains essentially horizontal and neglect damping. Find the natural circular frequency ω_0 of motion of the mass in terms of L, k, m, and the gravitational constant g.

FIGURE 3.6.10. The pendulum-and-spring system of Problem 21.

22. A mass m hangs on the end of a cord around a pulley of radius a and moment of inertia I, as shown in Fig. 3.6.11. The rim of the pulley is attached to a spring (with constant k). Assume small oscillations so that the spring remains essentially horizontal and neglect friction. Find the natural circular frequency of the system in terms of m, a, k, I, and g.

FIGURE 3.6.11. The mass-spring-pulley system of Problem 22.

23. A building consists of two floors. The first floor is attached rigidly to the ground, and the second floor is of mass $m = 1000$ slugs (fps units) and weighs 16 tons (32,000 lb). The elastic frame of the building behaves as a spring that resists horizontal displacements of the second floor; it requires a horizontal force of 5 tons to displace the second floor a distance of 1 ft. Assume that in an earthquake the ground oscillates horizontally with amplitude A_0 and circular frequency ω, resulting in an external horizontal force $F(t) = mA_0\omega^2 \sin \omega t$ on the second floor. **(a)** What is the natural frequency (in hertz) of oscillations of the second floor? **(b)** If the ground undergoes one oscillation every 2.25 s with an amplitude of 3 in., what is the amplitude of the resulting forced oscillations of the second floor?

24. A mass on a spring without damping is acted on by the external force $F(t) = F_0 \cos^3 \omega t$. Show that there are *two* values of ω for which resonance occurs, and find both.

25. Derive the steady periodic solution of

$$mx'' + cx' + kx = F_0 \sin \omega t.$$

In particular, show that it is what one would expect—the same as the formula in (20) with the same values of C and ω, except with $\sin(\omega t - \alpha)$ in place of $\cos(\omega t - \alpha)$.

26. Given the differential equation

$$mx'' + cx' + kx = E_0 \cos \omega t + F_0 \sin \omega t$$

—with both cosine and sine forcing terms—derive the steady periodic solution

$$x_{\text{sp}}(t) = \frac{\sqrt{E_0^2 + F_0^2}}{\sqrt{(k - m\omega^2)^2 + (c\omega)^2}} \cos(\omega t - \alpha - \beta),$$

where α is defined in Eq. (22) and $\beta = \tan^{-1}(F_0/E_0)$. (*Suggestion:* Add the steady periodic solutions separately corresponding to $E_0 \cos \omega t$ and $F_0 \sin \omega t$ (see Problem 25).)

27. According to Eq. (21), the amplitude of forced steady periodic oscillations for the system $mx'' + cx' + kx = F_0 \cos \omega t$ is given by

$$C(\omega) = \frac{F_0}{\sqrt{(k - m\omega^2)^2 + (c\omega)^2}}.$$

(a) If $c \geq c_{cr}/\sqrt{2}$, where $c_{cr} = \sqrt{4km}$, show that C steadily decreases as ω increases. **(b)** If $c < c_{cr}/\sqrt{2}$, show that C attains a maximum value (practical resonance) when

$$\omega = \omega_m = \sqrt{\frac{k}{m} - \frac{c^2}{2m^2}} < \omega_0 = \sqrt{\frac{k}{m}}.$$

28. As indicated by the cart-with-flywheel example discussed in this section, an unbalanced rotating machine part typically results in a force having amplitude proportional to the *square* of the frequency ω. **(a)** Show that the amplitude of the steady periodic solution of the differential equation

$$mx'' + cx' + kx = mA\omega^2 \cos \omega t$$

(with a forcing term similar to that in Eq. (17)) is given by

$$C(\omega) = \frac{mA\omega^2}{\sqrt{(k - m\omega^2)^2 + (c\omega)^2}}.$$

(b) Suppose that $c^2 < 2mk$. Show that the maximum amplitude occurs at the frequency ω_m given by

$$\omega_m = \sqrt{\frac{k}{m}\left(\frac{2mk}{2mk - c^2}\right)}.$$

Thus the resonance frequency in this case is *larger* (in contrast with the result of Problem 27) than the natural frequency $\omega_0 = \sqrt{k/m}$. (*Suggestion*: Maximize the *square* of C.)

Automobile Vibrations

Problems 29 and 30 deal further with the car of Example 5. Its upward displacement function satisfies the equation

$mx'' + cx' + kx = cy' + ky$ *when the shock absorber is connected (so that $c > 0$). With $y = a \sin \omega t$ for the road surface, this differential equation becomes*

$$mx'' + cx' + kx = E_0 \cos \omega t + F_0 \sin \omega t$$

where $E_0 = c\omega a$ and $F_0 = ka$.

29. Apply the result of Problem 26 to show that the amplitude C of the resulting steady periodic oscillation for the car is given by

$$C = \frac{a\sqrt{k^2 + (c\omega)^2}}{\sqrt{(k - m\omega^2)^2 + (c\omega)^2}}.$$

Because $\omega = 2\pi v/L$ when the car is moving with velocity v, this gives C as a function of v.

30. Figure 3.6.12 shows the graph of the amplitude function $C(\omega)$ using the numerical data given in Example 5 (including $c = 3000$ N·s/m). It indicates that, as the car accelerates gradually from rest, it initially oscillates with amplitude slightly over 5 cm. Maximum resonance vibrations with amplitude about 14 cm occur around 32 mi/h, but then subside to more tolerable levels at high speeds. Verify these graphically based conclusions by analyzing the function $C(\omega)$. In particular, find the practical resonance frequency and the corresponding amplitude.

FIGURE 3.6.12. Amplitude of vibrations of the car on a washboard surface.

3.6 Application: Forced Vibrations

Here we investigate forced vibrations of the mass-spring-dashpot system with equation

$$mx'' + cx' + kx = F(t). \tag{1}$$

To simplify the notation, let's take $m = p^2$, $c = 2p$, and $k = p^2q^2 + 1$ where $p > 0$ and $q > 0$. Then the complementary function of Eq. (1) is

$$x_c(t) = e^{-t/p}(c_1 \cos qt + c_2 \sin qt). \tag{2}$$

We will take $p = 5$, $q = 3$, and thus investigate the transient and steady periodic solutions corresponding to

$$25x'' + 10x' + 226x = F(t), \quad x(0) = 0, \quad x'(0) = 0 \tag{3}$$

with several illustrative possibilities for the external force $F(t)$. For your personal investigations to carry out similarly, you might select integers p and q with $6 \leq p \leq 9$ and $2 \leq q \leq 5$.

INVESTIGATION 1: With periodic external force $F(t) = 901 \cos 3t$, the MAT-LAB commands

```
x = dsolve('25*D2x+10*Dx+226*x=901*cos(3*t)',
    'x(0)=0, Dx(0)=0');
x = simple(x);
syms t, xsp = cos(3*t) + 30*sin(3*t);
ezplot(x, [0  6*pi]),hold on
ezplot(xsp, [0  6*pi])
```

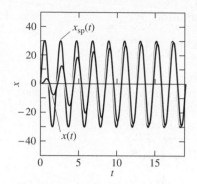

FIGURE 3.6.13. The solution $x(t) = x_{\mathrm{tr}}(t) + x_{\mathrm{sp}}(t)$ and the steady periodic solution $x(t) = x_{\mathrm{sp}}(t)$ with periodic external force $F(t) = 901 \cos 3t$.

produce the plot shown in Fig. 3.6.13. We see the (transient plus steady periodic) solution

$$x(t) = \cos 3t + 30 \sin 3t + e^{-t/5}\left(-\cos 3t - \tfrac{151}{15} \sin 3t\right)$$

rapidly "building up" to the steady periodic oscillation $x_{\mathrm{sp}}(t) = \cos 3t + 30 \sin 3t$.

INVESTIGATION 2: With damped oscillatory external force

$$F(t) = 900e^{-t/5} \cos 3t,$$

we have duplication with the complementary function in (2). The *Maple* commands

```
de2 := 25*diff(x(t),t,t)+10*diff(x(t),t)+226*x(t) =
    900*exp(-t/5)*cos(3*t);
dsolve({de2,x(0)=0,D(x)(0)=0}, x(t));
x := simplify(combine(rhs(%),trig));
C := 6*t*exp(-t/5);
plot({x,C,-C},t=0..8*Pi);
```

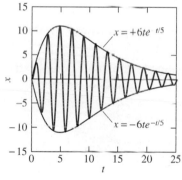

FIGURE 3.6.14. The solution $x(t) = 6te^{-t/5} \sin 3t$ and the envelope curves $x(t) = \pm 6te^{-t/5}$ with damped oscillatory force $F(t) = 900e^{-t/5} \cos 3t$.

produce the plot shown in Fig. 3.6.14. We see the solution

$$x(t) = 6te^{-t/5} \sin 3t$$

oscillating up-and-down between the envelope curves $x = \pm 6te^{-t/5}$. (Note the factor of t that signals a resonance situation.)

INVESTIGATION 3: With damped oscillatory external force

$$F(t) = 2700te^{-t/5} \cos 3t,$$

we have a still more complicated resonance situation. The *Mathematica* commands

```
de3 = 25 x''[t] + 10 x'[t] + 226 x[t] ==
    2700 t Exp[-t/5] Cos[3t]
soln = DSolve[{de3, x[0] == 0, x'[0] == 0}, x[t], t]
x = First[x[t]  /.  soln]
```

```
amp = Exp[-t/5] Sqrt[(3t)^2 + (9t^2 - 1)^2]
Plot[{x, amp, -amp}, {t, 0, 10 Pi}];
```

produce the plot shown in Fig. 3.6.15. We see the solution

$$x(t) = e^{-t/5}\left[3t\cos t + (9t^2 - 1)\sin 3t\right]$$

oscillating up-and-down between the envelope curves

$$x = \pm e^{-t/5}\sqrt{(3t)^2 + (9t^2 + 1)^2}.$$

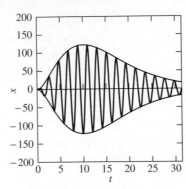

FIGURE 3.6.15. The solution $x(t) = e^{-t/5}[3t\cos t + (9t^2 - 1)\sin 3t]$ and the envelope curves $x(t) = \pm e^{-t/5}\sqrt{(3t)^2 + (9t^2 + 1)^2}$ with external force $F(t) = 2700te^{-t/5}\cos 3t$

3.7 Electrical Circuits

FIGURE 3.7.1. The series RLC circuit.

Circuit Element	Voltage Drop
Inductor	$L\dfrac{dI}{dt}$
Resistor	RI
Capacitor	$\dfrac{1}{C}Q$

FIGURE 3.7.2. Table of voltage drops.

Here we examine the RLC circuit that is a basic building block in more complicated electrical circuits and networks. As shown in Fig. 3.7.1, it consists of

A **resistor** with a resistance of R *ohms*,

An **inductor** with an inductance of L *henries*, and

A **capacitor** with a capacitance of C *farads*

in series with a source of electromotive force (such as a battery or a generator) that supplies a voltage of $E(t)$ *volts* at time t. If the switch shown in the circuit of Fig. 3.7.1 is closed, this results in a current of $I(t)$ *amperes* in the circuit and a charge of $Q(t)$ *coulombs* on the capacitor at time t. The relation between the functions Q and I is

$$\frac{dQ}{dt} = I(t). \tag{1}$$

We will always use mks electric units, in which time is measured in seconds.

According to elementary principles of electricity, the **voltage drops** across the three circuit elements are those shown in the table in Fig. 3.7.2. We can analyze the behavior of the series circuit of Fig. 3.7.1 with the aid of this table and one of Kirchhoff's laws:

The (algebraic) sum of the voltage drops across the elements in a simple loop of an electrical circuit is equal to the applied voltage.

As a consequence, the current and charge in the simple RLC circuit of Fig. 3.7.1 satisfy the basic circuit equation

$$L\frac{dI}{dt} + RI + \frac{1}{C}Q = E(t). \tag{2}$$

If we substitute (1) in Eq. (2), we get the second-order linear differential equation

$$LQ'' + RQ' + \frac{1}{C}Q = E(t) \tag{3}$$

for the charge $Q(t)$, under the assumption that the voltage $E(t)$ is known.

In most practical problems it is the current I rather than the charge Q that is of primary interest, so we differentiate both sides of Eq. (3) and substitute I for Q' to obtain

$$LI'' + RI' + \frac{1}{C}I = E'(t). \tag{4}$$

We do *not* assume here a prior familiarity with electrical circuits. It suffices to regard the resistor, inductor, and capacitor in an electrical circuit as "black boxes" that are calibrated by the constants R, L, and C. A battery or generator is described by the voltage $E(t)$ that it supplies. When the switch is open, no current flows in the circuit; when the switch is closed, there is a current $I(t)$ in the circuit and a charge $Q(t)$ on the capacitor. All we need to know about these constants and functions is that they satisfy Eqs. (1) through (4), our mathematical model for the RLC circuit. We can then learn a good deal about electricity by studying this mathematical model.

The Mechanical-Electrical Analogy

It is striking that Eqs. (3) and (4) have precisely the same form as the equation

$$mx'' + cx' + kx = F(t) \tag{5}$$

of a mass-spring-dashpot system with external force $F(t)$. The table in Fig. 3.7.3 details this important **mechanical-electrical analogy.** As a consequence, most of the results derived in Section 3.6 for mechanical systems can be applied at once to electrical circuits. The fact that the same differential equation serves as a mathematical model for such different physical systems is a powerful illustration of the unifying role of mathematics in the investigation of natural phenomena. More concretely, the correspondences in Fig. 3.7.3 can be used to construct an electrical model of a given mechanical system, using inexpensive and readily available circuit elements. The performance of the mechanical system can then be predicted by means of accurate

Mechanical System	Electrical System
Mass m	Inductance L
Damping constant c	Resistance R
Spring constant k	Reciprocal capacitance $1/C$
Position x	Charge Q (using (3) (or current I using (4)))
Force F	Electromotive force E (or its derivative E')

FIGURE 3.7.3. Mechanical-electrical analogies.

but simple measurements in the electrical model. This is especially useful when the actual mechanical system would be expensive to construct or when measurements of displacements and velocities would be inconvenient, inaccurate, or even dangerous. This idea is the basis of *analog computers*—electrical models of mechanical systems. Analog computers modeled the first nuclear reactors for commercial power and submarine propulsion before the reactors themselves were built.

In the typical case of an alternating current voltage $E(t) = E_0 \sin \omega t$, Eq. (4) takes the form

$$LI'' + RI' + \frac{1}{C}I = \omega E_0 \cos \omega t. \tag{6}$$

As in a mass-spring-dashpot system with a simple harmonic external force, the solution of Eq. (6) is the sum of a **transient current** I_{tr} that approaches zero as $t \to +\infty$ (under the assumption that the coefficients in Eq. (6) are all positive, so the roots of the characteristic equation have negative real parts), and a **steady periodic current** I_{sp}; thus

$$I = I_{tr} + I_{sp}. \tag{7}$$

Recall from Section 3.6 (Eqs. (19) through (21) there) that the steady periodic solution of Eq. (5) with $F(t) = F_0 \cos \omega t$ is

$$x_{sp}(t) = \frac{F_0 \cos(\omega t - \alpha)}{\sqrt{(k - m\omega^2)^2 + (c\omega)^2}},$$

where

$$\alpha = \tan^{-1} \frac{c\omega}{k - m\omega^2}, \quad 0 \leqq \alpha \leqq \pi.$$

If we make the substitutions L for m, R for c, $1/C$ for k, and ωE_0 for F_0, we get the steady periodic current

$$I_{sp}(t) = \frac{E_0 \cos(\omega t - \alpha)}{\sqrt{R^2 + \left(\omega L - \frac{1}{\omega C}\right)^2}} \tag{8}$$

with the phase angle

$$\alpha = \tan^{-1} \frac{\omega R C}{1 - L C \omega^2}, \quad 0 \leqq \alpha \leqq \pi. \tag{9}$$

Reactance and Impedance

The quantity in the denominator in (8),

$$Z = \sqrt{R^2 + \left(\omega L - \frac{1}{\omega C}\right)^2} \quad \text{(ohms)}, \tag{10}$$

is called the **impedance** of the circuit. Then the steady periodic current

$$I_{sp}(t) = \frac{E_0}{Z} \cos(\omega t - \alpha) \tag{11}$$

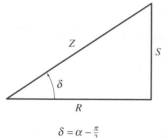

$$\delta = \alpha - \tfrac{\pi}{2}$$

FIGURE 3.7.4. Reactance and delay angle.

has amplitude

$$I_0 = \frac{E_0}{Z}, \tag{12}$$

reminiscent of Ohm's law $I = E/R$.

Equation (11) gives the steady periodic current as a cosine function, whereas the input voltage $E(t) = E_0 \sin \omega t$ was a sine function. To convert I_{sp} to a sine function, we first introduce the **reactance**

$$S = \omega L - \frac{1}{\omega C}. \tag{13}$$

Then $Z = \sqrt{R^2 + S^2}$, and we see from Eq. (9) that α is as in Fig. 3.7.4, with delay angle $\delta = \alpha - \frac{1}{2}\pi$. Equation (11) now yields

$$
\begin{aligned}
I_{sp}(t) &= \frac{E_0}{Z}(\cos \alpha \cos \omega t + \sin \alpha \sin \omega t) \\
&= \frac{E_0}{Z}\left(-\frac{S}{Z}\cos \omega t + \frac{R}{Z}\sin \omega t\right) \\
&= \frac{E_0}{Z}(\cos \delta \sin \omega t - \sin \delta \cos \omega t)
\end{aligned}
$$

Therefore,

$$I_{sp}(t) = \frac{E_0}{Z}\sin(\omega t - \delta), \tag{14}$$

where

$$\delta = \tan^{-1}\frac{S}{R} = \tan^{-1}\frac{LC\omega^2 - 1}{\omega R C}. \tag{15}$$

This finally gives the **time lag** δ/ω (in seconds) of the steady periodic current I_{sp} behind the input voltage (Fig. 3.7.5).

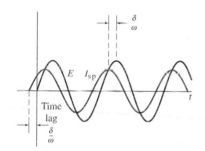

FIGURE 3.7.5. Time lag of current behind imposed voltage.

Initial Value Problems

When we want to find the transient current, we are usually given the initial values $I(0)$ and $Q(0)$. So we must first find $I'(0)$. To do so, we substitute $t = 0$ in Eq. (2) to obtain the equation

$$LI'(0) + RI(0) + \frac{1}{C}Q(0) = E(0) \tag{16}$$

to determine $I'(0)$ in terms of the initial values of current, charge, and voltage.

| Example 1 | Consider an RLC circuit with $R = 50$ ohms (Ω), $L = 0.1$ henry (H), and $C = 5 \times 10^{-4}$ farad (F). At time $t = 0$, when both $I(0)$ and $Q(0)$ are zero, the circuit is connected to a 110-V, 60-Hz alternating current generator. Find the current in the circuit and the time lag of the steady periodic current behind the voltage. |

Solution A frequency of 60 Hz means that $\omega = (2\pi)(60)$ rad/s, approximately 377 rad/s. So we take $E(t) = 110 \sin 377t$ and use equality in place of the symbol for approximate equality in this discussion. The differential equation in (6) takes the form

$$(0.1)I'' + 50I' + 2000I = (377)(110)\cos 377t.$$

We substitute the given values of R, L, C, and $\omega = 377$ in Eq. (10) to find that the impedance is $Z = 59.58 \ \Omega$, so the steady periodic amplitude is

$$I_0 = \frac{110 \text{ (volts)}}{59.58 \text{ (ohms)}} = 1.846 \text{ amperes (A)}.$$

With the same data, Eq. (15) gives the sine phase angle:

$$\delta = \tan^{-1}(0.648) = 0.575.$$

Thus the time lag of current behind voltage is

$$\frac{\delta}{\omega} = \frac{0.575}{377} = 0.0015 \text{ s,}$$

and the steady periodic current is

$$I_{\text{sp}} = (1.846)\sin(377t - 0.575).$$

The characteristic equation $(0.1)r^2 + 50r + 2000 = 0$ has the two roots $r_1 \approx -44$ and $r_2 \approx -456$. With these approximations, the general solution is

$$I(t) = c_1 e^{-44t} + c_2 e^{-456t} + (1.846)\sin(377t - 0.575),$$

with derivative

$$I'(t) = -44c_1 e^{-44t} - 456c_2 e^{-456t} + 696\cos(377t - 0.575).$$

Because $I(0) = Q(0) = 0$, Eq. (16) gives $I'(0) = 0$ as well. With these initial values substituted, we obtain the equations

$$I(0) = c_1 + c_2 - 1.004 = 0,$$
$$I'(0) = -44c_1 - 456c_2 + 584 = 0;$$

their solution is $c_1 = -0.307$, $c_2 = 1.311$. Thus the transient solution is

$$I_{\text{tr}}(t) = (-0.307)e^{-44t} + (1.311)e^{-456t}.$$

The observation that after one-fifth of a second we have $|I_{\text{tr}}(0.2)| < 0.000047$ A (comparable to the current in a single human nerve fiber) indicates that the transient solution dies out very rapidly, indeed. ∎

Example 2 Suppose that the RLC circuit of Example 1, still with $I(0) = Q(0) = 0$, is connected at time $t = 0$ to a battery supplying a constant 110 V. Now find the current in the circuit.

Solution We now have $E(t) \equiv 110$, so Eq. (16) gives

$$I'(0) = \frac{E(0)}{L} = \frac{110}{0.1} = 1100 \ (\text{A/s}),$$

and the differential equation is

$$(0.1)I'' + 50I' + 2000I = E'(t) = 0.$$

Its general solution is the complementary function we found in Example 1:

$$I(t) = c_1 e^{-44t} + c_2 e^{-456t}.$$

We solve the equations

$$I(0) = c_1 + c_2 = 0,$$
$$I'(0) = -44c_1 - 456c_2 = 1100$$

for $c_1 = -c_2 = 2.670$. Therefore,

$$I(t) = (2.670)(e^{-44t} - e^{-456t}).$$

Note that $I(t) \to 0$ as $t \to +\infty$ even though the voltage is constant. ∎

Electrical Resonance

Consider again the current differential equation in (6) corresponding to a sinusoidal input voltage $E(t) = E_0 \sin \omega t$. We have seen that the amplitude of its steady periodic current is

$$I_0 = \frac{E_0}{Z} = \frac{E_0}{\sqrt{R^2 + \left(\omega L - \dfrac{1}{\omega C}\right)^2}}. \tag{17}$$

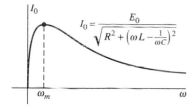

FIGURE 3.7.6. The effect of frequency on I_0.

For typical values of the constants R, L, C, and E_0, the graph of I_0 as a function of ω resembles the one shown in Fig. 3.7.6. It reaches a maximum value at $\omega_m = 1/\sqrt{LC}$ and then approaches zero as $\omega \to +\infty$; the critical frequency ω_m is the **resonance frequency** of the circuit.

In Section 3.6 we emphasized the importance of avoiding resonance in most mechanical systems (the cello is an example of a mechanical system in which resonance is *sought*). By contrast, many common electrical devices could not function properly without taking advantage of the phenomenon of resonance. The radio is a familiar example. A highly simplified model of its tuning circuit is the RLC circuit we have discussed. Its inductance L and resistance R are constant, but its capacitance C is varied as one operates the tuning dial.

Suppose that we wanted to pick up a particular radio station that is broadcasting at frequency ω, and thereby (in effect) provides an input voltage $E(t) = E_0 \sin \omega t$ to the tuning circuit of the radio. The resulting steady periodic current I_{sp} in the tuning circuit drives its amplifier, and in turn its loudspeaker, with the volume of sound we hear roughly proportional to the amplitude I_0 of I_{sp}. To hear our preferred station (of frequency ω) the loudest—and simultaneously tune out stations broadcasting at other frequencies—we therefore want to choose C to maximize I_0.

But examine Eq. (17), thinking of ω as a constant with C the only variable. We see at a glance—no calculus required—that I_0 is maximal when

$$\omega L - \frac{1}{\omega C} = 0;$$

that is, when

$$C = \frac{1}{L\omega^2}. \tag{18}$$

So we merely turn the dial to set the capacitance to this value.

This is the way that old crystal radios worked, but modern AM radios have a more sophisticated design. A *pair* of variable capacitors are used. The first controls the frequency selected as described earlier; the second controls the frequency of a signal that the radio itself generates, kept close to 455 kilohertz (kHz) above the desired frequency. The resulting *beat* frequency of 455 kHz, known as the *intermediate frequency*, is then amplified in several stages. This technique has the advantage that the several RLC circuits used in the amplification stages easily can be designed to resonate at 455 kHz and reject other frequencies, resulting in far more selectivity of the receiver as well as better amplification of the desired signal.

3.7 Problems

Problems 1 through 6 deal with the RL circuit of Fig. 3.7.7, a series circuit containing an inductor with an inductance of L henries, a resistor with a resistance of R ohms, and a source of electromotive force (emf), but no capacitor. In this case Eq. (2) reduces to the linear first-order equation

$$LI' + RI = E(t).$$

FIGURE 3.7.7. The circuit for Problems 1 through 6.

1. In the circuit of Fig. 3.7.7, suppose that $L = 5$ H, $R = 25$ Ω, and that the source E of emf is a battery supplying 100 V to the circuit. Suppose also that the switch has been in position 1 for a long time, so that a steady current of 4 A is flowing in the circuit. At time $t = 0$, the switch is thrown to position 2, so that $I(0) = 4$ and $E = 0$ for $t \geq 0$. Find $I(t)$.

2. Given the same circuit as in Problem 1, suppose that the switch is initially in position 2, but is thrown to position 1 at time $t = 0$, so that $I(0) = 0$ and $E = 100$ for $t \geq 0$. Find $I(t)$ and show that $I(t) \to 4$ as $t \to +\infty$.

3. Suppose that the battery in Problem 2 is replaced with an alternating current generator that supplies a voltage of $E(t) = 100 \cos 60t$ volts. With everything else the same, now find $I(t)$.

4. In the circuit of Fig. 3.7.7, with the switch in position 1, suppose that $L = 2$, $R = 40$, $E(t) = 100e^{-10t}$, and $I(0) = 0$. Find the maximum current in the circuit for $t \geq 0$.

5. In the circuit of Fig. 3.7.7, with the switch in position 1, suppose that $E(t) = 100e^{-10t} \cos 60t$, $R = 20$, $L = 2$, and $I(0) = 0$. Find $I(t)$.

6. In the circuit of Fig. 3.7.7, with the switch in position 1, take $L = 1$, $R = 10$, and $E(t) = 30 \cos 60t + 40 \sin 60t$. **(a)** Substitute $I_{sp}(t) = A \cos 60t + B \sin 60t$ and then determine A and B to find the steady-state current I_{sp} in the circuit. **(b)** Write the solution in the form $I_{sp}(t) = C \cos(\omega t - \alpha)$.

Problems 7 through 10 deal with the RC circuit in Fig. 3.7.8, containing a resistor (R ohms), a capacitor (C farads), a switch, a source of emf, but no inductor. Substitution of L = 0 in Eq. (3) gives the linear first-order differential equation

$$R\frac{dQ}{dt} + \frac{1}{C}Q = E(t)$$

for the charge Q = Q(t) on the capacitor at time t. Note that I(t) = Q'(t).

FIGURE 3.7.8. The circuit for Problems 7 through 10.

7. (a) Find the charge $Q(t)$ and current $I(t)$ in the RC circuit if $E(t) \equiv E_0$ (a constant voltage supplied by a battery) and the switch is closed at time $t = 0$, so that $Q(0) = 0$. **(b)** Show that

$$\lim_{t \to +\infty} Q(t) = E_0 C \quad \text{and that} \quad \lim_{t \to +\infty} I(t) = 0.$$

8. Suppose that in the circuit of Fig. 3.7.8, we have $R = 10$, $C = 0.02$, $Q(0) = 0$, and $E(t) = 100e^{-5t}$ (volts). **(a)** Find $Q(t)$ and $I(t)$. **(b)** What is the maximum charge on the capacitor for $t \geq 0$ and when does it occur?

9. Suppose that in the circuit of Fig. 3.7.8, $R = 200$, $C = 2.5 \times 10^{-4}$, $Q(0) = 0$, and $E(t) = 100 \cos 120t$. **(a)** Find $Q(t)$ and $I(t)$. **(b)** What is the amplitude of the steady-state current?

10. An emf of voltage $E(t) = E_0 \cos \omega t$ is applied to the RC circuit of Fig. 3.7.8 at time $t = 0$ (with the switch closed), and $Q(0) = 0$. Substitute $Q_{sp}(t) = A \cos \omega t + B \sin \omega t$ in the differential equation to show that the steady periodic charge on the capacitor is

$$Q_{sp}(t) = \frac{E_0 C}{\sqrt{1 + \omega^2 R^2 C^2}} \cos(\omega t - \beta)$$

where $\beta = \tan^{-1}(\omega RC)$.

In Problems 11 through 16, the parameters of an RLC circuit with input voltage $E(t)$ are given. Substitute

$$I_{sp}(t) = A \cos \omega t + B \sin \omega t$$

in Eq. (4), using the appropriate value of ω, to find the steady periodic current in the form $I_{sp}(t) = I_0 \sin(\omega t - \delta)$.

11. $R = 30 \; \Omega$, $L = 10$ H, $C = 0.02$ F; $E(t) = 50 \sin 2t$ V

12. $R = 200 \; \Omega$, $L = 5$ H, $C = 0.001$ F; $E(t) = 100 \sin 10t$ V

13. $R = 20 \; \Omega$, $L = 10$ H, $C = 0.01$ F; $E(t) = 200 \cos 5t$ V

14. $R = 50 \; \Omega$, $L = 5$ H, $C = 0.005$ F; $E(t) = 300 \cos 100t + 400 \sin 100t$ V

15. $R = 100 \; \Omega$, $L = 2$ H, $C = 5 \times 10^{-6}$ F; $E(t) = 110 \sin 60\pi t$ V

16. $R = 25 \; \Omega$, $L = 0.2$ H, $C = 5 \times 10^{-4}$ F; $E(t) = 120 \cos 377t$ V

In Problems 17 through 22, an RLC circuit with input voltage $E(t)$ is described. Find the current $I(t)$ using the given initial current (in amperes) and charge on the capacitor (in coulombs).

17. $R = 16 \; \Omega$, $L = 2$ H, $C = 0.02$ F; $E(t) = 100$ V; $I(0) = 0$, $Q(0) = 5$

18. $R = 60 \; \Omega$, $L = 2$ H, $C = 0.0025$ F; $E(t) = 100e^{-t}$ V; $I(0) = 0$, $Q(0) = 0$

19. $R = 60 \; \Omega$, $L = 2$ H, $C = 0.0025$ F; $E(t) = 100e^{-10t}$ V; $I(0) = 0$, $Q(0) = 1$

In each of Problems 20 through 22, plot both the steady periodic current $I_{sp}(t)$ and the total current $I(t) = I_{sp}(t) + I_{tr}(t)$.

20. The circuit and input voltage of Problem 11 with $I(0) = 0$ and $Q(0) = 0$

21. The circuit and input voltage of Problem 13 with $I(0) = 0$ and $Q(0) = 3$

22. The circuit and input voltage of Problem 15 with $I(0) = 0$ and $Q(0) = 0$

23. Consider an LC circuit—that is, an RLC circuit with $R = 0$—with input voltage $E(t) = E_0 \sin \omega t$. Show that unbounded oscillations of current occur for a certain resonance frequency; express this frequency in terms of L and C.

24. It was stated in the text that, if R, L, and C are positive, then any solution of $LI'' + RI' + I/C = 0$ is a transient solution—it approaches zero as $t \to +\infty$. Prove this.

25. Prove that the amplitude I_0 of the steady periodic solution of Eq. (6) is maximal at frequency $\omega = 1/\sqrt{LC}$.

3.8 Endpoint Problems and Eigenvalues

You are now familiar with the fact that a solution of a second-order linear differential equation is uniquely determined by two initial conditions. In particular, the only solution of the initial value problem

$$y'' + p(x)y' + q(x)y = 0; \quad y(a) = 0, \quad y'(a) = 0 \tag{1}$$

is the trivial solution $y(x) \equiv 0$. Most of Chapter 3 has been based, directly or indirectly, on the uniqueness of solutions of linear initial value problems (as guaranteed by Theorem 2 of Section 3.2).

In this section we will see that the situation is radically different for a problem such as

$$y'' + p(x)y' + q(x)y = 0; \quad y(a) = 0, \quad y(b) = 0. \tag{2}$$

The difference between the problems in Eqs. (1) and (2) is that in (2) the two conditions are imposed at two *different* points a and b with (say) $a < b$. In (2) we are to find a solution of the differential equation on the interval (a, b) that satisfies the conditions $y(a) = 0$ and $y(b) = 0$ at the endpoints of the interval. Such a problem is called an **endpoint** or **boundary value problem.** Examples 1 and 2 illustrate the sorts of complications that can arise in endpoint problems.

Example 1 Consider the endpoint problem

$$y'' + 3y = 0; \quad y(0) = 0, \quad y(\pi) = 0. \tag{3}$$

The general solution of the differential equation is

$$y(x) = A \cos x\sqrt{3} + B \sin x\sqrt{3}.$$

The endpoint conditions give

$$y(0) = A = 0 \quad \text{and} \quad y(\pi) = B \sin \pi \sqrt{3} \approx (-0.7458)B = 0,$$

so $B = 0$ as well. Thus the only solution of the endpoint problem in (3) is the trivial solution $y(x) \equiv 0$; there is no surprise here. ∎

Example 2 Consider the endpoint problem

$$y'' + 4y = 0; \quad y(0) = 0, \quad y(\pi) = 0. \tag{4}$$

The general solution of the differential equation is

$$y(x) = A \cos 2x + B \sin 2x.$$

The endpoint conditions yield $y(0) = A = 0$ and

$$y(\pi) = A \cos 2\pi + B \sin 2\pi = B \cdot 0 = 0.$$

No matter what the value of B, the condition $y(\pi) = 0$ is automatically satisfied because $\sin 2\pi = 0$. Thus the function

$$y(x) = B \sin 2x$$

satisfies the endpoint problem in (4) for *every* value of B. This is an example of an endpoint problem having infinitely many nontrivial solutions. ∎

Eigenvalue Problems

Rather than being the exceptional cases, Examples 1 and 2 illustrate the typical situation for an endpoint problem as in (2): It may have no nontrivial solutions, or it may have infinitely many nontrivial solutions. Note that the problems in (3) and (4) both can be written in the form

➤ $$y'' + p(x)y' + \lambda q(x)y = 0; \quad y(a) = 0, \quad y(b) = 0, \tag{5}$$

with $p(x) \equiv 0$, $q(x) \equiv 1$, $a = 0$, and $b = \pi$. The number λ is a parameter in the problem (nothing to do with the parameters that were varied in Section 3.5). If we take $\lambda = 3$, we get the equations in (3); with $\lambda = 4$, we obtain the equations in (4). Examples 1 and 2 show that the situation in an endpoint problem containing a parameter can (and generally will) depend strongly on the specific numerical value of the parameter.

An endpoint problem containing a parameter λ, such as the endpoint problem in (5), is called an **eigenvalue problem.** The question we ask in an eigenvalue problem is this: For what (real) values of the parameter λ does there exist a *nontrivial* (i.e., nonzero) solution of the endpoint problem? Such a value of λ is called an **eigenvalue** or **characteristic value** of the problem. Thus we saw in Example 2 that $\lambda = 4$ is an eigenvalue of the endpoint problem

$$y'' + \lambda y = 0; \quad y(0) = 0, \quad y(\pi) = 0. \tag{6}$$

In Example 1 we saw that $\lambda = 3$ is not an eigenvalue of this problem.

Suppose that λ_\star is an eigenvalue of the problem in (5) and that y_\star is a nontrivial solution of the problem with this value of λ inserted; that is, that

$$y_\star'' + p(x)y_\star' + \lambda_\star q(x)y_\star = 0 \quad \text{and} \quad y_\star(a) = 0, \quad y_\star(b) = 0.$$

Then we call y_\star an **eigenfunction** associated with the eigenvalue λ_\star. Thus we saw in Example 2 that $y_\star(x) = \sin 2x$ is an eigenfunction associated with the eigenvalue $\lambda_\star = 4$, as is any constant multiple of $\sin 2x$.

More generally, note that the problem in (5) is *homogeneous* in the sense that any constant multiple of an eigenfunction is again an eigenfunction; indeed, one associated with the same eigenvalue. That is, if $y = y_\star(x)$ satisfies the problem in (5) with $\lambda = \lambda_\star$, then so does any constant multiple $cy_\star(x)$. It can be proved (under mild restrictions on the coefficient functions p and q) that any two eigenfunctions associated with the same eigenvalue must be linearly dependent.

Example 3 Determine the eigenvalues and associated eigenfunctions for the endpoint problem

$$y'' + \lambda y = 0; \quad y(0) = 0, \quad y(L) = 0 \quad (L > 0). \tag{7}$$

Solution We must consider all possible (real) values of λ—positive, zero, and negative. If $\lambda = 0$, then the equation is simply $y'' = 0$ and its general solution is

$$y(x) = Ax + B.$$

Then the endpoint conditions $y(0) = 0 = y(L)$ immediately imply that $A = B = 0$, so the only solution in this case is the trivial function $y(x) \equiv 0$. Therefore, $\lambda = 0$ is *not* an eigenvalue of the problem in (7).

If $\lambda < 0$, let us then write $\lambda = -\alpha^2$ (with $\alpha > 0$) to be specific. Then the differential equation takes the form

$$y'' - \alpha^2 y = 0,$$

and its general solution is

$$y(x) = c_1 e^{\alpha x} + c_2 e^{-\alpha x} = A \cosh \alpha x + B \sinh \alpha x,$$

where $A = c_1 + c_2$ and $B = c_1 - c_2$. (Recall that $\cosh \alpha x = (e^{\alpha x} + e^{-\alpha x})/2$ and that $\sinh \alpha x = (e^{\alpha x} - e^{-\alpha x})/2$.) The condition $y(0) = 0$ then gives

$$y(0) = A \cosh 0 + B \sinh 0 = A = 0,$$

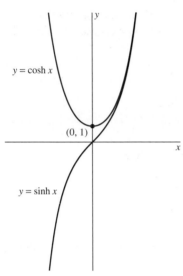

$y = \cosh x$

$(0, 1)$

$y = \sinh x$

FIGURE 3.8.1. The hyperbolic sine and cosine graphs.

so that $y(x) = B \sinh \alpha x$. But now the second endpoint condition, $y(L) = 0$, gives $y(L) = B \sinh \alpha L = 0$. This implies that $B = 0$, because $\alpha \neq 0$, and $\sinh x = 0$ only for $x = 0$ (examine the graphs of $y = \sinh x$ and $y = \cosh x$ in Fig. 3.8.1.) Thus the only solution of the problem in (7) in the case $\lambda < 0$ is the trivial solution $y \equiv 0$, and we may therefore conclude that the problem has *no* negative eigenvalues.

The only remaining possibility is that $\lambda = \alpha^2 > 0$ with $\alpha > 0$. In this case the differential equation is

$$y'' + \alpha^2 y = 0,$$

with general solution

$$y(x) = A \cos \alpha x + B \sin \alpha x.$$

The condition $y(0) = 0$ implies that $A = 0$, so $y(x) = B \sin \alpha x$. The condition $y(L) = 0$ then gives

$$y(L) = B \sin \alpha L = 0.$$

Can this occur if $B \neq 0$? Yes, but only provided that αL is a (positive) integral multiple of π:

$$\alpha L = \pi, \quad 2\pi, \quad 3\pi, \quad \ldots, \quad n\pi, \quad \ldots;$$

that is, if

$$\lambda = \alpha^2 = \frac{\pi^2}{L^2}, \quad \frac{4\pi^2}{L^2}, \quad \frac{9\pi^2}{L^2}, \quad \ldots, \quad \frac{n^2\pi^2}{L^2}, \quad \ldots .$$

Thus we have discovered that the problem in (7) has an *infinite sequence* of positive eigenvalues,

$$\lambda_n = \frac{n^2\pi^2}{L^2}, \quad n = 1, 2, 3, \ldots . \tag{8}$$

With $B = 1$, the eigenfunction associated with the eigenvalue λ_n is

$$y_n(x) = \sin \frac{n\pi x}{L}, \quad n = 1, 2, 3, \ldots . \tag{9}$$

■

Example 3 illustrates the general situation. According to a theorem whose precise statement we will defer until Section 10.1, under the assumption that $q(x) > 0$ on the interval $[a, b]$, any eigenvalue of the form in (5) has a divergent increasing sequence

$$\lambda_1 < \lambda_2 < \lambda_3 < \cdots < \lambda_n < \cdots \to +\infty$$

of eigenvalues, each with an associated eigenfunction. This is also true of the following slightly more general type of eigenvalue problem, in which the endpoint conditions involve values of the derivative y' as well as values of y:

$$y'' + p(x)y' + \lambda q(x)y = 0;$$
$$a_1 y(a) + a_2 y'(a) = 0, \quad b_1 y(b) + b_2 y'(b) = 0, \tag{10}$$

where a_1, a_2, b_1, and b_2 are given constants. With $a_1 = 1 = b_2$ and $a_2 = 0 = b_1$, we get the problem of Example 4 (in which $p(x) \equiv 0$ and $q(x) \equiv 1$, as in the previous example).

Example 4 Determine the eigenvalues and eigenfunctions of the problem

$$y'' + \lambda y = 0; \quad y(0) = 0, \quad y'(L) = 0. \tag{11}$$

Solution Virtually the same argument as that used in Example 3 shows that the only possible eigenvalues are positive, so we take $\lambda = \alpha^2 > 0$ $(\alpha > 0)$ to be specific. Then the differential equation is

$$y'' + \alpha^2 y = 0,$$

with general solution

$$y(x) = A \cos \alpha x + B \sin \alpha x.$$

The condition $y(0) = 0$ immediately gives $A = 0$, so

$$y(x) = B \sin \alpha x \quad \text{and} \quad y'(x) = B\alpha \cos \alpha x.$$

The second endpoint condition $y'(L) = 0$ now gives

$$y'(L) = B\alpha \cos \alpha L = 0.$$

This will hold with $B \neq 0$ provided that αL is an odd positive integral multiple of $\pi/2$:

$$\alpha L = \frac{\pi}{2}, \quad \frac{3\pi}{2}, \quad \dots, \quad \frac{(2n-1)\pi}{2}, \quad \dots;$$

that is, if

$$\lambda = \frac{\pi^2}{4L^2}, \quad \frac{9\pi^2}{4L^2}, \quad \dots, \quad \frac{(2n-1)^2\pi^2}{4L^2}, \quad \dots.$$

Thus the nth eigenvalue λ_n and associated eigenfunction of the problem in (11) are given by

$$\lambda_n = \frac{(2n-1)^2\pi^2}{4L^2} \quad \text{and} \quad y_n(x) = \sin \frac{(2n-1)\pi x}{2L} \tag{12}$$

for $n = 1, 2, 3, \dots$. ∎

A general procedure for determining the eigenvalues of the problem in (10) can be outlined as follows. We first write the general solution of the differential equation in the form

$$y = Ay_1(x, \lambda) + By_2(x, \lambda).$$

We write $y_i(x, \lambda)$ because y_1 and y_2 will depend on λ, as in Examples 3 and 4, in which

$$y_1(x) = \cos \alpha x = \cos x\sqrt{\lambda} \quad \text{and} \quad y_2(x) = \sin \alpha x = \sin x\sqrt{\lambda}.$$

Equilibrium position

(a)

Whirling string

(b)

String

(c)

FIGURE 3.8.2. The whirling string.

FIGURE 3.8.3. Forces on a short segment of the whirling string.

Then we impose the two endpoint conditions, noting that each is linear in y and y', and hence also linear in A and B. When we collect coefficients of A and B in the resulting pair of equations, we therefore get a system of the form

$$\alpha_1(\lambda)A + \beta_1(\lambda)B = 0,$$
$$\alpha_2(\lambda)A + \beta_2(\lambda)B = 0. \tag{13}$$

Now λ is an eigenvalue if and only if the system in (13) has a nontrivial solution (one with A and B not both zero). But such a homogeneous system of linear equations has a nontrivial solution if and only if the determinant of its coefficients vanishes. We therefore conclude that the eigenvalues of the problem in (10) are the (real) solutions of the equation

$$D(\lambda) = \alpha_1(\lambda)\beta_2(\lambda) - \alpha_2(\lambda)\beta_1(\lambda) = 0. \tag{14}$$

This can be a formidable equation to solve and may require a numerical approximation technique such as Newton's method.

Most of the interest in eigenvalue problems is due to their very diverse physical applications. The remainder of this section is devoted to three such applications. Numerous additional applications are included in Chapters 9 and 10 (on partial differential equations and boundary value problems).

The Whirling String

Who of us has not wondered about the shape of a quickly spinning jump rope? Let us consider the shape assumed by a tightly stretched flexible string of length L and constant linear density ρ (mass per unit length) if it is rotated or whirled (like a jump rope) with constant angular speed ω (in radians per second) around its equilibrium position along the x-axis. We assume that the portion of the string to one side of any point exerts a constant tension force T on the portion of the string to the other side of the point, with the direction of T tangential to the string at that point. We further assume that, as the string whirls around the x-axis, each point moves in a circle centered at that point's equilibrium position on the x-axis. Thus the string is elastic, so that as it whirls it also stretches to assume a curved shape. Denote by $y(x)$ the displacement of the string from the point x on the axis of rotation. Finally, we assume that the deflection of the string is so slight that $\sin\theta \approx \tan\theta = y'(x)$ in Fig. 3.8.2(c).

We plan to derive a differential equation for $y(x)$ by application of Newton's law $F = ma$ to the piece of string of mass $\rho\,\Delta x$ corresponding to the interval $[x, x + \Delta x]$. The only forces acting on this piece are the tension forces at its two ends. From Fig. 3.8.3 we see that the net vertical force in the positive y-direction is

$$F = T\sin(\theta + \Delta\theta) - T\sin\theta \approx T\tan(\theta + \Delta\theta) - T\tan\theta,$$

so that

$$F \approx Ty'(x + \Delta x) - Ty'(x). \tag{15}$$

Next we recall from elementary calculus or physics the formula $a = r\omega^2$ for the (inward) centripetal acceleration of a body in uniform circular motion (r is the radius of the circle and ω is the angular velocity of the body). Here we have $r = y$, so the vertical acceleration of our piece of string is $a = -\omega^2 y$, the minus sign because

the inward direction is the negative y-direction. Because $m = \rho \, \Delta x$, substitution of this and (15) in $F = ma$ yields

$$Ty'(x + \Delta x) - Ty'(x) \approx -\rho \omega^2 y \Delta x,$$

so that

$$T \cdot \frac{y'(x + \Delta x) - y'(x)}{\Delta x} \approx -\rho \omega^2 y.$$

We now take the limit as $\Delta x \to 0$ to get the differential equation of motion of the string:

$$Ty'' + \rho \omega^2 y = 0. \tag{16}$$

If we write

$$\lambda = \frac{\rho \omega^2}{T} \tag{17}$$

and impose the condition that the ends of the string are fixed, we finally get the eigenvalue problem

$$y'' + \lambda y = 0; \quad y(0) = 0, \quad y(L) = 0 \tag{7}$$

that we considered in Example 3. We found there that the eigenvalues of the problem in (7) are

$$\lambda_n = \frac{n^2 \pi^2}{L^2}, \quad n = 1, 2, 3, \ldots, \tag{8}$$

with the eigenfunction $y_n(x) = \sin(n\pi x/L)$ associated with λ_n.

But what does all this mean in terms of the whirling string? It means that unless λ in (17) is one of the eigenvalues in (8), then the only solution of the problem in (7) is the trivial solution $y(x) \equiv 0$. In this case the string remains in its equilibrium position with zero deflection. But if we equate (17) and (8) and solve for the value ω_n corresponding to λ_n,

$$\omega_n - \sqrt{\frac{\lambda_n T}{\rho}} = \frac{n\pi}{L} \sqrt{\frac{T}{\rho}} \tag{18}$$

for $n = 1, 2, 3, \ldots$, we get a sequence of **critical speeds** of angular rotation. Only at these critical angular speeds can the string whirl up out of its equilibrium position. At angular speed ω_n it assumes a shape of the form $y_n = c_n \sin(n\pi x/L)$; our mathematical model is not sufficiently complete to determine the coefficient c_n.

Suppose that we start the string rotating at speed

$$\omega < \omega_1 = \frac{\pi}{L} \sqrt{\frac{T}{\rho}},$$

then gradually increase its speed of rotation. So long as $\omega < \omega_1$, the string remains in its undeflected position $y \equiv 0$. But when $\omega = \omega_1$, the string pops into a whirling position $y = c_1 \sin(\pi x/L)$. And when ω is increased still further, the string pops back into its undeflected position along the axis of rotation!

FIGURE 3.8.4. Distortion of a horizontal beam.

The Deflection of a Uniform Beam

We include now an example of the use of a relatively simple endpoint value problem to explain a complicated physical phenomenon—the shape of a horizontal beam on which a vertical force is acting.

Consider the horizontal beam shown in Fig. 3.8.4, uniform both in cross section and in material. If it is supported only at its ends, then the force of its own weight distorts its longitudinal axis of symmetry into the curve shown as a dashed line in the figure. We want to investigate the shape $y = y(x)$ of this curve, the **deflection curve** of the beam. We will use the coordinate system indicated in Fig. 3.8.5, with the positive y-axis directed downward.

A consequence of the theory of elasticity is that for relatively small deflections of such a beam (so small that $[y'(x)]^2$ is negligible in comparison with unity), an adequate mathematical model of the deflection curve is the fourth-order differential equation

FIGURE 3.8.5. The deflection curve.

$$EIy^{(4)} = F(x), \tag{19}$$

where

- E denotes the Young's modulus of the material of the beam,
- I denotes the moment of inertia of the cross section of the beam around a horizontal line through the centroid of the cross section, and
- $F(x)$ denotes the density of *downward* force acting vertically on the beam at the point x.

Density of force? Yes; this means that the force acting downward on a very short segment $[x, x + \Delta x]$ of the beam is approximately $F(x)\Delta x$. The units of $F(x)$ are those of force per unit length, such as pounds per foot. We will consider here the case in which the only force distributed along the beam is its own weight, w pounds per foot, so that $F(x) \equiv w$. Then Eq. (19) takes the form

$$EIy^{(4)} = w \tag{20}$$

where E, I, and w are all constant.

Note: We assume no previous familiarity with the theory of elasticity or with Eq. (19) or (20) here. It is important to be able to begin with a differential equation that arises in a specific applied discipline and then analyze its implications; thus we develop an understanding of the equation by examining its solutions. Observe that, in essence, Eq. (20) implies that the fourth derivative $y^{(4)}$ is proportional to the weight density w. This proportionality involves, however, *two* constants: E, which depends only on the material of the beam, and I, which depends only on the shape of the cross section of the beam. Values of the Young's modulus E of various materials can be found in handbooks of physical constants; $I = \frac{1}{4}\pi a^4$ for a circular cross section of radius a.

Although Eq. (20) is a fourth-order differential equation, its solution involves only the solution of simple first-order equations by successive simple integrations. One integration of Eq. (20) yields

$$EIy^{(3)} = wx + C_1;$$

a second yields

$$EIy'' = \tfrac{1}{2}wx^2 + C_1x + C_2;$$

another yields

$$EIy' = \tfrac{1}{6}wx^3 + \tfrac{1}{2}C_1x^2 + C_2x + C_3;$$

a final integration gives

$$EIy = \tfrac{1}{24}wx^4 + \tfrac{1}{6}C_1x^3 + \tfrac{1}{2}C_2x^2 + C_3x + C_4,$$

where C_1, C_2, C_3, and C_4 are arbitrary constants. Thus we obtain a solution of Eq. (20) of the form

$$y(x) = \frac{w}{24EI}x^2 + Ax^3 + Bx^2 + Cx + D, \qquad (21)$$

where A, B, C, and D are constants resulting from the four integrations.

Simply supported or hinged

Built in

FIGURE 3.8.6. Two ways of supporting a beam.

These last four constants are determined by the way in which the beam is supported at its ends, where $x = 0$ and $x = L$. Figure 3.8.6 shows two common types of support. A beam might also be supported one way at one end but another way at the other end. For instance, Fig. 3.8.7 shows a **cantilever**—a beam firmly fastened at $x = 0$ but *free* (no support whatsoever) at $x = L$. The following table shows the **boundary** or **endpoint conditions** corresponding to the three most common cases. We will see that these conditions are applied readily in beam problems, although a discussion here of their origin would take us too far afield.

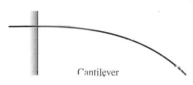

Cantilever

FIGURE 3.8.7. The cantilever.

Support	Endpoint Condition
Simply supported	$y = y'' = 0$
Built-in or fixed end	$y = y' = 0$
Free end	$y'' = y^{(3)} = 0$

For example, the deflection curve of the cantilever in Fig. 3.8.7 would be given by Eq. (21), with the coefficients A, B, C, and D determined by the conditions

$$y(0) = y'(0) = 0 \quad \text{and} \quad y''(L) = y^{(3)}(L) = 0, \qquad (22)$$

corresponding to the fixed end at $x = 0$ and the free end at $x = L$. The conditions in (22) together with the differential equation in (21) constitute an **endpoint value problem.**

Example 5

Determine the shape of the deflection curve of a uniform horizontal beam of length L and weight w per unit length and simply supported at each end.

Solution

We have the endpoint conditions

$$y(0) = y''(0) = 0 = y(L) = y''(L).$$

Rather than imposing then directly on Eq. (21), let us begin with the differential equation $EIy^{(4)} = w$ and determine the constants as we proceed with the four successive integrations. The first two integrations yield

$$EIy^{(3)} = wx + A; \quad EIy'' = \tfrac{1}{2}wx^2 + Ax + B.$$

Hence $y''(0) = 0$ implies that $B = 0$, and then $y''(L) = 0$ gives

$$0 = \tfrac{1}{2}wL^2 + AL.$$

It follows that $A = -wL/2$ and thus that

$$EIy'' = \tfrac{1}{2}x^2 - \tfrac{1}{2}wLx.$$

Then two more integrations give

$$EIy' = \tfrac{1}{6}wx^3 - \tfrac{1}{4}wLx^2 + C,$$

and finally,

$$EIy(x) = \tfrac{1}{24}wx^4 - \tfrac{1}{12}wLx^3 + Cx + D. \tag{23}$$

Now $y(0) = 0$ implies that $D = 0$; then, because $y(L) = 0$,

$$0 = \tfrac{1}{24}wL^4 - \tfrac{1}{12}wL^4 + CL.$$

It follows that $C = wL^3/24$. Hence from Eq. (23) we obtain

$$y(x) = \frac{w}{24EI}(x^4 - 2Lx^3 + L^3x) \tag{24}$$

as the shape of the simply supported beam. It is apparent from symmetry (see also Problem 17) that the *maximum deflection* y_{max} of the beam occurs at its midpoint $x = L/2$, and thus has the value

$$y_{\text{max}} = y\left(\frac{L}{2}\right) = \frac{w}{24EI}\left(\frac{1}{16}L^4 - \frac{2}{8}L^4 + \frac{1}{2}L^4\right);$$

that is,

$$y_{\text{max}} = \frac{5wL^4}{384EI}. \tag{25}$$

∎

Example 6 For instance, suppose that we want to calculate the maximum deflection of a simply supported steel rod 20 ft long with a circular cross section 1 in. in diameter. From a handbook we find that typical steel has density $\delta = 7.75$ g/cm^3 and that its Young's modulus is $E = 2 \times 10^{12}$ g/cm·s^2, so it will be more convenient to work in cgs units. Thus our rod has

$$\text{length:} \quad L = (20 \text{ ft})\left(30.48 \,\frac{\text{cm}}{\text{ft}}\right) = 609.60 \text{ cm}$$

and

$$\text{radius:} \quad a = \left(\frac{1}{2}\text{ in.}\right)\left(2.54 \,\frac{\text{cm}}{\text{in.}}\right) = 1.27 \text{ cm}.$$

Its *linear* mass density (that is, its mass per unit length) is

$$\rho = \pi a^2 \delta = \pi(1.27)^2(7.75) \approx 39.27 \,\frac{\text{g}}{\text{cm}},$$

so

$$w = \rho g = \left(39.27 \, \frac{\text{g}}{\text{cm}}\right)\left(980 \, \frac{\text{cm}}{\text{s}^2}\right) \approx 38484.6 \, \frac{\text{dyn}}{\text{cm}}.$$

The area moment of inertia of a circular disk of radius a around a diameter is $I = \frac{1}{4}\pi a^4$, so

$$I = \frac{1}{4}\pi(1.27)^4 \approx 2.04 \, \text{cm}^4.$$

Therefore Eq. (25) yields

$$y_{\text{max}} \approx \frac{(5)(38484.6)(609.60)^4}{(384)(2 \times 10^{12})(2.04)} \approx 16.96 \, \text{cm},$$

about 6.68 in., as the maximum deflection of the rod at its midpoint. It is interesting to note that y_{max} is proportional to L^4, so if the rod were only 10 ft long, its maximum deflection would be only one-sixteenth as much—only about 0.42 in. Because $I = \frac{1}{4}\pi a^4$, we see from Eq. (25) that the same reduction in maximum deflection could be achieved by doubling the radius a of the rod. ∎

The Buckled Rod

FIGURE 3.8.8. The buckled rod.

Figure 3.8.8 shows a uniform rod of length L, hinged at each end, that has been "buckled" by an axial force of compression P applied at one end. We assume this buckling to be so slight that the deflection curve $y = y(x)$ of the rod may be regarded as defined on the interval $0 \le x \le L$.

In the theory of elasticity the linear endpoint boundary value problem

$$EIy'' + Py = 0, \quad y(0) = y(L) = 0 \tag{26}$$

is used to model the actual (nonlinear) behavior of the rod. As in our discussion of the deflection of a uniform beam, E denotes the Young's modulus of the material of the beam and I denotes the moment of inertia of each cross section of the beam around a horizontal line through its centroid.

If we write

$$\lambda = \frac{P}{EI}, \tag{27}$$

then the problem in (26) becomes the eigenvalue problem

$$y'' + \lambda y = 0; \quad y(0) = y(L) = 0 \tag{7}$$

that we considered in Example 3. We found that its eigenvalues $\{\lambda_n\}$ are given by

$$\lambda_n = \frac{n^2\pi^2}{L^2}, \quad n = 1, 2, 3, \ldots \tag{8}$$

with the eigenfunction $y_n = \sin(n\pi x/L)$ associated with λ_n. (Thus whirling strings and buckled rods lead to the same eigenvalues and eigenfunctions.)

To interpret this result in terms of the buckled rod, recall from Eq. (27) that $P = \lambda EI$. The forces

$$P_n = \lambda_n EI = \frac{n^2\pi^2 EI}{L^2}, \quad n = 1, 2, 3, \ldots \tag{28}$$

are the *critical buckling forces* of the rod. Only when the compressive force P is one of these critical forces should the rod "buckle" out of its straight (undeflected) shape. The smallest compressive force for which this occurs is

$$P_1 = \frac{\pi^2 EI}{L^2}. \tag{29}$$

This smallest critical force P_1 is called the *Euler buckling force* for the rod; it is the upper bound for those compressive forces to which the rod can safely be subjected without buckling. (In practice a rod may fail at a significantly smaller force due to a contribution of factors not taken into account by the mathematical model discussed here.)

| Example 7 | For instance, suppose that we want to compute the Euler buckling force for a steel rod 10 ft long having a circular cross section 1 in. in diameter. In cgs units we have

$$E = 2 \times 10^{12} \text{ g/cm·s}^2,$$

$$L = (10 \text{ ft}) \left(30.48 \frac{\text{cm}}{\text{ft}}\right) = 304.8 \text{ cm}, \quad \text{and}$$

$$I = \frac{\pi}{4}\left[(0.5 \text{ in.})\left(2.54 \frac{\text{cm}}{\text{in.}}\right)\right]^4 \approx 2.04 \text{ cm}^4.$$

Upon substituting these values in Eq. (29) we find that the critical force for this rod is

$$P_1 \approx 4.34 \times 10^8 \text{ dyn} \approx 976 \text{ lb},$$

using the conversion factor 4.448×10^5 dyn/lb. ∎

3.8 Problems

The eigenvalues in Problems 1 through 5 are all nonnegative. First determine whether $\lambda = 0$ is an eigenvalue; then find the positive eigenvalues and associated eigenfunctions.

1. $y'' + \lambda y = 0$; $y'(0) = 0$, $y(1) = 0$
2. $y'' + \lambda y = 0$; $y'(0) = 0$, $y'(\pi) = 0$
3. $y'' + \lambda y = 0$; $y(-\pi) = 0$, $y(\pi) = 0$
4. $y'' + \lambda y = 0$; $y'(-\pi) = 0$, $y'(\pi) = 0$
5. $y'' + \lambda y = 0$; $y(-2) = 0$, $y'(2) = 0$
6. Consider the eigenvalue problem

$$y'' + \lambda y = 0; \quad y'(0) = 0, \quad y(1) + y'(1) = 0.$$

All the eigenvalues are nonnegative, so write $\lambda = \alpha^2$ where $\alpha \geq 0$. **(a)** Show that $\lambda = 0$ is not an eigenvalue. **(b)** Show that $y = A\cos\alpha x + B\sin\alpha x$ satisfies the endpoint conditions if and only if $B = 0$ and α is a positive root of the equation $\tan z = 1/z$. These roots $\{\alpha_n\}_1^\infty$ are the abscissas of the points of intersection of the curves $y = \tan z$ and $y = 1/z$, as indicated in Fig. 3.8.9. Thus the eigenvalues and eigenfunctions of this problem are the numbers $\{\alpha_n^2\}_1^\infty$ and the functions $\{\cos\alpha_n x\}_1^\infty$, respectively.

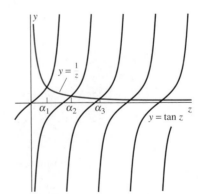

FIGURE 3.8.9. The eigenvalues are determined by the intersections of the graphs of $y = \tan z$ and $y = 1/z$.

7. Consider the eigenvalue problem

$$y'' + \lambda y = 0; \quad y(0) = 0, \quad y(1) + y'(1) = 0;$$

all its eigenvalues are nonnegative. **(a)** Show that $\lambda = 0$ is not an eigenvalue. **(b)** Show that the eigenfunctions

are the functions $\{\sin \alpha_n x\}_1^\infty$, where α_n is the nth positive root of the equation $\tan z = -z$. **(c)** Draw a sketch indicating the roots $\{\alpha_n\}_1^\infty$ as the points of intersection of the curves $y = \tan z$ and $y = -z$. Deduce from this sketch that $\alpha_n \approx (2n-1)\pi/2$ when n is large.

8. Consider the eigenvalue problem

$$y'' + \lambda y = 0; \quad y(0) = 0, \quad y(1) = y'(1);$$

all its eigenvalues are nonnegative. **(a)** Show that $\lambda = 0$ is an eigenvalue with associated eigenfunction $y_0(x) = x$. **(b)** Show that the remaining eigenfunctions are given by $y_n(x) = \sin \beta_n x$, where β_n is the nth positive root of the equation $\tan z = z$. Draw a sketch showing these roots. Deduce from this sketch that $\beta_n \approx (2n+1)\pi/2$ when n is large.

9. Prove that the eigenvalue problem of Example 4 has no negative eigenvalues.

10. Prove that the eigenvalue problem

$$y'' + \lambda y = 0; \quad y(0) = 0, \quad y(1) + y'(1) = 0$$

has no negative eigenvalues. (*Suggestion*: Show graphically that the only root of the equation $\tanh z = -z$ is $z = 0$.)

11. Use a method similar to that suggested in Problem 10 to show that the eigenvalue problem in Problem 6 has no negative eigenvalues.

12. Consider the eigenvalue problem

$$y'' + \lambda y = 0; \quad y(-\pi) = y(\pi), \quad y'(-\pi) = y'(\pi),$$

which is not of the type in (10) because the two endpoint conditions are not "separated" between the two endpoints. **(a)** Show that $\lambda_0 = 0$ is an eigenvalue with associated eigenfunction $y_0(x) \equiv 1$. **(b)** Show that there are no negative eigenvalues. **(c)** Show that the nth positive eigenvalue is n^2 and that it has two linearly independent associated eigenfunctions, $\cos nx$ and $\sin nx$.

13. Consider the eigenvalue problem

$$y'' + 2y' + \lambda y = 0; \quad y(0) = y(1) = 0.$$

(a) Show that $\lambda = 1$ is not an eigenvalue. **(b)** Show that there is no eigenvalue λ such that $\lambda < 1$. **(c)** Show that the nth positive eigenvalue is $\lambda_n = n^2\pi^2 + 1$, with associated eigenfunction $y_n(x) = e^{-x} \sin n\pi x$.

14. Consider the eigenvalue problem

$$y'' + 2y' + \lambda y = 0; \quad y(0) = 0, \quad y'(1) = 0.$$

Show that the eigenvalues are all positive and that the nth positive eigenvalue is $\lambda_n = \alpha_n^2 + 1$ with associated eigenfunction $y_n(x) = e^{-x} \sin \alpha_n x$, where α_n is the nth positive root of $\tan z = z$.

15. **(a)** A uniform cantilever beam is fixed at $x = 0$ and free at its other end, where $x = L$. Show that its shape is given by

$$y(x) = \frac{w}{24EI}(x^4 - 4Lx^3 + 6L^2x^2).$$

(b) Show that $y'(x) = 0$ only at $x = 0$, and thus that it follows (why?) that the maximum deflection of the cantilever is $y_{max} = y(L) = wL^4/(8EI)$.

16. **(a)** Suppose that a beam is fixed at its ends $x = 0$ and $x = L$. Show that its shape is given by

$$y(x) = \frac{w}{24EI}(x^4 - 2Lx^3 + L^2x^2).$$

(b) Show that the roots of $y'(x) = 0$ are $x = 0$, $x = L$, and $x = L/2$, so it follows (why?) that the maximum deflection of the beam is

$$y_{max} = y\left(\frac{L}{2}\right) = \frac{wL^4}{384EI},$$

one-fifth that of a beam with simply supported ends.

17. For the simply supported beam whose deflection curve is given by Eq. (24), show that the only root of $y'(x) = 0$ in $[0, L]$ is $x = L/2$, so it follows (why?) that the maximum deflection is indeed that given in Eq. (25).

18. **(a)** A beam is fixed at its left end $x = 0$ but is simply supported at the other end $x = L$. Show that its deflection curve is

$$y(x) = \frac{w}{48EI}(2x^4 - 5Lx^3 + 3L^2x^2).$$

(b) Show that its maximum deflection occurs where $x = (15 - \sqrt{33})L/16$ and is about 41.6% of the maximum deflection that would occur if the beam were simply supported at each end.

C H A P T E R

4

Introduction to Systems of Differential Equations

4.1 First-Order Systems and Applications

In the preceding chapters we have discussed methods for solving an ordinary differential equation that involves only one dependent variable. Many applications, however, require the use of two or more dependent variables, each a function of a single independent variable (typically time). Such a problem leads naturally to a *system* of simultaneous ordinary differential equations. We will usually denote the independent variable by t and the dependent variables (the unknown functions of t) by x_1, x_2, x_3, \ldots, or by x, y, z, \ldots. Primes will indicate derivatives with respect to t.

We will restrict our attention to systems in which the number of equations is the same as the number of dependent variables (unknown functions). For instance, a system of two first-order equations in the dependent variables x and y has the general form

$$\begin{aligned} f(t, x, y, x', y') &= 0, \\ g(t, x, y, x', y') &= 0, \end{aligned} \tag{1}$$

where the functions f and g are given. A **solution** of this system is a pair $x(t)$, $y(t)$ of functions of t that satisfy both equations identically over some interval of values of t.

For an example of a second-order system, consider a particle of mass m that moves in space under the influence of a force field \mathbf{F} that depends on time t, the position $(x(t), y(t), z(t))$ of the particle, and its velocity $(x'(t), y'(t), z'(t))$. Applying Newton's law $m\mathbf{a} = \mathbf{F}$ componentwise, we get the system

$$\begin{aligned} mx'' &= F_1(t, x, y, z, x', y', z'), \\ my'' &= F_2(t, x, y, z, x', y', z'), \\ mz'' &= F_3(t, x, y, z, x', y', z') \end{aligned} \tag{2}$$

of three second-order equations with independent variable t and dependent variables x, y, z; the three right-hand side functions F_1, F_2, F_3 are the components of the vector-valued function \mathbf{F}.

Initial Applications

Examples 1 through 3 further illustrate how systems of differential equations arise naturally in scientific problems.

Example 1

$x(t)$

Equilibrium positions

FIGURE 4.1.1. The mass-and-spring system of Example 1.

$k_1 x$ m_1 $k_2(y-x)$

$k_2(y-x)$ m_2 $f(t)$

FIGURE 4.1.2. The "free body diagrams" for the system of Example 1.

Consider the system of two masses and two springs shown in Fig. 4.1.1, with a given external force $f(t)$ acting on the right-hand mass m_2. We denote by $x(t)$ the displacement (to the right) of the mass m_1 from its static equilibrium position (when the system is motionless and in equilibrium and $f(t) = 0$), and by $y(t)$ the displacement of the mass m_2 from its static position. Thus the two springs are neither stretched nor compressed when x and y are zero.

In the configuration in Fig. 4.1.1, the first spring is stretched x units and the second by $y - x$ units. We apply Newton's law of motion to the two "free body diagrams" shown in Fig. 4.1.2; we thereby obtain the system

$$m_1 x'' = -k_1 x + k_2(y - x),$$
$$m_2 y'' = -k_2(y - x) + f(t) \tag{3}$$

of differential equations that the position functions $x(t)$ and $y(t)$ must satisfy. For instance, if $m_1 = 2$, $m_2 = 1$, $k_1 = 4$, $k_2 = 2$, and $f(t) = 40 \sin 3t$ in appropriate physical units, then the system in (3) reduces to

$$2x'' = -6x + 2y,$$
$$y'' = 2x - 2y + 40 \sin 3t \tag{4}$$

∎

Example 2

20 gal/min
Fresh water

10 gal/min

$x(t)$ lb
100 gal

$y(t)$ lb
200 gal

Tank 1 30 gal/min Tank 2

20 gal/min

FIGURE 4.1.3. The two brine tanks of Example 2.

Consider two brine tanks connected as shown in Fig. 4.1.3. Tank 1 contains $x(t)$ pounds of salt in 100 gal of brine and tank 2 contains $y(t)$ pounds of salt in 200 gal of brine. The brine in each tank is kept uniform by stirring, and brine is pumped from each tank to the other at the rates indicated in Fig. 4.1.3. In addition, fresh water flows into tank 1 at 20 gal/min, and the brine in tank 2 flows out at 20 gal/min (so the total volume of brine in the two tanks remains constant). The salt concentrations in the two tanks are $x/100$ pounds per gallon and $y/200$ pounds per gallon, respectively. When we compute the rates of change of the amount of salt in the two tanks, we therefore get the system of differential equations that $x(t)$ and $y(t)$ must satisfy:

$$x' = -30 \cdot \frac{x}{100} + 10 \cdot \frac{y}{200} = -\frac{3}{10}x + \frac{1}{20}y,$$
$$y' = 30 \cdot \frac{x}{100} - 10 \cdot \frac{y}{200} - 20 \cdot \frac{y}{200} = \frac{3}{10}x - \frac{3}{20}y;$$

that is,

$$20x' = -6x + y,$$
$$20y' = 6x - 3y. \tag{5}$$

∎

FIGURE 4.1.4. The electrical network of Example 3.

Circuit Element	Voltage Drop
Inductor	$L\dfrac{dI}{dt}$
Resistor	RI
Capacitor	$\dfrac{1}{C}Q$

FIGURE 4.1.5. Voltage drops across common circuit elements.

Example 3

Consider the electrical network shown in Fig. 4.1.4, where $I_1(t)$ denotes the current in the indicated direction through the inductor L and $I_2(t)$ denotes the current through the resistor R_2. The current through the resistor R_1 is $I = I_1 - I_2$ in the direction indicated. We recall Kirchhoff's voltage law to the effect that the (algebraic) sum of the voltage drops around any closed loop of such a network is zero. As in Section 3.7, the voltage drops across the three types of circuit elements are those shown in Fig. 4.1.5. We apply Kirchhoff's law to the left-hand loop of the network to obtain

$$2\frac{dI_1}{dt} + 50(I_1 - I_2) - 100 = 0, \tag{6}$$

because the voltage drop from the negative to the positive pole of the battery is -100. The right-hand loop yields the equation

$$125Q_2 + 25I_2 + 50(I_2 - I_1) = 0, \tag{7}$$

where $Q_2(t)$ is the charge on the capacitor. Because $dQ_2/dt = I_2$, differentiation of each side of Eq. (7) yields

$$125I_2 + 75\frac{dI_2}{dt} - 50\frac{dI_1}{dt} = 0. \tag{8}$$

After dividing Eqs. (6) and (8) by the factors 2 and -25, respectively, we get the system

$$\frac{dI_1}{dt} + 25I_1 - 25I_2 = 50,$$

$$2\frac{dI_1}{dt} - 3\frac{dI_2}{dt} - 5I_2 = 0 \tag{9}$$

of differential equations that the currents $I_1(t)$ and $I_2(t)$ must satisfy. ∎

First-Order Systems

Consider a system of differential equations that can be solved for the highest-order derivatives of the dependent variables that appear, as explicit functions of t and lower-order derivatives of the dependent variables. For instance, in the case of a system of two second-order equations, our assumption is that it can be written in the form

$$
\begin{aligned}
x_1'' &= f_1(t, x_1, x_2, x_1', x_2'), \\
x_2'' &= f_2(t, x_1, x_2, x_1', x_2').
\end{aligned}
\tag{10}
$$

It is of both practical and theoretical importance that any such higher-order system can be transformed into an equivalent system of *first-order* equations.

To describe how such a transformation is accomplished, we consider first the "system" consisting of the single nth-order equation

➤
$$
x^{(n)} = f(t, x, x', \ldots, x^{(n-1)}).
\tag{11}
$$

We introduce the dependent variables x_1, x_2, \ldots, x_n defined as follows:

$$
x_1 = x, \quad x_2 = x', \quad x_3 = x'', \quad \ldots, \quad x_n = x^{(n-1)}.
\tag{12}
$$

Note that $x_1' = x' = x_2$, $x_2' = x'' = x_3$, and so on. Hence the substitution of (12) in Eq. (11) yields the system

$$
\begin{aligned}
x_1' &= x_2, \\
x_2' &= x_3, \\
&\ \ \vdots \\
x_{n-1}' &= x_n, \\
x_n' &= f(t, x_1, x_2, \ldots, x_n)
\end{aligned}
\tag{13}
$$

➤

of n *first-order* equations. Evidently, this system is equivalent to the original nth-order equation in (11), in the sense that $x(t)$ is a solution of Eq. (11) if and only if the functions $x_1(t), x_2(t), \ldots, x_n(t)$ defined in (12) satisfy the system of equations in (13).

Example 4

The third-order equation

$$
x^{(3)} + 3x'' + 2x' - 5x = \sin 2t
$$

is of the form in (11) with

$$
f(t, x, x', x'') = 5x - 2x' - 3x'' + \sin 2t.
$$

Hence the substitutions

$$
x_1 = x, \quad x_2 = x' = x_1', \quad x_3 = x'' = x_2'
$$

yield the system

$$
\begin{aligned}
x_1' &= x_2, \\
x_2' &= x_3, \\
x_3' &= 5x_1 - 2x_2 - 3x_3 + \sin 2t
\end{aligned}
$$

of three first-order equations. ∎

It may appear that the first-order system obtained in Example 4 offers little advantage because we could use the methods of Chapter 3 to solve the original (linear) third-order equation. But suppose that we were confronted with the nonlinear equation

$$x'' = x^3 + (x')^3,$$

to which none of our earlier methods can be applied. The corresponding first-order system is

$$\begin{aligned} x_1' &= x_2, \\ x_2' &= (x_1)^3 + (x_2)^3, \end{aligned} \tag{14}$$

and we will see in Section 4.3 that there exist effective numerical techniques for approximating the solution of essentially any first-order system. So in this case the transformation to a first-order system *is* advantageous. From a practical viewpoint, large systems of higher-order differential equations typically are solved numerically with the aid of the computer, and the first step is to transform such a system into a first-order system for which a standard computer program is available.

Example 5 The system

$$\begin{aligned} 2x'' &= -6x + 2y, \\ y'' &= 2x - 2y + 40\sin 3t \end{aligned} \tag{4}$$

of second-order equations was derived in Example 1. Transform this system into an equivalent first-order system.

Solution Motivated by the equations in (12), we define

$$x_1 = x, \quad x_2 = x' = x_1', \quad y_1 = y, \quad y_2 = y' = y_1'.$$

Then the system in (4) yields the system

$$\begin{aligned} x_1' &= x_2, \\ 2x_2' &= -6x_1 + 2y_1, \\ y_1' &= y_2, \\ y_2' &= 2x_1 - 2y_1 + 40\sin 3t \end{aligned} \tag{15}$$

of four first-order equations in the dependent variables x_1, x_2, y_1, and y_2. ∎

Simple Two-Dimensional Systems

The linear second-order differential equation

$$x'' + px' + qx = 0 \tag{16}$$

(with constant coefficients and independent variable t) transforms via the substitutions $x' = y$, $x'' = y'$ into the two-dimensional linear system

$$\begin{aligned} x' &= y, \\ y' &= -qx - py. \end{aligned} \tag{17}$$

Conversely, we can solve this system in (17) by solving the familiar single equation in (16).

| Example 6 | To solve the two-dimensional system |

$$x' = -2y,$$
$$y' = \tfrac{1}{2}x,$$

(18)

we begin with the observation that

$$x'' = -2y' = -2\left(\tfrac{1}{2}x\right) = -x.$$

This gives the single second-order equation $x'' + x = 0$ with general solution

$(x^2 + x) = 0 \qquad x(x+1) = 0$

$$x(t) = A\cos t + B\sin t = C\cos(t - \alpha) \qquad x = 0$$
$$x = -1$$

where $A = C\cos\alpha$ and $B = C\sin\alpha$. Then

$$y(t) = -\tfrac{1}{2}x'(t) = -\tfrac{1}{2}(-A\sin t + B\cos t)$$
$$= \tfrac{1}{2}C\sin(t - \alpha).$$

The identity $\cos^2\theta + \sin^2\theta = 1$ therefore implies that, for each value of t, the point $(x(t), y(t))$ lies on the ellipse

$$\frac{x^2}{C^2} + \frac{y^2}{(C/2)^2} = 1$$

with semiaxes C and $C/2$. Figure 4.1.6 shows several such ellipses in the xy-plane.

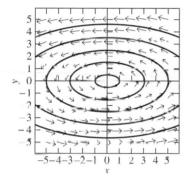

FIGURE 4.1.6. Direction field and solution curves for the system $x' = -2y$, $y' = \tfrac{1}{2}x$ of Example 6.

A solution $(x(t), y(t))$ of a two-dimensional system

$$x' = f(t, x, y),$$
$$y' = g(t, x, y)$$

may be regarded as a parametrization of a **solution curve** or **trajectory** of the system in the xy-plane. Thus the trajectories of the system in (18) are the ellipses of Fig. 4.1.6. The choice of an initial point $(x(0), y(0))$ determines which one of these trajectories a particular solution parametrizes.

The picture showing a system's trajectories in the xy-plane—its so-called *phase plane portrait*—fails to reveal precisely how the point $(x(t), y(t))$ moves along its trajectory. If the functions f and g do not involve the independent variable t, then a **direction field**—showing typical arrows representing vectors with components (proportional to) the derivatives $x' = f(x, y)$ and $y' = g(x, y)$—can be plotted. Because the moving point $(x(t), y(t))$ has velocity vector $(x'(t), y'(t))$, this direction field indicates the point's direction of motion along its trajectory. For instance, the direction field plotted in Fig. 4.1.6 indicates that each such point moves counterclockwise around its elliptical trajectory. Additional information can be shown in the separate graphs of $x(t)$ and $y(t)$ as functions of t.

Example 6
Continued

With initial values $x(0) = 2$, $y(0) = 0$, the general solution in Example 6 yields

$$x(0) = A = 2, \quad y(0) = -\tfrac{1}{2}B = 0.$$

The resulting particular solution is given by

$$x(t) = 2\cos t, \quad y(t) = \sin t.$$

The graphs of the two functions are shown in Fig. 4.1.7. We see that $x(t)$ initially decreases while $y(t)$ increases. It follows that, as t increases, the solution point $(x(t), y(t))$ traverses the trajectory $\frac{1}{4}x^2 + y^2 = 1$ in the counterclockwise direction, as indicated by the direction field vectors in Fig. 4.1.6. ■

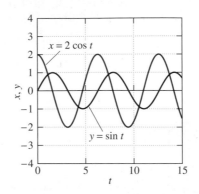

FIGURE 4.1.7. x- and y-solution curves for the initial value problem $x' = -2y$, $y' = \frac{1}{2}x$, $x(0) = 2$, $y(0) = 0$.

Example 7

To find a general solution of the system

$$\begin{aligned} x' &= y, \\ y' &= 2x + y, \end{aligned} \tag{19}$$

we begin with the observation that

$$x'' = y' = 2x + y = x' + 2x.$$

This gives the single linear second-order equation

$$x'' - x' - 2x = 0$$

with characteristic equation

$$r^2 - r - 2 = (r + 1)(r - 2) = 0$$

and general solution

$$x(t) = Ae^{-t} + Be^{2t}. \tag{20}$$

Then

$$y(t) = x'(t) = -Ae^{-t} + 2Be^{2t}. \tag{21}$$

FIGURE 4.1.8. Direction field and solution curves for the system $x' = y$, $y' = 2x + y$ of Example 7.

Typical phase plane trajectories of the system in (19) parametrized by Eqs. (20) and (21) are shown in Fig. 4.1.8. See Problem 23 for further discussion of these trajectories. ■

Example 8 To solve the initial value problem

$$x' = -y,$$
$$y' = (1.01)x - (0.2)y, \tag{22}$$
$$x(0) = 0, \quad y(0) = -1,$$

we begin with the observation that

$$x'' = -y' = -[(1.01)x - (0.2)y] = (-1.01)x - (0.2)x'.$$

This gives the single linear second-order equation

$$x'' + (0.2)x' + (1.01)x = 0$$

with characteristic equation

$$r^2 + (0.2)r + 1.01 = (r + 0.1)^2 + 1 = 0,$$

characteristic roots $-0.1 \pm i$, and general solution

$$x(t) = e^{-t/10}(A \cos t + B \sin t).$$

Then $x(0) = A = 0$, so

$$x(t) = Be^{-t/10} \sin t,$$
$$y(t) = -x'(t) = \tfrac{1}{10}Be^{-t/10} \sin t - Be^{-t/10} \cos t.$$

Finally, $y(0) = -B = -1$, so the desired solution of the system in (22) is

$$x(t) = e^{-t/10} \sin t,$$
$$y(t) = \tfrac{1}{10}e^{-t/10}(\sin t - 10 \cos t). \tag{23}$$

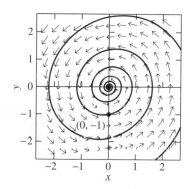

FIGURE 4.1.9. Direction field and solution curve for the system $x' = -y$, $y' = (1.01)x - (0.2)y$ of Example 8.

These equations parametrize the spiral trajectory in Fig. 4.1.9; the trajectory approaches the origin as $t \to +\infty$. Figure 4.1.10 shows the x- and y-solution curves given in (23). ∎

When we study linear systems in Chapter 5, we will learn why the superficially similar systems in Examples 6 through 8 have the markedly different trajectories shown in Figs. 4.1.6, 4.1.8, and 4.1.9.

Linear Systems

FIGURE 4.1.10. x- and y-solution curves for the initial value problem of Example 8.

In addition to practical advantages for numerical computation, the general theory of systems and systematic solution techniques are more easily and more concisely described for first-order systems than for higher-order systems. For instance, consider a *linear* first-order system of the form

$$x_1' = p_{11}(t)x_1 + p_{12}(t)x_2 + \cdots + p_{1n}x_n + f_1(t),$$
$$x_2' = p_{21}(t)x_1 + p_{22}(t)x_2 + \cdots + p_{2n}x_n + f_2(t),$$
$$\vdots$$
$$x_n' = p_{n1}(t)x_1 + p_{n2}(t)x_2 + \cdots + p_{nn}x_n + f_n(t). \tag{24}$$

We say that this system is **homogeneous** if the functions f_1, f_2, \ldots, f_n are all identically zero; otherwise, it is **nonhomogeneous.** Thus the linear system in (5) is homogeneous, whereas the linear system in (15) is nonhomogeneous. The system in (14) is nonlinear because the right-hand side of the second equation is not a linear function of the dependent variables x_1 and x_2.

A **solution** of the system in (24) is an n-tuple of functions $x_1(t), x_2(t), \ldots,$ $x_n(t)$ that (on some interval) identically satisfy each of the equations in (24). We will see that the general theory of a system of n linear first-order equations shares many similarities with the general theory of a single nth-order linear differential equation. Theorem 1 (proved in the Appendix) is analogous to Theorem 2 of Section 3.2. It tells us that if the coefficient functions p_{ij} and f_j in (24) are continuous, then the system has a unique solution satisfying given initial conditions.

THEOREM 1: Existence and Uniqueness for Linear Systems

Suppose that the functions $p_{11}, p_{12}, \ldots, p_{nn}$ and the functions f_1, f_2, \ldots, f_n are continuous on the open interval I containing the point a. Then, given the n numbers b_1, b_2, \ldots, b_n, the system in (24) has a unique solution on the entire interval I that satisfies the n initial conditions

$$x_1(a) = b_1, \quad x_2(a) = b_2, \quad \ldots, \quad x_n(a) = b_n. \tag{25}$$

▲

Thus n initial conditions are needed to determine a solution of a system of n linear first-order equations, and we therefore expect a general solution of such a system to involve n arbitrary constants. For instance, we saw in Example 5 that the second-order linear system

$$
\begin{aligned}
2x'' &= -6x + 2y, \\
y'' &= 2x - 2y + 40\sin 3t,
\end{aligned}
$$

which describes the position functions $x(t)$ and $y(t)$ of Example 1, is equivalent to the system of *four* first-order linear equations in (15). Hence four initial conditions would be needed to determine the subsequent motions of the two masses in Example 1. Typical initial values would be the initial positions $x(0)$ and $y(0)$ and the initial velocities $x'(0)$ and $y'(0)$. On the other hand, we found that the amounts $x(t)$ and $y(t)$ of salt in the two tanks of Example 2 are described by the system

$$
\begin{aligned}
20x' &= -6x + y, \\
20y' &= 6x - 3y
\end{aligned}
$$

of *two* first-order linear equations. Hence the two initial values $x(0)$ and $y(0)$ should suffice to determine the solution. Given a higher-order system, we often must transform it into an equivalent first-order system to discover how many initial conditions are needed to determine a unique solution. Theorem 1 tells us that the number of such conditions is precisely the same as the number of equations in the equivalent first-order system.

4.1 Problems

In Problems 1 through 10, transform the given differential equation or system into an equivalent system of first-order differential equations.

1. $x'' + 3x' + 7x = t^2$
2. $x^{(4)} + 6x'' - 3x' + x = \cos 3t$
3. $t^2 x'' + tx' + (t^2 - 1)x = 0$
4. $t^3 x^{(3)} - 2t^2 x'' + 3tx' + 5x = \ln t$
5. $x^{(3)} = (x')^2 + \cos x$
6. $x'' - 5x + 4y = 0, \ y'' + 4x - 5y = 0$
7. $x'' = -\dfrac{kx}{(x^2 + y^2)^{3/2}}, \ y'' = -\dfrac{ky}{(x^2 + y^2)^{3/2}}$
8. $x'' + 3x' + 4x - 2y = 0, \ y'' + 2y' - 3x + y = \cos t$
9. $x'' = 3x - y + 2z, \ y'' = x + y - 4z, \ z'' = 5x - y - z$
10. $x'' = (1 - y)x, \ y'' = (1 - x)y$

Use the method of Examples 6, 7, and 8 to find general solutions of the systems in Problems 11 through 20. If initial conditions are given, find the corresponding particular solution. For each problem, use a computer system or graphing calculator to construct a direction field and typical solution curves for the given system.

11. $x' = y, \ y' = -x$
12. $x' = y, \ y' = x$
13. $x' = -2y, \ y' = 2x; \ x(0) = 1, \ y(0) = 0$
14. $x' = 10y, \ y' = -10x; \ x(0) = 3, \ y(0) = 4$
15. $x' = \frac{1}{2}y, \ y' = -8x$
16. $x' = 8y, \ y' = -2x$
17. $x' = y, \ y' = 6x - y; \ x(0) = 1, \ y(0) = 2$
18. $x' = -y, \ y' = 10x - 7y; \ x(0) = 2, \ y(0) = -7$
19. $x' = -y, \ y' = 13x + 4y; \ x(0) = 0, \ y(0) = 3$
20. $x' = y, \ y' = -9x + 6y$

21. **(a)** Calculate $[x(t)]^2 + [y(t)]^2$ to show that the trajectories of the system $x' = y, \ y' = -x$ of Problem 11 are circles. **(b)** Calculate $[x(t)]^2 - [y(t)]^2$ to show that the trajectories of the system $x' = y, \ y' = x$ of Problem 12 are hyperbolas.

22. **(a)** Beginning with the general solution of the system $x' = -2y, \ y' = 2x$ of Problem 13, calculate $x^2 + y^2$ to show that the trajectories are circles. **(b)** Show similarly that the trajectories of the system $x' = \frac{1}{2}y, \ y' = -8x$ of Problem 15 are ellipses with equations of the form $16x^2 + y^2 = C^2$.

23. First solve Eqs. (20) and (21) for e^{-t} and e^{2t} in terms of $x(t), y(t)$, and the constants A and B. Then substitute the results in $(e^{2t})(e^{-t})^2 = 1$ to show that the trajectories of the system $x' = y, \ y' = 2x + y$ in Example 7 satisfy an equation of the form

$$4x^3 - 3xy^2 + y^3 = C \quad \text{(constant)}.$$

Then show that $C = 0$ yields the straight lines $y = -x$ and $y = 2x$ that are visible in Fig. 4.1.8.

24. Derive the equations

$$m_1 x_1'' = -(k_1 + k_2)x_1 + k_2 x_2,$$
$$m_2 x_2'' = k_2 x_1 - (k_2 + k_3)x_2$$

for the displacements (from equilibrium) of the two masses shown in Fig. 4.1.11.

FIGURE 4.1.11. The system of Problem 24.

25. Two particles each of mass m are attached to a string under (constant) tension T, as indicated in Fig. 4.1.12. Assume that the particles oscillate vertically (that is, parallel to the y-axis) with amplitudes so small that the sines of the angles shown are accurately approximated by their tangents. Show that the displacements y_1 and y_2 satisfy the equations

$$ky_1'' = -2y_1 + y_2, \quad ky_2'' = y_1 - 2y_2$$

where $k = mL/T$.

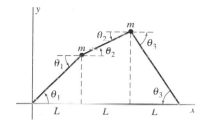

FIGURE 4.1.12. The mechanical system of Problem 25.

26. Three 100-gal fermentation vats are connected as indicated in Fig. 4.1.13, and the mixtures in each tank are kept uniform by stirring. Denote by $x_i(t)$ the amount (in pounds) of alcohol in tank T_i at time t ($i = 1, 2, 3$). Suppose that the mixture circulates between the tanks at the rate of 10 gal/min. Derive the equations

$$10x_1' = -x_1 + x_3$$
$$10x_2' = x_1 - x_2$$
$$10x_3' = x_2 - x_3.$$

FIGURE 4.1.13. The fermentation tanks of Problem 26.

27. Set up a system of first-order differential equations for the currents in the electrical circuit shown in Fig. 4.1.14.

FIGURE 4.1.14. The electrical circuit of Problem 27.

28. Repeat Problem 27, except with the generator replaced with a battery supplying an emf of 100 V and with the inductor replaced with a 1-millifarad (mF) capacitor.

29. A particle of mass m moves in the plane with coordinates $(x(t), y(t))$ under the influence of a force that is directed toward the origin and has magnitude $k/(x^2 + y^2)$—an inverse-square central force field. Show that

$$mx'' = -\frac{kx}{r^3} \quad \text{and} \quad my'' = -\frac{ky}{r^3},$$

where $r = \sqrt{x^2 + y^2}$.

30. Suppose that a projectile of mass m moves in a vertical plane in the atmosphere near the surface of the earth under the influence of two forces: a downward gravitational force of magnitude mg, and a resistive force \mathbf{F}_R that is directed opposite to the velocity vector \mathbf{v} and has magnitude kv^2 (where $v = |\mathbf{v}|$ is the speed of the projectile; see Fig. 4.1.15). Show that the equations of motion of the projectile are

$$mx'' = -kvx', \quad my'' = -kvy' - mg,$$

where $v = \sqrt{(x')^2 + (y')^2}$.

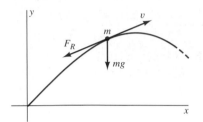

FIGURE 4.1.15. The trajectory of the projectile of Problem 30.

31. Suppose that a particle with mass m and electrical charge q moves in the xy-plane under the influence of the magnetic field $\mathbf{B} = B\mathbf{k}$ (thus a uniform field parallel to the z-axis), so the force on the particle is $\mathbf{F} = q\mathbf{v} \times \mathbf{B}$ if its velocity is \mathbf{v}. Show that the equations of motion of the particle are

$$mx'' = +qBy', \quad my'' = -qBx'.$$

4.1 Application: Gravitation and Kepler's Laws of Planetary Motion

Around the turn of the 17th century, Johannes Kepler analyzed a lifetime of planetary observations by the astronomer Tycho Brahe. Kepler concluded that the motion of the planets around the sun is described by the following three propositions, now known as **Kepler's laws of planetary motion:**

1. The orbit of each planet is an ellipse with the sun at one focus.
2. The radius vector from the sun to each planet sweeps out area at a constant rate.
3. The square of the planet's period of revolution is proportional to the cube of the major semiaxis of its elliptical orbit.

In his *Principia Mathematica* (1687) Isaac Newton deduced the inverse-square law of gravitation from Kepler's laws. In this application we lead you through a derivation of Kepler's first two laws from Newton's law of gravitation.

Assume that the sun is located at the origin in the plane of motion of a planet, and write the position vector of the planet in the form

$$\mathbf{r}(t) = (x(t), y(t)) = x\mathbf{i} + y\mathbf{j}, \tag{1}$$

where $\mathbf{i} = (1, 0)$ and $\mathbf{j} = (0, 1)$ denote the unit vectors in the positive x- and y-directions. Then the inverse-square law of gravitation implies (Problem 29) that the

acceleration vector $\mathbf{r}''(t)$ of the planet is given by

$$\mathbf{r}'' = -\frac{k\mathbf{r}}{r^3}, \tag{2}$$

where $r = \sqrt{x^2 + y^2}$ is the distance from the sun to the planet. If the polar coordinates of the planet at time t are $(r(t), \theta(t))$, then the radial and transverse unit vectors shown in Fig. 4.1.16 are given by

$$\mathbf{u}_r = \mathbf{i}\cos\theta + \mathbf{j}\sin\theta \quad \text{and} \quad \mathbf{u}_\theta = -\mathbf{i}\sin\theta + \mathbf{j}\cos\theta. \tag{3}$$

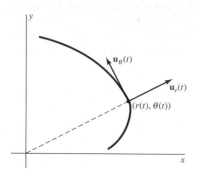

FIGURE 4.1.16. The radial and transverse unit vectors \mathbf{u}_r and \mathbf{u}_θ.

The radial unit vector \mathbf{u}_r (when located at the planet's position) always points directly away from the origin, and the transverse unit vector \mathbf{u}_θ is obtained from \mathbf{u}_r by a 90° counterclockwise rotation.

STEP 1: Differentiate the equations in (3) componentwise to show that

$$\frac{d\mathbf{u}_r}{dt} = \mathbf{u}_\theta \frac{d\theta}{dt} \quad \text{and} \quad \frac{d\mathbf{u}_\theta}{dt} = -\mathbf{u}_r \frac{d\theta}{dt}. \tag{4}$$

STEP 2: Use the equations in (4) to differentiate the planet's position vector $\mathbf{r} = r\mathbf{u}_r$ and thereby show that its velocity vector is given by

$$\mathbf{v} = \frac{d\mathbf{r}}{dt} = \mathbf{u}_r \frac{dr}{dt} + r\frac{d\theta}{dt}\mathbf{u}_\theta. \tag{5}$$

STEP 3: Differentiate again to show that the planet's acceleration vector $\mathbf{a} = d\mathbf{v}/dt$ is given by

$$\mathbf{a} = \left[\frac{d^2r}{dt^2} - r\left(\frac{d\theta}{dt}\right)^2\right]\mathbf{u}_r + \left[\frac{1}{r}\frac{d}{dt}\left(r^2\frac{d\theta}{dt}\right)\right]\mathbf{u}_\theta. \tag{6}$$

STEP 4: The radial and transverse components on the right-hand sides in Eqs. (2) and (6) must agree. Equating the transverse components, we get

$$\frac{1}{r}\frac{d}{dt}\left(r^2\frac{d\theta}{dt}\right) = 0, \tag{7}$$

so it follows that

$$r^2\frac{d\theta}{dt} = h, \tag{8}$$

where h is a constant. Because the polar-coordinate area element—for computation of the area $A(t)$ in Fig. 4.1.17—is given by $dA = \frac{1}{2}r^2 d\theta$, Eq. (8) implies that the derivative $A'(t)$ is *constant*, which is a statement of Kepler's second law.

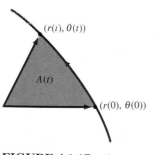

FIGURE 4.1.17. Area swept out by the radius vector.

STEP 5: Equate radial components in (2) and (6) and then use the result in (8) to show that the planet's radial coordinate function $r(t)$ satisfies the second-order differential equation

$$\frac{d^2r}{dt^2} - \frac{h^2}{r^3} = -\frac{k}{r^2}. \tag{9}$$

STEP 6: ⋅Assuming that the orbit can be written in the polar-coordinate form $r = r(\theta)$, use the chain rule and Eq. (8) to show that if $r = 1/z$, then

$$\frac{dr}{dt} = -h\frac{dz}{d\theta}.$$

Differentiate again to deduce from Eq. (9) that the function $z(\theta) = 1/r(\theta)$ satisfies the second-order equation

$$\frac{d^2z}{d\theta^2} + z = \frac{k}{h^2}. \tag{10}$$

STEP 7: Show that the general solution of Eq. (10) is

$$z(\theta) = A\sin\theta + B\cos\theta + \frac{k}{h^2}. \tag{11}$$

STEP 8: Finally, deduce from Eq. (11) that $r(\theta) = 1/z(\theta)$ is given by

$$r(\theta) = \frac{L}{1 + e\cos(\theta - \alpha)} \tag{12}$$

with $e = Ch^2/k$, $C\cos\alpha = A$, $C\sin\alpha = B$, and $L = h^2/k$. The polar-coordinate graph of Eq. (12) is a conic section of eccentricity e—an ellipse if $0 \leqq e < 1$, a parabola if $e = 1$, and a hyperbola if $e > 1$—with focus at the origin. Planetary orbits are bounded and therefore are ellipses with eccentricity $e < 1$. As indicated in Fig. 4.1.18, the major axis of the ellipse lies along the radial line $\theta = \alpha$.

STEP 9: Plot some typical elliptical orbits as described by (12) with different eccentricities, sizes, and orientations. In rectangular coordinates you can write

$$x(t) = r(t)\cos t, \quad y(t) = r(t)\sin t, \quad 0 \leqq t \leqq 2\pi$$

to plot an elliptical orbit with eccentricity e, semilatus rectum L (Fig. 4.1.18), and rotation angle α. The eccentricity of the earth's orbit is $e \approx 0.0167$, so close to zero that the orbit looks nearly circular (though with the sun off center), and the eccentricities of the other planetary orbits range from 0.0068 for Venus and 0.0933 for Mars to 0.2056 for Mercury and 0.2486 for Pluto. But many comets have highly eccentric orbits, like Halley's comet with $e \approx 0.97$ (Fig. 4.1.19).

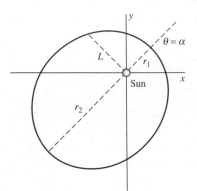

FIGURE 4.1.18. The elliptical orbit

$$r = \frac{L}{1 + e\cos(\theta - \alpha)}$$

with perihelion distance $r_1 = L/(1 + e)$ and aphelion distance $r_2 = L/(1 - e)$.

FIGURE 4.1.19. The shape of the orbit of Halley's comet.

4.2 The Method of Elimination

The most elementary approach to linear systems of differential equations involves the elimination of dependent variables by appropriately combining pairs of equations. The object of this procedure is to eliminate dependent variables in succession until there remains only a single equation containing only one dependent variable. This remaining equation will usually be a linear equation of high order and can frequently be solved by the methods of Chapter 3. After its solution has been found, the other dependent variables can be found in turn, using either the original differential equations or those that have appeared in the elimination process.

The **_method of elimination_** for linear differential systems is similar to the solution of a linear system of algebraic equations by a process of eliminating the

unknowns one at a time until only a single equation with a single unknown remains. It is most convenient in the case of manageably small systems: those containing no more than two or three equations. For such systems the method of elimination provides a simple and concrete approach that requires little preliminary theory or formal machinery. But for larger systems of differential equations, as well as for theoretical discussion, the matrix methods of Chapter 5 are preferable.

Example 1 Find the particular solution of the system

$$x' = 4x - 3y, \quad y' = 6x - 7y \tag{1}$$

that satisfies the initial conditions $x(0) = 2$, $y(0) = -1$.

Solution If we solve the second equation in (1) for x, we get

$$x = \tfrac{1}{6}y' + \tfrac{7}{6}y, \tag{2}$$

so that

$$x' = \tfrac{1}{6}y'' + \tfrac{7}{6}y'. \tag{3}$$

We then substitute these expressions for x and x' in the first equation of the system in (1); this yields

$$\tfrac{1}{6}y'' + \tfrac{7}{6}y' = 4\left(\tfrac{1}{6}y' + \tfrac{7}{6}y\right) - 3y,$$

which we simplify to

$$y'' + 3y' - 10y = 0.$$

This second-order linear equation has characteristic equation

$$r^2 + 3r - 10 = (r - 2)(r + 5) = 0,$$

so its general solution is

$$y(t) = c_1 e^{2t} + c_2 e^{-5t}. \tag{4}$$

Next, substitution of (4) in (2) gives

$$x(t) = \tfrac{1}{6}\left(2c_1 e^{2t} - 5c_2 e^{-5t}\right) + \tfrac{7}{6}\left(c_1 e^{2t} + c_2 e^{-5t}\right);$$

that is,

$$x(t) = \tfrac{3}{2}c_1 e^{2t} + \tfrac{1}{3}c_2 e^{-5t}. \tag{5}$$

Thus Eqs. (4) and (5) constitute the general solution of the system in (1).

The given initial conditions imply that

$$x(0) = \tfrac{3}{2}c_1 + \tfrac{1}{3}c_2 = 2$$

and that

$$y(0) = c_1 + c_2 = -1;$$

these equations are readily solved for $c_1 = 2$ and $c_2 = -3$. Hence the desired solution is

$$x(t) = 3e^{2t} - e^{-5t}, \quad y(t) = 2e^{2t} - 3e^{-5t}.$$

FIGURE 4.2.1. Direction field and solution curves for the system $x' = 4x - 3y$, $y' = 6x - 7y$ of Example 1.

Figure 4.2.1 shows this and other typical solution curves parametrized by the equations $x(t) = \tfrac{3}{2}c_1 e^{2t} + \tfrac{1}{3}c_2 e^{-5t}$, $y(t) = c_1 e^{2t} + c_2 e^{-5t}$ with different values of the arbitrary constants c_1 and c_2. We see two families of curves resembling hyperbolas sharing the same pair of (oblique) asymptotes.

Remark: The general solution defined by Eqs. (4) and (5) may be regarded as the pair or vector $(x(t), y(t))$. Recalling the componentwise addition of vectors (and multiplication of vectors by scalars), we can write the general solution in (4) and (5) in the form

$$(x(t), y(t)) = \left(\tfrac{3}{2}c_1 e^{2t} + \tfrac{1}{3}c_2 e^{-5t}, c_1 e^{2t} + c_2 e^{-5t}\right)$$

$$= c_1 \left(\tfrac{3}{2}e^{2t}, e^{2t}\right) + c_2 \left(\tfrac{1}{3}e^{-5t}, e^{-5t}\right).$$

This expression presents the general solution of the system in (1) as a linear combination of the two particular solutions

$$(x_1, y_1) = \left(\tfrac{3}{2}e^{2t}, e^{2t}\right) \quad \text{and} \quad (x_2, y_2) = \left(\tfrac{1}{3}e^{-5t}, e^{-5t}\right). \qquad \blacksquare$$

Linear Operators

In Example 1 we used an ad hoc procedure to eliminate one of the independent variables by expressing it in terms of the other. We now describe a systematic elimination procedure. Operator notation is most convenient for these purposes. A **linear differential operator** of order n with *constant* coefficients is one of the form

$$L = a_n D^n + a_{n-1} D^{n-1} + \cdots + a_1 D + a_0, \qquad (6)$$

where D denotes differentiation with respect to the independent variable t. If L_1 and L_2 are two such operators, then their product $L_1 L_2$ is defined in this way:

$$\blacktriangleright \qquad\qquad L_1 L_2[x] = L_1[L_2 x]. \qquad (7)$$

For instance, if $L_1 = D + a$ and $L_2 = D + b$, then

$$L_1 L_2[x] = (D+a)[(D+b)x] = D(Dx + bx) + a(Dx + bx)$$

$$= [D^2 + (a+b)D + ab]x.$$

This illustrates the fact that two linear operators with constant coefficients can be multiplied as if they were ordinary polynomials in the "variable" D. Because the multiplication of such polynomials is commutative, it follows that

$$L_1 L_2[x] = L_2 L_1[x] \qquad (8)$$

if the necessary derivatives of $x(t)$ exist. By contrast, this property of commutativity generally fails for linear operators with variable coefficients—see Problems 21 and 22.

Any system of two linear differential equations with constant coefficients can be written in the form

$$\begin{aligned} L_1 x + L_2 y &= f_1(t), \\ L_3 x + L_4 y &= f_2(t), \end{aligned} \qquad (9)$$

where L_1, L_2, L_3, and L_4 are linear differential operators (perhaps of different orders) as in Eq. (6), and $f_1(t)$ and $f_2(t)$ are given functions. For instance, the system in (1) (Example 1) can be written in the form

$$\begin{aligned} (D-4)x + \qquad\quad 3y &= 0, \\ -6x + (D+7)y &= 0, \end{aligned} \qquad (10)$$

with $L_1 = D - 4$, $L_2 = 3$, $L_3 = -6$, and $L_4 = D + 7$.

To eliminate the dependent variable x from the system in (9), we operate with L_3 on the first equation and with L_1 on the second. Thus we obtain the system

$$L_3 L_1 x + L_3 L_2 y = L_3 f_1(t),$$
$$L_1 L_3 x + L_1 L_4 y = L_1 f_2(t). \tag{11}$$

Subtraction of the first from the second of these equations yields the single equation

$$(L_1 L_4 - L_2 L_3) y = L_1 f_2(t) - L_3 f_1(t) \tag{12}$$

in the single dependent variable y. After solving for $y = y(t)$ we can substitute the result into either of the original equations in (9) and then solve for $x = x(t)$.

Alternatively, we could eliminate in like manner the dependent variable y from the original system in (9). If so, we would get the equation

$$(L_1 L_4 - L_2 L_3) x = L_4 f_1(t) - L_2 f_2(t), \tag{13}$$

which can now be solved for $x = x(t)$.

Note that the same operator $L_1 L_4 - L_2 L_3$ appears on the left-hand side in both Eq. (12) and Eq. (13). This is the **operational determinant**

$$\begin{vmatrix} L_1 & L_2 \\ L_3 & L_4 \end{vmatrix} = L_1 L_4 - L_2 L_3 \tag{14}$$

of the system in (9). In determinant notation Eqs. (12) and (13) can be rewritten as

$$\begin{vmatrix} L_1 & L_2 \\ L_3 & L_4 \end{vmatrix} x = \begin{vmatrix} f_1(t) & L_2 \\ f_2(t) & L_4 \end{vmatrix},$$
$$\begin{vmatrix} L_1 & L_2 \\ L_3 & L_4 \end{vmatrix} y = \begin{vmatrix} L_1 & f_1(t) \\ L_3 & f_2(t) \end{vmatrix}. \tag{15}$$

It is important to note that the determinants on the right-hand side in (15) are evaluated by means of the operators operating on the functions. The equations in (15) are strongly reminiscent of Cramer's rule for the solution of two linear equations in two (algebraic) variables and are thereby easy to remember. Indeed, you can solve a system of two linear differential equations either by carrying out the systematic elimination procedure described here or by directly employing the determinant notation in (15). Either process is especially simple if the system is homogeneous ($f_1(t) \equiv 0$ and $f_2(t) \equiv 0$), because in this case the right-hand sides of the equations in (12), (13), and (15) are zero.

Example 2 Find a general solution of the system

$$(D - 4)x + 3y = 0,$$
$$-6x + (D + 7)y = 0. \tag{10}$$

Solution The operational determinant of this system is

$$(D-4)(D+7) - 3 \cdot (-6) = D^2 + 3D - 10. \tag{16}$$

Hence Eqs. (13) and (12) are

$$x'' + 3x' - 10x = 0,$$
$$y'' + 3y' - 10y = 0.$$

The characteristic equation of each is

$$r^2 + 3r - 10 = (r-2)(r+5) = 0,$$

so their (separate) general solutions are

$$x(t) = a_1 e^{2t} + a_2 e^{-5t},$$
$$y(t) = b_1 e^{2t} + b_2 e^{-5t}. \tag{17}$$

At this point we appear to have *four* arbitrary constants a_1, a_2, b_1, and b_2. But it follows from Theorem 1 in Section 4.1 that the general solution of a system of two first-order equations involves only two arbitrary constants. This apparent difficulty demands a resolution.

The explanation is simple: There must be some hidden relations among our four constants. We can discover them by substituting the solutions in (17) into either of the original equations in (10). On substitution in the first equation, we get

$$0 = x' - 4x + 3y$$
$$= (2a_1 e^{2t} - 5a_2 e^{-5t}) - 4(a_1 e^{2t} + a_2 e^{-5t}) + 3(b_1 e^{2t} + b_2 e^{-5t});$$

that is,

$$0 = (-2a_1 + 3b_1)e^{2t} + (-9a_2 + 3b_2)e^{-5t}.$$

But e^{2t} and e^{-5t} are linearly independent functions, so it follows that $a_1 = \frac{3}{2}b_1$ and $a_2 = \frac{1}{3}b_2$. Therefore, the desired general solution is given by

$$x(t) = \tfrac{3}{2}b_1 e^{2t} + \tfrac{1}{3}b_2 e^{-5t}, \quad y(t) = b_1 e^{2t} + b_2 e^{-5t}.$$

Note that this result is in accord with the general solution (Eqs. (4) and (5)) that we obtained by a different method in Example 1. ∎

As illustrated by Example 2, the elimination procedure ordinarily will introduce a number of interdependent—thus unnecessary and undesirable—arbitrary constants. These must be eliminated by substitution of the proposed general solution into one or more of the original differential equations. If the operational determinant defined in (15) is *not identically zero*, then the number of independent arbitrary constants in a general solution of the system in (9) is equal to the order of the operational determinant—that is, its degree as a polynomial in D. (For a proof of this fact, see pages 144–150 of E. L. Ince's *Ordinary Differential Equations* (New York: Dover, 1956).) Thus the general solution of the system in (10) of Example 2 involves two arbitrary constants because its operational determinant $D^2 + 3D - 10$ is of order 2.

If the operational determinant is identically zero, then the system is said to be **degenerate**. A degenerate system may have either no solution or infinitely many independent solutions. For instance, the equations

$$Dx - Dy = 0,$$
$$2Dx - 2Dy = 1$$

with operational determinant zero are obviously inconsistent and thus have no solutions. On the other hand, the equations

$$Dx + Dy = t,$$
$$2Dx + 2Dy = 2t$$

with operational determinant zero are obviously redundant; we can substitute *any* (continuously differentiable) function for $x(t)$ and then integrate to obtain $y(t)$. Roughly speaking, every degenerate system is equivalent to either an inconsistent system or a redundant system.

Although the aforementioned procedures and results are described for the case of a system of two equations, they can be generalized readily to systems of three or more equations. For the system

$$L_{11}x + L_{12}y + L_{13}z = f_1(t),$$
$$L_{21}x + L_{22}y + L_{23}z = f_2(t),$$
$$L_{31}x + L_{32}y + L_{33}z = f_3(t)$$

(18)

of three linear equations, the dependent variable $x(t)$ satisfies the single linear equation

$$\begin{vmatrix} L_{11} & L_{12} & L_{13} \\ L_{21} & L_{22} & L_{23} \\ L_{31} & L_{32} & L_{33} \end{vmatrix} x = \begin{vmatrix} f_1(t) & L_{12} & L_{13} \\ f_2(t) & L_{22} & L_{23} \\ f_3(t) & L_{32} & L_{33} \end{vmatrix}$$

(19)

with analogous equations for $y = y(t)$ and $z = z(t)$. For most systems of more than three equations, however, the method of operational determinants is too tedious to be practical.

Mechanical Vibrations

A mechanical system typically vibrates or oscillates periodically in one or more specific ways. The methods of this section often can be applied to analyze the "natural modes of oscillation" of a given system. Example 3 illustrates this approach.

Example 3 In Example 1 of Section 4.1, we derived the equations

$$(D^2 + 3)x + (-1)y = 0,$$
$$-2x + (D^2 + 2)y = 0$$

(20)

for the displacements of the two masses in Fig. 4.2.2. Here $f(t) \equiv 0$ because we assume that there is no external force. Find the general solution of the system in (20).

Solution The operational determinant of the system in (20) is

$$(D^2 + 3)(D^2 + 2) - (-1)(-2) = D^4 + 5D^2 + 4 = (D^2 + 1)(D^2 + 4).$$

Hence the equations for $x(t)$ and $y(t)$ are

$$(D^2 + 1)(D^2 + 4)x = 0,$$
$$(D^2 + 1)(D^2 + 4)y = 0. \tag{21}$$

$k_1 = 4$ $k_2 = 2$
$m_1 = 2$ $m_2 = 1$
$x(t)$ $y(t)$

Equilibrium positions

FIGURE 4.2.2. The mass-and-spring system of Example 3.

The characteristic equation $(r^2 + 1)(r^2 + 4) = 0$ has roots i, $-i$, $2i$, and $-2i$. So the general solutions of the equations in (21) are

$$x(t) = a_1 \cos t + a_2 \sin t + b_1 \cos 2t + b_2 \sin 2t,$$
$$y(t) = c_1 \cos t + c_2 \sin t + d_1 \cos 2t + d_2 \sin 2t. \tag{22}$$

Because the operational determinant is of order 4, the general solution should contain four (rather than eight) arbitrary constants. When we substitute $x(t)$ and $y(t)$ from (22) in the first equation in (20), we get

$$\begin{aligned} 0 &= x'' + 3x - y \\ &= (-a_1 \cos t - a_2 \sin t - 4b_1 \cos 2t - 4b_2 \sin 2t) \\ &\quad + 3(a_1 \cos t + a_2 \sin t + b_1 \cos 2t + b_2 \sin 2t) \\ &\quad - (c_1 \sin t + c_2 \sin t + d_1 \cos 2t + d_2 \sin 2t); \end{aligned}$$

thus

$$\begin{aligned} 0 &= (2a_1 - c_1) \cos t + (2a_2 - c_2) \sin t \\ &\quad + (-b_1 - d_1) \cos 2t + (-b_2 - d_2) \sin 2t. \end{aligned}$$

Because $\cos t$, $\cos 2t$, $\sin t$, and $\sin 2t$ are linearly independent, it follows that their coefficients in the last equation are zero. Thus

$$c_1 = 2a_1, \quad c_2 = 2a_2, \quad d_1 = -b_1, \quad \text{and} \quad d_2 = -b_2.$$

Therefore

$$x(t) = a_1 \cos t + a_2 \sin t + b_1 \cos 2t + b_2 \sin 2t \ ,$$
$$y(t) = 2a_1 \cos t + 2a_2 \sin t - b_1 \cos 2t - b_2 \sin 2t \tag{23}$$

is the desired general solution of the system in (20). ∎

The equations in (23) describe **free oscillations** of the mass-and-spring system of Fig. 4.2.2—motion subject to *no* external forces. Four initial conditions (typically, initial displacements and velocities) would be required to determine the values of a_1, a_2, b_1, and b_2. The expression

$$\begin{aligned} (x(t), y(t)) &= a_1(\cos t, 2\cos t) + a_2(\sin t, 2\sin t) \\ &\quad + b_1(\cos 2t, -\cos 2t) + b_2(\sin 2t, -\sin 2t) \end{aligned} \tag{24}$$

then presents the general solution of the system in (20) as a linear combination of four particular solutions. Moreover, the first two of these particular solutions represent physically similar oscillations of the masses, as do the latter two.

Indeed, we can (by the usual trigonometric machinations) write

$$a_1 \cos t + a_2 \sin t = A \cos(t - \alpha),$$
$$2a_1 \cos t + 2a_2 \sin t = 2A \cos(t - \alpha)$$

and

$$b_1 \cos 2t + b_2 \sin 2t = B \cos(2t - \beta),$$
$$-b_1 \cos 2t - b_2 \sin 2t = -B \cos(2t - \beta)$$

with $A = \sqrt{a_1^2 + a_2^2}$, $\tan \alpha = a_2/a_1$, $B = \sqrt{b_1^2 + b_2^2}$, and $\tan \beta = b_2/b_1$. Then Eq. (24) takes the form

$$(x, y) = A(x_1, y_1) + B(x_2, y_2), \tag{25}$$

where the particular solutions

$$(x_1(t), y_1(t)) = (\cos(t - \alpha), 2\cos(t - \alpha)) \tag{26}$$

and

$$(x_2(t), y_2(t)) = (\cos(2t - \beta), -\cos(2t - \beta)) \tag{27}$$

describe the two **natural modes of oscillation** of the mass-and-spring system. Moreover, they exhibit its two (circular) **natural frequencies** $\omega_1 = 1$ and $\omega_2 = 2$.

The linear combination in Eq. (25) represents an arbitrary free oscillation of the mass-and-spring system as a superposition of its two natural modes of oscillation, with the constants A, α, B, and β determined by the initial conditions. Figure 4.2.3 (where $\alpha = 0$) illustrates the natural mode (x_1, y_1) of Eq. (26), in which the two masses move in synchrony in the same direction with the same frequency of oscillation $\omega_1 = 1$, but with the amplitude of m_2 twice that of m_1 (because $y_1 = 2x_1$). Figure 4.2.4 (where $\beta = 0$) illustrates the natural mode (x_2, y_2) of Eq. (27), in which the masses move in synchrony in opposite directions, with the same frequency $\omega_2 = 2$ and with equal amplitudes of oscillation (because $y_2 = -x_2$).

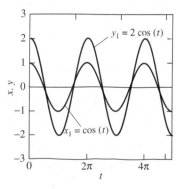

FIGURE 4.2.3. The two masses move in the same direction, each with frequency $\omega_1 = 1$.

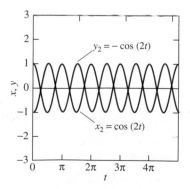

FIGURE 4.2.4. The two masses move in opposite directions with frequency $\omega_2 = 2$.

4.2 Problems

Find general solutions of the linear systems in Problems 1 through 20. If initial conditions are given, find the particular solution that satisfies them. In Problems 1 through 6, use a computer system or graphing calculator to construct a direction field and typical solution curves for the given system.

1. $x' = -x + 3y$, $y' = 2y$
2. $x' = x - 2y$, $y' = 2x - 3y$
3. $x' = -3x + 2y$, $y' = -3x + 4y$; $x(0) = 0$, $y(0) = 2$
4. $x' = 3x - y$, $y' = 5x - 3y$; $x(0) = 1$, $y(0) = -1$
5. $x' = -3x - 4y$, $y' = 2x + y$
6. $x' = x + 9y$, $y' = -2x - 5y$; $x(0) = 3$, $y(0) = 2$
7. $x' = 4x + y + 2t$, $y' = -2x + y$
8. $x' = 2x + y$, $y' = x + 2y - e^{2t}$
9. $x' = 2x - 3y + 2\sin 2t$, $y' = x - 2y - \cos 2t$
10. $x' + 2y' = 4x + 5y$, $2x' - y' = 3x$; $x(0) = 1$, $y(0) = -1$
11. $2y' - x' = x + 3y + e^t$, $3x' - 4y' = x - 15y + e^{-t}$
12. $x'' = 6x + 2y$, $y'' = 3x + 7y$
13. $x'' = -5x + 2y$, $y'' = 2x - 8y$
14. $x'' = -4x + \sin t$, $y'' = 4x - 8y$
15. $x'' - 3y' - 2x = 0$, $y'' + 3x' - 2y = 0$
16. $x'' + 13y' - 4x = 6\sin t$, $y'' - 2x' - 9y = 0$
17. $x'' + y'' - 3x' - y' - 2x + 2y = 0$,
 $2x'' + 3y'' - 9x' - 2y' - 4x + 6y = 0$
18. $x' = x + 2y + z$, $y' = 6x - y$, $z' = -x - 2y - z$
19. $x' = 4x - 2y$, $y' = -4x + 4y - 2z$, $z' = -4y + 4z$
20. $x' = y + z + e^{-t}$, $y' = x + z$, $z' = x + y$ (*Suggestion*: Solve the characteristic equation by inspection.)
21. Suppose that $L_1 = a_1 D^2 + b_1 D + C_1$ and $L_2 = a_2 D^2 + b_2 D + c_2$, where the coefficients are all constants, and that $x(t)$ is a twice differentiable function. Verify that $L_1 L_2 x = L_2 L_1 x$.
22. Suppose that $L_1 x = t Dx + x$ and that $L_2 x = Dx + tx$. Show that $L_1 L_2 x \neq L_2 L_1 x$. Thus linear operators with *variable* coefficients generally do not commute.

Show that the systems in Problems 23 through 25 are degenerate. In each problem determine—by attempting to solve the system—whether it has infinitely many solutions or no solutions.

23. $(D + 2)x + (D + 2)y = e^{-3t}$
 $(D + 3)x + (D + 3)y = e^{-2t}$
24. $(D + 2)x + (D + 2)y = t$
 $(D + 3)x + (D + 3)y = t^2$
25. $(D^2 + 5D + 6)x + D(D + 2)y = 0$
 $(D + 3)x + Dy = 0$

In Problems 26 through 29, first calculate the operational determinant, then attempt to solve the given system in order to determine the number of arbitrary constants that appear in its general solution.

26. $(D^2 + 1)x + D^2 y = 2e^{-t}$
 $(D^2 - 1)x + D^2 y = 0$
27. $(D^2 + 1)x + (D^2 + 2)y = 2e^{-t}$
 $(D^2 - 1)x + D^2 y = 0$
28. $(D^2 + D)x + D^2 y = 2e^{-t}$
 $(D^2 - 1)x + (D^2 - D)y = 0$
29. $(D^2 + 1)x - D^2 y = 2e^{-t}$
 $(D^2 - 1)x + D^2 y = 0$
30. Suppose that the salt concentration in each of the two brine tanks of Example 2 of Section 4.1 initially ($t = 0$) is 0.5 lb/gal. Then solve the system in Eq. (5) there to find the amounts $x(t)$ and $y(t)$ of salt in the two tanks at time t.
31. Suppose that the electrical network of Example 3 of Section 4.1 is initially open—no currents are flowing. Assume that it is closed at time $t = 0$; solve the system in Eq. (9) there to find $I_1(t)$ and $I_2(t)$.
32. Repeat Problem 31, except use the electrical network of Problem 27 of Section 4.1.
33. Repeat Problem 31, except use the electrical network of Problem 28 of Section 4.1. Assume that $I_1(0) = 2$ and $Q(0) = 0$, so that at time $t = 0$ there is no charge on the capacitor.
34. Three 100-gal brine tanks are connected as indicated in Fig. 4.1.13 of Section 4.1. Assume that the first tank initially contains 100 lb of salt, whereas the other two are filled with fresh water. Find the amounts of salt in each of the three tanks at time t. (*Suggestion*: Examine the equations to be derived in Problem 26 of Section 4.1.)
35. From Problem 31 of Section 4.1, recall the equations of motion

$$mx'' = qBy', \qquad my'' = -qBx'$$

for a particle of mass m and electrical charge q under the influence of the uniform magnetic field $\mathbf{B} = B\mathbf{k}$. Suppose that the initial conditions are $x(0) = r_0$, $y(0) = 0$, $x'(0) = 0$, and $y'(0) = -\omega r_0$ where $\omega = qB/m$. Show that the trajectory of the particle is a circle of radius r_0.

36. If, in addition to the magnetic field $\mathbf{B} = B\mathbf{k}$, the charged particle of Problem 35 moves with velocity \mathbf{v} under the influence of a uniform electric field $\mathbf{E} = E\mathbf{i}$, then the force acting on it is $\mathbf{F} = q(\mathbf{E} + \mathbf{v} \times \mathbf{B})$. Assume that the particle starts from rest at the origin. Show that its trajectory is the cycloid

$$x = a(1 - \cos \omega t), \qquad y = -a(\omega t - \sin \omega t)$$

where $a = E/(\omega B)$ and $\omega = qB/m$. The graph of such a cycloid is shown in Fig. 4.2.5.

FIGURE 4.2.5. The cycloidal path of the particle of Problem 36.

37. In the mass-and-spring system of Example 3, suppose instead that $m_1 = 2$, $m_2 = 0.5$, $k_1 = 75$, and $k_2 = 25$. **(a)** Find the general solution of the equations of motion of the system. In particular, show that its natural frequencies are $\omega_1 = 5$ and $\omega_2 = 5\sqrt{3}$. **(b)** Describe the natural modes of oscillation of the system.

38. Consider the system of two masses and three springs shown in Fig. 4.2.6. Derive the equations of motion

$$m_1 x'' = -(k_1 + k_2)x + k_2 y$$
$$m_2 y'' = k_2 x - (k_2 + k_3)y.$$

FIGURE 4.2.6. The mechanical system of Problem 38.

In Problems 39 through 46, find the general solution of the system in Problem 38 with the given masses and spring constants. Find the natural frequencies of the mass-and-spring system and describe its natural modes of oscillation. Use a computer system or graphing calculator to illustrate the two natural modes graphically (as in Figs. 4.2.3 and 4.2.4).

39. $m_1 = 4$, $m_2 = 2$, $k_1 = 8$, $k_2 = 4$, $k_3 = 0$
40. $m_1 = 2$, $m_2 = 1$, $k_1 = 100$, $k_2 = 50$, $k_3 = 0$
41. $m_1 = 1$, $m_2 = 1$, $k_1 = 1$, $k_2 = 4$, $k_3 = 1$
42. $m_1 = 1$, $m_2 = 2$, $k_1 = 1$, $k_2 = 2$, $k_3 = 2$
43. $m_1 = 1$, $m_2 = 1$, $k_1 = 1$, $k_2 = 2$, $k_3 = 1$
44. $m_1 = 1$, $m_2 = 1$, $k_1 = 2$, $k_2 = 1$, $k_3 = 2$
45. $m_1 = 1$, $m_2 = 2$, $k_1 = 2$, $k_2 = 4$, $k_3 = 4$
46. $m_1 = 1$, $m_2 = 1$, $k_1 = 4$, $k_2 = 6$, $k_3 = 4$

47. (a) For the system shown in Fig. 4.2.7, derive the equations of motion

$$mx'' = -2kx + ky$$
$$my'' = kx - 2ky + kz$$
$$mz'' = ky - 2kz.$$

(b) Assume that $m = k = 1$. Show that the natural frequencies of oscillation of the system are

$$\omega_1 = \sqrt{2}, \quad \omega_2 = \sqrt{2 - \sqrt{2}}, \quad \text{and} \quad \omega_3 = \sqrt{2 + \sqrt{2}}.$$

FIGURE 4.2.7. The mechanical system of Problem 47.

48. Suppose that the trajectory $(x(t), y(t))$ of a particle moving in the plane satisfies the initial value problem

$$x'' - 2y' + 3x = 0,$$
$$y'' + 2x' + 3y = 0;$$
$$x(0) = 4, \quad y(0) = x'(0) = y'(0) = 0.$$

Solve this problem. You should obtain

$$x(t) = 3\cos t + \cos 3t,$$
$$y(t) = 3\sin t - \sin 3t.$$

Verify that these equations describe the *hypocycloid* traced by a point $P(x, y)$ fixed on the circumference of a circle of radius $b = 1$ that rolls around inside a circle of radius $a = 4$. If P begins at $A(a, 0)$ when $t = 0$, then the parameter t represents the angle AOC shown in Fig. 4.2.8.

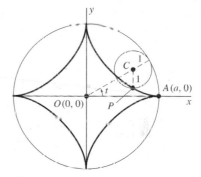

FIGURE 4.2.8. The hypocycloid of Problem 48.

4.2 Application: Computer Algebra Solution of Systems

Computer algebra systems can be used to solve systems as well as single differential equations. For instance, consider the system

$$\frac{dx}{dt} = 4x - 3y, \quad \frac{dy}{dt} = 6x - 7y \tag{1}$$

of Example 1. The *Maple* command

```
dsolve({diff(x(t),t) = 4*x(t) - 3*y(t),
```

```
                    diff(y(t),t) = 6*x(t) - 7*y(t)}, {x(t),y(t)});
```

yields

$$x(t) = \tfrac{1}{7}(3a_1 - 2a_2)e^{-5t} + \tfrac{1}{7}(-3a_1 + 9a_2)e^{2t},$$

$$y(t) = \tfrac{1}{7}(9a_1 - 6a_2)e^{-5t} + \tfrac{1}{7}(-2a_1 + 6a_2)e^{2t} \tag{2}$$

(after a bit of simplification), and the *Mathematica* command

```
    DSolve[ {x'[t] == 4 x[t] - 3 y[t],
             y'[t] == 6 x[t] - 7 y[t]}, {x[t],y[t]}, t ]
```

yields

$$x(t) = b_1 e^{-5t} + 3b_2 e^{2t}, \quad y(t) = 3b_1 e^{-5t} + 2b_2 e^{2t}. \tag{3}$$

Is it clear that the general solutions in (2) and (3) are equivalent to each other and to the general solution

$$x(t) = \tfrac{3}{2}c_1 e^{2t} + \tfrac{1}{3}c_2 e^{-5t}, \quad y(t) = c_1 e^{2t} + c_2 e^{-5t} \tag{4}$$

found in the text? What is the relation between the constants a_1, a_2 in (2), the constants b_1, b_2 in (3), and the constants c_1, c_2 in (4)?

Now consider the initial value problem

$$\begin{aligned} x'' &= -3x + y, & x(0) &= 0, & x'(0) &= 6, \\ y'' &= 2x - 2y, & y(0) &= 0, & y'(0) &= 6 \end{aligned} \tag{5}$$

for the mass-and-spring system of Example 3 in the text. Then the *Maple* command

```
    dsolve({diff(x(t),t,t) = -3*x(t) + y(t),
            diff(y(t),t,t) = 2*x(t) - 2*y(t),
            x(0) = 0, y(0) = 0, D(x)(0) = 6, D(y)(0) = 6},
            {x(t),y(t)});
```

and the *Mathematica* command

```
    DSolve[{x''[t] == -3 x[t] + y[t],
            y''[t] == 2 x[t] - 2 y[t],
            x[0] == 0, y[0] == 0, x'[0] == 6, y'[0] == 6},
            {x[t], y[t]}, t ] // ExpToTrig // Simplify
```

both yield the solution

$$x(t) = 4\sin t + \sin 2t, \quad y(t) = 8\sin t - \sin 2t \tag{6}$$

where we see a linear combination of

- an oscillation with frequency 1 in which the two masses move synchronously with the amplitude of motion of the second mass twice that of the first, and
- an oscillation of frequency 2 in which the two masses move in opposite directions with the same amplitude of motion.

You can apply similarly an available computer algebra system to solve Problems 1 through 20 and 39 through 46 in this section (supplying initial conditions for the latter ones if you like).

4.3 Numerical Methods for Systems

We now discuss the numerical approximation of solutions of systems of differential equations. Our goal is to apply the methods of Sections 2.4 through 2.6 to the initial value problem

➤ $$\mathbf{x}' = \mathbf{f}(t, \mathbf{x}), \quad \mathbf{x}(t_0) = \mathbf{x}_0 \tag{1}$$

for a system of m first-order differential equations. In (1) the independent variable is the scalar t, and

$$\mathbf{x} = (x_1, x_2, \ldots, x_m) \quad \text{and} \quad \mathbf{f} = (f_1, f_2, \ldots, f_m)$$

are vector-valued functions. If the component functions of \mathbf{f} and their first-order partial derivatives are all continuous in a neighborhood of the point (t_0, \mathbf{x}_0), then Theorems 3 and 4 of the Appendix guarantee the existence and uniqueness of a solution $\mathbf{x} = \mathbf{x}(t)$ of (1) on some subinterval [of the t-axis] containing t_0. With this assurance we can proceed to discuss the numerical approximation of this solution.

Beginning with step size h, we want to approximate the values of $\mathbf{x}(t)$ at the points t_1, t_2, t_3, \ldots, where $t_{n+1} = t_n + h$ for $n \geq 0$. Suppose that we have already computed the *approximations*

$$\mathbf{x}_1, \quad \mathbf{x}_2, \quad \mathbf{x}_3, \quad \ldots, \quad \mathbf{x}_n$$

to the *actual values*

$$\mathbf{x}(t_1), \quad \mathbf{x}(t_2), \quad \mathbf{x}(t_3), \quad \ldots, \quad \mathbf{x}(t_n)$$

of the exact solution of the system in (1). We can then make the step from \mathbf{x}_n to the next approximation $\mathbf{x}_{n+1} \approx \mathbf{x}(t_{n+1})$ by any one of the methods of Sections 2.4 through 2.6. Essentially all that is required is to write the iterative formula of the selected method in the vector notation of the present discussion.

Euler Methods for Systems

For example, the iterative formula of Euler's method for systems is

➤ $$\mathbf{x}_{n+1} = \mathbf{x}_n + h\mathbf{f}(t, \mathbf{x}_n). \tag{2}$$

To examine the case $m = 2$ of a pair of first-order differential equations, let us write

$$\mathbf{x} = \begin{bmatrix} x \\ y \end{bmatrix} \quad \text{and} \quad \mathbf{f} = \begin{bmatrix} f \\ g \end{bmatrix}.$$

Then the initial value problem in (1) is

$$\begin{aligned} x' &= f(t, x, y), \quad x(t_0) = x_0, \\ y' &= g(t, x, y), \quad y(t_0) = y_0, \end{aligned} \tag{3}$$

and the scalar components of the vector formula in (2) are

$$\begin{aligned} x_{n+1} &= x_n + hf(t_n, x_n, y_n), \\ y_{n+1} &= y_n + hg(t_n, x_n, y_n). \end{aligned} \tag{4}$$

Note that each iterative formula in (4) has the form of a single Euler iteration, but with y_n inserted like a parameter in the first formula (for x_{n+1}) and with x_n inserted like a parameter in the second formula (for y_{n+1}). The generalization to the system in (3) of each of the other methods in Sections 2.4 through 2.6 follows a similar pattern.

The improved Euler method for systems consists at each step of calculating first the predictor

$$\mathbf{u}_{n+1} = \mathbf{x}_n + h\mathbf{f}(t_n, \mathbf{x}_n) \tag{5}$$

and then the corrector

$$\mathbf{x}_{n+1} = \mathbf{x}_n + \frac{h}{2}[\mathbf{f}(t_n, \mathbf{x}_n) + \mathbf{f}(t_{n+1}, \mathbf{u}_{n+1})]. \tag{6}$$

For the case of the two-dimensional initial value problem in (3), the scalar components of the formulas in (5) and (6) are

$$\begin{aligned} u_{n+1} &= x_n + hf(t_n, x_n, y_n), \\ v_{n+1} &= y_n + hg(t_n, x_n, y_n) \end{aligned} \tag{7}$$

and

$$x_{n+1} = x_n + \frac{h}{2}[f(t_n, x_n, y_n) + f(t_{n+1}, u_{n+1}, v_{n+1})],$$
$$y_{n+1} = y_n + \frac{h}{2}[g(t_n, x_n, y_n) + g(t_{n+1}, u_{n+1}, v_{n+1})]. \tag{8}$$

Example 1 Consider the initial value problem

$$\begin{aligned} x' &= 3x - 2y, & x(0) &= 3; \\ y' &= 5x - 4y, & y(0) &= 6. \end{aligned} \tag{9}$$

The exact solution of the system in (9) is

$$x(t) = 2e^{-2t} + e^t, \qquad y(t) = 5e^{-2t} + e^t. \tag{10}$$

Here we have $f(x, y) = 3x - 2y$ and $g(x, y) = 5x - 4y$ in (3), so the Euler iterative formulas in (4) are

$$x_{n+1} = x_n + h \cdot (3x_n - 2y_n), \quad y_{n+1} = y_n + h \cdot (5x_n - 4y_n).$$

With step size $h = 0.1$ we calculate

$$x_1 = 3 + (0.1) \cdot [3 \cdot 3 - 2 \cdot 6] = 2.7,$$
$$y_1 = 6 + (0.1) \cdot [5 \cdot 3 - 4 \cdot 6] = 5.1$$

and

$$x_2 = 2.7 + (0.1) \cdot [3 \cdot (2.7) - 2 \cdot (5.1)] = 2.49,$$
$$y_2 = 5.1 + (0.1) \cdot [5 \cdot (2.7) - 4 \cdot (5.1)] = 4.41.$$

The actual values at $t_2 = 0.2$ given by (10) are $x(0.2) \approx 2.562$ and $y(0.2) \approx 4.573$.

To compute the improved Euler approximations to $x(0.2)$ and $y(0.2)$ with a single step of size $h = 0.2$, we first calculate the predictors

$$u_1 = 3 + (0.2) \cdot [3 \cdot 3 - 2 \cdot 6] = 2.4,$$
$$v_1 = 6 + (0.2) \cdot [5 \cdot 3 - 4 \cdot 6] = 4.2.$$

Then the corrector formulas in (8) yield

$$x_1 = 3 + (0.1) \cdot ([3 \cdot 3 - 2 \cdot 6] + [3 \cdot (2.4) - 2 \cdot (4.2)]) = 2.58,$$
$$y_1 = 6 + (0.1) \cdot ([5 \cdot 3 - 4 \cdot 6] + [5 \cdot (2.4) - 4 \cdot (4.2)]) = 4.62.$$

As we would expect, a single improved Euler step gives better accuracy than two ordinary Euler steps. ∎

The Runge-Kutta Method and Second-Order Equations

The vector version of the iterative formula for the Runge-Kutta method is

➤ $$\mathbf{x}_{n+1} = \mathbf{x}_n + \frac{h}{6}(\mathbf{k}_1 + 2\mathbf{k}_2 + 2\mathbf{k}_3 + \mathbf{k}_4) \tag{11}$$

where the vectors \mathbf{k}_1, \mathbf{k}_2, \mathbf{k}_3, and \mathbf{k}_4 are defined (by analogy with Eqs. (5a)–(5d) of Section 2.6) as follows:

➤ $$\begin{aligned} \mathbf{k}_1 &= \mathbf{f}(t_n, \mathbf{x}_n), \\ \mathbf{k}_2 &= \mathbf{f}\left(t_n + \tfrac{1}{2}h, \mathbf{x}_n + \tfrac{1}{2}h\mathbf{k}_1\right), \\ \mathbf{k}_3 &= \mathbf{f}\left(t_n + \tfrac{1}{2}h, \mathbf{x}_n + \tfrac{1}{2}h\mathbf{k}_2\right), \\ \mathbf{k}_4 &= \mathbf{f}(t_n + h, \mathbf{x}_n + h\mathbf{k}_3). \end{aligned} \tag{12}$$

To describe in scalar notation the Runge-Kutta method for the two-dimensional initial value problem

$$\begin{aligned} x' &= f(t, x, y), & x(t_0) &= x_0, \\ y' &= g(t, x, y), & y(t_0) &= y_0, \end{aligned} \tag{3}$$

let us write

$$\mathbf{x} = \begin{bmatrix} x \\ y \end{bmatrix}, \quad \mathbf{f} = \begin{bmatrix} f \\ g \end{bmatrix}, \quad \text{and} \quad \mathbf{k}_i = \begin{bmatrix} F_i \\ G_i \end{bmatrix}.$$

Then the Runge-Kutta iterative formulas for the step from (x_n, y_n) to the next approximation $(x_{n+1}, y_{n+1}) \approx (x(t_{n+1}), y(t_{n+1}))$ are

$$x_{n+1} = x_n + \frac{h}{6}(F_1 + 2F_2 + 2F_3 + F_4),$$

$$y_{n+1} = y_n + \frac{h}{6}(G_1 + 2G_2 + 2G_3 + G_4), \tag{13}$$

where the values F_1, F_2, F_3, and F_4 of the function f are

$$\begin{aligned}
F_1 &= f(t_n, x_n, y_n), \\
F_2 &= f\left(t_n + \tfrac{1}{2}h, x_n + \tfrac{1}{2}hF_1, y_n + \tfrac{1}{2}hG_1\right), \\
F_3 &= f\left(t_n + \tfrac{1}{2}h, x_n + \tfrac{1}{2}hF_2, y_n + \tfrac{1}{2}hG_2\right), \\
F_4 &= f(t_n + h, x_n + hF_3, y_n + hG_3);
\end{aligned} \tag{14}$$

G_1, G_2, G_3, and G_4 are the similarly defined values of the function g.

Perhaps the most common application of the two-dimensional Runge-Kutta method is to the numerical solution of second-order initial value problems of the form

$$\begin{aligned}
x'' &= g(t, x, x'), \\
x(t_0) &= x_0, \quad x'(t_0) = y_0.
\end{aligned} \tag{15}$$

If we introduce the auxiliary variable $y = x'$, then the problem in (15) translates into the two-dimensional first-order problem

$$\begin{aligned}
x' &= y, & x(t_0) &= x_0, \\
y' &= g(t, x, y), & y(t_0) &= y_0.
\end{aligned} \tag{16}$$

This is a problem of the form in (3) with $f(t, x, y) = y$.

If the functions f and g are not too complicated, then it is feasible to carry out manually a reasonable number of steps of the two-dimensional Runge-Kutta method described here. But the first operating electronic computers were constructed (during World War II) specifically to implement methods similar to the Runge-Kutta method for the numerical computation of trajectories of artillery projectiles. The application material for this section lists TI-85 and BASIC versions of Program RK2DIM that can be used with two-dimensional systems.

Example 2 The exact solution of the initial value problem

$$x'' = -x; \quad x(0) = 0, \quad x'(0) = 1 \tag{17}$$

is $x(t) = \sin t$. The substitution $y = x'$ translates (17) into the two-dimensional problem

$$\begin{aligned}
x' &= y, & x(0) &= 0; \\
y' &= -x, & y(0) &= 1,
\end{aligned} \tag{18}$$

which has the form in (3) with $f(t, x, y) = y$ and $g(t, x, y) = -x$. The table in Fig. 4.3.1 shows the results produced for $0 \leq t \leq 5$ (radians) using Program RK2DIM with step size $h = 0.05$. The values shown for $x = \sin t$ and $y = \cos t$ are all accurate to five decimal places. ■

Example 3 In Example 4 of Section 2.3 we considered a lunar lander that initially is falling freely toward the surface of the moon. Its retrorockets, when fired, provide a deceleration of $T = 4$ m/s^2. We found that a soft touchdown at the lunar surface is achieved by igniting these retrorockets when the lander is at a height of 41,870 meters (just over 26 miles) above the surface, and is then descending at the rate of 450 m/s.

t	$x = \sin t$	$y = \cos t$
0.5	+0.47943	+0.87758
1.0	+0.84147	+0.54030
1.5	+0.99749	+0.07074
2.0	+0.90930	−0.41615
2.5	+0.59847	−0.80114
3.0	+0.14112	−0.98999
3.5	−0.35078	−0.93646
4.0	−0.75680	−0.65364
4.5	−0.97753	−0.21080
5.0	−0.95892	+0.28366

FIGURE 4.3.1. Runge-Kutta values (with $h = 0.05$) for the problem in Eq. (18).

Now we want to compute the *descent time* of the lunar lander. Let the distance $x(t)$ of the lander from the center of the moon be measured in meters and measure time t in seconds. According to the analysis in Section 2.3 (where we used $r(t)$ instead of $x(t)$), $x(t)$ satisfies the initial value problem

$$\frac{d^2x}{dt^2} = T - \frac{GM}{x^2} = 4 - \frac{4.9044 \times 10^{12}}{x^2},$$

$$x(0) = R + 41870 = 1781870, \quad x'(0) = -450 \tag{19}$$

where $G \approx 6.6726 \times 10^{-11}$ N·(m/kg)2 is the universal gravitational constant and $M = 7.35 \times 10^{22}$ kg and $R = 1.74 \times 10^6$ m are the mass and radius of the moon. We seek the value of t when $x(t) = R = 1740000$.

The problem in (19) is equivalent to the first-order system

$$\frac{dx}{dt} = y, \quad x(0) = 1781870;$$

$$\frac{dy}{dx} = 4 - \frac{4.9044 \times 10^{12}}{x^2}, \quad y(0) = -450. \tag{20}$$

The table in Fig. 4.3.2 shows the result of a Runge-Kutta approximation with step size $h = 1$ (the indicated data agreeing with those obtained with step size $h = 2$). Evidently, touchdown on the lunar surface ($x = 1740000$) occurs at some time between $t = 180$ and $t = 190$ seconds. The table in Fig. 4.3.3 shows a second Runge-Kutta approximation with $t(0) = 180$, $x(0) = 1740059$, $y(0) = -16.83$, and $h = 0.1$. Now it is apparent that the lander's time of descent to the lunar surface is very close to 187 seconds; that is, 3 min 7 s. (The final velocity terms in these two tables are positive because the lander would begin to ascend if its retrorockets were not turned off at touchdown.) ∎

t (s)	x (m)	v (m/s)
0	1781870	−450.00
20	1773360	−401.04
40	1765826	−352.37
60	1759264	−303.95
80	1753667	−255.74
100	1749033	−207.73
120	1745357	−159.86
140	1742637	−112.11
160	1740872	−64.45
180	1740059	−16.83
200	1740199	30.77

FIGURE 4.3.2. The lander's descent to the lunar surface.

t (s)	x (m)	v (m/s)
180	1740059	−16.83
181	1740044	−14.45
182	1740030	−12.07
183	1740019	−9.69
184	1740011	−7.31
185	1740005	−4.93
186	1740001	−2.55
187	1740000	−0.17
188	1740001	2.21
189	1740004	4.59
190	1740010	6.97

FIGURE 4.3.3. Focusing on the lunar lander's soft touchdown.

Higher-Order Systems

As we saw in Section 4.1, any system of higher-order differential equations can be replaced with an equivalent system of first-order differential equations. For exam-

ple, consider the system

$$x'' = F(t, x, y, x', y'),$$
$$y'' = G(t, x, y, x', y') \qquad (21)$$

of second-order equations. If we substitute

$$x = x_1, \quad y = x_2, \quad x' = x_3 = x_1', \quad y' = x_4 = x_2',$$

then we get the equivalent system

$$x_1' = x_3,$$
$$x_2' = x_4,$$
$$x_3' = F(t, x_1, x_2, x_3, x_4), \qquad (22)$$
$$x_4' = G(t, x_1, x_2, x_3, x_4)$$

of four first-order equations in the unknown functions $x_1(t) = x(t)$, $x_2(t) = y(t)$, $x_3(t)$, and $x_4(t)$. It would be a routine (if slightly tedious) matter to write a four-dimensional version of program RK2DIM for the purpose of solving such a system. But in a programming language that accommodates vectors, an n-dimensional Runge-Kutta program is scarcely more complicated than a one-dimensional program. For instance, the application material for this section lists the n-dimensional MATLAB program **rkn** that closely resembles the one-dimensional program **rk** of Fig. 2.6.11.

Example 4 Suppose that a batted ball starts at $x_0 = 0$, $y_0 = 0$ with initial velocity $v_0 = 160$ ft/s and with initial angle of inclination $\theta = 30°$. If air resistance is ignored, we find by the elementary methods of Section 1.2 that the baseball travels a [horizontal] distance of $400\sqrt{3}$ ft (approximately 693 ft) in 5 s before striking the ground. Now suppose that in addition to a downward gravitational acceleration ($g = 32$ ft/s^2), the baseball experiences an acceleration due to air resistance of $(0.0025)v^2$ feet per second per second, directed opposite to its instantaneous direction of motion. Determine how far the baseball will travel horizontally under these conditions.

Solution According to Problem 30 of Section 4.1, the equations of motion of the baseball are

$$\frac{d^2x}{dt^2} = -cv\frac{dx}{dt}, \quad \frac{d^2y}{dt^2} = -cv\frac{dy}{dt} - g \qquad (23)$$

where $v = \sqrt{(x')^2 + (y')^2}$ is the speed of the ball, and where $c = 0.0025$ and $g = 32$ in fps units. We convert to a first-order system as in (22) and thereby obtain the system

$$x_1' = x_3,$$
$$x_2' = x_4,$$
$$x_3' = -cx_3\sqrt{x_3^2 + x_4^2}, \qquad (24)$$
$$x_4' = -cx_4\sqrt{x_3^2 + x_4^2} - g$$

of four first-order differential equations with

$$x_1(0) = x_2(0) = 0,$$
$$x_3(0) = 80\sqrt{3}, \quad x_4(0) = 80. \tag{25}$$

Note that $x_3(t)$ and $x_4(t)$ are simply the x- and y-components of the baseball's velocity vector, so $v = \sqrt{x_3^2 + x_4^2}$.

WITHOUT AIR RESISTANCE: Figure 4.3.4 shows the numerical results obtained when a Runge-Kutta program such as **rkn** is applied with step size $k = 0.1$ and with $c = 0$ (no air resistance). For convenience in interpreting the results, the printed output at each selected step consists of the horizontal and vertical coordinates x and y of the baseball, its velocity v, and the angle of inclination α of its velocity vector (in degrees measured from the horizontal). These results agree with the exact solution when $c = 0$. The ball travels a horizontal distance of $400\sqrt{3} \approx 692.82$ ft in exactly 5 s, having reached a maximum height of 100 ft after 2.5 s. Note also that the ball strikes the ground at the same angle and with the same speed as its initial angle and speed.

t	x	y	v	α
0.0	0.00	0.00	160.00	+30
0.5	69.28	36.00	152.63	+25
1.0	138.56	64.00	146.64	+19
1.5	207.85	84.00	142.21	+13
2.0	277.13	96.00	139.48	+7
2.5	346.41	100.00	138.56	+0
3.0	415.69	96.00	139.48	−7
3.5	484.97	84.00	142.21	−13
4.0	554.26	64.00	146.64	−19
4.5	623.54	36.00	152.63	−25
5.0	692.82	0.00	160.00	−30

FIGURE 4.3.4. The batted baseball with no air resistance ($c = 0$).

t	x	y	v	α
0.0	0.00	0.00	160.00	+30
0.5	63.25	32.74	127.18	+24
1.0	117.11	53.20	104.86	+17
1.5	164.32	63.60	89.72	+8
2.0	206.48	65.30	80.17	−3
2.5	244.61	59.22	75.22	−15
3.0	279.29	46.05	73.99	−27
3.5	310.91	26.41	75.47	−37
4.0	339.67	0.91	78.66	−46

FIGURE 4.3.5. The batted baseball with air resistance ($c = 0.0025$).

WITH AIR RESISTANCE: Figure 4.3.5 shows the results obtained with the fairly realistic value of $c = 0.0025$ for the air resistance for a batted baseball. To within a hundredth of a foot in either direction, the same results are obtained with step sizes $h = 0.05$ and $h = 0.025$. We now see that with air resistance the ball travels a distance well under 400 ft in just over 4 s. The more refined data in Fig. 4.3.6 show that the ball travels horizontally only about 340 ft and that its maximum height is only about 66 ft. As illustrated in Fig. 4.3.7, air resistance has converted a massive home run into a routine fly ball (if hit straightaway to center field). Note also that when the ball strikes the ground, it has slightly under *half* its initial speed (only about 79 ft/s) and is falling at a steeper angle (about 46°). Every baseball fan has observed empirically these aspects of the trajectory of a fly ball.

t	x	y	v	α
1.5	164.32	63.60	89.72	+8
1.6	173.11	64.60	87.40	+5
1.7	181.72	65.26	85.29	+3
1.8	190.15	65.60	83.39	+1
1.9	198.40	65.61	81.68	−1 ← Apex
2.0	206.48	65.30	80.17	−3
⋮	⋮	⋮	⋮	⋮
3.8	328.50	11.77	77.24	−42
3.9	334.14	6.45	77.93	−44
4.0	339.67	0.91	78.66	−46
4.1	345.10	−4.84	79.43	−47 ← Impact
4.2	350.41	−10.79	80.22	−49

FIGURE 4.3.6. The batted ball's apex and its impact with the ground.

Variable Step Size Methods

The Runge-Kutta method for a large system requires an appreciable amount of computational labor, even when a computer is employed. Therefore, just as the step size h should not be so large that the resulting error in the solution is unacceptable, h ought not to be so small that too many steps are needed, hence requiring an unacceptable amount of computation. Thus the practical numerical solution of differential equations involves a tradeoff between accuracy and efficiency.

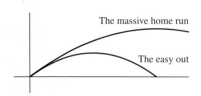

The massive home run

The easy out

FIGURE 4.3.7. An "easy out" or a home run?

To facilitate this tradeoff, modern *variable step size methods* vary the step size h as the solution process proceeds. Large steps are taken in regions where the dependent variables are changing slowly; smaller steps are taken when these variables are changing rapidly, in order to prevent large errors.

An *adaptable* or *variable step size* Runge-Kutta method employs both a pre-assigned minimum error tolerance *MinTol* and a maximum error tolerance *MaxTol* to attempt to ensure that the error made in the typical step from \mathbf{x}_n to \mathbf{x}_{n+1} is neither too large (and hence inaccurate) nor too small (and hence inefficient). A fairly simple scheme for doing this may be outlined as follows:

- Having reached \mathbf{x}_n with a Runge-Kutta step of length $t_n - t_{n-1} = h$, let $\mathbf{x}^{(1)}$ denote the result of a further Runge-Kutta step of length h and let $\mathbf{x}^{(2)}$ denote the result of *two* successive Runge-Kutta steps each of length $h/2$.
- On the grounds that $\mathbf{x}^{(2)}$ should be a more accurate approximation to $\mathbf{x}(t_n + h)$ than is $\mathbf{x}^{(1)}$, take

$$Err = |\mathbf{x}^{(1)} - \mathbf{x}^{(2)}|$$

as an estimate of the error in $\mathbf{x}^{(1)}$.

- If $MinTol \leq Err \leq MaxTol$, then let $\mathbf{x}_{n+1} = \mathbf{x}^{(1)}$, $t_{n+1} = t_n + h$, and proceed to the next step.
- If $Err < MinTol$, then the error is too small! Hence let $\mathbf{x}_{n+1} = \mathbf{x}^{(1)}$, $t_{n+1} = t_n + h$, but *double* the step size to $2h$ before making the next step.
- If $Err > MaxTol$, then the error is too large. Hence reject $\mathbf{x}^{(1)}$ and start afresh at \mathbf{x}_n with the *halved* step size $h/2$.

The detailed implementation of such a scheme can be complicated. For a much more complete but readable discussion of adaptive Runge-Kutta methods, see Section 15.2 of William H. Press et al., *Numerical Recipes*: *The Art of Scientific Computing* (Cambridge University Press, 1986).

Several widely available scientific computing packages (such as *Maple, Mathematica*, and MATLAB) include sophisticated variable step size programs that will accommodate an essentially arbitrary number of simultaneous differential equations. Such a general-purpose program might be used, for example, to model numerically the major components of the solar system: the sun and the nine (known) major planets. If m_i denotes the mass and $\mathbf{r}_i = (x_i, y_i, z_i)$ denotes the position vector of the ith one of these 10 bodies, then—by Newton's laws—the equation of motion of m_i is

$$m_i \mathbf{r}_i'' = \sum_{j \neq i} \frac{G m_i m_j}{(r_{ij})^3} (\mathbf{r}_j - \mathbf{r}_i), \tag{26}$$

where $r_{ij} = |\mathbf{r}_j - \mathbf{r}_i|$ denotes the distance between m_i and m_j. For each $i = 1$, 2, ..., 10, the summation in Eq. (26) is over all values of $j \neq i$ from 1 to 10. The 10 vector equations in (26) constitute a system of 30 second-order scalar equations, and the equivalent first-order system consists of 60 differential equations in the coordinates and velocity components of the 10 major bodies in the solar system. Mathematical models that involve this many (or more) differential equations—and that require sophisticated software and hardware for their numerical analysis—are quite common in science, engineering, and applied technology.

Earth-Moon Satellite Orbits

For an example of a program whose efficient solution requires adaptive step size methods, we consider an Apollo satellite in orbit about the Earth E and Moon M. Figure 4.3.8 shows an x_1x_2-coordinate system whose origin lies at the center of mass of the Earth and the Moon and which rotates at the rate of one revolution per "moon month" of approximately $\tau = 27.32$ days, so the Earth and Moon remain fixed in their positions on the x_1-axis. If we take as unit distance the distance (about 384,000 kilometers, assumed constant) between the Earth and Moon centers, then their coordinates are $E(-\mu, 0)$ and $M(1 - \mu, 0)$, where $\mu = m_M/(m_E + m_M)$ in terms of the Earth mass m_E and Moon mass m_M. If we take the total mass $m_E + m_M$ as the unit of mass and $\tau/(2\pi) \approx 4.35$ days as the unit of time, then the gravitational constant is $G = 1$ in Eq. (26), and the equations of motion of the satellite position $S(x_1, x_2)$ are

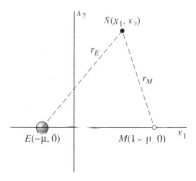

FIGURE 4.3.8. The Earth-Moon center-of-mass coordinate system.

$$x_1'' = x_1 + 2x_2' - \frac{(1 - \mu)(x_1 + \mu)}{(r_E)^3} - \frac{\mu(x_1 - 1 + \mu)}{(r_M)^3},$$

$$x_2'' = x_2 - 2x_1' - \frac{(1 - \mu)x_2}{(r_E)^3} - \frac{\mu x_2}{(r_M)^3}, \tag{27}$$

where r_E and r_M denote the satellite's distance to the Earth and Moon (indicated in Fig. 4.3.8). The initial two terms on the right-hand side of each equation result from the rotation of the coordinate system. In the system of units described here, the lunar mass is approximately $m_M = 0.012277471$. The second-order system in (27) can be converted to an equivalent first-order system (of four differential equations) by substituting

$$x_1' = x_3, \quad x_2' = x_4, \quad \text{so that} \quad x_1'' = x_3', \quad x_2'' = x_4'.$$

Suppose that the satellite initially is in a clockwise circular orbit of radius about 2400 kilometers about the Moon. At its farthest point from the Earth ($x_1 = 0.994$) it is "launched" into Earth-Moon orbit with initial velocity v_0. The corresponding initial conditions are

$$x_1(0) = 0.994, \quad x_2(0) = 0, \quad x_3(0) = 0, \quad x_4(0) = -v_0.$$

An adaptive step size method (**ode45**) in the MATLAB software system was used to solve numerically the system in (27). The orbits in Figs. 4.3.9 and 4.3.10 were obtained with

$$v_0 = 2.031732629557 \quad \text{and} \quad v_0 = 2.001585106379,$$

respectively. [In the system of units used here, the unit of velocity is approximately 3680 km/h.] In each case a closed but multilooped periodic trajectory about the Earth and the Moon—a so-called bus orbit—is obtained, but a relatively small change in the initial velocity changes the number of loops! For more information, see NASA Contractor Report CR-61139, "Study of the Methods for the Numerical Solution of Ordinary Differential Equations," prepared by O. B. Francis, Jr. et al. for the NASA–George C. Marshall Space Flight Center, June 7, 1966.

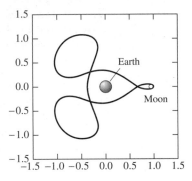

FIGURE 4.3.9. Apollo Moon-Earth bus orbit with insertion velocity $v_0 = 7476$ km/h.

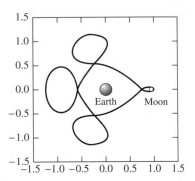

FIGURE 4.3.10. Apollo Moon-Earth bus orbit with insertion velocity $v_0 = 7365$ km/h.

4.3 Problems

*A hand-held calculator will suffice for Problems 1 through 8. In each problem an initial value problem and its exact solution are given. Approximate the values of $x(0.2)$ and $y(0.2)$ in three ways: **(a)** by the Euler method with two steps of size $h = 0.1$; **(b)** by the improved Euler method with a single step of size $h = 0.2$; and **(c)** by the Runge-Kutta method with a single step of size $h = 0.2$. Compare the approximate values with the actual values $x(0.2)$ and $y(0.2)$.*

1. $x' = x + 2y$, $x(0) = 0$,
 $y' = 2x + y$, $y(0) = 2$;
 $x(t) = e^{3t} - e^{-t}$, $y(t) = e^{3t} + e^{-t}$

2. $x' = 2x + 3y$, $x(0) = 1$,
 $y' = 2x + y$, $y(0) = -1$;
 $x(t) = e^{-t}$, $y(t) = -e^{-t}$

3. $x' = 3x + 4y$, $x(0) = 1$,
 $y' = 3x + 2y$, $y(0) = 1$;
 $x(t) = \frac{1}{7}(8e^{6t} - e^{-t})$, $y(t) = \frac{1}{7}(6e^{6t} + e^{-t})$

4. $x' = 9x + 5y$, $x(0) = 1$,
 $y' = -6x - 2y$, $y(0) = 0$;
 $x(t) = -5e^{3t} + 6e^{4t}$, $y(t) = 6e^{3t} - 6e^{4t}$

5. $x' = 2x - 5y$, $x(0) = 2$,
 $y' = 4x - 2y$, $y(0) = 3$;
 $x(t) = 2\cos 4t - \frac{11}{4}\sin 4t$, $y(t) = 3\cos 4t + \frac{1}{2}\sin 4t$

6. $x' = x - 2y$, $x(0) = 0$,
 $y' = 2x + y$, $y(0) = 4$;
 $x(t) = -4e^t \sin 2t$, $y(t) = 4e^t \cos 2t$

7. $x' = 3x - y$, $x(0) = 2$,
 $y' = x + y$, $y(0) = 1$;
 $x(t) = (t + 2)e^{2t}$, $y(t) = (t + 1)e^{2t}$

8. $x' = 5x - 9y$, $x(0) = 0$,
$\quad y' = 2x - y$, $y(0) = -1$;
$\quad x(t) = 3e^{2t} \sin 3t$, $y(t) = e^{2t}(\sin 3t - \cos 3t)$

A computer will be required for the remaining problems in this section. In Problems 9 through 12, an initial value problem and its exact solution are given. In each of these four problems, use the Runge-Kutta method with step sizes $h = 0.1$ and $h = 0.05$ to approximate to five decimal places the values $x(1)$ and $y(1)$. Compare the approximations with the actual values.

9. $x' = 2x - y$, $\quad x(0) = 1$,
$\quad y' = x + 2y$, $\quad y(0) = 0$;
$\quad x(t) = e^{2t} \cos t$, $\quad y(t) = e^{2t} \sin t$
10. $x' = x + 2y$, $\quad x(0) = 0$,
$\quad y' = x + e^{-t}$, $\quad y(0) = 0$;
$\quad x(t) = \frac{1}{9}(2e^{2t} - 2e^{-t} + 6te^{-t})$,
$\quad y(t) = \frac{1}{9}(e^{2t} - e^{-t} + 6te^{-t})$
11. $x' = -x - y - (1 + t^3)e^{-t}$, $\quad x(0) = 0$,
$\quad y' = -x - y - (t - 3t^2)e^{-t}$, $\quad y(0) = 1$;
$\quad x(t) = e^{-t}(\sin t - t)$, $\quad y(t) = e^{-t}(\cos t + t^3)$
12. $x'' + x = \sin t$, $\quad x(0) = 0$;
$\quad x(t) = \frac{1}{2}(\sin t - t \cos t)$
13. Suppose that a crossbow bolt is shot straight upward with initial velocity 288 ft/s. If its deceleration due to air resistance is $(0.04)v$, then its height $x(t)$ satisfies the initial value problem

$$x'' = -32 - (0.04)x'; \quad x(0) = 0, \quad x'(0) = 288.$$

Find the maximum height that the bolt attains and the time required for it to reach this height.
14. Repeat Problem 13, but assume instead that the deceleration of the bolt due to air resistance is $(0.0002)v^2$.
15. Suppose that a projectile is fired straight upward with initial velocity v_0 from the surface of the earth. If air resistance is not a factor, then its height $x(t)$ at time t satisfies the initial value problem

$$\frac{d^2x}{dt^2} = -\frac{gR^2}{(x + R)^2}; \quad x(0) = 0, \quad x'(0) = v_0.$$

Use the values $g = 32.15$ ft/s$^2 \approx 0.006089$ mi/s^2 for the gravitational acceleration of the earth at its surface and $R = 3960$ mi as the radius of the earth. If $v_0 = 1$ mi/s, find the maximum height attained by the projectile and its time of ascent to this height.

Problems 16 through 18 deal with the batted baseball of Example 4, having initial velocity 160 ft/s and air resistance coefficient $c = 0.0025$.

16. Find the *range*—the horizontal distance the ball travels before it hits the ground—and its total time of flight with initial inclination angles 40°, 45°, and 50°.
17. Find (to the nearest degree) the initial inclination that maximizes the range. If there were no air resistance it would be exactly 45°, but your answer should be less than 45°.
18. Find (to the nearest half degree) the initial inclination angle greater than 45° for which the range is 300 ft.
19. Find the initial velocity of a baseball hit by Babe Ruth (with $c = 0.0025$ and initial inclination 40°) if it hit the bleachers at a point 50 ft high and 500 horizontal feet from home plate.
20. Consider the crossbow bolt of Problem 14, fired with the same initial velocity of 288 ft/s and with the air resistance deceleration $(0.0002)v^2$ directed opposite its direction of motion. Suppose that this bolt is fired from ground level at an initial angle of 45°. Find how high vertically and how far horizontally it goes, and how long it remains in the air.
21. Suppose that an artillery projectile is fired from ground level with initial velocity 3000 ft/s and initial inclination angle 40°. Assume that its air resistance deceleration is $(0.0001)v^2$. **(a)** What is the range of the projectile and what is its total time of flight? What is its speed at impact with the ground? **(b)** What is the maximum altitude of the projectile, and when is that altitude attained? **(c)** You will find that the projectile is still losing speed at the apex of its trajectory. What is the *minimum* speed that it attains during its descent?

4.3 Application: Comets and Spacecraft

Figure 4.3.11 lists TI-85 and BASIC versions of the two-dimensional Runge-Kutta program RK2DIM. You should note that it closely parallels the one-dimensional Runge-Kutta program listed in Fig. 2.6.11, with a single line there replaced (where appropriate) with two lines here to calculate a *pair* of x- and y-values or slopes. Note also that the notation used is essentially that of Eqs. (13) and (14) in this section. The first several lines define the functions and initial data needed for Example 1.

Figure 4.3.12 exhibits an n-dimensional MATLAB implementation of the Runge-Kutta method. The MATLAB function **f** defines the vector of right-hand sides of the differential equations in the system $\mathbf{x}' = \mathbf{f}(t, \mathbf{x})$ to be solved. The **rkn** function then takes as input the initial t-value **t**, the column vector **x** of initial x-values, the final t-value **t1**, and the desired number **n** of subintervals. As output

TI-85	BASIC	Comment
PROGRAM:RK2DIM	Program RK2DIM	Program title
:F=Y	DEF FN F(T,X,Y) = Y	Define function f
:G=-X	DEF FN G(T,X,Y) = -X	Define function g
:50→N	N = 50	No. of steps
:0→T	T = 0	Initial t
:0→X	X = 0	Initial x
:1→Y	Y = 1	Initial y
:5→T1	T1 = 5	Final t
:(T1-T)/N→H	H = (T1-T)/N	Step size
:For(I,1,N)	FOR I=1 TO N	Begin loop
:T→T0	T0 = T	Save previous t
:X→X0	X0 = X	Save previous x
:Y→Y0	Y0 = Y	Save previous y
:F→F1	F1 = FNF(T,X,Y)	First f-slope
:G→G1	G1 = FNG(T,X,Y)	First g-slope
:T0+H/2→T	T = T0 + H/2	Midpoint t
:X0+H*F1/2→X	X = X0 + H*F1/2	Midpt x-predictor
:Y0+H*G1/2→Y	Y = Y0 + H*G1/2	Midpt y-predictor
:F→F2	F2 = FNF(T,X,Y)	Second f-slope
:G→G2	G2 = FNG(T,X,Y)	Second g-slope
:X0+H*F2/2→X	X = X0 + H*F2/2	Midpt x-predictor
:Y0+H*G2/2→Y	Y = Y0 + H*G2/2	Midpt y-predictor
:F→F3	F3 = FNF(T,X,Y)	Third f-slope
:G→G3	G3 = FNG(T,X,Y)	Third g-slope
:T0+H→T	T = T0 + H	New t
:X0+H*F3→X	X = X0 + H*F3	Endpt x-predictor
:Y0+H*G3→Y	Y = Y0 + H*G3	Endpt y-predictor
:F→F4	F4 = FNF(T,X,Y)	Fourth f-slope
:G→G4	G4 = FNG(T,X,Y)	Fourth g-slope
:(F1+2*F2+2*F3 +F4)/6→FA	FA = (F1+2*F2+2*F3 +F4)/6	Average f-slope
:(G1+2*G2+2*G3 +G4)/6→GA	GA = (G1+2*G2+2*G3 +G4)/6	Average g-slope
:X0+H*FA→X	X = Y0 + H*FA	x-corrector
:Y0+H*GA→Y	Y = Y0 + H*GA	y-corrector
:Disp T,X,Y	PRINT T,X,Y	Display results
:End	NEXT I	End loop

FIGURE 4.3.11. TI-85 and BASIC two-dimensional Runge-Kutta programs.

it produces the resulting column vector **T** of *t*-values and the matrix **X** whose rows give the corresponding *x*-values. For instance, with **f** as indicated in the figure, the MATLAB command

```
function xp = f(t,x)
xp = x;
xp(1) = x(2);
xp(2) = -x(1);

function [T,Y] = rkn(t,x,t1,n)
h = (t1 - t)/n;                 % step size
T = t;                          % initial t
X = x';                         % initial x-vector
for i = 1:n                     % begin loop
    k1 = f(t,x);                % first k-vector
    k2 = f(t+h/2,x+h*k1/2);     % second k-vector
    k3 = f(t+h/2,x+h*k2/2);     % third k-vector
    k4 = f(t+h  ,x+h*k3  );     % fourth k-vector
    k  = (k1+2*k2+2*k3+k4)/6;   % average k-vector
    t  = t + h;                 % new t
    x  = x + h*k;               % new x
    T = [T;t];                  % update t-column
    X = [X;x'];                 % update x-matrix
end                             % end loop
```

FIGURE 4.3.12. MATLAB implementation of the Runge-Kutta method.

```
[T,X] = rkn(0, [0;1], 5, 50)
```

then generates the data shown in the table of Fig. 4.3.1 (which lists only every fifth value of each variable).

You can use Examples 1 through 3 in this section to test your own implementation of the Runge-Kutta method. Then investigate the comet and spacecraft problems described next. Additional application material at the Web site www.prenhall.com/edwards describes additional numerical ODE investigations ranging from batted baseballs to the Apollo orbits shown in Figs. 4.3.9 and 4.3.10.

Your Spacecraft Landing

Your spacecraft is traveling at constant velocity V, approaching a distant earthlike planet with mass M and radius R. When activated, your deceleration system provides a constant thrust T until impact with the surface of the planet. During the period of deceleration, your distance $x(t)$ from the center of the planet satisfies the differential equation

$$\frac{d^2x}{dt^2} = T - \frac{GM}{x^2}, \tag{1}$$

where $G \approx 6.6726 \times 10^{-11}$ N·(m/kg)2 as in Example 3. Your question is this: At what altitude above the surface should your deceleration system be activated in order to achieve a soft touchdown? For a reasonable problem, you can take

$$M = 5.97 \times 10^{24} \text{ (kg)},$$
$$R = 6.38 \times 10^6 \text{ (m)},$$
$$V = p \times 10^4 \text{ (km/h)},$$
$$T = g + q \text{ (m/s}^2)$$

where $g = GM/R^2$ is the surface gravitational acceleration of the planet. Choose p to be the smallest nonzero digit and q the next-to-smallest nonzero digit in your ID number. Find the "ignition altitude" accurate to the nearest meter and the resulting "descent time" accurate to the nearest tenth of a second.

Kepler's Law of Planetary (or Satellite) Motion

Consider a satellite in elliptical orbit around a planet of mass M, and suppose that physical units are so chosen that $GM = 1$ (where G is the gravitational constant). If the planet is located at the origin in the xy-plane, then the equations of motion of the satellite are

$$\frac{d^2x}{dt^2} = -\frac{x}{(x^2 + y^2)^{3/2}}, \quad \frac{d^2y}{dt^2} = -\frac{y}{(x^2 + y^2)^{3/2}}. \tag{2}$$

Let T denote the period of revolution of the satellite. Kepler's third law says that the *square* of T is proportional to the *cube* of the major semiaxis a of its elliptical orbit. In particular, if $GM = 1$, then

$$T^2 = 4\pi^2 a^3. \tag{3}$$

(For details, see Section 12.6 of Edwards and Penney, *Calculus*, 6th ed. (Upper Saddle River, N.J.: Prentice Hall, 2002).) If the satellite's x- and y-components of velocity, $x_3 = x' = x_1'$ and $x_4 = y' = x_2'$, are introduced, then the system in (2) translates into a system of four first-order differential equations having the form of those in Eq. (22) of this section.

(a) Solve this 4×4 system numerically with the initial conditions

$$x(0) = 1, \quad y(0) = 0, \quad x'(0) = 0, \quad y'(0) = 1$$

that correspond theoretically to a circular orbit of radius $a = 1$, so Eq. (3) gives $T = 2\pi$. Is this what you get?

(b) Now solve the system numerically with the initial conditions

$$x(0) = 1, \quad y(0) = 0, \quad x'(0) = 0, \quad y'(0) = \tfrac{1}{2}\sqrt{6}$$

that correspond theoretically to an elliptical orbit with major semiaxis $a = 2$, so Eq. (3) gives $T = 4\pi\sqrt{2}$. Is this what you get?

Halley's Comet

Halley's comet last reached perihelion (its point of closest approach to the sun at the origin) on February 9, 1986. Its position and velocity components at that time were

$$\mathbf{p}_0 = (0.325514, -0.459460, 0.166229) \quad \text{and}$$
$$\mathbf{v}_0 = (-9.096111, -6.916686, -1.305721)$$

(respectively), with position in AU (astronomical units, in which the unit of distance is the major semiaxis of the earth's orbit) and time in years. In this system, the three-dimensional equations of motion of the comet are

$$\frac{d^2x}{dt^2} = -\frac{\mu x}{r^3}, \quad \frac{d^2y}{dt^2} = -\frac{\mu y}{r^3}, \quad \frac{d^2z}{dt^2} = -\frac{\mu z}{r^3} \tag{4}$$

FIGURE 4.3.13. yz-projection of the orbit of Halley's comet.

where

$$\mu = 4\pi^2 \quad \text{and} \quad r = \sqrt{x^2 + y^2 + z^2}.$$

Solve the equations in (4) numerically to verify the appearance of the yz-projection of the orbit of Halley's comet shown in Fig. 4.3.13. Plot the xy- and xz-projections as well.

Figure 4.3.14 shows the graph of the distance $r(t)$ of Halley's comet from the sun. Inspection of this graph indicates that Halley's comet reaches a maximum distance (at aphelion) of about 35 AU in a bit less than 40 years and returns to perihelion after about three-quarters of a century. The closer look in Fig. 4.3.15 indicates that the period of revolution of Halley's comet is about 76 years. Use your numerical solution to refine these observations. What is your best estimate of the calendar date of the comet's next perihelion passage?

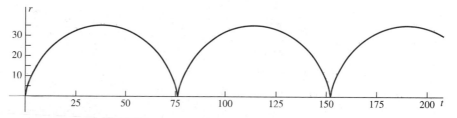

FIGURE 4.3.14. 200-year plot of the distance $r(t)$ of Halley's comet from the sun. Is there a cusp near $t = 75$?

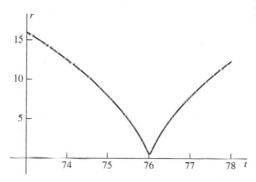

FIGURE 4.3.15. A closer look at Halley's perihelion passage after about 76 years.

Your Own Comet

The night before your birthday in 1997 you set up your telescope on a nearby mountaintop. It was a clear night, and you had a stroke of luck: At 12:30 A.M. you spotted a new comet. After repeating the observation on successive nights, you were able to calculate its solar system coordinates $\mathbf{p}_0 = (x_0, y_0, z_0)$ and its velocity vector $\mathbf{v}_0 = (vx_0, vy_0, vz_0)$ on that first night. Using this information, determine the following:

- the comet's perihelion (point nearest the sun) and aphelion (point farthest from the sun),
- the comet's velocity at perihelion and at aphelion,
- the comet's period of revolution around the sun, and
- the comet's next two dates of perihelion passage.

Using units of length in AU and time in earth years, the equations of motion of your comet are given in (4). For your personal comet, begin with random initial position and velocity vectors with the same order of magnitude as those of Halley's comet. Repeat the random selection of initial position and velocity vectors, if necessary, until you get a plausible eccentric orbit that ranges well outside the earth's orbit (as most real comets do).

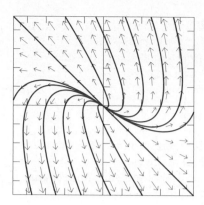

C H A P T E R

5

Linear Systems of Differential Equations

5.1 Matrices and Linear Systems

Although the simple elimination techniques of Section 4.2 suffice for the solution of small linear systems containing only two or three equations with constant coefficients, the general properties of linear systems—as well as solution methods suitable for larger systems—are most easily and concisely described using the language and notation of vectors and matrices. For ready reference and review, this section begins with a complete and self-contained account of the matrix notation and terminology that is needed. Special techniques of linear algebra—specifically, those associated with eigenvalues and eigenvectors—are introduced as needed in subsequent sections of this chapter.

Review of Matrix Notation and Terminology

An $m \times n$ **matrix A** is a rectangular array of mn numbers (or **elements**) arranged in m (horizontal) **rows** and n (vertical) **columns**:

$$\mathbf{A} = \begin{bmatrix} a_{11} & a_{12} & a_{13} & \cdots & a_{1j} & \cdots & a_{1n} \\ a_{21} & a_{22} & a_{23} & \cdots & a_{2j} & \cdots & a_{2n} \\ a_{31} & a_{32} & a_{33} & \cdots & a_{3j} & \cdots & a_{3n} \\ \vdots & \vdots & \vdots & & \vdots & & \vdots \\ a_{i1} & a_{i2} & a_{i3} & \cdots & a_{ij} & \cdots & a_{in} \\ \vdots & \vdots & \vdots & & \vdots & & \vdots \\ a_{m1} & a_{m2} & a_{m3} & \cdots & a_{mj} & \cdots & a_{mn} \end{bmatrix}. \tag{1}$$

We will ordinarily denote matrices by **boldface** capital letters. Sometimes we use the abbreviation $\mathbf{A} = \begin{bmatrix} a_{ij} \end{bmatrix}$ for the matrix with the element a_{ij} in the ith row and jth column, as in Eq. (1). We denote the **zero matrix,** each entry of which is zero,

by

$$\mathbf{0} = \begin{bmatrix} 0 & 0 & \cdots & 0 \\ 0 & 0 & \cdots & 0 \\ \vdots & \vdots & & \vdots \\ 0 & 0 & \cdots & 0 \end{bmatrix}. \tag{2}$$

Actually for each pair of positive integers m and n there is an $m \times n$ zero matrix, but the single symbol $\mathbf{0}$ will suffice for all these zero matrices.

Two $m \times n$ matrices $\mathbf{A} = \begin{bmatrix} a_{ij} \end{bmatrix}$ and $\mathbf{B} = \begin{bmatrix} b_{ij} \end{bmatrix}$ are said to be **equal** if corresponding elements are equal; that is, if $a_{ij} = b_{ij}$ for $1 \leqq i \leqq m$ and $1 \leqq j \leqq n$. We **add** \mathbf{A} and \mathbf{B} by adding corresponding entries:

➤ $$\mathbf{A} + \mathbf{B} = \begin{bmatrix} a_{ij} \end{bmatrix} + \begin{bmatrix} b_{ij} \end{bmatrix} = \begin{bmatrix} a_{ij} + b_{ij} \end{bmatrix}. \tag{3}$$

Thus the element in row i and column j of $\mathbf{C} = \mathbf{A} + \mathbf{B}$ is $c_{ij} = a_{ij} + b_{ij}$. To multiply the matrix \mathbf{A} by the number c, we simply multiply each of its elements by c:

➤ $$c\mathbf{A} = \mathbf{A}c = [ca_{ij}]. \tag{4}$$

Example 1 If

$$\mathbf{A} = \begin{bmatrix} 2 & -3 \\ 4 & 7 \end{bmatrix}, \quad \mathbf{B} = \begin{bmatrix} -13 & 10 \\ 7 & -5 \end{bmatrix}, \quad \text{and} \quad \mathbf{C} = \begin{bmatrix} 3 & 0 \\ 5 & -7 \end{bmatrix},$$

then

$$\mathbf{A} + \mathbf{B} = \begin{bmatrix} 2 & -3 \\ 4 & 7 \end{bmatrix} + \begin{bmatrix} -13 & 10 \\ 7 & -5 \end{bmatrix} = \begin{bmatrix} -11 & 7 \\ 11 & 2 \end{bmatrix}$$

and

$$6\mathbf{C} = 6 \cdot \begin{bmatrix} 3 & 0 \\ 5 & -7 \end{bmatrix} = \begin{bmatrix} 18 & 0 \\ 30 & -42 \end{bmatrix}. \qquad ■$$

We denote $(-1)\mathbf{A}$ by $-\mathbf{A}$ and define **subtraction** of matrices as follows:

$$\mathbf{A} - \mathbf{B} = \mathbf{A} + (-\mathbf{B}). \tag{5}$$

The matrix operations just defined have the following properties, each of which is analogous to a familiar algebraic property of the real number system:

$$\mathbf{A} + \mathbf{0} = \mathbf{0} + \mathbf{A} = \mathbf{A}, \qquad \mathbf{A} - \mathbf{A} = \mathbf{0}; \tag{6}$$

$$\mathbf{A} + \mathbf{B} = \mathbf{B} + \mathbf{A} \qquad \text{(commutativity)}; \tag{7}$$

$$\mathbf{A} + (\mathbf{B} + \mathbf{C}) = (\mathbf{A} + \mathbf{B}) + \mathbf{C} \qquad \text{(associativity)}; \tag{8}$$

$$c(\mathbf{A} + \mathbf{B}) = c\mathbf{A} + c\mathbf{B},$$
$$(c + d)\mathbf{A} = c\mathbf{A} + d\mathbf{A}. \qquad \text{(distributivity)} \tag{9}$$

Each of these properties is readily verified by elementwise application of a corresponding property of the real numbers. For example, $a_{ij} + b_{ij} = b_{ij} + a_{ij}$ for all i and j because addition of real numbers is commutative. Consequently,

$$\mathbf{A} + \mathbf{B} = \begin{bmatrix} a_{ij} + b_{ij} \end{bmatrix} = \begin{bmatrix} b_{ij} + a_{ij} \end{bmatrix} = \mathbf{B} + \mathbf{A}.$$

The **transpose** \mathbf{A}^T of the $m \times n$ matrix $\mathbf{A} = [a_{ij}]$ is the $n \times m$ (note!) matrix whose jth column is the jth row of \mathbf{A} (and consequently, whose ith row is the ith column of \mathbf{A}). Thus $\mathbf{A}^T = \begin{bmatrix} a_{ji} \end{bmatrix}$, although this is not notationally perfect; you must remember that \mathbf{A}^T will not have the same shape as \mathbf{A} unless \mathbf{A} is a **square** matrix—that is, unless $m = n$.

An $m \times 1$ matrix—one having only a single column—is called a **column vector,** or simply a **vector.** We often denote column vectors by **boldface** lowercase letters, as in

$$\mathbf{b} = \begin{bmatrix} 3 \\ -7 \\ 0 \end{bmatrix} \quad \text{or} \quad \mathbf{x} = \begin{bmatrix} x_1 \\ x_2 \\ \vdots \\ x_m \end{bmatrix}.$$

Similarly, a **row vector** is a $1 \times n$ matrix—one having only a single row, such as $\mathbf{c} = \begin{bmatrix} 5 & 17 & 0 & -3 \end{bmatrix}$. For aesthetic and typographical reasons, we will frequently write a column vector as the transpose of a row vector; for example, the two preceding column vectors may be written in the forms

$$\mathbf{b} = \begin{bmatrix} 3 & -7 & 0 \end{bmatrix}^T \quad \text{and} \quad \mathbf{x} = \begin{bmatrix} x_1 & x_2 & \cdots & x_m \end{bmatrix}^T.$$

Sometimes it is convenient to describe an $m \times n$ matrix in terms of either its m row vectors or its n column vectors. Thus if we write

$$\mathbf{A} = \begin{bmatrix} \mathbf{a}_1 \\ \mathbf{a}_2 \\ \vdots \\ \mathbf{a}_m \end{bmatrix} \quad \text{and} \quad \mathbf{B} = \begin{bmatrix} \mathbf{b}_1 & \mathbf{b}_2 & \cdots & \mathbf{b}_n \end{bmatrix},$$

it is understood that $\mathbf{a}_1, \mathbf{a}_2, \ldots,$ and \mathbf{a}_m are the *row* vectors of the matrix \mathbf{A} and that $\mathbf{b}_1, \mathbf{b}_2, \ldots,$ and \mathbf{b}_n are the *column* vectors of the matrix \mathbf{B}.

Matrix Multiplication

The properties listed in Eqs. (6) through (9) are quite natural and expected. The first surprises in the realm of matrix arithmetic come with multiplication. We define first the **scalar product** of a row vector \mathbf{a} and a column vector \mathbf{b}, each having the same number p of elements. If

$$\mathbf{a} = \begin{bmatrix} a_1 & a_2 & \cdots & a_p \end{bmatrix} \quad \text{and} \quad \mathbf{b} = \begin{bmatrix} b_1 & b_2 & \cdots & b_p \end{bmatrix}^T,$$

then $\mathbf{a} \cdot \mathbf{b}$ is defined as follows:

$$\blacktriangleright \qquad \mathbf{a} \cdot \mathbf{b} = \sum_{k=1}^{p} a_k b_k = a_1 b_1 + a_2 b_2 + \cdots + a_p b_p, \qquad (10)$$

exactly as in the scalar or *dot* product of two vectors—a familiar topic from elementary calculus.

The product \mathbf{AB} of two matrices is defined only if the number of columns of \mathbf{A} is equal to the number of rows of \mathbf{B}. If \mathbf{A} is an $m \times p$ matrix and \mathbf{B} is a $p \times n$ matrix, then their **product AB** is the $m \times n$ matrix $\mathbf{C} = \begin{bmatrix} c_{ij} \end{bmatrix}$, where c_{ij} is the scalar product of the ith row vector \mathbf{a}_i of \mathbf{A} and the jth column vector \mathbf{b}_j of \mathbf{B}. Thus

$$\blacktriangleright \qquad \mathbf{C} = \mathbf{AB} = \begin{bmatrix} \mathbf{a}_i \cdot \mathbf{b}_j \end{bmatrix}. \qquad (11)$$

In terms of the individual entries of $\mathbf{A} = \begin{bmatrix} a_{ij} \end{bmatrix}$ and $\mathbf{B} = \begin{bmatrix} b_{ij} \end{bmatrix}$, Eq. (11) can be recast in the form

$$c_{ij} = \sum_{k=1}^{p} a_{ik}b_{kj}. \tag{12}$$

For purposes of hand computation, the definition in Eqs. (11) and (12) is easy to remember by visualizing the picture

$$\mathbf{a}_i \longrightarrow \begin{bmatrix} a_{11} & a_{12} & \cdots & a_{1p} \\ a_{21} & a_{22} & \cdots & a_{2p} \\ \vdots & \vdots & & \vdots \\ a_{i1} & a_{i2} & \cdots & a_{ip} \\ \vdots & \vdots & & \vdots \\ a_{m1} & a_{m2} & \cdots & a_{mp} \end{bmatrix} \begin{bmatrix} b_{11} & b_{12} & \cdots & b_{1j} & \cdots & b_{1n} \\ b_{21} & b_{22} & \cdots & b_{2j} & \cdots & b_{2n} \\ \vdots & \vdots & & \vdots & & \vdots \\ b_{p1} & b_{p2} & \cdots & b_{pj} & \cdots & b_{pn} \end{bmatrix},$$

$$\begin{array}{c} \uparrow \\ \mathbf{b}_j \end{array}$$

which shows that one forms the dot product of the row vector \mathbf{a}_i with the column vector \mathbf{b}_j to obtain the element c_{ij} in the ith row and the jth column of \mathbf{AB}. It may help to think of "pouring the rows of \mathbf{A} down the columns of \mathbf{B}." This also reminds us that the number of columns of \mathbf{A} must be equal to the number of rows of \mathbf{B}.

Example 2

Check your understanding of the definition of matrix multiplication by verifying that if

$$\mathbf{A} = \begin{bmatrix} 2 & -3 \\ -1 & 5 \end{bmatrix} \quad \text{and} \quad \mathbf{B} = \begin{bmatrix} 13 & 9 \\ 4 & 0 \end{bmatrix},$$

then

$$\mathbf{AB} = \begin{bmatrix} 2 & -3 \\ -1 & 5 \end{bmatrix} \begin{bmatrix} 13 & 9 \\ 4 & 0 \end{bmatrix} = \begin{bmatrix} 14 & 18 \\ 7 & -9 \end{bmatrix}.$$

Similarly, verify that

$$\begin{bmatrix} 2 & -3 & 1 \\ 4 & 5 & -2 \\ 6 & -7 & 0 \end{bmatrix} \begin{bmatrix} x \\ y \\ z \end{bmatrix} = \begin{bmatrix} 2x - 3y + z \\ 4x + 5y - 2z \\ 6x - 7y \end{bmatrix}$$

and that

$$\begin{bmatrix} 1 & 2 \\ 3 & 4 \\ 5 & 6 \\ 7 & 8 \end{bmatrix} \begin{bmatrix} 2 & 1 & 3 \\ -1 & 3 & -2 \end{bmatrix} = \begin{bmatrix} 0 & 7 & -1 \\ 2 & 15 & 1 \\ 4 & 23 & 3 \\ 6 & 31 & 5 \end{bmatrix}. \qquad \blacksquare$$

It can be shown by direct (though lengthy) computation based on its definition that matrix multiplication is associative and is also distributive with respect to matrix addition; that is,

$$\mathbf{A}(\mathbf{BC}) = (\mathbf{AB})\mathbf{C} \tag{13}$$

and

$$A(B + C) = AB + AC, \tag{14}$$

provided that the matrices are of such sizes that the indicated multiplications and additions are possible.

But matrix multiplication is not commutative. That is, if A and B are both $n \times n$ matrices (so that both the products AB and BA are defined and have the same dimensions—$n \times n$), then, in general,

$$AB \neq BA. \tag{15}$$

Moreover, it can happen that

$$AB = 0 \quad \text{even though} \quad A \neq 0 \quad \text{and} \quad B \neq 0. \tag{16}$$

Examples illustrating the phenomena in (15) and (16) may be found in the problems, although you can easily construct your own examples using 2×2 matrices with small integral elements.

Inverse Matrices

A square $n \times n$ matrix is said to have **order** n. The **identity** matrix of order n is the square matrix

$$I = \begin{bmatrix} 1 & 0 & 0 & 0 & \cdots & 0 \\ 0 & 1 & 0 & 0 & \cdots & 0 \\ 0 & 0 & 1 & 0 & \cdots & 0 \\ 0 & 0 & 0 & 1 & \cdots & 0 \\ \vdots & \vdots & \vdots & \vdots & & \vdots \\ 0 & 0 & 0 & 0 & \cdots & 1 \end{bmatrix} \tag{17}$$

for which each entry on the **principal diagonal** is 1 and all off-diagonal entries are zero. It is quite easy to verify that

➤
$$AI = A = IA \tag{18}$$

for every square matrix A of the same order as I.

If A is a square matrix, then an **inverse** of A is a square matrix B of the same order as A such that *both*

$$AB = I \quad \text{and} \quad BA = I.$$

It is not difficult to show that if the matrix A has an inverse, then this inverse is unique. Consequently, we may speak of *the* inverse of A, and we will denote it by A^{-1}. Thus

➤
$$AA^{-1} = I = A^{-1}A, \tag{19}$$

given the existence of A^{-1}. It is clear that some square matrices do not have inverses—consider any square zero matrix. It is also easy to show that if A^{-1} exists, then $(A^{-1})^{-1}$ exists and $(A^{-1})^{-1} = A$.

In linear algebra it is proved that A^{-1} exists if and only the determinant $\det(A)$ of the square matrix A is nonzero. If so, the matrix A is said to be **nonsingular**; if $\det(A) = 0$, then A is called a **singular** matrix.

Determinants

We assume that the student has computed 2×2 and 3×3 determinants in earlier courses. If $\mathbf{A} = \left[\, a_{ij} \,\right]$ is a 2×2 matrix, then its **determinant** $\det(\mathbf{A}) = |\mathbf{A}|$ is defined as

$$|\mathbf{A}| = \begin{vmatrix} a_{11} & a_{12} \\ a_{21} & a_{22} \end{vmatrix} = a_{11}a_{22} - a_{12}a_{21}.$$

Determinants of higher order may be defined by induction, as follows. If $\mathbf{A} = \left[\, a_{ij} \,\right]$ is an $n \times n$ matrix, let \mathbf{A}_{ij} denote the $(n-1) \times (n-1)$ matrix obtained from \mathbf{A} by deleting its ith row and its jth column. The *expansion* of the determinant $|\mathbf{A}|$ along its ith row is given by

$$|\mathbf{A}| = \sum_{j=1}^{n} (-1)^{i+j} a_{ij} |\mathbf{A}_{ij}| \qquad (i \text{ fixed}), \tag{20a}$$

and its expansion along its jth column is given by

$$|\mathbf{A}| = \sum_{i=1}^{n} (-1)^{i+j} a_{ij} |\mathbf{A}_{ij}| \qquad (j \text{ fixed}). \tag{20b}$$

It is shown in linear algebra that whichever row we use in Eq. (20a) and whichever column we use in Eq. (20b), the results are the same in all $2n$ cases. Hence $|\mathbf{A}|$ is well defined by these formulas.

Example 3 If

$$\mathbf{A} = \begin{bmatrix} 3 & 1 & -2 \\ 4 & 2 & 1 \\ -2 & 3 & 5 \end{bmatrix},$$

then the expansion of $|\mathbf{A}|$ along its second row is

$$|\mathbf{A}| = -4 \cdot \begin{vmatrix} 1 & -2 \\ 3 & 5 \end{vmatrix} + 2 \cdot \begin{vmatrix} 3 & -2 \\ -2 & 5 \end{vmatrix} - 1 \cdot \begin{vmatrix} 3 & 1 \\ -2 & 3 \end{vmatrix}$$

$$= -4 \cdot 11 + 2 \cdot 11 - 1 \cdot 11 = -33.$$

And the expansion of $|\mathbf{A}|$ along its third column is

$$|\mathbf{A}| = -2 \cdot \begin{vmatrix} 4 & 2 \\ -2 & 3 \end{vmatrix} - 1 \cdot \begin{vmatrix} 3 & 1 \\ -2 & 3 \end{vmatrix} + 5 \cdot \begin{vmatrix} 3 & 1 \\ 4 & 2 \end{vmatrix}$$

$$= -2 \cdot 16 - 1 \cdot 11 + 5 \cdot 2 = -33. \qquad \blacksquare$$

Calculators and computers are convenient for the calculation of higher-dimensional determinants and inverse matrices, but determinants and inverses of 2×2 matrices are easy to compute by hand. For instance, if the 2×2 matrix

$$\mathbf{A} = \begin{bmatrix} a & b \\ c & d \end{bmatrix}$$

has nonzero determinant $|\mathbf{A}| = ad - bc \neq 0$, then its inverse matrix is

$$\mathbf{A}^{-1} = \frac{1}{|\mathbf{A}|} \begin{bmatrix} d & -b \\ -c & a \end{bmatrix}. \tag{21}$$

Note that the matrix on the right-hand side of Eq. (21) is obtained from **A** by interchanging the diagonal elements and changing the signs of the off-diagonal elements.

Example 4 If

$$\mathbf{A} = \begin{bmatrix} 6 & 8 \\ 5 & 7 \end{bmatrix},$$

then $|\mathbf{A}| = 6 \cdot 7 - 5 \cdot 8 = 2$. Hence Eq. (21) gives

$$\mathbf{A}^{-1} = \frac{1}{2} \begin{bmatrix} 7 & -8 \\ -5 & 6 \end{bmatrix} = \begin{bmatrix} \frac{7}{2} & -4 \\ -\frac{5}{2} & 3 \end{bmatrix}.$$

You should pause to verify that

$$\mathbf{A}^{-1}\mathbf{A} = \begin{bmatrix} \frac{7}{2} & -4 \\ -\frac{5}{2} & 3 \end{bmatrix} \begin{bmatrix} 6 & 8 \\ 5 & 7 \end{bmatrix} = \begin{bmatrix} 1 & 0 \\ 0 & 1 \end{bmatrix}. \qquad ■$$

Matrix-Valued Functions

A **matrix-valued function,** or simply **matrix function,** is a matrix such as

$$\mathbf{x}(t) = \begin{bmatrix} x_1(t) \\ x_2(t) \\ \vdots \\ x_n(t) \end{bmatrix} \qquad (22a)$$

or

$$\mathbf{A}(t) = \begin{bmatrix} a_{11}(t) & a_{12}(t) & \cdots & a_{1n}(t) \\ a_{21}(t) & a_{22}(t) & \cdots & a_{2n}(t) \\ \vdots & \vdots & & \vdots \\ a_{m1}(t) & a_{m2}(t) & \cdots & a_{mn}(t) \end{bmatrix}, \qquad (22b)$$

in which each entry is a function of t. We say that the matrix function $\mathbf{A}(t)$ is **continuous** (or **differentiable**) at a point (or on an interval) if each of its elements has the same property. The **derivative** of a differentiable matrix function is defined by elementwise differentiation; that is,

➤
$$\mathbf{A}'(t) = \frac{d\mathbf{A}}{dt} = \left[\frac{da_{ij}}{dt} \right]. \qquad (23)$$

Example 5 If

$$\mathbf{x}(t) = \begin{bmatrix} t \\ t^2 \\ e^{-t} \end{bmatrix} \quad \text{and} \quad \mathbf{A}(t) = \begin{bmatrix} \sin t & 1 \\ t & \cos t \end{bmatrix},$$

then

$$\frac{d\mathbf{x}}{dt} = \begin{bmatrix} 1 \\ 2t \\ -e^{-t} \end{bmatrix} \quad \text{and} \quad \mathbf{A}'(t) = \begin{bmatrix} \cos t & 0 \\ 1 & -\sin t \end{bmatrix}. \qquad \blacksquare$$

The differentiation rules

$$\frac{d}{dt}(\mathbf{A} + \mathbf{B}) = \frac{d\mathbf{A}}{dt} + \frac{d\mathbf{B}}{dt} \tag{24}$$

and

$$\frac{d}{dt}(\mathbf{A}\mathbf{B}) = \mathbf{A}\frac{d\mathbf{B}}{dt} + \frac{d\mathbf{A}}{dt}\mathbf{B} \tag{25}$$

follow readily by elementwise application of the analogous differentiation rules of elementary calculus for real-valued functions. If c is a (constant) real number and \mathbf{C} is a constant matrix, then

$$\frac{d}{dt}(c\mathbf{A}) = c\frac{d\mathbf{A}}{dt}, \quad \frac{d}{dt}(\mathbf{C}\mathbf{A}) = \mathbf{C}\frac{d\mathbf{A}}{dt}, \quad \text{and} \quad \frac{d}{dt}(\mathbf{A}\mathbf{C}) = \frac{d\mathbf{A}}{dt}\mathbf{C}. \tag{26}$$

Because of the noncommutativity of matrix multiplication, it is important not to reverse the order of the factors in Eqs. (25) and (26).

First-Order Linear Systems

The notation and terminology of matrices and vectors may seem rather elaborate when first encountered, but it is readily assimilated with practice. Our main use for matrix notation will be the simplification of computations with systems of differential equations, especially those computations that would be burdensome in scalar notation.

We discuss here the general system of n first-order linear equations

$$
\begin{aligned}
x_1' &= p_{11}(t)x_1 + p_{12}(t)x_2 + \cdots + p_{1n}(t)x_n + f_1(t), \\
x_2' &= p_{21}(t)x_1 + p_{22}(t)x_2 + \cdots + p_{2n}(t)x_n + f_2(t), \\
x_3' &= p_{31}(t)x_1 + p_{32}(t)x_2 + \cdots + p_{3n}(t)x_n + f_3(t), \\
&\qquad\qquad\qquad \vdots \\
x_n' &= p_{n1}(t)x_1 + p_{n2}(t)x_2 + \cdots + p_{nn}(t)x_n + f_n(t).
\end{aligned}
\tag{27}
$$

If we introduce the *coefficient matrix*

$$\mathbf{P}(t) = \begin{bmatrix} p_{ij}(t) \end{bmatrix}$$

and the column vectors

$$\mathbf{x} = \begin{bmatrix} x_i \end{bmatrix} \quad \text{and} \quad \mathbf{f}(t) = \begin{bmatrix} f_i(t) \end{bmatrix},$$

then the system in (27) takes the form of a single matrix equation

$$\frac{d\mathbf{x}}{dt} = \mathbf{P}(t)\mathbf{x} + \mathbf{f}(t). \tag{28}$$

We will see that the general theory of the linear system in (27) closely parallels that of a single nth-order equation. The matrix notation used in Eq. (28) not only emphasizes this analogy, but also saves a great deal of space.

A **solution** of Eq. (28) on the open interval I is a column vector function $\mathbf{x}(t) = \left[\, x_i(t)\, \right]$ such that the component functions of \mathbf{x} satisfy the system in (27) identically on I. If the functions $p_{ij}(t)$ and $f_i(t)$ are all continuous on I, then Theorem 1 of Section 4.1 guarantees the existence on I of a unique solution $\mathbf{x}(t)$ satisfying preassigned initial conditions $\mathbf{x}(a) = \mathbf{b}$.

| **Example 6** | The first-order system |

$$x_1' = 4x_1 - 3x_2,$$
$$x_2' = 6x_1 - 7x_2$$

can be written as the single matrix equation

$$\frac{d\mathbf{x}}{dt} = \begin{bmatrix} 4 & -3 \\ 6 & -7 \end{bmatrix} \mathbf{x} = \mathbf{P}\mathbf{x}.$$

To verify that the vector functions

$$\mathbf{x}_1(t) = \begin{bmatrix} 3e^{2t} \\ 2e^{2t} \end{bmatrix} \quad \text{and} \quad \mathbf{x}_2(t) = \begin{bmatrix} e^{-5t} \\ 3e^{-5t} \end{bmatrix}$$

are both solutions of the matrix differential equation with coefficient matrix \mathbf{P}, we need only calculate

$$\mathbf{P}\mathbf{x}_1 = \begin{bmatrix} 4 & -3 \\ 6 & -7 \end{bmatrix} \begin{bmatrix} 3e^{2t} \\ 2e^{2t} \end{bmatrix} = \begin{bmatrix} 6e^{2t} \\ 4e^{2t} \end{bmatrix} = \mathbf{x}_1'$$

and

$$\mathbf{P}\mathbf{x}_2 = \begin{bmatrix} 4 & -3 \\ 6 & -7 \end{bmatrix} \begin{bmatrix} e^{-5t} \\ 3e^{-5t} \end{bmatrix} = \begin{bmatrix} -5e^{-5t} \\ -15e^{-5t} \end{bmatrix} = \mathbf{x}_2'. \qquad \blacksquare$$

To investigate the general nature of the solutions of Eq. (28), we consider first the **associated homogeneous equation**

$$\frac{d\mathbf{x}}{dt} = \mathbf{P}(t)\mathbf{x}, \tag{29}$$

which has the form shown in Eq. (28), but with $\mathbf{f}(t) \equiv \mathbf{0}$. We expect it to have n solutions $\mathbf{x}_1, \mathbf{x}_2, \ldots, \mathbf{x}_n$ that are independent in some appropriate sense, and such

that every solution of Eq. (29) is a linear combination of these n particular solutions. Given n solutions $\mathbf{x}_1, \mathbf{x}_2, \ldots, \mathbf{x}_n$ of Eq. (29), let us write

$$\mathbf{x}_j(t) = \begin{bmatrix} x_{1j}(t) \\ \vdots \\ x_{ij}(t) \\ \vdots \\ x_{nj}(t) \end{bmatrix}. \tag{30}$$

Thus $x_{ij}(t)$ denotes the ith component of the vector $\mathbf{x}_j(t)$, so the second subscript refers to the vector function $\mathbf{x}_j(t)$, whereas the first subscript refers to a component of this function. Theorem 1 is analogous to Theorem 1 of Section 3.2.

THEOREM 1: Principle of Superposition

Let $\mathbf{x}_1, \mathbf{x}_2, \ldots, \mathbf{x}_n$ be n solutions of the homogeneous linear equation in (29) on the open interval I. If c_1, c_2, \ldots, c_n are constants, then the linear combination

$$\mathbf{x}(t) = c_1\mathbf{x}_1(t) + c_2\mathbf{x}_2(t) + \cdots + c_n\mathbf{x}_n(t) \tag{31}$$

is also a solution of Eq. (29) on I.

***Proof*:** We know that $\mathbf{x}'_i = \mathbf{P}(t)\mathbf{x}_i$ for each i ($1 \leqq i \leqq n$), so it follows immediately that

$$\begin{aligned} \mathbf{x}' &= c_1\mathbf{x}'_1 + c_2\mathbf{x}'_2 + \cdots + c_n\mathbf{x}'_n \\ &= c_1\mathbf{P}(t)\mathbf{x}_1 + c_2\mathbf{P}(t)\mathbf{x}_2 + \cdots + c_n\mathbf{P}(t)\mathbf{x}_n \\ &= \mathbf{P}(t)(c_1\mathbf{x}_1 + c_2\mathbf{x}_2 + \cdots + c_n\mathbf{x}_n). \end{aligned}$$

That is, $\mathbf{x}' = \mathbf{P}(t)\mathbf{x}$, as desired. The remarkable simplicity of this proof demonstrates clearly one advantage of matrix notation. ▲

Example 6

Continued

If \mathbf{x}_1 and \mathbf{x}_2 are the two solutions of

$$\frac{d\mathbf{x}}{dt} = \begin{bmatrix} 4 & -3 \\ 6 & -7 \end{bmatrix} \mathbf{x}$$

discussed in Example 6, then the linear combination

$$\mathbf{x}(t) = c_1\mathbf{x}_1(t) + c_2\mathbf{x}_2(t) = c_1 \begin{bmatrix} 3e^{2t} \\ 2e^{2t} \end{bmatrix} + c_2 \begin{bmatrix} e^{-5t} \\ 3e^{-5t} \end{bmatrix}$$

is also a solution. In scalar form with $\mathbf{x} = [x_1 \quad x_2]^T$, this gives the solution

$$\begin{aligned} x_1(t) &= 3c_1e^{2t} + c_2e^{-5t}, \\ x_2(t) &= 2c_1e^{2t} + 3c_2e^{-5t}, \end{aligned}$$

which is equivalent to the general solution we found by the method of elimination in Example 2 of Section 4.2. ■

Independence and General Solutions

Linear independence is defined in the same way for vector-valued functions as for real-valued functions (Section 3.2). The vector-valued functions \mathbf{x}_1, \mathbf{x}_2, ..., \mathbf{x}_n are **linearly dependent** on the interval I provided that there exist constants $c_1, c_2, \ldots,$ c_n *not all zero* such that

➤
$$c_1\mathbf{x}_1(t) + c_2\mathbf{x}_2(t) + \cdots + c_n\mathbf{x}_n(t) = \mathbf{0} \tag{32}$$

for all t in I. Otherwise, they are **linearly independent.** Equivalently, they are linearly independent provided that no one of them is a linear combination of the others. For instance, the two solutions \mathbf{x}_1 and \mathbf{x}_2 of Example 6 are linearly independent because, clearly, neither is a scalar multiple of the other.

Just as in the case of a single nth-order equation, there is a Wronskian determinant that tells us whether or not n given solutions of the homogeneous equation in (29) are linearly dependent. If \mathbf{x}_1, \mathbf{x}_2, ..., \mathbf{x}_n are such solutions, then their **Wronskian** is the $n \times n$ determinant

➤
$$W(t) = \begin{vmatrix} x_{11}(t) & x_{12}(t) & \cdots & x_{1n}(t) \\ x_{21}(t) & x_{22}(t) & \cdots & x_{2n}(t) \\ \vdots & \vdots & & \vdots \\ x_{n1}(t) & x_{n2}(t) & \cdots & x_{nn}(t) \end{vmatrix}, \tag{33}$$

using the notation in (30) for the components of the solutions. We may write either $W(t)$ or $W(\mathbf{x}_1, \mathbf{x}_2, \ldots, \mathbf{x}_n)$. Note that W is the determinant of the matrix that has as its *column* vectors the solutions $\mathbf{x}_1, \mathbf{x}_2, \ldots, \mathbf{x}_n$. Theorem 2 is analogous to Theorem 3 of Section 3.2. Moreover, its proof is essentially the same, with the definition of $W(\mathbf{x}_1, \mathbf{x}_2, \ldots, \mathbf{x}_n)$ in Eq. (33) substituted for the definition of the Wronskian of n solutions of a single nth-order equation (see Problems 42 through 44).

THEOREM 2: Wronskians of Solutions

Suppose that \mathbf{x}_1, \mathbf{x}_2, ..., \mathbf{x}_n are n solutions of the homogeneous linear equation $\mathbf{x}' = \mathbf{P}(t)\mathbf{x}$ on an open interval I. Suppose also that $\mathbf{P}(t)$ is continuous on I. Let

$$W = W(\mathbf{x}_1, \mathbf{x}_2, \ldots, \mathbf{x}_n).$$

Then:

- If \mathbf{x}_1, \mathbf{x}_2, ..., \mathbf{x}_n are linearly dependent on I, then $W = 0$ at every point of I.
- If \mathbf{x}_1, \mathbf{x}_2, ..., \mathbf{x}_n are linearly independent on I, then $W \neq 0$ at each point of I.

Thus there are only two possibilities for solutions of homogeneous systems: Either $W = 0$ at *every* point of I, or $W = 0$ at *no* point of I. ▲

| Example 7 | It is readily verified (as in Example 6) that |

$$\mathbf{x}_1(t) = \begin{bmatrix} 2e^t \\ 2e^t \\ e^t \end{bmatrix}, \quad \mathbf{x}_2(t) = \begin{bmatrix} 2e^{3t} \\ 0 \\ -e^{3t} \end{bmatrix}, \quad \text{and} \quad \mathbf{x}_3(t) = \begin{bmatrix} 2e^{5t} \\ -2e^{5t} \\ e^{5t} \end{bmatrix}$$

are solutions of the equation

$$\frac{d\mathbf{x}}{dt} = \begin{bmatrix} 3 & -2 & 0 \\ -1 & 3 & -2 \\ 0 & -1 & 3 \end{bmatrix} \mathbf{x}. \tag{34}$$

The Wronskian of these solutions is

$$W = \begin{vmatrix} 2e^t & 2e^{3t} & 2e^{5t} \\ 2e^t & 0 & -2e^{5t} \\ e^t & -e^{3t} & e^{5t} \end{vmatrix} = e^{9t} \begin{vmatrix} 2 & 2 & 2 \\ 2 & 0 & -2 \\ 1 & -1 & 1 \end{vmatrix} = -16e^{9t},$$

which is never zero. Hence Theorem 2 implies that the solutions \mathbf{x}_1, \mathbf{x}_2, and \mathbf{x}_3 are linearly independent (on any open interval). ■

Theorem 3 is analogous to Theorem 4 of Section 3.2. It says that a **general solution** of the *homogeneous* $n \times n$ system $\mathbf{x}' = \mathbf{P}(t)\mathbf{x}$ is a linear combination

$$\blacktriangleright \qquad \mathbf{x} = c_1\mathbf{x}_1 + c_2\mathbf{x}_2 + \cdots + c_n\mathbf{x}_n \tag{35}$$

of any n given linearly independent solutions \mathbf{x}_1, \mathbf{x}_2, ..., \mathbf{x}_n.

THEOREM 3: General Solutions of Homogeneous Systems

Let \mathbf{x}_1, \mathbf{x}_2, ..., \mathbf{x}_n be n linearly independent solutions of the homogeneous linear equation $\mathbf{x}' = \mathbf{P}(t)\mathbf{x}$ on an open interval I where $\mathbf{P}(t)$ is continuous. If $\mathbf{x}(t)$ is any solution whatsoever of the equation $\mathbf{x}' = \mathbf{P}(t)\mathbf{x}$ on I, then there exist numbers c_1, c_2, ..., c_n such that

$$\mathbf{x}(t) = c_1\mathbf{x}_1(t) + c_2\mathbf{x}_2(t) + \cdots + c_n\mathbf{x}_n(t) \tag{35}$$

for all t in I.

Proof: Let a be a fixed point of I. We show first that there exist numbers c_1, c_2, ..., c_n such that the solution

$$\mathbf{y}(t) = c_1\mathbf{x}_1(t) + c_2\mathbf{x}_2(t) + \cdots + c_n\mathbf{x}_n(t) \tag{36}$$

has the same initial values at $t = a$ as does the given solution $\mathbf{x}(t)$; that is, such that

$$c_1\mathbf{x}_1(a) + c_2\mathbf{x}_2(a) + \cdots + c_n\mathbf{x}_n(a) = \mathbf{x}(a). \tag{37}$$

Let $\mathbf{X}(t)$ be the $n \times n$ matrix with column vectors \mathbf{x}_1, \mathbf{x}_2, ..., \mathbf{x}_n, and let \mathbf{c} be the column vector with components c_1, c_2, ..., c_n. Then Eq. (37) may be written in the form

$$\mathbf{X}(a)\mathbf{c} = \mathbf{x}(a). \tag{38}$$

The Wronskian determinant $W(a) = |\mathbf{X}(a)|$ is nonzero because the solutions \mathbf{x}_1, \mathbf{x}_2, ..., \mathbf{x}_n are linearly independent. Hence the matrix $\mathbf{X}(a)$ has an inverse matrix $\mathbf{X}(a)^{-1}$. Therefore the vector $\mathbf{c} = \mathbf{X}(a)^{-1}\mathbf{x}(a)$ satisfies Eq. (38), as desired.

Finally, note that the given solution $\mathbf{x}(t)$ and the solution $\mathbf{y}(t)$ of Eq. (36)—with the values of c_i determined by the equation $\mathbf{c} = \mathbf{X}(a)^{-1}\mathbf{x}(a)$—have the same initial values (at $t = a$). It follows from the existence-uniqueness theorem of Section 4.1 that $\mathbf{x}(t) = \mathbf{y}(t)$ for all t in I. This establishes Eq. (35). ▲

Remark: Every $n \times n$ system $\mathbf{x}' = \mathbf{P}(t)\mathbf{x}$ with continuous coefficient matrix does have a set of n linearly independent solutions $\mathbf{x}_1, \mathbf{x}_2, \ldots, \mathbf{x}_n$ as in the hypotheses of Theorem 3. It suffices to choose for $\mathbf{x}_j(t)$ the unique solution such that

$$\mathbf{x}_j(a) = \begin{bmatrix} 0 \\ 0 \\ 0 \\ \vdots \\ 0 \\ 1 \\ 0 \\ \vdots \\ 0 \end{bmatrix} \longleftarrow \text{position } j$$

—that is, the column vector with all elements zero except for a 1 in row j. (In other words, $\mathbf{x}_j(a)$ is merely the jth column of the identity matrix.) Then

$$W(\mathbf{x}_1, \mathbf{x}_2, \ldots, \mathbf{x}_n)\big|_{t=a} = |\mathbf{I}| \neq 0,$$

so the solutions $\mathbf{x}_1, \mathbf{x}_2, \ldots, \mathbf{x}_n$ are linearly independent by Theorem 2. How actually to find these solutions explicitly is another matter—one that we address in Section 5.2 (for the case of constant coefficient matrices). ∎

Initial Value Problems and Elementary Row Operations

The general solution in Eq. (35) of the homogeneous linear system $\mathbf{x}' = \mathbf{P}(t)\mathbf{x}$ can be written in the form

$$\mathbf{x}(t) = \mathbf{X}(t)\mathbf{c}, \tag{39}$$

where

$$\mathbf{X}(t) = \begin{bmatrix} \mathbf{x}_1(t) & \mathbf{x}_2(t) & \cdots & \mathbf{x}_n(t) \end{bmatrix} \tag{40}$$

is the $n \times n$ matrix whose *column vectors* are the linearly independent solutions \mathbf{x}_1, $\mathbf{x}_2, \ldots, \mathbf{x}_n$, and $\mathbf{c} = \begin{bmatrix} c_1 & c_2 & \cdots & c_n \end{bmatrix}^T$ is the vector of coefficients in the linear combination

$$\mathbf{x}(t) = c_1\mathbf{x}_1(t) + c_2\mathbf{x}_2(t) + \cdots + c_n\mathbf{x}_n(t). \tag{35}$$

Suppose now that we wish to solve the *initial value problem*

$$\frac{d\mathbf{x}}{dt} = \mathbf{P}\mathbf{x}, \quad \mathbf{x}(a) = \mathbf{b}, \tag{41}$$

where the initial vector $\mathbf{b} = \begin{bmatrix} b_1 & b_2 & \cdots & b_n \end{bmatrix}^T$ is given. Then, according to Eq. (39), it suffices to solve the system

$$\mathbf{X}(a)\mathbf{c} = \mathbf{b} \tag{42}$$

to find the coefficients c_1, c_2, \ldots, c_n in Eq. (35).

We therefore review briefly the elementary technique of *row reduction* to solve an $n \times n$ *algebraic* linear system

$$
\begin{aligned}
a_{11}x_1 + a_{12}x_2 + \cdots + a_{1n}x_n &= b_1, \\
a_{21}x_1 + a_{22}x_2 + \cdots + a_{2n}x_n &= b_2, \\
&\ \ \vdots \\
a_{n1}x_1 + a_{n2}x_2 + \cdots + a_{nn}x_n &= b_n
\end{aligned}
\tag{43}
$$

with nonsingular coefficient matrix $\mathbf{A} = [a_{ij}]$, constant vector $\mathbf{b} = [b_i]$, and unknowns x_1, x_2, \ldots, x_n. The basic idea is to transform the system in (43) into the simpler *upper triangular form*

$$
\begin{aligned}
\overline{a}_{11}x_1 + \overline{a}_{12}x_2 + \cdots + \overline{a}_{1n}x_n &= \overline{b}_1, \\
\overline{a}_{22}x_2 + \cdots + \overline{a}_{2n}x_n &= \overline{b}_2, \\
&\ \ \vdots \\
\overline{a}_{nn}x_n &= \overline{b}_n
\end{aligned}
\tag{44}
$$

in which only the unknowns $x_j, x_{j+1}, \ldots, x_n$ appear explicitly in the jth equation $(j = 1, 2, \ldots, n)$. The transformed system is then easily solved by the process of *back substitution*. First the last equation in (44) is solved for x_n, then the next-to-last is solved for x_{n-1}, and so forth, until the first equation is finally solved for x_1.

The transformation of the system in (43) to upper triangular form is most easily described in terms of elementary row operations on the *augmented coefficient matrix*

$$
\left[\, \mathbf{A} \mid \mathbf{b} \,\right] =
\begin{bmatrix}
a_{11} & a_{12} & \cdots & a_{1n} & b_1 \\
a_{21} & a_{22} & \cdots & a_{2n} & b_2 \\
\vdots & \vdots & & \vdots & \vdots \\
a_{n1} & a_{n2} & \cdots & a_{nn} & b_n
\end{bmatrix}
\tag{45}
$$

that is obtained by adjoining the vector \mathbf{b} to the matrix \mathbf{A} as an additional column. The admissible **elementary row operations** are of the following three types:

1. Multiply any (single) row of the matrix by a nonzero constant.
2. Interchange any two rows of the matrix.
3. Subtract a constant multiple of one row from any other row.

The goal is to use a sequence of such operations (one by one, in turn) to transform $\left[\, \mathbf{A} \mid \mathbf{b} \,\right]$ into an upper triangular matrix, one that has only zeros beneath its principal diagonal. This upper triangular augmented coefficient matrix then corresponds to an upper triangular system as in (44). The process of transforming $\left[\, \mathbf{A} \mid \mathbf{b} \,\right]$ is carried out one column at a time, from left to right, as in the next example.

Example 8 Use the solution vectors given in Example 7 to solve the initial value problem

$$
\frac{d\mathbf{x}}{dt} =
\begin{bmatrix}
3 & -2 & 0 \\
-1 & 3 & -2 \\
0 & -1 & 3
\end{bmatrix} \mathbf{x}, \qquad
\mathbf{x}(0) =
\begin{bmatrix}
0 \\
2 \\
6
\end{bmatrix}.
\tag{46}
$$

Solution It follows from Theorem 3 that the linear combination

$$\mathbf{x}(t) = c_1\mathbf{x}_1(t) + c_2\mathbf{x}_2(t) + c_3\mathbf{x}_3(t)$$

$$= c_1 \begin{bmatrix} 2e^t \\ 2e^t \\ e^t \end{bmatrix} + c_2 \begin{bmatrix} 2e^{3t} \\ 0 \\ -e^{3t} \end{bmatrix} + c_3 \begin{bmatrix} 2e^{5t} \\ -2e^{5t} \\ e^{5t} \end{bmatrix}$$

is a general solution of the 3×3 linear system in (46). In scalar form, this gives the general solution

$$\begin{aligned} x_1(t) &= 2c_1e^t + 2c_2e^{3t} + 2c_3e^{5t}, \\ x_2(t) &= 2c_1e^t \qquad\quad - 2c_3e^{5t}, \\ x_3(t) &= \ c_1e^t - \ c_2e^{3t} + \ c_3e^{5t}. \end{aligned}$$

We seek the particular solution satisfying the initial conditions

$$x_1(0) = 0, \quad x_2(0) = 2, \quad x_3(0) = 6.$$

When we substitute these values in the three preceding scalar equations, we get the algebraic linear system

$$\begin{aligned} 2c_1 + 2c_2 + 2c_3 &= 0, \\ 2c_1 \qquad\quad - 2c_3 &= 2, \\ c_1 - \ c_2 + \ c_3 &= 6 \end{aligned}$$

with augmented coefficient matrix

$$\begin{bmatrix} 2 & 2 & 2 & \vdots & 0 \\ 2 & 0 & -2 & \vdots & 2 \\ 1 & -1 & 1 & \vdots & 6 \end{bmatrix}.$$

Multiplication of each of the first two rows by $\frac{1}{2}$ gives

$$\begin{bmatrix} 1 & 1 & 1 & \vdots & 0 \\ 1 & 0 & -1 & \vdots & 1 \\ 1 & -1 & 1 & \vdots & 6 \end{bmatrix},$$

then subtraction of the first row both from the second row and from the third row gives the matrix

$$\begin{bmatrix} 1 & 1 & 1 & \vdots & 0 \\ 0 & -1 & -2 & \vdots & 1 \\ 0 & -2 & 0 & \vdots & 6 \end{bmatrix}.$$

The first column of this matrix now has the desired form.

Now we multiply the second row by -1, then add twice the result to the third row. Thereby we get the upper triangular augmented coefficient matrix

$$\begin{bmatrix} 1 & 1 & 1 & \vdots & 0 \\ 0 & 1 & 2 & \vdots & -1 \\ 0 & 0 & 4 & \vdots & 4 \end{bmatrix}$$

that corresponds to the transformed system

$$\begin{aligned} c_1 + c_2 + \ c_3 &= \ 0, \\ c_2 + 2c_3 &= -1, \\ 4c_3 &= \ 4. \end{aligned}$$

We finally solve in turn for $c_3 = 1$, $c_2 = -3$, and $c_1 = 2$. Thus the desired particular solution is given by

$$\mathbf{x}(t) = 2\mathbf{x}_1(t) - 3\mathbf{x}_2(t) + \mathbf{x}_3(t) = \begin{bmatrix} 4e^t - 6e^{3t} + 2e^{5t} \\ 4e^t \qquad\quad - 2e^{5t} \\ 2e^t + 3e^{3t} + \ e^{5t} \end{bmatrix}. \qquad\blacksquare$$

Nonhomogeneous Solutions

We finally turn our attention to a *nonhomogeneous* linear system of the form

$$\frac{d\mathbf{x}}{dt} = \mathbf{P}(t)\mathbf{x} + \mathbf{f}(t). \qquad (47)$$

The following theorem is analogous to Theorem 5 of Section 3.2 and is proved in precisely the same way, substituting the preceding theorems in this section for the analogous theorems of Section 3.2. In brief, Theorem 4 means that the general solution of Eq. (47) has the form

$$\mathbf{x}(t) = \mathbf{x}_c(t) + \mathbf{x}_p(t), \qquad (48)$$

where $\mathbf{x}_p(t)$ is a single particular solution of Eq. (47) and the **complementary function** $\mathbf{x}_c(t)$ is a general solution of the associated homogeneous equation $\mathbf{x}' = \mathbf{P}(t)\mathbf{x}$.

THEOREM 4: Solutions of Nonhomogeneous Systems

Let \mathbf{x}_p be a particular solution of the nonhomogeneous linear equation in (47) on an open interval I on which the functions $\mathbf{P}(t)$ and $\mathbf{f}(t)$ are continuous. Let $\mathbf{x}_1, \mathbf{x}_2, \ldots,$ \mathbf{x}_n be linearly independent solutions of the associated homogeneous equation on I. If $\mathbf{x}(t)$ is any solution whatsoever of Eq. (47) on I, then there exist numbers $c_1, c_2,$ \ldots, c_n such that

$$\mathbf{x}(t) = c_1\mathbf{x}_1(t) + c_2\mathbf{x}_2(t) + \cdots + c_n\mathbf{x}_n(t) + \mathbf{x}_p(t) \qquad (49)$$

for all t in I. ▲

Thus finding a general solution of a nonhomogeneous linear system involves two separate steps:

1. Finding the general solution $\mathbf{x}_c(t)$ of the associated homogeneous system;
2. Finding a single particular solution $\mathbf{x}_p(t)$ of the nonhomogeneous system.

The sum $\mathbf{x}(t) = \mathbf{x}_c(t) + \mathbf{x}_p(t)$ will then be a general solution of the nonhomogeneous system.

Example 9 The nonhomogeneous linear system

$$\begin{aligned} x_1' &= \ 3x_1 - 2x_2 \qquad\quad - 9t + 13\,, \\ x_2' &= -x_1 + 3x_2 - 2x_3 + 7t - 15, \\ x_3' &= \qquad\ -\ x_2 + 3x_3 - 6t + \ 7 \end{aligned}$$

is of the form in (47) with

$$\mathbf{P}(t) = \begin{bmatrix} 3 & -2 & 0 \\ -1 & 3 & -2 \\ 0 & -1 & 3 \end{bmatrix}, \quad \mathbf{f}(t) = \begin{bmatrix} -9t + 13 \\ 7t - 15 \\ -6t + 7 \end{bmatrix}.$$

In Example 7 we saw that a general solution of the associated homogeneous linear system

$$\frac{d\mathbf{x}}{dt} = \begin{bmatrix} 3 & -2 & 0 \\ -1 & 3 & -2 \\ 0 & -1 & 3 \end{bmatrix} \mathbf{x}$$

is given by

$$\mathbf{x}_c(t) = \begin{bmatrix} 2c_1 e^t + 2c_2 e^{3t} + 2c_3 e^{5t} \\ 2c_1 e^t \qquad\qquad - 2c_3 e^{5t} \\ c_1 e^t - c_2 e^{3t} + c_2 e^{5t} \end{bmatrix},$$

and we can verify by substitution that the function

$$\mathbf{x}_p(t) = \begin{bmatrix} 3t \\ 5 \\ 2t \end{bmatrix}$$

(found using a computer algebra system) is a particular solution of the original non-homogeneous system. Consequently, Theorem 4 implies that a general solution of the nonhomogeneous system is given by

$$\mathbf{x}(t) = \mathbf{x}_c(t) + \mathbf{x}_p(t);$$

that is, by

$$x_1(t) = 2c_1 e^t + 2c_2 e^{3t} + 2c_3 e^{5t} + 3t,$$
$$x_2(t) = 2c_1 e^t \qquad\qquad - 2c_3 e^{5t} + 5,$$
$$x_3(t) = c_1 e^t - c_2 e^{3t} + c_3 e^{5t} + 2t.$$

5.1 Problems

1. Let

$$\mathbf{A} = \begin{bmatrix} 2 & -3 \\ 4 & 7 \end{bmatrix} \quad \text{and} \quad \mathbf{B} = \begin{bmatrix} 3 & -4 \\ 5 & 1 \end{bmatrix}.$$

Find **(a)** $2\mathbf{A} + 3\mathbf{B}$; **(b)** $3\mathbf{A} - 2\mathbf{B}$; **(c)** \mathbf{AB}; **(d)** \mathbf{BA}.

2. Verify that **(a)** $\mathbf{A}(\mathbf{BC}) = (\mathbf{AB})\mathbf{C}$ and that **(b)** $\mathbf{A}(\mathbf{B}+\mathbf{C}) = \mathbf{AB} + \mathbf{AC}$, where \mathbf{A} and \mathbf{B} are the matrices given in Problem 1 and

$$\mathbf{C} = \begin{bmatrix} 0 & 2 \\ 3 & -1 \end{bmatrix}.$$

3. Find \mathbf{AB} and \mathbf{BA} given

$$\mathbf{A} = \begin{bmatrix} 2 & 0 & -1 \\ 3 & -4 & 5 \end{bmatrix} \quad \text{and} \quad \mathbf{B} = \begin{bmatrix} 1 & 3 \\ -7 & 0 \\ 3 & -2 \end{bmatrix}.$$

4. Let \mathbf{A} and \mathbf{B} be the matrices given in Problem 3 and let

$$\mathbf{x} = \begin{bmatrix} 2t \\ e^{-t} \end{bmatrix} \quad \text{and} \quad \mathbf{y} = \begin{bmatrix} t^2 \\ \sin t \\ \cos t \end{bmatrix}.$$

Find \mathbf{Ay} and \mathbf{Bx}. Are the products \mathbf{Ax} and \mathbf{By} defined? Explain your answer.

5. Let

$$\mathbf{A} = \begin{bmatrix} 3 & 2 & -1 \\ 0 & 4 & 3 \\ -5 & 2 & 7 \end{bmatrix} \quad \text{and} \quad \mathbf{B} = \begin{bmatrix} 0 & -3 & 2 \\ 1 & 4 & -3 \\ 2 & 5 & -1 \end{bmatrix}.$$

Find **(a)** $7\mathbf{A} + 4\mathbf{B}$; **(b)** $3\mathbf{A} - 5\mathbf{B}$; **(c)** \mathbf{AB}; **(d)** \mathbf{BA}; **(e)** $\mathbf{A} - t\mathbf{I}$.

6. Let

$$\mathbf{A}_1 = \begin{bmatrix} 2 & 1 \\ -3 & 2 \end{bmatrix}, \quad \mathbf{A}_2 = \begin{bmatrix} 1 & 3 \\ -1 & -2 \end{bmatrix},$$

$$\mathbf{B} = \begin{bmatrix} 2 & 4 \\ 1 & 2 \end{bmatrix}.$$

(a) Show that $\mathbf{A}_1\mathbf{B} = \mathbf{A}_2\mathbf{B}$ and note that $\mathbf{A}_1 \neq \mathbf{A}_2$. Thus the cancellation law does not hold for matrices; that is, if $\mathbf{A}_1\mathbf{B} = \mathbf{A}_2\mathbf{B}$ and $\mathbf{B} \neq \mathbf{0}$, it does not follow that $\mathbf{A}_1 = \mathbf{A}_2$.
(b) Let $\mathbf{A} = \mathbf{A}_1 - \mathbf{A}_2$ and show that $\mathbf{AB} = \mathbf{0}$. Thus the product of two nonzero matrices may be the zero matrix.

7. Compute the determinants of the matrices \mathbf{A} and \mathbf{B} in Problem 6. Are your results consistent with the theorem to the effect that

$$\det(\mathbf{AB}) = \det(\mathbf{A}) \cdot \det(\mathbf{B})$$

for any two square matrices \mathbf{A} and \mathbf{B} of the same order?

8. Suppose that \mathbf{A} and \mathbf{B} are the matrices of Problem 5. Verify that $\det(\mathbf{AB}) = \det(\mathbf{BA})$.

In Problems 9 and 10, verify the product law for differentiation, $(\mathbf{AB})' = \mathbf{A}'\mathbf{B} + \mathbf{AB}'$.

9. $\mathbf{A}(t) = \begin{bmatrix} t & 2t-1 \\ t^3 & \dfrac{1}{t} \end{bmatrix}$ and $\mathbf{B}(t) = \begin{bmatrix} 1-t & 1+t \\ 3t^2 & 4t^3 \end{bmatrix}$

10. $\mathbf{A}(t) = \begin{bmatrix} e^t & t & t^2 \\ -t & 0 & 2 \\ 8t & -1 & t^3 \end{bmatrix}$ and $\mathbf{B}(t) = \begin{bmatrix} 3 \\ 2e^{-t} \\ 3t \end{bmatrix}$

In Problems 11 through 20, write the given system in the form $\mathbf{x}' = \mathbf{P}(t)\mathbf{x} + \mathbf{f}(t)$.

11. $x' = -3y, \ y' = 3x$
12. $x' = 3x - 2y, \ y' = 2x + y$
13. $x' = 2x + 4y + 3e^t, \ y' = 5x - y - t^2$
14. $x' = tx - e^t y + \cos t, \ y' = e^{-t}x + t^2 y - \sin t$
15. $x' = y + z, \ y' = z + x, \ z' = x + y$
16. $x' = 2x - 3y, \ y' = x + y + 2z, \ z' = 5y - 7z$
17. $x' = 3x - 4y + z + t, \ y' = x - 3z + t^2, \ z' = 6y - 7z + t^3$
18. $x' = tx - y + e^t z, \ y' = 2x + t^2 y - z, \ z' = e^{-t}x + 3ty + t^3 z$
19. $x'_1 = x_2, \ x'_2 = 2x_3, \ x'_3 = 3x_4, \ x'_4 = 4x_1$
20. $x'_1 = x_2 + x_3 + 1, \ x'_2 = x_3 + x_4 + t,$
$\quad x'_3 = x_1 + x_4 + t^2, \ x'_4 = x_1 + x_2 + t^3$

In Problems 21 through 30, first verify that the given vectors are solutions of the given system. Then use the Wronskian to show that they are linearly independent. Finally, write the general solution of the system.

21. $\mathbf{x}' = \begin{bmatrix} 4 & 2 \\ -3 & -1 \end{bmatrix} \mathbf{x}; \ \mathbf{x}_1 = \begin{bmatrix} 2e^t \\ -3e^t \end{bmatrix}, \ \mathbf{x}_2 = \begin{bmatrix} e^{2t} \\ -e^{2t} \end{bmatrix}$

22. $\mathbf{x}' = \begin{bmatrix} -3 & 2 \\ -3 & 4 \end{bmatrix} \mathbf{x}; \ \mathbf{x}_1 = \begin{bmatrix} e^{3t} \\ 3e^{3t} \end{bmatrix}, \ \mathbf{x}_2 = \begin{bmatrix} 2e^{-2t} \\ e^{-2t} \end{bmatrix}$

23. $\mathbf{x}' = \begin{bmatrix} 3 & -1 \\ 5 & -3 \end{bmatrix} \mathbf{x}; \ \mathbf{x}_1 = e^{2t}\begin{bmatrix} 1 \\ 1 \end{bmatrix}, \ \mathbf{x}_2 = e^{-2t}\begin{bmatrix} 1 \\ 5 \end{bmatrix}$

24. $\mathbf{x}' = \begin{bmatrix} 4 & 1 \\ -2 & 1 \end{bmatrix} \mathbf{x}; \ \mathbf{x}_1 = e^{3t}\begin{bmatrix} 1 \\ -1 \end{bmatrix}, \ \mathbf{x}_2 = e^{2t}\begin{bmatrix} 1 \\ -2 \end{bmatrix}$

25. $\mathbf{x}' = \begin{bmatrix} 4 & -3 \\ 6 & -7 \end{bmatrix} \mathbf{x}; \ \mathbf{x}_1 = \begin{bmatrix} 3e^{2t} \\ 2e^{2t} \end{bmatrix}, \ \mathbf{x}_2 = \begin{bmatrix} e^{-5t} \\ 3e^{-5t} \end{bmatrix}$

26. $\mathbf{x}' = \begin{bmatrix} 3 & -2 & 0 \\ -1 & 3 & -2 \\ 0 & -1 & 3 \end{bmatrix} \mathbf{x}; \ \mathbf{x}_1 = e^t\begin{bmatrix} 2 \\ 2 \\ 1 \end{bmatrix},$

$\mathbf{x}_2 = e^{3t}\begin{bmatrix} -2 \\ 0 \\ 1 \end{bmatrix}, \ \mathbf{x}_3 = e^{5t}\begin{bmatrix} 2 \\ -2 \\ 1 \end{bmatrix}$

27. $\mathbf{x}' = \begin{bmatrix} 0 & 1 & 1 \\ 1 & 0 & 1 \\ 1 & 1 & 0 \end{bmatrix} \mathbf{x}; \ \mathbf{x}_1 = e^{2t}\begin{bmatrix} 1 \\ 1 \\ 1 \end{bmatrix},$

$\mathbf{x}_2 = e^{-t}\begin{bmatrix} 1 \\ 0 \\ -1 \end{bmatrix}, \ \mathbf{x}_3 = e^{-t}\begin{bmatrix} 0 \\ 1 \\ -1 \end{bmatrix}$

28. $\mathbf{x}' = \begin{bmatrix} 1 & 2 & 1 \\ 6 & -1 & 0 \\ -1 & -2 & -1 \end{bmatrix} \mathbf{x}; \ \mathbf{x}_1 = \begin{bmatrix} 1 \\ 6 \\ -13 \end{bmatrix},$

$\mathbf{x}_2 = e^{3t}\begin{bmatrix} 2 \\ 3 \\ -2 \end{bmatrix}, \ \mathbf{x}_3 = e^{-4t}\begin{bmatrix} -1 \\ 2 \\ 1 \end{bmatrix}$

29. $\mathbf{x}' = \begin{bmatrix} -8 & -11 & -2 \\ 6 & 9 & 2 \\ -6 & -6 & 1 \end{bmatrix} \mathbf{x}; \ \mathbf{x}_1 = e^{-2t}\begin{bmatrix} 3 \\ -2 \\ 2 \end{bmatrix},$

$\mathbf{x}_2 = e^t\begin{bmatrix} 1 \\ -1 \\ 1 \end{bmatrix}, \ \mathbf{x}_3 = e^{3t}\begin{bmatrix} 1 \\ -1 \\ 0 \end{bmatrix}$

30. $\mathbf{x}' = \begin{bmatrix} 1 & -4 & 0 & -2 \\ 0 & 1 & 0 & 0 \\ 6 & -12 & -1 & -6 \\ 0 & -4 & 0 & -1 \end{bmatrix} \mathbf{x}; \ \mathbf{x}_1 = e^{-t}\begin{bmatrix} 1 \\ 0 \\ 0 \\ 1 \end{bmatrix},$

$\mathbf{x}_2 = e^{-t}\begin{bmatrix} 0 \\ 0 \\ 1 \\ 0 \end{bmatrix}, \ \mathbf{x}_3 = e^t\begin{bmatrix} 0 \\ 1 \\ 0 \\ -2 \end{bmatrix}, \ \mathbf{x}_4 = e^t\begin{bmatrix} 1 \\ 0 \\ 3 \\ 0 \end{bmatrix}$

In Problems 31 through 40, find a particular solution of the indicated linear system that satisfies the given initial conditions.

31. The system of Problem 22: $x_1(0) = 0, \ x_2(0) = 5$
32. The system of Problem 23: $x_1(0) = 5, \ x_2(0) = -3$
33. The system of Problem 24: $x_1(0) = 11, \ x_2(0) = -7$
34. The system of Problem 25: $x_1(0) = 8, \ x_2(0) = 0$
35. The system of Problem 26: $x_1(0) = 0, \ x_2(0) = 0, \ x_3(0) = 4$
36. The system of Problem 27: $x_1(0) = 10, \ x_2(0) = 12, \ x_3(0) = -1$
37. The system of Problem 29: $x_1(0) = 1, \ x_2(0) = 2, \ x_3(0) = 3$
38. The system of Problem 29: $x_1(0) = 5, \ x_2(0) = -7, \ x_3(0) = 11$

39. The system of Problem 30: $x_1(0) = x_2(0) = x_3(0) = x_4(0) = 1$

40. The system of Problem 30: $x_1(0) = 1$, $x_2(0) = 3$, $x_3(0) = 4$, $x_4(0) = 7$

41. (a) Show that the vector functions

$$\mathbf{x}_1(t) = \begin{bmatrix} t \\ t^2 \end{bmatrix} \quad \text{and} \quad \mathbf{x}_2 = \begin{bmatrix} t^2 \\ t^3 \end{bmatrix}$$

are linearly independent on the real line. (b) Why does it follow from Theorem 2 that there is *no* continuous matrix $\mathbf{P}(t)$ such that \mathbf{x}_1 and \mathbf{x}_2 are both solutions of $\mathbf{x}' = \mathbf{P}(t)\mathbf{x}$?

42. Suppose that one of the vector functions

$$\mathbf{x}_1(t) = \begin{bmatrix} x_{11}(t) \\ x_{21}(t) \end{bmatrix} \quad \text{and} \quad \mathbf{x}_2(t) = \begin{bmatrix} x_{12}(t) \\ x_{22}(t) \end{bmatrix}$$

is a constant multiple of the other on the open interval I. Show that their Wronskian $W(t) = |[x_{ij}(t)]|$ must vanish identically on I. This proves part (a) of Theorem 2 in the case $n = 2$.

43. Suppose that the vectors $\mathbf{x}_1(t)$ and $\mathbf{x}_2(t)$ of Problem 42 are solutions of the equation $\mathbf{x}' = \mathbf{P}(t)\mathbf{x}$, where the 2×2 matrix $\mathbf{P}(t)$ is continuous on the open interval I. Show that if there exists a point a of I at which their Wronskian $W(a)$ is zero, then there exist numbers c_1 and c_2 not both zero such that $c_1\mathbf{x}_1(a) + c_2\mathbf{x}_2(a) = \mathbf{0}$. Then conclude from the uniqueness of solutions of the equation $\mathbf{x}' = \mathbf{P}(t)\mathbf{x}$ that

$$c_1\mathbf{x}_1(t) + c_2\mathbf{x}_2(t) = \mathbf{0}$$

for all t in I; that is, that \mathbf{x}_1 and \mathbf{x}_2 are linearly dependent. This proves part (b) of Theorem 2 in the case $n = 2$.

44. Generalize Problems 42 and 43 to prove Theorem 2 for n an arbitrary positive integer.

45. Let $\mathbf{x}_1(t)$, $\mathbf{x}_2(t)$, ..., $\mathbf{x}_n(t)$ be vector functions whose ith components (for some fixed i) $x_{i1}(t)$, $x_{i2}(t)$, ..., $x_{in}(t)$ are linearly independent real-valued functions. Conclude that the vector functions are themselves linearly independent.

5.1 Application: Automatic Solution of Linear Systems

Linear systems with more than two or three equations are most frequently solved with the aid of calculators or computers. For instance, recall that in Example 8 we needed to solve the linear system

$$\begin{aligned} 2c_1 + 2c_2 + 2c_3 &= 0, \\ 2c_1 \qquad\quad - 2c_3 &= 2, \\ c_1 - c_2 + c_3 &= 6 \end{aligned} \qquad (1)$$

FIGURE 5.1.1. TI-86 solution of the system $\mathbf{AC} = \mathbf{B}$ in (1).

that can be written in the form $\mathbf{AC} = \mathbf{B}$ with 3×3 coefficient matrix \mathbf{A}, right-hand side the 3×1 column vector $\mathbf{B} = \begin{bmatrix} 0 & 2 & 6 \end{bmatrix}^T$, and unknown column vector $\mathbf{C} = \begin{bmatrix} c_1 & c_2 & c_3 \end{bmatrix}^T$. Figure 5.1.1 shows a TI calculator solution for $\mathbf{C} = \mathbf{A}^{-1}\mathbf{B}$, with the result that $c_1 = 2$, $c_2 = -3$, and $c_3 = 1$. The same result can be found using the *Maple* commands

```
with(linalg):
A := array([[2, 2, 2], [2, 0, -2], [1, -1, 1]]):
B := array([[0], [2], [6] ]):
C := multiply(inverse(A),B);
```

the *Mathematica* commands

```
A = {{2, 2, 2}, {2, 0, -2}, {1, -1, 1}};
B = {{0}, {2}, {6}};
C = Inverse[A].B
```

or the MATLAB commands

```
A = [[2    2    2]; [2    0    -2]; [1    -1    1]];
B = [0; 2; 6];
C = inv(A)*B
```

Use your own calculator or available computer algebra system to solve "automatically" Problems 31 through 40 in this section.

5.2 The Eigenvalue Method for Homogeneous Systems

We now introduce a powerful alternative to the method of elimination for constructing the general solution of a *homogeneous* first-order linear system with *constant* coefficients,

$$
\begin{aligned}
x_1' &= a_{11}x_1 + a_{12}x_2 + \cdots + a_{1n}x_n, \\
x_2' &= a_{21}x_1 + a_{22}x_2 + \cdots + a_{2n}x_n, \\
&\vdots \\
x_n' &= a_{n1}x_1 + a_{n2}x_2 + \cdots + a_{nn}x_n.
\end{aligned} \tag{1}
$$

By Theorem 3 of Section 5.1, we know that it suffices to find n linearly independent solution vectors $\mathbf{x}_1, \mathbf{x}_2, \ldots, \mathbf{x}_n$; the linear combination

$$
\mathbf{x}(t) = c_1\mathbf{x}_1 + c_2\mathbf{x}_2 + \cdots + c_n\mathbf{x}_n \tag{2}
$$

with arbitrary coefficients will then be a general solution of the system in (1).

To search for the n needed linearly independent solution vectors, we proceed by analogy with the characteristic root method for solving a single homogeneous equation with constant coefficients (Section 3.3). It is reasonable to anticipate solution vectors of the form

$$
\mathbf{x}(t) = \begin{bmatrix} x_1 \\ x_2 \\ x_3 \\ \vdots \\ x_n \end{bmatrix} = \begin{bmatrix} v_1 e^{\lambda t} \\ v_2 e^{\lambda t} \\ v_3 e^{\lambda t} \\ \vdots \\ v_n e^{\lambda t} \end{bmatrix} = \begin{bmatrix} v_1 \\ v_2 \\ v_3 \\ \vdots \\ v_n \end{bmatrix} e^{\lambda t} = \mathbf{v}e^{\lambda t} \tag{3}
$$

where $\lambda, v_1, v_2, v_3, \ldots, v_n$ are appropriate scalar constants. For if we substitute

$$
x_i = v_i e^{\lambda t}, \qquad x_i' = \lambda v_i e^{\lambda t}
$$

($i = 1, 2, \ldots, n$) in (1), then the factor $e^{\lambda t}$ will cancel throughout. This will leave us with n linear equations which—for appropriate values of λ—we can hope to solve for values of the coefficients v_1, v_2, \ldots, v_n in Eq. (3) so that $\mathbf{x}(t) = \mathbf{v}e^{\lambda t}$ is, indeed, a solution of the system in (1).

To investigate this possibility, it is more efficient to write the system in (1) in the matrix form

$$
\mathbf{x}' = \mathbf{Ax} \tag{4}
$$

where $\mathbf{A} = \begin{bmatrix} a_{ij} \end{bmatrix}$. When we substitute the trial solution $\mathbf{x} = \mathbf{v}e^{\lambda t}$ with derivative $\mathbf{x}' = \lambda\mathbf{v}e^{\lambda t}$ in Eq. (4), the result is

$$
\lambda\mathbf{v}e^{\lambda t} = \mathbf{Av}e^{\lambda t}.
$$

We cancel the nonzero scalar factor $e^{\lambda t}$ to get

$$
\mathbf{Av} = \lambda\mathbf{v}. \tag{5}
$$

This means that $\mathbf{x} = \mathbf{v}e^{\lambda t}$ will be a nontrivial solution of Eq. (4) provided that \mathbf{v} is a *nonzero* vector and λ is a constant such that Eq. (5) holds; that is, the *matrix product* \mathbf{Av} *is a scalar multiple of the vector* \mathbf{v}. The question now is this: How do we find \mathbf{v} and λ?

To answer this question, we rewrite Eq. (5) in the form

$$\mathbf{(A - \lambda I)v = 0}. \tag{6}$$

Given λ, this is a system of n homogeneous linear equations in the unknowns v_1, v_2, ..., v_n. By a standard theorem of linear algebra, it has a nontrivial solution if and only if the determinant of its coefficient matrix vanishes; that is, if and only if

$$|\mathbf{A} - \lambda\mathbf{I}| = \det(\mathbf{A} - \lambda\mathbf{I}) = 0. \tag{7}$$

In its simplest formulation, the **eigenvalue method** for solving the system $\mathbf{x}' = \mathbf{Ax}$ consists of finding λ so that Eq. (7) holds and next solving Eq. (6) with this value of λ to obtain v_1, v_2, ..., v_n. Then $\mathbf{x} = \mathbf{v}e^{\lambda t}$ will be a solution vector. The name of the method comes from the following definition.

Definition: Eigenvalues and Eigenvectors

The number λ (either zero or nonzero) is called an **eigenvalue** of the $n \times n$ matrix \mathbf{A} provided that

$$|\mathbf{A} - \lambda\mathbf{I}| = 0. \tag{7}$$

An **eigenvector** associated with the eigenvalue λ is a *nonzero* vector \mathbf{v} such that $\mathbf{Av} = \lambda\mathbf{v}$, so that

$$\mathbf{(A - \lambda I)v = 0}. \tag{6}$$

Note that if \mathbf{v} is an eigenvector associated with the eigenvalue λ, then so is any nonzero constant scalar multiple $c\mathbf{v}$ of \mathbf{v}—this follows upon multiplication of each side in Eq. (6) by $c \neq 0$.

The prefix *eigen* is a German word with the approximate translation *characteristic* in this context; the terms *characteristic value* and *characteristic vector* are in common use. For this reason, the equation

$$|\mathbf{A} - \lambda\mathbf{I}| = \begin{vmatrix} a_{11} - \lambda & a_{12} & \cdots & a_{1n} \\ a_{21} & a_{22} - \lambda & \cdots & a_{2n} \\ \vdots & \vdots & & \vdots \\ a_{n1} & a_{n2} & \cdots & a_{nn} - \lambda \end{vmatrix} = 0 \tag{8}$$

is called the **characteristic equation** of the matrix \mathbf{A}; its roots are the eigenvalues of \mathbf{A}. Upon expanding the determinant in (8), we evidently get an nth-degree polynomial of the form

$$(-1)^n\lambda^n + b_{n-1}\lambda^{n-1} + \cdots + b_1\lambda + b_0 = 0. \tag{9}$$

By the fundamental theorem of algebra, this equation has n roots—possibly some are complex, possibly some are repeated—and thus an $n \times n$ matrix has n eigenvalues (counting repetitions, if any). Although we assume that the elements of \mathbf{A} are

real numbers, we allow the possibility of complex eigenvalues and complex-valued eigenvectors.

Our discussion of Eqs. (4) through (7) provides a proof of the following theorem, which is the basis for the eigenvalue method of solving a first-order linear system with constant coefficients.

THEOREM 1: Eigenvalue Solutions of $\mathbf{x}' = \mathbf{A}\mathbf{x}$

Let λ be an eigenvalue of the [constant] coefficient matrix \mathbf{A} of the first-order linear system

$$\frac{d\mathbf{x}}{dt} = \mathbf{A}\mathbf{x}.$$

If \mathbf{v} is an eigenvector associated with λ, then

$$\mathbf{x}(t) = \mathbf{v}e^{\lambda t}$$

is a nontrivial solution of the system. ▲

The Eigenvalue Method

In outline, this method for solving the $n \times n$ homogeneous constant-coefficient system $\mathbf{x}' = \mathbf{A}\mathbf{x}$ proceeds as follows:

1. We first solve the characteristic equation in (8) for the eigenvalues $\lambda_1, \lambda_2, \ldots, \lambda_n$ of the matrix \mathbf{A}.
2. Next we attempt to find n *linearly independent* eigenvectors $\mathbf{v}_1, \mathbf{v}_2, \ldots, \mathbf{v}_n$ associated with these eigenvalues.
3. Step 2 is not always possible, but when it is, we get n linearly independent solutions

$$\mathbf{x}_1(t) = \mathbf{v}_1 e^{\lambda_1 t}, \quad \mathbf{x}_2(t) = \mathbf{v}_2 e^{\lambda_2 t}, \quad \ldots, \quad \mathbf{x}_n(t) = \mathbf{v}_n e^{\lambda_n t}. \quad (10)$$

In this case the general solution of $\mathbf{x}' = \mathbf{A}\mathbf{x}$ is a linear combination

$$\mathbf{x}(t) = c_1 \mathbf{x}_1(t) + c_2 \mathbf{x}_2(t) + \cdots + c_n \mathbf{x}_n(t)$$

of these n solutions.

We will discuss separately the various cases that can occur, depending on whether the eigenvalues are distinct or repeated, real or complex. The case of repeated eigenvalues—multiple roots of the characteristic equation—will be deferred to Section 5.4.

Distinct Real Eigenvalues

If the eigenvalues $\lambda_1, \lambda_2, \ldots, \lambda_n$ are real and distinct, then we substitute each of them in turn in Eq. (6) and solve for the associated eigenvectors $\mathbf{v}_1, \mathbf{v}_2, \ldots, \mathbf{v}_n$. In this case it can be proved that the particular solution vectors given in (10) are always linearly independent. (For instance, see Section 6.2 of Edwards and Penney, *Elementary Linear Algebra* (Englewood Cliffs, N.J.: Prentice Hall, 1988).) In any particular example such linear independence can always be verified by using the Wronskian determinant of Section 5.1. The following example illustrates the procedure.

Example 1 Find a general solution of the system

$$x_1' = 4x_1 + 2x_2,$$
$$x_2' = 3x_1 - x_2. \tag{11}$$

Solution The matrix form of the system in (11) is

$$\mathbf{x}' = \begin{bmatrix} 4 & 2 \\ 3 & -1 \end{bmatrix} \mathbf{x}. \tag{12}$$

The characteristic equation of the coefficient matrix is

$$\begin{vmatrix} 4 - \lambda & 2 \\ 3 & -1 - \lambda \end{vmatrix} = (4 - \lambda)(-1 - \lambda) - 6$$
$$= \lambda^2 - 3\lambda - 10 = (\lambda + 2)(\lambda - 5) = 0,$$

so we have the distinct real eigenvalues $\lambda_1 = -2$ and $\lambda_2 = 5$.

For the coefficient matrix \mathbf{A} in Eq. (12) the eigenvector equation $(\mathbf{A} - \lambda \mathbf{I})\mathbf{v} = \mathbf{0}$ takes the form

$$\begin{bmatrix} 4 - \lambda & 2 \\ 3 & -1 - \lambda \end{bmatrix} \begin{bmatrix} a \\ b \end{bmatrix} = \begin{bmatrix} 0 \\ 0 \end{bmatrix} \tag{13}$$

for the associated eigenvector $\mathbf{v} = \begin{bmatrix} a & b \end{bmatrix}^T$.

CASE 1: $\lambda_1 = -2$. Substitution of the first eigenvalue $\lambda_1 = -2$ in Eq. (13) yields the system

$$\begin{bmatrix} 6 & 2 \\ 3 & 1 \end{bmatrix} \begin{bmatrix} a \\ b \end{bmatrix} = \begin{bmatrix} 0 \\ 0 \end{bmatrix};$$

that is, the two scalar equations

$$6a + 2b = 0,$$
$$3a + b = 0. \tag{14}$$

In contrast with the nonsingular (algebraic) linear systems whose solutions we discussed in Section 5.1, the homogeneous linear system in (14) is *singular*—the two scalar equations obviously are equivalent (each being a multiple of the other). Therefore, Eq. (14) has infinitely many nonzero solutions—we can choose a arbitrary (but nonzero) and then solve for b.

Substitution of an eigenvalue λ in the eigenvector equation $(\mathbf{A} - \lambda \mathbf{I})\mathbf{v} = \mathbf{0}$ always yields a singular homogeneous linear system, and among its infinity of solutions we generally seek a "simple" solution with small integer values (if possible). Looking at the second equation in (14), the choice $a = 1$ yields $b = -3$, and thus

$$\mathbf{v}_1 = \begin{bmatrix} 1 \\ -3 \end{bmatrix}$$

is an eigenvector associated with $\lambda_1 = -2$ (as is any nonzero constant multiple of \mathbf{v}_1).

Remark: If instead of the "simplest" choice $a = 1$, $b = -3$, we had made another choice $a = c \neq 0$, $b = -3c$, we would have obtained the eigenvector

$$\mathbf{v}_1 = \begin{bmatrix} c \\ -3c \end{bmatrix} = c \begin{bmatrix} 1 \\ -3 \end{bmatrix}.$$

Because this is a constant multiple of our previous result, any choice we make leads to [a constant multiple of] the same solution

$$\mathbf{x}_1(t) = \begin{bmatrix} 1 \\ -3 \end{bmatrix} e^{-2t}.$$

CASE 2: $\lambda_2 = 5$. Substitution of the second eigenvalue $\lambda = 5$ in (13) yields the pair

$$\begin{aligned} -a + 2b &= 0, \\ 3a - 6b &= 0 \end{aligned} \tag{15}$$

of equivalent scalar equations. With $b = 1$ in the first equation we get $a = 2$, so

$$\mathbf{v}_2 = \begin{bmatrix} 2 \\ 1 \end{bmatrix}$$

is an eigenvector associated with $\lambda_2 = 5$. A different choice $a = 2c$, $b = c \neq 0$ would merely give a [constant] multiple of \mathbf{v}_2.

These two eigenvalues and associated eigenvectors yield the two solutions

$$\mathbf{x}_1(t) = \begin{bmatrix} 1 \\ -3 \end{bmatrix} e^{-2t} \quad \text{and} \quad \mathbf{x}_2(t) = \begin{bmatrix} 2 \\ 1 \end{bmatrix} e^{5t}.$$

They are linearly independent because their Wronskian

$$\begin{vmatrix} e^{-2t} & 2e^{5t} \\ -3e^{-2t} & e^{5t} \end{vmatrix} = 7e^{3t}$$

is nonzero. Hence a general solution of the system in (11) is

$$\mathbf{x}(t) = c_1 \mathbf{x}_1(t) + c_2 \mathbf{x}_2(t) = c_1 \begin{bmatrix} 1 \\ -3 \end{bmatrix} e^{-2t} + c_2 \begin{bmatrix} 2 \\ 1 \end{bmatrix} e^{5t};$$

in scalar form,

$$\begin{aligned} x_1(t) &= c_1 e^{-2t} + 2c_2 e^{5t}, \\ x_2(t) &= -3c_1 e^{-2t} + c_2 e^{5t}. \end{aligned}$$

Fig. 5.2.1 shows some typical solution curves of the system in (11). We see two families of hyperbolas sharing the same pair of asymptotes: the line $x_1 = 2x_2$ obtained from the general solution with $c_1 = 0$ and the line $x_2 = -3x_1$ obtained with $c_2 = 0$. Given initial values $x_1(0) = b_1$, $x_2(0) = b_2$, it is apparent from the figure that

- If (b_1, b_2) lies to the right of the line $x_2 = -3x_1$, then $x_1(t)$ and $x_2(t)$ both tend to $+\infty$ as $t \to +\infty$;
- If (b_1, b_2) lies to the left of the line $x_2 = -3x_1$, then $x_1(t)$ and $x_2(t)$ both tend to $-\infty$ as $t \to +\infty$. ∎

FIGURE 5.2.1. Direction field and solution curves for the linear system $x_1' = 4x_1 + 2x_2$, $x_2' = 3x_1 - x_2$ of Example 1.

Remark: As in Example 1, it is convenient when discussing a linear system $\mathbf{x}' = \mathbf{A}\mathbf{x}$ to use vectors $\mathbf{x}_1, \mathbf{x}_2, \ldots, \mathbf{x}_n$ to denote different vector-valued solutions of the system, whereas the *scalars* x_1, x_2, \ldots, x_n denote the components of a single vector-valued solution \mathbf{x}. ∎

r (gal/min)

Tank 1
V_1 (gal)

r

Tank 2
V_2

r

Tank 3
V_3

r

FIGURE 5.2.2. The three brine tanks of Example 2.

Compartmental Analysis

Frequently a complex process or system can be broken down into simpler subsystems or "compartments" that can be analyzed separately. The whole system can then be modeled by describing the interactions between the various compartments. Thus a chemical plant may consist of a succession of separate stages (or even physical compartments) in which various reactants and products combine or are mixed. It may happen that a single differential equation describes each compartment of the system, and then the whole physical system is modeled by a system of differential equations.

As a simple example of a three-stage system, Fig. 5.2.2 shows three brine tanks containing V_1, V_2, and V_3 gallons of brine, respectively. Fresh water flows into tank 1, while mixed brine flows from tank 1 into tank 2, from tank 2 into tank 3, and out of tank 3. Let $x_i(t)$ denote the amount (in pounds) of salt in tank i at time t for $i = 1$, 2, and 3. If each flow rate is r gallons per minute, then a simple accounting of salt concentrations, as in Example 2 of Section 4.1, yields the first-order system

$$\begin{aligned} x_1' &= -k_1 x_1, \\ x_2' &= k_1 x_1 - k_2 x_2, \\ x_3' &= k_2 x_2 - k_3 x_3, \end{aligned} \tag{16}$$

where

$$k_i = \frac{r}{V_i}, \quad i = 1, 2, 3. \tag{17}$$

Example 2 If $V_1 = 20$, $V_2 = 40$, $V_3 = 50$, $r = 10$ (gal/min), and the initial amounts of salt in the three brine tanks, in pounds, are

$$x_1(0) = 15, \quad x_2(0) = x_3(0) = 0,$$

find the amount of salt in each tank at time $t \geq 0$.

Solution Substituting the given numerical values in (16) and (17), we get the initial value problem

$$\mathbf{x}'(t) = \begin{bmatrix} -0.5 & 0.0 & 0.0 \\ 0.5 & -0.25 & 0.0 \\ 0.0 & 0.25 & -0.2 \end{bmatrix} \mathbf{x}, \quad \mathbf{x}(0) = \begin{bmatrix} 15 \\ 0 \\ 0 \end{bmatrix} \tag{18}$$

for the vector $\mathbf{x}(t) = \begin{bmatrix} x_1(t) & x_2(t) & x_3(t) \end{bmatrix}^T$. The simple form of the matrix

$$\mathbf{A} - \lambda \mathbf{I} = \begin{bmatrix} -0.5 - \lambda & 0.0 & 0.0 \\ 0.5 & -0.25 - \lambda & 0.0 \\ 0.0 & 0.25 & -0.2 - \lambda \end{bmatrix} \tag{19}$$

leads readily to the characteristic equation

$$|\mathbf{A} - \lambda \mathbf{I}| = (-0.5 - \lambda)(-0.25 - \lambda)(-0.2 - \lambda) = 0.$$

Thus the coefficient matrix \mathbf{A} in (18) has the distinct eigenvalues $\lambda_1 = -0.5$, $\lambda_2 = -0.25$, and $\lambda_3 = -0.2$.

CASE 1: $\lambda_1 = -0.5$. Substituting $\lambda = -0.5$ in (19), we get the equation

$$\left[\, \mathbf{A} + (0.5) \cdot \mathbf{I}\,\right] \mathbf{v} = \begin{bmatrix} 0.0 & 0.0 & 0.0 \\ 0.5 & 0.25 & 0.0 \\ 0.0 & 0.25 & 0.3 \end{bmatrix} \begin{bmatrix} a \\ b \\ c \end{bmatrix} = \begin{bmatrix} 0 \\ 0 \\ 0 \end{bmatrix}$$

for the associated eigenvector $\mathbf{v} = \begin{bmatrix} a & b & c \end{bmatrix}^T$. The last two rows, after division by 0.25 and 0.05, respectively, yield the scalar equations

$$\begin{aligned} 2a + \; b \quad\;\;\; &= 0, \\ 5b + 6c &= 0. \end{aligned}$$

The second equation is satisfied by $b = -6$ and $c = 5$, and then the first equation gives $a = 3$. Thus the eigenvector

$$\mathbf{v}_1 = \begin{bmatrix} 3 & -6 & 5 \end{bmatrix}^T$$

is associated with the eigenvalue $\lambda_1 = -0.5$.

CASE 2: $\lambda_2 = -0.25$. Substituting $\lambda = -0.25$ in (19), we get the equation

$$\left[\, \mathbf{A} + (0.25) \cdot \mathbf{I}\,\right] \mathbf{v} = \begin{bmatrix} -0.25 & 0 & 0 \\ 0.5 & 0 & 0 \\ 0 & 0.25 & 0.05 \end{bmatrix} \begin{bmatrix} a \\ b \\ c \end{bmatrix} = \begin{bmatrix} 0 \\ 0 \\ 0 \end{bmatrix}$$

for the associated eigenvector $\mathbf{v} = \begin{bmatrix} a & b & c \end{bmatrix}^T$. Each of the first two rows implies that $a = 0$, and division of the third row by 0.05 gives the equation

$$5b + c = 0,$$

which is satisfied by $b = 1$, $c = -5$. Thus the eigenvector

$$\mathbf{v}_2 = \begin{bmatrix} 0 & 1 & -5 \end{bmatrix}^T$$

is associated with the eigenvalue $\lambda_2 = -0.25$.

CASE 3: $\lambda_3 = -0.2$. Substituting $\lambda = -0.2$ in (19), we get the equation

$$\left[\, \mathbf{A} + (0.2) \cdot \mathbf{I}\,\right] \mathbf{v} = \begin{bmatrix} -0.3 & 0.0 & 0.0 \\ 0.5 & -0.05 & 0.0 \\ 0.0 & 0.25 & 0.0 \end{bmatrix} \begin{bmatrix} a \\ b \\ c \end{bmatrix} = \begin{bmatrix} 0 \\ 0 \\ 0 \end{bmatrix}$$

for the eigenvector \mathbf{v}. The first and third rows imply that $a = 0$, and $b = 0$, respectively, but the all-zero third column leaves c arbitrary (but nonzero). Thus

$$\mathbf{v}_3 = \begin{bmatrix} 0 & 0 & 1 \end{bmatrix}^T$$

is an eigenvector associated with $\lambda_3 = -0.2$.

 The general solution

$$\mathbf{x}(t) = c_1 \mathbf{v}_1 e^{\lambda_1 t} + c_2 \mathbf{v}_2 e^{\lambda_2 t} + c_3 \mathbf{v}_3 e^{\lambda_3 t}$$

therefore takes the form

$$\mathbf{x}(t) = c_1 \begin{bmatrix} 3 \\ -6 \\ 5 \end{bmatrix} e^{(-0.5)t} + c_2 \begin{bmatrix} 0 \\ 1 \\ -5 \end{bmatrix} e^{(-0.25)t} + c_3 \begin{bmatrix} 0 \\ 0 \\ 1 \end{bmatrix} e^{(-0.2)t}.$$

The resulting scalar equations are

$$\begin{aligned} x_1(t) &= 3c_1 e^{(-0.5)t}, \\ x_2(t) &= -6c_1 e^{(-0.5)t} + c_2 e^{(-0.25)t}, \\ x_3(t) &= 5c_1 e^{(-0.5)t} - 5c_2 e^{(-0.25)t} + c_3 e^{(-0.2)t}. \end{aligned}$$

When we impose the initial conditions $x_1(0) = 15$, $x_2(0) = x_3(0) = 0$, we get the equations

$$\begin{aligned} 3c_1 &= 15, \\ -6c_1 + c_2 &= 0, \\ 5c_1 - 5c_2 + c_3 &= 0 \end{aligned}$$

that are readily solved (in turn) for $c_1 = 5$, $c_2 = 30$, and $c_3 = 125$. Thus, finally, the amounts of salt at time t in the three brine tanks are given by

$$\begin{aligned} x_1(t) &= 15e^{(-0.5)t}, \\ x_2(t) &= -30e^{(-0.5)t} + 30e^{(-0.25)t}, \\ x_3(t) &= 25e^{(-0.5)t} - 150e^{(-0.25)t} + 125e^{(-0.2)t}. \end{aligned}$$

Figure 5.2.3 shows the graphs of $x_1(t)$, $x_2(t)$, and $x_3(t)$. As we would expect, tank 1 is rapidly "flushed" by the incoming fresh water, and $x_1(t) \to 0$ as $t \to +\infty$. The amounts $x_2(t)$ and $x_3(t)$ of salt in tanks 2 and 3 peak in turn and then approach zero as the whole three-tank system is purged of salt as $t \to +\infty$. ∎

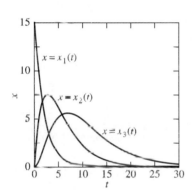

FIGURE 5.2.3. The salt content functions of Example 2.

Complex Eigenvalues

Even if some of the eigenvalues are complex, so long as they are distinct the method described previously still yields n linearly independent solutions. The only complication is that the eigenvectors associated with complex eigenvalues are ordinarily complex-valued, so we will have complex-valued solutions.

To obtain real-valued solutions, we note that—because we are assuming that the matrix \mathbf{A} has only real entries—the coefficients in the characteristic equation in (8) will all be real. Consequently any complex eigenvalues must appear in complex conjugate pairs. Suppose then that $\lambda = p + qi$ and $\bar{\lambda} = p - qi$ are such a pair of eigenvalues. If \mathbf{v} is an eigenvector associated with λ, so that

$$(\mathbf{A} - \lambda\mathbf{I})\mathbf{v} = \mathbf{0},$$

then taking complex conjugates in this equation yields

$$(\mathbf{A} - \bar{\lambda}\mathbf{I})\bar{\mathbf{v}} = \mathbf{0}.$$

Thus the conjugate $\bar{\mathbf{v}}$ of \mathbf{v} is an eigenvector associated with $\bar{\lambda}$. Of course the conjugate of a vector is defined componentwise; if

$$\mathbf{v} = \begin{bmatrix} a_1 + b_1 i \\ a_2 + b_2 i \\ \vdots \\ a_n + b_n i \end{bmatrix} = \begin{bmatrix} a_1 \\ a_2 \\ \vdots \\ a_n \end{bmatrix} + \begin{bmatrix} b_1 \\ b_2 \\ \vdots \\ b_n \end{bmatrix} i = \mathbf{a} + \mathbf{b}i, \tag{20}$$

then $\bar{\mathbf{v}} = \mathbf{a} - \mathbf{b}i$. The complex-valued solution associated with λ and \mathbf{v} is then

$$\mathbf{x}(t) = \mathbf{v}e^{\lambda t} = \mathbf{v}e^{(p+qi)t} = (\mathbf{a} + \mathbf{b}i)e^{pt}(\cos qt + i \sin qt);$$

that is,

$$\mathbf{x}(t) = e^{pt}(\mathbf{a}\cos qt - \mathbf{b}\sin qt) + ie^{pt}(\mathbf{b}\cos qt + \mathbf{a}\sin qt). \tag{21}$$

Because the real and imaginary parts of a complex-valued solution are also solutions, we thus get the two *real-valued* solutions

$$\begin{aligned} \mathbf{x}_1(t) &= \mathrm{Re}(\mathbf{x}(t)) = e^{pt}(\mathbf{a}\cos qt - \mathbf{b}\sin qt), \\ \mathbf{x}_2(t) &= \mathrm{Im}(\mathbf{x}(t)) = e^{pt}(\mathbf{b}\cos qt + \mathbf{a}\sin qt) \end{aligned} \tag{22}$$

associated with the complex conjugate eigenvalues $p \pm qi$. It is easy to check that the same two real-valued solutions result from taking real and imaginary parts of $\bar{\mathbf{v}}e^{\bar{\lambda}t}$. Rather than memorizing the formulas in (22), it is preferable in a specific example to proceed as follows:

- First find explicitly a single complex-valued solution $\mathbf{x}(t)$ associated with the complex eigenvalue λ;
- Then find the real and imaginary parts $\mathbf{x}_1(t)$ and $\mathbf{x}_2(t)$ to get two independent real-valued solutions corresponding to the two complex conjugate eigenvalues λ and $\bar{\lambda}$.

Example 3 Find a general solution of the system

$$\begin{aligned} \frac{dx_1}{dt} &= 4x_1 - 3x_2, \\[6pt] \frac{dx_2}{dt} &= 3x_1 + 4x_2. \end{aligned} \tag{23}$$

Solution The coefficient matrix

$$\mathbf{A} = \begin{bmatrix} 4 & -3 \\ 3 & 4 \end{bmatrix}$$

has characteristic equation

$$|\mathbf{A} - \lambda\mathbf{I}| = \begin{vmatrix} 4 - \lambda & -3 \\ 3 & 4 - \lambda \end{vmatrix} = (4 - \lambda)^2 + 9 = 0,$$

and hence has the complex conjugate eigenvalues $\lambda = 4 - 3i$ and $\bar{\lambda} = 4 + 3i$.

Substituting $\lambda = 4 - 3i$ in the eigenvector equation $(\mathbf{A} - \lambda\mathbf{I})\mathbf{v} = \mathbf{0}$, we get the equation

$$[\mathbf{A} - (4 - 3i) \cdot \mathbf{I}]\mathbf{v} = \begin{bmatrix} 3i & -3 \\ 3 & 3i \end{bmatrix}\begin{bmatrix} a \\ b \end{bmatrix} = \begin{bmatrix} 0 \\ 0 \end{bmatrix}$$

for an associated eigenvalue $\mathbf{v} = \begin{bmatrix} a & b \end{bmatrix}^T$. Division of each row by 3 yields the two scalar equations

$$ia - b = 0,$$
$$a + ib = 0,$$

each of which is satisfied by $a = 1$ and $b = i$. Thus $\mathbf{v} = \begin{bmatrix} 1 & i \end{bmatrix}^T$ is a complex eigenvector associated with the complex eigenvalue $\lambda = 4 - 3i$.

The corresponding complex-valued solution $\mathbf{x}(t) = \mathbf{v}e^{\lambda t}$ of $\mathbf{x}' = \mathbf{A}\mathbf{x}$ is then

$$\mathbf{x}(t) = \begin{bmatrix} 1 \\ i \end{bmatrix}e^{(4-3i)t} = \begin{bmatrix} 1 \\ i \end{bmatrix}e^{4t}(\cos 3t - i\sin 3t) = e^{4t}\begin{bmatrix} \cos 3t - i\sin 3t \\ i\cos 3t + \sin 3t \end{bmatrix}.$$

The real and imaginary parts of $\mathbf{x}(t)$ are the real-valued solutions

$$\mathbf{x}_1(t) = e^{4t}\begin{bmatrix} \cos 3t \\ \sin 3t \end{bmatrix} \quad \text{and} \quad \mathbf{x}_2(t) = e^{4t}\begin{bmatrix} -\sin 3t \\ \cos 3t \end{bmatrix}.$$

A real-valued general solution of $\mathbf{x}' = \mathbf{A}\mathbf{x}$ is then given by

$$\mathbf{x}(t) = c_1\mathbf{x}_1(t) + c_2\mathbf{x}_2(t) = e^{4t}\begin{bmatrix} c_1\cos 3t - c_2\sin 3t \\ c_1\sin 3t + c_2\cos 3t \end{bmatrix}.$$

Finally, a general solution of the system in (23) in scalar form is

$$x_1(t) = e^{4t}(c_1\cos 3t - c_2\sin 3t),$$
$$x_2(t) = e^{4t}(c_1\sin 3t + c_2\cos 3t).$$

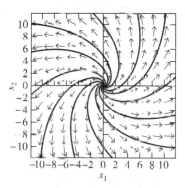

FIGURE 5.2.4. Direction field and solution curves for the linear system $x_1' = 4x_1 - 3x_2$, $x_2' = 3x_1 + 4x_2$ of Example 3.

Figure 5.2.4 shows some typical solution curves of the system in (23). Each appears to spiral counterclockwise as it emanates from the origin in the x_1x_2-plane. Actually, because of the factor e^{4t} in the general solution, we see that

- Along each solution curve, the point $(x_1(t), x_2(t))$ approaches the origin as $t \to -\infty$, whereas
- The absolute values of $x_1(t)$ and $x_2(t)$ both increase without bound as $t \to +\infty$. ∎

FIGURE 5.2.5. The three brine tanks of Example 4.

Figure 5.2.5 shows a "closed" system of three brine tanks with volumes V_1, V_2, and V_3. The difference between this system and the "open" system of Fig. 5.2.2 is that now the inflow to tank 1 is the outflow from tank 3. With the same notation as in Example 2, the appropriate modification of Eq. (16) is

$$\frac{dx_1}{dt} = -k_1 x_1 \qquad\qquad + k_3 x_3,$$

$$\frac{dx_2}{dt} = \quad k_1 x_1 - k_2 x_2, \qquad\qquad (24)$$

$$\frac{dx_3}{dt} = \qquad\qquad k_2 x_2 - k_3 x_3,$$

where $k_i = r/V_i$ as in (17).

Example 4 Find the amounts $x_1(t)$, $x_2(t)$, and $x_3(t)$ of salt at time t in the three brine tanks of Fig. 5.2.5 if $V_1 = 50$ gal, $V_2 = 25$ gal, $V_3 = 50$ gal, and $r = 10$ gal/min.

Solution With the given numerical values, (24) takes the form

$$\frac{d\mathbf{x}}{dt} = \begin{bmatrix} -0.2 & 0 & 0.2 \\ 0.2 & -0.4 & 0 \\ 0 & 0.4 & -0.2 \end{bmatrix} \mathbf{x} \tag{25}$$

with $\mathbf{x} = \begin{bmatrix} x_1 & x_2 & x_3 \end{bmatrix}^T$ as usual. When we expand the determinant of the matrix

$$\mathbf{A} - \lambda \cdot \mathbf{I} = \begin{bmatrix} -0.2 - \lambda & 0.0 & 0.2 \\ 0.2 & -0.4 - \lambda & 0.0 \\ 0.0 & 0.4 & -0.2 - \lambda \end{bmatrix} \tag{26}$$

along its first row, we find that the characteristic equation of \mathbf{A} is

$$(-0.2 - \lambda)(-0.4 - \lambda)(-0.2 - \lambda) + (0.2)(0.2)(0.4)$$
$$= -\lambda^3 - (0.8) \cdot \lambda^2 - (0.2) \cdot \lambda$$
$$= -\lambda \left[(\lambda + 0.4)^2 + (0.2)^2 \right] = 0.$$

Thus \mathbf{A} has the zero eigenvalue $\lambda_0 = 0$ and the complex conjugate eigenvalues λ, $\overline{\lambda} = -0.4 \pm (0.2)i$.

CASE 1: $\lambda_0 = 0$. Substitution of $\lambda = 0$ in Eq. (26) gives the eigenvector equation

$$(\mathbf{A} - 0 \cdot \mathbf{I})\mathbf{v} = \begin{bmatrix} -0.2 & 0.0 & 0.2 \\ 0.2 & -0.4 & 0.0 \\ 0.0 & 0.4 & -0.2 \end{bmatrix} \begin{bmatrix} a \\ b \\ c \end{bmatrix} = \begin{bmatrix} 0 \\ 0 \\ 0 \end{bmatrix}$$

for $\mathbf{v} = \begin{bmatrix} a & b & c \end{bmatrix}^T$. The first row gives $a = c$ and the second row gives $a = 2b$, so $\mathbf{v}_0 = \begin{bmatrix} 2 & 1 & 2 \end{bmatrix}^T$ is an eigenvector associated with the eigenvalue $\lambda_0 = 0$. The corresponding solution $\mathbf{x}_0(t) = \mathbf{v}_0 e^{\lambda_0 t}$ of Eq. (25) is the constant solution

$$\mathbf{x}_0(t) = \begin{bmatrix} 2 \\ 1 \\ 2 \end{bmatrix}. \tag{27}$$

CASE 2: $\lambda = -0.4 - (0.2)i$. Substitution of $\lambda = -0.4 - (0.2)i$ in Eq. (26) gives the eigenvector equation

$$[\mathbf{A} - (-0.4 - (0.2)i)\mathbf{I}] \mathbf{v} = \begin{bmatrix} 0.2 + (0.2)i & 0.0 & 0.2 \\ 0.2 & (0.2)i & 0.0 \\ 0.0 & 0.4 & 0.2 + (0.2)i \end{bmatrix} \begin{bmatrix} a \\ b \\ c \end{bmatrix}$$
$$= \begin{bmatrix} 0 \\ 0 \\ 0 \end{bmatrix}.$$

The second equation $(0.2)a + (0.2)ib = 0$ is satisfied by $a = 1$ and $b = i$. Then the first equation

$$[0.2 + (0.2)i]a + (0.2)c = 0$$

gives $c = -1 - i$. Thus $\mathbf{v} = \begin{bmatrix} 1 & i & (-1-i) \end{bmatrix}^T$ is a complex eigenvector associated with the complex eigenvalue $\lambda = -0.4 - (0.2)i$.

The corresponding complex-valued solution $\mathbf{x}(t) = \mathbf{v}e^{\lambda t}$ of (25) is

$$
\begin{aligned}
\mathbf{x}(t) &= \begin{bmatrix} 1 & i & -1-i \end{bmatrix}^T e^{(-0.4-0.2i)t} \\
&= \begin{bmatrix} 1 & i & -1-i \end{bmatrix}^T e^{(-0.4)t}(\cos 0.2t - i \sin 0.2t) \\
&= e^{(-0.4)t}\begin{bmatrix} \cos 0.2t - i \sin 0.2t \\ \sin 0.2t + i \cos 0.2t \\ -\cos 0.2t - \sin 0.2t - i \cos 0.2t + i \sin 0.2t \end{bmatrix}.
\end{aligned}
$$

The real and imaginary parts of $\mathbf{x}(t)$ are the real-valued solutions

$$
\mathbf{x}_1(t) = e^{(-0.4)t}\begin{bmatrix} \cos 0.2t \\ \sin 0.2t \\ -\cos 0.2t - \sin 0.2t \end{bmatrix},
$$

$$
\mathbf{x}_2(t) = e^{(-0.4)t}\begin{bmatrix} -\sin 0.2t \\ \cos 0.2t \\ -\cos 0.2t + \sin 0.2t \end{bmatrix}. \tag{28}
$$

The general solution

$$
\mathbf{x}(t) = c_0\mathbf{x}_0(t) + c_1\mathbf{x}_1(t) + c_2\mathbf{x}_2(t)
$$

has scalar components

$$
\begin{aligned}
x_1(t) &= 2c_0 + e^{(-0.4)t}(c_1 \cos 0.2t - c_2 \sin 0.2t), \\
x_2(t) &= c_0 + e^{(-0.4)t}(c_1 \sin 0.2t + c_2 \cos 0.2t), \\
x_3(t) &= 2c_0 + e^{(-0.4)t}[(-c_1 - c_2) \cos 0.2t + (-c_1 + c_2) \sin 0.2t]
\end{aligned} \tag{29}
$$

giving the amounts of salt in the three tanks at time t.

Observe that

$$
x_1(t) + x_2(t) + x_3(t) \equiv 5c_0. \tag{30}
$$

FIGURE 5.2.6. The salt content functions of Example 4.

Of course the total amount of salt in the closed system is constant; the constant c_0 in (30) is one-fifth the total amount of salt. Because of the factors of $e^{(-0.4)t}$ in (29), we see that

$$
\lim_{t\to\infty} x_1(t) = 2c_0, \quad \lim_{t\to\infty} x_2(t) = c_0, \quad \text{and} \quad \lim_{t\to\infty} x_3(t) = 2c_0.
$$

Thus as $t \to +\infty$ the salt in the system approaches a *steady-state* distribution with 40% of the salt in each of the two 50-gallon tanks and 20% in the 25-gallon tank. So whatever the initial distribution of salt among the three tanks, the limiting distribution is one of uniform concentration throughout the system. Figure 5.2.6 shows the graphs of the three solution functions with $c_0 = 10$, $c_1 = 30$, and $c_2 = -10$, in which case

$$
x_1(0) = 50 \quad \text{and} \quad x_2(0) = x_3(0) = 0. \qquad \blacksquare
$$

5.2 Problems

In Problems 1 through 16, apply the eigenvalue method of this section to find a general solution of the given system. If initial values are given, find also the corresponding particular solution. For each problem, use a computer system or graphing calculator to construct a direction field and typical solution curves for the given system.

1. $x_1' = x_1 + 2x_2, \quad x_2' = 2x_1 + x_2$
2. $x_1' = 2x_1 + 3x_2, \quad x_2' = 2x_1 + x_2$
3. $x_1' = 3x_1 + 4x_2, \quad x_2' = 3x_1 + 2x_2; \quad x_1(0) = x_2(0) = 1$
4. $x_1' = 4x_1 + x_2, \quad x_2' = 6x_1 - x_2$
5. $x_1' = 6x_1 - 7x_2, \quad x_2' = x_1 - 2x_2$
6. $x_1' = 9x_1 + 5x_2, \quad x_2' = -6x_1 - 2x_2; \quad x_1(0) = 1, \quad x_2(0) = 0$
7. $x_1' = -3x_1 + 4x_2, \quad x_2' = 6x_1 - 5x_2$
8. $x_1' = x_1 - 5x_2, \quad x_2' = x_1 - x_2$
9. $x_1' = 2x_1 - 5x_2, \quad x_2' = 4x_1 - 2x_2; \quad x_1(0) = 2, \quad x_2(0) = 3$
10. $x_1' = -3x_1 - 2x_2, \quad x_2' = 9x_1 + 3x_2$
11. $x_1' = x_1 - 2x_2, \quad x_2' = 2x_1 + x_2; \quad x_1(0) = 0, \quad x_2(0) = 4$
12. $x_1' = x_1 - 5x_2, \quad x_2' = x_1 + 3x_2$
13. $x_1' = 5x_1 - 9x_2, \quad x_2' = 2x_1 - x_2$
14. $x_1' = 3x_1 - 4x_2, \quad x_2' = 4x_1 + 3x_2$
15. $x_1' = 7x_1 - 5x_2, \quad x_2' = 4x_1 + 3x_2$
16. $x_1' = -50x_1 + 20x_2, \quad x_2' = 100x_1 - 60x_2$

In Problems 17 through 25, the eigenvalues of the coefficient matrix can be found by inspection and factoring. Apply the eigenvalue method to find a general solution of each system.

17. $x_1' = 4x_1 + x_2 + 4x_3, \quad x_2' = x_1 + 7x_2 + x_3,$
 $x_3' = 4x_1 + x_2 + 4x_3$
18. $x_1' = x_1 + 2x_2 + 2x_3, \quad x_2' = 2x_1 + 7x_2 + x_3,$
 $x_3' = 2x_1 + x_2 + 7x_3$
19. $x_1' = 4x_1 + x_2 + x_3, \quad x_2' = x_1 + 4x_2 + x_3, \quad x_3' = x_1 + x_2 + 4x_3$
20. $x_1' = 5x_1 + x_2 + 3x_3, \quad x_2' = x_1 + 7x_2 + x_3,$
 $x_3' = 3x_1 + x_2 + 5x_3$
21. $x_1' = 5x_1 - 6x_3, \quad x_2' = 2x_1 - x_2 - 2x_3, \quad x_3' = 4x_1 - 2x_2 - 4x_3$
22. $x_1' = 3x_1 + 2x_2 + 2x_3, \quad x_2' = -5x_1 - 4x_2 - 2x_3,$
 $x_3' = 5x_1 + 5x_2 + 3x_3$
23. $x_1' = 3x_1 + x_2 + x_3, \quad x_2' = -5x_1 - 3x_2 - x_3,$
 $x_3' = 5x_1 + 5x_2 + 3x_3$
24. $x_1' = 2x_1 + x_2 - x_3, \quad x_2' = -4x_1 - 3x_2 - x_3,$
 $x_3' = 4x_1 + 4x_2 + 2x_3$
25. $x_1' = 5x_1 + 5x_2 + 2x_3, \quad x_2' = -6x_1 - 6x_2 - 5x_3,$
 $x_3' = 6x_1 + 6x_2 + 5x_3$
26. Find the particular solution of the system

$$\frac{dx_1}{dt} = 3x_1 + x_3,$$

$$\frac{dx_2}{dt} = 9x_1 - x_2 + 2x_3,$$

$$\frac{dx_3}{dt} = -9x_1 + 4x_2 - x_3$$

that satisfies the initial conditions $x_1(0) = 0$, $x_2(0) = 0$, $x_3(0) = 17$.

The amounts $x_1(t)$ and $x_2(t)$ of salt in the two brine tanks of Fig. 5.2.7 satisfy the differential equations

$$\frac{dx_1}{dt} = -k_1 x_1, \quad \frac{dx_2}{dt} = k_1 x_1 - k_2 x_2,$$

where $k_i = r/V_i$ for $i = 1, 2$. In Problems 27 and 28 the volumes V_1 and V_2 are given. First solve for $x_1(t)$ and $x_2(t)$, assuming that $r = 10$ (gal/min), $x_1(0) = 15$ (lb), and $x_2(0) = 0$. Then find the maximum amount of salt ever in tank 2. Finally, construct a figure showing the graphs of $x_1(t)$ and $x_2(t)$.

Fresh water
Flow rate r

Tank 1
Volume V_1
Salt $x_1(t)$

r

Tank 2
Volume V_2
Salt $x_2(t)$

r

FIGURE 5.2.7. The two brine tanks of Problems 27 and 28.

27. $V_1 = 50$ (gal), $V_2 = 25$ (gal)
28. $V_1 = 25$ (gal), $V_2 = 40$ (gal)

The amounts $x_1(t)$ and $x_2(t)$ of salt in the two brine tanks of Fig. 5.2.8 satisfy the differential equations

$$\frac{dx_1}{dt} = -k_1 x_1 + k_2 x_2, \quad \frac{dx_2}{dt} = k_1 x_1 - k_2 x_2,$$

where $k_i = r/V_i$ as usual. In Problems 29 and 30, solve for $x_1(t)$ and $x_2(t)$, assuming that $r = 10$ (gal/min), $x_1(0) = 15$ (lb), and $x_2(0) = 0$. Then construct a figure showing the graphs of $x_1(t)$ and $x_2(t)$.

r

Tank 1 Tank 2

r

FIGURE 5.2.8. The two brine tanks of Problems 29 and 30.

29. $V_1 = 50$ (gal), $V_2 = 25$ (gal)
30. $V_1 = 25$ (gal), $V_2 = 40$ (gal)

Problems 31 through 34 deal with the open three-tank system of Fig. 5.2.2. Fresh water flows into tank 1; mixed brine flows from tank 1 into tank 2, from tank 2 into tank 3, and out of tank 3; all at the given flow rate r gallons per minute. The initial amounts $x_1(0) = x_0$ (lb), $x_2(0) = 0$, and $x_3(0) = 0$ of salt in the three tanks are given, as are their volumes V_1, V_2, and V_3 (in gallons). First solve for the amounts of salt in the three tanks at time t, then determine the maximal amount of salt that tank 3 ever contains. Finally, construct a figure showing the graphs of $x_1(t)$, $x_2(t)$, and $x_3(t)$.

31. $r = 30$, $x_0 = 27$, $V_1 = 30$, $V_2 = 15$, $V_3 = 10$
32. $r = 60$, $x_0 = 45$, $V_1 = 20$, $V_2 = 30$, $V_3 = 60$
33. $r = 60$, $x_0 = 45$, $V_1 = 15$, $V_2 = 10$, $V_3 = 30$
34. $r = 60$, $x_0 = 40$, $V_1 = 20$, $V_2 = 12$, $V_3 = 60$

Problems 35 through 37 deal with the closed three-tank system of Fig. 5.2.5, which is described by the equations in (24). Mixed brine flows from tank 1 into tank 2, from tank 2 into tank 3, and from tank 3 into tank 1, all at the given flow rate r gallons per minute. The initial amounts $x_1(0) = x_0$ (pounds), $x_2(0) = 0$, and $x_3(0) = 0$ of salt in the three tanks are given, as are their volumes V_1, V_2, and V_3 (in gallons). First solve for the amounts of salt in the three tanks at time t, then determine the limiting amount (as $t \rightarrow +\infty$) of salt in each tank. Finally, construct a figure showing the graphs of $x_1(t)$, $x_2(t)$, and $x_3(t)$.

35. $r = 120$, $x_0 = 33$, $V_1 = 20$, $V_2 = 6$, $V_3 = 40$
36. $r = 10$, $x_0 = 18$, $V_1 = 20$, $V_2 = 50$, $V_3 = 20$
37. $r = 60$, $x_0 = 55$, $V_1 = 60$, $V_2 = 20$, $V_3 = 30$

For each matrix A given in Problems 38 through 40, the zeros in the matrix make its characteristic polynomial easy to calculate. Find the general solution of $x' = Ax$.

38. $A = \begin{bmatrix} 1 & 0 & 0 & 0 \\ 2 & 2 & 0 & 0 \\ 0 & 3 & 3 & 0 \\ 0 & 0 & 4 & 4 \end{bmatrix}$

39. $A = \begin{bmatrix} -2 & 0 & 0 & 9 \\ 4 & 2 & 0 & -10 \\ 0 & 0 & -1 & 8 \\ 0 & 0 & 0 & 1 \end{bmatrix}$

40. $A = \begin{bmatrix} 2 & 0 & 0 & 0 \\ -21 & -5 & -27 & -9 \\ 0 & 0 & 5 & 0 \\ 0 & 0 & -21 & -2 \end{bmatrix}$

41. The coefficient matrix A of the 4×4 system

$$x_1' = 4x_1 + x_2 + x_3 + 7x_4,$$
$$x_2' = x_1 + 4x_2 + 10x_3 + x_4,$$
$$x_3' = x_1 + 10x_2 + 4x_3 + x_4,$$
$$x_4' = 7x_1 + x_2 + x_3 + 4x_4$$

has eigenvalues $\lambda_1 = -3$, $\lambda_2 = -6$, $\lambda_3 = 10$, and $\lambda_4 = 15$. Find the particular solution of this system that satisfies the initial conditions

$$x_1(0) = 3, \quad x_2(0) = x_3(0) = 1, \quad x_4(0) = 3.$$

In Problems 42 through 50, use a calculator or computer system to calculate the eigenvalues and eigenvectors (as illustrated in the 5.2 Application below) in order to find a general solution of the linear system $x' = Ax$ with the given coefficient matrix A.

42. $A = \begin{bmatrix} -40 & -12 & 54 \\ 35 & 13 & -46 \\ -25 & -7 & 34 \end{bmatrix}$

43. $A = \begin{bmatrix} -20 & 11 & 13 \\ 12 & -1 & -7 \\ -48 & 21 & 31 \end{bmatrix}$

44. $A = \begin{bmatrix} 147 & 23 & -202 \\ -90 & -9 & 129 \\ 90 & 15 & -123 \end{bmatrix}$

45. $A = \begin{bmatrix} 9 & -7 & -5 & 0 \\ -12 & 7 & 11 & 9 \\ 24 & -17 & -19 & -9 \\ -18 & 13 & 17 & 9 \end{bmatrix}$

46. $A = \begin{bmatrix} 13 & -42 & 106 & 139 \\ 2 & -16 & 52 & 70 \\ 1 & 6 & -20 & -31 \\ -1 & -6 & 22 & 33 \end{bmatrix}$

47. $A = \begin{bmatrix} 23 & -18 & -16 & 0 \\ -8 & 6 & 7 & 9 \\ 34 & -27 & -26 & -9 \\ -26 & 21 & 25 & 12 \end{bmatrix}$

48. $A = \begin{bmatrix} 47 & -8 & 5 & -5 \\ -10 & 32 & 18 & -2 \\ 139 & -40 & -167 & -121 \\ -232 & 64 & 360 & 248 \end{bmatrix}$

49. $A = \begin{bmatrix} 139 & -14 & -52 & -14 & 28 \\ -22 & 5 & 7 & 8 & -7 \\ 370 & -38 & -139 & -38 & 76 \\ 152 & -16 & -59 & -13 & 35 \\ 95 & -10 & -38 & -7 & 23 \end{bmatrix}$

50. $A = \begin{bmatrix} 9 & 13 & 0 & 0 & 0 & -13 \\ -14 & 19 & -10 & -20 & 10 & 4 \\ -30 & 12 & -7 & -30 & 12 & 18 \\ -12 & 10 & -10 & -9 & 10 & 2 \\ 6 & 9 & 0 & 6 & 5 & -15 \\ -14 & 23 & -10 & -20 & 10 & 0 \end{bmatrix}$

5.2 Application: Automatic Calculation of Eigenvalues and Eigenvectors

Most computational systems offer the capability to find eigenvalues and eigenvectors readily. For instance, Fig. 5.2.9 shows a graphing calculator computation of the eigenvalues and eigenvectors of the matrix

$$\mathbf{A} = \begin{bmatrix} -0.5 & 0.0 & 0.0 \\ 0.5 & -0.25 & 0.0 \\ 0.0 & 0.25 & -0.2 \end{bmatrix}$$

of Example 2. We see each eigenvector displayed as a column vector beneath its eigenvalue. Note that with results presented in decimal form, it is up to us to guess (and verify by matrix multiplication) that the exact eigenvector associated with the third eigenvalue $\lambda = -\frac{1}{2}$ is $\mathbf{v} = \begin{bmatrix} 1 & -2 & \frac{5}{3} \end{bmatrix}^T$. The *Maple* commands

```
with(linalg)
A := matrix(3,3,[-0.5,0,0,0.5,-0.25,0,0,0.25,-0.2]);
eigenvects(A);
```

the *Mathematica* commands

```
A = {{-0.5,0,0},{0.5,-0.25,0},{0,0.25,-0.2}}
Eigensystem[A]
```

and the MATLAB commands

```
A = [-0.5,0,0; 0.5,-0.25,0; 0,0.25,-0.2]
[V,D] = eig(A)
```

FIGURE 5.2.9. TI-86 calculation of the eigenvalues and eigenvectors of the matrix \mathbf{A}.

(where **D** will be a diagonal matrix displaying the eigenvalues of **A** and the column vectors of **V** are the corresponding eigenvectors) produce similar results. You can use these commands to find the eigenvalues and eigenvectors needed for any of the problems in this section.

For a more substantial investigation, choose a positive integer $n < 10$ ($n = 5$, for instance) and let q_1, q_2, \ldots, q_n denote the first n nonzero digits in your student ID number. Now consider an open system of brine tanks as in Fig. 5.2.2, except with n rather than three successive tanks having volumes $V_i = 10q_i$ ($i = 1, 2, \ldots, n$) in gallons. If each flow rate is $r = 10$ gallons per minute, then the salt amounts $x_1(t), x_2(t), \ldots, x_n(t)$ satisfy the linear system

$$\begin{aligned} x_1' &= -k_1 x_1, \\ x_i' &= k_{i-1}x_{i-1} - k_i x_i \quad (i = 2, 3, \ldots, n), \end{aligned}$$

where $k_i = r/V_i$. Apply the eigenvalue method to solve this system with initial conditions

$$x_1(0) = 10, \quad x_2(0) = x_3(0) = \cdots = x_n(0) = 0.$$

Graph the solution functions and estimate graphically the maximum amount of salt that each tank ever contains.

For an alternative investigation, suppose that the system of n tanks is closed as in Fig. 5.2.5, so that tank 1 receives as inflow the outflow from tank n (rather than fresh water). Then the first equation should be replaced with $x_1' = k_n x_n - k_1 x_1$. Now show that, in this *closed* system, as $t \to +\infty$ the salt originally in tank 1 distributes itself with constant density throughout the various tanks. A plot like Fig. 5.2.6 should make this fairly obvious.

5.3 Second-Order Systems and Mechanical Applications*

FIGURE 5.3.1. Three spring-coupled masses.

In this section we apply the matrix methods of Sections 5.1 and 5.2 to investigate the oscillations of typical mass-and-spring systems having two or more degrees of freedom. Our examples are chosen to illustrate phenomena that are generally characteristic of complex mechanical systems.

Figure 5.3.1 shows three masses connected to each other and to two walls by the four indicated springs. We assume that the masses slide without friction and that each spring obeys Hooke's law—its extension or compression x and force F of reaction are related by the formula $F = -kx$. If the rightward displacements x_1, x_2, and x_3 of the three masses (from their respective equilibrium positions) are all positive, then

- The first spring is stretched the distance x_1;
- The second spring is stretched the distance $x_2 - x_1$;
- The third spring is stretched the distance $x_3 - x_2$;
- The fourth spring is compressed the distance x_3.

Therefore, application of Newton's law $F = ma$ to the three masses (as in Example 1 of Section 4.1) yields their equations of motion:

$$
\begin{aligned}
m_1 x_1'' &= -k_1 x_1 &&+ k_2(x_2 - x_1), \\
m_2 x_2'' &= -k_2(x_2 - x_1) &&+ k_3(x_3 - x_2), \\
m_3 x_3'' &= -k_3(x_3 - x_2) - k_4 x_3. &&
\end{aligned}
\tag{1}
$$

In terms of the **displacement vector** $\mathbf{x} = \begin{bmatrix} x_1 & x_2 & x_3 \end{bmatrix}^T$, the **mass matrix**

$$
\mathbf{M} = \begin{bmatrix} m_1 & 0 & 0 \\ 0 & m_2 & 0 \\ 0 & 0 & m_3 \end{bmatrix}
\tag{2}
$$

and the **stiffness matrix**

$$
\mathbf{K} = \begin{bmatrix} -(k_1 + k_2) & k_2 & 0 \\ k_2 & -(k_2 + k_3) & k_3 \\ 0 & k_3 & -(k_3 + k_4) \end{bmatrix},
\tag{3}
$$

FIGURE 5.3.2. A system of n spring-coupled masses.

the system in (1) takes the matrix form

$$
\blacktriangleright \qquad \mathbf{M}\mathbf{x}'' = \mathbf{K}\mathbf{x}.
\tag{4}
$$

The notation in Eqs. (1) through (4) generalizes in a natural way to the system of n spring-coupled masses shown in Fig. 5.3.2. We need only write

$$
\mathbf{M} = \begin{bmatrix} m_1 & 0 & \cdots & 0 \\ 0 & m_2 & \cdot & 0 \\ \vdots & \vdots & & \vdots \\ 0 & 0 & \cdots & m_n \end{bmatrix}
\tag{5}
$$

*This optional section may be omitted without loss of continuity. It provides a sample of the more technical applications of eigenvalues to physics and engineering problems.

and

$$\mathbf{K} = \begin{bmatrix} -(k_1 + k_2) & k_2 & 0 & \cdots & 0 \\ k_2 & -(k_2 + k_3) & k_3 & \cdots & 0 \\ 0 & k_3 & -(k_3 + k_4) & \cdots & 0 \\ 0 & 0 & k_4 & \cdots & 0 \\ \vdots & \vdots & & & \vdots \\ 0 & 0 & \cdots & -(k_{n-1} + k_n) & k_n \\ 0 & 0 & \cdots & k_n & -(k_n + k_{n+1}) \end{bmatrix} \tag{6}$$

for the mass and stiffness matrices in Eq. (4).

The diagonal matrix \mathbf{M} is obviously nonsingular; to get its inverse \mathbf{M}^{-1} we need only replace each diagonal element with its reciprocal. Hence multiplication of each side in Eq. (4) by \mathbf{M}^{-1} yields the homogeneous **second-order system**

➤
$$\mathbf{x}'' = \mathbf{A}\mathbf{x}, \tag{7}$$

where $\mathbf{A} = \mathbf{M}^{-1}\mathbf{K}$. There is a wide variety of *frictionless* mechanical systems for which a displacement or position vector \mathbf{x}, a nonsingular mass matrix \mathbf{M}, and a stiffness matrix \mathbf{K} satisfying Eq. (4) can be defined.

Solution of Second-Order Systems

To seek a solution of Eq. (7), we substitute (as in Section 5.2 for a first-order system) a trial solution of the form

$$\mathbf{x}(t) = \mathbf{v}e^{\alpha t}, \tag{8}$$

where \mathbf{v} is a constant vector. Then $\mathbf{x}'' = \alpha^2 \mathbf{v}e^{\alpha t}$, so substitution of Eq. (8) in (7) gives

$$\alpha^2 \mathbf{v}e^{\alpha t} = \mathbf{A}\mathbf{v}e^{\alpha t},$$

which implies that

$$\mathbf{A}\mathbf{v} = \alpha^2 \mathbf{v}. \tag{9}$$

Therefore $\mathbf{x}(t) = \mathbf{v}e^{\alpha t}$ is a solution of $\mathbf{x}'' = \mathbf{A}\mathbf{x}$ if and only if $\alpha^2 = \lambda$, an eigenvalue of the matrix \mathbf{A}, and \mathbf{v} is an associated eigenvector.

If $\mathbf{x}'' = \mathbf{A}\mathbf{x}$ models a mechanical system, then it is typical that the eigenvalues of \mathbf{A} are *negative* real numbers. If

$$\alpha^2 = \lambda = -\omega^2 < 0,$$

then $\alpha = \pm\omega i$. In this case the solution given by Eq. (8) is

$$\mathbf{x}(t) = \mathbf{v}e^{i\omega t} = \mathbf{v}(\cos \omega t + i \sin \omega t).$$

The real and imaginary parts

$$\mathbf{x}_1(t) = \mathbf{v}\cos \omega t \quad \text{and} \quad \mathbf{x}_2(t) = \mathbf{v}\sin \omega t \tag{10}$$

of $\mathbf{x}(t)$ are then linearly independent *real-valued* solutions of the system. This analysis leads to the following theorem.

THEOREM 1: Second-Order Homogeneous Linear Systems

If the $n \times n$ matrix \mathbf{A} has distinct negative eigenvalues $-\omega_1^2, -\omega_2^2, \ldots, -\omega_n^2$ with associated [real] eigenvectors $\mathbf{v}_1, \mathbf{v}_2, \ldots, \mathbf{v}_n$, then a general solution of

➤
$$\mathbf{x}'' = \mathbf{A}\mathbf{x}$$

is given by

➤
$$\mathbf{x}(t) = \sum_{i=1}^{n} (a_i \cos \omega_i t + b_i \sin \omega_i t)\mathbf{v}_i \tag{11}$$

with a_i and b_i arbitrary constants. In the special case of a nonrepeated zero eigenvalue λ_0 with associated eigenvector \mathbf{v}_0,

$$\mathbf{x}_0(t) = (a_0 + b_0 t)\mathbf{v}_0 \tag{12}$$

is the corresponding part of the general solution. ▲

Remark: The nonzero vector \mathbf{v}_0 is an eigenvector corresponding to $\lambda_0 = 0$ provided that $\mathbf{A}\mathbf{v}_0 = \mathbf{0}$. If $\mathbf{x}(t) = (a_0 + b_0 t)\mathbf{v}_0$, then

$$\mathbf{x}'' = 0 \cdot \mathbf{v}_0 = (a_0 + b_0 t) \cdot \mathbf{0} = (a_0 + b_0 t) \cdot (\mathbf{A}\mathbf{v}_0) = \mathbf{A}\mathbf{x},$$

thus verifying the form in Eq. (12). ■

Example 1

Consider the mass-and-spring system with $n = 2$ shown in Fig. 5.3.3. Because there is no third spring connected to a right-hand wall, we set $k_3 = 0$. If $m_1 = 2$, $m_2 = 1$, $k_1 = 100$, and $k_2 = 50$, then the equation $\mathbf{M}\mathbf{x}'' = \mathbf{K}\mathbf{x}$ is

$$\begin{bmatrix} 2 & 0 \\ 0 & 1 \end{bmatrix} \mathbf{x}'' = \begin{bmatrix} -150 & 50 \\ 50 & -50 \end{bmatrix} \mathbf{x}, \tag{13}$$

which reduces to $\mathbf{x}'' = \mathbf{A}\mathbf{x}$ with

$$\mathbf{A} = \begin{bmatrix} -75 & 25 \\ 50 & -50 \end{bmatrix}.$$

Equilibrium positions

FIGURE 5.3.3. The mass-and-spring system of Example 1.

The characteristic equation of \mathbf{A} is

$$(-75 - \lambda)(-50 - \lambda) - 50 \cdot 25 = \lambda^2 + 125\lambda + 2500$$
$$= (\lambda + 25)(\lambda + 100) = 0,$$

so \mathbf{A} has the negative eigenvalues $\lambda_1 = -25$ and $\lambda_2 = -100$. By Theorem 1, the system in (13) therefore has solutions with [circular] frequencies $\omega_1 = 5$ and $\omega_2 = 10$.

CASE 1: $\lambda_1 = -25$. The eigenvector equation $(\mathbf{A} - \lambda\mathbf{I})\mathbf{v} = \mathbf{0}$ is

$$\begin{bmatrix} -50 & 25 \\ 50 & -25 \end{bmatrix} \begin{bmatrix} a \\ b \end{bmatrix} = \begin{bmatrix} 0 \\ 0 \end{bmatrix},$$

so an eigenvector associated with $\lambda_1 = -25$ is $\mathbf{v}_1 = \begin{bmatrix} 1 & 2 \end{bmatrix}^T$.

CASE 2: $\lambda_2 = -100$. The eigenvector equation $(\mathbf{A} - \lambda\mathbf{I})\mathbf{v} = \mathbf{0}$ is

$$\begin{bmatrix} 25 & 25 \\ 50 & 50 \end{bmatrix} \begin{bmatrix} a \\ b \end{bmatrix} = \begin{bmatrix} 0 \\ 0 \end{bmatrix},$$

so an eigenvector associated with $\lambda_2 = -100$ is $\mathbf{v}_2 = \begin{bmatrix} 1 & -1 \end{bmatrix}^T$.

By Eq. (11) it follows that a general solution of the system in (13) is given by

$$\mathbf{x}(t) = (a_1 \cos 5t + b_1 \sin 5t)\mathbf{v}_1 + (a_2 \cos 10t + b_2 \sin 10t)\mathbf{v}_2. \qquad (14)$$

As in the discussion of Example 3 of Section 4.2, the two terms on the right in Eq. (14) represent **free oscillations** of the mass-and-spring system. They describe the physical system's two **natural modes of oscillation** at its two [circular] **natural frequencies** $\omega_1 = 5$ and $\omega_2 = 10$. The natural mode

$$\mathbf{x}_1(t) = (a_1 \cos 5t + b_1 \sin 5t)\mathbf{v}_1 = c_1 \cos(5t - \alpha_1) \begin{bmatrix} 1 \\ 2 \end{bmatrix}$$

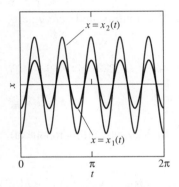

FIGURE 5.3.4. Oscillations in the same direction with frequency $\omega_1 = 5$; the amplitude of motion of mass 2 is twice that of mass 1.

(with $c_1 = \sqrt{a_1^2 + b_1^2}$, $\cos \alpha_1 = a_1/c_1$, and $\sin \alpha_1 = b_1/c_1$) has the scalar component equations

$$\begin{aligned} x_1(t) &= c_1 \cos(5t - \alpha_1), \\ x_2(t) &= 2c_1 \cos(5t - \alpha_1), \end{aligned} \qquad (15)$$

and therefore describes a free oscillation in which the two masses move in synchrony in the same direction and with the same frequency $\omega_1 = 5$, but with the amplitude of motion of m_2 twice that of m_1 (see Fig. 5.3.4). The natural mode

$$\mathbf{x}_2(t) = (a_2 \cos 10t + b_2 \sin 10t)\mathbf{v}_2 = c_2 \cos(10t - \alpha_2) \begin{bmatrix} 1 \\ -1 \end{bmatrix}$$

has the scalar component equations

$$\begin{aligned} x_1(t) &= c_2 \cos(10t - \alpha_2), \\ x_2(t) &= -c_2 \cos(10t - \alpha_2), \end{aligned} \qquad (16)$$

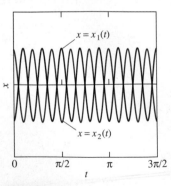

FIGURE 5.3.5. Oscillations in opposite directions with frequency $\omega_2 = 10$; the amplitudes of motion of the two masses are the same.

and therefore describes a free oscillation in which the two masses move in synchrony in opposite directions with the same frequency $\omega_2 = 10$ and with equal amplitudes of oscillation (see Fig. 5.3.5). ■

Example 2

FIGURE 5.3.6. The three railway cars of Example 2.

Figure 5.3.6 shows three railway cars connected by buffer springs that react when compressed, but disengage instead of stretching. With $n = 3$, $k_2 = k_3 = k$, and $k_1 = k_4 = 0$ in Eqs. (2) through (4), we get the system

$$\begin{bmatrix} m_1 & 0 & 0 \\ 0 & m_2 & 0 \\ 0 & 0 & m_3 \end{bmatrix} \mathbf{x}'' = \begin{bmatrix} -k & k & 0 \\ k & -2k & k \\ 0 & k & -k \end{bmatrix} \mathbf{x}, \tag{17}$$

which is equivalent to

$$\mathbf{x}'' = \begin{bmatrix} -c_1 & c_1 & 0 \\ c_2 & -2c_2 & c_2 \\ 0 & c_3 & -c_3 \end{bmatrix} \mathbf{x} \tag{18}$$

with

$$c_i = \frac{k}{m_i} \quad (i = 1, 2, 3). \tag{19}$$

If we assume further that $m_1 = m_3$, so that $c_1 = c_3$, then a brief computation gives

$$-\lambda(\lambda + c_1)(\lambda + c_1 + 2c_2) = 0 \tag{20}$$

for the characteristic equation of the coefficient matrix \mathbf{A} in Eq. (18). Hence the matrix \mathbf{A} has eigenvalues

$$\lambda_1 = 0, \quad \lambda_2 = -c_1, \quad \lambda_3 = -c_1 - 2c_2 \tag{21a}$$

corresponding to the natural frequencies

$$\omega_1 = 0, \quad \omega_2 = \sqrt{c_1}, \quad \omega_3 = \sqrt{c_1 + 2c_2} \tag{21b}$$

of the physical system.

For a numerical example, suppose that the first and third railway cars weigh 12 tons each, that the middle car weighs 8 tons, and that the spring constant is $k = 1.5$ tons/ft; i.e., $k = 3000$ lb/ft. Then, using fps units with mass measured in slugs (a weight of 32 pounds has a mass of 1 slug), we have

$$m_1 = m_3 = 750, \quad m_2 = 500,$$

and

$$c_1 = \frac{3000}{750} = 4, \quad c_2 = \frac{3000}{500} = 6.$$

Hence the coefficient matrix \mathbf{A} is

$$\mathbf{A} = \begin{bmatrix} -4 & 4 & 0 \\ 6 & -12 & 6 \\ 0 & 4 & -4 \end{bmatrix}, \tag{22}$$

and the eigenvalue-frequency pairs given by (21a) and (21b) are $\lambda_1 = 0$, $\omega_1 = 0$; $\lambda_2 = -4$, $\omega_2 = 2$; and $\lambda_3 = -16$, $\omega_3 = 4$.

CASE 1: $\lambda_1 = 0$, $\omega_1 = 0$. The eigenvector equation $(\mathbf{A} - \lambda\mathbf{I})\mathbf{v} = \mathbf{0}$ is

$$\mathbf{Av} = \begin{bmatrix} -4 & 4 & 0 \\ 6 & -12 & 6 \\ 0 & 4 & -4 \end{bmatrix} \begin{bmatrix} a \\ b \\ c \end{bmatrix} = \begin{bmatrix} 0 \\ 0 \\ 0 \end{bmatrix},$$

so it is clear that $\mathbf{v}_1 = \begin{bmatrix} 1 & 1 & 1 \end{bmatrix}^T$ is an eigenvector associated with $\lambda_1 = 0$. According to Theorem 1, the corresponding part of a general solution of $\mathbf{x}'' = \mathbf{Ax}$ is

$$\mathbf{x}_1(t) = (a_1 + b_1 t)\mathbf{v}_1.$$

CASE 2: $\lambda_2 = -4$, $\omega_2 = 2$. The eigenvector equation $(\mathbf{A} - \lambda\mathbf{I})\mathbf{v} = \mathbf{0}$ is

$$(\mathbf{A} + 4\mathbf{I})\mathbf{v} = \begin{bmatrix} 0 & 4 & 0 \\ 6 & -8 & 6 \\ 0 & 4 & 0 \end{bmatrix} \begin{bmatrix} a \\ b \\ c \end{bmatrix} = \begin{bmatrix} 0 \\ 0 \\ 0 \end{bmatrix}$$

so it is clear that $\mathbf{v}_2 = \begin{bmatrix} 1 & 0 & -1 \end{bmatrix}^T$ is an eigenvector associated with $\lambda_2 = -4$. According to Theorem 1, the corresponding part of a general solution of $\mathbf{x}'' = \mathbf{Ax}$ is

$$\mathbf{x}_2(t) = (a_2 \cos 2t + b_2 \sin 2t)\mathbf{v}_2.$$

CASE 3: $\lambda_3 = -16$, $\omega_3 = 4$. The eigenvector equation $(\mathbf{A} - \lambda\mathbf{I})\mathbf{v} = \mathbf{0}$ is

$$(\mathbf{A} + 16\mathbf{I})\mathbf{v} = \begin{bmatrix} 12 & 4 & 0 \\ 6 & 4 & 6 \\ 0 & 4 & 12 \end{bmatrix} \begin{bmatrix} a \\ b \\ c \end{bmatrix} = \begin{bmatrix} 0 \\ 0 \\ 0 \end{bmatrix},$$

so it is clear that $\mathbf{v}_3 = \begin{bmatrix} 1 & -3 & 1 \end{bmatrix}^T$ is an eigenvector associated with $\lambda_3 = -16$. According to Theorem 1, the corresponding part of a general solution of $\mathbf{x}'' = \mathbf{Ax}$ is

$$\mathbf{x}_3(t) = (a_3 \cos 4t + b_3 \sin 4t)\mathbf{v}_3.$$

The general solution $\mathbf{x} = \mathbf{x}_1 + \mathbf{x}_2 + \mathbf{x}_3$ of $\mathbf{x}'' = \mathbf{Ax}$ is therefore given by

$$\mathbf{x}(t) = a_1 \begin{bmatrix} 1 \\ 1 \\ 1 \end{bmatrix} + b_1 t \begin{bmatrix} 1 \\ 1 \\ 1 \end{bmatrix} + a_2 \begin{bmatrix} 1 \\ 0 \\ -1 \end{bmatrix} \cos 2t$$

$$+ b_2 \begin{bmatrix} 1 \\ 0 \\ -1 \end{bmatrix} \sin 2t + a_3 \begin{bmatrix} 1 \\ -3 \\ 1 \end{bmatrix} \cos 4t + b_3 \begin{bmatrix} 1 \\ -3 \\ 1 \end{bmatrix} \sin 4t. \quad (23)$$

To determine a particular solution, let us suppose that the leftmost car is moving to the right with velocity v_0 and at time $t = 0$ strikes the other two cars, which are together but at rest. The corresponding initial conditions are

$$x_1(0) = x_2(0) = x_3(0) = 0, \quad (24a)$$
$$x_1'(0) = v_0, \quad x_2'(0) = x_3'(0) = 0. \quad (24b)$$

Then substitution of (24a) in (23) gives the scalar equations

$$
\begin{aligned}
a_1 + a_2 + \ a_3 &= 0, \\
a_1 \quad\ \ - 3a_3 &= 0, \\
a_1 - a_2 + \ a_3 &= 0,
\end{aligned}
$$

which readily yield $a_1 = a_2 = a_3 = 0$. Hence the position functions of the three cars are

$$
\begin{aligned}
x_1(t) &= b_1 t + b_2 \sin 2t + \ b_3 \sin 4t, \\
x_2(t) &= b_1 t \qquad\qquad - 3b_3 \sin 4t, \\
x_3(t) &= b_1 t - b_2 \sin 2t + \ b_3 \sin 4t,
\end{aligned}
\tag{25}
$$

and their velocity functions are

$$
\begin{aligned}
x_1'(t) &= b_1 + 2b_2 \cos 2t + 4b_3 \cos 4t, \\
x_2'(t) &= b_1 \qquad\qquad\ - 12b_3 \cos 4t, \\
x_3'(t) &= b_1 - 2b_2 \cos 2t + 4b_3 \cos 4t.
\end{aligned}
\tag{26}
$$

Substitution of (24b) in (26) gives the equations

$$
\begin{aligned}
b_1 + 2b_2 + \ 4b_3 &= v_0, \\
b_1 \qquad\ - 12b_3 &= 0, \\
b_1 - 2b_2 + \ 4b_3 &= 0
\end{aligned}
$$

that readily yield $b_1 = \frac{3}{8} v_0$, $b_2 = \frac{1}{4} v_0$, and $b_3 = \frac{1}{32} v_0$. Finally, the position functions in (25) are

$$
\begin{aligned}
x_1(t) &= \tfrac{1}{32} v_0 (12t + 8 \sin 2t + \ \sin 4t), \\
x_2(t) &= \tfrac{1}{32} v_0 (12t \qquad\qquad - 3 \sin 4t), \\
x_3(t) &= \tfrac{1}{32} v_0 (12t - 8 \sin 2t + \ \sin 4t).
\end{aligned}
\tag{27}
$$

But these equations hold only so long as the two buffer springs remain compressed; that is, while both

$$
x_2 - x_1 < 0 \quad \text{and} \quad x_3 - x_2 < 0.
$$

To discover what this implies about t, we compute

$$
\begin{aligned}
x_2(t) - x_1(t) &= \tfrac{1}{32} v_0 (-8 \sin 2t - 4 \sin 4t) \\
&= -\tfrac{1}{32} v_0 (8 \sin 2t + 8 \sin 2t \cos 2t) \\
&= -\tfrac{1}{4} v_0 (\sin 2t)(1 + \cos 2t)
\end{aligned}
$$

and, similarly,

$$
x_3(t) - x_2(t) = -\tfrac{1}{4} v_0 (\sin 2t)(1 - \cos 2t).
$$

It follows that $x_2 - x_1 < 0$ and $x_3 - x_2 < 0$ until $t = \pi/2 \approx 1.57$ (seconds), at which time the equations in (26) and (27) give the values

$$x_1\left(\frac{\pi}{2}\right) = x_2\left(\frac{\pi}{2}\right) = x_3\left(\frac{\pi}{2}\right) = \frac{3\pi v_0}{16},$$

$$x_1'\left(\frac{\pi}{2}\right) = x_2'\left(\frac{\pi}{2}\right) = 0, \quad x_3'\left(\frac{\pi}{2}\right) = v_0.$$

We conclude that the three railway cars remain engaged and moving to the right until disengagement occurs at time $t = \pi/2$. Thereafter, cars 1 and 2 remain at rest (!), while car 3 continues to the right with speed v_0. If, for instance, $v_0 = 48$ feet per second (about 33 miles per hour), then the three cars travel a distance of $9\pi \approx 28.27$ (ft) during their 1.57 seconds of engagement, and

$$x_1(t) = x_2(t) = 9\pi, \quad x_3(t) = 48t - 15\pi \tag{27'}$$

for $t > \pi/2$. Figure 5.3.7 illustrates the "before" and "after" situations, and Fig. 5.3.8 shows the graphs of the functions $x_1(t)$, $x_2(t)$, and $x_3(t)$ in Eqs. (27) and (27'). ∎

FIGURE 5.3.7. (a) Before; (b) after.

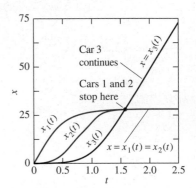

FIGURE 5.3.8. Position functions of the three railway cars of Example 2.

Forced Oscillations and Resonance

Suppose now that the ith mass of the mass-and-spring system in Fig. 5.3.2 is subject to an external force F_i ($i = 1, 2, \ldots, n$) in addition to the forces exerted by the springs attached to it. Then the homogeneous equation $\mathbf{Mx''} = \mathbf{Kx}$ is replaced with the nonhomogeneous equation

$$\mathbf{Mx''} = \mathbf{Kx} + \mathbf{F} \tag{28}$$

where $\mathbf{F} = \begin{bmatrix} F_1 & F_2 & \ldots & F_n \end{bmatrix}^T$ is the **external force vector** for the system. Multiplication by \mathbf{M}^{-1} yields

$$\mathbf{x''} = \mathbf{Ax} + \mathbf{f} \tag{29}$$

where \mathbf{f} is the external force vector *per unit mass*. We are especially interested in the case of a *periodic external force*

$$\mathbf{f}(t) = \mathbf{F}_0 \cos \omega t \tag{30}$$

(where \mathbf{F}_0 is a constant vector). We then anticipate a periodic particular solution

$$\mathbf{x}_p(t) = \mathbf{c} \cos \omega t \tag{31}$$

with the known external frequency ω and with a coefficient vector \mathbf{c} yet to be determined. Because $\mathbf{x}_p'' = -\omega^2 \mathbf{c} \cos \omega t$, substitution of (30) and (31) in (29), followed by cancellation of the common factor $\cos \omega t$, gives the linear system

$$(\mathbf{A} + \omega^2 \mathbf{I})\mathbf{c} = -\mathbf{F}_0 \tag{32}$$

to be solved for \mathbf{c}.

Observe that the matrix $\mathbf{A} + \omega^2 \mathbf{I}$ is nonsingular—in which case Eq. (32) can be solved for \mathbf{c}—unless $-\omega^2 = \lambda$, an eigenvalue of \mathbf{A}. Thus a periodic particular solution of the form in Eq. (31) exists provided that the external forcing frequency does *not* equal one of the natural frequencies $\omega_1, \omega_2, \ldots, \omega_n$ of the system. The case in which ω *is* a natural frequency corresponds to the phenomenon of **resonance** discussed in Section 3.6.

<table>
<tr><td>**Example 3**</td></tr>
</table>

Suppose that the second mass in Example 1 is subjected to the external periodic force $50 \cos \omega t$. Then with $m_1 = 2$, $m_2 = 1$, $k_1 = 100$, $k_2 = 50$, and $F_0 = 50$ in Fig. 5.3.9, Eq. (29) takes the form

$$\mathbf{x}'' = \begin{bmatrix} -75 & 25 \\ 50 & -50 \end{bmatrix} \mathbf{x} + \begin{bmatrix} 0 \\ 50 \end{bmatrix} \cos \omega t, \tag{33}$$

and the substitution $\mathbf{x} = \mathbf{c} \cos \omega t$ leads to the equation

$$\begin{bmatrix} \omega^2 - 75 & 25 \\ 50 & \omega^2 - 50 \end{bmatrix} \mathbf{c} = \begin{bmatrix} 0 \\ -50 \end{bmatrix} \tag{34}$$

for the coefficient vector $\mathbf{c} = \begin{bmatrix} c_1 & c_2 \end{bmatrix}^T$. This system is readily solved for

$$c_1 = \frac{1250}{(\omega^2 - 25)(\omega^2 - 100)}, \quad c_2 = -\frac{50(\omega^2 - 75)}{(\omega^2 - 25)(\omega^2 - 100)}. \tag{35}$$

For instance, if the external squared frequency is $\omega^2 = 50$, then (35) yields $c_1 = -1$, $c_2 = -1$. The resulting forced periodic oscillation is described by

$$x_1(t) = -\cos \omega t, \quad x_2(t) = -\cos \omega t.$$

Thus the two masses oscillate in synchrony with equal amplitudes and in the same direction.

If the external squared frequency is $\omega^2 = 125$, then (35) yields $c_1 = \frac{1}{2}$, $c_2 = -1$. The resulting forced periodic oscillation is described by

$$x_1(t) = \tfrac{1}{2} \cos \omega t, \quad x_2(t) = -\cos \omega t,$$

and now the two masses oscillate in synchrony in opposite directions, but with the amplitude of motion of m_2 twice that of m_1.

It is evident from the denominators in (35) that c_1 and c_2 approach $+\infty$ as ω approaches either of the two natural frequencies $\omega_1 = 5$ and $\omega_2 = 10$ (found in Example 1). Figure 5.3.10 shows a plot of the amplitude $\sqrt{c_1^2 + c_2^2}$ of the forced periodic solution $\mathbf{x}(t) = \mathbf{c} \cos \omega t$ as a function of the forced frequency ω. The peaks at $\omega_2 = 5$ and $\omega_2 = 10$ exhibit visually the phenomenon of resonance. ∎

FIGURE 5.3.9. The forced mass-and-spring system of Example 3.

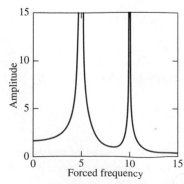

FIGURE 5.3.10. Frequency-amplitude plot for Example 3.

Periodic and Transient Solutions

It follows from Theorem 4 of Section 5.1 that a particular solution of the forced system

$$\mathbf{x}'' = \mathbf{A}\mathbf{x} + \mathbf{F}_0 \cos \omega t \tag{36}$$

will be of the form

➤ $$\mathbf{x}(t) = \mathbf{x}_c(t) + \mathbf{x}_p(t), \tag{37}$$

where $\mathbf{x}_p(t)$ is a particular solution of the nonhomogeneous system and $\mathbf{x}_c(t)$ is a solution of the corresponding homogeneous system. It is typical for the effects of frictional resistance in mechanical systems to damp out the complementary function solution $\mathbf{x}_c(t)$, so that

$$\mathbf{x}_c(t) \to \mathbf{0} \quad \text{as} \quad t \to +\infty. \tag{38}$$

Hence $\mathbf{x}_c(t)$ is a **transient solution** that depends only on the initial conditions; it dies out with time, leaving the **steady periodic solution $\mathbf{x}_p(t)$** resulting from the external driving force:

$$\mathbf{x}(t) \to \mathbf{x}_p(t) \quad \text{as} \quad t \to +\infty. \tag{39}$$

As a practical matter, every physical system includes frictional resistance (however small) that damps out transient solutions in this manner.

5.3 Problems

Problems 1 through 7 deal with the mass-and-spring system shown in Fig. 5.3.11 with stiffness matrix

$$\mathbf{K} = \begin{bmatrix} -(k_1 + k_2) & k_2 \\ k_2 & -(k_2 + k_3) \end{bmatrix}$$

and with the given mks values for the masses and spring constants. Find the two natural frequencies of the system and describe its two natural modes of oscillation.

FIGURE 5.3.11. The mass-and-spring system for Problems 1 through 6.

1. $m_1 = m_2 = 1$; $k_1 = 0, k_2 = 2, k_3 = 0$ (no walls)
2. $m_1 = m_2 = 1$; $k_1 = 1, k_2 = 4, k_3 = 1$
3. $m_1 = 1, m_2 = 2$; $k_1 = 1, k_2 = k_3 = 2$
4. $m_1 = m_2 = 1$; $k_1 = 1, k_2 = 2, k_3 = 1$
5. $m_1 = m_2 = 1$; $k_1 = 2, k_2 = 1, k_3 = 2$
6. $m_1 = 1, m_2 = 2$; $k_1 = 2, k_2 = k_3 = 4$
7. $m_1 = m_2 = 1$; $k_1 = 4, k_2 = 6, k_3 = 4$

In Problems 8 through 10 the indicated mass-and-spring system is set in motion from rest ($x_1'(0) = x_2'(0) = 0$) in its equilibrium position ($x_1(0) = x_2(0) = 0$) with the given external forces $F_1(t)$ and $F_2(t)$ acting on the masses m_1 and m_2, respectively. Find the resulting motion of the system and describe it as a superposition of oscillations at three different frequencies.

8. The mass-and-spring system of Problem 2, with $F_1(t) = 96 \cos 5t$, $F_2(t) \equiv 0$
9. The mass-and-spring system of Problem 3, with $F_1(t) \equiv 0$, $F_2(t) = 120 \cos 3t$
10. The mass-and-spring system of Problem 7, with $F_1(t) = 30 \cos t$, $F_2(t) = 60 \cos t$
11. Consider a mass-and-spring system containing two masses $m_1 = 1$ and $m_2 = 1$ whose displacement functions $x(t)$ and $y(t)$ satisfy the differential equations

$$\begin{aligned} x'' &= -40x + 8y, \\ y'' &= 12x - 60y. \end{aligned}$$

(a) Describe the two fundamental modes of free oscillation of the system. **(b)** Assume that the two masses start in motion with the initial conditions

$$x(0) = 19, \quad x'(0) = 12$$

and

$$y(0) = 3, \quad y'(0) = 6$$

and are acted on by the same force, $F_1(t) = F_2(t) = -195 \cos 7t$. Describe the resulting motion as a superposition of oscillations at three different frequencies.

In Problems 12 and 13, find the natural frequencies of the three-mass system of Fig. 5.3.1, using the given masses and spring constants. For each natural frequency ω, give the ratio $a_1:a_2:a_3$ of amplitudes for a corresponding natural mode $x_1 = a_1 \cos \omega t$, $x_2 = a_2 \cos \omega t$, $x_3 = a_3 \cos \omega t$.

12. $m_1 = m_2 = m_3 = 1$; $\quad k_1 = k_2 = k_3 = k_4 = 1$
13. $m_1 = m_2 = m_3 = 1$; $\quad k_1 = k_2 = k_3 = k_4 = 2$
 (*Hint:* One eigenvalue is $\lambda = -4$.)
14. In the system of Fig. 5.3.12, assume that $m_1 = 1$, $k_1 = 50$, $k_2 = 10$, and $F_0 = 5$ in mks units, and that $\omega = 10$. Then find m_2 so that in the resulting steady periodic oscillations, the mass m_1 will remain at rest(!). Thus the effect of the second mass-and-spring pair will be to neutralize the effect of the force on the first mass. This is an example of a *dynamic damper*. It has an electrical analogy that some cable companies use to prevent your reception of certain cable channels.

FIGURE 5.3.12. The mechanical system of Problem 14.

15. Suppose that $m_1 = 2$, $m_2 = \frac{1}{2}$, $k_1 = 75$, $k_2 = 25$, $F_0 = 100$, and $\omega = 10$ (all in mks units) in the forced mass-and-spring system of Fig. 5.3.9. Find the solution of the system $\mathbf{Mx}'' = \mathbf{Kx} + \mathbf{F}$ that satisfies the initial conditions $\mathbf{x}(0) = \mathbf{x}'(0) = \mathbf{0}$

16. Figure 5.3.13 shows two railway cars with a buffer spring. We want to investigate the transfer of momentum that occurs after car 1 with initial velocity v_0 impacts car 2 at rest. The analogue of Eq. (18) in the text is

$$\mathbf{x}'' = \begin{bmatrix} -c_1 & c_1 \\ c_2 & -c_2 \end{bmatrix} \mathbf{x}$$

with $c_i = k/m_i$ for $i = 1, 2$. Show that the eigenvalues of the coefficient matrix \mathbf{A} are $\lambda_1 = 0$ and $\lambda_2 = -c_1 - c_2$, with associated eigenvectors $\mathbf{v}_1 = \begin{bmatrix} 1 & 1 \end{bmatrix}^T$ and $\mathbf{v}_2 = \begin{bmatrix} c_1 & -c_2 \end{bmatrix}^T$.

FIGURE 5.3.13. The two railway cars of Problems 16 through 19.

17. If the two cars of Problem 16 both weigh 16 tons (so that $m_1 = m_2 = 1000$ (slugs)) and $k = 1$ ton/ft (that is, 2000 lb/ft), show that the cars separate after $\pi/2$ seconds, and that $x_1'(t) = 0$ and $x_2'(t) = v_0$ thereafter. Thus the original momentum of car 1 is completely transferred to car 2.

18. If cars 1 and 2 weigh 8 and 16 tons, respectively, and $k = 3000$ lb/ft, show that the two cars separate after $\pi/3$ seconds, and that

$$x_1'(t) = -\tfrac{1}{3}v_0 \quad \text{and} \quad x_2'(t) = +\tfrac{2}{3}v_0$$

thereafter. Thus the two cars rebound in opposite directions.

19. If cars 1 and 2 weigh 24 and 8 tons, respectively, and $k = 1500$ lb/ft, show that the cars separate after $\pi/2$ seconds, and that

$$x_1'(t) = +\tfrac{1}{2}v_0 \quad \text{and} \quad x_2'(t) = +\tfrac{3}{2}v_0$$

thereafter. Thus both cars continue in the original direction of motion, but with different velocities.

20. In the three-railway-car system of Fig. 5.3.6, suppose that cars 1 and 3 each weigh 32 tons, that car 2 weighs 8 tons, and that each spring constant is 4 tons/ft. If $x_1'(0) = v_0$ and $x_2'(0) = x_3'(0) = 0$, show that the two springs are compressed until $t = \pi/2$ and that

$$x_1'(t) = -\tfrac{1}{9}v_0 \quad \text{and} \quad x_2'(t) = x_3'(t) = +\tfrac{8}{9}v_0$$

thereafter. Thus car 1 rebounds, but cars 2 and 3 continue with the same velocity.

The Two-Axle Automobile

In Example 4 of Section 3.6 we investigated the vertical oscillations of a one-axle car—actually a unicycle. Now we can analyze a more realistic model: a car with two axles and with separate front and rear suspension systems. Figure 5.3.14 represents the suspension system of such a car. We assume that the car body acts as would a solid bar of mass m and length $L = L_1 + L_2$. It has moment of inertia I about its center of mass C, which is at distance L_1 from the front of the car. The car has front and back suspension springs with Hooke's constants k_1 and k_2, respectively. When the car is in motion, let $x(t)$ denote the vertical displacement of the center of mass of the car from equilibrium; let $\theta(t)$ denote its angular displacement (in radians) from the horizontal. Then Newton's laws of motion for linear and angular acceleration can be used to derive the equations

$$mx'' = -(k_1 + k_2)x + (k_1 L_1 - k_2 L_2)\theta,$$

$$I\theta'' = (k_1 L_1 - k_2 L_2)x - (k_1 L_1^2 + k_2 L_2^2)\theta. \tag{40}$$

FIGURE 5.3.14. Model of the two-axle automobile.

21. Suppose that $m = 75$ slugs (the car weighs 2400 lb), $L_1 = 7$ ft, $L_2 = 3$ ft (it's a rear-engine car), $k_1 = k_2 = 2000$ lb/ft, and $I = 1000$ ft·lb·s². Then the equations in (40) take the form

$$75x'' + 4000x - 8000\theta = 0,$$

$$1000\theta'' - 8000x + 116,000\theta = 0.$$

(a) Find the two natural frequencies ω_1 and ω_2 of the car.
(b) Now suppose that the car is driven at a speed of v feet per second along a washboard surface shaped like a sine curve with a wavelength of 40 ft. The result is a periodic force on the car with frequency $\omega = 2\pi v/40 = \pi v/20$.

Resonance occurs when with $\omega = \omega_1$ or $\omega = \omega_2$. Find the corresponding two critical speeds of the car (in feet per second and in miles per hour).

22. Suppose that $k_1 = k_2 = k$ and $L_1 = L_2 = \frac{1}{2}L$ in Fig. 5.3.14 (the symmetric situation). Then show that every free oscillation is a combination of a vertical oscillation with frequency

$$\omega_1 = \sqrt{2k/m}$$

and an angular oscillation with frequency

$$\omega_2 = \sqrt{kL^2/(2I)}.$$

*In Problems 23 through 25, the system of Fig. 5.3.14 is taken as a model for an undamped car with the given parameters in fps units. **(a)** Find the two natural frequencies of oscillation (in hertz). **(b)** Assume that this car is driven along a sinusoidal washboard surface with a wavelength of 40 ft. Find the two critical speeds.*

23. $m = 100$, $I = 800$, $L_1 = L_2 = 5$, $k_1 = k_2 = 2000$
24. $m = 100$, $I = 1000$, $L_1 = 6$, $L_2 = 4$, $k_1 = k_2 = 2000$
25. $m = 100$, $I = 800$, $L_1 = L_2 = 5$, $k_1 = 1000$, $k_2 = 2000$

5.3 Application: Earthquake-Induced Vibrations of Multistory Buildings

FIGURE 5.3.15. The seven-story building.

FIGURE 5.3.16. Forces on the ith floor.

In this application you are to investigate the response to transverse earthquake ground oscillations of the seven-story building illustrated in Fig. 5.3.15. Suppose that each of the seven [aboveground] floors weighs 16 tons, so the mass of each is $m = 1000$ (slugs). Also assume a horizontal restoring force of $k = 5$ (tons per foot) between adjacent floors. That is, the internal forces in response to horizontal displacements of the individual floors are those shown in Fig. 5.3.16. It follows that the free transverse oscillations indicated in Fig. 5.3.15 satisfy the equation $\mathbf{Mx''} = \mathbf{Kx}$ with $n = 7$, $m_i = 1000$ (for each i), and $k_i = 10000$ (lb/ft) for $1 \leq i \leq 7$. The system then reduces to the form $\mathbf{x''} = \mathbf{Ax}$ with

$$\mathbf{A} = \begin{bmatrix} -20 & 10 & 0 & 0 & 0 & 0 & 0 \\ 10 & -20 & 10 & 0 & 0 & 0 & 0 \\ 0 & 10 & -20 & 10 & 0 & 0 & 0 \\ 0 & 0 & 10 & -20 & 10 & 0 & 0 \\ 0 & 0 & 0 & 10 & -20 & 10 & 0 \\ 0 & 0 & 0 & 0 & 10 & -20 & 10 \\ 0 & 0 & 0 & 0 & 0 & 10 & -10 \end{bmatrix}. \qquad (1)$$

Once the matrix \mathbf{A} has been entered, the TI-86 command `eigVl A` takes about 15 seconds to calculate the seven eigenvalues shown in the λ-column of the table in Fig. 5.3.17. Alternatively, you can use the *Maple* command `eigenvals(A)`, the MATLAB command `eig(A)`, or the *Mathematica* command `Eigenvalues[A]`. Then calculate the entries in the remaining columns of the table showing the natural frequencies and periods of oscillation of the seven-story building. Note that a typical earthquake producing ground oscillations with a period of 2 seconds is un-

comfortably close to the fifth natural frequency (with period 1.9869 seconds) of the building.

i	Eigenvalue λ	Frequency $\omega = \sqrt{-\lambda}$	Period $P = \dfrac{2\pi}{\omega}$ (sec)
1	−38.2709	6.1863	1.0157
2	−33.3826	5.7778	1.0875
3	−26.1803	5.1167	1.2280
4	−17.9094	4.2320	1.4847
5	−10.0000	3.1623	1.9869
6	−3.8197	1.9544	3.2149
7	−0.4370	0.6611	9.5042

FIGURE 5.3.17. Frequencies and periods of natural oscillations of the seven-story building.

A horizontal earthquake oscillation $E \cos \omega t$ of the ground, with amplitude E and acceleration $a = -E\omega^2 \cos \omega t$, produces an opposite inertial force $F = ma = mE\omega^2 \sin \omega t$ on each floor of the building. The resulting nonhomogeneous system is

$$\mathbf{x}'' = \mathbf{A}\mathbf{x} + (E\omega^2 \cos \omega t)\mathbf{b}, \tag{2}$$

where $\mathbf{b} = \begin{bmatrix} 1 & 1 & 1 & 1 & 1 & 1 & 1 \end{bmatrix}^T$ and \mathbf{A} is the matrix of Eq. (1). Figure 5.3.18 shows a plot of *maximal amplitude* (for the forced oscillations of any single floor) versus the *period* of the earthquake vibrations. The spikes correspond to the first six of the seven resonant frequencies. We see, for instance, that whereas an earthquake with period 2 (s) likely would produce destructive resonance vibrations in the building, it probably would be unharmed by an earthquake with period 2.5 (s). Different buildings have different natural frequencies of vibration, and so a given earthquake may demolish one building but leave untouched the one next door. This seeming anomaly was observed in Mexico City after the devastating earthquake of September 19, 1985.

For your personal seven-story building to investigate, let the weight (in tons) of each story equal the largest digit of your student ID number and let k (in tons/ft) equal the smallest nonzero digit. Produce numerical and graphical results like those illustrated in Figs. 5.3.17 and 5.3.18. Is your building susceptible to likely damage from an earthquake with period in the 2- to 3-second range?

You might like to begin by working manually the following warm-up problems.

FIGURE 5.3.18. Resonance vibrations of a seven-story building—maximal amplitude as a function of period.

1. Find the periods of the natural vibrations of a building with two aboveground floors, each weighing 16 tons and with each restoring force being $k = 5$ tons/ft.

2. Find the periods of the natural vibrations of a building with three aboveground floors, with each weighing 16 tons and with each restoring force being $k = 5$ tons/ft.

3. Find the natural frequencies and natural modes of vibration of a building with three aboveground floors as in Problem 2, except that the upper two floors

weigh 8 tons each instead of 16. Give the ratios of the amplitudes A, B, and C of oscillations of the three floors in the form $A:B:C$ with $A = 1$.

4. Suppose that the building of Problem 3 is subject to an earthquake in which the ground undergoes horizontal sinusoidal oscillations with a period of 3 s and an amplitude of 3 in. Find the amplitudes of the resulting steady periodic oscillations of the three aboveground floors. Assume the fact that a motion $E \sin \omega t$ of the ground, with acceleration $a = -E\omega^2 \sin \omega t$, produces an opposite inertial force $F = -ma = mE\omega^2 \sin \omega t$ on a floor of mass m.

5.4 Multiple Eigenvalue Solutions

In Section 5.2 we saw that if the $n \times n$ matrix \mathbf{A} has n *distinct* (real or complex) eigenvalues $\lambda_1, \lambda_2, \ldots, \lambda_n$ with respective associated eigenvectors $\mathbf{v}_1, \mathbf{v}_2, \ldots, \mathbf{v}_n$, then a general solution of the system

$$\frac{d\mathbf{x}}{dt} = \mathbf{A}\mathbf{x} \tag{1}$$

is given by

$$\mathbf{x}(t) = c_1 \mathbf{v}_1 e^{\lambda_1 t} + c_2 \mathbf{v}_2 e^{\lambda_2 t} + \cdots + c_n \mathbf{v}_n e^{\lambda_n t} \tag{2}$$

with arbitrary constants c_1, c_2, \ldots, c_n. In this section we discuss the situation when the characteristic equation

$$|\mathbf{A} - \lambda \mathbf{I}| = 0 \tag{3}$$

does *not* have n distinct roots, and thus has at least one repeated root.

An eigenvalue is of **multiplicity** k if it is a k-fold root of Eq. (3). For each eigenvalue λ, the eigenvector equation

$$(\mathbf{A} - \lambda \mathbf{I})\mathbf{v} = \mathbf{0} \tag{4}$$

has at least one nonzero solution \mathbf{v}, so there is at least one eigenvector associated with λ. But an eigenvalue of multiplicity $k > 1$ may have *fewer* than k linearly independent associated eigenvectors. In this case we are unable to find a "complete set" of n linearly independent eigenvectors of \mathbf{A}, as needed to form the general solution in (2).

Let us call an eigenvalue of multiplicity k **complete** if it has k linearly independent associated eigenvectors. If every eigenvalue of the matrix \mathbf{A} is complete, then—because eigenvectors associated with different eigenvalues are linearly independent—it follows that \mathbf{A} does have a complete set of n linearly independent eigenvectors $\mathbf{v}_1, \mathbf{v}_2, \ldots, \mathbf{v}_n$ associated with the eigenvalues $\lambda_1, \lambda_2, \ldots, \lambda_n$ (each repeated with its multiplicity). In this case a general solution of $\mathbf{x}' = \mathbf{A}\mathbf{x}$ is still given by the usual combination in (2).

Example 1 Find a general solution of the system

$$\mathbf{x}' = \begin{bmatrix} 9 & 4 & 0 \\ -6 & -1 & 0 \\ 6 & 4 & 3 \end{bmatrix} \mathbf{x}. \tag{5}$$

Solution The characteristic equation of the coefficient matrix \mathbf{A} in Eq. (5) is

$$|\mathbf{A} - \lambda\mathbf{I}| = \begin{vmatrix} 9 - \lambda & 4 & 0 \\ -6 & -1 - \lambda & 0 \\ 6 & 4 & 3 - \lambda \end{vmatrix}$$

$$= (3 - \lambda)[(9 - \lambda)(-1 - \lambda) + 24]$$
$$= (3 - \lambda)(15 - 8\lambda + \lambda^2)$$
$$= (5 - \lambda)(3 - \lambda)^2 = 0.$$

Thus \mathbf{A} has the distinct eigenvalue $\lambda_1 = 5$ and the repeated eigenvalue $\lambda_2 = 3$ of multiplicity $k = 2$.

CASE 1: $\lambda_1 = 5$. The eigenvector equation $(\mathbf{A} - \lambda\mathbf{I})\mathbf{v} = \mathbf{0}$, where $\mathbf{v} = [a \quad b \quad c]^T$, is

$$(\mathbf{A} - 5\mathbf{I})\mathbf{v} = \begin{bmatrix} 4 & 4 & 0 \\ -6 & -6 & 0 \\ 6 & 4 & -2 \end{bmatrix}\begin{bmatrix} a \\ b \\ c \end{bmatrix} = \begin{bmatrix} 0 \\ 0 \\ 0 \end{bmatrix}.$$

Each of the first two equations, $4a + 4b = 0$ and $-6a - 6b = 0$, yields $b = -a$. Then the third equation reduces to $2a - 2c = 0$, so that $c = a$. The choice $a = 1$ then yields the eigenvector

$$\mathbf{v}_1 = \begin{bmatrix} 1 & -1 & 1 \end{bmatrix}^T$$

associated with the eigenvalue $\lambda_1 = 5$.

CASE 2: $\lambda_2 = 3$. Now the eigenvector equation is

$$(\mathbf{A} - 3\mathbf{I})\mathbf{v} = \begin{bmatrix} 6 & 4 & 0 \\ -6 & -4 & 0 \\ 6 & 4 & 0 \end{bmatrix}\begin{bmatrix} a \\ b \\ c \end{bmatrix} = \begin{bmatrix} 0 \\ 0 \\ 0 \end{bmatrix},$$

so the nonzero vector $\mathbf{v} = \begin{bmatrix} a & b & c \end{bmatrix}^T$ is an eigenvector if and only if

$$6a + 4b = 0; \tag{6}$$

that is, $b = -\frac{3}{2}a$. The fact that Eq. (6) does not involve c means that c is arbitrary, subject to the condition $\mathbf{v} \neq \mathbf{0}$. If $c = 1$, then we may choose $a = b = 0$; this gives the eigenvector

$$\mathbf{v}_2 = \begin{bmatrix} 0 & 0 & 1 \end{bmatrix}^T$$

associated with $\lambda_2 = 3$. If $c = 0$, then we must choose a to be nonzero. For instance, if $a = 2$ (to avoid fractions), then $b = -3$, so

$$\mathbf{v}_3 = \begin{bmatrix} 2 & -3 & 0 \end{bmatrix}^T$$

is a second linearly independent eigenvector associated with the multiplicity 2 eigenvalue $\lambda_2 = 3$.

Thus we have found a complete set $\mathbf{v}_1, \mathbf{v}_2, \mathbf{v}_3$ of three eigenvectors associated with the eigenvalues 5, 3, 3. The corresponding general solution of Eq. (5) is

$$\mathbf{x}(t) = c_1\mathbf{v}_1 e^{5t} + c_2\mathbf{v}_2 e^{3t} + c_3\mathbf{v}_3 e^{3t} \tag{7}$$

$$= c_1 \begin{bmatrix} 1 \\ -1 \\ 1 \end{bmatrix} e^{5t} + c_2 \begin{bmatrix} 0 \\ 0 \\ 1 \end{bmatrix} e^{3t} + c_3 \begin{bmatrix} 2 \\ -3 \\ 0 \end{bmatrix} e^{3t},$$

with scalar component functions given by

$$\begin{aligned} x_1(t) &= c_1 e^{5t} + 2c_3 e^{3t}, \\ x_2(t) &= -c_1 e^{5t} - 3c_3 e^{3t}, \\ x_3(t) &= c_1 e^{5t} + c_2 e^{3t}. \end{aligned}$$ ∎

Remark: Our choice in Example 1 of the two eigenvectors

$$\mathbf{v}_2 = \begin{bmatrix} 0 & 0 & 1 \end{bmatrix}^T \quad \text{and} \quad \mathbf{v}_3 = \begin{bmatrix} 2 & -3 & 0 \end{bmatrix}^T$$

associated with the repeated eigenvalue $\lambda_2 = 3$ bears comment. The fact that $b = -\frac{3}{2}a$ for any eigenvector associated with $\lambda_2 = 3$ means that any such eigenvector can be written as

$$\mathbf{v} = \begin{bmatrix} a \\ -\frac{3}{2}a \\ c \end{bmatrix} = c \begin{bmatrix} 0 \\ 0 \\ 1 \end{bmatrix} + \frac{1}{2}a \begin{bmatrix} 2 \\ -3 \\ 0 \end{bmatrix} = c\mathbf{v}_2 + \frac{1}{2}a\mathbf{v}_3,$$

and thus is a linear combination of \mathbf{v}_2 and \mathbf{v}_3. Therefore, given a and c not both zero, we could choose \mathbf{v} rather than \mathbf{v}_3 as our third eigenvector, and the new general solution

$$\mathbf{x}(t) = c_1\mathbf{v}_1 e^{5t} + c_2\mathbf{v}_2 e^{3t} + c_3\mathbf{v} e^{3t}$$

would be equivalent to the one in Eq. (7). Thus we need not worry about making the "right" choice of independent eigenvectors associated with a multiple eigenvalue. Any choice will do; we generally make the simplest one we can. ∎

Defective Eigenvalues

The following example shows that—unfortunately—not all multiple eigenvalues are complete.

Example 2 The matrix

$$\mathbf{A} = \begin{bmatrix} 1 & -3 \\ 3 & 7 \end{bmatrix} \tag{8}$$

has characteristic equation

$$\begin{aligned} |\mathbf{A} - \lambda\mathbf{I}| &= \begin{vmatrix} 1 - \lambda & -3 \\ 3 & 7 - \lambda \end{vmatrix} \\ &= (1 - \lambda)(7 - \lambda) + 9 \\ &= \lambda^2 - 8\lambda + 16 = (\lambda - 4)^2 = 0. \end{aligned}$$

Thus \mathbf{A} has the single eigenvalue $\lambda_1 = 4$ of multiplicity 2. The eigenvector equation

$$(\mathbf{A} - 4\mathbf{I})\mathbf{v} = \begin{bmatrix} -3 & -3 \\ 3 & 3 \end{bmatrix} \begin{bmatrix} a \\ b \end{bmatrix} = \begin{bmatrix} 0 \\ 0 \end{bmatrix}$$

then amounts to the equivalent scalar equations

$$-3a - 3b = 0, \quad 3a + 3b = 0.$$

Hence $b = -a$ if $\mathbf{v} = \begin{bmatrix} a & b \end{bmatrix}^T$ is to be an eigenvector of \mathbf{A}. Therefore any eigenvector associated with $\lambda_1 = 4$ is a nonzero multiple of $\mathbf{v} = \begin{bmatrix} 1 & -1 \end{bmatrix}^T$. Thus the multiplicity 2 eigenvalue $\lambda_1 = 4$ has only one independent eigenvector, and hence is incomplete. ∎

An eigenvalue λ of multiplicity $k > 1$ is called **defective** if it is not complete. If λ has only $p < k$ linearly independent eigenvectors, then the number

$$d = k - p \tag{9}$$

of "missing" eigenvectors is called the **defect** of the defective eigenvalue λ. Thus the defective eigenvalue $\lambda_1 = 4$ in Example 2 has multiplicity $k = 2$ and defect $d = 1$, because we saw that it has only $p = 1$ associated eigenvector.

If the eigenvalues of the $n \times n$ matrix \mathbf{A} are not all complete, then the eigenvalue method as yet described will produce *fewer* than the needed n linearly independent solutions of the system $\mathbf{x}' = \mathbf{A}\mathbf{x}$. We therefore need to discover how to find the "missing solutions" corresponding to a defective eigenvalue λ of multiplicity $k > 1$.

The Case of Defect $k = 2$

Let us begin with the case $k = 2$, and suppose that we have found (as in Example 2) that there is only a single eigenvector \mathbf{v}_1 associated with the defective eigenvalue λ. Then at this point we have found only the single solution

$$\blacktriangleright \qquad \mathbf{x}_1(t) = \mathbf{v}_1 e^{\lambda t} \tag{10}$$

of $\mathbf{x}' = \mathbf{A}\mathbf{x}$. By analogy with the case of a repeated characteristic root for a single linear differential equation (Section 3.3), we might hope to find a second solution of the form

$$\mathbf{x}_2(t) = (\mathbf{v}_2 t)e^{\lambda t} = \mathbf{v}_2 t e^{\lambda t}. \tag{11}$$

When we substitute $\mathbf{x} = \mathbf{v}_2 t e^{\lambda t}$ in $\mathbf{x}' = \mathbf{A}\mathbf{x}$, we get the equation

$$\mathbf{v}_2 e^{\lambda t} + \lambda \mathbf{v}_2 t e^{\lambda t} = \mathbf{A}\mathbf{v}_2 t e^{\lambda t}.$$

But because the coefficients of both $e^{\lambda t}$ and $t e^{\lambda t}$ must balance, it follows that $\mathbf{v}_2 = \mathbf{0}$, and hence that $\mathbf{x}_2(t) \equiv \mathbf{0}$. This means that—contrary to our hope—the system $\mathbf{x}' = \mathbf{A}\mathbf{x}$ does not have a *nontrivial* solution of the form assumed in (11).

Instead of simply giving up on the idea behind Eq. (11), let us extend it slightly and replace $\mathbf{v}_2 t$ with $\mathbf{v}_1 t + \mathbf{v}_2$. Thus we explore the possibility of a second solution of the form

$$\blacktriangleright \qquad \mathbf{x}_2(t) = (\mathbf{v}_1 t + \mathbf{v}_2)e^{\lambda t} = \mathbf{v}_1 t e^{\lambda t} + \mathbf{v}_2 e^{\lambda t} \tag{12}$$

where \mathbf{v}_1 and \mathbf{v}_2 are nonzero constant vectors. When we substitute $\mathbf{x} = \mathbf{v}_1 t e^{\lambda t} + \mathbf{v}_2 e^{\lambda t}$ in $\mathbf{x}' = \mathbf{A}\mathbf{x}$, we get the equation

$$\mathbf{v}_1 e^{\lambda t} + \lambda \mathbf{v}_1 t e^{\lambda t} + \lambda \mathbf{v}_2 e^{\lambda t} = \mathbf{A}\mathbf{v}_1 t e^{\lambda t} + \mathbf{A}\mathbf{v}_2 e^{\lambda t}. \tag{13}$$

We equate coefficients of $e^{\lambda t}$ and $t e^{\lambda t}$ here, and thereby obtain the two equations

$$(\mathbf{A} - \lambda \mathbf{I})\mathbf{v}_1 = \mathbf{0} \tag{14}$$

and

$$(\mathbf{A} - \lambda \mathbf{I})\mathbf{v}_2 = \mathbf{v}_1 \tag{15}$$

that the vectors \mathbf{v}_1 and \mathbf{v}_2 must satisfy in order for (12) to give a solution of $\mathbf{x}' = \mathbf{A}\mathbf{x}$.

Note that Eq. (14) merely confirms that \mathbf{v}_1 is an eigenvector of \mathbf{A} associated with the eigenvalue λ. Then Eq. (15) says that the vector \mathbf{v}_2 satisfies the equation

$$(\mathbf{A} - \lambda \mathbf{I})^2 \mathbf{v}_2 = (\mathbf{A} - \lambda \mathbf{I})[(\mathbf{A} - \lambda \mathbf{I})\mathbf{v}_2] = (\mathbf{A} - \lambda \mathbf{I})\mathbf{v}_1 = \mathbf{0}.$$

It follows that, in order to solve simultaneously the two equations in (14) and (15), it suffices to find a solution \mathbf{v}_2 of the single equation $(\mathbf{A} - \lambda \mathbf{I})^2 \mathbf{v}_2 = \mathbf{0}$ such that the resulting vector $\mathbf{v}_1 = (\mathbf{A} - \lambda \mathbf{I})\mathbf{v}_2$ is *nonzero*. It turns out that this is always possible if the defective eigenvalue λ of \mathbf{A} is of multiplicity 2. Consequently, the procedure described in the following algorithm always succeeds in finding two independent solutions associated with such an eigenvalue.

ALGORITHM: Defective Multiplicity 2 Eigenvalues

1. First find a nonzero solution \mathbf{v}_2 of the equation

$$(\mathbf{A} - \lambda \mathbf{I})^2 \mathbf{v}_2 = \mathbf{0} \tag{16}$$

such that

$$(\mathbf{A} - \lambda \mathbf{I})\mathbf{v}_2 = \mathbf{v}_1 \tag{17}$$

is nonzero, and therefore is an eigenvector \mathbf{v}_1 associated with λ.

2. Then form the two independent solutions

$$\mathbf{x}_1(t) = \mathbf{v}_1 e^{\lambda t} \tag{18}$$

and

$$\mathbf{x}_2(t) = (\mathbf{v}_1 t + \mathbf{v}_2) e^{\lambda t} \tag{19}$$

of $\mathbf{x}' = \mathbf{A}\mathbf{x}$ corresponding to λ. ▲

Example 3 Find a general solution of the system

$$\mathbf{x}' = \begin{bmatrix} 1 & -3 \\ 3 & 7 \end{bmatrix} \mathbf{x}. \tag{20}$$

Solution In Example 2 we found that the coefficient matrix \mathbf{A} in Eq. (20) has the defective eigenvalue $\lambda = 4$ of multiplicity 2. We therefore begin by calculating

$$(\mathbf{A} - 4\mathbf{I})^2 = \begin{bmatrix} -3 & -3 \\ 3 & 3 \end{bmatrix} \begin{bmatrix} -3 & -3 \\ 3 & 3 \end{bmatrix} = \begin{bmatrix} 0 & 0 \\ 0 & 0 \end{bmatrix}.$$

Hence Eq. (16) is

$$\begin{bmatrix} 0 & 0 \\ 0 & 0 \end{bmatrix} \mathbf{v}_2 = \mathbf{0},$$

and therefore is satisfied by *any* choice of \mathbf{v}_2. In principle, it could happen that $(\mathbf{A} - 4\mathbf{I})\mathbf{v}_2$ is nonzero (as desired) for some choices of \mathbf{v}_2 though not for others. If we try $\mathbf{v}_2 = \begin{bmatrix} 1 & 0 \end{bmatrix}^T$ we find that

$$(\mathbf{A} - 4\mathbf{I})\mathbf{v}_2 = \begin{bmatrix} -3 & -3 \\ 3 & 3 \end{bmatrix} \begin{bmatrix} 1 \\ 0 \end{bmatrix} = \begin{bmatrix} -3 \\ 3 \end{bmatrix} = \mathbf{v}_1$$

is nonzero, and therefore is an eigenvector associated with $\lambda = 4$. (It is -3 times the eigenvector found in Example 2.) Therefore the two solutions of Eq. (20) given by Eqs. (18) and (19) are

$$\mathbf{x}_1(t) = \mathbf{v}_1 e^{4t} = \begin{bmatrix} -3 \\ 3 \end{bmatrix} e^{4t},$$

$$\mathbf{x}_2(t) = (\mathbf{v}_1 t + \mathbf{v}_2) e^{4t} = \begin{bmatrix} -3t + 1 \\ 3t \end{bmatrix} e^{4t}.$$

The resulting general solution

$$\mathbf{x}(t) = c_1 \mathbf{x}_1(t) + c_2 \mathbf{x}_2(t)$$

has scalar component functions

$$x_1(t) = (-3c_2 t + c_2 - 3c_1) e^{4t},$$
$$x_2(t) = (3c_2 t + 3c_1) e^{4t}.$$

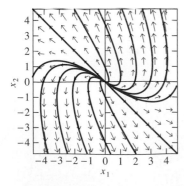

FIGURE 5.4.1. Direction field and solution curves for the linear system $x_1' = x_1 - 3x_2$, $x_2' = 3x_1 + 7x_2$ of Example 3.

With $c_2 = 0$ these solution equations reduce to the equations $x_1(t) = -3c_1 e^{4t}$, $x_2(t) = 3c_1 e^{4t}$, which parametrize the line $x_1 = -x_2$ in the $x_1 x_2$-plane. The point $(x_1(t), x_2(t))$ then recedes along this line away from the origin as $t \to +\infty$, to the northwest if $c_1 > 0$ and to the southeast if $c_1 < 0$. As indicated in Fig. 5.4.1, each solution curve with $c_2 \neq 0$ is tangent to the line $x_1 = -x_2$ at the origin; the point $(x_1(t), x_2(t))$ approaches the origin as $t \to -\infty$ and approaches $+\infty$ along the solution curve as $t \to +\infty$. ∎

Generalized Eigenvectors

The vector \mathbf{v}_2 in Eq. (16) is an example of a generalized eigenvector. If λ is an eigenvalue of the matrix \mathbf{A}, then a **rank r generalized eigenvector** associated with λ is a vector \mathbf{v} such that

$$\blacktriangleright \qquad (\mathbf{A} - \lambda\mathbf{I})^r\mathbf{v} = \mathbf{0} \quad \text{but} \quad (\mathbf{A} - \lambda\mathbf{I})^{r-1}\mathbf{v} \neq \mathbf{0}. \qquad (21)$$

If $r = 1$, then (21) simply means that \mathbf{v} is an eigenvector associated with λ. Thus a rank 1 generalized eigenvector is an ordinary eigenvector. The vector \mathbf{v}_2 in (16) is a rank 2 generalized eigenvector (and not an ordinary eigenvector).

The multiplicity 2 method described earlier boils down to finding a pair $\{\mathbf{v}_1, \mathbf{v}_2\}$ of generalized eigenvectors, one of rank 1 and one of rank 2, such that $(\mathbf{A} - \lambda\mathbf{I})\mathbf{v}_2 = \mathbf{v}_1$. Higher multiplicity methods involve longer "chains" of generalized eigenvectors. A **length k chain of generalized eigenvectors based on the eigenvector \mathbf{v}_1** is a set $\{\mathbf{v}_1, \mathbf{v}_2, \ldots, \mathbf{v}_k\}$ of k generalized eigenvectors such that

$$\blacktriangleright \qquad \begin{aligned} (\mathbf{A} - \lambda\mathbf{I})\mathbf{v}_k &= \mathbf{v}_{k-1}, \\ (\mathbf{A} - \lambda\mathbf{I})\mathbf{v}_{k-1} &= \mathbf{v}_{k-2}, \\ &\vdots \\ (\mathbf{A} - \lambda\mathbf{I})\mathbf{v}_2 &= \mathbf{v}_1. \end{aligned} \qquad (22)$$

Because \mathbf{v}_1 is an ordinary eigenvector, $(\mathbf{A} - \lambda\mathbf{I})\mathbf{v}_1 = \mathbf{0}$. Therefore, it follows from (22) that

$$(\mathbf{A} - \lambda\mathbf{I})^k\mathbf{v}_k = \mathbf{0}. \qquad (23)$$

If $\{\mathbf{v}_1, \mathbf{v}_2, \mathbf{v}_3\}$ is a length 3 chain of generalized eigenvectors associated with the multiple eigenvalue λ of the matrix \mathbf{A}, then it is easy to verify that three linearly independent solutions of $\mathbf{x}' = \mathbf{A}\mathbf{x}$ are given by

$$\blacktriangleright \qquad \begin{aligned} \mathbf{x}_1(t) &= \mathbf{v}_1 e^{\lambda t}, \\ \mathbf{x}_2(t) &= (\mathbf{v}_1 t + \mathbf{v}_2) e^{\lambda t}, \\ \mathbf{x}_3(t) &= \left(\tfrac{1}{2}\mathbf{v}_1 t^2 + \mathbf{v}_2 t + \mathbf{v}_3\right) e^{\lambda t}. \end{aligned} \qquad (24)$$

For instance, the equations in (22) give

$$\mathbf{A}\mathbf{v}_3 = \mathbf{v}_2 + \lambda\mathbf{v}_3, \quad \mathbf{A}\mathbf{v}_2 = \mathbf{v}_1 + \lambda\mathbf{v}_2, \quad \mathbf{A}\mathbf{v}_1 = \lambda\mathbf{v}_1,$$

so

$$\begin{aligned} \mathbf{A}\mathbf{x}_3 &= \left[\tfrac{1}{2}\mathbf{A}\mathbf{v}_1 t^2 + \mathbf{A}\mathbf{v}_2 t + \mathbf{A}\mathbf{v}_3\right] e^{\lambda t} \\ &= \left[\tfrac{1}{2}\lambda\mathbf{v}_1 t^2 + (\mathbf{v}_1 + \lambda\mathbf{v}_2)t + (\mathbf{v}_2 + \lambda\mathbf{v}_3)\right] e^{\lambda t} \\ &= (\mathbf{v}_1 t + \mathbf{v}_2)e^{\lambda t} + \lambda\left(\tfrac{1}{2}\mathbf{v}_1 t^2 + \mathbf{v}_2 t + \mathbf{v}_3\right) e^{\lambda t} \\ &= \mathbf{x}_3'. \end{aligned}$$

Therefore, $\mathbf{x}_3(t)$ in (24) does, indeed, define a solution of $\mathbf{x}' = \mathbf{A}\mathbf{x}$.

Consequently, in order to "handle" a multiplicity 3 eigenvalue λ, it suffices to find a length 3 chain $\{\mathbf{v}_1, \mathbf{v}_2, \mathbf{v}_3\}$ of generalized eigenvalues associated with λ. Looking at Eq. (23), we see that we need only find a solution \mathbf{v}_3 of

$$(\mathbf{A} - \lambda\mathbf{I})^3 \mathbf{v}_3 = \mathbf{0}$$

such that the vectors

$$\mathbf{v}_2 = (\mathbf{A} - \lambda\mathbf{I})\mathbf{v}_3 \quad \text{and} \quad \mathbf{v}_1 = (\mathbf{A} - \lambda\mathbf{I})\mathbf{v}_2$$

are both nonzero (although, as we will see, this is not always possible).

| **Example 4** | Find three linearly independent solutions of the system

$$\mathbf{x}' = \begin{bmatrix} 0 & 1 & 2 \\ -5 & -3 & -7 \\ 1 & 0 & 0 \end{bmatrix} \mathbf{x}. \tag{25}$$

Solution The characteristic equation of the coefficient matrix in Eq. (25) is

$$|\mathbf{A} - \lambda\mathbf{I}| = \begin{bmatrix} -\lambda & 1 & 2 \\ -5 & -3-\lambda & -7 \\ 1 & 0 & -\lambda \end{bmatrix}$$

$$= 1 \cdot [-7 - 2 \cdot (-3-\lambda)] + (-\lambda)[(-\lambda)(-3-\lambda) + 5]$$

$$= -\lambda^3 - 3\lambda^2 - 3\lambda - 1 = -(\lambda + 1)^3 = 0,$$

and thus \mathbf{A} has the eigenvalue $\lambda = -1$ of multiplicity 3. The eigenvector equation $(\mathbf{A} - \lambda\mathbf{I})\mathbf{v} = \mathbf{0}$ for an eigenvector $\mathbf{v} = \begin{bmatrix} a & b & c \end{bmatrix}^T$ is

$$(\mathbf{A} + \mathbf{I})\mathbf{v} = \begin{bmatrix} 1 & 1 & 2 \\ -5 & -2 & -7 \\ 1 & 0 & 1 \end{bmatrix} \begin{bmatrix} a \\ b \\ c \end{bmatrix} = \begin{bmatrix} 0 \\ 0 \\ 0 \end{bmatrix}.$$

The third row $a + c = 0$ gives $c = -a$, then the first row $a + b + 2c = 0$ gives $b = a$. Thus, to within a constant multiple, the eigenvalue $\lambda = -1$ has only the single associated eigenvector $\mathbf{v} = \begin{bmatrix} a & a & -a \end{bmatrix}^T$ with $a \neq 0$, and so the defect of $\lambda = -1$ is 2.

To apply the method described here for triple eigenvalues, we first calculate

$$(\mathbf{A} + \mathbf{I})^2 = \begin{bmatrix} 1 & 1 & 2 \\ -5 & -2 & -7 \\ 1 & 0 & 1 \end{bmatrix} \begin{bmatrix} 1 & 1 & 2 \\ -5 & -2 & -7 \\ 1 & 0 & 1 \end{bmatrix} = \begin{bmatrix} -2 & -1 & -3 \\ -2 & -1 & -3 \\ 2 & 1 & 3 \end{bmatrix}$$

and

$$(\mathbf{A} + \mathbf{I})^3 = \begin{bmatrix} 1 & 1 & 2 \\ -5 & -2 & -7 \\ 1 & 0 & 1 \end{bmatrix} \begin{bmatrix} -2 & -1 & -3 \\ -2 & -1 & -3 \\ 2 & 1 & 3 \end{bmatrix} = \begin{bmatrix} 0 & 0 & 0 \\ 0 & 0 & 0 \\ 0 & 0 & 0 \end{bmatrix}.$$

Thus *any* nonzero vector \mathbf{v}_3 will be a solution of the equation $(\mathbf{A} + \mathbf{I})^3 \mathbf{v}_3 = \mathbf{0}$. Beginning with $\mathbf{v}_3 = \begin{bmatrix} 1 & 0 & 0 \end{bmatrix}^T$, for instance, we calculate

$$\mathbf{v}_2 = (\mathbf{A} + \mathbf{I})\mathbf{v}_3 = \begin{bmatrix} 1 & 1 & 2 \\ -5 & -2 & -7 \\ 1 & 0 & 1 \end{bmatrix} \begin{bmatrix} 1 \\ 0 \\ 0 \end{bmatrix} = \begin{bmatrix} 1 \\ -5 \\ 1 \end{bmatrix},$$

$$\mathbf{v}_1 = (\mathbf{A} + \mathbf{I})\mathbf{v}_2 = \begin{bmatrix} 1 & 1 & 2 \\ -5 & -2 & -7 \\ 1 & 0 & 1 \end{bmatrix} \begin{bmatrix} 1 \\ -5 \\ 1 \end{bmatrix} = \begin{bmatrix} -2 \\ -2 \\ 2 \end{bmatrix}.$$

Note that \mathbf{v}_1 is the previously found eigenvector \mathbf{v} with $a = -2$; this agreement serves as a check of the accuracy of our matrix computations.

Thus we have found a length 3 chain $\{\mathbf{v}_1, \mathbf{v}_2, \mathbf{v}_3\}$ of generalized eigenvectors associated with the triple eigenvalue $\lambda = -1$. Substitution in (24) now yields the linearly independent solutions

$$\mathbf{x}_1(t) = \mathbf{v}_1 e^{-t} = \begin{bmatrix} -2 \\ -2 \\ 2 \end{bmatrix} e^{-t},$$

$$\mathbf{x}_2(t) = (\mathbf{v}_1 t + \mathbf{v}_2)e^{-t} = \begin{bmatrix} -2t + 1 \\ -2t - 5 \\ 2t + 1 \end{bmatrix} e^{-t},$$

$$\mathbf{x}_3(t) = \left(\tfrac{1}{2}\mathbf{v}_1 t^2 + \mathbf{v}_2 t + \mathbf{v}_3\right) e^{-t} = \begin{bmatrix} -t^2 + t + 1 \\ -t^2 - 5t \\ t^2 + t \end{bmatrix} e^{-t}$$

of the system $\mathbf{x}' = \mathbf{A}\mathbf{x}$. ∎

The General Case

A fundamental theorem of linear algebra states that every $n \times n$ matrix \mathbf{A} has n linearly independent generalized eigenvectors. These n generalized eigenvectors may be arranged in chains, with the sum of the lengths of the chains associated with a given eigenvalue λ equal to the multiplicity of λ. But the structure of these chains depends on the defect of λ, and can be quite complicated. For instance, a multiplicity 4 eigenvalue can correspond to

- Four length 1 chains (defect 0);
- Two length 1 chains and a length 2 chain (defect 1);
- Two length 2 chains (defect 2);
- A length 1 chain and a length 3 chain (defect 2); or
- A length 4 chain (defect 3).

Note that, in each of these cases, the length of the longest chain is at most $d + 1$, where d is the defect of the eigenvalue. Consequently, once we have found all the ordinary eigenvectors associated with a multiple eigenvalue λ, and therefore know the defect d of λ, we can begin with the equation

$$\blacktriangleright \qquad (\mathbf{A} - \lambda\mathbf{I})^{d+1}\mathbf{u} = \mathbf{0} \qquad\qquad (26)$$

to start building the chains of generalized eigenvectors associated with λ.

ALGORITHM: Chains of Generalized Eigenvalues

Begin with a nonzero solution \mathbf{u}_1 of Eq. (26) and successively multiply by the matrix $\mathbf{A} - \lambda\mathbf{I}$ until the zero vector is obtained. If

$$(\mathbf{A} - \lambda\mathbf{I})\mathbf{u}_1 = \mathbf{u}_2 \neq \mathbf{0},$$

$$\vdots$$

$$(\mathbf{A} - \lambda\mathbf{I})\mathbf{u}_{k-1} = \mathbf{u}_k \neq \mathbf{0},$$

but $(\mathbf{A} - \lambda\mathbf{I})\mathbf{u}_k = \mathbf{0}$, then the vectors

$$\{\mathbf{v}_1, \mathbf{v}_2, \ldots, \mathbf{v}_k\} = \{\mathbf{u}_k, \mathbf{u}_{k-1}, \ldots, \mathbf{u}_2, \mathbf{u}_1\}$$

(listed in reverse order of their appearance) form a length k chain of generalized eigenvectors based on the (ordinary) eigenvector \mathbf{v}_1. ▲

Each length k chain $\{\mathbf{v}_1, \mathbf{v}_2, \ldots, \mathbf{v}_k\}$ of generalized eigenvectors (with \mathbf{v}_1 an ordinary eigenvector associated with λ) determines a set of k independent solutions of $\mathbf{x}' = \mathbf{A}\mathbf{x}$ corresponding to the eigenvalue λ:

$$
\begin{aligned}
\mathbf{x}_1(t) &= \mathbf{v}_1 e^{\lambda t}, \\
\mathbf{x}_2(t) &= (\mathbf{v}_1 t + \mathbf{v}_2)e^{\lambda t}, \\
\mathbf{x}_3(t) &= \left(\tfrac{1}{2}\mathbf{v}_1 t^2 + \mathbf{v}_2 t + \mathbf{v}_3\right)e^{\lambda t}, \\
&\vdots \\
\mathbf{x}_k(t) &= \left(\frac{\mathbf{v}_1 t^{k-1}}{(k-1)!} + \cdots + \frac{\mathbf{v}_{k-2}t^2}{2!} + \mathbf{v}_{k-1}t + \mathbf{v}_k\right)e^{\lambda t}.
\end{aligned}
\tag{27}
$$

Note that (27) reduces to Eqs. (18) through (19) and (24) in the cases $k = 2$ and $k = 3$, respectively.

To ensure that we obtain n generalized eigenvectors of the $n \times n$ matrix \mathbf{A} that are actually linearly independent, and therefore produce a complete set of n linearly independent solutions of $\mathbf{x}' = \mathbf{A}\mathbf{x}$ when we amalgamate all the "chains of solutions" corresponding to different chains of generalized eigenvectors, we may rely on the following two facts:

- Any chain of generalized eigenvectors constitutes a linearly independent set of vectors.
- If two chains of generalized eigenvectors are based on linearly independent eigenvectors, then the union of these two chains is a linearly independent set of vectors (whether the two base eigenvectors are associated with different eigenvalues or with the same eigenvalue).

Example 5　Suppose that the 6×6 matrix \mathbf{A} has two multiplicity 3 eigenvalues $\lambda_1 = -2$ and $\lambda_2 = 3$ with defects 1 and 2, respectively. Then λ_1 must have an associated eigenvector \mathbf{u}_1 and a length 2 chain $\{\mathbf{v}_1, \mathbf{v}_2\}$ of generalized eigenvectors (with the eigenvectors \mathbf{u}_1 and \mathbf{v}_1 being linearly independent), whereas λ_2 must have a length 3 chain $\{\mathbf{w}_1, \mathbf{w}_2, \mathbf{w}_3\}$ of generalized eigenvectors based on its single eigenvector \mathbf{w}_1. The six

generalized eigenvectors \mathbf{u}_1, \mathbf{v}_1, \mathbf{v}_2, \mathbf{w}_1, \mathbf{w}_2, and \mathbf{w}_3 are then linearly independent and yield the following six independent solutions of $\mathbf{x}' = \mathbf{A}\mathbf{x}$:

$$\mathbf{x}_1(t) = \mathbf{u}_1 e^{-2t},$$
$$\mathbf{x}_2(t) = \mathbf{v}_1 e^{-2t},$$
$$\mathbf{x}_3(t) = (\mathbf{v}_1 t + \mathbf{v}_2) e^{-2t},$$
$$\mathbf{x}_4(t) = \mathbf{w}_1 e^{3t},$$
$$\mathbf{x}_5(t) = (\mathbf{w}_1 t + \mathbf{w}_2) e^{3t},$$
$$\mathbf{x}_6(t) = \left(\tfrac{1}{2} \mathbf{w}_1 t^2 + \mathbf{w}_2 t + \mathbf{w}_3\right) e^{3t}. \qquad \blacksquare$$

As Example 5 illustrates, the computation of independent solutions corresponding to different eigenvalues and chains of generalized eigenvalues is a routine matter. The determination of the chain structure associated with a given multiple eigenvalue can be more interesting (as in Example 6).

An Application

FIGURE 5.4.2. The railway cars of Example 6.

Figure 5.4.2 shows two railway cars that are connected with a spring (permanently attached to both cars) and with a damper that exerts opposite forces on the two cars, of magnitude $c(x_1' - x_2')$ proportional to their relative velocity. The two cars are also subject to frictional resistance forces $c_1 x_1'$ and $c_2 x_2'$ proportional to their respective velocities. An application of Newton's law $ma = F$ (as in Example 1 of Section 4.1) yields the equations of motion

$$m_1 x_1'' = k(x_2 - x_1) - c_1 x_1' - c(x_1' - x_2'),$$
$$m_2 x_2'' = k(x_1 - x_2) - c_2 x_2' - c(x_2' - x_1'). \qquad (28)$$

In terms of the position vector $\mathbf{x}(t) = \begin{bmatrix} x_1(t) & x_2(t) \end{bmatrix}^T$, these equations can be written in the matrix form

$$\mathbf{M}\mathbf{x}'' = \mathbf{K}\mathbf{x} + \mathbf{R}\mathbf{x}', \qquad (29)$$

where \mathbf{M} and \mathbf{K} are mass and stiffness matrices (as in Eqs. (2) and (3) of Section 5.3), and

$$\mathbf{R} = \begin{bmatrix} -(c + c_1) & c \\ c & -(c + c_2) \end{bmatrix}$$

is the *resistance matrix*. Unfortunately, because of the presence of the term involving \mathbf{x}', the methods of Section 5.3 cannot be used.

Instead, we write (28) as a first-order system in the four unknown functions $x_1(t)$, $x_2(t)$, $x_3(t) = x_1'(t)$, and $x_4(t) = x_2'(t)$. If $m_1 = m_2 = 1$ we get

$$\mathbf{x}' = \mathbf{A}\mathbf{x} \qquad (30)$$

where now $\mathbf{x} = \begin{bmatrix} x_1 & x_2 & x_3 & x_4 \end{bmatrix}^T$ and

$$\mathbf{A} = \begin{bmatrix} 0 & 0 & 1 & 0 \\ 0 & 0 & 0 & 1 \\ -k & k & -(c + c_1) & c \\ k & -k & c & -(c + c_2) \end{bmatrix}. \qquad (31)$$

Example 6 With $m_1 = m_2 = c = 1$ and $k = c_1 = c_2 = 2$, the system in Eq. (30) is

$$\mathbf{x}' = \begin{bmatrix} 0 & 0 & 1 & 0 \\ 0 & 0 & 0 & 1 \\ -2 & 2 & -3 & 1 \\ 2 & -2 & 1 & -3 \end{bmatrix} \mathbf{x}. \tag{32}$$

It is not too tedious to calculate manually—although a computer algebra system such as *Maple, Mathematica*, or MATLAB is useful here—the characteristic equation

$$\lambda^4 + 6\lambda^3 + 12\lambda^2 + 8\lambda = \lambda(\lambda + 2)^3 = 0$$

of the coefficient matrix \mathbf{A} in Eq. (32). Thus \mathbf{A} has the distinct eigenvalue $\lambda_0 = 0$ and the triple eigenvalue $\lambda_1 = -2$.

CASE 1: $\lambda_0 = 0$. The eigenvalue equation $(\mathbf{A} - \lambda\mathbf{I})\mathbf{v} = \mathbf{0}$ for the eigenvector $\mathbf{v} = \begin{bmatrix} a & b & c & d \end{bmatrix}^T$ is

$$\mathbf{Av} = \begin{bmatrix} 0 & 0 & 1 & 0 \\ 0 & 0 & 0 & 1 \\ -2 & 2 & -3 & 1 \\ 2 & -2 & 1 & -3 \end{bmatrix} \begin{bmatrix} a \\ b \\ c \\ d \end{bmatrix} = \begin{bmatrix} 0 \\ 0 \\ 0 \\ 0 \end{bmatrix}.$$

The first two rows give $c = d = 0$, then the last two rows yield $a = b$. Thus

$$\mathbf{v}_0 = \begin{bmatrix} 1 & 1 & 0 & 0 \end{bmatrix}^T$$

is an eigenvector associated with $\lambda_0 = 0$.

CASE 2: $\lambda_1 = -2$. The eigenvalue equation $(\mathbf{A} - \lambda\mathbf{I})\mathbf{v} = \mathbf{0}$ is

$$(\mathbf{A} + 2\mathbf{I})\mathbf{v} = \begin{bmatrix} 2 & 0 & 1 & 0 \\ 0 & 2 & 0 & 1 \\ -2 & 2 & -1 & 1 \\ 2 & -2 & 1 & -1 \end{bmatrix} \begin{bmatrix} a \\ b \\ c \\ d \end{bmatrix} = \begin{bmatrix} 0 \\ 0 \\ 0 \\ 0 \end{bmatrix}.$$

The third and fourth scalar equations here are the differences of the first and second equations, and therefore are redundant. Hence \mathbf{v} is determined by the first two equations,

$$2a + c = 0 \quad \text{and} \quad 2b + d = 0.$$

We can choose a and b independently, then solve for c and d. Thereby we obtain *two* eigenvectors associated with the triple eigenvalue $\lambda_1 = -2$. The choice $a = 1$, $b = 0$ yields $c = -2$, $d = 0$ and thereby the eigenvector

$$\mathbf{u}_1 = \begin{bmatrix} 1 & 0 & -2 & 0 \end{bmatrix}^T.$$

The choice $a = 0$, $b = 1$ yields $c = 0$, $d = -2$ and thereby the eigenvector

$$\mathbf{u}_2 = \begin{bmatrix} 0 & 1 & 0 & -2 \end{bmatrix}^T.$$

Because $\lambda_1 = -2$ has defect 1, we need a generalized eigenvector of rank 2, and hence a nonzero solution \mathbf{v}_2 of the equation

$$(\mathbf{A} + 2\mathbf{I})^2 \mathbf{v}_2 = \begin{bmatrix} 2 & 2 & 1 & 1 \\ 2 & 2 & 1 & 1 \\ 0 & 0 & 0 & 0 \\ 0 & 0 & 0 & 0 \end{bmatrix} \mathbf{v}_2 = \mathbf{0}.$$

Obviously,

$$\mathbf{v}_2 = \begin{bmatrix} 0 & 0 & 1 & -1 \end{bmatrix}^T$$

is such a vector, and we find that

$$(\mathbf{A} + 2\mathbf{I})\mathbf{v}_2 = \begin{bmatrix} 2 & 0 & 1 & 0 \\ 0 & 2 & 0 & 1 \\ -2 & 2 & -1 & 1 \\ 2 & -2 & 1 & -1 \end{bmatrix} \begin{bmatrix} 0 \\ 0 \\ 1 \\ -1 \end{bmatrix} = \begin{bmatrix} 1 \\ -1 \\ -2 \\ 2 \end{bmatrix} = \mathbf{v}_1$$

is nonzero, and therefore is an eigenvector associated with $\lambda_1 = -2$. Then $\{\mathbf{v}_1, \mathbf{v}_2\}$ is the length 2 chain we need.

The eigenvector \mathbf{v}_1 just found is neither of the two eigenvectors \mathbf{u}_1 and \mathbf{u}_2 found previously, but we observe that $\mathbf{v}_1 = \mathbf{u}_1 - \mathbf{u}_2$. For a length 1 chain \mathbf{w}_1 to complete the picture we can choose any linear combination of \mathbf{u}_1 and \mathbf{u}_2 that is independent of \mathbf{v}_1. For instance, we could choose either $\mathbf{w}_1 = \mathbf{u}_1$ or $\mathbf{w}_1 = \mathbf{u}_2$. However, we will see momentarily that the particular choice

$$\mathbf{w}_1 = \mathbf{u}_1 + \mathbf{u}_2 = \begin{bmatrix} 1 & 1 & -2 & -2 \end{bmatrix}^T$$

yields a solution of the system that is of physical interest.

Finally, the chains $\{\mathbf{v}_0\}$, $\{\mathbf{w}_1\}$, and $\{\mathbf{v}_1, \mathbf{v}_2\}$ yield the four independent solutions

$$\begin{aligned} \mathbf{x}_1(t) &= \mathbf{v}_0 e^{0 \cdot t} = \begin{bmatrix} 1 & 1 & 0 & 0 \end{bmatrix}^T, \\ \mathbf{x}_2(t) &= \mathbf{w}_1 e^{-2t} = \begin{bmatrix} 1 & 1 & -2 & -2 \end{bmatrix}^T e^{-2t}, \\ \mathbf{x}_3(t) &= \mathbf{v}_1 e^{-2t} = \begin{bmatrix} 1 & -1 & -2 & 2 \end{bmatrix}^T e^{-2t}, \\ \mathbf{x}_4(t) &= (\mathbf{v}_1 t + \mathbf{v}_2) e^{-2t} \\ &= \begin{bmatrix} t & -t & -2t+1 & 2t-1 \end{bmatrix}^T e^{-2t} \end{aligned} \tag{33}$$

of the system $\mathbf{x}' = \mathbf{A}\mathbf{x}$ in (32).

The four scalar components of the general solution

$$\mathbf{x}(t) = c_1 \mathbf{x}_1(t) + c_2 \mathbf{x}_2(t) + c_3 \mathbf{x}_3(t) + c_4 \mathbf{x}_4(t)$$

are described by the equations

$$\begin{aligned} x_1(t) &= c_1 + e^{-2t}(c_2 + c_3 + c_4 t), \\ x_2(t) &= c_1 + e^{-2t}(c_2 - c_3 - c_4 t), \\ x_3(t) &= e^{-2t}(-2c_2 - 2c_3 + c_4 - 2c_4 t), \\ x_4(t) &= e^{-2t}(-2c_2 + 2c_3 - c_4 + 2c_4 t). \end{aligned} \tag{34}$$

Recall that $x_1(t)$ and $x_2(t)$ are the position functions of the two masses, whereas $x_3(t) = x_1'(t)$ and $x_4(t) = x_2'(t)$ are their respective velocity functions.

For instance, suppose that $x_1(0) = x_2(0) = 0$ and that $x_1'(0) = x_2'(0) = v_0$. Then the equations

$$
\begin{aligned}
x_1(0) &= c_1 + c_2 + c_3 &&= 0, \\
x_2(0) &= c_1 + c_2 - c_3 &&= 0, \\
x_1'(0) &= -2c_2 - 2c_3 + c_4 &&= v_0, \\
x_2'(0) &= -2c_2 + 2c_3 - c_4 &&= v_0
\end{aligned}
$$

are readily solved for $c_1 = \frac{1}{2}v_0$, $c_2 = -\frac{1}{2}v_0$, and $c_3 = c_4 = 0$, so

$$
\begin{aligned}
x_1(t) &= x_2(t) = \tfrac{1}{2}v_0\left(1 - e^{-2t}\right), \\
x_1'(t) &= x_2'(t) = v_0 e^{-2t}.
\end{aligned}
$$

In this case the two railway cars continue in the same direction with equal but exponentially damped velocities, approaching the displacements $x_1 = x_2 = \frac{1}{2}v_0$ as $t \to +\infty$.

It is of interest to interpret physically the individual generalized eigenvector solutions given in (33). The degenerate ($\lambda_0 = 0$) solution

$$
\mathbf{x}_1(t) = \begin{bmatrix} 1 & 1 & 0 & 0 \end{bmatrix}^T
$$

describes the two masses at rest with position functions $x_1(t) \equiv 1$ and $x_2(t) \equiv 1$. The solution

$$
\mathbf{x}_2(t) = \begin{bmatrix} 1 & 1 & -2 & -2 \end{bmatrix}^T e^{-2t}
$$

corresponding to the carefully chosen eigenvector \mathbf{w}_1 describes damped motions $x_1(t) = e^{-2t}$ and $x_2(t) = e^{-2t}$ of the two masses, with equal velocities in the same direction. Finally, the solutions $\mathbf{x}_3(t)$ and $\mathbf{x}_4(t)$ resulting from the length 2 chain $\{\mathbf{v}_1, \mathbf{v}_2\}$ both describe damped motion with the two masses moving in opposite directions. ∎

The methods of this section apply to complex multiple eigenvalues just as to real multiple eigenvalues (although the necessary computations tend to be somewhat lengthy). Given a complex conjugate pair $\alpha \pm \beta i$ of eigenvalues of multiplicity k, we work with one of them (say, $\alpha - \beta i$) as if it were real to find k independent complex-valued solutions. The real and imaginary parts of these complex-valued solutions then provide $2k$ real-valued solutions associated with the two eigenvalues $\lambda = \alpha - \beta i$ and $\bar{\lambda} = \alpha + \beta i$ each of multiplicity k. See Problems 33 and 34.

5.4 Problems

Find general solutions of the systems in Problems 1 through 22. In Problems 1 through 6, use a computer system or graphing calculator to construct a direction field and typical solution curves for the given system.

1. $\mathbf{x}' = \begin{bmatrix} -2 & 1 \\ -1 & -4 \end{bmatrix} \mathbf{x}$

2. $\mathbf{x}' = \begin{bmatrix} 3 & -1 \\ 1 & 1 \end{bmatrix} \mathbf{x}$

3. $\mathbf{x}' = \begin{bmatrix} 1 & -2 \\ 2 & 5 \end{bmatrix} \mathbf{x}$

4. $\mathbf{x}' = \begin{bmatrix} 3 & -1 \\ 1 & 5 \end{bmatrix} \mathbf{x}$

5. $\mathbf{x}' = \begin{bmatrix} 7 & 1 \\ -4 & 3 \end{bmatrix} \mathbf{x}$

6. $\mathbf{x}' = \begin{bmatrix} 1 & -4 \\ 4 & 9 \end{bmatrix} \mathbf{x}$

7. $\mathbf{x}' = \begin{bmatrix} 2 & 0 & 0 \\ -7 & 9 & 7 \\ 0 & 0 & 2 \end{bmatrix} \mathbf{x}$

8. $\mathbf{x}' = \begin{bmatrix} 25 & 12 & 0 \\ -18 & -5 & 0 \\ 6 & 6 & 13 \end{bmatrix} \mathbf{x}$

9. $\mathbf{x}' = \begin{bmatrix} -19 & 12 & 84 \\ 0 & 5 & 0 \\ -8 & 4 & 33 \end{bmatrix} \mathbf{x}$

10. $\mathbf{x}' = \begin{bmatrix} -13 & 40 & -48 \\ -8 & 23 & -24 \\ 0 & 0 & 3 \end{bmatrix} \mathbf{x}$

11. $\mathbf{x}' = \begin{bmatrix} -3 & 0 & -4 \\ -1 & -1 & -1 \\ 1 & 0 & 1 \end{bmatrix} \mathbf{x}$

12. $\mathbf{x}' = \begin{bmatrix} -1 & 0 & 1 \\ 0 & -1 & 1 \\ 1 & -1 & -1 \end{bmatrix} \mathbf{x}$

13. $\mathbf{x}' = \begin{bmatrix} -1 & 0 & 1 \\ 0 & 1 & -4 \\ 0 & 1 & -3 \end{bmatrix} \mathbf{x}$

14. $\mathbf{x}' = \begin{bmatrix} 0 & 0 & 1 \\ -5 & -1 & -5 \\ 4 & 1 & -2 \end{bmatrix} \mathbf{x}$

15. $\mathbf{x}' = \begin{bmatrix} -2 & -9 & 0 \\ 1 & 4 & 0 \\ 1 & 3 & 1 \end{bmatrix} \mathbf{x}$

16. $\mathbf{x}' = \begin{bmatrix} 1 & 0 & 0 \\ -2 & -2 & -3 \\ 2 & 3 & 4 \end{bmatrix} \mathbf{x}$

17. $\mathbf{x}' = \begin{bmatrix} 1 & 0 & 0 \\ 18 & 7 & 4 \\ -27 & -9 & -5 \end{bmatrix} \mathbf{x}$

18. $\mathbf{x}' = \begin{bmatrix} 1 & 0 & 0 \\ 1 & 3 & 1 \\ -2 & -4 & -1 \end{bmatrix} \mathbf{x}$

19. $\mathbf{x}' = \begin{bmatrix} 1 & -4 & 0 & -2 \\ 0 & 1 & 0 & 0 \\ 6 & -12 & -1 & -6 \\ 0 & -4 & 0 & -1 \end{bmatrix} \mathbf{x}$

20. $\mathbf{x}' = \begin{bmatrix} 2 & 1 & 0 & 1 \\ 0 & 2 & 1 & 0 \\ 0 & 0 & 2 & 1 \\ 0 & 0 & 0 & 2 \end{bmatrix} \mathbf{x}$

21. $\mathbf{x}' = \begin{bmatrix} -1 & -4 & 0 & 0 \\ 1 & 3 & 0 & 0 \\ 1 & 2 & 1 & 0 \\ 0 & 1 & 0 & 1 \end{bmatrix} \mathbf{x}$

22. $\mathbf{x}' = \begin{bmatrix} 1 & 3 & 7 & 0 \\ 0 & -1 & -4 & 0 \\ 0 & 1 & 3 & 0 \\ 0 & -6 & -14 & 1 \end{bmatrix} \mathbf{x}$

In Problems 23 through 32 the eigenvalues of the coefficient matrix A are given. Find a general solution of the indicated system $\mathbf{x}' = \mathbf{Ax}$. Especially in Problems 29 through 32, use of a computer algebra system (as in the application material for this section) may be useful.

23. $\mathbf{x}' = \begin{bmatrix} 39 & 8 & -16 \\ -36 & -5 & 16 \\ 72 & 16 & -29 \end{bmatrix} \mathbf{x}; \quad \lambda = -1, 3, 3$

24. $\mathbf{x}' = \begin{bmatrix} 28 & 50 & 100 \\ 15 & 33 & 60 \\ -15 & -30 & -57 \end{bmatrix} \mathbf{x}; \quad \lambda = -2, 3, 3$

25. $\mathbf{x}' = \begin{bmatrix} -2 & 17 & 4 \\ -1 & 6 & 1 \\ 0 & 1 & 2 \end{bmatrix} \mathbf{x}; \quad \lambda = 2, 2, 2$

26. $\mathbf{x}' = \begin{bmatrix} 5 & -1 & 1 \\ 1 & 3 & 0 \\ -3 & 2 & 1 \end{bmatrix} \mathbf{x}; \quad \lambda = 3, 3, 3$

27. $\mathbf{x}' = \begin{bmatrix} -3 & 5 & -5 \\ 3 & -1 & 3 \\ 8 & -8 & 10 \end{bmatrix} \mathbf{x}; \quad \lambda = 2, 2, 2$

28. $\mathbf{x}' = \begin{bmatrix} -15 & -7 & 4 \\ 34 & 16 & -11 \\ 17 & 7 & 5 \end{bmatrix} \mathbf{x}; \quad \lambda = 2, 2, 2$

29. $\mathbf{x}' = \begin{bmatrix} -1 & 1 & 1 & -2 \\ 7 & -4 & -6 & 11 \\ 5 & -1 & 1 & 3 \\ 6 & -2 & -2 & 6 \end{bmatrix} \mathbf{x}; \quad \lambda = -1, -1, 2, 2$

30. $\mathbf{x}' = \begin{bmatrix} 2 & 1 & -2 & 1 \\ 0 & 3 & -5 & 3 \\ 0 & -13 & 22 & -12 \\ 0 & -27 & 45 & -25 \end{bmatrix} \mathbf{x}; \quad \lambda = -1, -1, 2, 2$

31. $\mathbf{x}' = \begin{bmatrix} 35 & -12 & 4 & 30 \\ 22 & -8 & 3 & 19 \\ -10 & 3 & 0 & -9 \\ -27 & 9 & -3 & -23 \end{bmatrix} \mathbf{x}; \quad \lambda = 1, 1, 1, 1$

32. $\mathbf{x}' = \begin{bmatrix} 11 & -1 & 26 & 6 & -3 \\ 0 & 3 & 0 & 0 & 0 \\ -9 & 0 & -24 & -6 & 3 \\ 3 & 0 & 9 & 5 & -1 \\ -48 & -3 & -138 & -30 & 18 \end{bmatrix} \mathbf{x};$
$\lambda = 2, 2, 3, 3, 3$

33. The characteristic equation of the coefficient matrix \mathbf{A} of the system

$$\mathbf{x}' = \begin{bmatrix} 3 & -4 & 1 & 0 \\ 4 & 3 & 0 & 1 \\ 0 & 0 & 3 & -4 \\ 0 & 0 & 4 & 3 \end{bmatrix} \mathbf{x}$$

is

$$\phi(\lambda) = (\lambda^2 - 6\lambda + 25)^2 = 0.$$

Therefore, \mathbf{A} has the repeated complex conjugate pair $3 \pm 4i$ of eigenvalues. First show that the complex vectors

$$\mathbf{v}_1 = \begin{bmatrix} 1 & i & 0 & 0 \end{bmatrix}^T \quad \text{and} \quad \mathbf{v}_2 = \begin{bmatrix} 9 & 0 & 1 & i \end{bmatrix}^T$$

form a length 2 chain $\{\mathbf{v}_1, \mathbf{v}_2\}$ associated with the eigenvalue $\lambda = 3 - 4i$. Then calculate the real and imaginary parts of the complex-valued solutions

$$\mathbf{v}_1 e^{\lambda t} \quad \text{and} \quad (\mathbf{v}_1 t + \mathbf{v}_2) e^{\lambda t}$$

to find four independent real-valued solutions of $\mathbf{x}' = \mathbf{Ax}$.

34. The characteristic equation of the coefficient matrix \mathbf{A} of the system

$$\mathbf{x}' = \begin{bmatrix} 2 & 0 & -8 & -3 \\ -18 & -1 & 0 & 0 \\ -9 & -3 & -25 & -9 \\ 33 & 10 & 90 & 32 \end{bmatrix} \mathbf{x}$$

is

$$\phi(\lambda) = (\lambda^2 - 4\lambda + 13)^2 = 0.$$

Therefore, \mathbf{A} has the repeated complex conjugate pair $2 \pm 3i$ of eigenvalues. First show that the complex vectors

$$\mathbf{v}_1 = \begin{bmatrix} -i & 3+3i & 0 & -i \end{bmatrix}^T,$$

$$\mathbf{v}_2 = \begin{bmatrix} 3 & -10+9i & -i & 0 \end{bmatrix}^T$$

form a length 2 chain $\{\mathbf{v}_1, \mathbf{v}_2\}$ associated with the eigenvalue $\lambda = 2 + 3i$. Then calculate (as in Problem 33) four independent real-valued solutions of $\mathbf{x}' = \mathbf{A}\mathbf{x}$.

35. Find the position functions $x_1(t)$ and $x_2(t)$ of the railway cars of Fig. 5.4.1 if the physical parameters are given by

$$m_1 = m_2 = c_1 = c_2 = c = k = 1$$

and the initial conditions are

$$x_1(0) = x_2(0) = 0, \quad x_1'(0) = x_2'(0) = v_0.$$

How far do the cars travel before stopping?

36. Repeat Problem 35 under the assumption that car 1 is shielded from air resistance by car 2, so now $c_1 = 0$. Show that, before stopping, the cars travel twice as far as those of Problem 35.

5.4 Application: Defective Eigenvalues and Generalized Eigenvectors

A typical computer algebra system can calculate both the eigenvalues of a given matrix \mathbf{A} and the linearly independent (ordinary) eigenvectors associated with each eigenvalue. For instance, consider the 4×4 matrix

$$\mathbf{A} = \begin{bmatrix} 35 & -12 & 4 & 30 \\ 22 & -8 & 3 & 19 \\ -10 & 3 & 0 & -9 \\ -27 & 9 & -3 & -23 \end{bmatrix} \tag{1}$$

of Problem 31 in this section. When the matrix \mathbf{A} has been entered, the *Maple* calculation

```
with(linalg): eigenvectors(A);
[1, 4, {[-1, 0, 1, 1], [0, 1, 3, 0]}]
```

or the *Mathematica* calculation

```
Eigensystem[A]
{{1,1,1,1},
{{-3,-1,0,3}, {0,1,3,0}, {0,0,0,0}, {0,0,0,0}}}
```

reveals that the matrix \mathbf{A} in Eq. (1) has the single eigenvalue $\lambda = 1$ of multiplicity 4 with only two independent associated eigenvectors \mathbf{v}_1 and \mathbf{v}_2. The MATLAB command

```
[V, D] = eig(sym(A))
```

provides the same information. The eigenvalue $\lambda = 1$ therefore has defect $d = 2$. If $\mathbf{B} = \mathbf{A} - (1)\mathbf{I}$, you should find that $\mathbf{B}^2 \neq \mathbf{0}$ but $\mathbf{B}^3 = \mathbf{0}$. If

$$\mathbf{u}_1 = \begin{bmatrix} 1 & 0 & 0 & 0 \end{bmatrix}^T, \quad \mathbf{u}_2 = \mathbf{B}\mathbf{u}_1, \quad \mathbf{u}_3 = \mathbf{B}\mathbf{u}_2,$$

then $\{\mathbf{u}_1, \mathbf{u}_2, \mathbf{u}_3\}$ should be a length 3 chain of generalized eigenvectors based on the ordinary eigenvector \mathbf{u}_3 (which should be a linear combination of the original

eigenvectors \mathbf{v}_1 and \mathbf{v}_2). Use your computer algebra system to carry out this construction, and finally write four linearly independent solutions of the linear system $\mathbf{x}' = \mathbf{A}\mathbf{x}$.

For a more exotic matrix to investigate, consider the MATLAB's `gallery(5)` example matrix

$$\mathbf{A} = \begin{bmatrix} -9 & 11 & -21 & 63 & -252 \\ 70 & -69 & 141 & -421 & 1684 \\ -575 & 575 & -1149 & 3451 & -13801 \\ 3891 & -3891 & 7782 & -23345 & 93365 \\ 1024 & -1024 & 2048 & -6144 & 24572 \end{bmatrix}. \tag{2}$$

Use appropriate commands like those illustrated here to show that \mathbf{A} has a single eigenvalue $\lambda = 0$ of multiplicity 5 and defect 4. Noting that $\mathbf{A} - (0)\mathbf{I} = \mathbf{A}$, you should find that $\mathbf{A}^4 \neq \mathbf{0}$ but that $\mathbf{A}^5 = \mathbf{0}$. Hence calculate the vectors

$$\mathbf{u}_1 = \begin{bmatrix} 1 & 0 & 0 & 0 & 0 \end{bmatrix}^T, \quad \mathbf{u}_2 = \mathbf{A}\mathbf{u}_1, \quad \mathbf{u}_3 = \mathbf{A}\mathbf{u}_2, \quad \mathbf{u}_4 = \mathbf{A}\mathbf{u}_3, \quad \mathbf{u}_5 = \mathbf{A}\mathbf{u}_4.$$

You should find that \mathbf{u}_5 is a nonzero vector such that $\mathbf{A}\mathbf{u}_5 = \mathbf{0}$, and is therefore an (ordinary) eigenvector of \mathbf{A} associated with the eigenvalue $\lambda = 0$. Thus $\{\mathbf{u}_1, \mathbf{u}_2, \mathbf{u}_3, \mathbf{u}_4, \mathbf{u}_5\}$ is a length 5 chain of generalized eigenvectors of the matrix \mathbf{A} in Eq. (2), and you can finally write five linearly independent solutions of the linear system $\mathbf{x}' = \mathbf{A}\mathbf{x}$.

5.5 Matrix Exponentials and Linear Systems

The solution vectors of a homogeneous linear system

$$\mathbf{x}' = \mathbf{A}\mathbf{x} \tag{1}$$

can be used to construct a matrix $\mathbf{X} = \boldsymbol{\Phi}(t)$ that satisfies the *matrix differential equation*

$$\mathbf{X}' = \mathbf{A}\mathbf{X} \tag{1'}$$

associated with Eq. (1). Suppose that $\mathbf{x}_1(t), \mathbf{x}_2(t), \ldots, \mathbf{x}_n(t)$ are n linearly independent solutions of Eq. (1). Then the $n \times n$ matrix

$$\boldsymbol{\Phi}(t) = \begin{bmatrix} | & | & & | \\ \mathbf{x}_1(t) & \mathbf{x}_2(t) & \cdots & \mathbf{x}_n(t) \\ | & | & & | \end{bmatrix}, \tag{2}$$

having these solution vectors as its column vectors, is called a **fundamental matrix** for the system in (1).

Fundamental Matrix Solutions

Because the column vector $\mathbf{x} = \mathbf{x}_j(t)$ of the fundamental matrix $\boldsymbol{\Phi}(t)$ in (2) satisfies the differential equation $\mathbf{x}' = \mathbf{A}\mathbf{x}$, it follows that the matrix $\mathbf{X} = \boldsymbol{\Phi}(t)$ itself satisfies the matrix differential equation $\mathbf{X}' = \mathbf{A}\mathbf{X}$. Because its column vectors are linearly independent, it also follows that the fundamental matrix $\boldsymbol{\Phi}(t)$ is nonsingular, and therefore has an inverse matrix $\boldsymbol{\Phi}(t)^{-1}$. Conversely, any nonsingular matrix solution $\boldsymbol{\Psi}(t)$ of Eq. (1') has linearly independent column vectors that satisfy Eq. (1), so $\boldsymbol{\Psi}(t)$ is a fundamental matrix for the system in (1).

In terms of the fundamental matrix $\boldsymbol{\Phi}(t)$ in (2), the general solution

$$\mathbf{x}(t) = c_1\mathbf{x}_1(t) + c_2\mathbf{x}_2(t) + \cdots + c_n\mathbf{x}_n(t) \tag{3}$$

of the system $\mathbf{X}' = \mathbf{A}\mathbf{X}$ can be written in the form

$$\mathbf{x}(t) = \boldsymbol{\Phi}(t)\mathbf{c} \tag{4}$$

where $\mathbf{c} = [c_1 \quad c_2 \quad \dots \quad c_n]^T$ is an arbitrary *constant* vector. If $\boldsymbol{\Psi}(t)$ is any other fundamental matrix for (1), then each column vector of $\boldsymbol{\Psi}(t)$ is a linear combination of the column vectors of $\boldsymbol{\Phi}(t)$, so it follows from Eq. (4) that

$$\boldsymbol{\Psi}(t) = \boldsymbol{\Phi}(t)\mathbf{C} \tag{4'}$$

for some $n \times n$ matrix \mathbf{C} of constants.

In order that the solution $\mathbf{x}(t)$ in (3) satisfy a given initial condition

$$\mathbf{x}(0) = \mathbf{x}_0, \tag{5}$$

it suffices that the coefficient vector \mathbf{c} in (4) be such that $\boldsymbol{\Phi}(0)\mathbf{c} = \mathbf{x}_0$; that is, that

$$\mathbf{c} = \boldsymbol{\Phi}(0)^{-1}\mathbf{x}_0. \tag{6}$$

When we substitute (6) in Eq. (4), we get the conclusion of the following theorem.

THEOREM 1: Fundamental Matrix Solutions

Let $\boldsymbol{\Phi}(t)$ be a fundamental matrix for the homogeneous linear system $\mathbf{x}' = \mathbf{A}\mathbf{x}$. Then the [unique] solution of the initial value problem

$$\mathbf{x}' = \mathbf{A}\mathbf{x}, \quad \mathbf{x}(0) = \mathbf{x}_0 \tag{7}$$

is given by

$$\mathbf{x}(t) = \boldsymbol{\Phi}(t)\boldsymbol{\Phi}(0)^{-1}\mathbf{x}_0. \tag{8}$$

▲

Section 5.2 tells us how to find a fundamental matrix for the system

$$\mathbf{x}' = \mathbf{A}\mathbf{x} \tag{9}$$

with constant $n \times n$ coefficient matrix \mathbf{A}, at least in the case where \mathbf{A} has a complete set of n linearly independent eigenvectors $\mathbf{v}_1, \mathbf{v}_2, \dots, \mathbf{v}_n$ associated with the

[not necessarily distinct] eigenvalues $\lambda_1, \lambda_2, \ldots, \lambda_n$, respectively. In this event the corresponding solution vectors of Eq. (9) are given by

$$\mathbf{x}_i(t) = \mathbf{v}_i e^{\lambda_i t}$$

for $i = 1, 2, \ldots, n$. Therefore, the $n \times n$ matrix

$$\blacktriangleright \qquad \Phi(t) = \begin{bmatrix} | & | & & | \\ \mathbf{v}_1 e^{\lambda_1 t} & \mathbf{v}_2 e^{\lambda_2 t} & \cdots & \mathbf{v}_n e^{\lambda_n t} \\ | & | & & | \end{bmatrix} \tag{10}$$

having the solutions $\mathbf{x}_1, \mathbf{x}_2, \ldots, \mathbf{x}_n$ as column vectors is a fundamental matrix for the system $\mathbf{x}' = \mathbf{A}\mathbf{x}$.

In order to apply Eq. (8), we must be able to compute the inverse matrix $\Phi(0)^{-1}$. The inverse of the nonsingular 2×2 matrix

$$\mathbf{A} = \begin{bmatrix} a & b \\ c & d \end{bmatrix}$$

is

$$\mathbf{A}^{-1} = \frac{1}{\Delta} \begin{bmatrix} d & -b \\ -c & a \end{bmatrix}, \tag{11}$$

where $\Delta = \det(\mathbf{A}) = ad - bc \neq 0$. The inverse of the nonsingular 3×3 matrix $\mathbf{A} = [a_{ij}]$ is given by

$$\mathbf{A}^{-1} = \frac{1}{\Delta} \begin{bmatrix} +A_{11} & -A_{12} & +A_{13} \\ -A_{21} & +A_{22} & -A_{23} \\ +A_{31} & -A_{32} & +A_{33} \end{bmatrix}^T, \tag{12}$$

where $\Delta = \det(\mathbf{A}) \neq 0$ and A_{ij} denotes the determinant of the 2×2 submatrix of \mathbf{A} obtained by deleting the ith row and jth column of \mathbf{A}. (Do not overlook the symbol T for *transpose* in Eq. (12).) The formula in (12) is also valid upon generalization to $n \times n$ matrices, but in practice inverses of larger matrices are usually computed instead by row reduction methods (see any linear algebra text) or by using a calculator or computer algebra system.

Example 1　Find a fundamental matrix for the system

$$\begin{aligned} x' &= 4x + 2y, \\ y' &= 3x - y, \end{aligned} \tag{13}$$

then use it to find the solution of (13) that satisfies the initial conditions $x(0) = 1$, $y(0) = -1$.

Solution The linearly independent solutions

$$\mathbf{x}_1(t) = \begin{bmatrix} e^{-2t} \\ -3e^{-2t} \end{bmatrix} \quad \text{and} \quad \mathbf{x}_2(t) = \begin{bmatrix} 2e^{5t} \\ e^{5t} \end{bmatrix}$$

found in Example 1 of Section 5.2 yield the fundamental matrix

$$\mathbf{\Phi}(t) = \begin{bmatrix} e^{-2t} & 2e^{5t} \\ -3e^{-2t} & e^{5t} \end{bmatrix}. \tag{14}$$

Then

$$\mathbf{\Phi}(0) = \begin{bmatrix} 1 & 2 \\ -3 & 1 \end{bmatrix},$$

and the formula in (11) gives the inverse matrix

$$\mathbf{\Phi}(0)^{-1} = \frac{1}{7}\begin{bmatrix} 1 & -2 \\ 3 & 1 \end{bmatrix}. \tag{15}$$

Hence the formula in (8) gives the solution

$$\mathbf{x}(t) = \begin{bmatrix} e^{-2t} & 2e^{5t} \\ -3e^{-2t} & e^{5t} \end{bmatrix}\left(\frac{1}{7}\right)\begin{bmatrix} 1 & 2 \\ 3 & 1 \end{bmatrix}\begin{bmatrix} 1 \\ -1 \end{bmatrix} = \left(\frac{1}{7}\right)\begin{bmatrix} e^{-2t} & 2e^{5t} \\ -3e^{-2t} & e^{5t} \end{bmatrix}\begin{bmatrix} 3 \\ 2 \end{bmatrix},$$

and so

$$\mathbf{x}(t) = \frac{1}{7}\begin{bmatrix} 3e^{-2t} + 4e^{5t} \\ -9e^{-2t} + 2e^{5t} \end{bmatrix}.$$

Thus the solution of the original initial value problem is given by

$$x(t) = \tfrac{3}{7}e^{-2t} + \tfrac{4}{7}e^{5t}, \quad y(t) = -\tfrac{9}{7}e^{-2t} + \tfrac{2}{7}e^{5t}. \qquad \blacksquare$$

Remark: An advantage of the fundamental matrix approach is this: Once we know the fundamental matrix $\mathbf{\Phi}(t)$ and the inverse matrix $\mathbf{\Phi}(0)^{-1}$, we can calculate rapidly by matrix multiplication the solutions corresponding to different initial conditions. For example, suppose that we seek the solution of the system in (13) satisfying the new initial conditions $x(0) = 77$, $y(0) = 49$. Then substitution of (14) and (15) in (8) gives the new particular solution

$$\mathbf{x}(t) = \frac{1}{7}\begin{bmatrix} e^{-2t} & 2e^{5t} \\ -3e^{-2t} & e^{5t} \end{bmatrix}\begin{bmatrix} 1 & -2 \\ 3 & 1 \end{bmatrix}\begin{bmatrix} 77 \\ 49 \end{bmatrix}$$

$$= \frac{1}{7}\begin{bmatrix} e^{-2t} & 2e^{5t} \\ -3e^{-2t} & e^{5t} \end{bmatrix}\begin{bmatrix} -21 \\ 280 \end{bmatrix} = \begin{bmatrix} -3e^{-2t} + 80e^{5t} \\ 9e^{-2t} + 40e^{5t} \end{bmatrix}. \qquad \blacksquare$$

Exponential Matrices

We now discuss the possibility of constructing a fundamental matrix for the constant-coefficient linear system $\mathbf{x}' = \mathbf{Ax}$ directly from the coefficient matrix \mathbf{A}—that is, without first applying the methods of earlier sections to find a linearly independent set of solution vectors.

We have seen that exponential functions play a central role in the solution of linear differential equations and systems, ranging from the scalar equation $x' = kx$ with solution $x(t) = x_0 e^{kt}$ to the vector solution $\mathbf{x}(t) = \mathbf{v}e^{\lambda t}$ of the linear system $\mathbf{x}' = \mathbf{Ax}$ whose coefficient matrix \mathbf{A} has eigenvalue λ with associated eigenvector \mathbf{v}. We now define exponentials of matrices in such a way that

$$\mathbf{X}(t) = e^{\mathbf{A}t}$$

is a matrix solution of the matrix differential equation

$$\mathbf{X}' = \mathbf{AX}$$

with $n \times n$ coefficient matrix \mathbf{A}.

The exponential e^z of the complex number z may be defined (as in Section 3.3) by means of the exponential series

$$e^z = 1 + z + \frac{z^2}{2!} + \frac{z^3}{3!} + \cdots + \frac{z^n}{n!} + \cdots . \tag{16}$$

Similarly, if \mathbf{A} is an $n \times n$ matrix, then the **exponential matrix** $e^{\mathbf{A}}$ is the $n \times n$ matrix defined by the series

$$e^{\mathbf{A}} = \mathbf{I} + \mathbf{A} + \frac{\mathbf{A}^2}{2!} + \cdots + \frac{\mathbf{A}^n}{n!} + \cdots , \tag{17}$$

where \mathbf{I} is the identity matrix. The meaning of the infinite series on the right in (17) is given by

$$\sum_{n=0}^{\infty} \frac{\mathbf{A}^n}{n!} = \lim_{k \to \infty} \left(\sum_{n=0}^{k} \frac{\mathbf{A}^n}{n!} \right) \tag{18}$$

where $\mathbf{A}^0 = \mathbf{I}$, $\mathbf{A}^2 = \mathbf{AA}$, $\mathbf{A}^3 = \mathbf{AA}^2$, and so on; inductively, $\mathbf{A}^{n+1} = \mathbf{AA}^n$ if $n \geq 0$. It can be shown that the limit in (18) exists for every $n \times n$ square matrix \mathbf{A}. That is, the exponential matrix $e^{\mathbf{A}}$ is defined (by Eq. (17)) for every square matrix \mathbf{A}.

Example 2 Consider the 2×2 diagonal matrix

$$\mathbf{A} = \begin{bmatrix} a & 0 \\ 0 & b \end{bmatrix}.$$

Then it is apparent that

$$\mathbf{A}^n = \begin{bmatrix} a^n & 0 \\ 0 & b^n \end{bmatrix}$$

for each integer $n \geqq 1$. It therefore follows that

$$e^{\mathbf{A}} = \mathbf{I} + \mathbf{A} + \cdots$$

$$= \begin{bmatrix} 1 & 0 \\ 0 & 1 \end{bmatrix} + \begin{bmatrix} a & 0 \\ 0 & b \end{bmatrix} + \begin{bmatrix} a^2/2! & 0 \\ 0 & b^2/2! \end{bmatrix} + \cdots$$

$$= \begin{bmatrix} 1 + a + a^2/2! + \cdots & 0 \\ 0 & 1 + b + b^2/2! + \cdots \end{bmatrix}.$$

Thus

$$e^{\mathbf{A}} = \begin{bmatrix} e^a & 0 \\ 0 & e^b \end{bmatrix},$$

so the exponential of the *diagonal* 2×2 matrix \mathbf{A} is obtained simply by exponentiating each diagonal element of \mathbf{A}. ∎

The $n \times n$ analogue of the 2×2 result in Example 2 is established in the same way. The exponential of the $n \times n$ diagonal matrix

$$\mathbf{D} = \begin{bmatrix} a_1 & 0 & \cdots & 0 \\ 0 & a_2 & \cdots & 0 \\ \vdots & \vdots & & \vdots \\ 0 & 0 & \cdots & a_n \end{bmatrix} \qquad (19)$$

is the $n \times n$ diagonal matrix

$$e^{\mathbf{D}} = \begin{bmatrix} e^{a_1} & 0 & \cdots & 0 \\ 0 & e^{a_2} & \cdots & 0 \\ \vdots & \vdots & & \vdots \\ 0 & 0 & \cdots & e^{a_n} \end{bmatrix} \qquad (20)$$

obtained by exponentiating each diagonal element of \mathbf{D}.

The exponential matrix $e^{\mathbf{A}}$ satisfies most of the exponential relations that are familiar in the case of scalar exponents. For instance, if $\mathbf{0}$ is the $n \times n$ zero matrix, then Eq. (17) yields

$$e^{\mathbf{0}} = \mathbf{I}, \qquad (21)$$

the $n \times n$ identity matrix. In Problem 31 we ask you to show that a useful law of exponents holds for $n \times n$ matrices that commute:

$$\text{If} \quad \mathbf{AB} = \mathbf{BA}, \quad \text{then} \quad e^{\mathbf{A}+\mathbf{B}} = e^{\mathbf{A}} e^{\mathbf{B}}. \qquad (22)$$

In Problem 32 we ask you to conclude that

$$\left(e^{\mathbf{A}} \right)^{-1} = e^{-\mathbf{A}}. \qquad (23)$$

In particular, the matrix $e^{\mathbf{A}}$ is nonsingular for every $n \times n$ matrix \mathbf{A} (reminiscent of the fact that $e^z \neq 0$ for all z). It follows from elementary linear algebra that the column vectors of $e^{\mathbf{A}}$ are always linearly independent.

If t is a scalar variable, then substitution of $\mathbf{A}t$ for \mathbf{A} in Eq. (17) gives

$$e^{\mathbf{A}t} = \mathbf{I} + \mathbf{A}t + \mathbf{A}^2 \frac{t^2}{2!} + \cdots + \mathbf{A}^n \frac{t^n}{n!} + \cdots. \qquad (24)$$

(Of course, $\mathbf{A}t$ is obtained simply by multiplying each element of \mathbf{A} by t.)

Example 3 If

$$\mathbf{A} = \begin{bmatrix} 0 & 3 & 4 \\ 0 & 0 & 6 \\ 0 & 0 & 0 \end{bmatrix},$$

then

$$\mathbf{A}^2 = \begin{bmatrix} 0 & 0 & 18 \\ 0 & 0 & 0 \\ 0 & 0 & 0 \end{bmatrix} \quad \text{and} \quad \mathbf{A}^3 = \begin{bmatrix} 0 & 0 & 0 \\ 0 & 0 & 0 \\ 0 & 0 & 0 \end{bmatrix},$$

so $\mathbf{A}^n = \mathbf{0}$ for $n \geq 3$. It therefore follows from Eq. (24) that

$$e^{\mathbf{A}t} = \mathbf{I} + \mathbf{A}\,t + \tfrac{1}{2}\mathbf{A}^2\,t^2$$

$$= \begin{bmatrix} 1 & 0 & 0 \\ 0 & 1 & 0 \\ 0 & 0 & 1 \end{bmatrix} + \begin{bmatrix} 0 & 3 & 4 \\ 0 & 0 & 6 \\ 0 & 0 & 0 \end{bmatrix} t + \tfrac{1}{2} \begin{bmatrix} 0 & 0 & 18 \\ 0 & 0 & 0 \\ 0 & 0 & 0 \end{bmatrix} t^2;$$

that is,

$$e^{\mathbf{A}t} = \begin{bmatrix} 1 & 3t & 4t + 9t^2 \\ 0 & 1 & 6t \\ 0 & 0 & 1 \end{bmatrix}. \qquad \blacksquare$$

Remark: If $\mathbf{A}^n = \mathbf{0}$ for some positive integer n, then the exponential series in (24) terminates after a finite number of terms, so the exponential matrix $e^{\mathbf{A}}$ (or $e^{\mathbf{A}t}$) is readily calculated as in Example 3. Such a matrix—with a vanishing power—is said to be **nilpotent**. $\qquad \blacksquare$

Example 4 If

$$\mathbf{A} = \begin{bmatrix} 2 & 3 & 4 \\ 0 & 2 & 6 \\ 0 & 0 & 2 \end{bmatrix},$$

then

$$\mathbf{A} = \begin{bmatrix} 2 & 0 & 0 \\ 0 & 2 & 0 \\ 0 & 0 & 2 \end{bmatrix} + \begin{bmatrix} 0 & 3 & 4 \\ 0 & 0 & 6 \\ 0 & 0 & 0 \end{bmatrix} = \mathbf{D} + \mathbf{B}$$

where $\mathbf{D} = 2\mathbf{I}$ is a diagonal matrix and \mathbf{B} is the nilpotent matrix of Example 3. Therefore, (20) and (22) give

$$e^{\mathbf{A}t} = e^{(\mathbf{D}+\mathbf{B})t} = e^{\mathbf{D}t}e^{\mathbf{B}t} = \begin{bmatrix} e^{2t} & 0 & 0 \\ 0 & e^{2t} & 0 \\ 0 & 0 & e^{2t} \end{bmatrix} \begin{bmatrix} 1 & 3t & 4t + 9t^2 \\ 0 & 1 & 6t \\ 0 & 0 & 1 \end{bmatrix};$$

thus

$$e^{\mathbf{A}t} = \begin{bmatrix} e^{2t} & 3te^{2t} & (4t + 9t^2)e^{2t} \\ 0 & e^{2t} & 6te^{2t} \\ 0 & 0 & e^{2t} \end{bmatrix}. \qquad \blacksquare$$

Matrix Exponential Solutions

It happens that term-by-term differentiation of the series in (24) is valid, with the result

$$\frac{d}{dt}\left(e^{\mathbf{A}t}\right) = \mathbf{A} + \mathbf{A}^2 t + \mathbf{A}^3 \frac{t^2}{2!} + \cdots = \mathbf{A}\left(\mathbf{I} + \mathbf{A}t + \mathbf{A}^2 \frac{t^2}{2!} + \cdots\right);$$

that is,

$$\frac{d}{dt}\left(e^{\mathbf{A}t}\right) = \mathbf{A}e^{\mathbf{A}t}, \tag{25}$$

in analogy to the formula $D_t\left(e^{kt}\right) = ke^{kt}$ from elementary calculus. Thus the matrix-valued function

$$\mathbf{X}(t) = e^{\mathbf{A}t}$$

satisfies the matrix differential equation

$$\mathbf{X}' = \mathbf{A}\mathbf{X}.$$

Because the matrix $e^{\mathbf{A}t}$ is nonsingular, it follows that *the matrix exponential $e^{\mathbf{A}t}$ is a fundamental matrix for the linear system $\mathbf{x}' = \mathbf{A}\mathbf{x}$*. In particular, it is *the* fundamental matrix $\mathbf{X}(t)$ such that $\mathbf{X}(0) = \mathbf{I}$. Therefore, Theorem 1 implies the following result.

THEOREM 2: Matrix Exponential Solutions

If \mathbf{A} is an $n \times n$ matrix, then the solution of the initial value problem

$$\mathbf{x}' = \mathbf{A}\mathbf{x}, \qquad \mathbf{x}(0) = \mathbf{x}_0 \tag{26}$$

is given by

$$\mathbf{x}(t) = e^{\mathbf{A}t}\mathbf{x}_0, \tag{27}$$

and this solution is unique. ▲

Thus the solution of homogeneous linear systems reduces to the task of computing exponential matrices. Conversely, if we already know a fundamental matrix $\mathbf{\Phi}(t)$ for the linear system $\mathbf{x}' = \mathbf{A}\mathbf{x}$, then the facts that $e^{\mathbf{A}t} = \mathbf{\Phi}(t)\mathbf{C}$ (by Eq. (4′)) and $e^{\mathbf{A}\cdot 0} = e^{\mathbf{0}} = \mathbf{I}$ (the identity matrix) yield

$$e^{\mathbf{A}t} = \mathbf{\Phi}(t)\mathbf{\Phi}(0)^{-1}. \tag{28}$$

So we can find the matrix exponential $e^{\mathbf{A}t}$ by solving the linear system $\mathbf{x}' = \mathbf{A}\mathbf{x}$.

Example 5 In Example 1 we found that the system $\mathbf{x}' = \mathbf{Ax}$ with

$$\mathbf{A} = \begin{bmatrix} 4 & 2 \\ 3 & -1 \end{bmatrix}$$

has fundamental matrix

$$\mathbf{\Phi}(t) = \begin{bmatrix} e^{-2t} & 2e^{5t} \\ -3e^{-2t} & e^{5t} \end{bmatrix} \quad \text{with} \quad \mathbf{\Phi}(0)^{-1} = \tfrac{1}{7} \begin{bmatrix} 1 & -2 \\ 3 & 1 \end{bmatrix}.$$

Hence Eq. (28) gives

$$e^{\mathbf{A}t} = \tfrac{1}{7} \begin{bmatrix} e^{-2t} & 2e^{5t} \\ -3e^{-2t} & e^{5t} \end{bmatrix} \begin{bmatrix} 1 & -2 \\ 3 & 1 \end{bmatrix}$$

$$= \tfrac{1}{7} \begin{bmatrix} e^{-2t} + 6e^{5t} & -2e^{-2t} + 2e^{5t} \\ -3e^{-2t} + 3e^{5t} & 6e^{-2t} + e^{5t} \end{bmatrix}. \qquad \blacksquare$$

Example 6 Use an exponential matrix to solve the initial value problem

$$\mathbf{x}' = \begin{bmatrix} 2 & 3 & 4 \\ 0 & 2 & 6 \\ 0 & 0 & 2 \end{bmatrix} \mathbf{x}, \quad \mathbf{x}(0) = \begin{bmatrix} 19 \\ 29 \\ 39 \end{bmatrix}. \qquad (29)$$

Solution The coefficient matrix \mathbf{A} in (29) evidently has characteristic equation $(2 - \lambda)^3 = 0$ and thus the triple eigenvalue $\lambda = 2, 2, 2$. It is easy to see that the eigenvector equation

$$(\mathbf{A} - 2\mathbf{I})\mathbf{v} = \begin{bmatrix} 0 & 3 & 4 \\ 0 & 0 & 6 \\ 0 & 0 & 0 \end{bmatrix} \begin{bmatrix} a \\ b \\ c \end{bmatrix} = \begin{bmatrix} 0 \\ 0 \\ 0 \end{bmatrix}$$

has (to within a constant multiple) the single solution $\mathbf{v} = \begin{bmatrix} 1 & 0 & 0 \end{bmatrix}^T$. Thus there is only a single eigenvector associated with the eigenvalue $\lambda = 2$, and so we do not yet have the three linearly independent solutions needed for a fundamental matrix. But we note that \mathbf{A} is the same matrix whose matrix exponential

$$e^{\mathbf{A}t} = \begin{bmatrix} e^{2t} & 3te^{2t} & (4t + 9t^2)e^{2t} \\ 0 & e^{2t} & 6te^{2t} \\ 0 & 0 & e^{2t} \end{bmatrix}$$

was calculated in Example 4. Hence, using Theorem 2, the solution of the initial value problem in (29) is given by

$$\mathbf{x}(t) = e^{\mathbf{A}t}\mathbf{x}(0) = \begin{bmatrix} e^{2t} & 3te^{2t} & (4t + 9t^2)e^{2t} \\ 0 & e^{2t} & 6te^{2t} \\ 0 & 0 & e^{2t} \end{bmatrix} \begin{bmatrix} 19 \\ 29 \\ 39 \end{bmatrix}$$

$$= \begin{bmatrix} (19 + 243t + 351t^2)e^{2t} \\ (29 + 234t)e^{2t} \\ 39e^{2t} \end{bmatrix}. \qquad \blacksquare$$

General Matrix Exponentials

The relatively simple calculation of $e^{\mathbf{A}t}$ carried out in Example 4 (and used in Example 6) was based on the observation that if

$$\mathbf{A} = \begin{bmatrix} 2 & 3 & 4 \\ 0 & 2 & 6 \\ 0 & 0 & 2 \end{bmatrix},$$

then $\mathbf{A} - 2\mathbf{I}$ is nilpotent:

$$(\mathbf{A} - 2\mathbf{I})^3 = \begin{bmatrix} 0 & 3 & 4 \\ 0 & 0 & 6 \\ 0 & 0 & 0 \end{bmatrix}^3 = \begin{bmatrix} 0 & 0 & 0 \\ 0 & 0 & 0 \\ 0 & 0 & 0 \end{bmatrix} = \mathbf{0}. \tag{30}$$

A similar result holds for any 3×3 matrix \mathbf{A} having a triple eigenvalue r, in which case its characteristic equation reduces to $(\lambda - r)^3 = 0$. For such a matrix, an explicit computation similar to that in Eq. (30) will show that

$$(\mathbf{A} - r\mathbf{I})^3 = \mathbf{0}. \tag{31}$$

(This particular result is a special case of the Cayley-Hamilton theorem of advanced linear algebra, according to which every matrix satisfies its own characteristic equation.) Thus the matrix $\mathbf{A} - r\mathbf{I}$ is nilpotent, and it follows that

$$e^{\mathbf{A}t} = e^{(r\mathbf{I}+\mathbf{A}-r\mathbf{I})t} = e^{r\mathbf{I}t} \cdot e^{(\mathbf{A}-r\mathbf{I})t} = e^{rt}\mathbf{1} \cdot \left[\mathbf{1} + (\mathbf{A} - r\mathbf{I})t + \tfrac{1}{2}(\mathbf{A} - r\mathbf{I})^2 t^2 \right], \quad (32)$$

the exponential series here terminating because of Eq. (31). In this way, we can rather easily calculate the matrix exponential $e^{\mathbf{A}t}$ for any square matrix having only a single eigenvalue.

The calculation in Eq. (32) motivates a method of calculating $e^{\mathbf{A}t}$ for any $n \times n$ matrix \mathbf{A} whatsoever. As we saw in Section 5.4, \mathbf{A} has n linearly independent generalized eigenvectors $\mathbf{u}_1, \mathbf{u}_2, \ldots, \mathbf{u}_n$. Each generalized eigenvector \mathbf{u} is associated with an eigenvalue λ of \mathbf{A} and has a *rank* $r \geq 1$ such that

$$(\mathbf{A} - \lambda\mathbf{I})^r \mathbf{u} = \mathbf{0} \qquad \text{but} \qquad (\mathbf{A} - \lambda\mathbf{I})^{r-1}\mathbf{u} \neq \mathbf{0}. \tag{33}$$

(If $r = 1$, then \mathbf{u} is an ordinary eigenvector such that $\mathbf{A}\mathbf{u} = \lambda\mathbf{u}$.)

Even if we do not yet know $e^{\mathbf{A}t}$ explicitly, we can consider the function $\mathbf{x}(t) = e^{\mathbf{A}t}\mathbf{u}$, which is a linear combination of the column vectors of $e^{\mathbf{A}t}$ and is therefore a solution of the linear system $\mathbf{x}' = \mathbf{A}\mathbf{x}$ with $\mathbf{x}(0) = \mathbf{u}$. Indeed, we can calculate \mathbf{x} explicitly in terms of \mathbf{A}, \mathbf{u}, λ, and r:

$$\mathbf{x}(t) = e^{\mathbf{A}t}\mathbf{u} = e^{(\lambda\mathbf{I}+\mathbf{A}-\lambda\mathbf{I})t}\mathbf{u} = e^{\lambda\mathbf{I}t}e^{(\mathbf{A}-\lambda\mathbf{I})t}\mathbf{u}$$

$$= e^{\lambda t}\mathbf{I}\left[\mathbf{I} + (\mathbf{A} - \lambda\mathbf{I})t + \cdots + (\mathbf{A} - \lambda\mathbf{I})^{r-1}\frac{t^{r-1}}{(r-1)!} + \cdots \right]\mathbf{u},$$

so

$$\mathbf{x}(t) = e^{\lambda t}\left[\mathbf{u} + (\mathbf{A} - \lambda\mathbf{I})\mathbf{u}t + (\mathbf{A} - \lambda\mathbf{I})^2\mathbf{u}\frac{t^2}{2!} + \cdots \right.$$

$$\left. + (\mathbf{A} - \lambda\mathbf{I})^{r-1}\mathbf{u}\frac{t^{r-1}}{(r-1)!} \right], \tag{34}$$

using (33) and the fact that $e^{\lambda\mathbf{I}t} = e^{\lambda t}\mathbf{I}$.

If the linearly independent solutions $\mathbf{x}_1(t)$, $\mathbf{x}_2(t)$, ..., $\mathbf{x}_n(t)$ of $\mathbf{x}' = \mathbf{Ax}$ are calculated using (34) with the linearly independent generalized eigenvectors \mathbf{u}_1, \mathbf{u}_2, ..., \mathbf{u}_n, then the $n \times n$ matrix

$$\boldsymbol{\Phi}(t) = \left[\begin{array}{cccc} \mathbf{x}_1(t) & \mathbf{x}_2(t) & \cdots & \mathbf{x}_n(t) \end{array} \right] \tag{35}$$

is a fundamental matrix for the system $\mathbf{x}' = \mathbf{Ax}$. Finally, the specific fundamental matrix $\mathbf{X}(t) = \boldsymbol{\Phi}(t)\boldsymbol{\Phi}(0)^{-1}$ satisfies the initial condition $\mathbf{X}(0) = \mathbf{I}$, and thus is the desired matrix exponential $e^{\mathbf{A}t}$. We have therefore outlined a proof of the following theorem.

THEOREM 3: Computation of $e^{\mathbf{A}t}$

Let \mathbf{u}_1, \mathbf{u}_2, ..., \mathbf{u}_n be n linearly independent generalized eigenvectors of the $n \times n$ matrix \mathbf{A}. For each i, $1 \leq i \leq n$, let $\mathbf{x}_i(t)$ be the solution of $\mathbf{x}' = \mathbf{Ax}$ given by (34), substituting $\mathbf{u} = \mathbf{u}_i$ and the associated eigenvalue λ and rank r of the generalized eigenvector \mathbf{u}_i. If the fundamental matrix $\boldsymbol{\Phi}(t)$ is defined by (35), then

$$e^{\mathbf{A}t} = \boldsymbol{\Phi}(t)\boldsymbol{\Phi}(0)^{-1}. \tag{36}$$

▲

Example 7 Find $e^{\mathbf{A}t}$ if

$$\mathbf{A} = \left[\begin{array}{ccc} 3 & 4 & 5 \\ 0 & 5 & 4 \\ 0 & 0 & 3 \end{array} \right]. \tag{37}$$

Solution Theorem 3 would apply even if the matrix \mathbf{A} were not upper triangular. But because \mathbf{A} is upper triangular, this fact enables us to see quickly that its characteristic equation is

$$(5 - \lambda)(3 - \lambda)^2 = 0.$$

Thus \mathbf{A} has the distinct eigenvalue $\lambda_1 = 5$ and the repeated eigenvalue $\lambda_2 = 3$.

CASE 1: $\lambda_1 = 5$. The eigenvector equation $(\mathbf{A} - \lambda\mathbf{I})\mathbf{u} = 0$ for $\mathbf{u} = \left[\begin{array}{ccc} a & b & c \end{array} \right]^T$ is

$$(\mathbf{A} - 5\mathbf{I})\mathbf{u} = \left[\begin{array}{ccc} -2 & 4 & 5 \\ 0 & 0 & 4 \\ 0 & 0 & -2 \end{array} \right] \left[\begin{array}{c} a \\ b \\ c \end{array} \right] = \left[\begin{array}{c} 0 \\ 0 \\ 0 \end{array} \right].$$

The last two scalar equations $4c = 0$ and $-2c = 0$ give $c = 0$. Then the first equation $-2a + 4b = 1$ is satisfied by $a = 2$ and $b = 1$. Thus the eigenvalue $\lambda_1 = 5$ has the (ordinary) eigenvector $\mathbf{u}_1 = \left[\begin{array}{ccc} 2 & 1 & 0 \end{array} \right]^T$. The corresponding solution of the system $\mathbf{x}' = \mathbf{Ax}$ is

$$\mathbf{x}_1(t) = e^{5t}\mathbf{u}_1 = e^{5t} \left[\begin{array}{ccc} 2 & 1 & 0 \end{array} \right]^T. \tag{38}$$

CASE 2: $\lambda_2 = 3$. The eigenvector equation $(\mathbf{A} - \lambda\mathbf{I})\mathbf{u} = 0$ for $\mathbf{u} = \begin{bmatrix} a & b & c \end{bmatrix}^T$ is

$$(\mathbf{A} - 3\mathbf{I})\mathbf{u} = \begin{bmatrix} 0 & 4 & 5 \\ 0 & 2 & 4 \\ 0 & 0 & 0 \end{bmatrix} \begin{bmatrix} a \\ b \\ c \end{bmatrix} = \begin{bmatrix} 0 \\ 0 \\ 0 \end{bmatrix}.$$

The first two equations $4b + 5c = 0$ and $2b + 4c = 0$ imply that $b = c = 0$, but leave a arbitrary. Thus the eigenvalue $\lambda_2 = 3$ has the single (ordinary) eigenvector $\mathbf{u}_2 = \begin{bmatrix} 1 & 0 & 0 \end{bmatrix}^T$. The corresponding solution of the system $\mathbf{x}' = \mathbf{A}\mathbf{x}$ is

$$\mathbf{x}_2(t) = e^{3t}\mathbf{u}_2 = e^{3t}\begin{bmatrix} 1 & 0 & 0 \end{bmatrix}^T. \tag{39}$$

To look for a generalized eigenvector of rank $r = 2$ in Eq. (33), we consider the equation

$$(\mathbf{A} - 3\mathbf{I})^2\mathbf{u} = \begin{bmatrix} 0 & 8 & 16 \\ 0 & 4 & 8 \\ 0 & 0 & 0 \end{bmatrix} \begin{bmatrix} a \\ b \\ c \end{bmatrix} = \begin{bmatrix} 0 \\ 0 \\ 0 \end{bmatrix}.$$

The first two equations $8b + 16c = 0$ and $4b + 8c = 0$ are satisfied by $b = 2$ and $c = -1$, but leave a arbitrary. With $a = 0$ we get the generalized eigenvector $\mathbf{u}_3 = \begin{bmatrix} 0 & 2 & -1 \end{bmatrix}^T$ of rank $r = 2$ associated with the eigenvalue $\lambda = 3$. Because $(\mathbf{A} - 3\mathbf{I})^2\mathbf{u} = \mathbf{0}$, Eq. (34) yields the third solution

$$\mathbf{x}_3(t) = e^{3t}[\mathbf{u}_3 + (\mathbf{A} - 3\mathbf{I})\mathbf{u}_3 t]$$

$$= e^{3t}\left(\begin{bmatrix} 0 \\ 2 \\ -1 \end{bmatrix} + \begin{bmatrix} 0 & 4 & 5 \\ 0 & 2 & 4 \\ 0 & 0 & 0 \end{bmatrix} \begin{bmatrix} 0 \\ 2 \\ -1 \end{bmatrix} t\right) = e^{3t}\begin{bmatrix} 3t \\ 2 \\ -1 \end{bmatrix}. \tag{40}$$

With the solutions listed in Eqs. (39) and (40), the fundamental matrix

$$\boldsymbol{\Phi}(t) = \begin{bmatrix} \mathbf{x}_1(t) & \mathbf{x}_2(t) & \mathbf{x}_3(t) \end{bmatrix}$$

defined by Eq. (35) is

$$\boldsymbol{\Phi}(t) = \begin{bmatrix} 2e^{5t} & e^{3t} & 3te^{3t} \\ e^{5t} & 0 & 2e^{3t} \\ 0 & 0 & e^{3t} \end{bmatrix} \quad \text{with} \quad \boldsymbol{\Phi}(0)^{-1} = \begin{bmatrix} 0 & 1 & 2 \\ 1 & -2 & -4 \\ 0 & 0 & -1 \end{bmatrix}.$$

Hence Theorem 3 finally yields

$$e^{\mathbf{A}t} = \boldsymbol{\Phi}(t)\boldsymbol{\Phi}(0)^{-1}$$

$$= \begin{bmatrix} 2e^{5t} & e^{3t} & 3te^{3t} \\ e^{5t} & 0 & 2e^{3t} \\ 0 & 0 & -e^{3t} \end{bmatrix} \begin{bmatrix} 0 & 1 & 2 \\ 1 & -2 & -4 \\ 0 & 0 & -1 \end{bmatrix}$$

$$= \begin{bmatrix} e^{3t} & 2e^{5t} - 2e^{3t} & 4e^{5t} - (4 + 3t)e^{3t} \\ 0 & e^{5t} & 2e^{5t} - 2e^{3t} \\ 0 & 0 & e^{3t} \end{bmatrix}.$$ ∎

Remark: As in Example 7, Theorem 3 suffices for the computation of $e^{\mathbf{A}t}$ provided that a basis consisting of generalized eigenvectors of \mathbf{A} can be found. Alternatively, a computer algebra system can be used as indicated in the project material for this section. ∎

5.5 Problems

Find a fundamental matrix of each of the systems in Problems 1 through 8, then apply Eq. (8) to find a solution satisfying the given initial conditions.

1. $\mathbf{x}' = \begin{bmatrix} 2 & 1 \\ 1 & 2 \end{bmatrix} \mathbf{x}, \quad \mathbf{x}(0) = \begin{bmatrix} 3 \\ -2 \end{bmatrix}$

2. $\mathbf{x}' = \begin{bmatrix} 2 & -1 \\ -4 & 2 \end{bmatrix} \mathbf{x}, \quad \mathbf{x}(0) = \begin{bmatrix} 2 \\ -1 \end{bmatrix}$

3. $\mathbf{x}' = \begin{bmatrix} 2 & -5 \\ 4 & -2 \end{bmatrix} \mathbf{x}, \quad \mathbf{x}(0) = \begin{bmatrix} 0 \\ 1 \end{bmatrix}$

4. $\mathbf{x}' = \begin{bmatrix} 3 & -1 \\ 1 & 1 \end{bmatrix} \mathbf{x}, \quad \mathbf{x}(0) = \begin{bmatrix} 1 \\ 0 \end{bmatrix}$

5. $\mathbf{x}' = \begin{bmatrix} -3 & -2 \\ 9 & 3 \end{bmatrix} \mathbf{x}, \quad \mathbf{x}(0) = \begin{bmatrix} 1 \\ -1 \end{bmatrix}$

6. $\mathbf{x}' = \begin{bmatrix} 7 & -5 \\ 4 & 3 \end{bmatrix} \mathbf{x}, \quad \mathbf{x}(0) = \begin{bmatrix} 2 \\ 0 \end{bmatrix}$

7. $\mathbf{x}' = \begin{bmatrix} 5 & 0 & -6 \\ 2 & -1 & -2 \\ 4 & -2 & -4 \end{bmatrix} \mathbf{x}, \quad \mathbf{x}(0) = \begin{bmatrix} 2 \\ 1 \\ 0 \end{bmatrix}$

8. $\mathbf{x}' = \begin{bmatrix} 3 & 2 & 2 \\ -5 & -4 & -2 \\ 5 & 5 & 3 \end{bmatrix} \mathbf{x}, \quad \mathbf{x}(0) = \begin{bmatrix} 1 \\ 0 \\ -1 \end{bmatrix}$

Compute the matrix exponential $e^{\mathbf{A}t}$ for each system $\mathbf{x}' = \mathbf{A}\mathbf{x}$ given in Problems 9 through 20.

9. $x_1' = 5x_1 - 4x_2, x_2' = 2x_1 - x_2$
10. $x_1' = 6x_1 - 6x_2, x_2' = 4x_1 - 4x_2$
11. $x_1' = 5x_1 - 3x_2, x_2' = 2x_1$
12. $x_1' = 5x_1 - 4x_2, x_2' = 3x_1 - 2x_2$
13. $x_1' = 9x_1 - 8x_2, x_2' = 6x_1 - 5x_2$
14. $x_1' = 10x_1 - 6x_2, x_2' = 12x_1 - 7x_2$
15. $x_1' = 6x_1 - 10x_2, x_2' = 2x_1 - 3x_2$
16. $x_1' = 11x_1 - 15x_2, x_2' = 6x_1 - 8x_2$
17. $x_1' = 3x_1 + x_2, x_2' = x_1 + 3x_2$
18. $x_1' = 4x_1 + 2x_2, x_2' = 2x_1 + 4x_2$
19. $x_1' = 9x_1 + 2x_2, x_2' = 2x_1 + 6x_2$
20. $x_1' = 13x_1 + 4x_2, x_2' = 4x_1 + 7x_2$

In Problems 21 through 24, show that the matrix \mathbf{A} is nilpotent and then use this fact to find (as in Example 3) the matrix exponential $e^{\mathbf{A}t}$.

21. $\mathbf{A} = \begin{bmatrix} 1 & -1 \\ 1 & -1 \end{bmatrix}$

22. $\mathbf{A} = \begin{bmatrix} 6 & 4 \\ -9 & -6 \end{bmatrix}$

23. $\mathbf{A} = \begin{bmatrix} 1 & -1 & -1 \\ 1 & -1 & 1 \\ 0 & 0 & 0 \end{bmatrix}$

24. $\mathbf{A} = \begin{bmatrix} 3 & 0 & -3 \\ 5 & 0 & 7 \\ 3 & 0 & -3 \end{bmatrix}$

Each coefficient matrix \mathbf{A} in Problems 25 through 30 is the sum of a nilpotent matrix and a multiple of the identity matrix. Use this fact (as in Example 6) to solve the given initial value problem.

25. $\mathbf{x}' = \begin{bmatrix} 2 & 5 \\ 0 & 2 \end{bmatrix} \mathbf{x}, \quad \mathbf{x}(0) = \begin{bmatrix} 4 \\ 7 \end{bmatrix}$

26. $\mathbf{x}' = \begin{bmatrix} 7 & 0 \\ 11 & 7 \end{bmatrix} \mathbf{x}, \quad \mathbf{x}(0) = \begin{bmatrix} 5 \\ -10 \end{bmatrix}$

27. $\mathbf{x}' = \begin{bmatrix} 1 & 2 & 3 \\ 0 & 1 & 2 \\ 0 & 0 & 1 \end{bmatrix} \mathbf{x}, \quad \mathbf{x}(0) = \begin{bmatrix} 4 \\ 5 \\ 6 \end{bmatrix}$

28. $\mathbf{x}' = \begin{bmatrix} 5 & 0 & 0 \\ 10 & 5 & 0 \\ 20 & 30 & 5 \end{bmatrix} \mathbf{x}, \quad \mathbf{x}(0) = \begin{bmatrix} 40 \\ 50 \\ 60 \end{bmatrix}$

29. $\mathbf{x}' = \begin{bmatrix} 1 & 2 & 3 & 4 \\ 0 & 1 & 6 & 3 \\ 0 & 0 & 1 & 2 \\ 0 & 0 & 0 & 1 \end{bmatrix} \mathbf{x}, \quad \mathbf{x}(0) = \begin{bmatrix} 1 \\ 1 \\ 1 \\ 1 \end{bmatrix}$

30. $\mathbf{x}' = \begin{bmatrix} 3 & 0 & 0 & 0 \\ 6 & 3 & 0 & 0 \\ 9 & 6 & 3 & 0 \\ 12 & 9 & 6 & 3 \end{bmatrix} \mathbf{x}, \quad \mathbf{x}(0) = \begin{bmatrix} 1 \\ 1 \\ 1 \\ 1 \end{bmatrix}$

31. Suppose that the $n \times n$ matrices \mathbf{A} and \mathbf{B} commute; that is, that $\mathbf{AB} = \mathbf{BA}$. Prove that $e^{\mathbf{A}+\mathbf{B}} = e^{\mathbf{A}}e^{\mathbf{B}}$. (*Suggestion*: Group the terms in the product of the two series on the right-hand side to obtain the series on the left.)

32. Deduce from the result of Problem 31 that, for every square matrix \mathbf{A}, the matrix $e^{\mathbf{A}}$ is nonsingular with $\left(e^{\mathbf{A}}\right)^{-1} = e^{-\mathbf{A}}$.

33. Suppose that

$$\mathbf{A} = \begin{bmatrix} 0 & 1 \\ 1 & 0 \end{bmatrix}.$$

Show that $\mathbf{A}^{2n} = \mathbf{I}$ and that $\mathbf{A}^{2n+1} = \mathbf{A}$ if n is a positive integer. Conclude that

$$e^{\mathbf{A}t} = \mathbf{I} \cosh t + \mathbf{A} \sinh t,$$

and apply this fact to find a general solution of $\mathbf{x}' = \mathbf{A}\mathbf{x}$. Verify that it is equivalent to the general solution found by the eigenvalue method.

34. Suppose that

$$\mathbf{A} = \begin{bmatrix} 0 & 2 \\ -2 & 0 \end{bmatrix}.$$

Show that $e^{\mathbf{A}t} = \mathbf{I} \cos 2t + \frac{1}{2}\mathbf{A} \sin 2t$. Apply this fact to find a general solution of $\mathbf{x}' = \mathbf{A}\mathbf{x}$, and verify that it is equivalent to the solution found by the eigenvalue method.

Apply Theorem 3 to calculate the matrix exponential $e^{\mathbf{A}t}$ for each of the matrices in Problems 35 through 40.

35. $\mathbf{A} = \begin{bmatrix} 3 & 4 \\ 0 & 3 \end{bmatrix}$

36. $\mathbf{A} = \begin{bmatrix} 1 & 2 & 3 \\ 0 & 1 & 4 \\ 0 & 0 & 1 \end{bmatrix}$

37. $\mathbf{A} = \begin{bmatrix} 2 & 3 & 4 \\ 0 & 1 & 3 \\ 0 & 0 & 1 \end{bmatrix}$

38. $\mathbf{A} = \begin{bmatrix} 5 & 20 & 30 \\ 0 & 10 & 20 \\ 0 & 0 & 5 \end{bmatrix}$

39. $\mathbf{A} = \begin{bmatrix} 1 & 3 & 3 & 3 \\ 0 & 1 & 3 & 3 \\ 0 & 0 & 2 & 3 \\ 0 & 0 & 0 & 2 \end{bmatrix}$

40. $\mathbf{A} = \begin{bmatrix} 2 & 4 & 4 & 4 \\ 0 & 2 & 4 & 4 \\ 0 & 0 & 2 & 4 \\ 0 & 0 & 0 & 3 \end{bmatrix}$

5.5 Application: Automated Matrix Exponential Solutions

If \mathbf{A} is an $n \times n$ matrix, then a computer algebra system can be used first to calculate the fundamental matrix $e^{\mathbf{A}t}$ for the system $\mathbf{x}' = \mathbf{A}\mathbf{x}$, then to calculate the matrix product $\mathbf{x}(t) = e^{\mathbf{A}t}\mathbf{x}_0$ to obtain a solution satisfying the initial condition $\mathbf{x}(0) = \mathbf{x}_0$. For instance, suppose that we want to solve the initial value problem

$$x_1' = 13x_1 + 4x_2,$$
$$x_2' = 4x_1 + 7x_2;$$
$$x_1(0) = 11, \quad x_2(0) = 23.$$

After the matrices

$$\mathbf{A} = \begin{bmatrix} 13 & 4 \\ 4 & 7 \end{bmatrix}, \quad \mathbf{x}_0 = \begin{bmatrix} 11 \\ 23 \end{bmatrix}$$

have been entered, either the *Maple* command

```
with(linalg): exponential(A*t)
```

the *Mathematica* command

```
MatrixExp[A t]
```

or the MATLAB command

```
syms t, expm(A*t)
```

yields the matrix exponential

$$\mathbf{expAt} = \frac{1}{5}\begin{bmatrix} e^{5t} + 4e^{15t} & -2e^{5t} + 2e^{15t} \\ -2e^{5t} + 2e^{15t} & 4e^{5t} + e^{15t} \end{bmatrix}.$$

Then either the *Maple* product `multiply(expAt,x0)`, the *Mathematica* product `expAt.x0`, or the MATLAB product `expAt*x0` gives the solution vector

$$\mathbf{x} = \begin{bmatrix} -7e^{5t} + 18e^{15t} \\ 14e^{5t} + 9e^{15t} \end{bmatrix}.$$

Obviously this, finally, is the way to do it!

For a three-dimensional example, solve the initial value problem

$$x_1' = -149x_1 - 50x_2 - 154x_3,$$
$$x_2' = 537x_1 + 180x_2 + 546x_3,$$
$$x_3' = -27x_1 - 9x_2 - 25x_3;$$
$$x_1(0) = 17, \quad x_2(0) = 43, \quad x_3(0) = 79.$$

And here's a four-dimensional problem:

$$x_1' = 4x_1 + x_2 + x_3 + 7x_4,$$
$$x_2' = x_1 + 4x_2 + 10x_3 + x_4,$$
$$x_3' = x_1 + 10x_2 + 4x_3 + x_4,$$
$$x_4' = 7x_1 + x_2 + x_3 + 4x_4;$$
$$x_1(0) = 15, \quad x_2(0) = 35, \quad x_3(0) = 55, \quad x_4(0) = 75.$$

If at this point you're having too much fun with matrix exponentials to stop, make up some problems of your own. For instance, choose any homogeneous linear system appearing in this chapter and experiment with different initial conditions. The exotic 5×5 matrix \mathbf{A} of the Section 5.4 application may suggest some interesting possibilities.

5.6 Nonhomogeneous Linear Systems

In Section 3.5 we exhibited two techniques for finding a single particular solution of a single nonhomogeneous nth-order linear differential equation—the method of undetermined coefficients and the method of variation of parameters. Each of these may be generalized to nonhomogeneous linear systems. In a linear system modeling a physical situation, nonhomogeneous terms typically correspond to external influences, such as inflow of liquid to a cascade of brine tanks or an external force acting on a mass-and-spring system.

Given the nonhomogeneous first-order linear system

$$\mathbf{x}' = \mathbf{A}\mathbf{x} + \mathbf{f}(t) \tag{1}$$

where \mathbf{A} is an $n \times n$ constant matrix and the "nonhomogeneous term" $\mathbf{f}(t)$ is a given continuous vector-valued function, we know from Theorem 4 of Section 5.1 that a general solution of Eq. (1) has the form

$$\mathbf{x}(t) = \mathbf{x}_c(t) + \mathbf{x}_p(t), \tag{2}$$

where

- $\mathbf{x}_c(t) = c_1\mathbf{x}_1(t) + c_2\mathbf{x}_2(t) + \cdots + c_n\mathbf{x}_n(t)$ is a general solution of the associated *homogeneous* system $\mathbf{x}' = \mathbf{A}\mathbf{x}$, and
- $\mathbf{x}_p(t)$ is a single particular solution of the original nonhomogeneous system in (1).

Preceding sections have dealt with $\mathbf{x}_c(t)$, so our task now is to find $\mathbf{x}_p(t)$.

Undetermined Coefficients

First we suppose that the nonhomogeneous term $\mathbf{f}(t)$ in (1) is a linear combination (with constant vector coefficients) of products of polynomials, exponential functions, and sines and cosines. Then the method of undetermined coefficients for systems is essentially the same as for a single linear differential equation. We make an intelligent guess as to the *general form* of a particular solution \mathbf{x}_p, then attempt to determine the coefficients in \mathbf{x}_p by substitution in Eq. (1). Moreover, the choice of this general form is essentially the same as in the case of a single equation (discussed in Section 3.5); we modify it only by using undetermined *vector* coefficients rather than undetermined scalars. We will therefore confine the present discussion to illustrative examples.

Example 1 Find a particular solution of the nonhomogeneous system

$$\mathbf{x}' = \begin{bmatrix} 3 & 2 \\ 7 & 5 \end{bmatrix} \mathbf{x} + \begin{bmatrix} 3 \\ 2t \end{bmatrix}. \tag{3}$$

Solution The nonhomogeneous term $\mathbf{f} = \begin{bmatrix} 3 & 2t \end{bmatrix}^T$ is linear, so it is reasonable to select a linear trial particular solution of the form

$$\mathbf{x}_p(t) = \mathbf{a}t + \mathbf{b} = \begin{bmatrix} a_1 \\ a_2 \end{bmatrix} t + \begin{bmatrix} b_1 \\ b_2 \end{bmatrix}. \tag{4}$$

Upon substitution of $\mathbf{x} = \mathbf{x}_p$ in Eq. (3), we get

$$\begin{bmatrix} a_1 \\ a_2 \end{bmatrix} = \begin{bmatrix} 3 & 2 \\ 7 & 5 \end{bmatrix} \begin{bmatrix} a_1 t + b_1 \\ a_2 t + b_2 \end{bmatrix} + \begin{bmatrix} 3 \\ 2t \end{bmatrix}$$

$$= \begin{bmatrix} 3a_1 + 2a_2 \\ 7a_1 + 5a_2 + 2 \end{bmatrix} t + \begin{bmatrix} 3b_1 + 2b_2 + 3 \\ 7b_1 + 5b_2 \end{bmatrix}.$$

We equate the coefficients of t and the constant terms (in both x_1- and x_2-components) and thereby obtain the equations

$$\begin{aligned} 3a_1 + 2a_2 &= 0, \\ 7a_1 + 5a_2 + 2 &= 0, \\ 3b_1 + 2b_2 + 3 &= a_1, \\ 7b_1 + 5b_2 &= a_2. \end{aligned} \tag{5}$$

We solve the first two equations in (5) for $a_1 = 4$ and $a_2 = -6$. With these values we can then solve the last two equations in (5) for $b_1 = 17$ and $b_2 = -25$. Substitution of these coefficients in Eq. (4) gives the particular solution $\mathbf{x} = \begin{bmatrix} x_1 & x_2 \end{bmatrix}^T$ of (3) described in scalar form by

$$\begin{aligned} x_1(t) &= 4t + 17, \\ x_2(t) &= -6t - 25. \end{aligned}$$ ∎

Example 2 Figure 5.6.1 shows the system of three brine tanks investigated in Example 2 of Section 5.2. The volumes of the three tanks are $V_1 = 20$, $V_2 = 40$, and $V_3 = 50$ (gal), and the common flow rate is $r = 10$ (gal/min). Suppose that all three tanks contain fresh water initially, but that the inflow to tank 1 is brine containing 2 pounds of salt per gallon, so that 20 pounds of salt flow into tank 1 per minute. Referring to Eq. (18) in Section 5.2, we see that the vector $\mathbf{x}(t) = \begin{bmatrix} x_1(t) & x_2(t) & x_3(t) \end{bmatrix}^T$ of amounts of salt (in pounds) in the three tanks at time t satisfies the nonhomogeneous initial value problem

$$\frac{d\mathbf{x}}{dt} = \begin{bmatrix} -0.5 & 0 & 0 \\ 0.5 & -0.25 & 0 \\ 0 & 0.25 & -0.2 \end{bmatrix} \mathbf{x} + \begin{bmatrix} 20 \\ 0 \\ 0 \end{bmatrix}, \quad \mathbf{x}(0) = \begin{bmatrix} 0 \\ 0 \\ 0 \end{bmatrix}. \tag{6}$$

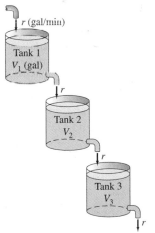
r (gal/min)

Tank 1
V_1 (gal)

Tank 2
V_2

Tank 3
V_3

FIGURE 5.6.1. The three brine tanks of Example 2.

The nonhomogeneous term $\mathbf{f} = \begin{bmatrix} 20 & 0 & 0 \end{bmatrix}^T$ here corresponds to the 20 lb/min inflow of salt to tank 1, with no (external) inflow of salt into tanks 2 and 3.

Because the nonhomogeneous term is constant, we naturally select a constant trial function $\mathbf{x}_p = \begin{bmatrix} a_1 & a_2 & a_3 \end{bmatrix}^T$, for which $\mathbf{x}'_p \equiv \mathbf{0}$. Then substitution of $\mathbf{x} = \mathbf{x}_p$ in (6) yields the system

$$\begin{bmatrix} 0 \\ 0 \\ 0 \end{bmatrix} = \begin{bmatrix} -0.5 & 0 & 0 \\ 0.5 & -0.25 & 0 \\ 0 & 0.25 & -0.2 \end{bmatrix} \begin{bmatrix} a_1 \\ a_2 \\ a_3 \end{bmatrix} + \begin{bmatrix} 20 \\ 0 \\ 0 \end{bmatrix}$$

that we readily solve for $a_1 = 40$, $a_2 = 80$, and $a_3 = 100$ in turn. Thus our particular solution is $\mathbf{x}_p(t) = \begin{bmatrix} 40 & 80 & 100 \end{bmatrix}^T$.

In Example 2 of Section 5.2 we found the general solution

$$\mathbf{x}_c(t) = c_1 \begin{bmatrix} 3 \\ -6 \\ 5 \end{bmatrix} e^{-t/2} + c_2 \begin{bmatrix} 0 \\ 1 \\ -5 \end{bmatrix} e^{-t/4} + c_3 \begin{bmatrix} 0 \\ 0 \\ 1 \end{bmatrix} e^{-t/5}$$

of the associated homogeneous system, so a general solution $\mathbf{x} = \mathbf{x}_c + \mathbf{x}_p$ of the nonhomogeneous system in (6) is given by

$$\mathbf{x}(t) = c_1 \begin{bmatrix} 3 \\ -6 \\ 5 \end{bmatrix} e^{-t/2} + c_2 \begin{bmatrix} 0 \\ 1 \\ -5 \end{bmatrix} e^{-t/4} + c_3 \begin{bmatrix} 0 \\ 0 \\ 1 \end{bmatrix} e^{-t/5} + \begin{bmatrix} 40 \\ 80 \\ 100 \end{bmatrix}. \tag{7}$$

When we apply the zero initial conditions in (6), we get the scalar equations

$$\begin{aligned} 3c_1 \quad\quad\quad\quad + 40 &= 0, \\ -6c_1 + c_2 \quad\quad + 80 &= 0, \\ 5c_1 - 5c_2 + c_3 + 100 &= 0 \end{aligned}$$

that are readily solved for $c_1 = -\frac{40}{3}$, $c_2 = -160$, and $c_3 = -\frac{2500}{3}$. Substituting these coefficients in Eq. (7), we find that the amounts of salt in the three tanks at time t are given by

$$\begin{aligned} x_1(t) &= 40 - 40e^{-t/2}, \\ x_2(t) &= 80 + 80e^{-t/2} - 160e^{-t/4}, \\ x_3(t) &= 100 + \tfrac{100}{3}\left(-2e^{-t/2} + 24e^{-t/4} - 25e^{-t/5}\right). \end{aligned} \tag{8}$$

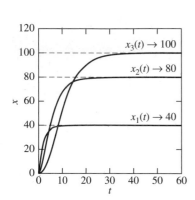

FIGURE 5.6.2. The salt amount solution curves defined in (8).

As illustrated in Fig. 5.6.2, we see the salt in each of the three tanks approaching, as $t \to +\infty$, a uniform density of 2 lb/gal—the same as the salt density in the inflow to tank 1. ∎

In the case of duplicate expressions in the complementary function and the nonhomogeneous terms, there is one difference between the method of undetermined coefficients for systems and for single equations (Rule 2 in Section 3.5). For a system, the usual first choice for a trial solution must be multiplied not only by the smallest integral power of t that will eliminate duplication, but also by all lower (nonnegative integral) powers of t as well, and all the resulting terms must be included in the trial solution.

Example 3 Consider the nonhomogeneous system

$$\mathbf{x}' = \begin{bmatrix} 4 & 2 \\ 3 & -1 \end{bmatrix} \mathbf{x} - \begin{bmatrix} 15 \\ 4 \end{bmatrix} te^{-2t}. \tag{9}$$

In Example 1 of Section 5.2 we found the solution

$$\mathbf{x}_c(t) = c_1 \begin{bmatrix} 1 \\ -3 \end{bmatrix} e^{-2t} + c_2 \begin{bmatrix} 2 \\ 1 \end{bmatrix} e^{5t} \tag{10}$$

of the associated homogeneous system. A preliminary trial solution $\mathbf{x}_p(t) = \mathbf{a}te^{-2t} + \mathbf{b}e^{-2t}$ exhibits duplication with the complementary function in (10). We would therefore select

$$\mathbf{x}_p(t) = \mathbf{a}t^2e^{-2t} + \mathbf{b}te^{-2t} + \mathbf{c}e^{-2t}$$

as our trial solution, and we would then have six scalar coefficients to determine. It is simpler to use the method of variation of parameters, our next topic. ∎

Variation of Parameters

Recall from Section 3.5 that the method of variation of parameters may be applied to a linear differential equation with variable coefficients and is not restricted to nonhomogeneous terms involving only polynomials, exponentials, and sinusoidal functions. The method of variation of parameters for systems enjoys the same flexibility and has a concise matrix formulation that is convenient for both practical and theoretical purposes.

We want to find a particular solution \mathbf{x}_p of the nonhomogeneous linear system

➤
$$\mathbf{x}' = \mathbf{P}(t)\mathbf{x} + \mathbf{f}(t), \tag{11}$$

given that we have already found a general solution

➤
$$\mathbf{x}_c(t) = c_1\mathbf{x}_1(t) + c_2\mathbf{x}_2(t) + \cdots + c_n\mathbf{x}_n(t) \tag{12}$$

of the associated homogeneous system

➤
$$\mathbf{x}' = \mathbf{P}(t)\mathbf{x}. \tag{13}$$

We first use the fundamental matrix $\mathbf{\Phi}(t)$ with column vectors $\mathbf{x}_1, \mathbf{x}_2, \ldots, \mathbf{x}_n$ to rewrite the complementary function in (12) as

$$\mathbf{x}_c(t) = \mathbf{\Phi}(t)\mathbf{c}. \tag{14}$$

Our idea is to replace the vector "parameter" \mathbf{c} with a variable vector $\mathbf{u}(t)$, and thus we seek a particular solution of the form

➤
$$\mathbf{x}_p(t) = \mathbf{\Phi}(t)\mathbf{u}(t). \tag{15}$$

We must determine $\mathbf{u}(t)$ so that \mathbf{x}_p does, indeed, satisfy Eq. (11).

The derivative of $\mathbf{x}_p(t)$ is (by the product rule)

$$\mathbf{x}_p'(t) = \mathbf{\Phi}'(t)\mathbf{u}(t) + \mathbf{\Phi}(t)\mathbf{u}'(t). \tag{16}$$

Hence substitution of Eqs. (15) and (16) in (11) yields

$$\mathbf{\Phi}'(t)\mathbf{u}(t) + \mathbf{\Phi}(t)\mathbf{u}'(t) = \mathbf{P}(t)\mathbf{\Phi}(t)\mathbf{u}(t) + \mathbf{f}(t). \tag{17}$$

But

$$\mathbf{\Phi}'(t) = \mathbf{P}(t)\mathbf{\Phi}(t) \tag{18}$$

because each column vector of $\mathbf{\Phi}(t)$ satisfies Eq. (13). Therefore, Eq. (17) reduces to

$$\mathbf{\Phi}(t)\mathbf{u}'(t) = \mathbf{f}(t). \tag{19}$$

Thus it suffices to choose $\mathbf{u}(t)$ so that

$$\mathbf{u}'(t) = \boldsymbol{\Phi}(t)^{-1}\mathbf{f}(t); \tag{20}$$

that is, so that

$$\mathbf{u}(t) = \int \boldsymbol{\Phi}(t)^{-1}\mathbf{f}(t)\,dt. \tag{21}$$

Upon substitution of (21) in (15), we finally obtain the desired particular solution, as stated in the following theorem.

THEOREM 1: Variation of Parameters

If $\boldsymbol{\Phi}(t)$ is a fundamental matrix for the homogeneous system $\mathbf{x}' = \mathbf{P}(t)\mathbf{x}$ on some interval where $\mathbf{P}(t)$ and $\mathbf{f}(t)$ are continuous, then a particular solution of the nonhomogeneous system

$$\mathbf{x}' = \mathbf{P}(t)\mathbf{x} + \mathbf{f}(t)$$

is given by

➤ $$\mathbf{x}_p(t) = \boldsymbol{\Phi}(t)\int \boldsymbol{\Phi}(t)^{-1}\mathbf{f}(t)\,dt. \tag{22}$$

▲

This is the **variation of parameters formula** for first-order linear systems. If we add this particular solution and the complementary function in (14), we get the general solution

➤ $$\mathbf{x}(t) = \boldsymbol{\Phi}(t)\mathbf{c} + \boldsymbol{\Phi}(t)\int \boldsymbol{\Phi}(t)^{-1}\mathbf{f}(t)\,dt \tag{23}$$

of the nonhomogeneous system in (11).

The choice of the constant of integration in Eq. (22) is immaterial, for we need only a single particular solution. In solving initial value problems it often is convenient to choose the constant of integration so that $\mathbf{x}_p(a) = \mathbf{0}$, and thus integrate from a to t:

$$\mathbf{x}_p(t) = \boldsymbol{\Phi}(t)\int_a^t \boldsymbol{\Phi}(s)^{-1}\mathbf{f}(s)\,ds. \tag{24}$$

If we add the particular solution of the nonhomogeneous problem

$$\mathbf{x}' = \mathbf{P}(t)\mathbf{x} + \mathbf{f}(t), \quad \mathbf{x}(a) = \mathbf{0}$$

in (24) to the solution $\mathbf{x}_c(t) = \boldsymbol{\Phi}(t)\boldsymbol{\Phi}(a)^{-1}\mathbf{x}_a$ of the associated homogeneous problem $\mathbf{x}' = \mathbf{P}(t)\mathbf{x}$, $\mathbf{x}(a) = \mathbf{x}_a$, we get the solution

$$\mathbf{x}(t) = \boldsymbol{\Phi}(t)\boldsymbol{\Phi}(a)^{-1}\mathbf{x}_a + \boldsymbol{\Phi}(t)\int_a^t \boldsymbol{\Phi}(s)^{-1}\mathbf{f}(s)\,ds \tag{25}$$

of the nonhomogeneous initial value problem

$$\mathbf{x}' = \mathbf{P}(t)\mathbf{x} + \mathbf{f}(t), \quad \mathbf{x}(a) = \mathbf{x}_a. \tag{26}$$

Equations (22) and (25) hold for any fundamental matrix $\Phi(t)$ of the homogeneous system $\mathbf{x}' = \mathbf{P}(t)\mathbf{x}$. In the constant-coefficient case $\mathbf{P}(t) \equiv \mathbf{A}$ we can use for $\Phi(t)$ the exponential matrix $e^{\mathbf{A}t}$—that is, the particular fundamental matrix such that $\Phi(0) = \mathbf{I}$. Then, because $(e^{\mathbf{A}t})^{-1} = e^{-\mathbf{A}t}$, substitution of $\Phi(t) = e^{\mathbf{A}t}$ in (22) yields the particular solution

$$\mathbf{x}_p(t) = e^{\mathbf{A}t} \int e^{-\mathbf{A}t}\mathbf{f}(t)\,dt \tag{27}$$

of the nonhomogeneous system $\mathbf{x}' = \mathbf{P}(t)\mathbf{x} + \mathbf{f}(t)$. Similarly, substitution of $\Phi(t) = e^{\mathbf{A}t}$ in Eq. (25) with $a = 0$ yields the solution

$$\blacktriangleright \qquad \mathbf{x}(t) = e^{\mathbf{A}t}\mathbf{x}_0 + e^{\mathbf{A}t} \int_0^t e^{-\mathbf{A}t}\mathbf{f}(t)\,dt \tag{28}$$

of the initial value problem

$$\blacktriangleright \qquad \mathbf{x}' = \mathbf{P}(t)\mathbf{x} + \mathbf{f}(t), \quad \mathbf{x}(0) = \mathbf{x}_0. \tag{29}$$

Remark: If we retain t as the independent variable but use s for the variable of integration, then the solutions in (27) and (28) can be rewritten in the forms

$$\mathbf{x}_p(t) = \int e^{-\mathbf{A}(s-t)}\mathbf{f}(s)\,ds \quad \text{and} \quad \mathbf{x}(t) = e^{\mathbf{A}t}\mathbf{x}_0 + \int_0^t e^{-\mathbf{A}(s-t)}\mathbf{f}(s)\,ds. \quad \blacksquare$$

Example 4 Solve the initial value problem

$$\mathbf{x}' = \begin{bmatrix} 4 & 2 \\ 3 & -1 \end{bmatrix}\mathbf{x} - \begin{bmatrix} 15 \\ 4 \end{bmatrix} te^{-2t}, \quad \mathbf{x}(0) = \begin{bmatrix} 7 \\ 3 \end{bmatrix}. \tag{30}$$

Solution The solution of the associated homogeneous system is displayed in Eq. (10). It gives the fundamental matrix

$$\Phi(t) = \begin{bmatrix} e^{-2t} & 2e^{5t} \\ -3e^{-2t} & e^{5t} \end{bmatrix} \quad \text{with} \quad \Phi(0)^{-1} = \tfrac{1}{7}\begin{bmatrix} 1 & -2 \\ 3 & 1 \end{bmatrix}.$$

It follows by Eq. (28) in Section 5.5 that the matrix exponential for the coefficient matrix \mathbf{A} in (30) is

$$e^{\mathbf{A}t} = \Phi(t)\Phi(0)^{-1} = \begin{bmatrix} e^{-2t} & 2e^{5t} \\ -3e^{-2t} & e^{5t} \end{bmatrix} \cdot \tfrac{1}{7}\begin{bmatrix} 1 & -2 \\ 3 & 1 \end{bmatrix}$$

$$= \tfrac{1}{7}\begin{bmatrix} e^{-2t} + 6e^{5t} & -2e^{-2t} + 2e^{5t} \\ -3e^{-2t} + 3e^{5t} & 6e^{-2t} + e^{5t} \end{bmatrix}.$$

Then the variation of parameters formula in Eq. (28) gives

$$e^{-\mathbf{A}t}\mathbf{x}(t) = \mathbf{x}_0 + \int_0^t e^{-\mathbf{A}s}\mathbf{f}(s)\,ds$$

$$= \begin{bmatrix} 7 \\ 3 \end{bmatrix} + \int_0^t \tfrac{1}{7}\begin{bmatrix} e^{2s} + 6e^{-5s} & -2e^{2s} + 2e^{-5s} \\ -3e^{2s} + 3e^{-5s} & 6e^{2s} + e^{-5s} \end{bmatrix}\begin{bmatrix} -15se^{-2s} \\ -4se^{-2s} \end{bmatrix}\,ds$$

$$= \begin{bmatrix} 7 \\ 3 \end{bmatrix} + \int_0^t \begin{bmatrix} -s - 14se^{-7s} \\ 3s - 7se^{-7s} \end{bmatrix}\,ds$$

$$= \begin{bmatrix} 7 \\ 3 \end{bmatrix} + \tfrac{1}{14}\begin{bmatrix} -4 - 7t^2 + 4e^{-7t} + 28te^{-7t} \\ -2 + 21t^2 + 2e^{-7t} + 14te^{-7t} \end{bmatrix}.$$

Therefore,

$$e^{-\mathbf{A}t}\mathbf{x}(t) = \tfrac{1}{14}\begin{bmatrix} 94 - 7t^2 + 4e^{-7t} + 28te^{-7t} \\ 40 + 21t^2 + 2e^{-7t} + 14te^{-7t} \end{bmatrix}.$$

Upon multiplication of the right-hand side here by $e^{\mathbf{A}t}$, we find that the solution of the initial value problem in (30) is given by

$$\mathbf{x}(t) = \tfrac{1}{7}\begin{bmatrix} e^{-2t} + 6e^{5t} & -2e^{-2t} + 2e^{5t} \\ -3e^{-2t} + 3e^{5t} & 6e^{-2t} + e^{5t} \end{bmatrix} \cdot \tfrac{1}{14}\begin{bmatrix} 94 - 7t^2 + 4e^{-7t} + 28te^{-7t} \\ 40 + 21t^2 + 2e^{-7t} + 14te^{-7t} \end{bmatrix}$$

$$= \tfrac{1}{14}\begin{bmatrix} (6 + 28t - 7t^2)e^{-2t} + 92e^{5t} \\ (-4 + 14t + 21t^2)e^{-2t} + 46e^{5t} \end{bmatrix}. \qquad\blacksquare$$

5.6 Problems

Apply the method of undetermined coefficients to find a particular solution of each of the systems in Problems 1 through 14. If initial conditions are given, find the particular solution that satisfies these conditions. Primes denote derivatives with respect to t.

1. $x' = x + 2y + 3$, $y' = 2x + y - 2$
2. $x' = 2x + 3y + 5$, $y' = 2x + y - 2t$
3. $x' = 3x + 4y$, $y' = 3x + 2y + t^2$; $x(0) = y(0) = 0$
4. $x' = 4x + y + e^t$, $y' = 6x - y - e^t$; $x(0) = y(0) = 1$
5. $x' = 6x - 7y + 10$, $y' = x - 2y - 2e^{-t}$
6. $x' = 9x + y + 2e^t$, $y' = -8x - 2y + te^t$
7. $x' = -3x + 4y + \sin t$, $y' = 6x - 5y$; $x(0) = 1$, $y(0) = 0$
8. $x' = x - 5y + 2\sin t$, $y' = x - y - 3\cos t$
9. $x' = x - 5y + \cos 2t$, $y' = x - y$
10. $x' = x - 2y$, $y' = 2x - y + e^t \sin t$
11. $x' = 2x + 4y + 2$, $y' = x + 2y + 3$; $x(0) = 1$, $y(0) = -1$
12. $x' = x + y + 2t$, $y' = x + y - 2t$
13. $x' = 2x + y + 2e^t$, $y' = x + 2y - 3e^t$
14. $x' = 2x + y + 1$, $y' = 4x + 2y + e^{4t}$

Problems 15 and 16 are similar to Example 2, but with two brine tanks (having volumes V_1 and V_2 gallons as in Fig. 5.6.2) instead of three tanks. Each tank initially contains fresh water, and the inflow to tank 1 at the rate of r gallons per minute has
a salt concentration of c_0 pounds per gallon. **(a)** *Find the amounts $x_1(t)$ and $x_2(t)$ of salt in the two tanks after t minutes.* **(b)** *Find the limiting (long-term) amount of salt in each tank.* **(c)** *Find how long it takes for each tank to reach a salt concentration of 1 lb/gal.*

15. $V_1 = 100$, $V_2 = 200$, $r = 10$, $c_0 = 2$
16. $V_1 = 200$, $V_2 = 100$, $r = 10$, $c_0 = 3$

In Problems 17 through 34, use the method of variation of parameters (and perhaps a computer algebra system) to solve the initial value problem

$$\mathbf{x}' = \mathbf{A}\mathbf{x} + \mathbf{f}(t), \qquad \mathbf{x}(a) = \mathbf{x}_a.$$

In each problem we provide the matrix exponential $e^{\mathbf{A}t}$ as provided by a computer algebra system.

17. $\mathbf{A} = \begin{bmatrix} 6 & -7 \\ 1 & -2 \end{bmatrix}$, $\mathbf{f}(t) = \begin{bmatrix} 60 \\ 90 \end{bmatrix}$, $\mathbf{x}(0) = \begin{bmatrix} 0 \\ 0 \end{bmatrix}$,

$$e^{\mathbf{A}t} = \tfrac{1}{6}\begin{bmatrix} -e^{-t} + 7e^{5t} & 7e^{-t} - 7e^{5t} \\ -e^{-t} + e^{5t} & 7e^{-t} - e^{5t} \end{bmatrix}$$

18. Repeat Problem 17, but with $\mathbf{f}(t)$ replaced with $\begin{bmatrix} 100t \\ -50t \end{bmatrix}$.

19. $A = \begin{bmatrix} 1 & 2 \\ 2 & -2 \end{bmatrix}$, $f(t) = \begin{bmatrix} 180t \\ 90 \end{bmatrix}$, $x(0) = \begin{bmatrix} 0 \\ 0 \end{bmatrix}$,

$e^{At} = \frac{1}{5} \begin{bmatrix} e^{-3t} + 4e^{2t} & -2e^{-3t} + 2e^{2t} \\ -2e^{-3t} + 2e^{2t} & 4e^{-3t} + e^{2t} \end{bmatrix}$

20. Repeat Problem 19, but with $f(t)$ replaced with $\begin{bmatrix} 75e^{2t} \\ 0 \end{bmatrix}$.

21. $A = \begin{bmatrix} 4 & -1 \\ 5 & -2 \end{bmatrix}$, $f(t) = \begin{bmatrix} 18e^{2t} \\ 30e^{2t} \end{bmatrix}$, $x(0) = \begin{bmatrix} 0 \\ 0 \end{bmatrix}$,

$e^{At} = \frac{1}{4} \begin{bmatrix} -e^{-t} + 5e^{3t} & e^{-t} - e^{3t} \\ -5e^{-t} + 5e^{3t} & 5e^{-t} - e^{3t} \end{bmatrix}$

22. Repeat Problem 21, but with $f(t)$ replaced with $\begin{bmatrix} 28e^{-t} \\ 20e^{3t} \end{bmatrix}$.

23. $A = \begin{bmatrix} 3 & -1 \\ 9 & -3 \end{bmatrix}$, $f(t) = \begin{bmatrix} 7 \\ 5 \end{bmatrix}$, $x(0) = \begin{bmatrix} 3 \\ 5 \end{bmatrix}$,

$e^{At} = \begin{bmatrix} 1 + 3t & -t \\ 9t & 1 - 3t \end{bmatrix}$

24. Repeat Problem 23, but with $f(t) = \begin{bmatrix} 0 \\ t^{-2} \end{bmatrix}$ and $x(1) = \begin{bmatrix} 3 \\ 7 \end{bmatrix}$.

25. $A = \begin{bmatrix} 2 & -5 \\ 1 & -2 \end{bmatrix}$, $f(t) = \begin{bmatrix} 4t \\ 1 \end{bmatrix}$, $x(0) = \begin{bmatrix} 0 \\ 0 \end{bmatrix}$,

$e^{At} = \begin{bmatrix} \cos t + 2\sin t & -5\sin t \\ \sin t & \cos t - 2\sin t \end{bmatrix}$

26. Repeat Problem 25, but with $f(t) = \begin{bmatrix} 4\cos t \\ 6\sin t \end{bmatrix}$ and $x(0) = \begin{bmatrix} 3 \\ 5 \end{bmatrix}$.

27. $A = \begin{bmatrix} 2 & -4 \\ 1 & -2 \end{bmatrix}$, $f(t) = \begin{bmatrix} 36t^2 \\ 6t \end{bmatrix}$, $x(0) = \begin{bmatrix} 0 \\ 0 \end{bmatrix}$,

$e^{At} = \begin{bmatrix} 1 + 2t & -4t \\ t & 1 - 2t \end{bmatrix}$

28. Repeat Problem 27, but with $f(t) = \begin{bmatrix} 4\ln t \\ t^{-1} \end{bmatrix}$ and $x(1) = \begin{bmatrix} 1 \\ -1 \end{bmatrix}$.

29. $A = \begin{bmatrix} 0 & -1 \\ 1 & 0 \end{bmatrix}$, $f(t) = \begin{bmatrix} \sec t \\ 0 \end{bmatrix}$, $x(0) = \begin{bmatrix} 0 \\ 0 \end{bmatrix}$,

$e^{At} = \begin{bmatrix} \cos t & -\sin t \\ \sin t & \cos t \end{bmatrix}$

30. $A = \begin{bmatrix} 0 & -2 \\ 2 & 0 \end{bmatrix}$, $f(t) = \begin{bmatrix} t\cos 2t \\ t\sin 2t \end{bmatrix}$, $x(0) = \begin{bmatrix} 0 \\ 0 \end{bmatrix}$,

$e^{At} = \begin{bmatrix} \cos 2t & -\sin 2t \\ \sin 2t & \cos 2t \end{bmatrix}$

31. $A = \begin{bmatrix} 1 & 2 & 3 \\ 0 & 1 & 2 \\ 0 & 0 & 1 \end{bmatrix}$, $f(t) = \begin{bmatrix} 0 \\ 0 \\ 6e^t \end{bmatrix}$, $x(0) = \begin{bmatrix} 0 \\ 0 \\ 0 \end{bmatrix}$,

$e^{At} = \begin{bmatrix} e^t & 2te^t & (3t + 2t^2)e^t \\ 0 & e^t & 2te^t \\ 0 & 0 & e^t \end{bmatrix}$

32. $A = \begin{bmatrix} 1 & 3 & 4 \\ 0 & 1 & 3 \\ 0 & 0 & 2 \end{bmatrix}$, $f(t) = \begin{bmatrix} 0 \\ 0 \\ 2e^{2t} \end{bmatrix}$, $x(0) = \begin{bmatrix} 0 \\ 0 \\ 0 \end{bmatrix}$,

$e^{At} = \begin{bmatrix} e^t & 3te^t & (-13 - 9t)e^t + 13e^{2t} \\ 0 & e^t & -3e^t + 3e^{2t} \\ 0 & 0 & e^{2t} \end{bmatrix}$

33. $A = \begin{bmatrix} 0 & 4 & 8 & 0 \\ 0 & 0 & 3 & 8 \\ 0 & 0 & 0 & 4 \\ 0 & 0 & 0 & 0 \end{bmatrix}$, $f(t) = 30\begin{bmatrix} t \\ t \\ t \\ t \end{bmatrix}$, $x(0) = \begin{bmatrix} 0 \\ 0 \\ 0 \\ 0 \end{bmatrix}$,

$e^{At} = \begin{bmatrix} 1 & 4t & 8t + 6t^2 & 32t^2 + 8t^3 \\ 0 & 1 & 3t & 8t + 6t^2 \\ 0 & 0 & 1 & 4t \\ 0 & 0 & 0 & 1 \end{bmatrix}$

34. $A = \begin{bmatrix} 0 & 4 & 8 & 0 \\ 0 & 0 & 0 & 8 \\ 0 & 0 & 2 & 4 \\ 0 & 0 & 0 & 2 \end{bmatrix}$, $f(t) = \begin{bmatrix} 0 \\ 6t \\ 0 \\ e^{2t} \end{bmatrix}$, $x(0) = \begin{bmatrix} 4 \\ 2 \\ 2 \\ 1 \end{bmatrix}$,

$e^{At} = \begin{bmatrix} 1 & 4t & 4(-1 + e^{2t}) & 16t(-1 + e^{2t}) \\ 0 & 1 & 0 & 4(-1 + e^{2t}) \\ 0 & 0 & e^{2t} & 4te^{2t} \\ 0 & 0 & 0 & e^{2t} \end{bmatrix}$

5.6 Application: Automated Variation of Parameters

The application of the variation of parameters formula in Eq. (28) encourages so mechanical an approach as to encourage especially the use of a computer algebra system. The following *Mathematica* commands were used to check the results in Example 4 of this section.

```
A = {{4,2}, {3,-1}};
x0 ={{7}, {3}};
f[t_] := {{-15 t Exp[-2t]},{-4 t Exp[-2t]}};
exp[A_] := MatrixExp[A]
x = exp[A*t].(x0 + Integrate[exp[-A*s].f[s], {s,0,t}])
```

The matrix exponential commands illustrated in the Section 5.5 application provide the basis for analogous *Maple* and MATLAB computations. You can then check routinely the answers for Problems 17 through 34 of this section.

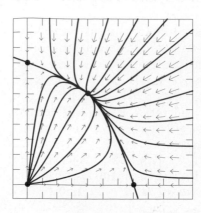

C H A P T E R 6

Nonlinear Systems and Phenomena

6.1 Stability and the Phase Plane

\mathbf{A} wide variety of natural phenomena are modeled by two-dimensional first-order systems of the form

$$\frac{dx}{dt} = F(x, y),$$

$$\frac{dy}{dt} = G(x, y) \tag{1}$$

in which the independent variable t does not appear explicitly. Such a system is called an **autonomous system.** The absence of t on the right-hand sides in (1) makes the system easier to analyze and its solutions easier to visualize. We assume that the functions F and G are continuously differentiable in some region R in the xy-plane, which is called the **phase plane** for the system in (1). Then, according to the existence and uniqueness theorems of the Appendix, given t_0 and any point (x_0, y_0) of R, there is a *unique* solution $x = x(t)$, $y = y(t)$ of (1) that is defined on some open interval (a, b) containing t_0 and satisfies the initial conditions

$$x(t_0) = x_0, \quad y(t_0) = y_0. \tag{2}$$

The equations $x = x(t)$, $y = y(t)$ then describe a parametrized solution curve in the phase plane. Any such solution curve is called a **trajectory** of the system in (1), and precisely one trajectory passes through each point of the region R (Problem 29). A **critical point** of the system in (1) is a point (x_\star, y_\star) such that

$$F(x_\star, y_\star) = G(x_\star, y_\star) = 0. \tag{3}$$

If (x_\star, y_\star) is a critical point of the system, then the constant-valued functions

$$x(t) \equiv x_\star, \quad y(t) \equiv y_\star \tag{4}$$

satisfy the equations in (1). Such a constant-valued solution is called an **equilibrium solution** of the system. Note that the trajectory of the equilibrium solution in (4) consists of the single point (x_\star, y_\star).

In some practical situations these very simple solutions and trajectories are the ones of greatest interest. For example, suppose that the system $x' = F(x, y)$, $y' = G(x, y)$ models two populations $x(t)$ and $y(t)$ of animals that cohabit the same environment, and perhaps compete for the same food or prey on one another; $x(t)$ might denote the number of rabbits and $y(t)$ the number of squirrels present at time t. Then a critical point (x_\star, y_\star) of the system specifies a *constant* population x_\star of rabbits and a *constant* population y_\star of squirrels that can coexist with one another in the environment. If (x_0, y_0) is *not* a critical point of the system, then it is *not* possible for constant populations of x_0 rabbits and y_0 squirrels to coexist; one or both must change with time.

Example 1 Find the critical points of the system

$$\frac{dx}{dt} = 14x - 2x^2 - xy,$$

$$\frac{dy}{dt} = 16y - 2y^2 - xy.$$

$$(5)$$

Solution When we look at the equations

$$14x - 2x^2 - xy - x(14 - 2x - y) = 0,$$
$$16y - 2y^2 - xy = y(16 - 2y - x) = 0$$

that a critical point (x, y) must satisfy, we see that either

$$x = 0 \quad \text{or} \quad 14 - 2x - y = 0, \qquad (6a)$$

and either

$$y = 0 \quad \text{or} \quad 16 - 2y - x = 0. \qquad (6b)$$

If $x = 0$ and $y \neq 0$, then the second equation in (6b) gives $y = 8$. If $y = 0$ and $x \neq 0$, then the second equation in (6a) gives $x = 7$. If both x and y are nonzero, then we solve the simultaneous equations

$$2x + y = 14, \quad x + 2y = 16$$

for $x = 4$, $y = 6$. Thus the system in (5) has the four critical points $(0, 0)$, $(0, 8)$, $(7, 0)$, and $(4, 6)$. If $x(t)$ and $y(t)$ denote the number of rabbits and the number of squirrels, respectively, and if both populations are *constant*, it follows that the equations in (5) allow only three nontrivial possibilities: either no rabbits and 8 squirrels, or 7 rabbits and no squirrels, or 4 rabbits and 6 squirrels. In particular, the critical point $(4, 6)$ describes the *only* possibility for the coexistence of constant nonzero populations of both species. ∎

Phase Portraits

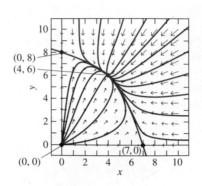

FIGURE 6.1.1. Direction field and phase portrait for the rabbit-squirrel system $x' = 14x - 2x^2 - xy$, $y' = 16y - 2y^2 - xy$ of Example 1.

If the initial point (x_0, y_0) is not a critical point, then the corresponding trajectory is a curve in the xy-plane along which the point $(x(t), y(t))$ moves as t increases. It turns out that any trajectory not consisting of a single point is a nondegenerate curve with no self-intersections (Problem 30). We can exhibit qualitatively the behavior of solutions of the system in (1) by sketching its **phase portrait**—a phase plane picture of its critical points and typical nondegenerate trajectories. We may construct a **slope field** by drawing typical line segments having slope

$$\frac{dy}{dx} = \frac{y'}{x'} = \frac{G(x, y)}{F(x, y)},$$

or a **direction field** by drawing typical vectors pointing in the same direction at each point as the vector $(F(x, y), G(x, y))$. Such a vector field then indicates which direction along a trajectory to travel in order to "go with the flow" described by the system.

Figure 6.1.1 shows a direction field and phase portrait for the rabbit-squirrel system of Example 1. The direction field arrows indicate the direction of motion of the point $(x(t), y(t))$. We see that, given any positive initial numbers $x_0 \neq 4$ and $y_0 \neq 6$ of rabbits and squirrels, this point moves along a trajectory approaching the critical point $(4, 6)$ as t increases.

Example 2 For the system

$$\begin{aligned} x' &= x - y, \\ y' &= 1 - x^2 \end{aligned} \tag{7}$$

we see from the first equation that $x = y$ and from the second that $x = \pm 1$ at each critical point. Thus this system has the two critical points $(-1, -1)$ and $(1, 1)$. The direction field in Fig. 6.1.2 suggests that trajectories somehow "circulate" counterclockwise around the critical point $(-1, -1)$, whereas it appears that some trajectories may approach, while others recede from, the critical point $(1, 1)$. These observations are corroborated by the phrase portrait in Fig. 6.1.3 for the system in (7). ∎

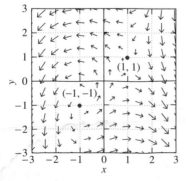

FIGURE 6.1.2. Direction field for the system in Eq. (7).

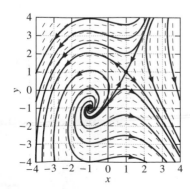

FIGURE 6.1.3. Phase portrait for the system in Eq. (7).

Critical Point Behavior

The behavior of the trajectories near an isolated critical point of an autonomous system is of particular interest. In the remainder of this section we illustrate with simple examples some of the most common possibilities.

Example 3 Consider the autonomous linear system

$$\frac{dx}{dt} = -x,$$

$$\frac{dy}{dt} = -ky \quad (k \text{ a nonzero constant}),$$

(8)

which has the origin $(0, 0)$ as its only critical point. The solution with initial point (x_0, y_0) is

$$x(t) = x_0 e^{-t}, \quad y(t) = y_0 e^{-kt}.$$

(9)

If $x_0 \neq 0$, we can write

$$y = y_0 e^{-kt} = \frac{y_0}{x_0^k}(x_0 e^{-t})^k = bx^k,$$

(10)

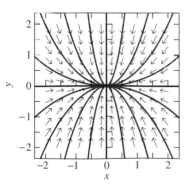

FIGURE 6.1.4. A proper node; the trajectories approach the origin, so it is a nodal sink.

where $b = y_0/x_0^k$. The nature of the critical point $(0, 0)$ depends on whether the nonzero parameter k is positive or negative.

CASE 1: $k > 0$. If k is positive, then we see from (9) that the point $(x(t), y(t))$ approaches the origin along the curve $y = bx^k$ as $t \to +\infty$. The appearance of this curve depends on the magnitude of k:

- If $k = 1$, then $y = bx$ with $b = y_0/x_0$ is a straight line through the point (x_0, y_0). These straight-line trajectories are illustrated by the phase portrait in Fig. 6.1.4.
- If $k > 1$ and neither x_0 nor y_0 in Eq. (10) is zero, then the curve $y = bx^k$ is tangent to the x-axis at the origin. This case is illustrated by the phase portrait in Fig. 6.1.5, where $k = 2$ and the trajectories are parabolas. More precisely, the trajectories are the *semi*axes and the right and left *halves* of these parabolas.
- If $0 < k < 1$ and neither x_0 nor y_0 is zero, then the phase portrait is similar to Fig. 6.1.5, except that each curve $y = bx^k$ is tangent to the y-axis (rather than to the x-axis) at the origin.

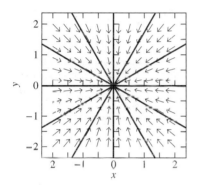

FIGURE 6.1.5. An improper node because all trajectories are tangent to a single line; they approach the origin, so it is a nodal sink.

The type of critical point illustrated in Figs. 6.1.4 and 6.1.5 is called a *node*. In general, the critical point (x_\star, y_\star) of the autonomous system in (1) is called a **node** provided that

- *Either* every trajectory approaches (x_\star, y_\star) as $t \to +\infty$ *or* every trajectory recedes from (x_\star, y_\star) as $t \to +\infty$, *and*
- Every trajectory is tangent at (x_\star, y_\star) to some straight line through the critical point.

A node is said to be **proper** provided that no two different pairs of "opposite" trajectories are tangent to the same straight line through the critical point. This is the situation in Fig. 6.1.4 (in which the trajectories *are* straight lines, not merely tangent to straight lines). A proper node might be called a "star point." In Fig. 6.1.5 all trajectories except for a single opposite pair are tangent to a single straight line through the critical point. This type of node is said to be **improper.**

A node is also called a **sink** if all trajectories approach the critical point, a **source** if all trajectories recede (or emanate) from it. Thus the origin in Fig. 6.1.4 is a *proper nodal sink,* whereas in Fig. 6.1.5 it is an *improper nodal sink.* If the direction field arrows in each figure were reversed, then the origin would be a *nodal source* in each figure.

CASE 2: $k < 0$. If k is negative, then the trajectories resemble those for the case $k = -1$, which is illustrated in Fig. 6.1.6. If neither x_0 nor y_0 is zero, then the corresponding trajectory in Fig. 6.1.6 is one branch of the rectangular hyperbola $xy = b$, and $|y(t)| \to +\infty$ as $t \to +\infty$. If either x_0 or y_0 *is* zero, then the trajectory is a semiaxis of the hyperbola. The point $(x(t), y(t))$ approaches the origin along the x-axis, but recedes from it along the y-axis, as $t \to +\infty$. Thus there are two trajectories that approach the critical point $(0, 0)$, but all others are unbounded as $t \to +\infty$. This type of critical point, illustrated in Fig. 6.1.6, is called a **saddle point.** ∎

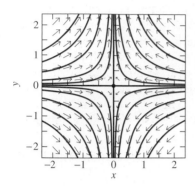

FIGURE 6.1.6. A saddle point with trajectories resembling the contour curves of a saddle point on a surface.

Stability

A critical point (x_\star, y_\star) of the autonomous system in (1) is said to be *stable* provided that if the initial point (x_0, y_0) is sufficiently close to (x_\star, y_\star), then $(x(t), y(t))$ remains close to (x_\star, y_\star) for all $t > 0$. In vector notation, with $\mathbf{x}(t) = (x(t), y(t))$, the distance between the initial point $\mathbf{x}_0 = (x_0, y_0)$ and the critical point $\mathbf{x}_\star = (x_\star, y_\star)$ is

$$|\mathbf{x}_0 - \mathbf{x}_\star| = \sqrt{(x_0 - x_\star)^2 + (y_0 - y_\star)^2}.$$

Thus the critical point \mathbf{x}_\star is **stable** provided that, for each $\epsilon > 0$, there exists $\delta > 0$ such that

➤ $$|\mathbf{x}_0 - \mathbf{x}_\star| < \delta \quad \text{implies that} \quad |\mathbf{x}(t) - \mathbf{x}_\star| < \epsilon \qquad (11)$$

for all $t > 0$. Note that the condition in (11) certainly holds if $\mathbf{x}(t) \to \mathbf{x}_\star$ as $t \to +\infty$, as in the case of a nodal sink. Thus the nodal sinks illustrated in Figs. 6.1.4 and 6.1.5 can also be described as *stable nodes.*

The critical point (x_\star, y_\star) is called **unstable** if it is not stable. The saddle point at $(0, 0)$ in Fig. 6.1.6 is an unstable critical point because the point $(x(t), y(t))$ goes to infinity as $t \to +\infty$ (assuming that $x_0 y_0 \neq 0$), and hence the condition in (11) is not satisfied.

Example 3

Continued

If the signs on the right-hand side in (8) are changed to obtain the system

$$\frac{dx}{dt} = x,$$

$$\frac{dy}{dt} = ky \quad (k \text{ a nonzero constant}), \qquad (12)$$

then the solution is $x(t) = x_0 e^t$, $y(t) = y_0 e^{kt}$. Then with $k = 1$ and $k = 2$, the trajectories are the same as those shown in Figs. 6.1.4 and 6.1.5, respectively, but

with the arrows reversed, so that the point $(x(t), y(t))$ goes to infinity as $t \to \infty$. The result in each case is a nodal source—that is, an *unstable node*—at $(0, 0)$. ∎

If (x_\star, y_\star) is a critical point, then the equilibrium solution $x(t) \equiv x_\star$, $y(t) \equiv y_\star$ is called **stable** or **unstable** depending on the nature of the critical point. In applications the stability of an equilibrium solution is often a crucial matter. For instance, suppose in Example 1 that $x(t)$ and $y(t)$ denote the rabbit and squirrel populations, respectively, *in hundreds.* We will see in Section 6.3 that the critical point $(4, 6)$ in Fig. 6.1.1 is stable. It follows that if we begin with *close to* 400 rabbits and 600 squirrels—rather than exactly these equilibrium values—then for all future time there will remain close to 400 rabbits and close to 600 squirrels. Thus the practical consequence of stability is that slight changes (perhaps due to random births and deaths) in the equilibrium populations will not so upset the equilibrium as to result in large deviations from the equilibrium solutions.

It is possible for trajectories to remain near a stable critical point without approaching it, as Example 4 shows.

Example 4 Consider a mass m that oscillates without damping on a spring with Hooke's constant k, so that its position function $x(t)$ satisfies the differential equation $x'' + \omega^2 x = 0$ (where $\omega^2 = k/m$). If we introduce the velocity $y = dx/dt$ of the mass, we get the system

$$\frac{dx}{dt} = y,$$

$$\frac{dy}{dt} = -\omega^2 x \qquad (13)$$

with general solution

$$x(t) = \quad A \cos \omega t + \quad B \sin \omega t, \qquad (14a)$$

$$y(t) = -A\omega \sin \omega t + B\omega \cos \omega t. \qquad (14b)$$

With $C = \sqrt{A^2 + B^2}$, $A = C \cos \alpha$, and $B = C \sin \alpha$, we can rewrite the solution in (14) in the form

$$x(t) = \quad C \cos(\omega t - \alpha), \qquad (15a)$$

$$y(t) = -\omega C \sin(\omega t - \alpha), \qquad (15b)$$

so it becomes clear that each trajectory other than the critical point $(0, 0)$ is an ellipse with equation of the form

$$\frac{x^2}{C^2} + \frac{y^2}{\omega^2 C^2} = 1. \qquad (16)$$

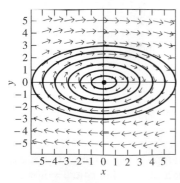

FIGURE 6.1.7. Direction field and elliptical trajectories for the system $x' = y$, $y' = -\frac{1}{4}x$. The origin is a stable center.

As illustrated by the phase portrait in Fig. 6.1.7 (where $\omega = \frac{1}{2}$), each point (x_0, y_0) other than the origin in the xy-plane lies on exactly one of these ellipses, and each solution $(x(t), y(t))$ traverses the ellipse through its initial point (x_0, y_0) in the clockwise direction with *period* $P = 2\pi/\omega$. (It is clear from (15) that $x(t + P) = x(t)$ and $y(t + P) = y(t)$ for all t.) Thus each nontrivial solution of the system in (13) is periodic and its trajectory is a simple closed curve enclosing the critical point at the origin. ∎

FIGURE 6.1.8. If the initial point (x_0, y_0) lies within distance δ of the origin, then the point $(x(t), y(t))$ stays within distance ϵ of the origin.

Figure 6.1.8 shows a typical elliptical trajectory in Example 4, with its minor semiaxis denoted by δ and its major semiaxis by ϵ. We see that if the initial point (x_0, y_0) lies within distance δ of the origin—so that its elliptical trajectory lies inside the one shown—then the point $(x(t), y(t))$ always remains within distance ϵ of the origin. Hence the origin $(0, 0)$ is a stable critical point of the system $x' = y$, $y' = -\omega^2 x$. Unlike the situation illustrated in Figs. 6.1.4 and 6.1.5, however, no single trajectory approaches the point $(0, 0)$. A stable critical point surrounded by simple closed trajectories representing periodic solutions is called a **(stable) center.**

Asymptotic Stability

The critical point (x_\star, y_\star) is called **asymptotically stable** if it is stable and, moreover, every trajectory that begins sufficiently close to (x_\star, y_\star) also approaches (x_\star, y_\star) as $t \to +\infty$. That is, there exists $\delta > 0$ such that

➤ $$|\mathbf{x} - \mathbf{x}_\star| < \delta \quad \text{implies that} \quad \lim_{t \to \infty} \mathbf{x}(t) = \mathbf{x}_\star, \qquad (17)$$

where $\mathbf{x}_0 = (x_0, y_0)$, $\mathbf{x}_\star = (x_\star, y_\star)$, and $\mathbf{x}(t) = (x(t), y(t))$ is a solution with $\mathbf{x}(0) = \mathbf{x}_0$.

Remark: The stable nodes shown in Figs. 6.1.4 and 6.1.5 are asymptotically stable because every trajectory approaches the critical point $(0, 0)$ as $t \to +\infty$. The center $(0, 0)$ shown in Fig. 6.1.7 is stable but not asymptotically stable, because however small an elliptical trajectory we consider, a point moving around this ellipse does not approach the origin. Thus asymptotic stability is a stronger condition than mere stability. ∎

Now suppose that $x(t)$ and $y(t)$ denote coexisting populations for which (x_\star, y_\star) is an asymptotically stable critical point. Then if the initial populations x_0 and y_0 are sufficiently close to x_\star and y_\star, respectively, it follows that both

$$\lim_{t \to \infty} x(t) = x_\star \quad \text{and} \quad \lim_{t \to \infty} y(t) = y_\star. \qquad (18)$$

That is, $x(t)$ and $y(t)$ actually approach the equilibrium populations x_\star and y_\star as $t \to +\infty$, rather than merely remaining close to those values.

For a mechanical system as in Example 4, a critical point represents an *equilibrium state* of the system—if the velocity $y = x'$ and the acceleration $y' = x''$ vanish simultaneously, then the mass remains at rest with no net force acting on it. Stability of a critical point concerns the question whether, when the mass is displaced slightly from its equilibrium, it

1. Moves back toward the equilibrium point as $t \to +\infty$,
2. Merely remains near the equilibrium point without approaching it, or
3. Moves farther away from equilibrium.

In Case 1 the critical [equilibrium] point is asymptotically stable; in Case 2 it is stable but not asymptotically so; in Case 3 it is an unstable critical point. A marble balanced on the top of a soccer ball is an example of an unstable critical point. A mass on a spring with damping illustrates the case of asymptotic stability of a mechanical system. The mass-and-spring without damping in Example 4 is an example of a system that is stable but not asymptotically stable.

Example 5

Suppose that $m = 1$ and $k = 2$ for the mass and spring of Example 4 and that the mass is attached also to a dashpot with damping constant $c = 2$. Then its displacement function $x(t)$ satisfies the second-order equation

$$x''(t) + 2x'(t) + 2x(t) = 0. \tag{19}$$

With $y = x'$ we obtain the equivalent first-order system

$$\frac{dx}{dt} = y,$$
$$\frac{dy}{dt} = -2x - 2y \tag{20}$$

with critical point $(0, 0)$. The characteristic equation $r^2 + 2r + 2 = 0$ of Eq. (19) has roots $-1 + i$ and $-1 - i$, so the general solution of the system in (20) is given by

$$x(t) = e^{-t}(A \cos t + B \sin t) = Ce^{-t} \cos(t - \alpha), \tag{21a}$$

$$y(t) = e^{-t}[(B - A) \cos t - (A + B) \sin t]$$
$$= -C\sqrt{2}e^{-t} \sin\left(t - \alpha + \tfrac{1}{4}\pi\right), \tag{21b}$$

FIGURE 6.1.9. A stable spiral point and one nearby trajectory.

where $C = \sqrt{A^2 + B^2}$ and $\alpha = \tan^{-1}(B/A)$. We see that $x(t)$ and $y(t)$ oscillate between positive and negative values and that both approach zero as $t \to +\infty$. Thus a typical trajectory spirals inward toward the origin, as illustrated by the spiral in Fig. 6.1.9. ∎

It is clear from (21) that the point $(x(t), y(t))$ approaches the origin as $t \to +\infty$, so it follows that $(0, 0)$ is an asymptotically stable critical point for the system $x' = y$, $y' = -2x - 2y$ of Example 5. Such an asymptotically stable critical point—around which the trajectories spiral as they approach it—is called a **stable spiral point** (or a **spiral sink**). In the case of a mass-spring-dashpot system, a spiral sink is the manifestation in the phase plane of the damped oscillations that occur because of resistance.

If the arrows in Fig. 6.1.9 were reversed, we would see a trajectory spiraling *outward* from the origin. An unstable critical point—around which the trajectories spiral as they emanate and recede from it—is called an **unstable spiral point** (or a **spiral source**). Example 6 shows that it also is possible for a trajectory to spiral into a **closed trajectory**—a simple closed solution curve that represents a periodic solution (like the elliptical trajectories in Fig. 6.1.7).

Example 6

Consider the system

$$\frac{dx}{dt} = -ky + x(1 - x^2 - y^2),$$
$$\frac{dy}{dt} = kx + y(1 - x^2 - y^2). \tag{22}$$

In Problem 21 we ask you to show that $(0, 0)$ is its only critical point. This system can be solved explicitly by introducing polar coordinates $x = r \cos\theta$, $y = r \sin\theta$, as follows. First note that

$$\frac{d\theta}{dt} = \frac{d}{dt}\left(\arctan\frac{y}{x}\right) = \frac{xy' - x'y}{x^2 + y^2}.$$

Then substitute the expressions given in (22) for x' and y' to obtain

$$\frac{d\theta}{dt} = \frac{k(x^2 + y^2)}{x^2 + y^2} = k.$$

It follows that

$$\theta(t) = kt + \theta_0, \qquad \text{where } \theta_0 = \theta(0). \tag{23}$$

Then differentiation of $r^2 = x^2 + y^2$ yields

$$2r\frac{dr}{dt} = 2x\frac{dx}{dt} + 2y\frac{dy}{dt}$$

$$= 2(x^2 + y^2)(1 - x^2 - y^2) = 2r^2(1 - r^2),$$

so $r = r(t)$ satisfies the differential equation

$$\frac{dr}{dt} = r(1 - r^2). \tag{24}$$

In Problem 22 we ask you to derive the solution

$$r(t) = \frac{r_0}{\sqrt{r_0^2 + (1 - r_0^2)e^{-2t}}}, \tag{25}$$

where $r_0 = r(0)$. Thus the typical solution of Eq. (22) may be expressed in the form

$$\begin{aligned} x(t) &= r(t)\cos(kt + \theta_0), \\ y(t) &= r(t)\sin(kt + \theta_0). \end{aligned} \tag{26}$$

FIGURE 6.1.10. Spiral trajectories of the system in Eq. (22) with $k = 5$.

If $r_0 = 1$, then Eq. (25) gives $r(t) \equiv 1$ (the unit circle). Otherwise, if $r_0 > 0$, then Eq. (25) implies that $r(t) \to 1$ as $t \to +\infty$. Hence the trajectory defined in (26) spirals in toward the unit circle if $r_0 > 1$ and spirals out toward this closed trajectory if $0 < r_0 < 1$. Figure 6.1.10 shows a trajectory spiraling outward from the origin and four trajectories spiraling inward, all approaching the closed trajectory $r(t) \equiv 1$. ∎

Under rather general hypotheses it can be shown that there are four possibilities for a nondegenerate trajectory of the autonomous system

$$\frac{dx}{dt} = F(x, y), \qquad \frac{dy}{dt} = G(x, y).$$

The four possibilities are these:

1. $(x(t), y(t))$ approaches a critical point as $t \to +\infty$.
2. $(x(t), y(t))$ is unbounded with increasing t.
3. $(x(t), y(t))$ is a periodic solution with a closed trajectory.
4. $(x(t), y(t))$ spirals toward a closed trajectory as $t \to +\infty$.

As a consequence, the qualitative nature of the phase plane picture of the trajectories of an autonomous system is determined largely by the locations of its critical points and by the behavior of its trajectories near its critical points. We will see in Section 6.2 that, subject to mild restrictions on the functions F and G, each isolated critical point of the system $x' = F(x, y)$, $y' = G(x, y)$ resembles qualitatively one of the examples of this section—it is either a node (proper or improper), a saddle point, a center, or a spiral point.

6.1 Problems

In Problems 1 through 8, find the critical point or points of the given autonomous system, and thereby match each system with its phase portrait among Figs. 6.1.11 through 6.1.18.

1. $\dfrac{dx}{dt} = 2x - y, \qquad \dfrac{dy}{dt} = x - 3y$

2. $\dfrac{dx}{dt} = x - y, \qquad \dfrac{dy}{dt} = x + 3y - 4$

3. $\dfrac{dx}{dt} = x - 2y + 3, \qquad \dfrac{dy}{dt} = x - y + 2$

4. $\dfrac{dx}{dt} = 2x - 2y - 4, \qquad \dfrac{dy}{dt} = x + 4y + 3$

5. $\dfrac{dx}{dt} = 1 - y^2, \qquad \dfrac{dy}{dt} = x + 2y$

FIGURE 6.1.11. Spiral point $(-2, 1)$ and saddle point $(2, -1)$.

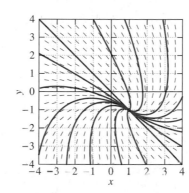

FIGURE 6.1.12. Spiral point $(1, -1)$.

FIGURE 6.1.13. Saddle point $(0, 0)$.

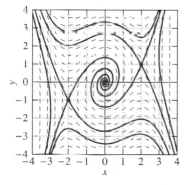

FIGURE 6.1.14. Spiral point $(0, 0)$; saddle points $(-2, -1)$ and $(2, 1)$.

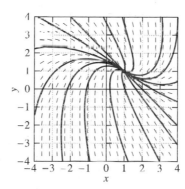

FIGURE 6.1.15. Node $(1, 1)$.

FIGURE 6.1.16. Spiral point $(-1, -1)$, saddle point $(0, 0)$, and node $(1, -1)$.

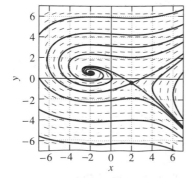

FIGURE 6.1.17. Spiral point $\left(-2, \frac{2}{3}\right)$ and saddle point $\left(2, -\frac{2}{5}\right)$.

FIGURE 6.1.18. Stable center $(-1, 1)$.

6. $\dfrac{dx}{dt} = 2 - 4x - 15y, \quad \dfrac{dy}{dt} = 4 - x^2$

7. $\dfrac{dx}{dt} = x - 2y, \quad \dfrac{dy}{dt} = 4x - x^3$

8. $\dfrac{dx}{dt} = x - y - x^2 + xy, \quad \dfrac{dy}{dt} = -y - x^2$

In Problems 9 through 12, find each equilibrium solution $x(t) \equiv x_0$ of the given second-order differential equation $x'' + f(x, x') = 0$. Use a computer system or graphing calculator to construct a phase portrait and direction field for the equivalent first-order system $x' = y$, $y' = -f(x, y)$. Thereby ascertain whether the critical point $(x_0, 0)$ looks like a center, a saddle point, or a spiral point of this system.

9. $x'' + 4x - x^3 = 0$
10. $x'' + 2x' + x + 4x^3 = 0$
11. $x'' + 3x' + 4\sin x = 0$
12. $x'' + (x^2 - 1)x' + x = 0$

Solve each of the linear systems in Problems 13 through 20 to determine whether the critical point $(0, 0)$ is stable, asymptotically stable, or unstable. Use a computer system or graphing calculator to construct a phase portrait and direction field for the given system. Thereby ascertain the stability or instability of each critical point, and identify it visually as a node, a saddle point, a center, or a spiral point.

13. $\dfrac{dx}{dt} = -2x, \quad \dfrac{dy}{dt} = -2y$

14. $\dfrac{dx}{dt} = 2x, \quad \dfrac{dy}{dt} = -2y$

15. $\dfrac{dx}{dt} = -2x, \quad \dfrac{dy}{dt} = -y$

16. $\dfrac{dx}{dt} = x, \quad \dfrac{dy}{dt} = 3y$

17. $\dfrac{dx}{dt} = y, \quad \dfrac{dy}{dt} = -x$

18. $\dfrac{dx}{dt} = -y, \quad \dfrac{dy}{dt} = 4x$

19. $\dfrac{dx}{dt} = 2y, \quad \dfrac{dy}{dt} = -2x$

20. $\dfrac{dx}{dt} = y, \quad \dfrac{dy}{dt} = -5x - 4y$

21. Verify that $(0, 0)$ is the only critical point of the system in Example 6.

22. Separate variables in Eq. (24) to derive the solution in (25).

In Problems 23 through 26, a system $dx/dt = F(x, y)$, $dy/dt = G(x, y)$ is given. Solve the equation

$$\frac{dy}{dx} = \frac{G(x, y)}{F(x, y)}$$

to find the trajectories of the given system. Use a computer system or graphing calculator to construct a phase portrait and

direction field for the system, and thereby identify visually the apparent character and stability of the critical point $(0, 0)$ of the given system.

23. $\dfrac{dx}{dt} = y, \quad \dfrac{dy}{dt} = -x$

24. $\dfrac{dx}{dt} = y(1 + x^2 + y^2), \quad \dfrac{dy}{dt} = x(1 + x^2 + y^2)$

25. $\dfrac{dx}{dt} = 4y(1 + x^2 + y^2), \quad \dfrac{dy}{dt} = -x(1 + x^2 + y^2)$

26. $\dfrac{dx}{dt} = y^3 e^{x+y}, \quad \dfrac{dy}{dt} = -x^3 e^{x+y}$

27. Let $(x(t), y(t))$ be a nontrivial solution of the nonautonomous system

$$\frac{dx}{dt} = y, \quad \frac{dy}{dt} = tx.$$

Suppose that $\phi(t) = x(t + \gamma)$ and $\psi(t) = y(t + \gamma)$, where $\gamma \neq 0$. Show that $(\phi(t), \psi(t))$ is *not* a solution of the system.

Problems 28 through 30 deal with the system

$$\frac{dx}{dt} = F(x, y), \quad \frac{dy}{dt} = G(x, y)$$

in a region where the functions F and G are continuously differentiable, so for each number a and point (x_0, y_0), there is a unique solution with $x(a) = x_0$ and $y(a) = y_0$.

28. Suppose that $(x(t), y(t))$ is a solution of the autonomous system and that $\gamma \neq 0$. Define $\phi(t) = x(t + \gamma)$ and $\psi(t) = y(t + \gamma)$. Then show (in contrast with the situation in Problem 27) that $(\phi(t), \psi(t))$ is also a solution of the system. Thus autonomous systems have the simple but important property that a "t-translate" of a solution is again a solution.

29. Let $(x_1(t), y_1(t))$ and $(x_2(t), y_2(t))$ be two solutions having trajectories that meet at the point (x_0, y_0); thus $x_1(a) = x_2(b) = x_0$ and $y_1(a) = y_2(b) = y_0$ for some values a and b of t. Define

$$x_3(t) = x_2(t + \gamma) \quad \text{and} \quad y_3(t) = y_2(t + \gamma),$$

where $\gamma = b - a$, so $(x_2(t), y_2(t))$ and $(x_3(t), y_3(t))$ have the same trajectory. Apply the uniqueness theorem to show that $(x_1(t), y_1(t))$ and $(x_3(t), y_3(t))$ are identical solutions. Hence the original two trajectories are identical. Thus no two different trajectories of an autonomous system can intersect.

30. Suppose that the solution $(x_1(t), y_1(t))$ is defined for all t and that its trajectory has an apparent self-intersection:

$$x_1(a) = x_1(a + P) = x_0, \quad y_1(a) = y_1(a + P) = y_0$$

for some $P > 0$. Introduce the solution

$$x_2(t) = x_1(t + P), \quad y_2(t) = y_1(t + P)$$

and then apply the uniqueness theorem to show that

$$x_1(t + P) = x_1(t) \quad \text{and} \quad y_1(t) = y_1(t + P)$$

for *all t*. Thus the solution $(x_1(t), y_1(t))$ is periodic with period P and has a closed trajectory. Consequently a solution of an autonomous system either is periodic with a closed trajectory, or else its trajectory never passes through the same point twice.

6.1 Application: Phase Plane Portraits and First-Order Equations

Consider a first-order differential equation of the form

$$\frac{dy}{dx} = \frac{G(x, y)}{F(x, y)}, \tag{1}$$

which may be difficult or impossible to solve explicitly. Its solution curves can nevertheless be plotted as trajectories of the corresponding autonomous two-dimensional system

$$\frac{dx}{dt} = F(x, y), \quad \frac{dy}{dt} = G(x, y). \tag{2}$$

Most ODE plotters can routinely generate phase portraits for autonomous systems. Many of those appearing in this chapter were plotted using (as illustrated in Fig. 6.1.19) John Polking's MATLAB-based `pplane` program that is available free for educational use (`math.rice.edu/~dfield`). Another freely available and user-friendly MATLAB-based ODE package with similar graphical capabilities is `Iode` (`www.math.uiuc.edu/iode`).

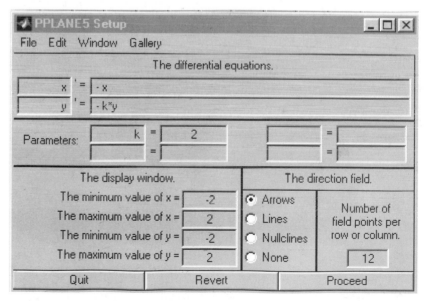

FIGURE 6.1.19. MATLAB `pplane` menu entries to plot a direction field and phase portrait for the system $x' = -x$, $y' = -2y$ (as shown in Fig. 6.1.5).

For example, to plot solution curves for the differential equation

$$\frac{dy}{dx} = \frac{2xy - y^2}{x^2 - 2xy}, \tag{3}$$

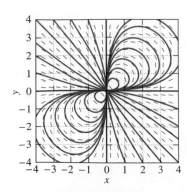

FIGURE 6.1.20. Phase portrait for the system in Eq. (4).

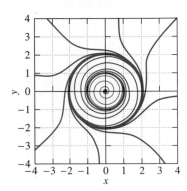

FIGURE 6.1.21. Phase portrait for the system corresponding to Eq. (5).

we plot trajectories of the system

$$\frac{dx}{dt} = x^2 - 2xy, \qquad \frac{dy}{dt} = 2xy - y^2. \qquad (4)$$

The result is shown in Fig. 6.1.20.

Plot similarly some solution curves for the following differential equations.

1. $\dfrac{dy}{dx} = \dfrac{4x - 5y}{2x + 3y}$

2. $\dfrac{dy}{dx} = \dfrac{4x - 5y}{2x - 3y}$

3. $\dfrac{dy}{dx} = \dfrac{4x - 3y}{2x - 5y}$

4. $\dfrac{dy}{dx} = \dfrac{2xy}{x^2 - y^2}$

5. $\dfrac{dy}{dx} = \dfrac{x^2 + 2xy}{y^2 + 2xy}$

Now construct some examples of your own. Homogeneous functions like those in Problems 1 through 5—rational functions with numerator and denominator of the same degree in x and y—work well. The differential equation

$$\frac{dy}{dx} = \frac{25x + y(1 - x^2 - y^2)(4 - x^2 - y^2)}{-25y + x(1 - x^2 - y^2)(4 - x^2 - y^2)} \qquad (5)$$

of this form generalizes Example 6 in this section but would be inconvenient to solve explicitly. Its phase portrait (Fig. 6.1.21) shows two periodic closed trajectories—the circles $r = 1$ and $r = 2$. Anyone want to try for three circles?

6.2 Linear and Almost Linear Systems

We now discuss the behavior of solutions of the autonomous system

$$\frac{dx}{dt} = f(x, y), \qquad \frac{dy}{dt} = g(x, y) \qquad (1)$$

near an isolated critical point (x_0, y_0) at which $f(x_0, y_0) = g(x_0, y_0) = 0$. A critical point is called **isolated** if some neighborhood of it contains no other critical point. We assume throughout that the functions f and g are continuously differentiable in a neighborhood of (x_0, y_0).

We can assume without loss of generality that $x_0 = y_0 = 0$. Otherwise, we make the substitutions $u = x - x_0$, $v = y - y_0$. Then $dx/dt = du/dt$ and $dy/dt = dv/dt$, so (1) is equivalent to the system

$$\frac{du}{dt} = f(u + x_0, v + y_0) = f_1(u, v),$$

$$\frac{dv}{dt} = g(u + x_0, v + y_0) = g_1(u, v) \qquad (2)$$

that has $(0, 0)$ as an isolated critical point.

Example 1 The system

$$\frac{dx}{dt} = 3x - x^2 - xy = x(3 - x - y),$$

$$\frac{dy}{dt} = y + y^2 - 3xy = y(1 - 3x + y)$$

(3)

has $(1, 2)$ as one of its critical points. We substitute $u = x - 1$, $v = y - 2$; that is, $x = u + 1$, $y = v + 2$. Then

$$3 - x - y = 3 - (u + 1) - (v + 2) = -u - v$$

and

$$1 - 3x + y = 1 - 3(u + 1) + (v + 2) = -3u + v,$$

so the system in (3) takes the form

$$\frac{du}{dt} = (u + 1)(-u - v) = -u - v - u^2 - uv,$$

$$\frac{dv}{dt} = (v + 2)(-3u + v) = -6u + 2v + v^2 - 3uv$$

(4)

and has $(0, 0)$ as a critical point. If we can determine the trajectories of the system in (4) near $(0, 0)$, then their translations under the rigid motion that carries $(0, 0)$ to $(1, 2)$ will be the trajectories near $(1, 2)$ of the original system in (3). This equivalence is illustrated by Fig. 6.2.1 (which shows computer-plotted trajectories of the system in (3) near the critical point $(1, 2)$ in the xy-plane) and Fig. 6.2.2 (which shows computer-plotted trajectories of the system in (4) near the critical point $(0, 0)$ in the uv-plane). ∎

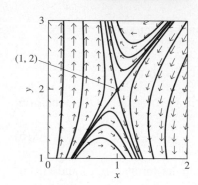

FIGURE 6.2.1. The saddle point $(1, 2)$ for the system $x' = 3x - x^2 - xy$, $y' = y + y^2 - 3xy$ of Example 1.

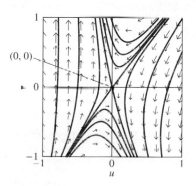

FIGURE 6.2.2. The saddle point $(0, 0)$ for the equivalent system $u' = -u - v - u^2 - uv$, $v' = -6u + 2v + v^2 - 3uv$.

Figures 6.2.1 and 6.2.2 illustrate the fact that the solution curves of the xy-system in (1) are simply the images under the translation $(u, v) \rightarrow (u + x_0, v + y_0)$ of the solution curves of the uv-system in (2). Near the two corresponding critical points—(x_0, y_0) in the xy-plane and $(0, 0)$ in the uv-plane—the two phase portraits therefore look precisely the same.

Linearization Near a Critical Point

Taylor's formula for functions of two variables implies that—if the function $f(x, y)$ is continuously differentiable near the fixed point (x_0, y_0)—then

$$f(x_0 + u, y_0 + v) = f(x_0, y_0) + f_x(x_0, y_0)u + f_y(x_0, y_0)v + r(u, v)$$

where the "remainder term" $r(u, v)$ satisfies the condition

$$\lim_{(u,v)\to(0,0)} \frac{r(u, v)}{\sqrt{u^2 + v^2}} = 0.$$

(Note that this condition would not be satisfied if $r(u, v)$ were a sum containing either constants or terms linear in u or v. In this sense, $r(u, v)$ consists of the "nonlinear part" of the function $f(x_0 + u, y_0 + v)$ of u and v.)

If we apply Taylor's formula to both f and g in (2) and assume that (x_0, y_0) is an isolated critical point so $f(x_0, y_0) = g(x_0, y_0) = 0$, the result is

$$\frac{du}{dt} = f_x(x_0, y_0)u + f_y(x_0, y_0)v + r(u, v),$$

$$\frac{dv}{dt} = g_x(x_0, y_0)u + g_y(x_0, y_0)v + s(u, v) \tag{5}$$

where $r(u, v)$ and the analogous remainder term $s(u, v)$ for g satisfy the condition

$$\lim_{(u,v)\to(0,0)} \frac{r(u, v)}{\sqrt{u^2 + v^2}} = \lim_{(u,v)\to(0,0)} \frac{s(u, v)}{\sqrt{u^2 + v^2}} = 0. \tag{6}$$

Then, when the values u and v are small, the remainder terms $r(u, v)$ and $s(u, v)$ are *very* small (being small even in comparison with u and v).

Consequently, when (u, v) is near $(0, 0)$, the (generally) nonlinear system $u' = f(x_0 + u, y_0 + v)$, $v' = g(x_0 + u, y_0 + v)$ in (5) is closely approximated by the **linearized system**

$$\frac{du}{dt} = f_x(x_0, y_0)u + f_y(x_0, y_0)v,$$

$$\frac{dv}{dt} = g_x(x_0, y_0)u + g_y(x_0, y_0)v \tag{7}$$

whose constant coefficients are the values $f_x(x_0, y_0)$, $f_y(x_0, y_0)$ and $g_x(x_0, y_0)$, $g_y(x_0, y_0)$ of the partial derivatives of the functions f and g at the critical point (x_0, y_0). Assuming that $(0, 0)$ is also an isolated critical point of this linear system, and that the remainder terms in (5) satisfy the condition in (6), the original system $x' = f(x, y)$, $y' = g(x, y)$ is said to be **almost linear** at the isolated critical point (x_0, y_0). In this case, its **linearization** at (x_0, y_0) is the linear system in (7). In short, this linearization is the linear system $\mathbf{u}' = \mathbf{Ju}$ (where $\mathbf{u} = \begin{bmatrix} u & v \end{bmatrix}^T$) whose coefficient matrix is the so-called **Jacobian matrix**

$$\mathbf{J}(x_0, y_0) = \begin{bmatrix} f_x(x_0, y_0) & f_y(x_0, y_0) \\ g_x(x_0, y_0) & g_y(x_0, y_0) \end{bmatrix} \tag{8}$$

of the functions f and g, evaluated at the point (x_0, y_0).

Example 1

Continued

In (3) we have $f(x, y) = 3x - x^2 - xy$ and $g(x, y) = y + y^2 - 3xy$. Then

$$\mathbf{J}(x, y) = \begin{bmatrix} 3 - 2x - y & -x \\ -3y & 1 + 2y - 3x \end{bmatrix}, \quad \text{so} \quad \mathbf{J}(1, 2) = \begin{bmatrix} -1 & -1 \\ -6 & 2 \end{bmatrix}.$$

Hence the linearization of the system $x' = 3x - x^2 - xy$, $y' = y + y^2 - 3xy$ at its critical point $(1, 2)$ is the linear system

$$u' = -u - v,$$
$$v' = -6u + 2v$$

that we get when we drop the nonlinear (quadratic) terms in (4). ∎

It turns out that in most (though not all) cases, the phase portrait of an almost linear system near an isolated critical point (x_0, y_0) strongly resembles—qualitatively—the phase portrait near the origin of its linearization. Consequently, the first step toward understanding general autonomous systems is to characterize the critical points of linear systems.

Critical Points of Linear Systems

We can use the eigenvalue-eigenvector method of Section 5.2 to investigate the critical point $(0, 0)$ of a linear system

$$\begin{bmatrix} x' \\ y' \end{bmatrix} = \begin{bmatrix} a & b \\ c & d \end{bmatrix} \begin{bmatrix} x \\ y \end{bmatrix} \tag{9}$$

with constant coefficient matrix \mathbf{A}. Recall that the eigenvalues λ_1 and λ_2 of \mathbf{A} are the solutions of the characteristic equation

$$\det(\mathbf{A} - \lambda \mathbf{I}) = \begin{vmatrix} a - \lambda & b \\ c & d - \lambda \end{vmatrix} = (a - \lambda)(d - \lambda) - bc = 0.$$

We assume that $(0, 0)$ is an *isolated* critical point of the system in (9), so it follows that the coefficient determinant $ad - bc$ of the system $ax + by = 0$, $cx + dy = 0$ is *nonzero*. This implies that $\lambda = 0$ is *not* a solution of (9), and hence that both eigenvalues of the matrix \mathbf{A} are nonzero.

The nature of the isolated critical point $(0, 0)$ then depends on whether the two nonzero eigenvalues λ_1 and λ_2 of \mathbf{A} are

- real and unequal with the same sign;
- real and unequal with opposite signs;
- real and equal;
- complex conjugates with nonzero real part; or
- pure imaginary numbers.

These five cases are discussed separately. In each case the critical point $(0, 0)$ resembles one of those we saw in the examples of Section 6.1—a node (proper or improper), a saddle point, a spiral point, or a center.

UNEQUAL REAL EIGENVALUES WITH THE SAME SIGN: In this case the matrix \mathbf{A} has linearly independent eigenvectors \mathbf{v}_1 and \mathbf{v}_2 and the general solution $\mathbf{x}(t) = \begin{bmatrix} x(t) & y(t) \end{bmatrix}^T$ of (9) takes the form

$$\mathbf{x}(t) = c_1 \mathbf{v}_1 e^{\lambda_1 t} + c_2 \mathbf{v}_2 e^{\lambda_2 t}. \tag{10}$$

This solution is most simply described in the oblique uv-coordinate system indicated in Fig. 6.2.3, in which the u- and v-axes are determined by the eigenvectors \mathbf{v}_1 and \mathbf{v}_2. Then the uv-coordinate functions $u(t)$ and $v(t)$ of the moving point $\mathbf{x}(t)$ are simply its distances from the origin measured in the directions parallel to the vectors \mathbf{v}_1 and \mathbf{v}_2, so it follows from Eq. (10) that a trajectory of the system is described by

$$u(t) = u_0 e^{\lambda_1 t}, \quad v(t) = v_0 e^{\lambda_2 t} \tag{11}$$

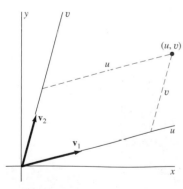

FIGURE 6.2.3. The oblique uv-coordinate system determined by the eigenvectors \mathbf{v}_1 and \mathbf{v}_2.

where $u_0 = u(0)$ and $v_0 = v(0)$. If $v_0 = 0$, then this trajectory lies on the u-axis, whereas if $u_0 = 0$, then it lies on the v-axis. Otherwise—if u_0 and v_0 are both nonzero—the parametric curve in (11) takes the explicit form $v = Cu^k$ where $k = \lambda_2/\lambda_1 > 0$. These solution curves are tangent at $(0, 0)$ to the u-axis if $k > 1$, to the v-axis if $0 < k < 1$. Thus we have in this case an **improper node** as in Example 3 of Section 6.1. If λ_1 and λ_2 are both positive, then we see from (10) and (11) that these solution curves "depart from the origin" as t increases, so $(0, 0)$ is a **nodal source**. But if λ_1 and λ_2 are both negative, then these solution curves approach the origin as t increases, so $(0, 0)$ is a **nodal sink**.

| Example 2 | **(a)** The matrix |

$$\mathbf{A} = \tfrac{1}{8}\begin{bmatrix} 7 & 3 \\ -3 & 17 \end{bmatrix}$$

has eigenvalues $\lambda_1 = 1$ and $\lambda_2 = 2$ with associated eigenvectors $\mathbf{v}_1 = \begin{bmatrix} 3 & 1 \end{bmatrix}^T$ and $\mathbf{v}_2 = \begin{bmatrix} 1 & 3 \end{bmatrix}^T$. Figure 6.2.4 shows a direction field and typical trajectories of the corresponding linear system $\mathbf{x}' = \mathbf{Ax}$. Note that the two eigenvectors point in the directions of the linear trajectories. As is typical of an improper node, all other trajectories are tangent to one of the oblique axes through the origin. In this example the two unequal real eigenvalues are both positive, so the critical point $(0, 0)$ is an improper nodal source.

(b) The matrix

$$\mathbf{B} = -\mathbf{A} = \tfrac{1}{8}\begin{bmatrix} -7 & -3 \\ 3 & -17 \end{bmatrix}$$

has eigenvalues $\lambda_1 = -1$ and $\lambda_2 = -2$ with the same associated eigenvectors $\mathbf{v}_1 = \begin{bmatrix} 3 & 1 \end{bmatrix}^T$ and $\mathbf{v}_2 = \begin{bmatrix} 1 & 3 \end{bmatrix}^T$. The new linear system $\mathbf{x}' = \mathbf{Bx}$ has the same direction field and trajectories as in Fig. 6.2.4 except with the direction field arrows now all reversed, so $(0, 0)$ is now an improper nodal sink. ∎

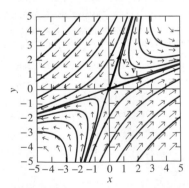

FIGURE 6.2.4. The improper nodal source of Example 2.

UNEQUAL REAL EIGENVALUES WITH OPPOSITE SIGNS: Here the situation is the same as in the previous case, except that $\lambda_2 < 0 < \lambda_1$ in (11). The trajectories with $u_0 = 0$ or $v_0 = 0$ lie on the u- and v-axes through the critical point $(0, 0)$. Those with u_0 and v_0 both nonzero are curves of the explicit form $v = Cu^k$, where $k = \lambda_2/\lambda_1 < 0$. As in the case $k < 0$ of Example 3 in Section 6.1, the nonlinear trajectories resemble hyperbolas, and the critical point $(0, 0)$ is therefore an unstable **saddle point.**

| Example 3 | The matrix |

$$\mathbf{A} = \tfrac{1}{4}\begin{bmatrix} 5 & -3 \\ 3 & -5 \end{bmatrix}$$

has eigenvalues $\lambda_1 = 1$ and $\lambda_2 = -1$ with associated eigenvectors $\mathbf{v}_1 = \begin{bmatrix} 3 & 1 \end{bmatrix}^T$ and $\mathbf{v}_2 = \begin{bmatrix} 1 & 3 \end{bmatrix}^T$. Figure 6.2.5 shows a direction field and typical trajectories of the corresponding linear system $\mathbf{x}' = \mathbf{Ax}$. Note that the two eigenvectors again point in the directions of the linear trajectories. Here $k = -1$ and the nonlinear trajectories are hyperbolas in the oblique uv-coordinate system, so we have the saddle point indicated in the figure. Note that the two eigenvectors point in the directions of the asymptotes to these hyperbolas. ∎

FIGURE 6.2.5. The saddle point of Example 3.

EQUAL REAL ROOTS: In this case, with $\lambda = \lambda_1 = \lambda_2 \neq 0$, the character of the critical point $(0, 0)$ depends on whether or not the coefficient matrix \mathbf{A} has two linearly independent eigenvectors \mathbf{v}_1 and \mathbf{v}_2. If so, then we have oblique uv-coordinates as in Fig. 6.2.3, and the trajectories are described by

$$u(t) = u_0 e^{\lambda t}, \quad v(t) = v_0 e^{\lambda t} \tag{12}$$

as in (11). But now $k = \lambda_2/\lambda_1 = 1$, so the trajectories with $u_0 \neq 0$ are all of the form $v = Cu$ and hence lie on straight lines through the origin. Therefore, $(0, 0)$ is a **proper node** (or **star**) as illustrated in Fig. 6.1.4, and is a source if $\lambda > 0$, a sink if $\lambda < 0$.

If the multiple eigenvalue $\lambda \neq 0$ has only a single associated eigenvector \mathbf{v}_1, then (as we saw in Section 5.4) there nevertheless exists a generalized eigenvector \mathbf{v}_2 such that $(\mathbf{A} - \lambda\mathbf{I})\mathbf{v}_2 = \mathbf{v}_1$, and the linear system $\mathbf{x}' = \mathbf{A}\mathbf{x}$ has the two linearly independent solutions

$$\mathbf{x}_1(t) = \mathbf{v}_1 e^{\lambda t} \quad \text{and} \quad \mathbf{x}_2(t) = (\mathbf{v}_1 t + \mathbf{v}_2)e^{\lambda t}. \tag{13}$$

We can still use the two vectors \mathbf{v}_1 and \mathbf{v}_2 to introduce oblique uv-coordinates as in Fig. 6.2.3. Then it follows from (13) that the coordinate functions $u(t)$ and $v(t)$ of the moving point $\mathbf{x}(t)$ on a trajectory are given by

$$u(t) = (u_0 + v_0 t)e^{\lambda t}, \quad v(t) = v_0 e^{\lambda t}, \tag{14}$$

where $u_0 = u(0)$ and $v_0 = v(0)$. If $v_0 = 0$ then this trajectory lies on the u-axis. Otherwise we have a nonlinear trajectory with

$$\frac{dv}{du} = \frac{dv/dt}{du/dt} = \frac{\lambda v_0 e^{\lambda t}}{v_0 e^{\lambda t} + \lambda(u_0 + v_0 t)e^{\lambda t}} = \frac{\lambda v_0}{v_0 + \lambda(u_0 + v_0 t)}.$$

We see that $dv/du \to 0$ as $t \to \pm\infty$, so it follows that each trajectory is tangent to the u-axis. Therefore, $(0, 0)$ is an **improper node.** If $\lambda < 0$, then we see from (14) that this node is a sink, but it is a source if $\lambda > 0$.

Example 4

The matrix

$$\mathbf{A} = \tfrac{1}{8}\begin{bmatrix} -11 & 9 \\ -1 & -5 \end{bmatrix}$$

has the multiple eigenvalue $\lambda = -1$ with the single associated eigenvector $\mathbf{v}_1 = \begin{bmatrix} 3 & 1 \end{bmatrix}^T$. It happens that $\mathbf{v}_2 = \begin{bmatrix} 1 & 3 \end{bmatrix}^T$ is a generalized eigenvector based on \mathbf{v}_1, but only the actual eigenvector shows up in a phase portrait for the linear system $\mathbf{x}' = \mathbf{A}\mathbf{x}$. As indicated in Fig. 6.2.6, the eigenvector \mathbf{v}_1 determines the u-axis through the improper nodal sink $(0, 0)$, this axis being tangent to each of the nonlinear trajectories. ∎

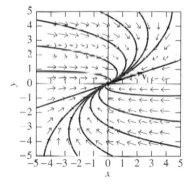

FIGURE 6.2.6. The improper nodal sink of Example 4.

COMPLEX CONJUGATE EIGENVALUES: Suppose that the matrix \mathbf{A} has eigenvalues $\lambda = p + qi$ and $\overline{\lambda} = p - qi$ (with p and q both nonzero) having associated complex conjugate eigenvectors $\mathbf{v} = \mathbf{a} + \mathbf{b}i$ and $\overline{\mathbf{v}} = \mathbf{a} - \mathbf{b}i$. Then we saw in Section 5.2—see Eq. (22) there—that the linear system $\mathbf{x}' = \mathbf{A}\mathbf{x}$ has the two independent real-valued solutions

$$\mathbf{x}_1(t) = e^{pt}(\mathbf{a}\cos qt - \mathbf{b}\sin qt) \quad \text{and} \quad \mathbf{x}_2(t) = e^{pt}(\mathbf{b}\cos qt + \mathbf{a}\sin qt). \tag{15}$$

Thus the components $x(t)$ and $y(t)$ of any solution $\mathbf{x}(t) = c_1\mathbf{x}_1(t) + c_2\mathbf{x}_2(t)$ oscillate between positive and negative values as t increases, so the critical point $(0, 0)$ is a **spiral point** as in Example 5 of Section 6.1. If the real part p of the eigenvalues is negative, then it is clear from (15) that $\mathbf{x}(t) \to \mathbf{0}$ as $t \to +\infty$, so the origin is a spiral sink. But if p is positive, then the critical point is a spiral source.

Example 5

The matrix

$$\mathbf{A} = \tfrac{1}{4}\begin{bmatrix} -10 & 15 \\ -15 & 8 \end{bmatrix}$$

has the complex conjugate eigenvalues $\lambda = -\tfrac{1}{4} \pm 3i$ with negative real part, so $(0, 0)$ is a spiral sink. Figure 6.2.7 shows a direction field and a typical spiral trajectory approaching the origin as $t \to +\infty$. ∎

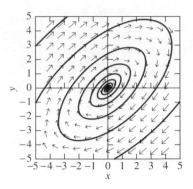

FIGURE 6.2.7. The spiral sink of Example 5.

PURE IMAGINARY EIGENVALUES: If the matrix \mathbf{A} has conjugate imaginary eigenvalues $\lambda = qi$ and $\bar{\lambda} = -qi$ with associated complex conjugate eigenvectors $\mathbf{v} = \mathbf{a} + \mathbf{b}i$ and $\bar{\mathbf{v}} = \mathbf{a} - \mathbf{b}i$, then (15) with $p = 0$ gives the independent solutions

$$\mathbf{x}_1(t) = \mathbf{a}\cos qt - \mathbf{b}\sin qt \quad \text{and} \quad \mathbf{x}_2(t) = \mathbf{b}\cos qt + \mathbf{a}\sin qt \qquad (16)$$

of the linear system $\mathbf{x}' = \mathbf{A}\mathbf{x}$. Just as in Example 4 of Section 6.1, it follows that any solution $\mathbf{x}(t) = c_1\mathbf{x}_1(t) + c_2\mathbf{x}_2(t)$ describes an ellipse centered at the origin in the xy-plane. Hence $(0, 0)$ is a **stable center** in this case.

Example 6

The matrix

$$\mathbf{A} = \frac{1}{4}\begin{bmatrix} -9 & 15 \\ -15 & 9 \end{bmatrix}$$

has the pure imaginary conjugate eigenvalues $\lambda = \pm 3i$, and therefore $(0, 0)$ is a stable center. Figure 6.2.8 shows a direction field and typical elliptical trajectories enclosing the critical point. ∎

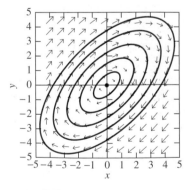

FIGURE 6.2.8. The stable center of Example 6.

Eigenvalues of A	Type of Critical Point
Real, unequal, same sign	Improper node
Real, unequal, opposite sign	Saddle point
Real and equal	Proper or improper node
Complex conjugate	Spiral point
Pure imaginary	Center

FIGURE 6.2.9. Classification of the critical point $(0, 0)$ of the two-dimensional system $\mathbf{x}' = \mathbf{A}\mathbf{x}$.

For the two-dimensional linear system $\mathbf{x}' = \mathbf{A}\mathbf{x}$ with $\det \mathbf{A} \neq 0$, the table in Fig. 6.2.9 lists the type of critical point at $(0, 0)$ found in the five cases discussed here, according to the nature of the eigenvalues λ_1 and λ_2 of the coefficient matrix \mathbf{A}. Our discussion of the various cases shows that the stability of the critical point $(0, 0)$ is determined by the *signs* of the real parts of these eigenvalues, as summarized in Theorem 1. Note that if λ_1 and λ_2 are real, then they are themselves their real parts.

> **THEOREM 1:** Stability of Linear Systems

Let λ_1 and λ_2 be the eigenvalues of the coefficient matrix **A** of the two-dimensional linear system

$$\frac{dx}{dt} = ax + by,$$

$$\frac{dy}{dt} = cx + dy \tag{17}$$

with $ad - bc \neq 0$. Then the critical point $(0, 0)$ is

1. Asymptotically stable if the real parts of λ_1 and λ_2 are both negative;
2. Stable but not asymptotically stable if the real parts of λ_1 and λ_2 are both zero (so that $\lambda_1, \lambda_2 = \pm qi$);
3. Unstable if either λ_1 or λ_2 has a positive real part. ▲

FIGURE 6.2.10. The effects of perturbation of pure imaginary roots.

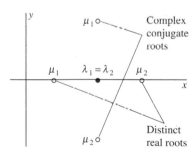

FIGURE 6.2.11. The effects of perturbation of real equal roots.

It is worthwhile to consider the effect of small perturbations in the coefficients a, b, c, and d of the linear system in (17), which result in small perturbations of the eigenvalues λ_1 and λ_2. If these perturbations are sufficiently small, then positive real parts (of λ_1 and λ_2) remain positive and negative real parts remain negative. Hence an asymptotically stable critical point remains asymptotically stable and an unstable critical point remains unstable. Part 2 of Theorem 1 is therefore the only case in which arbitrarily small perturbations can affect the stability of the critical point $(0, 0)$. In this case pure imaginary roots $\lambda_1, \lambda_2 = \pm qi$ of the characteristic equation can be changed to nearby complex roots $\mu_1, \mu_2 = r \pm si$, with r either positive or negative (see Fig. 6.2.10). Consequently, a small perturbation of the coefficients of the linear system in (7) can change a stable center to a spiral point that is either unstable or asymptotically stable.

There is one other exceptional case in which the type, though not the stability, of the critical point $(0, 0)$ can be altered by a small perturbation of its coefficients. This is the case with $\lambda_1 = \lambda_2$, equal roots that (under a small perturbation of the coefficients) can split into two roots μ_1 and μ_2, which are either complex conjugates or unequal real roots (see Fig. 6.2.11). In either case, the sign of the real parts of the roots is preserved, so the stability of the critical point is unaltered. Its nature may change, however; the table in Fig. 6.2.9 shows that a node with $\lambda_1 = \lambda_2$ can either remain a node (if μ_1 and μ_2 are real) or change to a spiral point (if μ_1 and μ_2 are complex conjugates).

Suppose that the linear system in (17) is used to model a physical situation. It is unlikely that the coefficients in (17) can be measured with total accuracy, so let the unknown precise linear model be

$$\frac{dx}{dt} = a^{\star}x + b^{\star}y,$$

$$\frac{dy}{dt} = c^{\star}x + d^{\star}y. \tag{17\star}$$

If the coefficients in (17) are sufficiently close to those in (17*), it then follows from the discussion in the preceding paragraph that the origin $(0, 0)$ is an asymptotically stable critical point for (17) if it is an asymptotically stable critical point for (17*), and is an unstable critical point for (17) if it is an unstable critical point for (17*).

Thus in this case the approximate model in (17) and the precise model in (17*) predict the same qualitative behavior (with respect to asymptotic stability versus instability).

Almost Linear Systems

We now return to the almost linear system

$$\frac{dx}{dt} = ax + by + r(x, y),$$

$$\frac{dy}{dt} = cx + dy + s(x, y) \tag{18}$$

having $(0, 0)$ as an isolated critical point with $ad - bc \neq 0$. Theorem 2, which we state without proof, essentially implies that—with regard to the type and stability of the critical point $(0, 0)$—the effect of the small nonlinear terms $r(x, y)$ and $s(x, y)$ is equivalent to the effect of a small perturbation in the coefficients of the associated *linear* system in (17).

THEOREM 2: Stability of Almost Linear Systems

Let λ_1 and λ_2 be the eigenvalues of the coefficient matrix of the linear system in (17) associated with the almost linear system in (18). Then

1. If $\lambda_1 = \lambda_2$ are equal real eigenvalues, then the critical point $(0, 0)$ of (18) is either a node or a spiral point, and is asymptotically stable if $\lambda_1 = \lambda_2 < 0$, unstable if $\lambda_1 = \lambda_2 > 0$.
2. If λ_1 and λ_2 are pure imaginary, then $(0, 0)$ is either a center or a spiral point, and may be either asymptotically stable, stable, or unstable.
3. Otherwise—that is, unless λ_1 and λ_2 are either real equal or pure imaginary— the critical point $(0, 0)$ of the almost linear system in (18) is of the same type and stability as the critical point $(0, 0)$ of the associated linear system in (17).

▲

Thus, if $\lambda_1 \neq \lambda_2$ and $\text{Re}(\lambda_1) \neq 0$, then the type and stability of the critical point of the almost linear system in (18) can be determined by analysis of its associated linear system in (17), and only in the case of pure imaginary eigenvalues is the stability of $(0, 0)$ not determined by the linear system. Except in the sensitive cases $\lambda_1 = \lambda_2$ and $\text{Re}(\lambda_i) = 0$, the trajectories near $(0, 0)$ will resemble qualitatively those of the associated linear system—they enter or leave the critical point in the same way, but may be "deformed" in a nonlinear manner. The table in Fig. 6.2.12 summarizes the situation.

An important consequence of the classification of cases in Theorem 2 is that a *critical point of an almost linear system is asymptotically stable if it is an asymptotically stable critical point of the linearization of the system.* Moreover, a critical point of the almost linear system is unstable if it is an unstable critical point of the linearized system. If an almost linear system is used to model a physical situation, then—apart from the sensitive cases mentioned earlier—it follows that the qualitative behavior of the system near a critical point can be determined by examining its linearization.

Eigenvalues λ_1, λ_2 for the Linearized System	Type of Critical Point of the Almost Linear System
$\lambda_1 < \lambda_2 < 0$	Stable improper node
$\lambda_1 = \lambda_2 < 0$	Stable node or spiral point
$\lambda_1 < 0 < \lambda_2$	Unstable saddle point
$\lambda_1 = \lambda_2 > 0$	Unstable node or spiral point
$\lambda_1 > \lambda_2 > 0$	Unstable improper node
$\lambda_1, \lambda_2 = a \pm bi \quad (a < 0)$	Stable spiral point
$\lambda_1, \lambda_2 = a \pm bi \quad (a > 0)$	Unstable spiral point
$\lambda_1, \lambda_2 = \pm bi$	Stable or unstable, center or spiral point

FIGURE 6.2.12. Classification of critical points of an almost linear system.

Example 7

Determine the type and stability of the critical point $(0, 0)$ of the almost linear system

$$\frac{dx}{dt} = 4x + 2y + 2x^2 - 3y^2,$$

$$\frac{dy}{dt} = 4x - 3y + 7xy. \tag{19}$$

Solution

The characteristic equation for the associated linear system (obtained simply by deleting the quadratic terms in (19)) is

$$(4 - \lambda)(-3 - \lambda) - 8 - (\lambda - 5)(\lambda + 4) = 0,$$

so the eigenvalues $\lambda_1 = 5$ and $\lambda_2 = -4$ are real, unequal, and have opposite signs. By our discussion of this case we know that $(0, 0)$ is an unstable saddle point of the linear system, and hence by Part 3 of Theorem 2, it is also an unstable saddle point of the almost linear system in (19). The trajectories of the linear system near $(0, 0)$ are shown in Fig. 6.2.13, and those of the nonlinear system in (19) are shown in Fig. 6.2.14. Figure 6.2.15 shows a phase portrait of the nonlinear system in (19) from a "wider view." In addition to the saddle point at $(0, 0)$, there are spiral points near the points $(0.279, 1.065)$ and $(0.933, -1.057)$, and a node near $(-2.354, -0.483)$. ∎

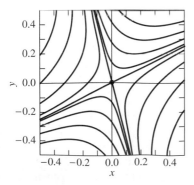

FIGURE 6.2.13. Trajectories of the linearized system of Example 7.

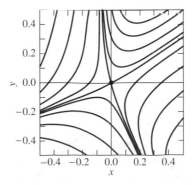

FIGURE 6.2.14. Trajectories of the original almost linear system of Example 7.

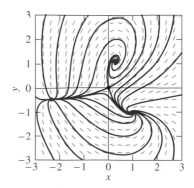

FIGURE 6.2.15. Phase portrait for the almost linear system in Eq. (19).

We have seen that the system $x' = f(x, y)$, $y' = g(x, y)$ with isolated critical point (x_0, y_0) transforms via the substitution $x = u + x_0$, $y = v + y_0$ to an equivalent uv-system with corresponding critical point $(0, 0)$ and linearization $\mathbf{u}' = \mathbf{J}\mathbf{u}$, whose coefficient matrix \mathbf{J} is the Jacobian in (8) of the functions f and g at (x_0, y_0). Consequently we need not carry out the substitution explicitly; instead, we can proceed directly to calculate the eigenvalues of \mathbf{J} preparatory to application of Theorem 2.

Example 8

Determine the type and stability of the critical point $(4, 3)$ of the almost linear system

$$\frac{dx}{dt} = 33 - 10x - 3y + x^2,$$

$$\frac{dy}{dt} = -18 + 6x + 2y - xy. \tag{20}$$

Solution

With $f(x, y) = 33 - 10x - 3y + x^2$, $g(x, y) = -18 + 6x + 2y - xy$ and $x_0 = 4$, $y_0 = 3$ we have

$$\mathbf{J}(x, y) = \begin{bmatrix} -10 + 2x & -3 \\ 6 - y & 2 - x \end{bmatrix}, \quad \text{so} \quad \mathbf{J}(4, 3) = \begin{bmatrix} -2 & -3 \\ 3 & -2 \end{bmatrix}.$$

The associated linear system

$$\frac{du}{dt} = -2u - 3v,$$

$$\frac{dv}{dt} = 3u - 2v \tag{21}$$

has characteristic equation $(\lambda + 2)^2 + 9 = 0$, with complex conjugate roots $\lambda = -2 \pm 3i$. Hence $(0, 0)$ is an asymptotically stable spiral point of the linear system in (21), so Theorem 2 implies that $(4, 3)$ is an asymptotically stable spiral point of the original almost linear system in (20). Figure 6.2.16 shows a typical trajectory of the linear system in (21), and Fig. 6.2.17 shows how this spiral point fits into the phase portrait for the original almost linear system in (20). ∎

FIGURE 6.2.16. Spiral trajectories of the linear system in Eq. (21).

FIGURE 6.2.17. Phase portrait for the almost linear system in Eq. (20).

6.2 Problems

In Problems 1 through 10, apply Theorem 1 to determine the type of the critical point $(0, 0)$ and whether it is asymtotically stable, stable, or unstable. Verify your conclusion by using a computer system or graphing calculator to construct a phase portrait for the given system.

1. $\dfrac{dx}{dt} = -2x + y, \quad \dfrac{dy}{dt} = x - 2y$

2. $\dfrac{dx}{dt} = 4x - y, \quad \dfrac{dy}{dt} = 2x + y$

3. $\dfrac{dx}{dt} = x + 2y, \quad \dfrac{dy}{dt} = 2x + y$

4. $\dfrac{dx}{dt} = 3x + y, \quad \dfrac{dy}{dt} = 5x - y$

5. $\dfrac{dx}{dt} = x - 2y, \quad \dfrac{dy}{dt} = 2x - 3y$

6. $\dfrac{dx}{dt} = 5x - 3y, \quad \dfrac{dy}{dt} = 3x - y$

7. $\dfrac{dx}{dt} = 3x - 2y, \quad \dfrac{dy}{dt} = 4x - y$

8. $\dfrac{dx}{dt} = x - 3y, \quad \dfrac{dy}{dt} = 6x - 5y$

9. $\dfrac{dx}{dt} = 2x - 2y, \quad \dfrac{dy}{dt} = 4x - 2y$

10. $\dfrac{dx}{dt} = x - 2y, \quad \dfrac{dy}{dt} = 5x - y$

Each of the systems in Problems 11 through 18 has a single critical point (x_0, y_0). Apply Theorem 2 to classify this critical point as to type and stability. Verify your conclusion by using a computer system or graphing calculator to construct a phase portrait for the given system

11. $\dfrac{dx}{dt} = x - 2y, \quad \dfrac{dy}{dt} = 3x - 4y - 2$

12. $\dfrac{dx}{dt} = x - 2y - 8, \quad \dfrac{dy}{dt} = x + 4y + 10$

13. $\dfrac{dx}{dt} = 2x - y - 2, \quad \dfrac{dy}{dt} = 3x - 2y - 2$

14. $\dfrac{dx}{dt} = x + y - 7, \quad \dfrac{dy}{dt} = 3x - y - 5$

15. $\dfrac{dx}{dt} = x - y, \quad \dfrac{dy}{dt} = 5x - 3y - 2$

16. $\dfrac{dx}{dt} = x - 2y + 1, \quad \dfrac{dy}{dt} = x + 3y - 9$

17. $\dfrac{dx}{dt} = x - 5y - 5, \quad \dfrac{dy}{dt} = x - y - 3$

18. $\dfrac{dx}{dt} = 4x - 5y + 3, \quad \dfrac{dy}{dt} = 5x - 4y + 6$

In Problems 19 through 28, investigate the type of the critical point $(0, 0)$ of the given almost linear system. Verify your conclusion by using a computer system or graphing calculator to construct a phase portrait. Also, describe the approximate locations and apparent types of any other critical points that are visible in your figure. Feel free to investigate these additional critical points; you can use the computational methods discussed in the application material for this section.

19. $\dfrac{dx}{dt} = x - 3y + 2xy, \quad \dfrac{dy}{dt} = 4x - 6y - xy$

20. $\dfrac{dx}{dt} = 6x - 5y + x^2, \quad \dfrac{dy}{dt} = 2x - y + y^2$

21. $\dfrac{dx}{dt} = x + 2y + x^2 + y^2, \quad \dfrac{dy}{dt} = 2x - 2y - 3xy$

22. $\dfrac{dx}{dt} = x + 4y - xy^2, \quad \dfrac{dy}{dt} = 2x - y + x^2y$

23. $\dfrac{dx}{dt} = 2x - 5y + x^3, \quad \dfrac{dy}{dt} = 4x - 6y + y^4$

24. $\dfrac{dx}{dt} = 5x - 3y + y(x^2 + y^2), \quad \dfrac{dy}{dt} = 5x + y(x^2 + y^2)$

25. $\dfrac{dx}{dt} = x - 2y + 3xy, \quad \dfrac{dy}{dt} = 2x - 3y - x^2 - y^2$

26. $\dfrac{dx}{dt} = 3x - 2y - x^2 - y^2, \quad \dfrac{dy}{dt} = 2x - y - 3xy$

27. $\dfrac{dx}{dt} = x - y + x^4 - y^2, \quad \dfrac{dy}{dt} = 2x - y + y^4 - x^2$

28. $\dfrac{dx}{dt} = 3x - y + x^3 + y^3, \quad \dfrac{dy}{dt} = 13x - 3y + 3xy$

In Problems 29 through 32, find all critical points of the given system, and investigate the type and stability of each. Verify your conclusions by means of a phase portrait constructed using a computer system or graphing calculator.

29. $\dfrac{dx}{dt} = x - y, \quad \dfrac{dy}{dt} = x^2 - y$

30. $\dfrac{dx}{dt} = y - 1, \quad \dfrac{dy}{dt} = x^2 - y$

31. $\dfrac{dx}{dt} = y^2 - 1, \quad \dfrac{dy}{dt} = x^3 - y$

32. $\dfrac{dx}{dt} = xy - 2, \quad \dfrac{dy}{dt} = x - 2y$

Bifurcations

The term bifurcation generally refers to something "splitting apart." With regard to differential equations or systems involving a parameter, it refers to abrupt changes in the character of the solutions as the parameter is changed continuously. Problems 33 through 36 illustrate sensitive cases in which small perturbations in the coefficients of a linear or almost linear system can change the type or stability (or both) of a critical point.

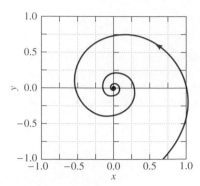

FIGURE 6.2.18(a). Stable spiral with $\epsilon = -0.2$.

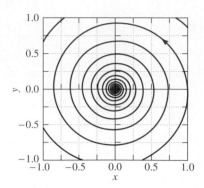

FIGURE 6.2.18(b). Stable spiral with $\epsilon = -0.05$.

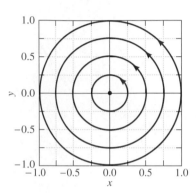

FIGURE 6.2.18(c). Stable center with $\epsilon = 0$.

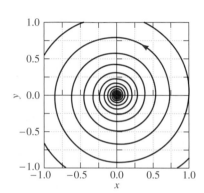

FIGURE 6.2.18(d). Unstable spiral with $\epsilon = 0.05$.

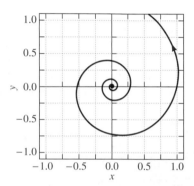

FIGURE 6.2.18(e). Unstable spiral with $\epsilon = 0.2$.

33. Consider the linear system

$$\frac{dx}{dt} = \epsilon x - y, \qquad \frac{dy}{dt} = x + \epsilon y.$$

Show that the critical point $(0, 0)$ is **(a)** a stable spiral point if $\epsilon < 0$; **(b)** a center if $\epsilon = 0$; **(c)** an unstable spiral point if $\epsilon > 0$. Thus small perturbations of the system $x' = -y$, $y' = x$ can change both the type and stability of the critical point. Figures 6.2.18(a)–(e) illustrate the loss of stability that occurs at $\epsilon = 0$ as the parameter increases from $\epsilon < 0$ to $\epsilon > 0$.

34. Consider the linear system

$$\frac{dx}{dt} = -x + \epsilon y, \qquad \frac{dy}{dt} = x - y.$$

Show that the critical point $(0, 0)$ is **(a)** a stable spiral point if $\epsilon < 0$; **(b)** a stable node if $0 \leqq \epsilon < 1$. Thus small perturbations of the system $x' = -x$, $y' = x - y$ can change the type of the critical point $(0, 0)$ without changing its stability.

35. This problem deals with the almost linear system

$$\frac{dx}{dt} = y + hx(x^2 + y^2), \qquad \frac{dy}{dt} = -x + hy(x^2 + y^2),$$

in illustration of the sensitive case of Theorem 2, in which the theorem provides no information about the stability of the critical point $(0, 0)$. **(a)** Show that $(0, 0)$ is a center of the linear system obtained by setting $h = 0$. **(b)** Suppose that $h \neq 0$. Let $r^2 = x^2 + y^2$, then apply the fact that

$$x\frac{dx}{dt} + y\frac{dy}{dt} = r\frac{dr}{dt}$$

to show that $dr/dt = hr^3$. **(c)** Suppose that $h = -1$. Integrate the differential equation in (b); then show that $r \to 0$ as $t \to +\infty$. Thus $(0, 0)$ is an asymptotically stable critical point of the almost linear system in this case. **(d)** Suppose that $h = +1$. Show that $r \to +\infty$ as t increases, so $(0, 0)$ is an unstable critical point in this case.

36. This problem presents the famous *Hopf bifurcation* for the almost linear system

$$\frac{dx}{dt} = \epsilon x + y - x(x^2 + y^2),$$

$$\frac{dy}{dt} = -x + \epsilon y - y(x^2 + y^2),$$

which has imaginary characteristic roots $\lambda = \pm i$ if $\epsilon = 0$. **(a)** Change to polar coordinates as in Example 6 of Section 6.1 to obtain the system $r' = r(\epsilon - r^2)$, $\theta' = -1$.

(b) Separate variables and integrate directly to show that if $\epsilon \leq 0$, then $r(t) \to 0$ as $t \to +\infty$, so in this case the origin is a stable spiral point. **(c)** Show similarly that if $\epsilon > 0$, then $r(t) \to \sqrt{\epsilon}$ as $t \to +\infty$, so in this case the origin is an unstable spiral point. The circle $r(t) \equiv \sqrt{\epsilon}$ itself is a closed periodic solution or *limit cycle.* Thus a limit cycle of increasing size is spawned as the parameter ϵ increases through the critical value 0.

37. In the case of a two-dimensional system that is *not* almost linear, the trajectories near an isolated critical point can exhibit a considerably more complicated structure than those near the nodes, centers, saddle points, and spiral points discussed in this section. For example, consider the system

$$\frac{dx}{dt} = x(x^3 - 2y^3),$$
$$\frac{dy}{dt} = y(2x^3 - y^3) \tag{22}$$

having $(0, 0)$ as an isolated critical point. This system is not almost linear because $(0, 0)$ is not an *isolated* critical point of the trivial associated linear system $x' = 0$, $y' = 0$. Solve the homogeneous first-order equation

$$\frac{dy}{dx} = \frac{y(2x^3 - y^3)}{x(x^3 - 2y^3)}$$

to show that the trajectories of the system in (22) are *folia of Descartes* of the form

$$x^3 + y^3 = 3cxy,$$

where c is an arbitrary constant (Fig. 6.2.19).

38. First note that the characteristic equation of the 2×2 matrix \mathbf{A} can be written in the form $\lambda^2 - T\lambda + D = 0$, where D is the determinant of \mathbf{A} and the trace T of the matrix \mathbf{A} is the sum of its two diagonal elements. Then apply Theorem 1 to show that the type of the critical point $(0, 0)$ of the system $\mathbf{x}' = \mathbf{A}\mathbf{x}$ is determined—as indicated in Fig. 6.2.20—by the location of the point (T, D) in the *trace-determinant plane* with horizontal T-axis and vertical D-axis.

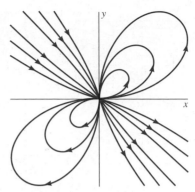

FIGURE 6.2.19. Trajectories of the system in Eq. (22).

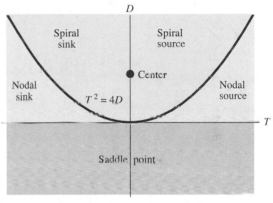

FIGURE 6.2.20. The critical point $(0, 0)$ of the system $\mathbf{x}' = \mathbf{A}\mathbf{x}$ is a

- spiral sink or source if the point (T, D) lies above the parabola $T^2 = 4D$ but off the D-axis;
- stable center if (T, D) lies on the positive D-axis;
- nodal sink or source if (T, D) lies between the parabola and the T-axis;
- saddle point if (T, D) lies beneath the T-axis.

6.2 Application: Phase Plane Portraits of Almost Linear Systems

Interesting and complicated phase portraits often result from simple nonlinear perturbations of linear systems. For instance, Fig. 6.2.21 shows a phase portrait for the almost linear system

$$\frac{dx}{dt} = -y\cos(x + y - 1),$$
$$\frac{dy}{dt} = x\cos(x - y + 1). \tag{1}$$

Among the seven critical points marked with dots, we see

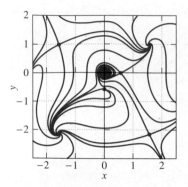

FIGURE 6.2.21. Phase plane portrait for the system in Eq. (1).

- Apparent spiral points in the first and third quadrants of the xy-plane;
- Apparent saddle points in the second and fourth quadrants, plus another one on the positive x-axis;
- A critical point of undetermined character on the negative y-axis; and
- An apparently "very weak" spiral point at the origin—meaning one that is approached very slowly as t increases or decreases (according as it is a sink or a source).

Some ODE software systems can automatically locate and classify critical points. For instance, Fig. 6.2.22 shows a screen produced by John Polking's MATLAB **pplane** program (cited in the Section 6.1 application). It shows that the fourth-quadrant critical point in Fig. 6.2.21 has approximate coordinates (1.5708, −2.1416), and that the coefficient matrix of the associated linear system has the positive eigenvalue $\lambda_1 \approx 2.8949$ and the negative eigenvalue $\lambda_2 \approx -2.3241$. It therefore follows from Theorem 2 that this critical point is, indeed, a saddle point of the almost linear system in (1).

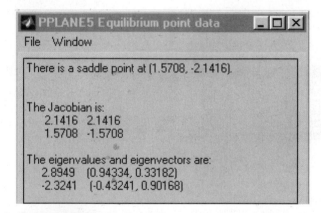

FIGURE 6.2.22. The fourth-quadrant saddle point revealed.

With a general computer algebra system such as *Maple* or *Mathematica*, you may have to do a bit of work yourself—or tell the computer precisely what to do—to find and classify a critical point. For instance, the *Maple* command

```
fsolve({-y*cos(x+y-1)=0,x*cos(x-y+1)=0},
       {x,y},{x=1..2,y=-3..-2});
```

or the *Mathematica* command

```
FindRoot[{-y*Cos[x+y-1] == 0, x*Cos[x-y+1] == 0},
        {x,1,2}, {y,-3,-2}]
```

will find the critical-point coordinates $a = 1.5708$, $b = -2.1416$ indicated earlier. Then the substitution $x = u + a$, $y = v + b$ yields the translated system

$$\frac{du}{dt} = (2.1416 - v)\cos(1.5708 - u - v) = f(u, v),$$

$$\frac{dv}{dt} = (1.5708 + u)\cos(4.7124 + u - v) = g(u, v). \tag{2}$$

If we substitute $u = v = 0$ in the Jacobian matrix $\partial(f, g)/\partial(u, v)$, we get the coefficient matrix

$$\mathbf{A} = \begin{bmatrix} 2.1416 & 2.1416 \\ 1.5708 & -1.5708 \end{bmatrix}$$

of the linear system corresponding to (2). Then the *Maple* command

```
evalf(Eigenvals(A))
```

or the *Mathematica* command

```
Eigenvalues[A]
```

yields the eigenvalues $\lambda_1 \approx 2.8949$ and $\lambda_2 \approx -2.3241$, thereby verifying that the critical point $(1.5708, -2.1416)$ of (1) is, indeed, a saddle point.

Use a computer algebra system to find and classify the other critical points of (1) indicated in Fig. 6.2.21. Then investigate similarly an almost linear system of your own construction. One convenient way to construct such a system is to begin with a linear system and insert sine or cosine factors resembling the ones in (1).

6.3 Ecological Models: Predators and Competitors

Some of the most interesting and important applications of stability theory involve the interactions between two or more biological populations occupying the same environment. We consider first a **predator-prey** situation involving two species. One species—the **predators**—feeds on the other species—the **prey**—which in turn feeds on some third food item readily available in the environment. A standard example is a population of foxes and rabbits in a woodland; the foxes (predators) eat rabbits (the prey), while the rabbits eat certain vegetation in the woodland. Other examples are sharks (predators) and food fish (prey), bass (predators) and sunfish (prey), ladybugs (predators) and aphids (prey), and beetles (predators) and scale insects (prey).

The classical mathematical model of a predator-prey situation was developed in the 1920s by the Italian mathematician Vito Volterra (1860–1940) in order to analyze the cyclic variations observed in the shark and food fish populations in the Adriatic Sea. To construct such a model, we denote the number of prey at time t by $x(t)$, the number of predators by $y(t)$, and make the following simplifying assumptions.

1. In the absence of predators, the prey population would grow at a natural rate, with $dx/dt = ax, a > 0$.

2. In the absence of prey, the predator population would decline at a natural rate, with $dy/dt = -by, b > 0$.

3. When both predators and prey are present, there occurs, in combination with these natural rates of growth and decline, a decline in the prey population and a growth in the predator population, each at a rate proportional to the frequency of encounters between individuals of the two species. We assume further that the frequency of such encounters is proportional to the product xy, reasoning that doubling either population alone should double the frequency of encounters, while doubling both populations ought to quadruple the frequency of encounters. Consequently, the consumption of prey by predators results in

 - an interaction rate of decline $-pxy$ in the prey population x, and

- an interaction rate of growth qxy in the predator population y.

When we combine the natural and interaction rates ax and $-pxy$ for the prey population x, as well as the natural and interaction rates $-by$ and qxy for the predator population y, we get the **predator-prey system**

$$\frac{dx}{dt} = ax - pxy = x(a - py),$$

$$\frac{dy}{dt} = -by + qxy = y(-b + qx), \tag{1}$$

with the constants a, b, p, and q all positive. *Note:* You may see the predator and prey equations written in either order in (1). It is important to recognize that the predator equation has negative linear term and positive interaction term, whereas the prey equation has positive linear term and negative interaction term.

Example 1 **The Critical Points** A critical point of the general predator-prey system in (1) is a solution (x, y) of the equations

$$x(a - py) = 0, \quad y(-b + qx) = 0. \tag{2}$$

The first of these two equations implies that either $x = 0$ or $y = a/p \neq 0$, and the second implies that either $y = 0$ or $x = b/q \neq 0$. It follows readily that this predator-prey system has the two critical points $(0, 0)$ and $(b/q, a/p)$.

THE CRITICAL POINT $(0, 0)$: The Jacobian matrix of the system in (1) is

$$\mathbf{J}(x, y) = \begin{bmatrix} a - py & -px \\ qy & -b + qx \end{bmatrix}, \quad \text{so} \quad \mathbf{J}(0, 0) = \begin{bmatrix} a & 0 \\ 0 & -b \end{bmatrix}. \tag{3}$$

The matrix $\mathbf{J}(0, 0)$ has characteristic equation $(a - \lambda)(-b - \lambda) = 0$ and the eigenvalues $\lambda_1 = a > 0$, $\lambda_2 = -b < 0$ with different signs. Hence it follows from Theorems 1 and 2 in Section 6.2 that the critical point $(0, 0)$ is an unstable saddle point, both of the predator-prey system and of its linearization at $(0, 0)$. The corresponding equilibrium solution $x(t) \equiv 0$, $y(t) \equiv 0$ merely describes simultaneous extinction of the prey (x) and predator (y) populations.

THE CRITICAL POINT $(b/q, a/p)$: The Jacobian matrix

$$\mathbf{J}(b/q, a/p) = \begin{bmatrix} 0 & -\dfrac{pb}{q} \\ \dfrac{aq}{p} & 0 \end{bmatrix} \tag{4}$$

has characteristic equation $\lambda^2 + ab = 0$ and the pure imaginary eigenvalues λ_1, $\lambda_2 = \pm i\sqrt{ab}$. It follows from Theorem 1 in Section 6.2 that the linearization of the predator-prey system at $(b/q, a/p)$ has a stable center at the origin. Thus we have the indeterminate case of Theorem 2 in Section 6.2, in which case the critical point can (aside from a stable center) also be either a stable spiral sink or an unstable spiral source of the predator-prey system itself. Hence further investigation is required to determine the actual character of the critical point $(b/q, a/p)$. The corresponding equilibrium solution $x(t) \equiv b/q$, $y(t) \equiv a/p$ describes the only *nonzero* constant prey (x) and predator (y) populations that coexist permanently.

THE PHASE PLANE PORTRAIT In Problem 1 we ask you to analyze numerically a typical predator-prey system and verify that the linearizations at its two critical points agree qualitatively with the phase plane portrait shown in Fig. 6.3.1—where the nontrivial critical point appears visually to be a stable center. Of course, only the first quadrant of this portrait corresponds to physically meaningful solutions describing nonnegative populations of prey and predators.

In Problem 2 we ask you to derive an exact implicit solution of the predator-prey system of Fig. 6.3.1—a solution that can be used to show that its phase plane trajectories in the first quadrant are, indeed, simple closed curves that encircle the critical point (75, 50) as indicated in the figure. It then follows from Problem 30 in Section 6.1 that the explicit solution functions $x(t)$ and $y(t)$ are both periodic functions of t—thus explaining the periodic fluctuations that are observed empirically in predator-prey populations. ∎

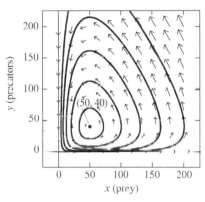

FIGURE 6.3.1. Phase plane portrait for the predator-prey system $x' = 200x - 4xy$, $y' = -150y + 2xy$ with critical points $(0, 0)$ and $(75, 50)$.

FIGURE 6.3.2. The predator prey phase portrait of Example 2.

Example 2

Oscillating Populations Figure 6.3.2 shows a computer-generated direction field and phase portrait for the predator-prey system

$$\frac{dx}{dt} = (0.2)x - (0.005)xy = (0.005)x(40 - y),$$

$$\frac{dy}{dt} = -(0.5)y + (0.01)xy = (0.01)y(-50 + x),$$

(5)

where $x(t)$ denotes the number of rabbits and $y(t)$ the number of foxes after t months. Evidently the critical point $(50, 40)$ is a stable center representing equilibrium populations of 50 rabbits and 40 foxes. Any other initial point lies on a closed trajectory enclosing this equilibrium point. The direction field indicates that the point $(x(t), y(t))$ traverses its trajectory in a counterclockwise direction, with the rabbit and fox populations oscillating periodically between their separate maximum and minimum values. A drawback is that the phase plane plot provides no indication as to the speed with which each trajectory is traversed.

This lost "sense of time" is recaptured by graphing the two individual population functions as functions of time t. In Fig. 6.3.3 we have graphed approximate

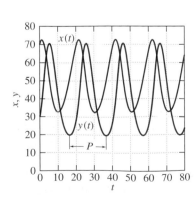

FIGURE 6.3.3. Periodic oscillations of the predator and prey populations in Example 2.

solution functions $x(t)$ and $y(t)$ calculated using the Runge-Kutta method of Section 4.3 with initial values $x(0) = 70$ and $y(0) = 40$. We see that the rabbit population oscillates between the extreme values $x_{max} \approx 72$ and $x_{min} \approx 33$, while the fox population oscillates (out of phase) between the extreme values $y_{max} \approx 70$ and $y_{min} \approx 20$. A careful measurement indicates that the period P of oscillation of each population is slightly over 20 months. One could "zoom in" on the maximum/minimum points on each graph in order to refine these estimates of the period and the maximum and minimum rabbit and fox populations.

Any positive initial conditions $x_0 = x(0)$ and $y_0 = y(0)$ yield a similar picture, with the rabbit and fox populations both surviving in coexistence with each other. ■

Competing Species

Now we consider two species (of animals, plants, or bacteria, for instance) with populations $x(t)$ and $y(t)$ at time t and which compete with each other for the food available in their common environment. This is in marked contrast to the case in which one species preys on the other. To construct a mathematical model that is as realistic as possible, let us assume that in the absence of either species, the other would have a bounded [logistic] population like those considered in Section 2.1. In the absence of any interaction or competition between the two species, their populations $x(t)$ and $y(t)$ would then satisfy the differential equations

$$\frac{dx}{dt} = a_1 x - b_1 x^2,$$

$$\frac{dy}{dt} = a_2 y - b_2 y^2,$$

(6)

each of the form of Eq. (2) of Section 2.1. But in addition, we assume that competition has the effect of a rate of decline in each population that is proportional to their product xy. We insert such terms with *negative* proportionality constants $-c_1$ and $-c_2$ in the equations in (6) to obtain the **competition system**

$$\frac{dx}{dt} = a_1 x - b_1 x^2 - c_1 xy = x(a_1 - b_1 x - c_1 y),$$

$$\frac{dy}{dt} = a_2 y - b_2 y^2 - c_2 xy = y(a_2 - b_2 y - c_2 x),$$

(7)

where the coefficients a_1, a_2, b_1, b_2, c_1, and c_2 are all positive.

The almost linear system in (7) has four critical points. Upon setting the right-hand sides of the two equations equal to zero, we see that if $x = 0$, then either $y = 0$ or $y = a_2/b_2$, whereas if $y = 0$, then either $x = 0$ or $x = a_1/b_1$. This gives the three critical points $(0, 0)$, $(0, a_2/b_2)$, and $(a_1/b_1, 0)$. The fourth critical point is obtained from the simultaneous solution of the equations

$$b_1 x + c_1 y = a_1, \quad b_2 x + c_2 y = a_2.$$

(8)

We assume that, as in most interesting applications, these equations have a single solution and that the corresponding critical point lies in the first quadrant of the xy-plane. This point (x_E, y_E) is then the fourth critical point of the system in (7), and it represents the possibility of coexistence of the two species, with constant nonzero equilibrium populations $x(t) \equiv x_E$ and $y(t) \equiv y_E$.

We are interested in the stability of the critical point (x_E, y_E). This turns out to depend on whether

$$c_1 c_2 < b_1 b_2 \quad \text{or} \quad c_1 c_2 > b_1 b_2. \tag{9}$$

Each inequality in (9) has a natural interpretation. Examining the equations in (6), we see that the coefficients b_1 and b_2 represent the inhibiting effect of each population on its own growth (possibly due to limitations of food or space). On the other hand, c_1 and c_2 represent the effect of competition between the two populations. Thus $b_1 b_2$ is a measure of *inhibition* while $c_1 c_2$ is a measure of *competition*. A general analysis of the system in (7) shows the following:

1. If $c_1 c_2 < b_1 b_2$, so that competition is small in comparison with inhibition, then (x_E, y_E) is an asymptotically stable critical point that is approached by each solution as $t \to +\infty$. Thus the two species can and do coexist in this case.

2. If $c_1 c_2 > b_1 b_2$, so that competition is large in comparison with inhibition, then (x_E, y_E) is an unstable critical point, and either $x(t)$ or $y(t)$ approaches zero as $t \to +\infty$. Thus the two species cannot coexist in this case; one survives and the other becomes extinct.

Rather than carrying out this general analysis, we present two examples that illustrate these two possibilities.

Example 3 **Survival of a Single Species** Suppose that the populations $x(t)$ and $y(t)$ satisfy the equations

$$\frac{dx}{dt} = 14x - \tfrac{1}{2}x^2 - xy,$$

$$\frac{dy}{dt} = 16y - \tfrac{1}{2}y^2 - xy, \tag{10}$$

in which $a_1 = 14$, $a_2 = 16$, $b_1 = b_2 = \tfrac{1}{2}$, and $c_1 = c_2 = 1$. Then $c_1 c_2 = 1 > \tfrac{1}{4} = b_1 b_2$, so we should expect survival of a single species as predicted in Case 2 previously. We find readily that the four critical points are $(0, 0)$, $(0, 32)$, $(28, 0)$, and $(12, 8)$. We shall investigate them individually.

THE CRITICAL POINT $(0, 0)$: The Jacobian matrix of the system in (10) is

$$\mathbf{J}(x, y) = \begin{bmatrix} 14 - x - y & -x \\ -y & 16 - y - x \end{bmatrix}, \quad \text{so} \quad \mathbf{J}(0, 0) = \begin{bmatrix} 14 & 0 \\ 0 & 16 \end{bmatrix}. \tag{11}$$

The matrix $\mathbf{J}(0, 0)$ has characteristic equation $(14 - \lambda)(16 - \lambda) = 0$ and has the eigenvalues

$$\lambda_1 = 14 \quad \text{with eigenvector} \quad \mathbf{v}_1 = \begin{bmatrix} 1 & 0 \end{bmatrix}^T$$

and

$$\lambda_2 = 16 \quad \text{with eigenvector} \quad \mathbf{v}_2 = \begin{bmatrix} 0 & 1 \end{bmatrix}^T.$$

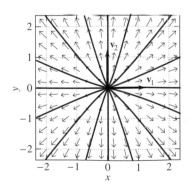

FIGURE 6.3.4. Phase plane portrait for the linear system $x' = 14x$, $y' = 16y$ corresponding to the critical point $(0, 0)$.

Both eigenvalues are positive, so it follows that $(0, 0)$ is a nodal source for the system's linearization $x' = 14x$, $y' = 16y$ at $(0, 0)$, and hence—by Theorem 2 in Section 6.2—is also an unstable nodal source for the original system in (10). Figure 6.3.4 shows a phase portrait for the linearized system near $(0, 0)$.

THE CRITICAL POINT (0, 32): The Jacobian matrix

$$\mathbf{J}(0, 32) = \begin{bmatrix} -18 & 0 \\ -32 & -16 \end{bmatrix} \tag{12}$$

corresponds to the linearization

$$\frac{du}{dt} = -18u,$$
$$\frac{dv}{dt} = -32u - 16v \tag{13}$$

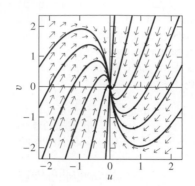

FIGURE 6.3.5. Phase plane portrait for the linear system in Eq. (13) corresponding to the critical point (0, 32).

of (10) at (0, 32) . The matrix $\mathbf{J}(0, 32)$ has characteristic equation $(-18-\lambda)(-16-\lambda) = 0$ and has the eigenvalues $\lambda_1 = -18$ with eigenvector $\mathbf{v}_1 = \begin{bmatrix} 1 & 16 \end{bmatrix}^T$ and $\lambda_2 = -16$ with eigenvector $\mathbf{v}_2 = \begin{bmatrix} 0 & 1 \end{bmatrix}^T$. Because both eigenvalues are negative, it follows that $(0, 0)$ is a nodal sink for the linearized system, and hence—by Theorem 2 in Section 6.2—that $(0, 32)$ is also a stable nodal sink for the original system in (10). Figure 6.3.5 shows a phase portrait for the linearized system near $(0, 0)$.

THE CRITICAL POINT (28, 0): The Jacobian matrix

$$\mathbf{J}(28, 0) = \begin{bmatrix} -14 & -28 \\ 0 & -12 \end{bmatrix} \tag{14}$$

corresponds to the linearization

$$\frac{du}{dt} = -14u - 28v,$$
$$\frac{dv}{dt} = -12v \tag{15}$$

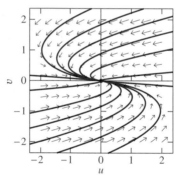

FIGURE 6.3.6. Phase plane portrait for the linear system in Eq. (15) corresponding to the critical point (28, 0).

of (10) at (28, 0) . The matrix $\mathbf{J}(28, 0)$ has characteristic equation $(-14-\lambda)(-12-\lambda) = 0$ and has the eigenvalues $\lambda_1 = -14$ with eigenvector $\mathbf{v}_1 = \begin{bmatrix} 1 & 0 \end{bmatrix}^T$ and $\lambda_2 = -12$ with eigenvector $\mathbf{v}_2 = \begin{bmatrix} -14 & 1 \end{bmatrix}^T$. Because both eigenvalues are negative, it follows that $(0, 0)$ is a nodal sink for the linearized system, and hence—by Theorem 2 in Section 6.2—that $(28, 0)$ is also a stable nodal sink for the original nonlinear system in (10). Figure 6.3.6 shows a phase portrait for the linearized system near $(0, 0)$.

THE CRITICAL POINT (12, 8): The Jacobian matrix

$$\mathbf{J}(12, 8) = \begin{bmatrix} -6 & -12 \\ -8 & -4 \end{bmatrix} \tag{16}$$

corresponds to the linearization

$$\frac{du}{dt} = -6u - 12v,$$
$$\frac{dv}{dt} = -8u - 4v \tag{17}$$

of (10) at (12, 8). The matrix $\mathbf{J}(12, 8)$ has characteristic equation

$$(-6 - \lambda)(-4 - \lambda) - (-8)(-12) = \lambda^2 + 10\lambda - 72 = 0$$

and has the eigenvalues

$$\lambda_1 = -5 - \sqrt{97} < 0 \quad \text{with eigenvector} \quad \mathbf{v}_1 = \left[\tfrac{1}{8}\left(1 + \sqrt{97}\right) \quad 1 \right]^T$$

and

$$\lambda_2 = -5 + \sqrt{97} > 0 \quad \text{with eigenvector} \quad \mathbf{v}_2 = \left[\tfrac{1}{8}\left(1 - \sqrt{97}\right) \quad 1 \right]^T.$$

Because the two eigenvalues have opposite signs, it follows that (0, 0) is a saddle point for the linearized system and hence—by Theorem 2 in Section 6.2—that (12, 8) is also an unstable saddle point for the original system in (10). Figure 6.3.7 shows a phase portrait for the linearized system near (0, 0).

Now that our local analysis of each of the four critical points is complete, it remains to assemble the information found into a coherent global picture. If we accept the facts that

- Near each critical point, the trajectories for the original system in (10) resemble qualitatively the linearized trajectories shown in Figs. 6.3.4–6.3.7, and
- As $t \to +\infty$ each trajectory either approaches a critical point or diverges toward infinity,

then it would appear that the phase plane portrait for the original system must resemble the rough sketch shown in Fig. 6.3.8. This sketch shows a few typical freehand trajectories connecting a nodal source at (0, 0), nodal sinks at (0, 32) and (28, 0), and a saddle point at (12, 8), with indicated directions of flow along these trajectories consistent with the known character of these critical points. Figure 6.3.9 shows a more precise computer-generated phase portrait and direction field for the nonlinear system in (10).

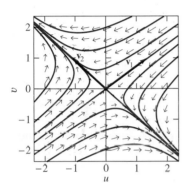

FIGURE 6.3.7. Phase plane portrait for the linear system in Eq. (17) corresponding to the critical point (12, 8).

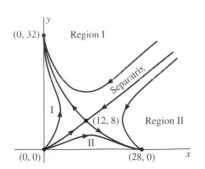

FIGURE 6.3.8. Rough sketch consistent with the analysis in Example 3.

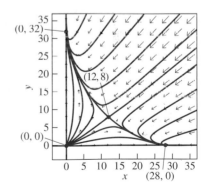

FIGURE 6.3.9. Phase plane portrait for the system in Example 3.

The two trajectories that approach the saddle point (12, 8), together with that saddle point, form a **separatrix** that separates regions I and II in Figure 6.3.9. It plays a crucial role in determining the long-term behavior of the two populations. If the initial point (x_0, y_0) lies precisely on the separatrix, then $(x(t), y(t))$

approaches $(12, 8)$ as $t \to +\infty$. Of course, random events make it extremely unlikely that $(x(t), y(t))$ will remain on the separatrix. If not, peaceful coexistence of the two species is impossible. If (x_0, y_0) lies in Region I above the separatrix, then $(x(t), y(t))$ approaches $(0, 32)$ as $t \to +\infty$, so the population $x(t)$ decreases to zero. Alternatively, if (x_0, y_0) lies in Region II below the separatrix, then $(x(t), y(t))$ approaches $(28, 0)$ as $t \to +\infty$, so the population $y(t)$ dies out. In short, whichever population has the initial competitive advantage survives, while the other faces extinction. ∎

Example 4　**Peaceful Coexistence of Two Species**　Suppose that the populations $x(t)$ and $y(t)$ satisfy the competition system

$$\frac{dx}{dt} = 14x - 2x^2 - xy,$$

$$\frac{dy}{dt} = 16y - 2y^2 - xy, \tag{18}$$

for which $a_1 = 14$, $a_2 = 16$, $b_1 = b_2 = 2$, and $c_1 = c_2 = 1$. Then $c_1 c_2 = 1 < 4 = b_1 b_2$, so now the effect of inhibition is greater than that of competition. We find readily that the four critical points are $(0, 0)$, $(0, 8)$, $(7, 0)$, and $(4, 6)$. We proceed as in Example 3.

THE CRITICAL POINT $(0,0)$:　When we drop the quadratic terms in (18), we get the same linearization $x' = 14x$, $y' = 16y$ at $(0, 0)$ as in Example 3. Thus its coefficient matrix has the two positive eigenvalues $\lambda_1 = 14$ and $\lambda_2 = 16$, and its phase portrait is the same as that shown in Fig. 6.3.3. Therefore, $(0, 0)$ is an unstable nodal source for the original system in (18).

THE CRITICAL POINT $(0,8)$:　The Jacobian matrix of the system in (18) is

$$\mathbf{J}(x, y) = \begin{bmatrix} 14 - 4x - y & -x \\ -y & 16 - 4y - x \end{bmatrix}, \quad \text{so} \quad \mathbf{J}(0, 8) = \begin{bmatrix} 6 & 0 \\ -8 & -16 \end{bmatrix}. \tag{19}$$

The matrix $\mathbf{J}(0, 8)$ corresponds to the linearization

$$\frac{du}{dt} = 6u,$$

$$\frac{dv}{dt} = -8u - 16v \tag{20}$$

FIGURE 6.3.10. Phase plane portrait for the linear system in Eq. (20) corresponding to the critical point $(0, 8)$.

of (18) at $(0, 8)$. It has characteristic equation $(6 - \lambda)(-16 - \lambda) = 0$ and has the positive eigenvalue $\lambda_1 = 6$ with eigenvector $\mathbf{v}_1 = \begin{bmatrix} 11 & -4 \end{bmatrix}^T$ and the negative eigenvalue $\lambda_2 = -16$ with eigenvector $\mathbf{v}_2 = \begin{bmatrix} 0 & 1 \end{bmatrix}^T$. It follows that $(0, 0)$ is a saddle point for the linearized system, and hence that $(0, 8)$ is an unstable saddle point for the original system in (18). Figure 6.3.10 shows a phase portrait for the linearized system near $(0, 0)$.

THE CRITICAL POINT (7, 0): The Jacobian matrix

$$\mathbf{J}(7,0) = \begin{bmatrix} -14 & -7 \\ 0 & 9 \end{bmatrix} \qquad (21)$$

corresponds to the linearization

$$\frac{du}{dt} = -14u - 7v,$$
$$\frac{dv}{dt} = 9v \qquad (22)$$

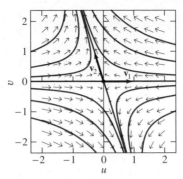

FIGURE 6.3.11. Phase plane portrait for the linear system in Eq. (22) corresponding to the critical point (7, 0).

of (18) at (7, 0). The matrix $\mathbf{J}(7, 0)$ has characteristic equation $(-14-\lambda)(9-\lambda) = 0$ and has the negative eigenvalue $\lambda_1 = -14$ with eigenvector $\mathbf{v}_1 = \begin{bmatrix} 1 & 0 \end{bmatrix}^T$ and the positive eigenvalue $\lambda_2 = 9$ with eigenvector $\mathbf{v}_2 = \begin{bmatrix} -7 & 23 \end{bmatrix}^T$. It follows that (0, 0) is a saddle point for the linearized system, and hence that (7, 0) is an unstable saddle point for the original system in (18). Figure 6.3.11 shows a phase portrait for the linearized system near (0, 0).

THE CRITICAL POINT (4,6): The Jacobian matrix

$$\mathbf{J}(4,6) = \begin{bmatrix} -8 & -4 \\ -6 & -12 \end{bmatrix} \qquad (23)$$

corresponds to the linearization

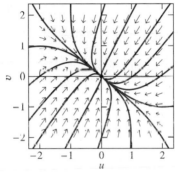

FIGURE 6.3.12. Phase plane portrait for the linear system in Eq. (24) corresponding to the critical point (4, 6).

$$\frac{du}{dt} = -8u - 4v,$$
$$\frac{dv}{dt} = 6u - 12v \qquad (24)$$

of (18) at (4, 6). The matrix $\mathbf{J}(4, 6)$ has characteristic equation

$$(-8 - \lambda)(-12 - \lambda) - (-6)(-4) = \lambda^2 + 20\lambda + 72 = 0$$

and has the two negative eigenvalues

$$\lambda_1 = 2(-5 - \sqrt{7}) \quad \text{with eigenvector} \quad \mathbf{v}_1 = \begin{bmatrix} \frac{1}{3}(-1 + \sqrt{7}) & 1 \end{bmatrix}^T$$

and

$$\lambda_2 = 2(-5 + \sqrt{7}) \quad \text{with eigenvector} \quad \mathbf{v}_2 = \begin{bmatrix} \frac{1}{3}(-1 - \sqrt{7}) & 1 \end{bmatrix}^T.$$

It follows that (0, 0) is a nodal sink for the linearized system, and hence that (4, 6) is a stable nodal sink for the original system in (18). Figure 6.3.12 shows a phase portrait for the linearized system near (0, 0).

Figure 6.3.13 assembles all this local information into a global phase plane portrait for the original system in (18). The notable feature of this system is that—for *any* positive initial population values x_0 and y_0—the point $(x(t), y(t))$ approaches the single critical point (4, 6) as $t \to +\infty$. It follows that the two species both survive in stable (peaceful) existence. ∎

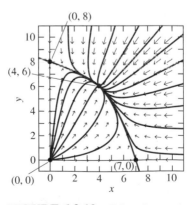

FIGURE 6.3.13. Direction field and phase portrait for the competition system
$x' = 14x - 2x^2 - xy,$
$y' = 16y - 2y^2 - xy$ of
Example 4.

Interactions of Logistic Populations

If the coefficients a_1, a_2, b_1, b_2 are positive but $c_1 = c_2 = 0$, then the equations

$$\frac{dx}{dt} = a_1 x - b_1 x^2 - c_1 xy,$$

$$\frac{dy}{dt} = a_2 y - b_2 y^2 - c_2 xy \tag{25}$$

describe two separate logistic populations $x(t)$ and $y(t)$ that have no effect on each other. Examples 3 and 4 illustrate cases in which the xy-coefficients c_1 and c_2 are both *positive*. The interaction between the two populations is then described as **competition**, because the effect of the xy-terms in (25) is to *decrease* the rates of growth of both populations—that is, each population is "hurt" by their mutual interaction.

Suppose, however, that the interaction coefficients c_1 and c_2 in (25) are both *negative*. Then the effect of the xy-terms is to increase the rates of growth of both populations—that is, each population is "helped" by their mutual interaction. This type of interaction is aptly described as **cooperation** between the two logistic populations.

Finally, the interaction between the two populations is one of **predation** if the interaction coefficients have *different signs*. For instance, if $c_1 > 0$ but $c_2 < 0$, then the x-population is hurt but the y-population is helped by their interaction. We may therefore describe $x(t)$ as a prey population and $y(t)$ as a predator population.

If either b_1 or b_2 is zero in (25), then the corresponding population would (in the absence of the other) exhibit exponential growth rather than logistic growth. For instance, suppose that $a_1 > 0$, $a_2 < 0$, $b_1 = b_2 = 0$, and $c_1 > 0$, $c_2 < 0$. Then $x(t)$ is a naturally growing prey population while $y(t)$ is a naturally declining predator population. This is the original predator-prey model with which we began this section.

Problems 26 through 34 illustrate a variety of the possibilities indicated here. The problems and examples in this section illustrate the power of elementary critical point analysis. But remember that ecological systems in nature are seldom so simple as in these examples. Frequently they involve more than two species, and the growth rates of these populations and the interactions among them often are more complicated than those discussed in this section. Consequently, the mathematical modeling of ecological systems remains an active area of current research.

6.3 Problems

Problems 1 and 2 deal with the predator-prey system

$$\frac{dx}{dt} = 200x - 4xy,$$

$$\frac{dy}{dt} = -150y + 2xy \tag{1}$$

that corresponds to Fig. 6.3.1.

1. Starting with the Jacobian matrix of the system in (1), derive its linearizations at the two critical points $(0, 0)$ and $(75, 50)$. Use a graphing calculator or computer system to construct phase plane portraits for these two lineariza-

tions that are consistent with the "big picture" shown in Fig. 6.3.1.

2. Separate the variables in the quotient

$$\frac{dy}{dx} = \frac{-150y + 2xy}{200x - 4xy}$$

of the two equations in (1), and thereby derive the exact implicit solution

$$200 \ln y + 150 \ln x - 2x - 4y = C$$

of the system. Use the contour plot facility of a graphing calculator or computer system to plot the contour curves

of this equation through the points (75, 100), (75, 150), (75, 200), (75, 250), and (75, 300) in the xy-plane. Are your results consistent with Fig. 6.3.1?

3. Let $x(t)$ be a harmful insect population (aphids?) that under natural conditions is held somewhat in check by a benign predator insect population $y(t)$ (ladybugs?). Assume that $x(t)$ and $y(t)$ satisfy the predator-prey equations in (1), so that the stable equilibrium populations are $x_E = b/q$ and $y_E = a/p$. Now suppose that an insecticide is employed that kills (per unit time) the same fraction $f < a$ of each species of insect. Show that the harmful population x_E is increased, while the benign population y_E is decreased, so the use of the insecticide is counterproductive. This is an instance in which mathematical analysis reveals undesirable consequences of a well-intentioned interference with nature.

Problems 4 through 7 deal with the competition system

$$\frac{dx}{dt} = 60x - 4x^2 - 3xy,$$

$$\frac{dy}{dt} = 42y - 2y^2 - 3xy,$$

(2)

in which $c_1 c_2 = 9 > 8 = b_1 b_2$, so the effect of competition should exceed that of inhibition. Problems 4 through 7 imply that the four critical points $(0, 0)$, $(0, 21)$, $(15, 0)$, and $(6, 12)$ of the system in (2) resemble those shown in Fig. 6.3.9—a nodal source at the origin, a nodal sink on each coordinate axis, and a saddle point interior to the first quadrant. In each of these problems use a graphing calculator or computer system to construct a phase plane portrait for the linearization at the indicated critical point. Finally, construct a first-quadrant phase plane portrait for the nonlinear system in (2). Do your local and global portraits look consistent?

4. Show that the coefficient matrix of the linearization $x' = 60x$, $y' = 42y$ of (2) at $(0, 0)$ has positive eigenvalues $\lambda_1 = 60$ and $\lambda_2 = 42$. Hence $(0, 0)$ is a nodal source for (2).

5. Show that the linearization of (2) at $(0, 21)$ is $u' = -3u$, $v' = -63u - 42v$. Then show that the coefficient matrix of this linear system has negative eigenvalues $\lambda_1 = -3$ and $\lambda_2 = -42$. Hence $(0, 21)$ is a nodal sink for the system in (2).

6. Show that the linearization of (2) at $(15, 0)$ is $u' = -60u - 45v$, $v' = -3v$. Then show that the coefficient matrix of this linear system has negative eigenvalues $\lambda_1 = -60$ and $\lambda_2 = -3$. Hence $(15, 0)$ is a nodal sink for the system in (2).

7. Show that the linearization of (2) at $(6, 12)$ is $u' = -24u - 18v$, $v' = -36u - 24v$. Then show that the coefficient matrix of this linear system has eigenvalues $\lambda_1 = -24 - 18\sqrt{2} < 0$ and $\lambda_2 = -24 + 18\sqrt{2} > 0$. Hence $(6, 12)$ is a saddle point for the system in (2).

Problems 8 through 10 deal with the competition system

$$\frac{dx}{dt} = 60x - 3x^2 - 4xy,$$

$$\frac{dy}{dt} = 42y - 3y^2 - 2xy,$$

(3)

in which $c_1 c_2 = 8 < 9 = b_1 b_2$, so the effect of inhibition should exceed that of competition. The linearization of the system in (3) at $(0, 0)$ is the same as that of (2). This observation and Problems 8 through 10 imply that the four critical points $(0, 0)$, $(0, 14)$, $(20, 0)$, and $(12, 6)$ of (3) resemble those shown in Fig. 6.3.13—a nodal source at the origin, a saddle point on each coordinate axis, and a nodal sink interior to the first quadrant. In each of these problems use a graphing calculator or computer system to construct a phase plane portrait for the linearization at the indicated critical point. Finally, construct a first-quadrant phase plane portrait for the nonlinear system in (3). Do your local and global portraits look consistent?

8. Show that the linearization of (3) at $(0, 14)$ is $u' = 4u$, $v' = -28u - 42v$. Then show that the coefficient matrix of this linear system has the positive eigenvalue $\lambda_1 = 4$ and the negative eigenvalue $\lambda_2 = -42$. Hence $(0, 14)$ is a saddle point for the system in (3).

9. Show that the linearization of (3) at $(20, 0)$ is $u' = -60u - 80v$, $v' = 2v$. Then show that the coefficient matrix of this linear system has the negative eigenvalue $\lambda_1 = -60$ and the positive eigenvalue $\lambda_2 = 2$. Hence $(20, 0)$ is a saddle point for the system in (3).

10. Show that the linearization of (3) at $(12, 6)$ is $u' = -36u - 48v$, $v' = -12u - 18v$. Then show that the coefficient matrix of this linear system has eigenvalues $\lambda_1 = -27 + 3\sqrt{73}$ and $\lambda_2 = -27 - 3\sqrt{73}$, both of which are negative. Hence $(12, 6)$ is a nodal sink for the system in (3).

Problems 11 through 13 deal with the predator-prey system

$$\frac{dx}{dt} = 5x - x^2 - xy,$$

$$\frac{dy}{dt} = -2y + xy,$$

(4)

in which the prey population $x(t)$ is logistic but the predator population $y(t)$ would (in the absence of any prey) decline naturally. Problems 11 through 13 imply that the three critical points $(0, 0)$, $(5, 0)$, and $(2, 3)$ of the system in (4) are as shown in Fig. 6.3.14—with saddle points at the origin and on the positive x-axis, and with a spiral sink interior to the first quadrant. In each of these problems use a graphing calculator or computer system to construct a phase plane portrait for the linearization at the indicated critical point. Do your local portraits look consistent with Fig. 6.3.14?

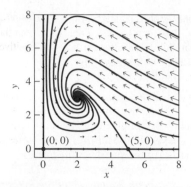

FIGURE 6.3.14. Direction field and phase portrait for the predator-prey system of Problems 11 through 13.

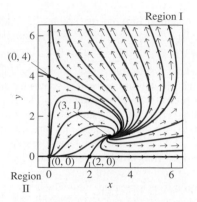

FIGURE 6.3.15. Direction field and phase portrait for the predator-prey system of Problems 14 through 17.

11. Show that the coefficient matrix of the linearization $x' = 5x$, $y' = -2y$ of (4) at $(0, 0)$ has the positive eigenvalue $\lambda_1 = 5$ and the negative eigenvalue $\lambda_2 = -2$. Hence $(0, 0)$ is a saddle point of the system in (4).

12. Show that the linearization of (4) at $(5, 0)$ is $u' = -5u - 5v$, $v' = 3v$. Then show that the coefficient matrix of this linear system has the negative eigenvalue $\lambda_1 = -5$ and the positive eigenvalue $\lambda_2 = 3$. Hence $(5, 0)$ is a saddle point for the system in (4).

13. Show that the linearization of (4) at $(2, 3)$ is $u' = -2u - 2v$, $v' = 3u$. Then show that the coefficient matrix of this linear system has the complex conjugate eigenvalues λ_1, $\lambda_2 = -1 \pm i\sqrt{5}$ with negative real part. Hence $(2, 3)$ is a spiral sink for the system in (4).

Problems 14 through 17 deal with the predator-prey system

$$\frac{dx}{dt} = x^2 - 2x - xy,$$

$$\frac{dy}{dt} = y^2 - 4y + xy. \tag{5}$$

Here each population—the prey population $x(t)$ and the predator population $y(t)$—is an unsophisticated population (like the alligators of Section 2.1) for which the only alternatives (in the absence of the other population) are doomsday and extinction. Problems 14 through 17 imply that the four critical points $(0, 0)$, $(0, 4)$, $(2, 0)$, and $(3, 1)$ of the system in (5) are as shown in Fig. 6.3.15—a nodal sink at the origin, a saddle point on each coordinate axis, and a spiral source interior to the first quadrant. This is a two-dimensional version of "doomsday versus extinction." If the initial point (x_0, y_0) lies in Region I, then both populations increase without bound (until doomsday), whereas if it lies in Region II, then both populations decrease to zero (and thus both become extinct). In each of these problems use a graphing calculator of computer system to construct a phase plane portrait for the linearization at the indicated critical point. Do your local portraits look consistent with Fig. 6.3.15?

14. Show that the coefficient matrix of the linearization $x' = -2x$, $y' = -4y$ of the system in (5) at $(0, 0)$ has the negative eigenvalues $\lambda_1 = -2$ and $\lambda_2 = -4$. Hence $(0, 0)$ is a nodal sink for (5).

15. Show that the linearization of (5) at $(0, 4)$ is $u' = -6u$, $v' = 4u + 4v$. Then show that the coefficient matrix of this linear system has the negative eigenvalue $\lambda_1 = -6$ and the positive eigenvalue $\lambda_2 = 4$. Hence $(0, 4)$ is a saddle point for the system in (5).

16. Show that the linearization of (5) at $(2, 0)$ is $u' = 2u - 2v$, $v' = -2v$. Then show that the coefficient matrix of this linear system has the positive eigenvalue $\lambda_1 = 2$ and the negative eigenvalue $\lambda_2 = -2$. Hence $(2, 0)$ is a saddle point for the system in (5).

17. Show that the linearization of (5) at $(3, 1)$ is $u' = 3u - 3v$, $v' = u + v$. Then show that the coefficient matrix of this linear system has complex conjugate eigenvalues λ_1, $\lambda_2 = 2 \pm i\sqrt{2}$ with positive real part. Hence $(3, 1)$ is a spiral source for (5).

Problems 18 through 25 deal with the predator-prey system

$$\frac{dx}{dt} = 2x - xy + \epsilon x(5 - x),$$

$$\frac{dy}{dt} = -5y + xy, \tag{6}$$

for which a bifurcation occurs at the value $\epsilon = 0$ of the parameter ϵ. Problems 18 and 19 deal with the case $\epsilon = 0$, in which case the system in (6) takes the form

$$\frac{dx}{dt} = 2x - xy, \qquad \frac{dy}{dt} = -5x + xy, \tag{7}$$

and these problems suggest that the two critical points $(0, 0)$ and $(5, 2)$ of the system in (7) are as shown in Fig. 6.3.16—a saddle point at the origin and a center at $(5, 2)$. In each problem use a graphing calculator of computer system to construct a phase plane portrait for the linearization at the indicated critical point. Do your local figures look portraits with Fig. 6.3.16?

FIGURE 6.3.16. The case $\epsilon = 0$
(Problems 18 and 19).

18. Show that the coefficient matrix of the linearization $x' = 2x$, $y' = -5y$ of (7) at $(0, 0)$ has the positive eigenvalue $\lambda_1 = 2$ and the negative eigenvalue $\lambda_2 = -5$. Hence $(0, 0)$ is a saddle point for the system in (7).

19. Show that the linearization of the system in (7) at $(5, 2)$ is $u' = -5v$, $v' = 2u$. Then show that the coefficient matrix of this linear system has conjugate imaginary eigenvalues λ_1, $\lambda_2 = \pm i\sqrt{10}$. Hence $(0, 0)$ is a stable center for the linear system. Although this is the indeterminate case of Theorem 2 in Section 6.2, Fig. 6.3.16 suggests that $(5, 2)$ also is a stable center for (7).

Problems 20 through 22 deal with the case $\epsilon = -1$, for which the system in (6) becomes

$$\frac{dx}{dt} = 3x + x^2 - xy, \qquad \frac{dy}{dt} = -5y + xy, \qquad (8)$$

and imply that the three critical points $(0, 0)$, $(3, 0)$, and $(5, 2)$ of (8) are as shown in Fig. 6.3.17—with a nodal sink at the origin, a saddle point on the positive x-axis, and a spiral source at $(5, 2)$. In each problem use a graphing calculator of computer system to construct a phase plane portrait for the linearization at the indicated critical point. Do your local portraits look consistent with Fig. 6.3.17?

FIGURE 6.3.17. The case $\epsilon = -1$
(Problems 20 through 22).

20. Show that the coefficient matrix of the linearization $x' = -3x$, $y' = -5y$ of the system in (8) at $(0, 0)$ has the negative eigenvalues $\lambda_1 = -3$ and $\lambda_2 = -5$. Hence $(0, 0)$ is a nodal sink for (8).

21. Show that the linearization of the system in (8) at $(3, 0)$ is $u' = 3u - 3v$, $v' = -2v$. Then show that the coefficient matrix of this linear system has the positive eigenvalue $\lambda_1 = 3$ and the negative eigenvalue $\lambda_2 = -2$. Hence $(3, 0)$ is a saddle point for (8).

22. Show that the linearization of (8) at $(5, 2)$ is $u' = 5u - 5v$, $v' = 2u$. Then show that the coefficient matrix of this linear system has complex conjugate eigenvalues λ_1, $\lambda_2 = \frac{1}{2}(5 \pm i\sqrt{15})$ with positive real part. Hence $(5, 2)$ is a spiral source for the system in (8).

Problems 23 through 25 deal with the case $\epsilon = 1$, so that the system in (6) takes the form

$$\frac{dx}{dt} = 7x - x^2 - xy, \qquad \frac{dy}{dt} = -5y + xy, \qquad (9)$$

and these problems imply that the three critical points $(0, 0)$, $(7, 0)$, and $(5, 2)$ of the system in (9) are as shown in Fig. 6.3.18—with saddle points at the origin and on the positive x-axis and with a spiral sink at $(5, 2)$. In each problem use a graphing calculator of computer system to construct a phase plane portrait for the linearization at the indicated critical point. Do your local portraits look consistent with Fig. 6.3.18?

FIGURE 6.3.18. The case $\epsilon = +1$
(Problems 23 through 25).

23. Show that the coefficient matrix of the linearization $x' = 7x$, $y' = -5y$ of (9) at $(0, 0)$ has the positive eigenvalue $\lambda_1 = 7$ and the negative eigenvalue $\lambda_2 = -5$. Hence $(0, 0)$ is a saddle point for the system in (9).

24. Show that the linearization of (9) at $(7, 0)$ is $u' = -7u - 7v$, $v' = 2v$. Then show that the coefficient matrix of this linear system has the negative eigenvalue $\lambda_1 = -7$ and the positive eigenvalue $\lambda_2 = 2$. Hence $(7, 0)$ is a saddle point for the system in (9).

25. Show that the linearization of (9) at $(5, 2)$ is $u' = -5u - 5v$, $v' = 2u$. Then show that the coefficient matrix of this linear system has the complex conjugate eigenvalues λ_1, $\lambda_2 = \frac{1}{2}(-5 \pm i\sqrt{15})$ with negative real part. Hence $(5, 2)$ is a spiral sink for the system in (9).

For each two-population system in Problems 26 through 34, first describe the type of x- and y-populations involved (exponential or logistic) and the nature of their interaction—competition, cooperation, or predation. Then find and characterize the system's critical points (as to type and stability). Determine what nonzero x- and y-populations can coexist? Finally, construct a phase plane portrait that enables you to describe the long-term behavior of the two populations in terms of their initial populations $x(0)$ and $y(0)$.

26. $\dfrac{dx}{dt} = 2x - xy, \quad \dfrac{dy}{dt} = 3y - xy$

27. $\dfrac{dx}{dt} = 2xy - 4x, \quad \dfrac{dy}{dt} = xy - 3y$

28. $\dfrac{dx}{dt} = 2xy - 16x, \quad \dfrac{dy}{dt} = 4y - xy$

29. $\dfrac{dx}{dt} = 3x - x^2 - \dfrac{1}{2}xy, \quad \dfrac{dy}{dt} = 4y - 2xy$

30. $\dfrac{dx}{dt} = 3x - x^2 + \dfrac{1}{2}xy, \quad \dfrac{dy}{dt} = \dfrac{1}{5}xy - y$

31. $\dfrac{dx}{dt} = 3x - x^2 - \dfrac{1}{4}xy, \quad \dfrac{dy}{dt} = xy - 2y$

32. $\dfrac{dx}{dt} = 30x - 3x^2 + xy, \quad \dfrac{dy}{dt} = 60y - 3y^2 + 4xy$

33. $\dfrac{dx}{dt} = 30x - 2x^2 - xy, \quad \dfrac{dy}{dt} = 80y - 4y^2 + 2xy$

34. $\dfrac{dx}{dt} = 30x - 2x^2 - xy, \quad \dfrac{dy}{dt} = 20y - 4y^2 + 2xy$

6.3 Application: Your Own Wildlife Conservation Preserve

You own a large wildlife conservation preserve that you originally stocked with F_0 foxes and R_0 rabbits on January 1, 1999. The following differential equations model the numbers $R(t)$ of rabbits and $F(t)$ foxes t months later:

$$\frac{dR}{dt} = (0.01)pR - (0.0001)aRF,$$

$$\frac{dF}{dt} = -(0.01)qF + (0.0001)bRF,$$

where p and q are the two largest digits (with $p < q$) and a and b are the two smallest nonzero digits (with $a < b$) in your student ID number.

The numbers of foxes and rabbits will oscillate periodically, out of phase with each other (like the functions $x(t)$ and $y(t)$ in Fig. 6.3.3). Choose your initial numbers F_0 of foxes and R_0 of rabbits—perhaps several hundred of each—so that the resulting solution curve in the RF-plane is a fairly eccentric closed curve. (The eccentricity may be increased if you begin with a relatively large number of rabbits and a small number of foxes, as any wildlife preserve owner would naturally do—because foxes prey on rabbits.)

Your task is to determine

1. The period of oscillation of the rabbit and fox populations
2. The maximum and minimum numbers of rabbits, and the calendar dates on which they first occur
3. The maximum and minimum numbers of foxes, and the calendar dates on which they first occur

With computer software that can plot both RF-trajectories and tR- and tF-solution curves like those in Figs. 6.3.2 and 6.3.3, you can "zoom in" graphically on the points whose coordinates provide the requested information.

6.4 Nonlinear Mechanical Systems

Now we apply the qualitative methods of Sections 6.1 and 6.2 to the analysis of simple mechanical systems like the mass-on-a-spring system shown in Fig. 6.4.1.

Let m denote the mass in a suitable system of units and let $x(t)$ denote the displacement of the mass at time t from its equilibrium position (in which the spring is unstretched). Previously we have always assumed that the force $F(x)$ exerted by the spring on the mass is a *linear* function of x: $F(x) = -kx$ (Hooke's law). In reality, however, every spring in nature actually is nonlinear (even if only slightly so). Moreover, springs in some automobile suspension systems deliberately are designed to be nonlinear. Here, then, we are interested specifically in the effects of nonlinearity.

So now we allow the force function $F(x)$ to be nonlinear. Because $F(0) = 0$ at the equilibrium position $x = 0$, we may assume that F has a power series expansion of the form

$$F(x) = -kx + \alpha x^2 + \beta x^3 + \cdots . \qquad (1)$$

We take $k > 0$ so that the reaction of the spring is directed opposite to the displacement when x is sufficiently small. If we assume also that the reaction of the spring is symmetric with respect to positive and negative displacements by the same distance, then $F(-x) = -F(x)$, so F is an *odd* function. In this case it follows that the coefficient of x^n in Eq. (1) is zero if n is even, so the first nonlinear term is the one involving x^3.

For a simple mathematical model of a nonlinear spring we therefore take

$$F(x) = -kx + \beta x^3, \qquad (2)$$

ignoring all terms in Eq. (1) of degree greater than 3. The equation of motion of the mass m is then

$$mx'' = -kx + \beta x^3. \qquad (3)$$

The Position-Velocity Phase Plane

If we introduce the *velocity*

$$y(t) = x'(t) \qquad (4)$$

of the mass with *position* $x(t)$, then we get from Eq. (3) the equivalent first-order system

$$\frac{dx}{dt} = y,$$
$$m\frac{dy}{dt} = -kx + \beta x^3. \qquad (5)$$

A phase plane trajectory of this system is a position-velocity plot that illustrates the motion of the mass on the spring. We can solve explicitly for the trajectories of this system by writing

$$\frac{dy}{dx} = \frac{dy/dt}{dx/dt} = \frac{-kx + \beta x^3}{my},$$

whence

$$my\, dy + (kx - \beta x^3)\, dx = 0.$$

Integration then yields

$$\tfrac{1}{2}my^2 + \tfrac{1}{2}kx^2 - \tfrac{1}{4}\beta x^4 = E \qquad (6)$$

Equilibrium
position

$x(t)$

FIGURE 6.4.1. The mass on a spring.

for the equation of a typical trajectory. We write E for the arbitrary constant of integration because $KE = \frac{1}{2}my^2$ is the kinetic energy of the mass with velocity y, and it is natural to define

$$PE = \frac{1}{2}kx^2 - \frac{1}{4}\beta x^4 \tag{7}$$

as the potential energy of the spring. Then Eq. (6) takes the form $KE + PE = E$, and thus expresses conservation of energy for the *undamped* motion of a mass on a spring.

The behavior of the mass depends on the sign of the nonlinear term in Eq. (2). The spring is called

- *hard* if $\beta < 0$,
- *soft* if $\beta > 0$.

We consider the two cases separately.

HARD SPRING OSCILLATIONS: If $\beta < 0$, then the second equation in (5) takes the form $my' = -x\left(|\beta|x^2 + k\right)$, so it follows that the only critical point of the system is the origin $(0, 0)$. Each trajectory

$$\frac{1}{2}my^2 + \frac{1}{2}kx^2 + \frac{1}{4}|\beta|x^4 = E > 0 \tag{8}$$

is an oval closed curve like those shown in Fig. 6.4.2, and thus $(0, 0)$ is a stable center. As the point $(x(t), y(t))$ traverses a trajectory in the clockwise direction, the position $x(t)$ and velocity $y(t)$ of the mass oscillate alternately as illustrated in Fig. 6.4.3. The mass is moving to the right (with x increasing) when $y > 0$, to the left when $y < 0$. Thus the behavior of a mass on a nonlinear hard spring resembles qualitatively that of a mass on a linear spring with $\beta = 0$ (as in Example 4 of Section 6.1). But one difference between the linear and nonlinear situations is that, whereas the period $T = 2\pi\sqrt{m/k}$ of oscillation of a mass on a linear spring is independent of the initial conditions, the period of a mass on a nonlinear spring depends on its initial position $x(0)$ and initial velocity $y(0)$ (Problems 17 through 22).

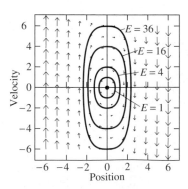

FIGURE 6.4.2. Position-velocity phase plane portrait for the hard mass-and-spring system with $m = k = 2$ and $\beta = -4 < 0$.

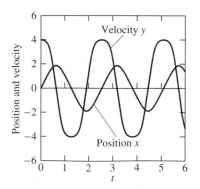

FIGURE 6.4.3. Position and velocity solution curves for the hard mass-and-spring system with $m = k = 2$ and $\beta = -4 < 0$.

Remark: The hard spring equation $mx'' = -kx - |\beta|x^3$ has equivalent first-order system

$$x' = y, \quad y' = -\frac{k}{m}x - \frac{|\beta|}{m}x^3$$

with Jacobian matrix

$$\mathbf{J}(x, y) = \begin{bmatrix} 0 & 1 \\ -\dfrac{k}{m} - \dfrac{3|\beta|}{m}x^2 & 0 \end{bmatrix}, \quad \text{so} \quad \mathbf{J}(0, 0) = \begin{bmatrix} 0 & 1 \\ -\omega^2 & 0 \end{bmatrix}$$

(writing $k/m = \omega^2$ as usual). The latter matrix has characteristic equation $\lambda^2 + \omega^2 = 0$ and pure imaginary eigenvalues $\lambda_1, \lambda_2 = \pm\omega i$. Thus the linearized system $x' = y$, $y' = -\omega^2 x$ has a stable center at the critical point $(0, 0)$—as we observed in Example 4 of Section 6.1. However, the nonlinear cubic term in the differential equation has (in effect) replaced although the elliptical trajectories (as in Fig. 6.1.7) of the linear system with the "flatter" quartic ovals we see in Fig. 6.4.2. ■

SOFT SPRING OSCILLATIONS: If $\beta > 0$, then the second equation in (5) takes the form $my' = x\left(\beta x^2 - k\right)$, so it follows that the system has the two critical points $\left(\pm\sqrt{k/\beta}, 0\right)$ in addition to the critical point $(0, 0)$. These three critical points yield the only solutions for which the mass can remain at rest. The following example illustrates the greater range of possible behaviors of a mass on a soft spring.

Example 1 If $m = 1$, $k = 4$, and $\beta = 1$, then the equation of motion of the mass is

$$\frac{d^2x}{dt^2} + 4x - x^3 = 0, \tag{9}$$

and Eq. (6) gives the trajectories in the form

$$\tfrac{1}{2}y^2 + 2x^2 - \tfrac{1}{4}x^4 = E. \tag{10}$$

After solving for

$$y = \pm\sqrt{2E - 4x^2 + \tfrac{1}{2}x^2}, \tag{10'}$$

we could select a fixed value of the constant energy E and plot manually a trajectory like one of those shown in the computer-generated position-velocity phase plane portrait in Fig. 6.4.4.

The different types of phase plane trajectories correspond to different values of the energy E. If we substitute $x = \pm\sqrt{k/\beta}$ and $y = 0$ into (6), we get the energy value $E = k^2/(4\beta) = 4$ (because $k = 4$ and $\beta = 1$) that corresponds to the trajectories that intersect the x-axis at the nontrivial critical points $(-2, 0)$ and $(2, 0)$. These emphasized trajectories are called **separatrices** because they separate phase plane regions of different behavior.

The nature of the motion of the mass is determined by which type of trajectory its initial conditions determine. The simple closed trajectories encircling $(0, 0)$ in the region bounded by the separatrices correspond to energies in the range $0 < E < 4$. These closed trajectories represent *periodic* oscillations of the mass back and forth around the equilibrium point $x = 0$.

The unbounded trajectories lying in the regions above and below the separatrices correspond to values of E greater than 4. These represent motions in which the mass approaches $x = 0$ with sufficient energy that it continues on through the equilibrium point, never to return again (as indicated in Fig. 6.4.5).

The unbounded trajectories opening to the right and left correspond to negative values of E. These represent motions in which the mass initially is headed toward

FIGURE 6.4.4. Position-velocity phase plane portrait for the soft mass-and-spring system with $m = 1, k = 4$, and $\beta = 1 > 0$. The separatrices are emphasized.

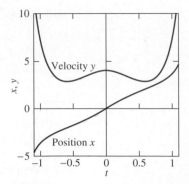

FIGURE 6.4.5. Position and velocity solution curves for the soft mass-and-spring system with $m = 1, k = 4, \beta = 1 > 0$, and energy $E = 8$—sufficiently great that the mass approaches the origin from the left and continues on indefinitely to the right.

the equilibrium point $x = 0$, but with insufficient energy to reach it. At some point the mass reverses direction and heads back whence it came.

In Fig. 6.4.4 it appears that the critical point $(0, 0)$ is a stable center, whereas the critical points $(\pm 2, 0)$ look like saddle points of the equivalent first-order system

$$x' = y, \quad y' = -4x + x^3 \tag{11}$$

with Jacobian matrix

$$\mathbf{J}(x, y) = \begin{bmatrix} 0 & 1 \\ -4 + 3x^2 & 0 \end{bmatrix}.$$

To check these observations against the usual critical point analysis, we note first that the Jacobian matrix

$$\mathbf{J}(0, 0) = \begin{bmatrix} 0 & 1 \\ -4 & 0 \end{bmatrix}$$

at the critical point $(0, 0)$ has characteristic equation $\lambda^2 + 4 = 0$ and pure imaginary eigenvalues $\lambda_1, \lambda_2 = \pm 2i$ consistent with a stable center. Moreover, the Jacobian matrix

$$\mathbf{J}(\pm 2, 0) = \begin{bmatrix} 0 & 1 \\ 8 & 0 \end{bmatrix}$$

corresponding to the other two critical points has characteristic equation $\lambda^2 - 8 = 0$ and real eigenvalues $\lambda_1, \lambda_2 = \pm\sqrt{8}$ of opposite sign, consistent with the saddle-point behavior that we observe near $(-2, 0)$ and $(+2, 0)$. ∎

Remark: Figures 6.4.2 and 6.4.4 illustrate a significant qualitative difference between hard springs with $\beta < 0$ and soft springs with $\beta > 0$ in the nonlinear equation $mx'' = kx + \beta x^3$. Whereas the phase plane trajectories for a hard spring are all bounded, a soft spring has unbounded phase plane trajectories (as well as bounded ones). However, we should realize that the unbounded soft-spring trajectories cease to represent physically realistic motions faithfully when they exceed the spring's capability of expansion without breaking. ∎

Damped Nonlinear Vibrations

Suppose now that the mass on a spring is connected also to a dashpot that provides a force of resistance proportional to the velocity $y = dx/dt$ of the mass. If the spring is still assumed nonlinear as in Eq. (2), then the equation of motion of the mass is

$$\blacktriangleright \qquad mx'' = -cx' - kx + \beta x^3, \qquad (12)$$

where $c > 0$ is the resistance constant. If $\beta > 0$, then the equivalent first-order system

$$\frac{dx}{dt} = y, \quad \frac{dy}{dt} = \frac{-kx - cy + \beta x^3}{m} = -\frac{c}{m}y - \frac{k}{m}x\left(1 - \frac{\beta}{k}x^2\right) \qquad (13)$$

has critical points $(0, 0)$ and $\left(\pm\sqrt{k/\beta}, 0\right)$ and Jacobian matrix

$$\mathbf{J}(x, y) = \begin{bmatrix} 0 & 1 \\ -\dfrac{k}{m} + \dfrac{3\beta}{m}x^2 & -\dfrac{c}{m} \end{bmatrix}.$$

Now the critical point at the origin is the most interesting one. The Jacobian matrix

$$\mathbf{J}(0, 0) = \begin{bmatrix} 0 & 1 \\ -\dfrac{k}{m} & -\dfrac{c}{m} \end{bmatrix}$$

has characteristic equation

$$(-\lambda)\left(-\frac{c}{m} - \lambda\right) + \frac{k}{m} = \frac{1}{m}(m\lambda^2 + c\lambda + k) = 0$$

and eigenvalues

$$\lambda_1, \lambda_2 = \frac{-c \pm \sqrt{c^2 - 4km}}{2m}.$$

It follows from Theorem 2 in Section 6.2 in that the critical point $(0, 0)$ of the system in (13) is

- a nodal sink if the resistance is so great that $c^2 > 4km$ (in which case the eigenvalues are negative and unequal), but is
- a spiral sink if $c^2 < 4km$ (in which case the eigenvalues are complex conjugates with negative real part).

The following example illustrates the latter case. (In the border-line case with equal negative eigenvalues, the origin may be either a nodal or a spiral sink.)

Example 2 Suppose that $m = 1, c = 2, k = 5$, and $\beta = \frac{5}{4}$. Then the nonlinear system in (13) is

$$\frac{dx}{dt} = y, \quad \frac{dy}{dt} = -5x - 2y + \frac{5}{4}x^3 = -2y - 5x(1 - \frac{1}{4}x^2). \qquad (14)$$

It has critical points $(0, 0)$, $(\pm 2, 0)$ and Jacobian matrix

$$\mathbf{J}(x, y) = \begin{bmatrix} 0 & 1 \\ -5 + \frac{15}{4}x^2 & -2 \end{bmatrix}.$$

At (0, 0): The Jacobian matrix

$$\mathbf{J}(0, 0) = \begin{bmatrix} 0 & 1 \\ -5 & -2 \end{bmatrix}$$

has characteristic equation $\lambda^2 + 2\lambda + 5 = 0$ and has complex conjugate eigenvalues λ_1, $\lambda_2 = -1 \pm 2i$ with negative real part. Hence $(0, 0)$ is a spiral sink of the nonlinear system in (14), and the linearized position function of the mass is of the form

$$x(t) = e^{-t}(A \cos 2t + B \sin 2t)$$

that corresponds to an exponentially damped oscillation about the equilibrium position $x = 0$.

At (\pm2, 0): The Jacobian matrix

$$\mathbf{J}(\pm 2, 0) = \begin{bmatrix} 0 & 1 \\ 10 & -2 \end{bmatrix}$$

has characteristic equation $\lambda^2 + 2\lambda - 10 = 0$ and real eigenvalues $\lambda_1 = -1 - \sqrt{11} < 0$ and $\lambda_2 = -1 + \sqrt{11} > 0$ with different signs. It follows that $(-2, 0)$ and $(+2, 0)$ are both saddle points of the system in (14).

The position-velocity phase plane portrait in Fig. 6.4.6 shows trajectories of (14) and the spiral sink at $(0, 0)$, as well as the unstable saddle points at $(-2, 0)$ and $(2, 0)$. The emphasized separatrices divide the phase plane into regions of different behavior. The behavior of the mass depends on the region in which its initial point (x_0, y_0) is located. If this initial point lies in

- Region I between the separatrices, then the trajectory spirals into the origin as $t \to +\infty$, and hence the periodic oscillations of the undamped case (Fig. 6.4.4) are now replaced with damped oscillations around the stable equilibrium position $x = 0$;
- Region II, then the mass passes through $x = 0$ moving from left to right (x increasing);
- Region III, then the mass passes through $x = 0$ moving from right to left (x decreasing);
- Region IV, then the mass approaches (but does not reach) the unstable equilibrium position $x = -2$ from the left, but stops and then returns to the left;
- Region V, then the mass approaches (but does not reach) the unstable equilibrium position $x = 2$ from the right, but stops and then returns to the right.

If the initial point (x_0, y_0) lies precisely on one of the separatrices, then the corresponding trajectory either approaches the stable spiral point or recedes to infinity from a saddle point as $t \to +\infty$. ∎

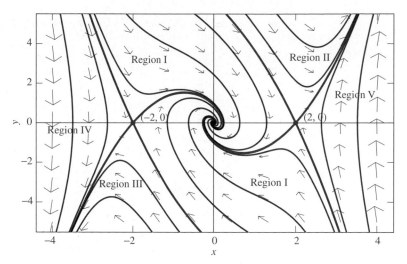

FIGURE 6.4.6. Position-velocity phase plane portrait for the soft mass-and-spring system with $m = 1$, $k = 5$, $\beta = \frac{5}{4}$, and resistance constant $c = 2$. The (black) separatrices are emphasized.

The Nonlinear Pendulum

In Section 3.4 we derived the equation

$$\frac{d^2\theta}{dt^2} + \frac{g}{L}\sin\theta = 0 \tag{15}$$

for the undamped oscillations of the simple pendulum shown in Fig. 6.4.7. There we used the approximation $\sin\theta \approx \theta$ for θ near zero to replace Eq. (15) with the linear model

$$\frac{d^2\theta}{dt^2} + \omega^2\theta = 0, \tag{16}$$

where $\omega^2 = g/L$. The general solution

$$\theta(t) = A\cos\omega t + B\sin\omega t \tag{17}$$

of Eq. (16) describes oscillations around the equilibrium position $\theta = 0$ with circular frequency ω and amplitude $C = (A^2 + B^2)^{1/2}$.

The linear model does not adequately describe the possible motions of the pendulum for large values of θ. For instance, the equilibrium solution $\theta(t) \equiv \pi$ of Eq. (15), with the pendulum standing straight up, does not satisfy the linear equation in (16). Nor does Eq. (17) include the situation in which the pendulum "goes over the top" repeatedly, so that $\theta(t)$ is a steadily increasing rather than an oscillatory function of t. To investigate these phenomena we must analyze the nonlinear equation $\theta'' + \omega^2\sin\theta = 0$ rather than merely its linearization $\theta'' + \omega^2\theta = 0$. We also want to include the possibility of resistance proportional to velocity, so we consider the general *nonlinear pendulum equation*

$$\frac{d^2\theta}{dt^2} + c\frac{d\theta}{dt} + \omega^2\sin\theta = 0. \tag{18}$$

FIGURE 6.4.7. The simple pendulum.

We examine first the undamped case, in which $c = 0$. With $x(t) = \theta(t)$ and $y(t) = \theta'(t)$ the equivalent first-order system is

$$\frac{dx}{dt} = y, \quad \frac{dy}{dt} = -\omega^2 \sin x. \tag{19}$$

We see that this system is almost linear by writing it in the form

$$\frac{dx}{dt} = y,$$

$$\frac{dy}{dt} = -\omega^2 x + g(x), \tag{20}$$

where

$$g(x) = -\omega^2(\sin x - x) = \omega^2 \left(\frac{x^3}{3!} - \frac{x^5}{5!} + \cdots \right)$$

has only higher-degree terms.

The critical points of the system in (19) are the points $(n\pi, 0)$ with n an integer, and its Jacobian matrix is given by

$$\mathbf{J}(x, y) = \begin{bmatrix} 0 & 1 \\ -\omega^2 \cos x & 0 \end{bmatrix}. \tag{21}$$

The nature of the critical point $(n\pi, 0)$ depends on whether n is even or odd.

EVEN CASE: If $n = 2m$ is an even integer, then $\cos n\pi = +1$, so (21) yields the matrix

$$\mathbf{J}(2m\pi, 0) = \begin{bmatrix} 0 & 1 \\ -\omega^2 & 0 \end{bmatrix}$$

with characteristic equation $\lambda^2 + \omega^2 = 0$ and pure imaginary eigenvalues $\lambda_1, \lambda_2 = \pm \omega i$. The linearization of (19) at $(n\pi, 0)$ is therefore the system

$$\frac{du}{dt} = v, \quad \frac{dv}{dt} = -\omega^2 u \tag{22}$$

for which $(0, 0)$ is the familiar stable center enclosed by elliptical trajectories (as in Example 4 of Section 6.1). Although this is the delicate case for which Theorem 2 of Section 6.2 does not settle the matter, we will see presently that $(2m\pi, 0)$ is also a stable center for the original nonlinear pendulum system in (19).

ODD CASE: If $n = 2m + 1$ is an odd integer, then $\cos n\pi = -1$, so (21) yields the matrix

$$\mathbf{J}((2m + 1)\pi, 0) = \begin{bmatrix} 0 & 1 \\ \omega^2 & 0 \end{bmatrix}$$

with characteristic equation $\lambda^2 - \omega^2 = 0$ and real eigenvalues $\lambda_1, \lambda_2 = \pm \omega$ with different signs. The linearization of (19) at $((2m + 1)\pi, 0)$ is therefore the system

$$\frac{du}{dt} = v, \quad \frac{dv}{dt} = \omega^2 u \tag{23}$$

for which $(0, 0)$ is a saddle point. It follows from Theorem 2 of Section 6.2 that the critical point $((2m + 1)\pi, 0)$ is a similar saddle point for the original nonlinear pendulum system in (19).

THE TRAJECTORIES: We can see how these "even centers" and "odd saddle points" fit together by solving the system in (19) explicitly for the phase plane trajectories. If we write

$$\frac{dy}{dx} = \frac{dy/dt}{dx/dt} = -\frac{\omega^2 \sin x}{y}$$

and separate the variables,

$$y\,dy + \omega^2 \sin x\,dx = 0,$$

then integration from $x = 0$ to $x = x$ yields

$$\tfrac{1}{2}y^2 + \omega^2(1 - \cos x) = E. \tag{24}$$

We write E for the arbitrary constant of integration because, if physical units are so chosen that $m = L = 1$, then the first term on the left is the kinetic energy and the second term the potential energy of the mass on the end of the pendulum. Then E is the total mechanical energy; Eq. (24) thus expresses conservation of mechanical energy for the undamped pendulum.

If we solve Eq. (24) for y and use a half-angle identity, we get the equation

$$y = \pm\sqrt{2E - 4\omega^2 \sin^2 \tfrac{1}{2}x} \tag{25}$$

that defines the phase plane trajectories. Note that the radicand in (25) remains positive if $E > 2\omega^2$. Figure 6.4.8 shows (along with a direction field) the results of plotting these trajectories for various values of the energy E.

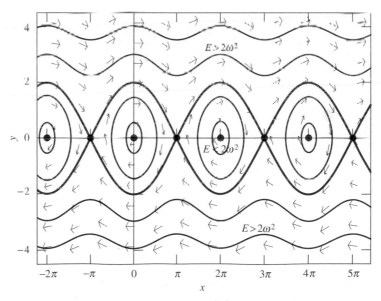

FIGURE 6.4.8. Position-velocity phase plane portrait for the undamped pendulum system $x' = y$, $y' = -\sin x$. The (black) separatrices are emphasized.

The emphasized separatrices in Fig. 6.4.8 correspond to the critical value $E = 2\omega^2$ of the energy; they enter and leave the unstable critical points $(n\pi, 0)$ with n an odd integer. Following the arrows along a separatrix, the pendulum theoretically approaches a balanced vertical position $\theta = x = (2m+1)\pi$ with just enough energy

to reach it but not enough to "go over the top." The instability of this equilibrium position indicates that this behavior may never be observed in practice!

The simple closed trajectories encircling the stable critical points—all of which correspond to the downward position $\theta = 2m\pi$ of the pendulum—represent periodic oscillations of the pendulum back and forth around the stable equilibrium position $\theta = 0$. These correspond to energies $E < 2\omega^2$ that are insufficient for the pendulum to ascend to the vertical upward position.

The unbounded trajectories with $E > 2\omega^2$ represent whirling motions of the pendulum in which it goes over the top repeatedly—in a clockwise direction if $y(t)$ remains positive, in a counterclockwise direction if $y(t)$ is negative.

Undamped Period of Oscillation

If the pendulum is released from rest with initial conditions

$$x(0) = \theta(0) = \alpha, \quad y(0) = \theta'(0) = 0, \tag{26}$$

then Eq. (24) with $t = 0$ reduces to

$$\omega^2(1 - \cos \alpha) = E. \tag{27}$$

Hence $E < 2\omega^2$ if $0 < \alpha < \pi$, so a periodic oscillation of the pendulum ensues. To determine the *period* of this oscillation, we subtract Eq. (27) from Eq. (24) and write the result (with $x = \theta$ and $y = d\theta/dt$) in the form

$$\frac{1}{2}\left(\frac{d\theta}{dt}\right)^2 = \omega^2(\cos \theta - \cos \alpha). \tag{28}$$

The period T of time required for one complete oscillation is four times the amount of time required for θ to decrease from $\theta = \alpha$ to $\theta = 0$, one-fourth of an oscillation. Hence we solve Eq. (28) for $dt/d\theta$ and integrate to get

$$T = \frac{4}{\omega\sqrt{2}} \int_0^\alpha \frac{d\theta}{\sqrt{\cos \theta - \cos \alpha}}. \tag{29}$$

To attempt to evaluate this integral we first use the identity $\cos \theta = 1 - 2\sin^2(\theta/2)$ and get

$$T = \frac{2}{\omega} \int_0^\alpha \frac{d\theta}{\sqrt{k^2 - \sin^2(\theta/2)}},$$

where

$$k = \sin \frac{\alpha}{2}.$$

Next, the substitution $u = (1/k)\sin(\theta/2)$ yields

$$T = \frac{4}{\omega} \int_0^1 \frac{du}{\sqrt{(1 - u^2)(1 - k^2 u^2)}}.$$

Finally, the substitution $u = \sin \phi$ gives

$$T = \frac{4}{\omega} \int_0^{\pi/2} \frac{d\phi}{\sqrt{1 - k^2 \sin^2 \phi}}. \tag{30}$$

The integral in (30) is the *elliptic integral of the first kind* that is often denoted by $F(k, \pi/2)$. It can be evaluated numerically as follows. First we use the binomial series

$$\frac{1}{\sqrt{1-x}} = 1 + \sum_{n=1}^{\infty} \frac{1 \cdot 3 \cdots (2n-1)}{2 \cdot 4 \cdots (2n)} x^n \tag{31}$$

with $x = k^2 \sin^2 \phi < 1$ to expand the integrand in (30). Then we integrate termwise using the tabulated integral formula

$$\int_0^{\pi/2} \sin^{2n} \phi \, d\phi = \frac{\pi}{2} \cdot \frac{1 \cdot 3 \cdots (2n-1)}{2 \cdot 4 \cdots (2n)}. \tag{32}$$

The final result is the formula

$$\begin{aligned}
T &= \frac{2\pi}{\omega} \left[1 + \sum_{n=1}^{\infty} \left(\frac{1 \cdot 3 \cdots (2n-1)}{2 \cdot 4 \cdots (2n)} \right)^2 k^{2n} \right] \\
&= T_0 \left[1 + \left(\frac{1}{2} \right)^2 k^2 + \left(\frac{1 \cdot 3}{2 \cdot 4} \right)^2 k^4 + \left(\frac{1 \cdot 3 \cdot 5}{2 \cdot 4 \cdot 6} \right)^2 k^6 + \cdots \right]
\end{aligned} \tag{33}$$

α	T/T_0
10°	1.0019
20°	1.0077
30°	1.0174
40°	1.0313
50°	1.0498
60°	1.0732
70°	1.1021
80°	1.1375
90°	1.1803

FIGURE 6.4.9. Dependence of the period T of a nonlinear pendulum on its initial angle α.

for the period T of the nonlinear pendulum released from rest with initial angle $\theta(0) = \alpha$, in terms of the linearized period $T_0 = 2\pi/\omega$ and $k = \sin(\alpha/2)$.

The infinite series within the second pair of brackets in Eq. (33) gives the factor T/T_0 by which the nonlinear period T is longer than the linearized period. The table in Fig. 6.4.9, obtained by summing this series numerically, shows how T/T_0 increases as α is increased. Thus T is 0.19% greater than T_0 if $\alpha = 10°$, whereas T is 18.03% greater than T_0 if $\alpha = 90°$. But even a 0.19% discrepancy is significant—the calculation

$$(0.0019) \times 3600 \frac{\text{seconds}}{\text{hour}} \times 24 \frac{\text{hours}}{\text{day}} \times 7 \frac{\text{days}}{\text{week}} \approx 1149 \text{ (seconds/week)}$$

shows that the linearized model is quite inadequate for a pendulum clock; a discrepancy of 19 min 9 s after only one week is unacceptable.

Damped Pendulum Oscillations

Finally, we discuss briefly the *damped* nonlinear pendulum. The almost linear first-order system equivalent to Eq. (19) is

$$\begin{aligned}
\frac{dx}{dt} &= y, \\
\frac{dy}{dt} &= -\omega^2 \sin x - cy,
\end{aligned} \tag{34}$$

and again the critical points are of the form $(n\pi, 0)$ where n is an integer. In Problems 9 through 11 we ask you to verify that

- If n is odd, then $(n\pi, 0)$ is an unstable saddle point of (34), just as in the undamped case; but
- If n is even and $c^2 > 4\omega^2$, then $(n\pi, 0)$ is a nodal sink; whereas
- If n is even and $c^2 < 4\omega^2$, then $(n\pi, 0)$ is a spiral sink.

Figure 6.4.10 illustrates the phase plane trajectories for the more interesting underdamped case, $c^2 < 4\omega^2$. Other than the physically unattainable separatrix trajectories that enter unstable saddle points, every trajectory eventually is "trapped" by one of the stable spiral points $(n\pi, 0)$ with n an even integer. What this means is that even if the pendulum starts with enough energy to go over the top, after a certain (finite) number of revolutions it has lost enough energy that thereafter it undergoes damped oscillations around its stable (lower) equilibrium position.

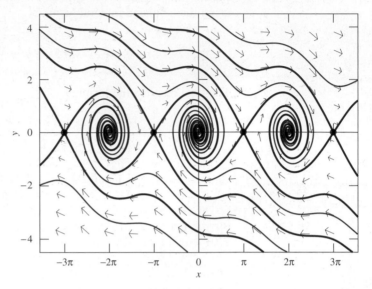

FIGURE 6.4.10. Position-velocity phase plane portrait for the damped pendulum system $x' = y$, $y' = -\sin x - \frac{1}{4}y$. The (black) separatrices are emphasized.

6.4 Problems

In Problems 1 through 4, show that the given system is almost linear with $(0, 0)$ as a critical point, and classify this critical point as to type and stability. Use a computer system or graphing calculator to construct a phase plane portrait that illustrates your conclusion.

1. $\dfrac{dx}{dt} = 1 - e^x + 2y$, $\dfrac{dy}{dt} = -x - 4\sin y$

2. $\dfrac{dx}{dt} = 2\sin x + \sin y$, $\dfrac{dy}{dt} = \sin x + 2\sin y$ (Fig. 6.4.11)

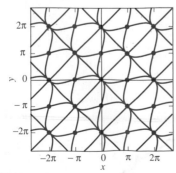

FIGURE 6.4.11. Trajectories of the system in Problem 2.

3. $\dfrac{dx}{dt} = e^x + 2y - 1$, $\dfrac{dy}{dt} = 8x + e^y - 1$

4. $\dfrac{dx}{dt} = \sin x \cos y - 2y$, $\dfrac{dy}{dt} = 4x - 3\cos x \sin y$

Find and classify each of the critical points of the almost linear systems in Problems 5 through 8. Use a computer system or graphing calculator to construct a phase plane portrait that illustrates your findings.

5. $\dfrac{dx}{dt} = -x + \sin y$, $\dfrac{dy}{dt} = 2x$

6. $\dfrac{dx}{dt} = y$, $\dfrac{dy}{dt} = \sin \pi x - y$

7. $\dfrac{dx}{dt} = 1 - e^{x-y}$, $\dfrac{dy}{dt} = 2\sin x$

8. $\dfrac{dx}{dt} = 3\sin x + y$, $\dfrac{dy}{dt} = \sin x + 2y$

Problems 9 through 11 deal with the damped pendulum system $x' = y$, $y' = -\omega^2 \sin x - cy$.

9. Show that if n is an odd integer, then the critical point $(n\pi, 0)$ is a saddle point for the damped pendulum system.

10. Show that if n is an even integer and $c^2 > 4\omega^2$, then the critical point $(n\pi, 0)$ is a nodal sink for the damped pendulum system.

11. Show that if n is an even integer and $c^2 < 4\omega^2$, then the critical point $(n\pi, 0)$ is a spiral sink for the damped pendulum system.

In each of Problems 12 through 16, a second-order equation of the form $x'' + f(x, x') = 0$, corresponding to a certain mass-and-spring system, is given. Find and classify the critical points of the equivalent first-order system.

12. $x'' + 20x - 5x^3 = 0$: Verify that the critical points resemble those shown in Fig. 6.4.4.

13. $x'' + 2x' + 20x - 5x^3 = 0$: Verify that the critical points resemble those shown in Fig. 6.4.6.

14. $x'' - 8x + 2x^3 = 0$: Here the linear part of the force is repulsive rather than attractive (as for an ordinary spring). Verify that the critical points resemble those shown in Fig. 6.4.12. Thus there are two stable equilibrium points and three types of periodic oscillations.

15. $x'' + 4x - x^2 = 0$: Here the force function is nonsymmetric. Verify that the critical points resemble those shown in Fig. 6.4.13.

16. $x'' + 4x - 5x^3 + x^5 = 0$: The idea here is that terms through the fifth degree in an odd force function have been retained. Verify that the critical points resemble those shown in Fig. 6.4.14.

In Problems 17 through 20, analyze the critical points of the indicated system, use a computer system to construct an illustrative position-velocity phase plane portrait, and describe the oscillations that occur.

17. Example 2 in this section illustrates the case of damped vibrations of a soft mass-spring system. Investigate an example of damped vibrations of a hard mass-spring system by using the same parameters as in Example 2, except now with $\beta = -\frac{5}{4} < 0$.

18. Example 2 illustrates the case of damped vibrations of a soft mass-spring system with the resistance proportional to the velocity. Investigate an example of resistance proportional to the *square* of the velocity by using the same parameters as in Example 2, but with resistance term $-cx'|x'|$ instead of $-cx'$ in Eq. (12).

19. Now repeat Example 2 with both the alterations corresponding to Problems 17 and 18. That is, take $\beta = -\frac{5}{4} < 0$ *and* replace the resistance term in Eq. (12) with $-cx'|x'|$.

20. The equations $x' = y$, $y' = -\sin x - \frac{1}{4}y|y|$ model a damped pendulum system as in Eqs. (34) and Fig. 6.4.10. But now the resistance is proportional to the *square* of the angular velocity of the pendulum. Compare the oscillations that occur with those that occur when the resistance is proportional to the angular velocity itself.

Problems 21 through 26 outline an investigation of the period T of oscillation of a mass on a nonlinear spring with equation of motion

$$\frac{d^2 x}{dt^2} + \phi(x) = 0. \qquad (35)$$

If $\phi(x) = kx$ with $k > 0$, then the spring actually is linear with period $T_0 = 2\pi/\sqrt{k}$.

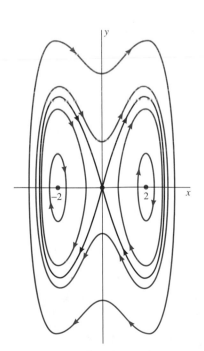

FIGURE 6.4.12. The phase portrait for Problem 14.

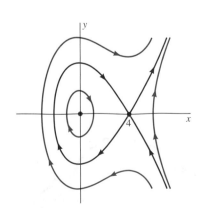

FIGURE 6.4.13. The phase portrait for Problem 15.

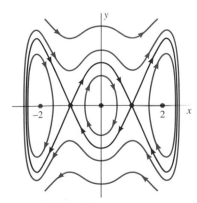

FIGURE 6.4.14. The phase portrait for Problem 16.

21. Integrate once (as in Eq. (6)) to derive the energy equation

$$\tfrac{1}{2}y^2 + V(x) = E, \tag{36}$$

where $y = dx/dt$ and

$$V(x) = \int_0^x \phi(u)\,du. \tag{37}$$

22. If the mass is released from rest with initial conditions $x(0) = x_0$, $y(0) = 0$ and periodic oscillations ensue, conclude from Eq. (36) that $E = V(x_0)$ and that the time T required for one complete oscillation is

$$T = \frac{4}{\sqrt{2}} \int_0^{x_0} \frac{du}{\sqrt{V(x_0) - V(u)}}. \tag{38}$$

23. If $\phi(x) = kx - \beta x^3$ as in the text, deduce from Eqs. (37) and (38) that

$$T = 4\sqrt{2} \int_0^{x_0} \frac{dx}{\sqrt{(x_0^2 - u^2)(2k - \beta x_0^2 - \beta u^2)}}. \tag{39}$$

24. Substitute $u = x_0 \cos\phi$ in (39) to show that

$$T = \frac{2T_0}{\pi\sqrt{1-\epsilon}} \int_0^{\pi/2} \frac{d\phi}{\sqrt{1 - \mu \sin^2\phi}}, \tag{40}$$

where $T_0 = 2\pi/\sqrt{k}$ is the linear period,

$$\epsilon = \frac{\beta}{k} x_0^2, \quad \text{and} \quad \mu = -\frac{1}{2} \cdot \frac{\epsilon}{1-\epsilon}. \tag{41}$$

25. Finally, use the binomial series in (31) and the integral formula in (32) to evaluate the elliptic integral in (40) and thereby show that the period T of oscillation is given by

$$T = \frac{T_0}{\sqrt{1-\epsilon}} \left(1 + \frac{1}{4}\mu + \frac{9}{64}\mu^2 + \frac{25}{256}\mu^3 + \cdots \right). \tag{42}$$

22. If $\epsilon = \beta x_0^2/k$ is sufficiently small that ϵ^2 is negligible, deduce from Eqs. (41) and (42) that

$$T \approx T_0\left(1 + \frac{3}{8}\epsilon\right) = T_0\left(1 + \frac{3\beta}{8k} x_0^2\right). \tag{43}$$

It follows that

- If $\beta > 0$, so the spring is *soft*, then $T > T_0$, and increasing x_0 increases T, so the larger ovals in Fig. 6.4.4 correspond to smaller frequencies.
- If $\beta < 0$, so the spring is *hard*, then $T < T_0$, and increasing x_0 decreases T, so the larger ovals in Fig. 6.4.2 correspond to larger frequencies.

6.4 Application: The Rayleigh and van der Pol Equations

The British mathematical physicist Lord Rayleigh (John William Strutt, 1842–1919) introduced an equation of the form

$$mx'' + kx = ax' - b(x')^3 \tag{1}$$

to model the oscillations of a clarinet reed. With $y = x'$ we get the autonomous system

$$x' = y,$$
$$y' = \frac{-kx + ay - by^3}{m}, \tag{2}$$

whose phase plane portrait is shown in Fig. 6.4.15 (for the case $m = k = a = b = 1$). The outward and inward spiral trajectories converge to a "limit cycle" solution that corresponds to periodic oscillations of the reed. The period T (and hence the frequency) of these oscillations can be measured on a tx-solution curve plotted as in Fig. 6.4.16. This period of oscillation depends only on the parameters m, k, a, and b in Eq. (1) and is independent of the initial conditions (why?).

Choose your own parameters m, k, a, and b (perhaps the least four nonzero digits in your student ID number), and use an available ODE plotting utility to plot trajectories and solution curves as in Figs. 6.4.15 and 6.4.16. Change *one* of your parameters to see how the amplitude and frequency of the resulting periodic oscillations are altered.

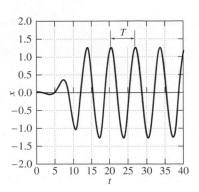

FIGURE 6.4.15. Phase plane portrait for the Rayleigh system in (2) with $m = k = a = b = 1$.

FIGURE 6.4.16. The tx-solution curve with initial conditions $x(0) = 0.01, x'(0) = 0$.

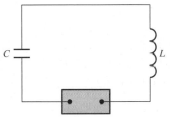

FIGURE 6.4.17. A simple circuit with an active element.

Van der Pol's Equation

Figure 6.4.17 shows a simple RLC circuit in which the usual (passive) resistance R has been replaced with an active element (such as a vacuum tube or semiconductor) across which the voltage drop V is given by a known function $f(I)$ of the current I. Of course, $V = f(I) = IR$ for a resistor. If we substitute $f(I)$ for IR in the familiar RLC-circuit equation $LI' + RI + Q/C = 0$ of Section 3.7, then differentiation gives the second-order equation

$$LI'' + f'(I)I' + \frac{I}{C} = 0. \tag{3}$$

In a 1924 study of oscillator circuits in early commercial radios, Balthasar van der Pol (1889–1959) assumed the voltage drop to be given by a nonlinear function of the form $f(I) = bI^3 - aI$, which with Eq. (3) becomes

$$LI'' + (3bI^2 - a)I' + \frac{I}{C} = 0. \tag{4}$$

This equation is closely related to Rayleigh's equation and has phase portraits resembling Fig. 6.4.15. Indeed, differentiation of the second equation in (2) and the resubstitution $x' = y$ yield the equation

$$my'' + (3by^2 - a)y' + ky = 0, \tag{5}$$

which has the same form as Eq. (4).

If we denote by τ the time variable in Eq. (4) and make the substitutions $I = px, t = \tau/\sqrt{LC}$, the result is

$$\frac{d^2x}{dt^2} + (3bp^2x^2 - a)\sqrt{\frac{C}{L}}\frac{dx}{dt} + x = 0.$$

With $p = \sqrt{a/(3b)}$ and $\mu = a\sqrt{C/L}$, this gives the standard form

$$x'' + \mu(x^2 - 1)x' + x = 0 \tag{6}$$

of *van der Pol's equation.*

For every nonnegative value of the parameter μ, the solution of van der Pol's equation with $x(0) = 2$, $x'(0) = 0$ is periodic, and the corresponding phase plane trajectory is a limit cycle to which the other trajectories converge (as in Fig. 6.4.15). It will be instructive for you to solve van der Pol's equation numerically and to plot this periodic trajectory for a selection of values from $\mu = 0$ to $\mu = 1000$ or more. With $\mu = 0$ it is a circle of radius 2 (why?). Figure 6.4.18 shows the periodic trajectory with $\mu = 1$, and Fig. 6.4.19 shows the corresponding $x(t)$ and $y(t)$ solution curves. When μ is large, van der Pol's equation is quite "stiff" and the periodic trajectory is more eccentric as in Fig. 6.4.20, which was plotted using MAT-LAB's stiff ODE solver **ode15s**. The corresponding $x(t)$ and $y(t)$ solution curves in Figs. 6.4.21 and 6.4.22 reveal surprising behavior of these component functions. Each alternates long intervals of very slow change with periods of abrupt change during very short time intervals that correspond to the "quasi-discontinuities" that are visible in Figs. 6.4.21 and 6.4.22. For instance, Fig. 6.4.23 shows that, between $t = 1614.28$ and $t = 1614.29$, the value of $y(t)$ zooms from near zero to over 1300 and back again. Perhaps you can measure the distance between x- or y-intercepts to show that the period of circuit around the cycle in Fig. 6.4.20 is approximately $T = 1614$. Indeed, this calculation and the construction of figures like those shown here may serve as a good test of the robustness of your computer system's ODE solver.

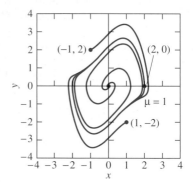

FIGURE 6.4.18. The phase plane trajectory of a periodic solution of van der Pol's equation with $\mu = 1$, as well as some trajectories spiraling in and out.

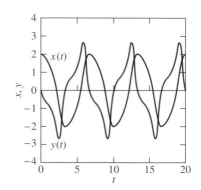

FIGURE 6.4.19. $x(t)$ and $y(t)$ solution curves defining the periodic solution of van der Pol's equation with $\mu = 1$.

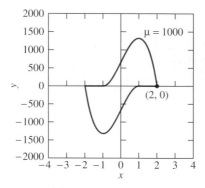

FIGURE 6.4.20. The phase plane trajectory of the periodic solution of van der Pol's equation with $\mu = 1000$.

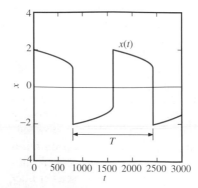

FIGURE 6.4.21. Graph of $x(t)$ with $\mu = 1000$.

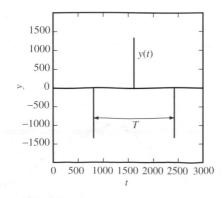

FIGURE 6.4.22. Graph of $y(t)$ with $\mu = 1000$.

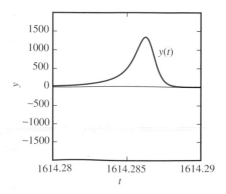

FIGURE 6.4.23. The upper spike in the graph of $y(t)$.

You might also plot other trajectories for $\mu = 10, 100$ or 1000 that (like the trajectories in Fig. 6.4.18) are "attracted" from within and without by the limit cycle. The origin looks like a spiral point in Fig. 6.4.18. Indeed, show that $(0, 0)$ is a spiral source for van der Pol's equation if $0 < \mu < 2$ but is a nodal source if $\mu \geq 2$.

6.5 Chaos in Dynamical Systems

In preceding sections we have looked at population growth and mechanical systems from a deterministic point of view—with the expectation that the initial state of a physical system fully determines its future evolution. But many common systems exhibit behavior that sometimes appears chaotic, in the sense that future states may not seem reliably predictable from a knowledge of initial conditions. This section includes project material illustrating the phenomenon of *chaos,* which is a topic of much current interest in science and engineering.

Population Growth and Period Doubling

In Section 2.1 we introduced the logistic differential equation

$$\frac{dP}{dt} = aP - bP^2 \qquad (a, b > 0) \tag{1}$$

that models a bounded (rather than exponentially growing) population. Indeed, if the population $P(t)$ satisfies Eq. (1), then as $t \to +\infty$, $P(t)$ approaches the [finite] limiting population $M = a/b$. We discuss here a "discrete" version of the logistic equation that has long been familiar to scientists, but only recently has been discovered to predict quite exotic and unexpected patterns of behavior for certain populations.

In order to solve Eq. (1) numerically as in Section 2.4, we first choose a fixed step size $h > 0$ and consider the sequence of discrete times

$$t_0, t_1, t_2, \ldots, t_n, t_{n+1}, \ldots, \tag{2}$$

where $t_{n+1} = t_n + h$ for each n. Beginning with the initial value $P_0 = P(t_0)$, we then calculate approximations

$$P_1, P_2, \ldots, P_n, P_{n+1}, \ldots \tag{3}$$

to the true values $P(t_1)$, $P(t_2)$, $P(t_3)$, \ldots of the actual population $P(t)$. For instance, Euler's method for the logistic equation in (1) consists of calculating the approximations in (3) iteratively by means of the formula

$$P_{n+1} = P_n + (aP_n - bP_n^2) \cdot h. \tag{4}$$

Now suppose that the population is one for which the step size h can be chosen so that the approximations calculated using Eq. (4) agree to acceptable accuracy with the actual population values. This might be the case, for instance, for an animal or insect population in which all reproduction takes place within short-duration breeding seasons that recur at regular intervals. If h is the interval between successive breeding seasons, then the population P_n during one breeding season may depend only on the population P_{n-1} during the previous season, and P_n may completely determine the population P_{n+1} during the next breeding season.

So let us assume for the sake of discussion that the successive values $P_n = P(t_n)$ of the population are given by the equation

$$P_{n+1} = P_n + (aP_n - bP_n^2) \cdot h. \tag{4}$$

Thus we replace the original differential equation in (1) with a "discrete" *difference equation*

$$\Delta P_n = (aP_n - bP_n^2)\,\Delta t \tag{5}$$

that gives the population difference $\Delta P_n = P_{n+1} - P_n$ in terms of the time difference $h = \Delta t$ and the preceding population P_n.

Equation (4) can be rewritten as the *logistic difference equation*

$$P_{n+1} = rP_n - sP_n^2, \tag{6}$$

where

$$r = 1 + ah \quad \text{and} \quad s = bh. \tag{7}$$

The substitution

$$P_n = \frac{r}{s}x_n \tag{8}$$

in Eq. (6) simplifies it still further to

$$x_{n+1} = rx_n(1 - x_n). \tag{9}$$

At this point we focus our attention on the final iterative formula in Eq. (9). Beginning with given values of x_0 and r, this formula generates a sequence x_1, x_2, x_3, ... of values corresponding to the successive times t_1, t_2, t_3, We may think of x_n, the value at time t_n, as the *fraction* of the maximum population that the environment can support. Assuming that the limiting fractional population

$$x_\infty = \lim_{n \to \infty} x_n \tag{10}$$

exists, we want to investigate the way in which x_∞ depends on the *growth parameter* r in Eq. (9). That is, if we regard r as the *input* to the process and x_∞ as the *output*, we ask how the output depends on the input.

The iteration in Eq. (9) is readily implemented in any available calculator or computer language. Figure 6.5.1 shows illustrative *Maple, Mathematica,* and MATLAB code for a simple program that begins with $x_1 = 0.5$ and calculates and assembles a list of the first couple of hundred ($k = 200$) iterates with $r = 1.5$.

Maple	*Mathematica*	MATLAB
`r := 1.5:`	`r = 1.5;`	`r = 1.5;`
`x = array(1..200):`	`x = Table[n,{n,1,200}];`	`x = 1:200;`
`x[1] := 0.5:`	`x[[1]] = 0.5;`	`x(1) = 0.5;`
`for n from 2 to 200 do`	`For[n=2, n<=200,`	`for n = 2:200`
` z := x[n-1]:`	` n=n+1,`	` z = x(n-1);`
` x[n] := r*z*(1-z):`	` z = x[[n-1]];`	` x(n) = r*z*(1-z);`
` od:`	` x[[n]] = r*z*(1-z)];`	`end`

FIGURE 6.5.1. *Maple, Mathematica,* and MATLAB versions of a simple iteration program.

Because $r = 1 + ah$ in (7), only values of r greater than 1 are pertinent to our idealized model of discrete population growth. It turns out that, for a typical such value of the growth parameter r entered at the first line, the results do not depend materially on the initial value x_1. After a reasonable number of iterations—the number required depends on the value of r—the value of x_n generally appears to "stabilize" on a limiting value x_∞ as in Eq. (10). For example, Fig. 6.5.2 shows results of runs of our simple iteration program with the values $r = 1.5$, 2.0, and 2.5 of the growth rate parameter, yielding limiting [fractional] populations

$$x_\infty = 0.333333, \qquad 0.500000, \qquad \text{and} \qquad 0.6000000,$$

respectively. Thus it appears that x_∞ exists and that its value grows moderately as r increases.

With	$r = 1.5$	$r = 2.0$	$r = 2.5$
x_1	0.5	0.5	0.5
x_2	0.3750	0.5000	0.6250
x_3	0.3516	0.5000	0.5859
\vdots	\vdots	\vdots	\vdots
x_{197}	0.3333	0.5000	0.6000
x_{198}	0.3333	0.5000	0.6000
x_{199}	0.3333	0.5000	0.6000
x_{200}	0.3333	0.5000	0.6000

FIGURE 6.5.2. Iterates with growth parameters $r = 1.5$, 2.0, and 2.5.

EXERCISE 1: Try several other values of the growth rate parameter in the range $1 < r < 3$. Do your results support the conjecture that the limiting population always exists and is an increasing function of r?

The results in Fig. 6.5.3 show that the conjecture stated in Exercise 1 is false! With growth rate parameters $r = 3.1$ and $r = 3.25$, the [fractional] population fails to stabilize on a single limiting population. (We calculated over a thousand iterates to make sure.) Instead, it oscillates between two different populations in alternate months (thinking of a month as our unit of time). For instance, with $r = 3.25$ we see that

$$x_{1001} = x_{1003} = x_{1005} = \cdots \approx 0.4953,$$

whereas

$$x_{1002} = x_{1004} = x_{1006} = \cdots \approx 0.8124.$$

Thus we have not a single limiting population, but rather a "limiting cycle" consisting of two distinct populations (as illustrated graphically in Fig. 6.5.4). Furthermore, when the growth rate is increased to $r = 3.5$, the *period* of the cycle doubles, and now we have a limiting cycle with a period of 4—the population cycles repeatedly through the four distinct values 0.5009, 0.8750, 0.3828, and 0.8269 (Fig. 6.5.5).

EXERCISE 2: Try values of the growth rate parameter in the range $2.9 < r < 3.1$ to determine as closely as possible just where the single limiting population splits (as r increases) into a cycle of period 2. Does this appear to happen just as r exceeds 3?

The results shown in Fig. 6.5.6 indicate that a cycle with period 8 is obtained with the growth rate parameter value $r = 3.55$. Events are now changing quite rapidly.

With	$r = 3.1$	$r = 3.25$	$r = 3.5$
x_1	0.5000	0.5000	0.5000
x_2	0.7750	0.8125	0.8750
x_3	0.5406	0.4951	0.3828
x_4	0.7699	0.8124	0.8269
⋮	⋮	⋮	⋮
x_{1001}	0.5580	0.4953	0.5009
x_{1002}	0.7646	0.8124	0.8750
x_{1003}	0.5580	0.4953	0.3828
x_{1004}	0.7646	0.8124	0.8269
x_{1005}	0.5580	0.4953	0.5009
x_{1006}	0.7646	0.8124	0.8750
x_{1007}	0.5580	0.4953	0.3828
x_{1008}	0.7646	0.8124	0.8269

FIGURE 6.5.3. Cycles with period 2 with $r = 3.1$ and $r = 3.25$; a cycle with period 4 with $r = 3.5$.

n	x_n
1001	0.5060
1002	0.8874
1003	0.3548
1004	0.8127
1005	0.5405
1006	0.8817
1007	0.3703
1008	0.8278
1009	0.5060
1010	0.8874
1011	0.3548
1012	0.8127
1013	0.5405
1014	0.8817
1015	0.3703
1016	0.8278

FIGURE 6.5.6. The period 8 cycle obtained with $r = 3.55$.

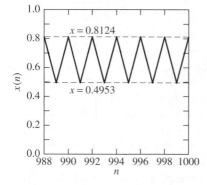

FIGURE 6.5.4. The graph of $x(n) = x_n$, showing the period 2 cycle of iterates obtained with $r = 3.25$.

FIGURE 6.5.5. The graph of $x(n) = x_n$, showing the period 4 cycle of iterates obtained with $r = 3.5$.

EXERCISE 3: Verify that a cycle with period 16 is obtained with the growth rate parameter value $r = 3.565$.

EXERCISE 4: See if you can find a cycle of period 32 somewhere between $r = 3.565$ and $r = 3.570$.

This is the phenomenon of **period doubling** for which the innocuous-looking iteration $x_{n+1} = rx_n(1 - x_n)$ has become famous in recent years. As the growth rate parameter is increased beyond $r = 3.56$, period doubling occurs so rapidly that utter chaos appears to break out somewhere near $r = 3.57$. Thus the graph shown in Fig. 6.5.7 indicates that, with $r = 3.57$, the earlier periodicity seems to have disappeared. No periodic cycle is evident, and the population appears to be changing (from one month to the next) in some essentially random fashion. Indeed, the deterministic population growth that is observed with smaller parameter val-

ues seems now to have degenerated into a nondeterministic process of apparently random change.

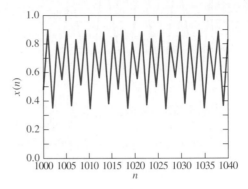

FIGURE 6.5.7. With $r = 3.57$: Chaos!

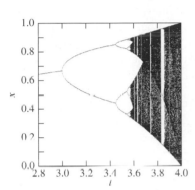

FIGURE 6.5.8. The pitchfork diagram with $2.8 \leq 4 \leq 4.0$, $0 \leq x \leq 1$.

FIGURE 6.5.9. The pitchfork diagram with $3.8 \leq r \leq 3.9$, $0 \leq x \leq 1$.

The solutions and applications manuals accompanying this text include MAT-LAB, *Mathematica*, and other versions of a program called PICHFORK. This program produces a visual presentation of the way in which the behavior of our iteration depends on the value of the growth parameter r. For each value of r in the input interval $a \leq r \leq b$ (the horizontal axis in the resulting diagram), 1000 iterations are first carried out to achieve "stability." Then the next 250 values of x generated by the iteration are plotted on the vertical axis—that is, the screen pixel at (r, x) is "turned on." The descriptively named "pitchfork diagram" that results then shows at a glance whether a given value of r corresponds to a cycle (with finite period) or to chaos. If the resolution in the picture suffices to make it clear that only finitely many values of x are placed above a given value of r, then we see that the iteration is "eventually periodic" for that specific value of the growth rate parameter.

Figure 6.5.8 shows the pitchfork diagram for the range $2.8 \leq r \leq 4.0$. Scanning from left to right, we see a single limiting population until $r \approx 3$, then a cycle with period 2 until $r \approx 3.45$, then a cycle of period 4, then one of period 8, and so forth, rapidly approaching the darkness of chaos. But note the vertical bands of "white space" that appear in the diagram between $r = 3.6$ and $r = 3.7$, between $r = 3.7$ and $r = 3.8$, and again between $r = 3.8$ and $r = 3.9$. These represent regions where [periodic] order returns from the preceding chaos.

For instance, Fig. 6.5.9 shows the interval $3.8 \leq r \leq 3.9$, where we observe a cycle of period 3 that emerges suddenly from the chaos near $r = 3.83$, and then splits successively into cycles of periods 6, 12, 24, ... (Figs. 6.5.10 and 6.5.11). This period doubling beginning with a cycle of period 3 is especially significant—a fundamental article by James Yorke and T.-Y. Li in the 1975 *American Mathematical Monthly* was entitled "Period Three Implies Chaos." According to this article, the existence of a cycle of period 3 (for an appropriate iteration) implies the existence of cycles of every other [finite] period, as well as chaotic "cycles" with no period at all.

PROJECT 1: Use Program PICHFORK to search for other interesting cycles, and verify their apparent periods by appropriate iterative computations. For instance, you should find a cycle with period 10 between $r = 3.60$ and $r = 3.61$, and one with period 14 between $r = 3.59$ and $r = 3.60$. Can you find cycles with period 5 and 7? If so, look for subsequent period doubling. A run of PICHFORK requires several hundred thousand iterations, so it will help if you have a fast computer (or one you can leave running overnight).

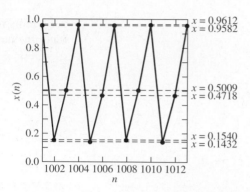

FIGURE 6.5.10. The graph of $x(n) = x_n$, showing the period 3 cycle of iterates obtained with $r = 3.84$.

FIGURE 6.5.11. The graph of $x(n) = x_n$, showing the period 6 cycle of iterates obtained with $r = 3.845$.

As we scan the pitchfork diagram (Fig. 6.5.8) from left to right, we spot the successive values r_1, r_2, r_3, \ldots of the growth rate parameter at which a *bifurcation* or qualitative change in the iteration $x_{n+1} = rx_n(1 - x_n)$ occurs as the value of r is increased further. These are the discrete values of r at which any sufficiently small increase in the growth parameter doubles the period of the iteration. In the 1970s the Los Alamos physicist Mitchell Feigenbaum discovered that a certain order underlies this period doubling toward chaos:

$$\lim_{k \to \infty} \frac{r_k - r_{k-1}}{r_{k+1} - r_k} = 4.66920160981\ldots. \tag{11}$$

The fraction on the left in Eq. (11) is the ratio of the lengths of successive constant-period "windows" in the pitchfork diagram. It is the fact that this ratio approaches a limit as $k \to +\infty$, rather than the specific value of this limit, that demonstrates a sort of order underlying the period doubling observed with the particular iteration $x_{n+1} = rx_n(1 - x_n)$. On the other hand, it is now known that precisely the same *Feigenbaum constant* 4.66920160981... plays exactly the same role for a wide variety of period-doubling phenomena arising in many different areas of science.

PROJECT 2: Feigenbaum used an HP-65 pocket calculator (rather than a powerful computer) to carry out the computations leading to the discovery of his famous constant. Perhaps you would like to use iterative computations and/or PICHFORK to isolate the first few bifurcation values r_1, r_2, r_3, \ldots with sufficient accuracy to verify that the limit in (11) is approximately 4.67. You can consult pages 124–126 of T. Gray and J. Glynn, *Exploring Mathematics with Mathematica* (Addison-Wesley, 1991) for a fancier approach.

Period Doubling in Mechanical Systems

In Section 6.4 we introduced the second-order differential equation

$$mx'' + cx' + kx + \beta x^3 = 0 \tag{12}$$

to model the free velocity-damped vibrations of a mass on a nonlinear spring. Recall that the term kx in Eq. (12) represents the force exerted on the mass by a *linear* spring, whereas the term βx^3 represents the nonlinearity of an actual spring.

We want now to discuss the *forced vibrations* that result when an external force $F(t) = F_0 \cos \omega t$ acts on the mass. With such a force adjoined to the system in Eq. (12), we obtain the **forced Duffing equation**

$$mx'' + cx' + kx + \beta x^3 = F_0 \cos \omega t \qquad (13)$$

for the displacement $x(t)$ of the mass from its equilibrium position. For most values of the parameters it is impossible to solve Eq. (13) explicitly for $x(t)$. Nevertheless, its solutions can be portrayed qualitatively by means of [numerically approximated] phase plane trajectories like those we used in Section 6.4 to describe free vibrations of nonlinear mechanical systems.

The Hooke's constant k is positive for a typical spring that resists displacement from equilibrium. But there do exist simple mechanical systems that emulate a spring having a negative Hooke's constant. For example, Fig. 6.5.12 shows a mass m atop a vertical metal filament. We assume that the thin metal filament can oscillate only in a vertical plane, and behaves like a flexible column that "buckles" or bends when the mass is displaced to either side of the vertical position. Then there is one stable equilibrium point to the left ($x < 0$) and another to the right ($x > 0$), but the vertical equilibrium position ($x = 0$) is unstable. When the mass is displaced slightly from this unstable equilibrium position, the internal force exerted on it is *repulsive* rather than attractive; this corresponds to a negative value of k in Eq. (13). If a periodic force is exerted on the mass by [say] an oscillating electromagnetic field, and air resistance damps its oscillations, then Eq. (13) with $k < 0$ but with $c > 0$ and $\beta > 0$ is a reasonable mathematical model for its horizontal displacement function $x(t)$.

In the absence of both damping and the external force, the phase plane trajectories of the free oscillations of the mass would resemble those shown in Fig. 6.4.12 (with Problem 14 in Section 6.4). The mass behaves as though it is repelled by the unstable critical point at $x = 0$ but is attracted by each of the two stable critical points symmetrically located on either side of the origin.

We saw in Section 3.6 that in the *linear* case a periodic external force $F(t) = F_0 \cos \omega t$ causes a steady periodic response $x(t) = C \cos(\omega t - \alpha)$ with the *same* frequency ω. The amplitude C of the steady periodic response is proportional to the amplitude F_0 of the external force. For instance, if the periodic external force is doubled in amplitude, then the only change in the response is that its amplitude is doubled as well.

To illustrate the quite different behavior of a nonlinear system, we take $k = -1$ and $m = c = \beta = \omega = 1$ in Eq. (13), so the differential equation is

$$x'' + x' - x + x^3 = F_0 \cos t. \qquad (14)$$

As an exercise you may verify that the two stable critical points are $(-1, 0)$ and $(1, 0)$. We want to examine the dependence of the (presumably steady periodic) response $x(t)$ on the amplitude F_0 of the periodic external force of period $2\pi/\omega = 2\pi$.

Figures 6.5.13 through 6.5.16 show the solutions of Eq. (14) obtained with the successive values $F_0 = 0.60, 0.70, 0.75$, and 0.80 of the amplitude of the external force. In each case the system was solved numerically with initial conditions $x(0) = 1, x'(0) = 0$ and the resulting solution plotted for the range $100 \leq t \leq 200$ (to show the steady periodic response remaining after the initial transient response has died out). Part (a) of each figure shows the phase plane trajectory $x = x(t)$, $y = x'(t)$, and part (b) shows the actual solution curve $x = x(t)$ in the tx-plane. Part (a)

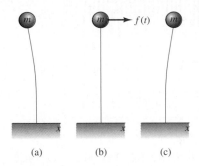

FIGURE 6.5.12. Equilibrium positions of a mass on a filament: (a) stable equilibrium with $x < 0$; (b) unstable equilibrium at $x = 0$; (c) stable equilibrium with $x > 0$.

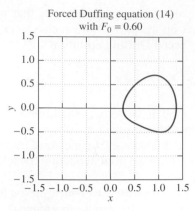

FIGURE 6.5.13(a). Period 2π response with $F_0 = 0.60$: phase plane trajectory.

FIGURE 6.5.13(b). Period 2π response with $F_0 = 0.60$: solution $x(t)$.

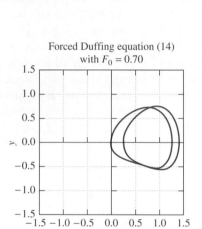

FIGURE 6.5.14(a). Period 4π response with $F_0 = 0.70$: phase plane trajectory.

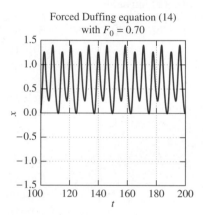

FIGURE 6.5.14(b). Period 4π response with $F_0 = 0.70$: solution $x(t)$.

exhibits the qualitative character of the solution more vividly, but part (b) is required to determine the period and frequency of the solution.

Figure 6.5.13 shows a simple oscillation of period 2π of the mass around the right-hand critical point. In the ensuing sequence of figures we see successive *period doubling* and finally *chaos* as the amplitude of the external force is increased in the range from $F_0 = 0.6$ to $F_0 = 0.8$. This *period doubling toward chaos* is a common characteristic of the behavior of a *nonlinear* mechanical system as an appropriate physical parameter (such as m, c, k, β, F_0, or ω in Eq. (13)) is increased or decreased. No such phenomenon occurs in linear systems.

PROJECT 3: Use an ODE plotting utility to see whether you can reproduce Figs. 6.5.13–6.5.16. Then investigate the parameter range $1.00 \leq F_0 \leq 1.10$ for the force constant in Eq. (14). With $F_0 = 1.00$ you should see a period 6π phase plane trajectory that encircles *both* stable critical points (as well as the unstable one). The period doubles around $F_0 = 1.07$ and chaos sets in around $F_0 = 1.10$. See whether you can spot a second period doubling somewhere between $F_0 = 1.07$ and $F_0 = 1.10$. Produce both phase plane trajectories and tx-solution curves on which you can measure the periods.

Forced Duffing equation (14)
with $F_0 = 0.75$

FIGURE 6.5.15(a). Period 8π response with $F_0 = 0.75$: phase plane trajectory.

Forced Duffing equation (14)
with $F_0 = 0.75$

FIGURE 6.5.15(b). Period 8π response with $F_0 = 0.75$: solution $x(t)$.

Forced Duffing equation (14)
with $F_0 = 0.80$

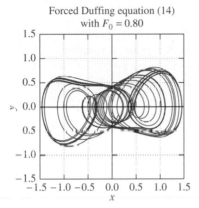

FIGURE 6.5.16(a). Chaotic response with $F_0 = 0.80$: phase plane trajectory.

Forced Duffing equation (14)
with $F_0 = 0.80$

FIGURE 6.5.16(b). Chaotic response with $F_0 = 0.80$: solution $x(t)$.

Warning: You should not expect your own hardware and ODE software to replicate the exact detail of the "chaotic tangle" shown in Fig. 6.5.16. To explain why, let's regard the forced Duffing equation (in (14)) with $F_0 = 0.80$ as an *input-output system* having the initial point $(x(0), x'(0))$ as input and the corresponding solution $x(t)$ as output. This input-output system is *chaotic* in the sense that very small changes in the input may cause very large changes in the output. For instance, the data shown in the table in Fig. 6.5.17 were generated using MATLAB's sophisticated numerical solver `ode45` with two nearby initial points and two different error tolerance settings. Solving the same numerical initial value problem repeatedly with different error tolerances can provide some indication of the reliability of the results; significant discrepancies certainly suggest little reliability. With initial conditions $x(0) = 1$, $x'(0) = 0$ it looks plausible (though hardly certain) that $x(100) \approx -1.1$ and $x(200) \approx -0.6$, but the value of $x(300)$ remains quite uncertain. By contrast, with initial conditions $x(0) = 1.000001$, $x'(0) = 0$ it appears that perhaps $x(200) \approx -0.3$ instead. If so, then a visually significant change in the solution results from a change of initial conditions no larger than might be expected from accumulated roundoff error in the course of a numerical approximation process that relies on machine arithmetic. In this event, any numerically computed solution is likely to diverge appreciably from the true solution over a long time interval. Hence we cannot be confident of the fine structure in a numerically generated trajectory

t	$x(t)$ with ErrTol $= 10^{-8}$	$x(t)$ with ErrTol $= 10^{-12}$	$x(t)$ with ErrTol $= 10^{-8}$	$x(t)$ with ErrTol $= 10^{-12}$
0	1	1	1.000001	1.000001
100	−1.1125	−1.1125	−1.1125	−1.1125
200	−0.5823	−0.5828	−0.2925	−0.2816
300	−1.2850	−0.1357	−0.0723	−0.1633

FIGURE 6.5.17. MATLAB attempts to approximate the solution of the forced Duffing equation $x'' + x' - x + x^3 = (0.80)\cos t$ on $0 \leq t \leq 300$ with $x'(0) = 0$ and two different values of $x(0)$ and two different error tolerances (ErrTol denoting the value used for both the absolute and the relative error tolerance in `ode45`).

such as that shown in Fig. 6.5.16. Investigations like this suggest only that the actual long-interval solution with initial conditions $x(0) = 1$, $x'(0) = 0$ is not periodic, but instead "wanders" back and forth in a seemingly unpredictable or chaotic fashion. Thus the qualitative character of the solution indicated in Fig. 6.5.16 may approximate reality without necessarily presenting an accurate picture of the precise detail of the trajectory. This behavior of solutions of the forced Duffing equation is not yet fully understood and remains a subject of current research. An interesting exposition with further references can be found in Chapter 15 of Dan Schwalbe and Stan Wagon, *VisualDSolve* (New York: Springer-Verlag, 1997).

The Lorenz Strange Attractor

The substitution of $x_1 = x$, $x_2 = x'$ in the forced Duffing equation in (13) yields a two-dimensional nonlinear system of first-order differential equations, and period-doubling phenomena are characteristic of such systems. But in higher dimensions even more exotic phenomena occur and are currently the subject of much active investigation. All this work stems ultimately from the original investigation of an extraordinary three-dimensional nonlinear system by the mathematical meteorologist E. N. Lorenz, who later described its discovery as follows.

> By the middle 1950s "numerical weather prediction;" i.e., forecasting by numerically integrating such approximations to the atmospheric equations as could feasibly be handled, was very much in vogue, despite the rather mediocre results which it was then yielding. A smaller but determined group favored statistical prediction. ... I was skeptical, and decided to test the idea by applying the statistical method to a set of artificial data, generated by solving a system of equations numerically. ... The first task was to find a suitable system of equations to solve. ... The system would have to be simple enough ... and the general solution would have to be aperiodic, since the statistical prediction of a periodic series would be a trivial matter, once the periodicity had been detected ... [In the course of talks with Dr. Barry Saltzman] he showed me some work on thermal convections, in which he used a system of seven ordinary differential equations. Most of his solutions soon acquired periodic behavior, but one solution refused to settle down. Moreover, in this solution four of the variables appeared to approach zero. Presumably the equations governing the remaining three variables, with the terms containing the four variables eliminated, would also possess aperiodic solutions. Upon my return I put the three equations on our computer, and

confirmed the aperiodicity which Saltzman had noted. We were finally in business. [Quoted in E. Hairer, S. P. Norsett, and G. Wanner, *Solving Ordinary Differential Equations I* (New York: Springer-Verlag, 1987).]

The famous *Lorenz system* of differential equations is given by

$$\frac{dx}{dt} = -sx + sy,$$

$$\frac{dy}{dt} = -xz + rx - y, \tag{15}$$

$$\frac{dz}{dt} = xy - bz.$$

The Lorenz strange attractor

FIGURE 6.5.18. The xz-projection of the Lorenz trajectory with $-20 \leq x \leq 20$, $0 \leq z \leq 50$.

A solution curve in xyz-space is best visualized by looking at its projection into some plane, typically one of the three coordinate planes. Figure 6.5.18 shows the projection into the xz-plane of the solution obtained by numerical integration with the parameter values $b = \frac{8}{3}$, $s = 10$, $r = 28$ and the initial values $x(0) = -8$, $y(0) = 8$, and $z(0) = 27$.

As the projection in Fig. 6.5.18 is traced in "real time," the moving solution point $P(x(t), y(t), z(t))$ appears to undergo a random number of oscillations on the right followed by a random number of oscillations on the left, then a random number on the right followed by a random number on the left, and so on. Given the meteorological origin of the Lorenz system, one naturally thinks of a random number of clear days followed by a random number of rainy days, then a random number of clear days followed by a random number of rainy days, and so on.

A close examination of such projections of the Lorenz trajectory shows that it is *not* simply oscillating back and forth around a pair of critical points (as Fig. 6.5.18 may initially suggest). Instead, as $t \to +\infty$, the solution point $P(t)$ on the trajectory wanders back and forth in space approaching closer and closer to a certain complicated set of points whose detailed structure is not yet fully understood. This elusive set that appears somehow to "attract" the solution point is the famous *Lorenz strange attractor.*

PROJECT 4: First use an ODE plotting utility to reproduce the xz-projection of the Lorenz trajectory shown in Fig. 6.5.18. Use the parameter values and initial conditions listed immediately after Eq. (15) and numerically integrate the Lorenz system on the interval $0 \leq t \leq 50$. Plot also the xy- and yz-projections of this same solution. Next, experiment with different parameter values and initial conditions. For instance, see if you can find a periodic solution with $r = 70$ (and $b = \frac{8}{3}$, $s = 10$ as before) and initial values $x_0 = -4$ and $z_0 = 64$. To get a trajectory that almost repeats itself, you will need to try different values of y_0 in the range $0 \leq y_0 \leq 10$ and look at xz-projections as in Fig. 6.5.18.

PROJECT 5: Another much-studied nonlinear three-dimensional system is the Rossler system

$$\frac{dx}{dt} = -y - z,$$

$$\frac{dy}{dt} = x + ay, \tag{16}$$

$$\frac{dz}{dt} = b + z(x - c).$$

The Rossler band

FIGURE 6.5.19. The xy-projection of the Rossler band.

Figure 6.5.19 shows an xy-projection of the *Rossler band,* a chaotic attractor obtained with the values $a = 0.398$, $b = 2$, and $c = 4$ of the parameters in (16). In the xy-plane the Rossler band looks "folded," but in space it appears twisted like a Möbius strip. Investigate the period doubling toward chaos that occurs with the Rossler system as the parameter a is increased, beginning with $a = 0.3$, $a = 0.35$, and $a = 0.375$ (take $b = 2$ and $c = 4$ in all cases).

In this section we have given just a taste of the ideas that are the focus of contemporary applications of nonlinear systems. To see how these ideas come full circle, consult the discussion of the Lorenz system on pages 117–123 of the book by Hairer et al. referenced previously. There you will see a certain aspect of the Lorenz trajectory described visually by means of a picture that looks very much like the pitchfork diagram shown in Fig. 6.5.9, together with the very same Feigenbaum constant 4.6692 . . . !

For an engaging account of the historical background to this final section of Chapter 6, see James Gleick, *Chaos: Making a New Science* (New York: Viking Press, 1987). For more detailed discussions of the forced Duffing, Lorenz, and Rossler equations, see J. M. T. Thompson and H. B. Stewart, *Nonlinear Dynamics and Chaos* (New York: John Wiley, 1986).

C H A P T E R

7

Laplace Transform Methods

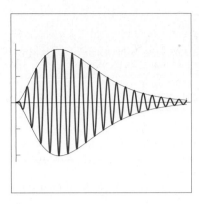

7.1 Laplace Transforms and Inverse Transforms

$D\{f(t)\} = f'(t)$

$\mathcal{L}\{f(t)\} = F(s)$

FIGURE 7.1.1. Transformation of a function: \mathcal{L} in analogy with D.

In Chapter 3 we saw that linear differential equations with constant coefficients have numerous applications and can be solved systematically. There are common situations, however, in which the alternative methods of this chapter are preferable. For example, recall the differential equations

$$mx'' + cx' + kx = F(t) \quad \text{and} \quad LI'' + RI' + \frac{1}{C}I = E'(t)$$

corresponding to a mass-spring-dashpot system and a series RLC circuit, respectively. It often happens in practice that the forcing term, $F(t)$ or $E'(t)$, has discontinuities—for example, when the voltage supplied to an electrical circuit is turned off and on periodically. In this case the methods of Chapter 3 can be quite awkward, and the Laplace transform method is more convenient.

The differentiation operator D can be viewed as a transformation which, when applied to the function $f(t)$, yields the new function $D\{f(t)\} = f'(t)$. The Laplace transformation \mathcal{L} involves the operation of integration and yields the new function $\mathcal{L}\{f(t)\} = F(s)$ of a new independent variable s. The situation is diagrammed in Fig. 7.1.1. After learning in this section how to compute the Laplace transform $F(s)$ of a function $f(t)$, we will see in Section 7.2 that the Laplace transform converts a *differential* equation in the unknown function $f(t)$ into an *algebraic* equation in $F(s)$. Because algebraic equations are generally easier to solve than differential equations, this is one method that simplifies the problem of finding the solution $f(t)$.

435

Definition: The Laplace Transform

Given a function $f(t)$ defined for all $t \geq 0$, the *Laplace transform* of f is the function F defined as follows:

➤ $$F(s) = \mathcal{L}\{f(t)\} = \int_0^\infty e^{-st} f(t) \, dt \tag{1}$$

for all values of s for which the improper integral converges.

Recall that an **improper integral** over an infinite interval is defined as a limit of integrals over bounded intervals; that is,

$$\int_a^\infty g(t) \, dt = \lim_{b \to \infty} \int_a^b g(t) \, dt. \tag{2}$$

If the limit in (2) exists, then we say that the improper integral **converges;** otherwise, it **diverges** or fails to exist. Note that the integrand of the improper integral in (1) contains the parameter s in addition to the variable of integration t. Therefore, when the integral in (1) converges, it converges not merely to a number, but to a *function* F of s. As in the following examples, it is typical for the improper integral in the definition of $\mathcal{L}\{f(t)\}$ to converge for some values of s and diverge for others.

Example 1 With $f(t) \equiv 1$ for $t \geq 0$, the definition of the Laplace transform in (1) gives

$$\mathcal{L}\{1\} = \int_0^\infty e^{-st} \, dt = \left[-\frac{1}{s} e^{-st} \right]_0^\infty = \lim_{b \to \infty} \left[-\frac{1}{s} e^{-bs} + \frac{1}{s} \right],$$

and therefore

$$\mathcal{L}\{1\} = \frac{1}{s} \quad \text{for} \quad s > 0. \tag{3}$$

As in (3), it's good practice to specify the domain of the Laplace transform—in problems as well as in examples. Also, in this computation we have used the common abbreviation

$$\left[g(t) \right]_a^\infty = \lim_{b \to \infty} \left[g(t) \right]_a^b. \tag{4}$$

∎

Remark: The limit we computed in Example 1 would not exist if $s < 0$, for then $(1/s)e^{-bs}$ would become unbounded as $b \to +\infty$. Hence $\mathcal{L}\{1\}$ is defined only for $s > 0$. This is typical of Laplace transforms; the domain of a transform is normally of the form $s > a$ for some number a. ∎

Example 2 With $f(t) = e^{at}$ for $t \geq 0$, we obtain

$$\mathcal{L}\{e^{at}\} = \int_0^\infty e^{-st} e^{at} \, dt = \int_0^\infty e^{-(s-a)t} \, dt = \left[-\frac{e^{-(s-a)t}}{s-a} \right]_{t=0}^\infty .$$

If $s - a > 0$, then $e^{-(s-a)t} \to 0$ as $t \to +\infty$, so it follows that

$$\mathcal{L}\{e^{at}\} = \frac{1}{s - a} \quad \text{for} \quad s > a. \tag{5}$$

Note here that the improper integral giving $\mathcal{L}\{e^{at}\}$ diverges if $s \leq a$. It is worth noting also that the formula in (5) holds if a is a complex number. For then, with $a = \alpha + i\beta$,

$$e^{-(s-a)t} = e^{i\beta t}e^{-(s-\alpha)t} \to 0$$

as $t \to +\infty$, provided that $s > \alpha = \mathrm{Re}(a)$; recall that $e^{i\beta t} = \cos \beta t + i \sin \beta t$. ∎

The Laplace transform $\mathcal{L}\{t^a\}$ of a power function is most conveniently expressed in terms of the **gamma function** $\Gamma(x)$, which is defined for $x > 0$ by the formula

$$\Gamma(x) = \int_0^\infty e^{-t}t^{x-1}\, dt. \tag{6}$$

For an elementary discussion of $\Gamma(x)$, see the subsection on the gamma function in Section 8.5, where it is shown that

$$\Gamma(1) = 1 \tag{7}$$

and that

$$\Gamma(x + 1) = x\Gamma(x) \tag{8}$$

for $x > 0$. It then follows that if n is a positive integer, then

$$\begin{aligned}
\Gamma(n + 1) &= n\Gamma(n) \\
&= n \cdot (n - 1)\Gamma(n - 1) \\
&= n \cdot (n - 1) \cdot (n - 2)\Gamma(n - 2) \\
&\qquad \vdots \\
&= n(n - 1)(n - 2) \cdots 2 \cdot \Gamma(2) \\
&= n(n - 1)(n - 2) \cdots 2 \cdot 1 \cdot \Gamma(1);
\end{aligned}$$

thus

$$\Gamma(n + 1) = n! \tag{9}$$

if n is a positive integer. Therefore, the function $\Gamma(x + 1)$, which is defined and continuous for all $x > -1$, agrees with the factorial function for $x = n$, a positive integer.

Example 3 Suppose that $f(t) = t^a$ where a is real and $a > -1$. Then

$$\mathcal{L}\{t^a\} = \int_0^\infty e^{-st}t^a\, dt.$$

If we substitute $u = st$, $t = u/s$, and $dt = du/s$ in this integral, we get

$$\mathcal{L}\{t^a\} = \frac{1}{s^{a+1}} \int_0^\infty e^{-u}u^a\, du = \frac{\Gamma(a + 1)}{s^{a+1}} \tag{10}$$

for all $s > 0$ (so that $u = st > 0$). Because $\Gamma(n + 1) = n!$ if n is a nonnegative integer, we see that

$$\mathcal{L}\{t^n\} = \frac{n!}{s^{n+1}} \quad \text{for} \quad s > 0. \tag{11}$$

For instance,

$$\mathcal{L}\{t\} = \frac{1}{s^2}, \quad \mathcal{L}\{t^2\} = \frac{2}{s^3}, \quad \text{and} \quad \mathcal{L}\{t^3\} = \frac{6}{s^4}.$$

As in Problems 1 and 2, these formulas can be derived immediately from the definition, without the use of the gamma function. ∎

Linearity of Transforms

It is not necessary for us to proceed much further in the computation of Laplace transforms directly from the definition. Once we know the Laplace transforms of several functions, we can combine them to obtain transforms of other functions. The reason is that the Laplace transformation is a *linear* operation.

THEOREM 1: Linearity of the Laplace Transform

If a and b are constants, then

$$\mathcal{L}\{af(t) + bg(t)\} = a\mathcal{L}\{f(t)\} + b\mathcal{L}\{g(t)\} \tag{12}$$

for all s such that the Laplace transforms of the functions f and g both exist. ▲

The proof of Theorem 1 follows immediately from the linearity of the operations of taking limits and of integration:

$$\mathcal{L}\{af(t) + bg(t)\} = \int_0^{\infty} e^{-st}[af(t) + bg(t)]\,dt$$

$$= \lim_{c \to \infty} \int_0^c e^{-st}[af(t) + bg(t)]\,dt$$

$$= a\left(\lim_{c \to \infty} \int_0^c e^{-st}f(t)\,dt\right) + b\left(\lim_{c \to \infty} \int_0^c e^{-st}g(t)\,dt\right)$$

$$= a\mathcal{L}\{f(t)\} + b\mathcal{L}\{g(t)\}.$$

Example 4 The computation of $\mathcal{L}\{t^{n/2}\}$ is based on the known special value

$$\Gamma\left(\frac{1}{2}\right) = \sqrt{\pi} \tag{13}$$

of the gamma function. For instance, it follows that

$$\Gamma\left(\frac{5}{2}\right) = \frac{3}{2}\Gamma\left(\frac{3}{2}\right) = \frac{3}{2} \cdot \frac{1}{2}\Gamma\left(\frac{1}{2}\right) = \frac{3}{4}\sqrt{\pi},$$

using the formula $\Gamma(x + 1) = x\Gamma(x)$ in (9), first with $x = \frac{3}{2}$ and then with $x = \frac{1}{2}$. Now the formulas in (10) through (12) yield

$$\mathcal{L}\{3t^2 + 4t^{3/2}\} = 3 \cdot \frac{2!}{s^3} + \frac{4\Gamma\left(\frac{5}{2}\right)}{s^{5/2}} = \frac{6}{s^3} + 3\sqrt{\frac{\pi}{s^5}}. \qquad \blacksquare$$

Example 5 Recall that $\cosh kt = (e^{kt} + e^{-kt})/2$. If $k > 0$, then Theorem 1 and Example 2 together give

$$\mathcal{L}\{\cosh kt\} = \frac{1}{2}\mathcal{L}\{e^{kt}\} + \frac{1}{2}\mathcal{L}\{e^{-kt}\} = \frac{1}{2}\left(\frac{1}{s - k} + \frac{1}{s + k}\right);$$

that is,

$$\mathcal{L}\{\cosh kt\} = \frac{s}{s^2 - k^2} \qquad \text{for } s > k > 0. \tag{14}$$

Similarly,

$$\mathcal{L}\{\sinh kt\} = \frac{k}{s^2 - k^2} \qquad \text{for } s > k > 0. \tag{15}$$

Because $\cos kt = (e^{ikt} + e^{-ikt})/2$, the formula in (5) (with $a = ik$) yields

$$\mathcal{L}\{\cos kt\} = \frac{1}{2}\left(\frac{1}{s - ik} + \frac{1}{s + ik}\right) = \frac{1}{2} \cdot \frac{2s}{s^2 - (ik)^2},$$

and thus

$$\mathcal{L}\{\cos kt\} = \frac{s}{s^2 + k^2} \qquad \text{for } s > 0. \tag{16}$$

(The domain follows from $s > \text{Re}(ik) = 0$.) Similarly,

$$\mathcal{L}\{\sin kt\} = \frac{k}{s^2 + k^2} \qquad \text{for } s > 0. \tag{17}$$

\blacksquare

Example 6 Applying linearity, the formula in (16), and a familiar trigonometric identity, we get

$$\mathcal{L}\{3e^{2t} + 2\sin^2 3t\} = \mathcal{L}\{3e^{2t} + 1 - \cos 6t\}$$

$$= \frac{3}{s - 2} + \frac{1}{s} - \frac{s}{s^2 + 36}$$

$$= \frac{3s^3 + 144s - 72}{s(s - 2)(s^2 + 36)} \qquad \text{for } s > 0. \qquad \blacksquare$$

Inverse Transforms

According to Theorem 3 of this section, no two different functions that are both continuous for all $t \geq 0$ can have the same Laplace transform. Thus if $F(s)$ is the transform of some continuous function $f(t)$, then $f(t)$ is uniquely determined. This observation allows us to make the following definition: If $F(s) = \mathcal{L}\{f(t)\}$, then we call $f(t)$ the **inverse Laplace transform** of $F(s)$ and write

➤
$$f(t) = \mathcal{L}^{-1}\{F(s)\}. \tag{18}$$

Example 7
$$\mathcal{L}^{-1}\left\{\frac{1}{s^3}\right\} = \frac{1}{2}t^2, \quad \mathcal{L}^{-1}\left\{\frac{1}{s+2}\right\} = e^{-2t}, \quad \mathcal{L}^{-1}\left\{\frac{2}{s^2+9}\right\} = \frac{2}{3}\sin 3t,$$

and so on. ∎

NOTATION: FUNCTIONS AND THEIR TRANSFORMS. Throughout this chapter we denote functions of t by lowercase letters. The transform of a function will always be denoted by that same letter capitalized. Thus $F(s)$ is the Laplace transform of $f(t)$ and $x(t)$ is the inverse Laplace transform of $X(s)$.

A table of Laplace transforms serves a purpose similar to that of a table of integrals. The table in Fig. 7.1.2 lists the transforms derived in this section; many additional transforms can be derived from these few, using various general properties of the Laplace transformation (which we will discuss in subsequent sections).

Piecewise Continuous Functions

As we remarked at the beginning of this section, we need to be able to handle certain types of discontinuous functions. The function $f(t)$ is said to be **piecewise continuous** on the bounded interval $a \leq t \leq b$ provided that $[a, b]$ can be subdivided into finitely many abutting subintervals in such a way that

1. f is continuous in the interior of each of these subintervals; and
2. $f(t)$ has a finite limit as t approaches each endpoint of each subinterval from its interior.

We say that f is piecewise continuous for $t \geq 0$ if it is piecewise continuous on every bounded subinterval of $[0, +\infty)$. Thus a piecewise continuous function has only simple discontinuities (if any) and only at isolated points. At such points the value of the function experiences a finite jump, as indicated in Fig. 7.1.3. The **jump in $f(t)$ at the point** c is defined to be $f(c+) - f(c-)$, where

$$f(c+) = \lim_{\epsilon \to 0^+} f(c + \epsilon) \quad \text{and} \quad f(c-) = \lim_{\epsilon \to 0^+} f(c - \epsilon).$$

Perhaps the simplest piecewise continuous (but discontinuous) function is the **unit step function**, whose graph appears in Fig. 7.1.4. It is defined as follows:

$$u(t) = \begin{cases} 0 & \text{for } t < 0, \\ 1 & \text{for } t \geq 0. \end{cases} \tag{19}$$

Because $u(t) = 1$ for $t \geq 0$ and because the Laplace transform involves only the values of a function for $t \geq 0$, we see immediately that

$$\mathcal{L}\{u(t)\} = \frac{1}{s} \quad (s > 0). \tag{20}$$

$f(t)$	$F(s)$			
1	$\dfrac{1}{s}$	$(s > 0)$		
t	$\dfrac{1}{s^2}$	$(s > 0)$		
$t^n \ (n \geq 0)$	$\dfrac{n!}{s^{n+1}}$	$(s > 0)$		
$t^a \ (a > -1)$	$\dfrac{\Gamma(a+1)}{s^{a+1}}$	$(s > 0)$		
e^{at}	$\dfrac{1}{s-a}$	$(s > 0)$		
$\cos kt$	$\dfrac{s}{s^2 + k^2}$	$(s > 0)$		
$\sin kt$	$\dfrac{k}{s^2 + k^2}$	$(s > 0)$		
$\cosh kt$	$\dfrac{s}{s^2 - k^2}$	$(s >	k)$
$\sinh kt$	$\dfrac{k}{s^2 - k^2}$	$(s >	k)$
$u(t - a)$	$\dfrac{e^{-as}}{s}$	$(s > 0)$		

FIGURE 7.1.2. A short table of Laplace transforms.

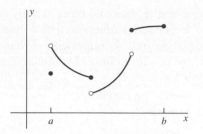

FIGURE 7.1.3. The graph of a piecewise continuous function; the solid dots indicate values of the function at discontinuities.

FIGURE 7.1.4. The graph of the unit step function.

FIGURE 7.1.5. The unit step function $u_a(t)$ has a jump at $t = a$.

The graph of the unit step function $u_a(t) = u(t - a)$ appears in Fig. 7.1.5. Its jump occurs at $t = a$ rather than at $t = 0$; equivalently,

$$u_a(t) = u(t - a) = \begin{cases} 0 & \text{for } t < a, \\ 1 & \text{for } t \geq a. \end{cases} \tag{21}$$

Example 8 Find $\mathcal{L}\{u_a(t)\}$ if $a > 0$.

Solution We begin with the definition of the Laplace transform. We obtain

$$\mathcal{L}\{u_a(t)\} = \int_0^\infty e^{-st} u_a(t)\, dt = \int_a^\infty e^{-st}\, dt = \lim_{b \to \infty} \left[-\frac{e^{-st}}{s} \right]_{t=a}^b ;$$

consequently,

$$\mathcal{L}\{u_a(t)\} = \frac{e^{-as}}{s} \quad (s > 0, \ a > 0). \tag{22}$$

∎

General Properties of Transforms

It is a familiar fact from calculus that the integral

$$\int_a^b g(t)\, dt$$

exists if g is piecewise continuous on the bounded interval $[a, b]$. Hence if f is piecewise continuous for $t \geq 0$, it follows that the integral

$$\int_0^b e^{-st} f(t)\, dt$$

exists for all $b < +\infty$. But in order for $F(s)$—the limit of this last integral as $b \to +\infty$—to exist, we need some condition to limit the rate of growth of $f(t)$ as $t \to +\infty$. The function f is said to be **of exponential order** as $t \to +\infty$ if there exist nonnegative constants M, c, and T such that

$$|f(t)| \leq Me^{ct} \quad \text{for } t \geq T. \tag{23}$$

Thus a function is of exponential order provided that it grows no more rapidly (as $t \to +\infty$) than a constant multiple of some exponential function with a linear exponent. The particular values of M, c, and T are not so important. What is important is that some such values exist so that the condition in (23) is satisfied.

Every polynomial $p(t)$ is of exponential order, with $M = c = 1$ in (23); this follows from the fact that $p(t)/e^t \to 0$ as $t \to +\infty$. The function $g(t) = e^{t^2}$ is an example of one that is *not* of exponential order—because

$$\lim_{t \to \infty} \frac{e^{t^2}}{e^{ct}} = \lim_{t \to \infty} e^{t^2 - ct} = +\infty,$$

the condition in (23) cannot hold for any [finite] value of M. The condition in (23) merely says that $f(t)/e^{ct}$ is *bounded* for t sufficiently large. In particular, any bounded function—such as $\cos kt$ or $\sin kt$—is of exponential order.

THEOREM 2: Existence of Laplace Transforms

If the function f is piecewise continuous for $t \geq 0$ and is of exponential order as $t \to +\infty$, then its Laplace transform $F(s) = \mathcal{L}\{f(t)\}$ exists. More precisely, if f is piecewise continuous and satisfies the condition in (23), then $F(s)$ exists for all $s > c$.

Proof: First we note that we can take $T = 0$ in (23). For by piecewise continuity, $|f(t)|$ is bounded on $[0, T]$. Increasing M in (23) if necessary, we can therefore assume that $|f(t)| \leq M$ if $0 \leq t \leq T$. Because $e^{ct} \geq 1$ for $t \geq 0$, it then follows that $|f(t)| \leq Me^{ct}$ for all $t \geq 0$.

By a standard theorem on convergence of improper integrals—the fact that absolute convergence implies convergence—it suffices for us to prove that the integral

$$\int_0^\infty |e^{-st} f(t)| \, dt$$

exists for $s > c$. To do this, it suffices in turn to show that the value of the integral

$$\int_0^b |e^{-st} f(t)| \, dt$$

remains bounded as $b \to +\infty$. But the fact that $|f(t)| \leq Me^{ct}$ for all $t \geq 0$ implies that

$$\int_0^b |e^{-st} f(t)| \, dt \leq \int_0^b |e^{-st} Me^{ct}| \, dt = M \int_0^b e^{-(s-c)t} \, dt$$

$$\leq M \int_0^\infty e^{-(s-c)t} \, dt = \frac{M}{s - c}$$

if $s > c$. This proves Theorem 2. ▲

We have shown, moreover, that

$$|F(s)| \leq \int_0^\infty |e^{-st} f(t)| \, dt \leq \frac{M}{s - c} \tag{24}$$

if $s > c$. When we take limits as $s \to +\infty$, we get the following result.

COROLLARY: *F(s) for s Large*

If $f(t)$ satisfies the hypotheses of Theorem 2, then

$$\lim_{s \to \infty} F(s) = 0. \tag{25}$$

▲

The condition in (25) severely limits the functions that can be Laplace transforms. For instance, the function $G(s) = s/(s+1)$ cannot be the Laplace transform of any "reasonable" function because its limit as $s \to +\infty$ is 1, not 0. More generally, a rational function—a quotient of two polynomials—can be (and is, as we shall see) a Laplace transform only if the degree of its numerator is less than that of its denominator.

On the other hand, the hypotheses of Theorem 2 are sufficient, but not necessary, conditions for existence of the Laplace transform of $f(t)$. For example, the function $f(t) = 1/\sqrt{t}$ fails to be piecewise continuous (at $t = 0$), but nevertheless (Example 3 with $a = -\frac{1}{2} > -1$) its Laplace transform

$$\mathcal{L}\{t^{-1/2}\} = \frac{\Gamma\left(\frac{1}{2}\right)}{s^{1/2}} = \sqrt{\frac{\pi}{s}}$$

both exists and violates the condition in (24), which would imply that $sF(s)$ remains bounded as $s \to +\infty$.

The remainder of this chapter is devoted largely to techniques for solving a differential equation by first finding the Laplace transform of its solution. It is then vital for us to know that this uniquely determines the solution of the differential equation; that is, that the function of s we have found has only one inverse Laplace transform that could be the desired solution. The following theorem is proved in Chapter 6 of Churchill's *Operational Mathematics*, 3rd ed. (New York: McGraw-Hill, 1972).

THEOREM 3: Uniqueness of Inverse Laplace Transforms

Suppose that the functions $f(t)$ and $g(t)$ satisfy the hypotheses of Theorem 2, so that their Laplace transforms $F(s)$ and $G(s)$ both exist. If $F(s) = G(s)$ for all $s > c$ (for some c), then $f(t) = g(t)$ wherever on $[0, +\infty)$ both f and g are continuous.

▲

Thus two piecewise continuous functions of exponential order with the same Laplace transform can differ only at their isolated points of discontinuity. This is of no importance in most practical applications, so we may regard inverse Laplace transforms as being essentially unique. In particular, two solutions of a differential equation must both be continuous, and hence must be the same solution if they have the same Laplace transform.

Historical Remark: Laplace transforms have an interesting history. The integral in the definition of the Laplace transform probably appeared first in the work of Euler. It is customary in mathematics to name a technique or theorem for the next person after Euler to discover it (else there will be several hundred different examples of "Euler's theorem"). In this case, the next person was the

French mathematician Pierre Simon de Laplace (1749–1827), who employed such integrals in his work on probability theory. The so-called operational techniques for solving differential equations, which are based on Laplace transforms, were not exploited by Laplace. Indeed, they were discovered and popularized by practicing engineers—notably the English electrical engineer Oliver Heaviside (1850–1925). These techniques were successfully and widely applied before they had been rigorously justified, and around the beginning of the twentieth century their validity was the subject of considerable controversy.

7.1 Problems

Apply the definition in (1) to find directly the Laplace transforms of the functions described (by formula or graph) in Problems 1 through 10.

1. $f(t) = t$
2. $f(t) = t^2$
3. $f(t) = e^{3t+1}$
4. $f(t) = \cos t$
5. $f(t) = \sinh t$
6. $f(t) = \sin^2 t$
7.

FIGURE 7.1.6.

8.

FIGURE 7.1.7.

9.

FIGURE 7.1.8.

10.

FIGURE 7.1.9.

Use the transforms in Fig. 7.1.2 to find the Laplace transforms of the functions in Problems 11 through 22. A preliminary integration by parts may be necessary.

11. $f(t) = \sqrt{t} + 3t$
12. $f(t) = 3t^{5/2} - 4t^3$
13. $f(t) = t - 2e^{3t}$
14. $f(t) = t^{3/2} - e^{-10t}$
15. $f(t) = 1 + \cosh 5t$
16. $f(t) = \sin 2t + \cos 2t$
17. $f(t) = \cos^2 2t$
18. $f(t) = \sin 3t \cos 3t$
19. $f(t) = (1 + t)^3$
20. $f(t) = te^t$
21. $f(t) = t \cos 2t$
22. $f(t) = \sinh^2 3t$

Use the transforms in Fig. 7.1.2 to find the inverse Laplace transforms of the functions in Problems 23 through 32.

23. $F(s) = \dfrac{3}{s^4}$
24. $F(s) = s^{-3/2}$
25. $F(s) = \dfrac{1}{s} - \dfrac{2}{s^{5/2}}$
26. $F(s) = \dfrac{1}{s+5}$
27. $F(s) = \dfrac{3}{s-4}$
28. $F(s) = \dfrac{3s+1}{s^2+4}$
29. $F(s) = \dfrac{5-3s}{s^2+9}$
30. $F(s) = \dfrac{9+s}{4-s^2}$
31. $F(s) = \dfrac{10s-3}{25-s^2}$
32. $F(s) = 2s^{-1}e^{-3s}$

33. Derive the transform of $f(t) = \sin kt$ by the method used in the text to derive the formula in (16).
34. Derive the transform of $f(t) = \sinh kt$ by the method used in the text to derive the formula in (14).
35. Use the tabulated integral

$$\int e^{ax} \cos bx \, dx = \frac{e^{ax}}{a^2+b^2}(a \cos bx + b \sin bx) + C$$

to obtain $\mathcal{L}\{\cos kt\}$ directly from the definition of the Laplace transform.

36. Show that the function $f(t) = \sin(e^{t^2})$ is of exponential order as $t \to +\infty$ but that its derivative is not.
37. Given $a > 0$, let $f(t) = 1$ if $0 \le t < a$, $f(t) = 0$ if $t \ge a$. First, sketch the graph of the function f, making clear its value at $t = a$. Then express f in terms of unit step functions to show that $\mathcal{L}\{f(t)\} = s^{-1}(1 - e^{-as})$.
38. Given that $0 < a < b$, let $f(t) = 1$ if $a \le t < b$, $f(t) = 0$ if either $t < a$ or $t \ge b$. First, sketch the graph of the function f, making clear its values at $t = a$ and $t = b$. Then express f in terms of unit step functions to show that $\mathcal{L}\{f(t)\} = s^{-1}(e^{-as} - e^{-bs})$.
39. The *unit staircase function* is defined as follows:

$$f(t) = n \quad \text{if} \quad n - 1 \le t < n, \quad n = 1, 2, 3, \dots.$$

(a) Sketch the graph of f to see why its name is appropriate. (b) Show that

$$f(t) = \sum_{n=0}^{\infty} u(t - n)$$

for all $t \ge 0$. (c) Assume that the Laplace transform of the infinite series in part (b) can be taken termwise (it can). Apply the geometric series to obtain the result

$$\mathcal{L}\{f(t)\} = \frac{1}{s(1 - e^{-s})}.$$

40. (a) The graph of the function f is shown in Fig. 7.1.10. Show that f can be written in the form

$$f(t) = \sum_{n=0}^{\infty}(-1)^n u(t-n).$$

(b) Use the method of Problem 39 to show that

$$\mathcal{L}\{f(t)\} = \frac{1}{s(1+e^{-s})}.$$

FIGURE 7.1.10. The graph of the function of Problem 40.

41. The graph of the *square wave function* $g(t)$ is shown in Fig. 7.1.11. Express g in terms of the function f of Problem 40 and hence deduce that

$$\mathcal{L}\{g(t)\} = \frac{1-e^{-s}}{s(1+e^{-s})} = \frac{1}{s}\tanh\frac{s}{2}.$$

FIGURE 7.1.11. The graph of the function of Problem 41.

42. Given constants a and b, define $h(t)$ for $t \geq 0$ by

$$h(t) = \begin{cases} a & \text{if } n-1 \leq t < n \text{ and } n \text{ is odd;} \\ b & \text{if } n-1 \leq t < n \text{ and } n \text{ is even.} \end{cases}$$

Sketch the graph of h and apply one of the preceding problems to show that

$$\mathcal{L}\{h(t)\} = \frac{a+be^{-s}}{s(1+e^{-s})}.$$

7.1 Application: Computer Algebra Transforms and Inverse Transforms

If $f(t) = t\cos 3t$, then the definition of the Laplace transform gives the improper integral

$$F(s) = \mathcal{L}\{f(t)\} = \int_0^{\infty} te^{-st}\cos 3t\, dt,$$

whose evaluation would appear to require a tedious integration by parts. Consequently a computer algebra Laplace transforms package is useful for the quick calculation of transforms. *Maple* contains the integral transforms package **int-trans** and *Mathematica* contains the package **Calculus:LaplaceTransform**. The *Maple* commands

```
with(inttrans):
f := t*cos(3*t):
F := laplace(f, t, s);
F := simplify(expand( F ));
```

or the *Mathematica* commands

```
Needs["Calculus`LaplaceTransform`"]
f = t*Cos[3*t];
F = LaplaceTransform[f, t, s]
Simplify[Expand[ F ]]
```

yield immediately the Laplace transform $F(s) = (s^2-9)/(s^2+9)^2$.

If you execute these commands you'll see why it's a good idea to routinely "expand and simplify" the result when calculating a Laplace transform. We can recover the original function $f(t)$ with the *Maple* command

```
invlaplace(F, s, t);
```

or the *Mathematica* command

```
InverseLaplaceTransform[F, s, t]
```

(although again we may want to expand and simplify before taking a close look at the result).

Remark: Note carefully the order of s and t in the preceding commands—first t, then s when transforming; first s, then t when inverse transforming. ■

You can use these computer algebra commands to check the answers to Problems 11 through 32 in this section, as well as a few interesting problems of your own selection.

7.2 Transformation of Initial Value Problems

We now discuss the application of Laplace transforms to solve a linear differential equation with constant coefficients, such as

$$ax''(t) + bx'(t) + cx(t) = f(t), \tag{1}$$

with given initial conditions $x(0) = x_0$ and $x'(0) = x_0'$. By the linearity of the Laplace transformation, we can transform Eq. (1) by separately taking the Laplace transform of each term in the equation. The transformed equation is

$$a\mathcal{L}\{x''(t)\} + b\mathcal{L}\{x'(t)\} + c\mathcal{L}\{x(t)\} = \mathcal{L}\{f(t)\}; \tag{2}$$

it involves the transforms of the derivatives x' and x'' of the unknown function $x(t)$. The key to the method is Theorem 1, which tells us how to express the transform of the *derivative* of a function in terms of the transform of the function itself.

THEOREM 1: Transforms of Derivatives

Suppose that the function $f(t)$ is continuous and piecewise smooth for $t \geqq 0$ and is of exponential order as $t \to +\infty$, so that there exist nonnegative constants M, c, and T such that

$$|f(t)| \leqq Me^{ct} \quad \text{for } t \geqq T. \tag{3}$$

Then $\mathcal{L}\{f'(t)\}$ exists for $s > c$, and

➤ $$\mathcal{L}\{f'(t)\} = s\mathcal{L}\{f(t)\} - f(0) = sF(s) - f(0). \tag{4}$$

▲

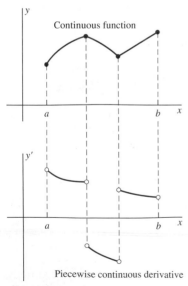

FIGURE 7.2.1. The discontinuities of f' correspond to "corners" on the graph of f.

The function f is called **piecewise smooth** on the bounded interval $[a, b]$ if it is piecewise continuous on $[a, b]$ and differentiable except at finitely many points, with $f'(t)$ being piecewise continuous on $[a, b]$. We may assign arbitrary values to $f(t)$ at the isolated points at which f is not differentiable. We say that f is piecewise smooth for $t \geqq 0$ if it is piecewise smooth on every bounded subinterval of $[0, +\infty)$. Figure 7.2.1 indicates how "corners" on the graph of f correspond to discontinuities in its derivative f'.

The main idea of the proof of Theorem 1 is exhibited best by the case in which $f'(t)$ is continuous (not merely piecewise continuous) for $t \geq 0$. Then, beginning with the definition of $\mathcal{L}\{f'(t)\}$ and integrating by parts, we get

$$\mathcal{L}\{f'(t)\} = \int_0^\infty e^{-st} f'(t) \, dt = \left[e^{-st} f(t) \right]_{t=0}^\infty + s \int_0^\infty e^{-st} f(t) \, dt.$$

Because of (3), the integrated term $e^{-st} f(t)$ approaches zero (when $s > c$) as $t \to +\infty$, and its value at the lower limit $t = 0$ contributes $-f(0)$ to the evaluation of the preceding expression. The integral that remains is simply $\mathcal{L}\{f(t)\}$; by Theorem 2 of Section 7.1, the integral converges when $s > c$. Then $\mathcal{L}\{f'(t)\}$ exists when $s > c$, and its value is that given in Eq. (4). We will defer the case in which $f'(t)$ has isolated discontinuities to the end of this section.

Solution of Initial Value Problems

In order to transform Eq. (1), we need the transform of the second derivative as well. If we assume that $g(t) = f'(t)$ satisfies the hypotheses of Theorem 1, then that theorem implies that

$$\begin{aligned}
\mathcal{L}\{f''(t)\} = \mathcal{L}\{g'(t)\} &= s\mathcal{L}\{g(t)\} - g(0) \\
&= s\mathcal{L}\{f'(t)\} - f'(0) \\
&= s\,[s\mathcal{L}\{f(t)\} - f(0)] - f'(0),
\end{aligned}$$

and thus

$$\mathcal{L}\{f''(t)\} = s^2 F(s) - sf(0) - f'(0). \tag{5}$$

A repetition of this calculation gives

$$\mathcal{L}\{f'''(t)\} = s\mathcal{L}\{f''(t)\} - f''(0) = s^3 F(s) - s^2 f(0) - sf'(0) - f''(0). \tag{6}$$

After finitely many such steps we obtain the following extension of Theorem 1.

COROLLARY: Transforms of Higher Derivatives

Suppose that the functions f, f', f'', ..., $f^{(n-1)}$ are continuous and piecewise smooth for $t \geq 0$, and that each of these functions satisfies the conditions in (3) with the same values of M and c. Then $\mathcal{L}\{f^{(n)}(t)\}$ exists when $s > c$, and

$$\begin{aligned}
\mathcal{L}\{f^{(n)}(t)\} &= s^n \mathcal{L}\{f(t)\} - s^{n-1} f(0) - s^{n-2} f'(0) - \cdots - f^{(n-1)}(0) \\
&= s^n F(s) - s^{n-1} f(0) - \cdots - sf^{(n-2)}(0) - f^{(n-1)}(0). \tag{7}
\end{aligned}$$

▲

Example 1 Solve the initial value problem

$$x'' - x' - 6x = 0; \quad x(0) = 2, \quad x'(0) = -1.$$

Solution With the given initial values, Eqs. (4) and (5) yield

$$\mathcal{L}\{x'(t)\} = s\mathcal{L}\{x(t)\} - x(0) = sX(s) - 2$$

and

$$\mathcal{L}\{x''(t)\} = s^2\mathcal{L}\{x(t)\} - sx(0) - x'(0) = s^2X(s) - 2s + 1,$$

where (according to our convention about notation) $X(s)$ denotes the Laplace transform of the (unknown) function $x(t)$. Hence the transformed equation is

$$\left[s^2X(s) - 2s + 1\right] - [sX(s) - 2] - 6[X(s)] = 0,$$

which we quickly simplify to

$$(s^2 - s - 6)X(s) - 2s + 3 = 0.$$

Thus

$$X(s) = \frac{2s - 3}{s^2 - s - 6} = \frac{2s - 3}{(s - 3)(s + 2)}.$$

By the method of partial fractions (of integral calculus), there exist constants A and B such that

$$\frac{2s - 3}{(s - 3)(s + 2)} = \frac{A}{s - 3} + \frac{B}{s + 2},$$

and multiplication of both sides of this equation by $(s - 3)(s + 2)$ yields the identity

$$2s - 3 = A(s + 2) + B(s - 3).$$

If we substitute $s = 3$, we find that $A = \frac{3}{5}$; substitution of $s = -2$ shows that $B = \frac{7}{5}$. Hence

$$X(s) = \mathcal{L}\{x(t)\} = \frac{\frac{3}{5}}{s - 3} + \frac{\frac{7}{5}}{s + 2}.$$

Because $\mathcal{L}^{-1}\{1/(s - a)\} = e^{at}$, it follows that

$$x(t) = \tfrac{3}{5}e^{3t} + \tfrac{7}{5}e^{-2t}$$

is the solution of the original initial value problem. Note that we did not first find the general solution of the differential equation. The Laplace transform method directly yields the desired particular solution, automatically taking into account—via Theorem 1 and its corollary—the given initial conditions. ∎

Example 2 Solve the initial value problem

$$x'' + 4x = \sin 3t; \quad x(0) = x'(0) = 0.$$

Such a problem arises in the motion of a mass-and-spring system with external force, as shown in Fig. 7.2.2.

Solution

FIGURE 7.2.2. A mass-and-spring system satisfying the initial value problem in Example 2. The mass is initially at rest in its equilibrium position.

Because both initial values are zero, Eq. (5) yields $\mathcal{L}\{x''(t)\} = s^2 X(s)$. We read the transform of $\sin 3t$ from the table in Fig. 7.1.2 (Section 7.1) and thereby get the transformed equation

$$s^2 X(s) + 4X(s) = \frac{3}{s^2 + 9}.$$

Therefore,

$$X(s) = \frac{3}{(s^2 + 4)(s^2 + 9)}.$$

The method of partial fractions calls for

$$\frac{3}{(s^2 + 4)(s^2 + 9)} = \frac{As + B}{s^2 + 4} + \frac{Cs + D}{s^2 + 9}.$$

The fact that there are no terms of odd degree on the left-hand side suggests that we set $A = C = 0$, because nonzero values for A or C would lead to terms of odd degree—but no such terms are present in the numerator on the left. So we replace A and C with zero, then multiply both sides by $(s^2 + 4)(s^2 + 9)$. The result is the identity

$$3 = B(s^2 + 9) + D(s^2 + 4) = (B + D)s^2 + (9B + 4D).$$

When we equate coefficients of like powers of s we get the linear equations

$$B + D = 0,$$
$$9B + 4D = 3,$$

which are readily solved for $B = \frac{3}{5}$ and $D = -\frac{3}{5}$. Hence

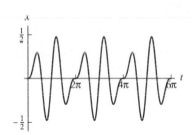

FIGURE 7.2.3. The position function $x(t)$ in Example 2.

$$X(s) = \mathcal{L}\{x(t)\} = \frac{3}{10} \cdot \frac{2}{s^2 + 4} - \frac{1}{5} \cdot \frac{3}{s^2 + 9}.$$

Because $\mathcal{L}\{\sin 2t\} = 2/(s^2 + 4)$ and $\mathcal{L}\{\sin 3t\} = 3/(s^2 + 9)$, it follows that

$$x(t) = \tfrac{3}{10} \sin 2t - \tfrac{1}{5} \sin 3t.$$

Figure 7.2.3 shows the graph of this period 2π position function of the mass. Note that the Laplace transform method again gives the solution directly, without the necessity of first finding the complementary function and a particular solution of the original nonhomogeneous differential equation. Thus nonhomogeneous equations are solved in exactly the same manner as are homogeneous equations. ■

Examples 1 and 2 illustrate the solution procedure that is outlined in Fig. 7.2.4.

Linear Systems

Laplace transforms are used frequently in engineering problems to solve linear systems in which the coefficients are all constants. When initial conditions are specified, the Laplace transform reduces such a linear system of differential equations to a linear system of algebraic equations in which the unknowns are the transforms of the solution functions. As Example 3 illustrates, the technique for a system is essentially the same as for a single linear differential equation with constant coefficients.

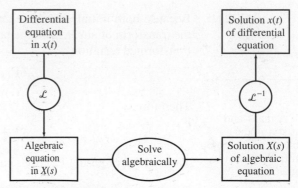

FIGURE 7.2.4. Using the Laplace transform to solve an initial value problem.

Example 3 Solve the system

$$2x'' = -6x + 2y,$$
$$y'' = 2x - 2y + 40\sin 3t, \tag{8}$$

subject to the initial conditions

$$x(0) = x'(0) = y(0) = y'(0) = 0. \tag{9}$$

Thus the force $f(t) = 40\sin 3t$ is suddenly applied to the second mass of Fig. 7.2.5 at the time $t = 0$ when the system is at rest in its equilibrium position.

FIGURE 7.2.5. A mass-and-spring system satisfying the initial value problem in Example 3. Both masses are initially at rest in their equilibrium positions.

Solution We write $X(s) = \mathcal{L}\{x(t)\}$ and $Y(s) = \mathcal{L}\{y(t)\}$. Then the initial conditions in (9) imply that

$$\mathcal{L}\{x''(t)\} = s^2 X(s) \quad \text{and} \quad \mathcal{L}\{y''(t)\} = s^2 Y(s).$$

Because $\mathcal{L}\{\sin 3t\} = 3/(s^2 + 9)$, the transforms of the equations in (8) are the equations

$$2s^2 X(s) = -6X(s) + 2Y(s),$$
$$s^2 Y(s) = 2X(s) - 2Y(s) + \frac{120}{s^2 + 9}.$$

Thus the transformed system is

$$(s^2 + 3)X(s) \qquad - Y(s) = 0,$$
$$-2X(s) + (s^2 + 2)Y(s) = \frac{120}{s^2 + 9}. \tag{10}$$

The determinant of this pair of linear equations in $X(s)$ and $Y(s)$ is

$$\begin{vmatrix} s^2 + 3 & -1 \\ -2 & s^2 + 2 \end{vmatrix} = (s^2 + 3)(s^2 + 2) - 2 = (s^2 + 1)(s^2 + 4),$$

and we readily solve—using Cramer's rule, for instance—the system in (10) for

$$X(s) = \frac{120}{(s^2 + 1)(s^2 + 4)(s^2 + 9)} = \frac{5}{s^2 + 1} - \frac{8}{s^2 + 4} + \frac{3}{s^2 + 9} \tag{11a}$$

and

$$Y(s) = \frac{120(s^2 + 3)}{(s^2 + 1)(s^2 + 4)(s^2 + 9)} = \frac{10}{s^2 + 1} + \frac{8}{s^2 + 4} - \frac{18}{s^2 + 9}. \tag{11b}$$

The partial fraction decompositions in Eqs. (11a) and (11b) are readily found using the method of Example 2. For instance, noting that the denominator factors are linear in s^2, we can write

$$\frac{120}{(s^2 + 1)(s^2 + 4)(s^2 + 9)} = \frac{A}{s^2 + 1} + \frac{B}{s^2 + 4} + \frac{C}{s^2 + 9},$$

and it follows that

$$120 = A(s^2 + 4)(s^2 + 9) + B(s^2 + 1)(s^2 + 9) + C(s^2 + 1)(s^2 + 4). \tag{12}$$

Substitution of $s^2 = -1$ (that is, $s = i$, a zero of the factor $s^2 + 1$) in Eq. (12) gives $120 = A \cdot 3 \cdot 8$, so $A = 5$. Similarly, substitution of $s^2 = -4$ in Eq. (12) yields $B = -8$, and substitution of $s^2 = -9$ yields $C = 3$. Thus we obtain the partial fraction decomposition shown in Eq. (11a).

At any rate, the inverse Laplace transforms of the expressions in Eqs. (11a) and (11b) give the solution

$$x(t) = 5 \sin t - 4 \sin 2t + \sin 3t,$$
$$y(t) = 10 \sin t + 4 \sin 2t - 6 \sin 3t.$$

Figure 7.2.6 shows the graphs of these two period 2π position functions of the two masses. ∎

FIGURE 7.2.6. The position functions $x(t)$ and $y(t)$ in Example 3.

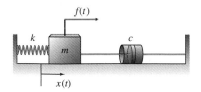

FIGURE 7.2.7. A mass-spring-dashpot system with external force $f(t)$.

The Transform Perspective

Let us regard the general constant-coefficient second-order equation as the equation of motion

$$mx'' + cx' + kx = f(t)$$

of the familiar mass-spring-dashpot system (Fig. 7.2.7). Then the transformed equation is

$$m\left[s^2 X(s) - sx(0) - x'(0)\right] + c\left[sX(s) - x(0)\right] + kX(s) = F(s). \tag{13}$$

Note that Eq. (13) is an *algebraic* equation—indeed, a linear equation—in the "unknown" $X(s)$. This is the source of the power of the Laplace transform method:

> Linear ***differential*** equations are transformed into readily solved ***algebraic*** equations.

If we solve Eq. (13) for $X(s)$, we get

$$X(s) = \frac{F(s)}{Z(s)} + \frac{I(s)}{Z(s)},$$ (14)

where

$$Z(s) = ms^2 + cs + k \quad \text{and} \quad I(s) = mx(0)s + mx'(0) + cx(0).$$

Note that $Z(s)$ depends only on the physical system itself. Thus Eq. (14) presents $X(s) = \mathcal{L}\{x(t)\}$ as the sum of a term depending only on the external force and one depending only on the initial conditions. In the case of an underdamped system, these two terms are the transforms

$$\mathcal{L}\{x_{\text{sp}}(t)\} = \frac{F(s)}{Z(s)} \quad \text{and} \quad \mathcal{L}\{x_{\text{tr}}(t)\} = \frac{I(s)}{Z(s)}$$

of the steady periodic solution and the transient solution, respectively. The only potential difficulty in finding these solutions is in finding the inverse Laplace transform of the right-hand side in Eq. (14). Much of the remainder of this chapter is devoted to finding Laplace transforms and inverse transforms. In particular, we seek those methods that are sufficiently powerful to enable us to solve problems that—unlike those in Examples 1 and 2—cannot be solved readily by the methods of Chapter 3.

Additional Transform Techniques

Example 4 Show that

$$\mathcal{L}\{te^{at}\} = \frac{1}{(s-a)^2}.$$

Solution If $f(t) = te^{at}$, then $f(0) = 0$ and $f'(t) = e^{at} + ate^{at}$. Hence Theorem 1 gives

$$\mathcal{L}\{e^{at} + ate^{at}\} = \mathcal{L}\{f'(t)\} = s\mathcal{L}\{f(t)\} = s\mathcal{L}\{te^{at}\}.$$

It follows from the linearity of the transform that

$$\mathcal{L}\{e^{at}\} + a\mathcal{L}\{te^{at}\} = s\mathcal{L}\{te^{at}\}.$$

Hence

$$\mathcal{L}\{te^{at}\} = \frac{\mathcal{L}\{e^{at}\}}{s-a} = \frac{1}{(s-a)^2}$$ (15)

because $\mathcal{L}\{e^{at}\} = 1/(s-a)$. ∎

Example 5 Find $\mathcal{L}\{t \sin kt\}$.

Solution Let $f(t) = t \sin kt$. Then $f(0) = 0$ and

$$f'(t) = \sin kt + kt \cos kt.$$

The derivative involves the new function $t \cos kt$, so we note that $f'(0) = 0$ and differentiate again. The result is

$$f''(t) = 2k \cos kt - k^2 t \sin kt.$$

But $\mathcal{L}\{f''(t)\} = s^2 \mathcal{L}\{f(t)\}$ by the formula in (5) for the transform of the second derivative, and $\mathcal{L}\{\cos kt\} = s/(s^2 + k^2)$, so we have

$$\frac{2ks}{s^2 + k^2} - k^2 \mathcal{L}\{t \sin kt\} = s^2 \mathcal{L}\{t \sin kt\}.$$

Finally, we solve this equation for

$$\mathcal{L}\{t \sin kt\} = \frac{2ks}{(s^2 + k^2)^2}. \tag{16}$$

This procedure is considerably more pleasant than the alternative of evaluating the integral

$$\mathcal{L}\{t \sin kt\} = \int_0^\infty t e^{-st} \sin kt \, dt. \qquad\blacksquare$$

Examples 4 and 5 exploit the fact that if $f(0) = 0$, then differentiation of f corresponds to multiplication of its transform by s. It is reasonable to expect the inverse operation of integration (antidifferentiation) to correspond to division of the transform by s.

THEOREM 2: Transforms of Integrals

If $f(t)$ is a piecewise continuous function for $t \geq 0$ and satisfies the condition of exponential order $|f(t)| \leq Me^{ct}$ for $t \geq T$, then

$$\mathcal{L}\left\{ \int_0^t f(\tau)\, d\tau \right\} = \frac{1}{s}\mathcal{L}\{f(t)\} = \frac{F(s)}{s} \tag{17}$$

for $s > c$. Equivalently,

$$\mathcal{L}^{-1}\left\{ \frac{F(s)}{s} \right\} = \int_0^t f(\tau)\, d\tau. \tag{18}$$

Proof: Because f is piecewise continuous, the fundamental theorem of calculus implies that

$$g(t) = \int_0^t f(\tau)\, d\tau$$

is continuous and that $g'(t) = f(t)$ where f is continuous; thus g is continuous and piecewise smooth for $t \geq 0$. Furthermore,

$$|g(t)| \leq \int_0^t |f(\tau)|\, d\tau \leq M \int_0^t e^{c\tau}\, d\tau = \frac{M}{c}(e^{ct} - 1) < \frac{M}{c}e^{ct},$$

so $g(t)$ is of exponential order as $t \to +\infty$. Hence we can apply Theorem 1 to g; this gives

$$\mathcal{L}\{f(t)\} = \mathcal{L}\{g'(t)\} = s\mathcal{L}\{g(t)\} - g(0).$$

Now $g(0) = 0$, so division by s yields

$$\mathcal{L}\left\{\int_0^t f(\tau)\, d\tau\right\} = \mathcal{L}\{g(t)\} = \frac{\mathcal{L}\{f(t)\}}{s},$$

which completes the proof. ▲

Example 6 Find the inverse Laplace transform of

$$G(s) = \frac{1}{s^2(s-a)}.$$

Solution In effect, Eq. (18) means that we can delete a factor of s from the denominator, find the inverse transform of the resulting simpler expression, and finally integrate from 0 to t (to "correct" for the missing factor s). Thus

$$\mathcal{L}^{-1}\left\{\frac{1}{s(s-a)}\right\} = \int_0^t \mathcal{L}^{-1}\left\{\frac{1}{s-a}\right\} d\tau = \int_0^t e^{a\tau}\, d\tau = \frac{1}{a}(e^{at} - 1).$$

We now repeat the technique to obtain

$$\mathcal{L}^{-1}\left\{\frac{1}{s^2(s-a)}\right\} = \int_0^t \mathcal{L}^{-1}\left\{\frac{1}{s(s-a)}\right\} d\tau = \int_0^t \frac{1}{a}(e^{a\tau} - 1)\, d\tau$$

$$= \left[\frac{1}{a}\left(\frac{1}{a}e^{a\tau} - \tau\right)\right]_0^t = \frac{1}{a^2}(e^{at} - at - 1).$$

This technique is often a more convenient way than the method of partial fractions for finding an inverse transform of a fraction of the form $P(s)/[s^n Q(s)]$. ∎

Proof of Theorem 1: We conclude this section with the proof of Theorem 1 in the general case in which f' is merely piecewise continuous. We need to prove that the limit

$$\lim_{b\to\infty} \int_0^b e^{-st} f'(t)\, dt$$

exists and also need to find its value. With b fixed, let $t_1, t_2, \ldots, t_{k-1}$ be the points interior to the interval $[0, b]$ at which f' is discontinuous. Let $t_0 = 0$ and $t_k = b$. Then we can integrate by parts on each interval (t_{n-1}, t_n) where f' is continuous. This yields

$$\int_0^b e^{-st} f'(t)\, dt = \sum_{n=1}^k \int_{t_{n-1}}^{t_n} e^{-st} f'(t)\, dt$$

$$= \sum_{n=1}^k \left(\left[e^{-st} f(t)\right]_{t_{n-1}}^{t_n} + s\int_{t_{n-1}}^{t_n} e^{-st} f(t)\, dt\right)$$

$$= -f(0) - \sum_{n=1}^{k-1} e^{-st_n} j_f(t_n) + e^{-sb} f(b) + s\int_0^b e^{-st} f(t)\, dt, \tag{19}$$

where

$$j_f(t_n) = f(t_n+) - f(t_n-) \tag{20}$$

is the jump in $f(t)$ at $t = t_n$. But because f is continuous, each jump is zero: $j_f(t_n) = 0$ for all n. Moreover, if $b > c$, then $e^{-sb} f(b) \to 0$ as $b \to +\infty$. Therefore when we take the limit in (19) as $b \to +\infty$, we get the desired result:

$$\mathcal{L}\{f'(t)\} = s\mathcal{L}\{f(t)\} - f(0). \qquad \blacktriangle$$

Extension of Theorem 1

Suppose that the original function $f(t)$ is itself only piecewise continuous (instead of continuous), with its (finite jump) discontinuities located at the points t_1, t_2, t_3, Assuming that $\mathcal{L}\{f'(t)\}$ exists, when we take the limit in (19) as $b \to +\infty$, we get

$$\mathcal{L}\{f'(t)\} = sF(s) - f(0) - \sum_{n=1}^{\infty} e^{-st_n} j_f(t_n). \tag{21}$$

\blacktriangle

| **Example 7** | Let $f(t) = 1 + [\![t]\!]$ be the unit staircase function; its graph is shown in Fig. 7.2.8. Then $f(0) = 1$, $f'(t) = 0$, and $j_f(n) = 1$ for each integer $n = 1, 2, 3, \ldots$. Hence Eq. (21) yields

$$0 = sF(s) - 1 - \sum_{n=1}^{\infty} e^{-ns},$$

so the Laplace transform of $f(t)$ is

$$F(s) = \frac{1}{s} \sum_{n=0}^{\infty} e^{-ns} = \frac{1}{s(1 - e^{-s})}.$$

FIGURE 7.2.8. The graph of the unit staircase function of Example 7.

In the last step we used the formula for the sum of a geometric series,

$$\sum_{n=0}^{\infty} x^n = \frac{1}{1 - x},$$

with $x = e^{-s} < 1$. \blacksquare

7.2 Problems

Use Laplace transforms to solve the initial value problems in Problems 1 through 16.

1. $x'' + 4x = 0$; $x(0) = 5$, $x'(0) = 0$

2. $x'' + 9x = 0$; $x(0) = 3$, $x'(0) = 4$

3. $x'' - x' - 2x = 0$; $x(0) = 0$, $x'(0) = 2$

4. $x'' + 8x' + 15x = 0$; $x(0) = 2$, $x'(0) = -3$

5. $x'' + x = \sin 2t$; $x(0) = 0 = x'(0)$

6. $x'' + 4x = \cos t$; $x(0) = 0 = x'(0)$

7. $x'' + x = \cos 3t$; $x(0) = 1$, $x'(0) = 0$

8. $x'' + 9x = 1$; $x(0) = 0 = x'(0)$

9. $x'' + 4x' + 3x = 1$; $x(0) = 0 = x'(0)$

10. $x'' + 3x' + 2x = t$; $x(0) = 0$, $x'(0) = 2$

11. $x' = 2x + y$, $y' = 6x + 3y$; $x(0) = 1$, $y(0) = -2$

12. $x' = x + 2y$, $y' = x + e^{-t}$; $x(0) = y(0) = 0$

13. $x' + 2y' + x = 0$, $x' - y' + y = 0$; $x(0) = 0$, $y(0) = 1$

14. $x'' + 2x + 4y = 0$, $y'' + x + 2y = 0$; $x(0) = y(0) = 0$, $x'(0) = y'(0) = -1$

15. $x'' + x' + y' + 2x - y = 0$, $y'' + x' + y' + 4x - 2y = 0$; $x(0) = y(0) = 1$, $x'(0) = y'(0) = 0$

16. $x' = x + z$, $y' = x + y$, $z' = -2x - z$; $x(0) = 1$, $y(0) = 0$, $z(0) = 0$

Apply Theorem 2 to find the inverse Laplace transforms of the functions in Problems 17 through 24.

17. $F(s) = \dfrac{1}{s(s-3)}$

18. $F(s) = \dfrac{3}{s(s+5)}$

19. $F(s) = \dfrac{1}{s(s^2+4)}$

20. $F(s) = \dfrac{2s+1}{s(s^2+9)}$

21. $F(s) = \dfrac{1}{s^2(s^2+1)}$

22. $F(s) = \dfrac{1}{s(s^2-9)}$

23. $F(s) = \dfrac{1}{s^2(s^2-1)}$

24. $F(s) = \dfrac{1}{s(s+1)(s+2)}$

25. Apply Theorem 1 to derive $\mathcal{L}\{\sin kt\}$ from the formula for $\mathcal{L}\{\cos kt\}$.

26. Apply Theorem 1 to derive $\mathcal{L}\{\cosh kt\}$ from the formula for $\mathcal{L}\{\sinh kt\}$.

27. (a) Apply Theorem 1 to show that

$$\mathcal{L}\{t^n e^{at}\} = \frac{n}{s-a}\mathcal{L}\{t^{n-1}e^{at}\}.$$

(b) Deduce that $\mathcal{L}\{t^n e^{at}\} = n!/(s-a)^{n+1}$ for $n = 1, 2, 3, \ldots$.

Apply Theorem 1 as in Example 5 to derive the Laplace transforms in Problems 28 through 30.

28. $\mathcal{L}\{t \cos kt\} = \dfrac{s^2 - k^2}{(s^2 + k^2)^2}$

29. $\mathcal{L}\{t \sinh kt\} = \dfrac{2ks}{(s^2 - k^2)^2}$

30. $\mathcal{L}\{t \cosh kt\} = \dfrac{s^2 + k^2}{(s^2 - k^2)^2}$

31. Apply the results in Example 5 and Problem 28 to show that

$$\mathcal{L}^{-1}\left\{\frac{1}{(s^2 + k^2)^2}\right\} = \frac{1}{2k^3}(\sin kt - kt \cos kt).$$

Apply the extension of Theorem 1 in Eq. (21) to derive the Laplace transforms given in Problems 32 through 37.

32. $\mathcal{L}\{u(t-a)\} = s^{-1}e^{-as}$ for $a > 0$.

33. If $f(t) = 1$ on the interval $[a, b]$ (where $0 < a < b$) and $f(t) = 0$ otherwise, then

$$\mathcal{L}\{f(t)\} = \frac{e^{-as} - e^{-bs}}{s}.$$

34. If $f(t) = (-1)^{[\![t]\!]}$ is the square wave function whose graph is shown in Fig. 7.2.9, then

$$\mathcal{L}\{f(t)\} = \frac{1}{s}\tanh\frac{s}{2}.$$

(*Suggestion*: Use the geometric series.)

FIGURE 7.2.9. The graph of the square wave function of Problem 34.

35. If $f(t)$ is the *unit on-off function* whose graph is shown in Fig. 7.2.10, then

$$\mathcal{L}\{f(t)\} = \frac{1}{s(1 + e^{-s})}.$$

FIGURE 7.2.10. The graph of the off-on function of Problem 35.

36. If $g(t)$ is the *triangular wave function* whose graph is shown in Fig. 7.2.11, then

$$\mathcal{L}\{g(t)\} = \frac{1}{s^2}\tanh\frac{s}{2}.$$

FIGURE 7.2.11. The graph of the triangular wave function of Problem 36.

37. If $f(t)$ is the *sawtooth function* whose graph is shown in Fig. 7.2.12, then

$$\mathcal{L}\{f(t)\} = \frac{1}{s^2} - \frac{e^{-s}}{s(1 - e^{-s})}.$$

(*Suggestion*: Note that $f'(t) \equiv 1$ where it is defined.)

FIGURE 7.2.12. The graph of the sawtooth function of Problem 37.

7.2 Application: Transforms of Initial Value Problems

The typical computer algebra system knows Theorem 1 and its corollary, and hence can transform not only functions (as in the Section 7.1 project), but also entire initial value problems. We illustrate the technique here with *Mathematica*, and in the Section 7.3 project with *Maple*. Consider the initial value problem

$$x'' + 4x = \sin 3t, \quad x(0) = x'(0) = 0$$

of Example 2. First we define the differential equation with its initial conditions, then load the Laplace transform package.

```
de = x''[t] + 4*x[t] == Sin[3*t]
inits = {x[0] -> 0, x'[0] -> 0}
Needs["Calculus`LaplaceTransform`"]
```

The Laplace transform of the differential equation is given by

```
DE = LaplaceTransform[ de, t, s ]
```

The result of this command—which we do not show explicitly here—is a linear (algebraic) equation in the as yet unknown `LaplaceTransform[x[t],t,s]`. We proceed to solve for this transform $X(s)$ of the unknown function $x(t)$ and substitute the initial conditions.

```
X = Solve[DE, LaplaceTransform[x[t],t,s]]
X = X // Last // Last // Last
X = X /. inits
```

$$\frac{3}{(s^2 + 4)(s^2 + 9)}$$

Finally we need only compute an inverse transform to find $x(t)$.

```
x = InverseLaplaceTransform[X,s,t] // Expand
```

$$\frac{3}{10} \sin(2t) - \frac{1}{5} \sin(3t)$$

Of course we could probably get this result immediately with `DSolve`, but the intermediate output generated by the steps shown here can be quite instructive. You can try it for yourself with the initial value problems in Problems 1 through 16.

7.3 Translation and Partial Fractions

As illustrated by Examples 1 and 2 of Section 7.2, the solution of a linear differential equation with constant coefficients can often be reduced to the matter of finding the inverse Laplace transform of a rational function of the form

$$R(s) = \frac{P(s)}{Q(s)} \tag{1}$$

where the degree of $P(s)$ is less than that of $Q(s)$. The technique for finding $\mathcal{L}^{-1}\{R(s)\}$ is based on the same method of partial fractions that we use in elementary calculus to integrate rational functions. The following two rules describe

the **partial fraction decomposition** of $R(s)$, in terms of the factorization of the denominator $Q(s)$ into linear factors and irreducible quadratic factors corresponding to the real and complex zeros, respectively, of $Q(s)$.

RULE 1: Linear Factor Partial Fractions

The part of the partial fraction decomposition of $R(s)$ corresponding to the linear factor $s - a$ of multiplicity n is a sum of n partial fractions, having the form

$$\frac{A_1}{s - a} + \frac{A_2}{(s - a)^2} + \cdots + \frac{A_n}{(s - a)^n}, \tag{2}$$

where $A_1, A_2, \ldots,$ and A_n are constants. ▲

RULE 2: Quadratic Factor Partial Fractions

The part of the partial fraction decomposition corresponding to the irreducible quadratic factor $(s - a)^2 + b^2$ of multiplicity n is a sum of n partial fractions, having the form

$$\frac{A_1 s + B_1}{(s - a)^2 + b^2} + \frac{A_2 s + B_2}{[(s - a)^2 + b^2]^2} + \cdots + \frac{A_n s + B_n}{[(s - a)^2 + b^2]^n}, \tag{3}$$

where $A_1, A_2, \ldots, A_n, B_1, B_2, \ldots,$ and B_n are constants. ▲

Finding $\mathcal{L}^{-1}\{R(s)\}$ involves two steps. First we must find the partial fraction decomposition of $R(s)$, and then we must find the inverse Laplace transform of each of the individual partial fractions of the types that appear in (2) and (3). The latter step is based on the following elementary property of Laplace transforms.

THEOREM 1: Translation on the s-Axis

If $F(s) = \mathcal{L}\{f(t)\}$ exists for $s > c$, then $\mathcal{L}\{e^{at} f(t)\}$ exists for $s > a + c$, and

➤ $$\mathcal{L}\{e^{at} f(t)\} = F(s - a). \tag{4}$$

Equivalently,

➤ $$\mathcal{L}^{-1}\{F(s - a)\} = e^{at} f(t). \tag{5}$$

Thus the translation $s \to s - a$ in the transform corresponds to multiplication of the original function of t by e^{at}.

Proof: If we simply replace s with $s - a$ in the definition of $F(s) = \mathcal{L}\{f(t)\}$, we obtain

$$F(s - a) = \int_0^\infty e^{-(s-a)t} f(t)\, dt = \int_0^\infty e^{-st}\left[e^{at} f(t)\right] dt = \mathcal{L}\{e^{at} f(t)\}.$$

This is Eq. (4), and it is clear that Eq. (5) is the same. ▲

If we apply the translation theorem to the formulas for the Laplace transforms of t^n, $\cos kt$, and $\sin kt$ that we already know—multiplying each of these functions

by e^{at} and replacing s with $s - a$ in the transforms—we get the following additions to the table in Fig. 7.1.2.

$f(t)$	$F(s)$		
$e^{at}t^n$	$\dfrac{n!}{(s-a)^{n+1}}$	$(s > a)$	(6)
$e^{at}\cos kt$	$\dfrac{s-a}{(s-a)^2 + k^2}$	$(s > a)$	(7)
$e^{at}\sin kt$	$\dfrac{k}{(s-a)^2 + k^2}$	$(s > a)$	(8)

For ready reference, all the Laplace transforms derived in this chapter are listed in the table of transforms that appears in the endpapers.

Example 1 Consider a mass-and-spring system with $m = \frac{1}{2}$, $k = 17$, and $c = 3$ in mks units (Fig. 7.3.1). As usual, let $x(t)$ denote the displacement of the mass m from its equilibrium position. If the mass is set in motion with $x(0) = 3$ and $x'(0) = 1$, find $x(t)$ for the resulting damped free oscillations.

Solution The differential equation is $\frac{1}{2}x'' + 3x' + 17x = 0$, so we need to solve the initial value problem

$$x'' + 6x' + 34x = 0; \quad x(0) = 3, \quad x'(0) = 1.$$

$k = 17$ $c = 3$ $m = \frac{1}{2}$

$x(t)$

FIGURE 7.3.1. The mass-spring-dashpot system of Example 1.

We take the Laplace transform of each term of the differential equation. Because [obviously] $\mathcal{L}\{0\} = 0$, we get the equation

$$\left[s^2 X(s) - 3s - 1\right] + 6\left[sX(s) - 3\right] + 34X(s) = 0,$$

which we solve for

$$X(s) = \frac{3s + 19}{s^2 + 6s + 34} = 3 \cdot \frac{s+3}{(s+3)^2 + 25} + 2 \cdot \frac{5}{(s+3)^2 + 25}.$$

Applying the formulas in (7) and (8) with $a = -3$ and $k = 5$, we now see that

$$x(t) = e^{-3t}\left(3\cos 5t + 2\sin 5t\right).$$

FIGURE 7.3.2. The position function $x(t)$ in Example 1.

Figure 7.3.2 shows the graph of this rapidly decaying damped oscillation. ■

Example 2 illustrates a useful technique for finding the partial fraction coefficients in the case of nonrepeated linear factors.

Example 2 Find the inverse Laplace transform of

$$R(s) = \frac{s^2 + 1}{s^3 - 2s^2 - 8s}.$$

Solution Note that the denominator of $R(s)$ factors as $Q(s) = s(s+2)(s-4)$. Hence

$$\frac{s^2+1}{s^3-2s^2-8s} = \frac{A}{s} + \frac{B}{s+2} + \frac{C}{s-4}.$$

Multiplication of each term of this equation by $Q(s)$ yields

$$s^2+1 = A(s+2)(s-4) + Bs(s-4) + Cs(s+2).$$

When we successively substitute the three zeros $s = 0$, $s = -2$, and $s = 4$ of the denominator $Q(s)$ in this equation, we get the results

$$-8A = 1, \quad 12B = 5, \quad \text{and} \quad 24C = 17.$$

Thus $A = -\frac{1}{8}$, $B = \frac{5}{12}$, and $C = \frac{17}{24}$, so

$$\frac{s^2+1}{s^3-2s^2-8s} = -\frac{\frac{1}{8}}{s} + \frac{\frac{5}{12}}{s+2} + \frac{\frac{17}{24}}{s-4},$$

and therefore

$$\mathcal{L}^{-1}\left\{\frac{s^2+1}{s^3-2s^2-8s}\right\} = -\frac{1}{8} + \frac{5}{12}e^{-2t} + \frac{17}{24}e^{4t}. \qquad \blacksquare$$

Example 3 illustrates a differentiation technique for finding the partial fraction coefficients in the case of repeated linear factors.

Example 3 Solve the initial value problem

$$y'' + 4y' + 4y = t^2; \quad y(0) = y'(0) = 0.$$

Solution The transformed equation is

$$s^2 Y(s) + 4sY(s) + 4Y(s) = \frac{2}{s^3}.$$

Thus

$$Y(s) = \frac{2}{s^3(s+2)^2} = \frac{A}{s^3} + \frac{B}{s^2} + \frac{C}{s} + \frac{D}{(s+2)^2} + \frac{E}{s+2}. \qquad (9)$$

To find A, B, and C, we multiply both sides by s^3 to obtain

$$\frac{2}{(s+2)^2} = A + Bs + Cs^2 + s^3 F(s), \qquad (10)$$

where $F(s) = D(s+2)^{-2} + E(s+2)^{-1}$ is the sum of the two partial fractions corresponding to $(s+2)^2$. Substitution of $s = 0$ in Eq. (10) yields $A = \frac{1}{2}$. To find B and C, we differentiate Eq. (10) twice to obtain

$$\frac{-4}{(s+2)^3} = B + 2Cs + 3s^2 F(s) + s^3 F'(s) \qquad (11)$$

and

$$\frac{12}{(s+2)^4} = 2C + 6sF(s) + 6s^2F'(s) + s^3F''(s). \qquad (12)$$

Now substitution of $s = 0$ in Eq. (11) yields $B = -\frac{1}{2}$, and substitution of $s = 0$ in Eq. (12) yields $C = \frac{3}{8}$.

To find D and E, we multiply each side in Eq. (9) by $(s+2)^2$ to get

$$\frac{2}{s^3} = D + E(s+2) + (s+2)^2 G(s), \qquad (13)$$

where $G(s) = As^{-3} + Bs^{-2} + Cs^{-1}$, and then differentiate to obtain

$$-\frac{6}{s^4} = E + 2(s+2)G(s) + (s+2)^2 G'(s). \qquad (14)$$

Substitution of $s = -2$ in Eqs. (13) and (14) now yields $D = -\frac{1}{4}$ and $E = -\frac{3}{8}$. Thus

$$Y(s) = \frac{\frac{1}{2}}{s^3} - \frac{\frac{1}{2}}{s^2} + \frac{\frac{3}{8}}{s} - \frac{\frac{1}{4}}{(s+2)^2} - \frac{\frac{3}{8}}{s+2},$$

so the solution of the given initial value problem is

$$y(t) = \tfrac{1}{4}t^2 - \tfrac{1}{2}t + \tfrac{3}{8} - \tfrac{1}{4}te^{-2t} - \tfrac{3}{8}e^{-2t}. \qquad \blacksquare$$

Examples 4, 5, and 6 illustrate techniques for dealing with quadratic factors in partial fraction decompositions.

Example 4 Consider the mass spring dashpot system as in Example 1, but with initial conditions $x(0) = x'(0) = 0$ and with the imposed external force $F(t) = 15 \sin 2t$. Find the resulting transient motion and steady periodic motion of the mass.

Solution The initial value problem we need to solve is

$$x'' + 6x' + 34x = 30 \sin 2t; \quad x(0) = x'(0) = 0.$$

The transformed equation is

$$s^2 X(s) + 6s X(s) + 34 X(s) = \frac{60}{s^2 + 4}.$$

Hence

$$X(s) = \frac{60}{(s^2 + 4)[(s+3)^2 + 25]} = \frac{As + B}{s^2 + 4} + \frac{Cs + D}{(s+3)^2 + 25}.$$

When we multiply both sides by the common denominator, we get

$$60 = (As + B)[(s+3)^2 + 25] + (Cs + D)(s^2 + 4). \qquad (15)$$

To find A and B, we substitute the zero $s = 2i$ of the quadratic factor $s^2 + 4$ in Eq. (15); the result is

$$60 = (2iA + B)[(2i+3)^2 + 25],$$

which we simplify to

$$60 = (-24A + 30B) + (60A + 12B)i.$$

We now equate real parts and imaginary parts on each side of this equation to obtain the two linear equations

$$-24A + 30B = 60 \quad \text{and} \quad 60A + 12B = 0,$$

which are readily solved for $A = -\frac{10}{29}$ and $B = \frac{50}{29}$.

To find C and D, we substitute the zero $s = -3 + 5i$ of the quadratic factor $(s + 3)^2 + 25$ in Eq. (15) and get

$$60 = [C(-3 + 5i) + D][(-3 + 5i)^2 + 4],$$

which we simplify to

$$60 = (186C - 12D) + (30C - 30D)i.$$

Again we equate real parts and imaginary parts; this yields the two linear equations

$$186C - 12D = 60 \quad \text{and} \quad 30C - 30D = 0,$$

and we readily find their solution to be $C = D = \frac{10}{29}$.

With these values of the coefficients A, B, C, and D, our partial fractions decomposition of $X(s)$ is

$$X(s) = \frac{1}{29}\left(\frac{-10s + 50}{s^2 + 4} + \frac{10s + 10}{(s + 3)^2 + 25}\right)$$

$$= \frac{1}{29}\left(\frac{-10s + 25 \cdot 2}{s^2 + 4} + \frac{10(s + 3) - 4 \cdot 5}{(s + 3)^2 + 25}\right).$$

After we compute the inverse Laplace transforms, we get the position function

$$x(t) = \tfrac{5}{29}(-2\cos 2t + 5\sin 2t) + \tfrac{2}{29}e^{-3t}(5\cos 5t - 2\sin 5t).$$

The terms of circular frequency 2 constitute the steady periodic forced oscillation of the mass, whereas the exponentially damped terms of circular frequency 5 constitute its transient motion, which disappears very rapidly (see Fig. 7.3.3). Note that the transient motion is nonzero even though both initial conditions are zero. ■

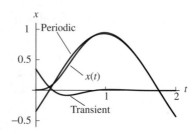

FIGURE 7.3.3. The periodic forced oscillation $x_{\mathrm{sp}}(t)$, damped transient motion $x_{\mathrm{tr}}(t)$, and solution $x(t) = x_{\mathrm{sp}}(t) + x_{\mathrm{tr}}(t)$ in Example 4.

Resonance and Repeated Quadratic Factors

The following two inverse Laplace transforms are useful in inverting partial fractions that correspond to the case of repeated quadratic factors:

$$\mathcal{L}^{-1}\left\{\frac{s}{(s^2 + k^2)^2}\right\} = \frac{1}{2k}t\sin kt, \tag{16}$$

$$\mathcal{L}^{-1}\left\{\frac{1}{(s^2 + k^2)^2}\right\} = \frac{1}{2k^3}(\sin kt - kt\cos kt). \tag{17}$$

These follow from Example 5 and Problem 31 of Section 7.2, respectively. Because of the presence in Eqs. (16) and (17) of the terms $t\sin kt$ and $t\cos kt$, a repeated quadratic factor ordinarily signals the phenomenon of resonance in an undamped mechanical or electrical system.

Example 5 Use Laplace transforms to solve the initial value problem

$$x'' + \omega_0^2 x = F_0 \sin \omega t; \quad x(0) = 0 = x'(0)$$

that determines the undamped forced oscillations of a mass on a spring.

Solution When we transform the differential equation, we get the equation

$$s^2 X(s) + \omega_0^2 X(s) = \frac{F_0 \omega}{s^2 + \omega^2}, \quad \text{so} \quad X(s) = \frac{F_0 \omega}{(s^2 + \omega^2)(s^2 + \omega_0^2)}.$$

If $\omega \neq \omega_0$, we find without difficulty that

$$X(s) = \frac{F_0 \omega}{\omega^2 - \omega_0^2} \left(\frac{1}{s^2 + \omega_0^2} - \frac{1}{s^2 + \omega^2} \right),$$

so it follows that

$$x(t) = \frac{F_0 \omega}{\omega^2 - \omega_0^2} \left(\frac{1}{\omega_0} \sin \omega_0 t - \frac{1}{\omega} \sin \omega t \right).$$

But if $\omega = \omega_0$, we have

$$X(s) = \frac{F_0 \omega_0}{(s^2 + \omega_0^2)^2},$$

so Eq. (17) yields the resonance solution

$$x(t) = \frac{F_0}{2\omega_0^2} (\sin \omega_0 t - \omega_0 t \cos \omega_0 t). \tag{18}$$

■

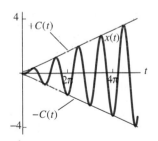

4 $C(t)$

$x(t)$

2π 4π t

$-C(t)$

-4

FIGURE 7.3.4. The resonance solution in (18) with $\omega_0 = \frac{1}{2}$ and $F_0 = 1$, together with its envelope curves $x = +C(t)$.

Remark: The solution curve defined in Eq. (18) bounces back and forth (see Fig. 7.3.4) between the "envelope curves" $x = \pm C(t)$ that are obtained by writing (18) in the form

$$x(t) = A(t) \cos \omega_0 t + B(t) \sin \omega_0 t$$

and then defining the usual "amplitude" $C = \sqrt{A^2 + B^2}$. In this case we find that

$$C(t) = \frac{F_0}{2\omega_0^2} \sqrt{\omega_0^2 t^2 + 1}.$$

This technique for constructing envelope curves of resonance solutions is illustrated further in the application material for this section. ■

Example 6 Solve the initial value problem

$$y^{(4)} + 2y'' + y = 4te^t; \quad y(0) = y'(0) = y''(0) = y^{(3)}(0) = 0.$$

Solution First we observe that

$$\mathcal{L}\{y''(t)\} = s^2 Y(s), \quad \mathcal{L}\{y^{(4)}(t)\} = s^4 Y(s), \quad \text{and} \quad \mathcal{L}\{te^t\} = \frac{1}{(s-1)^2}.$$

Hence the transformed equation is

$$(s^4 + 2s^2 + 1)Y(s) = \frac{4}{(s-1)^2}.$$

Thus our problem is to find the inverse transform of

$$Y(s) = \frac{4}{(s-1)^2(s^2+1)^2}$$

$$= \frac{A}{(s-1)^2} + \frac{B}{s-1} + \frac{Cs+D}{(s^2+1)^2} + \frac{Es+F}{s^2+1}. \tag{19}$$

If we multiply by the common denominator $(s-1)^2(s^2+1)^2$, we get the equation

$$A(s^2+1)^2 + B(s-1)(s^2+1)^2 + Cs(s-1)^2$$
$$+ D(s-1)^2 + Es(s-1)^2(s^2+1) + F(s-1)^2(s^2+1) = 4. \tag{20}$$

Upon substituting $s = 1$ we find that $A = 1$.

Equation (20) is an identity that holds for all values of s. To find the values of the remaining coefficients, we substitute in succession the values $s = 0$, $s = -1$, $s = 2$, $s = -2$, and $s = 3$ in Eq. (20). This yields the system

$$\begin{aligned}
-B & & + D & & + F &= 3, \\
-8B - & 4C + 4D - & 8E + & 8F &= 0, \\
25B + & 2C + & D + & 10E + & 5F &= -21, \\
-75B - & 18C + 9D - & 90E + 45F &= -21, \\
200B + & 12C + 4D + & 120E + 40F &= -96
\end{aligned} \tag{21}$$

of five linear equations in B, C, D, E, and F. With the aid of a calculator programmed to solve linear systems, we find that $B = -2$, $C = 2$, $D = 0$, $E = 2$, and $F = 1$.

We now substitute in Eq. (18) the coefficients we have found, and thus obtain

$$Y(s) = \frac{1}{(s-1)^2} - \frac{2}{s-1} + \frac{2s}{(s^2+1)^2} + \frac{2s+1}{s^2+1}.$$

Recalling Eq. (16), the translation property, and the familiar transforms of $\cos t$ and $\sin t$, we see finally that the solution of the given initial value problem is

$$y(t) = (t-2)e^t + (t+1)\sin t + 2\cos t. \quad \blacksquare$$

7.3 Problems

Apply the translation theorem to find the Laplace transforms of the functions in Problems 1 through 4.

1. $f(t) = t^4 e^{\pi t}$ **2.** $f(t) = t^{3/2} e^{-4t}$

3. $f(t) = e^{-2t} \sin 3\pi t$ **4.** $f(t) = e^{-t/2} \cos 2\left(t - \frac{1}{8}\pi\right)$

Apply the translation theorem to find the inverse Laplace transforms of the functions in Problems 5 through 10.

5. $F(s) = \dfrac{3}{2s - 4}$ **6.** $F(s) = \dfrac{s - 1}{(s + 1)^3}$

7. $F(s) = \dfrac{1}{s^2 + 4s + 4}$ **8.** $F(s) = \dfrac{s + 2}{s^2 + 4s + 5}$

9. $F(s) = \dfrac{3s + 5}{s^2 - 6s + 25}$ **10.** $F(s) = \dfrac{2s - 3}{9s^2 - 12s + 20}$

Use partial fractions to find the inverse Laplace transforms of the functions in Problems 11 through 22.

11. $F(s) = \dfrac{1}{s^2 - 4}$ **12.** $F(s) = \dfrac{5s - 6}{s^2 - 3s}$

13. $F(s) = \dfrac{5 - 2s}{s^2 + 7s + 10}$ **14.** $F(s) = \dfrac{5s - 4}{s^3 - s^2 - 2s}$

15. $F(s) = \dfrac{1}{s^3 - 5s^2}$ **16.** $F(s) = \dfrac{1}{(s^2 + s - 6)^2}$

17. $F(s) = \dfrac{1}{s^4 - 16}$ **18.** $F(s) = \dfrac{s^3}{(s - 4)^4}$

19. $F(s) = \dfrac{s^2 - 2s}{s^4 + 5s^2 + 4}$ **20.** $F(s) = \dfrac{1}{s^4 - 8s^2 + 16}$

21. $F(s) = \dfrac{s^2 + 3}{(s^2 + 2s + 2)^2}$ **22.** $F(s) = \dfrac{2s^3 - s^2}{(4s^2 - 4s + 5)^2}$

Use the factorization

$$s^4 + 4a^4 = (s^2 - 2as + 2a^2)(s^2 + 2as + 2a^2)$$

to derive the inverse Laplace transforms listed in Problems 23 through 26.

23. $\mathcal{L}^{-1}\left\{\dfrac{s^3}{s^4 + 4a^4}\right\} = \cosh at \cos at$

24. $\mathcal{L}^{-1}\left\{\dfrac{s}{s^4 + 4a^4}\right\} = \dfrac{1}{2a^2} \sinh at \sin at$

25. $\mathcal{L}^{-1}\left\{\dfrac{s^2}{s^4 + 4a^4}\right\} = \dfrac{1}{2a}(\cosh at \sin at + \sinh at \cos at)$

26. $\mathcal{L}^{-1}\left\{\dfrac{1}{s^4 + 4a^4}\right\} = \dfrac{1}{4a^3}(\cosh at \sin at - \sinh at \cos at)$

Use Laplace transforms to solve the initial value problems in Problems 27 through 38.

27. $x'' + 6x' + 25x = 0$; $x(0) = 2, x'(0) = 3$

28. $x'' - 6x' + 8x = 2$; $x(0) = x'(0) = 0$

29. $x'' - 4x = 3t$; $x(0) = x'(0) = 0$

30. $x'' + 4x' + 8x = e^{-t}$; $x(0) = x'(0) = 0$

31. $x^{(3)} + x'' - 6x' = 0$; $x(0) = 0, x'(0) = x''(0) = 1$

32. $x^{(4)} - x = 0$; $x(0) = 1, x'(0) = x''(0) = x^{(3)}(0) = 0$

33. $x^{(4)} + x = 0$; $x(0) = x'(0) = x''(0) = 0, x^{(3)}(0) = 1$

34. $x^{(4)} + 13x'' + 36x = 0$; $x(0) = x''(0) = 0, x'(0) = 2,$ $x^{(3)}(0) = -13$

35. $x^{(4)} + 8x'' + 16x = 0$; $x(0) = x'(0) = x''(0) = 0,$ $x^{(3)}(0) = 1$

36. $x^{(4)} + 2x'' + x = e^{2t}$; $x(0) = x'(0) = x''(0) = x^{(3)}(0) = 0$

37. $x'' + 4x' + 13x = te^{-t}$; $x(0) = 0, x'(0) = 2$

38. $x'' + 6x' + 18x = \cos 2t$; $x(0) = 1, x'(0) = -1$

Problems 39 and 40 illustrate two types of resonance in a mass-spring-dashpot system with given external force $f(t)$ and with the initial conditions $x(0) = x'(0) = 0$.

39. Suppose that $m = 1, k = 9, c = 0$, and $f(t) = 6 \cos 3t$. Use the inverse transform given in Eq. (16) to derive the solution $x(t) = t \sin 3t$. Construct a figure that illustrates the resonance that occurs.

40. Suppose that $m = 1, k = 9.04, c = 0.4$, and $f(t) = 6e^{-t/3} \cos 3t$. Derive the solution

$$x(t) = te^{-t/5} \sin 3t.$$

Show that the maximum value of the amplitude function $A(t) = te^{-t/5}$ is $A(5) = 5/e$. Thus (as indicated in Fig. 7.3.5) the oscillations of the mass increase in amplitude during the first 5 s before being damped out as $t \to +\infty$.

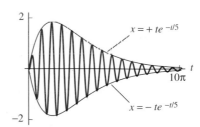

FIGURE 7.3.5. The graph of the damped oscillation in Problem 40.

7.3 Application: Damping and Resonance Investigations

Here we outline a *Maple* investigation of the behavior of the mass-spring-dashpot system

$$mx'' + cx' + kx = F(t), \quad x(0) = x'(0) = 0 \tag{1}$$

with parameter values

```
m := 25;      c := 10;      k := 226;
```

in response to a variety of possible external forces:

1. $F(t) \equiv 0$:

This is the case of free damped oscillations, similar to those illustrated in Fig. 3.4.12.

2. $F(t) = 901 \cos 3t$:

With this periodic external force you should see a steady periodic oscillation with an exponentially damped transient motion (as illustrated in Fig. 3.6.13).

3. $F(t) = 900e^{-t/5} \cos 3t$:

Now the periodic external force is exponentially damped, and the transform $X(s)$ includes a repeated quadratic factor that signals the presence of a resonance phenomenon. The response $x(t)$ is a constant multiple of that shown in Fig. 7.3.5.

4. $F(t) = 900te^{-t/5} \cos 3t$:

We have inserted the factor t to make it a bit more interesting. The solution in this case is illustrated below.

5. $F(t) = 162t^3 e^{-t/5} \cos 3t$:

In this case you'll find that the transform $X(s)$ involves the *fifth* power of a quadratic factor, and its inverse transform by manual methods would be impossibly tedious.

To illustrate the *Maple* approach, we first set up the differential equation corresponding to Case 4.

```
F := 900*t*exp(-t/5)*cos(3*t);
de := m*diff(x(t),t$2) + c*diff(x(t),t) + k*x(t) = F;
```

Then we apply the Laplace transform and substitute the initial conditions.

```
with(inttrans):
DE := laplace(de, t, s):
X(s) := solve(DE, laplace(x(t), t, s)):
X(s) := simplify(subs(x(0)=0, D(x)(0)=0, X(s)));
```

At this point the command `factor(denom(X(s)))` shows that

$$X(s) = \frac{22500(25s^2 + 10s - 224)}{(25s^2 + 10s + 226)^3}.$$

The cubed quadratic factor would be difficult to handle manually, but the command

```
x(t) := invlaplace(X(s), s, t);
```

soon yields

$$x(t) = e^{-t/5} \left(t \cos 3t + \left(3t^2 - \tfrac{1}{3} \right) \sin 3t \right).$$

The amplitude function for these damped oscillations is defined by

```
C(t) := exp(-t/5)*sqrt(t^2 + (3*t^2 - 1/3)^2);
```

and finally the command

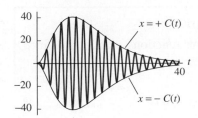

FIGURE 7.3.6. The resonance solution and its envelope curves in Case 4.

```
plot({x(t), C(t), -C(t)}, t=0..40);
```

produces the plot shown in Fig. 7.3.6. The resonance resulting from the repeated quadratic factor consists of a temporary buildup before the oscillations are damped out.

For each of Cases 1 through 5 you can enter the appropriate external force and re-execute the commands listed previously. To see the advantage of using Laplace transforms, set up the differential equation de for Case 5 and examine the result of the command

```
dsolve({de, x(0)=0, D(x)(0)=0}, x(t));
```

Of course you can substitute your own favorite mass-spring-dashpot parameters for those used here. But it will simplify the calculations if you choose m, c, and k so that

$$mr^2 + cr + k = (pr + a)^2 + b^2 \tag{2}$$

where p, a, and b are integers. One way is to select the latter integers first, then use Eq. (2) to determine m, c, and k.

7.4 Derivatives, Integrals, and Products of Transforms

The Laplace transform of the (initially unknown) solution of a differential equation is sometimes recognizable as the product of the transforms of two *known* functions. For example, when we transform the initial value problem

$$x'' + x = \cos t; \quad x(0) = x'(0) = 0,$$

we get

$$X(s) = \frac{s}{(s^2+1)^2} = \frac{s}{s^2+1} \cdot \frac{1}{s^2+1} = \mathcal{L}\{\cos t\} \cdot \mathcal{L}\{\sin t\}.$$

This strongly suggests that there ought to be a way of combining the two functions $\sin t$ and $\cos t$ to obtain a function $x(t)$ whose transform is the *product* of *their* transforms. But obviously $x(t)$ is *not* simply the product of $\cos t$ and $\sin t$, because

$$\mathcal{L}\{\cos t \sin t\} = \mathcal{L}\left\{\tfrac{1}{2}\sin 2t\right\} = \frac{1}{s^2+4} \neq \frac{s}{(s^2+1)^2}.$$

Thus $\mathcal{L}\{\cos t \sin t\} \neq \mathcal{L}\{\cos t\} \cdot \mathcal{L}\{\sin t\}$.

Theorem 1 of this section will tell us that the function

$$h(t) = \int_0^t f(\tau)g(t-\tau)\,d\tau \tag{1}$$

has the desired property that

$$\mathcal{L}\{h(t)\} = H(s) = F(s) \cdot G(s). \tag{2}$$

The new function of t defined as the integral in (1) depends only on f and g and is called the *convolution* of f and g. It is denoted by $f * g$, the idea being that it is a new type of product of f and g, so tailored that its transform is the product of the transforms of f and g.

Definition: The Convolution of Two Functions

The **convolution** $f * g$ of the piecewise continuous functions f and g is defined for $t \geq 0$ as follows:

$$(f * g)(t) = \int_0^t f(\tau)g(t - \tau)\,d\tau. \qquad (3)$$

We will also write $f(t) * g(t)$ when convenient. In terms of the convolution product, Theorem 1 of this section says that

$$\mathcal{L}\{f * g\} = \mathcal{L}\{f\} \cdot \mathcal{L}\{g\}.$$

If we make the substitution $u = t - \tau$ in the integral in (3), we see that

$$f(t) * g(t) = \int_0^t f(\tau)g(t - \tau)\,d\tau = \int_t^0 f(t - u)g(u)(-du)$$

$$= \int_0^t g(u)f(t - u)\,du = g(t) * f(t).$$

Thus the convolution is *commutative*: $f * g = g * f$.

Example 1

The convolution of $\cos t$ and $\sin t$ is

$$(\cos t) * (\sin t) = \int_0^t \cos \tau \sin(t - \tau)\,d\tau.$$

We apply the trigonometric identity

$$\cos A \sin B = \tfrac{1}{2}\left[\sin(A + B) - \sin(A - B)\right]$$

to obtain

$$(\cos t) * (\sin t) = \int_0^t \tfrac{1}{2}\left[\sin t - \sin(2\tau - t)\right]\,d\tau$$

$$= \tfrac{1}{2}\left[\tau \sin t + \tfrac{1}{2}\cos(2\tau - t)\right]_{\tau=0}^t;$$

that is,

$$(\cos t) * (\sin t) = \tfrac{1}{2}t \sin t.$$

And we recall from Example 5 of Section 7.2 that the Laplace transform of $\frac{1}{2}t \sin t$ is indeed $s/(s^2 + 1)^2$. ∎

Theorem 1 is proved at the end of this section.

THEOREM 1: The Convolution Property

Suppose that $f(t)$ and $g(t)$ are piecewise continuous for $t \geq 0$ and that $|f(t)|$ and $|g(t)|$ are bounded by Me^{ct} as $t \to +\infty$. Then the Laplace transform of the convolution $f(t) * g(t)$ exists for $s > c$; moreover,

$$\mathcal{L}\{f(t) * g(t)\} = \mathcal{L}\{f(t)\} \cdot \mathcal{L}\{g(t)\} \qquad (4)$$

and

$$\mathcal{L}^{-1}\{F(s)\cdot G(s)\} = f(t) * g(t).\qquad(5)$$

Thus we can find the inverse transform of the product $F(s)\cdot G(s)$, provided that we can evaluate the integral

$$\mathcal{L}^{-1}\{F(s)\cdot G(s)\} = \int_0^t f(\tau)g(t-\tau)\,d\tau.\qquad(5')$$

Example 2 illustrates the fact that convolution often provides a convenient alternative to the use of partial fractions for finding inverse transforms.

Example 2 With $f(t) = \sin 2t$ and $g(t) = e^t$, convolution yields

$$\mathcal{L}^{-1}\left\{\frac{2}{(s-1)(s^2+4)}\right\} = (\sin 2t) * e^t = \int_0^t e^{t-\tau}\sin 2\tau\,d\tau$$

$$= e^t \int_0^t e^{-\tau}\sin 2\tau\,d\tau = e^t\left[\frac{e^{-\tau}}{5}(-\sin 2\tau - 2\cos 2\tau)\right]_0^t,$$

so

$$\mathcal{L}^{-1}\left\{\frac{2}{(s-1)(s^2+4)}\right\} = \frac{2}{5}e^t - \frac{1}{5}\sin 2t - \frac{2}{5}\cos 2t. \qquad\blacksquare$$

Differentiation of Transforms

According to Theorem 1 of Section 7.2, if $f(0) = 0$ then differentiation of $f(t)$ corresponds to multiplication of its transform by s. Theorem 2, proved at the end of this section, tells us that differentiation of the transform $F(s)$ corresponds to multiplication of the original function $f(t)$ by $-t$.

THEOREM 2: Differentiation of Transforms

If $f(t)$ is piecewise continuous for $t \geqq 0$ and $|f(t)| \leqq Me^{ct}$ as $t \to +\infty$, then

$$\mathcal{L}\{-tf(t)\} = F'(s)\qquad(7)$$

for $s > c$. Equivalently,

$$f(t) = \mathcal{L}^{-1}\{F(s)\} = -\frac{1}{t}\mathcal{L}^{-1}\{F'(s)\}.\qquad(8)$$

Repeated application of Eq. (6) gives

$$\mathcal{L}\{t^n f(t)\} = (-1)^n F^{(n)}(s)\qquad(9)$$

for $n = 1, 2, 3, \dots$.

Example 3 Find $\mathcal{L}\{t^2 \sin kt\}$.

Solution Equation (8) gives

$$\mathcal{L}\{t^2 \sin kt\} = (-1)^2 \frac{d^2}{ds^2}\left(\frac{k}{s^2 + k^2}\right)$$

$$= \frac{d}{ds}\left[\frac{-2ks}{(s^2 + k^2)^2}\right] = \frac{6ks^2 - 2k^3}{(s^2 + k^2)^3}. \qquad (10)$$

∎

The form of the differentiation property in Eq. (7) is often helpful in finding an inverse transform when the *derivative* of the transform is easier to work with than the transform itself.

Example 4 Find $\mathcal{L}^{-1}\{\tan^{-1}(1/s)\}$.

Solution The derivative of $\tan^{-1}(1/s)$ is a simple rational function, so we apply Eq. (7):

$$\mathcal{L}^{-1}\left\{\tan^{-1}\frac{1}{s}\right\} = -\frac{1}{t}\mathcal{L}^{-1}\left\{\frac{d}{ds}\tan^{-1}\frac{1}{s}\right\} = -\frac{1}{t}\mathcal{L}^{-1}\left\{\frac{-1/s^2}{1 + (1/s)^2}\right\}$$

$$= -\frac{1}{t}\mathcal{L}^{-1}\left\{\frac{-1}{s^2 + 1}\right\} = -\frac{1}{t}(-\sin t).$$

Therefore,

$$\mathcal{L}^{-1}\left\{\tan^{-1}\frac{1}{s}\right\} = \frac{\sin t}{t}.$$

∎

Equation (8) can be applied to transform a linear differential equation having polynomial, rather than constant, coefficients. The result will be a differential equation involving the transform; whether this procedure leads to success depends, of course, on whether we can solve the new equation more readily than the old one.

Example 5 Let $x(t)$ be the solution of Bessel's equation of order zero,

$$tx'' + x' + tx = 0,$$

such that $x(0) = 1$ and $x'(0) = 0$. This solution of Bessel's equation is customarily denoted by $J_0(t)$. Because

$$\mathcal{L}\{x'(t)\} = sX(s) - 1 \quad \text{and} \quad \mathcal{L}\{x''(t)\} = s^2 X(s) - s,$$

and because x and x'' are each multiplied by t, application of Eq. (7) yields the transformed equation

$$-\frac{d}{ds}\left[s^2 X(s) - s\right] + [sX(s) - 1] - \frac{d}{ds}[X(s)] = 0.$$

The result of differentiation and simplification is the differential equation

$$(s^2 + 1)X'(s) + sX(s) = 0.$$

This equation is separable—

$$\frac{X'(s)}{X(s)} = -\frac{s}{s^2 + 1};$$

its general solution is

$$X(s) = \frac{C}{\sqrt{s^2 + 1}}.$$

In Problem 39 we outline the argument that $C = 1$. Because $X(s) = \mathcal{L}\{J_0(t)\}$, it follows that

$$\mathcal{L}\{J_0(t)\} = \frac{1}{\sqrt{s^2 + 1}}. \tag{11}$$

■

Integration of Transforms

Differentiation of $F(s)$ corresponds to multiplication of $f(t)$ by t (together with a change of sign). It is therefore natural to expect that integration of $F(s)$ will correspond to division of $f(t)$ by t. Theorem 3, proved at the end of this section, confirms this, provided that the resulting quotient $f(t)/t$ remains well behaved as $t \to 0$ from the right; that is, provided that

$$\lim_{t \to 0^+} \frac{f(t)}{t} \quad \text{exists and is finite.} \tag{12}$$

THEOREM 3: Integration of Transforms

Suppose that $f(t)$ is piecewise continuous for $t \geq 0$, that $f(t)$ satisfies the condition in (11), and that $|f(t)| \leq Me^{ct}$ as $t \to +\infty$. Then

$$\mathcal{L}\left\{\frac{f(t)}{t}\right\} = \int_s^\infty F(\sigma)\,d\sigma \tag{13}$$

for $s > c$. Equivalently,

$$f(t) = \mathcal{L}^{-1}\{F(s)\} = t\mathcal{L}^{-1}\left\{\int_s^\infty F(\sigma)\,d\sigma\right\}. \tag{14}$$

▲

Example 6 Find $\mathcal{L}\{(\sinh t)/t\}$.

Solution We first verify that the condition in (11) holds:

$$\lim_{t \to 0} \frac{\sinh t}{t} = \lim_{t \to 0} \frac{e^t - e^{-t}}{2t} = \lim_{t \to 0} \frac{e^t + e^{-t}}{2} = 1,$$

with the aid of l'Hôpital's rule. Then Eq. (12), with $f(t) = \sinh t$, yields

$$\mathcal{L}\left\{\frac{\sinh t}{t}\right\} = \int_s^\infty \mathcal{L}\{\sinh t\}\, d\sigma = \int_s^\infty \frac{d\sigma}{\sigma^2 - 1}$$

$$= \frac{1}{2}\int_s^\infty \left(\frac{1}{\sigma - 1} - \frac{1}{\sigma + 1}\right) d\sigma = \frac{1}{2}\left[\ln\frac{\sigma - 1}{\sigma + 1}\right]_s^\infty.$$

Therefore,

$$\mathcal{L}\left\{\frac{\sinh t}{t}\right\} = \frac{1}{2}\ln\frac{s + 1}{s - 1},$$

because $\ln 1 = 0$. ∎

The form of the integration property in Eq. (13) is often helpful in finding an inverse transform when the indefinite *integral* of the transform is easier to handle than the transform itself.

Example 7 Find $\mathcal{L}^{-1}\{2s/(s^2 - 1)^2\}$.

Solution We could use partial fractions, but it is much simpler to apply Eq. (13). This gives

$$\mathcal{L}^{-1}\left\{\frac{2s}{(s^2 - 1)^2}\right\} = t\mathcal{L}^{-1}\left\{\int_s^\infty \frac{2\sigma}{(\sigma^2 - 1)^2}\, d\sigma\right\}$$

$$= t\mathcal{L}^{-1}\left\{\left[\frac{-1}{\sigma^2 - 1}\right]_s^\infty\right\} = t\mathcal{L}^{-1}\left\{\frac{1}{s^2 - 1}\right\},$$

and therefore

$$\mathcal{L}^{-1}\left\{\frac{2s}{(s^2 - 1)^2}\right\} = t\sinh t. \qquad ∎$$

*Proofs of Theorems

Proof of Theorem 1: The transforms $F(s)$ and $G(s)$ exist when $s > c$ by Theorem 2 of Section 7.1. The definition of the Laplace transform gives

$$G(s) = \int_0^\infty e^{-su} g(u)\, du = \int_{-\tau}^\infty e^{-s(t-\tau)} g(t - \tau)\, dt \quad (u = t - \tau),$$

and therefore

$$G(s) = e^{s\tau}\int_0^\infty e^{-st} g(t - \tau)\, dt,$$

because we may *define* $f(t)$ and $g(t)$ to be zero for $t < 0$. Then

$$F(s)G(s) = G(s)\int_0^\infty e^{-s\tau} f(\tau)\, d\tau = \int_0^\infty e^{-s\tau} f(\tau)G(s)\, d\tau$$

$$= \int_0^\infty e^{-s\tau} f(\tau)\left(e^{s\tau}\int_0^\infty e^{-st} g(t - \tau)\, dt\right) d\tau$$

$$= \int_0^\infty \left(\int_0^\infty e^{-st} f(\tau)g(t - \tau)\, dt\right) d\tau.$$

Now our hypotheses on f and g imply that the order of integration may be reversed. (The proof of this requires a discussion of uniform convergence of improper integrals, and can be found in Chapter 2 of Churchill's *Operational Mathematics*, 3rd ed. (New York: McGraw-Hill, 1972).) Hence

$$F(s)G(s) = \int_0^\infty \left(\int_0^\infty e^{-st} f(\tau) g(t - \tau) \, d\tau \right) dt$$

$$= \int_0^\infty e^{-st} \left(\int_0^t f(\tau) g(t - \tau) \, d\tau \right) dt$$

$$= \int_0^\infty e^{-st} \left[f(t) * g(t) \right] \, dt,$$

and therefore,

$$F(s)G(s) = \mathcal{L}\{f(t) * g(t)\}.$$

We replace the upper limit of the inner integral with t because $g(t-\tau) = 0$ whenever $\tau > t$. This completes the proof of Theorem 1. ▲

Proof of Theorem 2: Because

$$F(s) = \int_0^\infty e^{-st} f(t) \, dt,$$

differentiation under the integral sign yields

$$F'(s) = \frac{d}{ds} \int_0^\infty e^{-st} f(t) \, dt$$

$$= \int_0^\infty \frac{d}{ds} \left[e^{-st} f(t) \right] \, dt = \int_0^\infty e^{-st} \left[-t f(t) \right] \, dt;$$

thus

$$F'(s) = \mathcal{L}\{-t f(t)\},$$

which is Eq. (6). We obtain Eq. (7) by applying \mathcal{L}^{-1} and then dividing by $-t$. The validity of differentiation under the integral sign depends on uniform convergence of the resulting integral; this is discussed in Chapter 2 of the book by Churchill just mentioned. ▲

Proof of Theorem 3: By definition,

$$F(\sigma) = \int_0^\infty e^{-\sigma t} f(t) \, dt.$$

So integration of $F(\sigma)$ from s to $+\infty$ gives

$$\int_s^\infty F(\sigma) \, d\sigma = \int_s^\infty \left(\int_0^\infty e^{-\sigma t} f(t) \, dt \right) d\sigma.$$

Under the hypotheses of the theorem, the order of integration may be reversed (see Churchill's book again); it follows that

$$\int_s^\infty F(\sigma)\, d\sigma = \int_0^\infty \left(\int_s^\infty e^{-\sigma t} f(t)\, d\sigma \right) dt$$

$$= \int_0^\infty \left[\frac{e^{-\sigma t}}{-t} \right]_{\sigma=s}^\infty f(t)\, dt$$

$$= \int_0^\infty e^{-st} \frac{f(t)}{t}\, dt = \mathcal{L}\left\{ \frac{f(t)}{t} \right\}.$$

This verifies Eq. (12), and Eq. (13) follows upon first applying \mathcal{L}^{-1} and then multiplying by t. ▲

7.4 Problems

*Find the convolution $f(t) * g(t)$ in Problems 1 through 6.*

1. $f(t) = t,\ g(t) \equiv 1$
2. $f(t) = t,\ g(t) = e^{at}$
3. $f(t) = g(t) = \sin t$
4. $f(t) = t^2,\ g(t) = \cos t$
5. $f(t) = g(t) = e^{at}$
6. $f(t) = e^{at},\ g(t) = e^{bt}\quad (a \neq b)$

Apply the convolution theorem to find the inverse Laplace transforms of the functions in Problems 7 through 14.

7. $F(s) = \dfrac{1}{s(s-3)}$
8. $F(s) = \dfrac{1}{s(s^2+4)}$

9. $F(s) = \dfrac{1}{(s^2+9)^2}$
10. $F(s) = \dfrac{1}{s^2(s^2+k^2)}$

11. $F(s) = \dfrac{s^2}{(s^2+4)^2}$
12. $F(s) = \dfrac{1}{s(s^2+4s+5)}$

13. $F(s) = \dfrac{s}{(s-3)(s^2+1)}$
14. $F(s) = \dfrac{s}{s^4+5s^2+4}$

In Problems 15 through 22, apply either Theorem 2 or Theorem 3 to find the Laplace transform of $f(t)$.

15. $f(t) = t\sin 3t$
16. $f(t) = t^2\cos 2t$
17. $f(t) = te^{2t}\cos 3t$
18. $f(t) = te^{-t}\sin^2 t$
19. $f(t) = \dfrac{\sin t}{t}$
20. $f(t) = \dfrac{1-\cos 2t}{t}$
21. $f(t) = \dfrac{e^{3t}-1}{t}$
22. $f(t) = \dfrac{e^t - e^{-t}}{t}$

Find the inverse transforms of the functions in Problems 23 through 28.

23. $F(s) = \ln\dfrac{s-2}{s+2}$
24. $F(s) = \ln\dfrac{s^2+1}{s^2+4}$

25. $F(s) = \ln\dfrac{s^2+1}{(s+2)(s-3)}$
26. $F(s) = \tan^{-1}\dfrac{3}{s+2}$

27. $F(s) = \ln\left(1 + \dfrac{1}{s^2}\right)$
28. $F(s) = \dfrac{s}{(s^2+1)^3}$

In Problems 29 through 34, transform the given differential equation to find a nontrivial solution such that $x(0) = 0$.

29. $tx'' + (t-2)x' + x = 0$
30. $tx'' + (3t-1)x' + 3x = 0$
31. $tx'' - (4t+1)x' + 2(2t+1)x = 0$
32. $tx'' + 2(t-1)x' - 2x = 0$
33. $tx'' - 2x' + tx = 0$
34. $tx'' + (4t-2)x' + (13t-4)x = 0$
35. Apply the convolution theorem to show that

$$\mathcal{L}^{-1}\left\{ \frac{1}{(s-1)\sqrt{s}} \right\} = \frac{2e^t}{\sqrt{\pi}} \int_0^{\sqrt{t}} e^{-u^2}\, du = e^t \operatorname{erf}\sqrt{t}.$$

(*Suggestion:* Substitute $u = \sqrt{t}$.)

In Problems 36 through 38, apply the convolution theorem to derive the indicated solution $x(t)$ of the given differential equation with initial conditions $x(0) = x'(0) = 0$.

36. $x'' + 4x = f(t);\ x(t) = \dfrac{1}{2}\displaystyle\int_0^t f(t-\tau)\sin 2\tau\, d\tau$

37. $x'' + 2x' + x = f(t);\ x(t) = \displaystyle\int_0^t \tau e^{-\tau} f(t-\tau)\, d\tau$

38. $x'' + 4x' + 13x = f(t);$

$$x(t) = \frac{1}{3}\int_0^t f(t-\tau)e^{-2\tau}\sin 3\tau\, d\tau$$

Termwise Inverse Transformation of Series

In Chapter 2 of Churchill's Operational Mathematics, *the following theorem is proved. Suppose that $f(t)$ is continuous for $t \geq 0$, that $f(t)$ is of exponential order as $t \to +\infty$, and that*

$$F(s) = \sum_{n=0}^\infty \frac{a_n}{s^{n+k+1}}$$

where $0 \leq k < 1$ and the series converges absolutely for $s > c$. Then

$$f(t) = \sum_{n=0}^{\infty} \frac{a_n t^{n+k}}{\Gamma(n+k+1)}.$$

Apply this result in Problems 39 through 41.

39. In Example 5 it was shown that

$$\mathcal{L}\{J_0(t)\} = \frac{C}{\sqrt{s^2+1}} = \frac{C}{s}\left(1 + \frac{1}{s^2}\right)^{-1/2}.$$

Expand with the aid of the binomial series and then compute the inverse transformation term by term to obtain

$$J_0(t) = C \sum_{n=0}^{\infty} \frac{(-1)^n t^{2n}}{2^{2n}(n!)^2}.$$

Finally, note that $J_0(0) = 1$ implies that $C = 1$.

40. Expand the function $F(s) = s^{-1/2}e^{-1/s}$ in powers of s^{-1} to show that

$$\mathcal{L}^{-1}\left\{\frac{1}{\sqrt{s}}e^{-1/s}\right\} = \frac{1}{\sqrt{\pi t}}\cos 2\sqrt{t}.$$

41. Show that

$$\mathcal{L}^{-1}\left\{\frac{1}{s}e^{-1/s}\right\} = J_0\left(2\sqrt{t}\right).$$

7.5 Periodic and Piecewise Continuous Input Functions

FIGURE 7.5.1. The graph of the unit step function at $t = a$.

Mathematical models of mechanical or electrical systems often involve functions with discontinuities corresponding to external forces that are turned abruptly on or off. One such simple off-on function is the **unit step function** at $t = a$; its formula is

$$u_a(t) = u(t-a) = \begin{cases} 0 & \text{if } t < a, \\ 1 & \text{if } t \geq a, \end{cases} \tag{1}$$

and its graph appears in Fig. 7.5.1.

In Example 8 of Section 7.1 we saw that if $a \geq 0$, then

$$\mathcal{L}\{u(t-a)\} = \frac{e^{-as}}{s}. \tag{2}$$

Because $\mathcal{L}\{u(t)\} = 1/s$, Eq. (2) implies that multiplication of the transform of $u(t)$ by e^{-as} corresponds to the translation $t \to t-a$ in the original independent variable. Theorem 1 tells us that this fact, when properly interpreted, is a general property of the Laplace transformation.

THEOREM 1: Translation on the t-Axis

If $\mathcal{L}\{f(t)\}$ exists for $s > c$, then

$$\mathcal{L}\{u(t-a)f(t-a)\} = e^{-as}F(s) \tag{3a}$$

and

$$\mathcal{L}^{-1}\{e^{-as}F(s)\} = u(t-a)f(t-a) \tag{3b}$$

for $s > c + a$. ▲

Note that

$$u(t-a)f(t-a) = \begin{cases} 0 & \text{if } t < a, \\ f(t-a) & \text{if } t \geq a. \end{cases} \tag{4}$$

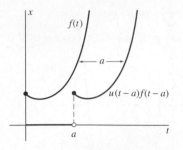

FIGURE 7.5.2. Translation of $f(t)$ a units to the right.

Thus Theorem 1 implies that $\mathcal{L}^{-1}\{e^{-as}F(s)\}$ is the function whose graph for $t \geq a$ is the translation a units to the right of the graph of $f(t)$. Note that the part of the graph of $f(t)$ to the left of $t = 0$ is cut off and not translated (Fig. 7.5.2).

Proof of Theorem 1: From the definition of $\mathcal{L}\{f(t)\}$, we get

$$e^{-as}F(s) = e^{-as}\int_0^\infty e^{-s\tau}f(\tau)\,d\tau = \int_0^\infty e^{-s(\tau+a)}f(\tau)\,d\tau.$$

The substitution $t = \tau + a$ then yields

$$e^{-as}F(s) = \int_a^\infty e^{-st}f(t-a)\,dt.$$

From Eq. (4) we see that this is the same as

$$e^{-as}F(s) = \int_0^\infty e^{-st}u(t-a)f(t-a)\,dt = \mathcal{L}\{u(t-a)f(t-a)\},$$

because $u(t-a)f(t-a) = 0$ for $t < a$. This completes the proof of Theorem 1. ▲

Example 1 With $f(t) = \frac{1}{2}t^2$, Theorem 1 gives

$$\mathcal{L}^{-1}\left\{\frac{e^{-as}}{s^3}\right\} = u(t-a)\frac{1}{2}(t-a)^2 = \begin{cases} 0 & \text{if } t < a, \\ \frac{1}{2}(t-a)^2 & \text{if } t \geq a \end{cases} \quad \text{(Fig. 7.5.3).} \blacksquare$$

Example 2 Find $\mathcal{L}\{g(t)\}$ if

$$g(t) = \begin{cases} 0 & \text{if } t < 3, \\ t^2 & \text{if } t \geq 3 \end{cases} \quad \text{(Fig. 7.5.4).}$$

Solution Before applying Theorem 1, we must first write $g(t)$ in the form $u(t-3)f(t-3)$. The function $f(t)$ whose translation 3 units to the right agrees (for $t \geq 3$) with $g(t) = t^2$ is $f(t) = (t+3)^2$ because $f(t-3) = t^2$. But then

$$F(s) = \mathcal{L}\{t^2 + 6t + 9\} = \frac{2}{s^3} + \frac{6}{s^2} + \frac{9}{s},$$

so now Theorem 1 yields

$$\mathcal{L}\{g(t)\} = \mathcal{L}\{u(t-3)f(t-3)\} = e^{-3s}F(s) = e^{-3s}\left(\frac{2}{s^3} + \frac{6}{s^2} + \frac{9}{s}\right). \blacksquare$$

Example 3 Find $\mathcal{L}\{f(t)\}$ if

$$f(t) = \begin{cases} \cos 2t & \text{if } 0 \leq t < 2\pi, \\ 0 & \text{if } t \geq 2\pi \end{cases} \quad \text{(Fig. 7.5.5).}$$

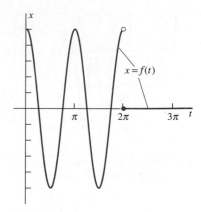

FIGURE 7.5.3. The graph of the inverse transform of Example 1.

FIGURE 7.5.4. The graph of the function $g(t)$ of Example 2.

FIGURE 7.5.5. The function $f(t)$ of Examples 3 and 4

Solution We note first that

$$f(t) = [1 - u(t - 2\pi)]\cos 2t = \cos 2t - u(t - 2\pi)\cos 2(t - 2\pi)$$

because of the periodicity of the cosine function. Hence Theorem 1 gives

$$\mathcal{L}\{f(t)\} = \mathcal{L}\{\cos 2t\} - e^{-2\pi s}\mathcal{L}\{\cos 2t\} = \frac{s(1 - e^{-2\pi s})}{s^2 + 4}.$$
∎

Example 4 A mass that weighs 32 lb (mass $m = 1$ slug) is attached to the free end of a long light spring that is stretched 1 ft by a force of 4 lb ($k = 4$ lb/ft). The mass is initially at rest in its equilibrium position. Beginning at time $t = 0$ (seconds), an external force $f(t) = \cos 2t$ is applied to the mass, but at time $t = 2\pi$ this force is turned off (abruptly discontinued) and the mass is allowed to continue its motion unimpeded. Find the resulting position function $x(t)$ of the mass.

Solution We need to solve the initial value problem

$$x'' + 4x = f(t); \quad x(0) = x'(0) = 0,$$

where $f(t)$ is the function of Example 3. The transformed equation is

$$(s^2 + 4)X(s) = F(s) = \frac{s(1 - e^{-2\pi s})}{s^2 + 4},$$

so

$$X(s) = \frac{s}{(s^2 + 4)^2} - e^{-2\pi s}\frac{s}{(s^2 + 4)^2}.$$

Because

$$\mathcal{L}^{-1}\left\{\frac{s}{(s^2 + 4)^2}\right\} = \tfrac{1}{4}t\sin 2t$$

by Eq. (16) of Section 7.3, it follows from Theorem 1 that

$$x(t) = \tfrac{1}{4}t\sin 2t - u(t - 2\pi)\cdot\tfrac{1}{4}(t - 2\pi)\sin 2(t - 2\pi)$$

$$= \tfrac{1}{4}[t - u(t - 2\pi)\cdot(t - 2\pi)]\sin 2t.$$

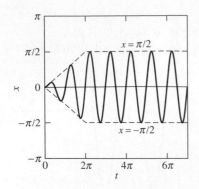

FIGURE 7.5.6. The graph of the function $x(t)$ of Example 4.

If we separate the cases $t < 2\pi$ and $t \geq 2\pi$, we find that the position function may be written in the form

$$x(t) = \begin{cases} \frac{1}{4}t \sin 2t & \text{if } t < 2\pi, \\ \frac{1}{2}\pi \sin 2t & \text{if } t \geq 2\pi. \end{cases}$$

As indicated by the graph of $x(t)$ shown in Fig. 7.5.6, the mass oscillates with circular frequency $\omega = 2$ and with linearly increasing amplitude until the force is removed at time $t = 2\pi$. Thereafter, the mass continues to oscillate with the same frequency but with constant amplitude $\pi/2$. The force $F(t) = \cos 2t$ would produce pure resonance if continued indefinitely, but we see that its effect ceases immediately at the moment it is turned off. ∎

If we were to attack Example 4 with the methods of Chapter 3, we would need to solve one problem for the interval $0 \leq t < 2\pi$ and then solve a new problem with different initial conditions for the interval $t \geq 2\pi$. In such a situation the Laplace transform method enjoys the distinct advantage of not requiring the solution of different problems on different intervals.

Example 5 Consider the RLC circuit shown in Fig. 7.5.7, with $R = 110\ \Omega$, $L = 1$ H, $C = 0.001$ F, and a battery supplying $E_0 = 90$ V. Initially there is no current in the circuit and no charge on the capacitor. At time $t = 0$ the switch is closed and left closed for 1 second. At time $t = 1$ it is opened and left open thereafter. Find the resulting current in the circuit.

Solution We recall from Section 3.7 the basic series circuit equation

$$L\frac{di}{dt} + Ri + \frac{1}{C}q = e(t); \tag{5}$$

we use lowercase letters for current, charge, and voltage and reserve uppercase letters for their transforms. With the given circuit elements, Eq. (5) is

$$\frac{di}{dt} + 110i + 1000q = e(t), \tag{6}$$

where $e(t) = 90[1 - u(t - 1)]$, corresponding to the opening and closing of the switch.

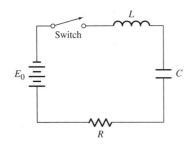

FIGURE 7.5.7. The series RLC circuit of Example 5.

In Section 3.7 our strategy was to differentiate both sides of Eq. (5), then apply the relation

$$i = \frac{dq}{dt} \tag{7}$$

to obtain the second-order equation

$$L\frac{d^2i}{dt^2} + R\frac{di}{dt} + \frac{1}{C}i = e'(t).$$

Here we do not use that method, because $e'(t) = 0$ except at $t = 1$, whereas the jump from $e(t) = 90$ when $t < 1$ to $e(t) = 0$ when $t > 1$ would seem to require that $e'(1) = -\infty$. Thus $e'(t)$ appears to have an infinite discontinuity at $t = 1$. This

phenomenon will be discussed in Section 7.6. For now, we will simply note that it is an odd situation and circumvent it rather than attempt to deal with it here.

To avoid the possible problem at $t = 1$, we observe that the initial value $q(0) = 0$ and Eq. (7) yield, upon integration,

$$q(t) = \int_0^t i(\tau)\,d\tau. \tag{8}$$

We substitute Eq. (8) in Eq. (5) to obtain

$$L\frac{di}{dt} + Ri + \frac{1}{C}\int_0^t i(\tau)\,d\tau = e(t). \tag{9}$$

This is the **integrodifferential equation** of a series RLC circuit; it involves both the integral and the derivative of the unknown function $i(t)$. The Laplace transform method works well with such an equation.

In the present example, Eq. (9) is

$$\frac{di}{dt} + 110i + 1000\int_0^t i(\tau)\,d\tau = 90\left[1 - u(t-1)\right]. \tag{10}$$

Because

$$\mathcal{L}\left\{\int_0^t i(\tau)\,d\tau\right\} = \frac{I(s)}{s}$$

by Theorem 2 of Section 7.2 on transforms of integrals, the transformed equation is

$$sI(s) + 110I(s) + 1000\frac{I(s)}{s} = \frac{90}{s}(1 - e^{-s}).$$

We solve this equation for $I(s)$ to obtain

$$I(s) = \frac{90(1 - e^{-s})}{s^2 + 110s + 1000}.$$

But

$$\frac{90}{s^2 + 110s + 1000} = \frac{1}{s + 10} - \frac{1}{s + 100},$$

so we have

$$I(s) = \frac{1}{s + 10} - \frac{1}{s + 100} - e^{-s}\left(\frac{1}{s + 10} - \frac{1}{s + 100}\right).$$

We now apply Theorem 1 with $f(t) = e^{-10t} - e^{-100t}$; thus the inverse transform is

$$i(t) = e^{-10t} - e^{-100t} - u(t-1)\left[e^{-10(t-1)} - e^{-100(t-1)}\right].$$

After we separate the cases $t < 1$ and $t \geq 1$, we find that the current in the circuit is given by

$$i(t) = \begin{cases} e^{-10t} - e^{-100t} & \text{if } t < 1, \\ e^{-10t} - e^{-10(t-1)} - e^{-100t} + e^{-100(t-1)} & \text{if } t \geq 1. \end{cases}$$

The portion $e^{-10t} - e^{-100t}$ of the solution would describe the current if the switch were left closed for all t rather than being open for $t \geq 1$. ∎

Transforms of Periodic Functions

FIGURE 7.5.8. The graph of a function with period p.

Periodic forcing functions in practical mechanical or electrical systems often are more complicated than pure sines or cosines. The nonconstant function $f(t)$ defined for $t \geq 0$ is said to be **periodic** if there is a number $p > 0$ such that

$$f(t + p) = f(t) \tag{11}$$

for all $t \geq 0$. The least positive value of p (if any) for which Eq. (11) holds is called the **period** of f. Such a function is shown in Fig. 7.5.8. Theorem 2 simplifies the computation of the Laplace transform of a periodic function.

THEOREM 2: Transforms of Periodic Functions

Let $f(t)$ be periodic with period p and piecewise continuous for $t \geq 0$. Then the transform $F(s) = \mathcal{L}\{f(t)\}$ exists for $s > 0$ and is given by

$$F(s) = \frac{1}{1 - e^{-ps}} \int_0^p e^{-st} f(t)\, dt. \tag{12}$$

Proof: The definition of the Laplace transform gives

$$F(s) = \int_0^\infty e^{-st} f(t)\, dt = \sum_{n=0}^\infty \int_{np}^{(n+1)p} e^{-st} f(t)\, dt.$$

The substitution $t = \tau + np$ in the nth integral following the summation sign yields

$$\int_{np}^{(n+1)p} e^{-st} f(t)\, dt = \int_0^p e^{-s(\tau+np)} f(\tau + np)\, d\tau = e^{-nps} \int_0^p e^{-s\tau} f(\tau)\, d\tau$$

because $f(\tau + np) = f(\tau)$ by periodicity. Thus

$$F(s) = \sum_{n=0}^\infty \left(e^{-nps} \int_0^p e^{-s\tau} f(\tau)\, d\tau \right)$$

$$= \left(1 + e^{-ps} + e^{-2ps} + \cdots \right) \int_0^p e^{-s\tau} f(\tau)\, d\tau.$$

Consequently,

$$F(s) = \frac{1}{1 - e^{-ps}} \int_0^p e^{-s\tau} f(\tau)\, d\tau.$$

We use the geometric series

$$\frac{1}{1 - x} = 1 + x + x^2 + x^3 + \cdots,$$

with $x = e^{-ps} < 1$ (for $s > 0$) to sum the series in the final step. Thus we have derived Eq. (12). ▲

The principal advantage of Theorem 2 is that it enables us to find the Laplace transform of a periodic function without the necessity of an explicit evaluation of an improper integral.

Example 6

FIGURE 7.5.9. The graph of the square wave function of Example 6.

Figure 7.5.9 shows the graph of the square wave function $f(t) = (-1)^{\llbracket t/a \rrbracket}$ of period $p = 2a$; $\llbracket x \rrbracket$ denotes the greatest integer not exceeding x. By Theorem 2 the Laplace transform of $f(t)$ is

$$F(s) = \frac{1}{1 - e^{-2as}} \int_0^{2a} e^{-st} f(t)\, dt$$

$$= \frac{1}{1 - e^{-2as}} \left(\int_0^a e^{-st}\, dt + \int_a^{2a} (-1)e^{-st}\, dt \right)$$

$$= \frac{1}{1 - e^{-2as}} \left(\left[-\frac{1}{s} e^{-st} \right]_0^a - \left[-\frac{1}{s} e^{-st} \right]_a^{2a} \right)$$

$$= \frac{(1 - e^{-as})^2}{s(1 - e^{-2as})} = \frac{1 - e^{-as}}{s(1 + e^{-as})}.$$

Therefore,

$$F(s) = \frac{1 - e^{-as}}{s(1 + e^{-as})} \tag{13a}$$

$$= \frac{e^{as/2}}{s(e^{as/2} + e^{-as/2})} = \frac{1}{s} \tanh \frac{as}{2}. \tag{13b}$$

■

Example 7

FIGURE 7.5.10. The graph of the triangular wave function of Example 7.

Figure 7.5.10 shows the graph of a triangular wave function $g(t)$ of period $p = 2a$. Because the derivative $g'(t)$ is the square wave function of Example 6, it follows from the formula in (13b) and Theorem 2 of Section 7.2 that the transform of this triangular wave function is

$$G(s) = \frac{F(s)}{s} = \frac{1}{s^2} \tanh \frac{as}{2}. \tag{14}$$

■

Example 8

Consider a mass-spring-dashpot system with $m = 1$, $c = 4$, and $k = 20$ in appropriate units. Suppose that the system is initially at rest at equilibrium ($x(0) = x'(0) = 0$) and that the mass is acted on beginning at time $t = 0$ by the external force $f(t)$ whose graph is shown in Fig. 7.5.11: the square wave with amplitude 20 and period 2π. Find the position function $f(t)$.

Solution The initial value problem is

$$x'' + 4x' + 20x = f(t); \quad x(0) = x'(0) = 0.$$

The transformed equation is

$$s^2 X(s) + 4s X(s) + 20 X(s) = F(s). \tag{15}$$

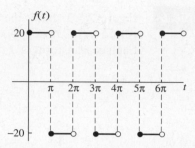

FIGURE 7.5.11. The graph of the external force function of Example 8.

From Example 6 with $a = \pi$ we see that the transform of $f(t)$ is

$$F(s) = \frac{20}{s} \cdot \frac{1 - e^{-\pi s}}{1 + e^{-\pi s}}$$

$$= \frac{20}{s} \left(1 - e^{-\pi s}\right) \left(1 - e^{-\pi s} + e^{-2\pi s} - e^{-3\pi s} + \cdots\right)$$

$$= \frac{20}{s} \left(1 - 2e^{-\pi s} + 2e^{-2\pi s} - 2e^{-3\pi s} + \cdots\right),$$

so that

$$F(s) = \frac{20}{s} + \frac{40}{s} \sum_{n=1}^{\infty} (-1)^n e^{-n\pi s}. \tag{16}$$

Substitution of Eq. (16) in Eq. (15) yields

$$X(s) = \frac{F(s)}{s^2 + 4s + 20}$$

$$= \frac{20}{s[(s+2)^2 + 16]} + 2 \sum_{n=1}^{\infty} (-1)^n \frac{20 e^{-n\pi s}}{s[(s+2)^2 + 16]}. \tag{17}$$

From the transform in Eq. (8) of Section 7.3, we get

$$\mathcal{L}^{-1} \left\{ \frac{20}{(s+2)^2 + 16} \right\} = 5 e^{-2t} \sin 4t,$$

so by Theorem 2 of Section 7.2 we have

$$g(t) = \mathcal{L}^{-1} \left\{ \frac{20}{s[(s+2)^2 + 16]} \right\} = \int_0^t 5 e^{-2\tau} \sin 4\tau \, d\tau.$$

Using a tabulated formula for $\int e^{at} \sin bt \, dt$, we get

$$g(t) = 1 - e^{-2t} \left(\cos 4t + \tfrac{1}{2} \sin 4t\right) = 1 - h(t), \tag{18}$$

where

$$h(t) = e^{-2t} \left(\cos 4t + \tfrac{1}{2} \sin 4t\right). \tag{19}$$

Now we apply Theorem 1 to find the inverse transform of the right-hand term in Eq. (17). The result is

$$x(t) = g(t) + 2\sum_{n=1}^{\infty} (-1)^n u(t - n\pi) g(t - n\pi),\qquad(20)$$

and we note that for any fixed value of t the sum in Eq. (20) is finite. Moreover,

$$g(t - n\pi) = 1 - e^{-2(t-n\pi)}\left[\cos 4(t - n\pi) + \tfrac{1}{2}\sin 4(t - n\pi)\right]$$

$$= 1 - e^{2n\pi}e^{-2t}\left(\cos 4t + \tfrac{1}{2}\sin 4t\right).$$

Therefore,

$$g(t - n\pi) = 1 - e^{2n\pi}h(t).\qquad(21)$$

Hence if $0 < t < \pi$, then

$$x(t) = 1 - h(t).$$

If $\pi < t < 2\pi$, then

$$x(t) = [1 - h(t)] - 2\left[1 - e^{2\pi}h(t)\right] = -1 + h(t) - 2h(t)\left[1 - e^{2\pi}\right].$$

If $2\pi < t < 3\pi$, then

$$x(t) = [1 - h(t)] - 2\left[1 - e^{2\pi}h(t)\right] + 2\left[1 - e^{4\pi}h(t)\right]$$

$$= 1 + h(t) - 2h(t)\left[1 - e^{2\pi} + e^{4\pi}\right].$$

The general expression for $n\pi < t < (n+1)\pi$ is

$$x(t) = h(t) + (-1)^n - 2h(t)\left[1 - e^{2\pi} + \cdots + (-1)^n e^{2n\pi}\right]$$

$$= h(t) + (-1)^n - 2h(t)\frac{1 + (-1)^n e^{2(n+1)\pi}}{1 + e^{2\pi}},\qquad(22)$$

which we obtained with the aid of the familiar formula for the sum of a finite geometric progression. A rearrangement of Eq. (22) finally gives, with the aid of Eq. (19),

$$x(t) = \frac{e^{2\pi} - 1}{e^{2\pi} + 1}e^{-2t}\left(\cos 4t + \tfrac{1}{2}\sin 4t\right) + (-1)^n$$

$$- \frac{2 \cdot (-1)^n e^{2\pi}}{e^{2\pi} + 1}e^{-2(t-n\pi)}\left(\cos 4t + \tfrac{1}{2}\sin 4t\right)\qquad(23)$$

for $n\pi < t < (n+1)\pi$. The first term in Eq. (23) is the transient solution

$$x_{\mathrm{tr}}(t) \approx (0.9963)e^{-2t}\left(\cos 4t + \tfrac{1}{2}\sin 4t\right) \approx (1.1139)e^{-2t}\cos(4t - 0.4636).\quad(24)$$

The last two terms in Eq. (23) give the steady periodic solution x_{sp}. To investigate it, we write $\tau = t - n\pi$ for t in the interval $n\pi < t < (n+1)\pi$. Then

$$x_{\mathrm{sp}}(t) = (-1)^n\left[1 - \frac{2e^{2\pi}}{e^{2\pi} + 1}e^{-2\tau}\left(\cos 4\tau + \tfrac{1}{2}\sin 4\tau\right)\right]$$

$$\approx (-1)^n\left[1 - (2.2319)e^{-2\tau}\cos(4\tau - 0.4636)\right].\qquad(25)$$

Figure 7.5.12 shows the graph of $x_{\mathrm{sp}}(t)$. Its most interesting feature is the appearance of periodically damped oscillations with a frequency *four times* that of the imposed force $f(t)$. In Chapter 9 (Fourier Series) we will see why a periodic external force sometimes excites oscillations at a higher frequency than the imposed frequency. ∎

FIGURE 7.5.12. The graph of the steady periodic solution for Example 8; note the "periodically damped" oscillations with frequency four times that of the imposed force.

7.5 Problems

Find the inverse Laplace transform $f(t)$ of each function given in Problems 1 through 10. Then sketch the graph of f.

1. $F(s) = \dfrac{e^{-3s}}{s^2}$

2. $F(s) = \dfrac{e^{-s} - e^{-3s}}{s^2}$

3. $F(s) = \dfrac{e^{-s}}{s+2}$

4. $F(s) = \dfrac{e^{-s} - e^{2-2s}}{s-1}$

5. $F(s) = \dfrac{e^{-\pi s}}{s^2+1}$

6. $F(s) = \dfrac{se^{-s}}{s^2+\pi^2}$

7. $F(s) = \dfrac{1 - e^{-2\pi s}}{s^2+1}$

8. $F(s) = \dfrac{s(1 - e^{-2s})}{s^2+\pi^2}$

9. $F(s) = \dfrac{s(1 + e^{-3s})}{s^2+\pi^2}$

10. $F(s) = \dfrac{2s(e^{-\pi s} - e^{-2\pi s})}{s^2+4}$

Find the Laplace transforms of the functions given in Problems 11 through 22.

11. $f(t) = 2$ if $0 \leq t < 3$; $f(t) = 0$ if $t \geq 3$
12. $f(t) = 1$ if $1 \leq t \leq 4$; $f(t) = 0$ if $t < 1$ or if $t > 4$
13. $f(t) = \sin t$ if $0 \leq t \leq 2\pi$; $f(t) = 0$ if $t > 2\pi$
14. $f(t) = \cos \pi t$ if $0 \leq t \leq 2$; $f(t) = 0$ if $t > 2$
15. $f(t) = \sin t$ if $0 \leq t \leq 3\pi$; $f(t) = 0$ if $t > 3\pi$
16. $f(t) = \sin 2t$ if $\pi \leq t \leq 2\pi$; $f(t) = 0$ if $t < \pi$ or if $t > 2\pi$
17. $f(t) = \sin \pi t$ if $2 \leq t \leq 3$; $f(t) = 0$ if $t < 2$ or if $t > 3$
18. $f(t) = \cos \frac{1}{2}\pi t$ if $3 \leq t \leq 5$; $f(t) = 0$ if $t < 3$ or if $t > 5$
19. $f(t) = 0$ if $t < 1$; $f(t) = t$ if $t \geq 1$
20. $f(t) = t$ if $t \leq 1$; $f(t) = 1$ if $t > 1$
21. $f(t) = t$ if $t \leq 1$; $f(t) = 2 - t$ if $1 \leq t \leq 2$; $f(t) = 0$ if $t > 2$
22. $f(t) = t^3$ if $1 \leq t \leq 2$; $f(t) = 0$ if $t < 1$ or if $t > 2$

23. Apply Theorem 2 with $p = 1$ to verify that $\mathcal{L}\{1\} = 1/s$.
24. Apply Theorem 2 to verify that $\mathcal{L}\{\cos kt\} = s/(s^2 + k^2)$.
25. Apply Theorem 2 to show that the Laplace transform of the square wave function of Fig. 7.5.13 is

$$\mathcal{L}\{f(t)\} = \frac{1}{s(1 + e^{-as})}.$$

FIGURE 7.5.13. The graph of the square wave function of Problem 25.

26. Apply Theorem 2 to show that the Laplace transform of the sawtooth function $f(t)$ of Fig. 7.5.14 is

$$F(s) = \frac{1}{as^2} - \frac{e^{-as}}{s(1 - e^{-as})}.$$

FIGURE 7.5.14. The graph of the sawtooth function of Problem 26.

27. Let $g(t)$ be the staircase function of Fig. 7.5.15. Show that $g(t) = (t/a) - f(t)$, where f is the sawtooth function of Fig. 7.5.14, and hence deduce that

$$\mathcal{L}\{g(t)\} = \frac{e^{-as}}{s(1 - e^{-as})}.$$

FIGURE 7.5.15. The graph of the staircase function of Problem 27.

28. Suppose that $f(t)$ is a periodic function of period $2a$ with $f(t) = t$ if $0 \le t < a$ and $f(t) = 0$ if $a \le t < 2a$. Find $\mathcal{L}\{f(t)\}$.

29. Suppose that $f(t)$ is the half-wave rectification of $\sin kt$, shown in Fig. 7.5.16. Show that

$$\mathcal{L}\{f(t)\} = \frac{k}{(s^2 + k^2)(1 - e^{-\pi s/k})}.$$

FIGURE 7.5.16. The half-wave rectification of $\sin kt$.

30. Let $g(t) = u(t - \pi/k)f(t - \pi/k)$, where $f(t)$ is the function of Problem 29 and $k > 0$. Note that $h(t) = f(t) + g(t)$ is the full-wave rectification of $\sin kt$ shown in Fig. 7.5.17. Hence deduce from Problem 29 that

$$\mathcal{L}\{h(t)\} = \frac{k}{s^2 + k^2} \coth \frac{\pi s}{2k}.$$

FIGURE 7.5.17. The full-wave rectification of $\sin kt$.

In Problems 31 through 35, the values of mass m, spring constant k, dashpot resistance c, and force $f(t)$ are given for a mass-spring-dashpot system with external forcing function. Solve the initial value problem

$$mx'' + cx' + kx = f(t), \quad x(0) = x'(0) = 0$$

and construct the graph of the position function $x(t)$.

31. $m = 1, k = 4, c = 0$; $f(t) = 1$ if $0 \le t < \pi$, $f(t) = 0$ if $t \ge \pi$

32. $m = 1, k = 4, c = 5$; $f(t) = 1$ if $0 \le t < 2$, $f(t) = 0$ if $t \ge 2$

33. $m = 1, k = 9, c = 0$; $f(t) = \sin t$ if $0 \le t \le 2\pi$, $f(t) = 0$ if $t > 2\pi$

34. $m = 1, k = 1, c = 0$; $f(t) = t$ if $0 \le t < 1$, $f(t) = 0$ if $t \ge 1$

35. $m = 1, k = 4, c = 4$; $f(t) = t$ if $0 \le t \le 2$, $f(t) = 0$ if $t > 2$

In Problems 36 through 40, the values of the elements of an RLC circuit are given. Solve the initial value problem

$$L\frac{di}{dt} + Ri + \frac{1}{C}\int_0^t i(\tau)\,d\tau = e(t); \quad i(0) = 0$$

with the given impressed voltage $e(t)$.

36. $L = 0, R = 100, C = 10^{-3}$; $e(t) = 100$ if $0 \le t < 1$; $e(t) = 0$ if $t \ge 1$

37. $L = 1, R = 0, C = 10^{-4}$; $e(t) = 100$ if $0 \le t < 2\pi$; $e(t) = 0$ if $t \ge 2\pi$

38. $L = 1, R = 0, C = 10^{-4}$; $e(t) = 100\sin 10t$ if $0 \le t < \pi$; $e(t) = 0$ if $t \ge \pi$

39. $L = 1, R = 150, C = 2 \times 10^{-4}$; $e(t) = 100t$ if $0 \le t < 1$; $e(t) = 0$ if $t \ge 1$

40. $L = 1, R = 100, C = 4 \times 10^{-4}$; $e(t) = 50t$ if $0 \le t < 1$; $e(t) = 0$ if $t \ge 1$

In Problems 41 and 42, a mass-spring-dashpot system with external force $f(t)$ is described. Under the assumption that $x(0) = x'(0) = 0$, use the method of Example 8 to find the transient and steady periodic motions of the mass. Then construct the graph of the position function $x(t)$. If you would like to check your graph using a numerical DE solver, it may be useful to note that the function

$$f(t) = A[2u((t - \pi)(t - 2\pi)(t - 3\pi) \cdot \\ (t - 4\pi)(t - 5\pi)(t - 6\pi)) - 1]$$

has the value $+A$ if $0 < t < \pi$, the value $-A$ if $\pi < t < 2\pi$, and so forth, and hence agrees on the interval $[0, 6\pi]$ with the square wave function that has amplitude A and period 2π. (See also the definition of a square wave function in terms of sawtooth and triangular wave functions in the application material for this section.)

41. $m = 1, k = 4, c = 0$; $f(t)$ is a square wave function with amplitude 4 and period 2π.

42. $m = 1, k = 10, c = 2$; $f(t)$ is a square wave function with amplitude 10 and period 2π.

7.5 Application: Engineering Functions

Periodic piecewise linear functions occur so frequently as input functions in engineering applications that they are sometimes called **engineering functions.** Computations with such functions are readily handled by computer algebra systems. In *Mathematica,* for instance, we can define

```
SawTooth[t_] := 2 t - 2 Floor[t] - 1
TriangularWave[t_] := Abs[2 SawTooth[(2 t - 1)/4]] - 1
SquareWave[t_] := Sign[ TriangularWave[t] ]
```

Plot each of the functions to verify that it has period 2 and that its name is aptly chosen. For instance, the result of

```
Plot[ SquareWave[t], {t, 0, 6}];
```

should look like Fig. 7.5.9. If $f(t)$ is one of these engineering functions and $p > 0$, then the function $f(2t/p)$ will have period p. To illustrate this, try

```
Plot[ TriangularWave[ 2 t/p ], {t, 0, 3 p}];
```

with various values of p.

Now let's consider the mass-spring-dashpot equation

```
diffEq = m x''[t] + c x'[t] + k x[t] == input
```

with selected parameter values and an input forcing function with period p and amplitude F_0.

```
m = 4; c = 8; k = 5; p = 1; F0 = 4;
input = F0 SquareWave[2 t/p];
```

You can plot this **input** function to verify that it has period 1:

```
Plot[ input, {t, 0, 10}];
```

FIGURE 7.5.18. Response $x(t)$ to period 1 square wave input.

Finally, let's suppose that the mass is initially at rest in its equilibrium position and solve numerically the resulting initial value problem.

```
response = NDSolve[ {diffEq, x[0] == 0, x'[0] == 0},
              x, {t, 0, 10}, MaxSteps -> 1000 ]
Plot[ x[t] /. response, {t, 0, 10}];
```

In the resulting Fig. 7.5.18 we see that after an initial transient dies out, the response function $x(t)$ settles down (as expected?) to a periodic oscillation with the same period as the input.

Investigate this initial value problem with several mass-spring-dashpot parameters—for instance, selected digits of your student ID number—and with input engineering functions having various amplitudes and periods.

7.6 Impulses and Delta Functions

Consider a force $f(t)$ that acts only during a very short time interval $a \leq t \leq b$, with $f(t) = 0$ outside this interval. A typical example would be the *impulsive force* of a bat striking a ball—the impact is almost instantaneous. A quick surge of voltage (resulting from a lightning bolt, for instance) is an analogous electrical phenomenon. In such a situation it often happens that the principal effect of the

force depends only on the value of the integral

$$p = \int_a^b f(t)\, dt \tag{1}$$

and does not depend otherwise on precisely how $f(t)$ varies with time t. The number p in Eq. (1) is called the **impulse** of the force $f(t)$ over the interval $[a, b]$.

In the case of a force $f(t)$ that acts on a particle of mass m in linear motion, integration of Newton's law

$$f(t) = mv'(t) = \frac{d}{dt}[mv(t)]$$

yields

$$p = \int_a^b \frac{d}{dt}[mv(t)]\, dt = mv(b) - mv(a). \tag{2}$$

Thus the impulse of the force is equal to the change in momentum of the particle. So if change in momentum is the only effect with which we are concerned, we need know only the impulse of the force; we need know neither the precise function $f(t)$ nor even the precise time interval during which it acts. This is fortunate, because in a situation such as that of a batted ball, we are unlikely to have such detailed information about the impulsive force that acts on the ball.

Our strategy for handling such a situation is to set up a reasonable mathematical model in which the unknown force $f(t)$ is replaced with a simple and explicit force that has the same impulse. Suppose for simplicity that $f(t)$ has impulse 1 and acts during some brief time interval beginning at time $t = a \geq 0$. Then we can select a fixed number $\epsilon > 0$ that approximates the length of this time interval and replace $f(t)$ with the specific function

$$d_{a,\epsilon}(t) = \begin{cases} \dfrac{1}{\epsilon} & \text{if } a \leq t < a + \epsilon, \\[2mm] 0 & \text{otherwise.} \end{cases} \tag{3}$$

FIGURE 7.6.1. The graph of the impulse function $d_{a,\epsilon}(t)$.

This is a function of t, with a and ϵ being parameters that specify the time interval $[a, a + \epsilon]$. If $b \geq a + \epsilon$, then we see (Fig. 7.6.1) that the impulse of $d_{a,\epsilon}$ over $[a, b]$ is

$$p = \int_a^b d_{a,\epsilon}(t)\, dt = \int_a^{a+\epsilon} \frac{1}{\epsilon}\, dt = 1.$$

Thus $d_{a,\epsilon}$ has a *unit* impulse, whatever the number ϵ may be. Essentially the same computation gives

$$\int_0^\infty d_{a,\epsilon}(t)\, dt = 1. \tag{4}$$

Because the precise time interval during which the force acts seems unimportant, it is tempting to think of an *instantaneous impulse* that occurs precisely at the instant $t = a$. We might try to model such an instantaneous unit impulse by taking the limit as $\epsilon \to 0$, thereby defining

$$\delta_a(t) = \lim_{\epsilon \to 0} d_{a,\epsilon}(t), \tag{5}$$

where $a \geq 0$. If we could also take the limit under the integral sign in Eq. (4), then it would follow that

$$\int_0^\infty \delta_a(t)\, dt = 1. \tag{6}$$

But the limit in Eq. (5) gives

$$\delta_a(t) = \begin{cases} +\infty & \text{if } t = a, \\ 0 & \text{if } t \neq a. \end{cases} \tag{7}$$

Obviously, no function can satisfy both (6) and (7)—if a function is zero except at a single point, then its integral is not 1 but zero. Nevertheless, the symbol $\delta_a(t)$ is very useful. However interpreted, it is called the **Dirac delta function** at a after the British theoretical physicist P. A. M. Dirac (1902–1984), who in the early 1930s introduced a "function" allegedly enjoying the properties in Eqs. (6) and (7).

Delta Functions as Operators

The following computation motivates the meaning that we will attach here to the symbol $\delta_a(t)$. If $g(t)$ is continuous function, then the mean value theorem for integrals implies that

$$\int_a^{a+\epsilon} g(t)\, dt = \epsilon g\left(\bar{t}\right)$$

for some point \bar{t} in $[a, a + \epsilon]$. It follows that

$$\lim_{\epsilon \to 0} \int_0^\infty g(t) d_{a,\epsilon}(t)\, dt = \lim_{\epsilon \to 0} \int_a^{a+\epsilon} g(t) \cdot \frac{1}{\epsilon}\, dt = \lim_{\epsilon \to 0} g\left(\bar{t}\right) = g(a) \tag{8}$$

by continuity of g at $t = a$. If $\delta_a(t)$ *were* a function in the strict sense of the definition, and if we could interchange the limit and the integral in Eq. (8), we therefore could conclude that

$$\int_0^\infty g(t)\delta_a(t)\, dt = g(a). \tag{9}$$

Function
$g(t)$

$$\int_0^\infty \cdots \delta_a(t)\,dt$$

Number
$g(a)$

FIGURE 7.6.2. A diagram illustrating how the delta function "sifts out" the value $g(a)$.

We take Eq. (9) as the *definition* (!) of the symbol $\delta_a(t)$. Although we call it the delta function, it is not a genuine function; instead, it specifies the *operation*

$$\int_0^\infty \cdots \delta_a(t)\, dt$$

which—when applied to a continuous function $g(t)$—sifts out or selects the value $g(a)$ of this function at the point $a \geq 0$. This idea is shown schematically in Fig. 7.6.2. Note that we will use the symbol $\delta_a(t)$ only in the context of integrals such as that in Eq. (9), or when it will appear subsequently in such an integral.

For instance, if we take $g(t) = e^{-st}$ in Eq. (9), the result is

$$\int_0^\infty e^{-st} \delta_a(t)\, dt = e^{-as}. \tag{10}$$

We therefore *define* the Laplace transform of the delta function to be

$$\mathcal{L}\{\delta_a(t)\} = e^{-as} \quad (a \geq 0). \tag{11}$$

If we write

$$\delta(t) = \delta_0(t) \quad \text{and} \quad \delta(t-a) = \delta_a(t), \tag{12}$$

then (11) with $a = 0$ gives

$$\mathcal{L}\{\delta(t)\} = 1. \tag{13}$$

Note that if $\delta(t)$ were an actual function, then Eq. (13) would contradict the corollary to Theorem 2 of Section 7.1. There is no problem here; $\delta(t)$ is not a function, and Eq. (13) is our *definition* of $\mathcal{L}\{\delta(t)\}$.

Delta Function Inputs

Now, finally, suppose that we are given a mechanical system whose response $x(t)$ to the external force $f(t)$ is determined by the differential equation

$$Ax'' + Bx' + Cx = f(t). \tag{14}$$

To investigate the response of this system to a unit impulse at the instant $t = a$, it seems reasonable to replace $f(t)$ with $\delta_a(t)$ and begin with the equation

$$Ax'' + Bx' + Cx = \delta_a(t). \tag{15}$$

But what is meant by the solution of such an equation? We will call $x(t)$ a solution of Eq. (15) provided that

$$x(t) = \lim_{\epsilon \to 0} x_\epsilon(t), \tag{16}$$

where $x_\epsilon(t)$ is a solution of

$$Ax'' + Bx' + Cx = d_{a,\epsilon}(t). \tag{17}$$

Because

$$d_{a,\epsilon}(t) = \frac{1}{\epsilon}\left[u_a(t) - u_{a+\epsilon}(t)\right] \tag{18}$$

is an ordinary function, Eq. (17) makes sense. For simplicity suppose the initial conditions to be $x(0) = x'(0) = 0$. When we transform Eq. (17), writing $X_\epsilon = \mathcal{L}\{x_\epsilon\}$, we get the equation

$$(As^2 + Bs + C)X_\epsilon(s) = \frac{1}{\epsilon}\left(\frac{e^{-as}}{s} - \frac{e^{-(a+\epsilon)s}}{s}\right) = (e^{-as})\frac{1 - e^{-s\epsilon}}{s\epsilon}.$$

If we take the limit in the last equation as $\epsilon \to 0$, and note that

$$\lim_{\epsilon \to 0} \frac{1 - e^{-s\epsilon}}{s\epsilon} = 1$$

by l'Hôpital's rule, we get the equation

$$(As^2 + Bs + C)X(s) = e^{-as}, \tag{19}$$

if

$$X(s) = \lim_{\epsilon \to 0} X_\epsilon(x).$$

Note that this is precisely the same result that we would obtain if we transformed Eq. (15) directly, using the fact that $\mathcal{L}\{\delta_a(t)\} = e^{-as}$.

On this basis it is reasonable to solve a differential equation involving a delta function by employing the Laplace transform method exactly as if $\delta_a(t)$ were an ordinary function. It is important to verify that the solution so obtained agrees with the one defined in Eq. (16), but this depends on a highly technical analysis of the limiting procedures involved; we consider it beyond the scope of the present discussion. The formal method is valid in all the examples of this section and will produce correct results in the subsequent problem set.

Example 1 A mass $m = 1$ is attached to a spring with constant $k = 4$; there is no dashpot. The mass is released from rest with $x(0) = 3$. At the instant $t = 2\pi$ the mass is struck with a hammer, providing an impulse $p = 8$. Determine the motion of the mass.

Solution We need to solve the initial value problem

$$x'' + 4x = 8\delta_{2\pi}(t); \quad x(0) = 3, \quad x'(0) = 0.$$

We apply the Laplace transform to get

$$s^2 X(s) - 3s + 4X(s) = 8e^{-2\pi s},$$

so

$$X(s) = \frac{3s}{s^2 + 4} + \frac{8e^{-2\pi s}}{s^2 + 4}.$$

Recalling the transforms of sine and cosine, as well as the theorem on translations on the t-axis (Theorem 1 of Section 7.5), we see that the inverse transform is

$$x(t) = 3\cos 2t + 4u(t - 2\pi)\sin 2(t - 2\pi)$$
$$= 3\cos 2t + 4u_{2\pi}(t)\sin 2t.$$

Because $3\cos 2t + 4\sin 2t = 5\cos(2t - \alpha)$ with $\alpha = \tan^{-1}(4/3) \approx 0.9273$, separation of the cases $t < 2\pi$ and $t \geq 2\pi$ gives

$$x(t) \approx \begin{cases} 3\cos 2t & \text{if } t \leq 2\pi, \\ 5\cos(2t - 0.9273) & \text{if } t \geq 2\pi. \end{cases}$$

The resulting motion is shown in Fig. 7.6.3. Note that the impulse at $t = 2\pi$ results in a visible discontinuity in the velocity at $t = 2\pi$, as it instantaneously increases the amplitude of the oscillations of the mass from 3 to 5. ∎

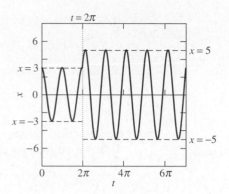

FIGURE 7.6.3. The motion of the mass of Example 1.

Delta Functions and Step Functions

It is useful to regard the delta function $\delta_a(t)$ as the derivative of the unit step function $u_a(t)$. To see why this is reasonable, consider the continuous approximation $u_{a,\epsilon}(t)$ to $u_a(t)$ shown in Fig. 7.6.4. We readily verify that

FIGURE 7.6.4. Approximation of $u_a(t)$ by $u_{a,\epsilon}(t)$.

$$\frac{d}{dt}u_{a,\epsilon}(t) = d_{a,\epsilon}(t).$$

Because

$$u_a(t) = \lim_{\epsilon \to 0} u_{a,\epsilon}(t) \quad \text{and} \quad \delta_a(t) = \lim_{\epsilon \to 0} d_{a,\epsilon}(t),$$

an interchange of limits and derivatives yields

$$\frac{d}{dt}u_a(t) = \lim_{\epsilon \to 0}\frac{d}{dt}u_{a,\epsilon}(t) = \lim_{\epsilon \to 0} d_{a,\epsilon}(t),$$

and therefore

$$\frac{d}{dt}u_a(t) = \delta_a(t) = \delta(t - a). \tag{20}$$

We may regard this as the *formal definition* of the derivative of the unit step function, although $u_a(t)$ is not differentiable in the ordinary sense at $t = a$.

Example 2

We return to the RLC circuit of Example 5 of Section 7.5, with $R = 110\ \Omega$, $L = 1$ H, $C = 0.001$ F, and a battery supplying $e_0 = 90$ V. Suppose that the circuit is initially passive—no current and no charge. At time $t = 0$ the switch is closed and at time $t = 1$ it is opened and left open. Find the resulting current $i(t)$ in the circuit.

Solution

In Section 7.5 we circumvented the discontinuity in the voltage by employing the integrodifferential form of the circuit equation. Now that delta functions are available, we may begin with the ordinary circuit equation

$$Li'' + Ri' + \frac{1}{C}i = e'(t).$$

In this example we have

$$e(t) = 90 - 90u(t - 1) = 90 - 90u_1(t),$$

so $e'(t) = -90\delta(t-1)$ by Eq. (20). Hence we want to solve the initial value problem

$$i'' + 110i' + 1000i = -90\delta(t-1); \quad i(0) = 0, \quad i'(0) = 90. \tag{21}$$

The fact that $i'(0) = 90$ comes from substitution of $t = 0$ in the equation

$$Li'(t) + Ri(t) + \frac{1}{C}q(t) = e(t)$$

with the numerical values $i(0) = q(0) = 0$ and $e(0) = 90$.

When we transform the problem in (21), we get the equation

$$s^2 I(s) - 90 + 110s I(s) + 1000I(s) = -90e^{-s}.$$

Hence

$$I(s) = \frac{90(1 - e^{-s})}{s^2 + 110s + 1000}.$$

This is precisely the same transform $I(s)$ we found in Example 5 of Section 7.5, so inversion of $I(s)$ yields the same solution $i(t)$ recorded there. ∎

Example 3 Consider a mass on a spring with $m = k = 1$ and $x(0) = x'(0) = 0$. At each of the instants $t = 0, \pi, 2\pi, 3\pi, \ldots, n\pi, \ldots$, the mass is struck a hammer blow with a unit impulse. Determine the resulting motion.

Solution We need to solve the initial value problem

$$x'' + x = \sum_{n=0}^{\infty} \delta_{n\pi}(t); \quad x(0) = 0 = x'(0).$$

Because $\mathcal{L}\{\delta_{n\pi}(t)\} = e^{-n\pi s}$, the transformed equation is

$$s^2 X(s) + X(s) = \sum_{n=0}^{\infty} e^{-n\pi s},$$

so

$$X(s) = \sum_{n=0}^{\infty} \frac{e^{-n\pi s}}{s^2 + 1}.$$

We compute the inverse Laplace transform term by term; the result is

$$x(t) = \sum_{n=0}^{\infty} u(t - n\pi) \sin(t - n\pi).$$

Because $\sin(t - n\pi) = (-1)^n \sin t$ and $u(t - n\pi) = 0$ for $t < n\pi$, we see that if $n\pi < t < (n+1)\pi$, then

$$x(t) = \sin t - \sin t + \sin t - \cdots + (-1)^n \sin t;$$

that is,

$$x(t) = \begin{cases} \sin t & \text{if } n \text{ is even,} \\ 0 & \text{if } n \text{ is odd.} \end{cases}$$

$x(t)$

$\pi \quad 2\pi \quad 3\pi \quad 4\pi \qquad t$

FIGURE 7.6.5. The half-wave rectification of $\sin t$.

Hence $x(t)$ is the half-wave rectification of $\sin t$ shown in Fig. 7.6.5. The physical explanation is that the first hammer blow (at time $t = 0$) starts the mass moving to the right; just as it returns to the origin, the second hammer blow stops it dead; it remains motionless until the third hammer blow starts it moving again, and so on. Of course, if the hammer blows are not perfectly synchronized then the motion of the mass will be quite different. ∎

Systems Analysis and Duhamel's Principle

Consider a physical system in which the *output* or *response* $x(t)$ to the *input* function $f(t)$ is described by the differential equation

$$ax'' + bx' + cx = f(t), \tag{22}$$

where the constant coefficients a, b, and c are determined by the physical parameters of the system and are independent of $f(t)$. The mass-spring-dashpot system and the series RLC circuit are familiar examples of this general situation.

For simplicity we assume that the system is initially passive: $x(0) = x'(0) = 0$. Then the transform of Eq. (22) is

$$as^2 X(s) + bs X(s) + c X(s) = F(s),$$

so

$$X(s) = \frac{F(s)}{as^2 + bs + c} = W(s)F(s). \tag{23}$$

The function

$$W(s) = \frac{1}{as^2 + bs + c} \tag{24}$$

is called the **transfer function** of the system. Thus the transform of the response to the input $f(t)$ is the product of $W(s)$ and the transform $F(s)$.

The function

$$w(t) = \mathcal{L}^{-1}\{W(s)\} \tag{25}$$

is called the **weight function** of the system. From Eq. (24) we see by convolution that

$$x(t) = \int_0^t w(\tau)f(t - \tau)\,d\tau. \tag{26}$$

This formula is **Duhamel's principle** for the system. What is important is that the weight function $w(t)$ is determined completely by the parameters of the system. Once $w(t)$ has been determined, the integral in (26) gives the response of the system to an arbitrary input function $f(t)$.

Example 4 Consider a mass-spring-dashpot system (initially passive) that responds to the external force $f(t)$ in accord with the equation $x'' + 6x' + 10x = f(t)$. Then

$$W(s) = \frac{1}{s^2 + 6s + 10} = \frac{1}{(s + 3)^2 + 1},$$

so the weight function is $w(t) = e^{-3t}\sin t$. Then Duhamel's principle implies that the response $x(t)$ to the force $f(t)$ is

$$x(t) = \int_0^t e^{-3\tau}(\sin \tau)f(t - \tau)\,d\tau. \qquad \blacksquare$$

Note that

$$W(s) = \frac{1}{as^2 + bs + c} = \frac{\mathcal{L}\{\delta(t)\}}{as^2 + bs + c}.$$

Consequently, it follows from Eq. (23) that the weight function is simply the response of the system to the delta function input $\delta(t)$. For this reason $w(t)$ is sometimes called the **unit impulse response**. A response that is usually easier to measure in practice is the response $h(t)$ to the unit step function $u(t)$; $h(t)$ is the **unit step response.** Because $\mathcal{L}\{u(t)\} = 1/s$, we see from Eq. (23) that the transform of $h(t)$ is

$$H(s) = \frac{W(s)}{s}.$$

It follows from the formula for transforms of integrals that

$$h(t) = \int_0^t w(\tau)\,d\tau, \quad \text{so that} \quad w(t) = h'(t). \tag{27}$$

Thus the weight function, or unit impulse response, is the derivative of the unit step response. Substitution of (27) in Duhamel's principle gives

$$x(t) = \int_0^t h'(t)f(t-\tau)\,d\tau \tag{28}$$

for the response of the system to the input $f(t)$.

APPLICATIONS: To describe a typical application of Eq. (28), suppose that we are given a complex series circuit containing many inductors, resistors, and capacitors. Assume that its circuit equation is a linear equation of the form in (22), but with i in place of x. What if the coefficients a, b, and c are unknown, perhaps only because they are too difficult to compute? We would still want to know the current $i(t)$ corresponding to any input $f(t) = e'(t)$. We connect the circuit to a linearly increasing voltage $e(t) = t$, so that $f(t) = e'(t) = 1 = u(t)$, and measure the response $h(t)$ with an ammeter. We then compute the derivative $h'(t)$, either numerically or graphically. Then according to Eq. (28), the output current $i(t)$ corresponding to the input voltage $e(t)$ will be given by

$$i(t) = \int_0^t h'(\tau)e'(t-\tau)\,d\tau$$

(using the fact that $f(t) = e'(t)$).

HISTORICAL REMARK: In conclusion, we remark that around 1950, after engineers and physicists had been using delta functions widely and fruitfully for about 20 years without rigorous justification, the French mathematician Laurent Schwartz developed a rigorous mathematical theory of *generalized functions* that supplied the missing logical foundation for delta function techniques. Every piecewise continuous ordinary function is a generalized function, but the delta function is an example of a generalized function that is not an ordinary function.

7.6 Problems

Solve the initial value problems in Problems 1 through 8, and graph each solution function $x(t)$.

1. $x'' + 4x = \delta(t);\ x(0) = x'(0) = 0$
2. $x'' + 4x = \delta(t) + \delta(t - \pi);\ x(0) = x'(0) = 0$
3. $x'' + 4x' + 4x = 1 + \delta(t - 2);\ x(0) = x'(0) = 0$
4. $x'' + 2x' + x = t + \delta(t);\ x(0) = 0,\ x'(0) = 1$
5. $x'' + 2x' + 2x = 2\delta(t - \pi);\ x(0) = x'(0) = 0$
6. $x'' + 9x = \delta(t - 3\pi) + \cos 3t;\ x(0) = x'(0) = 0$
7. $x'' + 4x' + 5x = \delta(t - \pi) + \delta(t - 2\pi);\ x(0) = 0,\ x'(0) = 2$
8. $x'' + 2x' + x = \delta(t) - \delta(t - 2);\ x(0) = x'(0) = 2$

Apply Duhamel's principle to write an integral formula for the solution of each initial value problem in Problems 9 through 12.

9. $x'' + 4x = f(t);\ x(0) = x'(0) = 0$
10. $x'' + 6x' + 9x = f(t);\ x(0) = x'(0) = 0$
11. $x'' + 6x' + 8x = f(t);\ x(0) = x'(0) = 0$
12. $x'' + 4x' + 8x = f(t);\ x(0) = x'(0) = 0$
13. This problem deals with a mass m, initially at rest at the origin, that receives an impulse p at time $t = 0$. **(a)** Find the solution $x_\epsilon(t)$ of the problem

 $$mx'' = pd_{0,\epsilon}(t);\quad x(0) = x'(0) = 0.$$

 (b) Show that $\lim_{\epsilon \to 0} x_\epsilon(t)$ agrees with the solution of the problem

 $$mx'' = p\delta(t);\quad x(0) = x'(0) = 0.$$

 (c) Show that $mv = p$ for $t > 0\ (v = dx/dt)$.
14. Verify that $u'(t - a) = \delta(t - a)$ by solving the problem

 $$x' = \delta(t - a);\quad x(0) = 0$$

 to obtain $x(t) = u(t - a)$.
15. This problem deals with a mass m on a spring (with constant k) that receives an impulse $p_0 = mv_0$ at time $t = 0$. Show that the initial value problems

 $$mx'' + kx = 0;\quad x(0) = 0,\quad x'(0) = v_0$$

 and

 $$mx'' + kx = p_0\delta(t);\quad x(0) = 0,\quad x'(0) = 0$$

 have the same solution. Thus the effect of $p_0\delta(t)$ is, indeed, to impart to the particle an initial momentum p_0.
16. This is a generalization of Problem 15. Show that the problems

 $$ax'' + bx' + cx = f(t);\quad x(0) = 0,\quad x'(0) = v_0$$

 and

 $$ax'' + bx' + cx = f(t) + av_0\delta(t);\quad x(0) = x'(0) = 0$$

 have the same solution for $t > 0$. Thus the effect of the term $av_0\delta(t)$ is to supply the initial condition $x'(0) = v_0$.

17. Consider an initially passive RC circuit (no inductance) with a battery supplying e_0 volts. **(a)** If the switch to the battery is closed at time $t = a$ and opened at time $t = b > a$ (and left open thereafter), show that the current in the circuit satisfies the initial value problem

 $$Ri' + \frac{1}{C}i = e_0\delta(t - a) - e_0\delta(t - b);\quad i(0) = 0.$$

 (b) Solve this problem if $R = 100\ \Omega$, $C = 10^{-4}$ F, $e_0 = 100$ V, $a = 1$ (s), and $b = 2$ (s). Show that $i(t) > 0$ if $1 < t < 2$ and that $i(t) < 0$ if $t > 2$.

18. Consider an initially passive LC circuit (no resistance) with a battery supplying e_0 volts. **(a)** If the switch is closed at time $t = 0$ and opened at time $t = a > 0$, show that the current in the circuit satisfies the initial value problem

 $$Li'' + \frac{1}{C}i = e_0\delta(t) - e_0\delta(t - a);$$

 $$i(0) = i'(0) = 0.$$

 (b) If $L = 1$ H, $C = 10^{-2}$ F, $e_0 = 10$ V, and $a = \pi$ (s), show that

 $$i(t) = \begin{cases} \sin 10t & \text{if } t < \pi, \\ 0 & \text{if } t > \pi. \end{cases}$$

 Thus the current oscillates through five cycles and then stops abruptly when the switch is opened (Fig. 7.6.6).

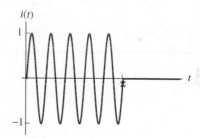

FIGURE 7.6.6. The current function of Problem 18.

19. Consider the LC circuit of Problem 18(b), except suppose that the switch is alternately closed and opened at times $t = 0, \pi/10, 2\pi/10, \ldots$. **(a)** Show that $i(t)$ satisfies the initial value problem

 $$i'' + 100i = 10\sum_{n=0}^{\infty}(-1)^n\delta\left(t - \frac{n\pi}{10}\right);\quad i(0) = i'(0) = 0.$$

 (b) Solve this initial value problem to show that

 $$i(t) = (n + 1)\sin 10t \quad \text{if} \quad \frac{n\pi}{10} < t < \frac{(n+1)\pi}{10}.$$

 Thus a resonance phenomenon occurs (see Fig. 7.6.7).

FIGURE 7.6.7. The current function of Problem 19.

20. Repeat Problem 19, except suppose that the switch is alternately closed and opened at times $t = 0, \pi/5, 2\pi/5, \ldots, n\pi/5, \ldots$. Now show that if

$$\frac{n\pi}{5} < t < \frac{(n+1)\pi}{5},$$

then

$$i(t) = \begin{cases} \sin 10t & \text{if } n \text{ is even;} \\ 0 & \text{if } n \text{ is odd.} \end{cases}$$

Thus the current in alternate cycles of length $\pi/5$ first executes a sine oscillation during one cycle, then is dormant during the next cycle, and so on (see Fig. 7.6.8).

FIGURE 7.6.8. The current function of Problem 20.

21. Consider an RLC circuit in series with a battery, with $L = 1$ H, $R = 60$ Ω, $C = 10^{-3}$ F, and $e_0 = 10$ V. **(a)** Suppose that the switch is alternately closed and opened at times $t = 0, \pi/10, 2\pi/10, \ldots$. Show that $i(t)$ satisfies the initial value problem

$$i'' + 60i' + 1000i = 10 \sum_{n=0}^{\infty} (-1)^n \delta\left(t - \frac{n\pi}{10}\right);$$

$$i(0) = i'(0) = 0.$$

(b) Solve this problem to show that if

$$\frac{n\pi}{10} < t < \frac{(n+1)\pi}{10},$$

then

$$i(t) = \frac{e^{3n\pi+3\pi} - 1}{e^{3\pi} - 1} e^{-30t} \sin 10t.$$

Construct a figure showing the graph of this current function.

22. Consider a mass $m = 1$ on a spring with constant $k = 1$, initially at rest, but struck with a hammer at each of the instants $t = 0, 2\pi, 4\pi, \ldots$. Suppose that each hammer blow imparts an impulse of $+1$. Show that the position function $x(t)$ of the mass satisfies the initial value problem

$$x'' + x = \sum_{n=0}^{\infty} \delta(t - 2n\pi); \quad x(0) = x'(0) = 0.$$

Solve this problem to show that if $2n\pi < t < 2(n+1)\pi$, then $x(t) = (n+1) \sin t$. Thus resonance occurs because the mass is struck each time it passes through the origin moving to the right—in contrast with Example 3, in which the mass was struck each time it returned to the origin. Finally, construct a figure showing the graph of this position function.

CHAPTER

Power Series Methods

8.1 Introduction and Review of Power Series

In Section 3.3 we saw that solving a homogeneous linear differential equation with constant coefficients can be reduced to the algebraic problem of finding the roots of its characteristic equation. There is no similar procedure for solving linear differential equations with *variable* coefficients, at least not routinely and in finitely many steps. With the exception of special types, such as the occasional equation that can be solved by inspection, linear equations with variable coefficients generally require the power series techniques of this chapter.

These techniques suffice for many of the nonelementary differential equations that appear most frequently in applications. Perhaps the most important (because of its applications in such areas as acoustics, heat flow, and electromagnetic radiation) is **Bessel's equation** of order n:

➤ $$x^2 y'' + x y' + (x^2 - n^2) y = 0.$$

Legendre's equation of order n is important in many applications. It has the form

➤ $$(1 - x^2) y'' - 2x y' + n(n+1) y = 0.$$

In this section we introduce the **power series method** in its simplest form and, along the way, state (without proof) several theorems that constitute a review of the basic facts about power series. Recall first that a **power series** in [powers of] $x - a$ is an infinite series of the form

➤ $$\sum_{n=0}^{\infty} c_n (x - a)^n = c_0 + c_1(x - a) + c_2(x - a)^2 + \cdots + c_n(x - a)^n + \cdots. \quad (1)$$

If $a = 0$, this is a power series in x:

➤ $$\sum_{n=0}^{\infty} c_n x^n = c_0 + c_1 x + c_2 x^2 + \cdots + c_n x^n + \cdots. \quad (2)$$

497

We will confine our review mainly to power series in x, but every general property of power series in x can be converted to a general property of power series in $x - a$ by replacement of x with $x - a$.

The power series in (2) **converges** on the interval I provided that the limit

$$\sum_{n=0}^{\infty} c_n x^n = \lim_{N \to \infty} \sum_{n=0}^{N} c_n x^n \tag{3}$$

exists for all x in I. In this case the sum

$$f(x) = \sum_{n=0}^{\infty} c_n x^n \tag{4}$$

is defined on I, and we call the series $\sum c_n x^n$ a **power series representation** of the function f on I. The following power series representations of elementary functions should be familiar to you from introductory calculus:

$$e^x = \sum_{n=0}^{\infty} \frac{x^n}{n!} = 1 + x + \frac{x^2}{2!} + \frac{x^3}{3!} + \cdots ; \tag{5}$$

$$\cos x = \sum_{n=0}^{\infty} \frac{(-1)^n x^{2n}}{(2n)!} = 1 - \frac{x^2}{2!} + \frac{x^4}{4!} - \cdots ; \tag{6}$$

$$\sin x = \sum_{n=0}^{\infty} \frac{(-1)^n x^{2n+1}}{(2n+1)!} = x - \frac{x^3}{3!} + \frac{x^5}{5!} - \cdots ; \tag{7}$$

$$\cosh x = \sum_{n=0}^{\infty} \frac{x^{2n}}{(2n)!} = 1 + \frac{x^2}{2!} + \frac{x^4}{4!} + \cdots ; \tag{8}$$

$$\sinh x = \sum_{n=0}^{\infty} \frac{x^{2n+1}}{(2n+1)!} = x + \frac{x^3}{3!} + \frac{x^5}{5!} + \cdots ; \tag{9}$$

$$\ln(1+x) = \sum_{n=1}^{\infty} \frac{(-1)^{n+1} x^n}{n} = x - \frac{x^2}{2} + \frac{x^3}{3} - \cdots ; \tag{10}$$

$$\frac{1}{1-x} = \sum_{n=0}^{\infty} x^n = 1 + x + x^2 + x^3 + \cdots ; \tag{11}$$

and

$$(1+x)^\alpha = 1 + \alpha x + \frac{\alpha(\alpha-1)x^2}{2!} + \frac{\alpha(\alpha-1)(\alpha-2)x^3}{3!} + \cdots . \tag{12}$$

In compact summation notation, we observe the usual conventions that $0! = 1$ and that $x^0 = 1$ for all x, including $x = 0$. The series in (5) through (9) converge to the indicated functions for all x. In contrast, the series in (10) and (11) converge if $|x| < 1$ but diverge if $|x| > 1$. (What if $|x| = 1$?) The series in (11) is the **geometric series**. The series in (12), with α an arbitrary real number, is the **binomial series**. If α is a nonnegative integer, then the binomial series converges for all x. Otherwise

it converges if $|x| < 1$ and diverges if $|x| > 1$; its behavior for $|x| = 1$ depends on the value of α.

Remark: Power series such as those listed in (5) through (12) are often derived as Taylor series. The **Taylor series** with **center** $x = a$ of the function f is the power series

$$\blacktriangleright \quad \sum_{n=0}^{\infty} \frac{f^{(n)}(a)}{n!}(x-a)^n = f(a) + f'(a)(x-a) + \frac{f''(a)}{2!}(x-a)^2 + \cdots \quad (13)$$

in powers of $x - a$, under the hypothesis that f is infinitely differentiable at $x = a$ (so that the coefficients in Eq. (13) are all defined). If $a = 0$, then the series in (13) is the **Maclaurin series**

$$\blacktriangleright \quad \sum_{n=0}^{\infty} \frac{f^{(n)}(0)}{n!}x^n = f(0) + f'(0)x + \frac{f''(0)}{2!}x^2 + \frac{f^{(3)}(0)}{3!}x^3 + \cdots . \quad (13')$$

For example, suppose that $f(x) = e^x$. Then $f^{(n)}(x) = e^x$, and hence $f^{(n)}(0) = 1$ for all $n \geq 0$. In this case Eq. (13') reduces to the **exponential series** in (5). \blacksquare

Power Series Operations

If the Taylor series of the function f converges to $f(x)$ for all x in some open interval containing a, then we say that the function f is **analytic** at $x = a$. For example, every polynomial is analytic everywhere and every rational function is analytic wherever its denominator is nonzero. More generally, if the two functions f and g are both analytic at $x = a$, then so are their sum $f + g$ and their product $f \cdot g$, as is their quotient f/g if $g(a) \neq 0$.

For instance, the function $h(x) = \tan x = (\sin x)/(\cos x)$ is analytic at $x = 0$ because $\cos 0 = 1 \neq 0$ and the sine and cosine functions are analytic (by virtue of their convergent power series representations in Eqs. (6) and (7)). It is rather awkward to compute the Taylor series of the tangent function using Eq. (13) because of the way in which its successive derivatives grow in complexity (try it!). Fortunately, power series may be manipulated algebraically in much the same way as polynomials. For example, if

$$f(x) = \sum_{n=0}^{\infty} a_n x^n \quad \text{and} \quad g(x) = \sum_{n=0}^{\infty} b_n x^n, \quad (14)$$

then

$$f(x) + g(x) = \sum_{n=0}^{\infty} (a_n + b_n)x^n \quad (15)$$

and

$$f(x)g(x) = \sum_{n=0}^{\infty} c_n x^n$$

$$= a_0 b_0 + (a_0 b_1 + a_1 b_0)x + (a_0 b_2 + a_1 b_1 + a_2 b_0)x^2 + \cdots , \quad (16)$$

where $c_n = a_0 b_n + a_1 b_{n-1} + \cdots + a_n b_0$. The series in (15) is the result of **termwise addition** and the series in (16) is the result of **formal multiplication**—multiplying each term of the first series by each term of the second and then collecting coefficients of like powers of x. (Thus the processes strongly resemble addition and multiplication of ordinary polynomials.) The series in (15) and (16) converge to $f(x) + g(x)$ and $f(x)g(x)$, respectively, on any open interval on which both the series in (14) converge. For example,

$$\sin x \cos x = \left(x - \frac{1}{6}x^3 + \frac{1}{120}x^5 - \cdots \right) \left(1 - \frac{1}{2}x^2 + \frac{1}{24}x^4 - \cdots \right)$$

$$= x + \left(-\frac{1}{6} - \frac{1}{2} \right) x^3 + \left(\frac{1}{24} + \frac{1}{12} + \frac{1}{120} \right) x^5 + \cdots$$

$$= x - \frac{4}{6}x^3 + \frac{16}{120}x^5 - \cdots$$

$$= \frac{1}{2}\left[(2x) - \frac{(2x)^3}{3!} + \frac{(2x)^5}{5!} - \cdots \right] = \frac{1}{2}\sin 2x$$

for all x.

Similarly, the quotient of two power series can be computed by long division, as illustrated by the computation shown in Fig. 8.1.1. This division of the Taylor series for $\cos x$ into that for $\sin x$ yields the first few terms of the series

$$\tan x = x + \frac{1}{3}x^3 + \frac{2}{15}x^5 + \frac{17}{315}x^7 + \cdots . \tag{17}$$

Division of power series is more treacherous than multiplication; the series thus obtained for f/g may fail to converge at some points where the series for f and g both converge. For example, the sine and cosine series converge for all x, but the tangent series in (17) converges only if $|x| < \pi/2$.

The Power Series Method

The **power series method** for solving a differential equation consists of substituting the power series

$$y = \sum_{n=0}^{\infty} c_n x^n \tag{18}$$

in the differential equation and then attempting to determine what the coefficients c_0, c_1, c_2, \ldots must be in order that the power series will satisfy the differential equation. This is reminiscent of the method of undetermined coefficients, but now we have infinitely many coefficients somehow to determine. This method is not always successful, but when it is we obtain an infinite series representation of a solution, in contrast to the "closed form" solutions that our previous methods have yielded.

Before we can substitute the power series in (18) in a differential equation, we must first know what to substitute for the derivatives y', y'', The following theorem (stated without proof) tells us that the derivative y' of $y = \sum c_n x^n$ is obtained

$$\begin{array}{r} x \;+\; \dfrac{x^3}{3} \;+\; \dfrac{2x^5}{15} \;+\; \dfrac{17x^7}{315} \;+\; \cdots \\[1em] 1 - \dfrac{x^2}{2} + \dfrac{x^4}{24} - \dfrac{x^6}{720} + \cdots \;\Big)\; x \;-\; \dfrac{x^3}{6} \;+\; \dfrac{x^5}{120} \;-\; \dfrac{x^7}{5040} \;+\; \cdots \end{array}$$

$$x \;-\; \frac{x^3}{2} \;+\; \frac{x^5}{24} \;-\; \frac{x^7}{720} \;+\; \cdots$$

$$\frac{x^3}{3} \;-\; \frac{x^5}{30} \;+\; \frac{x^7}{840} \;+\; \cdots$$

$$\frac{x^3}{3} \;-\; \frac{x^5}{6} \;+\; \frac{x^7}{72} \;-\; \cdots$$

$$\frac{2x^5}{15} \;-\; \frac{4x^7}{315} \;+\; \cdots$$

$$\frac{2x^5}{15} \;-\; \frac{x^7}{15} \;+\; \cdots$$

$$\frac{17x^7}{315} \;+\; \cdots$$

$$\vdots$$

FIGURE 8.1.1. Obtaining the series for $\tan x$ by division of series.

by the simple procedure of writing the sum of the derivatives of the individual terms in the series for y.

THEOREM 1: Termwise Differentiation of Power Series

If the power series representation

$$f(x) = \sum_{n=0}^{\infty} c_n x^n = c_0 + c_1 x + c_2 x^2 + c_3 x^3 + \cdots \tag{19}$$

of the function f converges on the open interval I, then f is differentiable on I, and

$$f'(x) = \sum_{n=1}^{\infty} n c_n x^{n-1} = c_1 + 2c_2 x + 3c_3 x^2 + \cdots \tag{20}$$

at each point of I. ▲

For example, differentiation of the geometric series

$$\frac{1}{1-x} = \sum_{n=0}^{\infty} x^n = 1 + x + x^2 + x^3 + \cdots \tag{11}$$

gives

$$\frac{1}{(1-x)^2} = \sum_{n=1}^{\infty} nx^{n-1} = 1 + 2x + 3x^2 + 4x^3 + \cdots .$$

The process of determining the coefficients in the series $y = \sum c_n x^n$ so that it will satisfy a given differential equation depends also on Theorem 2. This theorem—stated without proof—tells us that if two power series represent the same function, then they are the same series. In particular, the Taylor series in (13) is the only power series (in powers of $x - a$) that represents the function f.

THEOREM 2: Identity Principle

If

$$\sum_{n=0}^{\infty} a_n x^n = \sum_{n=0}^{\infty} b_n x^n$$

for every point x in some open interval I, then $a_n = b_n$ for all $n \geq 0$. ▲

In particular, if $\sum a_n x^n = 0$ for all x in some open interval, it follows from Theorem 2 that $a_n = 0$ for all $n \geq 0$.

Example 1 Solve the equation $y' + 2y = 0$.

Solution We substitute the series

➤ $$y = \sum_{n=0}^{\infty} c_n x^n \quad \text{and} \quad y' = \sum_{n=1}^{\infty} n c_n x^{n-1},$$

and obtain

$$\sum_{n=1}^{\infty} n c_n x^{n-1} + 2 \sum_{n=0}^{\infty} c_n x^n = 0. \tag{21}$$

To compare coefficients here, we need the general term in each sum to be the term containing x^n. To accomplish this, we shift the index of summation in the first sum. To see how to do this, note that

$$\sum_{n=1}^{\infty} n c_n x^{n-1} = c_1 + 2c_2 x + 3c_3 x^2 + \cdots = \sum_{n=0}^{\infty} (n+1) c_{n+1} x^n.$$

Thus we can replace n with $n + 1$ if, at the same time, we start counting one step lower; that is, at $n = 0$ rather than at $n = 1$. This is a shift of $+1$ in the index of summation. The result of making this shift in Eq. (21) is the identity

$$\sum_{n=0}^{\infty} (n+1) c_{n+1} x^n + 2 \sum_{n=0}^{\infty} c_n x^n = 0;$$

that is,

$$\sum_{n=0}^{\infty} [(n+1) c_{n+1} + 2c_n] x^n = 0.$$

If this equation holds on some interval, then it follows from the identity principle that $(n + 1)c_{n+1} + 2c_n = 0$ for all $n \geq 0$; consequently,

$$c_{n+1} = -\frac{2c_n}{n + 1} \qquad (22)$$

for all $n \geq 0$. Equation (22) is a **recurrence relation** from which we can successively compute c_1, c_2, c_3, \ldots in terms of c_0; the latter will turn out to be the arbitrary constant that we expect to find in a general solution of a first-order differential equation.

With $n = 0$, Eq. (22) gives

$$c_1 = -\frac{2c_0}{1}.$$

With $n = 1$, Eq. (22) gives

$$c_2 = -\frac{2c_1}{2} = +\frac{2^2 c_0}{1 \cdot 2} = \frac{2^2 c_0}{2!}.$$

With $n = 2$, Eq. (22) gives

$$c_3 = -\frac{2c_2}{3} = -\frac{2^3 c_0}{1 \cdot 2 \cdot 3} - -\frac{2^3 c_0}{3!}.$$

By now it should be clear that after n such steps, we will have

$$c_n = (-1)^n \frac{2^n c_0}{n!}, \qquad n \geq 1.$$

(This is easy to prove by induction on n.) Consequently, our solution takes the form

$$y(x) = \sum_{n=0}^{\infty} c_n x^n = \sum_{n=0}^{\infty} (-1)^n \frac{2^n c_0}{n!} x^n = c_0 \sum_{n=0}^{\infty} \frac{(-2x)^n}{n!} = c_0 e^{-2x}.$$

In the final step we have used the familiar exponential series in Eq. (5) to identify our power series solution as the same solution $y(x) = c_0 e^{-2x}$ we could have obtained by the method of separation of variables. ∎

Shift of Index of Summation

In the solution of Example 1 we wrote

$$\sum_{n=1}^{\infty} n c_n x^{n-1} = \sum_{n=0}^{\infty} (n + 1) c_{n+1} x^n \qquad (23)$$

by shifting the index of summation by $+1$ in the series on the left. That is, we simultaneously *increased* the index of summation by 1 (replacing n with $n + 1$, $n \to n + 1$) and *decreased* the starting point by 1, from $n = 1$ to $n = 0$, thereby obtaining the series on the right. This procedure is valid because each infinite series in (23) is simply a compact notation for the single series

$$c_1 + 2c_2 x + 3c_3 x^2 + 4c_4 x^3 + \cdots . \qquad (24)$$

More generally, we can shift the index of summation by k in an infinite series by simultaneously *increasing* the summation index by k ($n \to n+k$) and *decreasing* the starting point by k. For instance, a shift by $+2$ ($n \to n+2$) yields

$$\sum_{n=3}^{\infty} a_n x^{n-1} = \sum_{n=1}^{\infty} a_{n+2} x^{n+1}.$$

If k is negative, we interpret a "decrease by k" as an increase by $-k = |k|$. Thus a shift by -2 ($n \to n-2$) in the index of summation yields

$$\sum_{n=1}^{\infty} n c_n x^{n-1} = \sum_{n=3}^{\infty} (n-2) c_{n-2} x^{n-3};$$

we have *decreased* the index of summation by 2 but *increased* the starting point by 2, from $n = 1$ to $n = 3$. You should check that the summation on the right is merely another representation of the series in (24).

We know that the power series obtained in Example 1 converges for all x because it is an exponential series. More commonly, a power series solution is not recognizable in terms of the familiar elementary functions. When we get an unfamiliar power series solution, we need a way of finding where it converges. After all, $y = \sum c_n x^n$ is merely an *assumed* form of the solution. The procedure illustrated in Example 1 for finding the coefficients $\{c_n\}$ is merely a formal process and may or may not be valid. Its validity—in applying Theorem 1 to compute y' and applying Theorem 2 to obtain a recurrence relation for the coefficients—depends on the convergence of the initially *unknown* series $y = \sum c_n x^n$. Hence this formal process is justified only if in the end we can show that the power series we obtain converges on some open interval. If so, it then represents a solution of the differential equation on that interval. The following theorem (which we state without proof) may be used for this purpose.

THEOREM 3: Radius of Convergence

Given the power series $\sum c_n x^n$, suppose that the limit

➤
$$\rho = \lim_{n \to \infty} \left| \frac{c_n}{c_{n+1}} \right| \tag{25}$$

exists (ρ is finite) or is infinite (in this case we will write $\rho = \infty$). Then

(a) If $\rho = 0$, then the series diverges for all $x \neq 0$.
(b) If $0 < \rho < \infty$, then $\sum c_n x^n$ converges if $|x| < \rho$ and diverges if $|x| > \rho$.
(c) If $\rho = \infty$, then the series converges for all x. ▲

The number ρ in (25) is called the **radius of convergence** of the power series $\sum c_n x^n$. For instance, for the power series obtained in Example 1, we have

$$\rho = \lim_{n \to \infty} \left| \frac{(-1)^n 2^n c_0 / n!}{(-1)^{n+1} 2^{n+1} c_0 / (n+1)!} \right| = \lim_{n \to \infty} \frac{n+1}{2} = \infty,$$

and consequently the series we obtained in Example 1 converges for all x. Even if the limit in (25) fails to exist, there always will be a number ρ such that exactly one of the three alternatives in Theorem 3 holds. This number may be difficult to

find, but for the power series we will consider in this chapter, Eq. (25) will be quite sufficient for computing the radius of convergence.

| **Example 2** | Solve the equation $(x - 3)y' + 2y = 0$. |

Solution As before, we substitute

$$y = \sum_{n=0}^{\infty} c_n x^n \quad \text{and} \quad y' = \sum_{n=1}^{\infty} n c_n x^{n-1}$$

to obtain

$$(x - 3) \sum_{n=1}^{\infty} n c_n x^{n-1} + 2 \sum_{n=0}^{\infty} c_n x^n = 0,$$

so that

$$\sum_{n-1}^{\infty} n c_n x^n - 3 \sum_{n=1}^{\infty} n c_n x^{n-1} + 2 \sum_{n=0}^{\infty} c_n x^n = 0.$$

In the first sum we can replace $n = 1$ with $n = 0$ with no effect on the sum. In the second sum we shift the index of summation by $+1$. This yields

$$\sum_{n=0}^{\infty} n c_n x^n - 3 \sum_{n=0}^{\infty} (n + 1) c_{n+1} x^n + 2 \sum_{n=0}^{\infty} c_n x^n = 0;$$

that is,

$$\sum_{n=0}^{\infty} \left[n c_n - 3(n + 1) c_{n+1} + 2 c_n \right] x^n = 0.$$

The identity principle then gives

$$n c_n - 3(n + 1) c_{n+1} + 2 c_n = 0,$$

from which we obtain the recurrence relation

$$c_{n+1} = \frac{n + 2}{3(n + 1)} c_n \quad \text{for} \quad n \geq 0.$$

We apply this formula with $n = 0$, $n = 1$, and $n = 2$, in turn, and find that

$$c_1 = \frac{2}{3} c_0, \quad c_2 = \frac{3}{3 \cdot 2} c_1 = \frac{3}{3^2} c_0, \quad \text{and} \quad c_3 = \frac{4}{3 \cdot 3} c_2 = \frac{4}{3^3} c_0.$$

This is almost enough to make the pattern evident; it is not difficult to show by induction on n that

$$c_n = \frac{n + 1}{3^n} c_0 \quad \text{if} \quad n \geq 1.$$

Hence our proposed power series solution is

$$y(x) = c_0 \sum_{n=0}^{\infty} \frac{n + 1}{3^n} x^n. \tag{26}$$

Its radius of convergence is

$$\rho = \lim_{n\to\infty} \left| \frac{c_n}{c_{n+1}} \right| = \lim_{n\to\infty} \frac{3n+3}{n+2} = 3.$$

Thus the series in (26) converges if $-3 < x < 3$ and diverges if $|x| > 3$. In this particular example we can explain why. An elementary solution (obtained by separation of variables) of our differential equation is $y = 1/(3-x)^2$. If we differentiate termwise the geometric series

$$\frac{1}{3-x} = \frac{\frac{1}{3}}{1 - \frac{x}{3}} = \frac{1}{3} \sum_{n=0}^{\infty} \frac{x^n}{3^n},$$

we get a constant multiple of the series in (26). Thus this series (with the arbitrary constant c_0 appropriately chosen) represents the solution

$$y(x) = \frac{1}{(3-x)^2}$$

on the interval $-3 < x < 3$, and the singularity at $x = 3$ is the reason why the radius of convergence of the power series solution turned out to be $\rho = 3$. ∎

Example 3 Solve the equation $x^2 y' = y - x - 1$.

Solution We make the usual substitutions $y = \sum c_n x^n$ and $y' = \sum n c_n x^{n-1}$, which yield

$$x^2 \sum_{n=1}^{\infty} n c_n x^{n-1} = -1 - x + \sum_{n=0}^{\infty} c_n x^n$$

so that

$$\sum_{n=1}^{\infty} n c_n x^{n+1} = -1 - x + \sum_{n=0}^{\infty} c_n x^n.$$

Because of the presence of the two terms -1 and $-x$ on the right-hand side, we need to split off the first two terms, $c_0 + c_1 x$, of the series on the right for comparison. If we also shift the index of summation on the left by -1 (replace $n = 1$ with $n = 2$ and n with $n - 1$), we get

$$\sum_{n=2}^{\infty} (n-1) c_{n-1} x^n = -1 - x + c_0 + c_1 x + \sum_{n=2}^{\infty} c_n x^n.$$

Because the left-hand side contains neither a constant term nor a term containing x to the first power, the identity principle now yields $c_0 = 1$, $c_1 = 1$, and $c_n = (n-1)c_{n-1}$ for $n \geq 2$. It follows that

$$c_2 = 1 \cdot c_1 = 1!, \quad c_3 = 2 \cdot c_2 = 2!, \quad c_4 = 3 \cdot c_3 = 3!,$$

and, in general, that

$$c_n = (n-1)! \quad \text{for} \quad n \geq 2.$$

Thus we obtain the power series

$$y(x) = 1 + x + \sum_{n=2}^{\infty} (n-1)!\, x^n.$$

But the radius of convergence of this series is

$$\rho = \lim_{n \to \infty} \frac{(n-1)!}{n!} = \lim_{n \to \infty} \frac{1}{n} = 0,$$

so the series converges only for $x = 0$. What does this mean? Simply that the given differential equation does not have a [convergent] power series solution of the assumed form $y = \sum c_n x^n$. This example serves as a warning that the simple act of writing $y = \sum c_n x^n$ involves an assumption that may be false. ■

Example 4 Solve the equation $y'' + y = 0$.

Solution If we assume a solution of the form

➤
$$y = \sum_{n=0}^{\infty} c_n x^n,$$

we find that

➤
$$y' = \sum_{n=1}^{\infty} n c_n x^{n-1} \quad \text{and} \quad y'' = \sum_{n-2}^{\infty} n(n-1) c_n x^{n-2}.$$

Substitution for y and y'' in the differential equation then yields

$$\sum_{n=2}^{\infty} n(n-1) c_n x^{n-2} + \sum_{n=0}^{\infty} c_n x^n = 0.$$

We shift the index of summation in the first sum by $+2$ (replace $n = 2$ with $n = 0$ and n with $n + 2$). This gives

$$\sum_{n=0}^{\infty} (n+2)(n+1) c_{n+2} x^n + \sum_{n=0}^{\infty} c_n x^n = 0.$$

The identity $(n+2)(n+1) c_{n+2} + c_n = 0$ now follows from the identity principle, and thus we obtain the recurrence relation

➤
$$c_{n+2} = -\frac{c_n}{(n+1)(n+2)} \tag{27}$$

for $n \geq 0$. It is evident that this formula will determine the coefficients c_n with even subscript in terms of c_0 and those of odd subscript in terms of c_1; c_0 and c_1 are not predetermined and thus will be the two arbitrary constants we expect to find in a general solution of a second-order equation.

When we apply the recurrence relation in (27) with $n = 0$, 2, and 4 in turn, we get

$$c_2 = -\frac{c_0}{2!}, \quad c_4 = \frac{c_0}{4!}, \quad \text{and} \quad c_6 = -\frac{c_0}{6!}.$$

Taking $n = 1, 3,$ and 5 in turn, we find that

$$c_3 = -\frac{c_1}{3!}, \quad c_5 = \frac{c_1}{5!}, \quad \text{and} \quad c_7 = -\frac{c_1}{7!}.$$

Again, the pattern is clear; we leave it for you to show (by induction) that for $k \geq 1$,

➤ $$c_{2k} = \frac{(-1)^k c_0}{(2k)!} \quad \text{and} \quad c_{2k+1} = \frac{(-1)^k c_1}{(2k+1)!}.$$

Thus we get the power series solution

$$y(x) = c_0 \left(1 - \frac{x^2}{2!} + \frac{x^4}{4!} - \frac{x^6}{6!} + \cdots \right) + c_1 \left(x - \frac{x^3}{3!} + \frac{x^5}{5!} - \frac{x^7}{7!} + \cdots \right);$$

that is, $y(x) = c_0 \cos x + c_1 \sin x$. Note that we have no problem with the radius of convergence here; the Taylor series for the sine and cosine functions converge for all x. ∎

The solution of Example 4 can bear further comment. Suppose that we had never heard of the sine and cosine functions, let alone their Taylor series. We would then have discovered the two power series solutions

$$C(x) = \sum_{n=0}^{\infty} \frac{(-1)^n x^{2n}}{(2n)!} = 1 - \frac{x^2}{2!} + \frac{x^4}{4!} - \cdots \tag{28}$$

and

$$S(x) = \sum_{n=0}^{\infty} \frac{(-1)^n x^{2n+1}}{(2n+1)!} = x - \frac{x^3}{3!} + \frac{x^5}{5!} - \cdots \tag{29}$$

of the differential equation $y'' + y = 0$. It is clear that $C(0) = 1$ and that $S(0) = 0$. After verifying that the two series in (28) and (29) converge for all x, we can differentiate them term by term to find that

$$C'(x) = -S(x) \quad \text{and} \quad S'(x) = C(x). \tag{30}$$

Consequently, $C'(0) = 0$ and $S'(0) = 1$. Thus with the aid of the power series method (all the while knowing nothing about the sine and cosine functions), we have discovered that $y = C(x)$ is the unique solution of

$$y'' + y = 0$$

that satisfies the initial conditions $y(0) = 1$ and $y'(0) = 0$, and that $y = S(x)$ is the unique solution that satisfies the initial conditions $y(0) = 0$ and $y'(0) = 1$. It follows that $C(x)$ and $S(x)$ are linearly independent, and—recognizing the importance of the differential equation $y'' + y = 0$—we can agree to call C the *cosine* function and S the *sine* function. Indeed, all the usual properties of these two functions can be established, using only their initial values (at $x = 0$) and the derivatives in (30); there is no need to refer to triangles or even to angles. (Can you use the series in (28) and (29) to show that $[C(x)]^2 + [S(x)]^2 = 1$ for all x?) This demonstrates that the cosine and sine functions are fully determined by the differential

equation $y'' + y = 0$ of which they are the natural linearly independent solutions. Figures 8.1.2 and 8.1.3 show how the geometric character of the graphs of $\cos x$ and $\sin x$ is revealed by the graphs of the Taylor polynomial approximations that we get by truncating the infinite series in (28) and (29).

This is by no means an uncommon situation. Many important special functions of mathematics occur in the first instance as power series solutions of differential equations and thus are in practice *defined* by means of these power series. In the remaining sections of this chapter we will see numerous examples of such functions.

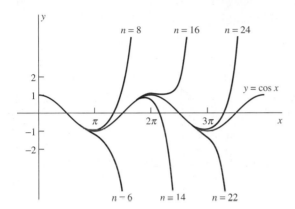

FIGURE 8.1.2. Taylor polynomial approximations to $\cos x$.

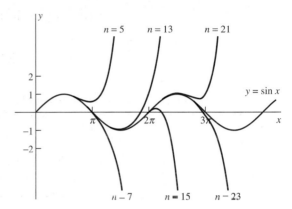

FIGURE 8.1.3. Taylor polynomial approximations to $\sin x$.

8.1 Problems

In Problems 1 through 10, find a power series solution of the given differential equation. Determine the radius of convergence of the resulting series, and use the series in Eqs. (5) through (12) to identify the series solution in terms of familiar elementary functions. (Of course, no one can prevent you from checking your work by also solving the equations by the methods of earlier chapters!)

1. $y' = y$
2. $y' = 4y$
3. $2y' + 3y = 0$
4. $y' + 2xy = 0$
5. $y' = x^2 y$
6. $(x - 2)y' + y = 0$
7. $(2x - 1)y' + 2y = 0$
8. $2(x + 1)y' = y$
9. $(x - 1)y' + 2y = 0$
10. $2(x - 1)y' = 3y$

In Problems 11 through 14, use the method of Example 4 to find two linearly independent power series solutions of the given differential equation. Determine the radius of convergence of each series, and identify the general solution in terms of familiar elementary functions.

11. $y'' = y$
12. $y'' = 4y$
13. $y'' + 9y = 0$
14. $y'' + y = x$

Show (as in Example 3) that the power series method fails to yield a power series solution of the form $y = \sum c_n x^n$ for the differential equations in Problems 15 through 18.

15. $xy' + y = 0$
16. $2xy' = y$
17. $x^2 y' + y = 0$
18. $x^3 y' = 2y$

In Problems 19 through 22, first derive a recurrence relation giving c_n for $n \geq 2$ in terms of c_0 or c_1 (or both). Then apply the given initial conditions to find the values of c_0 and c_1. Next determine c_n (in terms of n, as in the text) and, finally, identify the particular solution in terms of familiar elementary functions.

19. $y'' + 4y = 0$; $y(0) = 0$, $y'(0) = 3$
20. $y'' - 4y = 0$; $y(0) = 2$, $y'(0) = 0$
21. $y'' - 2y' + y = 0$; $y(0) = 0$, $y'(0) = 1$
22. $y'' + y' - 2y = 0$; $y(0) = 1$, $y'(0) = -2$
23. Show that the equation

$$x^2 y'' + x^2 y' + y = 0$$

has no power series solution of the form $y = \sum c_n x^n$.
24. Establish the binomial series in (12) by means of the following steps. **(a)** Show that $y = (1 + x)^\alpha$ satisfies the initial value problem $(1 + x)y' = \alpha y$, $y(0) = 1$. **(b)** Show that the power series method gives the binomial series in (12) as the solution of the initial value problem in part (a), and that this series converges if $|x| < 1$. **(c)** Explain why the validity of the binomial series given in (12) follows from parts (a) and (b).
25. For the initial value problem

$$y'' = y' + y, \quad y(0) = 0, \quad y(1) = 1$$

derive the power series solution

$$y(x) = \sum_{n=1}^{\infty} \frac{F_n}{n!} x^n$$

where $\{F_n\}_{n=0}^{\infty}$ is the sequence 0, 1, 1, 2, 3, 5, 8, 13, ... of *Fibonacci numbers* defined by $F_0 = 0$, $F_1 = 1$, $F_n = F_{n-2} + F_{n-1}$ for $n > 1$.

26. (a) Show that the solution of the initial value problem

$$y' = 1 + y^2, \quad y(0) = 0$$

is $y(x) = \tan x$. **(b)** Because $y(x) = \tan x$ is an odd function with $y'(0) = 1$, its Taylor series is of the form

$$y = x + c_3 x^3 + c_5 x^5 + c_7 x^7 + \cdots.$$

Substitute this series in $y' = 1 + y^2$ and equate like powers of x to derive the following relations:

$$3c_3 = 1, \qquad\qquad 5c_5 = 2c_3,$$
$$7c_7 = 2c_5 + (c_3)^2, \qquad 9c_9 = 2c_7 + 2c_3 c_5,$$
$$11c_{11} = 2c_9 + 2c_3 c_7 + (c_5)^2.$$

(c) Conclude that

$$\tan x = x + \frac{1}{3}x^3 + \frac{2}{15}x^5 + \frac{17}{315}x^7$$

$$+ \frac{62}{2835}x^9 + \frac{1382}{155925}x^{11} + \cdots.$$

27. This section introduces the use of infinite series to solve differential equations. Conversely, differential equations can sometimes be used to sum infinite series. For example, consider the infinite series

$$1 + \frac{1}{1!} - \frac{1}{2!} + \frac{1}{3!} + \frac{1}{4!} - \frac{1}{5!} + \cdots;$$

note the $++-++-\cdots$ pattern of signs superimposed on the terms of the series for the number e. We could evaluate this series if we could obtain a formula for the function

$$f(x) = 1 + x - \frac{1}{2!}x^2 + \frac{1}{3!}x^3 + \frac{1}{4!}x^4 - \frac{1}{5!}x^5 + \cdots,$$

because the sum of the numerical series in question is simply $f(1)$. **(a)** It's possible to show that the power series given here converges for all x and that termwise differentiation is valid. Given these facts, show that $f(x)$ satisfies the initial value problem

$$y^{(3)} = y; \quad y(0) = y'(0) = 1, \quad y''(0) = -1.$$

(b) Solve this initial value problem to show that

$$f(x) = \frac{1}{3}e^x + \frac{2}{3}e^{-x/2}\left(\cos\frac{\sqrt{3}}{2}x + \sqrt{3}\sin\frac{\sqrt{3}}{2}x\right).$$

For a suggestion, see Problem 48 of Section 3.3. **(c)** Evaluate $f(1)$ to find the sum of the numerical series given here.

8.2 Series Solutions Near Ordinary Points

The power series method introduced in Section 8.1 can be applied to linear equations of any order, but its most important applications are to homogeneous second-order linear differential equations of the form

$$A(x)y'' + B(x)y' + C(x)y = 0, \tag{1}$$

where the coefficients A, B, and C are analytic functions of x. Indeed, in most applications these coefficient functions are simple polynomials.

We saw in Example 3 of Section 8.1 that the series method does not always yield a series solution. To discover when it does succeed, we rewrite Eq. (1) in the form

$$y'' + P(x)y' + Q(x)y = 0 \tag{2}$$

with leading coefficient 1, and with $P = B/A$ and $Q = C/A$. Note that $P(x)$ and $Q(x)$ will generally fail to be analytic at points where $A(x)$ vanishes. For instance, consider the equation

$$xy'' + y' + xy = 0. \tag{3}$$

The coefficient functions in (3) are continuous everywhere. But in the form of (2) it is the equation

$$y'' + \frac{1}{x}y' + y = 0 \tag{4}$$

with $P(x) = 1/x$ not analytic at $x = 0$.

The point $x = a$ is called an **ordinary point** of Eq. (2)—and of the equivalent Eq. (1)—provided that the functions $P(x)$ and $Q(x)$ are both analytic at $x = a$. Otherwise, $x = a$ is a **singular point.** Thus the only singular point of Eqs. (3) and (4) is $x = 0$. Recall that a quotient of analytic functions is analytic wherever the denominator is nonzero. It follows that, if $A(a) \neq 0$ in Eq. (1) with analytic coefficients, then $x = a$ is an ordinary point. If $A(x)$, $B(x)$, and $C(x)$ are *polynomials* with no common factors, then $x = a$ is an ordinary point if and only if $A(a) \neq 0$.

Example 1 The point $x = 0$ is an ordinary point of the equation

$$xy'' + (\sin x)y' + x^2 y = 0,$$

despite the fact that $A(x) = x$ vanishes at $x = 0$. The reason is that

$$P(x) = \frac{\sin x}{x} = \frac{1}{x}\left(x - \frac{x^3}{3!} + \frac{x^5}{5!} - \cdots\right) = 1 - \frac{x^2}{3!} + \frac{x^4}{5!} - \cdots$$

is nevertheless analytic at $x = 0$ because the division by x yields a convergent power series. ∎

Example 2 The point $x = 0$ is *not* an ordinary point of the equation

$$y'' + x^2 y' + x^{1/2} y = 0.$$

For while $P(x) = x^2$ is analytic at the origin, $Q(x) = x^{1/2}$ is not. The reason is that $Q(x)$ is not differentiable at $x = 0$ and hence is not analytic there. (Theorem 1 of Section 8.1 implies that an analytic function must be differentiable.) ∎

Example 3 The point $x = 0$ is an ordinary point of the equation

$$(1 - x^3)y'' + (7x^2 + 3x^5)y' + (5x - 13x^4)y = 0$$

because the coefficient functions $A(x)$, $B(x)$, and $C(x)$ are polynomials with $A(0) \neq 0$. ∎

Theorem 2 of Section 3.1 implies that Eq. (2) has two linearly independent solutions on any open interval where the coefficient functions $P(x)$ and $Q(x)$ are continuous. The basic fact for our present purpose is that near an *ordinary* point a, these solutions will be power series in powers of $x - a$. A proof of the following theorem can be found in Chapter 3 of Coddington, *An Introduction to Ordinary Differential Equations* (Englewood Cliffs, N.J.: Prentice Hall, 1961).

THEOREM 1: Solutions Near an Ordinary Point

Suppose that a is an ordinary point of the equation

$$A(x)y'' + B(x)y' + C(x)y = 0; \tag{1}$$

that is, the functions $P = B/A$ and $Q = C/A$ are analytic at $x = a$. Then Eq. (1) has two linearly independent solutions, each of the form

$$y(x) = \sum_{n=0}^{\infty} c_n (x - a)^n. \tag{5}$$

The radius of convergence of any such series solution is at least as large as the distance from a to the nearest (real or complex) singular point of Eq. (1). The coefficients in the series in (5) can be determined by its substitution in Eq. (1). ▲

Example 4　Determine the radius of convergence guaranteed by Theorem 1 of a series solution of

$$(x^2 + 9)y'' + xy' + x^2 y = 0 \tag{6}$$

in powers of x. Repeat for a series in powers of $x - 4$.

Solution　This example illustrates the fact that we must take into account complex singular points as well as real ones. Because

$$P(x) = \frac{x}{x^2 + 9} \quad \text{and} \quad Q(x) = \frac{x^2}{x^2 + 9},$$

the only singular points of Eq. (6) are $\pm 3i$. The distance (in the complex plane) of each of these from 0 is 3, so a series solution of the form $\sum c_n x^n$ has radius of convergence at least 3. The distance of each singular point from 4 is 5, so a series solution of the form $\sum c_n (x - 4)^n$ has radius of convergence at least 5 (see Fig. 8.2.1). ■

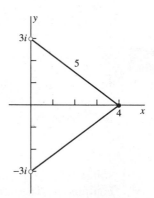

FIGURE 8.2.1. Radius of convergence as distance to nearest singularity.

Example 5　Find the general solution in powers of x of

$$(x^2 - 4)y'' + 3xy' + y = 0. \tag{7}$$

Then find the particular solution with $y(0) = 4$, $y'(0) = 1$.

Solution The only singular points of Eq. (7) are ± 2, so the series we get will have radius of convergence at least 2. Substitution of

$$\blacktriangleright \qquad y = \sum_{n=0}^{\infty} c_n x^n, \quad y' = \sum_{n=1}^{\infty} n c_n x^{n-1}, \quad \text{and} \quad y'' = \sum_{n=2}^{\infty} n(n-1) c_n x^{n-2}$$

in Eq. (7) yields

$$\sum_{n=2}^{\infty} n(n-1) c_n x^n - 4 \sum_{n=2}^{\infty} n(n-1) c_n x^{n-2} + 3 \sum_{n=1}^{\infty} n c_n x^n + \sum_{n=0}^{\infty} c_n x^n = 0.$$

We can begin the first and third summations at $n = 0$ as well, because no nonzero terms are thereby introduced. We shift the index of summation in the second sum by $+2$, replacing n with $n + 2$ and using the initial value $n = 0$. This gives

$$\sum_{n=0}^{\infty} n(n-1) c_n x^n - 4 \sum_{n=0}^{\infty} (n+2)(n+1) c_{n+2} x^n + 3 \sum_{n=0}^{\infty} n c_n x^n + \sum_{n=0}^{\infty} c_n x^n = 0.$$

After collecting coefficients of c_n and c_{n+2}, we obtain

$$\sum_{n=0}^{\infty} \left[(n^2 + 2n + 1) c_n - 4(n+2)(n+1) c_{n+2} \right] x^n = 0.$$

The identity principle yields

$$(n+1)^2 c_n - 4(n+2)(n+1) c_{n+2} = 0,$$

which leads to the recurrence relation

$$\blacktriangleright \qquad c_{n+2} = \frac{(n+1) c_n}{4(n+2)} \qquad (8)$$

for $n \geq 0$. With $n = 0, 2,$ and 4 in turn, we get

$$c_2 = \frac{c_0}{4 \cdot 2}, \quad c_4 = \frac{3 c_2}{4 \cdot 4} = \frac{3 c_0}{4^2 \cdot 2 \cdot 4}, \quad \text{and} \quad c_6 = \frac{5 c_4}{4 \cdot 6} = \frac{3 \cdot 5 c_0}{4^3 \cdot 2 \cdot 4 \cdot 6}.$$

Continuing in this fashion, we evidently would find that

$$c_{2n} = \frac{1 \cdot 3 \cdot 5 \cdots (2n-1)}{4^n \cdot 2 \cdot 4 \cdots (2n)} c_0.$$

With the common notation

$$(2n+1)!! = 1 \cdot 3 \cdot 5 \cdots (2n+1) = \frac{(2n+1)!}{2^n \cdot n!}$$

and the observation that $2 \cdot 4 \cdot 6 \cdots (2n) = 2^n \cdot n!$, we finally obtain

$$\blacktriangleright \qquad c_{2n} = \frac{(2n-1)!!}{2^{3n} \cdot n!} c_0. \qquad (9)$$

(We also used the fact that $4^n \cdot 2^n = 2^{3n}$.)

With $n = 1, 3$, and 5 in Eq. (8), we get

$$c_3 = \frac{2c_1}{4 \cdot 3}, \quad c_5 = \frac{4c_3}{4 \cdot 5} = \frac{2 \cdot 4c_1}{4^2 \cdot 3 \cdot 5}, \quad \text{and} \quad c_7 = \frac{6c_5}{4 \cdot 7} = \frac{2 \cdot 4 \cdot 6c_1}{4^3 \cdot 3 \cdot 5 \cdot 7}.$$

It is apparent that the pattern is

$$\blacktriangleright \qquad c_{2n+1} = \frac{2 \cdot 4 \cdot 6 \cdots (2n)}{4^n \cdot 1 \cdot 3 \cdot 5 \cdots (2n+1)} c_1 = \frac{n!}{2^n \cdot (2n+1)!!} c_1. \qquad (10)$$

The formula in (9) gives the coefficients of even subscript in terms of c_0; the formula in (10) gives the coefficients of odd subscript in terms of c_1. After we separately collect the terms of the series of even and odd degree, we get the general solution

$$y(x) = c_0 \left(1 + \sum_{n=1}^{\infty} \frac{(2n-1)!!}{2^{3n} \cdot n!} x^{2n} \right) + c_1 \left(x + \sum_{n=1}^{\infty} \frac{n!}{2^n \cdot (2n+1)!!} x^{2n+1} \right). \qquad (11)$$

Alternatively,

$$y(x) = c_0 \left(1 + \frac{1}{8} x^2 + \frac{3}{128} x^4 + \frac{5}{1024} x^6 + \cdots \right)$$

$$+ c_1 \left(x + \frac{1}{6} x^3 + \frac{1}{30} x^5 + \frac{1}{140} x^7 + \cdots \right). \qquad (11')$$

Because $y(0) = c_0$ and $y'(0) = c_1$, the given initial conditions imply that $c_0 = 4$ and $c_1 = 1$. Using these values in Eq. (11'), the first few terms of the particular solution satisfying $y(0) = 4$ and $y'(0) = 1$ are

$$y(x) = 4 + x + \frac{1}{2} x^2 + \frac{1}{6} x^3 + \frac{3}{32} x^4 + \frac{1}{30} x^5 + \cdots. \qquad (12)$$

■

Remark: As in Example 5, substitution of $y = \sum c_n x^n$ in a linear second-order equation with $x = 0$ an ordinary point typically leads to a recurrence relation that can be used to express each of the successive coefficients c_2, c_3, c_4, \ldots in terms of the first two, c_0 and c_1. In this event two linearly independent solutions are obtained as follows. Let $y_0(x)$ be the solution obtained with $c_0 = 1$ and $c_1 = 0$, and let $y_1(x)$ be the solution obtained with $c_0 = 0$ and $c_1 = 1$. Then

$$y_0(0) = 1, \quad y_0'(0) = 0 \quad \text{and} \quad y_1(0) = 0, \quad y_1'(0) = 1,$$

so it is clear that y_0 and y_1 are linearly independent. In Example 5, $y_0(x)$ and $y_1(x)$ are defined by the two series that appear on the right-hand side in Eq. (11), which expresses the general solution in the form $y = c_0 y_0 + c_1 y_1$. ■

Translated Series Solutions

If in Example 5 we had sought a particular solution with given initial values $y(a)$ and $y'(a)$, we would have needed the general solution in the form

$$y(x) = \sum_{n=0}^{\infty} c_n(x-a)^n; \tag{13}$$

that is, in powers of $x - a$ rather than in powers of x. For only with a solution of the form in (13) is it true that the initial conditions

$$y(a) = c_0 \quad \text{and} \quad y'(a) = c_1$$

determine the arbitrary constants c_0 and c_1 in terms of the initial values of y and y'. Consequently, to solve an initial value problem, we need a series expansion of the general solution centered at the point where the initial conditions are specified.

Example 6 Solve the initial value problem

$$(t^2 - 2t - 3)\frac{d^2y}{dt^2} + 3(t-1)\frac{dy}{dt} + y = 0; \quad y(1) = 4, \quad y'(1) = -1. \tag{14}$$

Solution We need a general solution of the form $\sum c_n(t-1)^n$. But instead of substituting this series in (14) to determine the coefficients, it simplifies the computations if we first make the substitution $x = t - 1$, so that we wind up looking for a series of the form $\sum c_n x^n$ after all. To transform Eq. (14) into one with the new independent variable x, we note that

$$t^2 - 2t - 3 = (x+1)^2 - 2(x+1) - 3 = x^2 - 4,$$

$$\frac{dy}{dt} = \frac{dy}{dx}\frac{dx}{dt} = \frac{dy}{dx} = y',$$

and

$$\frac{d^2y}{dt^2} = \left[\frac{d}{dx}\left(\frac{dy}{dx}\right)\right]\frac{dx}{dt} = \frac{d}{dx}(y') = y'',$$

where primes denote differentiation with respect to x. Hence we transform Eq. (14) into

$$(x^2 - 4)y'' + 3xy' + y = 0$$

with initial conditions $y = 4$ and $y' = 1$ at $x = 0$ (corresponding to $t = 1$). This is the initial value problem we solved in Example 5, so the particular solution in (12) is available. We substitute $t - 1$ for x in Eq. (12) and thereby obtain the desired particular solution

$$y(t) = 4 + (t-1) + \frac{1}{2}(t-1)^2 + \frac{1}{6}(t-1)^3$$

$$+ \frac{3}{32}(t-1)^4 + \frac{1}{30}(t-1)^5 + \cdots.$$

This series converges if $-1 < t < 3$. (Why?) A series such as this can be used to estimate numerical values of the solution. For instance,

$$y(0.8) = 4 + (-0.2) + \frac{1}{2}(-0.2)^2 + \frac{1}{6}(-0.2)^3$$

$$+ \frac{3}{32}(-0.2)^4 + \frac{1}{30}(-0.2)^5 + \cdots,$$

so that $y(0.8) \approx 3.8188$. ■

The last computation in Example 6 illustrates the fact that series solutions of differential equations are useful not only for establishing general properties of a solution, but also for numerical computations when an expression of the solution in terms of elementary functions is unavailable.

Types of Recurrence Relations

The formula in Eq. (8) is an example of a **two-term** recurrence relation; it expresses each coefficient in the series in terms of *one* of the preceding coefficients. A **many-term** recurrence relation expresses each coefficient in the series in terms of two or more preceding coefficients. In the case of a many-term recurrence relation, it is generally inconvenient or even impossible to find a formula that gives the typical coefficient c_n in terms of n. The next example shows what we sometimes can do with a three-term recurrence relation.

Example 7 Find two linearly independent solutions of

$$y'' - xy' - x^2 y = 0. \tag{15}$$

Solution We make the usual substitution of the power series $y = \sum c_n x^n$. This results in the equation

$$\sum_{n=2}^{\infty} n(n-1)c_n x^{n-2} - \sum_{n=1}^{\infty} nc_n x^n - \sum_{n=0}^{\infty} c_n x^{n+2} = 0.$$

We can start the second sum at $n = 0$ without changing anything else. To make each term include x^n in its general term, we shift the index of summation in the first sum by $+2$ (replace n with $n + 2$), and we shift it by -2 in the third sum (replace n with $n - 2$). These shifts yield

$$\sum_{n=0}^{\infty}(n+2)(n+1)c_{n+2}x^n - \sum_{n=0}^{\infty} nc_n x^n - \sum_{n=2}^{\infty} c_{n-2}x^n = 0.$$

The common range of these three summations is $n \geq 2$, so we must separate the terms corresponding to $n = 0$ and $n = 1$ in the first two sums before collecting coefficients of x^n. This gives

$$2c_2 + 6c_3 x - c_1 x + \sum_{n=2}^{\infty}\left[(n+2)(n+1)c_{n+2} - nc_n - c_{n-2}\right]x^n = 0.$$

The identity principle now implies that $2c_2 = 0$, that $c_3 = \frac{1}{6}c_1$, and the three-term recurrence relation

$$c_{n+2} = \frac{nc_n + c_{n-2}}{(n+2)(n+1)} \tag{16}$$

for $n \geq 2$. In particular,

$$c_4 = \frac{2c_2 + c_0}{12}, \quad c_5 = \frac{3c_3 + c_1}{20}, \quad c_6 = \frac{4c_4 + c_2}{30},$$

$$c_7 = \frac{5c_5 + c_3}{42}, \quad c_8 = \frac{6c_6 + c_4}{56}. \tag{17}$$

Thus all values of c_n for $n \geq 4$ are given in terms of the arbitrary constants c_0 and c_1 because $c_2 = 0$ and $c_3 = \frac{1}{6}c_1$.

To get our first solution y_1 of Eq. (15), we choose $c_0 = 1$ and $c_1 = 0$, so that $c_2 = c_3 = 0$. Then the formulas in (17) yield

$$c_4 = \frac{1}{12}, \quad c_5 = 0, \quad c_6 = \frac{1}{90}, \quad c_7 = 0, \quad c_8 = \frac{3}{1120};$$

thus

$$y_1(x) = 1 + \frac{1}{12}x^4 + \frac{1}{90}x^6 + \frac{3}{1120}x^8 + \cdots. \tag{18}$$

Because $c_1 = c_3 = 0$, it is clear from Eq. (16) that this series contains only terms of even degree.

To obtain a second linearly independent solution y_2 of Eq. (15), we take $c_0 = 0$ and $c_1 = 1$, so that $c_2 = 0$ and $c_3 = \frac{1}{6}$. Then the formulas in (17) yield

$$c_4 = 0, \quad c_5 = \frac{3}{40}, \quad c_6 = 0, \quad c_7 = \frac{13}{1008},$$

so that

$$y_2(x) = x + \frac{1}{6}x^3 + \frac{3}{40}x^5 + \frac{13}{1008}x^7 + \cdots. \tag{19}$$

Because $c_0 = c_2 = 0$, it is clear from Eq. (16) that this series contains only terms of odd degree. The solutions $y_1(x)$ and $y_2(x)$ are linearly independent because $y_1(0) = 1$ and $y_1'(0) = 0$, whereas $y_2(0) = 0$ and $y_2'(0) = 1$. A general solution of Eq. (15) is a linear combination of the power series in (18) and (19). Equation (15) has no singular points, so the power series representing $y_1(x)$ and $y_2(x)$ converge for all x. ∎

The Legendre Equation

The **Legendre equation** of order α is the second-order linear differential equation

$$\blacktriangleright \qquad (1 - x^2)y'' - 2xy' + \alpha(\alpha + 1)y = 0, \tag{20}$$

where the real number α satisfies the inequality $\alpha > -1$. This differential equation has extensive applications, ranging from numerical integration formulas (such as Gaussian quadrature) to the problem of determining the steady-state temperature within a solid spherical ball when the temperatures at points of its boundary are known. The only singular points of the Legendre equation are at $+1$ and -1, so it has two linearly independent solutions that can be expressed as power series in

powers of x with radius of convergence at least 1. The substitution $y = \sum c_m x^m$ in Eq. (20) leads (see Problem 31) to the recurrence relation

$$\blacktriangleright \qquad c_{m+2} = -\frac{(\alpha - m)(\alpha + m + 1)}{(m+1)(m+2)} c_m \qquad (21)$$

for $m \geq 0$. We are using m as the index of summation because we have another role for n to play.

In terms of the arbitrary constants c_0 and c_1, Eq. (21) yields

$$c_2 = -\frac{\alpha(\alpha+1)}{2!} c_0,$$

$$c_3 = -\frac{(\alpha-1)(\alpha+2)}{3!} c_1,$$

$$c_4 = \frac{\alpha(\alpha-2)(\alpha+1)(\alpha+3)}{4!} c_0,$$

$$c_5 = \frac{(\alpha-1)(\alpha-3)(\alpha+2)(\alpha+4)}{5!} c_1.$$

We can show without much trouble that for $m > 0$,

$$c_{2m} = (-1)^m \frac{\alpha(\alpha-2)(\alpha-4)\cdots(\alpha-2m+2)(\alpha+1)(\alpha+3)\cdots(\alpha+2m-1)}{(2m)!} c_0 \qquad (22)$$

and

$$c_{2m+1} = (-1)^m \frac{(\alpha-1)(\alpha-3)\cdots(\alpha-2m+1)(\alpha+2)(\alpha+4)\cdots(\alpha+2m)}{(2m+1)!} c_1. \qquad (23)$$

Alternatively,

$$c_{2m} = (-1)^m a_{2m} c_0 \quad \text{and} \quad c_{2m+1} = (-1)^m a_{2m+1} c_1,$$

where a_{2m} and a_{2m+1} denote the fractions in Eqs. (22) and (23), respectively. With this notation, we get two linearly independent power series solutions

$$y_1(x) = c_0 \sum_{m=0}^{\infty} (-1)^m a_{2m} x^{2m} \quad \text{and} \quad y_2(x) = c_1 \sum_{m=0}^{\infty} (-1)^m a_{2m+1} x^{2m+1} \qquad (24)$$

of Legendre's equation of order α.

Now suppose that $\alpha = n$, a nonnegative *integer*. If $\alpha = n$ is even, we see from Eq. (22) that $a_{2m} = 0$ when $2m > n$. In this case, $y_1(x)$ is a *polynomial* of degree n and y_2 is a (nonterminating) infinite series. If $\alpha = n$ is an odd positive integer, we see from Eq. (23) that $a_{2m+1} = 0$ when $2m + 1 > n$. In this case, $y_2(x)$ is a *polynomial* of degree n and y_1 is a (nonterminating) infinite series. Thus in either case, one of the two solutions in (24) is a polynomial and the other is a nonterminating series.

With an appropriate choice (made separately for each n) of the arbitrary constants c_0 (n even) or c_1 (n odd), the nth-degree polynomial solution of Legendre's equation of order n,

$$(1 - x)^2 y'' - 2xy' + n(n+1)y = 0, \tag{25}$$

is denoted by $P_n(x)$ and is called the **Legendre polynomial** of degree n. It is customary (for a reason indicated in Problem 32) to choose the arbitrary constant so that the coefficient of x^n in $P_n(x)$ is $(2n)! / \left[2^n (n!)^2 \right]$. It then turns out that

$$P_n(x) = \sum_{k=0}^{N} \frac{(-1)^k (2n - 2k)!}{2^n k! \, (n-k)! \, (n-2k)!} x^{n-2k}, \tag{26}$$

where $N = [\![n/2]\!]$, the integral part of $n/2$. The first six Legendre polynomials are

$$P_0(x) \equiv 1, \qquad\qquad P_1(x) = x,$$

$$P_2(x) = \frac{1}{2}(3x^2 - 1), \qquad\qquad P_3(x) = \frac{1}{2}(5x^3 - 3x),$$

$$P_4(x) - \frac{1}{8}(35x^4 - 30x^2 + 3), \qquad P_5(x) = \frac{1}{8}(63x^5 - 70x^3 + 15x),$$

and their graphs are shown in Fig. 8.2.2.

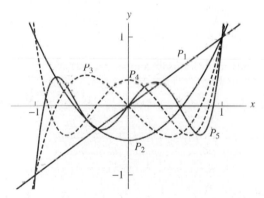

FIGURE 8.2.2. Graphs $y = P_n(x)$ of the Legendre polynomials for $n - 1, 2, 3, 4,$ and 5. The graphs are distinguished by the fact that all n zeros of $P_n(x)$ lie in the interval $-1 < x < 1$.

8.2 Problems

Find general solutions in powers of x of the differential equations in Problems 1 through 15. State the recurrence relation and the guaranteed radius of convergence in each case.

1. $(x^2 - 1)y'' + 4xy' + 2y = 0$
2. $(x^2 + 2)y'' + 4xy' + 2y = 0$
3. $y'' + xy' + y = 0$
4. $(x^2 + 1)y'' + 6xy' + 4y = 0$
5. $(x^2 - 3)y'' + 2xy' = 0$
6. $(x^2 - 1)y'' - 6xy' + 12y = 0$

7. $(x^2 + 3)y'' - 7xy' + 16y = 0$
8. $(2 - x^2)y'' - xy' + 16y = 0$
9. $(x^2 - 1)y'' + 8xy' + 12y = 0$
10. $3y'' + xy' - 4y = 0$
11. $5y'' - 2xy' + 10y = 0$
12. $y'' - x^2 y' - 3xy = 0$
13. $y'' + x^2 y' + 2xy = 0$
14. $y'' + xy = 0$ (an *Airy equation*)
15. $y'' + x^2 y = 0$

Use power series to solve the initial value problems in Problems 16 and 17.

16. $(1 + x^2)y'' + 2xy' - 2y = 0$; $y(0) = 0$, $y'(0) = 1$
17. $y'' + xy' - 2y = 0$; $y(0) = 1$, $y'(0) = 0$

Solve the initial value problems in Problems 18 through 22. First make a substitution of the form $t = x - a$, then find a solution $\sum c_n t^n$ of the transformed differential equation. State the interval of values of x for which Theorem 1 of this section guarantees convergence.

18. $y'' + (x - 1)y' + y = 0$; $y(1) = 2$, $y'(1) = 0$
19. $(2x - x^2)y'' - 6(x - 1)y' - 4y = 0$; $y(1) = 0$, $y'(1) = 1$
20. $(x^2 - 6x + 10)y'' - 4(x - 3)y' + 6y = 0$; $y(3) = 2$, $y'(3) = 0$
21. $(4x^2 + 16x + 17)y'' = 8y$; $y(-2) = 1$, $y'(-2) = 0$
22. $(x^2 + 6x)y'' + (3x + 9)y' - 3y = 0$; $y(-3) = 0$, $y'(-3) = 2$

In Problems 23 through 26, find a three-term recurrence relation for solutions of the form $y = \sum c_n x^n$. Then find the first three nonzero terms in each of two linearly independent solutions.

23. $y'' + (1 + x)y = 0$
24. $(x^2 - 1)y'' + 2xy' + 2xy = 0$
25. $y'' + x^2 y' + x^2 y = 0$
26. $(1 + x^3)y'' + x^4 y = 0$
27. Solve the initial value problem

$$y'' + xy' + (2x^2 + 1)y = 0; \quad y(0) = 1, \, y'(0) = -1.$$

Determine sufficiently many terms to compute $y(1/2)$ accurate to four decimal places.

In Problems 28 through 30, find the first three nonzero terms in each of two linearly independent solutions of the form $y = \sum c_n x^n$. Substitute known Taylor series for the analytic functions and retain enough terms to compute the necessary coefficients.

28. $y'' + e^{-x} y = 0$
29. $(\cos x)y'' + y = 0$
30. $xy'' + (\sin x)y' + xy = 0$
31. Derive the recurrence relation in (21) for the Legendre equation.
32. Follow the steps outlined in this problem to establish **Rodrigues's formula**

$$P_n(x) = \frac{1}{n! \, 2^n} \frac{d^n}{dx^n}(x^2 - 1)^n$$

for the nth-degree Legendre polynomial. **(a)** Show that $v = (x^2 - 1)^n$ satisfies the differential equation

$$(1 - x^2)v' + 2nxv = 0.$$

Differentiate each side of this equation to obtain

$$(1 - x^2)v'' + 2(n - 1)xv' + 2nv = 0.$$

(b) Differentiate each side of the last equation n times in succession to obtain

$$(1 - x^2)v^{(n+2)} - 2xv^{(n+1)} + n(n + 1)v^{(n)} = 0.$$

Thus $u = v^{(n)} = D^n(x^2 - 1)^n$ satisfies Legendre's equation of order n. **(c)** Show that the coefficient of x^n in u is $(2n)!/n!$; then state why this proves Rodrigues' formula. (Note that the coefficient of x^n in $P_n(x)$ is $(2n)! / \left[2^n (n!)^2\right]$.)

33. The **Hermite equation** of order α is

$$y'' - 2xy' + 2\alpha y = 0.$$

(a) Derive the two power series solutions

$$y_1 = 1 + \sum_{m=1}^{\infty} (-1)^m \frac{2^m \alpha(\alpha - 2) \cdots (\alpha - 2m + 2)}{(2m)!} x^{2m}$$

and

$$y_2 = x$$

$$+ \sum_{m=1}^{\infty} (-1)^m \frac{2^m (\alpha - 1)(\alpha - 3) \cdots (\alpha - 2m + 1)}{(2m + 1)!} x^{2m+1}.$$

Show that y_1 is a polynomial if α is an even integer, whereas y_2 is a polynomial if α is an odd integer. **(b)** The **Hermite polynomial** of degree n is denoted by $H_n(x)$. It is the nth-degree polynomial solution of Hermite's equation, multiplied by a suitable constant so that the coefficient of x^n is 2^n. Show that the first six Hermite polynomials are

$H_0(x) \equiv 1$,	$H_1(x) = 2x$,
$H_2(x) = 4x^2 - 2$,	$H_3(x) = 8x^3 - 12x$,
$H_4(x) = 16x^4 - 48x^2 + 12$,	
$H_5(x) = 32x^5 - 160x^3 + 120x$.	

A general formula for the Hermite polynomials is

$$H_n(x) = (-1)^n e^{x^2} \frac{d^n}{dx^n}\left(e^{-x^2}\right).$$

Verify that this formula does in fact give an nth-degree polynomial. It is interesting to use a computer algebra system to investigate the conjecture that (for each n) the zeros of the Hermite polynomials $H_n(x)$ and H_{n+1} are "interlaced"—that is, the n zeros of H_n line in the n bounded open intervals whose endpoints are successive pairs of zeros of H_{n+1}.

34. The discussion following Example 4 in Section 8.1 suggests that the differential equation $y'' + y = 0$ could be used to introduce and define the familiar sine and cosine functions. In a similar fashion, the *Airy equation*

$$y'' = xy$$

serves to introduce two new special functions that appear in applications ranging from radio waves to molecular vibrations. Derive the first three or four terms of two different power series solutions of the Airy equation. Then verify that your results agree with the formulas

$$y_1(x) = 1 + \sum_{k=1}^{\infty} \frac{1 \cdot 4 \cdots (3k-2)}{(3k)!} x^{3k}$$

and

$$y_2(x) = x + \sum_{k=1}^{\infty} \frac{2 \cdot 5 \cdots (3k-1)}{(3k+1)!} x^{3k+1}$$

for the solutions that satisfy the initial conditions $y_1(0) = 1$, $y_1'(0) = 0$ and $y_2(0) = 0$, $y_2'(0) = 1$, respectively. The special combinations

$$\text{Ai}(x) = \frac{y_1(x)}{3^{2/3}\Gamma(\frac{2}{3})} - \frac{y_2(x)}{3^{1/3}\Gamma(\frac{1}{3})}$$

and

$$\text{Bi}(x) = \frac{y_1(x)}{3^{1/6}\Gamma(\frac{2}{3})} + \frac{y_2(x)}{3^{-1/6}\Gamma(\frac{1}{3})}$$

define the standard *Airy functions* that appear in mathematical tables and computer algebra systems. Their graphs shown in Fig. 8.2.3 exhibit trigonometric-like oscillatory behavior for $x < 0$, whereas $\text{Ai}(x)$ decreases exponentially and $\text{Bi}(x)$ increases exponentially as $x \to +\infty$. It is interesting to use a computer algebra system to investigate how many terms must be retained in the y_1- and y_2-series above to produce a figure that is visually indistinguishable from Fig. 8.2.3 (which is based on high-precision approximations to the Airy functions).

FIGURE 8.2.3. The Airy function graphs $y = \text{Ai}(x)$ and $y = \text{Bi}(x)$.

8.2 Application: Automatic Computation of Series Coefficients

Repeated application of a recurrence relation to grind out successive coefficients is —especially in the case of a recurrence relation with three or more terms—a tedious aspect of the infinite series method. Here we illustrate the use of a computer algebra system for this task. In Example 7 we saw that the coefficients in the series solution $y = \sum c_n x^n$ of the differential equation $y'' - xy' - x^2 y = 0$ are given in terms of the two arbitrary coefficients c_0 and c_1 by

$$c_2 = 0, \quad c_3 = \frac{c_1}{6}, \quad \text{and} \quad c_{n+2} = \frac{nc_n + c_{n-2}}{(n+2)(n+1)} \quad \text{for} \quad n \geq 2. \tag{1}$$

It would appear to be a routine matter to implement such a recurrence relation, but a twist results from the fact that a typical computer system array is indexed by the subscripts 1, 2, 3, ..., rather than by the subscripts 0, 1, 2, ... that match the exponents in the successive terms of a power series that begins with a constant term. For this reason we first rewrite our proposed power series solution in the form

$$y = \sum_{n=0}^{\infty} c_n x^n = \sum_{n=1}^{\infty} b_n x^{n-1} \tag{2}$$

where $b_n = c_{n-1}$ for each $n \geq 1$. Then the first two conditions in (1) imply that $b_3 = 0$ and $b_4 = \frac{1}{6}b_2$; also, the recurrence relation (with n replaced with $n-1$) yields the new recurrence relation

$$b_{n+2} = c_{n+1} = \frac{(n-1)c_{n-1} + c_{n-3}}{(n+1)n} = \frac{(n-1)b_n + b_{n-2}}{n(n+1)}. \tag{3}$$

Now we are ready to begin. Suppose that we want to calculate the terms through the 10th degree in (2) with the initial conditions $b_1 = b_2 = 1$. Then either the *Maple* commands

```
k := 11:              # k terms
b := array(1..k):
b[1] := 1:            # arbitrary
b[2] := 1:            # arbitrary
b[3] := 0:
b[4] := b[2]/6:
for n from 3 by 1 to k - 2 do
    b[n+2] := ((n-1)*b[n] + b[n-2])/(n*(n+1));
    od;
```

or the *Mathematica* commands

```
k = 11;               (* k terms *)
b = Table[0, {n,1,k}];
b[[1]] = 1;           (* arbitrary *)
b[[2]] = 1;           (* arbitrary *)
b[[3]] = 0;
b[[4]] = b[[2]]/6;
For[n=3, n<=k-2,
    b[[n+2]]=((n-1)*b[[n]] + b[[n-2]])/(n*(n+1)); n=n+1];
```

quickly yield the coefficients $\{b_n\}$ corresponding to the solution

$$y(x) =$$

$$1 + x + \frac{x^3}{6} + \frac{x^4}{12} + \frac{3x^5}{40} + \frac{x^6}{90} + \frac{13x^7}{1008} + \frac{3x^8}{1120} + \frac{119x^9}{51840} + \frac{41x^{10}}{113400} + \cdots. \quad (4)$$

You might note that the even- and odd-degree terms here agree with those shown in Eqs. (18) and (19), respectively, of Example 7.

The MATLAB commands

```
k = 11;               % k terms
b = 0*(1:k);
b(1) = 1;             % arbitrary
b(2) = 1;             % arbitrary
b(3) = 0;
b(4) = b(2)/6;
for n = 3:k-2
    b(n+2) = ((n-1)*b(n) + b(n-2))/(n*(n+1));
    end
format rat, b
```

give the same results, *except that* the coefficient b_{10} of x^9 is shown as $73/31801$ rather than the correct value $119/51840$ shown in Eq. (4). It happens that

$$\frac{73}{31801} \approx 0.0022955253 \quad \text{while} \quad \frac{119}{51840} \approx 0.0022955247,$$

so the two rational fractions agree when rounded to 9 decimal places. The explanation is that (unlike *Mathematica* and *Maple*) MATLAB works internally with decimal rather than exact arithmetic. But at the end its **format rat** algorithm converts a correct 14-place approximation for b_{10} into an incorrect rational fraction that's "close but no cigar."

You can substitute $b_1 = 1$, $b_2 = 0$ and $b_1 = 0$, $b_2 = 1$ separately (instead of $b_1 = b_2 = 1$) in the commands shown here to derive partial sums of the two linearly independent solutions displayed in Eqs. (18) and (19) of Example 7. This technique can be applied to any of the examples and problems in this section.

8.3 Regular Singular Points

We now investigate the solution of the homogeneous second-order linear equation

➤
$$A(x)y'' + B(x)y' + C(x)y = 0 \tag{1}$$

near a singular point. Recall that if the functions A, B, and C are polynomials having no common factors, then the singular points of Eq. (1) are simply those points where $A(x) = 0$. For instance, $x = 0$ is the only singular point of the Bessel equation of order n,

$$x^2 y'' + xy' + (x^2 - n^2)y = 0,$$

whereas the Legendre equation of order n,

$$(1 - x^2)y'' - 2xy' + n(n + 1)y = 0,$$

has the two singular points $x = -1$ and $x = 1$. It turns out that some of the features of the solutions of such equations of the most importance for applications are largely determined by their behavior near their singular points.

We will restrict our attention to the case in which $x = 0$ is a singular point of Eq. (1). A differential equation having $x = a$ as a singular point is easily transformed by the substitution $t = x - a$ into one having a corresponding singular point at 0. For example, let us substitute $t = x - 1$ into the Legendre equation of order n. Because

$$y' = \frac{dy}{dx} = \frac{dy}{dt}\frac{dt}{dx} = \frac{dy}{dt},$$

$$y'' = \frac{d^2y}{dx^2} = \left[\frac{d}{dt}\left(\frac{dy}{dx}\right)\right]\frac{dt}{dx} = \frac{d^2y}{dt^2},$$

and $1 - x^2 = 1 - (t + 1)^2 = -2t - t^2$, we get the equation

$$-t(t + 2)\frac{d^2y}{dt^2} - 2(t + 1)\frac{dy}{dt} + n(n + 1)y = 0.$$

This new equation has the singular point $t = 0$ corresponding to $x = 1$ in the original equation; it has also the singular point $t = -2$ corresponding to $x = -1$.

Types of Singular Points

A differential equation having a singular point at 0 ordinarily will *not* have power series solutions of the form $y(x) = \sum c_n x^n$, so the straightforward method of Section 8.2 fails in this case. To investigate the form that a solution of such an equation might take, we assume that Eq. (1) has analytic coefficient functions and rewrite it in the standard form

➤
$$y'' + P(x)y' + Q(x)y = 0, \tag{2}$$

where $P = B/A$ and $Q = C/A$. Recall that $x = 0$ is an ordinary point (as opposed to a singular point) of Eq. (2) if the functions $P(x)$ and $Q(x)$ are analytic at $x = 0$; that is, if $P(x)$ and $Q(x)$ have convergent power series expansions in powers of x on some open interval containing $x = 0$. Now it can be proved that each of the functions $P(x)$ and $Q(x)$ *either* is analytic *or* approaches $\pm\infty$ as $x \to 0$.

Consequently, $x = 0$ is a singular point of Eq. (2) provided that either $P(x)$ or $Q(x)$ (or both) approaches $\pm\infty$ as $x \to 0$. For instance, if we rewrite the Bessel equation of order n in the form

$$y'' + \frac{1}{x}y' + \left(1 - \frac{n^2}{x^2}\right)y = 0,$$

we see that $P(x) = 1/x$ and $Q(x) = 1 - (n/x)^2$ both approach infinity as $x \to 0$.

We will see presently that the power series method can be generalized to apply near the singular point $x = 0$ of Eq. (2), provided that $P(x)$ approaches infinity no more rapidly than $1/x$, and $Q(x)$ no more rapidly than $1/x^2$, as $x \to 0$. This is a way of saying that $P(x)$ and $Q(x)$ have only "weak" singularities at $x = 0$. To state this more precisely, we rewrite Eq. (2) in the form

$$\blacktriangleright \qquad\qquad y'' + \frac{p(x)}{x}y' + \frac{q(x)}{x^2}y = 0, \qquad\qquad (3)$$

where

$$\blacktriangleright \qquad\qquad p(x) = xP(x) \quad \text{and} \quad q(x) = x^2 Q(x). \qquad\qquad (4)$$

Definition: Regular Singular Point

The singular point $x = 0$ of Eq. (3) is a **regular singular point** if the functions $p(x)$ and $q(x)$ are both analytic at $x = 0$. Otherwise it is an **irregular singular point**.

In particular, the singular point $x = 0$ is a *regular* singular point if $p(x)$ and $q(x)$ are both polynomials. For instance, we see that $x = 0$ is a regular singular point of Bessel's equation of order n by writing that equation in the form

$$y'' + \frac{1}{x}y' + \frac{x^2 - n^2}{x^2}y = 0,$$

noting that $p(x) \equiv 1$ and $q(x) = x^2 - n^2$ are both polynomials in x.

By contrast, consider the equation

$$2x^3 y'' + (1 + x)y' + 3xy = 0,$$

which has the singular point $x = 0$. If we write this equation in the form of (3), we get

$$y'' + \frac{(1+x)/(2x^2)}{x}y' + \frac{\frac{3}{2}}{x^2}y = 0.$$

Because

$$p(x) = \frac{1+x}{2x^2} = \frac{1}{2x^2} + \frac{1}{2x} \to \infty$$

as $x \to 0$ (although $q(x) \equiv \frac{3}{2}$ is a polynomial), we see that $x = 0$ is an irregular singular point. We will not discuss the solution of differential equations near irregular singular points; this is a considerably more advanced topic than the solution of differential equations near regular singular points.

Example 1 Consider the differential equation

$$x^2(1+x)y'' + x(4-x^2)y' + (2+3x)y = 0.$$

In the standard form $y'' + Py' + Qy = 0$ it is

$$y'' + \frac{4-x^2}{x(1+x)}y' + \frac{2+3x}{x^2(1+x)}y = 0.$$

Because

$$P(x) = \frac{4-x^2}{x(1+x)} \quad \text{and} \quad Q(x) = \frac{2+3x}{x^2(1+x)}$$

both approach ∞ as $x \to 0$, we see that $x = 0$ is a singular point. To determine the nature of this singular point we write the differential equation in the form of Eq. (3):

$$y'' + \frac{(4-x^2)/(1+x)}{x}y' + \frac{(2+3x)/(1+x)}{x^2}y = 0.$$

Thus

$$p(x) = \frac{4-x^2}{1+x} \quad \text{and} \quad q(x) = \frac{2+3x}{1+x}.$$

Because a quotient of polynomials is analytic wherever the denominator is nonzero, we see that $p(x)$ and $q(x)$ are both analytic at $x = 0$. Hence $x = 0$ is a *regular singular point* of the given differential equation. ∎

It may happen that when we begin with a differential equation in the general form in Eq. (1) and rewrite it in the form in (3), the functions $p(x)$ and $q(x)$ as given in (4) are indeterminate forms at $x = 0$. In this case the situation is determined by the limits

$$p_0 = p(0) = \lim_{x \to 0} p(x) = \lim_{x \to 0} xP(x) \tag{5}$$

and

$$q_0 = q(0) = \lim_{x \to 0} q(x) = \lim_{x \to 0} x^2Q(x). \tag{6}$$

If $p_0 = 0 = q_0$, then $x = 0$ may be an ordinary point. Otherwise, if the limits in both (5) and (6) exist and are *finite*, then $x = 0$ is a regular singular point. If either limit fails to exist or is infinite, then $x = 0$ is an irregular singular point.

Remark: The most common case in applications, for the differential equation written in the form

$$y'' + \frac{p(x)}{x}y' + \frac{q(x)}{x^2}y = 0, \tag{3}$$

is that the functions $p(x)$ and $q(x)$ are *polynomials*. In this case $p_0 = p(0)$ and $q_0 = q(0)$ are simply the constant terms of these polynomials, so there is no need to evaluate the limits in Eqs. (5) and (6). ∎

Example 2 To investigate the nature of the point $x = 0$ for the differential equation

$$x^4 y'' + (x^2 \sin x) y' + (1 - \cos x) y = 0,$$

we first write it in the form in (3):

$$y'' + \frac{(\sin x)/x}{x} y' + \frac{(1 - \cos x)/x^2}{x^2} y = 0.$$

Then l'Hôpital's rule gives the values

$$p_0 = \lim_{x \to 0} \frac{\sin x}{x} = \lim_{x \to 0} \frac{\cos x}{1} = 1$$

and

$$q_0 = \lim_{x \to 0} \frac{1 - \cos x}{x^2} = \lim_{x \to 0} \frac{\sin x}{2x} = \frac{1}{2}$$

for the limits in Eqs. (5) and (6). While not both are zero, we see that $x = 0$ is a regular singular point because each limit is finite. Alternatively, we could write

$$p(x) = \frac{\sin x}{x} = \frac{1}{x} \left(x - \frac{x^3}{3!} + \frac{x^5}{5!} - \cdots \right) = 1 - \frac{x^2}{3!} + \frac{x^4}{5!} - \cdots$$

and

$$q(x) = \frac{1 - \cos x}{x^2} = \frac{1}{x^2} \left[1 - \left(1 - \frac{x^2}{2!} + \frac{x^4}{4!} - \frac{x^6}{6!} + \cdots \right) \right]$$

$$= \frac{1}{2!} - \frac{x^2}{4!} + \frac{x^4}{6!} - \cdots.$$

These (convergent) power series show explicitly that $p(x)$ and $q(x)$ are analytic and moreover that $p_0 = p(0) = 1$ and $q_0 = q(0) = \frac{1}{2}$. ∎

The Method of Frobenius

We now approach the task of actually finding solutions of a second-order linear differential equation near the regular singular point $x = 0$. The simplest such equation is the constant-coefficient *equidimensional equation*

$$x^2 y'' + p_0 x y' + q_0 y = 0 \tag{7}$$

to which Eq. (3) reduces when $p(x) \equiv p_0$ and $q(x) \equiv q_0$ are constants. In this case we can verify by direct substitution that the simple power function $y(x) = x^r$ is a solution of Eq. (7) if and only if r is a root of the quadratic equation

$$r(r - 1) + p_0 r + q_0 = 0. \tag{8}$$

In the general case, in which $p(x)$ and $q(x)$ are power series rather than constants, it is a reasonable conjecture that our differential equation might have a solution of the form

$$\blacktriangleright \qquad y(x) = x^r \sum_{n=0}^{\infty} c_n x^n = \sum_{n=0}^{\infty} c_n x^{n+r} = c_0 x^r + c_1 x^{r+1} + c_2 x^{r+2} + \cdots \tag{9}$$

—the product of x^r and a power series. This turns out to be a very fruitful conjecture; according to Theorem 1 (soon to be stated formally), every equation of the form in (1) having $x = 0$ as a regular singular point does, indeed, have at least one such solution. This fact is the basis for the **method of Frobenius**, named for the German mathematician Georg Frobenius (1848–1917), who discovered the method in the 1870s.

An infinite series of the form in (9) is called a **Frobenius series**. Note that a Frobenius series is generally *not* a power series. For instance, with $r = -\frac{1}{2}$ the series in (9) takes the form

$$y = c_0 x^{-1/2} + c_1 x^{1/2} + c_2 x^{3/2} + c_3 x^{5/2} + \cdots ;$$

it is not a series in *integral* powers of x.

To investigate the possible existence of Frobenius series solutions, we begin with the equation

➤ $$x^2 y'' + x p(x) y' + q(x) y = 0 \tag{10}$$

obtained by multiplying the equation in (3) by x^2. If $x = 0$ is a regular singular point, then $p(x)$ and $q(x)$ are analytic at $x = 0$, so

$$\begin{aligned}
p(x) &= p_0 + p_1 x + p_2 x^2 + p_3 x^3 + \cdots , \\
q(x) &= q_0 + q_1 x + q_2 x^2 + q_3 x^3 + \cdots .
\end{aligned} \tag{11}$$

Suppose that Eq. (10) has the Frobenius series solution

➤ $$y = \sum_{n=0}^{\infty} c_n x^{n+r}. \tag{12}$$

We may (and always do) assume that $c_0 \neq 0$ because the series must have a first nonzero term. Termwise differentiation in Eq. (12) leads to

➤ $$y' = \sum_{n=0}^{\infty} c_n (n + r) x^{n+r-1} \tag{13}$$

and

➤ $$y'' = \sum_{n=0}^{\infty} c_n (n + r)(n + r - 1) x^{n+r-2}. \tag{14}$$

Substitution of the series in Eqs. (11) through (14) in Eq. (10) now yields

$$\left[r(r-1)c_0 x^r + (r+1)r c_1 x^{r+1} + \cdots \right]$$
$$+ \left[p_0 x + p_1 x^2 + \cdots \right] \cdot \left[r c_0 x^{r-1} + (r+1)c_1 x^r + \cdots \right]$$
$$+ \left[q_0 + q_1 x + \cdots \right] \cdot \left[c_0 x^r + c_1 x^{r+1} + \cdots \right] = 0. \tag{15}$$

The lowest power of x that appears in Eq. (15) is x^r. If Eq. (15) is to be satisfied identically, then the coefficient $r(r-1)c_0 + p_0 r c_0 + q_0 c_0$ of x^r must vanish. Because $c_0 \neq 0$, it follows that r must satisfy the quadratic equation

➤ $$r(r-1) + p_0 r + q_0 = 0 \tag{16}$$

of precisely the same form as that obtained with the equidimensional equation in (7). Equation (16) is called the **indicial equation** of the differential equation in (10), and its two roots (possibly equal) are the **exponents** of the differential equation (at the regular singular point $x = 0$).

Our derivation of Eq. (16) shows that *if* the Frobenius series $y = x^r \sum c_n x^n$ is to be a solution of the differential equation in (10), *then* the exponent r must be one of the roots r_1 and r_2 of the indicial equation in (16). If $r_1 \neq r_2$, it follows that there are two possible Frobenius series solutions, whereas if $r_1 = r_2$ there is only one possible Frobenius series solution; the second solution cannot be a Frobenius series. The exponents r_1 and r_2 in the possible Frobenius series solutions are determined (using the indicial equation) by the values $p_0 = p(0)$ and $q_0 = q(0)$ that we have discussed. In practice, particularly when the coefficients in the differential equation in the original form in (1) are polynomials, the simplest way of finding p_0 and q_0 is often to write the equation in the form

$$y'' + \frac{p_0 + p_1 x + p_2 x^2 + \cdots}{x} y' + \frac{q_0 + q_1 x + q_2 x^2 + \cdots}{x^2} y = 0. \tag{17}$$

Then inspection of the series that appear in the two numerators reveals the constants p_0 and q_0.

Example 3 Find the exponents in the possible Frobenius series solutions of the equation

$$2x^2(1 + x)y'' + 3x(1 + x)^3 y' - (1 - x^2)y = 0.$$

Solution We divide each term by $2x^2(1 + x)$ to recast the differential equation in the form

$$y'' + \frac{\frac{3}{2}(1 + 2x + x^2)}{x} y' + \frac{-\frac{1}{2}(1 - x)}{x^2} y = 0,$$

and thus see that $p_0 = \frac{3}{2}$ and $q_0 = -\frac{1}{2}$. Hence the indicial equation is

$$r(r - 1) + \tfrac{3}{2}r - \tfrac{1}{2} = r^2 + \tfrac{1}{2}r - \tfrac{1}{2} = (r + 1)(r - \tfrac{1}{2}) = 0,$$

with roots $r_1 = \frac{1}{2}$ and $r_2 = -1$. The two possible Frobenius series solutions are then of the forms

$$y_1(x) = x^{1/2} \sum_{n=0}^{\infty} a_n x^n \quad \text{and} \quad y_2(x) = x^{-1} \sum_{n=0}^{\infty} b_n x^n. \qquad ■$$

Frobenius Series Solutions

Once the exponents r_1 and r_2 are known, the coefficients in a Frobenius series solution are determined by substitution of the series in Eqs. (12) through (14) in the differential equation, essentially the same method as was used to determine coefficients in power series solutions in Section 8.2. If the exponents r_1 and r_2 are complex conjugates, then there always exist two linearly independent Frobenius series solutions. We will restrict our attention here to the case in which r_1 and r_2 are both real. We also will seek solutions only for $x > 0$. Once such a solution has been found, we need only replace x^{r_1} with $|x|^{r_1}$ to obtain a solution for $x < 0$. The following theorem is proved in Chapter 4 of Coddington's *An Introduction to Ordinary Differential Equations*.

THEOREM 1: Frobenius Series Solutions

Suppose that $x = 0$ is a regular singular point of the equation

$$x^2 y'' + xp(x)y' + q(x)y = 0. \tag{10}$$

Let $\rho > 0$ denote the minimum of the radii of convergence of the power series

$$p(x) = \sum_{n=0}^{\infty} p_n x^n \quad \text{and} \quad q(x) = \sum_{n=0}^{\infty} q_n x^n.$$

Let r_1 and r_2 be the roots, with $r_1 \geqq r_2$, of the indicial equation $r(r-1) + p_0 r + q_0 = 0$. Then

(a) For $x > 0$, there exists a solution of Eq. (10) of the form

$$y_1(x) = x^{r_1} \sum_{n=0}^{\infty} a_n x^n \qquad (a_0 \neq 0) \tag{18}$$

corresponding to the larger root r_1.

(b) If $r_1 - r_2$ is neither zero nor a positive integer, then there exists a second linearly independent solution for $x > 0$ of the form

$$y_2(x) = x^{r_2} \sum_{n=0}^{\infty} b_n x^n \qquad (b_0 \neq 0) \tag{19}$$

corresponding to the smaller root r_2.

The radii of convergence of the power series in Eqs. (18) and (19) are each at least ρ. The coefficients in these series can be determined by substituting the series in the differential equation

$$x^2 y'' + xp(x)y' + q(x)y = 0. \qquad \blacktriangle$$

We have already seen that if $r_1 = r_2$, then there can exist only one Frobenius series solution. It turns out that, if $r_1 - r_2$ is a positive integer, there may or may not exist a second Frobenius series solution of the form in Eq. (19) corresponding to the smaller root r_2. These exceptional cases are discussed in Section 8.4. Examples 4 through 6 illustrate the process of determining the coefficients in those Frobenius series solutions that are guaranteed by Theorem 1.

Example 4

Find the Frobenius series solutions of

$$2x^2 y'' + 3xy' - (x^2 + 1)y = 0. \tag{20}$$

Solution

First we divide each term by $2x^2$ to put the equation in the form in (17):

$$y'' + \frac{\frac{3}{2}}{x} y' + \frac{-\frac{1}{2} - \frac{1}{2}x^2}{x^2} y = 0. \tag{21}$$

We now see that $x = 0$ is a regular singular point, and that $p_0 = \frac{3}{2}$ and $q_0 = -\frac{1}{2}$. Because $p(x) \equiv \frac{3}{2}$ and $q(x) = -\frac{1}{2} - \frac{1}{2}x^2$ are polynomials, the Frobenius series we obtain will converge for all $x > 0$. The indicial equation is

$$r(r - 1) + \tfrac{3}{2}r - \tfrac{1}{2} = \left(r - \tfrac{1}{2}\right)(r + 1) = 0,$$

so the exponents are $r_1 = \frac{1}{2}$ and $r_2 = -1$. They do not differ by an integer, so Theorem 1 guarantees the existence of two linearly independent Frobenius series solutions. Rather than separately substituting

$$y_1 = x^{1/2} \sum_{n=0}^{\infty} a_n x^n \quad \text{and} \quad y_2 = x^{-1} \sum_{n=0}^{\infty} b_n x^n$$

in Eq. (20), it is more efficient to begin by substituting $y = x^r \sum c_n x^n$. We will then get a recurrence relation that depends on r. With the value $r_1 = \frac{1}{2}$ it becomes a recurrence relation for the series for y_1, whereas with $r_2 = -1$ it becomes a recurrence relation for the series for y_2.

When we substitute

$$\blacktriangleright \qquad y = \sum_{n=0}^{\infty} c_n x^{n+r}, \quad y' = \sum_{n=0}^{\infty} (n+r)c_n x^{n+r-1},$$

and

$$\blacktriangleright \qquad y'' = \sum_{n=0}^{\infty} (n+r)(n+r-1)c_n x^{n+r-2}$$

in Eq. (20)—the original differential equation, rather than Eq. (21)—we get

$$2\sum_{n=0}^{\infty} (n+r)(n+r-1)c_n x^{n+r} + 3\sum_{n=0}^{\infty} (n+r)c_n x^{n+r}$$

$$- \sum_{n=0}^{\infty} c_n x^{n+r+2} - \sum_{n=0}^{\infty} c_n x^{n+r} = 0. \qquad (22)$$

At this stage there are several ways to proceed. A good standard practice is to shift indices so that each exponent will be the same as the smallest one present. In this example, we shift the index of summation in the third sum by -2 to reduce its exponent from $n + r + 2$ to $n + r$. This gives

$$2\sum_{n=0}^{\infty} (n+r)(n+r-1)c_n x^{n+r} + 3\sum_{n=0}^{\infty} (n+r)c_n x^{n+r}$$

$$- \sum_{n=2}^{\infty} c_{n-2} x^{n+r} - \sum_{n=0}^{\infty} c_n x^{n+r} = 0. \qquad (23)$$

The common range of summation is $n \geq 2$, so we must treat $n = 0$ and $n = 1$ separately. Following our standard practice, the terms corresponding to $n = 0$ will always give the indicial equation

$$[2r(r-1) + 3r - 1]c_0 = 2\left(r^2 + \tfrac{1}{2}r - \tfrac{1}{2}\right)c_0 = 0.$$

The terms corresponding to $n = 1$ yield

$$[2(r+1)r + 3(r+1) - 1]c_1 = (2r^2 + 5r + 2)c_1 = 0.$$

Because the coefficient $2r^2 + 5r + 2$ of c_1 is nonzero whether $r = \frac{1}{2}$ or $r = -1$, it follows that

$$c_1 = 0 \tag{24}$$

in either case.

The coefficient of x^{n+r} in Eq. (23) is

$$2(n+r)(n+r-1)c_n + 3(n+r)c_n - c_{n-2} - c_n = 0.$$

We solve for c_n and simplify to obtain the recurrence relation

➤ $$c_n = \frac{c_{n-2}}{2(n+r)^2 + (n+r) - 1} \qquad \text{for } n \geqq 2. \tag{25}$$

CASE 1: $r_1 = \frac{1}{2}$. We now write a_n in place of c_n and substitute $r = \frac{1}{2}$ in Eq. (25). This gives the recurrence relation

➤ $$a_n = \frac{a_{n-2}}{2n^2 + 3n} \qquad \text{for } n \geqq 2. \tag{26}$$

With this formula we can determine the coefficients in the first Frobenius solution y_1. In view of Eq. (24) we see that $a_n = 0$ whenever n is odd. With $n = 2, 4$, and 6 in Eq. (26), we get

$$a_2 = \frac{a_0}{14}, \quad a_4 = \frac{a_2}{44} = \frac{a_0}{616}, \quad \text{and} \quad a_6 = \frac{a_4}{90} = \frac{a_0}{55440}.$$

Hence the first Frobenius solution is

$$y_1(x) = a_0 x^{1/2} \left(1 + \frac{x^2}{14} + \frac{x^4}{616} + \frac{x^6}{55440} + \cdots \right).$$

CASE 2: $r_2 = -1$. We now write b_n in place of c_n and substitute $r = -1$ in Eq. (25). This gives the recurrence relation

➤ $$b_n = \frac{b_{n-2}}{2b^2 - 3n} \qquad \text{for } n \geqq 2. \tag{27}$$

Again, Eq. (24) implies that $b_n = 0$ for n odd. With $n = 2, 4$, and 6 in (27), we get

$$b_2 = \frac{b_0}{2}, \quad b_4 = \frac{b_2}{20} = \frac{b_0}{40}, \quad \text{and} \quad b_6 = \frac{b_4}{54} = \frac{b_0}{2160}.$$

Hence the second Frobenius solution is

$$y_2(x) = b_0 x^{-1} \left(1 + \frac{x^2}{2} + \frac{x^4}{40} + \frac{x^6}{2160} + \cdots \right). \qquad \blacksquare$$

Example 5 Find a Frobenius solution of Bessel's equation of order zero,

$$x^2 y'' + x y' + x^2 y = 0. \tag{28}$$

Solution In the form of (17), Eq. (28) becomes

$$y'' + \frac{1}{x} y' + \frac{x^2}{x^2} y = 0.$$

Hence $x = 0$ is a regular singular point with $p(x) \equiv 1$ and $q(x) = x^2$, so our series will converge for all $x > 0$. Because $p_0 = 1$ and $q_0 = 0$, the indicial equation is

$$r(r-1) + r = r^2 = 0.$$

Thus we obtain only the single exponent $r = 0$, and so there is only one Frobenius series solution

$$y(x) = x^0 \sum_{n=0}^{\infty} c_n x^n$$

of Eq. (28); it is in fact a power series.

Thus we substitute $y = \sum c_n x^n$ in (28); the result is

$$\sum_{n=0}^{\infty} n(n-1) c_n x^n + \sum_{n=0}^{\infty} n c_n x^n + \sum_{n=0}^{\infty} c_n x^{n+2} = 0.$$

We combine the first two sums and shift the index of summation in the third by -2 to obtain

$$\sum_{n=0}^{\infty} n^2 c_n x^n + \sum_{n=2}^{\infty} c_{n-2} x^n = 0.$$

The term corresponding to x^0 gives $0 = 0$: no information. The term corresponding to x^1 gives $c_1 = 0$, and the term for x^n yields the recurrence relation

$$c_n = -\frac{c_{n-2}}{n^2} \qquad \text{for } n \geq 2. \tag{29}$$

Because $c_1 = 0$, we see that $c_n = 0$ whenever n is odd. Substituting $n = 2, 4,$ and 6 in Eq. (29), we get

$$c_2 = -\frac{c_0}{2^2}, \quad c_4 = -\frac{c_2}{4^2} = \frac{c_0}{2^2 \cdot 4^2}, \quad \text{and} \quad c_6 = -\frac{c_4}{6^2} = -\frac{c_0}{2^2 \cdot 4^2 \cdot 6^2}.$$

Evidently, the pattern is

$$c_{2n} = \frac{(-1)^n c_0}{2^2 \cdot 4^2 \cdots (2n)^2} = \frac{(-1)^n c_0}{2^{2n} (n!)^2}.$$

The choice $c_0 = 1$ gives us one of the most important special functions in mathematics, the **Bessel function of order zero of the first kind**, denoted by $J_0(x)$. Thus

$$J_0(x) = \sum_{n=0}^{\infty} \frac{(-1)^n x^{2n}}{2^{2n} (n!)^2} = 1 - \frac{x^2}{4} + \frac{x^4}{64} - \frac{x^6}{2304} + \cdots. \tag{30}$$

In this example we have not been able to find a second linearly independent solution of Bessel's equation of order zero. We will derive that solution in Section 8.4; it will not be a Frobenius series. ∎

When $r_1 - r_2$ Is an Integer

Recall that, if $r_1 - r_2$ is a positive integer, then Theorem 1 guarantees only the existence of the Frobenius series solution corresponding to the larger exponent r_1. Example 6 illustrates the fortunate case in which the series method nevertheless yields a second Frobenius series solution. The case in which the second solution is not a Frobenius series will be discussed in Section 8.4.

Example 6 Find the Frobenius series solutions of

$$xy'' + 2y' + xy = 0. \tag{31}$$

Solution In standard form the equation becomes

$$y'' + \frac{2}{x}y' + \frac{x^2}{x^2}y = 0,$$

so we see that $x = 0$ is a regular singular point with $p_0 = 2$ and $q_0 = 0$. The indicial equation

$$r(r-1) + 2r = r(r+1) = 0$$

has roots $r_1 = 0$ and $r_2 = -1$, which differ by an integer. In this case when $r_1 - r_2$ is an integer, it is better to depart from the standard procedure of Example 4 and begin our work with the *smaller* exponent. As you will see, the recurrence relation will then tell us whether or not a second Frobenius series solution exists. If it does exist, our computations will simultaneously yield *both* Frobenius series solutions. If the second solution does not exist, we begin anew with the larger exponent $r = r_1$ to obtain the one Frobenius series solution guaranteed by Theorem 1.

Hence we begin by substituting

$$y - x^{-1}\sum_{n=0}^{\infty} c_n x^n - \sum_{n=0}^{\infty} c_n x^{n-1}$$

in Eq. (31). This gives

$$\sum_{n=0}^{\infty}(n-1)(n-2)c_n x^{n-2} + 2\sum_{n=0}^{\infty}(n-1)c_n x^{n-2} + \sum_{n=0}^{\infty} c_n x^n = 0.$$

We combine the first two sums and shift the index by -2 in the third to obtain

$$\sum_{n=0}^{\infty} n(n-1)c_n x^{n-2} + \sum_{n=2}^{\infty} c_{n-2} x^{n-2} = 0. \tag{32}$$

The cases $n = 0$ and $n = 1$ reduce to

$$0 \cdot c_0 = 0 \quad \text{and} \quad 0 \cdot c_1 = 0.$$

Hence we have *two* arbitrary constants c_0 and c_1 and therefore can expect to find a general solution incorporating two linearly independent Frobenius series solutions. If, for $n = 1$, we had obtained an equation such as $0 \cdot c_1 = 3$, which can be satisfied for *no* choice of c_1, this would have told us that no second Frobenius series solution could exist.

Now knowing that all is well, from (32) we read the recurrence relation

$$c_n = -\frac{c_{n-2}}{n(n-1)} \qquad \text{for } n \geq 2. \tag{33}$$

The first few values of n give

$$c_2 = -\frac{1}{2 \cdot 1}c_0, \qquad\qquad c_3 = -\frac{1}{3 \cdot 2}c_1,$$

$$c_4 = -\frac{1}{4 \cdot 3}c_2 = \frac{c_0}{4!}, \qquad c_5 = -\frac{1}{5 \cdot 4}c_3 = \frac{c_1}{5!},$$

$$c_6 = -\frac{1}{6 \cdot 5}c_4 = -\frac{c_0}{6!}, \qquad c_7 = -\frac{1}{7 \cdot 6}c_6 = -\frac{c_1}{7!};$$

evidently the pattern is

$$c_{2n} = \frac{(-1)^n c_0}{(2n)!}, \qquad c_{2n+1} = \frac{(-1)^n c_1}{(2n+1)!}$$

for $n \geq 1$. Therefore, a general solution of Eq. (31) is

$$y(x) = x^{-1} \sum_{n=0}^{\infty} c_n x^n$$

$$= \frac{c_0}{x}\left(1 - \frac{x^2}{2!} + \frac{x^4}{4!} - \cdots\right) + \frac{c_1}{x}\left(x - \frac{x^3}{3!} + \frac{x^5}{5!} - \cdots\right)$$

$$= \frac{c_0}{x}\sum_{n=0}^{\infty}\frac{(-1)^n x^{2n}}{(2n)!} + \frac{c_1}{x}\sum_{n=0}^{\infty}\frac{(-1)^n x^{2n+1}}{(2n+1)!}.$$

Thus

$$y(x) = \frac{1}{x}(c_0 \cos x + c_1 \sin x).$$

We have thus found a general solution expressed as a linear combination of the two Frobenius series solutions

$$y_1(x) = \frac{\cos x}{x} \quad \text{and} \quad y_2(x) = \frac{\sin x}{x}. \tag{34}$$

As indicated in Fig. 8.3.1, one of these Frobenius series solutions is bounded but the other is unbounded near the regular singular point $x = 0$—a common occurrence in the case of exponents differing by an integer. ∎

FIGURE 8.3.1. The solutions $y_1(x) = \dfrac{\cos x}{x}$ and $y_2(x) = \dfrac{\sin x}{x}$ in Example 6.

Summary

When confronted with a linear second-order differential equation

$$A(x)y'' + B(x)y' + C(x)y = 0$$

with analytic coefficient functions, in order to investigate the possible existence of series solutions we first write the equation in the standard form

➤
$$y'' + P(x)y' + Q(x)y = 0.$$

If $P(x)$ and $Q(x)$ are both analytic at $x = 0$, then $x = 0$ is an ordinary point, and the equation has two linearly independent power series solutions.

Otherwise, $x = 0$ is a singular point, and we next write the differential equation in the form

➤
$$y'' + \frac{p(x)}{x}y' + \frac{q(x)}{x^2}y = 0.$$

If $p(x)$ and $q(x)$ are both analytic at $x = 0$, then $x = 0$ is a regular singular point. In this case we find the two exponents r_1 and r_2 (assumed real, and with $r_1 \geq r_2$) by solving the indicial equation

➤
$$r(r - 1) + p_0 r + q_0 = 0,$$

where $p_0 = p(0)$ and $q_0 = q(0)$. There always exists a Frobenius series solution $y = x^{r_1} \sum a_n x^n$ associated with the larger exponent r_1, and if $r_1 - r_2$ is not an integer, the existence of a second Frobenius series solution $y_2 = x^{r_2} \sum b_n x^n$ is also guaranteed.

8.3 Problems

In Problems 1 through 8, determine whether $x = 0$ is an ordinary point, a regular singular point, or an irregular singular point. If it is a regular singular point, find the exponents of the differential equation at $x = 0$.

1. $xy'' + (x - x^3)y' + (\sin x)y = 0$
2. $xy'' + x^2 y' + (e^x - 1)y = 0$
3. $x^2 y'' + (\cos x)y' + xy = 0$
4. $3x^3 y'' + 2x^2 y' + (1 - x^2)y = 0$
5. $x(1 + x)y'' + 2y' + 3xy = 0$
6. $x^2(1 - x^2)y'' + 2xy' - 2y = 0$
7. $x^2 y'' + (6 \sin x)y' + 6y = 0$
8. $(6x + 2x^3)y'' + 21xy' + 9(x^2 - 1)y = 0$

If $x = a \neq 0$ is a singular point of a second-order linear differential equation, then the substitution $t = x - a$ transforms it into a differential equation having $t = 0$ as a singular point. We then attribute to the original equation at $x = a$ the behavior of the new equation at $t = 0$. Classify (as regular or irregular) the singular points of the differential equations in Problems 9 through 16.

9. $(1 - x)y'' + xy' + x^2 y = 0$
10. $(1 - x)^2 y'' + (2x - 2)y' + y = 0$
11. $(1 - x^2)y'' - 2xy' + 12y = 0$
12. $(x - 2)^3 y'' + 3(x - 2)^2 y' + x^3 y = 0$
13. $(x^2 - 4)y'' + (x - 2)y' + (x + 2)y = 0$
14. $(x^2 - 9)^2 y'' + (x^2 + 9)y' + (x^2 + 4)y = 0$
15. $(x - 2)^2 y'' - (x^2 - 4)y' + (x + 2)y = 0$

16. $x^3(1 - x)y'' + (3x + 2)y' + xy = 0$

Find two linearly independent Frobenius series solutions (for $x > 0$) of each of the differential equations in Problems 17 through 26.

17. $4xy'' + 2y' + y = 0$
18. $2xy'' + 3y' - y = 0$
19. $2xy'' - y' - y = 0$
20. $3xy'' + 2y' + 2y = 0$
21. $2x^2 y'' + xy' - (1 + 2x^2)y = 0$
22. $2x^2 y'' + xy' - (3 - 2x^2)y = 0$
23. $6x^2 y'' + 7xy' - (x^2 + 2)y = 0$
24. $3x^2 y'' + 2xy' + x^2 y = 0$
25. $2xy'' + (1 + x)y' + y = 0$
26. $2xy'' + (1 - 2x^2)y' - 4xy = 0$

Use the method of Example 6 to find two linearly independent Frobenius series solutions of the differential equations in Problems 27 through 31. Then construct a graph showing their graphs for $x > 0$.

27. $xy'' + 2y' + 9xy = 0$
28. $xy'' + 2y' - 4xy = 0$
29. $4xy'' + 8y' + xy = 0$
30. $xy'' - y' + 4x^3 y = 0$
31. $4x^2 y'' - 4xy' + (3 - 4x^2)y = 0$

In Problems 32 through 34, find the first three nonzero terms of each of two linearly independent Frobenius series solutions.

32. $2x^2y'' + x(x+1)y' - (2x+1)y = 0$

33. $(2x^2 + 5x^3)y'' + (3x - x^2)y' - (1+x)y = 0$

34. $2x^2y'' + (\sin x)y' - (\cos x)y = 0$

35. Note that $x = 0$ is an irregular point of the equation

$$x^2y'' + (3x - 1)y' + y = 0.$$

(a) Show that $y = x^r \sum c_n x^n$ can satisfy this equation only if $r = 0$. **(b)** Substitute $y = \sum c_n x^n$ to derive the "formal" solution $y = \sum n! x^n$. What is the radius of convergence of this series?

36. (a) Suppose that A and B are nonzero constants. Show that the equation $x^2y'' + Ay' + By = 0$ has at most one solution of the form $y = x^r \sum c_n x^n$. **(b)** Repeat part (a) with the equation $x^3y'' + Axy' + By = 0$. **(c)** Show that the equation $x^3y'' + Ax^2y' + By = 0$ has no Frobenius series solution. (*Suggestion:* In each case substitute $y = x^r \sum c_n x^n$ in the given equation to determine the possible values of r.)

37. (a) Use the method of Frobenius to derive the solution $y_1 = x$ of the equation $x^3y'' - xy' + y = 0$. **(b)** Verify by substitution the second solution $y_2 = xe^{-1/x}$. Does y_2 have a Frobenius series representation?

38. Apply the method of Frobenius to Bessel's equation of order $\frac{1}{2}$,

$$x^2y'' + xy' + \left(x^2 - \tfrac{1}{4}\right)y = 0,$$

to derive its general solution for $x > 0$,

$$y(x) = c_0 \frac{\cos x}{\sqrt{x}} + c_1 \frac{\sin x}{\sqrt{x}}.$$

Figure 8.3.2 shows the graphs of the two indicated solutions.

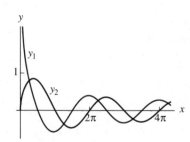

FIGURE 8.3.2. The solutions

$$y_1(x) = \frac{\cos x}{\sqrt{x}} \text{ and } y_2(x) = \frac{\sin x}{\sqrt{x}} \text{ in}$$

Problem 38.

39. (a) Show that Bessel's equation of order 1,

$$x^2y'' + xy' + (x^2 - 1)y = 0,$$

has exponents $r_1 = 1$ and $r_2 = -1$ at $x = 0$, and that the Frobenius series corresponding to $r_1 = 1$ is

$$J_1(x) = \frac{x}{2} \sum_{n=0}^{\infty} \frac{(-1)^n x^{2n}}{n!\,(n+1)!\,2^{2n}}.$$

(b) Show that there is no Frobenius solution corresponding to the smaller exponent $r_2 = -1$; that is, show that it is impossible to determine the coefficients in

$$y_2(x) = x^{-1} \sum_{n=0}^{\infty} c_n x^n.$$

40. Consider the equation $x^2y'' + xy' + (1-x)y = 0$. **(a)** Show that its exponents are $\pm i$, so it has complex-valued Frobenius series solutions

$$y_+ = x^i \sum_{n=0}^{\infty} p_n x^n \quad \text{and} \quad y_- = x^{-i} \sum_{n=0}^{\infty} q_n x^n$$

with $p_0 = q_0 = 1$. **(b)** Show that the recursion formula is

$$c_n = \frac{c_{n-1}}{n^2 + 2rn}.$$

Apply this formula with $r = i$ to obtain $p_n = c_n$, then with $r = -i$ to obtain $q_n = c_n$. Conclude that p_n and q_n are complex conjugates: $p_n = a_n + ib_n$ and $q_n = a_n - ib_n$, where the numbers $\{a_n\}$ and $\{b_n\}$ are real. **(c)** Deduce from part (b) that the differential equation given in this problem has real-valued solutions of the form

$$y_1(x) = A(x)\cos(\ln x) - B(x)\sin(\ln x),$$
$$y_2(x) = A(x)\sin(\ln x) + B(x)\cos(\ln x),$$

where $A(x) = \sum a_n x^n$ and $B(x) = \sum b_n x^n$.

41. Consider the differential equation

$$x(x-1)(x+1)^2 y'' + 2x(x-3)(x+1)y' - 2(x-1)y = 0$$

that appeared in an advertisement for a symbolic algebra program in the March 1984 issue of the *American Mathematical Monthly*. **(a)** Show that $x = 0$ is a regular singular point with exponents $r_1 = 1$ and $r_2 = 0$. **(b)** It follows from Theorem 1 that this differential equation has a power series solution of the form

$$y_1(x) = x + c_2 x^2 + c_3 x^3 + \cdots.$$

Substitute this series (with $c_1 = 1$) in the differential equation to show that $c_2 = -2$, $c_3 = 3$, and

$$c_{n+2} = $$
$$\frac{(n^2 - n)c_{n-1} + (n^2 - 5n - 2)c_n - (n^2 + 7n + 4)c_{n+1}}{(n+1)(n+2)}$$

for $n \geq 2$. **(c)** Use the recurrence relation in part (b) to prove by induction that $c_n = (-1)^{n+1} n$ for $n \geq 1$ (!). Hence deduce (using the geometric series) that

$$y_1(x) = \frac{x}{(1+x)^2}$$

for $0 < x < 1$.

42. This problem is a brief introduction to Gauss's **hypergeometric equation**

$$x(1-x)y'' + [\gamma - (\alpha + \beta + 1)x]y' - \alpha\beta y = 0, \quad (35)$$

where α, β, and γ are constants. This famous equation has wide-ranging applications in mathematics and physics. **(a)** Show that $x = 0$ is a regular singular point of Eq. (35), with exponents 0 and $1 - \gamma$. **(b)** If γ is not zero or a negative integer, it follows (why?) that Eq. (35) has a power series solution

$$y(x) = x^0 \sum_{n=0}^{\infty} c_n x^n = \sum_{n=0}^{\infty} c_n x^n$$

with $c_0 \neq 0$. Show that the recurrence relation for this series is

$$c_{n+1} = \frac{(\alpha + n)(\beta + n)}{(\gamma + n)(1 + n)} c_n$$

for $n \geq 0$. **(c)** Conclude that with $c_0 = 1$ the series in part (b) is

$$y(x) = 1 + \sum_{n=0}^{\infty} \frac{\alpha_n \beta_n}{n! \, \gamma_n} x^n \quad (36)$$

where $\alpha_n = \alpha(\alpha + 1)(\alpha + 2) \cdots (\alpha + n - 1)$ for $n \geq 1$, and β_n and γ_n are defined similarly. **(d)** The series in (36) is known as the **hypergeometric series** and is commonly denoted by $F(\alpha, \beta, \gamma, x)$. Show that

(i) $F(1, 1, 1, x) = \dfrac{1}{1 - x}$ (the geometric series);

(ii) $xF(1, 1, 2, -x) = \ln(1 + x)$;

(iii) $xF\left(\frac{1}{2}, 1, \frac{3}{2}, -x^2\right) = \tan^{-1} x$;

(iv) $F(-k, 1, 1, -x) = (1 + x)^k$ (the binomial series).

8.3 Application: Automating the Frobenius Series Method

Here we illustrate the use of a computer algebra system such as *Maple* to apply the method of Frobenius. More complete versions of this application—illustrating the use of *Maple*, *Mathematica*, and MATLAB—can be found in the computing applications manual that accompanies this text. We consider the differential equation

$$2x^2 y'' + 3xy' - (x^2 + 1)y = 0 \quad (1)$$

of Example 4 in this section, where we found the two indicial roots $r_1 = \frac{1}{2}$ and $r_2 = -1$.

Beginning with the indicial root $r_1 = \frac{1}{2}$, we first write the initial seven terms of a proposed Frobenius series solution:

```
a := array(0..6):
y := x^(1/2)*sum( a[n]*x^(n), n = 0..6);
```

$$y := \sqrt{x} \left(a_0 + a_1 x + a_2 x^2 + a_3 x^3 + a_4 x^4 + a_5 x^5 + a_6 x^6 \right)$$

Then we substitute this series (actually, partial sum) into the left-hand side of Eq. (1).

```
deq1 := 2*x^2*diff(y,x$2) + 3*x*diff(y,x) - (x^2 + 1)*y:
```

Evidently $x^{3/2}$ will factor out upon simplification, so we multiply by $x^{-3/2}$ and then collect coefficients of like powers of x.

```
deq2 := collect( x^(-3/2)*simplify(deq1), x);
```

$$deq2 := -x^7 a_6 - x^6 a_5 + (90a_6 - a_4)x^5 + (-a_3 + 65a_5)x^4 + (-a_2 + 44a_4)x^3 + (-a_1 + 27a_3)x^2 + (14a_2 - a_0)x + 5a_1$$

We see here the equations that the successive coefficients must satisfy. We can select them automatically by defining an array, then filling the elements of this array by equating to zero (in turn) each of the coefficients in the series.

```
eqs := array(0..5):
for n from 0 to 5 do
   eqs[n] := coeff(deq1,x,n) = 0: od:
coeffEqs := convert(eqs, set);
```

$$coeffEqs := \{5a_1 = 0, \ -a_2 + 44a_4 = 0, \ -a_3 + 65a_5 = 0,$$
$$90a_6 - a_4 = 0, \ 14a_2 - a_0 = 0, \ -a_1 + 27a_3 = 0\}$$

We now have a collection of six linear equations relating the seven coefficients (a_0 through a_6). Hence we can proceed to solve for the successive coefficients in terms of a_0.

```
succCoeffs := convert([seq(a[n], n=1..6)], set);
ourCoeffs := solve(coeffEqs, succCoeffs);
```

$$ourCoeffs := \left\{ a_1 = 0, \ a_6 = \frac{1}{55440}a_0, \ a_4 = \frac{1}{616}a_0, \right.$$
$$\left. a_2 = \frac{1}{14}a_0, \ a_5 = 0, \ a_3 = 0 \right\}$$

Thus we get the first particular solution

$$y_1(x) = a_0 x^{1/2} \left(1 + \frac{x^2}{14} + \frac{x^4}{616} + \frac{x^6}{55440} + \cdots \right)$$

found in Example 4. You can now repeat this process, beginning with the indicial root $r_2 = -1$, to derive similarly the second particular solution.

In the following problems, use this method to derive Frobenius series solutions that can be checked against the given known general solutions.

1. $xy'' - y' + 4x^3 y = 0$, $\quad y(x) = A \cos x^2 + B \sin x^2$

2. $xy'' - 2y' + 9x^5 y = 0$, $\quad y(x) = A \cos x^3 + B \sin x^3$

3. $4xy'' - 2y' + y = 0$, $\quad y(x) = A \cos \sqrt{x} + B \sin \sqrt{x}$

4. $xy'' + 2y' + xy = 0$, $\quad y(x) = \dfrac{1}{x}(A \cos x + B \sin x)$

5. $4xy'' + 6y' + y = 0$, $\quad y(x) = \dfrac{1}{\sqrt{x}}\left(A \cos \sqrt{x} + B \sin \sqrt{x}\right)$

6. $x^2 y'' + xy' + (4x^4 - 1)y = 0$, $\quad y(x) = \dfrac{1}{x}(A \cos x^2 + B \sin x^2)$

7. $xy'' + 3y' + 4x^3 y = 0$, $\quad y(x) = \dfrac{1}{x^2}(A \cos x^2 + B \sin x^2)$

8. $x^2 y'' + x^2 y' - 2y = 0$, $\quad y(x) = \dfrac{1}{x}[A(2 - x) + B(2 + x)e^{-x}]$

Problems 9 through 11 involve the arctangent series

$$\tan^{-1} x = x - \frac{x^3}{3} + \frac{x^5}{5} - \frac{x^7}{7} + \cdots .$$

9. $(x + x^3)y'' + (2 + 4x^2)y' + 2xy = 0$, $\quad y(x) = \dfrac{1}{x}(A + B \tan^{-1} x)$

10. $(2x + 2x^2)y'' + (3 + 5x)y' + y = 0$, $\quad y(x) = \dfrac{1}{\sqrt{x}}\left(A + B \tan^{-1} \sqrt{x}\right)$

11. $(x + x^5)y'' + (3 + 7x^4)y' + 8x^3 y = 0$, $\quad y(x) = \dfrac{1}{x^2}(A + B \tan^{-1} x^2)$

8.4 Method of Frobenius: The Exceptional Cases

We continue our discussion of the equation

$$y'' + \frac{p(x)}{x}y' + \frac{q(x)}{x^2}y = 0 \tag{1}$$

where $p(x)$ and $q(x)$ are analytic at $x = 0$, and $x = 0$ is a regular singular point. If the roots r_1 and r_2 of the indicial equation

$$\phi(r) = r(r-1) + p_0 r + q_0 = 0 \tag{2}$$

do not differ by an integer, then Theorem 1 of Section 8.3 guarantees that Eq. (1) has two linearly independent Frobenius series solutions. We consider now the more complex situation in which $r_1 - r_2$ is an integer. If $r_1 = r_2$, then there is only one exponent available, and thus there can be only one Frobenius series solution. But we saw in Example 6 of Section 8.3 that if $r_1 = r_2 + N$, with N a positive integer, then it is possible that a second Frobenius series solution exists. We will also see that it is possible that such a solution does not exist. In fact, the second solution involves $\ln x$ when it is not a Frobenius series. As you will see in Examples 3 and 4, these exceptional cases occur in the solution of Bessel's equation. For applications, this is the most important second-order linear differential equation with variable coefficients.

The Nonlogarithmic Case with $r_1 = r_2 + N$

In Section 8.3 we derived the indicial equation by substituting the power series $p(x) = \sum p_n x^n$ and $q(x) = \sum q_n x^n$ and the Frobenius series

$$y(x) = x^r \sum_{n=0}^{\infty} c_n x^n = \sum_{n=0}^{\infty} c_n x^{n+r} \qquad (c_0 \neq 0) \tag{3}$$

in the differential equation in the form

$$x^2 y'' + xp(x)y' + q(x)y = 0. \tag{4}$$

The result of this substitution, after collection of the coefficients of like powers of x, is an equation of the form

$$\sum_{n=0}^{\infty} F_n(r)x^{n+r} = 0 \tag{5}$$

in which the coefficients depend on r. It turns out that the coefficient of x^r is

$$F_0(r) = [r(r-1) + p_0 r + q_0]c_0 = \phi(r)c_0, \tag{6}$$

which gives the indicial equation because $c_0 \neq 0$ by assumption; also, for $n \geq 1$, the coefficient of x^{n+r} has the form

$$F_n(r) = \phi(r+n)c_n + L_n(r; c_0, c_1, \ldots, c_{n-1}). \tag{7}$$

Here L_n is a certain linear combination of $c_0, c_1, \ldots, c_{n-1}$. Although the exact formula is not necessary for our purposes, it happens that

$$L_n = \sum_{k=0}^{n-1} [(r+k)p_{n-k} + q_{n-k}]c_k. \tag{8}$$

Because all the coefficients in (5) must vanish for the Frobenius series to be a solution of Eq. (4), it follows that the exponent r and the coefficients c_0, c_1, \ldots, c_n must satisfy the equation

$$\phi(r+n)c_n + L_n(r; c_0, c_1, \ldots, c_{n-1}) = 0. \tag{9}$$

This is a *recurrence relation* for c_n in terms of $c_0, c_1, \ldots, c_{n-1}$.

Now suppose that $r_1 = r_2 + N$ with N a positive integer. If we use the larger exponent r_1 in Eq. (9), then the coefficient $\phi(r_1 + n)$ of c_n will be nonzero for every $n \geq 1$ because $\phi(r) = 0$ only when $r = r_1$ and when $r = r_2 < r_1$. Once $c_0, c_1, \ldots, c_{n-1}$ have been determined, we therefore can solve Eq. (9) for c_n and continue to compute successive coefficients in the Frobenius series solution corresponding to the exponent r_1.

But when we use the smaller exponent r_2, there is a potential difficulty in computing c_N. For in this case $\phi(r_2 + N) = 0$, so Eq. (9) becomes

$$0 \cdot c_N + L_N(r_2; c_0, c_1, \ldots, c_{N-1}) = 0. \tag{10}$$

At this stage $c_0, c_1, \ldots, c_{N-1}$ have already been determined. If it happens that

$$L_N(r_2; c_0, c_1, \ldots, c_{N-1}) = 0,$$

then we can choose c_N arbitrarily and continue to determine the remaining coefficients in a second Frobenius series solution. But if it happens that

$$L_N(r_2; c_0, c_1, \ldots, c_{N-1}) \neq 0,$$

then Eq. (10) is not satisfied with any choice of c_N; in this case there cannot exist a second Frobenius series solution corresponding to the smaller exponent r_2. Examples 1 and 2 illustrate these two possibilities.

Example 1 Consider the equation

$$x^2 y'' + (6x + x^2)y' + xy = 0. \tag{11}$$

Here $p_0 = 6$ and $q_0 = 0$, so the indicial equation is

$$\phi(r) = r(r-1) + 6r = r^2 + 5r = 0 \tag{12}$$

with roots $r_1 = 0$ and $r_2 = -5$; the roots differ by the integer $N = 5$. We substitute the Frobenius series $y = \sum c_n x^{n+r}$ and get

$$\sum_{n=0}^{\infty} (n+r)(n+r-1)c_n x^{n+r} + 6\sum_{n=0}^{\infty} (n+r)c_n x^{n+r}$$

$$+ \sum_{n=0}^{\infty} (n+r)c_n x^{n+r+1} + \sum_{n=0}^{\infty} c_n x^{n+r+1} = 0.$$

When we combine the first two and also the last two sums, and in the latter shift the index by -1, the result is

$$\sum_{n=0}^{\infty}[(n+r)^2 + 5(n+r)]c_n x^{n+r} + \sum_{n=1}^{\infty}(n+r)c_{n-1}x^{n+r} = 0.$$

The terms corresponding to $n = 0$ give the indicial equation in (12), whereas for $n \geq 1$ we get the equation

$$[(n+r)^2 + 5(n+r)]c_n + (n+r)c_{n-1} = 0, \qquad (13)$$

which in this example corresponds to the general solution in (9). Note that the coefficient of c_n is $\phi(n+r)$.

We now follow the recommendation in Section 8.3 for the case $r_1 = r_2 + N$: We begin with the smaller root $r_2 = -5$. With $r_2 = -5$, Eq. (13) reduces to

➤ $$n(n-5)c_n + (n-5)c_{n-1} = 0. \qquad (14)$$

If $n \neq 5$, we can solve this equation for c_n to obtain the recurrence relation

➤ $$c_n = -\frac{c_{n-1}}{n} \qquad \text{for } n \neq 5. \qquad (15)$$

This yields

$$c_1 = -c_0, \qquad\qquad c_2 = -\frac{c_1}{2} = \frac{c_0}{2},$$

$$c_3 = \frac{c_2}{3} = -\frac{c_0}{6}, \quad \text{and} \quad c_4 = -\frac{c_3}{4} = \frac{c_0}{24}. \qquad (16)$$

In the case $r_1 = r_2 + N$, it is always the coefficient c_N that requires special consideration. Here $N = 5$, and for $n = 5$ Eq. (14) takes the form $0 \cdot c_5 + 0 = 0$. Hence c_5 is a second arbitrary constant, and we can compute additional coefficients, still using the recursion formula in (15):

$$c_6 = -\frac{c_5}{6}, \quad c_7 = -\frac{c_6}{7} = \frac{c_5}{6\cdot 7}, \quad c_8 = -\frac{c_7}{8} = -\frac{c_5}{6\cdot 7\cdot 8}, \qquad (17)$$

and so on.

When we combine the results in (16) and (17), we get

$$y = x^{-5}\sum_{n=0}^{\infty}c_n x^n$$

$$= c_0 x^{-5}\left(1 - x + \frac{x^2}{2} - \frac{x^3}{6} + \frac{x^4}{24}\right)$$

$$+ c_5 x^{-5}\left(x^5 - \frac{x^6}{6} + \frac{x^7}{6\cdot 7} - \frac{x^8}{6\cdot 7\cdot 8} + \cdots\right)$$

in terms of the two arbitrary constants c_0 and c_5. Thus we have found the two Frobenius series solutions

$$y_1(x) = x^{-5}\left(1 - x + \frac{x^2}{2} - \frac{x^3}{6} + \frac{x^4}{24}\right)$$

and

$$y_2(x) = 1 + \sum_{n=1}^{\infty} \frac{(-1)^n x^n}{6 \cdot 7 \cdots (n+5)} = 1 + 120 \sum_{n=1}^{\infty} \frac{(-1)^n x^n}{(n+5)!}$$

of Eq. (11). ∎

Example 2 Determine whether or not the equation

$$x^2 y'' - xy' + (x^2 - 8)y = 0 \tag{18}$$

has two linearly independent Frobenius series solutions.

Solution Here $p_0 = -1$ and $q_0 = -8$, so the indicial equation is

$$\phi(r) = r(r-1) - r - 8 = r^2 - 2r - 8 = 0$$

with roots $r_1 = 4$ and $r_2 = -2$ differing by $N = 6$. On substitution of $y = \sum c_n x^{n+r}$ in Eq. (18), we get

$$\sum_{n=0}^{\infty} (n+r)(n+r-1)c_n x^{n+r} - \sum_{n=0}^{\infty} (n+r)c_n x^{n+r}$$

$$+ \sum_{n=0}^{\infty} c_n x^{n+r+2} - 8 \sum_{n=0}^{\infty} c_n x^{n+r} = 0.$$

If we shift the index by -2 in the third sum and combine the other three sums, we get

$$\sum_{n=0}^{\infty} [(n+r)^2 - 2(n+r) - 8] c_n x^{n+r} + \sum_{n=2}^{\infty} c_{n-2} x^{n+r} = 0.$$

The coefficient of x^r gives the indicial equation, and the coefficient of x^{r+1} gives

$$\left[(r+1)^2 - 2(r+1) - 8 \right] c_1 = 0.$$

Because the coefficient of c_1 is nonzero both for $r = 4$ and for $r = -2$, it follows that $c_1 = 0$ in each case. For $n \geq 2$ we get the equation

$$\left[(n+r)^2 - 2(n+r) - 8 \right] c_n + c_{n-2} = 0, \tag{19}$$

which corresponds in this example to the general equation in (9); note that the coefficient of c_n is $\phi(n+r)$.

We work first with the smaller root $r = r_2 = -2$. Then Eq. (19) becomes

$$n(n-6)c_n + c_{n-2} = 0 \tag{20}$$

for $n \geq 2$. For $n \neq 6$ we can solve for the recurrence relation

$$c_n = -\frac{c_{n-2}}{n(n-6)} \qquad (n \geq 2, \ n \neq 6). \tag{21}$$

Because $c_1 = 0$, this formula gives

$$c_2 = \frac{c_0}{8}, \qquad\qquad c_3 = 0,$$

$$c_4 = \frac{c_2}{8} = \frac{c_0}{64}, \qquad \text{and} \qquad c_5 = 0.$$

Now Eq. (20) with $n = 6$ reduces to

$$0 \cdot c_6 + \frac{c_0}{64} = 0.$$

But $c_0 \neq 0$ by assumption, and hence there is no way to choose c_6 so that this equation holds. Thus there is *no* Frobenius series solution corresponding to the smaller root $r_2 = -2$.

To find the single Frobenius series solution corresponding to the larger root $r_1 = 4$, we substitute $r = 4$ in Eq. (19) to obtain the recurrence relation

➤
$$c_n = -\frac{c_{n-2}}{n(n+6)} \qquad (n \geq 2). \tag{22}$$

This gives

$$c_2 = -\frac{c_0}{2 \cdot 8}, \quad c_4 = -\frac{c_2}{4 \cdot 10} = \frac{c_0}{2 \cdot 4 \cdot 8 \cdot 10}.$$

The general pattern is

$$c_{2n} = \frac{(-1)^n c_0}{2 \cdot 4 \cdots (2n) \cdot 8 \cdot 10 \cdots (2n+6)} = \frac{(-1)^n 6 c_0}{2^{2n} n! \, (n+3)!}.$$

This yields the Frobenius series solution

$$y_1(x) = x^4 \left(1 + 6 \sum_{n=1}^{\infty} \frac{(-1)^n x^{2n}}{2^{2n} n! \, (n+3)!} \right)$$

of Eq. (18). ∎

Reduction of Order

When only a single Frobenius series solution exists, we need an additional technique. We discuss here the method of **reduction of order**, which enables us to use one known solution y_1 of a second-order homogeneous linear differential equation to find a second linearly independent solution y_2. Consider the second-order equation

➤
$$y'' + P(x)y' + Q(x)y = 0 \tag{23}$$

on an open interval I on which P and Q are continuous. Suppose that we know one solution y_1 of Eq. (23). By Theorem 2 of Section 3.1, there exists a second linearly independent solution y_2; our problem is to find y_2. Equivalently, we would like to find the quotient

$$v(x) = \frac{y_2(x)}{y_1(x)}. \tag{24}$$

Once we know $v(x)$, y_2 will then be given by

$$y_2(x) = v(x)y_1(x). \tag{25}$$

We begin by substituting the expression in (25) in Eq. (23), using the derivatives

$$y_2' = vy_1' + v'y_1 \quad \text{and} \quad y_2'' = vy_1'' + 2v'y_1' + v''y_1.$$

We get

$$\left[vy_1'' + 2v'y_1' + v''y_1\right] + P\left[vy_1' + v'y_1\right] + Qvy_1 = 0,$$

and rearrangement gives

$$v\left[y_1'' + Py_1' + Qy_1\right] + v''y_1 + 2v'y_1' + Pv'y_1 = 0.$$

But the bracketed expression in this last equation vanishes because y_1 is a solution of Eq. (23). This leaves the equation

$$v''y_1 + (2y_1' + Py_1)v' = 0. \tag{26}$$

The key to the success of this method is that Eq. (26) is *linear* in v'. Thus the substitution in (25) has reduced the second-order linear equation in (23) to the first-order (in v') linear equation in (26). If we write $u = v'$ and assume that $y_1(x)$ never vanishes on I, then Eq. (26) yields

$$u' + \left(2\frac{y_1'}{y_1} + P(x)\right)u = 0. \tag{27}$$

An integrating factor for Eq. (27) is

$$\rho = \exp\left(\int\left(2\frac{y_1'}{y_1} + P(x)\right)dx\right) = \exp\left(2\ln|y_1| + \int P(x)\,dx\right);$$

thus

$$\rho(x) = y_1^2 \exp\left(\int P(x)\,dx\right).$$

We now integrate the equation in (27) to obtain

$$uy_1^2 \exp\left(\int P(x)\,dx\right) = C, \quad \text{so} \quad v' = u = \frac{C}{y_1^2}\exp\left(-\int P(x)\,dx\right).$$

Another integration now gives

$$\frac{y_2}{y_1} = v = C\int \frac{\exp\left(-\int P(x)\,dx\right)}{y_1^2}\,dx + K.$$

With the particular choices $C = 1$ and $K = 0$, we get

$$y_2 = y_1 \int \frac{\exp\left(-\int P(x)\,dx\right)}{y_1^2}\,dx. \tag{28}$$

This formula provides a second solution $y_2(x)$ of Eq. (23) on any interval where $y_1(x)$ is never zero. Note that because an exponential function never vanishes, $y_2(x)$ is a nonconstant multiple of $y_1(x)$, so y_1 and y_2 are linearly independent solutions.

The Logarithmic Cases

We now investigate the general form of the second solution of the equation

➤
$$y'' + \frac{p(x)}{x}y' + \frac{q(x)}{x^2}y = 0, \tag{1}$$

under the assumption that its exponents r_1 and $r_2 = r_1 - N$ differ by the integer $N \geq 0$. We assume that we have already found the Frobenius series solution

➤
$$y_1(x) = x^{r_1}\sum_{n=0}^{\infty} a_n x^n \qquad (a_0 \neq 0) \tag{29}$$

for $x > 0$ corresponding to the larger exponent r_1. Let us write $P(x)$ for $p(x)/x$ and $Q(x)$ for $q(x)/x^2$. Thus we can rewrite Eq. (1) in the form $y'' + Py' + Qy = 0$ of Eq. (23).

Because the indicial equation has roots r_1 and $r_2 = r_1 - N$, it can be factored easily:

$$r^2 + (p_0 - 1)r + q_0 = (r - r_1)(r - r_1 + N)$$
$$= r^2 + (N - 2r_1)r + (r_1^2 - r_1 N) = 0,$$

so we see that

$$p_0 - 1 = N - 2r_1;$$

that is,

$$-p_0 - 2r_1 = -1 - N. \tag{30}$$

In preparation for use of the reduction of order formula in (28), we write

$$P(x) = \frac{p_0 + p_1 x + p_2 x^2 + \cdots}{x} = \frac{p_0}{x} + p_1 + p_2 x + \cdots$$

Then

$$\exp\left(-\int P(x)\,dx\right) = \exp\left(-\int\left[\frac{p_0}{x} + p_1 + p_2 x + \cdots\right]dx\right)$$
$$= \exp\left(-p_0\ln x - p_1 x - \tfrac{1}{2}p_2 x^2 - \cdots\right)$$
$$= x^{-p_0}\exp\left(-p_1 x - \tfrac{1}{2}p_2 x^2 - \cdots\right),$$

so that

$$\exp\left(-\int P(x)\,dx\right) = x^{-p_0}\left(1 + A_1 x + A_2 x^2 + \cdots\right). \tag{31}$$

In the last step we have used the fact that a composition of analytic functions is analytic and therefore has a power series representation; the initial coefficient of that series in (31) is 1 because $e^0 = 1$.

We now substitute (29) and (31) in (28); with the choice $a_0 = 1$ in (29), this yields

$$y_2 = y_1 \int \frac{x^{-p_0}\left(1 + A_1 x + A_2 x^2 + \cdots\right)}{x^{2r_1}\left(1 + a_1 x + a_2 x^2 + \cdots\right)^2}\,dx.$$

We expand the denominator and simplify:

$$y_2 = y_1 \int \frac{x^{-p_0 - 2r_1} \left(1 + A_1 x + A_2 x^2 + \cdots\right)}{1 + B_1 x + B_2 x^2 + \cdots}\, dx$$

$$= y_1 \int x^{-1-N} \left(1 + C_1 x + C_2 x^2 + \cdots\right)\, dx \tag{32}$$

(we used long division and (30) in the last step). We now consider separately the cases $N = 0$ and $N > 0$. We want to ascertain the general form of y_2 without keeping track of specific coefficients.

CASE 1: EQUAL EXPONENTS ($r_1 = r_2$). With $N = 0$, Eq. (32) gives

$$y_2 = y_1 \int \left(\frac{1}{x} + C_1 + C_2 x^2 + \cdots\right) dx$$

$$= y_1 \ln x + y_1 \left(C_1 x + \tfrac{1}{2} C_2 x^2 + \cdots\right)$$

$$= y_1 \ln x + x^{r_1} \left(1 + a_1 x + \cdots\right)\left(C_1 x + \tfrac{1}{2} C_2 x^2 + \cdots\right)$$

$$= y_1 \ln x + x^{r_1} \left(b_0 x + b_1 x^2 + b_2 x^3 + \cdots\right).$$

Consequently, in the case of equal exponents, the general form of y_2 is

➤ $$y_2(x) = y_1(x) \ln x + x^{1+r_1} \sum_{n=0}^{\infty} b_n x^n. \tag{33}$$

Note the logarithmic term; it is always present when $r_1 = r_2$.

CASE 2: POSITIVE INTEGRAL DIFFERENCE ($r_1 = r_2 + N$). With $N > 0$, Eq. (32) gives

$$y_2 = y_1 \int x^{-1-N} \left(1 + C_1 x + C_2 x^2 + \cdots + C_N x^N + \cdots\right) dx$$

$$= y_1 \int \left(\frac{C_N}{x} + \frac{1}{x^{N+1}} + \frac{C_1}{x^N} + \cdots\right) dx$$

$$= C_N y_1 \ln x + y_1 \left(\frac{x^{-N}}{-N} + \frac{C_1 x^{-N+1}}{-N+1} + \cdots\right)$$

$$= C_N y_1 \ln x + x^{r_2+N} \left(\sum_{n=0}^{\infty} a_n x^n\right) x^{-N} \left(-\frac{1}{N} + \frac{C_1 x}{-N+1} + \cdots\right),$$

so that

➤ $$y_2(x) = C_N y_1(x) \ln x + x^{r_2} \sum_{n=0}^{\infty} b_n x^n, \tag{34}$$

where $b_0 = -a_0/N \neq 0$. This gives the general form of y_2 in the case of exponents differing by a positive integer. Note the coefficient C_N that appears in (34) but not

in (33). If it happens that $C_N = 0$, then there is no logarithmic term; if so, Eq. (1) has a second Frobenius series solution (as in Example 1).

In our derivation of Eqs. (33) and (34)—which exhibit the general form of the second solution in the cases $r_1 = r_2$ and $r_1 - r_2 = N > 0$, respectively—we have said nothing about the radii of convergence of the various power series that appear. Theorem 1 (next) is a summation of the preceding discussion and also tells where the series in (33) and (34) converge. As in Theorem 1 of Section 8.3, we restrict our attention to solutions for $x > 0$.

THEOREM 1: The Exceptional Cases

Suppose that $x = 0$ is a regular singular point of the equation

$$x^2 y'' + x p(x) y' + q(x) y = 0. \tag{4}$$

Let $\rho > 0$ denote the minimum of the radii of convergence of the power series

$$p(x) = \sum_{n=0}^{\infty} p_n x^n \quad \text{and} \quad q(x) = \sum_{n=0}^{\infty} q_n x^n.$$

Let r_1 and r_2 be the roots, with $r_1 \geq r_2$, of the indicial equation

$$r(r - 1) + p_0 r + q_0 = 0.$$

(a) If $r_1 = r_2$, then Eq. (4) has two solutions y_1 and y_2 of the forms

$$\blacktriangleright \qquad y_1(x) = x^{r_1} \sum_{n=0}^{\infty} a_n x^n \qquad (a_0 \neq 0) \tag{35a}$$

and

$$\blacktriangleright \qquad y_2(x) = y_1(x) \ln x + x^{r_1+1} \sum_{n=0}^{\infty} b_n x^n. \tag{35b}$$

(b) If $r_1 - r_2 = N$, a positive integer, then Eq. (4) has two solutions y_1 and y_2 of the forms

$$\blacktriangleright \qquad y_1(x) = x^{r_1} \sum_{n=0}^{\infty} a_n x^n \qquad (a_0 \neq 0) \tag{36a}$$

and

$$\blacktriangleright \qquad y_2(x) = C y_1(x) \ln x + x^{r_2} \sum_{n=0}^{\infty} b_n x^n. \tag{36b}$$

In Eq. (36b), $b_0 \neq 0$ but C may be either zero or nonzero, so the logarithmic term may or may not actually be present in this case. The radii of convergence of the power series of this theorem are all at least ρ. The coefficients in these series (and the constant C in Eq. (36b)) may be determined by direct substitution of the series in the differential equation in (4). ▲

Example 3 We will illustrate the case $r_1 = r_2$ by deriving the second solution of Bessel's equation of order zero,

$$x^2 y'' + x y' + x^2 y = 0, \tag{37}$$

for which $r_1 = r_2 = 0$. In Example 5 of Section 8.3 we found the first solution

$$y_1(x) = J_0(x) = \sum_{n=0}^{\infty} \frac{(-1)^n x^{2n}}{2^{2n}(n!)^2} \tag{38}$$

According to Eq. (35b) the second solution will have the form

$$y_2 = y_1 \ln x + \sum_{n=1}^{\infty} b_n x^n. \tag{39}$$

The first two derivatives of y_2 are

$$y_2' = y_1' \ln x + \frac{y_1}{x} + \sum_{n=1}^{\infty} n b_n x^{n-1}$$

and

$$y_2'' = y_1'' \ln x + \frac{2 y_1'}{x} - \frac{y_1}{x^2} + \sum_{n=2}^{\infty} n(n-1) b_n x^{n-2}.$$

We substitute these in Eq. (37) and use the fact that $J_0(x)$ also satisfies this equation to obtain

$$
\begin{aligned}
0 &= x^2 y_2'' + x y_2' + x^2 y_2 \\
&= \left[x^2 y_1'' + x y_1' + x^2 y_1 \right] \ln x + 2 x y_1' \\
&\quad + \sum_{n=2}^{\infty} n(n-1) b_n x^n + \sum_{n=1}^{\infty} n b_n x^n + \sum_{n=1}^{\infty} b_n x^{n+2},
\end{aligned}
$$

and it follows that

$$0 = 2 \sum_{n=1}^{\infty} \frac{(-1)^n 2n x^{2n}}{2^{2n}(n!)^2} + b_1 x + 2^2 b_2 x^2 + \sum_{n=3}^{\infty} (n^2 b_n + b_{n-2}) x^n. \tag{40}$$

The only term involving x in Eq. (40) is $b_1 x$, so $b_1 = 0$. But $n^2 b_n + b_{n-2} = 0$ if n is odd, and it follows that all the coefficients of odd subscript in y_2 vanish.

Now we examine the coefficients with even subscripts in Eq. (40). First we see that

$$b_2 = -2 \cdot \frac{(-1)(2)}{2^2 \cdot 2^2 \cdot (1!)^2} = \frac{1}{4}. \tag{41}$$

For $n \geq 2$, we read the recurrence relation

$$(2n)^2 b_{2n} + b_{2n-2} = -\frac{(2)(-1)^n (2n)}{2^{2n}(n!)^2} \tag{42}$$

from (40). Nonhomogeneous recurrence relations such as (42) are typical of the exceptional cases of the method of Frobenius, and their solution often requires a bit of ingenuity. The usual strategy depends on detecting the most conspicuous dependence of b_{2n} on n. We note the presence of $2^{2n}(n!)^2$ on the right-hand side in (42); in conjunction with the coefficient $(2n)^2$ on the left-hand side, we are induced to think of b_{2n} as *something* divided by $2^{2n}(n!)^2$. Noting also the alternation of sign, we make the substitution

$$b_{2n} = \frac{(-1)^{n+1}c_{2n}}{2^{2n}(n!)^2}, \tag{43}$$

in the expectation that the recurrence relation for c_{2n} will be simpler than the one for b_{2n}. We chose $(-1)^{n+1}$ rather than $(-1)^n$ because $b_2 = \frac{1}{4} > 0$; with $n = 1$ in (43), we get $c_2 = 1$. Substitution of (43) in (42) gives

$$(2n)^2 \frac{(-1)^{n+1}c_{2n}}{2^{2n}(n!)^2} + \frac{(-1)^n c_{2n-2}}{2^{2n-2}[(n-1)!]^2} = \frac{(-2)(-2)^n(2n)}{2^{2n}(n!)^2},$$

which boils down to the extremely simple recurrence relation

$$c_{2n} = c_{2n-2} + \frac{1}{n}.$$

Thus

$$c_4 = c_2 + \frac{1}{2} = 1 + \frac{1}{2},$$

$$c_6 = c_4 + \frac{1}{3} = 1 + \frac{1}{2} + \frac{1}{3},$$

$$c_8 = c_6 + \frac{1}{4} = 1 + \frac{1}{2} + \frac{1}{3} + \frac{1}{4},$$

and so on. Evidently,

$$c_{2n} = 1 + \frac{1}{2} + \frac{1}{3} + \cdots + \frac{1}{n} = H_n, \tag{44}$$

where we denote by H_n the nth partial sum of the harmonic series $\sum(1/n)$.

Finally, keeping in mind that the coefficients of odd subscript are all zero, we substitute (43) and (44) in (39) to obtain the second solution

$$y_2(x) = J_0(x)\ln x + \sum_{n=1}^{\infty} \frac{(-1)^{n+1}H_n x^{2n}}{2^{2n}(n!)^2}$$

$$= J_0(x)\ln x + \frac{x^2}{4} - \frac{3x^4}{128} + \frac{11x^6}{13824} - \cdots \tag{45}$$

of Bessel's equation of order zero. The power series in (45) converges for all x. The most commonly used linearly independent [of $J_0(x)$] second solution is

$$Y_0(x) = \frac{2}{\pi}(\gamma - \ln 2)y_1 + \frac{2}{\pi}y_2;$$

that is,

$$Y_0(x) = \frac{2}{\pi} \left[\left(\gamma + \ln \frac{x}{2} \right) J_0(x) + \sum_{n=1}^{\infty} \frac{(-1)^{n+1} H_n x^{2n}}{2^{2n}(n!)^2} \right], \tag{46}$$

where γ denotes *Euler's constant*:

$$\gamma = \lim_{n \to \infty} (H_n - \ln n) \approx 0.57722. \tag{47}$$

This particular combination $Y_0(x)$ is chosen because of its nice behavior as $x \to +\infty$; it is called the **Bessel function of order zero of the second kind.** ∎

Example 4 As an alternative to the method of substitution, we illustrate the case $r_1 - r_2 = N$ by employing the technique of reduction of order to derive a second solution of Bessel's equation of order 1,

$$x^2 y'' + xy' + (x^2 - 1)y = 0; \tag{48}$$

the associated indicial equation has roots $r_1 = 1$ and $r_2 = -1$. According to Problem 39 of Section 8.3, one solution of Eq. (48) is

$$y_1(x) = J_1(x) = \frac{x}{2} \sum_{n=0}^{\infty} \frac{(-1)^n x^{2n}}{2^{2n} n! \, (n+1)!} = \frac{x}{2} - \frac{x^3}{16} + \frac{x^5}{384} - \frac{x^7}{18432} + \cdots. \tag{49}$$

With $P(x) = 1/x$ from (48), the reduction of order formula in (28) yields

$$y_2 = y_1 \int \frac{1}{xy_1^2} \, dx$$

$$= y_1 \int \frac{1}{x(x/2 - x^3/16 + x^5/384 - x^7/18432 + \cdots)^2} \, dx$$

$$= y_1 \int \frac{4}{x^3(1 - x^2/8 + x^4/192 - x^6/9216 + \cdots)^2} \, dx$$

$$= 4y_1 \int \frac{1}{x^3(1 - x^2/4 + 5x^4/192 - 7x^6/4608 + \cdots)} \, dx$$

$$= 4y_1 \int \frac{1}{x^3} \left(1 + \frac{x^2}{4} + \frac{7x^4}{192} + \frac{19x^6}{4608} + \cdots \right) dx$$

$$= 4y_1 \int \left(\frac{1}{4x} + \frac{1}{x^3} + \frac{7x}{192} + \frac{19x^3}{4608} + \cdots \right) dx$$

$$= y_1 \ln x + 4y_1 \left(-\frac{1}{2x^2} + \frac{7x^2}{384} + \frac{19x^4}{18432} + \cdots \right).$$

Thus

$$y_2(x) = y_1(x) \ln x - \frac{1}{x} + \frac{x}{8} + \frac{x^3}{32} - \frac{11x^5}{4608} + \cdots. \tag{50}$$

Note that the technique of reduction of order readily yields the first several terms of the series, but does not provide a recurrence relation that can be used to determine the general term of the series.

With a computation similar to that shown in Example 3 (but more complicated —see Problem 21), the method of substitution can be used to derive the solution

$$y_3(x) = y_1(x) \ln x - \frac{1}{x} + \sum_{n=1}^{\infty} \frac{(-1)^n (H_n + H_{n-1}) x^{2n-1}}{2^{2n} n! (n-1)!}, \tag{51}$$

where H_n is defined in (44) for $n \geq 1$; $H_0 = 0$. The reader can verify that the terms shown in Eq. (50) agree with

$$y_2(x) = \frac{3}{4} J_1(x) + y_3(x). \tag{52}$$

The most commonly used linearly independent [of J_1] solution of Bessel's equation of order 1 is the combination

$$Y_1(x) = \frac{2}{\pi} (\gamma - \ln 2) y_1(x) + \frac{2}{\pi} y_3(x)$$

$$= \frac{2}{\pi} \left[\left(\gamma + \ln \frac{x}{2} \right) J_1(x) - \frac{1}{x} + \sum_{n-1}^{\infty} \frac{(-1)^n (H_n + H_{n-1}) x^{2n-1}}{2^{2n} n! (n-1)!} \right]. \tag{53}$$

∎

Examples 3 and 4 illustrate two methods of finding the solution in the logarithmic cases— direct substitution and reduction of order. A third alternative is outlined in Problem 19.

8.4 Problems

In Problems 1 through 8, either apply the method of Example 1 to find two linearly independent Frobenius series solutions, or find one such solution and show (as in Example 2) that a second such solution does not exist.

1. $xy'' + (3 - x)y' - y = 0$
2. $xy'' + (5 - x)y' - y = 0$
3. $xy'' + (5 + 3x)y' + 3y = 0$
4. $5xy'' + (30 + 3x)y' + 3y = 0$
5. $xy'' - (4 + x)y' + 3y = 0$
6. $2xy'' - (6 + 2x)y' + y = 0$
7. $x^2 y'' + (2x + 3x^2)y' - 2y = 0$
8. $x(1 - x)y'' - 3y' + 2y = 0$

In Problems 9 through 14, first find the first four nonzero terms in a Frobenius series solution of the given differential equation. Then use the reduction of order technique (as in Example 4) to find the logarithmic term and the first three nonzero terms in a second linearly independent solution.

9. $xy'' + y' - xy = 0$
10. $x^2 y'' - xy' + (x^2 + 1)y = 0$
11. $x^2 y'' + (x^2 - 3x)y' + 4y = 0$

12. $x^2 y'' + x^2 y' - 2y = 0$
13. $x^2 y'' + (2x^2 - 3x)y' + 3y = 0$
14. $x^2 y'' + x(1 + x)y' - 4y = 0$
15. Begin with

$$J_0(x) = 1 - \frac{x^2}{4} + \frac{x^4}{64} - \frac{x^6}{2304} + \cdots.$$

Using the method of reduction of order, derive the second linearly independent solution

$$y_2(x) = J_0(x) \ln x + \frac{x^2}{4} - \frac{3x^4}{128} + \frac{11x^6}{13284} - \cdots$$

of Bessel's equation of order zero.

16. Find two linearly independent Frobenius series solutions of Bessel's equation of order $\frac{3}{2}$,

$$x^2 y'' + xy' + \left(x^2 - \frac{9}{4} \right) y = 0.$$

17. (a) Verify that $y_1(x) = xe^x$ is one solution of

$$x^2 y'' - x(1 + x)y' + y = 0.$$

(b) Note that $r_1 = r_2 = 1$. Substitute

$$y_2 = y_1 \ln x + \sum_{n=1}^{\infty} b_n x^{n+1}$$

in the differential equation to deduce that $b_1 = -1$ and that

$$nb_n - b_{n-1} = -\frac{1}{n!} \quad \text{for } n \geq 2.$$

(c) Substitute $b_n = c_n/n!$ in this recurrence relation and conclude from the result that $c_n = -H_n$. Thus the second solution is

$$y_2(x) = xe^x \ln x - \sum_{n=1}^{\infty} \frac{H_n x^{n+1}}{n!}.$$

18. Consider the equation $xy'' - y = 0$, which has exponents $r_1 = 1$ and $r_2 = 0$ at $x = 0$. **(a)** Derive the Frobenius series solution

$$y_1(x) = \sum_{n=1}^{\infty} \frac{x^n}{n!\,(n-1)!}.$$

(b) Substitute

$$y_2 = Cy_1 \ln x + \sum_{n=0}^{\infty} b_n x^n$$

in the equation $xy'' - y = 0$ to derive the recurrence relation

$$n(n+1)b_{n+1} - b_n = -\frac{2n+1}{(n+1)!\,n!}C.$$

Conclude from this result that a second solution is

$$y_2(x) = y_1(x) \ln x + 1 - \sum_{n=1}^{\infty} \frac{H_n + H_{n-1}}{n!\,(n-1)!}x^n.$$

19. Suppose that the differential equation

$$L[y] = x^2 y'' + xp(x)y' + q(x)y = 0 \tag{54}$$

has equal exponents $r_1 = r_2$ at the regular singular point $x = 0$, so that its indicial equation is

$$\phi(r) = (r - r_1)^2 = 0.$$

Let $c_0 = 1$ and define $c_n(r)$ for $n \geq 1$ by using Eq. (9); that is,

$$c_n(r) = -\frac{L_n(r; c_0, c_1, \ldots, c_{n-1})}{\phi(r+n)}. \tag{55}$$

Then define the function $y(x, r)$ of x and r to be

$$y(x, r) = \sum_{n=0}^{\infty} c_n(r) x^{n+r}. \tag{56}$$

(a) Deduce from the discussion preceding Eq. (9) that

$$L[y(x, r)] = x^r (r - r_1)^2. \tag{57}$$

Hence deduce that

$$y_1 = y(x, r_1) = \sum_{n=0}^{\infty} c_n(r_1) x^{n+r_1} \tag{58}$$

is one solution of Eq. (54). **(b)** Differentiate Eq. (57) with respect to r to show that

$$L[y_r(x, r_1)] = \frac{\partial}{\partial r}\left[x^r (r - r_1)^2\right]\Bigg|_{r=r_1} = 0.$$

Deduce that $y_2 = y_r(x, r_1)$ is a second solution of Eq. (54). **(c)** Differentiate Eq. (58) with respect to r to show that

$$y_2 = y_1 \ln x + x^{r_1} \sum_{n=1}^{\infty} c_n'(r_1) x^n. \tag{59}$$

20. Use the method of Problem 19 to derive both the solutions in (38) and (45) of Bessel's equation of order zero. The following steps outline this computation. **(a)** Take $c_0 = 1$; show that Eq. (55) reduces in this case to

$$(r+1)^2 c_1(r) = 0 \quad \text{and}$$
$$c_n(r) = -\frac{c_{n-2}(r)}{(n+r)^2} \quad \text{for } n \geqq 2. \tag{60}$$

(b) Next show that $c_1(0) = c_1'(0) = 0$, and then deduce from (60) that $c_n(0) = c_n'(0) = 0$ for n odd. Hence you need to compute $c_n(0)$ and $c_n'(0)$ only for n even. **(c)** Deduce from (60) that

$$c_{2n}(r) = \frac{(-1)^n}{(r+2)^2(r+4)^2 \cdots (r+2n)^2}. \tag{61}$$

With $r = r_1 = 0$ in (58), this gives $J_0(x)$. **(d)** Differentiate (61) to show that

$$c_{2n}'(0) = \frac{(-1)^{n+1} H_n}{2^{2n}(n!)^2}.$$

Substitution of this result in (59) gives the second solution in (45).

21. Derive the logarithmic solution in (51) of Bessel's equation of order 1 by the method of substitution. The following steps outline this computation. **(a)** Substitute

$$y_2 = CJ_1(x) \ln x + x^{-1}\left(1 + \sum_{n=1}^{\infty} b_n x^n\right)$$

in Bessel's equation to obtain

$$-b_1 + x + \sum_{n=2}^{\infty}[(n^2 - 1)b_{n+1} + b_{n-1}]x^n$$

$$+ C\left[x + \sum_{n=1}^{\infty} \frac{(-1)^n (2n+1)x^{2n+1}}{2^{2n}(n+1)!\,n!}\right] = 0. \tag{62}$$

(b) Deduce from Eq. (62) that $C = -1$ and that $b_n = 0$ for n odd. **(c)** Next deduce the recurrence relation

$$\left[(2n+1)^2 - 1\right] b_{2n+2} + b_{2n} = \frac{(-1)^n (2n+1)}{2^{2n}(n+1)!\,n!} \qquad (63)$$

for $n \geq 1$. Note that if b_2 is chosen arbitrarily, then b_{2n} is determined for all $n > 1$. **(d)** Take $b_2 = \frac{1}{4}$ and substitute

$$b_{2n} = \frac{(-1)^n c_{2n}}{2^{2n}(n-1)!\,n!}$$

in Eq. (63) to obtain

$$c_{2n+2} - c_{2n} = \frac{1}{n+1} + \frac{1}{n}.$$

(e) Note that $c_2 = 1 = H_1 + H_0$ and deduce that

$$c_{2n} = H_n + H_{n-1}.$$

8.4 Application: The Exceptional Case by Reduction of Order

Here we illustrate the use of a computer algebra system such as *Mathematica* to implement the reduction of order formula in Eq. (28) of this section. More complete versions of this application—using *Maple, Mathematica*, and MATLAB—can be found in the computing applications manual that accompanies this text. To illustrate the method, we will derive a second solution of Bessel's equation of order zero, beginning with the known power series solution

$$J_0(x) = \sum_{n=0}^{\infty} (-1)^n \frac{x^{2n}}{2^{2n}(n!)^2},$$

which we enter in the form

```
y1 = Sum[((-x^2/4)^n) / (n!)^2, {n, 0, 5}] + O[x]^12
```

Then with

```
P = 1/x;
```

we need only substitute in the integral in Eq. (28):

```
integral = Integrate[ Exp[ -Integrate[P,x]]/y1^2, x ]
```

$$\log(x) + \frac{x^2}{4} + \frac{5x^4}{128} + \frac{23x^6}{3456} + \frac{677x^8}{589824} + \frac{7313x^{10}}{36864000} + O(x^{12})$$

Then—multiplying J0 = y1 times the logarithm and the series terms separately—the computation

```
y2 = J0*Log[x] + y1*(integral - Log[x])
```

yields the second solution

$$J_0(x)\log(x) + \frac{x^2}{4} - \frac{3x^4}{128} + \frac{11x^6}{13824} - \frac{25x^8}{1769472} + \frac{137x^{10}}{884736000} + O(x^{12})$$

of Bessel's equation of order zero (as we see it in Eq. (45) of the text).

After verifying (with your computer algebra system) the computations we present here, you can begin with the power series for $J_1(x)$ in Eq. (49) of this section and derive similarly the second solution in (50) of Bessel's equation of order 1. Problems 9 through 14 can also be partially automated in this way.

8.5 Bessel's Equation

We have already seen several cases of Bessel's equation of order $p \geq 0$,

$$x^2 y'' + xy' + (x^2 - p^2)y = 0. \tag{1}$$

Its solutions are now called Bessel functions of order p. Such functions first appeared in the 1730s in the work of Daniel Bernoulli and Euler on the oscillations of a vertically suspended chain. The equation itself appears in a 1764 article by Euler on the vibrations of a circular drumhead, and Fourier used Bessel functions in his classical treatise on heat (1822). But their general properties were first studied systematically in an 1824 memoir by the German astronomer and mathematician Friedrich W. Bessel (1784–1846), who was investigating the motion of planets. The standard source of information on Bessel functions is G. N. Watson's *A Treatise on the Theory of Bessel Functions*, 2nd ed. (Cambridge: Cambridge University Press, 1944). Its 36 pages of references, which cover only the period up to 1922, give some idea of the vast literature of this subject.

Bessel's equation in (1) has indicial equation $r^2 - p^2 = 0$, with roots $r = \pm p$. If we substitute $y = \sum c_m x^{m+r}$ in Eq. (1), we find in the usual manner that $c_1 = 0$ and that

$$\left[(m+r)^2 - p^2 \right] c_m + c_{m-2} = 0 \tag{2}$$

for $m \geq 2$. The verification of Eq. (2) is left to the reader (Problem 6).

The Case $r = p > 0$

If we use $r = p$ and write a_m in place of c_m, then Eq. (2) yields the recursion formula

$$a_m = -\frac{a_{m-2}}{m(2p+m)}. \tag{3}$$

Because $a_1 = 0$, it follows that $a_m = 0$ for all odd values of m. The first few even coefficients are

$$a_2 = -\frac{a_0}{2(2p+2)} = -\frac{a_0}{2^2(p+1)},$$

$$a_4 = -\frac{a_2}{4(2p+4)} = \frac{a_0}{2^4 \cdot 2(p+1)(p+2)},$$

$$a_6 = -\frac{a_4}{6(2p+6)} = -\frac{a_0}{2^6 \cdot 2 \cdot 3(p+1)(p+2)(p+3)}.$$

The general pattern is

$$a_{2m} = \frac{(-1)^m a_0}{2^{2m} m! (p+1)(p+2) \cdots (p+m)},$$

so with the larger root $r = p$ we get the solution

$$y_1(x) = a_0 \sum_{m=0}^{\infty} \frac{(-1)^m x^{2m+p}}{2^{2m} m! (p+1)(p+2) \cdots (p+m)}. \tag{4}$$

If $p = 0$ this is the only Frobenius series solution; with $a_0 = 1$ as well, it is the function $J_0(x)$ we have seen before.

The Case $r = -p < 0$

If we use $r = -p$ and write b_m in place of c_m, Eq. (2) takes the form

$$m(m - 2p)b_m + b_{m-2} = 0 \tag{5}$$

for $m \geq 2$, whereas $b_1 = 0$. We see that there is a potential difficulty if it happens that $2p$ is a positive integer—that is, if p is either a positive integer or an odd positive integral multiple of $\frac{1}{2}$. For then when $m = 2p$, Eq. (5) is simply $0 \cdot b_m + b_{m-2} = 0$. Thus if $b_{m-2} \neq 0$, then no value of b_m can satisfy this equation.

But if p is an odd positive integral multiple of $\frac{1}{2}$, we can circumvent this difficulty. For suppose that $p = k/2$ where k is an odd positive integer. Then we need only choose $b_m = 0$ for all odd values of m. The crucial step is the kth step,

$$k(k - k)b_k + b_{k-2} = 0;$$

and this equation will hold because $b_k = b_{k-2} = 0$.

Hence if p is *not* a positive integer, we take $b_m = 0$ for m odd and define the coefficients of even subscript in terms of b_0 by means of the recursion formula

$$b_m = -\frac{b_{m-2}}{m(m - 2p)}, \qquad m \geq 2. \tag{6}$$

In comparing (6) with (3), we see that (6) will lead to the same result as that in (4), except with p replaced with $-p$. Thus in this case we obtain the second solution

$$y_2(x) = b_0 \sum_{m=0}^{\infty} \frac{(-1)^m x^{2m-p}}{2^{2m} m! \, (-p + 1)(-p + 2) \cdots (-p + m)}. \tag{7}$$

The series in (4) and (7) converge for all $x > 0$ because $x = 0$ is the only singular point of Bessel's equation. If $p > 0$, then the leading term in y_1 is $a_0 x^p$, whereas the leading term in y_2 is $b_0 x^{-p}$. Hence $y_1(0) = 0$, but $y_2(x) \to \pm\infty$ as $x \to 0$, so it is clear that y_1 and y_2 are linearly independent solutions of Bessel's equation of order $p > 0$.

The Gamma Function

The formulas in (4) and (7) can be simplified by use of the **gamma function** $\Gamma(x)$, which is defined for $x > 0$ by

$$\Gamma(x) = \int_0^{\infty} e^{-t} t^{x-1} \, dt. \tag{8}$$

It is not difficult to show that this improper integral converges for each $x > 0$. The gamma function is a generalization for $x > 0$ of the factorial function $n!$, which is defined only if n is a nonnegative integer. To see the way in which $\Gamma(x)$ is a generalization of $n!$, we note first that

$$\Gamma(1) = \int_0^{\infty} e^{-t} \, dt = \lim_{b \to \infty} \left[-e^{-t}\right]_0^b = 1. \tag{9}$$

Then we integrate by parts with $u = t^x$ and $dv = e^{-t}\, dt$:

$$\Gamma(x+1) = \lim_{b \to \infty} \int_0^b e^{-t} t^x \, dt = \lim_{b \to \infty} \left(\left[-e^{-t} t^x \right]_0^b + \int_0^b x e^{-t} t^{x-1} \, dt \right)$$

$$= x \left(\lim_{b \to \infty} \int_0^b e^{-t} t^{x-1} \, dt \right);$$

that is,

> $$\Gamma(x+1) = x\Gamma(x). \tag{10}$$

This is the most important property of the gamma function.

If we combine Eqs. (9) and (10), we see that

$$\Gamma(2) = 1 \cdot \Gamma(1) = 1!, \quad \Gamma(3) = 2 \cdot \Gamma(2) = 2!, \quad \Gamma(4) = 3 \cdot \Gamma(3) = 3!,$$

and in general that

> $$\Gamma(n+1) = n! \qquad \text{for } n \geqq 0 \text{ an integer.} \tag{11}$$

An important special value of the gamma function is

$$\Gamma\left(\tfrac{1}{2}\right) = \int_0^\infty e^{-t} t^{-1/2} \, dt = 2 \int_0^\infty e^{-u^2} \, du = \sqrt{\pi}, \tag{12}$$

where we have substituted u^2 for t in the first integral; the fact that

$$\int_0^\infty e^{-u^2} \, du = \frac{\sqrt{\pi}}{2}$$

is known, but is far from obvious.

Although $\Gamma(x)$ is defined in (8) only for $x > 0$, we can use the recursion formula in (10) to define $\Gamma(x)$ whenever x is neither zero nor a negative integer. If $-1 < x < 0$, then

$$\Gamma(x) = \frac{\Gamma(x+1)}{x};$$

the right-hand side is defined because $0 < x + 1 < 1$. The same formula may then be used to extend the definition of $\Gamma(x)$ to the open interval $(-2, -1)$, then to the open interval $(-3, -2)$, and so on. The graph of the gamma function thus extended is shown in Fig. 8.5.1. The student who would like to pursue this fascinating topic further should consult Artin's *The Gamma Function* (New York: Holt, Rinehart and Winston, 1964). In only 39 pages, this is one of the finest expositions in the entire literature of mathematics.

FIGURE 8.5.1. The graph of the extended gamma function.

Bessel Functions of the First Kind

If we choose $a_0 = 1/[2^p\Gamma(p + 1)]$ in (4), where $p > 0$, and note that

$$\Gamma(p + m + 1) = (p + m)(p + m - 1) \cdots (p + 2)(p + 1)\Gamma(p + 1)$$

by repeated application of Eq. (10), we can write the **Bessel function of the first kind of order** p very concisely with the aid of the gamma function:

$$\blacktriangleright \qquad J_p(x) = \sum_{m=0}^{\infty} \frac{(-1)^m}{m!\,\Gamma(p + m + 1)} \left(\frac{x}{2}\right)^{2m+p}. \qquad (13)$$

Similarly, if $p > 0$ is not an integer, we choose $b_0 = 1/[2^{-p}\Gamma(-p + 1)]$ in (7) to obtain the linearly independent second solution

$$\blacktriangleright \qquad J_{-p}(x) = \sum_{m=0}^{\infty} \frac{(-1)^m}{m!\,\Gamma(-p + m + 1)} \left(\frac{x}{2}\right)^{2m-p} \qquad (14)$$

of Bessel's equation of order p. If p is not an integer, we have the general solution

$$y(x) = c_1 J_p(x) + c_2 J_{-p}(x) \qquad (15)$$

for $x > 0$; x^p must be replaced with $|x|^p$ in Eqs. (13) through (15) to get the correct solutions for $x < 0$.

If $p = n$, a nonnegative integer, then Eq. (13) gives

$$\blacktriangleright \qquad J_n(x) = \sum_{m=0}^{\infty} \frac{(-1)^m}{m!\,(m + n)!} \left(\frac{x}{2}\right)^{2m+n} \qquad (16)$$

for the Bessel functions of the first kind of integral order. Thus

$$J_0(x) = \sum_{m=0}^{\infty} \frac{(-1)^m x^{2m}}{2^{2m}(m!)^2} = 1 - \frac{x^2}{2^2} + \frac{x^4}{2^2 \cdot 4^2} - \frac{x^6}{2^2 \cdot 4^2 \cdot 6^2} + \cdots \qquad (17)$$

and

$$J_1(x) = \sum_{m=0}^{\infty} \frac{(-1)^m 2^{2m+1}}{2^{2m+1} m!\,(m + 1)!} = \frac{x}{2} - \frac{1}{2!}\left(\frac{x}{2}\right)^3 + \frac{1}{2! \cdot 3!}\left(\frac{x}{2}\right)^5 - \cdots. \qquad (18)$$

The graphs of $J_0(x)$ and $J_1(x)$ are shown in Fig. 8.5.2. In a general way they resemble damped cosine and sine oscillations, respectively (see Problem 27). Indeed, if you examine the series in (17), you can see part of the reason why $J_0(x)$ and $\cos x$ *might* be similar—only minor changes in the denominators in (17) are needed to produce the Taylor series for $\cos x$. As suggested by Fig. 8.5.2, the zeros of the functions $J_0(x)$ and $J_1(x)$ are *interlaced*—between any two consecutive zeros of $J_0(x)$ there is precisely one zero of $J_1(x)$ (Problem 26) and vice versa. The first four zeros of $J_0(x)$ are approximately 2.4048, 5.5201, 8.6537, and 11.7915. For n large, the nth zero of $J_0(x)$ is approximately $\left(n - \frac{1}{4}\right)\pi$; the nth zero of $J_1(x)$ is approximately $\left(n + \frac{1}{4}\right)\pi$. Thus the interval between consecutive zeros of either $J_0(x)$ or $J_1(x)$ is approximately π—another similarity with $\cos x$ and $\sin x$. You can see the way the accuracy of these approximations increases with increasing n by rounding the entries in the table in Fig. 8.5.3 to two decimal places.

FIGURE 8.5.2. The graphs of the Bessel functions $J_0(x)$ and $J_1(x)$.

n	nth zero of $J_0(x)$	$\left(n - \frac{1}{4}\right)\pi$	nth zero of $J_1(x)$	$\left(n + \frac{1}{4}\right)\pi$
1	2.4048	2.3562	3.8317	3.9270
2	5.5201	5.4978	7.0156	7.0686
3	8.6537	8.6394	10.1735	10.2102
4	11.7915	11.7810	13.3237	13.3518
5	14.9309	14.9226	16.4706	16.4934

FIGURE 8.5.3. Zeros of $J_0(x)$ and $J_1(x)$.

It turns out that $J_p(x)$ is an elementary function if the order p is half an odd integer. For instance, on substitution of $p = \frac{1}{2}$ and $p = -\frac{1}{2}$ in Eqs. (13) and (14), respectively, the results can be recognized (Problem 2) as

$$J_{1/2}(x) = \sqrt{\frac{2}{\pi x}}\,\sin x \qquad \text{and} \qquad J_{-1/2}(x) = \sqrt{\frac{2}{\pi x}}\,\cos x. \qquad (19)$$

Bessel Functions of the Second Kind

The methods of Section 8.4 must be used to find linearly independent second solutions of integral order. A very complicated generalization of Example 3 in that section gives the formula

$$Y_n(x) = \frac{2}{\pi}\left(\gamma + \ln\frac{x}{2}\right) J_n(x) - \frac{1}{\pi}\sum_{m=0}^{n-1}\frac{2^{n-2m}(n-m-1)!}{m!\,x^{n-2m}}$$

$$- \frac{1}{\pi}\sum_{m=0}^{\infty}\frac{(-1)^m(H_m + H_{m+n})}{m!\,(m+n)!}\left(\frac{x}{2}\right)^{n+2m}, \qquad (20)$$

with the notation used there. If $n = 0$ then the first sum in (20) is taken to be zero. Here, $Y_n(x)$ is called the **Bessel function of the second kind of integral order** $n \geq 0$.

The general solution of Bessel's equation of integral order n is

➤ $$y(x) = c_1 J_n(x) + c_2 Y_n(x). \qquad (21)$$

It is important to note that $Y_n(x) \to -\infty$ as $x \to 0$ (Fig. 8.5.4). Hence $c_2 = 0$ in Eq. (21) if $y(x)$ is continuous at $x = 0$. Thus if $y(x)$ is a *continuous* solution of Bessel's equation of order n, it follows that

$$y(x) = c\,J_n(x)$$

for some constant c. Because $J_0(0) = 1$, we see in addition that if $n = 0$, then $c = y(0)$. In Section 10.4 we will see that this single fact regarding Bessel functions has numerous physical applications.

Figure 8.5.5 illustrates the fact that for $n > 1$ the graphs of $J_n(x)$ and $Y_n(x)$ look generally like those of $J_1(x)$ and $Y_1(x)$. In particular, $J_n(0) = 0$ while $Y_n(x) \to -\infty$ as $x \to 0^+$, and both functions undergo damped oscillation as $x \to +\infty$.

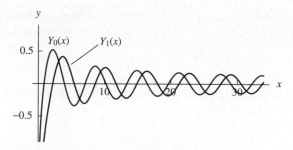

FIGURE 8.5.4. The graphs of the Bessel functions $Y_0(x)$ and $Y_1(x)$.

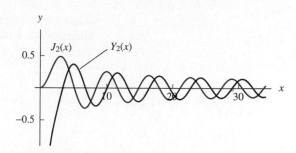

FIGURE 8.5.5. The graphs of the Bessel functions $J_2(x)$ and $Y_2(x)$.

Bessel Function Identities

Bessel functions are analogous to trigonometric functions in that they satisfy a large number of standard identities of frequent utility, especially in the evaluation of integrals involving Bessel functions. Differentiation of

$$J_p(x) = \sum_{m=0}^{\infty} \frac{(-1)^m}{m!\,\Gamma(p+m+1)} \left(\frac{x}{2}\right)^{2m+p} \tag{13}$$

in the case that p is a nonnegative integer gives

$$\frac{d}{dx}\left[x^p J_p(x)\right] = \frac{d}{dx} \sum_{m=0}^{\infty} \frac{(-1)^m x^{2m+2p}}{2^{2m+p} m!\,(p+m)!}$$

$$= \sum_{m=0}^{\infty} \frac{(-1)^m x^{2m+2p-1}}{2^{2m+p-1} m!\,(p+m-1)!}$$

$$= x^p \sum_{m=0}^{\infty} \frac{(-1)^m x^{2m+p-1}}{2^{2m+p-1} m!\,(p+m-1)!},$$

and thus we have shown that

$$\frac{d}{dx}\left[x^p J_p(x)\right] = x^p J_{p-1}(x). \tag{22}$$

Similarly,

$$\frac{d}{dx}\left[x^{-p} J_p(x)\right] = -x^{-p} J_{p+1}(x). \tag{23}$$

If we carry out the differentiations in Eqs. (22) and (23) and then divide the resulting identities by x^p and x^{-p}, respectively, we obtain (Problem 8) the identities

$$J_p'(x) = J_{p-1}(x) - \frac{p}{x} J_p(x) \tag{24}$$

and

$$J_p'(x) = \frac{p}{x} J_p(x) - J_{p+1}(x). \tag{25}$$

Thus we may express the derivatives of Bessel functions in terms of Bessel functions themselves. Subtraction of Eq. (25) from Eq. (24) gives the recursion formula

$$J_{p+1}(x) = \frac{2p}{x} J_p(x) - J_{p-1}(x), \tag{26}$$

which can be used to express Bessel functions of high order in terms of Bessel functions of lower orders. In the form

$$J_{p-1}(x) = \frac{2p}{x} J_p(x) - J_{p+1}(x), \tag{27}$$

it can be used to express Bessel functions of large negative order in terms of Bessel functions of numerically smaller negative orders.

The identities in Eqs. (22) through (27) hold wherever they are meaningful—that is, whenever no Bessel functions of negative integral order appear. In particular, they hold for all nonintegral values of p.

Example 1 With $p = 0$, Eq. (22) gives

$$\int x J_0(x)\, dx = x J_1(x) + C.$$

Similarly, with $p = 0$, Eq. (23) gives

$$\int J_1(x)\, dx = -J_0(x) + C. \qquad \blacksquare$$

Example 2 Using first $p = 2$ and then $p = 1$ in Eq. (26), we get

$$J_3(x) = \frac{4}{x} J_2(x) - J_1(x) = \frac{4}{x} \left[\frac{2}{x} J_1(x) - J_0(x) \right] - J_1(x),$$

so that

$$J_3(x) = -\frac{4}{x} J_0(x) + \left(\frac{8}{x^2} - 1 \right) J_1(x).$$

With similar manipulations every Bessel function of positive integral order can be expressed in terms of $J_0(x)$ and $J_1(x)$. $\qquad \blacksquare$

Example 3 To antidifferentiate $x J_2(x)$, we first note that

$$\int x^{-1} J_2(x)\, dx = -x^{-1} J_1(x) + C$$

by Eq. (23) with $p = 1$. We therefore write

$$\int x J_2(x)\, dx = \int x^2 \left[x^{-1} J_2(x) \right] dx$$

and integrate by parts with

$$u = x^2, \qquad\qquad dv = x^{-1} J_2(x)\, dx,$$
$$du = 2x\, dx, \qquad \text{and} \qquad v = -x^{-1} J_1(x).$$

This gives

$$\int x J_2(x)\, dx = -x J_1(x) + 2 \int J_1(x)\, dx = -x J_1(x) - 2 J_0(x) + C,$$

with the aid of the second result of Example 1. $\qquad \blacksquare$

The Parametric Bessel Equation

The **parametric Bessel equation of order** n is

$$\blacktriangleright \qquad x^2 y'' + x y' + (\alpha^2 x^2 - n^2) y = 0, \tag{28}$$

where α is a positive parameter. As we will see in Chapter 10, this equation appears in the solution of Laplace's equation in polar coordinates. It is easy to see (Problem 9) that the substitution $t = \alpha x$ transforms Eq. (28) into the (standard) Bessel equation

$$t^2 \frac{d^2 y}{dt^2} + t \frac{dy}{dt} + (t^2 - n^2) y = 0 \tag{29}$$

with general solution $y(t) = c_1 J_n(t) + c_2 Y_n(t)$. Hence the general solution of Eq. (28) is

$$\blacktriangleright \qquad y(x) = c_1 J_n(\alpha x) + c_2 Y_n(\alpha x). \tag{30}$$

Now consider the eigenvalue problem

$$x^2 y'' + x y' + (\lambda x^2 - n^2) = 0,$$
$$y(L) = 0 \tag{31}$$

on the interval $[0, L]$. We seek the *positive* values of λ for which there exists a nontrivial solution of (31) that is *continuous* on $[0, L]$. If we write $\lambda = \alpha^2$, then the differential equation in (31) is that in Eq. (28), so its general solution is given in Eq. (30). Because $Y_n(x) \to -\infty$ as $x \to 0$ but $J_n(0)$ is finite, the continuity of $y(x)$ requires that $c_2 = 0$. Thus $y(x) = c_1 J_n(\alpha x)$. The endpoint condition $y(L) = 0$ now implies that $z = \alpha L$ must be a (positive) root of the equation

$$J_n(z) = 0. \tag{32}$$

For $n > 1$, $J_n(x)$ oscillates rather like $J_1(x)$ in Fig. 8.5.2 and hence has an infinite sequence of positive zeros $\gamma_{n1}, \gamma_{n2}, \gamma_{n3}, \ldots$ (see Fig 8.5.6). It follows that the kth positive eigenvalue of the problem in (31) is

$$\lambda_k = (\alpha_k)^2 = \frac{(\gamma_{nk})^2}{L^2}, \tag{33}$$

and that its associated eigenfunction is

$$y_k(x) = J_n\left(\frac{\gamma_{nk}}{L} x\right). \tag{34}$$

The roots γ_{nk} of Eq. (32) for $n \leq 8$ and $k \leq 20$ are tabulated in Table 9.5 of M. Abramowitz and I. A. Stegun, *Handbook of Mathematical Functions* (New York: Dover, 1965).

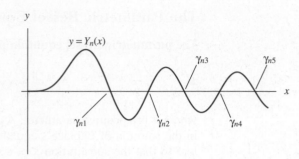

FIGURE 8.5.6. The positive zeros $\gamma_{n1}, \gamma_{n2}, \gamma_{n3}, \ldots$ of the Bessel function $J_n(x)$.

8.5 Problems

1. Differentiate termwise the series for $J_0(x)$ to show directly that $J_0'(x) = -J_1(x)$ (another analogy with the cosine and sine functions).

2. (a) Deduce from Eqs. (10) and (12) that

$$\Gamma\left(n + \tfrac{1}{2}\right) = \frac{1 \cdot 3 \cdot 5 \cdots (2n-1)}{2^n}\sqrt{\pi}.$$

(b) Use the result of part (a) to verify the formulas in Eq. (19) for $J_{1/2}(x)$ and $J_{-1/2}(x)$, and construct a figure showing the graphs of these functions.

3. (a) Suppose that m is a positive integer. Show that

$$\Gamma\left(m + \tfrac{2}{3}\right) = \frac{2 \cdot 5 \cdot 8 \cdots (3m-1)}{3^m}\Gamma\left(\tfrac{2}{3}\right).$$

(b) Conclude from part (a) and Eq. (13) that

$$J_{-1/3}(x) =$$
$$\frac{(x/2)^{-1/3}}{\Gamma\left(\tfrac{2}{3}\right)}\left(1 + \sum_{m=1}^{\infty} \frac{(-1)^m 3^m x^{2m}}{2^{2m}\, m! \cdot 2 \cdot 5 \cdots (3m-1)}\right).$$

4. Apply Eqs. (19), (26), and (27) to show that

$$J_{3/2}(x) = \sqrt{\frac{2}{\pi x^3}}\,(\sin x - x\cos x)$$

and

$$J_{-3/2}(x) = -\sqrt{\frac{2}{\pi x^3}}\,(\cos x + x\sin x).$$

Construct a figure showing the graphs of these two functions.

5. Express $J_4(x)$ in terms of $J_0(x)$ and $J_1(x)$.

6. Derive the recursion formula in Eq. (2) for Bessel's equation.

7. Verify the identity in (23) by termwise differentiation.

8. Deduce the identities in Eqs. (24) and (25) from those in Eqs. (22) and (23).

9. Verify that the substitution $t = \alpha x$ transforms the parametric Bessel equation in (28) into the equation in (29).

10. Show that

$$4J_p''(x) = J_{p-2}(x) - 2J_p(x) + J_{p+2}(x).$$

11. Use the relation $\Gamma(x+1) = x\Gamma(x)$ to deduce from Eqs. (13) and (14) that if p is not a negative integer, then

$$J_p(x) =$$
$$\frac{(x/2)^p}{\Gamma(p+1)}\left[1 + \sum_{m=1}^{\infty} \frac{(-1)^m (x/2)^{2m}}{m!\,(p+1)(p+2)\cdots(p+m)}\right].$$

This form is more convenient for the computation of $J_p(x)$ because only the single value $\Gamma(p+1)$ of the gamma function is required.

12. Use the series of Problem 11 to find $y(0) = \lim_{x\to 0} y(x)$ if

$$y(x) = x^2\left[\frac{J_{5/2}(x) + J_{-5/2}(x)}{J_{1/2}(x) + J_{-1/2}(x)}\right].$$

Use a computer algebra system to graph $y(x)$ for x near 0. Does the graph corroborate your value of $y(0)$?

Any integral of the form $\int x^m J_n(x)\,dx$ can be evaluated in terms of Bessel functions and the indefinite integral $\int J_0(x)\,dx$. The latter integral cannot be simplified further, but the function $\int_0^x J_0(t)\,dt$ is tabulated in Table 11.1 of Abramowitz and Stegun. Use the identities in Eqs. (22) and (23) to evaluate the integrals in Problems 13 through 21.

13. $\displaystyle\int x^2 J_0(x)\,dx$

14. $\displaystyle\int x^3 J_0(x)\,dx$

15. $\displaystyle\int x^4 J_0(x)\,dx$

16. $\displaystyle\int x J_1(x)\,dx$

17. $\displaystyle\int x^2 J_1(x)\,dx$

18. $\displaystyle\int x^3 J_1(x)\,dx$

19. $\displaystyle\int x^4 J_1(x)\,dx$

20. $\displaystyle\int J_2(x)\,dx$

21. $\displaystyle\int J_3(x)\,dx$

22. Prove that

$$J_0(x) = \frac{1}{\pi} \int_0^\pi \cos(x \sin \theta) \, d\theta$$

by showing that the right-hand side satisfies Bessel's equation of order zero and has the value $J_0(0)$ when $x = 0$. Explain why this constitutes a proof.

23. Prove that

$$J_1(x) = \frac{1}{\pi} \int_0^\pi \cos(\theta - x \sin \theta) \, d\theta$$

by showing that the right-hand side satisfies Bessel's equation of order 1 and that its derivative has the value $J_1'(0)$ when $x = 0$. Explain why this constitutes a proof.

24. It can be shown that

$$J_n(x) = \frac{1}{\pi} \int_0^\pi \cos(n\theta - x \sin \theta) \, d\theta.$$

With $n \geq 2$, show that the right-hand side satisfies Bessel's equation of order n and also agrees with the values $J_n(0)$ and $J_n'(0)$. Explain why this does *not* suffice to prove the preceding assertion.

25. Deduce from Problem 22 that

$$J_0(x) = \frac{1}{2\pi} \int_0^{2\pi} e^{ix \sin \theta} \, d\theta.$$

(*Suggestion*: Show first that

$$\int_0^{2\pi} e^{ix \sin \theta} \, d\theta = \int_0^\pi \left(e^{ix \sin \theta} + e^{-ix \sin \theta} \right) d\theta;$$

then use Euler's formula.)

26. Use Eqs. (22) and (23) and Rolle's theorem to prove that between any two consecutive zeros of $J_n(x)$ there is precisely one zero of $J_{n+1}(x)$. Use a computer algebra system to construct a figure illustrating this fact with $n = 10$ (for instance).

27. (a) Show that the substitution $y = x^{-1/2}z$ in Bessel's equation of order p,

$$x^2 y'' + x y' + (x^2 - p^2)y = 0,$$

yields

$$z'' + \left(1 - \frac{p^2 - \frac{1}{4}}{x^2} \right) z = 0.$$

(b) If x is so large that $\left(p^2 - \frac{1}{4} \right) / x^2$ is negligible, then the latter equation reduces to $z'' + z \approx 0$. Explain why this suggests (without proving it) that if $y(x)$ is a solution of Bessel's equation, then

$$\begin{aligned} y(x) &\approx x^{-1/2}(A \cos x + B \sin x) \\ &= Cx^{-1/2} \cos(x - \alpha) \end{aligned} \tag{35}$$

with C and α constants, and x large.

Asymptotic Approximations

It is known that the choices $C = \sqrt{2/\pi}$ and $\alpha = (2n + 1)\pi/4$ in (35) yield the best approximation to $J_n(x)$ for x large:

$$J_n(x) \approx \sqrt{\frac{2}{\pi x}} \cos \left(x - \tfrac{1}{4}(2n + 1)\pi \right). \tag{36}$$

Similarly,

$$Y_n(x) \approx \sqrt{\frac{2}{\pi x}} \sin \left(x - \tfrac{1}{4}(2n + 1)\pi \right). \tag{37}$$

In particular,

$$J_0(x) \approx \sqrt{\frac{2}{\pi x}} \cos \left(x - \tfrac{1}{4}\pi \right)$$

and

$$Y_0(x) \approx \sqrt{\frac{2}{\pi x}} \sin \left(x - \tfrac{1}{4}\pi \right)$$

if x is large. These are *asymptotic approximations* in that the ratio of the two sides in each approximation approaches unity as $x \to +\infty$.

8.6 Applications of Bessel Functions

The importance of Bessel functions stems not only from the frequent appearance of Bessel's equation in applications, but also from the fact that the solutions of many other second-order linear differential equations can be expressed in terms of Bessel functions. To see how this comes about, we begin with Bessel's equation of order p in the form

$$z^2 \frac{d^2 w}{dz^2} + z \frac{dw}{dz} + (z^2 - p^2)w = 0, \tag{1}$$

and substitute

$$w = x^{-\alpha} y, \qquad z = kx^\beta. \tag{2}$$

Then a routine (but lengthy) transformation of Eq. (1) yields

$$x^2 y'' + (1 - 2\alpha)xy' + (\alpha^2 - \beta^2 p^2 + \beta^2 k^2 x^{2\beta})y = 0;$$

that is,

$$x^2 y'' + Axy' + (B + Cx^q)y = 0, \tag{3}$$

where the constants A, B, C, and q are given by

$$A = 1 - 2\alpha, \quad B = \alpha^2 - \beta^2 p^2, \quad C = \beta^2 k^2, \quad \text{and} \quad q = 2\beta. \tag{4}$$

It is a simple matter to solve the equations in (4) for

$$\alpha = \frac{1 - A}{2}, \qquad \beta = \frac{q}{2},$$

$$k = \frac{2\sqrt{C}}{q}, \quad \text{and} \quad p = \frac{\sqrt{(1 - A)^2 - 4B}}{q}. \tag{5}$$

Under the assumption that the square roots in (5) are real, it follows that the general solution of Eq. (3) is

$$y(x) = x^\alpha w(z) = x^\alpha w(kx^\beta),$$

where

$$w(z) = c_1 J_p(z) + c_2 Y_{-p}(z)$$

(assuming that p is not an integer) is the general solution of the Bessel equation in (1). This establishes the following result.

THEOREM 1: Solutions in Bessel Functions

If $C > 0$, $q \neq 0$, and $(1 - A)^2 \geq 4B$, then the general solution (for $x > 0$) of Eq. (3) is

$$y(x) = x^\alpha \left[c_1 J_p(kx^\beta) + c_2 J_{-p}(kx^\beta) \right], \tag{6}$$

where α, β, k, and p are given by the equations in (5). If p is an integer, then J_{-p} is to be replaced with Y_p. ▲

Example 1 Solve the equation

$$4x^2 y'' + 8xy' + (x^4 - 3)y = 0. \tag{7}$$

Solution To compare Eq. (7) with Eq. (3), we rewrite the former as

$$x^2 y'' + 2xy' + \left(-\tfrac{3}{4} + \tfrac{1}{4}x^4 \right) y = 0$$

and see that $A = 2$, $B = -\tfrac{3}{4}$, $C = \tfrac{1}{4}$, and $q = 4$. Then the equations in (5) give $\alpha = -\tfrac{1}{2}$, $\beta = 2$, $k = \tfrac{1}{4}$, and $p = \tfrac{1}{2}$. Thus the general solution in (6) of Eq. (7) is

$$y(x) = x^{-1/2} \left[c_1 J_{1/2} \left(\tfrac{1}{4}x^2 \right) + c_2 J_{-1/2} \left(\tfrac{1}{4}x^2 \right) \right].$$

If we recall from Eq. (19) of Section 8.5 that

$$J_{1/2}(z) = \sqrt{\frac{2}{\pi z}} \sin z \quad \text{and} \quad J_{-1/2}(z) = \sqrt{\frac{2}{\pi z}} \cos z,$$

we see that a general solution of Eq. (7) can be written in the elementary form

$$y(x) = x^{-3/2} \left(A \cos \frac{x^2}{4} + B \sin \frac{x^2}{4} \right). \qquad \blacksquare$$

Example 2 Solve the Airy equation

$$y'' + 9xy = 0. \qquad (8)$$

Solution First we rewrite the given equation in the form

$$x^2 y'' + 9x^3 y = 0.$$

This is the special case of Eq. (3) with $A = B = 0$, $C = 9$, and $q = 3$. It follows from the equations in (5) that $\alpha = \frac{1}{2}$, $\beta = \frac{3}{2}$, $k = 2$, and $p = \frac{1}{3}$. Thus the general solution of Eq. (8) is

$$y(x) = x^{1/2} \left[c_1 J_{1/3}(2x^{3/2}) + c_2 J_{-1/3}(2x^{3/2}) \right]. \qquad \blacksquare$$

Buckling of a Vertical Column

For a practical application, we now consider the problem of determining when a uniform vertical column will buckle under its own weight. We take $x = 0$ at the free top end of the column and $x = L > 0$ at its bottom; we assume that the bottom is rigidly imbedded in the ground, perhaps in concrete; see Fig. 8.6.1. Denote the angular deflection of the column at the point x by $\theta(x)$. From the theory of elasticity it follows that

$$EI \frac{d^2\theta}{dx^2} + g\rho x\theta = 0, \qquad (9)$$

where E is the Young's modulus of the material of the column, I is its cross-sectional moment of inertia, ρ is the linear density of the column, and g is gravitational acceleration. The boundary conditions are

$$\theta'(0) = 0, \quad \theta(L) = 0. \qquad (10)$$

We will accept (9) and (10) as an appropriate statement of the problem and attempt to solve it in this form. With

$$\lambda = \gamma^2 = \frac{g\rho}{EI}, \qquad (11)$$

we have the eigenvalue problem

$$\theta'' + \gamma^2 x\theta = 0; \quad \theta'(0) = 0, \quad \theta(L) = 0. \qquad (12)$$

FIGURE 8.6.1. The buckling column.

The column can buckle only if there is a nontrivial solution of (12); otherwise the column will remain in its undeflected vertical position.

The differential equation in (12) is an Airy equation similar to the one in Example 2. It has the form of Eq. (3) with $A = B = 0$, $C = \gamma^2$, and $q = 3$. The equations in (5) give $\alpha = \frac{1}{2}$, $\beta = \frac{3}{2}$, $k = \frac{2}{3}\gamma$, and $p = \frac{1}{3}$. So the general solution is

$$\theta(x) = x^{1/2}\left[c_1 J_{1/3}\left(\tfrac{2}{3}\gamma x^{3/2}\right) + c_2 J_{-1/3}\left(\tfrac{2}{3}\gamma x^{3/2}\right)\right]. \tag{13}$$

In order to apply the initial conditions, we substitute $p = \pm\frac{1}{3}$ in

$$J_p(x) = \sum_{m=0}^{\infty} \frac{(-1)^m}{m!\,\Gamma(p+m+1)}\left(\frac{x}{2}\right)^{2m+p},$$

and find after some simplifications that

$$\theta(x) = \frac{c_1\gamma^{1/3}}{3^{1/3}\Gamma\left(\frac{4}{3}\right)}\left(x - \frac{\gamma^2 x^4}{12} + \frac{\gamma^4 x^7}{504} - \cdots\right)$$

$$+ \frac{c_2 3^{1/3}}{\gamma^{1/3}\Gamma\left(\frac{2}{3}\right)}\left(1 - \frac{\gamma^2 x^3}{6} + \frac{\gamma^4 x^6}{180} - \cdots\right).$$

From this it is clear that the endpoint condition $\theta'(0) = 0$ implies that $c_1 = 0$, so

$$\theta(x) = c_2 x^{1/2} J_{-1/3}\left(\tfrac{2}{3}\gamma x^{3/2}\right). \tag{14}$$

The endpoint condition $\theta(L) = 0$ now gives

$$J_{-1/3}\left(\tfrac{2}{3}\gamma L^{3/2}\right) = 0. \tag{15}$$

Thus the column will buckle only if $z = \frac{2}{3}\gamma L^{3/2}$ is a root of the equation $J_{-1/3}(z) = 0$. The graph of

$$J_{-1/3}(z) = \frac{(z/2)^{-1/3}}{\Gamma\left(\frac{2}{3}\right)}\left(1 + \sum_{m=1}^{\infty}\frac{(-1)^m 3^m z^{2m}}{2^{2m}\,m!\cdot 2\cdot 5\cdots(3m-1)}\right) \tag{16}$$

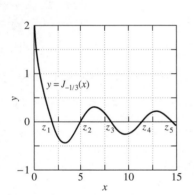

FIGURE 8.6.2. The graph of $J_{-1/3}(x)$.

(see Problem 3 of Section 8.5) is shown in Fig. 8.6.2, where we see that the smallest positive zero z_1 is a bit less than 2. Most technical computing systems can find roots like this one. For instance, each of the computer system commands

```
fsolve(BesselJ(-1/3,x)=0, x, 1..2)        (Maple)
FindRoot[BesselJ[-1/3,x]==0, {x,2}]        (Mathematica)
fzero('besselj(-1/3,x)', 2)                (MATLAB)
```

yield the value $z_1 = 1.86635$ (rounded accurate to five decimal places).

The shortest length L_1 for which the column will buckle under its own weight is

$$L_1 = \left(\frac{3z_1}{2\gamma}\right)^{2/3} = \left[\frac{3z_1}{2}\left(\frac{EI}{\rho g}\right)^{1/2}\right]^{2/3}.$$

If we substitute $z_1 \approx 1.86635$ and $\rho = \delta A$, where δ is the volumetric density of the material of the column and A is its cross-sectional area, we finally get

$$L_1 \approx (1.986)\left(\frac{EI}{g\delta A}\right)^{1/3} \tag{17}$$

for the critical buckling length. For example, with a steel column or rod for which $E = 2.8 \times 10^7$ lb/in.2 and $g\delta = 0.28$ lb/in.3, the formula in (17) gives the results shown in the table in Fig. 8.6.3.

Cross Section of Rod	Shortest Buckling Length L_1
Circular with $r = 0.5$ in.	30 ft 6 in.
Circular with $r = 1.5$ in.	63 ft 5 in.
Annular with $r_{inner} = 1.25$ in. and $r_{outer} = 1.5$ in.	75 ft 7 in.

FIGURE 8.6.3.

We have used the familiar formulas $A = \pi r^2$ and $I = \frac{1}{4}\pi r^4$ for a circular disk. The data in the table show why flagpoles are hollow.

8.6 Problems

In Problems 1 through 12, express the general solution of the given differential equation in terms of Bessel functions.

1. $x^2 y'' - xy' + (1 + x^2)y = 0$
2. $xy'' + 3y' + xy = 0$
3. $xy'' - y' + 36x^3 y = 0$
4. $x^2 y'' - 5xy' + (8 + x)y = 0$
5. $36x^2 y'' + 60xy' + (9x^3 - 5)y = 0$
6. $16x^2 y'' + 24xy' + (1 + 144x^3)y = 0$
7. $x^2 y'' + 3xy' + (1 + x^2)y = 0$
8. $4x^2 y'' - 12xy' + (15 + 16x)y = 0$
9. $16x^2 y'' - (5 - 144x^3)y = 0$
10. $2x^2 y'' - 3xy' - 2(14 - x^5)y = 0$
11. $y'' + x^4 y = 0$
12. $y'' + 4x^3 y = 0$
13. Apply Theorem 1 to show that the general solution of

$$xy'' + 2y' + xy = 0$$

is $y(x) = x^{-1}(A \cos x + B \sin x)$.

14. Verify that the substitutions in (2) in Bessel's equation (Eq. (1)) yield Eq. (3).
15. **(a)** Show that the substitution

$$y = -\frac{1}{u}\frac{du}{dx}$$

transforms the Riccati equation $dy/dx = x^2 + y^2$ into $u'' + x^2 u = 0$. **(b)** Show that the general solution of $dy/dx = x^2 + y^2$ is

$$y(x) = x\frac{J_{3/4}\left(\frac{1}{2}x^2\right) - cJ_{-3/4}\left(\frac{1}{2}x^2\right)}{cJ_{1/4}\left(\frac{1}{2}x^2\right) + J_{-1/4}\left(\frac{1}{2}x^2\right)}.$$

(*Suggestion*: Apply the identities in Eqs. (22) and (23) of Section 8.5.)

16. **(a)** Substitute the series of Problem 11 of Section 8.5 in the result of Problem 15 here to show that the solution of the initial value problem

$$\frac{dy}{dx} = x^2 + y^2, \quad y(0) = 0$$

is

$$y(x) = x\frac{J_{3/4}\left(\frac{1}{2}x^2\right)}{J_{-1/4}\left(\frac{1}{2}x^2\right)}.$$

(b) Deduce similarly that the solution of the initial value problem

$$\frac{dy}{dx} = x^2 + y^2, \quad y(0) = 1$$

is

$$y(x) = x\frac{2\Gamma\left(\frac{3}{4}\right)J_{3/4}\left(\frac{1}{2}x^2\right) + \Gamma\left(\frac{1}{4}\right)J_{-3/4}\left(\frac{1}{2}x^2\right)}{2\Gamma\left(\frac{3}{4}\right)J_{-1/4}\left(\frac{1}{2}x^2\right) - \Gamma\left(\frac{1}{4}\right)J_{1/4}\left(\frac{1}{2}x^2\right)}.$$

Some solution curves of the equation $dy/dx = x^2 + y^2$ are shown in Fig. 8.6.4. The location of the asymptotes where $y(x) \to +\infty$ can be found by using Newton's method to find the zeros of the denominators in the formulas for the solutions as listed here.

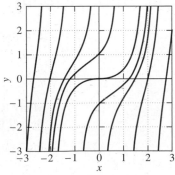

FIGURE 8.6.4. Solution curves of $\dfrac{dy}{dx} = x^2 + y^2$.

17. Figure 8.6.5 shows a linearly tapered rod with circular cross section, subject to an axial force P of compression. As in Section 3.8, its deflection curve $y = y(x)$ satisfies the endpoint value problem

$$EIy'' + Py = 0; \quad y(a) = y(b) = 0. \quad (18)$$

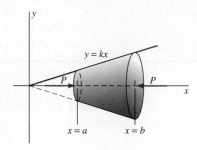

FIGURE 8.6.5. The tapered rod of Problem 17.

Here, however, the moment of inertia $I = I(x)$ of the cross section at x is given by

$$I(x) = \frac{1}{4}\pi (kx)^4 = I_0 \cdot \left(\frac{x}{b}\right)^4,$$

where $I_0 = I(b)$, the value of I at $x = b$. Substitution of $I(x)$ in the differential equation in (18) yields the eigenvalue problem

$$x^4 y'' + \lambda y = 0, \quad y(a) = y(b) = 0,$$

where $\lambda = \mu^2 = Pb^4/EI_0$. **(a)** Apply the theorem of this section to show that the general solution of $x^4 y'' + \mu^2 y = 0$ is

$$y(x) = x \left(A \cos \frac{\mu}{x} + B \sin \frac{\mu}{x}\right).$$

(b) Conclude that the nth eigenvalue is given by $\mu_n = n\pi ab/L$, where $L = b - a$ is the length of the rod, and hence that the nth buckling force is

$$P_n = \frac{n^2 \pi^2}{L^2} \left(\frac{a}{b}\right)^2 EI_0.$$

Note that if $a = b$, this result reduces to Eq. (28) of Section 3.8.

18. Consider a variable-length pendulum as indicated in Fig. 8.6.6. Assume that its length is increasing linearly with time, $L(t) = a + bt$. It can be shown that the oscillations of this pendulum satisfy the differential equation

$$L\theta'' + 2L'\theta' + g\theta = 0$$

under the usual condition that θ is so small that $\sin\theta$ is very well approximated by θ: $\theta \approx \sin\theta$. Substitute $L = a + bt$ to derive the general solution

$$\theta(t) = \frac{1}{\sqrt{L}} \left[A J_1\left(\frac{2}{b}\sqrt{gL}\right) + B Y_1\left(\frac{2}{b}\sqrt{gL}\right) \right].$$

For the application of this solution to a discussion of the steadily descending pendulum ("its nether extremity was formed of a crescent of glittering steel, about a foot in length from horn to horn; the horns upward, and the under edge as keen as that of a razor ... and the whole hissed as it swung through the air ... down and still down it came") of Edgar Allan Poe's macabre classic "The Pit and the Pendulum," see the article by Borrelli, Coleman, and Hobson in the March 1985 issue of *Mathematics Magazine* (Vol. 58, pp. 78–83).

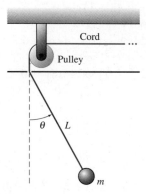

FIGURE 8.6.6. A variable length pendulum.

8.6 Application: Riccati Equations and Modified Bessel Functions

A Riccati equation is one of the form

$$\frac{dy}{dx} = A(x)y^2 + B(x)y + C(x).$$

Many Riccati equations like the ones listed next can be solved explicitly in terms of Bessel functions.

$$\frac{dy}{dx} = x^2 + y^2; \quad (1)$$

$$\frac{dy}{dx} = x^2 - y^2; \quad (2)$$

$$\frac{dy}{dx} = y^2 - x^2; \tag{3}$$

$$\frac{dy}{dx} = x + y^2; \tag{4}$$

$$\frac{dy}{dx} = x - y^2; \tag{5}$$

$$\frac{dy}{dx} = y^2 - x. \tag{6}$$

For example, Problem 15 in this section says that the general solution of Eq. (1) is given by

$$y(x) = x \frac{J_{3/4}\left(\frac{1}{2}x^2\right) - cJ_{-3/4}\left(\frac{1}{2}x^2\right)}{cJ_{1/4}\left(\frac{1}{2}x^2\right) + J_{-1/4}\left(\frac{1}{2}x^2\right)}. \tag{7}$$

See whether the symbolic DE solver command in your computer algebra system, such as the *Maple* command

```
dsolve(diff(y(x),x) = x^2 + y(x)^2, y(x))
```

or the *Mathematica* command

```
DSolve[ y'[x] == x^2 + y[x]^2, y[x], x ]
```

agrees with Eq. (7). If Bessel functions other than those appearing in Eq. (7) are involved, you may need to apply the identities in (26) and (27) of Section 8.5 to transform the computer's "answer" to (7). Then see whether your system can take the limit as $x \to 0$ in (7) to show that the arbitrary constant c is given in terms of the initial value $y(0)$ by

$$c = -\frac{y(0)\Gamma\left(\frac{1}{4}\right)}{2\Gamma\left(\frac{3}{4}\right)}. \tag{8}$$

Now you should be able to use built-in Bessel functions to plot typical solution curves like those shown in Fig. 8.6.4.

Next, investigate similarly one of the other equations in (2) through (6). Each has a general solution of the same general form in (7)—a quotient of linear combinations of Bessel functions. In addition to $J_p(x)$ and $Y_p(x)$, these solutions may involve the *modified Bessel functions*

$$I_p(x) = i^{-p} J_p(ix)$$

and

$$K_p(x) = \frac{\pi}{2} i^{-p} \left[J_p(ix) + Y_p(ix) \right]$$

that satisfy the *modified Bessel equation*

$$x^2 y'' + xy' - (x^2 + p^2)y = 0$$

of order p. For instance, the general solution of Eq. (5) is given for $x > 0$ by

$$y(x) = x^{1/2} \frac{I_{2/3}\left(\frac{2}{3}x^{3/2}\right) - cI_{-2/3}\left(\frac{2}{3}x^{3/2}\right)}{I_{-1/3}\left(\frac{2}{3}x^{3/2}\right) - cI_{1/3}\left(\frac{2}{3}x^{3/2}\right)}, \tag{9}$$

where

$$c = -\frac{y(0)\Gamma\left(\frac{1}{3}\right)}{3^{1/3}\Gamma\left(\frac{2}{3}\right)}. \tag{10}$$

FIGURE 8.6.7. Solution curves of $\dfrac{dy}{dx} = x - y^2$.

Figure 8.6.7 shows some typical solution curves, together with the parabola $y^2 = x$ that appears to bear an interesting relation to Eq. (6)—we see a funnel near $y = +\sqrt{x}$ and a spout near $y = -\sqrt{x}$.

The Bessel functions with imaginary argument that appear in the definitions of $I_p(x)$ and $K_p(x)$ may look exotic, but the power series of the modified function $I_n(x)$ is simply that of the unmodified function $J_n(x)$ except without the alternating minus signs. For instance,

$$I_0(x) = 1 + \frac{x^2}{4} + \frac{x^4}{64} + \frac{x^6}{2304} + \cdots$$

and

$$I_1(x) = \frac{x}{2} + \frac{x^3}{16} + \frac{x^5}{384} + \frac{x^7}{18432} + \cdots .$$

Check these power series expansions using your computer algebra system—look at **BesselI** in either *Maple* or *Mathematica*—and compare them with Eqs. (17) and (18) in Section 8.5.

The second-order differential equations with the same right-hand sides as in Eqs. (1) through (6) have interesting solutions which, however, cannot be expressed in terms of elementary functions and/or "known" special functions such as Bessel functions. Nevertheless, they can be investigated using an ODE plotter. For instance, the interesting pattern in Fig. 8.6.8 shows solution curves of the second-order equation

$$y'' = y^2 - x \tag{11}$$

with the same initial value $y(0) = 0$ but different slopes $y'(0) = -3.3, -3.1, \ldots,$ 0.7. Equation (11) is a form of the *first Painlevé transcendant,* an equation that arose historically in the classification of nonlinear second-order differential equations in terms of their critical points (see Chapter 14 of E. L. Ince, *Ordinary Differential Equations*, New York: Dover Publications, 1956). Figure 8.6.8 was suggested by an article by Anne Noonburg containing a similar figure in the Spring 1993 issue of the *C • ODE • E Newsletter.*

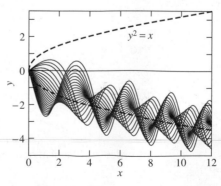

FIGURE 8.6.8. The first Painlevé transcendant $y'' = y^2 - x$, $y(0) = 0$, $y'(0) = -3.3, -3.1, \ldots, 0.7$.

Finally, here's a related example that was inspired by a *Maple* demonstration package. The *Maple* `dsolve` command yields the general solution

$$y(x) = x^{-1}(c_1 J_{10}(x) + c_2 Y_{10}(x))$$
$$+ x^{-11}(1857945600 + 51609600x^2 + 806400x^4 + 9600x^6 + 100x^8 + x^{10}) \quad (12)$$

of the nonhomogeneous second-order equation

$$x^2 y'' + 3xy' + (x^2 - 99)y = x. \quad (13)$$

Show that Theorem 1 in this section explains the "Bessel part" of the alleged solution in Eq. (12). Can you explain where the rational function part comes from, or at least verify it? For further examples of this sort, you can replace the coefficient 99 in Eq. (13) with $r^2 - 1$, where r is an *even* integer, and/or replace the x on the right-hand side with x^s, where s is an *odd* integer. (With parities other than these, more exotic special functions are involved.)

CHAPTER 9

Fourier Series Methods

9.1 Periodic Functions and Trigonometric Series

As motivation for the subject of Fourier series, we consider the differential equation

$$\frac{d^2x}{dt^2} + \omega_0^2 x = f(t) \tag{1}$$

that models the behavior of a mass-and-spring system with natural [circular] frequency ω_0, moving under the influence of an external force of magnitude $f(t)$ per unit mass. As we saw in Section 3.6, a particular solution of Eq. (1) can easily be found by the method of undetermined coefficients if $f(t)$ is a simple harmonic function—a sine or cosine function. For instance, the equation

$$\frac{d^2x}{dt^2} + \omega_0^2 x = A \cos \omega t \tag{2}$$

with $\omega^2 \neq \omega_0^2$ has the particular solution

$$x_p(t) = \frac{A}{\omega_0^2 - \omega^2} \cos \omega t, \tag{3}$$

which is readily found by beginning with the trial solution $x_p(t) = a \cos \omega t$.

Now suppose, more generally, that the force function $f(t)$ in Eq. (1) is a linear combination of simple harmonic functions. Then, on the basis of Eq. (3) and the analogous formula with sine in place of cosine, we can apply the principle of superposition to construct a particular solution of Eq. (1). For example, consider

the equation

$$\frac{d^2x}{dt^2} + \omega_0^2 x = \sum_{n=1}^{N} A_n \cos \omega_n t, \tag{4}$$

in which ω_0^2 is equal to none of the ω_n^2. Equation (4) has the particular solution

$$x_p(t) = \sum_{n=1}^{N} \frac{A_n}{\omega_0^2 - \omega_n^2} \cos \omega_n t, \tag{5}$$

obtained by adding the solutions given in Eq. (3) corresponding to the N terms on the right-hand side in Eq. (4).

Mechanical (and electrical) systems often involve periodic forcing functions that are not [simply] finite linear combinations of sines and cosines. Nevertheless, as we will soon see, any reasonably nice periodic function $f(t)$ has a representation as an *infinite series* of trigonometric terms. This fact opens the way toward solving Eq. (1) by superposition of trigonometric "building blocks," with the finite sum in Eq. (5) replaced with an infinite series.

Definition: Periodic Function

The function $f(t)$ defined for all t is said to be **periodic** provided that there exists a positive number p such that

$$f(t + p) = f(t) \tag{6}$$

for all t. The number p is then called a **period** of the function f.

Note that the period of a periodic function is not unique; for example, if p is a period of $f(t)$, then so are the numbers $2p$, $3p$, and so on. Indeed, every positive number is a period of any constant function.

If there exists a smallest positive number P such that $f(t)$ is periodic with period P, then we call P *the* period of f. For instance, the period of the functions $g(t) = \cos nt$ and $h(t) = \sin nt$ (where n is a positive integer) is $2\pi/n$ because

$$\cos n \left(t + \frac{2\pi}{n} \right) = \cos(nt + 2\pi) = \cos nt$$

and (7)

$$\sin n \left(t + \frac{2\pi}{n} \right) = \sin(nt + 2\pi) = \sin nt.$$

Moreover, 2π itself is a period of the functions $g(t)$ and $h(t)$. Ordinarily we will have no need to refer to the smallest period of a function $f(t)$ and will simply say that $f(t)$ has period p if p is any period of $f(t)$.

In Section 7.5 we saw several examples of piecewise continuous periodic functions. For instance, the square wave function having the graph shown in Fig. 9.1.1 has period 2π.

FIGURE 9.1.1. A square wave function.

Because $g(t) = \cos nt$ and $h(t) = \sin nt$ each have period 2π, any linear combination of sines and cosines of integral multiples of t, such as

$$f(t) = 3 + \cos t - \sin t + 5 \cos 2t + 17 \sin 3t,$$

has period 2π. But every such linear combination is continuous, so the square wave function cannot be expressed in such manner. In his celebrated treatise *The Analytic Theory of Heat* (1822), the French scientist Joseph Fourier (1768–1830) made the remarkable assertion that every function $f(t)$ with period 2π can be represented by an *infinite* trigonometric series of the form

$$\frac{a_0}{2} + \sum_{n=1}^{\infty} (a_n \cos nt + b_n \sin nt). \tag{8}$$

We will see in Section 9.2 that under rather mild restrictions on the function $f(t)$, this is so! An infinite series of the form in (8) is called a *Fourier series*, and the representation of functions by Fourier series is one of the most widely used techniques in applied mathematics, especially for the solution of partial differential equations (see Sections 9.5 through 9.7).

Fourier Series of Period 2π Functions

In this section we will confine our attention to functions of period 2π. We want to determine what the coefficients in the Fourier series in (8) must be if it is to converge to a given function $f(t)$ of period 2π. For this purpose we need the following integrals, in which m and n denote positive integers (Problems 27 through 29):

$$\int_{-\pi}^{\pi} \cos mt \cos nt \, dt = \begin{cases} 0 & \text{if } m \neq n, \\ \pi & \text{if } m = n. \end{cases} \tag{9}$$

$$\int_{-\pi}^{\pi} \sin mt \sin nt \, dt = \begin{cases} 0 & \text{if } m \neq n, \\ \pi & \text{if } m = n. \end{cases} \tag{10}$$

$$\int_{-\pi}^{\pi} \cos mt \sin nt \, dt = 0 \quad \text{for all } m \text{ and } n. \tag{11}$$

These formulas imply that the functions $\cos nt$ and $\sin nt$ for $n = 1, 2, 3, \ldots$ constitute a *mutually orthogonal* set of functions on the interval $[-\pi, \pi]$. Two real-valued functions $u(t)$ and $v(t)$ are said to be **orthogonal** on the interval $[a, b]$ provided that

$$\int_a^b u(t)v(t) \, dt = 0. \tag{12}$$

Suppose now that the piecewise continuous function $f(t)$ of period 2π has a Fourier series representation

$$f(t) = \frac{a_0}{2} + \sum_{m=1}^{\infty} (a_m \cos mt + b_m \sin mt), \tag{13}$$

in the sense that the infinite series on the right converges to the value $f(t)$ for every t. We assume in addition that, when the infinite series in Eq. (13) is multiplied by

any continuous function, the resulting series can be integrated term by term. Then the result of termwise integration of Eq. (13) itself from $t = -\pi$ to $t = \pi$ is

$$\int_{-\pi}^{\pi} f(t)\,dt = \frac{a_0}{2} \int_{-\pi}^{\pi} 1\,dt$$

$$+ \sum_{m=1}^{\infty} \left(a_m \int_{-\pi}^{\pi} \cos mt\,dt \right) + \sum_{m=1}^{\infty} \left(b_m \int_{-\pi}^{\pi} \sin mt\,dt \right) = \pi a_0$$

because all the trigonometric integrals vanish. Hence

$$a_0 = \frac{1}{\pi} \int_{-\pi}^{\pi} f(t)\,dt. \tag{14}$$

If we first multiply each side in Eq. (13) by $\cos nt$ and then integrate termwise, the result is

$$\int_{-\pi}^{\pi} f(t) \cos nt\,dt = \frac{a_0}{2} \int_{-\pi}^{\pi} \cos nt\,dt$$

$$+ \sum_{m=1}^{\infty} \left(a_m \int_{-\pi}^{\pi} \cos mt\,\cos nt\,dt \right) + \sum_{m=1}^{\infty} \left(b_m \int_{\pi}^{\pi} \sin mt\,\cos nt\,dt \right);$$

it then follows from Eq. (11) that

$$\int_{-\pi}^{\pi} f(t) \cos nt\,dt = \sum_{m=1}^{\infty} a_m \left(\int_{-\pi}^{\pi} \cos mt\,\cos nt\,dt \right). \tag{15}$$

But of all the integrals (for $m = 1, 2, 3, \ldots$) on the right-hand side in (15), only the one for which $m = n$ is nonzero. Thus Eq. (9) yields

$$\int_{-\pi}^{\pi} f(t) \cos nt\,dt = a_n \int_{-\pi}^{\pi} \cos^2 nt\,dt = \pi a_n,$$

so the value of the coefficient a_n is

$$a_n = \frac{1}{\pi} \int_{-\pi}^{\pi} f(t) \cos nt\,dt. \tag{16}$$

Note that with $n = 0$, the formula in (16) reduces to Eq. (14); this explains why we denote the constant term in the original Fourier series by $\frac{1}{2}a_0$ (rather than simply a_0). If we multiply each side in Eq. (13) by $\sin nt$ and then integrate termwise, we find in a similar way that

$$b_n = \frac{1}{\pi} \int_{-\pi}^{\pi} f(t) \sin nt\,dt. \tag{17}$$

In short, we have found that if the series in (13) converges to $f(t)$ and if the termwise integrations carried out here are valid, then the coefficients in the series must have the values given in Eqs. (16) and (17). This motivates us to *define* the Fourier series of a periodic function by means of these formulas, whether or not the resulting series converges to the function (or even converges at all).

Definition: Fourier Series and Fourier Coefficients

Let $f(t)$ be a piecewise continuous function of period 2π that is defined for all t. Then the **Fourier series** of $f(t)$ is the series

$$\blacktriangleright \qquad \frac{a_0}{2} + \sum_{n=1}^{\infty}(a_n \cos nt + b_n \sin nt), \qquad (18)$$

where the **Fourier coefficients** a_n and b_n are defined by means of the formulas

$$\blacktriangleright \qquad a_n = \frac{1}{\pi}\int_{-\pi}^{\pi} f(t)\cos nt\, dt \qquad (16)$$

for $n = 0, 1, 2, 3, \ldots$ and

$$\blacktriangleright \qquad b_n = \frac{1}{\pi}\int_{-\pi}^{\pi} f(t)\sin nt\, dt \qquad (17)$$

for $n = 1, 2, 3, \ldots.$

It can (and often does) happen that the Fourier series of a function fails to converge to the function at certain points in the domain of the function. We will therefore write

$$f(t) \sim \frac{a_0}{2} + \sum_{n=1}^{\infty}(a_n \cos nt + b_n \sin nt), \qquad (19)$$

not using an equals sign between the function and its Fourier series until we have discussed convergence of Fourier series in Section 9.2.

Suppose that the piecewise continuous function $f(t)$ as given initially is defined only on the interval $[-\pi, \pi]$, and assume that $f(-\pi) = f(\pi)$. Then we can extend f so that its domain includes all real numbers by means of the periodicity condition $f(t + 2\pi) = f(t)$ for all t. We continue to denote this extension of the original function by f, and note that it automatically has period 2π. Its graph looks the same on every interval of the form

$$(2n - 1)\pi \leqq t \leqq (2n + 1)\pi$$

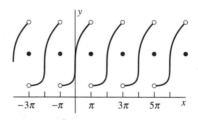

FIGURE 9.1.2. Extending a function to produce a periodic function.

where n is an integer (Fig. 9.1.2). For instance, the square wave function of Fig. 9.1.1 can be described as the period 2π function such that

$$f(t) = \begin{cases} -1 & \text{if } -\pi < t < 0; \\ +1 & \text{if } 0 < t < \pi; \\ \;\;0 & \text{if } t = -\pi, 0, \text{ or } \pi. \end{cases} \qquad (20)$$

Thus the square wave function is the period 2π function defined on one full period by means of Eq. (20).

We need to consider Fourier series of piecewise continuous functions because many functions that appear in applications are only piecewise continuous, not continuous. Note that the integrals in Eqs. (16) and (17) exist if $f(t)$ is piecewise continuous, so every piecewise continuous function has a Fourier series.

Example 1 Find the Fourier series of the square wave function defined in Eq. (20).

Solution It is always a good idea to calculate a_0 separately, using Eq. (14). Thus

$$a_0 = \frac{1}{\pi} \int_{-\pi}^{\pi} f(t)\, dt = \frac{1}{\pi} \int_{-\pi}^{0} (-1)\, dt + \frac{1}{\pi} \int_{0}^{\pi} (+1)\, dt$$

$$= \frac{1}{\pi}(-\pi) + \frac{1}{\pi}(\pi) = 0.$$

We split the first integral into two integrals because $f(t)$ is defined by different formulas on the intervals $(-\pi, 0)$ and $(0, \pi)$; the values of $f(t)$ at the endpoints of these intervals do not affect the values of the integrals.

Equation (16) yields (for $n > 0$)

$$a_n = \frac{1}{\pi} \int_{-\pi}^{\pi} f(t) \cos nt\, dt = \frac{1}{\pi} \int_{-\pi}^{0} (-\cos nt)\, dt + \frac{1}{\pi} \int_{0}^{\pi} \cos nt\, dt$$

$$= \frac{1}{\pi}\left[-\frac{1}{n} \sin nt \right]_{-\pi}^{0} + \frac{1}{\pi}\left[\frac{1}{n} \sin nt \right]_{0}^{\pi} = 0.$$

And Eq. (17) yields

$$b_n = \frac{1}{\pi} \int_{-\pi}^{\pi} f(t) \sin nt\, dt = \frac{1}{\pi} \int_{-\pi}^{0} (-\sin nt)\, dt + \frac{1}{\pi} \int_{0}^{\pi} \sin nt\, dt$$

$$= \frac{1}{\pi}\left[\frac{1}{n} \cos nt \right]_{-\pi}^{0} + \frac{1}{\pi}\left[-\frac{1}{n} \cos nt \right]_{0}^{\pi}$$

$$= \frac{2}{n\pi}(1 - \cos n\pi) = \frac{2}{n\pi}[1 - (-1)^n].$$

Thus $a_n = 0$ for all $n \geqq 0$, and

$$b_n = \begin{cases} \dfrac{4}{n\pi} & \text{for } n \text{ odd;} \\[2mm] 0 & \text{for } n \text{ even.} \end{cases}$$

The last result follows because $\cos(-n\pi) = \cos(n\pi) = (-1)^n$. With these values of the Fourier coefficients, we obtain the Fourier series

$$f(t) \sim \frac{4}{\pi} \sum_{n \text{ odd}} \frac{\sin nt}{n} = \frac{4}{\pi}\left(\sin t + \frac{1}{3} \sin 3t + \frac{1}{5} \sin 5t + \cdots \right). \qquad (21)$$

Here we have introduced the useful abbreviation

$$\sum_{n \text{ odd}} \quad \text{for} \quad \sum_{\substack{n=1 \\ n \text{ odd}}}^{\infty}$$

—for example,

$$\sum_{n \text{ odd}} \frac{1}{n} = 1 + \frac{1}{3} + \frac{1}{5} + \cdots.$$

Figure 9.1.3 shows the graphs of several of the partial sums

$$S_N(t) = \frac{4}{\pi} \sum_{n=1}^{N} \frac{\sin(2n-1)t}{2n-1}$$

of the Fourier series in (21). Note that as t approaches a discontinuity of $f(t)$ from either side, the value of $S_n(t)$ tends to overshoot the limiting value of $f(t)$—either $+1$ or -1 in this case. This behavior of a Fourier series near a point of discontinuity of its function is typical and is known as **Gibbs's phenomenon**. ∎

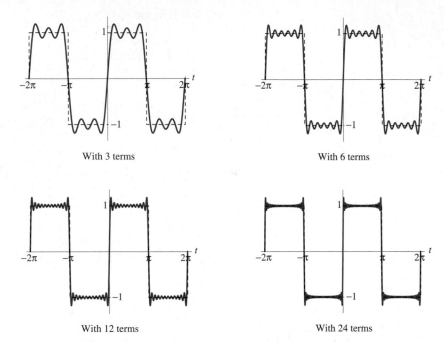

With 3 terms With 6 terms

With 12 terms With 24 terms

FIGURE 9.1.3. Graphs of partial sums of the Fourier series of the square wave function (Example 1) with $N = 3, 6, 12,$ and 24 terms.

The following integral formulas, easily derived by integration by parts, are useful in computing Fourier series of polynomial functions:

$$\int u \cos u \, du = \cos u + u \sin u + C; \tag{22}$$

$$\int u \sin u \, du = \sin u - u \cos u + C; \tag{23}$$

$$\int u^n \cos u \, du = u^n \sin u - n \int u^{n-1} \sin u \, du; \tag{24}$$

$$\int u^n \sin u \, du = -u^n \cos u + n \int u^{n-1} \cos u \, du. \tag{25}$$

Example 2 Find the Fourier series of the period 2π function that is defined in one period to be

$$f(t) = \begin{cases} 0 & \text{if } -\pi < t \leqq 0; \\ t & \text{if } 0 \leqq t < \pi; \\ \dfrac{\pi}{2} & \text{if } t = \pm\pi. \end{cases} \tag{26}$$

The graph of f is shown in Fig. 9.1.4.

Solution The values of $f(\pm\pi)$ are irrelevant because they have no effect on the values of the integrals that yield the Fourier coefficients. Because $f(t) \equiv 0$ on the interval $(-\pi, 0)$, each integral from $t = -\pi$ to $t = \pi$ may be replaced with an integral from $t = 0$ to $t = \pi$. Equations (14), (16), and (17) therefore give

FIGURE 9.1.4. The periodic function of Example 2.

$$a_0 = \frac{1}{\pi} \int_0^\pi t \, dt = \frac{1}{\pi}\left[\frac{1}{2}t^2\right]_0^\pi = \frac{\pi}{2};$$

$$a_n = \frac{1}{\pi} \int_0^\pi t \cos nt \, dt = \frac{1}{n^2\pi} \int_0^{n\pi} u \cos u \, du \qquad \left(u = nt, \ t = \frac{u}{n}\right)$$

$$= \frac{1}{n^2\pi}\Big[\cos u + u \sin u\Big]_0^{n\pi} \qquad \text{(by Eq. (22))}$$

$$= \frac{1}{n^2\pi}[(-1)^n - 1].$$

Consequently, $a_n = 0$ if n is even and $n \geqq 2$;

$$a_n = -\frac{2}{n^2\pi} \qquad \text{if } n \text{ is odd.}$$

Next,

$$b_n = \frac{1}{\pi} \int_0^\pi t \sin nt \, dt = \frac{1}{n^2\pi} \int_0^{n\pi} u \sin u \, du$$

$$= \frac{1}{n^2\pi}\Big[\sin u - u \cos u\Big]_0^{n\pi} \qquad \text{(by Eq. (20))}$$

$$= -\frac{1}{n}\cos n\pi.$$

Thus

$$b_n = \frac{(-1)^{n+1}}{n} \qquad \text{for all } n \geqq 1.$$

Therefore, the Fourier series of $f(t)$ is

$$f(t) \sim \frac{\pi}{4} - \frac{2}{\pi} \sum_{n \text{ odd}} \frac{\cos nt}{n^2} + \sum_{n=1}^{\infty} \frac{(-1)^{n+1} \sin nt}{n}. \tag{27}$$

∎

If $f(t)$ is a function of period 2π, it is readily verified (Problem 30) that

$$\int_{-\pi}^{\pi} f(t)\,dt = \int_{a}^{a+2\pi} f(t)\,dt \tag{28}$$

for all a. That is, the integral of $f(t)$ over one interval of length 2π is equal to its integral over any other such interval. In case $f(t)$ is given explicitly on the interval $[0, 2\pi]$ rather than on $[-\pi, \pi]$, it may be more convenient to compute its Fourier coefficients as

$$a_n = \frac{1}{\pi} \int_{0}^{2\pi} f(t) \cos nt\,dt \tag{29a}$$

and

$$b_n = \frac{1}{\pi} \int_{0}^{2\pi} f(t) \sin nt\,dt. \tag{29b}$$

9.1 Problems

In Problems 1 through 10, sketch the graph of the function f defined for all t by the given formula, and determine whether it is periodic. If so, find its smallest period.

1. $f(t) = \sin 3t$

2. $f(t) = \cos 2\pi t$

3. $f(t) = \cos \dfrac{3t}{2}$

4. $f(t) = \sin \dfrac{\pi t}{3}$

5. $f(t) = \tan t$

6. $f(t) = \cot 2\pi t$

7. $f(t) = \cosh 3t$

8. $f(t) = \sinh \pi t$

9. $f(t) = |\sin t|$

10. $f(t) = \cos^2 3t$

In Problems 11 through 26, the values of a period 2π function $f(t)$ in one full period are given. Sketch several periods of its graph and find its Fourier series.

11. $f(t) \equiv 1 , -\pi \leqq t \leqq \pi$

12. $f(t) = \begin{cases} +3, & -\pi < t \leqq 0; \\ -3, & 0 < t \leqq \pi \end{cases}$

13. $f(t) = \begin{cases} 0, & -\pi < t \leqq 0; \\ 1, & 0 < t \leqq \pi \end{cases}$

14. $f(t) = \begin{cases} 3, & -\pi < t \leqq 0; \\ -2, & 0 < t \leqq \pi \end{cases}$

15. $f(t) = t, -\pi < t \leqq \pi$

16. $f(t) = t, 0 < t < 2\pi$

17. $f(t) = |t|, -\pi \leqq t \leqq \pi$

18. $f(t) = \begin{cases} \pi + t, & -\pi < t \leqq 0; \\ \pi - t, & 0 < t \leqq \pi \end{cases}$

19. $f(t) = \begin{cases} \pi + t, & -\pi \leqq t < 0; \\ 0, & 0 \leqq t \leqq \pi \end{cases}$

20. $f(t) = \begin{cases} 0, & -\pi \leqq t < -\pi/2; \\ 1, & -\pi/2 \leqq t \leqq \pi/2; \\ 0, & \pi/2 < t \leqq \pi \end{cases}$

21. $f(t) = t^2 , -\pi \leqq t \leqq \pi$

22. $f(t) = t^2 , 0 \leqq t < 2\pi$

23. $f(t) = \begin{cases} 0, & -\pi \leqq t \leqq 0; \\ t^2, & 0 \leqq t < \pi \end{cases}$

24. $f(t) = |\sin t| , -\pi \leqq t \leqq \pi$

25. $f(t) = \cos^2 2t , -\pi \leqq t \leqq \pi$

26. $f(t) = \begin{cases} 0, & -\pi \leqq t \leqq 0; \\ \sin t, & 0 \leqq t \leqq \pi \end{cases}$

27. Verify Eq. (9). (*Suggestion*: Use the trigonometric identity

$$\cos A \cos B = \tfrac{1}{2}[\cos(A + B) + \cos(A - B)].)$$

28. Verify Eq. (10).

29. Verify Eq. (11).

30. Let $f(t)$ be a piecewise continuous function with period P. **(a)** Suppose that $0 \leqq a < P$. Substitute $u = t - P$ to show that

$$\int_{P}^{a+P} f(t)\,dt = \int_{0}^{a} f(t)\,dt.$$

Conclude that

$$\int_{a}^{a+P} f(t)\,dt = \int_{0}^{P} f(t)\,dt.$$

(b) Given A, choose n so that $A = nP + a$ with $0 \leqq a < P$. Then substitute $v = t - nP$ to show that

$$\int_{A}^{A+P} f(t)\,dt = \int_{a}^{a+P} f(t)\,dt = \int_{0}^{P} f(t)\,dt.$$

31. Multiply each side in Eq. (13) by $\sin nt$ and then integrate term by term to derive Eq. (17).

9.2 General Fourier Series and Convergence

In Section 9.1 we defined the Fourier series of a periodic function of period 2π. Now let $f(t)$ be a function that is piecewise continuous for all t and has arbitrary period $P > 0$. We write

$$P = 2L, \tag{1}$$

so L is the **half-period** of the function f. Let us define the function g as follows:

$$g(u) = f\left(\frac{Lu}{\pi}\right) \tag{2}$$

for all u. Then

$$g(u + 2\pi) = f\left(\frac{Lu}{\pi} + 2L\right) = f\left(\frac{Lu}{\pi}\right) = g(u),$$

and hence $g(u)$ is also periodic and has period 2π. Consequently, g has the Fourier series

$$g(u) \sim \frac{a_0}{2} + \sum_{n=1}^{\infty}(a_n \cos nu + b_n \sin nu) \tag{3}$$

with Fourier coefficients

$$a_n = \frac{1}{\pi}\int_{-\pi}^{\pi} g(u) \cos nu\, du \tag{4a}$$

and

$$b_n = \frac{1}{\pi}\int_{-\pi}^{\pi} g(u) \sin nu\, du. \tag{4b}$$

If we now write

$$t = \frac{Lu}{\pi}, \quad u = \frac{\pi t}{L}, \quad f(t) = g(u), \tag{5}$$

then

$$f(t) = g\left(\frac{\pi t}{L}\right) \sim \frac{a_0}{2} + \sum_{n=1}^{\infty}\left(a_n \cos \frac{n\pi t}{L} + b_n \frac{n\pi t}{L}\right), \tag{6}$$

and then substitution of (5) in (4) yields

$$a_n = \frac{1}{\pi}\int_{-\pi}^{\pi} g(u) \cos nu\, du \quad \left(u = \frac{\pi t}{L},\ du = \frac{\pi}{L}dt\right)$$

$$= \frac{1}{L}\int_{-L}^{L} g\left(\frac{\pi t}{L}\right) \cos \frac{n\pi t}{L}\, dt.$$

Therefore,

$$a_n = \frac{1}{L} \int_{-L}^{L} f(t) \cos \frac{n\pi t}{L} \, dt; \tag{7}$$

similarly,

$$b_n = \frac{1}{L} \int_{-L}^{L} f(t) \sin \frac{n\pi t}{L} \, dt. \tag{8}$$

This computation motivates the following definition of the Fourier series of a periodic function of period $2L$.

Definition: Fourier Series and Fourier Coefficients

Let $f(t)$ be a piecewise continuous function of period $2L$ that is defined for all t. Then the **Fourier series** of $f(t)$ is the series

➤
$$f(t) \sim \frac{a_0}{2} + \sum_{n=1}^{\infty} \left(a_n \cos \frac{n\pi t}{L} + b_n \sin \frac{n\pi t}{L} \right), \tag{6}$$

where the **Fourier coefficients** $\{a_n\}_0^\infty$ and $\{b_n\}_1^\infty$ are defined to be

➤
$$a_n = \frac{1}{L} \int_{-L}^{L} f(t) \cos \frac{n\pi t}{L} \, dt \tag{7}$$

and

➤
$$b_n = \frac{1}{L} \int_{-L}^{L} f(t) \sin \frac{n\pi t}{L} \, dt. \tag{8}$$

With $n = 0$, Eq. (7) takes the simple form

$$a_0 = \frac{1}{L} \int_{-L}^{L} f(t) \, dt, \tag{9}$$

which demonstrates that the constant term $\frac{1}{2} a_0$ in the Fourier series of f is simply the average value of $f(t)$ on the interval $[-L, L]$.

As a consequence of Problem 30 of Section 9.1, we may evaluate the integrals in (7) and (8) over any other interval of length $2L$. For instance, if $f(t)$ is given by a single formula for $0 < t < 2L$, it may be more convenient to compute the integrals

$$a_n = \frac{1}{L} \int_{0}^{2L} f(t) \cos \frac{n\pi t}{L} \, dt \tag{10a}$$

and

$$b_n = \frac{1}{L} \int_{0}^{2L} f(t) \sin \frac{n\pi t}{L} \, dt. \tag{10b}$$

Example 1 Figure 9.2.1 shows the graph of a square wave function with period 4. Find its Fourier series.

Solution Here, $L = 2$; also, $f(t) = -1$ if $-2 < t < 0$, while $f(t) = 1$ if $0 < t < 2$. Hence Eqs. (7), (8), and (9) yield

$$a_0 = \frac{1}{2} \int_{-2}^{2} f(t)\, dt = \frac{1}{2} \int_{-2}^{0} (-1)\, dt + \frac{1}{2} \int_{0}^{2} (+1)\, dt = 0,$$

$$a_n = \frac{1}{2} \int_{-2}^{0} (-1) \cos \frac{n\pi t}{2}\, dt + \frac{1}{2} \int_{0}^{2} (+1) \cos \frac{n\pi t}{2}\, dt$$

$$= \frac{1}{2} \left[-\frac{2}{n\pi} \sin \frac{n\pi t}{2} \right]_{-2}^{0} + \frac{1}{2} \left[\frac{2}{n\pi} \sin \frac{n\pi t}{2} \right]_{0}^{2} = 0,$$

FIGURE 9.2.1. The square wave of Example 1.

and

$$b_n = \frac{1}{2} \int_{-2}^{0} (-1) \sin \frac{n\pi t}{2}\, dt + \frac{1}{2} \int_{0}^{2} (+1) \sin \frac{n\pi t}{2}\, dt$$

$$+ \frac{1}{2} \left[\frac{2}{n\pi} \cos \frac{n\pi t}{2} \right]_{-2}^{0} + \frac{1}{2} \left[-\frac{2}{n\pi} \cos \frac{n\pi t}{2} \right]_{0}^{2}$$

$$= \frac{2}{n\pi} \left| 1 - (-1)^n \right| = \begin{cases} \dfrac{4}{n\pi} & \text{if } n \text{ is odd,} \\[2mm] 0 & \text{if } n \text{ is even.} \end{cases}$$

Thus the Fourier series is

$$f(t) \sim \frac{4}{\pi} \sum_{n \text{ odd}} \frac{1}{n} \sin \frac{n\pi t}{2} \tag{11a}$$

$$= \frac{4}{\pi} \left(\sin \frac{\pi t}{2} + \frac{1}{3} \sin \frac{3\pi t}{2} + \frac{1}{5} \sin \frac{5\pi t}{2} + \cdots \right). \tag{11b}$$

∎

The Convergence Theorem

We want to impose conditions on the periodic function f that are enough to guarantee that its Fourier series actually converges to $f(t)$ at least at those values of t at which f is continuous. Recall that the function f is said to be *piecewise continuous* on the interval $[a, b]$ provided that there is a finite partition of $[a, b]$ with endpoints

$$a = t_0 < t_1 < t_2 < \cdots < t_{n-1} < t_n = b$$

such that

1. f is continuous on each open interval $t_{i-1} < t < t_i$; and
2. At each endpoint t_i of such a subinterval the limit of $f(t)$, as t approaches t_i from within the subinterval, exists and is finite.

FIGURE 9.2.2. A finite jump discontinuity.

The function f is called *piecewise continuous* for all t if it is piecewise continuous on every bounded interval. It follows that a piecewise continuous function is continuous except possibly at isolated points, and that at each such point of discontinuity, the one-sided limits

$$f(t+) = \lim_{u \to t^+} f(u) \quad \text{and} \quad f(t-) = \lim_{u \to t^-} f(u) \qquad (12)$$

both exist and are finite. Thus a piecewise continuous function has only isolated "finite jump" discontinuities like the one shown in Fig. 9.2.2.

The square wave and sawtooth functions that we saw in Chapter 7 are typical examples of periodic piecewise continuous functions. The function $f(t) = \tan t$ is a periodic function (of period π) that is not piecewise continuous because it has infinite discontinuities. The function $g(t) = \sin(1/t)$ is not piecewise continuous on $[-1, 1]$ because its one-sided limits at $t = 0$ do not exist. The function

$$h(t) = \begin{cases} t & \text{if } t = \dfrac{1}{n} \quad (n \text{ an integer}), \\ 0 & \text{otherwise} \end{cases}$$

on $[-1, 1]$ has one-sided limits everywhere, but is not piecewise continuous because its discontinuities are not isolated—it has the infinite sequence $\{1/n\}_1^\infty$ of discontinuities; a piecewise continuous function can have only finitely many discontinuities in any bounded interval.

Note that a piecewise continuous function need not be defined at its isolated points of discontinuity. Alternatively, it can be defined arbitrarily at such points. For instance, the square wave function f of Fig. 9.2.1 is piecewise continuous no matter what its values might be at the points $\ldots, -4, -2, 0, 2, 4, 6, \ldots$ at which it is discontinuous. Its derivative f' is also piecewise continuous; $f'(t) = 0$ unless t is an even integer, in which case $f'(t)$ is undefined.

The piecewise continuous function f is said to be **piecewise smooth** provided that its derivative f' is piecewise continuous. Theorem 1 (next) tells us that the Fourier series of a piecewise smooth function converges everywhere. More general Fourier convergence theorems—with weaker hypotheses on the periodic function f—are known. But the hypothesis that f is piecewise smooth is easy to check and is satisfied by most functions encountered in practical applications. A proof of the following theorem may be found in G. P. Tolstov, *Fourier Series* (New York: Dover, 1976).

THEOREM 1: Convergence of Fourier Series

Suppose that the periodic function f is piecewise smooth. Then its Fourier series in (6) converges

(a) to the value $f(t)$ at each point where f is continuous, and

(b) to the value $\frac{1}{2}[f(t+) + f(t-)]$ at each point where f is discontinuous. ▲

Note that $\frac{1}{2}[f(t+) + f(t-)]$ is the *average* of the right-hand and left-hand limits of f at the point t. If f is continuous at t, then $f(t) = f(t+) = f(t-)$, so

$$f(t) = \frac{f(t+) + f(t-)}{2}. \qquad (13)$$

Hence Theorem 1 could be rephrased as follows: The Fourier series of a piecewise smooth function f converges for *every* t to the average value in (13). For this reason it is customary to write

$$f(t) = \frac{a_0}{2} + \sum_{n=1}^{\infty} \left(a_n \cos \frac{n\pi t}{L} + b_n \sin \frac{n\pi t}{L} \right), \tag{14}$$

with the understanding that the piecewise smooth function f has been redefined (if necessary) at each of its points of discontinuity in order to satisfy the average value condition in (13).

Example 1

Continued

Figure 9.2.1 shows us at a glance that if t_0 is an even integer, then

$$\lim_{t \to t_0^+} f(t) = +1 \quad \text{and} \quad \lim_{t \to t_0^-} f(t) = -1.$$

Hence

$$\frac{f(t_0+) + f(t_0-)}{2} = 0.$$

Note that, in accord with Theorem 1, the Fourier series of $f(t)$ in (11) clearly converges to zero if n is an even integer (because $\sin n\pi = 0$). ∎

Example 2

Let $f(t)$ be a function of period 2 with $f(t) = t^2$ if $0 < t < 2$. We define $f(t)$ for t an even integer by the average value condition in (13); consequently, $f(t) = 2$ if t is an even integer. The graph of the function f appears in Fig. 9.2.3. Find its Fourier series.

Solution

Here $L = 1$, and it is most convenient to integrate from $t = 0$ to $t = 2$. Then

$$a_0 = \frac{1}{1} \int_0^2 t^2 \, dt = \left[\frac{1}{3} t^3 \right]_0^2 = \frac{8}{3}.$$

With the aid of the integral formulas in Eqs. (22) through (25) of Section 9.1, we obtain

$$a_n = \int_0^2 t^2 \cos n\pi t \, dt$$

$$= \frac{1}{n^3 \pi^3} \int_0^{2n\pi} u^2 \cos u \, du \qquad \left(u = n\pi t, \ t = \frac{u}{n\pi} \right)$$

$$= \frac{1}{n^3 \pi^3} \left[u^2 \sin u - 2 \sin u + 2u \cos u \right]_0^{2n\pi} = \frac{4}{n^2 \pi^2};$$

$$b_n = \int_0^2 t^2 \sin n\pi t \, dt = \frac{1}{n^3 \pi^3} \int_0^{2n\pi} u^2 \sin u \, du$$

$$= \frac{1}{n^3 \pi^3} \left[-u^2 \cos u + 2 \cos u + 2u \sin u \right]_0^{2n\pi} = -\frac{4}{n\pi}.$$

FIGURE 9.2.3. The period 2 function of Example 2.

Hence the Fourier series of f is

$$f(t) = \frac{4}{3} + \frac{4}{\pi^2} \sum_{n=1}^{\infty} \frac{\cos n\pi t}{n^2} - \frac{4}{\pi} \sum_{n=1}^{\infty} \frac{\sin n\pi t}{n}, \tag{15}$$

and Theorem 1 assures us that this series converges to $f(t)$ for all t. ∎

We can draw some interesting consequences from the Fourier series in (15). If we substitute $t = 0$ on each side, we find that

$$f(0) = 2 = \frac{4}{3} + \frac{4}{\pi^2} \sum_{n=1}^{\infty} \frac{1}{n^2}.$$

On solving for the series, we obtain the lovely summation

$$\blacktriangleright \qquad \sum_{n=1}^{\infty} \frac{1}{n^2} = 1 + \frac{1}{2^2} + \frac{1}{3^2} + \frac{1}{4^2} + \cdots = \frac{\pi^2}{6} \tag{16}$$

that was discovered by Euler. If we substitute $t = 1$ in Eq. (15), we get

$$f(1) = 1 = \frac{4}{3} + \frac{4}{\pi^2} \sum_{n=1}^{\infty} \frac{(-1)^n}{n^2},$$

which yields

$$\sum_{n=1}^{\infty} \frac{(-1)^{n+1}}{n^2} = 1 - \frac{1}{2^2} + \frac{1}{3^2} - \frac{1}{4^2} + \cdots = \frac{\pi^2}{12}. \tag{17}$$

If we add the series in Eqs. (16) and (17) and then divide by 2, the "even" terms cancel and the result is

$$\sum_{n \text{ odd}} \frac{1}{n^2} = 1 + \frac{1}{3^2} + \frac{1}{5^2} + \frac{1}{7^2} + \cdots = \frac{\pi^2}{8}. \tag{18}$$

9.2 Problems

In Problems 1 through 14, the values of a periodic function $f(t)$ in one full period are given; at each discontinuity the value of $f(t)$ is that given by the average value condition in (13). Sketch the graph of f and find its Fourier series.

1. $f(t) = \begin{cases} -2, & -3 < t < 0; \\ 2, & 0 < t < 3 \end{cases}$

2. $f(t) = \begin{cases} 0, & -5 < t < 0; \\ 1, & 0 < t < 5 \end{cases}$

3. $f(t) = \begin{cases} 2, & -2\pi < t < 0; \\ -1, & 0 < t < 2\pi \end{cases}$

4. $f(t) = t, \ -2 < t < 2$

5. $f(t) = t, \ -2\pi < t < 2\pi$

6. $f(t) = t, \ 0 < t < 3$

7. $f(t) = |t|, \ -1 < t < 1$

8. $f(t) = \begin{cases} 0, & 0 < t < 1; \\ 1, & 1 < t < 2; \\ 0, & 2 < t < 3 \end{cases}$

9. $f(t) = t^2, \ -1 < t < 1$

10. $f(t) = \begin{cases} 0, & -2 < t < 0; \\ t^2, & 0 < t < 2 \end{cases}$

11. $f(t) = \cos \dfrac{\pi t}{2}, \ -1 < t < 1$

12. $f(t) = \sin \pi t, \ 0 < t < 1$

13. $f(t) = \begin{cases} 0, & -1 < t < 0; \\ \sin \pi t, & 0 < t < 1 \end{cases}$

14. $f(t) = \begin{cases} 0, & -2\pi < t < 0; \\ \sin t, & 0 < t < 2\pi \end{cases}$

15. (a) Suppose that f is a function of period 2π with $f(t) = t^2$ for $0 < t < 2\pi$. Show that

$$f(t) = \frac{4\pi^2}{3} + 4 \sum_{n=1}^{\infty} \frac{\cos nt}{n^2} - 4\pi \sum_{n=1}^{\infty} \frac{\sin nt}{n}$$

and sketch the graph of f, indicating the value at each discontinuity. **(b)** Deduce the series summations in Eqs. (16) and (17) from the Fourier series in part (a).

16. (a) Suppose that f is a function of period 2 such that $f(t) = 0$ if $-1 < t < 0$ and $f(t) = t$ if $0 < t < 1$. Show that

$$f(t) = \frac{1}{4} - \frac{2}{\pi^2} \sum_{n \text{ odd}} \frac{\cos n\pi t}{n^2} + \frac{1}{\pi} \sum_{n=1}^{\infty} \frac{(-1)^{n+1} \sin n\pi t}{n}$$

and sketch the graph of f, indicating the value at each discontinuity. **(b)** Deduce the series summation in Eq. (18) from the Fourier series in part (a).

17. (a) Suppose that f is a function of period 2 with $f(t) = t$ for $0 < t < 2$. Show that

$$f(t) = 1 - \frac{2}{\pi} \sum_{n=1}^{\infty} \frac{\sin n\pi t}{n}$$

and sketch the graph of f, indicating the value at each discontinuity. **(b)** Substitute an appropriate value of t to deduce **Leibniz's series**

$$1 - \frac{1}{3} + \frac{1}{5} - \frac{1}{7} + \cdots = \frac{\pi}{4}.$$

Derive the Fourier series listed in Problems 18 through 21, and graph the period 2π function to which each series converges.

18. $\displaystyle\sum_{n=1}^{\infty} \frac{\sin nt}{n} = \frac{\pi - t}{2} \quad (0 < t < 2\pi)$

19. $\displaystyle\sum_{n=1}^{\infty} \frac{(-1)^{n+1} \sin nt}{n} = \frac{t}{2} \quad (-\pi < t < \pi)$

20. $\displaystyle\sum_{n=1}^{\infty} \frac{\cos nt}{n^2} = \frac{3t^2 - 6\pi t + 2\pi^2}{12} \quad (0 < t < 2\pi)$

21. $\displaystyle\sum_{n=1}^{\infty} \frac{(-1)^{n+1} \cos nt}{n^2} = \frac{\pi^2 - 3t^2}{12} \quad (-\pi < t < \pi)$

22. Suppose that $p(t)$ is a polynomial of degree n. Show by repeated integration by parts that

$$\int p(t)g(t) \, dt = p(t)G_1(t) - p'(t)G_2(t)$$
$$+ p''(t)G_3(t) - \cdots + (-1)^n p^{(n)}(t)G_{n+1}(t)$$

where $G_k(t)$ denotes the kth iterated antiderivative $G_k(t) = (D^{-1})^k g(t)$. This formula is useful in computing Fourier coefficients of polynomials.

23. Apply the integral formula of Problem 22 to show that

$$\int t^4 \cos t \, dt = t^4 \sin t + 4t^3 \cos t$$
$$- 12t^2 \sin t - 24t \cos t + 24 \sin t + C$$

and that

$$\int t^4 \sin t \, dt = -t^4 \cos t + 4t^3 \sin t$$
$$+ 12t^2 \cos t - 24t \sin t - 24 \cos t + C.$$

24. (a) Show that for $0 < t < 2\pi$,

$$t^4 = \frac{16\pi^4}{5} + 16 \sum_{n=1}^{\infty} \left(\frac{2\pi^2}{n^2} - \frac{3}{n^4} \right) \cos nt$$
$$+ 16\pi \sum_{n=1}^{\infty} \left(\frac{3}{n^3} - \frac{\pi^2}{n} \right) \sin nt$$

and sketch the graph of f, indicating the value at each discontinuity. **(b)** From the Fourier series in part (a), deduce the summations

$$\sum_{n=1}^{\infty} \frac{1}{n^4} = \frac{\pi^4}{90}, \qquad \sum_{n=1}^{\infty} \frac{(-1)^{n+1}}{n^4} = \frac{7\pi^4}{720},$$

and

$$\sum_{n \text{ odd}} \frac{1}{n^4} = \frac{\pi^4}{96}.$$

25. (a) Find the Fourier series of the period 2π function f with $f(t) = t^3$ if $-\pi < t < \pi$. **(b)** Use the series of part (a) to derive the summation

$$1 - \frac{1}{3^3} + \frac{1}{5^3} - \frac{1}{7^3} + \cdots = \frac{\pi^3}{32}$$

and sketch the graph of f, indicating the value at each discontinuity. **(c)** *Attempt* to evaluate the series

$$\sum_{n=1}^{\infty} \frac{1}{n^3} = 1 + \frac{1}{2^3} + \frac{1}{3^3} + \frac{1}{4^3} + \cdots$$

by substituting an appropriate value of t in the Fourier series of part (a). Is your attempt successful? Explain.

9.2 Application: Computer Algebra Calculation of Fourier Coefficients

A computer algebra system can greatly ease the burden of calculation of the Fourier coefficients of a given function $f(t)$. In the case of a function that is defined "piecewise," we must take care to "split" the integral according to the different intervals of definition of the function. We illustrate the method by deriving the Fourier series of the period 2π square wave function defined on $(-\pi, \pi)$ by

$$f(t) = \begin{cases} -1 & \text{if } -\pi < t < 0, \\ +1 & \text{if } 0 < t < \pi. \end{cases} \tag{1}$$

In this case the function is defined by different formulas on two different intervals, so each Fourier coefficient integral from $-\pi$ to π must be calculated as the sum of two integrals:

$$a_n = \frac{1}{\pi} \int_{-\pi}^{0} (-1) \cos nt \, dt + \frac{1}{\pi} \int_{0}^{\pi} (+1) \cos nt \, dt,$$

$$b_n = \frac{1}{\pi} \int_{-\pi}^{0} (-1) \sin nt \, dt + \frac{1}{\pi} \int_{0}^{\pi} (+1) \sin nt \, dt. \tag{2}$$

We can define the coefficients in (2) as functions of n by the *Maple* commands

```
a := n -> (1/Pi)*(int(-cos(n*t), t=-Pi..0) +
                    int(+cos(n*t), t=0..Pi)):
b := n -> (1/Pi)*(int(-sin(n*t), t=-Pi..0) +
                    int(+sin(n*t), t=0..Pi)):
```

or by the *Mathematica* commands

```
a[n_] := (1/Pi)*(Integrate[-Cos[n*t], {t, -Pi, 0}] +
                    Integrate[+Cos[n*t], {t, 0, Pi}])
b[n_] := (1/Pi)*(Integrate[-Sin[n*t], {t, -Pi, 0}] +
                    Integrate[+Sin[n*t], {t, 0, Pi}])
```

Because the function $f(t)$ in Eq. (1) is odd, we naturally find that $a_n \equiv 0$. Hence the *Maple* commands

```
fourierSum := sum(b(n)*sin(n*t), n=1..9);
plot(fourierSum, t=-2*Pi..4*Pi);
```

or the *Mathematica* commands

```
fourierSum = Sum[b[n]*Sin[n*t], {n,1,9}]
Plot[fourierSum, {t, -2*Pi, 4*Pi}];
```

yield the partial sum

$$\sum_{n=1}^{9} b_n \sin nt = \frac{4}{\pi} \left(\sin t + \frac{\sin 3t}{3} + \frac{\sin 5t}{5} + \frac{\sin 7t}{7} + \frac{\sin 9t}{9} \right)$$

and generate a graph like one of those in Fig. 9.1.3. The corresponding MATLAB commands are entirely analogous and can be found in the applications manual that accompanies this text.

To practice the symbolic derivation of Fourier series in this manner, you can begin by verifying the Fourier series calculated manually in Examples 1 and 2 of this section. Then Problems 1 through 21 are fair game. Finally, the period 2π triangular wave and trapezoidal wave functions illustrated in Figs. 9.2.4 and 9.2.5 have especially interesting Fourier series that we invite you to discover for yourself.

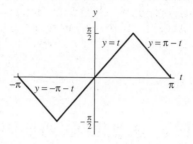

FIGURE 9.2.4. The triangular wave.

FIGURE 9.2.5. The trapezoidal wave.

9.3 Fourier Sine and Cosine Series

Certain properties of functions are reflected prominently in their Fourier series. The function f defined for all t is said to be **even** if

$$f(-t) = f(t) \tag{1}$$

for all t; f is **odd** if

$$f(-t) = -f(t) \tag{2}$$

for all t. The first condition implies that the graph of $y = f(t)$ is symmetric with respect to the y-axis, whereas the condition in (2) implies that the graph of an odd function is symmetric with respect to the origin (see Fig. 9.3.1). The functions $f(t) = t^{2n}$ and $g(t) = \cos t$ are even functions, whereas the functions $f(t) = t^{2n+1}$ and $g(t) = \sin t$ are odd. We will see that the Fourier series of an even periodic function has only cosine terms and that the Fourier series of an odd periodic function has only sine terms.

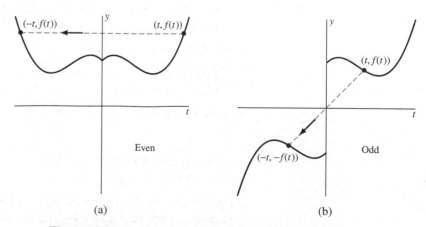

FIGURE 9.3.1. (a) An even function; (b) an odd function.

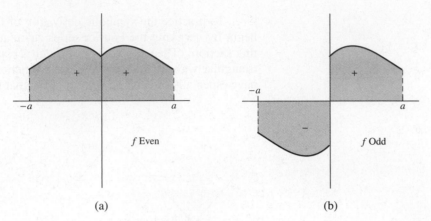

(a) (b)

FIGURE 9.3.2. Area under the graph of (a) an even function and (b) an odd.

Addition and cancellation of areas as indicated in Fig. 9.3.2 reminds us of the following basic facts about integrals of even and odd functions over an interval $[-a, a]$ that is symmetric around the origin.

$$\text{If } f \text{ is even:} \quad \int_{-a}^{a} f(t)\, dt = 2 \int_{0}^{a} f(t)\, dt. \tag{3}$$

$$\text{If } f \text{ is odd:} \quad \int_{-a}^{a} f(t)\, dt = 0. \tag{4}$$

These facts are easy to verify analytically (Problem 17).

It follows immediately from Eqs. (1) and (2) that the product of two even functions is even, as is the product of two odd functions; the product of an even function and an odd function is odd. In particular, if $f(t)$ is an *even* periodic function of period $2L$, then $f(t)\cos(n\pi t/L)$ is even, whereas $f(t)\sin(n\pi t/L)$ is odd, because the cosine function is even and the sine function is odd. When we compute the Fourier coefficients of f, we therefore get

$$a_n = \frac{1}{L} \int_{-L}^{L} f(t) \cos \frac{n\pi t}{L}\, dt = \frac{2}{L} \int_{0}^{L} f(t) \cos \frac{n\pi t}{L}\, dt \tag{5a}$$

and

$$b_n = \frac{1}{L} \int_{-L}^{L} f(t) \sin \frac{n\pi t}{L}\, dt = 0 \tag{5b}$$

because of (3) and (4). Hence the Fourier series of the *even* function f of period $2L$ has only *cosine* terms:

$$f(t) = \frac{a_0}{2} + \sum_{n=1}^{\infty} a_n \cos \frac{n\pi t}{L} \qquad (f \text{ even}) \tag{6}$$

with the values of a_n given by Eq. (5a). If $f(t)$ is *odd*, then $f(t)\cos(n\pi t/L)$ is odd, whereas $f(t)\sin(n\pi t/L)$ is even, so

$$a_n = \frac{1}{L} \int_{-L}^{L} f(t) \cos \frac{n\pi t}{L}\, dt = 0 \tag{7a}$$

and

$$b_n = \frac{1}{L} \int_{-L}^{L} f(t) \sin \frac{n\pi t}{L} \, dt = \frac{2}{L} \int_{0}^{L} f(t) \sin \frac{n\pi t}{L} \, dt. \tag{7b}$$

Hence the Fourier series of the *odd* function f of period $2L$ has only *sine* terms:

$$f(t) = \sum_{n=1}^{\infty} b_n \sin \frac{n\pi t}{L} \qquad (f \text{ odd}) \tag{8}$$

with the coefficients b_n given in Eq. (7b).

Even and Odd Extensions

In all our earlier discussion and examples, we began with a periodic function defined *for all* t; the Fourier series of such a function is uniquely determined by the Fourier coefficient formulas. In many practical situations, however, we begin with a function f defined only on an interval of the form $0 < t < L$, and we want to represent its values on this interval by a Fourier series of period $2L$. The first step is the necessary extension of f to the interval $-L < t < 0$. Granted this, we may extend f to the entire real line by the periodicity condition $f(t + 2L) = f(t)$ (and use the average value property should any discontinuities arise). But *how* we define f for $-L < t < 0$ is our choice, and the Fourier series representation for $f(t)$ on $(0, L)$ that we obtain will depend on that choice. Specifically, different choices of the extension of f to the interval $(-L, 0)$ will yield different Fourier series that converge to the same function $f(t)$ in the original interval $0 < t < L$, but converge to the different extensions of f on the interval $-L < t < 0$.

In practice, given $f(t)$ defined for $0 < t < L$, we generally make one of two natural choices—we extend f to obtain either an even function or an odd function. The **even period** $2L$ **extension** of f is the function f_E defined as

$$f_E(t) = \begin{cases} f(t) & \text{if } 0 < t < L; \\ f(-t) & \text{if } -L < t < 0 \end{cases} \tag{9}$$

and by $f_E(t+2L) = f_E(t)$ for all t. The values of $f_E(t)$ for t an integral multiple of L are not important because they cannot affect the Fourier series of f_E. The Fourier series of the even extension f_E, given by Eqs. (5) and (6), will contain only cosine terms and is called the *Fourier cosine series* of the original function f.

The **odd period** $2L$ **extension** of f is the function f_O defined as

$$f_O(t) = \begin{cases} f(t) & \text{if } 0 < t < L; \\ -f(-t) & \text{if } -L < t < 0 \end{cases} \tag{10}$$

and by $f_O(t + 2L) = f_O(t)$ for all t. The Fourier series of the odd extension f_O, given by Eqs. (7) and (8), will contain only sine terms and is called the *Fourier sine series* of the original function f.

Definition: Fourier Cosine and Sine Series

Suppose that the function $f(t)$ is piecewise continuous on the interval $[0, L]$. Then the **Fourier cosine series** of f is the series

$$f(t) = \frac{a_0}{2} + \sum_{n=1}^{\infty} a_n \cos \frac{n\pi t}{L} \qquad (11)$$

with

$$a_n = \frac{2}{L} \int_0^L f(t) \cos \frac{n\pi t}{L} \, dt. \qquad (12)$$

The **Fourier sine series** of f is the series

$$f(t) = \sum_{n=1}^{\infty} b_n \sin \frac{n\pi t}{L} \qquad (13)$$

with

$$b_n = \frac{2}{L} \int_0^L f(t) \sin \frac{n\pi t}{L} \, dt. \qquad (14)$$

Assuming that f is piecewise smooth and satisfies the average value condition $f(t) = \frac{1}{2}[f(t+) + f(t-)]$ at each of its isolated discontinuities, Theorem 1 of Section 9.2 implies that each of the two series in (11) and (13) converges to $f(t)$ for all t in the interval $0 < t < L$. Outside this interval, the cosine series in (11) converges to the even period $2L$ extension of f, whereas the sine series in (13) converges to the odd period $2L$ extension of f. In many cases of interest we have no concern with the values of f outside the original interval $(0, L)$, and therefore the choice between (11) and (12) or (13) and (14) is determined by whether we prefer to represent $f(t)$ in the interval $(0, L)$ by a cosine series or a sine series. (See Example 2 for a situation that dictates our choice between a Fourier cosine series and a Fourier sine series to represent a given function.)

Example 1 Suppose that $f(t) = t$ for $0 < t < L$. Find both the Fourier cosine series and the Fourier sine series for f.

Solution Equation (12) gives

$$a_0 = \frac{2}{L} \int_0^L t \, dt = \frac{2}{L} \left[\frac{1}{2} t^2 \right]_0^L = L$$

and

$$a_n = \frac{2}{L} \int_0^L t \cos \frac{n\pi t}{L} \, dt = \frac{2L}{n^2 \pi^2} \int_0^{n\pi} u \cos u \, du$$

$$= \frac{2L}{n^2 \pi^2} \left[u \sin u + \cos u \right]_0^{n\pi} = \begin{cases} -\dfrac{4L}{n^2 \pi^2} & \text{for } n \text{ odd;} \\ \quad 0 & \text{for } n \text{ even.} \end{cases}$$

Thus the Fourier cosine series of f is

$$t = \frac{L}{2} - \frac{4L}{\pi^2}\left(\cos\frac{\pi t}{L} + \frac{1}{3^2}\cos\frac{3\pi t}{L} + \frac{1}{5^2}\cos\frac{5\pi t}{L} + \cdots\right) \tag{15}$$

for $0 < t < L$. Next, Eq. (14) gives

$$b_n = \frac{2}{L}\int_0^L t\sin\frac{n\pi t}{L}\,dt = \frac{2L}{n^2\pi^2}\int_0^{n\pi} u\sin u\,du$$

$$= \frac{2L}{n^2\pi^2}\Big[-u\cos u + \sin u\Big]_0^{n\pi} = \frac{2L}{n\pi}(-1)^{n+1}.$$

Thus the Fourier sine series of f is

$$t = \frac{2L}{\pi}\left(\sin\frac{\pi t}{L} - \frac{1}{2}\sin\frac{2\pi t}{L} + \frac{1}{3}\sin\frac{3\pi t}{L} - \cdots\right) \tag{16}$$

for $0 < t < L$. The series in Eq. (15) converges to the even period $2L$ extension of f shown in Fig. 9.3.3; the series in Eq. (16) converges to the odd period $2L$ extension shown in Fig. 9.3.4. ∎

FIGURE 9.3.3. The even period $2L$ extension of f.

FIGURE 9.3.4. The odd period $2L$ extension of f.

Termwise Differentiation of Fourier Series

In this and in subsequent sections, we want to consider Fourier series as possible solutions of differential equations. In order to substitute a Fourier series for the unknown dependent variable in a differential equation to check whether it is a solution, we first need to differentiate the series in order to compute the derivatives that appear in the equation. Care is required here; term by term differentiation of an infinite series of variable terms is not always valid. Theorem 1 gives sufficient conditions for the validity of termwise differentiation of a Fourier series.

THEOREM 1: Termwise Differentiation of Fourier Series

Suppose that the function f is continuous for all t, periodic with period $2L$, and that its derivative f' is piecewise smooth for all t. Then the Fourier series of f' is the series

$$f'(t) = \sum_{n=1}^{\infty}\left(-\frac{n\pi}{L}a_n\sin\frac{n\pi t}{L} + \frac{n\pi}{L}b_n\cos\frac{n\pi t}{L}\right) \tag{17}$$

obtained by termwise differentiation of the Fourier series

$$f(t) = \frac{a_0}{2} + \sum_{n=1}^{\infty}\left(a_n\cos\frac{n\pi t}{L} + b_n\sin\frac{n\pi t}{L}\right). \tag{18}$$

Proof: The point of the theorem is that the differentiated series in Eq. (17) actually converges to $f'(t)$ (with the usual proviso about average values). But because f' is periodic and piecewise smooth, we know from Theorem 1 of Section 9.2 that the Fourier series of f' converges to $f'(t)$:

$$f'(t) = \frac{\alpha_0}{2} + \sum_{n=1}^{\infty}\left(\alpha_n\cos\frac{n\pi t}{L} + \beta_n\sin\frac{n\pi t}{L}\right). \tag{19}$$

In order to prove Theorem 1, it therefore suffices to show that the series in Eqs. (17) and (19) are identical. We will do so under the additional hypothesis that f' is continuous everywhere. Then

$$\alpha_0 = \frac{1}{L} \int_{-L}^{L} f'(t)\, dt = \frac{1}{L}\Big[f(t)\Big]_{-L}^{L} = 0$$

because $f(L) = f(-L)$ by periodicity, and

$$\alpha_n = \frac{1}{L} \int_{-L}^{L} f'(t) \cos \frac{n\pi t}{L}\, dt$$

$$= \frac{1}{L}\left[f(t) \cos \frac{n\pi t}{L} \right]_{-L}^{L} + \frac{n\pi}{L} \cdot \frac{1}{L} \int_{-L}^{L} f(t) \sin \frac{n\pi t}{L}\, dt$$

—integration by parts. It follows that

$$\alpha_n = \frac{n\pi}{L} b_n.$$

Similarly, we find that

$$\beta_n = -\frac{n\pi}{L} a_n,$$

and therefore the series in Eqs. (17) and (19) are, indeed, identical. ▲

Whereas the assumption that the derivative f' is continuous is merely a convenience—the proof of Theorem 1 can be strengthened to allow isolated discontinuities in f'—it is important to note that the conclusion of Theorem 1 generally fails when f itself is discontinuous. For example, consider the Fourier series

$$t = \frac{2L}{\pi}\left(\sin \frac{\pi t}{L} - \frac{1}{2} \sin \frac{2\pi t}{L} + \frac{1}{3} \sin \frac{3\pi t}{L} - \cdots \right), \tag{16}$$

$-L < t < L$, of the discontinuous sawtooth function having the graph shown in Fig. 9.3.4. All the hypotheses of Theorem 1 are satisfied apart from the continuity of f, and f has only isolated jump discontinuities. But the series

$$2\left(\cos \frac{\pi t}{L} - \cos \frac{2\pi t}{L} + \cos \frac{3\pi t}{L} - \cdots \right) \tag{20}$$

obtained by differentiating the series in Eq. (16) term by term diverges (for instance, when $t = 0$ and when $t = L$), and therefore termwise differentiation of the series in Eq. (16) is not valid.

By contrast, consider the [continuous] triangular wave function $f(t)$ having the graph shown in Fig. 9.3.3, with $f(t) = |t|$ for $-L < t < L$. This function satisfies all the hypotheses of Theorem 1, so its Fourier series

$$f(t) = \frac{L}{2} - \frac{4L}{\pi^2}\left(\cos \frac{\pi t}{L} + \frac{1}{3^2} \cos \frac{3\pi t}{L} + \frac{1}{5^2} \cos \frac{5\pi t}{L} + \cdots \right) \tag{15}$$

can be differentiated termwise. The result is

$$f'(t) = \frac{4}{\pi}\left(\sin \frac{\pi t}{L} + \frac{1}{3} \sin \frac{3\pi t}{L} + \frac{1}{5} \sin \frac{5\pi t}{L} + \cdots \right), \tag{21}$$

which is the Fourier series of the period $2L$ square wave function that takes the value -1 for $-L < t < 0$ and $+1$ for $0 < t < L$.

Fourier Series Solutions of Differential Equations

In the remainder of this chapter and in Chapter 10, we will frequently need to solve endpoint value problems of the general form

$$ax'' + bx' + cx = f(t) \quad (0 < t < L); \tag{22}$$
$$x(0) = x(L) = 0, \tag{23}$$

where the function $f(t)$ is given. We might consider applying the techniques of Chapter 3, solving the problem by

1. First finding the general solution $x_c = c_1 x_1 + c_2 x_2$ of the associated homogeneous differential equation;
2. Then finding a single particular solution x_p of the nonhomogeneous equation in (22); and
3. Finally, determining the constants c_1 and c_2 so that $x = x_c + x_p$ satisfies the endpoint conditions in (23).

In many problems, the following Fourier series method is more convenient and more useful. We first extend the definition of the function $f(t)$ to the interval $L < t < 0$ in an appropriate way, and then to the entire real line by the periodicity conditions $f(t + 2L) = f(t)$. Then the function f, if piecewise smooth, has a Fourier series

$$f(t) = \frac{A_0}{2} + \sum_{n=1}^{\infty} \left(A_n \cos \frac{n\pi t}{L} + B_n \sin \frac{n\pi t}{L} \right), \tag{24}$$

which has coefficients $\{A_n\}$ and $\{B_n\}$ that we can and do compute. We then assume that the differential equation in (22) has a solution $x(t)$ with a Fourier series

$$x(t) = \frac{a_0}{2} + \sum_{n=1}^{\infty} \left(a_n \cos \frac{n\pi t}{L} + b_n \sin \frac{n\pi t}{L} \right) \tag{25}$$

that may validly be differentiated twice termwise. We attempt to determine the coefficients in Eq. (25) by first substituting the series in Eqs. (24) and (25) into the differential equation in (22) and then equating coefficients of like terms. If this procedure is carried out in such a way that the resulting series in Eq. (25) also satisfies the endpoint conditions in (23), then we have a "formal Fourier series solution" of the original endpoint value problem; that is, a solution subject to verification of the assumed termwise differentiability. Example 2 illustrates this process.

Example 2 Find a formal Fourier series solution of the endpoint value problem

$$x'' + 4x = 4t, \tag{26}$$
$$x(0) = x(1) = 0. \tag{27}$$

Solution Here $f(t) = 4t$ for $0 < t < 1$. A crucial first step—which we did not make explicit in the preceding outline—is to choose a periodic extension $f(t)$ so that each term in its Fourier series satisfies the endpoint conditions in (27). For this purpose we

choose the odd period 2 extension, because each term of the form $\sin n\pi t$ satisfies (27). Then from the series in Eq. (16) with $L = 1$, we get the Fourier series

$$4t = \frac{8}{\pi} \sum_{n=1}^{\infty} \frac{(-1)^{n+1}}{n} \sin n\pi t \tag{28}$$

for $0 < t < 1$. We therefore anticipate a sine series solution

$$x(t) = \sum_{n=1}^{\infty} b_n \sin n\pi t, \tag{29}$$

noting that any such series will satisfy the endpoint conditions in (27). When we substitute the series in (28) and (29) in Eq. (26), the result is

$$\sum_{n=1}^{\infty} (-n^2\pi^2 + 4)b_n \sin n\pi t = \frac{8}{\pi} \sum_{n=1}^{\infty} \frac{(-1)^{n+1}}{n} \sin n\pi t. \tag{30}$$

We next equate coefficients of like terms in Eq. (30). This yields

$$b_n = \frac{8 \cdot (-1)^{n+1}}{n\pi(4 - n^2\pi^2)},$$

so our formal Fourier series solution is

$$x(t) = \frac{8}{\pi} \sum_{n=1}^{\infty} \frac{(-1)^{n+1} \sin n\pi t}{n(4 - n^2\pi^2)}. \tag{31}$$

In Problem 16 we ask you to derive the exact solution

$$x(t) = t - \frac{\sin 2t}{\sin 2} \qquad (0 \leq t \leq 1), \tag{32}$$

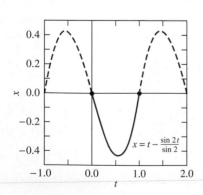

FIGURE 9.3.5. Graph of the solution in Example 2.

and to verify that (31) is the Fourier series of the odd period 2 extension of this solution.

The dashed curve in Fig. 9.3.5 was plotted by summing 10 terms of the Fourier series in (31). The solid curve for $0 \leq t \leq 2$ is the graph of the exact solution in (32). ∎

Termwise Integration of Fourier Series

Theorem 2 guarantees that the Fourier series of a piecewise continuous periodic function can always be integrated term by term, whether or not it converges! A proof is outlined in Problem 25.

THEOREM 2: Termwise Integration of Fourier Series

Suppose that f is a piecewise continuous periodic function with period $2L$ and Fourier series

$$f(t) \sim \frac{a_0}{2} + \sum_{n=1}^{\infty} \left(a_n \cos \frac{n\pi t}{L} + b_n \sin \frac{n\pi t}{L} \right), \tag{33}$$

which may not converge. Then

$$\int_0^t f(s)\, ds = \frac{a_0 t}{2} + \sum_{n=1}^{\infty} \frac{L}{n\pi} \left[a_n \sin \frac{n\pi t}{L} - b_n \left(\cos \frac{n\pi t}{L} - 1 \right) \right], \tag{34}$$

with the series on the right-hand side convergent for all t. Note that the series in Eq. (34) is the result of term-by-term integration of the series in (33). ▲

Example 3 Let us attempt to verify the conclusion of Theorem 2 in the case that $f(t)$ is the period 2π function such that

$$f(t) = \begin{cases} -1, & -\pi < t < 0; \\ +1, & 0 < t < \pi. \end{cases} \tag{35}$$

By Example 1 of Section 9.1, the Fourier series of f is

$$f(t) = \frac{4}{\pi} \left(\sin t + \frac{1}{3} \sin 3t + \frac{1}{5} \sin 5t + \cdots \right). \tag{36}$$

Theorem 2 then implies that

$$F(t) = \int_0^t f(s)\, ds$$

$$= \int_0^t \frac{4}{\pi} \left(\sin s + \frac{1}{3} \sin 3s + \frac{1}{5} \sin 5s + \cdots \right) ds$$

$$= \frac{4}{\pi} \left[(1 - \cos t) + \frac{1}{3^2}(1 - \cos 3t) + \frac{1}{5^2}(1 - \cos 5t) + \cdots \right].$$

Thus

$$F(t) = \frac{4}{\pi} \left(1 + \frac{1}{3^2} + \frac{1}{5^2} + \cdots \right)$$

$$- \frac{4}{\pi} \left(\cos t + \frac{1}{3^2} \cos 3t + \frac{1}{5^2} \cos 5t + \cdots \right). \tag{37}$$

On the other hand, direct integration of (35) yields

$$F(t) = \int_0^t f(s)\, ds = |t| = \begin{cases} -t, & -\pi < t < 0, \\ t, & 0 < t < \pi. \end{cases}$$

We know from Example 1 in this section (with $L = \pi$) that

$$|t| = \frac{\pi}{2} - \frac{4}{\pi}\left(\cos t + \frac{1}{3^2}\cos 3t + \frac{1}{5^2}\cos 5t + \cdots\right). \tag{38}$$

We also know from Eq. (18) of Section 9.2 that

$$1 + \frac{1}{3^2} + \frac{1}{5^2} + \frac{1}{7^2} + \cdots = \frac{\pi^2}{8},$$

so it follows that the two series in Eqs. (37) and (38) are indeed identical. ∎

9.3 Problems

In Problems 1 through 10, a function $f(t)$ defined on an interval $0 < t < L$ is given. Find the Fourier cosine and sine series of f and sketch the graphs of the two extensions of f to which these two series converge.

1. $f(t) = 1, 0 < t < \pi$
2. $f(t) = 1 - t, 0 < t < 1$
3. $f(t) = 1 - t, 0 < t < 2$
4. $f(t) = \begin{cases} t, & 0 < t \leq 1; \\ 2 - t, & 1 \leq t < 2 \end{cases}$
5. $f(t) = \begin{cases} 0, & 0 < t < 1; \\ 1, & 1 < t < 2; \\ 0, & 2 < t < 3 \end{cases}$
6. $f(t) = t^2, 0 < t < \pi$
7. $f(t) = t(\pi - t), 0 < t < \pi$
8. $f(t) = t - t^2, 0 < t < 1$
9. $f(t) = \sin t, 0 < t < \pi$
10. $f(t) = \begin{cases} \sin t, & 0 < t \leq \pi \\ 0, & \pi \leq t < 2\pi \end{cases}$

Find formal Fourier series solutions of the endpoint value problems in Problems 11 through 14.

11. $x'' + 2x = 1, x(0) = x(\pi) = 0$
12. $x'' - 4x = 1, x(0) = x(\pi) = 0$
13. $x'' + x = t, x(0) = x(1) = 0$
14. $x'' + 2x = t, x(0) = x(2) = 0$
15. Find a formal Fourier series solution of the endpoint value problem

$$x'' + 2x = t, \quad x'(0) = x'(\pi) = 0.$$

(*Suggestion:* Use a Fourier cosine series in which each term satisfies the endpoint conditions.)

16. (a) Derive the solution $x(t) = t - (\sin 2t)/(\sin 2)$ of the endpoint value problem

$$x'' + 4x = 4t, \quad x(0) = x(1) = 0.$$

(b) Show that the series in Eq. (31) is the Fourier sine series of the solution in part (a).

17. (a) Suppose that f is an even function. Show that

$$\int_{-a}^0 f(t)\, dt = \int_0^a f(t)\, dt.$$

(b) Suppose that f is an odd function. Show that

$$\int_{-a}^0 f(t)\, dt = -\int_0^a f(t)\, dt.$$

18. By Example 2 of Section 9.2, the Fourier series of the period 2 function f with $f(t) = t^2$ for $0 < t < 2$ is

$$f(t) = \frac{4}{3} + \frac{4}{\pi^2}\sum_{n=1}^{\infty}\frac{\cos n\pi t}{n^2} - \frac{4}{\pi}\sum_{n=1}^{\infty}\frac{\sin n\pi t}{n}.$$

Show that the termwise derivative of this series does not converge to $f'(t)$.

19. Begin with the Fourier series

$$t = 2\sum_{n=1}^{\infty}\frac{(-1)^{n+1}}{n}\sin nt, \quad -\pi < t < \pi,$$

and integrate termwise three times in succession to obtain the series

$$\frac{1}{24}t^4 = \frac{\pi^2 t^2}{12} - 2\sum_{n=1}^{\infty}\frac{(-1)^n}{n^4}\cos nt + 2\sum_{n=1}^{\infty}\frac{(-1)^n}{n^4}.$$

20. Substitute $t = \pi/2$ and $t = \pi$ in the series of Problem 19 to obtain the summations

$$\sum_{n=1}^{\infty} \frac{1}{n^4} = \frac{\pi^4}{90}, \quad \sum_{n=1}^{\infty} \frac{(-1)^{n+1}}{n^4} = \frac{7\pi^4}{720},$$

and

$$1 + \frac{1}{3^4} + \frac{1}{5^4} + \frac{1}{7^4} + \cdots = \frac{\pi^4}{96}.$$

21. (*Odd half-multiple sine series*) Let $f(t)$ be given for $0 < t < L$, and define $F(t)$ for $0 < t < 2L$ as follows:

$$F(t) = \begin{cases} f(t), & 0 < t < L; \\ f(2L - t), & L < t < 2L. \end{cases}$$

Thus the graph of $F(t)$ is symmetric around the line $t = L$ (Fig. 9.3.6). Then the period $4L$ Fourier sine series of F is

$$F(t) = \sum_{n=1}^{\infty} b_n \sin \frac{n\pi t}{2L},$$

where

$$b_n = \frac{1}{L} \int_0^L f(t) \sin \frac{n\pi t}{2L} \, dt$$
$$+ \frac{1}{L} \int_L^{2L} f(2L - t) \sin \frac{n\pi t}{2L} \, dt.$$

Substitute $s = 2L - t$ in the second integral to derive the series (for $0 < t < L$)

$$f(t) = \sum_{n \text{ odd}} b_n \sin \frac{n\pi t}{2L},$$

where

$$b_n = \frac{2}{L} \int_0^L f(t) \sin \frac{n\pi t}{2L} \, dt \quad (n \text{ odd}).$$

FIGURE 9.3.6. Construction of F from f in Problem 21.

22. (*Odd half-multiple cosine series*) Let $f(t)$ be given for $0 < t < L$, and define $G(t)$ for $0 < t < 2L$ as follows:

$$G(t) = \begin{cases} f(t), & 0 < t < L; \\ -f(2L - t), & L < t < 2L. \end{cases}$$

Use the period $4L$ Fourier cosine series of $G(t)$ to derive the series (for $0 < t < L$)

$$f(t) = \sum_{n \text{ odd}} a_n \cos \frac{n\pi t}{2L},$$

where

$$a_n = \frac{2}{L} \int_0^L f(t) \cos \frac{n\pi t}{2L} \, dt \quad (n \text{ odd}).$$

23. Given: $f(t) = t, 0 < t < \pi$. Derive the odd half-multiple sine series (Problem 21)

$$f(t) = \frac{8}{\pi} \sum_{n \text{ odd}} \frac{(-1)^{(n-1)/2}}{n^2} \sin \frac{nt}{2}.$$

24. Given the endpoint value problem

$$x'' - x = t, \quad x(0) = 0, \quad x'(\pi) = 0,$$

note that any constant multiple of $\sin(nt/2)$ with n odd satisfies the endpoint conditions. Hence use the odd half-multiple sine series of Problem 23 to derive the formal Fourier series solution

$$x(t) = \frac{32}{\pi} \sum_{n \text{ odd}} \frac{(-1)^{(n+1)/2}}{n^2(n^2 + 4)} \sin \frac{nt}{2}.$$

25. In this problem we outline the proof of Theorem 2. Suppose that $f(t)$ is a piecewise continuous period $2L$ function. Define

$$F(t) = \int_0^t \left[f(s) - \tfrac{1}{2}a_0 \right] \, ds,$$

where $\{a_n\}$ and $\{b_n\}$ denote the Fourier coefficients of $f(t)$. **(a)** Show directly that $F(t + 2L) = F(t)$, so that F is a continuous period $2L$ function and therefore has a convergent Fourier series

$$F(t) = \frac{A_0}{2} + \sum_{n=1}^{\infty} \left(A_n \cos \frac{n\pi t}{L} + B_n \sin \frac{n\pi t}{L} \right).$$

(b) Suppose that $n \geq 1$. Show by direct computation that

$$A_n = -\frac{L}{n\pi} b_n \quad \text{and} \quad B_n = \frac{L}{n\pi} a_n.$$

(c) Thus

$$\int_0^t f(s) \, ds = \frac{t}{2} a_0 + \frac{1}{2} A_0$$
$$+ \sum_{n=1}^{\infty} \frac{L}{n\pi} \left(a_n \sin \frac{n\pi t}{L} - b_n \cos \frac{n\pi t}{L} \right).$$

Finally, substitute $t = 0$ to see that

$$\frac{1}{2} A_0 = \sum_{n=1}^{\infty} \frac{L}{n\pi} b_n.$$

where

$$a_n = \frac{2}{L} \int_0^L f(t) \cos \frac{n\pi t}{2L} \, dt \quad (n \text{ odd}).$$

9.3 Application: Fourier Series of Piecewise Smooth Functions

Most computer algebra systems permit the use of unit step functions for the efficient derivation of Fourier series of "piecewise-defined" functions. Here we illustrate the use of *Maple* for this purpose. *Mathematica* and MATLAB versions can be found in the applications manual that accompanies this text.

Let the "unit function" unit(t, a, b) have the value 1 on the interval $a \leqq t < b$ and the value 0 otherwise. Then we can define a given piecewise smooth function $f(t)$ as a "linear combination" of different unit functions corresponding to the separate intervals on which the function is smooth, with the unit function for each interval multiplied by the formula defining $f(t)$ on that interval. For example, consider the even period 2π function whose graph is shown in Fig. 9.3.7. This "trapezoidal wave function" is defined for $0 < t < \pi$ by

$$f(t) = \frac{\pi}{3}\text{unit}\left(t, 0, \frac{\pi}{6}\right)$$

$$+ \left(\frac{\pi}{2} - t\right)\text{unit}\left(t, \frac{\pi}{6}, \frac{5\pi}{6}\right) + \left(-\frac{\pi}{3}\right)\text{unit}\left(t, \frac{5\pi}{6}, \pi\right). \qquad (1)$$

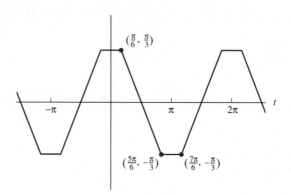

FIGURE 9.3.7. Even period 2π trapezoidal wave function.

The unit step function (with values 0 for $t < 0$ and 1 for $t > 0$) is available in *Maple* as the "Heaviside function." For instance, Heaviside$(-2) = 0$ and Heaviside$(3) = 1$. The unit function on the interval $[a, b]$ can be defined by

```
unit := (t,a,b) -> Heaviside(t-a) - Heaviside(t-b):
```

Then the trapezoidal wave function in Eq. (1) is defined for $0 \leqq t \leqq \pi$ by

```
f := t -> (Pi/3)*unit(t, 0, Pi/6) +
          (Pi/2 - t)*unit(t, Pi/6, 5*Pi/6) +
          (-Pi/3)*unit(t, 5*Pi/6, Pi):
```

We can now calculate the Fourier coefficients in the cosine series $f(t) = \frac{1}{2}a_0 + \sum a_n \cos nt$.

```
a := n -> (2/Pi)*int(f(t)*cos(n*t), t=0..Pi);
```

We then find that a typical partial sum of the series is given by

```
fourierSum := a(0)/2 + sum(a(n)*cos(n*t), n=1..25);
```

$$fourierSum := 2\frac{\sqrt{3}\cos(t)}{\pi} - \frac{2}{25}\frac{\sqrt{3}\cos(5t)}{\pi} - \frac{2}{49}\frac{\sqrt{3}\cos(7t)}{\pi}$$

$$+ \frac{2}{121}\frac{\sqrt{3}\cos(11t)}{\pi} + \frac{2}{169}\frac{\sqrt{3}\cos(13t)}{\pi} - \frac{2}{289}\frac{\sqrt{3}\cos(17t)}{\pi}$$

$$- \frac{2}{361}\frac{\sqrt{3}\cos(19t)}{\pi} + \frac{2}{529}\frac{\sqrt{3}\cos(23t)}{\pi} + \frac{2}{625}\frac{\sqrt{3}\cos(25t)}{\pi}$$

Thus we discover the lovely Fourier series

$$f(t) = \frac{2\sqrt{3}}{\pi}\sum \frac{(\pm)\cos nt}{n^2} \tag{2}$$

with a $+--++--++$ pattern of signs, and where the summation is taken over all odd positive integers n that are *not* multiples of 3. You can enter the command

```
plot(fourierSum, t=-2*Pi..3*Pi);
```

to verify that this Fourier series is consistent with Fig. 9.3.7.

You can then apply this method to find the Fourier series of the following period 2π functions.

1. The even square wave function whose graph is shown in Fig. 9.3.8.
2. The even and odd triangular wave functions whose graphs are shown in Figs. 9.2.4 and 9.3.9.
3. The odd trapezoidal wave function whose graph is shown in Fig. 9.2.5.

Then find similarly the Fourier series of some piecewise smooth functions of your own choice, perhaps ones that have periods other than 2π and are neither even nor odd.

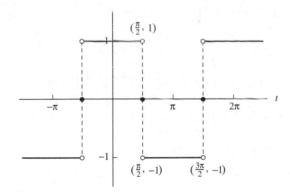

FIGURE 9.3.8. Even period 2π square wave function.

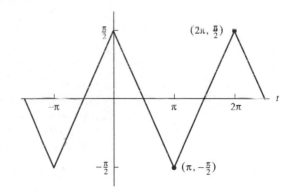

FIGURE 9.3.9. Even period 2π triangular wave function.

9.4 Applications of Fourier Series

We consider first the motion of a mass m on a spring with Hooke's constant k under the influence of a *periodic* external force $F(t)$, as indicated in Fig. 9.4.1. Its displacement $x(t)$ from equilibrium satisfies the familiar equation

$$mx'' + kx = F(t). \tag{1}$$

FIGURE 9.4.1. A mass-and-spring system with external force.

The general solution of Eq. (1) is of the form

$$x(t) = c_1 \cos \omega_0 t + c_2 \sin \omega_0 t + x_p(t), \tag{2}$$

where $\omega_0 = \sqrt{k/m}$ is the natural frequency of the system and $x_p(t)$ is a particular solution of Eq. (1). The values c_1 and c_2 would be determined by the initial conditions. Here we want to use Fourier series to find a *periodic* particular solution of Eq. (1). We will denote it by $x_{sp}(t)$ and call it a **steady periodic solution.**

We assume for simplicity that $F(t)$ is an odd function with period $2L$, so its Fourier series has the form

$$F(t) = \sum_{n=1}^{\infty} B_n \sin \frac{n\pi t}{L}. \tag{3}$$

If $n\pi/L$ is not equal to ω_0 for any positive integer n, we can determine a steady periodic solution of the form

$$x_{sp}(t) = \sum_{n=1}^{\infty} b_n \sin \frac{n\pi t}{L} \tag{4}$$

by substituting the series in Eqs. (3) and (4) in Eq. (1) to find the coefficients in Eq. (4). Example 1 illustrates this procedure.

Example 1 Suppose that $m = 2$ kg, $k = 32$ N/m, and that $F(t)$ is an odd periodic force with a period of 2 s given in one period by

$$F(t) = \begin{cases} +10\,\text{N} & \text{if } 0 < t < 1; \\ -10\,\text{N} & \text{if } 1 < t < 2. \end{cases} \tag{5}$$

Find the steady periodic motion $x_{sp}(t)$.

Solution The graph of the periodic forcing function $F(t)$ is shown in Fig. 9.4.2. By essentially the same computation as in Example 1 of Section 9.1, the Fourier series of $F(t)$ is

$$F(t) = \frac{40}{\pi} \sum_{n \text{ odd}} \frac{\sin n\pi t}{n}; \tag{6}$$

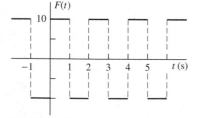

FIGURE 9.4.2. The graph of the forcing function of Example 1.

note that it contains only terms corresponding to n odd. When we substitute this series and

$$x_{sp}(t) = \sum_{n \text{ odd}} b_n \sin n\pi t, \tag{7}$$

a trial solution likewise containing only odd terms, into Eq. (1) with $m = 2$ and $k = 32$, we get

$$\sum_{n \text{ odd}} b_n(-2n^2\pi^2 + 32) \sin n\pi t = \frac{40}{\pi} \sum_{n \text{ odd}} \frac{\sin n\pi t}{n}.$$

FIGURE 9.4.3. The graph of $x_{sp}(t)$.

We equate coefficients of like terms; the result is that

$$b_n = \frac{20}{n\pi(16 - n^2\pi^2)} \qquad \text{for } n \text{ odd.}$$

Hence

$$x_{sp}(t) = \frac{20}{\pi} \sum_{n \text{ odd}} \frac{\sin n\pi t}{n(16 - n^2\pi^2)}. \tag{8}$$

The fact that each term in Eq. (8) is symmetric around $t = \frac{1}{2}$ suggests that $x_{sp}(t)$ is maximal when $t = \frac{1}{2}$, as does the graph in Fig. 9.4.3. Assuming this, we find the amplitude of the steady periodic motion to be

$$x_{sp}\left(\tfrac{1}{2}\right) = \frac{20}{\pi} \sum_{n \text{ odd}} \frac{1}{n(16 - n^2\pi^2)} \sin \frac{n\pi}{2}.$$

The first 10 terms yield $x\left(\frac{1}{2}\right) \approx 1.063$ m. ∎

If there is a nonzero term $B_N \sin(N\pi t/L)$ in the Fourier series of $F(t)$ for which $N\pi/L = \omega_0$, then this term causes **pure resonance**. The reason is that the equation

$$mx'' + kx = B_N \sin \omega_0 t$$

has the resonance solution

$$x(t) = -\frac{B_N}{2m\omega_0} t \cos \omega_0 t$$

if $\omega_0 = \sqrt{k/m}$. The solution corresponding to Eq. (4) in this case is then

$$x(t) = -\frac{B_N}{2m\omega_0} t \cos \omega_0 t + \sum_{n \neq N} \frac{B_N}{m(\omega_0^2 - n^2\pi^2/L^2)} \sin \frac{n\pi t}{L}. \tag{9}$$

Example 2 Suppose that $m = 2$ kg and $k = 32$ N/m as in Example 1. Determine whether pure resonance will occur if $F(t)$ is the odd periodic function defined in one period to be

(a) $F(t) = \begin{cases} +10, & 0 < t < \pi; \\ -10, & \pi < t < 2\pi. \end{cases}$

(b) $F(t) = 10t, \quad -\pi < t < \pi.$

Solution (a) The natural frequency is $\omega_0 = 4$, and the Fourier series of $F(t)$ is

$$F(t) = \frac{40}{\pi} \left(\sin t + \frac{1}{3} \sin 3t + \frac{1}{5} \sin 5t + \cdots \right).$$

Because this series contains no $\sin 4t$ term, no resonance occurs.

(b) In this case the Fourier series is

$$F(t) = 20 \sum_{n=1}^{\infty} \frac{(-1)^{n+1}}{n} \sin nt.$$

Pure resonance occurs because of the presence of the term containing the factor $\sin 4t$. ∎

Example 3 illustrates the *near resonance* that can occur when a single term in the solution is magnified because its frequency is close to the natural frequency ω_0.

Example 3 Find a steady periodic solution of

$$x'' + 10x = F(t), \tag{10}$$

where $F(t)$ is the period 4 function with $F(t) = 5t$ for $-2 < t < 2$ and Fourier series

$$F(t) = \frac{20}{\pi} \sum_{n=1}^{\infty} \frac{(-1)^{n+1}}{n} \sin \frac{n\pi t}{2}. \tag{11}$$

Solution When we substitute Eq. (11) and

$$x_{\text{sp}}(t) = \sum_{n=1}^{\infty} b_n \sin \frac{n\pi t}{2}$$

in (10), we obtain

$$\sum_{n=1}^{\infty} b_n \left(-\frac{n^2\pi^2}{4} + 10 \right) \sin \frac{n\pi t}{2} = \frac{20}{\pi} \sum_{n=1}^{\infty} \frac{(-1)^{n+1}}{n} \sin \frac{n\pi t}{2}.$$

We equate coefficients of like terms and then solve for b_n to get the steady periodic solution

$$x_{\text{sp}}(t) = \frac{80}{\pi} \sum_{n=1}^{\infty} \frac{(-1)^{n+1}}{n(40 - n^2\pi^2)} \sin \frac{n\pi t}{2}$$

$$\approx (0.8452) \sin \frac{\pi t}{2} - (24.4111) \sin \frac{2\pi t}{2} - (0.1738) \sin \frac{3\pi t}{2} + \cdots .$$

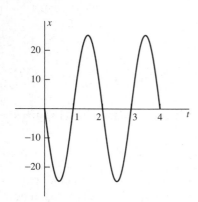

FIGURE 9.4.4. The graph of $x_{\text{sp}}(t)$.

The very large magnitude of the second term results from the fact that $\omega_0 = \sqrt{10} \approx \pi = 2\pi/2$. Thus the dominant motion of a spring with the differential equation in (10) would be an oscillation with frequency π radians per second, period 2 s, and amplitude about 24, consistent with the graph of $x_{\text{sp}}(t)$ shown in Fig. 9.4.4. ∎

Damped Forced Oscillations

FIGURE 9.4.5. A damped mass-and-spring system with external force.

Now we consider the motion of a mass m attached both to a spring with Hooke's constant k and to a dashpot with damping constant c, under the influence of a *periodic* external force $F(t)$ (Fig. 9.4.5). The displacement $x(t)$ of the mass from equilibrium satisfies the equation

$$mx'' + cx' + kx = F(t). \tag{12}$$

We recall from Problem 25 of Section 3.6 that the steady periodic solution of Eq. (12) with $F(t) = F_0 \sin \omega t$ is

$$x(t) = \frac{F_0}{\sqrt{(k - m\omega^2)^2 + (c\omega)^2}} \sin(\omega t - \alpha), \tag{13}$$

where

$$\alpha = \tan^{-1} \frac{c\omega}{k - m\omega^2}, \quad 0 \leqq \alpha \leqq \pi. \tag{14}$$

If $F(t)$ is an odd period $2L$ function with Fourier series

$$F(t) = \sum_{n=1}^{\infty} B_n \sin \frac{n\pi t}{L}, \tag{15}$$

then the preceding formulas yield, by superposition, the steady periodic solution

$$x_{\text{sp}}(t) = \sum_{n=1}^{\infty} \frac{B_n \sin(\omega_n t - \alpha_n)}{\sqrt{(k - m\omega_n^2)^2 + (c\omega_n)^2}}, \tag{16}$$

where $\omega_n = n\pi/L$ and α_n is the angle determined by Eq. (14) with this value of ω. Example 4 illustrates the interesting fact that the dominant frequency of the steady periodic solution can be an *integral multiple* of the frequency of the force $F(t)$.

Example 4 Suppose that $m = 3$ kg, $c = 0.02$ N/m/s, $k = 27$ N/m, and $F(t)$ is the odd period 2π function with $F(t) = \pi t - t^2$ if $0 < t < \pi$. Find the steady periodic motion $x_{\text{sp}}(t)$.

Solution We find that the Fourier series of $F(t)$ is

$$F(t) = \frac{8}{\pi} \left(\sin t + \frac{1}{3^3} \sin 3t + \frac{1}{5^3} \sin 5t + \cdots \right). \tag{17}$$

Thus $B_n = 0$ for n even, $B_n = 8/(\pi n^3)$ for n odd, and $\omega_n = n$. Equation (16) gives

$$x_{\text{sp}}(t) = \frac{8}{\pi} \sum_{n \text{ odd}} \frac{\sin(nt - \alpha_n)}{n^3 \sqrt{(27 - 3n^2)^2 + (0.02n)^2}} \tag{18}$$

with

$$\alpha_n = \tan^{-1} \frac{(0.02)n}{27 - 3n^2}, \quad 0 \leqq \alpha_n \leqq \pi. \tag{19}$$

With the aid of a programmable calculator, we find that

$$x_{\text{sp}}(t) \approx (0.1061) \sin(t - 0.0008) + (1.5719) \sin\left(3t - \tfrac{1}{2}\pi\right)$$
$$+ (0.0004) \sin(5t - 3.1437) + (0.0001) \sin(7t - 3.1428) + \cdots. \tag{20}$$

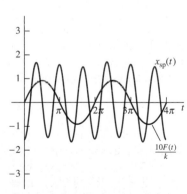

FIGURE 9.4.6. The imposed force and the resulting steady periodic motion in Example 4.

Because the coefficient corresponding to $n = 3$ is much larger than the others, the response of this system is approximately a sinusoidal motion with frequency *three times* that of the input force. Figure 9.4.6 shows $x_{\text{sp}}(t)$ in comparison with the scaled force $10F(t)/k$ that has the appropriate dimension of distance.

What is happening here is this: The mass $m = 3$ on a spring with $k = 27$ has (if we ignore the small effect of the dashpot) a natural frequency $\omega_0 = \sqrt{k/m} = 3$ rad/s. The imposed external force $F(t)$ has a [smallest] period of 2π s and hence a fundamental frequency of 1 rad/s. Consequently, the term corresponding to $n = 3$ in the Fourier series of $F(t)$ (in Eq. (17)) has the same frequency as the natural

frequency of the system. Thus near resonance vibrations occur, with the mass completing essentially *three* oscillations for every single oscillation of the external force. This is the physical effect of the dominant $n = 3$ term on the right-hand side in Eq. (20). For instance, you can push a friend in a swing quite high even if you push the swing only every *third* time it returns to you. This also explains why some transformers "hum" at a frequency much higher than 60 Hz.

This is a general phenomenon that must be taken into account in the design of mechanical systems. To avoid the occurrence of abnormally large and potentially destructive near resonance vibrations, the system must be so designed that it is not subject to any external periodic force, some *integral multiple* of whose fundamental frequency is close to a natural frequency of vibration. ∎

9.4 Problems

Find the steady periodic solution $x_{sp}(t)$ of each of the differential equations in 1 through 6. Use a computer algebra system to plot enough terms of the series to determine the visual appearance of the graph of $x_{sp}(t)$.

1. $x'' + 5x = F(t)$, where $F(t)$ is the function of period 2π such that $F(t) = 3$ if $0 < t < \pi$, $F(t) = -3$ if $\pi < t < 2\pi$.

2. $x'' + 10x = F(t)$, where $F(t)$ is the even function of period 4 such that $F(t) = 3$ if $0 < t < 1$, $F(t) = -3$ if $1 < t < 2$.

3. $x'' + 3x = F(t)$, where $F(t)$ is the odd function of period 2π such that $F(t) = 2t$ if $0 < t < \pi$.

4. $x'' + 4x = F(t)$, where $F(t)$ is the even function of period 4 such that $F(t) = 2t$ if $0 < t < 2$.

5. $x'' + 10x = F(t)$, where $F(t)$ is the odd function of period 2 such that $F(t) = t - t^2$ if $0 < t < 1$.

6. $x'' + 2x = F(t)$, where $F(t)$ is the even function of period 2π such that $F(t) = \sin t$ if $0 < t < \pi$.

In each of Problems 7 through 12, the mass m and Hooke's constant k for a mass-and-spring system are given. Determine whether or not pure resonance will occur under the influence of the given external periodic force $F(t)$.

7. $m = 1, k = 9$; $F(t)$ is the odd function of period 2π with $F(t) = 1$ for $0 < t < \pi$.

8. $m = 2, k = 10$; $F(t)$ is the odd function of period 2 with $F(t) = 1$ for $0 < t < 1$.

9. $m = 3, k = 12$; $F(t)$ is the odd function of period 2π with $F(t) = 3$ for $0 < t < \pi$.

10. $m = 1, k = 4\pi^2$; $F(t)$ is the odd function of period 2 with $F(t) = 2t$ for $0 < t < 1$.

11. $m = 3, k = 48$; $F(t)$ is the even function of period 2π with $F(t) = t$ for $0 < t < \pi$.

12. $m = 2, k = 50$; $F(t)$ is the odd function of period 2π with $F(t) = \pi t - t^2$ for $0 < t < \pi$.

In each of Problems 13 through 16, the values of m, c, and k for a damped mass-and-spring system are given. Find the steady periodic motion—in the form of Eq. (16)—of the mass under the influence of the given external force $F(t)$. Compute the coefficients and phase angles for the first three nonzero terms in the series for $x_{sp}(t)$.

13. $m = 1, c = 0.1, k = 4$; $F(t)$ is the force of Problem 1.

14. $m = 2, c = 0.1, k = 18$; $F(t)$ is the force of Problem 3.

15. $m = 3, c = 1, k = 30$; $F(t)$ is the force of Problem 5.

16. $m = 1, c = 0.01, k = 4$; $F(t)$ is the force of Problem 4.

17. Consider a forced damped mass-and-spring system with $m = \frac{1}{4}$ slug, $c = 0.6$ lb/ft/s, $k = 36$ lb/ft. The force $F(t)$ is the period 2 (s) function with $F(t) = 15$ if $0 < t < 1$, $F(t) = -15$ if $1 < t < 2$. (a) Find the steady periodic solution in the form

$$x_{sp}(t) = \sum_{n=1}^{\infty} b_n \sin(n\pi t - \alpha_n).$$

(b) Find the location—to the nearest tenth of an inch—of the mass when $t = 5$ s.

18. Consider a forced damped mass-and-spring system with $m = 1, c = 0.01$, and $k = 25$. The force $F(t)$ is the odd function of period 2π with $F(t) = t$ if $0 < t < \pi/2$, $F(t) = \pi - t$ if $\pi/2 < t < \pi$. Find the steady periodic motion; compute enough terms of its series to see that the dominant frequency of the motion is *five* times that of the external force.

9.5 Heat Conduction and Separation of Variables

The most important applications of Fourier series are to the solution of partial differential equations by means of the method of separation of variables that we introduce in this section. Recall that a *partial differential equation* is one containing one or

more *partial* derivatives of a dependent variable that is a function of at least two independent variables. An example is the **one-dimensional heat equation**

$$\frac{\partial u}{\partial t} = k\frac{\partial^2 u}{\partial x^2}, \qquad (1)$$

in which the dependent variable u is an unknown function of x and t, and k is a given constant.

The Heated Rod

Equation (1) models the variation of temperature u with position x and time t in a heated rod that extends along the x-axis. We assume that the rod has uniform cross section with area A perpendicular to the axis and that it is made of a homogeneous material. We assume further that the cross section of the rod is so small that u is constant on each cross section, and that the lateral surface of the rod is insulated so that no heat can pass through it. Then u will, indeed, be a function of x and t, and heat will flow along the rod in only the x-direction. In general, we envision heat as flowing like a fluid from the warmer to the cooler parts of a body.

The **heat flux** $\phi(x,\,t)$ in the rod is the rate of flow of heat at time t across a unit area of the cross section at x. Typical units for ϕ are calories per second per square centimeter. The derivation of Eq. (1) is based on the empirical principle that

$$\phi = -K\frac{\partial u}{\partial x}, \qquad (2)$$

where the proportionality constant K is called the **thermal conductivity** of the material of the rod. The minus sign in Eq. (2) corresponds to the fact that heat flows in the negative x-direction when $u_x > 0$, whereas ϕ is positive for heat flow in the positive x-direction; in short, heat flows from a hot place to a cold place, not vice versa.

Now consider a small segment of the rod corresponding to the interval $[x,\, x + \Delta x]$, as shown in Fig. 9.5.1. The rate of flow R (in calories per second) of heat *into* this segment through its two ends is

$$R = A\phi(x,\,t) - A\phi(x + \Delta x,\,t) = KA[u_x(x + \Delta x,\,t) - u_x(x,\,t)]. \qquad (3)$$

$A\phi(x,\,t)$ $A\phi(x+\Delta x,\,t)$

x $x+\Delta x$

FIGURE 9.5.1. Net flow of heat into a short segment of the rod.

The resulting time rate of change u_t of the temperature in the segment depends on its density δ (grams per cubic centimeter) and specific heat c (both assumed constant). The **specific heat** c is the amount of heat (in calories) required to raise by $1°$ the temperature of 1 g of material. Consequently $c\delta u$ calories of heat are required to raise 1 cm^3 of the material from temperature zero to temperature u. A short slice of the rod of length dx has volume $A\,dx$, so $c\delta u A\,dx$ calories of heat are required to raise the temperature of this slice from 0 to u. The **heat content**

$$Q(t) = \int_x^{x+\Delta x} c\delta A u(x,\,t)\,dx \qquad (4)$$

of the segment $[x,\, x + \Delta x]$ of the rod is the amount of heat needed to raise it from zero to the given temperature $u(x,\,t)$. Because heat enters and leaves the segment only through its ends, we see from Eq. (3) that

$$Q'(t) = KA[u_x(x + \Delta x,\,t) - u_x(x,\,t)], \qquad (5)$$

because $R = dQ/dt$. Thus by differentiating Eq. (4) within the integral and applying the mean value theorem for integrals, we see that

$$Q'(t) = \int_x^{x+\Delta x} c\delta A u_t(x, t)\, dx = c\delta A u_t(\overline{x}, t)\, \Delta x \tag{6}$$

for some \overline{x} in $(x, x + \Delta x)$. Upon equating the values in Eqs. (5) and (6), we get

$$c\delta A u_t(\overline{x}, t)\, \Delta x = K A [u_x(x + \Delta x, t) - u_x(x, t)], \tag{7}$$

so

$$u_t(\overline{x}, t) = k \frac{u_x(x + \Delta x, t) - u_x(x, t)}{\Delta x}, \tag{8}$$

where

$$k = \frac{K}{c\delta} \tag{9}$$

is the **thermal diffusivity** of the material. Finally, by taking the limit in Eq. (8) as $\Delta x \to 0$ and $\overline{x} \to x$, we obtain the one-dimensional heat equation

$$\frac{\partial u}{\partial t} = k \frac{\partial^2 u}{\partial x^2}. \tag{1}$$

Thus the temperature $u(x, t)$ in the thin rod with insulated sides must satisfy this partial differential equation.

Boundary Conditions

Now suppose that the rod has finite length L, extending from $x = 0$ to $x = L$. Its temperature function $u(x, t)$ will be determined among all possible solutions of Eq. (1) by appropriate subsidiary conditions. In fact, whereas a solution of an ordinary differential equation involves arbitrary constants, a solution of a partial differential equation generally involves arbitrary *functions*. In the case of the heated rod, we can specify its temperature function $f(x)$ at time $t = 0$. This gives the *initial condition*

$$u(x, 0) = f(x). \tag{10}$$

We may also specify fixed temperatures at the two ends of the rod. For instance, if each end were clamped against a large block of ice at temperature zero, we would have the *endpoint conditions*

$$u(0, t) = u(L, t) = 0 \qquad \text{(for all } t\text{)}. \tag{11}$$

Combining all this, we get the **boundary value problem**

$$\frac{\partial u}{\partial t} = k \frac{\partial^2 u}{\partial x^2} \qquad (0 < x < L, \ t > 0); \tag{12a}$$

$$u(0, t) = u(L, t) = 0, \tag{12b}$$

$$u(x, 0) = f(x). \tag{12c}$$

FIGURE 9.5.2. A geometric interpretation of the boundary value problem in Eqs. (12a)–(12c).

Figure 9.5.2 gives a geometric interpretation of the boundary value problem in (12): We are to find a function $u(x, t)$ that is continuous on the unbounded strip (including its boundary) shaded in the xt-plane. This function must satisfy the differential equation in (12a) at each interior point of the strip, and on the boundary of the strip must have the values prescribed by the *boundary conditions* in (12b) and (12c). Physical intuition suggests that if $f(x)$ is a reasonable function, then there will exist one and only one such function $u(x, t)$.

Instead of having fixed temperatures, the two ends of the rod might be insulated. In this case no heat would flow through either end, so we see from Eq. (2) that the conditions in (12b) would be replaced in the boundary value problem by the endpoint conditions

$$u_x(0, t) = u_x(L, t) = 0 \tag{13}$$

(for all t). Alternatively, the rod could be insulated at one end and have a fixed temperature at the other. This and other endpoint possibilities are discussed in the Problems.

Superposition of Solutions

Note that the heat equation in (12a) is *linear*. That is, any linear combination $u = c_1 u_1 + c_2 u_2$ of two solutions of (12a) is also a solution of (12a); this follows immediately from the linearity of partial differentiation. It is also true that if u_1 and u_2 each satisfy the conditions in (12b), then so does any linear combination $u = c_1 u_1 + c_2 u_2$. The conditions in (12b) are therefore called **homogeneous** boundary conditions (though a more descriptive term might be *linear* boundary conditions). By contrast, the final boundary condition in (12c) is not homogeneous; it is a **nonhomogeneous** boundary condition.

Our overall strategy for solving the boundary value problem in (12) will be to find functions u_1, u_2, u_3, ... that satisfy both the partial differential equation in (12a) and the homogeneous boundary conditions in (12b), and then attempt to combine these functions by superposition, as if they were building blocks, in the hope of obtaining a solution $u = c_1 u_1 + c_2 u_2 + \cdots$ that satisfies the nonhomogeneous condition in (12c) as well. Example 1 illustrates this approach.

Example 1

It is easy to verify by direct substitution that each of the functions

$$u_1(x, t) = e^{-t} \sin x, \quad u_2(x, t) = e^{-4t} \sin 2x, \quad \text{and} \quad u_3(x, t) = e^{-9t} \sin 3x$$

satisfies the equation $u_t = u_{xx}$. Use these functions to construct a solution of the boundary value problem

$$\frac{\partial u}{\partial t} = \frac{\partial^2 u}{\partial x^2} \qquad (0 < x < \pi, \quad t > 0); \tag{14a}$$

$$u(0, t) = u(\pi, t) = 0, \tag{14b}$$

$$u(x, 0) = 80 \sin^3 x = 60 \sin x - 20 \sin 3x. \tag{14c}$$

Solution Any linear combination of the form

$$u(x, t) = c_1 u_1(x, t) + c_2 u_2(x, t) + c_3 u_3(x, t)$$
$$= c_1 e^{-t} \sin x + c_2 e^{-4t} \sin 2x + c_3 e^{-9t} \sin 3x$$

satisfies both the differential equation in (14a) and the homogeneous conditions in (14b). Because

$$u(x, 0) = c_1 \sin x + c_2 \sin 2x + c_3 \sin 3x,$$

we see that we can also satisfy the nonhomogeneous condition in (14c) simply by choosing $c_1 = 60$, $c_2 = 0$, and $c_3 = -20$. Thus a solution of the given boundary value problem is

$$u(x, t) = 60e^{-t} \sin x - 20e^{-9t} \sin 3x. \qquad \blacksquare$$

The boundary value problem in Example 1 is exceptionally simple in that only a finite number of homogeneous solutions are needed to satisfy by superposition the nonhomogeneous boundary condition. It is more usual that an infinite sequence u_1, u_2, u_3, ... of functions satisfying (12a) and (12b) is required. If so, we write the infinite series

$$\blacktriangleright \qquad u(x, t) = \sum_{n=1}^{\infty} c_n u_n(x, t) \qquad (15)$$

and then attempt to determine the coefficients c_1, c_2, c_3, \ldots in order to satisfy (12c) as well. The following principle summarizes the properties of this infinite series that must be verified to ensure that we have a solution of the boundary value problem in (12).

PRINCIPLE: Superposition of Solutions

Suppose that each of the functions u_1, u_2, u_3, ... satisfies both the differential equation in (12a) (for $0 < x < L$ and $t > 0$) and the homogeneous conditions in (12b). Suppose also that the coefficients in Eq. (15) are chosen to meet the following three criteria:

1. For $0 < x < L$ and $t > 0$, the function determined by the series in (15) is continuous and termwise differentiable (once with respect to t and twice with respect to x).

2. $\displaystyle\sum_{n=1}^{\infty} c_n u_n(x, 0) = f(x).$

3. The function $u(x, t)$ determined by Eq. (15) interior to the strip $0 \leqq x \leqq L$ and $t \geqq 0$, and by the boundary conditions in (12b) and (12c) on its boundary, is continuous.

Then $u(x, t)$ is a solution of the boundary value problem in (12). \blacktriangle

In the method of separation of variables described next, we concentrate on finding the solutions u_1, u_2, u_3, \ldots satisfying the homogeneous conditions and on determining the coefficients so that the series in Eq. (15) satisfies the nonhomogeneous conditions upon substitution of $t = 0$. At this point we have only a *formal* series solution of the boundary value problem—one that is subject to verification of the continuity and differentiability conditions given in part (1) of the superposition principle stated here. If the function $f(x)$ in (12c) is piecewise smooth, it can be proved that a formal series solution always satisfies the restrictions and, moreover, is the unique solution of the boundary value problem. For a proof, see Chapter 6 of R. V. Churchill and J. W. Brown, *Fourier Series and Boundary Value Problems*, 3rd ed. (New York: McGraw-Hill, 1978).

Separation of Variables

This method of solving the boundary value problem in (12) for the heated rod was introduced by Fourier in his study of heat cited in Section 9.1. We first search for the *building block* functions u_1, u_2, u_3, ... that satisfy the differential equation $u_t = k u_{xx}$ and the homogeneous conditions $u(0, t) = u(L, t) = 0$, with each of these functions being of the special form

➤
$$u(x, t) = X(x)T(t) \tag{16}$$

in which the variables x and t are "separated." Substitution of Eq. (16) in $u_t = k u_{xx}$ yields $XT' = kX''T$, where for brevity we write T' for $T'(t)$ and X'' for $X''(x)$. Division of both sides by kXT then gives

➤
$$\frac{X''}{X} = \frac{T'}{kT}. \tag{17}$$

The left-hand side of Eq. (17) is a function of x alone, but the right-hand side is a function of t alone. If t is held constant on the right-hand side, then the left-hand side X''/X must remain constant as x varies. Similarly, if x is held constant on the left-hand side, then the right-hand side T'/kT must remain constant as t varies. Consequently, equality can hold only if each of these two expressions is the same *constant*, which for convenience we denote by $-\lambda$. Thus Eq. (17) becomes

$$\frac{X''}{X} = \frac{T'}{kT} = -\lambda, \tag{18}$$

which consists of the two equations

$$X''(x) + \lambda X(x) = 0, \tag{19}$$
$$T'(t) + \lambda k T(t) = 0. \tag{20}$$

It follows that the product function $u(x, t) = X(x)T(t)$ satisfies the partial differential equation $u_t = k u_{xx}$ if $X(x)$ and $T(t)$ separately satisfy the ordinary differential equations in (19) and (20) for some value of the constant λ.

We focus first on $X(x)$. The homogeneous endpoint conditions are

$$u(x, 0) = X(0)T(t) = 0, \qquad u(L, t) = X(L)T(t) = 0. \tag{21}$$

If $T(t)$ is to be a nontrivial function of t, then (21) can hold only if $X(0) = X(L) = 0$. Thus $X(x)$ must satisfy the endpoint value problem

$$X'' + \lambda X = 0,$$
$$X(0) = 0, \quad X(L) = 0. \tag{22}$$

This is actually an eigenvalue problem of the type we discussed in Section 3.8. Indeed, we saw in Example 3 of that section that (22) has a nontrivial solution if and only if λ is one of the eigenvalues

$$\lambda_n = \frac{n^2 \pi^2}{L^2}, \qquad n = 1, 2, 3, \ldots, \tag{23}$$

and that an eigenfunction associated with λ_n is

$$X_n(x) = \sin\frac{n\pi x}{L}, \qquad n = 1, 2, 3, \ldots. \tag{24}$$

Recall that the reasoning behind (23) and (24) is as follows. If $\lambda = 0$, then (22) obviously implies that $X(x) \equiv 0$. If $\lambda = -\alpha^2 < 0$, then

$$X(x) = A\cosh\alpha x + B\sinh\alpha x,$$

and then the conditions $X(0) = 0 = X(L)$ imply that $A = B = 0$. Hence the only possibility for a nontrivial eigenfunction is that $\lambda = \alpha^2 > 0$. Then

$$X(x) = A\cos\alpha x + B\sin\alpha x,$$

and the conditions $X(0) = 0 = X(L)$ then imply that $A = 0$ and that $\alpha = n\pi/L$ for some positive integer n. Whenever separation of variables leads to an unfamiliar eigenvalue problem, we generally must consider separately the cases $\lambda = 0$, $\lambda = -\alpha^2 < 0$, and $\lambda = \alpha^2 > 0$.

Now we turn our attention to Eq. (20), knowing that the constant λ must be one of the eigenvalues listed in (23). For the nth of these possibilities we write Eq. (20) as

$$T_n' + \frac{n^2\pi^2 k}{L^2}T_n = 0, \tag{25}$$

in anticipation of a different solution $T_n(t)$ for each different positive integer n. A nontrivial solution of this equation is

$$T_n(t) = \exp\left(-n^2\pi^2 kt/L^2\right). \tag{26}$$

We omit the arbitrary constant of integration because it will (in effect) be inserted later.

To summarize our progress, we have discovered the two associated sequences $\{X_n\}_1^\infty$ and $\{T_n\}_1^\infty$ of functions given in (24) and (26). Together they yield the sequence of product functions

$$u_n(x, t) = X_n(x)T_n(t) = \exp\left(-n^2\pi^2 kt/L^2\right)\sin\frac{n\pi x}{L}, \tag{27}$$

$n = 1, 2, 3, \ldots$. Each of these functions satisfies both the heat equation $u_t = ku_{xx}$ and the homogeneous conditions $u(0, t) = u(L, t) = 0$. Now we combine these functions (*superposition*) to attempt to satisfy the nonhomogeneous condition $u(x, 0) = f(x)$ as well. We therefore form the infinite series

$$u(x, t) = \sum_{n=1}^\infty c_n u_n(x, t) = \sum_{n=1}^\infty c_n \exp\left(-n^2\pi^2 kt/L^2\right)\sin\frac{n\pi x}{L}. \tag{28}$$

It remains only to determine the constant coefficients $\{c_n\}_1^\infty$ so that

$$u(x, 0) = \sum_{n=1}^\infty c_n \sin\frac{n\pi x}{L} = f(x). \tag{29}$$

But this will be the Fourier sine series of $f(x)$ on $[0, L]$ provided that we choose

$$\blacktriangleright \qquad c_n = b_n = \frac{2}{L} \int_0^L f(x) \sin \frac{n\pi x}{L} \, dx \qquad (30)$$

for each $n = 1, 2, 3, \ldots$. Thus we have the following result.

THEOREM 1: The Heated Rod with Zero Endpoint Temperatures

The boundary value problem in (12) for a heated rod with zero endpoint temperatures has the formal series solution

$$\blacktriangleright \qquad u(x, t) = \sum_{n=1}^{\infty} b_n \exp\left(-n^2\pi^2 kt/L^2\right) \sin \frac{n\pi x}{L}, \qquad (31)$$

where the $\{b_n\}$ are the Fourier sine coefficients in Eq. (30) of the rod's initial temperature function $f(x) = u(x, 0)$. ▲

Remark: By taking the limit in (31) termwise as $t \to \infty$, we get $u(x, \infty) \equiv 0$, as we expect because the two ends of the rod are held at temperature zero. ■

The series solution in Eq. (31) usually converges quite rapidly, unless t is quite small, because of the presence of the negative exponential factors. Therefore it is practical for numerical computations. For use in problems and examples, values of the thermal diffusivity constant k for some common materials are listed in the table in Fig. 9.5.3.

Material	k (cm²/s)
Silver	1.70
Copper	1.15
Aluminum	0.85
Iron	0.15
Concrete	0.005

FIGURE 9.5.3. Some thermal diffusivity constants.

Example 2

Suppose that a rod of length $L = 50$ cm is immersed in steam until its temperature is $u_0 = 100°C$ throughout. At time $t = 0$, its lateral surface is insulated and its two ends are imbedded in ice at $0°C$. Calculate the rod's temperature at its midpoint after half an hour if it is made of (a) iron; (b) concrete.

Solution

The boundary value problem for this rod's temperature function $u(x, t)$ is

$$u_t = ku_{xx},$$
$$u(0, t) = u(L, t) = 0;$$
$$u(x, 0) = u_0.$$

Recall the square-wave series

$$f(t) = \frac{4}{\pi} \sum_{n \text{ odd}} \frac{1}{n} \sin \frac{n\pi t}{L} = \begin{cases} +1 & \text{if } 0 < t < L, \\ -1 & \text{if } -L < t < 0 \end{cases}$$

that we derived in Example 1 of Section 9.2. It follows that the Fourier sine series of $f(x) \equiv u_0$ is

$$f(x) = \frac{4u_0}{\pi} \sum_{n \text{ odd}} \frac{1}{n} \sin \frac{n\pi x}{L}$$

for $0 < x < L$. Hence the Fourier coefficients in Eq. (31) are given by

$$b_n = \begin{cases} \dfrac{4u_0}{n\pi} & \text{for } n \text{ odd,} \\[2mm] 0 & \text{for } n \text{ even,} \end{cases}$$

and therefore the rod's temperature function is given by

$$u(x,t) = \frac{4u_0}{\pi} \sum_{n \text{ odd}} \frac{1}{n} \exp\left(-\frac{n^2\pi^2 kt}{L^2}\right) \sin\frac{n\pi x}{L}.$$

Figure 9.5.4 shows the graph $y = u(x,t)$ with $u_0 = 100$ and $L = 50$. As t increases we see the maximum temperature of the rod (evidently at its midpoint) steadily decreasing. The temperature at the midpoint $x = 25$ after $t = 1800$ seconds is

$$u(25, 1800) = \frac{400}{\pi} \sum_{n \text{ odd}} \frac{(-1)^{n+1}}{n} \exp\left(-\frac{18n^2\pi^2 k}{25}\right).$$

(a) With $k = 0.15$ for iron, this series gives

$$u(25, 1800) \approx 43.8519 - 0.0029 + 0.0000 - \cdots \approx 43.85°\text{C}.$$

(b) With $k = 0.005$ for concrete, it gives

$$\begin{aligned} u(25, 1800) \approx\ & 122.8795 - 30.8257 + 10.4754 - 3.1894 \\ & + 0.7958 - 0.1572 + 0.0242 - 0.0029 \\ & + 0.0003 - 0.0000 + \cdots \approx 100.00°\text{C}. \end{aligned}$$

Thus concrete is a very effective insulator. ∎

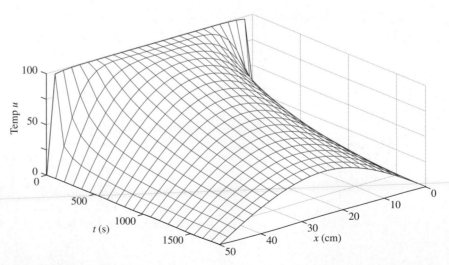

FIGURE 9.5.4. The graph of the temperature function $u(x,t)$ in Example 2.

Insulated Endpoint Conditions

We now consider the boundary value problem

$$\frac{\partial u}{\partial t} = k\frac{\partial^2 u}{\partial x^2} \qquad (0 < x < L, \;\; t > 0); \tag{32a}$$

➤

$$u_x(0, t) = u_x(L, t) = 0, \tag{32b}$$

$$u(x, 0) = f(x), \tag{32c}$$

which corresponds to a rod of length L with initial temperature $f(x)$, but with its two ends insulated. The separation of variables $u(x, t) = X(x)T(t)$ proceeds as in Eqs. (16) through (20) without change. But the homogeneous endpoint conditions in (32b) yield $X'(0) = X'(L) = 0$. Thus $X(x)$ must satisfy the endpoint value problem

$$\begin{aligned} &X'' + \lambda X = 0; \\ &X'(0) = 0, \quad X'(L) = 0. \end{aligned} \tag{33}$$

We must consider separately the possibilities $\lambda = 0$, $\lambda = -\alpha^2 < 0$, and $\lambda = \alpha^2 > 0$ for the eigenvalues.

With $\lambda = 0$, the general solution of $X'' = 0$ is $X(x) = Ax + B$, so $X'(x) = A$. Hence the endpoint conditions in (33) require $A = 0$, but B may be nonzero. Because a constant multiple of an eigenfunction is an eigenfunction, we can choose any constant value we wish for B. Thus, with $B = 1$, we have the zero eigenvalue and associated eigenfunction

$$\lambda_0 = 0, \quad X_0(x) \equiv 1. \tag{34}$$

With $\lambda = 0$ in Eq. (20), we get $T'(t) = 0$, so we may take $T_0(t) \equiv 1$ as well.

With $\lambda = -\alpha^2 < 0$, the general solution of the equation $X'' - \alpha^2 X = 0$ is

$$X(x) = A\cosh \alpha x + B\sinh \alpha x,$$

and we readily verify that $X'(0) = X'(L) = 0$ only if $A = B = 0$. Thus there are no negative eigenvalues.

With $\lambda = \alpha^2 > 0$, the general solution of $X'' + \alpha^2 X = 0$ is

$$\begin{aligned} X(x) &= A\cos \alpha x + B\sin \alpha x, \\ X'(x) &= -A\alpha \sin \alpha x + B\alpha \cos \alpha x. \end{aligned}$$

Hence $X'(0) = 0$ implies that $B = 0$, and then

$$X'(L) = -A\alpha \sin \alpha L = 0$$

requires that αL be an integral multiple of π, because $\alpha \neq 0$ and $A \neq 0$ if we are to have a nontrivial solution. Thus we have the infinite sequence of eigenvalues and associated eigenfunctions

$$\lambda_n = \alpha_n^2 = \frac{n^2\pi^2}{L^2}, \quad X_n(x) = \cos\frac{n\pi x}{L} \tag{35}$$

for $n = 1, 2, 3, \ldots$. Just as before, the solution of Eq. (20) with $\lambda = n^2\pi^2/L^2$ is $T_n(t) = \exp\left(-n^2\pi^2kt/L^2\right)$.

Therefore, the product functions satisfying the homogeneous conditions are

$$u_0(x, t) \equiv 1; \quad u_n(x, t) = \exp\left(n^2\pi^2kt/L^2\right)\cos\frac{n\pi x}{L} \tag{36}$$

for $n = 1, 2, 3, \ldots$. Hence the trial solution is

$$u(x, t) = c_0 + \sum_{n=1}^{\infty} c_n \exp\left(-n^2\pi^2kt/L^2\right)\cos\frac{n\pi x}{L}. \tag{37}$$

To satisfy the nonhomogeneous condition $u(x, 0) = f(x)$, we obviously want Eq. (37) to reduce when $t = 0$ to the Fourier cosine series

$$f(x) = \frac{a_0}{2} + \sum_{n=1}^{\infty} a_n \cos\frac{n\pi x}{L}, \tag{38}$$

where

$$a_n = \frac{2}{L}\int_0^L f(x)\cos\frac{n\pi x}{L}\,dx \tag{39}$$

for $n = 0, 1, 2, \ldots$. Thus we have the following result.

THEOREM 2: Heated Rod with Insulated Ends

The boundary value problem in (32) for a heated rod with insulated ends has the formal series solution

$$u(x, t) = \frac{a_0}{2} + \sum_{n=1}^{\infty} a_n \exp\left(-n^2\pi^2kt/L^2\right)\cos\frac{n\pi x}{L} \tag{40}$$

where the $\{a_n\}$ are the Fourier cosine coefficients in (39) of the rod's initial temperature function $f(x) = u(x, 0)$. ▲

Remark: Note that

$$\lim_{t\to\infty} u(x, t) = \frac{a_0}{2} = \frac{1}{L}\int_0^L f(x)\,dx, \tag{41}$$

the average value of the initial temperature. With both the lateral surface and the ends of the rod insulated, its original heat content ultimately distributes itself uniformly throughout the rod. ■

Example 3

We consider the same 50-cm rod as in Example 2, but now suppose that its initial temperature is given by the "triangular function" graphed in Fig. 9.5.5. At time $t = 0$, the rod's lateral surface *and* its two ends are insulated. Then its temperature function $u(x, t)$ satisfies the boundary value problem

$$u_t = ku_{xx},$$
$$u_x(0, t) = u_x(50, t) = 0,$$
$$u(x, 0) = f(x).$$

FIGURE 9.5.5. The graph of the initial temperature function $u(x, 0) = f(x)$ in Example 3.

Now substitution of $L = 25$ in the even triangular-wave series of Eq. (15) in Section 9.3 (where the length of the interval is denoted by $2L$), followed by multiplication by 4, yields the Fourier cosine series

$$f(x) = 50 - \frac{400}{\pi^2} \sum_{n \text{ odd}} \frac{1}{n^2} \cos \frac{n\pi x}{25}$$

(for $0 < x < 50$) of our initial temperature function. But in order to match terms with the series in (40) with $L = 50$, we need to exhibit terms of the form $\cos(n\pi x/50)$ rather than terms of the form $\cos(n\pi x/25)$. Hence we replace n with $n/2$ throughout and thereby rewrite the series in the form

$$f(x) = 50 - \frac{1600}{\pi^2} \sum_{n=2,6,10,\ldots} \frac{1}{n^2} \cos \frac{n\pi x}{50};$$

note that the summation runs through all positive integers of the form $4m - 2$. Then Theorem 2 implies that the rod's temperature function is given by

$$u(x, t) = 50 - \frac{1600}{n^2} \sum_{n=2,6,10,\ldots} \frac{1}{n^2} \exp\left(-\frac{n^2\pi^2 kt}{2500}\right) \cos \frac{n\pi x}{50}.$$

Figure 9.5.6 shows the graph $u = u(x, t)$ for the first 1200 seconds, and we see the temperature in the rod beginning with a sharp maximum at the midpoint $x = 25$, but rapidly "averaging out" as the heat in the rod is redistributed with increasing t ∎

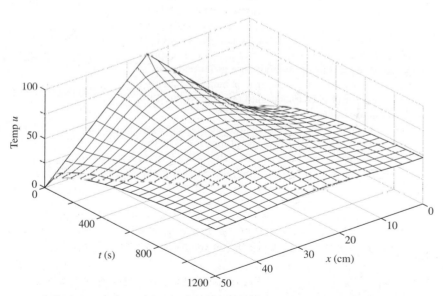

FIGURE 9.5.6. The graph of the temperature function $u(x, t)$ in Example 3.

Finally, we point out that, although we set up the boundary value problems in (12) and (32) for a rod of length L, they also model the temperature $u(x, t)$ within the infinite slab $0 \leq x \leq L$ in three-dimensional space if its initial temperature $f(x)$ depends only on x and its two faces $x = 0$ and $x = L$ are either both insulated or both held at temperature zero.

9.5 Problems

Solve the boundary value problems in Problems 1 through 12.

1. $u_t = 3u_{xx}$, $0 < x < \pi$, $t > 0$; $u(0, t) = u(\pi, t) = 0$,
 $u(x, 0) = 4 \sin 2x$

2. $u_t = 10u_{xx}$, $0 < x < 5$, $t > 0$; $u_x(0, t) = u_x(5, t) = 0$,
 $u(x, 0) = 7$

3. $u_t = 2u_{xx}$, $0 < x < 1$, $t > 0$; $u(0, t) = u(1, t) = 0$,
 $u(x, 0) = 5 \sin \pi x - \frac{1}{5} \sin 3\pi x$

4. $u_t = u_{xx}$, $0 < x < \pi$, $t > 0$; $u(0, t) = u(\pi, t) = 0$,
 $u(x, 0) = 4 \sin 4x \cos 2x$

5. $u_t = 2u_{xx}$, $0 < x < 3$, $t > 0$; $u_x(0, t) = u_x(3, t) = 0$,
 $u(x, 0) = 4 \cos \frac{2}{3} \pi x - 2 \cos \frac{4}{3} \pi x$

6. $2u_t = u_{xx}$, $0 < x < 1$, $t > 0$; $u(0, t) = u(1, t) = 0$,
 $u(x, 0) = 4 \sin \pi x \cos^3 \pi x$

7. $3u_t = u_{xx}$, $0 < x < 2$, $t > 0$; $u_x(0, t) = u_x(2, t) = 0$,
 $u(x, 0) = \cos^2 2\pi x$

8. $u_t = u_{xx}$, $0 < x < 2$, $t > 0$; $u_x(0, t) = u_x(2, t) = 0$,
 $u(x, 0) = 10 \cos \pi x \cos 3\pi x$

9. $10u_t = u_{xx}$, $0 < x < 5$, $t > 0$; $u(0, t) = u(5, t) = 0$,
 $u(x, 0) = 25$

10. $5u_t = u_{xx}$, $0 < x < 10$, $t > 0$; $u(0, t) = u(10, t) = 0$,
 $u(x, 0) = 4x$

11. $5u_t = u_{xx}$, $0 < x < 10$, $t > 0$; $u_x(0, t) = u_x(10, t) = 0$,
 $u(x, 0) = 4x$

12. $u_t = u_{xx}$, $0 < x < 100$, $t > 0$; $u(0, t) = u(100, t) = 0$,
 $u(x, 0) = x(100 - x)$

13. Suppose that a rod 40 cm long with insulated lateral surface is heated to a uniform temperature of $100°C$, and that at time $t = 0$ its two ends are embedded in ice at $0°C$. **(a)** Find the formal series solution for the temperature $u(x, t)$ of the rod. **(b)** In the case the rod is made of copper, show that after 5 min the temperature at its midpoint is about $15°C$. **(c)** In the case the rod is made of concrete, use the first term of the series to find the time required for its midpoint to cool to $15°C$.

14. A copper rod 50 cm long with insulated lateral surface has initial temperature $u(x, 0) = 2x$, and at time $t = 0$ its two ends are insulated. **(a)** Find $u(x, t)$. **(b)** What will its temperature be at $x = 10$ after 1 min? **(c)** After approximately how long will its temperature at $x = 10$ be $45°C$?

15. The two faces of the slab $0 \leq x \leq L$ are kept at temperature zero, and the initial temperature of the slab is given by $u(x, 0) = A$ (a constant) for $0 < x < L/2$, $u(x, 0) = 0$ for $L/2 < x < L$. Derive the formal series solution

$$u(x, t) =$$
$$\frac{4A}{\pi} \sum_{n=1}^{\infty} \frac{\sin^2(n\pi/4)}{n} \exp\left(-n^2\pi^2kt/L^2\right) \sin \frac{n\pi x}{L}.$$

16. Two iron slabs are each 25 cm thick. Initially one is at temperature $100°C$ throughout and the other is at temperature $0°C$. At time $t = 0$ they are placed face to face, and their outer faces are kept at $0°C$. **(a)** Use the result of Problem 15 to verify that after a half hour the temperature

of their common face is approximately $22°C$. **(b)** Suppose that the two slabs are instead made of concrete. How long will it be until their common face reaches a temperature of $22°C$?

17. (Steady-state and transient temperatures) Let a laterally insulated rod with initial temperature $u(x, 0) = f(x)$ have fixed endpoint temperatures $u(0, t) = A$ and $u(L, t) = B$. **(a)** It is observed empirically that as $t \to +\infty$, $u(x, t)$ approaches a **steady-state temperature** $u_{ss}(x)$ that corresponds to setting $u_t = 0$ in the boundary value problem. Thus $u_{ss}(x)$ is the solution of the endpoint value problem

$$\frac{\partial^2 u_{ss}}{\partial x^2} = 0; \quad u_{ss}(0) = A, \quad u_{ss}(L) = B.$$

Find $u_{ss}(x)$. **(b)** The **transient temperature** $u_{tr}(x, t)$ is defined to be

$$u_{tr}(x, t) = u(x, t) - u_{ss}(x).$$

Show that u_{tr} satisfies the boundary value problem

$$\frac{\partial u_{tr}}{\partial t} = k \frac{\partial^2 u_{tr}}{\partial x^2};$$
$$u_{tr}(0, t) = u_{tr}(L, t) = 0,$$
$$u_{tr}(x, 0) = g(x) = f(x) - u_{ss}(x).$$

(c) Conclude from the formulas in (30) and (31) that

$$u(x, t) = u_{ss}(x) + u_{tr}(x, t)$$
$$= u_{ss}(x) + \sum_{n=1}^{\infty} c_n \exp\left(-n^2\pi^2kt/L^2\right) \sin \frac{n\pi x}{L},$$

where

$$c_n = \frac{2}{L} \int_0^L [f(x) - u_{ss}(x)] \sin \frac{n\pi x}{L} dx.$$

18. Suppose that a laterally insulated rod with length $L = 50$ and thermal diffusivity $k = 1$ has initial temperature $u(x, 0) = 0$ and endpoint temperatures $u(0, t) = 0$, $u(50, t) = 100$. Apply the result of Problem 17 to show that

$$u(x, t) =$$
$$2x - \frac{200}{\pi} \sum_{n=1}^{\infty} \frac{(-1)^{n+1}}{n} \exp\left(-n^2\pi^2kt/2500\right) \sin \frac{n\pi x}{50}.$$

19. Suppose that heat is generated within a laterally insulated rod at the rate of $q(x, t)$ calories per second per cubic centimeter. Extend the derivation of the heat equation in this section to derive the equation

$$\frac{\partial u}{\partial t} = k \frac{\partial^2 u}{\partial x^2} + \frac{q(x, t)}{c\delta}.$$

20. Suppose that current flowing through a laterally insulated rod generates heat at a constant rate; then Problem 19 yields the equation

$$\frac{\partial u}{\partial t} = k\frac{\partial^2 u}{\partial x^2} + C.$$

Assume the boundary conditions $u(0, t) = u(L, t) = 0$ and $u(x, 0) = f(x)$. **(a)** Find the steady-state temperature $u_{ss}(x)$ determined by

$$0 = k\frac{d^2 u_{ss}}{dx^2} + C, \quad u_{ss}(0) = u_{ss}(L) = 0.$$

(b) Show that the transient temperature

$$u_{tr}(x, t) = u(x, t) - u_{ss}(x)$$

satisfies the boundary value problem

$$\frac{\partial u_{tr}}{\partial t} = k\frac{\partial^2 u_{tr}}{\partial x^2};$$

$$u_{tr}(0, t) = u_{tr}(L, t) = 0,$$

$$u_{tr}(x, 0) = g(x) = f(x) - u_{ss}(x).$$

Hence conclude from the formulas in (34) and (35) that

$$u(x, t) = u_{ss}(x) + \sum_{n=1}^{\infty} c_n \exp\left(-n^2\pi^2 kt/L^2\right) \sin\frac{n\pi x}{L},$$

where

$$c_n = \frac{2}{L}\int_0^L [f(x) - u_{ss}(x)]\sin\frac{n\pi x}{L}\,dx.$$

21. The answer to part (a) of Problem 20 is $u_{ss}(x) = Cx(L - x)/2k$. If $f(x) \equiv 0$ in Problem 20, so the rod being heated is initially at temperature zero, deduce from the result of part (b) that

$$u(x, t) = \frac{Cx}{2k}(L - x)$$

$$- \frac{4CL^2}{k\pi^3}\sum_{n \text{ odd}}\frac{1}{n^3}\exp\left(-n^2\pi^2 kt/L^2\right)\sin\frac{n\pi x}{L}.$$

22. Consider the temperature $u(x, t)$ in a bare slender wire with $u(0, t) = u(L, t) = 0$ and $u(x, 0) = f(x)$. Instead of being laterally insulated, the wire loses heat to a

surrounding medium (at fixed temperature zero) at a rate proportional to $u(x, t)$. **(a)** Conclude from Problem 19 that

$$\frac{\partial u}{\partial t} = k\frac{\partial^2 u}{\partial x^2} - hu,$$

where h is a positive constant. **(b)** Then substitute

$$u(x, t) = e^{-ht}v(x, t)$$

to show that $v(x, t)$ satisfies the boundary value problem having the solution given in (30) and (31). Hence conclude that

$$u(x, t) = e^{-ht}\sum_{n=1}^{\infty} c_n \exp\left(-n^2\pi^2 kt/L^2\right)\sin\frac{n\pi x}{L},$$

where

$$c_n = \frac{2}{L}\int_0^L f(x)\sin\frac{n\pi x}{L}\,dx.$$

23. Consider a slab with thermal conductivity K occupying the region $0 \leq x \leq L$. Suppose that, in accord with Newton's law of cooling, each face of the slab loses heat to the surrounding medium (at temperature zero) at the rate of Hu calories per second per square centimeter. Deduce from Eq. (2) that the temperature $u(x, t)$ in the slab satisfies the boundary conditions

$$hu(0, t) - u_x(0, t) = 0 = hu(L, t) + u_x(L, t)$$

where $h = H/K$.

24. Suppose that a laterally insulated rod with length L, thermal diffusivity k, and initial temperature $u(x, 0) = f(x)$ is insulated at the end $x = L$ and held at temperature zero at $x = 0$. **(a)** Separate the variables to show that the eigenfunctions are

$$X_n(x) = \sin\frac{n\pi x}{2L}$$

for n odd. **(b)** Use the odd half-multiple sine series of Problem 21 in Section 9.3 to derive the solution

$$u(x, t) = \sum_{n \text{ odd}} c_n \exp\left(-n^2\pi^2 kt/4L^2\right)\sin\frac{n\pi x}{2L},$$

where

$$c_n = \frac{2}{L}\int_0^L f(x)\sin\frac{n\pi x}{2L}\,dx.$$

9.5 Application: Heated Rod Investigations

First let's investigate numerically the temperature function

$$u(x, t) = \frac{4u_0}{\pi}\sum_{n \text{ odd}}\frac{1}{n}\exp\left(-\frac{n^2\pi^2 kt}{L^2}\right)\sin\frac{n\pi x}{L}$$

of the heated rod of Example 2, having length $L = 50$ cm, uniform initial temperature $u_0 = 100°C$, and thermal diffusivity $k = 0.15$ (for iron). The following

MATLAB function sums the first N nonzero terms of this series.

```
function u = u(x,t)
k = 0.15;              % diffusivity of iron
L = 50;                % length of rod
u0 = 100;              % initial temperature
S = 0;                 % initial sum
N = 50;                % number of terms
for n = 1:2:2*N+1;
    S = S +(1/n)*exp(-n^2*pi^2*k*t/L^2).*sin(n*pi*x/L);
end
u = 4*u0*S/pi;
```

This function was used to plot Figs. 9.5.7 through 9.5.10. The corresponding *Maple* and *Mathematica* functions are provided in the applications manual that accompanies this text. As a practical matter, $N = 50$ terms suffice to give the value $u(x, t)$ after 10 seconds (or longer) with two decimal places of accuracy throughout the interval $0 \leq x \leq 50$. (How might you check this assertion?)

FIGURE 9.5.7. The graph of $u(x, 30)$ giving rod temperatures after 30 seconds.

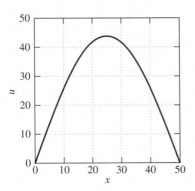

FIGURE 9.5.8. The graph of $u(x, 1800)$ giving rod temperature after 30 minutes.

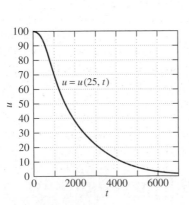

FIGURE 9.5.9. The graph of $u(25, t)$ giving the midpoint temperatures of the rod.

FIGURE 9.5.10. Magnification of the graph of $u(25, t)$ giving the midpoint temperatures of the rod.

The graph of $u(x, 30)$ in Fig. 9.5.7 shows that after 30 seconds the rod has cooled appreciably only near its two ends and still has temperature near $100°C$ for $10 \leq x \leq 40$. Figure 9.5.8 shows the graph of $u(x, 1800)$ after 30 minutes and

illustrates the fact (?) that the rod's maximum temperature is always at its midpoint, where $x = 25$.

The graph of $u(25, t)$ for a two-hour period shown in Fig. 9.5.9 indicates that the midpoint temperature takes something more than 1500 seconds (25 minutes) to fall to 50°. Figure 9.5.10 shows a magnification of the graph near its intersection point with the horizontal line $u = 50$ and indicates that this actually takes about 1578 seconds (26 min 18 s).

For your very own rod with constant initial temperature $f(x) = T_0 = 100$ to investigate in this manner, let

$$L = 100 + 10p \quad \text{and} \quad k = 1 + (0.1)q,$$

where p is the largest and q is the smallest nonzero digit of your student ID number.

1. If the two ends of the rod are both held at temperature zero, then determine how long (to the nearest second) it will take for the rod's midpoint temperature to fall to 50°.

2. If the end $x = L$ of the rod is insulated, but the end $x = 0$ is held at temperature zero, then the temperature function $u(x, t)$ is given by the Fourier series in Problem 24 of this section. Determine how long it will be until the maximum temperature anywhere in the rod is 50°.

9.6 Vibrating Strings and the One-Dimensional Wave Equation

Although Fourier systematized the method of separation of variables, trigonometric series solutions of partial differential equations had appeared earlier in eighteenth-century investigations of vibrating strings by Euler, d'Alembert, and Daniel Bernoulli. To derive the partial differential equation that models the vibrations of a string, we begin with a flexible uniform string with linear density ρ (in grams per centimeter or slugs per foot) stretched under a tension of T (dynes or pounds) between the fixed points $x = 0$ and $x = L$. Suppose that, as the string vibrates in the xy-plane around its equilibrium position, each point moves parallel to the y-axis, so we can denote by $y(x, t)$ the displacement at time t of the point x of the string. Then, for any fixed value of t, the shape of the string at time t is the curve $y = y(x, t)$. We assume also that the deflection of the string remains so slight that the approximation $\sin \theta \approx \tan \theta = y_x(x, t)$ is quite accurate (Fig. 9.6.1). Finally, we assume that in addition to the internal forces of tension acting tangentially to the string, it is acted on by an external vertical force with linear density $F(x)$ in such units as dynes per centimeter or pounds per foot.

We want to apply Newton's second law $F = ma$ to the short segment of string of mass $\rho \Delta x$ corresponding to the interval $[x, x + \Delta x]$, with a being the vertical acceleration $y_{tt}(\overline{x}, t)$ of its midpoint. Reading the vertical components of the force shown in Fig. 9.6.1, we get

$$(\rho \Delta x) \cdot y_{tt}(\overline{x}, t) \approx T \sin(\theta + \Delta\theta) - T \sin\theta + F(\overline{x}) \Delta x$$
$$\approx T y_x(x + \Delta x, t) - T y_x(x, t) + F(\overline{x}) \Delta x,$$

so division by Δx yields

$$\rho y_{tt}(\overline{x}, t) \approx T \frac{y_x(x + \Delta x, t) - y_x(x, t)}{\Delta x} + F(\overline{x}).$$

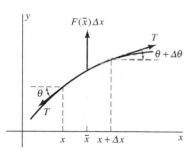

FIGURE 9.6.1. Forces on a short segment of the vibrating string.

When we take the limit as $\Delta x \to 0$ and $\overline{x} \to x$, we obtain the partial differential equation

$$\rho \frac{\partial^2 y}{\partial t^2} = T \frac{\partial^2 y}{\partial x^2} + F(x) \tag{1}$$

that describes the vertical vibrations of a flexible string with constant linear density ρ and tension T under the influence of an external vertical force with linear density $F(x)$.

If we set

$$a^2 = \frac{T}{\rho} \tag{2}$$

and set $F(x) \equiv 0$ in Eq. (1), we get the **one-dimensional wave equation**

$$\frac{\partial^2 y}{\partial t^2} = a^2 \frac{\partial^2 y}{\partial x^2} \tag{3}$$

that models the *free* vibrations of a uniform flexible string.

The fixed ends of the string at the points $x = 0$ and $x = L$ on the x-axis correspond to the *endpoint conditions*

$$y(0, t) = y(L, t) = 0. \tag{4}$$

Our intuition about the physics of the situation suggests that the motion of the string will be determined if we specify both its **initial position function**

$$y(x, 0) = f(x) \qquad (0 < x < L) \tag{5}$$

and its **initial velocity function**

$$y_t(x, 0) = g(x) \qquad (0 < x < L). \tag{6}$$

On combining Eqs. (3) through (6), we get the boundary value problem

$$\frac{\partial^2 y}{\partial t^2} = a^2 \frac{\partial^2 y}{\partial x^2} \qquad (0 < x < L, \ t > 0); \tag{7a}$$

$$y(0, t) = y(L, t) = 0, \tag{7b}$$

$$y(x, 0) = f(x) \qquad (0 < x < L), \tag{7c}$$

$$y_t(x, 0) = g(x) \qquad (0 < x < L) \tag{7d}$$

for the displacement function $y(x, t)$ of a freely vibrating string with fixed ends, initial position $f(x)$, and initial velocity $g(x)$.

Solution by Separation of Variables

Like the heat equation, the wave equation in (7a) is linear: Any linear combination of two solutions is again a solution. Another similarity is that the endpoint conditions in (7b) are homogeneous. Unfortunately, the conditions in both (7c) and (7d) are nonhomogeneous; we must deal with *two* nonhomogeneous boundary conditions.

As we described in Section 9.5, the method of separation of variables involves superposition of solutions satisfying the homogeneous conditions to obtain a solution that also satisfies a single nonhomogeneous boundary condition. To adapt the

technique to the situation at hand, we adopt the "divide and conquer" strategy of splitting the problem in (7) into the following two separate boundary value problems, each involving only a single nonhomogeneous boundary condition:

Problem A	Problem B
$y_{tt} = a^2 y_{xx};$	$y_{tt} = a^2 y_{xx};$
$y(0, t) = y(L, t) = 0,$	$y(0, t) = y(L, t) = 0,$
$y(x, 0) = f(x),$	$y(x, 0) = 0,$
$y_t(x, 0) = 0.$	$y_t(x, 0) = g(x).$

If we can separately find a solution $y_A(x, t)$ of Problem A and a solution $y_B(x, t)$ of Problem B, then their sum $y(x, t) = y_A(x, t) + y_B(x, t)$ will be a solution of the original problem in (7) because

$$y(x, 0) = y_A(x, 0) + y_B(x, 0) = f(x) + 0 = f(x)$$

and

$$y_t(x, 0) = \{y_A\}_t(x, 0) + \{y_B\}_t(x, 0) = 0 + g(x) = g(x).$$

So let us attack Problem A with the method of separation of variables. Substitution of

$$y(x, t) = X(x)T(t) \tag{8}$$

in $y_{tt} = a^2 y_{xx}$ yields $XT'' = a^2 X''T$ where (as before) we write X'' for $X''(x)$ and T'' for $T''(t)$. Therefore,

$$\frac{X''}{X} = \frac{T''}{a^2 T}. \tag{9}$$

The functions X''/X of x and $T''/a^2 T$ of t can agree for all x and t only if each is equal to the same constant. Consequently, we may conclude that

$$\frac{X''}{X} = \frac{T''}{a^2 T} = -\lambda \tag{10}$$

for some constant λ; the minus sign is inserted here merely to facilitate recognition of the eigenvalue problem in (13). Thus our partial differential equation separates into the two ordinary differential equations

$$X'' + \lambda X = 0, \tag{11}$$
$$T'' + \lambda a^2 T = 0. \tag{12}$$

The endpoint conditions

$$y(0, t) = X(0)T(t) = 0, \quad y(L, t) = X(L)T(t) = 0$$

require that $X(0) = X(L) = 0$ if $T(t)$ is nontrivial. Hence $X(x)$ must satisfy the now familiar eigenvalue problem

$$X'' + \lambda X = 0, \quad X(0) = X(L) = 0. \tag{13}$$

As in Eqs. (23) and (24) of Section 9.5, the eigenvalues of this problem are the numbers

$$\lambda_n = \frac{n^2\pi^2}{L^2}, \quad n = 1, 2, 3, \ldots, \tag{14}$$

and the eigenfunction associated with λ_n is

$$X_n(x) = \sin\frac{n\pi x}{L}, \quad n = 1, 2, 3, \ldots. \tag{15}$$

Now we turn to Eq. (12). The homogeneous initial condition

$$y_t(x, 0) = X(x)T'(0) = 0$$

implies that $T'(0) = 0$. Therefore, the solution $T_n(t)$ associated with the eigenvalue $\lambda_n = n^2\pi^2/L^2$ must satisfy the conditions

$$T_n'' + \frac{n^2\pi^2 a^2}{L^2}T_n = 0, \quad T_n'(0) = 0. \tag{16}$$

The general solution of the differential equation in (16) is

$$T_n(t) = A_n\cos\frac{n\pi at}{L} + B_n\sin\frac{n\pi at}{L}. \tag{17}$$

Its derivative

$$T_n'(t) = \frac{n\pi a}{L}\left(-A_n\sin\frac{n\pi at}{L} + B_n\cos\frac{n\pi at}{L}\right)$$

satisfies the condition $T_n'(0) = 0$ if $B_n = 0$. Thus a nontrivial solution of (16) is

$$T_n(t) = \cos\frac{n\pi at}{L}. \tag{18}$$

We combine the results in Eqs. (15) and (18) to obtain the infinite sequence of product functions

$$y_n(x, t) = X_n(x)T_n(t) = \cos\frac{n\pi at}{L}\sin\frac{n\pi x}{L}, \tag{19}$$

$n = 1, 2, 3, \ldots$. Each of these *building block* functions satisfies both the wave equation $y_{tt} = a^2 y_{xx}$ and the homogeneous boundary conditions in Problem A. By superposition we get the infinite series

$$y_n(x, t) = \sum_{n=1}^{\infty} A_n X_n(x)T_n(t) = \sum_{n=1}^{\infty} A_n\cos\frac{n\pi at}{L}\sin\frac{n\pi x}{L}. \tag{20}$$

It remains only to choose the coefficients $\{A_n\}$ to satisfy the nonhomogeneous boundary condition

$$y(x, 0) = \sum_{n=1}^{\infty} A_n\sin\frac{n\pi x}{L} = f(x) \tag{21}$$

for $0 < x < L$. But this will be the Fourier sine series of $f(x)$ on $[0, L]$ provided that we choose

➤
$$A_n = \frac{2}{L} \int_0^L f(x) \sin \frac{n\pi x}{L} \, dx. \tag{22}$$

Thus we see finally that a formal series solution of Problem A is

➤
$$y(x, t) = \sum_{n=1}^{\infty} A_n \cos \frac{n\pi a t}{L} \sin \frac{n\pi x}{L}, \tag{23}$$

with the coefficients $\{A_n\}_1^{\infty}$ computed using Eq. (22). Note that the series in (23) is obtained from the Fourier sine series of $f(x)$ simply by inserting the factor $\cos(n\pi a t/L)$ in the nth term. Note also that this term has [circular] frequency $\omega_n = n\pi a/L$.

Example 1 It follows immediately that the solution of the boundary value problem

$$\frac{\partial^2 y}{\partial t^2} = 4 \frac{\partial^2 y}{\partial x^2} \qquad (0 < x < \pi, \ t > 0);$$

$$y(0, t) = y(\pi, t) = 0,$$

$$y(x, 0) = \tfrac{1}{10} \sin^3 x = \tfrac{3}{40} \sin x - \tfrac{1}{40} \sin 3x,$$

$$y_t(x, 0) = 0,$$

for which $L = \pi$ and $a = 2$, is

$$y(x, t) = \tfrac{3}{40} \cos 2t \, \sin x - \tfrac{1}{40} \cos 6t \, \sin 3x.$$

The reason is that we are given explicitly the Fourier sine series of $f(x)$ with $A_1 = \tfrac{3}{40}$, $A_3 = -\tfrac{1}{40}$, and $A_n = 0$ otherwise. ∎

Example 2 **A plucked string** Figure 9.6.2 shows the initial position function $f(x)$ for a stretched string (of length L) that is set in motion by moving its midpoint $x = L/2$ aside the distance $\tfrac{1}{2}bL$ and then releasing it from rest at time $t = 0$. The corresponding boundary value problem is

$$y_{tt} = a^2 y_{xx} \qquad (0 < x < L, \ t > 0);$$

$$y(0, t) = y(L, t) = 0,$$

$$y(x, 0) = f(x),$$

$$y_t(x, 0) = 0,$$

where $f(x) = bx$ for $0 \leqq x \leqq L/2$ and $f(x) = b(L - x)$ for $L/2 \leqq x \leqq L$. Find $y(x, t)$.

FIGURE 9.6.2. The initial position of the plucked string of Example 2.

Solution The nth Fourier sine coefficient of $f(x)$ is

$$A_n = \frac{2}{L} \int_0^L f(x) \sin \frac{n\pi x}{L} \, dx$$

$$= \frac{2}{L} \int_0^{L/2} bx \sin \frac{n\pi x}{L} \, dx + \frac{2}{L} \int_{L/2}^L b(L - x) \sin \frac{n\pi x}{L} \, dx;$$

it follows that

$$A_n = \frac{4bL}{n^2\pi^2} \sin \frac{n\pi}{2}.$$

Hence Eq. (23) yields the formal series solution

$$y(x, t) = \frac{4bL}{\pi^2} \sum_{n=1}^{\infty} \frac{1}{n^2} \sin \frac{n\pi}{2} \cos \frac{n\pi a t}{L} \sin \frac{n\pi x}{L}$$

$$= \frac{4bL}{\pi^2} \left(\cos \frac{\pi a t}{L} \sin \frac{\pi x}{L} - \frac{1}{3^2} \cos \frac{3\pi a t}{L} \sin \frac{3\pi x}{L} + \cdots \right). \qquad (24)$$

■

Music

Numerous familiar musical instruments employ vibrating strings to generate the sounds they produce. When a string vibrates with a given frequency, vibrations at this frequency are transmitted through the air—in the form of periodic variations in air density called **sound waves**—to the ear of the listener. For example, middle C is a *tone* with a frequency of approximately 256 Hz. When several tones are heard simultaneously, the combination is perceived as harmonious if the ratios of their frequencies are nearly ratios of small whole numbers; otherwise many perceive the combinations as dissonant.

The series in Eq. (23) represents the motion of a string as a superposition of infinitely many vibrations with different frequencies. The *n*th term

$$A_n \cos \frac{n\pi a t}{L} \sin \frac{n\pi x}{L}$$

represents a vibration with frequency

$$\nu_n = \frac{\omega_n}{2\pi} = \frac{n\pi a/L}{2\pi} = \frac{n}{2L} \sqrt{\frac{T}{\rho}} \qquad (\text{Hz}). \qquad (25)$$

The lowest of these frequencies,

$$\nu_1 = \frac{1}{2L} \sqrt{\frac{T}{\rho}} \quad (\text{Hz}), \qquad (26)$$

is called the **fundamental frequency,** and it is ordinarily predominant in the sound we hear. The frequency $\nu_n = n\nu_1$ of the *n*th **overtone** or **harmonic** is an integral multiple of ν_1, and this is why the sound of a single vibrating string is harmonious rather than dissonant.

Note in Eq. (26) that the fundamental frequency ν_1 is proportional to \sqrt{T} and inversely proportional to L and to $\sqrt{\rho}$. Thus we can double this frequency—and hence get a fundamental tone one **octave** higher—either by halving the length L or by quadrupling the tension T. The initial conditions do *not* affect ν_1; instead, they determine the coefficients in (23) and hence the extent to which the higher harmonics contribute to the sound produced. Therefore the initial conditions affect the **timbre,** or overall frequency mixture, rather than the fundamental frequency. (Technically this is true only for relatively small displacements of the string; if you strike a piano key rather forcefully you can detect a slight and brief initial deviation from the usual frequency of the note.)

According to one (rather simplistic) theory of hearing, the loudness of the sound produced by a vibrating string is proportional to its total [kinetic plus potential] energy, which is given by

$$E = \frac{1}{2} \int_0^L \left[\rho \left(\frac{\partial y}{\partial t} \right)^2 + T \left(\frac{\partial y}{\partial x} \right)^2 \right] dx. \tag{27}$$

In Problem 17 we ask you to show that substitution of the series in (23) in the formula in (27) yields

$$E = \frac{\pi^2 T}{4L} \sum_{n=1}^{\infty} n^2 A_n^2. \tag{28}$$

The ratio of the nth term $n^2 A_n^2$ to the sum $\sum n^2 A_n^2$ is then the portion of the whole sound attributable to the nth harmonic.

We illustrate this theory with the series in (24), which describes the motion of the plucked string of Example 2. Note that the even harmonics are missing, and that $A_n = 4bL/(\pi^2 n^2)$ for n odd. Hence Eq. (28) gives

$$E = \frac{\pi^2 T}{4L} \sum_{n \text{ odd}} n^2 \frac{16 b^2 L^2}{\pi^4 n^4} = \frac{4 b^2 L T}{\pi^2} \sum_{n \text{ odd}} \frac{1}{n^2},$$

so the proportion of the sound associated with the nth harmonic (for n odd) is

$$\frac{1/n^2}{\displaystyle\sum_{n \text{ odd}} 1/n^2} = \frac{1/n^2}{\pi^2/8} = \frac{8}{\pi^2 n^2}.$$

Substituting $n = 1$ and $n = 3$, we find that 81.06% of the sound of the string of Example 2 is associated with the fundamental tone and 9.01% with the harmonic corresponding to $n = 3$.

The d'Alembert Solution

In contrast with series solutions of the heat equation, formal series solutions of the wave equation ordinarily do *not* possess sufficient termwise differentiability to permit verification of the solution by application of a superposition theorem analogous to the one stated in Section 9.5. For instance, termwise differentiation of the series in (24) would yield the series

$$\frac{\partial^2 y}{\partial x^2} = -\frac{4b}{L} \sum_{n=1}^{\infty} \sin \frac{n\pi}{2} \cos \frac{n\pi a t}{L} \sin \frac{n\pi x}{L},$$

which generally fails to converge, because the "convergence factor" $1/n^2$ has disappeared after the second differentiation.

There is, however, an alternative approach that both verifies the solution in (23) and yields valuable additional information about it. If we apply the trigonometric identity

$$2 \sin A \cos B = \sin(A + B) + \sin(A - B)$$

with $A = n\pi x/L$ and $B = n\pi at/L$, then (23) gives

$$y(x, t) = \sum_{n=1}^{\infty} A_n \sin \frac{n\pi x}{L} \cos \frac{n\pi at}{L}$$

$$= \frac{1}{2} \sum_{n=1}^{\infty} A_n \sin \frac{n\pi}{L}(x + at) + \frac{1}{2} \sum_{n=1}^{\infty} A_n \sin \frac{n\pi}{L}(x - at). \qquad (29)$$

But by definition of the coefficients,

$$\sum_{n=1}^{\infty} A_n \sin \frac{n\pi x}{L} = F(x)$$

for all x, where $F(x)$ is the *odd* extension of period $2L$ of the initial position function $f(x)$. Hence Eq. (29) means that

$$\blacktriangleright \qquad y(x, t) = \tfrac{1}{2}\left[F(x + at) + F(x - at)\right]. \qquad (30)$$

Therefore, the series in (23) converges to the expression on the right-hand side in (30), which is known as the **d'Alembert form** of the solution of Problem A for the vibrating string. Moreover, using the chain rule it is easy to verify (Problems 13 and 14) that the function $y(x, t)$ defined in (30) does, indeed, satisfy the equation $y_{tt} = a^2 y_{xx}$ (under the assumption that F is twice differentiable), as well as the boundary conditions $y(0, t) = y(L, t) = 0$ and $y(x, 0) = f(x)$.

For any function $F(x)$, the functions $F(x + at)$ and $F(x - at)$ represent "waves" moving to the left and right, respectively, along the x-axis with speed a. This fact is illustrated in Fig. 9.6.3, which shows the graphs of these two functions for successive values of t, in a typical case where the initial position function $F(x)$

(a) At time $t = 0$ (b) At time $t = \pi/8$

(c) At time $t = \pi/4$ (d) At time $t = 3\pi/8$

FIGURE 9.6.3. The pulse $F(x)$ produces two waves—one moving to the left, one to the right.

is a pulse function centered at the midpoint $x = \pi/2$ of a string of length $L = \pi$ with $a = 1$ (so the fundamental period of vibration is 2π). Thus the d'Alembert solution in (30) expresses $y(x, t)$ as a superposition of two waves moving in opposite directions with speed a. This is why the equation $y_{tt} = a^2 y_{xx}$ is called the *wave* equation.

Strings with Initial Velocity

The separation of variables solution of Problem B, with initial conditions $y(x, 0) = 0$ and $y_t(x, 0) = g(x)$, proceeds precisely as did that of Problem A until Eq. (16) is reached. But then $y(x, 0) = X(x)T(0) = 0$ implies that $T(0) = 0$, so instead of (16) we have

$$\frac{d^2 T_n}{dt^2} + \frac{n^2 \pi^2 a^2}{L^2} T_n = 0, \quad T_n(0) = 0. \tag{31}$$

From Eq. (17) we see that a nontrivial solution of (31) is

$$T_n(t) = \sin \frac{n\pi a t}{L}. \tag{32}$$

The resulting formal power series solution is therefore of the form

$$\blacktriangleright \qquad y(x, t) = \sum_{n=1}^{\infty} B_n \sin \frac{n\pi a t}{L} \sin \frac{n\pi x}{L}, \tag{33}$$

so we want to choose the coefficients $\{B_n\}$ so that

$$y_t(x, 0) = \sum_{n=1}^{\infty} B_n \frac{n\pi a}{L} \sin \frac{n\pi x}{L} = g(x). \tag{34}$$

Thus we want $B_n \cdot n\pi a/L$ to be the Fourier sine coefficient b_n of $g(x)$ on $[0, L]$.

$$B_n \frac{n\pi a}{L} = b_n = \frac{2}{L} \int_0^L g(x) \sin \frac{n\pi x}{L}\, dx.$$

Hence we choose

$$\blacktriangleright \qquad B_n = \frac{2}{n\pi a} \int_0^L g(x) \sin \frac{n\pi x}{L}\, dx \tag{35}$$

in order for $y(x, t)$ in (33) to be a formal series solution of Problem B.

Example 3 Consider a string on a guitar lying crosswise in the back of a pickup trick that at time $t = 0$ slams into a brick wall with speed v_0. Then $g(x) \equiv v_0$, so

$$B_n = \frac{2}{n\pi a} \int_0^L v_0 \sin \frac{n\pi x}{L}\, dx = \frac{2v_0 L}{n^2 \pi^2 a} \left[1 - (-1)^n\right].$$

Hence the series in (33) gives

$$y(x, t) = \frac{4v_0 L}{\pi^2 a} \sum_{n \text{ odd}} \frac{1}{n^2} \sin \frac{n\pi a t}{L} \sin \frac{n\pi x}{L}. \qquad \blacksquare$$

If we differentiate the series in (33) termwise with respect to t, we get

$$y_t(x, t) = \sum_{n=1}^{\infty} b_n \sin \frac{n\pi x}{L} \cos \frac{n\pi a t}{L} = \frac{1}{2}[G(x + at) + G(x - at)], \qquad (36)$$

where G is the odd period $2L$ extension of the initial velocity function $g(x)$, using the same device as in the derivation of Eq. (30). In Problem 15 we ask you to deduce that

$$y(x, t) = \frac{1}{2a}[H(x + at) + H(x - at)], \qquad (37)$$

where the function $H(x)$ is defined to be

$$H(x) = \int_0^x G(s)\, ds. \qquad (38)$$

If, finally, a string has both a nonzero initial position function $y(x, 0) = f(x)$ and a nonzero initial velocity function $y_t(x, 0) = g(x)$, then we can obtain its displacement function by adding the d'Alembert solutions of Problems A and B given in Eqs. (30) and (37), respectively. Hence the vibrations of this string with general initial conditions are described by

$$y(x, t) = \frac{1}{2}[F(x + at) + F(x - at)] + \frac{1}{2a}[H(x + at) + H(x - at)], \qquad (39)$$

a superposition of four waves moving along the x-axis with speed a, two moving to the left and two moving to the right.

9.6 Problems

Solve the boundary value problems in Problems 1 through 10.

1. $y_{tt} = 4y_{xx}$, $0 < x < \pi$, $t > 0$; $y(0, t) = y(\pi, t) = 0$,
 $y(x, 0) = \frac{1}{10} \sin 2x$, $y_t(x, 0) = 0$

2. $y_{tt} = y_{xx}$, $0 < x < 1$, $t > 0$; $y(0, t) = y(1, t) = 0$,
 $y(x, 0) = \frac{1}{10} \sin \pi x - \frac{1}{20} \sin 3\pi x$, $y_t(x, 0) = 0$

3. $4y_{tt} = y_{xx}$, $0 < x < \pi$, $t > 0$; $y(0, t) = y(\pi, t) = 0$,
 $y(x, 0) = y_t(x, 0) = \frac{1}{10} \sin x$

4. $4y_{tt} = y_{xx}$, $0 < x < 2$, $t > 0$; $y(0, t) = y(2, t) = 0$,
 $y(x, 0) = \frac{1}{5} \sin \pi x \cos \pi x$, $y_t(x, 0) = 0$

5. $y_{tt} = 25y_{xx}$, $0 < x < 3$, $t > 0$; $y(0, t) = y(3, t) = 0$,
 $y(x, 0) = \frac{1}{4} \sin \pi x$, $y_t(x, 0) = 10 \sin 2\pi x$

6. $y_{tt} = 100y_{xx}$, $0 < x < \pi$, $t > 0$; $y(0, t) = y(\pi, t) = 0$,
 $y(x, 0) = x(\pi - x)$, $y_t(x, 0) = 0$

7. $y_{tt} = 100y_{xx}$, $0 < x < 1$, $t > 0$; $y(0, t) = y(1, t) = 0$,
 $y(x, 0) = 0$, $y_t(x, 0) = x$

8. $y_{tt} = 4y_{xx}$, $0 < x < \pi$, $t > 0$; $y(0, t) = y(\pi, t) = 0$,
 $y(x, 0) = \sin x$, $y_t(x, 0) = 1$

9. $y_{tt} = 4y_{xx}$, $0 < x < 1$, $t > 0$; $y(0, t) = y(1, t) = 0$,
 $y(x, 0) = 0$, $y_t(x, 0) = x(1 - x)$

10. $y_{tt} = 25y_{xx}$, $0 < x < \pi$, $t > 0$; $y(0, t) = y(\pi, t) = 0$,
 $y(x, 0) = y_t(x, 0) = \sin^2 x$

11. Suppose that a string 2 ft long weighs $\frac{1}{32}$ oz and is subjected to a tension of 32 lb. Find the fundamental frequency with which it vibrates and the velocity with which the vibration waves travel along it.

12. Show that the amplitude of the oscillations of the midpoint of the string of Example 3 is

$$y\left(\frac{L}{2}, \frac{L}{2a}\right) = \frac{4v_0 L}{\pi^2 a} \sum_{n \text{ odd}} \frac{1}{n^2} = \frac{v_0 L}{2a}.$$

If the string is the string of Problem 11 and the impact speed of the pickup truck is 60 mi/h, show that this amplitude is approximately 1 in.

13. Suppose that the function $F(x)$ is twice differentiable for all x. Use the chain rule to verify that the functions

$$y(x, t) = F(x + at) \quad \text{and} \quad y(x, t) = F(x - at)$$

satisfy the equation $y_{tt} = a^2 y_{xx}$.

14. Given the differentiable odd period $2L$ function $F(x)$, show that the function

$$y(x, t) = \frac{1}{2}[F(x + at) + F(x - at)]$$

satisfies the conditions $y(0, t) = y(L, t) = 0$, $y(x, 0) = F(x)$, and $y_t(x, 0) = 0$.

15. If $y(x, 0) = 0$, then Eq. (36) implies (why?) that

$$y(x, t) = \frac{1}{2} \int_0^t G(x + a\tau) \, d\tau + \frac{1}{2} \int_0^t G(x - a\tau) \, d\tau.$$

Make appropriate substitutions in these integrals to derive Eqs. (37) and (38).

16. (a) Show that the substitutions $u = x + at$ and $v = x - at$ transform the equation $y_{tt} = a^2 y_{xx}$ into the equation $y_{uv} = 0$. **(b)** Conclude that every solution of $y_{tt} = a^2 y_{xx}$ is of the form

$$y(x, t) = F(x + at) + G(x - at),$$

which represents two waves traveling in opposite directions, each with speed a.

17. Suppose that

$$y(x, t) = \sum_{n=1}^{\infty} A_n \cos \frac{n\pi at}{L} \sin \frac{n\pi x}{L}.$$

Square the derivatives y_t and y_x and then integrate termwise—applying the orthogonality of the sine and cosine functions—to verify that

$$E = \frac{1}{2} \int_0^L (\rho y_t^2 + T y_x^2) \, dx = \frac{\pi^2 T}{4L} \sum_{n=1}^{\infty} n^2 A_n^2.$$

18. Consider a stretched string, initially at rest; its end at $x = 0$ is fixed, but its end at $x = L$ is partially free—it is allowed to slide without friction along the vertical line $x = L$. The corresponding boundary value problem is

$$y_{tt} = a^2 y_{xx} \qquad (0 < x < L, \ t > 0);$$
$$y(0, t) = y_x(L, t) = 0,$$
$$y(x, 0) = f(x),$$
$$y_t(x, 0) = 0.$$

Separate the variables and use the odd half-multiple sine series of $f(x)$, as in Problem 24 of Section 9.5, to derive the solution

$$y(x, t) = \sum_{n \text{ odd}} A_n \cos \frac{n\pi at}{2L} \sin \frac{n\pi x}{2L},$$

where

$$A_n = \frac{2}{L} \int_0^L f(x) \sin \frac{n\pi x}{2L} \, dx.$$

Problems 19 and 20 deal with the vibrations of a string under the influence of the downward force $F(x) = -\rho g$ of gravity. According to Eq. (1), its displacement function satisfies the partial differential equation

$$\frac{\partial^2 y}{\partial t^2} = a^2 \frac{\partial^2 y}{\partial x^2} - g \tag{40}$$

with endpoint conditions $y(0, t) = y(L, t) = 0$.

19. Suppose first that the string hangs in a stationary position, so that $y = y(x)$ and $y_{tt} = 0$, and hence its differential equation of motion takes the simple form $a^2 y'' = g$. Derive the stationary solution

$$y(x) = \phi(x) = \frac{gx}{2a^2}(x - L).$$

20. Now suppose that the string is released from rest in equilibrium; consequently the initial conditions are $y(x, 0) = 0$ and $y_t(x, 0) = 0$. Define

$$v(x, t) = y(x, t) - \phi(x),$$

where $\phi(x)$ is the stationary solution of Problem 19. Deduce from Eq. (40) that $v(x, t)$ satisfies the boundary value problem

$$v_{tt} = a^2 v_{xx};$$
$$v(0, t) = v(L, t) = 0,$$
$$v(x, 0) = -\phi(x),$$
$$v_t(x, 0) = 0.$$

Conclude from Eqs. (22) and (23) that

$$y(x, t) - \phi(x) = \sum_{n=1}^{\infty} A_n \cos \frac{n\pi at}{L} \sin \frac{n\pi x}{L},$$

where the coefficients $\{A_n\}$ are the Fourier sine coefficients of $f(x) = -\phi(x)$. Finally, explain why it follows that the string oscillates between the positions $y = 0$ and $y = 2\phi(x)$.

21. For a string vibrating in air with resistance proportional to velocity, the boundary value problem is

$$y_{tt} = a^2 y_{xx} - 2h y_t;$$
$$y(0, t) = y(L, t) = 0,$$
$$y(x, 0) = f(x),$$
$$y_t(x, 0) = 0. \tag{41}$$

Assume that $0 < h < \pi a/L$. **(a)** Substitute

$$y(x, t) = X(x) T(t)$$

in (41) to obtain the equations

$$X'' + \lambda X = 0, \quad X(0) = X(L) = 0 \tag{42}$$

and

$$T'' + 2hT' + a^2 \lambda T = 0, \quad T'(0) = 0. \tag{43}$$

(b) The eigenvalues and eigenfunctions of (42) are

$$\lambda_n = \frac{n^2 \pi^2}{L^2} \quad \text{and} \quad X_n(x) = \sin \frac{n\pi x}{L}$$

(as usual). Show that the general solution with $\lambda = n^2 \pi^2 / L^2$ of the differential equation in (43) is

$$T_n(t) = e^{-ht} (A_n \cos \omega_n t + B_n \sin \omega_n t),$$

where $\omega_n = \sqrt{(n^2\pi^2 a^2/L^2) - h^2} < n\pi a/L$. **(c)** Show that $T_n'(0) = 0$ implies that $B_n = hA_n/\omega_n$, and hence that to within a constant multiplicative coefficient,

$$T_n(t) = e^{-ht}\cos(\omega_n t - \alpha_n)$$

where $\alpha_n = \tan^{-1}(h/\omega_n)$. **(d)** Finally, conclude that

$$y(x, t) = e^{-ht}\sum_{n=1}^{\infty} c_n \cos(\omega_n t - \alpha_n)\sin\frac{n\pi x}{L},$$

where

$$c_n = \frac{2}{L\cos\alpha_n}\int_0^L f(x)\sin\frac{n\pi x}{L}\,dx.$$

From this formula we see that the air resistance has three main effects: exponential damping of amplitudes, decreased frequencies $\omega_n < n\pi a/L$, and the introduction of the phase delay angles α_n.

22. Rework Problem 21 as follows: First substitute $y(x, t) = e^{-ht}v(x, t)$ in Eq. (41) and then show that the boundary value problem for $v(x, t)$ is

$$v_{tt} = a^2 v_{xx} + h^2 v;$$
$$v(0, t) = v(L, t) = 0,$$
$$v(x, 0) = f(x),$$
$$v_t(x, 0) = hf(x).$$

Next show that the substitution $v(x, t) = X(x)T(t)$ leads to the equations

$$X'' + \lambda X = 0, \quad X(0) = X(L) = 0,$$
$$T'' + (\lambda a^2 - h^2)T = 0.$$

Proceed in this manner to derive the solution $y(x, t)$ given in part (d) of Problem 21.

The snapshots in Fig. 9.6.4 show successive positions of a vibrating string with length $L = \pi$ and $a = 1$ (so its period of oscillation is 2π). The string is initially at rest with fixed endpoints, and at time $t = 0$ it is set in motion with initial position function

$$f(x) = 2\sin^2 x = 1 - \cos 2x. \tag{44}$$

23. The most interesting snapshot is the one in Fig. 9.6.4(e), where it appears that the string exhibits a momentary "flat spot" at the instant $t = \pi/4$. Indeed, apply the d'Alembert formula in (30) to prove that the string's position function $y(x, t)$ satisfies the condition

$$y\left(x, \frac{\pi}{4}\right) = 1 \quad \text{for} \quad \frac{\pi}{4} \leq x \leq \frac{3\pi}{4}.$$

24. **(a)** Show that the position function $f(x)$ defined in Eq. (44) has inflection points $[f''(x) = 0]$ at $x = \pi/4$ and at $x = 3\pi/4$. **(b)** In snapshots (a)–(e) of Fig. 9.6.4 it appears that these two inflection points may remain fixed during some initial portion of the string's vibration. Indeed, apply the d'Alembert formula to show that if either $x = \pi/4$ or $x = 3\pi/4$, then $y(x, t) = 1$ for $0 \leq t \leq \pi/4$.

9.6 Application: Vibrating String Investigations

Here we describe a *Mathematica* implementation of the d'Alembert solution

$$y(x, t) = \tfrac{1}{2}[F(x + at) + F(x - at)] \tag{1}$$

of the vibrating string problem, and apply it to investigate graphically the motions that result from a variety of different initial positions of the string. *Maple* and MAT-LAB versions of this implementation are included in the applications manual that accompanies this text.

To plot the snapshots shown in Fig. 9.6.4, we began with the initial position function

```
f[x_] := 2*Sin[x]^2
```

To define the odd period 2π extension $F(x)$ of $f(x)$, we need the function $s(x)$ that shifts the point x by an integral multiple of π into the interval $[0, \pi]$.

```
s[x_] := Block[{k}, k = Floor[N[x/Pi]];
            If[EvenQ[k], (* k is even *)
            (* then *) N[x - k*Pi],
            (* else *) N[x - k*Pi - Pi]]]
```

Then the desired odd extension is defined by

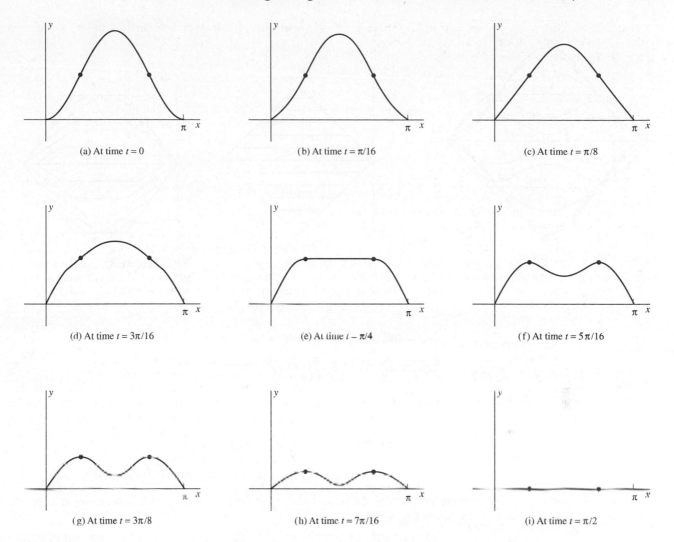

FIGURE 9.6.4. Successive snapshots of the vibrating string of Problems 23 and 24.

```
F[x_] := If[s[x] > 0, (* then *) f[s[x]],
                      (* else *)-f[-s[x]]]
```

Finally, the d'Alembert solution in (1) is

```
G[x_, t_] := ( F[x + t] + F[x - t] )/2
```

A snapshot of the position of the string at time t is defined by

```
stringAt[t_] := Plot[g[x,t], {x, 0, Pi} ];
```

For example, the command `stringAt[Pi/4]` plots the snapshot corresponding to $t = \pi/4$ shown in Fig. 9.6.4(e), which exhibits the apparent flat spot discussed in Problem 23. We can plot a whole sequence of snapshots at once:

```
snapshots = Table[ stringAt[t], {t, 0, Pi, Pi/12}]
```

These snapshots can be animated to show the vibrating string in motion, or we can exhibit simultaneously the successive positions of the string in a single composite figure (see Fig. 9.6.5) with the command `Show[snapshots]`.

FIGURE 9.6.5. Successive positions of the vibrating string with initial position $f(x) = 2\sin^2 x.$

FIGURE 9.6.6. Successive positions of the vibrating string with triangular initial position.

FIGURE 9.6.7. Successive positions of the vibrating string with trapezoidal initial position.

The initial position function

```
f[x_] := If[x < Pi/2, (* then *) x,
            (* else *) Pi - x] // N
```

(corresponding to the triangular wave function of Fig. 9.2.4) generates in this way the composite picture shown in Fig. 9.6.6. Similarly, the initial position function

```
f[x_] := Which[ 0 <= x < Pi/3, x,
               Pi/3 <= x < 2*Pi/3, Pi/3,
               2*Pi/3 <= x <= Pi, Pi - x ] // N
```

(corresponding to the trapezoidal wave function of Fig. 9.2.5) produces the picture shown in Fig. 9.6.7.

You can test your implementation of d'Alembert's method by attempting to generate Figs. 9.6.6 and 9.6.7 for yourself. The initial position "bump function"

$$f(x) = \sin^{200} x, \quad 0 \leqq x \leqq \pi$$

generates traveling waves traveling (initially) in opposite directions, as indicated in Fig. 9.6.3 in the text. The initial position function

$$f(x) = \begin{cases} \sin^{200}\left(x + \frac{3}{2}\right) & \text{for } 0 < x < \pi/2, \\ 0 & \text{for } \pi/2 < x < \pi \end{cases}$$

generates a single wave that begins at $x = 0$ and (initially) travels to the right. (Think of a jump rope tied to a tree, whose free end is given a "snap" at time $t = 0$.)

After exploring some of the possibilities indicated previously, try some initial position functions of your own choice. Any continuous function f such that $f(0) = f(\pi) = 0$ is fair game. The more exotic the resulting vibration of the string, the better.

9.7 Steady-State Temperature and Laplace's Equation

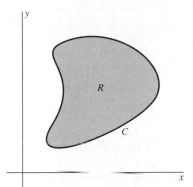

FIGURE 9.7.1. A plane region R and its bounding curve C.

We now consider the temperature in a two-dimensional plate, or lamina, that occupies a region R in the xy-plane bounded by a piecewise smooth curve C, as shown in Fig. 9.7.1. We assume that the faces of the plate are insulated, and that it is so thin that the temperature within it does not vary in the direction perpendicular to the xy-plane. We want to determine, under various conditions, the temperature $u(x, y, t)$ at the point (x, y) at time t.

Let the plate consist of material with density δ (mass per unit volume), specific heat c (per unit mass), and thermal conductivity K, all assumed constant throughout the plate. Under these assumptions, it can be shown (by a generalization of the derivation of the one-dimensional heat equation in Section 9.5) that the temperature function $u(x, y, t)$ satisfies the **two-dimensional heat equation**

$$\frac{\partial u}{\partial t} = k\left(\frac{\partial^2 u}{\partial x^2} + \frac{\partial^2 u}{\partial y^2}\right). \tag{1}$$

As in Section 9.5, k denotes the thermal diffusivity,

$$k = \frac{K}{c\delta}, \tag{2}$$

of the material of the plate. The sum of second derivatives on the right-hand side in (1) is the **Laplacian** of the function u, commonly denoted by

$$\nabla^2 u = \frac{\partial^2 u}{\partial x^2} + \frac{\partial^2 u}{\partial y^2}, \tag{3}$$

so the two-dimensional heat equation may be written as

$$\frac{\partial u}{\partial t} = k\nabla^2 u. \tag{1'}$$

In comparing Eq. (1′) with the one-dimensional heat equation $u_t = ku_{xx}$, we see that in passing from one to two dimensions, the second-order space derivative u_{xx} is replaced with the Laplacian $\nabla^2 u$. This is an instance of a general phenomenon. For instance, if a flexible stretched membrane occupies in equilibrium a region in the xy-plane and vibrates in the (perpendicular) z-direction, then its displacement function $z = z(x, y, t)$ satisfies the **two-dimensional wave equation**

$$\frac{\partial^2 z}{\partial t^2} = a^2\left(\frac{\partial^2 z}{\partial x^2} + \frac{\partial^2 z}{\partial y^2}\right) = a^2\nabla^2 z. \tag{4}$$

This equation has the same relation to the one-dimensional wave equation $z_{tt} = a^2 z_{xx}$ (here writing $z(x, y)$ for the displacement of a string) as Eq. (1′) has to the one-dimensional heat equation.

In this section we confine our attention to the steady-state situation in which the temperature u does not vary with time, and so is a function only of x and y. Thus we are interested in the *steady-state temperature* of a plate. In this case $u_t = 0$, so Eq. (1) becomes the **two-dimensional Laplace equation**

$$\nabla^2 u = \frac{\partial^2 u}{\partial x^2} + \frac{\partial^2 u}{\partial y^2} = 0. \tag{5}$$

This important partial differential equation is also known as the **potential equation.** The three-dimensional Laplace equation $u_{xx} + u_{yy} + u_{zz} = 0$ is satisfied (in empty space) by electric and gravitational potential functions. It is also satisfied by the velocity potential function for the steady irrotational flow of an incompressible and inviscid liquid.

Dirichlet Problems

A particular solution of Laplace's equation in a bounded plane region R is determined by appropriate boundary conditions. For example, it is plausible on physical grounds that the steady-state temperature $u(x, y)$ in a plate is determined if we know that $u(x, y)$ agrees with a given function $f(x, y)$ at each point of the boundary curve C of the plate. To find the steady-state temperature in a plate with assigned boundary values, we need to solve the boundary value problem

$$\frac{\partial^2 u}{\partial x^2} + \frac{\partial^2 u}{\partial y^2} = 0 \qquad \text{(within } R\text{);}$$

$$u(x, y) = f(x, y) \qquad \text{(if } (x, y) \text{ is on } C\text{).}$$

(6)

Such a problem, finding a solution of Laplace's equation in a region R with given boundary values, is called a **Dirichlet problem.** It is known that if the boundary curve C and the boundary value function f are reasonably well behaved, then there exists a unique solution of the Dirichlet problem in (6).

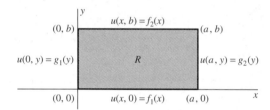

FIGURE 9.7.2. A rectangular plate with given boundary values.

Figure 9.7.2 shows a rectangular plate with indicated boundary values along its four edges. The corresponding Dirichlet problem is

$$u_{xx} + u_{yy} = 0 \qquad \text{(within } R\text{);}$$
$$u(x, 0) = f_1(x), \quad u(x, b) = f_2(x) \qquad (0 < x < a),$$
$$u(0, y) = g_1(y), \quad u(a, y) = g_2(y) \qquad (0 < y < b).$$

(7)

Because there are four nonhomogeneous conditions (rather than one), this boundary value problem is not directly susceptible to the method of separation of variables. In Section 9.6, when confronted with this difficulty, we split the problem of the vibrating string with both nonzero initial position and nonzero initial velocity into problems with a single nonhomogeneous condition each. In a similar way, the boundary value problem in (7) can be split into *four* problems, each with a single nonhomogeneous boundary condition. For each of these problems, $u(x, y)$ will be zero along three edges of the rectangle and will have assigned values on the fourth. Each of these four boundary values can be solved by the method of separation of variables, and the sum of the four solutions is the solution of the original problem in (7). In the following example we solve one of these four problems—the one illustrated in Fig. 9.7.3.

Example 1 Solve the boundary value problem

$$u_{xx} + u_{yy} = 0;$$
$$u(0, y) = u(a, y) = u(x, b) = 0, \tag{8}$$
$$u(x, 0) = f(x)$$

for the rectangle in Fig. 9.7.3.

Solution Substitution of $u(x, y) = X(x)Y(y)$ gives $X''Y + XY'' = 0$, so

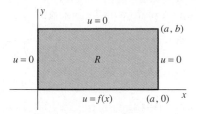

$$\frac{X''}{X} = -\frac{Y''}{Y} = -\lambda \tag{9}$$

for some constant λ. Thus $X(x)$ must satisfy the familiar eigenvalue problem

$$X'' + \lambda X = 0,$$
$$X(0) = X(a) = 0.$$

FIGURE 9.7.3. The boundary value problem of Example 1.

The eigenvalues and associated eigenfunctions are

$$\lambda_n = \frac{n^2 \pi^2}{a^2}, \quad X_n(x) = \sin \frac{n\pi x}{a} \tag{10}$$

for $n = 1, 2, 3, \dots$ From Eq. (9) with $\lambda = n^2\pi^2/a^2$ and the remaining homogeneous boundary conditions in (8), we get

$$Y_n'' - \frac{n^2\pi^2}{a^2} Y_n = 0, \quad Y_n(b) = 0. \tag{11}$$

The general solution of the differential equation in (11) is

$$Y_n(y) = A_n \cosh \frac{n\pi y}{a} + B_n \sinh \frac{n\pi y}{a},$$

and the condition

$$Y_n(b) = A_n \cosh \frac{n\pi b}{a} + B_n \sinh \frac{n\pi b}{a} = 0$$

implies that $B_n = -[A_n \cosh(n\pi b/a)]/[\sinh(n\pi b/a)]$, so

$$Y_n(y) = A_n \cosh \frac{n\pi y}{a} - \frac{A_n \cosh(n\pi b/a)}{\sinh(n\pi b/a)} \sinh \frac{n\pi y}{a}$$

$$= \frac{A_n}{\sinh(n\pi b/a)} \left(\sinh \frac{n\pi b}{a} \cosh \frac{n\pi y}{a} - \cosh \frac{n\pi b}{a} \sinh \frac{n\pi y}{a} \right).$$

Therefore,

$$Y_n(y) = c_n \sinh \frac{n\pi (b - y)}{a} \tag{12}$$

where $c_n = A_n / \sinh(n\pi b/a)$. From (10) and (12) we obtain the formal series solution

$$\blacktriangleright \qquad u(x, y) = \sum_{n=1}^{\infty} X_n(x) Y_n(y) = \sum_{n=1}^{\infty} c_n \sin \frac{n\pi x}{a} \sinh \frac{n\pi (b - y)}{a}. \qquad (13)$$

It remains only to choose the coefficients $\{c_n\}$ to satisfy the nonhomogeneous condition

$$u(x, 0) = \sum_{n=1}^{\infty} \left(c_n \sinh \frac{n\pi b}{a} \right) \sin \frac{n\pi x}{a} = f(x).$$

For this purpose we want

$$c_n \sinh \frac{n\pi b}{a} = b_n = \frac{2}{a} \int_0^a f(x) \sin \frac{n\pi x}{a} \, dx,$$

so

$$\blacktriangleright \qquad c_n = \frac{2}{a \sinh(n\pi b/a)} \int_0^a f(x) \sin \frac{n\pi x}{a} \, dx. \qquad (14)$$

With this choice of coefficients, the series in (13) is a formal series solution of the Dirichlet problem in (8). ∎

For instance, suppose that

$$f(x) \equiv T_0 = \frac{4T_0}{\pi} \sum_{n \text{ odd}} \frac{1}{n} \sin \frac{n\pi x}{a}$$

if $0 < x < a$, so that $b_n = 4T_0/(\pi n)$ for n odd and $b_n = 0$ for n even. Then the formulas in (13) and (14) yield

$$u(x, y) = \frac{4T_0}{\pi} \sum_{n \text{ odd}} \frac{\sin(n\pi x/a) \, \sinh(n\pi [b - y]/a)}{n \sinh(n\pi b/a)} \qquad (15)$$

for the steady-state temperature in a rectangular plate with its base held at temperature T_0 and its other three edges at temperature zero. In particular, suppose that $T_0 = 100$ and that $a = b = 10$. Then from Eq. (15) we find that the temperature at the center of the plate is

$$u(5, 5) = \frac{400}{\pi} \sum_{n \text{ odd}} \frac{\sin(n\pi/2) \, \sinh(n\pi/2)}{n \sinh n\pi}$$

$$\approx 25.3716 - 0.3812 + 0.0099 - 0.0003;$$

that is, $u(5, 5)$ is approximately 25.00.

Remark 1: Can you supply an argument—perhaps using symmetry—that proves (without using the series solution) that $u(5, 5)$ is exactly 25?

Remark 2: In practice, engineers use solutions like the one in Eq. (15) to plot the variation of temperature within a plate or piece of machinery; they are particularly interested in any "hot spots" that may appear. Such information is often presented by constructing *level curves* on which the temperature function $u(x, y)$ is constant. Figure 9.7.4 shows typical level curves for the function $u(x, y)$ in (15) with $a = b = 10$ and $T_0 = 100$. ∎

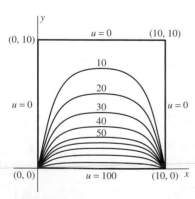

FIGURE 9.7.4. Typical level curves of the function $u(x, y)$ in Eq. (15).

Example 2
Let R be the "semi-infinite" strip shown in Fig. 9.7.5. Solve the boundary value problem

$$u_{xx} + u_{yy} = 0 \qquad \text{(in } R\text{)};$$
$$u(x, 0) = u(x, b) = 0 \qquad (0 < x < \infty),$$
$$u(x, y) \text{ is bounded as } x \to +\infty,$$
$$u(0, y) = g(y). \tag{16}$$

Solution

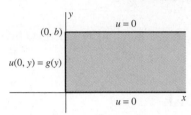

FIGURE 9.7.5. The "semi-infinite" strip of Example 2.

The condition that $u(x, y)$ is bounded as $x \to +\infty$ plays the role of a homogeneous boundary condition associated with the "missing" right edge of the "rectangle"; this is typical of Dirichlet problems for unbounded regions. With $u(x, y) = X(x)Y(y)$, the actual homogeneous conditions give $Y(0) = Y(b) = 0$. Hence it is $Y(y)$ that will satisfy an eigenvalue problem. We therefore write the separation in (9) as

$$\frac{Y''}{Y} = -\frac{X''}{X} = -\lambda, \tag{17}$$

changing the sign of λ to get the familiar eigenvalue problem

$$Y'' + \lambda Y = 0, \quad Y(0) = Y(b) = 0$$

with eigenvalues and associated eigenfunctions

$$\lambda_n = \frac{n^2 \pi^2}{b^2}, \quad Y_n(y) = \sin \frac{n\pi y}{b}. \tag{18}$$

The general solution of

$$X_n'' - \frac{n^2 \pi^2}{b^2} X_n = 0$$

is $X_n(x) = A_n \exp(n\pi x/b) + B_n \exp(-n\pi x/b)$. We write the solution in exponential form because then the condition that $u(x, y)$, and hence $X(x)$, is bounded as $x \to +\infty$ immediately implies that $A_n = 0$ for all n. Suppressing the constant, this leaves

$$X_n(x) = \exp\left(-\frac{n\pi x}{b}\right). \tag{19}$$

From Eqs. (18) and (19) we get the formal series solution

$$\blacktriangleright \qquad u(x, y) = \sum_{n=1}^{\infty} b_n X_n(x) Y_n(y) = \sum_{n=1}^{\infty} b_n \exp\left(-\frac{n\pi x}{b}\right) \sin \frac{n\pi y}{b} \tag{20}$$

that satisfies the homogeneous boundary conditions and the boundedness condition. In order also to satisfy the nonhomogeneous condition

$$u(0, y) = \sum_{n=1}^{\infty} b_n \sin \frac{n\pi y}{b} = g(y),$$

we choose b_n to be the Fourier sine coefficient

$$\blacktriangleright \qquad b_n = \frac{2}{b} \int_0^b g(y) \sin \frac{n\pi y}{b} \, dy \tag{21}$$

of the function $g(y)$ on $[0, b]$.

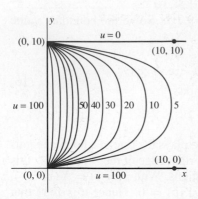

FIGURE 9.7.6. Typical level curves of the function $u(x, y)$ of Example 2.

For instance, if

$$g(y) = T_0 = \frac{4T_0}{\pi} \sum_{n \text{ odd}} \frac{1}{n} \sin \frac{n\pi y}{b}$$

for $0 < y < b$, then Eqs. (20) and (21) yield

$$u(x, y) = \frac{4T_0}{\pi} \sum_{n \text{ odd}} \frac{1}{n} \exp\left(-\frac{n\pi x}{b}\right) \sin \frac{n\pi y}{b}.$$

Figure 9.7.6 shows some typical level curves of the temperature function defined by this formula with $b = 10$ and $T_0 = 100$. ∎

The Dirichlet Problem for a Circular Disk

We now investigate the steady-state temperature in a circular disk of radius a with insulated faces and with given boundary temperatures. Obviously we should accommodate the geometry of the circle by expressing $u = u(r, \theta)$ in terms of polar coordinates r and θ, with $x = r\cos\theta$ and $y = r\sin\theta$. When these equations are used to transform the Laplacian, the result is **Laplace's equation in polar coordinates:**

$$\blacktriangleright \qquad \nabla^2 u = \frac{\partial^2 u}{\partial r^2} + \frac{1}{r}\frac{\partial u}{\partial r} + \frac{1}{r^2}\frac{\partial^2 u}{\partial \theta^2} = 0. \qquad (22)$$

We prescribe assigned boundary temperatures (as indicated in Fig. 9.7.7) by requiring that

$$\blacktriangleright \qquad u(a, \theta) = f(\theta), \qquad (23)$$

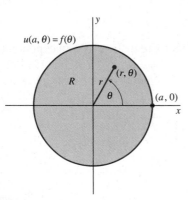

FIGURE 9.7.7. The Dirichlet problem for a circular disk.

where the function $f(\theta)$ of period 2π is given. In addition, because (r, θ) and $(r, \theta + 2\pi)$ are the polar coordinates of the same point, we impose on u the periodicity condition

$$\blacktriangleright \qquad u(r, \theta) = u(r, \theta + 2\pi) \qquad (24)$$

for $r < a$ and all θ; this will play the role of a homogeneous boundary condition.

To solve the boundary value problem in Eqs. (22) through (24), we substitute $u(r, \theta) = R(r)\Theta(\theta)$ in Eq. (22); the result is

$$R''\Theta + \frac{1}{r}R'\Theta + \frac{1}{r^2}R\Theta'' = 0.$$

Division by $R\Theta/r^2$ gives

$$\frac{r^2 R'' + r R'}{R} + \frac{\Theta''}{\Theta} = 0,$$

so it follows that

$$\frac{r^2 R'' + r R'}{R} = -\frac{\Theta''}{\Theta} = \lambda$$

for some constant λ. This yields the two ordinary differential equations

$$r^2 R'' + r R' - \lambda R = 0 \tag{25}$$

and

$$\Theta'' + \lambda \Theta = 0. \tag{26}$$

The general solution of Eq. (26) is

$$\Theta(\theta) = A \cos \alpha \theta + B \sin \alpha \theta \quad \text{if } \lambda = \alpha^2 > 0, \tag{27a}$$
$$\Theta(\theta) = A + B\Theta \quad \text{if } \lambda = 0, \tag{27b}$$
$$\Theta(\theta) = A e^{\alpha \theta} + B e^{-\alpha \theta} \quad \text{if } \lambda = -\alpha^2 < 0. \tag{27c}$$

Now we apply the condition in (24), which implies that $\Theta(\theta) = \Theta(\theta + 2\pi)$; that is, that $\Theta(\theta)$ has period 2π. In Problems 22 and 23 we ask you to show that this is so only if $\lambda = n^2$ (n an integer) in (27a) or if $\lambda = 0$ and $B = 0$ in (27b). Thus we have the eigenvalues and associated eigenfunctions

$$\lambda_0 = 0, \quad \Theta_0(\theta) = 1; \tag{28a}$$
$$\lambda_n = n^2, \quad \Theta_n(\theta) = A_n \cos n\theta + B_n \sin n\theta \tag{28b}$$

for $n = 1, 2, 3, \dots$.

Next we turn our attention to Eq. (25). With $\lambda_0 = 0$ it reduces to the equation $r^2 R_0'' + r R_0' = 0$ with general solution

$$R_0(r) = C_0 + D_0 \ln r. \tag{29}$$

Of course we want $u(r, \theta)$, and hence $R(r)$, to be continuous at $r = 0$; this requires that $D_0 = 0$ in Eq. (29), so $R_0(r) = C_0$. Then, with $\lambda = \lambda_n = n^2$, Eq. (25) is

$$r^2 R_n'' + r R_n' - n^2 R_n = 0.$$

On substituting the trial solution $R(r) = r^k$, we find that $k = \pm n$, so the general solution of the previous equation is

$$R_n(r) = C_n r^n + \frac{D_n}{r^n}. \tag{30}$$

Continuity at $r = 0$ then requires that $D_n = 0$, so $R_n(r) = C_n r^n$.

Finally we combine the results in Eqs. (28) through (30) with $D_n \equiv 0$ to obtain a formal series solution of the form

$$u(r, \theta) = \sum_{n=0}^{\infty} R_n(r) \Theta_n(\theta);$$

that is,

$$u(r, \theta) = \frac{a_0}{2} + \sum_{n=1}^{\infty} (a_n \cos n\theta + b_n \sin n\theta) r^n. \tag{31}$$

In order that the boundary condition $u(a, \theta) = f(\theta)$ be satisfied, we want

$$u(a, \theta) = \frac{a_0}{2} + \sum_{n=1}^{\infty} (a_n a^n \cos n\theta + b_n a^n \sin n\theta)$$

to be the Fourier series of $f(\theta)$ on $[0, 2\pi]$. Hence we choose

$$a_n = \frac{1}{\pi a^n} \int_0^{2\pi} f(\theta) \cos n\theta \, d\theta \quad (n = 0, 1, 2, \dots) \tag{32a}$$

and

$$b_n = \frac{1}{\pi a^n} \int_0^{2\pi} f(\theta) \sin n\theta \, d\theta \qquad (n = 1, 2, 3, \dots). \tag{32b}$$

Example 3 For instance, if $F(\theta) = T_0$ for $0 < \theta < \pi$ and $F(\theta) = -T_0$ for $\pi < \theta < 2\pi$, so that

$$F(\theta) = \frac{4T_0}{\pi} \sum_{n \text{ odd}} \frac{1}{n} \sin n\theta,$$

then $a_n = 0$ for all $n \geq 0$, $b_n = 0$ for n even, and $b_n = 4T_0/(\pi n a^n)$ for n odd. So Eqs. (31) and (32) yield

$$u(r, \theta) = \frac{4T_0}{\pi} \sum_{n \text{ odd}} \frac{r^n}{na^n} \sin n\theta. \qquad \blacksquare$$

Note: In the Dirichlet problems of this section, we have concentrated on temperature functions with preassigned boundary values. In the case of a plate with an *insulated* edge—across which no heat can flow—the boundary condition is that the derivative of $u(x, y)$ in the direction normal to this edge vanishes. For instance, consider the rectangular plate in Fig. 9.7.2. If the edge $x = 0$ is insulated, then $u_x(0, y) \equiv 0$, whereas if the edge $y = b$ is insulated, then $u_y(x, b) \equiv 0$.

9.7 Problems

In Problems 1 through 3, solve the Dirichlet problem for the rectangle $0 < x < a$, $0 < y < b$ consisting of Laplace's equation $u_{xx} + u_{yy} = 0$ and the given boundary value conditions.

1. $u(x, 0) = u(x, b) = u(0, y) = 0$, $u(a, y) = g(y)$
2. $u(x, 0) = u(x, b) = u(a, y) = 0$, $u(0, y) = g(y)$
3. $u(x, 0) = u(0, y) = u(a, y) = 0$, $u(x, b) = f(x)$
4. Consider the boundary value problem

$$u_{xx} + u_{yy} = 0;$$
$$u_x(0, y) = u_x(a, y) = u(x, 0) = 0,$$
$$u(x, b) = f(x)$$

corresponding to a rectangular plate $0 < x < a$, $0 < y < b$ with the edges $x = 0$ and $x = a$ insulated. Derive the solution

$$u(x, y) = \frac{a_0 y}{2b} + \sum_{n=1}^{\infty} a_n \left(\cos \frac{n\pi x}{a} \right) \left(\frac{\sinh(n\pi y/a)}{\sinh(n\pi b/a)} \right),$$

where

$$a_n = \frac{2}{a} \int_0^a f(x) \cos \frac{n\pi x}{a} \, dx \qquad (n = 0, 1, 2, \dots).$$

(*Suggestion:* Show first that λ_0 is an eigenvalue with $X_0(x) \equiv 1$ and $Y_0(y) = y$.)

In Problems 5 and 6, find a solution of Laplace's equation $u_{xx} + u_{yy} = 0$ in the rectangle $0 < x < a$, $0 < y < b$ that satisfies the given boundary conditions (see Problem 4).

5. $u_y(x, 0) = u_y(x, b) = u(a, y) = 0$, $u(0, y) = g(y)$
6. $u_x(0, y) = u_x(a, y) = u_y(x, 0) = 0$, $u(x, b) = f(x)$

In Problems 7 and 8, find a solution of Laplace's equation in the semi-infinite strip $0 < x < a$, $y > 0$ that satisfies the given boundary conditions and the additional condition that $u(x, y)$ is bounded as $y \to +\infty$.

7. $u(0, y) = u(a, y) = 0$, $u(x, 0) = f(x)$
8. $u_x(0, y) = u_x(a, y) = 0$, $u(x, 0) = f(x)$
9. Suppose that $a = 10$ and $f(x) = 10x$ in Problem 8. Show that

$$u(x, y) = 50 - \frac{400}{\pi^2} \sum_{n \text{ odd}} \frac{1}{n^2} e^{-n\pi y/10} \cos \frac{n\pi x}{10}.$$

Then compute (with two-decimal-place accuracy) the values $u(0, 5)$, $u(5, 5)$, and $u(10, 5)$.

10. The edge $x = a$ of the rectangular plate $0 < x < a$, $0 < y < b$ is insulated, the edges $x = 0$ and $y = 0$ are held at temperature zero, and $u(x, b) = f(x)$. Use the odd half-multiple sine series of Problem 21 of Section 9.3 to derive a solution of the form

$$u(x, y) = \sum_{n \text{ odd}} c_n \sin \frac{n\pi x}{2a} \sinh \frac{n\pi y}{2a}.$$

Then give a formula for c_n.

11. The edge $y = 0$ of the rectangular plate $0 < x < a$, $0 < y < b$ is insulated, the edges $x = a$ and $y = b$ are held at temperature zero, and $u(0, y) = g(y)$. Use the odd half-multiple cosine series of Problem 22 of Section 9.3 to find $u(x, y)$.

12. A vertical cross section of a long high wall 30 cm thick has the shape of the semi-infinite strip $0 < x < 30$, $y > 0$. The face $x = 0$ is held at temperature zero, but the face $x = 30$ is insulated. Given $u(x, 0) = 25$, derive the formula

$$u(x, y) = \frac{100}{\pi} \sum_{n \text{ odd}} \frac{1}{n} e^{-n\pi y/60} \sin \frac{n\pi x}{60}$$

for the steady-state temperature within the wall.

Problems 13 through 15 deal with the semicircular plate of radius a shown in Fig. 9.7.8. The circular edge has a given temperature $u(a, \theta) = f(\theta)$. In each problem, derive the given series for the steady-state temperature $u(r, \theta)$ satisfying the given boundary conditions along $\theta = 0$ and $\theta - \pi$, and give the formula for the coefficients c_n.

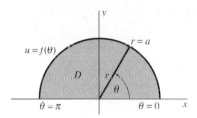

FIGURE 9.7.8. The semicircular plate of Problems 13 through 15.

13. $u(r, 0) = u(r, \pi) = 0$;

$$u(r, \theta) - \sum_{n=1}^{\infty} c_n r^n \sin n\theta$$

14. $u_\theta(r, 0) = u_\theta(r, \pi) = 0$;

$$u(r, \theta) = \frac{c_0}{2} + \sum_{n=1}^{\infty} c_n r^n \cos n\theta$$

15. $u(r, 0) = u_\theta(r, \pi) = 0$;

$$u(r, \theta) = \sum_{n \text{ odd}} c_n r^{n/2} \sin \frac{n\theta}{2}$$

16. Consider Dirichlet's problem for the region *exterior* to the circle $r = a$. You want to find a solution of

$$r^2 u_{rr} + r u_r + u_{\theta\theta} = 0$$

such that $u(a, \theta) = f(\theta)$ and $u(r, \theta)$ is bounded as $r \to +\infty$. Derive the series

$$u(r, \theta) = \frac{a_0}{2} + \sum_{n=1}^{\infty} \frac{1}{r^n} (a_n \cos n\theta + b_n \sin n\theta),$$

and give formulas for the coefficients $\{a_n\}$ and $\{b_n\}$.

17. The velocity potential function $u(r, \theta)$ for steady flow of an ideal fluid around a cylinder of radius $r = a$ satisfies the boundary value problem

$$r^2 u_{rr} + r u_r + u_{\theta\theta} = 0 \qquad (r > a);$$
$$u_r(a, \theta) = 0, \qquad u(r, \theta) = u(r, -\theta),$$
$$\lim_{r \to \infty} [u(r, \theta) - U_0 r \cos \theta] = 0.$$

(a) By separation of variables, derive the solution

$$u(r, \theta) = \frac{U_0}{r}(r^2 + a^2) \cos \theta.$$

(b) Hence show that the velocity components of the flow are

$$u_x = \frac{\partial u}{\partial x} = \frac{U_0}{r^2}(r^2 - a^2 \cos 2\theta)$$

and

$$u_y = \frac{\partial u}{\partial y} = -\frac{U_0}{r^2}a^2 \sin 2\theta.$$

The streamlines for this fluid flow around the cylinder are shown in Fig. 9.7.9.

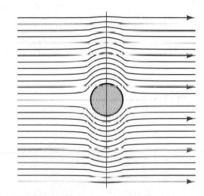

FIGURE 9.7.9. Streamlines for ideal fluid flow around a cylinder.

Comment: The streamlines in Fig. 9.7.9 are the level curves of the function $\psi(x, y)$ given in part (b) of Problem 18. It is troublesome to write a computer program to plot such level curves. It is far easier to plot some solutions of the differential equation $dy/dx = u_y/u_x$. We did so with initial conditions $x_0 = -5$, $y_0 = -2.7, -2.5, -2.3, \ldots, 2.5, 2.7$. The solutions were obtained numerically using the improved Euler method (Section 2.5) with step size 0.02.

18. (a) Show that the velocity potential in part (a) of Problem 17 can be written in rectangular coordinates as

$$u(x, y) = U_0 x \left(1 + \frac{a^2}{x^2 + y^2} \right).$$

(b) The stream function for the flow is

$$\psi(x, y) = U_0 y \left(\frac{a^2}{x^2 + y^2} - 1 \right).$$

Show that $\nabla u \cdot \nabla \psi \equiv 0$. Because $\mathbf{v} = \nabla u$ is the velocity vector, this shows that the streamlines of the flow are the level curves of $\psi(x, y)$.

Problems 19 through 21 deal with a solid spherical ball of radius $r = a$ with initial temperature T_0 throughout. At time $t = 0$ it is packed in ice, so its temperature at the surface $r = a$ is zero thereafter. Its temperature then depends only on time t and the distance r from the center of the ball, so we write $u = u(r, t)$.

19. **(a)** The heat content of the spherical shell with inner radius r and outer radius $r + \Delta r$ is

$$Q(t) = \int_r^{r+\Delta r} c\delta u(x, t) \cdot 4\pi s^2 \, ds.$$

Show that $Q'(t) = 4\pi c\delta r^2 u_t(\bar{r}, t)$ for some \bar{r} in the interval $(r, r + \Delta r)$. **(b)** The radial heat flux is $\phi = -Ku_r$ across the bounding spherical surfaces of the shell in part (a). Conclude that

$$Q'(t) = 4\pi K \left[(r + \Delta r)^2 u_r(r + \Delta r, t) - r^2 u_r(r, t) \right].$$

(c) Equate the values of $Q'(t)$ in parts (a) and (b); then take the limit as $\Delta r \to 0$ to get the equation

$$\frac{\partial u}{\partial t} = \frac{k}{r^2} \frac{\partial}{\partial r} \left(r^2 \frac{\partial u}{\partial r} \right).$$

(d) Finally, show that this last equation is equivalent to

$$\frac{\partial}{\partial t}(ru) = k \frac{\partial^2}{\partial r^2}(ru).$$

20. It follows from part (d) of Problem 19 that $u(r, t)$ satisfies the boundary value problem

$$\frac{\partial}{\partial t}(ru) = k \frac{\partial^2}{\partial r^2}(ru) \qquad (r < a);$$

$$u(a, t) = 0, \quad u(r, 0) = T_0.$$

Show that the new function $v(r, t) = ru(r, t)$ satisfies a familiar boundary value problem (in Section 9.5), and thus

derive the solution

$$u(r, t) = \frac{2aT_0}{\pi} \sum_{n=1}^{\infty} \frac{(-1)^{n+1}}{nr} \exp\left(-\frac{n^2\pi^2 kt}{a^2} \right) \sin \frac{n\pi r}{a}.$$

21. **(a)** Deduce from the solution to Problem 20 that the temperature at the center of the ball at time t is

$$u(0, t) = 2T_0 \sum_{n=1}^{\infty} (-1)^{n+1} \exp\left(-\frac{n^2\pi^2 kt}{a^2} \right).$$

(b) Let $a = 30$ cm and $T_0 = 100°$C. Compute $u(0, t)$ after 15 min if the ball is made of iron with $k = 0.15$ cgs. (*Answer:* Approximately $45°$C.) **(c)** If you have access to a programmable calculator, repeat part (b) for a ball made of concrete with $k = 0.005$ in cgs units. About 15 terms are required to show that $u = 100.00°$C accurate to two decimal places.

22. In the discussion of the Dirichlet problem for a circular disk in this section, we obtained the ordinary differential equation $\Theta'' + \lambda\Theta = 0$ with the periodicity condition $\Theta(\theta) = \Theta(\theta + 2\pi)$. **(a)** Suppose that $\lambda = \alpha^2 > 0$. Show that the general solution

$$\Theta(\theta) = A \cos \alpha\theta + B \sin \alpha\theta$$

has period 2π only if $\lambda = n^2$ with n an integer. **(b)** In the case $\lambda = 0$, show that the general solution

$$\Theta(\theta) = A\theta + B$$

is periodic only if $A = 0$.

23. If $\lambda = -\alpha^2 < 0$ in Problem 22, then the general solution of the differential equation is

$$\Theta(\theta) = Ae^{\alpha\theta} + Be^{-\alpha\theta}.$$

Show that this function is not periodic unless $A = B = 0$.

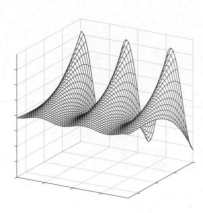

CHAPTER 10

Eigenvalues and Boundary Value Problems

10.1 Sturm-Liouville Problems and Eigenfunction Expansions

In the last three sections of Chapter 9, we saw that numerous different boundary value problems all lead —by separation of variables—to the same ordinary differential equation

➤ $$X''(x) + \lambda X(x) = 0 \qquad (0 < x < L), \tag{1}$$

containing an eigenvalue λ, but with different endpoint conditions, such as

$$X(0) = X(L) = 0 \tag{2}$$

or

$$X'(0) = X'(L) = 0 \tag{3}$$

or

$$X(0) = X'(L) = 0, \tag{4}$$

depending on the original boundary conditions.

For example, recall the problem (in Section 9.5) of finding the temperature $u(x, t)$ in a rod $0 \leq x \leq L$ with given initial temperature $u(x, 0) = f(x)$. As a boundary value problem, this problem of the rod is the same as the problem of finding the temperature within a large slab that occupies the region $0 \leq x \leq L$ in xyz-space. If its initial temperature depends only on x and is independent of y and z—that is, if $u(x, 0) = f(x)$—then the same will be true of its temperature $u = u(x, t)$ at time t. When we substitute

$$u(x, t) = X(x)T(t)$$

645

in the heat equation

$$\frac{\partial u}{\partial t} = k\frac{\partial^2 u}{\partial x^2},$$

we find that $X(x)$ satisfies the endpoint conditions in (2) if the faces $x = 0$ and $x = L$ of the slab are held at temperature zero, those in (3) if both faces are insulated, and those in (4) if one face is insulated and the other is held at temperature zero. But if each face loses heat to a surrounding medium (at temperature zero) in accord with Newton's law of cooling, then (according to Problem 23 of Section 9.5) the endpoint conditions take the form

$$hX(0) - X'(0) = 0 = hX(L) + X'(L), \tag{5}$$

where h is a nonnegative heat transfer coefficient.

The point is that when we impose different endpoint conditions on the solution of Eq. (1), we get different eigenvalue problems, and hence different eigenvalues $\{\lambda_n\}$ and different eigenfunctions $\{X_n(x)\}$ to use in constructing a formal power series solution

$$u(x, t) = \sum c_n X_n(x) T_n(t) \tag{6}$$

of the boundary value problem. The final step in this construction is to choose the coefficient $\{c_n\}$ in Eq. (6) so that

$$u(x, 0) = \sum c_n T_n(0) X_n(x) = f(x). \tag{7}$$

Thus we need finally an *eigenfunction expansion* of the given function $f(x)$ in terms of the eigenfunctions of the pertinent endpoint value problem.

Sturm-Liouville Problems

In order to unify and generalize the method of separation of variables, it is useful to formulate a general type of eigenvalue problem that includes as special cases each of those mentioned previously. Equation (1), with y instead of X as the dependent variable, can be written in the form

$$\frac{d}{dx}\left[p(x)\frac{dy}{dx}\right] - q(x)y + \lambda r(x)y = 0, \tag{8}$$

where $p(x) = r(x) \equiv 1$ and $q(x) \equiv 0$. Indeed, we indicate in Problem 16 that essentially any linear second-order differential equation of the form

$$A(x)y'' + B(x)y' + C(x)y + \lambda D(x)y = 0$$

takes the form in (8) after multiplication by a suitable factor.

Example 1 If we multiply the parametric Bessel equation

$$x^2 y'' + xy' + (\lambda x^2 - n^2)y = 0 \qquad (x > 0)$$

of order n by $1/x$, the result may be written as

$$\frac{d}{dx}\left[x\frac{dy}{dx}\right] - \frac{n^2}{x}y + \lambda xy = 0,$$

which is of the form in (8) with $p(x) = r(x) = x$ and $q(x) = n^2/x$. ∎

Now, on solutions of Eq. (8) in a bounded open interval (a, b), let us impose homogeneous [linear] endpoint conditions of the form

$$\alpha_1 y(a) - \alpha_2 y'(a) = 0, \quad \beta_1 y(b) + \beta_2 y'(b) = 0, \tag{9}$$

where the coefficients α_1, α_2, β_1, and β_2 are constants. In addition to being homogeneous, these two conditions are *separated*, in that one of them involves the values of $y(x)$ and $y'(x)$ at one endpoint $x = a$, whereas the other involves values at the other endpoint $x = b$. For instance, note that the conditions $y(a) = y'(b) = 0$ are of the form in (9) with $\alpha_1 = \beta_2 = 1$ and $\alpha_2 = \beta_1 = 0$.

Definition: Sturm-Liouville Problem

A **Sturm-Liouville problem** is an endpoint value problem of the form

$$\frac{d}{dx}\left[p(x)\frac{dy}{dx} \right] - q(x)y + \lambda r(x)y = 0 \qquad (a < x < b); \tag{8}$$

$$\alpha_1 y(a) - \alpha_2 y'(a) = 0, \qquad \beta_1 y(b) + \beta_2 y'(b) = 0, \tag{9}$$

with neither α_1 and α_2 both zero nor β_1 and β_2 both zero. The parameter λ in (8) is the "eigenvalue" whose possible (constant) values are sought.

Sturm-Liouville problems generalize the endpoint value problems of Section 3.8 with sine and cosine solutions. They are named for the French mathematicians Charles Sturm (1803–1855) and Joseph Liouville (1809–1882), who investigated such problems in the 1830s.

Example 2 We get different Sturm-Liouville problems by pairing the same differential equation

$$y'' + \lambda y = 0 \qquad (\text{for } 0 < x < L)$$

with the different homogeneous endpoint conditions

- $y(0) = y(L) = 0$ (where $\alpha_1 = \beta_1 = 1$ and $\alpha_2 = \beta_2 = 0$),
- $y'(0) = y'(L) = 0$ (where $\alpha_1 = \beta_1 = 0$ and $\alpha_2 = \beta_2 = 1$), and
- $y(0) = y'(L) = 0$ (where $\alpha_1 = \beta_2 = 1$ and $\alpha_2 = \beta_1 = 0$). ∎

Note that the Sturm-Liouville problem in (8)–(9) always has the trivial solution $y(x) \equiv 0$. We seek the values of λ—the **eigenvalues**—for which this problem has a *nontrivial* real-valued solution (an **eigenfunction**), and for each eigenvalue its associated eigenfunction (or eigenfunctions). Note that any constant [nonzero] multiple of an eigenfunction will also be an eigenfunction. The following theorem provides sufficient conditions under which the problem in (8)–(9) has an infinite sequence $\{\lambda_n\}_1^\infty$ of nonnegative eigenvalues, with each eigenvalue λ_n having (to within a constant multiple) exactly one associated eigenfunction $y_n(x)$. A proof of this theorem is outlined in Chapter 9 of G. P. Tolstov, *Fourier Series* (New York: Dover, 1976).

THEOREM 1: Sturm-Liouville Eigenvalues

Suppose that the functions $p(x)$, $p'(x)$, $q(x)$, and $r(x)$ in Eq. (8) are continuous on the interval $[a, b]$ and that $p(x) > 0$ and $r(x) > 0$ at each point of $[a, b]$. Then the eigenvalues of the Sturm-Liouville problem in (8)–(9) constitute an increasing sequence

$$\lambda_1 < \lambda_2 < \lambda_3 < \cdots < \lambda_{n-1} < \lambda_n < \cdots \tag{10}$$

of real numbers with

$$\lim_{n \to \infty} \lambda_n = +\infty. \tag{11}$$

To within a constant factor, only a single eigenfunction $y_n(x)$ is associated with each eigenvalue λ_n. Moreover, if $q(x) \geqq 0$ on $[a, b]$ and the coefficients α_1, α_2, β_1, and β_2 in (9) are all nonnegative, then the eigenvalues are all nonnegative. ▲

Note: It is important to observe the signs in (8) and (9) when verifying the hypotheses of Theorem 1. Sometimes the Sturm-Liouville problem in (8)–(9) is called **regular** if it satisfies the hypotheses of Theorem 1; otherwise it is **singular**. In this section we will confine our attention to regular Sturm-Liouville problems. Singular Sturm-Liouville problems associated with Bessel's equation will appear in Section 10.4.

Example 3 In Example 3 of Section 3.8, we saw that the Sturm-Liouville problem

$$\begin{aligned} y'' + \lambda y &= 0 \qquad (0 < x < L); \\ y(0) &= 0, \qquad y(L) = 0 \end{aligned} \tag{12}$$

has eigenvalues $\lambda_n = n^2\pi^2/L^2$ and associated eigenfunctions $y_n(x) = \sin(n\pi x/L)$ ($n = 1, 2, 3, \ldots$). There we had to consider the cases $\lambda = -\alpha^2 < 0$, $\lambda = 0$, and $\lambda = \alpha^2 > 0$ separately. Here $p(x) = r(x) \equiv 1$, $q(x) \equiv 0$, $\alpha_1 = \beta_1 = 1$, and $\alpha_2 = \beta_2 = 0$, so we know from Theorem 1 that the problem in (12) has only nonnegative eigenvalues. Therefore, only the two cases $\lambda = 0$ and $\lambda = \alpha^2 > 0$ would need to be considered if we were starting afresh. ■

We are also familiar (from the problems in Section 9.5) with the Sturm-Liouville problem

$$\begin{aligned} y'' + \lambda y &= 0 \qquad (0 < x < L); \\ y'(0) &= 0, \qquad y'(L) = 0, \end{aligned} \tag{13}$$

which has eigenvalues $\lambda_0 = 0$, $\lambda_n = n^2\pi^2/L^2$ ($n = 1, 2, 3, \ldots$) and associated eigenfunctions $y_0(x) \equiv 1$, $y_n(x) = \cos(n\pi x/L)$. We will customarily write $\lambda_0 = 0$ if 0 is an eigenvalue and otherwise write λ_1 for the smallest eigenvalue; thus λ_1 always denotes the smallest positive eigenvalue.

Example 4 Find the eigenvalues and associated eigenfunctions of the Sturm-Liouville problem

$$\begin{aligned} y'' + \lambda y &= 0 \qquad (0 < x < L); \\ y'(0) &= 0, \qquad y(L) = 0. \end{aligned} \tag{14}$$

Solution This is a Sturm-Liouville problem satisfying the hypotheses of Theorem 1 with $\alpha_1 = \beta_2 = 0$ and $\alpha_2 = \beta_1 = 1$, so there are no negative eigenvalues. If $\lambda = 0$, then $y(x) = Ax + B$, and thus $y'(0) = A = 0$ and $y(L) = B = 0$. Therefore 0 is not an eigenvalue. If $\lambda = \alpha^2$, then

$$y(x) = A \cos \alpha x + B \sin \alpha x \quad \text{and} \quad y'(x) = -A\alpha \sin \alpha x + B\alpha \cos \alpha x.$$

Hence $y'(0) = 0$ implies that $B = 0$, and then $y(L) = A \cos \alpha L = 0$, so it follows that αL is an *odd* integral multiple of $\pi/2$: $\alpha L = (2n - 1)\pi/2$. Thus we have the eigenvalues and associated eigenfunctions

$$\blacktriangleright \qquad \lambda_n = \alpha_n^2 = \frac{(2n - 1)^2 \pi^2}{4L^2} \quad \text{and} \quad y_n(x) = \cos \frac{(2n - 1)\pi x}{2L}$$

for $n = 1, 2, 3, \ldots$. ∎

Example 5 Find the eigenvalues and associated eigenfunctions of the Sturm-Liouville problem

$$\blacktriangleright \qquad \begin{aligned} y'' + \lambda y &= 0 \qquad (0 < x < L); \\ y(0) = 0, \qquad hy(L) &+ y'(L) = 0 \quad (h > 0). \end{aligned} \qquad (15)$$

Solution This problem satisfies the hypotheses of Theorem 1 with $\alpha_1 = \beta_2 = 1$, $\alpha_2 = 0$, and $\beta_1 = h > 0$, so there are no negative eigenvalues. If $\lambda = 0$, then $y(x) = Ax + B$, thus $y(0) = B = 0$. Then

$$hy(L) + y'(L) = h \cdot AL + A = A \cdot (hL + 1) = 0,$$

and it follows that $A = 0$ as well. Thus 0 is not an eigenvalue.
If $\lambda = \alpha^2$, then

$$y(x) = A \cos \alpha x + B \sin \alpha x.$$

Hence $y(0) = A = 0$, so

$$y(x) = B \sin \alpha x \quad \text{and} \quad y'(x) = B\alpha \cos \alpha x.$$

Therefore,

$$hy(L) + y'(L) = hB \sin \alpha L + B\alpha \cos \alpha L = 0.$$

If $B \neq 0$, it follows that

$$\tan \alpha L = -\frac{\alpha}{h} = -\frac{\alpha L}{hL}. \qquad (16)$$

Thus $\beta = \alpha L$ is a [positive] solution of the equation

$$\tan x = -\frac{x}{hL}. \qquad (17)$$

FIGURE 10.1.1. The solution of Eq. (17) by geometric methods.

The solutions of Eq. (17) are the points of intersection of the graphs of $y(x) = -\tan x$ and $y(x) = x/hL$, as indicated in Fig. 10.1.1. It is apparent from the figure that there is an infinite sequence of positive roots $\beta_1, \beta_2, \beta_3, \ldots$, and that when n is large, β_n is only slightly larger than $(2n - 1)\pi/2$. See the table in Fig. 10.1.2, in which the first eight solutions of Eq. (17) are listed for the case $hL = 1$. At any

n	x_n	$(2n-1)\pi/2$
1	2.0288	1.5708
2	4.9132	4.7124
3	7.9787	7.8540
4	11.0855	10.9956
5	14.2074	14.1372
6	17.3364	17.2788
7	20.4692	20.4204
8	23.6043	23.5619

FIGURE 10.1.2. Approximate values of the first eight positive solutions of $\tan x = -x$.

rate, the eigenvalues and associated eigenfunctions of the problem in (15) are given by

$$\lambda_n = \alpha_n^2 = \frac{\beta_n^2}{L^2}, \quad y_n(x) = \sin\alpha_n x = \sin\frac{\beta_n x}{L} \tag{18}$$

for $n = 1, 2, 3, \ldots$. Equation (17) appears frequently in certain applications (mechanical vibrations and heat conduction are only two of many examples), and its solutions for various values of hL are tabulated in Table 4.19 of Abramowitz and Stegun, *Handbook of Mathematical Functions* (New York: Dover, 1965). ∎

Orthogonal Eigenfunctions

Recall from Section 9.1 that the familiar formulas for the coefficients in a Fourier series follow from the orthogonality of the sine and cosine functions. Similarly, the expansion of a given function in terms of the eigenfunctions of a Sturm-Liouville problem depends on a crucial orthogonality property of these eigenfunctions. The functions $\phi(x)$ and $\psi(x)$ are said to be **orthogonal** on the interval $[a, b]$ with respect to the **weight function** $r(x)$ provided that

$$\int_a^b \phi(x)\psi(x)r(x)\,dx = 0. \tag{19}$$

The following theorem implies that any two eigenfunctions of a regular Sturm-Liouville problem that are associated with distinct eigenvalues are orthogonal with respect to the weight function $r(x)$ in Eq. (8).

THEOREM 2: Orthogonality of Eigenfunctions

Suppose that the functions p, q, and r in the Sturm-Liouville problem in Eqs. (8)–(9) satisfy the hypotheses of Theorem 1, and let $y_i(x)$ and $y_j(x)$ be eigenfunctions associated with the distinct eigenvalues λ_i and λ_j. Then

$$\int_a^b y_i(x)y_j(x)r(x)\,dx = 0. \tag{20}$$

Proof: We begin with the equations

$$\frac{d}{dx}\left[p(x)\frac{dy_i}{dx}\right] - q(x)y_i + \lambda_i r(x)y_i = 0,$$

$$\frac{d}{dx}\left[p(x)\frac{dy_j}{dx}\right] - q(x)y_j + \lambda_j r(x)y_j = 0. \tag{21}$$

These equations imply that λ_i, y_i and λ_j, y_j are eigenvalue-eigenfunction pairs. If we multiply the first equation by y_j and the second by y_i, then subtract the results, we get

$$y_j\frac{d}{dx}\left[p(x)\frac{dy_i}{dx}\right] - y_i\frac{d}{dx}\left[p(x)\frac{dy_j}{dx}\right] + (\lambda_i - \lambda_j)r(x)y_i y_j = 0.$$

Hence

$$(\lambda_i - \lambda_j) y_i y_j r(x) = y_i \frac{d}{dx}\left[p(x) \frac{dy_j}{dx}\right] - y_j \frac{d}{dx}\left[p(x) \frac{dy_i}{dx}\right]$$

$$= \frac{d}{dx}\left[p(x) \left(y_i \frac{dy_j}{dx} - y_j \frac{dy_i}{dx}\right)\right],$$

the latter of the two equalities being verifiable by direct differentiation. Therefore, integration from $x = a$ to $x = b$ yields

$$(\lambda_i - \lambda_j) \int_a^b y_i(x) y_j(x) r(x)\, dx = \left[p(x) \left(y_i(x) y_j'(x) - y_j(x) y_i'(x)\right)\right]_a^b. \quad (22)$$

From the endpoint condition in (9), we have

$$\alpha_1 y_i(a) - \alpha_2 y_i'(a) = 0 \quad \text{and} \quad \alpha_1 y_j(a) - \alpha_2 y_j'(a) = 0.$$

Because α_1 and α_2 are not both zero, it follows that the determinant of coefficients must be zero:

$$y_i(a) y_j'(a) - y_j(a) y_i'(a) = 0.$$

Similarly, the second endpoint condition in (9) implies that

$$y_i(b) y_j'(b) - y_j(b) y_i'(b) = 0.$$

Thus the right-hand side in Eq. (22) vanishes. Inasmuch as $\lambda_i \neq \lambda_j$, the result in (20) follows, and the proof is complete. ▲

Eigenfunction Expansions

Now suppose that the function $f(x)$ can be represented in the interval $[a, b]$ by an **eigenfunction series**

➤
$$f(x) = \sum_{m=1}^{\infty} c_m y_m(x), \quad (23)$$

where the functions y_1, y_2, y_3, ... are the eigenfunctions of the regular Sturm-Liouville problem in Eqs. (8)–(9). To determine the coefficients c_1, c_2, c_3, ..., we generalize the technique by which we determined the ordinary Fourier coefficients in Section 9.1. First we multiply each side of Eq. (23) by $y_n(x) r(x)$, then integrate from $x = a$ to $x = b$. Under the assumption that termwise integration is valid, we obtain

$$\int_a^b f(x) y_n(x) r(x)\, dx = \sum_{m=1}^{\infty} c_m \int_a^b y_m(x) y_n(x) r(x)\, dx. \quad (24)$$

But because of the orthogonality in (20), the only nonzero term on the right-hand side in Eq. (24) is the one for which $m = n$. Thus Eq. (24) takes the form

$$\int_a^b f(x) y_n(x) r(x)\, dx = c_n \int_a^b [y_n(x)]^2 r(x)\, dx,$$

and therefore

$$c_n = \frac{\displaystyle\int_a^b f(x)y_n(x)r(x)\,dx}{\displaystyle\int_a^b [y_n(x)]^2\, r(x)\,dx}. \tag{25}$$

We therefore *define* the eigenfunction series in (23)—representing $f(x)$ in terms of the eigenfunctions of the given Sturm-Liouville problem—by means of the choice of the coefficients specified by Eq. (25).

For instance, suppose that the Sturm-Liouville problem is the familiar

$$\begin{aligned} y'' + \lambda y &= 0 \qquad (0 < x < \pi); \\ y(0) = y(\pi) &= 0, \end{aligned} \tag{26}$$

for which $r(x) \equiv 1$ and the eigenfunctions are $y_n(x) = \sin nx$ for $n = 1, 2, 3, \ldots$. Then Eq. (25) yields

$$c_n = \frac{\displaystyle\int_0^\pi f(x)\sin nx\,dx}{\displaystyle\int_0^\pi \sin^2 nx\,dx} = \frac{2}{\pi}\int_0^\pi f(x)\sin nx\,dx,$$

because

$$\int_0^\pi \sin^2 x\,dx = \frac{\pi}{2}.$$

This is the familiar formula for the Fourier sine coefficients, and so the eigenfunction series in (23) is simply the familiar Fourier sine series

$$f(x) = \sum_{n=1}^\infty c_n \sin nx$$

of $f(x)$ on $[0, \pi]$.

The following theorem, stated without proof, generalizes the Fourier convergence theorem of Section 9.2.

THEOREM 3: Convergence of Eigenfunction Series

Let y_1, y_2, y_3, \ldots be the eigenfunctions of a regular Sturm-Liouville problem on $[a, b]$. If the function $f(x)$ is piecewise smooth on $[a, b]$, then the eigenfunction series in Eq. (23) converges for $a < x < b$ to the value $f(x)$ wherever f is continuous, and to the average $\frac{1}{2}[f(x+) + f(x-)]$ of right-hand and left-hand limits at each point of discontinuity. ▲

Example 6 For the Sturm-Liouville problem $y'' + \lambda y = 0$ $(0 < x < L)$, $y'(0) = y(L) = 0$ of Example 4, we found the eigenfunctions $y_n(x) = \cos(2n-1)\pi x/(2L)$, $n = 1, 2, 3, \ldots$. The corresponding eigenfunction series for a function $f(x)$ is

$$f(x) = \sum_{n=1}^\infty c_n \cos\frac{(2n-1)\pi x}{2L} \tag{27}$$

with

$$c_n = \frac{\displaystyle\int_0^L f(x) \cos\left([2n-1]\pi x/2L\right)\, dx}{\displaystyle\int_0^L \cos^2\left([2n-1]\pi x/2L\right)\, dx}$$

$$= \frac{2}{L}\int_0^L f(x)\cos\frac{(2n-1)\pi x}{2L}\, dx, \tag{28}$$

because

$$\int_0^L \cos^2\frac{(2n-1)\pi x}{2L}\, dx = \frac{L}{2}.$$

Thus the series in Eq. (27) is the odd half-multiple cosine series of Problem 22 of Section 9.3. Similarly, the Sturm-Liouville problem $y'' + \lambda y = 0$, $y(0) = y'(L) = 0$ leads to the odd half-multiple sine series

$$f(x) = \sum_{n=1}^{\infty} c_n \sin\frac{(2n-1)\pi x}{2L}, \quad c_n = \frac{2}{L}\int_0^L f(x)\sin\frac{(2n-1)\pi x}{2L}\, dx \tag{29}$$

of Problem 21 of Section 9.3. ∎

Example 7 Represent the function $f(x) = A$ (a constant) for $0 < x < 1$ as a series of eigenfunctions of the Sturm-Liouville problem

$$\begin{aligned} y'' + \lambda y &= 0 & (0 < x < 1); \\ y(0) &= 0, & y(1) + 2y'(1) = 0. \end{aligned} \tag{30}$$

Solution Comparing (30) with (15), we see that this is the Sturm-Liouville problem of Example 5 with $L = 1$ and $h = \frac{1}{2}$. From (17) and (18) we see that the eigenfunctions of the problem are therefore $y_n(x) = \sin\beta_n x$ where $\beta_1, \beta_2, \beta_3, \ldots$ are the positive roots of the equation $\tan x = -2x$. Hence the coefficients in the desired series are given by

$$c_n = \frac{\displaystyle\int_0^1 A\sin\beta_n x\, dx}{\displaystyle\int_0^1 \sin^2\beta_n x\, dx}. \tag{31}$$

Now $\displaystyle\int_0^1 A\sin\beta_n x\, dx = A(1-\cos\beta_n)/\beta_n$ and

$$\int_0^1 \sin^2\beta_n x\, dx = \int_0^1 \frac{1}{2}(1-\cos 2\beta_n x)\, dx = \frac{1}{2}\left[x - \frac{1}{2\beta_n}\sin 2\beta_n x\right]_0^1$$

$$= \frac{1}{2}\left(1 - \frac{\sin\beta_n \cos\beta_n}{\beta_n}\right).$$

Consequently,

$$\int_0^1 \sin^2\beta_n x\, dx = \frac{1}{2}\left(1 + 2\cos^2\beta_n\right).$$

In the last step we used the fact from Eq. (17) that $(\sin \beta_n)/\beta_n = -2\cos\beta_n$. Substituting these values for the integrals in (31), we get the eigenfunction series

$$f(x) = \sum_{n=1}^{\infty} \frac{2A(1-\cos\beta_n)}{\beta_n(1+2\cos^2\beta_n)} \sin\beta_n x. \tag{32}$$

∎

Remark: A numerical investigation of the eigenfunction series in Eq. (32) is outlined in the computing project for this section. ∎

Summary

According to the theorems in this section, every regular Sturm-Liouville problem has the following three properties:

- It has an infinite sequence of eigenvalues diverging to infinity (Theorem 1).
- The eigenfunctions are orthogonal with appropriate weight function (Theorem 2).
- Any piecewise smooth function can be represented by an eigenfunction series (Theorem 3).

There are other types of eigenvalue problems in applied mathematics that also enjoy these three important properties. Some isolated examples will appear in subsequent sections, though we will confine our attention largely to applications of Sturm-Liouville problems.

10.1 Problems

The problems for Section 3.8 deal with eigenvalues and eigenfunctions and may be used here as well. In Problems 1 through 5, verify that the eigenvalues and eigenfunctions for the indicated Sturm-Liouville problem are those listed.

1. $y'' + \lambda y = 0$, $y'(0) = y'(L) = 0$; $\lambda_0 = 0$, $y_0(x) \equiv 1$ and $\lambda_n = n^2\pi^2/L^2$, $y_n(x) = \cos n\pi x/L$

2. $y'' + \lambda y = 0$, $y(0) = y'(L) = 0$; $\lambda_n = (2n-1)^2\pi^2/4L^2$, $y_n(x) = \sin(2n-1)\pi x/2L$, $n \geq 1$

3. $y'' + \lambda y = 0$, $y'(0) = hy(L) + y'(L) = 0$ $(h > 0)$; $\lambda_n = \beta_n^2/L^2$, $y_n(x) = \cos\beta_n x/L$ $(n \geq 1)$, where β_n is the nth positive root of $\tan x = hL/x$. Sketch $y = \tan x$ and $y = hL/x$ to estimate the value of β_n for n large.

4. $y'' + \lambda y = 0$, $hy(0) - y'(0) = y(L) = 0$ $(h > 0)$; $\lambda_n = \beta_n^2/L^2$, $y_n(x) = \beta_n \cos\beta_n x/L + hL\sin\beta_n x/L$ $(n \geq 1)$, where β_n is the nth positive root of $\tan x = -x/hL$.

5. $y'' + \lambda y = 0$, $hy(0) - y'(0) = hy(L) + y'(L) = 0$ $(h > 0)$; $\lambda_n = \beta_n^2/L^2$, $y_n(x) = \beta_n \cos\beta_n x/L + hL\sin\beta_n x/L$ $(n \geq 1)$, where β_n is the nth positive root of $\tan x = 2hLx/(x^2 - h^2L^2)$. Estimate β_n for n large by sketching the graphs of $y = 2hL\cot x$ and the hyperbola $y = (x^2 - h^2L^2)/x$.

6. Show that the Sturm-Liouville problem $y'' + \lambda y = 0$, $y(0) = y'(L) = 0$ leads to the odd half-multiple sine series in Eq. (29) (see Problem 2).

In Problems 7 through 10, represent the given function $f(x)$ as a series of eigenfunctions of the indicated Sturm-Liouville problem.

7. $f(x) \equiv 1$; the Sturm-Liouville problem of Example 5
8. $f(x) \equiv 1$; the Sturm-Liouville problem of Problem 3
9. $f(x) = x$; the Sturm-Liouville problem of Example 5 with $L = 1$
10. $f(x) = x$; the Sturm-Liouville problem of Problem 3 with $L = 1$

Problems 11 through 14 deal with the regular Sturm-Liouville problem

$$\begin{aligned} y'' + \lambda y &= 0 \qquad (0 < x < L); \\ y(0) &= 0, \qquad hy(L) - y'(L) = 0, \end{aligned} \tag{33}$$

where $h > 0$. Note that Theorem 1 does not exclude the possibility of negative eigenvalues.

11. Show that $\lambda_0 = 0$ is an eigenvalue if and only if $hL = 1$, in which case the associated eigenfunction is $y_0(x) = x$.

12. Show that the problem in (33) has a single negative eigenvalue λ_0 if and only if $hL > 1$, in which case $\lambda_0 = -\beta_0^2/L^2$ and $y_0(x) = \sinh\beta_0 x/L$, where β_0 is the positive root of the equation $\tanh x = x/hL$. (*Suggestion*: Sketch the graphs of $y = \tanh x$ and $y = x/hL$.)

13. Show that the positive eigenvalues and associated eigenfunctions of the problem in (33) are $\lambda_n = \beta_n^2/L^2$ and $y_n(x) = \sin \beta_n x/L$ $(n \geq 1)$, where β_n is the nth positive root of $\tan x = x/hL$.

14. Suppose that $hL = 1$ in (33) and that $f(x)$ is piecewise smooth. Show that

$$f(x) = c_0 x + \sum_{n=1}^{\infty} c_n \sin \frac{\beta_n x}{L},$$

where $\{\beta_n\}_1^{\infty}$ are the positive roots of $\tan x = x$, and

$$c_0 = \frac{3}{L^3} \int_0^L x f(x)\, dx,$$

$$c_n = \frac{2}{L \sin^2 \beta_n} \int_0^L f(x) \sin \frac{\beta_n x}{L}\, dx.$$

15. Show that the eigenvalues and eigenfunctions of the Sturm-Liouville problem

$$y'' + \lambda y = 0 \qquad (0 < x < 1);$$
$$y(0) + y'(0) = 0, \qquad y(1) = 0$$

are given by $\lambda_0 = 0$, $y_0(x) = x - 1$ and

$$\lambda_n = \beta_n^2, \quad y_n(x) = \beta_n \cos \beta_n x - \sin \beta_n x$$

for $n \geq 1$, where $\{\beta_n\}_1^{\infty}$ are the positive roots of $\tan x = x$.

16. Beginning with the equation

$$A(x)y'' + B(x)y' + C(x)y + \lambda D(x)y = 0,$$

first divide by $A(x)$ and then multiply by

$$p(x) = \exp\left(\int \frac{B(x)}{A(x)}\, dx\right).$$

Show that the resulting equation can be written in the Sturm-Liouville form

$$\frac{d}{dx}\left[p(x)\frac{dy}{dx}\right] - q(x)y + \lambda r(x)y = 0$$

with $q(x) = -p(x)C(x)/A(x)$ and

$$r(x) = p(x)D(x)/A(x).$$

Loaded Uniform Beam

Consider a uniform beam with [downward] load $w(x)$, whose deflection function $y(x)$ satisfies the fourth-order equation

$$EI y^{(4)} = w(x)$$

for $0 < x < L$ and the endpoint conditions

- $y = y'' = 0$ at a hinged (or simply supported) end;
- $y = y' = 0$ at a fixed (built-in) end;
- $y'' = y^{(3)} = 0$ at a free end.

In the case in which each end is hinged, $y(x)$ can be found by the Fourier series method discussed in Section 9.4—that is, by substituting the Fourier series

$$y(x) = \sum b_n \sin \frac{n\pi x}{L}$$

in the differential equation

$$EI y^{(4)} = w(x) = \sum_{n=1}^{\infty} c_n \sin \frac{n\pi x}{L} \tag{34}$$

(where c_n is the nth Fourier sine coefficient of $w(x)$) to determine the coefficients $\{b_n\}$.

17. Suppose that w is constant in Eq. (34). Apply the method described here to obtain the deflection function

$$y(x) = \frac{4wL^4}{EI\pi^5} \sum_{n \text{ odd}} \frac{1}{n^5} \sin \frac{n\pi x}{L}.$$

18. Suppose that $w = bx$ in (34). Derive the deflection function

$$y(x) = \frac{2bL^5}{EI\pi^5} \sum_{n=1}^{\infty} \frac{(-1)^{n+1}}{n^5} \sin \frac{n\pi x}{L}.$$

The method used in Problems 17 and 18 succeeds because the functions $\sin(n\pi x/L)$ satisfy the hinged/hinged conditions $y(0) = y''(0) = y(L) = y''(L) = 0$, so that $y(x)$ does also.

If, instead, both ends of the beam are fixed, in place of the sine functions we can use the eigenfunctions of the problem

$$y^{(4)} - \lambda y = 0 \qquad (0 < x < L),$$
$$y(0) = y'(0) = 0, \qquad y(L) = y'(L) = 0, \tag{35}$$

because these eigenfunctions satisfy the fixed/fixed endpoint conditions. The eigenvalues of this problem are all positive, and by Problem 22 the associated eigenfunctions are orthogonal with weight function $r(x) \equiv 1$. Hence we can write

$$w(x) = \sum_{n=1}^{\infty} c_n y_n(x), \qquad c_n = \frac{\int_0^L w(x)y_n(x)\, dx}{\int_0^L [y_n(x)]^2\, dx}, \tag{36}$$

according to the analogue of Theorem 3 that holds for the problem in (35). If we write $\lambda = \alpha^4$, then $y_n(x)$ is of the form

$$y(x) = A \cosh \alpha x$$
$$+ B \sinh \alpha x + C \cos \alpha x + D \sin \alpha x, \tag{37}$$

with $\alpha = \alpha_n$, so it follows that $y_n^{(4)}(x) = \alpha_n^4 y_n(x)$. When we substitute the series $y(x) = \sum b_n y_n(x)$—which evidently satisfies the fixed/fixed endpoint conditions—in (35), we obtain

$$EI \sum_{n=1}^{\infty} b_n \alpha_n^4 y_n(x) = \sum_{n=1}^{\infty} c_n y_n(x).$$

Hence $EI b_n \alpha_n^4 = c_n$, so the deflection function of the beam is

$$y(x) = \sum_{n=1}^{\infty} \frac{c_n}{EI \alpha_n^4} y_n(x). \tag{38}$$

The following problems deal with the eigenvalues and eigenfunctions of the problem in (35) and similar problems.

19. Begin with the general solution in (37) of $y^{(4)} - \alpha^4 y = 0$. First note that $y(0) = y'(0) = 0$ implies that $C = -A$ and $D = -B$. Then impose the conditions $y(L) = y'(L) = 0$ to get two homogeneous linear equations in A and B. Hence the determinant of coefficients of A and B must vanish; deduce from this that $\cosh \alpha L \cos \alpha L = 1$. Conclude that the nth eigenvalue is $\lambda_n = \beta_n^4 / L^4$ where $\{\beta_n\}_1^\infty$ are the positive roots of the equation $\cosh x \cos x = 1$ (see Fig. 10.1.3). Finally, show that an associated eigenfunction is

$$y_n(x) = (\sinh \beta_n - \sin \beta_n) \left(\cosh \frac{\beta_n x}{L} - \cos \frac{\beta_n x}{L} \right)$$

$$- (\cosh \beta_n - \cos \beta_n) \left(\sinh \frac{\beta_n x}{L} - \sin \frac{\beta_n x}{L} \right).$$

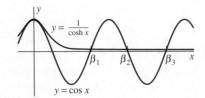

FIGURE 10.1.3. The solutions of $\cosh x \cos x = 1$ (the graph is not drawn to scale).

20. For the case of a cantilever (fixed/free) beam, we need to

solve the eigenvalue problem

$$y^{(4)} - \lambda y = 0 \qquad (0 < x < L);$$
$$y(0) = y'(0) = 0, \qquad y''(L) = y^{(3)}(L) = 0.$$

Proceeding as in Problem 19, show that the nth eigenvalue is $\lambda_n = \beta_n^4 / L^4$, where $\{\beta_n\}_1^\infty$ are the positive roots of the equation $\cosh x \cos x = -1$. Then find the associated eigenfunction.

21. For the eigenvalue problem

$$y^{(4)} - \lambda y = 0, \qquad y(0) = y'(0) = 0 = y(L) = y''(L)$$

corresponding to a fixed/hinged beam, show that the nth eigenvalue is $\lambda_n = \beta_n^4 / L^4$, where $\{\beta_n\}_1^\infty$ are the positive roots of the equation $\tanh x = \tan x$.

22. Suppose that $y_m^{(4)} = \lambda_m y_m$ and $y_n^{(4)} = \lambda_n y_n$. Apply the method of proof of Theorem 2 and integrate by parts twice to show that

$$(\lambda_m - \lambda_n) \int_0^L y_m(x) y_n(x) \, dx$$

$$= \left[y_n(x) \frac{d^3 y_m}{dx^3} - y_m(x) \frac{d^3 y_n}{dx^3} \right.$$

$$\left. - \frac{dy_n}{dx} \frac{d^2 y_m}{dx^2} + \frac{dy_m}{dx} \frac{d^2 y_n}{dx^2} \right]_0^L.$$

Conclude that if each endpoint condition is either $y = y' = 0$, $y = y'' = 0$, or $y'' = y^{(3)} = 0$, then y_m and y_n are orthogonal if $\lambda_m \neq \lambda_n$.

10.1 Application: Numerical Eigenfunction Expansions

We outline a MATLAB-based numerical investigation of the eigenfunction series

$$f(x) = \sum_{n=1}^\infty \frac{2(1 - \cos \beta_n)}{\beta_n(1 + 2\cos^2 \beta_n)} \sin \beta_n x \qquad (1)$$

for $0 < x < 1$. (We have taken $A = 1$ in Eq. (32) of this section.) *Maple* and *Mathematica* versions of this investigation are included in the applications manual that accompanies this text.

We saw in the solution of Example 7 that the values $\{\beta_n\}_1^\infty$ in Eq. (1) are the positive solutions of the equation $\tan x = -2x$. Just as indicated in Fig. 10.1.1 (for the similar equation $\tan x = -x$), the value of β_n is (for n large) just slightly larger than $(2n - 1)\pi/2$. But because of the discontinuities in the tangent function near these solutions, it is best for automatic root-finding to rewrite our eigenvalue equation in the form

$$\sin x + 2x \cos x = 0. \qquad (2)$$

Then the MATLAB command

```
b = fsolve('sin(x)+2*x.*cos(x)', pi/2+0.1+(0:pi:99*pi));
```

n	β_n
1	1.8366
2	4.8158
3	7.9171
4	11.0408
5	14.1724
6	17.3076
7	20.4448
8	23.5831
9	26.7222
10	29.8619

FIGURE 10.1.4. The first ten positive solutions of Eq. (2).

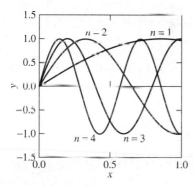

FIGURE 10.1.5. The eigenfunctions $y_n(x) = \sin \beta_n x$ for $n = 1, 2, 3, 4$.

quickly generates the first 100 values of β_n (see Fig. 10.1.4). (We have provided a vector

$$\mathbf{x}_0 = \frac{\pi}{2} + 0.1 + \begin{bmatrix} 0 & \pi & 2\pi & 3\pi & \dots & 99\pi \end{bmatrix}$$

of 100 initial guesses.) The commands

```
x = 0 : 1/200 : 1;
for n = 1 : 4
    plot(x, sin(b(n)*x)
end
```

graph the first four eigenfunctions (Fig. 10.1.5).

You may find it interesting to apply the MATLAB command **diff(b)** to verify that $\beta_n - \beta_{n-1} \approx 3.1416$ for $n \geq 65$. Thus successive eigenvalues do, indeed, appear to be separated by π. The command

```
c = 2*(1 - cos(b))./(b.*(1 + 2*cos(b).^2));
```

calculates the first 100 coefficients in the eigenfunction series in (1). We find that

$$f(x) \approx (1.2083) \sin \beta_1 x + (0.3646) \sin \beta_2 x + (0.2664) \sin \beta_3 x$$
$$+ (0.1722) \sin \beta_4 x + \cdots + (0.0066) \sin \beta_{97} x + (0.0065) \sin \beta_{98} x$$
$$+ (0.0065) \sin \beta_{99} x + (0.0064) \sin \beta_{100} x$$

with the numerical coefficients evidently decreasing quite slowly. Finally, the commands

```
x = 0 : 1/500 : 1;
y = zeros(size(x));
for n = 1 : 100
    y = y + c(n)*sin(b(n)*x);
end
plot(x,y)
```

graph this partial sum (Fig. 10.1.6). Note the Gibbs phenomenon that appears at $x = 0$.

You can check your understanding of eigenfunction expansions by reworking Example 7, except on the interval $0 < x < L$ and with the right-endpoint condition

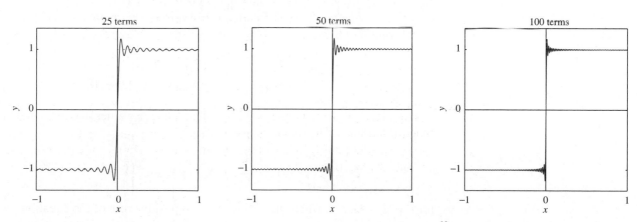

FIGURE 10.1.6. Graphs with $N = 25$, $N = 50$, $N = 100$ of the partial sum $\sum_{n=1}^{N} c_n \sin \beta_n x$ of the eigenfunction series in Eq. (1).

in (30) replaced with $y(L) + ky'(L) = 0$. Select your own values $L > 1$ and $k > 2$. Then calculate the resulting values of β_n and construct graphs like those in Figs. 10.1.5 and 10.1.6. In Fig. 10.1.6 we see convergence to $f(x) \equiv 1$ within the interval $0 < x < 1$, but a typical Gibbs phenomenon at $x = 0$.

10.2 Applications of Eigenfunction Series

This section is devoted to three examples that illustrate the application of the eigenfunction series of Section 10.1 to boundary value problems. In each, the method of separation of variables leads to a Sturm-Liouville problem in which the eigenfunctions are used as building blocks to construct a solution satisfying the nonhomogeneous boundary condition in the original problem.

Example 1 A uniform slab of material with thermal diffusivity k occupies the space region $0 \leqq x \leqq L$ and initially has temperature U_0 throughout. Beginning at time $t = 0$, the face $x = 0$ is held at temperature zero; at the face $x = L$, heat exchange takes place with a surrounding medium at temperature zero, so $hu + u_x = 0$ there (by Problem 23 of Section 9.5; the constant h is an appropriate heat transfer coefficient). We want to find the temperature $u(x, t)$ of the slab at position x at time t; $u(x, t)$ satisfies the boundary value problem

$$u_t = ku_{xx} \qquad (0 < x < L,\ t > 0); \tag{1}$$

$$u(0, t) = 0, \tag{2}$$

➤
$$hu(L, t) + u_x(L, t) = 0, \tag{3}$$

$$u(x, 0) = U_0. \tag{4}$$

Solution As in Section 9.5, we begin by substituting $u(x, t) = X(x)T(t)$ in Eq. (1) and thereby obtain

$$\frac{X''}{X} = \frac{T'}{kT} = -\lambda$$

as usual, so

$$X'' + \lambda X = T' + \lambda kT = 0. \tag{5}$$

Now Eq. (2) gives $X(0) = 0$, and Eq. (3) yields $hX(L)T(t) + X'(L)T(t) = 0$. We naturally assume that $T(t)$ is not identically zero (because we are not looking for a trivial solution); it follows that $X(x)$ and λ satisfy the Sturm-Liouville problem

$$X'' + \lambda X = 0 \qquad (0 < x < L);$$
$$X(0) = 0, \qquad hX(L) + X'(L) = 0. \tag{6}$$

In Example 3 of Section 10.1, we found that the eigenvalues and associated eigenfunctions of this problem are

$$\lambda_n = \frac{\beta_n^2}{L^2}, \qquad X_n(x) = \sin\frac{\beta_n x}{L} \tag{7}$$

for $n = 1, 2, 3, \ldots$, where β_n denotes the nth positive root of the equation

$$\tan x = -\frac{x}{hL}. \tag{8}$$

When we substitute $\lambda = \beta_n^2/L^2$ in the right-hand equation in (5), we get the first-order equation

$$T_n' = -\frac{\beta_n^2 k}{L^2} T_n$$

with solution (to within a multiplicative constant)

$$T_n(t) = \exp\left(-\frac{\beta_n^2 kt}{L^2}\right). \tag{9}$$

Consequently, each of the functions

$$u_n(x, t) = X_n(x)T_n(t) = \exp\left(-\frac{\beta_n^2 kt}{L^2}\right)\sin\frac{\beta_n x}{L}$$

satisfies the homogeneous conditions in (1)–(3) of the boundary value problem. It remains only for us to choose the coefficients so that the formal series

$$u(x, t) = \sum_{n=1}^{\infty} c_n \exp\left(-\frac{\beta_n^2 kt}{L^2}\right)\sin\frac{\beta_n x}{L} \tag{10}$$

also satisfies the nonhomogeneous condition

$$u(x, 0) = \sum_{n=1}^{\infty} c_n \sin\frac{\beta_n x}{L} = U_0. \tag{11}$$

Now $r(x) \equiv 1$ in the Sturm-Liouville problem in (6), so by Theorem 3 and Eq. (25) of Section 10.1, we can satisfy Eq. (11) by choosing

$$c_n = \frac{\displaystyle\int_0^L U_0 \sin\frac{\beta_n x}{L}\, dx}{\displaystyle\int_0^L \sin^2\frac{\beta_n x}{L}\, dx}.$$

But

$$\int_0^L U_0 \sin\frac{\beta_n x}{L}\, dx = \frac{U_0 L}{\beta_n}\left(1 - \cos\beta_n\right),$$

and by essentially the same computations as in Example 5 of Section 10.1 we find that

$$\int_0^L \sin^2\frac{\beta_n x}{L}\, dx = \frac{1}{2h}\left(hL + \cos^2\beta_n\right).$$

Hence

$$c_n = \frac{2U_0 hL(1 - \cos^2\beta_n)}{\beta_n(hL + \cos^2\beta_n)},$$

and substitution of this value in (10) yields the formal series solution

$$u(x, t) = 2U_0 hL \sum_{n=1}^{\infty} \frac{1 - \cos\beta_n}{\beta_n(hL + \cos^2\beta_n)} \exp\left(-\frac{\beta_n^2 kt}{L^2}\right)\sin\frac{\beta_n x}{L}. \tag{12}$$

∎

Remark: A numerical investigation of the eigenfunction series in (12) is outlined in the computing project for this section. ∎

Longitudinal Vibrations of Bars

Suppose that a uniform elastic bar has length L, cross-sectional area A, and density δ (mass per unit volume), and occupies the interval $0 \leqq x \leqq L$ when it is unstretched. We consider longitudinal vibrations of the bar in which each cross section (normal to the x-axis) moves only in the x-direction. We can then describe the motion of the bar in terms of the displacement $u(x, t)$ at time t of the cross section whose unstretched position is x. It follows (see Problem 13) from Hooke's law and the definition of the Young's modulus E of the material of the bar that the force $F(x, t)$ exerted on the cross section x by the part of the bar to the *left* of this section is

$$F(x, t) = -AEu_x(x, t), \tag{13}$$

FIGURE 10.2.1. A small segment of the bar.

the minus sign signifying that F acts to the left when $u_x > 0$ (so the bar is then stretched near x rather than compressed) (Fig. 10.2.1).

We take Eq. (13) as the starting point for our derivation of the partial differential equation that the displacement function $u(x, t)$ satisfies when the displacements are sufficiently small that Hooke's law may be applied. If we apply Newton's second law of motion to the segment of the bar between cross section x and cross section $x + \Delta x$, we get

$$(\delta A\, \Delta x)u_{tt}(\overline{x}, t) \approx -F(x + \Delta x, t) + F(x, t)$$
$$= AE\left[u_x(x + \Delta x, t) - u_x(x, t)\right], \tag{14}$$

where \overline{x} denotes the midpoint of $[x, x + \Delta x]$, because this segment has mass $\delta A \Delta x$ and approximate acceleration $u_{tt}(\overline{x}, t)$. When we divide the expression in (14) by $\delta A\, \Delta x$ and then take the limit as $\Delta x \to 0$, the result is the one-dimensional wave equation

➤
$$\frac{\partial^2 u}{\partial t^2} = a^2 \frac{\partial^2 u}{\partial x^2}, \tag{15}$$

where

$$a^2 = \frac{E}{\delta}. \tag{16}$$

Because Eq. (15) is identical to the equation of the vibrating string, it follows from our discussion of the d'Alembert solution in Section 9.6 that the [free] longitudinal vibrations of a bar with fixed ends are represented by waves of the form $u(x, t) = F(x \pm at)$. The velocity $a = \sqrt{E/\delta}$ with which these waves travel is the velocity of sound in the material of the bar. Indeed, the wave equation in (15) also describes ordinary one-dimensional sound waves in a gas in a pipe. In this case, E in Eq. (16) denotes the bulk modulus (fractional increase in density per unit increase in pressure) of the gas and δ is its equilibrium density.

Example 2

A bar has length L, density δ, cross-sectional area A, Young's modulus E, and total mass $M = \delta AL$. Its end $x = 0$ is fixed and a mass m is attached to its free end (Fig. 10.2.2). The bar initially is stretched linearly by moving m a distance $d = bL$ to the right; at time $t = 0$ the system is released from rest. To determine the subsequent vibrations of the bar, we must solve the boundary value problem

FIGURE 10.2.2. The bar-and-mass system of Example 2.

➤
$$u_{tt} = a^2 u_{xx} \qquad (0 < x < L,\ t > 0); \tag{17a}$$
$$u(0, t) = 0, \tag{17b}$$
$$mu_{tt}(L, t) = -AEu_x(L, t), \tag{17c}$$
$$u(x, 0) = bx, \qquad u_t(x, 0) = 0. \tag{17d}$$

Solution The endpoint condition at $x = L$ in (17c) comes from equating $ma = mu_{tt}$ for the mass with the force $F = -AEu_x$ given in (13)—the mass being acted on only by the bar. Substitution of $u(x, t) = X(x)T(t)$ in $u_{tt} = a^2 u_{xx}$ leads, as usual, to the equations

$$X'' + \lambda X = 0, \quad T'' + \lambda a^2 T = 0. \tag{18}$$

The boundary condition in (17b) yields $u(0, t) = X(0)T(t) = 0$, so one endpoint condition is $X(0) = 0$. Because $u_{tt} = XT''$ and $u_x = X'T$, Eq. (17c) yields

$$mX(L)T''(t) = -AEX'(L)T(t)$$

as the other endpoint condition. Substitution of

$$T''(t) = -\lambda a^2 T(t) = -\frac{\lambda E}{\delta}T(t),$$

followed by division by $-ET(t)/\delta$, gives $m\lambda X(L) = A\delta X'(L)$. Thus the eigenvalue problem for $X(x)$ is

$$X'' + \lambda X = 0; \tag{19}$$
$$X(0) = 0, \quad m\lambda X(L) = A\delta X'(L).$$

It is important to note that—because of the presence of λ in the right-endpoint condition—this is *not* a Sturm-Liouville problem, so Theorems 1 through 3 of Section 10.1 do not apply. Nevertheless, all eigenvalues of (19) are positive (Problem 9). We therefore write $\lambda = \alpha^2$, and note that $X(x) = \sin \alpha x$ satisfies $X(0) = 0$. The right-endpoint condition in (19) then yields

$$m\alpha^2 \sin \alpha L = A\delta \alpha \cos \alpha L,$$

and thus

$$\tan \alpha L = \frac{A\delta}{m\alpha} = \frac{M/m}{\alpha L}, \tag{20}$$

where $M = A\delta L$. We put $\beta = \alpha L$; it follows that the eigenvalues of (19) and associated eigenfunctions are

$$\lambda_n = \frac{\beta_n^2}{L^2}, \quad X_n(x) = \sin \frac{\beta_n x}{L} \tag{21}$$

for $n = 1, 2, 3, \ldots$, where $\{\beta_n\}_1^\infty$ are the positive roots of the equation

$$\tan x = \frac{M/m}{x}, \tag{22}$$

indicated in Fig. 10.2.3.

Next, we deduce in the usual way from

$$T_n'' + \frac{\beta_n^2 a^2}{L^2} T_n = 0, \quad T_n'(0) = 0 \tag{23}$$

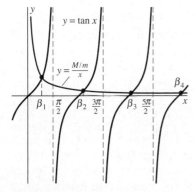

FIGURE 10.2.3. The positive roots $\{\beta_n\}_1^\infty$ of Eq. (22).

that $T_n(t) = \cos(\beta_n at/L)$ to within a multiplicative constant. Thus it remains only

to find coefficients $\{c_n\}_1^\infty$ so that the series

$$u(x,t) = \sum_{n=1}^{\infty} c_n \cos \frac{\beta_n at}{L} \sin \frac{\beta_n x}{L} \qquad (24)$$

satisfies the nonhomogeneous condition

$$u(x,0) = \sum_{n=1}^{\infty} c_n \sin \frac{\beta_n x}{L} = f(x) = bx. \qquad (25)$$

Caution is required because (19) is not a Sturm-Liouville problem. Indeed, we ask you to show in Problems 14 and 15 that the eigenfunctions $\{\sin(\beta_n x/L)\}_1^\infty$ are *not* orthogonal on $[0, L]$, so the coefficient formula in Eq. (25), Section 10.1, is not applicable here.

But in the case of longitudinal vibrations one ordinarily is not so much concerned with the displacement function $u(x,t)$ itself. Of greater interest is the question of how the bar's natural frequencies of vibration are affected by the mass m on its free end. Whatever the coefficients in Eq. (25) may be, we see from Eq. (24) that the nth circular frequency is

$$\omega_n = \frac{\beta_n a}{L} = \frac{\beta_n}{L} \sqrt{\frac{E}{\delta}}, \qquad (26)$$

where β_n is the nth positive solution of Eq. (22), which can be rewritten in the form

$$\cot x = \frac{mx}{M}. \qquad (27)$$

Thus the natural frequencies are determined by the ratio of the mass m to the total mass M of the bar. ∎

Remark: The case of *no* mass on the free end of the bar corresponds to $m = 0$. Then Eq. (27) reduces to the equation $\cos x = 0$, whose nth positive solution is $\beta_n = (2n - 1)\pi/2$. Then the nth circular frequency is given by

$$\omega_n = \frac{(2n - 1)\pi}{2L} \sqrt{\frac{E}{\delta}}. \qquad (28)$$

In Problem 8 we ask you to derive this result directly, beginning with the boundary value problem for a bar with one end fixed and the other end (completely) free. ∎

Transverse Vibrations of Bars

FIGURE 10.2.4. A bar undergoing transverse vibrations.

We now discuss vibrations of a uniform elastic bar in which the motion of each point is not longitudinal, but instead is perpendicular to the x-axis (the axis of the bar in its equilibrium position). Let $y(x,t)$ denote the transverse displacement of the cross section at x at time t, as indicated in Fig. 10.2.4. We merely want to outline a derivation of the partial differential equation that the deflection function $y(x,t)$ satisfies. Recall first that in Section 3.8 we introduced the equation $EIy^{(4)} = F$ for the static deflection of a bar or beam under the influence of a transverse external force F (per unit length). According to a general dynamical principle, we can transform the static equation $EIy^{(4)} = F$ into a dynamical equation (with no external force)

by replacing F with the reversed inertial force $F = -\rho y_{tt}$, where ρ is the linear density (mass/length) of the bar, and also by replacing $y^{(4)}$ with y_{xxxx}. This gives

$$\blacktriangleright \qquad EI\frac{\partial^4 y}{\partial x^4} = -\rho\frac{\partial^2 y}{\partial t^2},$$

which may be written in the form

$$\blacktriangleright \qquad \frac{\partial^2 y}{\partial t^2} + a^4\frac{\partial^4 y}{\partial x^4} = 0, \tag{29}$$

where

$$a^4 = \frac{EI}{\rho}. \tag{30}$$

The following example illustrates the solution of this fourth-order partial differential equation by the method of separation of variables.

Example 3 A uniform bar with linear density ρ, Young's modulus E, and cross-sectional moment of inertia I is simply supported (or hinged) at its two ends $x = 0$ and $x = L$. If the bar is set in motion from rest with given initial position $f(x)$, then its displacement function $y(x, t)$ satisfies the boundary value problem

$$y_{tt} + a^4 y_{xxxx} = 0 \qquad (0 < x < L,\ t > 0); \tag{31a}$$
$$\blacktriangleright \qquad y(0, t) = y_{xx}(0, t) = y(L, t) = y_{xx}(L, t) = 0, \tag{31b}$$
$$y(x, 0) = f(x), \qquad y_t(x, 0) = 0. \tag{31c}$$

Find $y(x, t)$. The boundary conditions in (31b) are the hinged end conditions that we accept without proof, and those in (31c) are the initial conditions.

Solution Substitution of $y(x, t) = X(x)T(t)$ in the differential equation yields $XT'' + a^4 X^{(4)}T = 0$, so

$$\frac{X^{(4)}}{X} = -\frac{T''}{a^4 T} = \lambda. \tag{32}$$

To determine the sign of λ, we reason that the equation

$$T'' + \lambda a^4 T = 0 \tag{33}$$

must have trigonometric rather than exponential solutions. The reason is that on the basis of practical experience—with tuning forks or xylophone bars, for instance—one expects periodic vibrations to take place. This could not occur if λ were negative and Eq. (33) had exponential solutions. Hence λ must be positive, and it is convenient to write $\lambda = \alpha^4 > 0$. Then $X(t)$ must satisfy the equation

$$X^{(4)}(x) - \alpha^4 X(x) = 0, \tag{34}$$

which has the general solution

$$X(x) = A\cos\alpha x + B\sin\alpha x + C\cosh\alpha x + D\sinh\alpha x,$$

with
$$X''(t) = \alpha^2(-A\cos\alpha x - B\sin\alpha x + C\cosh\alpha x + D\sinh\alpha x).$$

The endpoint conditions in (31) yield

$$X(0) = X''(0) = X(L) = X''(L) = 0. \tag{35}$$

Hence
$$X(0) = A + C = 0 \quad\text{and}\quad X''(0) = -A + C = 0,$$

and these two equations imply that $A = C = 0$. Therefore,

$$X(L) = B\sin\alpha L + D\sinh\alpha L = 0$$

and
$$X''(L) = \alpha^2(-B\sin\alpha L + D\sinh\alpha L) = 0.$$

It follows that
$$B\sin\alpha L = 0 \quad\text{and}\quad D\sinh\alpha L = 0.$$

But $\sinh\alpha L \neq 0$ because $\alpha \neq 0$; consequently, $D = 0$. Hence $B \neq 0$ if we are to have a nontrivial solution, so $\sin\alpha L = 0$. Thus α must be an integral multiple of π/L. Therefore, the eigenvalues and associated eigenfunctions of the problem determined by Eqs. (34) and (35) are

$$\lambda_n = \alpha_n^4 = \frac{n^4\pi^4}{L^4}, \quad X_n(x) = \sin\frac{n\pi x}{L} \tag{36}$$

for $n = 1, 2, 3, \ldots$.
 With $\lambda = n^4\pi^4/L^4$ in (33) we get the equation

$$T_n'' + \frac{n^4\pi^4 a^4}{L^4}T_n = 0. \tag{37}$$

Because the initial condition $y_t(x, 0) = 0$ yields $T_n'(0) = 0$, we take

$$T_n(t) = \cos\frac{n^2\pi^2 a^2 t}{L^2}. \tag{38}$$

Combining the results in (36) and (38), we construct the series

$$\blacktriangleright \qquad y(x, t) = \sum_{n=1}^{\infty} c_n \cos\frac{n^2\pi^2 a^2 t}{L^2}\sin\frac{n\pi x}{L} \tag{39}$$

that formally satisfies the partial differential equation in (31) and the homogeneous boundary conditions. The nonhomogeneous condition is

$$y(x, 0) = \sum_{n=1}^{\infty} c_n \sin\frac{n\pi x}{L} = f(x),$$

so we choose the Fourier sine coefficients, given by

$$\blacktriangleright \qquad c_n = \frac{2}{L}\int_0^L f(x)\sin\frac{n\pi x}{L}\,dx, \tag{40}$$

in order for Eq. (39) to provide a formal series solution. ∎

Note that the circular frequency of the nth term in (39) is

$$\omega_n = \frac{n^2 \pi^2 a^2}{L^2} = n^2 \omega_1, \tag{41}$$

where the fundamental frequency of the bar is

$$\omega_1 = \frac{\pi^2 a^2}{L^2} = \frac{\pi^2}{L^2}\sqrt{\frac{EI}{\rho}}. \tag{42}$$

Because the frequency of each of the higher harmonics is an integral multiple of ω_1, the sound of a vibrating bar with simply supported ends is musical. Because the higher frequencies $\{n^2 \omega_1\}_2^\infty$ are sparser than the higher frequencies $\{n \omega_1\}_2^\infty$ of a vibrating string, the tone of a vibrating bar is purer than that of a vibrating string. This partly accounts for the silkiness of the sound of the vibraphone as played by Milt Jackson, former member of the Modern Jazz Quartet.

10.2 Problems

Find formal series solutions of the boundary value problems in Problems 1 through 6. Express each answer in the form given in Problem 1.

1. $u_t = k u_{xx}$ $(0 < x < L, \quad t > 0); \; u_x(0, t) = h u(L, t) + u_x(L, t) = 0, u(x, 0) = f(x)$
Answer:

$$u(x, t) = \sum_{n=1}^\infty c_n \exp\left(-\frac{\beta_n^2 kt}{L^2}\right)\cos\frac{\beta_n x}{L},$$

where $\{\beta_n\}_1^\infty$ are the positive roots of the equation $\tan x = hL/x$ and

$$c_n = \frac{2h}{hL + \sin^2 \beta_n}\int_0^L f(x)\cos\frac{\beta_n x}{L}\,dx.$$

2. $u_{xx} + u_{yy} = 0$ $(0 < x < 1, 0 < y < L); u(0, y) = h u(L, y) + u_x(L, y) = 0, \quad u(x, L) = 0, u(x, 0) = f(x)$
3. $u_{xx} + u_{yy} = 0$ $(0 < x < L, \quad 0 < y < L); u_y(x, 0) = h u(x, L) + u_y(x, L) = 0, \; u(L, y) = 0, u(0, y) = g(y)$
4. $u_{xx} + u_{yy} = 0$ $(0 < x < L, \quad y > 0); u(0, y) = h u(L, y) + u_x(L, y) = 0, u(x, y)$ bounded as $y \to +\infty$, $u(x, 0) = f(x)$
5. $u_t = k u_{xx}$ $(0 < x < L, \quad t > 0); h u(0, t) - u_x(0, t) = u(L, t) = 0, u(x, 0) = f(x)$
6. $u_t = k u_{xx}$ $(0 < x < L, \quad t > 0); h u(0, t) - u_x(0, t) = h u(L, t) + u_x(L, t) = 0, u(x, 0) = f(x)$
7. Let $u(x, y)$ denote the bounded steady-state temperature in an infinitely high wall with base $y = 0$ and faces $x = 0$ and $x = 1$. The face $x = 0$ is insulated, the base $y = 0$ is kept at temperature $100°C$, and heat transfer with $h = 1$ takes place at the face $x = 1$. Derive the solution

$$u(x, y) = 200\sum_{n=1}^\infty \frac{\exp(-\alpha_n y)\sin\alpha_n\cos\alpha_n x}{\alpha_n + \sin\alpha_n\cos\alpha_n},$$

where $\{\alpha_n\}_1^\infty$ are the positive roots of the equation $\cot x = x$. Given $\alpha_1 = 0.860$, $\alpha_2 = 3.426$, $\alpha_3 = 6.437$, and $\alpha_4 = 9.529$, calculate the temperature $u(1, 1)$ accurate to $0.1°C$.

8. If the bar in Example 2 has *no* mass attached to the end $x = L$, then Eq. (17c) is replaced with the free end condition $u_x(L, t) = 0$. Separate variables in the resulting boundary value problem to derive the series solution

$$u(x, t) = \sum_{n=1}^\infty c_n \cos\frac{(2n - 1)\pi at}{2L}\sin\frac{(2n - 1)\pi x}{2L},$$

where

$$c_n = \frac{2}{L}\int_0^L bx\sin\frac{(2n - 1)\pi x}{2L}\,dx = \frac{8bL(-1)^{n+1}}{(2n - 1)^2\pi^2}.$$

In particular, the bar's natural frequencies of longitudinal vibration are given by Eq. (28).

9. (a) Show that $\lambda = 0$ is not an eigenvalue of the problem in (19). **(b)** Show that this problem has no negative eigenvalues. (*Suggestion:* Sketch the graphs $y = \tanh x$ and $y = -k/x$ with $k > 0$.)

10. Calculate the speed (in miles per hour) of longitudinal sound waves in each case. **(a)** Steel, with $\delta = 7.75$ g/cm³ and $E = 2 \times 10^{12}$ in cgs units. **(b)** Water, with $\delta = 1$ g/cm³ and bulk modulus $K = 2.25 \times 10^{10}$ in cgs units.

11. Consider a mass $m = nm_0$ of an ideal gas of molecular weight m_0 whose pressure p and volume V satisfy the law $pV = nRT_K$, where n is the number of moles of the gas, $R = 8314$ in mks units, and $T_K = T_C + 273$, where T_C is the Celsius temperature. The bulk modulus of the gas

is $K = \gamma p$, where the value of the dimensionless constant γ is 1.4 for air having molecular weight $m_0 = 29$. **(a)** Show that the velocity of sound in this gas is

$$a = \sqrt{\frac{K}{\delta}} = \sqrt{\frac{\gamma R T_K}{m_0}}.$$

(b) Use this formula to show that the speed of sound in air at Celsius temperature T_C is approximately $740 + (1.36)T_C$ miles per hour.

12. Suppose that the free end of the bar of Example 2 is attached to a spring (rather than to a mass), as shown in Fig. 10.2.5. The endpoint condition then becomes $ku(L, t) + AEu_x(L, t) = 0$. (Why?) Assume that $u(x, 0) = f(x)$ and that $u_t(x, 0) = 0$. Derive a solution of the form

$$u(x, t) = \sum_{n=1}^{\infty} c_n \cos \frac{\beta_n at}{L} \sin \frac{\beta_n x}{L},$$

where $\{\beta_n\}_1^{\infty}$ are the positive roots of the equation $\tan x = -AEx/kL$.

FIGURE 10.2.5. The bar of Problem 12.

13. If a bar has natural length L, cross-sectional area A, and Young's modulus E, then (as a consequence of Hooke's law) the axial force at each end required to stretch it the small amount ΔL is $F = (AE \, \Delta L)/L$. Apply this result to a segment of the bar of natural length $L = \Delta x$ between the cross sections at x and $x + \Delta x$ that is stretched by the amount $\Delta L = u(x + \Delta x, t) - u(x, t)$. Then let $\Delta x \to 0$ to derive Eq. (13).

14. Show that the eigenfunctions $\{X_n(x)\}_1^{\infty}$ of the problem in (19) are not orthogonal. (*Suggestion*: Apply Eq. (22) of Section 10.1 to show that if $m \neq n$, then

$$\int_0^L X_m(x)X_n(x) \, dx = -\frac{m}{A\delta} X_m(L)X_n(L).)$$

15. Show that the eigenfunctions $\{\sin \beta_n x/L\}_1^{\infty}$ of the problem in (19) are not orthogonal; do so by obtaining the explicit value of the integral

$$\int_0^L \sin \frac{\beta_m x}{L} \sin \frac{\beta_n x}{L} \, dx.$$

(*Suggestion*: Use the fact that $\{\beta_n\}_1^{\infty}$ are the roots of the equation $x \tan x = M/m$.)

16. According to Problem 19 of Section 9.7, the temperature $u(r, t)$ in a uniform solid spherical ball of radius a satisfies the partial differential equation $(ru)_t = k(ru)_{rr}$. Suppose that the ball has initial temperature $u(r, 0) = f(r)$ and that its surface $r = a$ is insulated, so that $u_r(a, t) = 0$. Substitute $v(r, t) = ru(r, t)$ to derive the solution

$$u(r, t) = c_0 + \sum_{n=1}^{\infty} \frac{c_n}{r} \exp\left(-\frac{\beta_n^2 kt}{a^2}\right) \sin \frac{\beta_n r}{a},$$

where $\{\beta_n\}_1^{\infty}$ are the positive roots of the equation $\tan x = x$ and

$$c_0 = \frac{3}{a^3} \int_0^a r^2 f(r) \, dr,$$

$$c_n = \frac{2}{a \sin^2 \beta_n} \int_0^a rf(r) \sin \frac{\beta_n r}{a} \, dr$$

(see Problem 14 of Section 10.1).

17. A problem concerning the diffusion of gas through a membrane leads to the boundary value problem

$$u_t = ku_{xx} \qquad (0 < x < L, \ t > 0);$$
$$u(0, t) = u_t(L, t) + hku_x(L, t) = 0,$$
$$u(x, 0) = 1.$$

Derive the solution

$$u(x, t) = 4 \sum_{n=1}^{\infty} \frac{1 - \cos \beta_n}{2\beta_n - \sin 2\beta_n} \exp\left(-\frac{\beta_n^2 kt}{L^2}\right) \sin \frac{\beta_n x}{L},$$

where $\{\beta_n\}_1^{\infty}$ are the positive roots of the equation $x \tan x = hL$.

18. Suppose that the simply supported uniform bar of Example 3 has, instead, initial position $y(x, 0) = 0$ and initial velocity $y_t(x, 0) = g(x)$. Then derive the solution

$$y(x, t) = \sum_{n=1}^{\infty} c_n \sin \frac{n^2 \pi^2 a^2 t}{L^2} \sin \frac{n\pi x}{L},$$

where

$$c_n = \frac{2L}{n^2 \pi^2 a^2} \int_0^L g(x) \sin \frac{n\pi x}{L} \, dx.$$

19. To approximate the effect of an initial momentum impulse P applied at the midpoint $x = L/2$ of a simply supported bar, substitute

$$g(x) = \begin{cases} \dfrac{P}{2\rho\epsilon} & \text{if } \dfrac{L}{2} - \epsilon < x < \dfrac{L}{2} + \epsilon, \\ 0 & \text{otherwise} \end{cases}$$

in the result of Problem 18. Then let $\epsilon \to 0$ to obtain the solution

$$y(x, t) = C \sum_{n=1}^{\infty} \frac{1}{n^2} \sin \frac{n\pi}{2} \sin \frac{n^2 \pi^2 a^2 t}{L^2} \sin \frac{n\pi x}{L},$$

where

$$C = \frac{2PL}{\pi^2 \sqrt{EI\rho}}.$$

20. **(a)** If $g(x) = v_0$ (a constant) in Problem 18, show that

$$y(x, t) = \frac{4v_0 L^2}{\pi^3 a^2} \sum_{n \text{ odd}} \frac{1}{n^3} \sin \frac{n^2 \pi^2 a^2 t}{L^2} \sin \frac{n\pi x}{L}.$$

This describes the vibrations of a hinged bar lying crosswise in the back of a pickup truck that hits a brick wall with speed v_0 at time $t = 0$. **(b)** Now suppose that the bar is made of steel ($E = 2 \times 10^{12}$ dyn/cm², $\delta = 7.75$ g/cm³), has a square cross section with edge $a = 1$ in. (so that $I = \frac{1}{12}a^4$), and length $L = 19$ in. What is its fundamental frequency (in hertz)?

10.2 Application: Numerical Heat Flow Investigations

n	β_n
1	2.6537
2	5.4544
3	8.3913
4	11.4086
5	14.4699
6	17.5562
7	20.6578
8	23.7693
9	26.8874
10	30.0102

FIGURE 10.2.6. The first 10 positive solutions of the eigenvalue equation

$$\tan x = -\frac{x}{hL}$$

with $L = 50$ and $h = 0.1$.

We outline a *Mathematica*-based numerical investigation of the temperature function

$$u(x,t) = 2u_0 hL \sum_{n=1}^{\infty} \frac{1 - \cos \beta_n}{\beta_n(hL + \cos^2 \beta_n)} \exp\left(-\frac{\beta_n^2 kt}{L^2}\right) \sin \frac{\beta_n x}{L} \tag{1}$$

of the heated slab of Example 1 in this section. *Maple* and MATLAB versions of this investigation are included in the applications manual that accompanies this text.

We suppose that the slab has thickness $L = 50$ cm, uniform initial temperature $u_0 = 100°$, fixed temperature $0°$ at the slab's left-hand boundary $x = 0$, thermal diffusivity $k = 0.15$ (for iron), and heat transfer coefficient $h = 0.1$ at the right-hand boundary $x = L$. According to Eq. (8) of this section, the eigenvalues $\{\beta_n\}_1^\infty$ in (1) are the positive solutions of the equation $\tan x = -x/(hL)$. As in Fig. 10.1.1, we see that if n is large then β_n is slightly larger than $(2n-1)\pi/2$. We can therefore approximate the first 20 of these eigenvalues using the commands

```
L = 50;   h = 0.1;
roots = Table[ FindRoot[ Tan[x] == -x/(h*L),
            {x, (2*n - 1)*Pi/2 + 0.1}], {n,1,20}];
beta = x /. solutions
```

The first ten of these eigenvalues are listed in the table of Fig. 10.2.6.

The following *Mathematica* function now sums the first 20 nonzero terms of the series in Eq. (1).

```
b = beta;   u0 = 100;    k = 0.15;
c = (1 - Cos[b])/(b*(h*L + Cos[b]^2));    (* coeffs *)
u[x_, t_] := 2*u0*h*L*Apply[ Plus,
        c*Exp[-b*b*k*t/L^2]*Sin[b*x/L]] // N
```

As a practical matter, this suffices to calculate the value $u(x, t)$ after 10 seconds (or more) with two-place accuracy throughout the interval $0 \leq x \leq 50$. (How might you verify this assertion?) The command

```
Plot[u[x,900], {x, 0, 50}];
```

then yields the graph $u = u(x, 900)$ for $0 \leq x \leq 50$ shown in Fig. 10.2.7. We see that after 15 minutes the right boundary $x = 50$ of the slab has cooled below $40°$, whereas interior temperatures near $x = 30$ remain above $80°$. The command

```
Plot[{u[50,t], 25}, {t, 0, 3600}];
```

yields the graph $u = u(50, t)$ for $0 \leq t \leq 3600$ shown in Fig. 10.2.8. Now we see that the right boundary temperature of the slab falls to $25°$ in a bit less than 2000 seconds. The calculation

```
t1 = t /. FindRoot[u[50,t] == 25, {t, 2000}]
```

shows that this actually occurs in about 1951 s; that is, about 32 min 31 s.

Figure 10.2.9 indicates that interior temperatures of the slab at time t_1 still remain above $50°$ at some points, with a maximum occurring near $x = 30$. We can find this maximum value of the temperature $u(x, t_1)$ by finding the minimum value of its negative $-u(x, t_1)$. The command

```
FindMinimum[-u[x,t1], {x, 30}]
```

yields the result that the maximum value is $u(29.36, t_1) \approx 53.80°$.

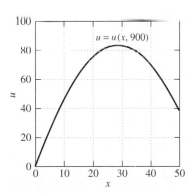

FIGURE 10.2.7. Temperatures in the slab after 15 minutes.

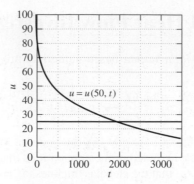

FIGURE 10.2.8. Right boundary temperatures of the slab.

FIGURE 10.2.9. Interior temperatures when the right boundary temperature is 25°.

For your own slab with uniform initial temperature $u(x, 0) = 100°$ to investigate in this manner, let $h = 0.1$, $L = 50(10 + p)$, and $k = q/10$, where p is the largest and q is the smallest nonzero digit of your student ID number. Then carry out the investigations outlined here to find

- When the right boundary temperature of the slab is 25°, and
- The maximum temperature within the slab at that instant.

As an alternative to finding roots and maximum values numerically, you can "zoom in" graphically on appropriate points in figures like 10.2.8 and 10.2.9.

10.3 Steady Periodic Solutions and Natural Frequencies

In Section 9.6 we derived the solution

$$y(x, t) = \sum_{n=1}^{\infty} \left(A_n \cos \frac{n\pi a t}{L} + B_n \sin \frac{n\pi a t}{L} \right) \sin \frac{n\pi x}{L}$$

$$= \sum_{n=1}^{\infty} C_n \cos \left(\frac{n\pi a t}{L} - \gamma_n \right) \sin \frac{n\pi x}{L} \tag{1}$$

of the vibrating string problem

$$\frac{\partial^2 y}{\partial t^2} = a^2 \frac{\partial^2 y}{\partial x^2} \qquad \left(a = \sqrt{\frac{T}{\rho}} \right); \tag{2}$$

$$y(0, t) = y(L, t) = 0, \tag{3}$$

$$y(x, 0) = f(x), \quad y_t(x, 0) = g(x). \tag{4}$$

The solution in Eq. (1) describes the *free* vibrations of a string with length L and linear density ρ under tension T; the constant coefficients in Eq. (1) are determined by the initial conditions in (4).

In particular, we see from the terms in Eq. (1) that the natural [circular] frequencies of vibration (in radians per second) of the string are given by

$$\omega_n = \frac{n\pi a}{L}, \tag{5}$$

$n = 1, 2, 3, \ldots$. These are the only values of ω for which Eq. (2) has a steady periodic solution of the form

$$y(x, t) = X(x) \cos(\omega t - \gamma) \tag{6}$$

that satisfies the endpoint conditions in (3). For if we substitute Eq. (6) in (2) and cancel the factor $\cos(\omega t - \gamma)$, we find that $X(x)$ must satisfy the equation

$$a^2 X''(x) + \omega^2 X(x) = 0,$$

whose general solution

$$X(x) = A \cos \frac{\omega x}{a} + B \sin \frac{\omega x}{a}$$

satisfies the conditions in (3) only if $A = 0$ and $\omega = n\pi a/L$ for some positive integer n.

Forced Vibrations and Resonance

Now suppose that the string is influenced by a periodic external force $F(t) = F_0 \cos \omega t$ (force per unit mass) that acts uniformly on the string along its length. Then, according to Eq. (1) of Section 9.6, the displacement $y(x, t)$ of the string will satisfy the nonhomogeneous partial differential equation

$$\frac{\partial^2 y}{\partial t^2} = a^2 \frac{\partial^2 y}{\partial x^2} + F_0 \cos \omega t \tag{7}$$

together with boundary conditions such as those in (3) and (4). For instance, if the string is initially at rest in equilibrium when the external force begins to act, we want to find a solution of Eq. (7) that satisfies the conditions

$$y(0, t) = y(L, t) = y(x, 0) = y_t(x, 0) = 0. \tag{8}$$

To do this, it suffices first to find a particular solution $y_p(x, t)$ of Eq. (7) that satisfies the fixed endpoint conditions in (3), and second to find a solution $y_c(x, t)$ like Eq. (1) of the familiar problem in (2)–(4) with

$$f(x) = -y_p(x, 0) \quad \text{and} \quad g(x) = -D_t y_p(x, 0).$$

Evidently,

$$y(x, t) = y_c(x, t) + y_p(x, t)$$

will then satisfy Eqs. (7) and (8).

So our new task is to find $y_p(x, t)$. Examination of the individual terms in Eq. (7) suggests that we try

$$y_p(x, t) = X(x) \cos \omega t. \tag{9}$$

Substitution of this in Eq. (7) and cancellation of the common factor $\cos \omega t$ yields the ordinary differential equation

$$a^2 X'' + \omega^2 X = -F_0$$

with general solution

$$X(x) = A \cos \frac{\omega x}{a} + B \sin \frac{\omega x}{a} - \frac{F_0}{\omega^2}. \tag{10}$$

The condition $x(0) = 0$ requires that $A = F_0/\omega^2$, and then $X(L) = 0$ requires that

$$X(L) = \frac{F_0}{\omega^2}\left(\cos \frac{\omega L}{a} - 1\right) + B \sin \frac{\omega L}{a} = 0. \tag{11}$$

Now suppose that the frequency ω of the periodic external force is *not* equal to any one of the natural frequencies $\omega_n = n\pi a/L$ of the string. Then $\sin(\omega L/a) \neq 0$, so we can solve Eq. (11) for B and then substitute the result in Eq. (10) to obtain

$$X(x) = \frac{F_0}{\omega^2}\left(\cos \frac{\omega x}{a} - 1\right) - \frac{F_0[\cos(\omega L/a) - 1]}{\omega^2 \sin(\omega L/a)} \sin \frac{\omega x}{a}. \tag{12}$$

Then Eq. (9), with this choice of $X(x)$, gives the desired particular solution $y_p(x, t)$.

Note, however, that as the value of ω approaches $\omega_n = n\pi a/L$ with n odd, the coefficient of $\sin(\omega x/a)$ in Eq. (12) approaches $\pm\infty$; thus *resonance* occurs. This explains the fact that when [only] one of two nearby identical strings is plucked, the other will begin to vibrate as well, due to its being acted on (through the medium of the air) by an external periodic force at its fundamental frequency. Observe also that if $\omega = \omega_n = n\pi a/L$ with n even, then we can choose $B = 0$ in Eq. (11), so resonance does not occur in this case. Problem 20 explains why some of the resonance possibilities are absent.

The vibrating string is typical of continuous systems that have an infinite sequence of natural frequencies of vibration. When a periodic external force acts on such a system, potentially destructive resonance vibrations may occur if the imposed frequency is close to one of the natural frequencies of the system. Hence an important aspect of proper structural design is the avoidance of such resonance vibrations.

Natural Frequencies of Beams

Figure 10.3.1 shows a uniform beam of length L, linear density ρ, and Young's modulus E, clamped at each end. For $0 < x < L$ and $t > 0$, its deflection function $y(x, t)$ satisfies the fourth-order equation

$$\frac{\partial^2 y}{\partial t^2} + a^4 \frac{\partial^4 y}{\partial x^4} = 0 \qquad \left(a^4 = \frac{EI}{\rho}\right) \tag{13}$$

that we discussed in Section 10.2; I denotes the moment of inertia of the cross section of the beam around its horizontal axis of symmetry. Because both the displacement and the slope are zero at each fixed end, the endpoint conditions are

$$y(0, t) = y_x(0, t) = 0 \tag{14}$$

and

$$y(L, t) = y_x(L, t) = 0. \tag{15}$$

$y(x, t)$

$x = 0$ $x = L$

FIGURE 10.3.1. A bar clamped at each end.

Here we want only to find the natural frequencies of vibration of the beam, so we will not be concerned with initial conditions. The natural frequencies are the values of ω for which Eq. (13) has a nontrivial solution of the form

$$y(x, t) = X(x) \cos(\omega t - \gamma) \tag{16}$$

that satisfies the conditions in Eqs. (14) and (15). When we substitute $y(x, t)$ from (16) into (13) and then cancel the common factor $\cos(\omega t - \gamma)$, we obtain the fourth-order ordinary differential equation $-\omega^2 X + a^4 X^{(4)} = 0$; that is,

$$X^{(4)} - \frac{\omega^2}{a^4} X = 0. \tag{17}$$

If we write $\alpha^4 = \omega^2/a^4$, we can express the general solution of Eq. (17) as

$$X(x) = A \cosh \alpha x + B \sinh \alpha x + C \cos \alpha x + D \sin \alpha x,$$

with

$$X'(x) = \alpha(A \sinh \alpha x + B \cosh \alpha x - C \sin \alpha x + D \cos \alpha x).$$

The conditions in (14) give

$$X(0) = A + C = 0 \quad \text{and} \quad X'(0) = \alpha(B + D) = 0,$$

so $C = -A$ and $D = -B$. Hence the conditions in (15) give

$$X(L) = A(\cosh \alpha L - \cos \alpha L) + B(\sinh \alpha L - \sin \alpha L) = 0$$

and

$$\frac{1}{\alpha} X'(L) = A(\sinh \alpha L + \sin \alpha L) + B(\cosh \alpha L - \cos \alpha L) = 0.$$

In order for these two linear homogeneous equations in A and B to have a nontrivial solution, the determinant of coefficients must be zero:

$$(\cosh \alpha L - \cos \alpha L)^2 - (\sinh^2 \alpha L - \sin^2 \alpha L) = 0;$$
$$(\cosh^2 \alpha L - \sinh^2 \alpha L) + (\cos^2 \alpha L + \sin^2 \alpha L) - 2 \cosh \alpha L \cos \alpha L = 0;$$
$$2 - 2 \cosh \alpha L \cos \alpha L = 0.$$

Then $\beta = \alpha L$ must be a nonzero root of the equation

$$\cosh x \cos x = 1. \tag{18}$$

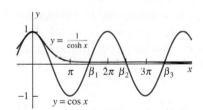

FIGURE 10.3.2. Solutions of $\cosh x \cos x = 1$.

From Fig. 10.3.2 we see that this equation has an increasing sequence of positive roots $\{\beta_n\}_1^\infty$. Now $\omega = \alpha^2 a^2 = \beta^2 a^2/L^2$ and $a^2 = \sqrt{EI/\rho}$, so it follows that the natural [circular] frequencies of vibration of the beam with clamped ends are given by

$$\omega_n = \frac{\beta_n^2}{L^2} \sqrt{\frac{EI}{\rho}} \quad \text{(rad/s)} \tag{19}$$

for $n = 1, 2, 3, \ldots$. The roots of Eq. (18) are $\beta_1 \approx 4.73004$, $\beta_2 \approx 7.85320$, $\beta_3 \approx 10.99561$, and $\beta_n \approx (2n + 1)\pi/2$ for $n \geq 4$ (as indicated in Fig. 10.3.2).

FIGURE 10.3.3. An idealized I beam.

Example 1

For example, suppose that the basic structural element of a bridge is a 120-ft-long steel I beam with the cross section indicated in Fig. 10.3.3, having moment of inertia $I = 9000$ cm^4. If we substitute the values

$$L = (120 \text{ ft}) \left(30.48 \frac{\text{cm}}{\text{ft}}\right), \quad \rho = \left(7.75 \frac{\text{g}}{\text{cm}^3}\right)(40 \text{ cm}^2), \quad E = 2 \times 10^{12} \frac{\text{dyn}}{\text{cm}^2},$$

and the values of β_1 and β_2 in Eq. (19), we find that the lowest two natural frequencies of the beam are

$$\omega_1 \approx 12.74 \frac{\text{rad}}{\text{s}} \quad \left(122 \frac{\text{cycles}}{\text{min}}\right) \quad \text{and} \quad \omega_2 \approx 35.13 \frac{\text{rad}}{\text{s}} \quad \left(335 \frac{\text{cycles}}{\text{min}}\right).$$

If a company of soldiers marching at about 120 steps per minute approaches this bridge, they therefore would be well advised to break cadence before crossing. From time to time bridges collapse because of resonance vibrations. Recall the Kansas City hotel disaster of July 17, 1981, in which a skywalk filled with dancers collapsed. Newspapers quoted investigators who speculated that the rhythmic movement of the dancers had set up destructive resonance vibrations in the steel I beams that supported the skywalk. ∎

Underground Temperature Oscillations

Let us assume that the underground temperature at a particular location is a function $u(x, t)$ of time t and the depth x beneath the surface. Then u satisfies the heat equation $u_t = ku_{xx}$, with k being the thermal diffusivity of the soil. We may regard the temperature $u(0, t)$ at the surface $x = 0$ as being known from weather records. In fact, the periodic seasonal variation of monthly average surface temperatures, with a maximum in midsummer (July in the northern hemisphere) and a minimum in midwinter (January), is very close to a sine or cosine oscillation. We shall therefore assume that

$$u(0, t) = T_0 + A_0 \cos \omega t, \tag{20}$$

where we take $t = 0$ at midsummer. Here T_0 is the average annual temperature, A_0 the amplitude of seasonal temperature variation, and ω is chosen to make the period of $u(0, t)$ exactly one year. (In cgs units, for instance, ω would be 2π divided by the number of seconds—31,557,341—in a year.)

It is reasonable to assume that the temperature at a fixed depth also varies periodically with t. If we introduce $U(x, t) = u(x, t) - T_0$ for convenience, then we are interested in periodic solutions of the heat equation of the form

$$U(x, t) = A(x) \cos(\omega t - \gamma) = V(x) \cos \omega t + W(x) \sin \omega t \tag{21}$$

of the problem

$$\frac{\partial U}{\partial t} = k \frac{\partial^2 U}{\partial x^2} \quad (x > 0, \ t > 0), \tag{22}$$

$$U(0, t) = A_0 \cos \omega t. \tag{23}$$

To solve this problem, let us regard $U(x, t)$ in Eq. (21) as the real part of the complex-valued function

$$\widetilde{U}(x, t) = X(x) e^{i\omega t}. \tag{24}$$

Then we want $\widetilde{U}(x, t)$ to satisfy the conditions

$$\widetilde{U}_t = k\widetilde{U}_{xx}, \tag{22'}$$

$$\widetilde{U}(0, t) = A_0 e^{i\omega t}. \tag{23'}$$

If we substitute Eq. (24) in Eq. (22'), we get $i\omega X = kX''$; that is,

$$X'' - \alpha^2 X = 0, \tag{25}$$

where

$$\alpha = \pm\sqrt{\frac{i\omega}{k}} = \pm(1 + i)\sqrt{\frac{\omega}{2k}} \tag{26}$$

because $\sqrt{i} = \pm(1 + i)/\sqrt{2}$. Hence the general solution of Eq. (25) is

$$X(x) = A \exp\left(-(1 + i)x\sqrt{\omega/2k}\right) + B \exp\left(+(1 + i)x\sqrt{\omega/2k}\right). \tag{27}$$

In order that $\widetilde{U}(x, t)$, and hence $X(x)$, be bounded as $x \to +\infty$, it is necessary that $B = 0$. Also, we see from Eqs. (23') and (24) that $A = X(0) = A_0$. So

$$X(x) = A_0 \exp\left(-(1 + i)x\sqrt{\omega/2k}\right). \tag{28}$$

Finally, the solution of our original problem in (22)–(23) is

$$U(x, t) = \operatorname{Re} \widetilde{U}(x, t) = \operatorname{Re} X(x)e^{i\omega t}$$

$$= \operatorname{Re}\left\{(A_0 \exp(i\omega t)) \exp\left(-(1 + i)x\sqrt{\omega/2k}\right)\right\}$$

$$= \operatorname{Re}\left\{A_0 \exp\left(-x\sqrt{\omega/2k}\right) \exp\left(i(\omega t - x\sqrt{\omega/2k})\right)\right\};$$

hence

$$U(x, t) = A_0 \exp\left(-x\sqrt{\omega/2k}\right) \cos\left(\omega t - x\sqrt{\omega/2k}\right). \tag{29}$$

Thus the amplitude $A(x)$ of the annual temperature is exponentially damped as a function of the depth x:

$$A(x) = A_0 \exp\left(-x\sqrt{\omega/2k}\right). \tag{30}$$

In addition, there is a phase delay $\gamma(x) = x\sqrt{\omega/2k}$ at the depth x.

Example 2 With $k = 0.005$ (a typical value for soil in cgs units) and the value of ω previously mentioned, we find that $\sqrt{\omega/2k} \approx 0.00446$ cm^{-1}. For instance, we then see from Eq. (30) that the amplitude is one-half the surface amplitude, $A(x) = \frac{1}{2}A_0$, when $(0.00446)x = \ln 2$; that is, when $x \approx 155.30$ cm ≈ 5.10 ft. If $A_0 = 16°$, it follows that at a depth of about 20 ft, the amplitude of annual temperature variation is only $1°$. ∎

Example 3 Another interesting consequence of Eq. (29) is the "reversal of seasons" that occurs when $\gamma(x) = (0.00446)x = \pi$; that is, at a depth of $x \approx 704.39$ cm, about 23.11 ft. Figure 10.3.4 shows the graph $U = U(x, t)$ for $0 \leqq x \leqq X$, $0 \leqq t \leqq T$, where X corresponds to 23 feet in centimeters and T to 2.5 years in seconds. Can you see by inspection of the right edge of the graph that the temperature at depth 23 feet appears to be at a maximum when the surface temperature is at a minimum? ■

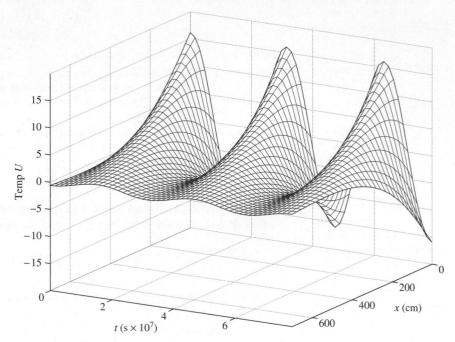

FIGURE 10.3.4. The graph of the temperature function $U(x, t)$ in Eq. (29).

10.3 Problems

A uniform bar of length L is made of material with density δ and Young's modulus E. In Problems 1 through 6, substitute $u(x, t) = X(x) \cos \omega t$ in $\delta u_{tt} = E u_{xx}$ to find the natural frequencies of longitudinal vibration of the bar with the two given conditions at its ends $x = 0$ and $x = L$.

1. Both ends are fixed.
2. Both ends are free.
3. The end at $x = 0$ is fixed; the end at $x = L$ is free.
4. The end at $x = 0$ is fixed; the free end at $x = L$ is attached to a mass m as in Example 2 of Section 10.2.
5. Each end is free, but the end at $x = L$ is attached to a spring with Hooke's constant k as in Problem 12 of Section 10.2.
6. The free ends are attached to masses m_0 and m_1.
7. Suppose that the mass on the free end at $x = L$ in Problem 4 is attached also to the spring of Problem 5. Show that the natural frequencies are given by $\omega_n = (\beta_n/L)\sqrt{E/\delta}$, where $\{\beta_n\}_1^\infty$ are the positive roots of the equation

$$(mEx^2 - k\delta L^2) \sin x = MEx \cos x.$$

(*Note*: The condition at $x = L$ is $mu_{tt} = -AEu_x - ku$.)

Problems 8 through 14 deal with transverse vibrations of the uniform beam of this section, but with various end conditions. In each case show that the natural frequencies are given by the formula in Eq. (19), with $\{\beta_n\}_1^\infty$ being the positive roots of the given frequency equation. Recall that $y = y' = 0$ at a fixed end, $y = y'' = 0$ at a hinged end, and $y'' = y^{(3)} = 0$ at a free end (primes denote derivatives with respect to x).

8. The ends at $x = 0$ and $x = L$ are both hinged; the frequency equation is $\sin x = 0$, so that $\beta_n = n\pi$.
9. The end at $x = 0$ is fixed and the end at $x = L$ is hinged; the frequency equation is $\tanh x = \tan x$.
10. The beam is a cantilever with the end at $x = 0$ fixed and the end at $x = L$ free; the frequency equation is $\cosh x \cos x = -1$.
11. The end at $x = 0$ is fixed and the end at $x = L$ is attached to a vertically sliding clamp, so $y' = y^{(3)} = 0$ there; the frequency equation is $\tanh x + \tan x = 0$.

12. The cantilever of Problem 10 has total mass $M = \rho L$ and has a mass m attached to its free end; the frequency equation is

$$M(1 + \cosh x \, \cos x) = mx(\cosh x \, \sin x - \sinh x \, \cos x).$$

The conditions at $x = L$ are $y_{xx} = 0$ and $my_{tt} = EIy_{xxx}$.

13. The free end at $x = L$ of the cantilever of Problem 10 is attached (as in Fig. 10.3.5) to a spring with Hooke's constant k; the frequency equation is

$$EIx^3(1 + \cosh x \, \cos x)$$
$$= kL^3(\sinh x \, \cos x - \cosh x \, \sin x).$$

The conditions at $x = L$ are $y_{xx} = 0$ and $ky = EIy_{xxx}$.

FIGURE 10.3.5. The cantilever of Problem 13.

14. Suppose that the mass m on the free end of the cantilever of Problem 12 is attached to the spring of Problem 13. The conditions at $x = L$ are $y_{xx} = 0$ and $my_{tt} = EIy_{xxx} - ky$. Derive the frequency equation

$$MEIx^3(1 + \cosh x \, \cos x)$$
$$= (kML^3 - mEIx^4)(\sinh x \, \cos x - \cosh x \, \sin x).$$

Note that the frequency equations in Problems 12 and 13 are the special cases $k = 0$ and $m = 0$, respectively.

15. Suppose that the cantilever of Problem 10 is a diving board made of steel with density $\delta = 7.75$ g/cm³. The diving board is 4 m long and its cross section is a rectangle of width $a = 30$ cm and thickness $b = 2$ cm. The moment of inertia of this rectangle around its horizontal axis of symmetry is $I = \frac{1}{12}ab^3$. Given that the least positive root of the equation $\cosh x \, \cos x = -1$ is $\beta_1 \approx 1.8751$, determine the frequency (in hertz) at which a person should bounce up and down at the tip of the diving board for maximal (resonant) effect.

16. If a uniform bar hinged at each end is subjected to an axial force of compression P, then its transverse vibrations satisfy the equation

$$\rho \frac{\partial^2 y}{\partial t^2} + P \frac{\partial^2 y}{\partial x^2} + EI \frac{\partial^4 y}{\partial x^4} = 0.$$

Show that its natural frequencies are given by

$$\omega_n = \frac{n^2 \pi^2}{L^2} \left(1 - \frac{PL^2}{n^2 \pi^2 EI}\right)^{1/2} \sqrt{\frac{EI}{\rho}}$$

for $n = 1, 2, 3, \ldots$. Note that with $P = 0$, this reduces to the result in Example 3 of Section 10.2 and that the effect of $P > 0$ is to increase the value of each of the ω_n.

17. A beam hinged at each end is sufficiently thick that its kinetic energy of rotation must be taken into account. Then the differential equation for its transverse vibrations is

$$\rho \frac{\partial^2 y}{\partial t^2} - \frac{I}{A} \frac{\partial^4 y}{\partial x^2 \partial t^2} + EI \frac{\partial^4 y}{\partial x^4} = 0.$$

Show that its natural frequencies are given by

$$\omega_n = \frac{n^2 \pi^2}{L^2} \left(1 + \frac{n^2 \pi^2 I}{\rho AL^2}\right)^{-1/2} \sqrt{\frac{EI}{\rho}}$$

for $n = 1, 2, 3, \ldots$.

18. Suppose that the end $x = 0$ of a uniform bar with cross-sectional area A and Young's modulus E is fixed, while the longitudinal force $F(t) = F_0 \sin \omega t$ acts on its end $x = L$, so that $AEu_x(L, t) = F_0 \sin \omega t$. Derive the steady periodic solution

$$u(x, t) = \frac{F_0 a \sin(\omega x / a) \sin \omega t}{AE\omega \cos(\omega L / a)}.$$

19. Repeat Problem 18, except here a transverse force $F(t) = F_0 \sin \omega t$ acts at the free end $x = L$, so that

$$y_{xx}(L, t) = EIy_{xxx}(L, t) + F_0 \sin \omega t = 0.$$

Determine the steady periodic transverse oscillations of the cantilever in the form $y(x, t) = X(x) \sin \omega t$.

20. A string with fixed ends is acted on by a periodic force $F(x, t) = F(x) \sin \omega t$ per unit mass, so

$$y_{tt} = a^2 y_{xx} + F(x) \sin \omega t.$$

Substitute

$$y(x, t) = \sum_{n=1}^{\infty} c_n \sin \frac{n\pi x}{L} \sin \omega t$$

and

$$F(x) = \sum_{n=1}^{\infty} F_n \sin \frac{n\pi x}{L}$$

to derive the steady periodic solution

$$y(x, t) = \sum_{n=1}^{\infty} \frac{F_n \sin(n\pi x / a) \sin \omega t}{\omega_n^2 - \omega^2},$$

where $\omega_n = n\pi a / L$. Hence resonance does not result if $\omega = \omega_n$ but $F_n = 0$.

21. Suppose that the string of Problem 20 is also subject to air resistance proportional to its velocity, so that

$$y_{tt} = a^2 y_{xx} - cy_t + F(x) \sin \omega t.$$

Generalize the method of Problem 20 to derive the steady periodic solution

$$y(x, t) = \sum_{n=1}^{\infty} \rho_n F_n \sin \frac{n\pi x}{L} \sin(\omega t - \alpha_n),$$

where
$$\rho_n = \left[(\omega_n^2 - \omega^2)^2 + \omega^2 c^2\right]^{-1/2}$$
and
$$\alpha_n = \tan^{-1} \frac{\omega c}{\omega_n^2 - \omega^2}.$$

Note the similarity to damped forced motion of a mass on a spring.

22. The **telephone equation** for the voltage $e(x, t)$ in a long transmission line at the point $x \geq 0$ at time t is

$$\frac{\partial^2 e}{\partial x^2} = LC \frac{\partial^2 e}{\partial t^2} + (RC + LG) \frac{\partial e}{\partial t} + RGe,$$

where R, L, G, and C denote resistance, inductance, conductance, and capacitance (all per unit length of line), respectively. The condition $e(0, t) = E_0 \cos \omega t$ represents a periodic signal voltage at the origin of transmission at $x = 0$. Assume that $e(x, t)$ is bounded as $x \to +\infty$. Substitute $\tilde{e}(x, t) = E(x)e^{i\omega t}$ to derive the steady periodic solution

$$e(x, t) = E_0 e^{-\alpha x} \cos(\omega t - \beta x),$$

where α and β are the real and imaginary parts, respectively, of the complex number

$$\left[(RG - LC\omega^2) + i\omega(RC + LG)\right]^{1/2}.$$

23. The temperature $u(x, t)$ at the point $x \geq 0$ at time t of water moving with velocity $\gamma \geq 0$ in a long pipe satisfies the equation

$$\frac{\partial u}{\partial t} = k \frac{\partial^2 u}{\partial x^2} - \gamma \frac{\partial u}{\partial x}.$$

Suppose that $u(0, t) = A_0 \cos \omega t$ and that $u(x, t)$ is bounded as $x \to +\infty$. Substitute $\tilde{u}(x, t) = X(x)e^{i\omega t}$ to derive the steady periodic solution

$$u(x, t) = A_0 e^{-\alpha x} \cos(\omega t - \beta x),$$

where

$$\alpha = \frac{1}{2k} \left(\gamma^4 + 16k^2\omega^2\right)^{1/4} \cos \frac{\phi}{2} - \frac{\gamma}{2k},$$

$$\beta = \frac{1}{2k} \left(\gamma^4 + 16k^2\omega^2\right)^{1/4} \sin \frac{\phi}{2}, \quad \text{and}$$

$$\phi = \tan^{-1} \frac{4k}{\gamma^2}.$$

Show also that when $\gamma = 0$, this solution reduces to that in Eq. (29).

10.3 Application: Vibrating Beams and Diving Boards

n	β_n
1	4.7300
2	7.8532
3	10.9956
4	14.1372
5	17.2788
6	20.4204
7	23.5619
8	26.7035
9	29.8451
10	32.9867

FIGURE 10.3.6. The first 10 positive solutions of the eigenvalue equation $\cosh x \, \cos x = 1$ for the vibrating beam with fixed ends.

In this project you are to investigate further the transverse vibrations of an elastic bar or beam of length L whose position function $y(x, t)$ satisfies the partial differential equation

$$\rho \frac{\partial^2 y}{\partial t^2} + EI \frac{\partial^4 y}{\partial x^4} = 0 \qquad (0 < x < L)$$

and the initial conditions $y(x, 0) = f(x)$, $y_t(x, 0) = 0$.

First separate the variables (as in Example 3 of Section 10.3) to derive the formal series solution

$$y(x, t) = \sum_{n=1}^{\infty} c_n \left(\cos \frac{\beta_n^2 a^2 t}{L^2}\right) X_n(x),$$

where $a^4 = EI/\rho$, the $\{c_n\}$ are the appropriate eigenfunction expansion coefficients of the initial position function $f(x)$, and $\{\beta_n\}_1^{\infty}$ and $X_n(x)$ are determined by the end conditions imposed on the bar. In a particular case, we want to find both the **frequency equation** whose positive roots are the $\{\beta_n\}_1^{\infty}$ and the eigenfunctions $\{X_n(x)\}$. In this section we saw that the frequency equation for the fixed/fixed case (with $y(0) = y'(0) = y(L) = y'(L) = 0$) is $\cosh x \, \cos x = 1$. This equation can be solved using computer system methods like those illustrated in the applications for Sections 10.1 and 10.2; the table in Fig. 10.3.6 lists the first ten positive solutions $\{\beta_n\}_1^{10}$. The calculations preceding Eq. (18) in this section imply that the nth eigenfunction is then given by

$$X_n(x) = A_n \left(\cosh \frac{\beta_n x}{L} - \cos \frac{\beta_n x}{L}\right) + B_n \left(\sinh \frac{\beta_n x}{L} - \sin \frac{\beta_n x}{L}\right)$$

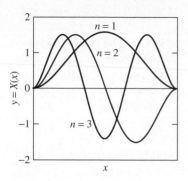

FIGURE 10.3.7. The first three eigenfunction graphs $y = X_n(x)$ for the fixed-fixed I beam.

where the ratio A_n/B_n is determined by the endpoint condition $X_n(L) = 0$. The first three of these eigenfunctions are graphed in Fig. 10.3.7, which was plotted with each $A_n = 1$ and with $L = 3657.6$ cm for the 120-ft steel I beam of this section. The fact that each (correctly) appears to have slope zero at both endpoints serves to corroborate the calculations that generated the figure.

The following cases describe other vibrating beam situations that you can investigate similarly.

Case 1: Hinged at each end

The boundary conditions are

$$y(0) = y''(0) = y(L) = y''(L) = 0.$$

According to Problem 8, the frequency equation is $\sin x = 0$, so $\beta_n = n\pi$ and $X_n(x) = \sin n\pi x$. Suppose that the bar is made of steel (with density $\delta = 7.75$ g/cm^3 and Young's modulus $E = 2 \times 10^{12}$ dyn/cm^2) and is 19 in. long with square cross section of edge $w = 1$ in. (so its moment of inertia is $I = \frac{1}{12}w^4$). Determine its first few natural frequencies of vibration (in Hz). How does this bar sound when it vibrates?

Case 2: Free at each end

The boundary conditions are

$$y''(0) = y^{(3)}(0) = y''(L) = y^{(3)}(L) = 0.$$

This case models, for example, a weightless bar suspended in space in an orbiting spacecraft. Now show that the frequency equation is $\cosh x \cos x = 1$ (although the eigenfunctions differ from those in the fixed/fixed case). Solve this equation (either graphically or numerically, as in the Section 10.2 application) to approximate the first several natural frequencies of vibration of the same bar considered in Case 1. How does it sound now?

Case 3: Fixed at $x = 0$, free at $x = L$.

Now the boundary conditions are

FIGURE 10.3.8. The most spectacular dives are determined by the frequency equation of Case 3.

$$y(0) = y'(0) = y''(L) = y^{(3)}(L) = 0.$$

This is a **cantilever** like the diving board illustrated in Fig. 10.3.8. According to Problem 15, the frequency equation is $\cosh x \cos x = -1$. Approximate the first several positive solutions of this equation and show graphically that $\beta_n \approx (2n - 1)\pi/2$ for n large. Thereby determine the first several natural frequencies (in Hz) of vibration of the particular diving board described in Problem 15. These are the frequencies at which a diver on this board should bounce up and down at the free end for maximal resonant effect.

10.4 Cylindrical Coordinate Problems

When the Laplacian $\nabla^2 u = u_{xx} + u_{yy} + u_{zz}$ of a function $u = y(x, y, z)$ is transformed by means of the substitution $x = r\cos\theta$, $y = r\sin\theta$, the result is the **Laplacian in cylindrical coordinates,**

$$\blacktriangleright \qquad \nabla^2 u = \frac{\partial^2 u}{\partial r^2} + \frac{1}{r}\frac{\partial u}{\partial r} + \frac{1}{r^2}\frac{\partial^2 u}{\partial \theta^2} + \frac{\partial^2 u}{\partial z^2}. \tag{1}$$

For a typical application, consider a very long uniform solid cylinder of radius c that lies along the z-axis. Suppose that it is heated to an initial temperature depending only on the distance r from the z-axis, and that at time $t = 0$ the condition

$$\left. (\beta_1 u + \beta_2 u_r)\right|_{r=c} = 0 \tag{2}$$

is imposed on its surface $r = c$. Note the following special cases for Eq. (2):

- It reduces to the condition $u = 0$ if $\beta_1 = 1$ and $\beta_2 = 0$.
- It reduces to the insulation condition $u_r = 0$ if $\beta_1 = 0$ and $\beta_2 = 1$.
- It reduces to the heat transfer condition $hu + u_r = 0$ if $\beta_1 = h$ and $\beta_2 = 1$.

It is reasonable to expect the temperature u within the cylinder at time t to depend only on r, so we write $u = u(r, t)$. Then $u_{\theta\theta} = u_{zz} = 0$, so substitution of (1) in the heat equation $u_t = k\nabla^2 u$ yields the boundary value problem

$$\frac{\partial u}{\partial t} = k\left(\frac{\partial^2 u}{\partial r^2} + \frac{1}{r}\frac{\partial u}{\partial r}\right) \qquad (r < c,\ t > 0); \tag{3}$$

$$\blacktriangleright \qquad \beta_1 u(c, t) + \beta_2 u_r(c, t) = 0, \tag{4}$$

$$u(r, 0) = f(r) \qquad \text{(initial temperature)}. \tag{5}$$

To solve this problem by separation of variables, we substitute

$$u(r, t) = R(r)T(t)$$

in Eq. (3); thus we obtain

$$RT' = k\left(R''T + \frac{1}{r}R'T\right). \tag{6}$$

Division by kRT yields

$$\frac{R'' + R'/r}{R} = \frac{T'}{kT} = -\lambda. \tag{7}$$

Hence $R(r)$ must satisfy the equation

$$R'' + \frac{1}{r}R' + \lambda R = 0, \tag{8}$$

as well as the equation $\beta_1 R(c) + \beta_2 R'(c) = 0$ that follows from Eq. (4). Moreover, the equation

$$T' = -\lambda k T$$

implies that $T(t) = e^{-\lambda k t}$ to within a constant multiple. Because the diffusivity k is positive, it follows that λ must be nonnegative if $T(t)$ is to remain bounded as $t \to +\infty$, as is required by the physical problem that Eqs. (3)–(5) model. We therefore write $\lambda = \alpha^2$, so Eq. (8) takes the form

$$r^2 R'' + r R' + \alpha^2 r^2 R = 0. \tag{8'}$$

A Singular Sturm-Liouville Problem

Equation (8′) is—with $x = r$ and $y(x) = R(r)$—the parametric Bessel equation

$$x^2 y'' + x y' + \alpha^2 x^2 y = 0 \tag{9}$$

of order zero that we discussed in Section 8.5. More generally, recall that the parametric Bessel equation

➤ $$x^2 y'' + x y' + (\alpha^2 x^2 - n^2) y = 0 \tag{10}$$

has general solution

➤ $$y(x) = A J_n(\alpha x) + B Y_n(\alpha x) \tag{11}$$

if $\alpha > 0$. Upon division by x, the Bessel equation in (10) takes the Sturm-Liouville form

➤ $$\frac{d}{dx}\left(x \frac{dy}{dx}\right) - \frac{n^2}{x} y + \lambda x y = 0 \tag{12}$$

with $p(x) = x$, $q(x) = n^2/x$, $r(x) = x$, and $\lambda = \alpha^2$. We want to determine the nonnegative values of λ for which there is a solution of Eq. (12) in $(0, c)$ that is *continuous* (along with its derivative dy/dx) on the closed interval $[0, c]$ and satisfies the endpoint condition

$$\beta_1 y(c) + \beta_2 y'(c) = 0, \tag{13}$$

where β_1 and β_2 are not both zero.

The Sturm-Liouville problem associated with Eqs. (12) and (13) is *singular* because $p(0) = r(0) = 0$ and $q(x) \to +\infty$ as $x \to 0^+$, whereas we assumed in Theorem 1 of Section 10.1 that $p(x)$ and $r(x)$ were positive and that $q(x)$ was continuous on the entire interval. This problem also fails to fit the pattern of Section 10.1 in that no condition like (13) is imposed at the left endpoint $x = 0$. Nevertheless, the requirement that $y(x)$ be continuous on $[0, c]$ plays the role of such a condition. Because $Y_n(x) \to -\infty$ as $x \to 0$, the solution in Eq. (11) for $\alpha > 0$ can be continuous at $x = 0$ only if $B = 0$, so

$$y(x) = J_n(\alpha x)$$

to within a constant multiple. It remains only to impose the condition in Eq. (13) at $x = c$.

It is convenient to distinguish the cases $\beta_2 = 0$ and $\beta_2 \neq 0$. If $\beta_2 = 0$, then Eq. (13) takes the simple form

➤
$$y(c) = 0. \tag{13a}$$

If $\beta_2 \neq 0$, we multiply each term in Eq. (13) by c/β_2 and then write $h = c\beta_1/\beta_2$ to obtain the equivalent condition

➤
$$hy(c) + cy'(c) = 0. \tag{13b}$$

We assume hereafter that $h \geqq 0$.

CASE 1: $\lambda = 0$. We first consider the possibility of a zero eigenvalue $\lambda = 0$. If both $\lambda = 0$ and $n = 0$, then Eq. (12) reduces to the equation $(xy')' = 0$ with general solution
$$y(x) = A \ln x + B,$$

and continuity on $[0, c]$ requires that $A = 0$. But then Eq. (13a) implies that $B = 0$ as well, as does Eq. (13b) unless $h = 0$, in which case $\lambda = 0$ is an eigenvalue with associated eigenfunction $y(x) \equiv 1$.

If $\lambda = 0$ but $n > 0$, then Eq. (12) is simply the equation

$$x^2 y'' + xy' - n^2 y = 0$$

with general solution (substituting the trial solution $y = x^k$ as in Section 9.7)

$$y(x) = Ax^n + Bx^{-n},$$

and continuity on $[0, c]$ requires that $B = 0$. But it is easy to check that $y(x) = Ax^n$ satisfies neither Eq. (13a) nor Eq. (13b) unless $A = 0$. Thus $\lambda = 0$ is not an eigenvalue if $n > 0$. We have therefore shown that $\lambda = 0$ *is an eigenvalue of the problem in* (12)–(13) *if and only if $n = h = 0$ and the endpoint condition at $x = c$ is $y'(c) = 0$, in which case an associated eigenfunction is $y(x) \equiv 1$.* In this case we write
$$\lambda_0 = 0 \quad \text{and} \quad y_0(x) \equiv 1.$$

CASE 2: $\lambda > 0$. If $\lambda > 0$, then $\lambda = \alpha^2 > 0$, and in this case the only solution of Eq. (12) that is continuous on $[0, c]$ is, to within a constant multiple,

$$y(x) = J_n(\alpha x).$$

Then Eq. (13a) implies that $J_n(\alpha c) = 0$; in other words, αc must be a positive root of the equation

➤
$$J_n(x) = 0. \tag{14a}$$

Recall from Section 8.5 that the graphs of $J_0(x)$ and $J_1(x)$ look as indicated in Fig. 10.4.1. The graph of $J_n(x)$ for $n > 1$ resembles that of $J_1(x)$, even in the detail that $J_n(0) = 0$. In particular, for each $n = 1, 2, 3, \ldots$, Eq. (14a) has an increasing infinite sequence $\{\gamma_{nk}\}_{k=1}^{\infty}$ of positive roots such that $\gamma_{nk} \to +\infty$ as $k \to +\infty$. These roots for $n \leq 8$ and $k \leq 20$ are given in Table 9.5 of Abramowitz and Stegun's *Handbook of Mathematical Functions*.

FIGURE 10.4.1. The graphs of the Bessel functions $J_0(x)$ and $J_1(x)$.

If $y(x) = J_n(\alpha x)$, so that $dy/dx = \alpha J_n'(\alpha x)$, then Eq. (13b) implies that

$$h J_n(\alpha c) + \alpha c J_n'(\alpha c) = 0$$

—that is, that αc be a positive root of the equation

$$h J_n(x) + x J_n'(x) = 0. \tag{14b}$$

It is known that this equation also has an increasing infinite sequence $\{\gamma_{nk}\}_{k=1}^{\infty}$ of positive roots diverging to $+\infty$. If $h = 0$, then Eq. (14b) reduces to the equation $J_n'(x) = 0$; the roots of this equation appear in Table 9.5 of Abramowitz and Stegun. In the important case $n = 0$, the first five roots of Eq. (14b) for various values of h can be found in Table 9.7 of the same reference.

If either of the boundary conditions in Eqs. (13a) and (13b) holds, then the kth positive eigenvalue is

$$\lambda_k = \frac{\gamma_k^2}{c^2},$$

where we write γ_k for the kth positive root of the appropriate one of Eqs. (14a) and (14b); the associated eigenfunction is

$$y_k(x) = J_n\left(\frac{\gamma_k x}{c}\right).$$

The table in Fig. 10.4.2 summarizes this situation for ready reference. The exceptional case $n = h = 0$ corresponding to the endpoint condition $y'(c) = 0$ is listed separately. We have discussed only nonnegative eigenvalues, but it can be proved that the problem in (12)–(13) has no negative eigenvalues. (See Section 78 of R. V. Churchill and J. W. Brown, *Fourier Series and Boundary Value Problems*, 3rd ed. (New York: McGraw-Hill, 1978).)

Fourier-Bessel Series

Now that we know that the singular Sturm-Liouville problem in (12)–(13) has an infinite sequence of eigenvalues and associated eigenfunctions similar to those of a regular Sturm-Liouville problem, we can discuss eigenfunction expansions. In either case 1 or case 2 of Fig. 10.4.2, we expect a piecewise smooth function $f(x)$ on $[0, c]$ to have an eigenfunction series of the form

$$f(x) = \sum_{k=1}^{\infty} c_k y_k(x) = \sum_{k=1}^{\infty} c_k J_n\left(\frac{\gamma_k x}{c}\right), \tag{15}$$

Endpoint Condition	Eigenvalues	Associated Eigenfunctions
Case 1: $y(c) = 0$	$\lambda_k = \gamma_k^2/c^2$; $\{\gamma_k\}_1^\infty$ the positive roots of $J_n(x) = 0$	$y_k(x) = J_n\left(\dfrac{\gamma_k x}{c}\right)$
Case 2: $hy(c) + cy'(c) = 0$; h and n not both zero	$\lambda_k = \gamma_k^2/c^2$; $\{\gamma_k\}_1^\infty$ the positive roots of $hJ_n(x) + xJ_n'(x) = 0$	$y_k(x) = J_n\left(\dfrac{\gamma_k x}{c}\right)$
Case 3: $y'(c) = 0$, $n = 0$	$\lambda_0 = 0,\ \lambda_k = \gamma_k^2/c^2$; $\{\gamma_k\}_1^\infty$ the positive roots of $J_0'(x) = 0$	$y_0(x) = 1$, $y_k(x) = J_0\left(\dfrac{\gamma_k x}{c}\right)$

FIGURE 10.4.2. Eigenvalues and eigenfunctions of the singular Sturm-Liouville problem

$$\frac{d}{dx}\left[x\frac{dy}{dx}\right] - \frac{n^2}{x}y + \lambda xy = 0, \quad \beta_1 y(c) + \beta_2 y'(c) = 0 \quad \text{on } [0, c].$$

while in the exceptional case 3, the series will also contain a constant term c_0 corresponding to $\lambda_0 = 0$, $y_0(x) \equiv 1$. If the conclusion of Theorem 2 of Section 10.1 is to hold (despite the fact that its hypotheses are not satisfied), then the eigenfunctions

$$J_n\left(\frac{\gamma_k x}{c}\right), \quad k = 1, 2, 3, \ldots$$

must be orthogonal on $[0, c]$ with weight function $r(x) = x$. Indeed, if we substitute $p(x) = r(x) = x$ and

$$y_k(x) = J_n\left(\frac{\gamma_k x}{c}\right), \quad y_k'(x) = \frac{\gamma_k}{c}J_n'\left(\frac{\gamma_k x}{c}\right)$$

in Eq. (22) of Section 10.1, the result is

$$(\lambda_i - \lambda_j)\int_0^c xJ_n\left(\frac{\gamma_i x}{c}\right)J_n\left(\frac{\gamma_j x}{c}\right)dx$$

$$= \left[x\left\{\frac{\gamma_j}{c}J_n\left(\frac{\gamma_i x}{c}\right)J_n'\left(\frac{\gamma_j x}{c}\right) - \frac{\gamma_i}{c}J_n\left(\frac{\gamma_j x}{c}\right)J_n'\left(\frac{\gamma_i x}{c}\right)\right\}\right]_0^c$$

$$= \gamma_j J_n(\gamma_i)J_n'(\gamma_j) - \gamma_i J_n(\gamma_j)J_n'(\gamma_i). \tag{16}$$

It is clear that the quantity in Eq. (16) is zero if γ_i and γ_j are both roots of Eq. (14a), $J_n(x) = 0$, whereas if both are roots of Eq. (14b) it reduces to

$$J_n(\gamma_i)\left[-hJ_n(\gamma_j)\right] - J_n(\gamma_j)\left[-hJ_n(\gamma_i)\right] = 0.$$

In either event we thus see that if $i \neq j$, then

$$\int_0^c xJ_n\left(\frac{\gamma_i x}{c}\right)J_n\left(\frac{\gamma_j x}{c}\right)dx = 0. \tag{17}$$

This orthogonality with weight function $r(x) = x$ is what we need to determine the coefficients in the eigenfunction series in Eq. (15). If we multiply each term in Eq. (15) by $x J_n(\gamma_k x/c)$ and then integrate termwise, we get

$$\int_0^c x f(x) J_n\left(\frac{\gamma_k x}{c}\right) dx = \sum_{j=1}^{\infty} c_j \int_0^c x J_n\left(\frac{\gamma_j x}{c}\right) J_n\left(\frac{\gamma_k x}{c}\right) dx$$

$$= c_k \int_0^c x \left[J_n\left(\frac{\gamma_k x}{c}\right)\right]^2 dx$$

by Eq. (17). Hence

➤
$$c_k = \frac{\int_0^c x f(x) J_n(\gamma_k x/c) \, dx}{\int_0^c x \left[J_n(\gamma_k x/c)\right]^2 dx}. \tag{18}$$

With these coefficients, series of the form in Eq. (15) are often called **Fourier-Bessel series**. It is known that a Fourier-Bessel series for a piecewise smooth function $f(x)$ satisfies the convergence conclusion of Theorem 3 of Section 10.1. That is, it converges to the average value $\frac{1}{2}[f(x+) + f(x-)]$ at each point of $(0, c)$, and hence to the value $f(x)$ at each interior point of continuity.

The Fourier-Bessel Coefficients

In spite of their appearance, the denominator integrals in Eq. (18) are not difficult to evaluate. Suppose that $y(x) = J_n(\alpha x)$, so that $y(x)$ satisfies the parametric Bessel equation of order n,

$$\frac{d}{dx}\left[x \frac{dy}{dx}\right] + \left(\alpha^2 x - \frac{n^2}{x}\right) y = 0. \tag{19}$$

By multiplying this equation by $2xy'(x)$ and integrating by parts (Problem 10), we can readily derive the formula

$$2\alpha^2 \int_0^c x \left[J_n(\alpha x)\right]^2 dx = \alpha^2 c^2 \left[J_n'(\alpha c)\right]^2 + (\alpha^2 c^2 - n^2) \left[J_n(\alpha c)\right]^2. \tag{20}$$

Now suppose that $\alpha = \gamma_k/c$, where γ_k is a root of the equation $J_n(x) = 0$. We apply Eq. (20) as well as the recurrence formula

$$x J_n'(x) = n J_n(x) - x J_{n+1}(x)$$

of Section 8.5, which implies that $J_n'(\gamma_k) = -J_{n+1}(\gamma_k)$. The result is

$$\int_0^c x \left[J_n\left(\frac{\gamma_k x}{c}\right)\right]^2 dx = \frac{c^2}{2}\left[J_n'(\gamma_k)\right]^2 = \frac{c^2}{2}\left[J_{n+1}(\gamma_k)\right]^2. \tag{21}$$

The other entries in the table in Fig. 10.4.3 follow similarly from Eq. (20). Fourier-Bessel series with $n = 0$ are the most common (Problem 9). The forms they take in our three cases are listed next.

$\{\gamma_k\}_1^\infty$ the Positive Roots of the Equation	Value of $\int_0^c x \left[J_n\left(\dfrac{\gamma_k x}{c}\right) \right]^2 dx$
Case 1: $J_n(x) = 0$	$\dfrac{c^2}{2} [J_{n+1}(\gamma_k)]^2$
Case 2: $hJ_n(x) + xJ_n'(x) = 0$ (*n* and *h* not both zero)	$\dfrac{c^2(\gamma_k^2 - n^2 + h^2)}{2\gamma_k^2} [J_n(\gamma_k)]^2$
Case 3: $J_0'(x) = 0$	$\dfrac{c^2}{2} [J_0(\gamma_k)]^2$

FIGURE 10.4.3.

CASE 1 WITH $n = 0$: If $\{\gamma_k\}_{k=1}^\infty$ are the positive roots of the equation $J_0(x) = 0$, then

$$f(x) = \sum_{k=1}^\infty c_k J_0\left(\frac{\gamma_k x}{c}\right);$$

$$c_k = \frac{2}{c^2 [J_1(\gamma_k)]^2} \int_0^c x f(x) J_0\left(\frac{\gamma_k x}{c}\right) dx. \tag{22}$$

CASE 2 WITH $n = 0$: If $\{\gamma_k\}_{k=1}^\infty$ are the positive roots of the equation $hJ_0(x) + xJ_0'(x) = 0$ with $h > 0$, then

$$f(x) = \sum_{k=1}^\infty c_k J_0\left(\frac{\gamma_k x}{c}\right);$$

$$c_k = \frac{2\gamma_k^2}{c^2(\gamma_k^2 + h^2)[J_0(\gamma_k)]^2} \int_0^c x f(x) J_0\left(\frac{\gamma_k x}{c}\right) dx. \tag{23}$$

CASE 3: If $\{\gamma_k\}_{k=1}^\infty$ are the positive roots of the equation $J_0'(x) = 0$, then (Problem 10)

$$f(x) = c_0 + \sum_{k=1}^\infty c_k J_0\left(\frac{\gamma_k x}{c}\right); \tag{24a}$$

$$c_0 = \frac{2}{c^2} \int_0^c x f(x)\, dx, \tag{24b}$$

$$c_k = \frac{2}{c^2 [J_0(\gamma_k)]^2} \int_0^c x f(x) J_0\left(\frac{\gamma_k x}{c}\right) dx. \tag{24c}$$

Applications

Example 1 Suppose that a long circular cylinder of radius c has initial temperature $u(r, 0) = u_0$, a constant. Find $u(r, t)$ in each of the following three cases:

(a) $u(c, t) = 0$ (zero boundary temperature);

(b) The boundary of the cylinder is insulated, so that $u_r(c, t) = 0$;

(c) Heat transfer takes place across the boundary of the cylinder, so that

$$Hu(c, t) + Ku_r(c, t) = 0, \quad \text{where } H > 0, K > 0.$$

Solution With $u(r, t) = R(r)T(t)$, we saw previously that $\lambda = \alpha^2 > 0$ in Eq. (7), so

$$r^2 R'' + r R' + \alpha^2 r^2 R = 0 \tag{25}$$

and

$$T' = -\alpha^2 kT. \tag{26}$$

Equation (25) is Bessel's parametric equation of order zero, and its only continuous nontrivial solutions on $[0, c]$ are of the form

$$R(r) = A J_0(\alpha r). \tag{27}$$

But the possible values of α depend on the boundary conditions that are imposed.

CASE (a): Zero Boundary Temperature. The condition $u(c, t) = 0$ yields $R(c) = A J_0(\alpha c) = 0$, so αc must be one of the roots $\{\gamma_k\}_{k=1}^{\infty}$ of the equation $J_0(x) = 0$. Thus the eigenvalues and eigenfunctions are

$$\lambda_n = \frac{\gamma_n^2}{c^2}, \quad R_n(r) = J_0\left(\frac{\gamma_n r}{c}\right) \tag{28}$$

for $n = 1, 2, 3, \ldots$. Then the equation

$$T_n' = -\left(\frac{\gamma_n^2}{c^2}\right) kT_n$$

yields

$$T_n(t) = \exp\left(-\frac{\gamma_n^2 kt}{c^2}\right)$$

to within a constant factor. Hence the series

$$u(r, t) = \sum_{n=1}^{\infty} c_n \exp\left(-\frac{\gamma_n^2 kt}{c^2}\right) J_0\left(\frac{\gamma_n r}{c}\right) \tag{29}$$

satisfies formally the heat equation and the boundary condition $u(c, t) = 0$. It remains to choose the coefficients so that

$$u(r, 0) = \sum_{n=1}^{\infty} c_n J_0\left(\frac{\gamma_n r}{c}\right) = u_0.$$

Because $J_0(\gamma_n) = 0$, Eq. (22) yields

$$c_n = \frac{2u_0}{c^2 [J_1(\gamma_n)]^2} \int_0^c r J_0\left(\frac{\gamma_n r}{c}\right) dr$$

$$= \frac{2u_0}{\gamma_n^2 [J_1(\gamma_n)]^2} \int_0^{\gamma_n} x J_0(x) \, dx$$

$$= \frac{2u_0}{\gamma_n^2 [J_1(\gamma_n)]^2} \left[x J_1(x)\right]_0^{\gamma_n} = \frac{2u_0}{\gamma_n J_1(\gamma_n)}.$$

Here we have used the integral

$$\int x J_0(x)\,dx = x J_1(x) + C.$$

On substitution of the coefficients $\{c_n\}$ in Eq. (29), we finally obtain

$$u(r, t) = 2u_0 \sum_{n=1}^{\infty} \frac{1}{\gamma_n J_1(\gamma_n)} \exp\left(\frac{-\gamma_n^2 kt}{c^2}\right) J_0\left(\frac{\gamma_n r}{c}\right). \tag{30}$$

Because of the presence of exponential factors, only a few terms are ordinarily needed for numerical computations. The data required for the first five terms are listed in Fig. 10.4.4. For instance, suppose that the cylinder has radius $c = 10$ cm, is made of iron with thermal diffusivity $k = 0.15$, and has initial temperature $u_0 = 100°C$ throughout. Then, because $J_0(0) = 1$, we find from Eq. (30) that the temperature at its axis $(r = 0)$ after two minutes $(t = 120)$ will be

$$u(0, 120) = 200 \sum_{n=1}^{\infty} \frac{1}{\gamma_n J_1(\gamma_n)} \exp(-[0.18]\gamma_n^2)$$

$$\approx 200 \cdot (0.28283 - 0.00221 + 0.00000 - \cdots);$$

thus $u(0, 120)$ will be approximately $56.12°C$.

n	γ_n	$J_1(\gamma_n)$
1	2.40483	+0.51915
2	5.52008	−0.34026
3	8.65373	+0.27145
4	11.79153	−0.23246
5	14.93092	+0.20655

FIGURE 10.4.4. Roots of $J_0(\gamma) = 0$.

CASE (b): Insulated Boundary. In this case no heat ever escapes from the heated cylinder, so we certainly should find that its temperature remains constant: $u(r, t) \equiv u_0$. Inasmuch as $u_r(c, t) = 0$ implies that $R'(c) = 0$, we have $n = h = 0$ in Eq. (14b), and hence case 3 in Fig. 10.4.2. Therefore, $\lambda_0 = 0$ is an eigenvalue with associated eigenfunction $R_0(r) \equiv 1$. A corresponding solution of Eq. (26) is $T_0(t) = 1$. The positive eigenvalues are again given by (28), except that now the numbers $\{\gamma_n\}_1^{\infty}$ are the positive roots of $J_0'(x) = 0$. The solution is therefore of the form

$$u(r, t) = c_0 + \sum_{n=1}^{\infty} c_n \exp\left(-\frac{\gamma_n^2 kt}{c^2}\right) J_0\left(\frac{\gamma_n r}{c}\right),$$

and we want

$$u(r, 0) = c_0 + \sum_{n=1}^{\infty} c_n J_0\left(\frac{\gamma_n r}{c}\right) = u_0.$$

But now the case 3 coefficient formulas in (24) yield

$$c_0 = \frac{2}{c^2} \int_0^c r u_0\,dr = \frac{2u_0}{c^2}\left[\frac{1}{2}r^2\right]_0^c = u_0;$$

$$c_n = \frac{2u_0}{c^2 [J_0(\gamma_n)]^2} \int_0^c r J_0\left(\frac{\gamma_n r}{c}\right) dr = \frac{2u_0}{\gamma_n^2 [J_0(\gamma_n)]^2} \int_0^{\gamma_n} x J_0(x)\,dx$$

$$= \frac{2u_0}{\gamma_n^2 [J_0(\gamma_n)]^2}\left[x J_1(x)\right]_0^{\gamma_n} = \frac{2u_0 J_1(\gamma_n)}{\gamma_n [J_0(\gamma_n)]^2} = 0$$

because $J_1(\gamma_n) = -J_0'(\gamma_n) = 0$. Thus we find that $u(r, t) \equiv u_0$, as expected.

CASE (c): Heat Transfer at the Boundary. Substitution of $R(r) = J_0(\alpha r)$ in the boundary condition $H R(c) + K R'(c) = 0$ gives the equation

$$H J_0(\alpha c) + K \alpha J_0'(\alpha c) = 0$$

that takes the form

$$h J_0(\alpha c) + \alpha c J_0'(\alpha c) = 0 \tag{31}$$

with $h = cH/K > 0$. Comparing Eq. (31) with Eq. (14b), we see that we have case 2 in Fig. 10.4.2 with $n = 0$. Hence the eigenvalues and associated eigenfunctions are given by

$$\lambda_n = \frac{\gamma_n^2}{c^2}, \quad R_n(r) = J_0\left(\frac{\gamma_n r}{c}\right), \tag{32}$$

where $\{\gamma_n\}_1^\infty$ are the positive roots of the equation $h J_0(x) + x J_0'(x) = 0$. Now $T' = -\lambda k T$ as before; consequently, the solution is of the form

$$u(r, t) = \sum_{n=1}^\infty c_n \exp\left(-\frac{\gamma_n^2 k t}{c^2}\right) J_0\left(\frac{\gamma_n r}{c}\right), \tag{33}$$

and Eq. (23) yields

$$c_n = \frac{2 u_0 \gamma_n^2}{c^2(\gamma_n^2 + h^2)\,[J_0(\gamma_n)]^2} \int_0^c r J_0\left(\frac{\gamma_n r}{c}\right) dr$$

$$= \frac{2 u_0}{(\gamma_n^2 + h^2)\,[J_0(\gamma_n)]^2} \int_0^{\gamma_n} x J_0(x)\,dx = \frac{2 u_0 \gamma_n J_1(\gamma_n)}{(\gamma_n^2 + h^2)\,[J_0(\gamma_n)]^2}.$$

Hence

$$u(x, t) = 2 u_0 \sum_{n=1}^\infty \frac{\gamma_n J_1(\gamma_n)}{(\gamma_n^2 + h^2)\,[J_0(\gamma_n)]^2} \exp\left(-\frac{\gamma_n^2 k t}{c^2}\right) J_0\left(\frac{\gamma_n r}{c}\right). \tag{34}$$

■

Example 2

Suppose that a flexible circular membrane of radius c vibrates under tension T in such a way that its [normal] displacement u depends only on time t and the distance r from its center. (This is the case of *radial vibrations*.) Then $u = u(r, t)$ satisfies the partial differential equation

$$\frac{\partial^2 u}{\partial t^2} = a^2 \nabla^2 u = a^2 \left(\frac{\partial^2 u}{\partial r^2} + \frac{1}{r}\frac{\partial u}{\partial r}\right) \tag{35}$$

where $a^2 = T/\rho$, and $u(c, t) = 0$ if the boundary of the membrane is fixed. Find the natural frequencies and normal modes of radial vibration of the membrane.

Solution We apply the method of Section 10.3: Substitution of the function $u(r, t) = R(r) \sin \omega t$ in Eq. (35) yields

$$-\omega^2 R \sin \omega t = a^2 \left(R'' \sin \omega t + \frac{1}{r} R' \sin \omega t \right).$$

Thus ω and $R(r)$ must satisfy the equation

$$r^2 R'' + r R' + \frac{\omega^2}{a^2} r^2 R = 0, \tag{36}$$

as well as the condition $R(c) = 0$ that follows from $u(c, t) = 0$. Equation (36) is the parametric Bessel equation of order zero with parameter $\alpha = \omega/a$, and its only nontrivial solution is (to within a constant multiple)

$$R(r) = J_0 \left(\frac{\omega r}{a} \right); \quad \text{hence} \quad R(c) = J_0 \left(\frac{\omega c}{a} \right) = 0,$$

so $\omega c/a$ must be one of the positive roots $\{\gamma_n\}_1^\infty$ of the equation $J_0(x) = 0$. Thus the nth natural [circular] frequency and a corresponding natural mode of vibration are

$$\omega_n = \frac{\gamma_n a}{c}, \quad u_n(r, t) = J_0 \left(\frac{\gamma_n r}{c} \right) \sin \frac{\gamma_n a t}{c}. \tag{37}$$

On inspection of the table of values of $\{\gamma_n\}$ in Fig. 10.4.4, we see that the higher natural frequencies ω_n are *not* integral multiples of $\omega_1 = \gamma_1 a/c$; this is why the sound of a vibrating circular drumhead is not perceived as being harmonious. The vibration of the membrane in the nth normal mode in (37) is illustrated in Fig. 10.4.5, which shows a vertical cross section through the center of the drumhead. In addition to the boundary $r = c$, there are $n - 1$ fixed circles, called *nodal circles*, with radii $r_i = \gamma_i c/\gamma_n$, $i = 1, 2, \ldots, n - 1$. The annular regions of the membrane between consecutive pairs of nodal circles move alternately up and down between the surfaces $u = \pm J_0(\gamma_n r/c)$. Figure 10.4.6 shows how the surface $u = J_0(\gamma_n r/c)$ looks for $n = 1, 2, 3,$ and 4. ∎

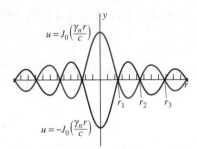

FIGURE 10.4.5. Cross section of the vibrating circular membrane.

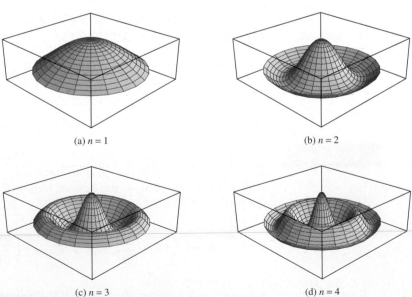

(a) $n = 1$ (b) $n = 2$

(c) $n = 3$ (d) $n = 4$

FIGURE 10.4.6. The surface $u = J_0(\gamma_n r/c)$: (a) $n = 1$; (b) $n = 2$; (c) $n = 3$; (d) $n = 4$.

10.4 Problems

1. Suppose that the circular membrane of Example 2 has initial position $u(r, 0) = f(r)$ and initial velocity $u_t(r, 0) = 0$. Derive by separation of variables the solution

$$u(r, t) = \sum_{n=1}^{\infty} c_n J_0\left(\frac{\gamma_n r}{c}\right) \cos \frac{\gamma_n a t}{c},$$

where $\{\gamma_n\}_1^{\infty}$ are the positive roots of $J_0(x) = 0$ and

$$c_n = \frac{2}{c^2 \left[J_1(\gamma_n)\right]^2} \int_0^c r f(r) J_0\left(\frac{\gamma_n r}{c}\right) dr.$$

2. Assume that the circular membrane of Example 2 has initial position $u(r, 0) = 0$ and initial velocity $u_t(r, 0) = v_0$ (constant). Derive the solution

$$u(r, t) = \frac{2cv_0}{a} \sum_{n=1}^{\infty} \frac{J_0(\gamma_n r/c) \sin(\gamma_n a t/c)}{\gamma_n^2 J_1(\gamma_n)},$$

where $\{\gamma_n\}_1^{\infty}$ are the positive roots of $J_0(x) = 0$.

3. (a) Find $u(r, t)$ in the case in which the circular membrane of Example 2 has initial position $u(r, 0) = 0$ and initial velocity

$$u_r(r, 0) = \begin{cases} \dfrac{P_0}{\rho \pi \epsilon^2} & \text{if } 0 \leqq r \leqq \epsilon, \\[2mm] 0 & \text{if } \epsilon < r \leqq c. \end{cases}$$

 (b) Use the fact that $[J_1(x)]/x \to \frac{1}{2}$ as $x \to 0$ to find the limiting value of the result in part (a) as $\epsilon \to 0$. You should obtain

$$u(r, t) = \frac{a P_0}{\pi c T} \sum_{n=1}^{\infty} \frac{J_0(\gamma_n r/c) \sin(\gamma_n a t/c)}{\gamma_n \left[J_1(\gamma_n)\right]^2},$$

where $\{\gamma_n\}_1^{\infty}$ are the positive roots of $J_0(x) = 0$. This function describes the motion of a drumhead resulting from an initial momentum impulse P_0 at its center.

4. (a) A circular plate of radius c has insulated faces and heat capacity s calories per degree per square centimeter. Find $u(r, t)$ given $u(c, t) = 0$ and

$$u(r, 0) = \begin{cases} \dfrac{q_0}{s \pi \epsilon^2} & \text{if } 0 \leqq r \leqq \epsilon, \\[2mm] 0 & \text{if } \epsilon < r \leqq c. \end{cases}$$

 (b) Take the limit as $\epsilon \to 0$ of the result in part (a) to obtain

$$u(r, t) = \frac{q_0}{s \pi c^2} \sum_{n=1}^{\infty} \frac{1}{[J_1(\gamma_n)]^2} \exp\left(-\frac{\gamma_n^2 k t}{c^2}\right) J_0\left(\frac{\gamma_n r}{c}\right),$$

(where $\{\gamma_n\}_1^{\infty}$ are the positive roots of $J_0(x) = 0$) for the temperature resulting from an injection of q_0 calories of heat at the center of the plate.

Problems 5 through 7 deal with the steady-state temperature $u = u(r, z)$ in a solid cylinder of radius $r = c$ with bottom $z = 0$ and top $z = L$, given that u satisfies Laplace's equation

$$\frac{\partial^2 u}{\partial r^2} + \frac{1}{r}\frac{\partial u}{\partial r} + \frac{\partial^2 u}{\partial z^2} = 0.$$

5. If $u(r, L) = u_0$ and the rest of the surface of the cylinder is held at temperature zero, use separation of variables to derive the solution

$$u(r, z) = 2u_0 \sum_{n=1}^{\infty} \frac{J_0(\gamma_n r/c)}{\gamma_n J_1(\gamma_n)} \cdot \frac{\sinh(\gamma_n z/c)}{\sinh(\gamma_n L/c)},$$

where $\{\gamma_n\}_1^{\infty}$ are the positive roots of $J_0(x) = 0$.

6. (a) If $u(r, L) = f(r)$, $u(r, 0) = 0$, and the cylindrical surface $r = c$ is insulated, derive a solution of the form

$$u(r, z) = c_0 z + \sum_{n=1}^{\infty} c_n J_0\left(\frac{\gamma_n r}{c}\right) \sinh \frac{\gamma_n z}{c},$$

where $\{\gamma_n\}_1^{\infty}$ are the positive roots of $J_0'(x) = 0$. (b) Suppose that $f(r) = u_0$ (constant). Deduce from the result in part (a) that $u(r, z) = u_0 z/L$.

7. Let $c = 1$ and $L \to +\infty$, so that the cylinder is semi-infinite. If $u(r, 0) = u_0$, $hu(1, z) + u_r(1, z) = 0$ (heat transfer on the cylindrical surface), and $u(r, z)$ is bounded as $z \to +\infty$, derive the solution

$$u(r, z) = 2hu_0 \sum_{n=1}^{\infty} \frac{\exp(-\gamma_n z) J_0(\gamma_n r)}{(\gamma_n^2 + h^2) J_0(\gamma_n)},$$

where $\{\gamma_n\}_1^{\infty}$ are the positive roots of the equation $h J_0(x) = x J_1(x)$.

8. Begin with the parametric Bessel equation of order n,

$$\frac{d}{dx}\left[x\frac{dy}{dx}\right] + \left(\alpha^2 x - \frac{n^2}{x}\right) y = 0. \tag{38}$$

Multiply each term by $2x\, dy/dx$; then write the result as

$$\frac{d}{dx}\left[x\frac{dy}{dx}\right]^2 + (\alpha^2 x^2 - n^2)\frac{d}{dx}(y^2) = 0.$$

Integrate each term, using integration by parts for the second term, to obtain

$$\left[\left(x\frac{dy}{dx}\right)^2 + (\alpha^2 x^2 - n^2)y^2\right]_0^c - 2\alpha^2 \int_0^c x y^2\, dx = 0.$$

Finally, substitute $y(x) = J_0(\alpha x)$, a solution of Eq. (38), to get the formula in Eq. (20) in the text.

9. This problem provides the coefficient integrals for Fourier-Bessel series with $n = 0$. **(a)** Substitute $n = 0$ in the result of Problem 8 to obtain the integral formula

$$\int_0^c x \left[J_0(\alpha x)\right]^2 dx = \frac{c^2}{2} \left([J_0(\alpha c)]^2 + [J_1(\alpha c)]^2\right). \quad (39)$$

(b) Suppose that $\alpha = \gamma_k/c$ where γ_k is a root of the equation $J_0(x) = 0$. Deduce that

$$\int_0^c x \left[J_0\left(\frac{\gamma_k x}{c}\right)\right]^2 dx = \frac{c^2}{2} [J_1(\gamma_k)]^2.$$

(c) Suppose that $\alpha = \gamma_k/c$ where γ_k is a root of the equation $J_0'(x) = 0$. Deduce that

$$\int_0^c x \left[J_0\left(\frac{\gamma_k x}{c}\right)\right]^2 dx = \frac{c^2}{2} [J_0(\gamma_k)]^2.$$

(d) Suppose that $\alpha = \gamma_k/c$ where γ_k is a root of the equation $h J_0(x) + x J_0'(x) = 0$. Deduce that

$$\int_0^c x \left[J_0\left(\frac{\gamma_k x}{c}\right)\right]^2 dx = \frac{c^2(\gamma_k^2 + h^2)}{2\gamma_k^2} [J_0(\gamma_k)]^2.$$

10. Suppose that $\{\gamma_m\}_1^\infty$ are the positive roots of the equation $J_0'(x) = 0$ and that

$$f(x) = c_0 + \sum_{m=1}^\infty c_m J_0\left(\frac{\gamma_m x}{c}\right). \quad (40)$$

(a) Multiply each side of Eq. (40) by x and then integrate termwise from $x = 0$ to $x = c$ to show that

$$c_0 = \frac{2}{c^2} \int_0^c x f(x)\, dx.$$

(b) Multiply each side of Eq. (40) by $x J_0(\gamma_n x/c)$ and then integrate termwise to show that

$$c_m = \frac{2}{c^2 [J_0(\gamma_m)]^2} \int_0^c x f(x) J_0\left(\frac{\gamma_m x}{c}\right) dx.$$

11. If a circular membrane with fixed boundary is subjected to a periodic force $F_0 \sin \omega t$ per unit mass uniformly distributed over the membrane, then its displacement function $u(r, t)$ satisfies the equation

$$\frac{\partial^2 u}{\partial t^2} = a^2 \left(\frac{\partial^2 u}{\partial r^2} + \frac{1}{r}\frac{\partial u}{\partial r}\right) + F_0 \sin \omega t.$$

Substitute $u(r, t) = R(r) \sin \omega t$ to find a steady periodic solution.

12. Consider a vertically hanging cable of length L and weight w per unit length, with fixed upper end at $x = L$ and free lower end at $x = 0$, as shown in Fig. 10.4.7. When the cable vibrates transversely, its displacement function $y(x, t)$ satisfies the equation

$$\frac{w}{g}\frac{\partial^2 y}{\partial t^2} = \frac{\partial}{\partial x}\left(wx\frac{\partial y}{\partial x}\right)$$

because the tension is $T(x) = wx$. Substitute the function $y(x, t) = X(x) \sin \omega t$, then apply the theorem of Section 8.6 to solve the ordinary differential equation that results. Deduce from the solution that the natural frequencies of vibration of the hanging cable are given by

$$\omega_n = \frac{\gamma_n}{2}\sqrt{\frac{g}{L}} \quad \text{(rad/s)},$$

where $\{\gamma_n\}_1^\infty$ are the roots of $J_0(x) = 0$. Historically, this problem was the first in which Bessel functions appeared.

FIGURE 10.4.7. The vertical hanging cable of Problem 12.

Problems 13 through 15 deal with a canal of length L off the ocean (see Fig. 10.4.8 for a top view). Its vertical cross section at x is rectangular with width w(x) and depth h(x); the latter is the equilibrium depth of the water at x. Consider a periodic canal tide such that the vertical displacement of the water surface is y(x, t) = X(x) cos ωt at time t. Then y(x, t) satisfies the equation

$$\frac{w(x)}{g}\frac{\partial^2 y}{\partial t^2} = \frac{\partial}{\partial x}\left[w(x)h(x)\frac{\partial y}{\partial x}\right].$$

Let $y_0 = X(L)$ be the amplitude of the tide at the mouth of the canal.

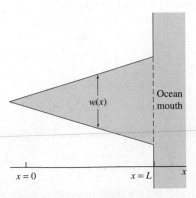

FIGURE 10.4.8. The canal of Problems 13 through 15.

13. Suppose that $w(x) = wx$ and $h(x) = h$ (a constant). Show that

$$y(x, t) = y_0 \frac{J_0\left(wx/\sqrt{gh}\right)}{J_0\left(wL\sqrt{gh}\right)} \cos \omega t.$$

Apply the theorem of Section 8.6.

14. Suppose that $w(x) = w$ (a constant) and that $h(x) = hx$. Show that

$$y(x, t) = y_0 \frac{J_0\left(2w\sqrt{x/gh}\right)}{J_0\left(2w\sqrt{L/gh}\right)} \cos \omega t.$$

15. Suppose that $w(x) = wx$ and $h(x) = hx$ with w and h both constant. Show that

$$y(x, t) = y_0 \sqrt{\frac{L}{x}} \frac{J_1\left(2w\sqrt{x/gh}\right)}{J_1\left(2w\sqrt{L/gh}\right)} \cos \omega t.$$

16. If $0 < a < b$, then the eigenvalue problem

$$\frac{d}{dx}\left[x\frac{dy}{dx}\right] + \lambda xy = 0, \quad y(a) = y(b) = 0$$

for the parametric Bessel equation of order zero is a regular Sturm-Liouville problem. By Problem 1 of Section 10.1, it therefore has an infinite sequence of nonnegative eigenvalues. **(a)** Prove that zero is not an eigenvalue. **(b)** Show that the nth eigenvalue is $\lambda_n = \gamma_n^2$, where $\{\gamma_n\}_1^\infty$ are the positive roots of the equation

$$J_0(ax)Y_0(bx) - J_0(bx)Y_0(ax) = 0. \qquad (41)$$

The first five roots of Eq. (41) for various values of a/b are listed in Table 9.7 of Abramowitz and Stegun's *Handbook*

of Mathematical Functions. **(c)** Show that an associated eigenfunction is

$$R_n(x) = Y_0(\gamma_n a) J_0(\gamma_n x) - J_0(\gamma_n a) Y_0(\gamma_n x). \qquad (42)$$

17. Suppose that an annular membrane with constant density ρ (per unit area) is stretched under constant tension T between the circles $r = a$ and $r = b > a$. Show that its nth natural [circular] frequency is $\omega_n = \gamma_n \sqrt{T/\rho}$, where $\{\gamma_n\}_1^\infty$ are the positive roots of Eq. (41).

18. Suppose that the infinite cylindrical shell $a \leq r \leq b$ has initial temperature $u(r, 0) = f(r)$, and thereafter $u(a, t) = u(b, t) = 0$. By separation of variables, derive the solution

$$u(r, t) = \sum_{n=1}^{\infty} c_n \exp\left(-\gamma_n^2 kt\right) R_n(r),$$

where $R_n(r)$ is the function in Eq. (42) and

$$c_n \int_a^b r \left[R_n(r)\right]^2 dr = \int_a^b rf(r)R_n(r)\, dr.$$

19. Consider the semi-infinite cylindrical shell

$$0 < a \leq r \leq b, \quad z \geq 0.$$

If $u(a, z) = u(b, z) = 0$ and $u(r, 0) = f(r)$, derive by separation of variables the steady-state temperature

$$u(r, z) = \sum_{n=1}^{\infty} c_n \exp\left(-\gamma_n z\right) R_n(r),$$

where $\{c_n\}$ and $\{R_n\}$ are as given in Problems 18 and 16.

10.4 Application: Bessel Functions and Heated Cylinders

Here we outline a *Maple*-based numerical investigation of the temperature function

$$u(r, t) = \sum_{n=1}^{\infty} a_n \exp\left(-\frac{\gamma_n^2 kt}{c^2}\right) J_0\left(\frac{\gamma_n r}{c}\right) \qquad (1)$$

of the heated cylindrical rod of Example 1 in this section. *Mathematica* and MATLAB versions of this investigation are included in the applications manual that accompanies this text.

We assume that the cylindrical rod has radius $c = 10$ cm, constant initial temperature $u_0 = 100°$, and thermal diffusivity $k = 0.15$ (for iron). The coefficients $\{a_n\}_1^\infty$ in Eq. (1) depend on the condition imposed at the boundary $r = c$ of the cylinder. In the case of the zero boundary condition $u(c, t) = 0$, Eq. (30) in this section implies that $a_n = 2u_0/ (\gamma_n J_1(\gamma_n))$, where the $\{\gamma_n\}_1^\infty$ are the positive solutions of the equation $J_0(x) = 0$.

Knowing that $\gamma_1 \approx 2.4$ and that successive roots differ approximately by π, we can we can approximate the first 20 values of γ_n by using the *Maple* commands

```
g := array(1..20):          #   g for gamma
for n from 1 to 20 do
```

```
g[n] := fsolve(BesselJ(0,x)=0, x = 2.4 + (n-1)*Pi):
od:
```

The table in Fig. 10.4.9 lists the first ten values of γ_n. Then the first 20 coefficients in (1) are calculated by the commands

```
a := array(1..20):
c := 10:    u0 := 100:    k := 0.15:
for n from 1 to 20 do
    a[n] := 2*u0/(g[n]*BesselJ(1, g[n])):
    od:
```

Finally, the following *Maple* function sums the corresponding terms of the series:

```
u := (r, t) -> sum(a[n]*exp(-g[n]^2*k*t/c^2)*
                  BesselJ(0, g[n]*r/c), n = 1..20);
```

The graph of $u(r, 120)$ in Fig. 10.4.10 shows how the temperature within the rod after two minutes varies with the distance r from its center, and we see that the centerline temperature has already fallen below $60°$. The graph of $u(0, t)$ for a five-minute period shown in Fig. 10.4.11 indicates that the centerline temperature takes a bit more than 200 seconds to fall to $25°$. Indeed, the computation

```
fsolve(u(0, t) = 25, t = 200..250);
```

reveals that this requires about 214 seconds.

For your own cylindrical rod with constant initial temperature $u_0 = 100°$ to investigate in the manner we illustrate, let $c = 2p$ and $k = q/10$, where p is the largest and q the smallest nonzero digit of your student ID number.

INVESTIGATION A: If the cylindrical boundary of the rod is held at zero temperature $u(c, t) = 0$, plot graphs like Figs. 10.4.10 and 10.4.11, then determine how long it will take for the rod's centerline temperature to fall to $25°$.

n	γ_n
1	2.4048
2	5.5201
3	8.6537
4	11.7915
5	14.9309
6	18.0711
7	21.2116
8	24.3525
9	27.4935
10	30.6346

FIGURE 10.4.9. The first 10 positive zeros of $J_0(x) = 0$.

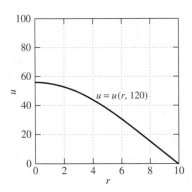

FIGURE 10.4.10. Temperature as a function of distance r from the rod's centerline.

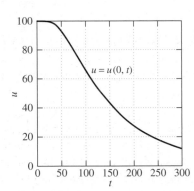

FIGURE 10.4.11. Centerline temperature as a function of time t.

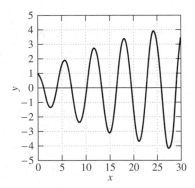

FIGURE 10.4.12. The graph $y = J_0(x) - xJ_0'(x)$.

INVESTIGATION B: Now suppose that heat transfer occurs at the rod's cylindrical boundary, so that the coefficients in the series in (1) are those appearing in Eq. (34) of the text. Assume that $h = 1$, so that the $\{\gamma_n\}_1^\infty$ are now the positive roots of the equation

$$J_0(x) + xJ_0'(x) = J_0(x) - xJ_1(x) = 0 \tag{2}$$

(because $J_0'(x) = -J_1(x)$). Figure 10.4.12 shows the graph of the left-hand side in Eq. (2) and indicates that $\gamma_1 \approx 1.25$, with successive roots differing (as usual)

by approximately π. Determine in this case *both* how long it takes the centerline temperature to fall to 25° *and* how long it takes the boundary temperature at $r = c$ to fall to 25°.

10.5 Higher-Dimensional Phenomena

In this section we discuss the application of the method of separation of variables to heat flow and vibration problems in which two or more space variables are involved. The section consists largely of illustrative examples, problems, and projects grouped according to whether rectangular, cylindrical, or spherical coordinates are most appropriate for their solutions.

Rectangular Coordinate Applications and Two-Dimensional Fourier Series

As indicated in Section 9.7, the Laplacian in two dimensions takes the form

$$\nabla^2 u = \frac{\partial^2 u}{\partial x^2} + \frac{\partial^2 u}{\partial y^2}.$$

For instance, consider a thin plate lying in the xy-plane with its two parallel faces insulated, so that heat flows within the plate only in the x- and y-directions. If $u(x, y, t)$ denotes the temperature at time t at the point (x, y) within the plate, then under standard assumptions it follows that u satisfies the **two-dimensional heat equation**

$$\frac{\partial u}{\partial t} = k\nabla^2 u = k\left(\frac{\partial^2 u}{\partial x^2} + \frac{\partial^2 u}{\partial y^2}\right). \tag{1}$$

Here k denotes the thermal diffusivity of the material comprising the plate. If the plate is rectangular and homogeneous boundary conditions are imposed along its four edges, then the method of separation of variables can be applied in a straight-forward manner.

Example 1 Suppose that a thin rectangular plate occupies the plane region $0 \leqq x \leqq a, 0 \leqq y \leqq b$, that its top and bottom faces are insulated, and that its four edges are held at temperature zero. If the plate has the initial temperature function $u(x, y, 0) = f(x, y)$, then its temperature function $u(x, y, t)$ satisfies the boundary value problem consisting of the heat equation in (1) and the boundary conditions

$$u(0, y, t) = u(a, y, t) = u(x, 0, t) = u(x, b, t) = 0, \tag{2}$$
$$u(x, y, 0) = f(x, y). \tag{3}$$

Find $u(x, y, t)$.

Solution We substitute $u(x, y, t) = X(x)Y(y)T(t)$ in Eq. (1). After division by $kXYT$, we obtain

$$\frac{T'}{kT} = \frac{X''}{X} + \frac{Y''}{Y}.$$

This relation can hold for all x, y, and t only if each term is a constant, so we write

$$\frac{X''}{X} = -\lambda, \quad \frac{Y''}{Y} = -\mu, \quad \frac{T'}{kT} = -(\lambda + \mu). \tag{4}$$

Taking into account the boundary conditions in (2), we see that $X(x)$ and $Y(y)$ satisfy the separate Sturm-Liouville problems

$$X'' + \lambda X = 0, \quad X(0) = X(a) = 0 \tag{5}$$

and

$$Y'' + \mu Y = 0, \quad Y(0) = Y(b) = 0. \tag{6}$$

The eigenvalues and eigenfunctions of the familiar problem in (5) are

$$\lambda_m = \frac{m^2 \pi^2}{a^2}, \quad X_m(x) = \sin \frac{m\pi x}{a} \tag{7}$$

for $m = 1, 2, 3, \dots$. Similarly, the eigenvalues and eigenfunctions of the problem in (6) are

$$\mu_n = \frac{n^2 \pi^2}{b^2}, \quad Y_n(y) = \sin \frac{n\pi y}{b} \tag{8}$$

for $n = 1, 2, 3, \dots$. We use distinct indices m and n in (7) and (8) because the two problems in Eqs. (5) and (6) are independent of each other.

For each pair m, n of positive integers, we must solve the third equation in (4),

$$T'_{mn} = -(\lambda_m + \mu_n)kT_{mn} = -\left(\frac{m^2}{a^2} + \frac{n^2}{b^2}\right)\pi^2 k T_{mn}. \tag{9}$$

To within a multiplicative constant, the solution of Eq. (9) is

$$T_{mn}(t) = \exp\left(-\gamma_{mn}^2 kt\right), \tag{10}$$

where

$$\gamma_{mn} = \pi \sqrt{\frac{m^2}{a^2} + \frac{n^2}{b^2}}. \tag{11}$$

Thus we have found a "doubly infinite" collection of building blocks, and it follows that the "doubly infinite" series

$$u(x, y, t) = \sum_{m=1}^{\infty} \sum_{n=1}^{\infty} c_{mn} \exp\left(-\gamma_{mn}^2 kt\right) \sin \frac{m\pi x}{a} \sin \frac{n\pi y}{b} \tag{12}$$

formally satisfies the heat equation in (1) and the homogeneous boundary conditions in (2).

It remains only to determine the coefficients $\{c_{mn}\}$ so that the series satisfies also the nonhomogeneous condition

$$u(x, y, 0) = \sum_{m=1}^{\infty} \sum_{n=1}^{\infty} c_{mn} \sin \frac{m\pi x}{a} \sin \frac{n\pi y}{b} = f(x, y). \tag{13}$$

To do so, we first group the terms in this *two-dimensional Fourier series* to display the total coefficient of $\sin(n\pi y/b)$ and write

$$f(x, y) = \sum_{n=1}^{\infty} \left(\sum_{m=1}^{\infty} c_{mn} \sin \frac{m\pi x}{a} \right) \sin \frac{n\pi y}{b}. \tag{14}$$

For each fixed x, we want the series in (14) to be the Fourier sine series of $f(x, y)$ on $0 \leq y \leq b$. This will be true provided that

$$\sum_{m=1}^{\infty} c_{mn} \sin \frac{m\pi x}{a} = \frac{2}{b} \int_0^b f(x, y) \sin \frac{n\pi y}{b} \, dy. \tag{15}$$

The right-hand side in (15) is, for each n, a function $F_n(x)$; viz.,

$$F_n(x) = \sum_{m=1}^{\infty} c_{mn} \sin \frac{m\pi x}{a}. \tag{16}$$

This requires that c_{mn} be the mth Fourier sine coefficient of $F_n(x)$ on $0 \leq x \leq a$; that is, that

$$c_{mn} = \frac{2}{a} \int_0^a F_n(x) \sin \frac{m\pi x}{a} \, dx. \tag{17}$$

Substituting the right-hand side in Eq. (15) for $F_n(x)$ in Eq. (17), we finally get

$$c_{mn} = \frac{4}{ab} \int_0^a \int_0^b f(x, y) \sin \frac{m\pi x}{a} \sin \frac{n\pi y}{b} \, dy \, dx \tag{18}$$

for $m, n = 1, 2, 3, \ldots$. With these coefficients, the series in (13) is the **double Fourier sine series** of $f(x, y)$ on the rectangle $0 \leq x \leq a$, $0 \leq y \leq b$, and the series in (12) formally satisfies the boundary value problem in (1)–(3). ■

PROBLEM 1 Suppose that $f(x, y) = u_0$, a constant. Compute the coefficients in (18) to obtain the solution

$$u(x, y, t) = \frac{16u_0}{\pi^2} \sum_{m \text{ odd}} \sum_{n \text{ odd}} \frac{\exp\left(-\gamma_{mn}^2 kt\right)}{mn} \sin \frac{m\pi x}{a} \sin \frac{n\pi y}{b}.$$

PROBLEM 2 Replace the boundary conditions (2) in Example 1 with

$$u(0, y, t) = u(x, 0, t) = 0,$$
$$hu(a, y, t) + u_x(a, y, t) = hu(x, b, t) + u_y(x, b, t) = 0.$$

Thus the edges $x = 0$ and $y = 0$ are still held at temperature zero, but now heat transfer takes place along the edges $x = a$ and $y = b$. Then derive the solution

$$u(x, y, t) = \sum_{m=1}^{\infty} \sum_{n=1}^{\infty} c_{mn} \exp\left(-\gamma_{mn}^2 kt\right) \sin \frac{\alpha_m x}{a} \sin \frac{\beta_n y}{b}$$

where $\gamma_{mn}^2 = (\alpha_m/a)^2 + (\beta_n/b)^2$, $\{\alpha_m\}$ are the positive roots of $ha \tan x = -x$, $\{\beta_n\}$ are the positive roots of $hb \tan x = -x$, and

$$c_{mn} = \frac{4}{A_m B_n} \int_0^a \int_0^b f(x, y) \sin \frac{\alpha_m x}{a} \sin \frac{\beta_n y}{b} \, dy \, dx,$$

where $A_m = (ha + \cos^2 \alpha_m)/h$ and $B_n = (hb + \cos^2 \beta_n)/h$.

PROJECT A Suppose that the three edges $x = 0$, $y = 0$, and $y = b$ of the plate of Example 1 are held at temperature zero, but the fourth edge $x = a$ is insulated, with corresponding boundary conditions

$$u(0, y, t) = u(x, 0, t) = u(x, b, t) = u_x(a, y, t) = 0.$$

If the plate's initial temperature is $u(x, y, 0) = f(x, y)$, show that its temperature function is given by

$$u(x, y, t) = \sum_{m=1}^{\infty} \sum_{n=1}^{\infty} c_{mn} \exp\left(-\gamma_{mn}^2 kt\right) \sin \frac{(2m-1)\pi x}{2a} \sin \frac{n\pi y}{b}, \qquad (19)$$

where

$$\left(\frac{\gamma_{mn}}{\pi}\right)^2 = \left(\frac{2m-1}{a}\right)^2 + \left(\frac{n}{b}\right)^2$$

and

$$c_{mn} = \frac{4}{ab} \int_0^a \int_0^b f(x, y) \sin \frac{(2m-1)\pi x}{2a} \sin \frac{n\pi y}{b} \, dy \, dx.$$

The result of Problem 21 in Section 9.3 may be useful. If $f(x, y) \equiv u_0$ (a constant), show that Eq. (2) yields

$$u(x, y, t) = \frac{16u_0}{\pi^2} \sum_{m=1}^{\infty} \sum_{n \text{ odd}} \frac{\exp\left(-\gamma_{mn}^2 kt\right)}{(2m-1)n} \sin \frac{(2m-1)\pi x}{2a} \sin \frac{n\pi y}{b}. \qquad (20)$$

For your personal plate to investigate numerically, take $u_0 = 100$, $a = 10p$, $b = 10q$, and $k = r/10$, where p and q are the two largest digits and r is the smallest nonzero digit in your student ID number. Plot $z = u(x, y, t)$ for typical values of t to verify that each such graph is symmetric with respect to the midline $y = b/2$, so it follows (why?) that the maximum temperature in the plate occurs at a point of this midline. Then determine (perhaps using the method of the Section 10.2 project)

- How long it takes for the maximum temperature on the edge $x = a$ to fall to $25°$;
- What is then the maximum interior temperature in the plate.

Vibrations of Rectangular Membranes

Now let's consider a two-dimensional flexible membrane whose equilibrium position occupies a region in the horizontal xy-plane. Suppose that this membrane vibrates up and down, with $u(x, y, t)$ denoting the vertical displacement of the point (x, y) of the membrane at time t. If T and ρ denote the membrane's tension and density (per unit area), respectively, then under standard assumptions it follows that its displacement function $u(x, y, t)$ satisfies the **two-dimensional wave equation**

$$\frac{\partial^2 u}{\partial t^2} = c^2 \nabla^2 u = c^2 \left(\frac{\partial^2 u}{\partial x^2} + \frac{\partial^2 u}{\partial y^2}\right), \qquad (21)$$

where $c^2 = T/\rho$.

PROBLEM 3 Suppose that the rectangular membrane $0 \leq x \leq a$, $0 \leq y \leq b$ is released from rest with given initial displacement $u(x, y, 0) = f(x, y)$. If the four edges of the membrane are held fixed thereafter with zero displacement, then the displacement function $u(x, y, t)$ satisfies the boundary value problem consisting of the wave equation in (21) and the boundary conditions

$$u(0, y, t) = u(a, y, t) = u(x, 0, t) = u(x, b, t) = 0,$$

$$u(x, y, 0) = f(x, y) \quad \text{(initial position)}, \tag{22}$$

$$u_t(x, y, 0) = 0 \qquad \text{(initial velocity)}.$$

Derive the solution

$$u(x, y, t) = \sum_{m=1}^{\infty} \sum_{n=1}^{\infty} c_{mn} \cos \gamma_{mn} c t \sin \frac{m\pi x}{a} \sin \frac{n\pi y}{b}, \tag{23}$$

where the numbers $\{\gamma_{mn}\}$ and the coefficients $\{c_{mn}\}$ are given by the formulas in (11) and (18), respectively.

The *mn*th term in Eq. (23) defines the rectangular membrane's *mn*th *natural mode of oscillation*, with displacement function

$$u_{mn}(x, y, t) = \sin \frac{m\pi x}{a} \sin \frac{n\pi y}{b} \cos \gamma_{mn} c t. \tag{24}$$

In this mode the membrane travels up and down between the (imaginary) surfaces

$$u = \pm \sin \frac{m\pi x}{a} \sin \frac{n\pi y}{b}$$

with circular frequency $\omega_{mn} = \gamma_{mn} c$ of oscillation. Figure 10.5.1 illustrates how these surfaces look for typical small values of m and n. If $c = 1$ and $a = b = \pi$ (for instance), then the successive frequencies

$$\omega_{12} = \omega_{21} = \sqrt{5} \approx 2.24, \qquad \omega_{22} = \sqrt{8} \approx 2.83,$$

$$\omega_{13} = \omega_{31} = \sqrt{10} \approx 3.16, \qquad \omega_{23} = \omega_{32} = \sqrt{13} \approx 3.61,$$

$$\omega_{33} = \sqrt{18} \approx 4.24, \qquad \cdots$$

are not integral multiples of the fundamental frequency $\omega_{11} = \sqrt{2} \approx 1.41$. This correctly suggests that the sound of a vibrating rectangular membrane is not harmonious, and therefore is usually perceived as noise rather than music.

PROBLEM 4 Suppose that the membrane in Problem 3 is a square tambourine lying upright and crosswise in a pickup truck that hits a brick wall at time $t = 0$. Then the membrane is set in motion with zero initial displacement and constant initial velocity, so the initial conditions are

$$u(x, y, 0) = 0, \quad u_t(x, y, 0) = v_0 \quad \text{(a constant)}.$$

Then derive the solution

$$u(x, y, t) = \frac{16 v_0}{\pi^2 c} \sum_{m \text{ odd}} \sum_{n \text{ odd}} \frac{\sin \gamma_{mn} c t}{mn \gamma_{mn}} \sin \frac{m\pi x}{a} \sin \frac{n\pi y}{b}.$$

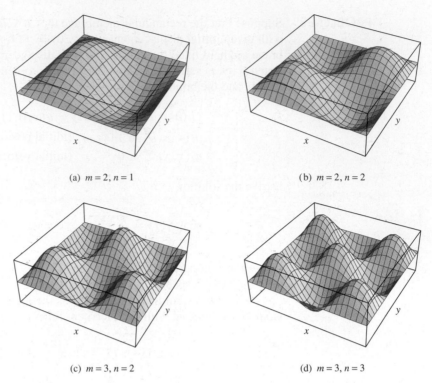

(a) $m = 2, n = 1$ (b) $m = 2, n = 2$

(c) $m = 3, n = 2$ (d) $m = 3, n = 3$

FIGURE 10.5.1. Surfaces of the form $u = \sin \dfrac{m\pi x}{a} \sin \dfrac{n\pi y}{b}$.

PROJECT B Suppose that the square membrane $0 \leq x \leq \pi$, $0 \leq y \leq \pi$ is plucked at its center point and set in motion from rest with the initial position function

$$u(x, y, 0) = f(x, y) = \min\{x, y, \pi - x, \pi - y\}, \tag{25}$$

whose graph over the square $0 \leq x \leq \pi$, $0 \leq y \leq \pi$ resembles a square-based tent or pyramid with height $\pi/2$ at its center. Thus the "tent function" $f(x, y)$ is the two-dimensional analogue of the familiar one-dimensional triangle function. It can be defined piecewise as indicated in Fig. 10.5.2.

Use a computer algebra system such as *Maple* or *Mathematica* to show that the double integral formula in (18) yields $c_{mn} = 0$ if $m \neq n$ and

$$c_{nn} = \frac{2[1 - (-1)^n]}{\pi n^2} = \begin{cases} \dfrac{4}{\pi n^2} & \text{for } n \text{ odd,} \\[2mm] 0 & \text{for } n \text{ even.} \end{cases}$$

It therefore follows from (23) that the membrane's resulting displacement function is given by

$$u(x, y, t) = \frac{4}{\pi} \sum_{n \text{ odd}} \frac{\sin nx \, \sin ny \, \cos nt\sqrt{2}}{n^2}. \tag{26}$$

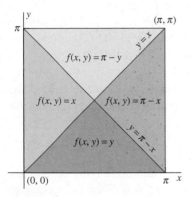

FIGURE 10.5.2. Piecewise definition of the tent function.

Is it clear that—because the summand in (26) contains no terms with $m \neq n$—the function $u(x, y, t)$ is periodic (in t) with period $\pi\sqrt{2}$? The fact that the tent function in (25) thus yields a "musical" vibration of a square membrane was first pointed out to us by John Polking. Figure 10.5.3 shows some typical snapshots of this vibration.

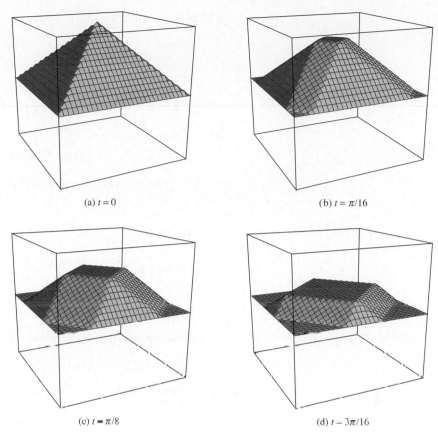

<div align="center">(a) $t = 0$ (b) $t = \pi/16$</div>

<div align="center">(c) $t = \pi/8$ (d) $t = 3\pi/16$</div>

FIGURE 10.5.3. Vibrations of the square membrane with period $P = \pi\sqrt{2}$.

Polar Coordinate Applications

FIGURE 10.5.4. The solid cylinder of Example 2.

In problems involving regions that enjoy circular symmetry around the origin in the plane (or the vertical z-axis in space), the use of polar (or cylindrical) coordinates is advantageous. In Section 9.7 we discussed the expression of the two-dimensional Laplacian in terms of the familiar plane polar coordinates (r, θ) for which $x = r\cos\theta$ and $y = r\sin\theta$. The three-dimensional Laplacian for a function $u(r, \theta, z)$ expressed in cylindrical coordinates is given by

$$\nabla^2 u = \frac{\partial^2 u}{\partial r^2} + \frac{1}{r}\frac{\partial u}{\partial r} + \frac{1}{r^2}\frac{\partial^2 u}{\partial \theta^2} + \frac{\partial^2 u}{\partial z^2}. \tag{27}$$

If u is independent of either θ or z then the corresponding second derivative is absent from the right-hand side in Eq. (27).

| Example 2 | Suppose that the solid cylinder $0 \leq r \leq c$, $0 \leq z \leq h$ of Fig. 10.5.4 is made of a uniform material with thermal diffusivity k and has an axially symmetric initial temperature $u(r, z, 0) = f(r, z)$ that is independent of θ. If the cylinder's boundary—consisting of the vertical cylindrical surface $r = c$ and its circular top and bottom—is thereafter held at temperature zero (perhaps because the cylinder is embedded in ice at time $t = 0$), then the resulting axially symmetric temperature function |

$u(r, z, t)$ satisfies the boundary value problem

$$\frac{\partial u}{\partial t} = k\nabla^2 u = k\left(\frac{\partial^2 u}{\partial r^2} + \frac{1}{r}\frac{\partial u}{\partial r} + \frac{\partial^2 u}{\partial z^2}\right), \tag{28}$$

$$u(c, z, t) = u(r, 0, t) = u(r, h, t) = 0, \tag{29}$$
$$\text{(boundary temperature zero)}$$

$$u(r, z, 0) = f(r, z). \tag{30}$$
$$\text{(given initial temperature)} \quad \blacksquare$$

The following two problems outline a solution of this heat problem for the case of constant initial temperature.

PROBLEM 5 Show that the substitution

$$u(r, z, t) = R(r)Z(z)T(t)$$

in Eq. (28) yields the separation of variables

$$r^2 R'' + r R' + \alpha^2 r^2 R = 0, \qquad R(c) = 0; \tag{31}$$
$$Z'' + \beta^2 Z = 0, \qquad\qquad Z(0) = Z(h) = 0; \tag{32}$$
$$T' = -(\alpha^2 + \beta^2)kT. \tag{33}$$

Note that the differential equation in (31) is the parametric Bessel equation of order zero for $R(r)$, and so (31) has the eigenvalues and eigenfunctions given in case 1 of the table in Fig. 10.4.2. Of course, the differential equations in (32) and (33) have familiar trigonometric and exponential solutions. Hence derive the formal series solution

$$u(r, z, t) = \sum_{m=1}^{\infty}\sum_{n=1}^{\infty} c_{mn} \exp\left(-\lambda_{mn}kt\right) J_0\left(\frac{\gamma_m r}{c}\right) \sin\frac{n\pi z}{h}, \tag{34}$$

where the $\{\gamma_m\}_1^{\infty}$ are the positive solutions of the equation $J_0(x) = 0$ (as listed in the Section 10.4 project),

$$\lambda_{mn} = \frac{\gamma_m^2}{c^2} + \frac{n^2\pi^2}{h^2}, \tag{35}$$

and the coefficients in Eq. (34) are given by the formula

$$c_{mn} = \frac{4}{hc^2\left[J_1(\gamma_m)\right]^2}\int_0^c\int_0^h rf(r, z)J_0\left(\frac{\gamma_m r}{c}\right)\sin\frac{n\pi z}{h}\,dz\,dr. \tag{36}$$

The derivation of Eq. (36) is similar to that of the double Fourier series coefficient formula in (18).

PROBLEM 6 If the initial temperature function is constant, $f(r, z) \equiv u_0$, deduce from Eqs. (34) and (36) that

$$u(r, z, t) = \frac{8u_0}{\pi}\sum_{m=1}^{\infty}\sum_{n\text{ odd}}\frac{\exp\left(-\lambda_{mn}kt\right)J_0\left(\frac{\gamma_m r}{c}\right)\sin\frac{n\pi z}{h}}{n\gamma_m J_1(\gamma_m)}, \tag{37}$$

where λ_{mn} is defined in Eq. (35).

PROJECT C For your personal heated cylinder to investigate numerically, take $u_0 = 100$, $h = 10p$, $c = 5q$, and $k = r/10$, where p and q are the two largest digits and r is the smallest nonzero digit in your student ID number. Assume zero boundary temperatures as in Problems 5 and 6. Plot the surface $z = u(r, z, t)$ for typical values of t to substantiate the physically plausible conjecture that, at any time t, the maximum temperature within the cylinder occurs at its center point, where $r = 0$ and $z = h/2$. It may help to plot also graphs of

1. u as a function of r with $z = h/2$ and t constant; and
2. u as a function of z with $r = 0$ and t constant.

Then determine how long it takes for the maximum temperature within the cylinder to fall to $25°$.

Repeat this investigation with the temperature held at zero on the cylinder's bottom and curved surface, but with its top now insulated, so that the boundary conditions in (29) are replaced with

$$u(c, z, t) = u(r, 0, t) = u_z(r, h, t) = 0.$$

Using (as in Project A) the result of Problem 21 in Section 9.3, you should find that (37) is replaced with

$$u(r, z, t) = \frac{8u_0}{\pi} \sum_{m=1}^{\infty} \sum_{n=1}^{\infty} \frac{\exp(-\lambda_{mn}kt) \, J_0\left(\frac{\gamma_m r}{c}\right) \sin\frac{(2n-1)\pi z}{2h}}{(2n-1)\gamma_m J_1(\gamma_m)}, \qquad (38)$$

where

$$\lambda_{mn} = \frac{\gamma_m^2}{c^2} + \frac{(2n-1)^2 \pi^2}{4h^2}.$$

Is it plausible that the cylinder's maximum temperature now always occurs at the center of its top? Substantiate this conjecture with appropriate graphical investigations. Also determine how long it takes for this maximum temperature to fall to $25°$.

PROJECT D This project deals with a vibrating uniform circular membrane of radius a. If the membrane's initial displacement and velocity functions depend on both polar coordinates r and θ, then the wave equation in (19) takes the polar coordinate form

$$\frac{\partial^2 u}{\partial t^2} = c^2 \nabla^2 u = c^2 \left(\frac{\partial^2 u}{\partial r^2} + \frac{1}{r}\frac{\partial u}{\partial r} + \frac{1}{r^2}\frac{\partial^2 u}{\partial \theta^2} \right). \qquad (39)$$

If the membrane is released from rest at time $t = 0$ and thereafter its boundary is held fixed (so on the circle $r = a$ we have the displacement u always zero), then the membrane's displacement function $u(r, \theta, t)$ satisfies both Eq. (39) and the boundary conditions

$$u(a, \theta, t) = 0 \qquad \text{(fixed boundary)}, \qquad (40)$$

$$u(r, \theta, 0) = f(r, \theta) \qquad \text{(given initial displacement)}, \qquad (41)$$

$$u_t(r, \theta, 0) = 0 \qquad \text{(zero initial velocity)}. \qquad (42)$$

Fill in the details in the following outline of a solution. Show first that the substitution

$$u(r, \theta, t) = R(r)\Theta(\theta)T(t)$$

in Eq. (39) yields the separation of variables

$$\frac{T''}{c^2 T} = \frac{R'' + \frac{1}{r}R'}{R} + \frac{\Theta''}{r^2\Theta} = -\alpha^2 \quad \text{(constant).} \tag{43}$$

Then

$$T'' + \alpha^2 c^2 T = 0, \quad T'(0) = 0 \tag{44}$$

implies that, to within a constant multiple,

$$T(t) = \cos \alpha c t. \tag{45}$$

Next, the right-hand equality in (43) yields the equation

$$\frac{r^2 R'' + r R'}{R} + \alpha^2 r^2 + \frac{\Theta''}{\Theta} = 0, \tag{46}$$

from which it follows that

$$\frac{\Theta''}{\Theta} = -\beta^2 \quad \text{(constant).} \tag{47}$$

In order that a solution $\Theta(\theta)$ of $\Theta'' + \beta^2\Theta = 0$ have the necessary periodicity of period 2π, the parameter β must be an integer, so we have the θ-solutions

$$\Theta_n(\theta) = \begin{cases} \cos n\theta, \\ \sin n\theta \end{cases} \tag{48}$$

for $n = 0, 1, 2, 3, \ldots$.

Substitution of $\Theta''/\Theta = -n^2$ in (46) now yields the parametric Bessel equation

$$r^2 R'' + r R' + (\alpha^2 r^2 - n^2)R = 0 \tag{49}$$

of order n, with bounded solution $J_n(\alpha r)$. Because the zero boundary condition in (40) yields $J_n(\alpha a) = 0$, case 1 in the table in Fig. 10.4.2 yields the r-eigenfunctions

$$R_{mn}(r) = J_n\left(\frac{\gamma_{mn}r}{a}\right) \qquad (m = 1, 2, 3, \ldots ; n = 0, 1, 2, \ldots), \tag{50}$$

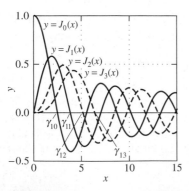

FIGURE 10.5.5. The graphs and initial zeros of the first several Bessel functions.

where γ_{mn} denotes the mth positive solution of the equation $J_n(x) = 0$. The numerical value of γ_{mn} can be approximated by first estimating the value $r_1 \approx \gamma_{1n}$ of the first root by inspecting a graph of $J_n(x)$ as in Fig. 10.5.5 and then using a command such as *Mathematica*'s

```
FindRoot[ BesselJ[n,x] == 0, {r, r1 + (m - 1)*Pi}]
```

Finally, substitution of $\alpha_{mn} = (\gamma_{mn})/a$ in (45) yields the t-function

$$T_{mn}(t) = \cos \frac{\gamma_{mn}ct}{a}. \tag{51}$$

Combining (48), (50), and (51), we see that our boundary value problem for the circular membrane released from rest has the formal series solution

$$u(r, \theta, t) = \sum_{m=1}^{\infty} \sum_{n=0}^{\infty} J_n\left(\frac{\gamma_{mn}r}{a}\right)(a_{mn}\cos n\theta + b_{mn}\sin n\theta)\cos\frac{\gamma_{mn}ct}{a}. \tag{52}$$

Thus the vibrating circular membrane's typical natural mode of oscillation with zero initial velocity is of the form

$$u_{mn}(r, \theta, t) = J_n\left(\frac{\gamma_{mn}r}{a}\right)\cos n\theta \cos\frac{\gamma_{mn}ct}{a} \tag{53}$$

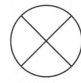

(a) $m = 2, n = 1$ (b) $m = 1, n = 2$

or the analogous form with $\sin n\theta$ in place of $\cos n\theta$. In this mode the membrane vibrates with $m - 1$ fixed *nodal circles* (in addition to its boundary $r = a$) with radii

$$r_{jn} = \frac{\gamma_{jn}a}{\gamma_{mn}} \qquad \text{for } j = 1, 2, \ldots, m - 1.$$

(c) $m = 2, n = 3$ (d) $m = 3, n = 2$

FIGURE 10.5.6. Typical nodal circles and radii for a vibrating circular membrane.

It also has $2n$ fixed *nodal radii* spaced at angles of π/n starting with $\theta = \pi/(2n)$. Figure 10.5.6 shows some typical configurations of these nodal circles and radii, which divide the circle into annular sectors that move alternately up and down as the membrane vibrates.

Formulas for the coefficients in (52) often are not required for practical applications involving vibrating circular membranes. We suggest that you explore vibrating membrane possibilities by graphing convenient linear combinations of eigenfunctions defined as in (53). For instance, Fig. 10.5.7 shows snapshots of the oscillation defined by

$$u(r, \theta, t) = J_1(\gamma_{21}r)\cos\theta\cos\gamma_{21}t + J_2(\gamma_{32}r)\cos 2\theta\cos\gamma_{32}t \tag{54}$$

for a circular membrane with $c = 1$ and radius $a = 1$. If multimedia hardware and software are available, you might enjoy constructing a movie showing such an oscillation.

Spherical Coordinate Applications

In problems involving regions that enjoy spherical symmetry around the origin in space, it is appropriate to use the spherical coordinates indicated in Fig. 10.5.8. The three-dimensional Laplacian for a function $u(\rho, \phi, \theta)$ expressed in spherical coordinates is given by

$$\nabla^2 u = \frac{1}{\rho^2}\left[\frac{\partial}{\partial \rho}\left(\rho^2\frac{\partial u}{\partial \rho}\right) + \frac{1}{\sin\phi}\frac{\partial}{\partial \phi}\left(\frac{\partial u}{\partial \phi}\sin\phi\right) + \frac{1}{\sin^2\phi}\frac{\partial^2 u}{\partial\theta^2}\right]. \tag{55}$$

Note that $\rho = \sqrt{x^2 + y^2 + z^2}$ denotes the distance of the point P from the origin O, ϕ is the angle down from the positive z-axis to OP, and θ is the ordinary polar coordinate angle in the xy-plane (although some texts reverse the roles of ϕ and θ). Observe also that if u is independent of either ρ, ϕ, or θ, then the corresponding second derivative term is missing on the right-hand side in Eq. (55).

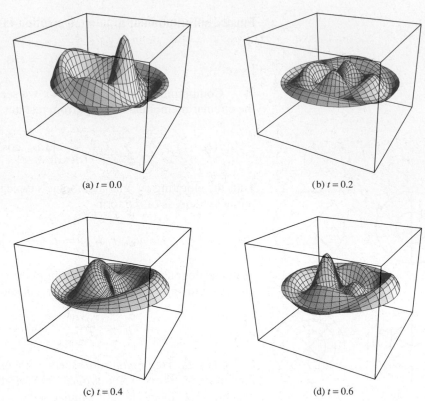

(a) $t = 0.0$ (b) $t = 0.2$

(c) $t = 0.4$ (d) $t = 0.6$

FIGURE 10.5.7. Snapshots of the circular membrane vibration defined by Eq. (54).

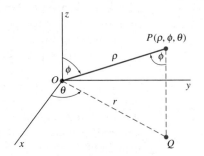

FIGURE 10.5.8. Finding the spherical coordinates of the point P.

Example 3	Suppose that a given axially symmetric temperature function $g(\phi)$ is imposed on the boundary sphere $\rho = a$ of the solid ball $0 \leqq \rho \leqq a$, and we seek the resulting axially symmetric steady-state temperature function $u(\rho, \phi)$ within the sphere. Because u is independent of θ, Laplace's equation $\nabla^2 u = 0$ takes (after multiplication by ρ^2 in Eq. (55)) the form

$$\frac{\partial}{\partial \rho}\left(\rho^2 \frac{\partial u}{\partial \rho}\right) + \frac{1}{\sin \phi}\frac{\partial}{\partial \phi}\left(\frac{\partial u}{\partial \phi}\sin \phi\right) = 0, \tag{56}$$

and we have the single boundary condition

$$u(a, \phi) = g(\phi) \qquad \text{(given boundary temperature)}. \tag{57}$$

∎

The following problems outline a solution of this boundary value problem to find the temperature function $u(\rho, \phi)$.

PROBLEM 7 Show that the substitution $u(\rho, \phi) = R(\rho)\Phi(\phi)$ in Eq. (56) yields the separation of variables

$$\rho^2 R'' + 2\rho R' - \lambda R = 0, \qquad (58)$$

$$[(\sin\phi)\Phi']' + \lambda(\sin\phi)\Phi = 0, \qquad (59)$$

where λ is the usual separation constant. We have no homogeneous boundary conditions to impose, but we do seek *continuous* functions $R(\rho)$ for $0 \le \rho \le a$ and $\Phi(\phi)$ for $0 \le \phi \le \pi$. Equation (58) is one for which the trial solution $R(\rho) = \rho^k$ suggests itself, but Eq. (59) appears to be completely unfamiliar.

PROBLEM 8 Show that the substitution

$$x = \cos\phi, \quad y(x) = \Phi(\cos^{-1} x) = \Phi(\phi) \qquad (-1 \le x \le 1)$$

in Eq. (59) yields the *Legendre equation*

$$(1 - x^2)y'' - 2xy + \lambda y = 0$$

that we discussed in Section 8.2. This equation has a solution $y(x)$ that is continuous for $1 \le x \le 1$ only if $\lambda = n(n+1)$, where n is a nonnegative integer. In this case $y(x)$ is a constant multiple of the nth *Legendre polynomial* $P_n(x)$. Thus we have eigenvalues and eigenfunctions of Eq. (59) given by

$$\lambda_n = n(n+1), \quad \Phi_n(\phi) - P_n(\cos\phi) \qquad (60)$$

for $n = 1, 2, 3, \ldots$. Recall from Section 8.2 that the first several Legendre polynomials are

$$P_0(x) \equiv 1, \qquad\qquad\qquad P_1(x) = x,$$

$$P_2(x) = \tfrac{1}{2}(3x^2 - 1), \qquad\qquad P_3(x) = \tfrac{1}{2}(5x^3 - 3x),$$

$$P_4(x) = \tfrac{1}{8}(35x^4 - 30x^2 + 3), \qquad\qquad \ldots.$$

PROBLEM 9 With $\lambda = n(n+1)$, Eq. (58) takes the form

$$\rho^2 R'' + 2\rho R' - n(n+1)R = 0.$$

Show that the trial solution $R(\rho) = \rho^k$ yields the general solution

$$R(\rho) = A\rho^n + \frac{B}{\rho^{n+1}}.$$

But continuity at $\rho = 0$ implies that $B = 0$ here, so it follows that the eigenfunction of Eq. (58) corresponding to $\lambda_n = n(n+1)$ is (a constant multiple of) $R_n(\rho) = \rho^n$. Thus we have found the building-block solutions

$$u_n(\rho, \phi) = \rho^n P_n(\cos\phi) \qquad (n = 0, 1, 2, \ldots)$$

of Laplace's equation in (56). In the usual fashion our next step is to write the formal series solution

$$u(\rho, \phi) = \sum_{n=0}^{\infty} b_n \rho^n P_n(\cos\phi). \qquad (61)$$

PROBLEM 10 It remains only to discuss the choice of coefficients in Eq. (61) to satisfy the nonhomogeneous condition

$$u(a, \phi) = g(\phi) = \sum_{n=0}^{\infty} b_n a^n P_n(\cos \phi).$$

With $x = \cos \phi$, $f(x) = g(\phi) = g(\cos^{-1} x)$, and $c_n = b_n a^n$, this equation takes the form of the *Fourier-Legendre series*

$$f(x) = \sum_{n=0}^{\infty} c_n P_n(x), \tag{62}$$

expressing the function $f(x)$ on $[-1, 1]$ in terms of Legendre polynomials. Given that the Legendre polynomials $\{P_n(x)\}_0^{\infty}$ are mutually orthogonal on $[-1, 1]$ with weight function $r(x) \equiv 1$, apply the formal eigenfunction series method of Section 10.1—multiplying each side in Eq. (62) by $P_k(x)$ and integrating termwise—to derive the *Fourier-Legendre coefficient formula*

$$c_n = \frac{\displaystyle\int_{-1}^{1} f(x) P_n(x) \, dx}{\displaystyle\int_{-1}^{1} [P_n(x)]^2 \, dx}.$$

But then the known integral

$$\int_{-1}^{1} [P_n(x)]^2 \, dx = \frac{2}{2n + 1}$$

gives

$$c_n = \frac{2n + 1}{2} \int_{-1}^{1} f(x) P_n(x) \, dx. \tag{63}$$

Show, finally, that this choice of coefficients yields the formal series solution

$$u(\rho, \phi) = \sum_{n=0}^{\infty} c_n \left(\frac{\rho}{a}\right)^n P_n(\cos \phi) \tag{64}$$

of the boundary value problem in Eqs. (56) and (57).

PROBLEM 11 Recall the notation
$$k!! = k(k - 2)(k - 4) \cdots 4 \cdot 2$$

for k even and
$$k!! = k(k - 2)(k - 4) \cdots 3 \cdot 1$$

for k odd. Use the integral

$$\int_0^1 P_n(x) \, dx = \begin{cases} 1 & \text{if } n = 0, \\[2mm] \frac{1}{2} & \text{if } n = 1, \\[2mm] 0 & \text{if } n = 2, 4, 6, \ldots, \\[2mm] (-1)^{(n-1)/2} \dfrac{(n - 2)!!}{(n + 1)!!} & \text{if } n = 3, 5, 7, \ldots \end{cases} \tag{65}$$

to derive the Fourier-Legendre series

$$s(x) = \tfrac{3}{2}P_1(x) - \tfrac{7}{8}P_3(x) + \tfrac{11}{16}P_5(x) - \tfrac{75}{128}P_7(x) + \cdots$$

$$= \tfrac{3}{2}P_1(x) + \sum_{\substack{n \text{ odd} \\ n \geq 3}} (-1)^{(n-1)/2}(2n+1)\frac{(n-2)!!}{(n+1)!!}P_n(x) \qquad (66)$$

of the square wave function

$$s(x) = \begin{cases} -1 & \text{if } -1 < x < 0, \\ +1 & \text{if } 0 < x < 1. \end{cases}$$

Figure 10.5.9 shows the graph of the partial sum with 25 terms of the series in Eq. (66), exhibiting a characteristic Gibbs phenomenon near $x = 0$.

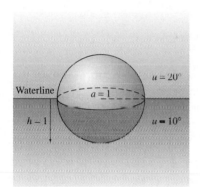

FIGURE 10.5.9. The graph of the 25-term partial sum of the rectangular wave series in Eq. (66).

PROJECT E

Figure 10.5.10 shows a solid spherical buoy of radius $a = 1$ meter floating in water. If the buoy has uniform specific gravity $\delta = 0.5$ (density half that of water), then it sinks to a depth $h = 1$ meter as shown. Suppose that the temperature of the water is $10°$ and that of the air is $20°$. Then we want to find the temperature function $u(\rho, \phi)$ within the buoy subject to the boundary condition

$$u(1, \phi) = g(\phi) = \begin{cases} 20 & \text{if } 0 < \phi < \pi/2, \\ 10 & \text{if } \pi/2 < \phi < \pi. \end{cases}$$

Then

$$f(x) = g(\cos^{-1}x) = \begin{cases} 10 & \text{if } -1 < x < 0 \\ 20 & \text{if } 0 < x < 1 \end{cases} = 15 + 5s(x)$$

in terms of the step function $s(x)$ of Problem 11. Hence use Eqs. (64) and (66) to show that

$$u(\rho, \phi) = 15 + \tfrac{15}{2}\rho P_1(\cos\phi)$$

$$+ \sum_{\substack{n \text{ odd} \\ n \geq 3}} 5 \cdot (-1)^{(n-1)/2}(2n+1)\frac{(n-2)!!}{(n+1)!!}\rho^n P_n(\cos\phi). \qquad (67)$$

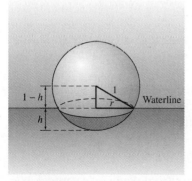

FIGURE 10.5.10. A floating spherical buoy with specific gravity $\delta = 0.5$.

Because the odd-degree Legendre polynomials have only terms of odd degree, it follow from Eq. (67) that the buoy has a constant average temperature of $15°$ on its water-level cross section (where $\phi = \pi/2$). As a first step in using the series in (67) to investigate the temperature distribution within the buoy, sum enough terms to plot u as a function $u(z)$ of height x ($-1 \leq z \leq 1$) on the vertical axis of symmetry of the buoy. Knowing that $u(0) = 15$, find the numerical values of z for which $u = 12.5°$ and $u = 17.5°$.

The buoy problem is more interesting if the specific gravity of the buoy is less than half that of water, $0 < \delta < 0.5$, so the buoy lies higher in the water as in Fig. 10.5.11. Archimedes' law of buoyancy implies that the buoy floats with the fraction δ of its total volume $4\pi/3$ submerged. Hence the part of the buoy lying beneath the waterline has volume $V = 4\pi\delta/3$. But this volume also is given by the sphere-segment volume formula

FIGURE 10.5.11. The floating buoy with $\delta < 0.5$.

$$V = \frac{\pi h}{6}(3r^2 + h^2).$$

Equate the two expressions for V to show that the depth h to which the buoy sinks in the water is a solution of the cubic equation

$$h^3 - 3h^2 + 4\delta = 0. \tag{68}$$

With your own choice of δ—perhaps one-tenth the smallest nonzero digit in your student ID number—graph this equation to see that it has a single root in the interval $0 < h < 1$. Approximate this root h accurate to at least two decimal places. Then the boundary condition (water temperature $10°$, air temperature $20°$) corresponds to

$$f(x) = \begin{cases} 10 & \text{if } -1 < x < h - 1, \\ 20 & \text{if } h - 1 < x < +1, \end{cases}$$

so (63) gives the formula

$$c_n = \int_{-1}^{h-1} 10 P_n(x)\, dx + \int_{h-1}^{1} 20 P_n(x)\, dx \tag{69}$$

for the coefficients $\{c_n\}_0^\infty$ in the series

$$u(\rho, \phi) = \sum_{n=0}^{\infty} c_n \rho^n P_n(\cos\phi). \tag{70}$$

Now the questions—as to the temperature at the waterline level within the buoy, and as to where within the buoy the temperature is exactly $12.5°$, $15°$, and $17.5°$—are more interesting. If $h \neq 0$ then numerical integration is necessary to evaluate the integrals in (69). For instance, the *Mathematica* command

```
c[n] = ((2*n + 1)/2)*
        (10*NIntegrate[LegendreP[n,x], {x, -1, h - 1}] +
        20*NIntegrate[LegendreP[n,x], {x, h - 1, 1}])
```

can be used. Explain why Eq. (70) implies that the temperature $u(z)$ at the point $(0, 0, z)$ (on the buoy's vertical axis of symmetry with $-1 \leqq z \leqq 1$) is then given by

$$u(z) = \sum_{n=0}^{\infty} c_n z^n. \tag{71}$$

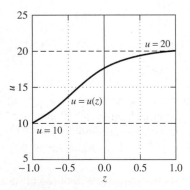

FIGURE 10.5.12. The temperature $u(z)$ at the point z on the vertical axis of the buoy with $h = 0.5$.

With δ chosen so that $h = 0.5$, the partial sum consisting of 50 terms in (71) was used to plot the graph of $u(z)$ shown in Fig. 10.5.12, where we see the temperature increasing nonlinearly from $u(-1) = 10$ to $u(1) = 20$. Using the *Mathematica* **FindRoot** function we find that $u(-0.6335) \approx 12.5$, $u(-0.3473) \approx 15$, and $u(0) = 17.5$.

Now the temperature at the center point of the buoy's cross section at the waterline is $u(-0.5) \approx 13.6603$ (rather than the average temperature of $15°$ that one might naively expect). In Fig. 10.5.13 we see that

$$\rho = \sqrt{x^2 + 0.25} \quad \text{and} \quad \phi = \pi - \arctan|2x|$$

at a point of the waterline cross section at distance x from the buoy's vertical axis. The graph of u (now as a function of x) shown in Fig. 10.5.14 was plotted by substituting these expressions for ρ and ϕ in a 50-term partial sum of the series in (67). Can you see intuitively why, as x increases starting at zero, the temperature $u(x)$ first decreases and then rises rapidly to $u = 15°$ at the boundary circle of the waterline cross section?

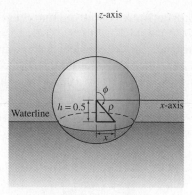

FIGURE 10.5.13. The waterline cross section with $z = -0.5$, $-\frac{\sqrt{3}}{2} \leq x \leq \frac{\sqrt{3}}{2}$.

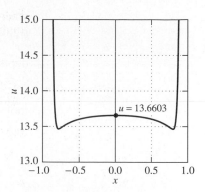

FIGURE 10.5.14. The temperature $u(x)$ in the waterline cross section of the buoy with $h = 0.5$.

Spherical Harmonics and the Book Cover

Equation (55) gives the three-dimensional Laplacian in spherical coordinates. For instance, consider radial vibrations of the surface of an elastic spherical planet of radius c. If $u(\phi, \theta, t)$ denotes the radial displacement at time t of the point (ϕ, θ) of the surface $\rho = c$ of the planet, then the wave equation $u_{tt} = a^2 \nabla^2 u$ (with $\rho = c$ and $u_\rho \equiv 0$) takes the form

$$\frac{\partial^2 u}{\partial t^2} = b^2 \nabla^2_{\phi\theta} u \tag{72}$$

where $b = a/c$ and

$$\nabla^2_{\phi\theta} u = \frac{1}{\sin\phi} \frac{\partial}{\partial\phi}\left((\sin\phi)\frac{\partial u}{\partial\phi}\right) + \frac{1}{\sin^2\phi}\frac{\partial^2 u}{\partial\theta^2}. \tag{73}$$

Alternatively, Eq. (72) models the oscillations of tidal waves on the surface of a spherical planet of radius c. In this case, $u(\phi, \theta, t)$ denotes the radial displacement (from equilibrium) of the water surface at the point (ϕ, θ) at time t and $b^2 = gh/c^2$, where h is the average depth of the water and g denotes gravitational acceleration at the surface of the planet.

In the applications manual that accompanies this text, we show that the separation of variables

$$u(\phi, \theta, t) = Y(\phi, \theta)T(t)$$

in Eq. (72) leads to eigenfunctions of the typical form

$$u_{mn}(\phi, \theta, t) = Y_{mn}(\phi, \theta)\cos\omega_n t \tag{74}$$

for $0 \leq m \leq n = 1, 2, 3, \ldots$. The frequency of this oscillation is $\omega_n = b\sqrt{n(n+1)}$ and Y_{mn} denotes the **spherical harmonic** defined by

$$Y_{mn}(\phi, \theta) = P_n^m(\cos\phi)\cos m\theta \tag{75}$$

in terms of the so-called **associated Legendre function**

$$P_n^m(x) = (1 - x^2)^{m/2} P_n^{(m)}(x), \tag{76}$$

where the mth derivative of the ordinary Legendre polynomial $P_n(x)$ appears on the right.

For instance, let us consider water waves with a (quite unrealistic) average depth of $h = 2$ sloshing about on the surface of a small spherical planet of radius $c = 5$. In Figs. 10.5.15 and 10.5.16 we show a couple of typical water surface shapes of the form $\rho = c + hY_{mn}(\phi, \theta)$. For the figure on the cover of this book we took $m = 3$ and $n = 9$.

FIGURE 10.5.15. With $m = 1$ and $n = 5$.

FIGURE 10.5.16. With $m = 2$ and $n = 6$.

REFERENCES FOR FURTHER STUDY

The literature of the theory and applications of differential equations is vast. The following list includes a selection of books that might be useful to readers who wish to pursue further the topics introduced in this book.

1. ABRAMOWITZ, M. and I. A. STEGUN, *Handbook of Mathematical Functions*. New York: Dover, 1965. The comprehensive collection of tables to which frequent reference is made in the text.

2. BIRKHOFF, G. and G.-C. ROTA, *Ordinary Differential Equations* (2nd ed.). New York: John Wiley, 1969. An intermediate-level text that includes more complete treatment of existence and uniqueness theorems, Sturm-Liouville problems, and eigenfunction expansions.

3. BRAUN, M., *Differential Equations and Their Applications* (3rd ed.). New York: Springer-Verlag, 1983. An introductory text at a slightly higher level than this book, it includes several interesting "case study" applications.

4. CHURCHILL, R. V., *Operational Mathematics* (3rd ed.). New York: McGraw-Hill, 1972. The standard reference for theory and applications of Laplace transforms, starting at about the same level as Chapter 7 of this book.

5. CHURCHILL, R. V. and J. W. BROWN, *Fourier Series and Boundary Value Problems* (3rd ed.). New York: McGraw-Hill, 1978. At about the same level as Chapters 9 and 10 of this book.

6. CODDINGTON, E. A., *An Introduction to Ordinary Differential Equations*. Englewood Cliffs, N.J.: Prentice Hall, 1961. An intermediate-level introduction; Chapters 3 and 4 include proofs of the theorems on power series and Frobenius series solutions stated in Chapter 8 of this book.

7. CODDINGTON, E. A. and N. LEVINSON, *Theory of Ordinary Differential Equations*. New York: McGraw-Hill, 1955. An advanced theoretical text; Chapter 5 discusses solutions near an irregular singular point.

8. DORMAND, J. R., *Numerical Methods for Differential Equations*. Boca Raton: CRC Press, 1996. More complete coverage of modern computational methods for approximate solution of differential equations.

9. HABERMAN, R., *Elementary Applied Partial Differential Equations* (3rd ed.). Upper Saddle River, N.J.: Prentice Hall, 1998. A next step beyond Chapters 9 and 10 of this book, but still quite accessible.

10. HUBBARD, J. H. and B. H. WEST, *Differential Equations: A Dynamical Systems Approach*. New York: Springer-Verlag, 1992 (part I) and 1995 (Higher-Dimensional Systems). Detailed treatment of qualitative phenomena, with a balanced combination of computational and theoretical viewpoints.

11. INCE, E. L., *Ordinary Differential Equations*. New York: Dover, 1956. First published in 1926, this is the classic older reference work on the subject.

12. LEBEDEV, N. N., *Special Functions and Their Applications*. New York: Dover, 1972. A comprehensive account of Bessel functions and the other special functions of mathematical physics.

13. LEBEDEV, N. N., I. P. SKALSKAYA, and Y. S. UFLYAND, *Worked Problems in Applied Mathematics*. New York: Dover, 1979. A large collection of applied examples and problems similar to those discussed in Chapter 10 of this book.

14. MCLACHLAN, N. W., *Bessel Functions for Engineers* (2nd ed.). London: Oxford University Press, 1955. Includes numerous physical applications of Bessel functions.

15. MCLACHLAN, N. W., *Ordinary Non-Linear Differential Equations in Engineering and Physical Sciences*. London: Oxford University Press, 1956. A concrete introduction to the effects of nonlinearity in physical systems.

16. POLKING, J. C. and D. Arnold, *Ordinary Differential Equations Using MATLAB* (2nd ed.). Upper Saddle River, N.J.: Prentice Hall, 1999. A manual for using MATLAB in an elementary differential equations course; based on the MATLAB programs `dfield` and `pplane` that are used and referenced in this text.

17. PRESS, W. H., B. P. FLANNERY, S. A. TEUKOLSKY, and W. T. VETTERLING, *Numerical Recipes: The Art of Scientific Computing*. Cambridge: Cambridge University Press, 1986. Chapter 15 discusses modern techniques for the numerical solution of differential equations. This edition include FORTRAN programs; Pascal and C editions have also been published.

18. RAINVILLE, E., *Intermediate Differential Equations* (2nd ed.). New York: Macmillan, 1964. Chapters 3 and 4 include proofs of the theorems on power series and Frobenius series solutions stated in Chapter 8 of this book.

19. SAGAN, H., *Boundary and Eigenvalue Problems in Mathematical Physics*. New York: John Wiley, 1961. Discusses the classical boundary value problems and the variational approach to Sturm-Liouville problems, eigenvalues, and eigenfunctions.

20. SIMMONS, G. F., *Differential Equations*. New York: McGraw-Hill, 1972. An introductory text with interesting historical notes and fascinating applications and with the most eloquent preface in any mathematics book currently in print.

21. TOLSTOV, G. P., *Fourier Series*. New York: Dover, 1976. An introductory text including detailed discussion of both convergence and applications of Fourier series.

22. THOMPSON, J. M. T. AND H. B. STEWART, *Nonlinear Dynamics and Chaos*. New York: John Wiley, 1986. Includes more detailed discussions of the forced Duffing, Lorenz, and Rossler systems (among others that exhibit nonlinear chaos phenomena).

23. WEINBERGER, H. F., *A First Course in Partial Differential Equations*. New York: Blaisdell, 1965. Includes separation of variables, Sturm-Liouville methods, and applications of Laplace transform methods to partial differential equations.

24. WEINSTOCK, R., *Calculus of Variations*. New York: Dover, 1974. Includes variational derivations of the partial differential equations of vibrating strings, membranes, rods, and bars.

APPENDIX

Existence and Uniqueness of Solutions

In Chapter 1 we saw that an initial value problem of the form

$$\frac{dy}{dx} = f(x, y), \quad y(a) = b \tag{1}$$

can fail (on a given interval containing the point $x = a$) to have a unique solution. For instance, the answer to Problem 33 in Section 1.3 says that the initial value problem

$$x^2 \frac{dy}{dx} + y^2 = 0, \quad y(0) = b \tag{2}$$

has no solutions at all unless $b = 0$, in which case there are infinitely many solutions. According to Problem 31 of Section 1.3, the initial value problem

$$\frac{dy}{dx} = -\sqrt{1 - y^2}, \quad y(0) = 1 \tag{3}$$

has the two distinct solutions $y_1(x) \equiv 1$ and $y_2(x) = \cos x$ on the interval $0 \leq x \leq \pi$. In this appendix we investigate conditions on the function $f(x, y)$ that suffice to guarantee that the initial value problem in (1) has one and only one solution, and then proceed to establish appropriate versions of the existence-uniqueness theorems that were stated without proof in Sections 1.3, 3.1, 3.2, and 4.1.

A.1 Existence of Solutions

The approach we employ is the **method of successive approximations**, which was developed by the French mathematician Emile Picard (1856–1941). This method is based on the fact that the function $y(x)$ satisfies the initial value problem in (1) on the open interval I containing $x = a$ if and only if it satisfies the integral equation

$$y(x) = b + \int_a^x f(t, y(t)) \, dt \tag{4}$$

for all x in I. In particular, if $y(x)$ satisfies Eq. (4), then clearly $y(a) = b$, and differentiation of both sides in (4)—using the fundamental theorem of calculus—yields the differential equation $y'(x) = f(x, y(x))$.

To attempt to solve Eq. (4), we begin with the initial function

$$y_0(x) \equiv b, \tag{5}$$

and then define iteratively a sequence y_1, y_2, y_3, ... of functions that we hope will converge to the solution. Specifically, we let

$$y_1(x) = b + \int_a^x f(t, y_0(t)) \, dt \quad \text{and} \quad y_2(x) = b + \int_a^x f(t, y_1(t)) \, dt. \tag{6}$$

In general, y_{n+1} is obtained by substitution of y_n for y in the right-hand side in Eq. (4):

$$y_{n+1}(x) = b + \int_a^x f(t, y_n(t)) \, dt. \tag{7}$$

Suppose we know that each of these functions $\{y_n(x)\}_0^\infty$ is defined on some open interval (the same for each n) containing $x = a$, and that the limit

$$y(x) = \lim_{n \to \infty} y_n(x) \tag{8}$$

exists at each point of this interval. Then it will follow that

$$y(x) = \lim_{n \to \infty} y_{n+1}(x) = \lim_{n \to \infty} \left[b + \int_a^x f(t, y_n(t)) \, dt \right]$$

$$= b + \lim_{n \to \infty} \int_a^x f(t, y_n(t)) \, dt \tag{9}$$

$$= b + \int_a^x f\left(t, \lim_{n \to \infty} y_n(t)\right) dt \tag{10}$$

and hence that

$$y(x) = b + \int_a^x f(t, y(t)) \, dt,$$

provided that we can validate the interchange of limit operations involved in passing from (9) to (10). It is therefore reasonable to expect that, under favorable conditions, the sequence $\{y_n(x)\}$ defined iteratively in Eqs. (5) and (7) will converge to a solution $y(x)$ of the integral equation in (4), and hence to a solution of the original initial value problem in (1).

| **Example 1** | To apply the method of successive approximations to the initial value problem |

$$\frac{dy}{dx} = y, \quad y(0) = 1, \tag{11}$$

we write Eqs. (5) and (7), thereby obtaining

$$y_0(x) \equiv 1, \quad y_{n+1}(x) = 1 + \int_0^x y_n(t) \, dt. \tag{12}$$

The iteration formula in (12) yields

$$y_1(x) = 1 + \int_0^x 1 \, dt = 1 + x,$$

$$y_2(x) = 1 + \int_0^x (1 + t) \, dt = 1 + x + \tfrac{1}{2}x^2,$$

$$y_3(x) = 1 + \int_0^x \left(1 + t + \tfrac{1}{2}t^2\right) dt = 1 + x + \tfrac{1}{2}x^2 + \tfrac{1}{6}x^3,$$

and

$$y_4(x) = 1 + \int_0^x \left(1 + t + \tfrac{1}{2}t^2 + \tfrac{1}{6}t^3\right) dt$$

$$= 1 + x + \tfrac{1}{2}x^2 + \tfrac{1}{6}x^3 + \tfrac{1}{24}x^4.$$

It is clear that we are generating the sequence of partial sums of a power series solution; indeed, we immediately recognize the series as that of $y(x) = e^x$. There is no difficulty in demonstrating that the exponential function is indeed the solution of the initial value problem in (11); moreover, a diligent student can verify (using a proof by induction on n) that $y_n(x)$, obtained in the aforementioned manner, is indeed the nth partial sum for the Taylor series with center zero for $y(x) = e^x$. ∎

Example 2 To apply the method of successive approximations to the initial value problem

$$\frac{dy}{dx} = 4xy, \qquad y(0) = 3, \tag{13}$$

we write Eqs. (5) and (7) as in Example 1. Now we obtain

$$y_0(x) \equiv 3, \quad y_{n+1}(x) = 3 + \int_0^x 4t y_n(t)\, dt. \tag{14}$$

The iteration formula in (14) yields

$$y_1(x) = 3 + \int_0^x (4t)(3)\, dt = 3 + 6x^2,$$

$$y_2(x) = 3 + \int_0^x (4t)(3 + 6t^2)\, dt = 3 + 6x^2 + 6x^4,$$

$$y_3(x) = 3 + \int_0^x (4t)(3 + 6t^2 + 6t^4)\, dt = 3 + 6x^2 + 6x^4 + 4x^6,$$

and

$$y_4(x) = 3 + \int_0^x (4t)(3 + 6t^2 + 6t^4 + 4t^6)\, dt$$

$$= 3 + 6x^2 + 6x^4 + 4x^6 + 2x^8.$$

It is again clear that we are generating partial sums of a power series solution. It is not quite so obvious what function has such a power series representation, but the initial value problem in (13) is readily solved by separation of variables:

$$y(x) = 3 \exp\left(2x^2\right) = 3 \sum_{n=0}^{\infty} \frac{(2x^2)^n}{n!}$$

$$= 3 + 6x^2 + 6x^4 + 4x^6 + 2x^8 + \tfrac{4}{5}x^{10} + \cdots. \quad ∎$$

In some cases it may be necessary to compute a much large number of terms, either in order to identify the solution or to use a partial sum of its series with large subscript to approximate the solution accurately for x near its initial value. Fortunately, computer algebra systems such as *Maple* and *Mathematica* can perform the

symbolic integrations (as opposed to numerical integrations) of the sort in Examples 1 and 2. If necessary, you could generate the first hundred terms in Example 2 in a matter of minutes.

In general, of course, we apply Picard's method because we cannot find a solution by elementary methods. Suppose that we have produced a large number of terms of what we believe to be the correct power series expansion of the solution. We *must* have conditions under which the sequence $\{y_n(x)\}$ provided by the method of successive approximations is guaranteed in advance to converge to a solution. It is just as convenient to discuss the initial value problem

$$\frac{d\mathbf{x}}{dt} = \mathbf{f}(\mathbf{x}, t), \quad \mathbf{x}(a) = \mathbf{b} \tag{15}$$

for a system of m first-order equations, where

$$\mathbf{x} = \begin{bmatrix} x_1 \\ x_2 \\ x_3 \\ \vdots \\ x_m \end{bmatrix}, \quad \mathbf{f} = \begin{bmatrix} f_1 \\ f_2 \\ f_3 \\ \vdots \\ f_m \end{bmatrix}, \quad \text{and} \quad \mathbf{b} = \begin{bmatrix} b_1 \\ b_2 \\ b_3 \\ \vdots \\ b_m \end{bmatrix}.$$

It turns out that with the aid of this vector notation (which we introduced in Section 5.1), most results concerning a single [scalar] equation $x' = f(x, t)$ can be generalized readily to analogous results for a system of m first-order equations, as abbreviated in (15). Consequently, the effort of using vector notation is amply justified by the generality it provides.

The method of successive approximations for the system in (15) calls for us to compute the sequence $\{\mathbf{x}_n(t)\}_0^\infty$ of vector-valued functions of t,

$$\mathbf{x}_n(t) = \begin{bmatrix} x_{1n}(t) \\ x_{2n}(t) \\ x_{3n}(t) \\ \vdots \\ x_{mn}(t) \end{bmatrix},$$

defined iteratively by

$$\mathbf{x}_0(a) \equiv \mathbf{b}, \quad \mathbf{x}_{n+1}(t) = \mathbf{b} + \int_a^t \mathbf{f}(\mathbf{x}_n(s), s)\, ds. \tag{16}$$

Recall that vector-valued functions are integrated componentwise.

Example 3 Consider the m-dimensional initial value problem

$$\frac{d\mathbf{x}}{dt} = \mathbf{A}\mathbf{x}, \quad \mathbf{x}(0) = \mathbf{b} \tag{17}$$

for a homogeneous linear system with $m \times m$ constant coefficient matrix \mathbf{A}. The equations in (16) take the form

$$\mathbf{x}_0(t) = \mathbf{b}, \quad \mathbf{x}_{n+1} = \mathbf{b} + \int_0^x \mathbf{A}\mathbf{x}_n(s)\, ds. \tag{18}$$

Thus

$$\mathbf{x}_1(t) = \mathbf{b} + \int_0^t \mathbf{A}\mathbf{b}\,ds = \mathbf{b} + \mathbf{A}\mathbf{b}t = (\mathbf{I} + \mathbf{A}t)\mathbf{b};$$

$$\mathbf{x}_2(t) = \mathbf{b} + \int_0^t \mathbf{A}(\mathbf{b} + \mathbf{A}\mathbf{b}s)\,ds = \mathbf{b} + \mathbf{A}\mathbf{b}t + \tfrac{1}{2}\mathbf{A}^2\mathbf{b}t^2 = (\mathbf{I} + \mathbf{A}t + \tfrac{1}{2}\mathbf{A}^2t^2)\mathbf{b}$$

and

$$\mathbf{x}_3(t) = \mathbf{b} + \int_0^t \mathbf{A}(\mathbf{b} + \mathbf{A}\mathbf{b}s + \tfrac{1}{2}\mathbf{A}^2\mathbf{b}s^2)\,ds = (\mathbf{I} + \mathbf{A}t + \tfrac{1}{2}\mathbf{A}^2t^2 + \tfrac{1}{6}\mathbf{A}^3t^3)\mathbf{b}.$$

We have therefore obtained the first several partial sums of the exponential series solution

$$\mathbf{x}(t) = e^{\mathbf{A}t}\mathbf{b} = \left(\sum_{n=0}^{\infty} \frac{(\mathbf{A}t)^n}{n!}\right)\mathbf{b} \tag{19}$$

of (17), which was derived earlier in Section 5.5. ∎

The key to establishing convergence in the method of successive approximations is an appropriate condition on the rate at which $\mathbf{f}(\mathbf{x}, t)$ changes when \mathbf{x} varies but t is held fixed. If R is a region in $(m + 1)$-dimensional (\mathbf{x}, t)-space, then the function $\mathbf{f}(\mathbf{x}, t)$ is said to be **Lipschitz continuous** on R if there exists a constant $k > 0$ such that

$$|\mathbf{f}(\mathbf{x}_1, t) - \mathbf{f}(\mathbf{x}_2, t)| \leqq k|\mathbf{x}_1 - \mathbf{x}_2| \tag{20}$$

if (\mathbf{x}_1, t) and (\mathbf{x}_2, t) are points of R. Recall that the **norm** of an m-dimensional point or vector \mathbf{x} is defined to be

$$|\mathbf{x}| = \sqrt{x_1^2 + x_2^2 + x_3^2 + \cdots + x_m^2}. \tag{21}$$

Then $|\mathbf{x}_1 - \mathbf{x}_2|$ is simply the Euclidean distance between the points \mathbf{x}_1 and \mathbf{x}_2.

Example 4 Let $f(x, t) = x^2 \exp\left(-t^2\right) \sin t$ and let R be the strip $0 \leqq x \leqq 2$ in the xy-plane. If (x_1, t) and (x_2, t) are both points of R, then

$$|f(x_1, t) - f(x_2, t)| = |\exp\left(-t^2\right) \sin t| \cdot |x_1 + x_2| \cdot |x_1 - x_2| \leqq 4|x_1 - x_2|,$$

because $\left|\exp\left(-t^2\right) \sin t\right| \leqq 1$ for all t and $|x_1 + x_2| \leqq 4$ if x_1 and x_2 are both in the interval $[0, 2]$. Thus f satisfies the Lipschitz condition in (20) with $k = 4$ and is therefore Lipschitz continuous in the strip R. ∎

Example 5 Let $f(x, t) = t\sqrt{x}$ on the rectangle R consisting of the points (x, t) in the xt-plane for which $0 \leqq x \leqq 1$ and $0 \leqq t \leqq 1$. Then, taking $x_1 = x$, $x_2 = 0$, and $t = 1$, we find that

$$|f(x, 1) - f(0, 1)| = \sqrt{x} = \frac{1}{\sqrt{x}}|x - 0|.$$

Because $x^{-1/2} \to +\infty$ as $x \to 0^+$, we see that the Lipschitz condition in (20) cannot be satisfied by any (finite) constant $k > 0$. Thus the function f, though obviously continuous on R, is *not* Lipschitz continuous on R. ∎

Suppose, however, that the function $f(x, t)$ has a continuous partial derivative $f_x(x, t)$ on the closed rectangle R in the xt-plane, and denote by k the maximum value of $|f_x(x, t)|$ on R. Then the mean value theorem of differential calculus yields

$$|f(x_1, t) - f(x_2, t)| = |f_x(\overline{x}, t) \cdot (x_1 - x_2)|$$

for some \overline{x} in (x_1, x_2), so it follows that

$$|f(x_1, t) - f(x_2, t)| \leqq k|x_1 - x_2|$$

because $|f_x(\overline{x}, t)| \leqq k$. Thus a continuously differentiable function $f(x, t)$ defined on a closed rectangle *is* Lipschitz continuous there. More generally, the multivariable mean value theorem of advanced calculus can be used similarly to prove that a *vector-valued function* $\mathbf{f}(\mathbf{x}, t)$ *with continuously differentiable component functions on a closed rectangular region R in (\mathbf{x}, t)-space is Lipschitz continuous on R.*

Example 6	The function $f(x, t) = x^2$ is Lipschitz continuous on any closed [bounded] region in the xt-plane. But consider this function on the infinite strip R consisting of the points (x, t) for which $0 \leqq t \leqq 1$ and x is arbitrary. Then

$$|f(x_1, t) - f(x_2, t)| = |x_1^2 \quad x_2^2| = |x_1 + x_2| \cdot |x_1 - x_2|.$$

Because $|x_1 + x_2|$ can be made arbitrarily large, it follows that f is *not* Lipschitz continuous on the infinite strip R. ∎

If I is an interval on the t-axis, then the set of all points (\mathbf{x}, t) with t in I is an infinite strip or slab in $(m + 1)$ space (as indicated in Fig. A.1). Example 6 shows that Lipschitz continuity of $\mathbf{f}(\mathbf{x}, t)$ on such an infinite slab is a very strong condition. Nevertheless, the existence of a solution of the initial value problem

$$\frac{d\mathbf{x}}{dt} = \mathbf{f}(\mathbf{x}, t), \quad \mathbf{x}(a) = \mathbf{b} \tag{15}$$

under the hypothesis of Lipschitz continuity of \mathbf{f} in such a slab is of considerable importance.

FIGURE A.1. An infinite slab in $(m + 1)$-space.

THEOREM 1: Global Existence of Solutions

Let \mathbf{f} be a vector-valued function (with m components) of $m + 1$ real variables, and let I be a [bounded or unbounded] open interval containing $t = a$. If $\mathbf{f}(\mathbf{x}, t)$ is continuous and satisfies the Lipschitz condition in (20) for all t in I and for all \mathbf{x}_1 and \mathbf{x}_2, then the initial value problem in (15) has a solution on the [entire] interval I.

Proof: We want to show that the sequence $\{\mathbf{x}_n(t)\}_0^\infty$ of successive approximations determined iteratively by

$$\mathbf{x}_0(a) = \mathbf{b}, \quad \mathbf{x}_{n+1}(t) = \mathbf{b} + \int_0^t \mathbf{f}(\mathbf{x}_n(s), s)\, ds \tag{16}$$

converges to a solution $\mathbf{x}(t)$ of (15). We see that each of these functions in turn is continuous on I, as each is an [indefinite] integral of a continuous function.

We may assume that $a = 0$, because the transformation $t \to t + a$ converts (15) into an equivalent problem with initial point $t = 0$. Also, we will consider only the portion $t \geqq 0$ of the interval I; the details for the case $t \leqq 0$ are very similar.

The main part of the proof consists in showing that if $[0, T]$ is a closed (and bounded) interval contained in I, then the sequence $\{\mathbf{x}_n(t)\}$ converges *uniformly* on $[0, T]$ to a limit function $\mathbf{x}(t)$. This means that, given $\epsilon > 0$, there exists an integer N such that

$$|\mathbf{x}_n(t) - \mathbf{x}(t)| < \epsilon \tag{22}$$

for all $n \geq N$ and all t in $[0, T]$. For ordinary (perhaps nonuniform) convergence the integer N, for which (22) holds for all $n \geq N$, may depend on t, with no single value of N working for all t in I. Once this uniform convergence of the sequence $\{\mathbf{x}_n(t)\}$ has been established, the following conclusions will follow from standard theorems of advanced calculus (see pages 620–622 of A. E. Taylor and W. R. Mann, *Advanced Calculus*, 3rd ed. (New York: John Wiley, 1983)):

1. The limit function $\mathbf{x}(t)$ is continuous on $[0, T]$.
2. If N is so chosen that the inequality in (22) holds for $n \geq N$, then the Lipschitz continuity of \mathbf{f} implies that

$$|\mathbf{f}(\mathbf{x}_n(t), t) - \mathbf{f}(\mathbf{x}(t), t)| \leq k|\mathbf{x}_n(t) - \mathbf{x}(t)| < k\epsilon$$

for all t in $[0, T]$ and $n \geq N$, so it follows that the sequence $\{\mathbf{f}(\mathbf{x}_n(t), t)\}_0^\infty$ converges uniformly to $\mathbf{f}(\mathbf{x}(t), t)$ on $[0, T]$.

3. But a uniformly convergent sequence or series can be integrated termwise, so it follows that, on taking limits in the iterative formula in (16),

$$\mathbf{x}(t) = \lim_{n \to \infty} \mathbf{x}_{n+1}(t) = \mathbf{b} + \lim_{n \to \infty} \int_0^t \mathbf{f}(\mathbf{x}_n(s), s)\, ds$$

$$= \mathbf{b} + \int_0^t \lim_{n \to \infty} \mathbf{f}(\mathbf{x}_n(s), s)\, ds;$$

thus

$$\mathbf{x}(t) = \mathbf{b} + \int_0^t \mathbf{f}(\mathbf{x}(s), s)\, ds. \tag{23}$$

4. Because the function $\mathbf{x}(t)$ is continuous on $[0, T]$, the integral equation in (23) (analogous to the one-dimensional case in (4)) implies that $\mathbf{x}'(t) = \mathbf{f}(\mathbf{x}(t), t)$ on $[0, T]$. But if this is true on every closed subinterval of the open interval I, then it is true on the entire interval I as well.

It therefore remains only to prove that the sequence $\{\mathbf{x}_n(t)\}_0^\infty$ converges uniformly on the closed interval $[0, T]$. Let M be the maximum value of $|\mathbf{f}(\mathbf{b}, t)|$ for t in $[0, T]$. Then

$$|\mathbf{x}_1(t) - \mathbf{x}_0(t)| = \left| \int_0^t \mathbf{f}(\mathbf{x}_0(s), s)\, ds \right| \leq \int_0^t |\mathbf{f}(\mathbf{b}, s)|\, ds \leq Mt. \tag{24}$$

Next,

$$|\mathbf{x}_2(t) - \mathbf{x}_1(t)| = \left| \int_0^t [\mathbf{f}(\mathbf{x}_1(s), s) - \mathbf{f}(\mathbf{x}_0(s), s)]\, ds \right| \leq k \int_0^t |\mathbf{x}_1(s) - \mathbf{x}_0(s)|\, ds,$$

and hence

$$|\mathbf{x}_2(t) - \mathbf{x}_1(t)| \leq k \int_0^t Ms\, ds = \tfrac{1}{2}kMt^2. \tag{25}$$

We now proceed by induction. Assume that

$$|\mathbf{x}_n(t) - \mathbf{x}_{n-1}(t)| \leqq \frac{M}{k} \cdot \frac{(kt)^n}{n!}. \tag{26}$$

It then follows that

$$|\mathbf{x}_{n+1}(t) - \mathbf{x}_n(t)| = \left| \int_0^t \mathbf{f}(\mathbf{x}_n(s), s) - \mathbf{f}(\mathbf{x}_{n-1}(s), s)] \, ds \right|$$

$$\leqq k \int_0^t |\mathbf{x}_n(s) - \mathbf{x}_{n-1}(s)| \, ds;$$

consequently,

$$|\mathbf{x}_{n+1}(t) - \mathbf{x}_n(t)| \leqq k \int_0^t \frac{M}{k} \cdot \frac{(ks)^n}{n!} \, ds.$$

It follows upon evaluating this integral that

$$|\mathbf{x}_{n+1}(t) - \mathbf{x}_n(t)| \leqq \frac{M}{k} \cdot \frac{(kt)^{n+1}}{(n+1)!}.$$

Thus (26) holds on $[0, T]$ for all $n \geqq 1$.

Hence the terms of the infinite series

$$\mathbf{x}_0(t) + \sum_{n=1}^{\infty} \left[\mathbf{x}_n(t) - \mathbf{x}_{n-1}(t) \right] \tag{27}$$

are dominated (in magnitude on the interval $[0, T]$) by the terms of the convergent series

$$\sum_{n=1}^{\infty} \frac{M}{k} \cdot \frac{(kT)^{n+1}}{(n+1)!} = \frac{M}{k}(e^{kT} - 1), \tag{28}$$

which is a series of positive constants. It therefore follows (from the Weierstrass M-test on pages 618–619 of Taylor and Mann) that the series in (27) converges uniformly on $[0, T]$. But the sequence of partial sums of this series is simply our original sequence $\{\mathbf{x}_n(t)\}_0^{\infty}$ of successive approximations, so the proof of Theorem 1 is finally complete. ▲

A.2 Linear Systems

An important application of the global existence theorem just given is to the initial value problem

$$\frac{d\mathbf{x}}{dt} = \mathbf{A}(t)\mathbf{x} + \mathbf{g}(t), \quad \mathbf{x}(a) = \mathbf{b} \tag{29}$$

for a linear system, where the $m \times m$ matrix-valued function $\mathbf{A}(t)$ and the vector-valued function $\mathbf{g}(t)$ are continuous on a [bounded or unbounded] open interval I containing the point $t = a$. In order to apply Theorem 1 to the linear system in (29), we note first that the proof of Theorem 1 requires only that, for each closed and bounded subinterval J of I, there exists a Lipschitz constant k such that

$$|\mathbf{f}(\mathbf{x}_1, t) - \mathbf{f}(\mathbf{x}_2, t)| \leqq k|\mathbf{x}_1 - \mathbf{x}_2| \tag{20}$$

for all t in J (and all \mathbf{x}_1 and \mathbf{x}_2). Thus we do not need a single Lipschitz constant for the entire open interval I.

In (29) we have $\mathbf{f}(\mathbf{x}, t) = \mathbf{A}(t)\mathbf{x} + \mathbf{g}$, so

$$\mathbf{f}(\mathbf{x}_1, t) - \mathbf{f}(\mathbf{x}_2, t) = \mathbf{A}(t)(\mathbf{x}_1 - \mathbf{x}_2). \tag{30}$$

It therefore suffices to show that, if $\mathbf{A}(t)$ is continuous on the closed and bounded interval J, then there is a constant k such that

$$|\mathbf{A}(t)\mathbf{x}| \leqq k|\mathbf{x}| \tag{31}$$

for all t in J. But this follows from the fact (Problem 17) that

$$|\mathbf{A}\mathbf{x}| \leqq \|\mathbf{A}\| \cdot |\mathbf{x}|, \tag{32}$$

where the **norm** $\|\mathbf{A}\|$ of the matrix \mathbf{A} is defined to be

$$\|\mathbf{A}\| = \left(\sum_{i,j=1}^{m} (a_{ij})^2 \right)^{1/2}. \tag{33}$$

Because $\mathbf{A}(t)$ is continuous on the closed and bounded interval J, the norm $\|\mathbf{A}\|$ is bounded on J, so Eq. (31) follows, as desired. Thus we have the following global existence theorem for the linear initial value problem in (29).

THEOREM 2: Existence for Linear Systems

Let the $m \times m$ matrix-valued function $\mathbf{A}(t)$ and the vector-valued function $\mathbf{g}(t)$ be continuous on the [bounded or unbounded] open interval I containing the point $t = a$. Then the initial value problem

$$\frac{d\mathbf{x}}{dt} = \mathbf{A}(t)\mathbf{x} + \mathbf{g}(t), \quad \mathbf{x}(a) = \mathbf{b} \tag{29}$$

has a solution on the [entire] interval I. ▲

As we saw in Section 4.1, the mth-order initial value problem

$$x^{(m)} + a_1(t)x^{(m-1)} + \cdots + a_{m-1}(t)x' + a_m(t)x = p(t),$$
$$x(a) = b_0, \quad x'(a) = b_1, \quad \ldots, \quad x^{(m-1)}(a) = b_{m-1} \tag{34}$$

is readily transformed into an equivalent $m \times m$ system of the form in (29). It therefore follows from Theorem 2 that if the functions $a_1(t), a_2(t), \ldots, a_m(t)$ and $p(t)$ in (34) are all continuous on the [bounded or unbounded] open interval I containing $t = a$, then the initial value problem in (34) has a solution on the [entire] interval I.

A.3 Local Existence

In the case of a *nonlinear* initial value problem

$$\frac{d\mathbf{x}}{dt} = \mathbf{f}(\mathbf{x}, t), \quad \mathbf{x}(a) = \mathbf{b}, \tag{35}$$

the hypothesis in Theorem 1 that \mathbf{f} satisfies a Lipschitz condition on a slab (\mathbf{x}, t) (t in I, all \mathbf{x}) is unrealistic and rarely satisfied. This is illustrated by the following simple example.

| Example 1 | Consider the initial value problem |

$$\frac{dy}{dx} = x^2, \quad x(0) = b > 0. \tag{36}$$

As we saw in Example 6, the equation $x' = x^2$ does not satisfy a "strip Lipschitz condition." When we solve (36) by separation of variables, we get

$$x(t) = \frac{b}{1 - bt}. \tag{37}$$

Because the denominator vanishes for $t = 1/b$, Eq. (37) provides a solution of the initial value problem in (36) only for $t < 1/b$, despite the fact that the differential equation $x' = x^2$ "looks nice" on the entire real line—certainly the function appearing on the right-hand side of the equation is continuous everywhere. In particular, if b is large, then we have a solution only on a very small interval to the right of $t = 0$. ∎

Although Theorem 2 assures us that *linear* equations have global solutions, Example 7 shows that, in general, even a "nice" nonlinear differential equation can be expected to have a solution only on a small interval around the initial point $t = a$, and that the length of this interval of existence can depend on the initial value $\mathbf{x}(a) = \mathbf{b}$, as well as on the differential equation itself. The reason is this. If $\mathbf{f}(\mathbf{x}, t)$ is continuously differentiable in a neighborhood of the point (\mathbf{b}, a) in $(m + 1)$-dimensional space, then—as indicated in the discussion preceding Example 6—we can conclude that $\mathbf{f}(\mathbf{x}, t)$ satisfies a Lipschitz condition on some rectangular region R centered at (\mathbf{b}, a), of the form

$$|t - a| < A, \quad |x_i - b_i| < B_i \tag{38}$$

($i = 1, 2, \ldots, m$). In the proof of Theorem 1, we need to apply the Lipschitz condition on the function \mathbf{f} in analyzing the iterative formula

$$\mathbf{x}_{n+1}(t) = \mathbf{b} + \int_a^t \mathbf{f}(\mathbf{x}_n(s), s) \, ds. \tag{39}$$

The potential difficulty is that unless the values of t are suitably restricted, then the point $(\mathbf{x}_n(t), t)$ appearing in the integrand in (39) may not lie in the region R where \mathbf{f} is known to satisfy a Lipschitz condition. On the other hand, it can be shown that—on a sufficiently small open interval J containing the point $t = a$—the graphs of the functions $\{\mathbf{x}_n(t)\}$ given iteratively by the formula in (39) remain within the region R, so the proof of convergence can then be carried out as in the proof of Theorem 1. A proof of the following *local* existence theorem can be found in Chapter 6 of G. Birkhoff and G.-C. Rota, *Ordinary Differential Equations*, 2nd ed. (New York: John Wiley, 1969).

THEOREM 3: Local Existence of Solutions

Let \mathbf{f} be a vector-valued function (with m components) of the $m + 1$ real variables x_1, x_2, \ldots, x_m, and t. If the first-order partial derivatives of \mathbf{f} all exist and are continuous in some neighborhood of the point $\mathbf{x} = \mathbf{b}$, $t = a$, then the initial value problem

$$\frac{d\mathbf{x}}{dt} = \mathbf{f}(\mathbf{x}, t), \quad \mathbf{x}(a) = \mathbf{b}, \tag{35}$$

has a solution on some open interval containing the point $t = a$. ▲

A.4 Uniqueness of Solutions

It is possible to establish the existence of solutions of the initial value problem in (35) under the much weaker hypothesis that $\mathbf{f}(\mathbf{x}, t)$ is merely continuous; techniques other than those used in this section are required. By contrast, the Lipschitz condition that we used in proving Theorem 1 is the key to *uniqueness* of solutions. In particular, the solution provided by Theorem 3 is unique near the point $t = a$.

THEOREM 4: Uniqueness of Solutions

Suppose that on some region R in $(m+1)$-space, the function \mathbf{f} in (35) is continuous and satisfies the Lipschitz condition

$$|\mathbf{f}(\mathbf{x}_1, t) - \mathbf{f}(\mathbf{x}_2, t)| \leqq k \cdot |\mathbf{x}_1 - \mathbf{x}_2|. \tag{20}$$

If $\mathbf{x}_1(t)$ and $\mathbf{x}_2(t)$ are two solutions of the initial problem in (35) on some open interval I containing $x = a$, such that the solution curves $(\mathbf{x}_1(t), t)$ and $(\mathbf{x}_2(t), t)$ both lie in R for all t in I, then $\mathbf{x}_1(t) = \mathbf{x}_2(t)$ for all t in I. ▲

We will outline the proof of Theorem 4 for the one-dimensional case in which x is a real variable. A generalization of this proof to the multivariable case can be found in Chapter 6 of Birkhoff and Rota.

Let us consider the function

$$\phi(t) = [x_1(t) - x_2(t)]^2 \tag{40}$$

for which $\phi(a) = 0$, because $x_1(a) = x_2(a) = b$. We want to show that $\phi(t) \equiv 0$, so that $x_1(t) \equiv x_2(t)$. We will consider only the case $t \geqq a$; the details are similar for the case $t \leqq a$.

If we differentiate each side in Eq. (40), we find that

$$\begin{aligned}
|\phi'(t)| &= |2[x_1(t) - x_2(t)] \cdot [x_1'(t) - x_2'(t)]| \\
&= |2[x_1(t) - x_2(t)] \cdot [f(x_1(t), t) - f(x_2(t), t)]| \\
&\leqq 2k|x_1(t) - x_2(t)|^2 = 2k\phi(t),
\end{aligned}$$

using the Lipschitz condition on f. Hence

$$\phi'(t) \leqq 2k\phi(t). \tag{41}$$

Now let us temporarily ignore the fact that $\phi(a) = 0$ and compare $\phi(t)$ with the solution of the differential equation

$$\Phi'(t) = 2k\Phi(t) \tag{42}$$

such that $\Phi(a) = \phi(a)$; clearly

$$\Phi(t) = \Phi(a)e^{2k(t-a)}. \tag{43}$$

In comparing (41) with (42), it seems inevitable that

$$\phi(t) \leqq \Phi(t) \qquad \text{for } t \geqq a, \tag{44}$$

and this is easily proved (Problem 18). Hence

$$0 \leqq [x_1(t) - x_2(t)]^2 \leqq [x_1(a) - x_2(a)]^2 e^{2k(t-a)}.$$

On taking square roots, we get

$$0 \leqq |x_1(t) - x_2(t)| \leqq |x_1(a) - x_2(a)|e^{k(t-a)}. \tag{45}$$

But $x_1(a) - x_2(a) = 0$, so (45) implies that $x_1(t) \equiv x_2(t)$.

Example 1 The initial value problem

$$\frac{dx}{dt} = 3x^{2/3}, \quad x(0) = 0 \tag{46}$$

has both the obvious solution $x_1(t) \equiv 0$ and the solution $x_2(t) = t^3$ that is readily found by separation of variables. Hence the function $f(x, t)$ must *fail* to satisfy a Lipschitz condition near $(0, 0)$. Indeed, the mean value theorem yields

$$|f(x, 0) - f(0, 0)| = |f_x(\overline{x}, 0)| \cdot |x - 0|$$

for some \overline{x} between 0 and x. But $f_x(x, 0) = 2x^{-1/3}$ is unbounded as $x \to 0$, so no Lipschitz condition can be satisfied. ∎

A.5 Well-Posed Problems and Mathematical Models

In addition to uniqueness, another consequence of the inequality in (45) is the fact that solutions of the differential equation

$$\frac{dx}{dt} = f(x, t) \tag{47}$$

depend *continuously* on the initial value $x(a)$; that is, if $x_1(t)$ and $x_2(t)$ are two solutions of (47) on the interval $a \leq t \leq T$ such that the initial values $x_1(a)$ and $x_2(a)$ are sufficiently close to one another, then the values of $x_1(t)$ and $x_2(t)$ remain close to one another. In particular, if $|x_1(a) - x_2(a)| \leqq \delta$, then (45) implies that

$$|x_1(t) - x_2(t)| \leq \delta e^{k(T-a)} = \epsilon \tag{48}$$

for all t with $a \leqq t \leqq T$. Obviously, we can make ϵ as small as we wish by choosing δ sufficiently close to zero.

This continuity of solutions of (47) with respect to initial values is important in practical applications where we are unlikely to know the initial value $x_0 = x(a)$ with absolute precision. For example, suppose that the initial value problem

$$\frac{dx}{dt} = f(x, t), \quad x(a) = x_0 \tag{49}$$

models a population for which we know only that the initial population is within $\delta > 0$ of the assumed value x_0. Then even if the function $f(x, t)$ is accurate, the solution $x(t)$ of (49) will be only an approximation to the actual population. But (45) implies that the actual population at time t will be within $\delta e^{k(T-a)}$ of the approximate population $x(t)$. Thus, on a given closed interval $[a, T]$, $x(t)$ will be a close approximation to the actual population provided that $\delta > 0$ is sufficiently small.

An initial value problem is usually considered *well posed* as a mathematical model for a real-world situation only if the differential equation has unique solutions

that are continuous with respect to initial values. Otherwise it is unlikely that the initial value problem adequately mirrors the real-world situation.

An even stronger "continuous dependence" of solutions is often desirable. In addition to possible inaccuracy in the initial value, the function $f(x, t)$ may not model precisely the physical situation. For instance, it may involve physical parameters (such as resistance coefficients) whose values cannot be measured with absolute precision. Birkhoff and Rota generalize the proof of Theorem 4 to establish the following result.

THEOREM 5: Continuous Dependence of Solutions

Let $\mathbf{x}(t)$ and $\mathbf{y}(t)$ be solutions of the equations

$$\frac{d\mathbf{x}}{dt} = \mathbf{f}(\mathbf{x}, t) \quad \text{and} \quad \frac{d\mathbf{y}}{dt} = \mathbf{g}(\mathbf{y}, t) \tag{50}$$

on the closed interval $[a, T]$. Let \mathbf{f} and \mathbf{g} be continuous for $a \leq t \leq T$ and for \mathbf{x} and \mathbf{y} in a common region D of n-space and assume that \mathbf{f} satisfies the Lipschitz condition in (20) on the region D. If

$$|\mathbf{f}(\mathbf{z}, t) - \mathbf{g}(\mathbf{z}, t)| \leq \mu \tag{51}$$

for all t in the interval $[a, T]$ and all \mathbf{z} in D, it then follows that

$$|\mathbf{x}(t) - \mathbf{y}(t)| \leq |\mathbf{x}(a) - \mathbf{y}(a)| \cdot e^{k(t-a)} + \frac{\mu}{k} \left[e^{k(t-a)} - 1 \right] \tag{52}$$

on the interval $[a, T]$. ▲

If $\mu > 0$ is small, then (51) implies that the functions \mathbf{f} and \mathbf{g} appearing in the two differential equations, though different, are "close" to each other. If $\epsilon > 0$ is given, then it is apparent from (52) that

$$|\mathbf{x}(t) - \mathbf{y}(t)| \leq \epsilon \tag{53}$$

for all t in $[a, T]$ if both $|\mathbf{x}(a) - \mathbf{y}(a)|$ and μ are sufficiently small. Thus Theorem 5 says (roughly) that if both the two initial values and the two differential equations in (50) are close to each other, then the two solutions remain close to each other for $a \leq t \leq T$.

For example, suppose that a falling body is subject both to constant gravitational acceleration g and to resistance proportional to some power of its velocity, so (with the positive axis directed downward) its velocity v satisfies the differential equation

$$\frac{dv}{dt} = g - cv^\rho. \tag{54}$$

Assume, however, that only an approximation \overline{c} to the actual resistance c and an approximation $\overline{\rho}$ to the actual exponent ρ are known. Then our mathematical model is based on the differential equation

$$\frac{du}{dt} = g - \overline{c}u^{\overline{\rho}} \tag{55}$$

instead of the actual equation in (54). Thus if we solve Eq. (55), we obtain only an approximation $u(t)$ to the actual velocity $v(t)$. But if the parameters \overline{c} and $\overline{\rho}$ are sufficiently close to the actual values c and ρ, then the right-hand sides in (54) and (55) will be close to each other. If so, then Theorem 5 implies that the actual and approximate velocity functions $v(t)$ and $u(t)$ are close to each other. In this case the approximation in (55) will be a good model of the actual physical situation.

Problems

In Problems 1 through 8, apply the successive approximation formula to compute $y_n(x)$ for $n \leq 4$. Then write the exponential series for which these approximations are partial sums (perhaps with the first term or two missing; for example,

$$e^x - 1 = x + \tfrac{1}{2}x^2 + \tfrac{1}{6}x^3 + \tfrac{1}{24}x^4 + \cdots).$$

1. $\dfrac{dy}{dx} = y,\ y(0) = 3$

2. $\dfrac{dy}{dx} = -2y,\ y(0) = 4$

3. $\dfrac{dy}{dx} = -2xy,\ y(0) = 1$

4. $\dfrac{dy}{dx} = 3x^2 y,\ y(0) = 2$

5. $\dfrac{dy}{dx} = 2y + 2,\ y(0) = 0$

6. $\dfrac{dy}{dx} = x + y,\ y(0) = 0$

7. $\dfrac{dy}{dx} = 2x(1 + y),\ y(0) = 0$

8. $\dfrac{dy}{dx} = 4x(y + 2x^2),\ y(0) = 0$

In Problems 9 through 12, compute the successive approximations $y_n(x)$ for $n \leq 3$; then compare them with the appropriate partial sums of the Taylor series of the exact solution.

9. $\dfrac{dy}{dx} = x + y,\ y(0) = 1$

10. $\dfrac{dy}{dx} = y + e^x,\ y(0) = 0$

11. $\dfrac{dy}{dx} = y^2,\ y(0) = 1$

12. $\dfrac{dy}{dx} = \tfrac{1}{2}y^3,\ y(0) = 1$

13. Apply the iterative formula in (16) to compute the first three successive approximations to the solution of the initial value problem

$$\frac{dx}{dt} = 2x - y, \qquad x(0) = 1;$$

$$\frac{dy}{dt} = 3x - 2y, \qquad y(0) = -1.$$

14. Apply the matrix exponential series in (19) to solve (in closed form) the initial value problem

$$\mathbf{x}'(t) = \begin{bmatrix} 1 & 1 \\ 0 & 1 \end{bmatrix} \mathbf{x}, \qquad \mathbf{x}(0) = \begin{bmatrix} 1 \\ 1 \end{bmatrix}.$$

(*Suggestion*: Show first that

$$\begin{bmatrix} 1 & 1 \\ 0 & 1 \end{bmatrix}^n = \begin{bmatrix} 1 & n \\ 0 & 1 \end{bmatrix}$$

for each positive integer n.)

15. For the initial value problem $dy/dx = 1 + y^3$, $y(1) = 1$, show that the second Picard approximation is

$$y_2(x) = 1 + 2(x-1) + 3(x-1)^2 + 4(x-1)^3 + 2(x-1)^4.$$

Then compute $y_2(1.1)$ and $y_2(1.2)$. The fourth-order Runge-Kutta method with step size $h = 0.005$ yields $y(1.1) \approx 1.2391$ and $y(1.2) \approx 1.6269$.

16. For the initial value problem $dy/dx = x^2 + y^2$, $y(0) = 0$, show that the third Picard approximation is

$$y_3(x) = \frac{1}{3}x^3 + \frac{1}{63}x^7 + \frac{2}{2079}x^{11} + \frac{1}{59535}x^{15}.$$

Compute $y_3(1)$. The fourth-order Runge-Kutta method yields $y(1) \approx 0.350232$, both with step size $h = 0.05$ and with step size $h = 0.025$.

17. Prove as follows the inequality $|\mathbf{Ax}| \leq \|\mathbf{A}\| \cdot |\mathbf{x}|$, where \mathbf{A} is an $m \times m$ matrix with row vectors $\mathbf{a}_1, \mathbf{a}_2, \ldots, \mathbf{a}_m$, and \mathbf{x} is an m-dimensional vector. First note that the components of the vector \mathbf{Ax} are $\mathbf{a}_1 \cdot \mathbf{x}, \mathbf{a}_2 \cdot \mathbf{x}, \ldots, \mathbf{a}_m \cdot \mathbf{x}$, so

$$|\mathbf{Ax}| = \left[\sum_{n=1}^{m} (\mathbf{a}_i \cdot \mathbf{x})^2 \right]^{1/2}.$$

Then use the Cauchy-Schwarz inequality $(\mathbf{a} \cdot \mathbf{x})^2 \leq |\mathbf{a}|^2 |\mathbf{x}|^2$ for the dot product.

18. Suppose that $\phi(t)$ is a differentiable function with

$$\phi'(t) \leq k\phi(t) \qquad (k > 0)$$

for $t \geq a$. Multiply both sides by e^{-kt}, then transpose to show that

$$\frac{d}{dt}\left[\phi(t)e^{-kt} \right] \leq 0$$

for $t \geq a$. Then apply the mean value theorem to conclude that

$$\phi(t) \leq \phi(a)e^{k(t-a)}$$

for $t \geq a$.

ANSWERS TO SELECTED PROBLEMS

Chapter 1

Section 1.1

11. If $y = y_1 = x^{-2}$, then $y'(x) = -2x^{-3}$ and $y''(x) = 6x^{-4}$, so
$x^2 y'' + 5xy' + 4y = x^2(6x^{-4}) + 5x(-2x^{-3}) + 4(x^{-2}) =$
$6x^{-2} - 10x^{-2} + 4x^{-2} = 0$. If $y = y_2 = x^{-2} \ln x$, then
$y'(x) = x^{-3} - 2x^{-3} \ln x$ and $y''(x) = -5x^{-4} + 6x^{-4} \ln x$, so
$x^2 y'' + 5xy' + 4y =$
$x^2(-5x^{-4} + 6x^{-4} \ln x) + 5x(x^{-3} - 2x^{-3} \ln x) + 4(x^{-2} \ln x) = 0$.

13. $r = \frac{2}{3}$ **14.** $r = \pm\frac{1}{2}$ **15.** $r = -2, 1$

16. $r = \frac{1}{6}(-3 \pm \sqrt{57})$

17. $C = 2$

18. $C = 3$

19. $C = 6$

20. $C = 11$

21. $C = 7$

22. $C = 1$

23. $C = -56$

24. $C = 17$

25. $C = \pi/4$

26. $C = -\pi$

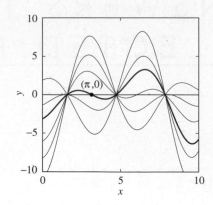

27. $y' = x + y$

28. $y' = 2y/x$

29. $y' = x/(1 - y)$

31. $y' = (y - x)/(y + x)$

32. $dP/dt = k\sqrt{P}$

33. $dv/dt = kv^2$

35. $dN/dt = k(P - N)$

37. $y \equiv 1$ or $y = x$

39. $y = x^2$

41. $y = \frac{1}{2}e^x$

42. $y = \cos x$ or $y = \sin x$

43. (b) The identically zero function $x(0) \equiv 0$

44. (a) The graphs (figure below) of typical solutions with $k = \frac{1}{2}$ suggest that (for each) the value $x(t)$ increases without bound as t increases.

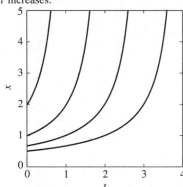

(b) The graphs (figure below) of typical solutions with $k = -\frac{1}{2}$ suggest that now the value $x(t)$ approaches 0 as t increases without bound.

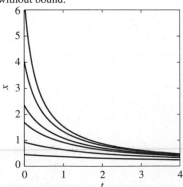

45. $P(t) = 100/(50 - t)$; $P = 100$ when $t = 49$, and $P = 1000$ when $t = 49.9$. Thus it appears that $P(t)$ grows without bound as t approaches 50.

46. $v(t) = 50/(5 + 2t)$; $v = 1$ when $t = 22.5$, and $v = \frac{1}{10}$ when $t = 247.5$. Thus it appears that $v(t)$ approaches 0 as t increases without bound.

47. (a) $C = 10.1$; (b) No such C, but the constant function $y(x) \equiv 0$ satisfies the conditions $y' = y^2$ and $y(0) = 0$.

Section 1.2

1. $y(x) = x^2 + x + 3$

2. $y(x) = \frac{1}{3}(x - 2)^3$

3. $y(x) = \frac{1}{3}(2x^{3/2} - 16)$

4. $y(x) = -1/x + 6$

5. $y(x) = 2\sqrt{x + 2} - 5$

6. $y(x) = \frac{1}{3}[(x^2 + 9)^{3/2} - 125]$

7. $y(x) = 10\tan^{-1} x$

8. $y(x) = \frac{1}{2}\sin 2x + 1$

9. $y(x) = \sin^{-1} x$

10. $y(x) = -(x + 1)e^{-x} + 2$

11. $x(t) = 25t^2 + 10t + 20$

12. $x(t) = -10t^2 - 15t + 5$

13. $x(t) = \frac{1}{2}t^3 + 5t$

14. $x(t) = \frac{1}{3}t^3 + \frac{1}{2}t - 7t + 4$

15. $x(t) = \frac{1}{3}(t + 3)^4 - 37t - 26$

16. $x(t) = \frac{4}{3}(t + 4)^{3/2} - 5t - \frac{29}{3}$

17. $x(t) - \frac{1}{2}\left[(t + 1)^{-1} + t - 1\right]$

19. $x(t) = \begin{cases} 5t & \text{if } 0 \le t \le 5, \\ 10t - \frac{1}{2}t^2 - \frac{25}{2} & \text{if } 5 \le t < 10. \end{cases}$

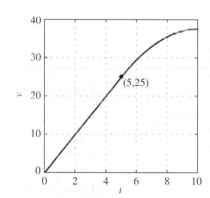

20. $x(t) = \begin{cases} \frac{1}{2}t^2 & \text{if } 0 \le t \le 5, \\ 5t - \frac{25}{2} & \text{if } 5 \le t \le 10. \end{cases}$

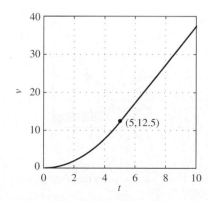

21. $x(t) = \begin{cases} \frac{1}{2}t^2 & \text{if } 0 \le t \le 5, \\ 10t - \frac{1}{2}t^2 - 25 & \text{if } 5 \le t \le 10. \end{cases}$

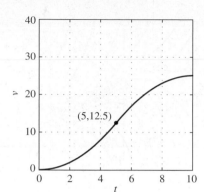

22. $x(t) = \begin{cases} \frac{5}{6}t^2 & \text{if } 0 \le t \le 3, \\ 5t - \frac{15}{2} & \text{if } 3 \le t \le 7, \\ \frac{1}{6}(-5t^2 + 100t - 290) & \text{if } 7 \le t \le 10. \end{cases}$

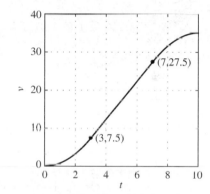

23. $v(t) = -(9.8)t + 49$, so the ball reaches its maximum height ($v = 0$) after $t = 5$ seconds. Its maximum height then is $y(5) = 122.5$ (m).

24. $v(5) = -160$ ft/s

25. The car stops when $t \approx 2.78$ (s), so the distance traveled before stopping is approximately $x(2.78) \approx 38.58$ (m).

26. (a) $y \approx 530$ m (b) $t \approx 20.41$ s (c) $t \approx 20.61$ s

27. $y_0 \approx 178.57$ (m)

28. $v(4.77) \approx -192.64$ ft/s

29. After 10 seconds the car has traveled 200 ft and is traveling at 70 ft/s.

30. $a = 22$ ft/s^2; it skids for 4 seconds.

31. $v_0 = 10\sqrt{30}$ (m/s), about 197.18 km/h

32. 60 m **33.** $20\sqrt{10} \approx 63.25$ (ft/s)

34. 460.8 ft **36.** About 13.6 ft

37. 25 (mi) **38.** 1:10 pm

39. 6 mph **40.** 2.4 mi

41. $\frac{544}{3} \approx 181.33$ ft/s **42.** 25 mi

43. Time: 6.12245×10^9 s ≈ 194 years; Distance: 1.8367×10^{17} m ≈ 19.4 light-years

44. About 54 mi/h

Section 1.3

1.

2.

3.

4.

5.

6.

7.

8.

9.

10.

11. A unique solution exists in some neighborhood of $x = 1$.

12. A unique solution exists in some neighborhood of $x = 1$.

13. A unique solution exists in some neighborhood of $x = 0$.

14. Existence but not uniqueness is guaranteed in some neighborhood of $x = 0$.

15. Neither existence nor uniqueness is guaranteed in any neighborhood of $x = 2$.

16. A unique solution exists in some neighborhood of $x = 2$.

17. A unique solution exists in some neighborhood of $x = 0$.

18. Neither existence nor uniqueness is guaranteed.

19. A unique solution exists in some neighborhood of $x = 0$,

20. A unique solution exists in some neighborhood of $x = 0$.

21. Your figure should suggest that $y(-4) \approx 3$; an exact solution of the differential equation gives $y(-4) = 3 + e^{-4} \approx 3.0183$.

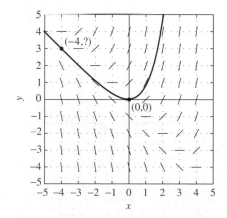

22. $y(-4) \approx -3$

23. Your figure should suggest that $y(2) \approx 1$; the actual value is closer to 1.004.

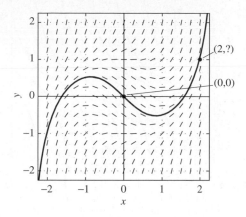

24. $y(2) \approx 1.5$

25. Your figure should suggest that the limiting velocity is about 20 ft/sec (quite survivable) and that the time required to reach 19 ft/sec is a little less than 2 seconds. An exact solution gives $v(t) = 19$ when $t = \frac{5}{8} \ln 20 \approx 1.8723$.

26. A figure suggests that there are 40 deer after about 60 months; a more accurate value is $t \approx 61.61$. The limiting population is 75 deer.

27. The initial value problem $y' = 2\sqrt{y}$, $y(0) = b$ has no solution if $b < 0$; a unique solution if $b > 0$; infinitely many solutions if $b = 0$.

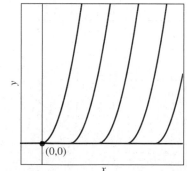

28. The initial value problem $xy' = y$, $y(a) = b$ has a unique solution if $a \neq 0$; infinitely many solutions if $a = b = 0$; no solution if $a = 0$ but $b \neq 0$.

29. The initial value problem $y' = 3y^{2/3}$, $y(a) = b$ always has infinitely many solutions defined for all x. However, if $b \neq 0$

then it has a unique solution near $x = a$.

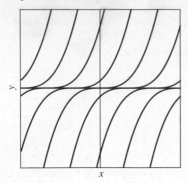

30. The initial value problem $y' = -\sqrt{1 - y^2}$, $y(a) = b$ has a unique solution if $|b| < 1$; no solution if $|b| > 1$, and infinitely many solutions (defined for all x) if $b = \pm 1$.

31. The initial value problem $y' = \sqrt{1 - y^2}$, $y(a) = b$ has a unique solution if $|b| < 1$; no solution if $|b| > 1$, and infinitely many solutions (defined for all x) if $b = \pm 1$.

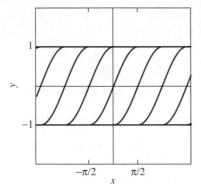

32. The initial value problem $y' = 4x\sqrt{y}$, $y(a) = b$ has infinitely many solutions (defined for all x) if $b \geqq 0$; no solutions if $b < 0$. However, if $b > 0$ then it has a unique solution near $x = a$.

33. The initial value problem $x^2 y' + y^2 = 0$, $y(a) = b$ has a unique solution with initial point (a, b) if $a \neq 0$, no solution if $a = 0$ but $b \neq 0$, infinitely many solutions if $a = b = 0$.

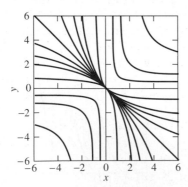

Section 1.4

1. $y(x) = C \exp(-x^2)$
2. $y(x) = 1/(x^2 + C)$
3. $y(x) = C \exp(-\cos x)$
4. $y(x) = C(1 + x)^4$
5. $y(x) = \sin\left(C + \sqrt{x}\right)$
6. $y(x) = (x^{3/2} + C)^2$
7. $y(x) = (2x^{4/3} + C)^{3/2}$
8. $y(x) = \sin^{-1}(x^2 + C)$
9. $y(x) = C(1 + x)/(1 - x)$

10. $y(x) = (1 + x)/[1 + C(1 + x)]$
11. $y(x) = (C - x^2)^{-1/2}$ **12.** $y^2 + 1 = Ce^{x^2}$
13. $\ln(y^4 + 1) = C + 4 \sin x$ **14.** $3y + 2y^{3/2} = 3x + 2x^{3/2} + C$
15. $1/(3y^3) - 2/y = 1/x + \ln|x| + C$
16. $y(x) = \sec^{-1}(C\sqrt{1 + x^2})$
17. $\ln|1 + y| = x + \frac{1}{2}x^2 + C$
18. $y(x) = \tan\left(C - \dfrac{1}{x} - x\right)$
19. $y(x) = 2\exp(e^x)$ **20.** $y(x) = \tan(x^3 + \pi/4)$
21. $y^2 = 1 + \sqrt{x^2 - 16}$ **22.** $y(x) = -3\exp(x^4 - x)$
23. $y(x) = \frac{1}{2}(1 + e^{2x-2})$ **24.** $y(x) = \frac{\pi}{2}\sin x$
25. $y(x) = x\exp(x^2 - 1)$ **26.** $y(x) = 1/(1 - x^2 - x^3)$
27. $y = \ln(3e^{2x} - 2)$ **28.** $y(x) = \tan^{-1}(\sqrt{x} - 1)$
29. (a) General solution $y(x) = -1/(x - C)$; (b) The singular solution $y(x) \equiv 0$. (c) In the following figure we see that there is a unique solution through every point of the xy- plane.

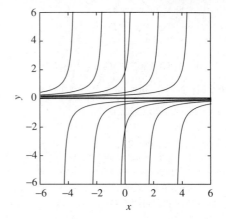

30. General solution $y(x) = (x - C)^2$; singular solution $y(x) \equiv 0$. (a) No solution if $b < 0$; (b) Infinitely many solutions (for all x) if $b \geqq 0$; (c) Two solutions near (a, b) if $b < 0$.

31. Separation of variables gives the same general solution $y = (x - C)^2$ as in Problem 30, but the restriction that $y' = 2\sqrt{y} \geqq 0$ implies that only the *right halves* of the parabolas qualify as solution curves. In the figure below we see that through the point (a, b) there passes (a) No solution curve if $b < 0$, (b) a unique solution curve if $b > 0$, (c) Infinitely many solution curves if $b = 0$.

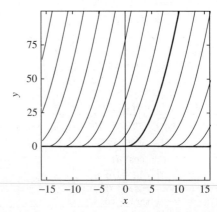

32. General solution $y(x) = \pm \sec(x - C)$; singular solutions $y(x) \equiv \pm 1$.
(a) No solution if $|b| < 1$; (b) A unique solution if $|b| > 1$;
(c) Infinitely many solutions if $b = \pm 1$.

33. About 51840 persons
34. $t \approx 3.87$ hr
35. About 14735 years
36. Age about 686 years
37. $21103.48
38. $44.52
39. 2585 mg
40. About 35 years
41. About 4.86×10^9 years ago
42. About 1.25 billion years
43. After a total of about 63 min have elapsed
44. About 2.41 minutes
45. (a) 0.495 m; (b $(8.32 \times 10^{-7})I_0$; (c) 3.29 m
46. (a) About 9.60 inches; (b) About 18,200 ft
47. After about 46 days
48. About 6 billion years
49. After about 66 min 40 s
50. (a) $A(t) = 10 \cdot 3^{2t/15}$; (b) About 20.80 pu; (c) About 15.72 years
51. (a) $A(t) = 15 \cdot (\frac{2}{3})^{t/5}$; (b) approximately 7.84 su; (c) After about 33.4 months
52. About 120 thousand years ago
53. About 74 thousand years ago
54. 3 hours
55. 972 s
56. At time $t = 2048/1562 \approx 1.31$ (in hours)
58. 1:20 P.M.
59. (a) $y(t) = (8 - 7t)^{2/3}$; (b) at 1:08:34 P.M.;
 (c) $r = \frac{1}{60}\sqrt{\frac{J}{12}} \approx 0.15$ (in.)
60. About 6 min 3 sec
61. Approximately 14 min 29 s
62. The tank is empty about 14 seconds after 2:00 P.M.
63. (a) 1:53:34 P.M.; (b) $r \approx 0.04442$ ft ≈ 0.53 in.
64. $r = \frac{1}{720}\sqrt{3}$ ft, about $\frac{1}{35}$ in.
65. At approximately 10:29 A.M.

Section 1.5

1. $y(x) = 2(1 - e^{-x})$
2. $y(x) = (3x + C)e^{2x}$
3. $y(x) = e^{-3x}(x^2 + C)$
4. $y(x) = (x + C)e^{x^2}$
5. $y(x) = x + 4x^{-2}$
6. $y(x) = x^2 + 32/x^5$
7. $y(x) = 5x^{1/2} + Cx^{-1/2}$
8. $y(x) = 3x + Cx^{-1/3}$
9. $y(x) = x(7 + \ln x)$
10. $y(x) = 3x^3 + Cx^{3/2}$
11. $y(x) \equiv 0$
12. $y(x) = \frac{1}{4}x^5 + 56x^{-3}$
13. $y(x) = (e^x + e^{-x})/2$
14. $y(x) = x^3 \ln x + 10x^3$
15. $y(x) = [1 - 5\exp(-x^2)]/2$
16. $y(x) = 1 + e^{-\sin x}$
17. $y(x) = (1 + \sin x)/(1 + x)$
18. $y(x) = x^2(\sin x + C)$
19. $y(x) = \frac{1}{2}\sin x + C\csc x$
20. $y(x) = -1 + \exp\left(-x - \frac{1}{2}x^2\right)$
21. $y(x) = x^3 \sin x$
22. $y(x) = (x^3 + 5)e^{-x^2}$
23. $y(x) = x^3(2 + Ce^{-2x})$
24. $y(x) = \frac{1}{3}[1 + 16(x^2+4)^{-3/2}]$
25. $y(x) = \left[\exp\left(-\frac{3}{2}x^2\right)\right]\left[3(x^2 + 1)^{3/2} - 2\right]$
26. $x(y) = 1/2y^2 + C/y^4$
27. $x(y) = e^y\left(C + \frac{1}{2}y^2\right)$
28. $x(y) = \frac{1}{2}[y + (1 + y^2)(\tan^{-1} y + C)]$
30. $y(x) = x^{1/2}\int_1^x t^{-1/2}\cos t\, dt$
29. $y(x) = \left[\exp(x^2)\right]\left[C + \frac{1}{2}\sqrt{\pi}\,\mathrm{erf}(x)\right]$
32. (a) $y(x) = \sin x - \cos x$; (b) $y(x) = Ce^{-x} + \sin x - \cos x$;
 (c) $y(x) = 2e^{-x} + \sin x - \cos x$
33. After about 7 min 41 s
34. About 22.2 days
35. About 5.5452 years
36. (a) $x(t) = (60 - t) - (60 - t)^3/3600$; (b) About 23.09 lb

37. 393.75 lb
38. (a) $x(t) = 50e^{-t/20}$; (b) $y(t) = 150e^{-t/40} - 100e^{-t/20}$;
 (c) 56.25 lb
39. (b) $y_{\max} = 100e^{-1} \approx 36.79$ (gal)
41. (b) Approximately $1,308,283
43. $-50.0529, -28.0265, -6.0000, 16.0265, 38.0529$
44. 3.99982, 4.00005, 4.00027, 4.00050, 4.00073
45. $x(t) = 20(1 - e^{-t/10})$; $x = 10$ after $t = 10\ln 2 \approx 6.93$ months.
46. $x(t) = \frac{20}{101}(101 - 102e^{-t/10} + \cos t + 10\sin t)$; $x = 10$ after $t = 6.47$ months.

Section 1.6

1. $x^2 - 2xy - y^2 = C$
2. $y^2 = x^2(\ln x + C)$
3. $y(x) = x(C + \ln|x|)^2$
4. $2\tan^{-1}(y/x) - \ln(y^2/x^2 + 1) = 2\ln x + C$
5. $\ln|xy| = C + xy^{-1}$
6. $2y\ln y = x + Cy$
7. $y^3 = 3x^3(C + \ln|x|)$
8. $y = -x\ln(C - \ln x)$
9. $y(x) = x/(C - \ln|x|)$
10. $x^2 + 2y^2 = Cx^6$
11. $y = C(x^2 + y^2)$
12. $4x^2 + y^2 = x^2(\ln x + C)^2$
13. $y + \sqrt{x^2 + y^2} = Cx^2$
14. $x - \sqrt{x^2 + y^2} = C$
15. $x^2(2xy + y^2) = C$
16. $x = 2\sqrt{x + y + 1} - 2\ln\left(1 + \sqrt{x + y + 1}\right) + C$
17. $y(x) = -4x + 2\tan(2x + C)$
18. $y = \ln(x + y + 1) + C$
19. $y^2 = x/(2 + Cx^5)$
20. $y^3 = 3 + Ce^{-3x^2}$
21. $y^2 = 1/(Ce^{-2x} - 1)$
22. $y^3 = 7x/(7Cx^7 + 15)$
23. $y(x) = (x + Cx^2)^{-3}$
24. $y^2 = e^{4x}/(C + \ln x)$
25. $2x^3y^3 = 3\sqrt{1 + x^4} + C$
26. $y^3 = e^{-x}(x + C)$
27. $y(x) = (x^4 + Cx)^{1/3}$
28. $y = \ln(Cx^2 + x^2e^{2x})$
29. $\sin^2 y = 4x^2 + Cx$
30. $x^2 - 2xe^y - e^{2y} = C$
31. $x^2 + 3xy + y^2 = C$
32. $x^2 - xy + 3y^2 = C$
33. $x^3 + 2xy^2 + 2y^3 = C$
34. $x^3 + x^2y^2 + y^4 = C$
35. $3x^4 + 4y^3 + 12y\ln x = C$
36. $x + e^{xy} + y^2 = C$
37. $\sin x + x\ln y + e^y = C$
38. $x^2 + 2x\tan^{-1} y + \ln(1 + y^2) = C$
39. $5x^3y^3 + 5xy^4 + y^5 = C$
40. $e^x\sin y + x\tan y = C$
41. $x^2y^{-1} + y^2x^{-3} + 2y^{1/2} = C$
42. $xy^{-2/3} + x^{-3/2}y = C$
43. $y(x) = Ax^2 + B$
44. $x(y) = Ay^2 + B$
45. $y(x) = A\cos 2x + B\sin 2x$
46. $y(x) = x^2 + A\ln x + B$
47. $y(x) = A - \ln|x + B|$
48. $y(x) = \ln x + Ax^{-2} + B$
49. $y(x) = \pm(A + Be^x)^{1/2}$
50. $y(x) = \ln|\sec(x + A)| - \frac{1}{2}x^2 + B$
51. $x(y) = -\frac{1}{3}(y^3 + Ay + B)$
52. $Ay^2 - (Ax + B)^2 = 1$
53. $y(x) = A\tan(Ax + B)$
54. $Ay^2(B - x) = 1$
58. $y = \exp(x^2 + C/x^2)$
59. $x^2 - 2xy - y^2 - 2x - 6y = C$
60. $(x + 3y + 3)^5 = C(y - x - 5)$
61. $x = \tan(x - y) + \sec(x - y) + C$
64. $y(x) = x + e^{-x^2}\left[C + \frac{1}{2}\sqrt{\pi}\,\mathrm{erf}(x)\right]^{-1}$
65. $y(x) = x + (C - x)^{-1}$
69. Approximately 3.68 mi

Chapter 1 Review Problems

1. Linear: $y(x) = x^3(C + \ln x)$
2. Separable: $y(x) = x/(3 - Cx - x \ln x)$
3. Homogeneous: $y(x) = x/(C - \ln x)$
4. Exact: $x^2 y^3 + e^x - \cos y = C$
5. Separable: $y(x) = C \exp(x^{-3} - x^{-2})$
6. Separable: $y(x) = x/(1 + Cx + 2x \ln x)$
7. Linear: $y(x) = x^{-2}(C + \ln x)$
8. Homogeneous: $y(x) = 3Cx/(C - x^3) = 3x/(1 + Kx^3)$
9. Bernoulli: $y(x) = (x^2 + Cx^{-1})^2$
10. Separable: $y(x) = \tan\left(C + x + \frac{1}{3}x^3\right)$
11. Homogeneous: $y(x) = x/(C - 3 \ln x)$
12. Exact: $3x^2 y^3 + 2xy^4 = C$
13. Separable: $y(x) = 1/(C + 2x^2 - x^5)$
14. Homogeneous: $y^2 = x^2/(C + 2 \ln x)$
15. Linear: $y(x) = (x^3 + C)e^{-3x}$
16. Substitution: $v = y - x$; solution: $y - x - 1 = Ce^{2x}(y - x + 1)$
17. Exact: $e^x + e^y + e^{xy} = C$
18. Homogeneous: $y^2 = Cx^2(x^2 - y^2)$
19. Separable: $y(x) = x^2/(x^5 + Cx^2 + 1)$
20. Linear: $y(x) = 2x^{-3/2} + Cx^{-3}$
21. Linear: $y(x) = [C + \ln(x - 1)]/(x + 1)$
22. Bernoulli: $y(x) = (2x^4 + Cx^2)^3$
23. Exact: $xe^y + y \sin x = C$
24. Separable: $y(x) = x^{1/2}/(6x^2 + Cx^{1/2} + 2)$
25. Linear: $y(x) = (x + 1)^{-2}(x^3 + 3x^2 + 3x + C)$
 $= x + 1 + K(x + 1)^{-2}$
26. Exact: $3x^{3/2}y^{4/3} - 5x^{6/5}y^{3/2} = C$
27. Bernoulli: $y(x) = x^{-1}(C + \ln x)^{-1/3}$
28. Linear: $y(x) = x^{-1}(C + e^{2x})$
29. Linear: $y(x) = (x^2 + x + C)(2x + 1)^{-1/2}$
30. Substitution: $v = x + y$; solution:
 $x = 2(x + y)^{1/2} - 2 \ln[1 + (x + y)^{1/2}] + C$
31. Separable and linear
32. Separable and Bernoulli
33. Exact and homogeneous
34. Exact and homogeneous
35. Separable and linear
36. Separable and Bernoulli

Chapter 2

Section 2.1

1. $x(t) = \dfrac{2}{2 - e^{-t}}$

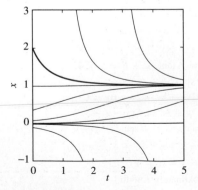

2. $x(t) = \dfrac{10}{1 + 9e^{-10t}}$

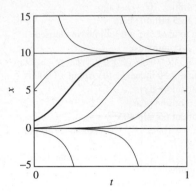

3. $x(t) = \dfrac{2 + e^{-2t}}{2 - e^{-2t}}$

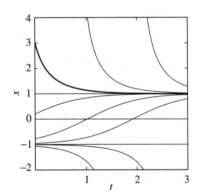

4. $x(t) = \dfrac{3(1 - e^{-12t})}{2(1 + e^{-12t})}$

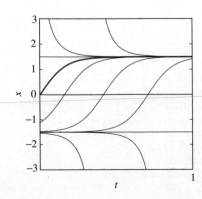

5. $x(t) = \dfrac{40}{8 - 3e^{-15t}}$

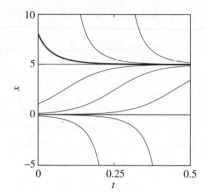

6. $x(t) = \dfrac{10}{2 + 3e^{15t}}$

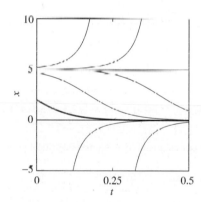

7. $x(t) = \dfrac{77}{11 - 4e^{-28t}}$

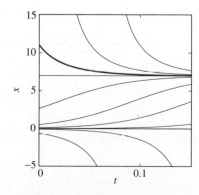

8. $x(t) = \dfrac{221}{17 - 4e^{91t}}$

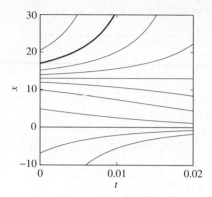

9. 484

10. 20 weeks

11. (b) $P(t) = \left(\frac{1}{2}t + 10\right)^2$

12. $P(t) = \dfrac{240}{20 - t}$

13. $P(t) = \dfrac{180}{30 - t}$

14. $P(t) = \dfrac{P_0}{1 + k P_0 t}$

16. About 27.69 months

17. About 44.22 months

19. About 24.41 months

20. About 42.12 months

21. $\dfrac{200}{1 + e^{-6/5}} \approx 153.7$ million

22. About 34.66 days

23. (a) $\lim\limits_{t \to \infty} x(t) = 200$ grams (b) $\frac{5}{4}\ln 3 \approx 1.37$ seconds

24. About 9.24 days

25. (a) $M = 100$ and $k = 0.0002$; (b) In the year 2033

26. $50 \ln \frac{9}{8} \approx 5.89$ months

27. (a) $100 \ln \frac{9}{5} \approx 58.78$ months; (b) $100 \ln 2 \approx 69.31$ months

28. (a) The alligators eventually die out. (b) Doomsday occurs after about 9 years 2 months.

29. (a) $P(140) \approx 127.008$ million; (b) About 210.544 million; (c) In 2000 we get $P \approx 196.169$, whereas the actual 2000 population was about 281.422 million.

31. $\alpha \approx 0.3915$; 2.15×10^6 cells

37. $k \approx 0.0000668717$, $M \approx 338.027$

38. $k \approx 0.000146679$, $M \approx 208.250$

39. $P(t) = P_0 \exp\left(kt + \dfrac{b}{2\pi}\sin 2\pi t\right)$; the colored curve in the figure below shows the graph with $P_0 = 100$, $k = 0.03$, and $b = 0.06$. It oscillates about the black curve which represents natural growth with $P_0 = 100$ and $k = 0.03$ We see that the two agree at the end of each full year.

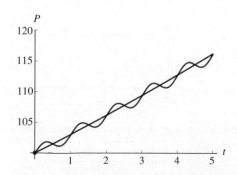

Section 2.2

1. Unstable critical point: $x = 4$;

$$x(t) = 4 + (x_0 - 4)e^t$$

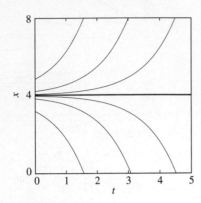

2. Stable critical point: $x = 3$;

$$x(t) = 3 + (x_0 - 3)e^{-t}$$

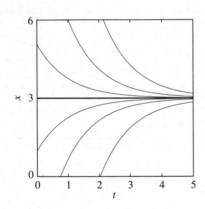

3. Stable critical point: $x = 0$; unstable critical point: $x = 4$;

$$x(t) = \frac{4x_0}{x_0 + (4 - x_0)e^{4t}}$$

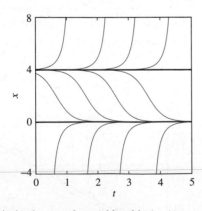

4. Stable critical point: $x = 3$; unstable critical point: $x = 0$;

$$x(t) = \frac{3x_0}{x_0 + (3 - x_0)e^{-3t}}$$

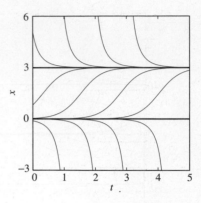

5. Stable critical point: $x = -2$; unstable critical point: $x = 2$;

$$x(t) = \frac{2[x_0 + 2 + (x_0 - 2)e^{4t}]}{x_0 + 2 - (x_0 - 2)e^{4t}}$$

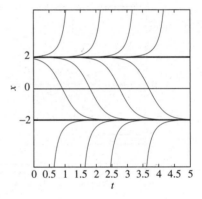

6. Stable critical point: $x = 3$; unstable critical point: $x = -3$;

$$x(t) = \frac{3[x_0 - 3 + (x_0 + 3)e^{6t}]}{3 - x_0 + (x_0 + 3)e^{6t}}$$

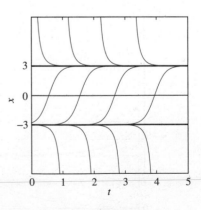

7. Semi-stable (see Problem 18) critical point: $x = 2$;

$$x(t) = \frac{(2t - 1)x_0 - 4t}{tx_0 - 2t - 1}$$

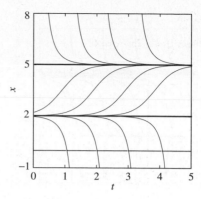

8. Semi-stable critical point: $x = 3$;

$$x(t) = \frac{(3t+1)x_0 - 9t}{tx_0 - 3t + 1}$$

11. Unstable critical point: $x = 1$;

$$\frac{1}{(x(t) - 1)^2} = \frac{1}{(x_0 - 1)^2} - 2t$$

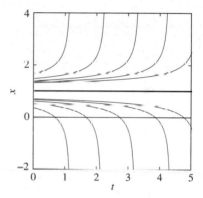

9. Stable critical point: $x = 1$; unstable critical point: $x = 4$;

$$x(t) = \frac{4(1 - x_0) + (x_0 - 4)e^{3t}}{1 - x_0 + (x_0 - 4)e^{3t}}$$

12. Stable critical point: $x = 2$;

$$\frac{1}{(2 - x(t))^2} = \frac{1}{(2 - x_0)^2} + 2t$$

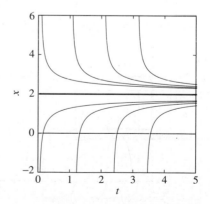

10. Stable critical point: $x = 5$; unstable critical point: $x = 2$;

$$x(t) = \frac{2(5 - x_0) + 5(x_0 - 2)e^{3t}}{5 - x_0 + (x_0 - 2)e^{3t}}$$

For each of Problems 13 through 18 we show a plot of slope field and typical solution curves. The equilibrium solutions of the given differential equation are labeled, and the stability or instability of each should be clear from the picture.

13.

$x' = (x + 2)(x - 2)^2$

14.

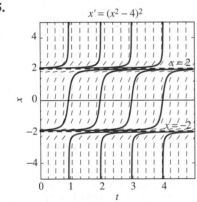

$x' = x(x^2 - 4)$

15.

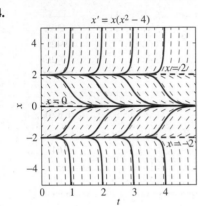

$x' = (x^2 - 4)^2$

16.

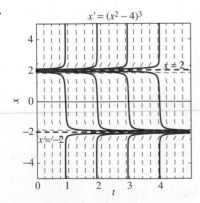

$x' = (x^2 - 4)^3$

17.

$x' = x^2(x^2 - 4)$

18.

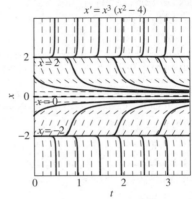

$x' = x^3(x^2 - 4)$

19. There are two critical points if $h < 2\frac{1}{2}$, one critical point if $h = 2\frac{1}{2}$, and no critical points if $h > 2\frac{1}{2}$. The bifurcation diagram is the parabola $(c - 5)^2 = 25 - 10h$ in the hc-plane.

20. There are two critical points if $s < \frac{1}{16}$, one critical point if $s = \frac{1}{16}$, and no critical points if $s > \frac{1}{16}$. The bifurcation diagram is the parabola $(2c - 5)^2 = 25(1 - 16s)$ in the sc-plane.

Section 2.3

1. Approximately 31.5 s

3. $400/(\ln 2) \approx 577$ ft

5. $400 \ln 7 \approx 778$ ft

7. (a) 100 ft/sec; (b) about 23 sec and 1403 ft to reach 90 ft/sec

8. (a) 100 ft/sec; (b) about 14.7 sec and 830 ft to reach 90 ft/sec

9. 50 ft/s

10. About 5 min 47 s

11. Time of fall: about 12.5 s

12. Approximately 648 ft

19. Approximately 30.46 ft/s; exactly 40 ft/s

20. Approximately 277.26 ft

22. Approximately 20.67 ft/s; about 484.57 s

23. Approximately 259.304 s

24. (a) About 0.88 cm; (b) about 2.91 km

25. (b) About 1.389 km/sec; (c) $r_{max} = 100R/19 \approx 5.26R$

26. Yes

28. (b) After about $8\frac{1}{2}$ minutes it hits the surface at about 4.116 km/sec.

29. About 51.427 km

30. Approximately 11.11 km/sec (as compared with the earth's escape velocity of about 11.18 km/sec).

Section 2.4

In Problems 1 through 10 we round off the indicated values to 3 decimal places.

1. Approximate values 1.125 and 1.181; true value 1.213
2. Approximate values 1.125 and 1.244; true value 1.359
3. Approximate values 2.125 and 2.221; true value 2.297
4. Approximate values 0.625 and 0.681; true value 0.713
5. Approximate values 0.938 and 0.889; true value 0.851
6. Approximate values 1.750 and 1.627; true value 1.558
7. Approximate values 2.859 and 2.737; true value 2.647
8. Approximate values 0.445 and 0.420; true value 0.405
9. Approximate values 1.267 and 1.278; true value 1.287
10. Approximate values 1.125 and 1.231; true value 1.333

Problems 11 through 24 call for tables of values that would occupy too much space for inclusion here. In Problems 11 through 16 we give first the final x-value, next the corresponding approximate y-values obtained with step sizes $h = 0.01$ and $h = 0.005$, and then the final true y-value. (All y-values rounded off accurate to 4 decimal places.)

11. 1.0, −0.7048, −0.7115, 0.7183
12. 1.0, 2.9864, 2.9931, 3.0000
13. 2.0, 4.8890, 4.8940, 4.8990
14. 2.0, 3.2031, 3.2304, 3.2589
15. 3.0, 3.4422, 3.4433, 3.4444
16. 3.0, 8.8440, 8.8445, 8.8451

In Problems 17 through 24 we give first the final x-value and then the corresponding approximate y-values obtained with step sizes $h = 0.1, h = 0.02, h = 0.004$, and $h = 0.0008$ respectively. (All y-values rounded off accurate to 4 decimal places.)

17. 1.0, 0.2925, 0.3379, 0.3477, 0.3497
18. 2.0, 1.6680, 1.6771, 1.6790, 1.6794
19. 2.0, 6.1831, 6.3653, 6.4022, 6.4096
20. 2.0, −1.3792, −1.2843, −1.2649, 1.2610
21. 2.0, 2.8508, 2.8681, 2.8716, 2.8723
22. 2.0, 6.9879, 7.2601, 7.3154, 7.3264
23. 1.0, 1.2262, 1.2300, 1.2306, 1.2307
24. 1.0, 0.9585, 0.9918, 0.9984, 0.9997

25. With both step sizes $h = 0.01$ and $h = 0.005$, the approximate velocity after 1 second is 16.0 ft/sec (80% of the limiting velocity of 20 ft/sec); after 2 seconds it is 19.2 ft/sec (96% of the limiting velocity).

26. With both step sizes $h = 1$ and $h = 0.5$, the approximate population after 5 years is 49 deer (65% of the limiting population of 75 deer); after 10 years it is 66 deer (88% of the limiting population).

27. With successive step sizes $h = 0.1, 0.01, 0.001, \ldots$ the first four approximations to $y(2)$ we obtain are 0.7772, 0.9777, 1.0017, and 1.0042. It therefore seems likely that $y(2) \approx 1.00$.

28. With successive step sizes $h = 0.1, 0.01, 0.001, \ldots$ the first four approximations to $y(2)$ we obtain are 1.2900, 1.4435, 1.4613, and 1.4631. It therefore seems likely that $y(2) \approx 1.46$.

29.

x	$h = 0.15$ y	$h = 0.03$ y	$h = 0.006$ y
−1.0	1.0000	1.0000	1.0000
−0.7	1.0472	1.0512	1.0521
−0.4	1.1213	1.1358	1.1390
−0.1	1.2826	1.3612	1.3835
0.2	0.8900	1.4711	0.8210
0.5	0.7460	1.2808	0.7192

30.

x	$h = 0.1$ y	$h = 0.01$ y
1.8	2.8200	4.3308
1.9	3.9393	7.9425
2.0	5.8521	28.3926

31.

x	$h = 0.1$ y	$h = 0.01$ y
0.7	4.3460	6.4643
0.8	5.8670	11.8425
0.9	8.3349	39.5010

Section 2.5

1.

x	Improved Euler y	Actual y
0.1	1.8100	1.8097
0.2	1.6381	1.6375
0.3	1.4824	1.4816
0.4	1.3416	1.3406
0.5	1.2142	1.2131

Note: In Problems 2 through 10, we give the value of x, the corresponding improved Euler value of y, and the true value of y.

2. 0.5, 1.3514, 1.3191
3. 0.5, 2.2949, 2.2974
4. 0.5, 0.7142, 0.7131
5. 0.5, 0.8526, 0.8513
6. 0.5, 1.5575, 1.5576
7. 0.5, 2.6405, 2.6475
8. 0.5, 0.4053, 0.4055
9. 0.5, 1.2873, 1.2874
10. 0.5, 1.3309, 1.3333

In Problems 11 through 16 we give the final value of x, the corresponding values of y with $h = 0.01$ and with $h = 0.005$, and the true value of y.

11. 1.0, −0.71824, −0.71827, −0.71828

12. 1.0, 2.99995, 2.99999, 3.00000
13. 2.0, 4.89901, 4.89899, 4.89898
14. 2.0, 3.25847, 3.25878, 3.25889
15. 3.0, 3.44445, 3.44445, 3.44444
16. 3.0, 8.84511, 8.84509, 8.84509

In Problems 17 through 24 we give the final value of x and the corresponding values of y for $h = 0.1, 0.02, 0.004$, and 0.0008.

17. 1.0, 0.35183, 0.35030, 0.35023, 0.35023
18. 2.0, 1.68043, 1.67949, 1.67946, 1.67946
19. 2.0, 6.40834, 6.41134, 6.41147, 6.41147
20. 2.0, -1.26092, -1.26003, -1.25999, -1.25999
21. 2.0, 2.87204, 2.87245, 2.87247, 2.87247
22. 2.0, 7.31578, 7.32841, 7.32916, 7.32920
23. 1.0, 1.22967, 1.23069, 1.23073, 1.23073
24. 1.0, 1.00006, 1.00000, 1.00000, 1.00000
25. With both step sizes $h = 0.01$ and $h = 0.005$ the approximate velocity after 1 second is 15.962 ft/sec (80% of the limiting velocity of 20 ft/sec); after 2 seconds it is 19.185 ft/sec (96% of the limiting velocity).
26. With both step sizes $h = 1$ and $h = 0.5$ the approximate population after 5 years is 49.391 deer (65% of the limiting population of 75 deer); after 10 seconds it is 66.113 deer (88% of the limiting population).
27. With successive step sizes $h = 0.1, 0.01, 0.001, \dots$ the first three approximations to $y(2)$ we obtain are 1.0109, 1.0045, and 1.0045. It therefore seems likely that $y(2) \approx 1.0045$.
28. With successive step sizes $h = 0.1, 0.01, 0.001, \dots$ the first four approximations to $y(2)$ we obtain are 1.4662, 1.4634, 1.4633, and 1.4633. It therefore seems likely that $y(2) \approx 1.4633$.
29. Impact speed approximately 43.22 m/s
30. Impact speed approximately 43.48 m/s

Section 2.6

1. $y(0.25) \approx 1.55762$; $y(0.25) = 1.55760$.
$y(0.5) \approx 1.21309$; $y(0.5) = 1.21306$.
Solution: $y = 2e^{-x}$

In Problems 2 through 10 we give the approximation to $y(0.5)$, its true value, and the solution.

2. 1.35867, 1.35914; $y = \frac{1}{2}e^{2x}$
3. 2.29740, 2.29744; $y = 2e^x - 1$
4. 0.71309, 0.71306; $y = 2e^{-x} + x - 1$
5. 0.85130, 0.85128; $y = -e^x + x + 2$
6. 1.55759, 1.55760; $u = 2\exp\left(-x^2\right)$
7. 2.64745, 2.64749; $y = 3\exp\left(-x^3\right)$
8. 0.40547, 0.40547; $y = \ln(x + 1)$
9. 1.28743, 1.28743; $y = \tan\frac{1}{4}(x + \pi)$
10. 1.33337, 1.33333; $y = (1 - x^2)^{-1}$
11. *Solution*: $y(x) = 2 - e^x$.

x	$h = 0.2$ y	$h = 0.1$ y	Exact y
0.0	1.00000	1.00000	1.00000
0.2	0.77860	0.77860	0.77860
0.4	0.50818	0.50818	0.50818
0.6	0.17789	0.17788	0.17788
0.8	-0.22552	-0.22554	-0.22554
1.0	-0.71825	-0.71828	-0.71828

In Problems 12 through 16 we give the final value of x, the corresponding Runge-Kutta approximations with $h = 0.2$ and with $h = 0.1$, the exact value of y, and the solution.

12. 1.0, 2.99996, 3.00000, 3.00000;
$y = 1 + 2/(2 - x)$
13. 2.0, 4.89900, 4.89898, 4.89898;
$y = \sqrt{8 + x^4}$
14. 2.0, 3.25795, 3.25882, 3.25889;
$y = 1/(1 - \ln x)$
15. 3.0, 3.44445, 3.44444, 3.44444;
$y = x + 4x^{-2}$
16. 3.0, 8.84515, 8.84509, 8.84509;
$y = (x^6 - 37)^{1/3}$

In Problems 17 through 24 we give the final value of x and the corresponding values of y with $h = 0.2, 0.1, 0.05$, and 0.025.

17. 1.0, 0.350258, 0.350234, 0.350232, 0.350232
18. 2.0, 1.679513, 1.679461, 1.679459, 1.679459
19. 2.0, 6.411464, 6.411474, 6.411474, 6.411474
20. 2.0, -1.259990, -1.259992, -1.259993, -1.259993
21. 2.0, 2.872467, 2.872468, 2.872468, 2.872468
22. 2.0, 7.326761, 7.328452, 7.328971, 7.329134
23. 1.0, 1.230735, 1.230731, 1.230731, 1.230731
24. 1.0, 1.000000, 1.000000, 1.000000, 1.000000
25. With both step sizes $h = 0.1$ and $h = 0.05$, the approximate velocity after 1 second is 15.962 ft/sec (80% of the limiting velocity of 20 ft/sec); after 2 seconds it is 19.185 ft/sec (96% of the limiting velocity).
26. With both step sizes $h = 6$ and $h = 3$, the approximate population after 5 years is 49.3915 deer (65% of the limiting population of 75 deer); after 10 years it is 66.1136 deer (88% of the limiting population).
27. With successive step sizes $h = 1, 0.1, 0.01, \dots$ the first four approximations to $y(2)$ we obtain are 1.05722, 1.00447, 1.00445 and 1.00445. Thus it seems likely that $y(2) \approx 1.00445$ accurate to 5 decimal places.
28. With successive step sizes $h = 1, 0.1, 0.01, \dots$ the first four approximations to $y(2)$ we obtain are 1.48990, 1.46332, 1.46331, and 1.46331. Thus it seems likely that $y(2) \approx 1.4633$ accurate to 5 decimal places.
29. Time aloft: approximately 9.41 seconds
30. Time aloft: approximately 9.41 seconds

Chapter 3

Section 3.1

1. $y(x) = \frac{5}{2}e^x - \frac{5}{2}e^{-x}$
2. $y(x) = 2e^{3x} - 3e^{-3x}$
3. $y(x) = 3\cos 2x + 4\sin 2x$
4. $y(x) = 10\cos 5x - 2\sin 5x$
5. $y(x) = 2e^x - e^{2x}$
6. $y(x) = 4e^{2x} + 3e^{-3x}$
7. $y(x) = 6 - 8e^{-x}$
8. $y(x) = \frac{1}{3}(14 - 2e^{3x})$
9. $y(x) = 2e^{-x} + xe^{-x}$
10. $y(x) = 3e^{5x} - 2xe^{5x}$
11. $y(x) = 5e^x \sin x$
12. $y(x) = e^{-3x}(2\cos 2x + 3\sin 2x)$

13. $y(x) = 5x - 2x^2$

14. $y(x) = 3x^2 - 16/x^3$

15. $y(x) = 7x - 5x \ln x$

16. $y(x) = 2\cos(\ln x) + 3\sin(\ln x)$

21. Linearly independent

22. Linearly independent

23. Linearly independent

24. Linearly dependent

25. Linearly independent

26. Linearly independent

28. $y(x) = 1 - 2\cos x - \sin x$

29. There is no contradiction because if the given differential equation is divided by x^2 to get the form in Eq. (8), then the resulting coefficient functions $p(x) = -4/x$ and $q(x) = 6/x^2$ are not continuous at $x = 0$.

33. $y(x) = c_1 e^x + c_2 e^{2x}$

34. $y(x) = c_1 e^{-5x} + c_2 e^{3x}$

35. $y(x) = c_1 + c_2 e^{-5x}$

36. $y(x) = c_1 + c_2 e^{-3x/2}$

37. $y(x) = c_1 e^{-x/2} + c_2 e^x$

38. $y(x) = c_1 e^{-x/2} + c_2 e^{-3x/2}$

39. $y(x) = (c_1 + c_2 x)e^{-x/2}$

40. $y(x) = (c_1 + c_2 x)e^{2x/3}$

41. $y(x) = c_1 e^{-4x/3} + c_2 e^{5x/2}$

42. $y(x) = c_1 e^{-4x/7} + c_2 e^{3x/5}$

43. $y'' + 10y' = 0$

44. $y'' - 100y = 0$

45. $y'' + 20y' + 100y = 0$

46. $y'' - 110y' + 1000y = 0$

47. $y'' = 0$

48. $y'' - 2y' - y = 0$

49. The high point is $\left(\ln \frac{7}{4}, \frac{16}{7}\right)$.

50. $(-\ln 2, -2)$

52. $y(x) = c_1 x + c_2/x$

53. $y(x) = c_1 x^{-4} + c_2 x^3$

54. $y(x) = c_1 x^{-3/2} + c_2 x^{1/2}$

55. $y(x) = c_1 + c_2 \ln x$

56. $y(x) = x^3(c_1 + c_2 \ln x)$

Section 3.2

1. $15 \cdot (2x) - 16 \cdot (3x^2) - 6 \cdot (5x - 8x^2) \equiv 0$

2. $(-4)(5) + (5)(2 - 3x^2) + (1)(10 + 15x^2) \equiv 0$

3. $1 \cdot 0 + 0 \cdot \sin x + 0 \cdot e^x \equiv 0$

4. $(6)(17) + (-51)(2\sin^2 x) + (-34)(3\cos^2 x) \equiv 0$

5. $1 \cdot 17 - 34 \cdot \cos^2 x + 17 \cdot \cos 2x \equiv 0$

6. $(-1)(e^x) + (1)(\cosh x) + (1)(\sinh x) \equiv 0$

13. $y(x) = \frac{4}{3}e^x - \frac{1}{3}e^{-2x}$

14. $y(x) = \frac{1}{2}(3e^x - 6e^{2x} + 3e^{3x})$

15. $y(x) = (2 - 2x + x^2)e^x$

16. $y(x) = -12e^x + 13e^{2x} - 10xe^{2x}$

17. $y(x) = \frac{1}{9}(29 - 2\cos 3x - 3\sin 3x)$

18. $y(x) = e^x(2 - \cos x - \sin x)$

19. $y(x) = x + 2x^2 + 3x^3$

20. $y(x) = 2x - x^{-2} + x^{-2}\ln x$

21. $y(x) = 2\cos x - 5\sin x + 3x$

22. $y(x) = 4e^{2x} - e^{-2x} - 3$

23. $y(x) = e^{-x} + 4e^{3x} - 2$

24. $y(x) = e^x(3\cos x + 4\sin x) + x + 1$

38. $y_2(x) = \dfrac{1}{x^3}$

39. $y_2(x) = xe^{x/2}$

40. $y_2(x) = xe^x$

41. $y_2(x) = x + 2$

42. $y_2(x) = 1 + x^2$

Section 3.3

1. $y(x) = c_1 e^{2x} + c_2 e^{-2x}$

2. $y(x) = c_1 + c_2 e^{3x/2}$

3. $y(x) = c_1 e^{2x} + c_2 e^{-5x}$

4. $y(x) = c_1 e^{x/2} + c_2 e^{3x}$

5. $y(x) = c_1 e^{-3x} + c_2 xe^{-3x}$

6. $y(x) = e^{-5x/2}\left[c_1 \exp\left(\frac{1}{2}x\sqrt{5}\right) + c_2 \exp\left(-\frac{1}{2}x\sqrt{5}\right)\right]$

7. $y(x) = c_1 e^{3x/2} + c_2 xe^{3x/2}$

8. $y(x) = e^{3x}(c_1 \cos 2x + c_2 \sin 2x)$

9. $y(x) = e^{-4x}(c_1 \cos 3x + c_2 \sin 3x)$

10. $y(x) = c_1 + c_2 x + c_3 x^2 + c_4 e^{-3x/5}$

11. $y(x) = c_1 + c_2 x + c_3 e^{4x} + c_4 xe^{4x}$

12. $y(x) = c_1 + c_2 e^x + c_3 xe^x + c_4 x^2 e^x$

13. $y(x) = c_1 + c_2 e^{-2x/3} + c_3 xe^{-2x/3}$

14. $y(x) = c_1 e^x + c_2 e^{-x} + c_3 \cos 2x + c_4 \sin 2x$

15. $y(x) = c_1 e^{2x} + c_2 xe^{2x} + c_3 e^{-2x} + c_4 xe^{-2x}$

16. $y(x) = (c_1 + c_2 x)\cos 3x + (c_3 + c_4 x)\sin 3x$

17. $y(x) = c_1 \cos\left(x/\sqrt{2}\right) + c_2 \sin\left(x/\sqrt{2}\right) + c_3 \cos\left(2x/\sqrt{3}\right) + c_4 \sin\left(2x/\sqrt{3}\right)$

18. $y(x) = c_1 e^{2x} + c_2 e^{-2x} + c_3 \cos 2x + c_4 \sin 2x$

19. $y(x) = c_1 e^x + c_2 e^{-x} + c_3 xe^{-x}$

20. $y(x) = e^{-x/2}\left[(c_1 + c_2 x)\cos\left(\frac{1}{2}x\sqrt{3}\right) + (c_3 + c_3 x)\sin\left(\frac{1}{2}x\sqrt{3}\right)\right]$

21. $y(x) = 5e^x + 2e^{3x}$

22. $y(x) = e^{-x/3}\left[3\cos\left(x/\sqrt{3}\right) + 5\sqrt{3}\sin\left(x/\sqrt{3}\right)\right]$

23. $y(x) = e^{3x}(3\cos 4x - 2\sin 4x)$

24. $y(x) = \frac{1}{2}(-7 + e^{9x} + 8e^{-x/2})$

25. $y(x) = \frac{1}{4}(-13 + 6x + 9e^{-2x/3})$

26. $y(x) = \frac{1}{5}(24 - 9e^{-5x} - 25xe^{-5x})$

27. $y(x) = c_1 e^x + c_2 e^{-2x} + c_3 xe^{-2x}$

28. $y(x) = c_1 e^{2x} + c_2 e^{-x} + c_3 e^{-x/2}$

29. $y(x) = c_1 e^{-3x} + e^{3x/2}\left[c_2 \cos\left(\frac{3}{2}x\sqrt{3}\right) + c_3 \sin\left(\frac{3}{2}x\sqrt{3}\right)\right]$

30. $y(x) = c_1 e^{-x} + c_2 e^x + c_3 \cos(x\sqrt{3}) + c_4 \sin(x\sqrt{3})$

31. $y(x) = c_1 e^x + e^{-2x}(c_2 \cos 2x + c_3 \sin 2x)$

32. $y(x) = c_1 e^{2x} + (c_2 + c_3 x + c_4 x^2)e^{-x}$

33. $y(x) = c_1 e^{3x} + e^{-3x}(c_2 \cos 3x + c_3 \sin 3x)$

34. $y(x) = c_1 e^{2x/3} + c_2 \cos 2x + c_3 \sin 2x$

35. $y(x) = c_1 e^{-x/2} + c_2 e^{-x/3} + c_3 \cos 2x + c_4 \sin 2x$

36. $y(x) = c_1 e^{7x/9} + e^{-x}(c_2 \cos x + c_3 \sin x)$

37. $y(x) = 11 + 5x + 3x^2 + 7e^x$

38. $y(x) = 2e^{5x} - 2\cos 10x$

39. $y^{(3)} - 6y'' + 12y' - 8y = 0$

40. $y^{(3)} - 2y'' + 4y' - 8y = 0$

41. $y^{(4)} - 16y = 0$

42. $y^{(6)} + 12y^{(4)} + 48y'' + 64y = 0$

44. (a) $x = i, -2i$ (b) $x = -i, 3i$

45. $y(x) = c_1 e^{-ix} + c_2 e^{3ix}$

46. $y(x) = c_1 e^{3ix} + c_2 e^{-2ix}$

47. $y(x) = c_1 \exp\left(\left[1 + i\sqrt{3}\right]x\right) + c_2 \exp\left(-\left[1 + i\sqrt{3}\right]x\right)$

48. $y(x) = \frac{1}{3}\left(e^x + \exp\left[\frac{1}{2}\left(-1 + i\sqrt{3}\right)x\right] + \exp\left[\frac{1}{2}\left(-1 - i\sqrt{3}\right)x\right]\right)$

49. $y(x) = 2e^{2x} - 5e^{-x} + 3\cos x - 9\sin x$

52. $y(x) = c_1 \cos(3\ln x) + c_2 \sin(3\ln x)$

53. $y(x) = x^{-3}[c_1 \cos(4\ln x) + c_2 \sin(4\ln x)]$

54. $y(x) = c_1 + c_2 \ln x + c_3 x^{-3}$

55. $y(x) = c_1 + x^2(c_2 + c_3 \ln x)$

56. $y(x) = c_1 + c_2 \ln x + c_3 (\ln x)^2$

57. $y(x) = c_1 + x^3 \left(c_2 x^{-\sqrt{3}} + c_3 x^{+\sqrt{3}} \right)$

58. $y(x) = x^{-1}[c_1 + c_2 \ln x + c_3 (\ln x)^2]$

Section 3.4

1. Frequency: 2 rad/s ($1/\pi$ Hz); period: π s

2. Frequency: 8 rad/sec ($4/\pi$ Hz); period: $\pi/4$ sec

3. Amplitude: 2 m; frequency: 5 rad/s; period: $2\pi/5$ s

4. (a) $x(t) = \frac{13}{12} \cos(12t - \alpha)$ with $\alpha = 2\pi - \tan^{-1}(5/12) \approx 5.8884$; (b) Amplitude: $\frac{13}{12}$ m; period: $\pi/6$ sec

6. About 7.33 mi

7. About 10450 ft

8. 29.59 in.

10. Amplitude: 100 cm; period: about 2.01 sec

11. About 3.8 in.

13. (a) $x(t) = 50(e^{-2t/5} - e^{-t/2})$; (b) 4.096 exactly

14. (a) $x(t) = 25e^{-t/5} \cos(3t - \alpha)$ with $\alpha = \tan^{-1}(3/4) \approx 0.6435$; (b) envelope curves $x = \pm 25e^{-t/5}$; pseudoperiod $2\pi/3$

15. $x(t) = 4e^{-2t} - 2e^{-4t}$, overdamped; $u(t) = 2\cos(2\sqrt{2}\,t)$

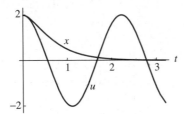

16. $x(t) = 4e^{-3t} - 2e^{-7t}$, overdamped; $u(t) \approx 2\sqrt{\frac{22}{21}} \cos(\sqrt{21}\,t - 0.2149)$

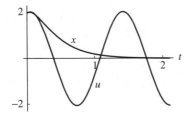

17. $x(t) = 5e^{-4t}(2t + 1)$, critically damped; $u(t) \approx \frac{5}{2}\sqrt{5} \cos(4t - 5.8195)$

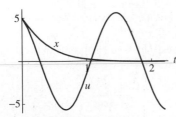

18. $x(t) = 2e^{-3t} \cos\left(4t - \frac{3\pi}{2}\right)$, underdamped; $u(t) = \frac{8}{5} \cos\left(5t - \frac{3\pi}{2}\right)$

19. $x(t) \approx \frac{1}{3}\sqrt{313}\, e^{-5t/2} \cos(6t - 0.8254)$, underdamped; $u(t) \approx \frac{4}{13}\sqrt{233} \cos\left(\frac{13}{2}t - 0.5517\right)$

20. $x(t) \approx 13e^{-4t} \cos(2t - 1.1760)$, underdamped; $u(t) \approx \sqrt{\frac{129}{5}} \cos(2\sqrt{5}\,t - 0.1770)$

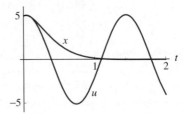

21. $x(t) \approx 10e^{-5t} \cos(10t - 0.9273)$, underdamped; $u(t) \approx 2\sqrt{14} \cos(5\sqrt{5}\,t - 0.6405)$

22. (b) The time-varying amplitude is $\frac{2}{3}\sqrt{3}$, the frequency is $4\sqrt{3}$ rad/s, and the phase angle is $\pi/6$.

23. (a) $k \approx 7018$ lb/ft; (b) After about 2.47 s

34. Damping constant: $c \approx 11.51$ lb/ft/s; spring constant: $k \approx 189.68$ lb/ft

Section 3.5

1. $y_p(x) = \frac{1}{25} e^{3x}$

2. $y_p(x) = -\frac{1}{4}(5 + 6x)$

3. $y_p(x) = \frac{1}{39}(\cos 3x - 5 \sin 3x)$

4. $y_p(x) = \frac{1}{9}(-4e^x + 3xe^x)$

5. $y_p(x) = \frac{1}{26}(13 + 3\cos 2x - 2\sin 2x)$

6. $y_p(x) = \frac{1}{343}(4 - 56x + 49x^2)$

7. $y_p(x) = -\frac{1}{6}(e^x - e^{-x}) = -\frac{1}{3} \sinh x$

8. $y_p(x) = \frac{1}{4}x \sinh 2x$

9. $y_p(x) = -\frac{1}{3} + \frac{1}{16}(2x^2 - x)e^x$

10. $y_p(x) = \frac{1}{6}(2x \sin 3x - 3x \cos 3x)$

11. $y_p(x) = \frac{1}{8}(3x^2 - 2x)$

12. $y_p(x) = 2x + \frac{1}{2}x \sin x$

13. $y_p(x) = \frac{1}{65}e^x(7 \sin x - 4 \cos x)$

14. $y_p(x) = \frac{1}{24}(-3x^2 e^x + x^3 e^x)$

15. $y_p(x) \equiv -17$

16. $y_p(x) = \frac{1}{81}(45 + e^{3x} - 6xe^{3x} + 9x^2 e^{3x})$

17. $y_p(x) = \frac{1}{4}(x^2 \sin x - x \cos x)$

18. $y_p(x) = -\frac{1}{144}(24xe^x - 19xe^{2x} + 6x^2 e^{2x})$

19. $y_p(x) = \frac{1}{8}(10x^2 - 4x^3 + x^4)$

20. $y_p(x) = -7 + \frac{1}{3}xe^x$

21. $y_p(x) = xe^x(A \cos x + B \sin x)$

22. $y_p(x) = Ax^3 + Bx^4 + Cx^5 + Dxe^x$

23. $y_p(x) = Ax \cos 2x + Bx \sin 2x + Cx^2 \cos 2x + Dx^2 \sin 2x$

24. $y_p(x) = Ax + Bx^2 + (Cx + Dx^2)e^{-3x}$

25. $y_p(x) = Axe^{-x} + Bx^2 e^{-x} + Cxe^{-2x} + Dx^2 e^{-2x}$

26. $y_p(x) = (Ax + Bx^2)e^{3x} \cos 2x + (Cx + Dx^2)e^{3x} \sin 2x$

27. $y_p(x) = Ax \cos x + Bx \sin x + Cx \sin x + Dx \sin x$

28. $y_p(x) = (Ax + Bx^2 + Cx^3) \cos 3x + (Dx + Ex^2 + Fx^3) \sin 3x$

29. $y_p(x) = Ax^3 e^x + Bx^4 e^x + Cxe^{2x} + Dxe^{-2x}$

30. $y_p(x) = (A + Bx + Cx^2) \cos x + (D + Ex + Fx^2) \sin x$

31. $y(x) = \cos 2x + \frac{3}{4} \sin 2x + \frac{1}{2}x$

32. $y(x) = \frac{1}{6}(15e^{-x} - 16e^{2x} + e^x)$

33. $y(x) = \cos 3x - \frac{2}{15} \sin 3x + \frac{1}{5} \sin 2x$

34. $y(x) = \cos x - \sin x + \frac{1}{2}x \sin x$

35. $y(x) = e^x(2 \cos x - \frac{5}{2} \sin x) + \frac{1}{2}x + 1$

36. $y(x) = \frac{1}{192}(234 + 240x - 9e^{-2x} - 33e^{2x} - 12x^2 - 4x^4)$

37. $y(x) = 4 - 4e^x + 3xe^x + x - \frac{1}{2}x^2 e^x + \frac{1}{6}x^3 e^x$

38. $y(x) = \frac{1}{85}[e^{-x}(176 \cos x + 197 \sin x) - (6 \cos 3x + 7 \sin 3x)]$

39. $y(x) = 3 + 3x - \frac{1}{2}x^2 + \frac{1}{6}x^3 + 4e^{-x} + xe^{-x}$

40. $y(x) = \frac{1}{4}(5e^{-x} + 5e^x + 10 \cos x - 20)$

41. $y_p(x) = 255 - 450x + 30x^2 + 20x^3 + 10x^4 - 4x^5$

42. $y(x) = 10e^{-x} + 35e^{2x} + 210 \cos x + 390 \sin x + y_p(x)$ where $y_p(x)$ is the particular solution of Problem 41.

43. (b) $y(x) = c_1 \cos 2x + c_2 \sin 2x + \frac{1}{4} \cos x - \frac{1}{20} \cos 3x$

44. $y(x) = e^{-x/2}\left[c_1 \cos\left(\frac{1}{2}x\sqrt{3}\right) + c_2 \sin\left(\frac{1}{2}x\sqrt{3}\right)\right] + \frac{1}{26}(-3 \cos 2x + 2 \sin 2x) + \frac{1}{482}(-15 \cos 4x + 4 \sin 4x)$

45. $y(x) = c_1 \cos 3x + c_2 \sin 3x + \frac{1}{24} - \frac{1}{10} \cos 2x - \frac{1}{56} \cos 4x$

46. $y(x) = c_1 \cos x + c_2 \sin x + \frac{1}{16}(3x \cos x + 3x^2 \sin x) + \frac{1}{128}(3 \sin 3x - 4x \cos 3x)$

47. $y_p(x) = \frac{2}{3}e^x$

48. $y_p(x) = \frac{1}{12}(6x + 1)e^{-2x}$

49. $y_p(x) = x^2 e^{2x}$

50. $y_p(x) = \frac{1}{16}(4x \cosh 2x - \sinh 2x)$

51. $y_p(x) = -\frac{1}{4}(\cos 2x \cos x - \sin 2x \sin x) + \frac{1}{20}(\cos 5x \cos 2x + \sin 5x \sin 2x) = -\frac{1}{5} \cos 3x$ (!)

52. $y_p(x) = -\frac{1}{6}x \cos 3x$

53. $y_p(x) = \frac{2}{3}x \sin 3x + \frac{2}{9}(\cos 3x) \ln|\cos 3x|$

54. $y_p(x) = -1 - (\cos x) \ln|\csc x - \cot x|$

55. $y_p(x) = \frac{1}{8}(1 - x \sin 2x)$

56. $y_p(x) = -\frac{1}{9}e^x(3x + 2)$

58. $y_p(x) = x^3(\ln x - 1)$

59. $y_p(x) = \frac{1}{4}x^4$

60. $y_p(x) = -\frac{72}{5}x^{4/3}$

61. $y_p(x) = \ln x$

62. $y_p(x) = -x^2 + x \ln\left|\frac{1 + x}{1 - x}\right| + \frac{1}{2}(1 + x^2) \ln|1 - x^2|$

Section 3.6

1. $x(t) = 2 \cos 2t - 2 \cos 3t$

2. $x(t) = \frac{3}{2} \sin 2t - \sin 3t$

3. $x(t) = \sqrt{138388} \cos(10t - \alpha) + 5 \cos(5t - \beta)$ with $\alpha = 2\pi - \tan^{-1}(1/186) \approx 6.2778$ and $\beta = \tan^{-1}(4/3) \approx 0.9273$.

4. $x(t) = 2\sqrt{106} \cos(5t - \alpha) + 10 \cos 4t$ with $\alpha = \pi - \tan^{-1}(9/5) \approx 2.0779$

5. $x(t) = (x_0 - C) \cos \omega_0 t + C \cos \omega t$, where $C = F_0/(k - m\omega^2)$

7. $x_{sp}(t) = \frac{10}{13} \cos(3t - \alpha)$ with $\alpha = \pi - \tan^{-1}(12/5) \approx 1.9656$

8. $x_{sp}(t) = \frac{4}{25} \cos(5t - \alpha)$ with $\alpha = 2\pi - \tan^{-1}(3/4) \approx 5.6397$

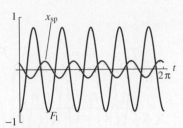

9. $x_{sp}(t) = \frac{3}{\sqrt{40001}} \cos(10t - \alpha)$ with
$\alpha = \pi + \tan^{-1}(199/20) \approx 4.6122$

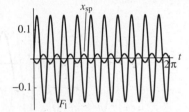

10. $x_{sp}(t) = \frac{10}{793} \sqrt{61} \cos(10t - \alpha)$ with
$\alpha = \pi + \tan^{-1}(171/478) \approx 3.4851$

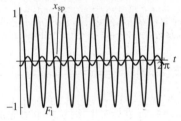

11. $x_{sp}(t) = \frac{\sqrt{10}}{4} \cos(3t - \alpha)$ with $\alpha = \pi - \tan^{-1}(3) \approx 1.8925$
$x_{tr}(t) = \frac{5}{4}\sqrt{2}\, e^{-2t} \cos(t - \beta)$ with $\beta = 2\pi - \tan^{-1}(7) \approx 4.8543$

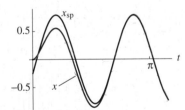

12. $x_{sp}(t) = \frac{5}{3\sqrt{29}} \cos(3t - \alpha)$ with $\alpha = \pi + \tan^{-1}(2/5) \approx 3.5221$
$x_{tr}(t) = \frac{25}{6\sqrt{29}} e^{-3t} \cos(2t - \beta)$ with $\beta = \tan^{-1}(5/2) \approx 1.1903$

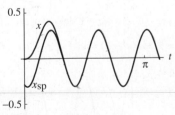

13. $x_{sp}(t) = \frac{300}{\sqrt{1469}} \cos(10t - \alpha)$ with
$\alpha = \pi - \tan^{-1}(10/37) \approx 2.9320$
$x_{tr}(t) = 2\sqrt{\frac{113314}{1469}}\, e^{-t} \cos(5t - \beta)$ with
$\beta = 2\pi - \tan^{-1}(421/12895) \approx 6.2505$

14. $x_{sp}(t) = \sqrt{485} \cos(t - \alpha)$ with $\alpha = \tan^{-1}(22) \approx 1.5254$
$x_{tr}(t) = \sqrt{3665}\, e^{-4t} \cos(3t - \beta)$ with
$\beta = \pi + \tan^{-1}(52/31) \approx 4.1748$

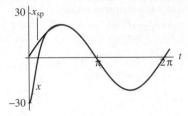

15. $C(\omega) = 2/\sqrt{4 + \omega^4}$; there is no practical resonance frequency.

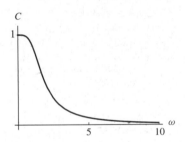

16. $C(\omega) = 10/\sqrt{25 + 6\omega^2 + \omega^4}$; there is no practical resonance frequency.

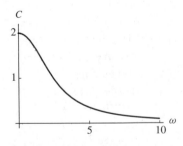

17. $C(\omega) = 50/\sqrt{2025 - 54\omega^2 + \omega^4}$; there is practical resonance at frequency $\omega = 3\sqrt{3}$.

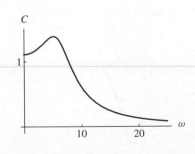

18. $C(\omega) = 100/\sqrt{422500 - 1200\omega^2 + \omega^4}$; there is practical resonance at frequency $\omega = 10\sqrt{6}$.

19. $\omega = \sqrt{384}$ rad/sec (approximately 3.12 Hz)
20. $\omega \approx 44.27$ rad/sec (approximately 7.05 Hz)
21. $\omega_0 = \sqrt{(g/L) + (k/m)}$
22. $\omega_0 = \sqrt{k/(m + I/a^2)}$
23. (a) Natural frequency: $\sqrt{10}$ rad/s (approximately 0.50 Hz); (b) amplitude: approximately 10.625 in.

Section 3.7

1. $I(t) = 4e^{-5t}$
2. $I(t) = 4(1 - e^{-5t})$
3. $I(t) = \frac{4}{145}(\cos 60t + 12\sin 60t - e^{-5t})$
4. $I(t) = 5(e^{-10t} - e^{-20t})$; $I_{max} = I(\frac{1}{10}\ln 2) = 5/4$.
5. $I(t) = \frac{5}{6}e^{-10t}\sin 60t$
6. $I_{sp}(t) = \frac{1}{37}(-21\cos 60t + 22\sin 60t) = (5/\sqrt{37})\cos(60t - \alpha)$, where $\alpha = \pi - \tan^{-1}(22/21) \approx 2.3329$.
7. (a) $Q(t) = E_0 C(1 - e^{-t/RC})$; $I(t) = (E_0/R)e^{-t/RC}$
8. (a) $Q(t) = 10te^{-5t}$; $I(t) = 10(1 - 5t)e^{-5t}$;
 (b) $Q_{max} = Q(1/5) = 2e^{-1}$.
9. (a) $Q(t) = (\cos 120t + 6\sin 120t - e^{-20t})/1480$
 $I(t) = (36\cos 120t - 6\sin 120t + e^{-20t})/74$
 (b) $I_{sp} = \frac{6}{74}(6\cos 120t - \sin 120t) = \frac{3}{\sqrt{37}}\cos(120t - \alpha)$ with
 $\alpha = 2\pi - \tan^{-1}\frac{1}{6}$.
11. $I_{sp}(t) = \frac{10}{\sqrt{37}}\sin(2t - \delta)$ with $\delta = 2\pi - \tan^{-1}(1/6) \approx 6.1180$
12. $I_{sp}(t) = \frac{2}{\sqrt{17}}\sin(10t - \delta)$ with $\delta = 2\pi - \tan^{-1}(1/4) \approx 6.0382$
13. $I_{sp}(t) = \frac{20}{\sqrt{13}}\sin(5t - \delta)$ with $\delta = 2\pi - \tan^{-1}(2/3) \approx 5.6952$
14. $I_{sp}(t) \approx 0.9990\sin(100t - 0.8272)$
15. $I_{sp}(t) \approx 0.1591\sin(60\pi t - 4.8576)$
16. $I_{sp}(t) \approx 1.6125\sin(377t - 1.2282)$
17. $I(t) = -25e^{-4t}\sin 3t$
18. $I(t) = \frac{50}{171}(19e^{-10t} - 18e^{-20t} - e^{-t})$
19. $I(t) = 10e^{-20t} - 10e^{-10t} - 50te^{-10t}$
20. $I(t) = -\frac{10}{37\sqrt{11}}e^{-3t/2}(\sqrt{11}\cos t\sqrt{11}/2 + 27\sin t\sqrt{11}/2) + \frac{10}{37}(\cos 2t + 6\sin 2t)$

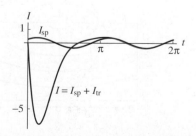

21. $I(t) = -\frac{10}{39}e^{-t}(12\cos 3t + 47\sin 3t) + \frac{20}{13}(2\cos 5t + 3\sin 5t)$

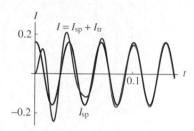

22. $I(t) \approx -e^{-25t}(0.1574\cos 25t\sqrt{159} + 0.0262\sin 25t\sqrt{159}) + (0.1574\cos 60\pi t + 0.0230\sin 60\pi t)$

Section 3.8

1. Only positive eigenvalues $\{n^2\pi^2/4\}$ and associated eigenfunctions $\{\cos(n\pi x/2)\}$ for $n = 1, 3, 5, \ldots$.
2. Eigenvalue $\lambda_0 = 0$ with eigenfunction $y_0(x) \equiv 1$, and positive eigenvalues $\{n^2\}$ with associated eigenfunctions $\{\cos nx\}$ for $n = 1, 2, 3, \ldots$.
3. Only positive eigenvalues $\{n^2/4\}$ for $n = 1, 2, 3, \ldots$. The nth eigenfunction $y_n(x)$ is $\cos(nx/2)$ if n is odd, $\sin(nx/2)$ if n is even.
4. Eigenvalue $\lambda_0 = 0$ with eigenfunction $y_0(x) \equiv 1$, and positive eigenvalues $\{n^2/4\}$ for $n = 1, 2, 3, \ldots$. The nth eigenfunction $y_n(x)$ is $\sin(nx/2)$ if n is odd, $\cos(nx/2)$ if n is even.
5. Only positive eigenvalues $\{n^2\pi^2/64\}$ for $n = 1, 2, 3, \ldots$. The nth eigenfunction $y_n(x)$ is $\cos(n\pi x/8) + \sin(n\pi x/8)$ if n is odd, $\cos(n\pi x/8) - \sin(n\pi x/8)$ if n is even.
7–8. In the figure below, points of intersection of the curve $y = \tan z$ with the lines $y = \pm z$ are labeled with their z-coordinates. We see that α_n lies just to the right of the vertical line $z = (2n - 1)\pi/2$, while β_n lies just to the left of the line $z = (2n + 1)\pi/2$.

Chapter 4

Section 4.1

1. $x_1' = x_2$, $x_2' = -7x_1 - 3x_2 + t^2$
2. $x_1' = x_2$, $x_2' = x_3$, $x_3' = x_4$, $x_4' = -x_1 + 3x_2 - 6x_3 + \cos 3t$
3. $x_1' = x_2$, $t^2x_2' = (1 - t^2)x_1 - tx_2$
4. $x_1' = x_2$, $x_2' = x_3$, $t^3x_3' = -5x_1 - 3tx_2 + 2t^2x_3 + \ln t$
5. $x_1' = x_2$, $x_2' = x_3$, $x_3' = x_2^2 + \cos x_1$

6. $x_1' = x_2, x_2' = 5x_1 - 4y_1, y_1' = y_2, y_2' = -4x_1 + 5y_1$

7. $x_1' = x_2, y_1' = y_2, x_2' = -kx_1 \cdot (x_1^2 + y_1^2)^{-3/2},$
$y_2' = -ky_1 \cdot (x_1^2 + y_1^2)^{-3/2}$

8. $x_1' = x_2, x_2' = -4x_1 + 2y_1 - 3x_2$
$y_1' = y_2, y_2' = 3x_1 - y_1 - 2y_2 + \cos t$

9. $x_1' = x_2, y_1' = y_2, z_1' = z_2, x_2' = 3x_1 - y_1 + 2z_1,$
$y_2' = x_1 + y_1 - 4z_1, z_2' = 5x_1 - y_1 - z_1$

10. $x_1' = x_2, x_2' = x_1(1 - y_1)$
$y_1' = y_2, y_2' = y_1(1 - x_1)$

11. $x(t) = A \cos t + B \sin t, \ y(t) = B \cos t - A \sin t$

12. $x(t) = Ae^t + Be^{-t}, \ y(t) = Ae^t - Be^{-t}$

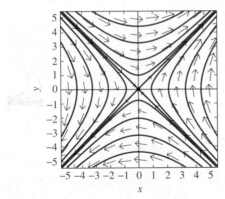

13. $x(t) = A \cos 2t + B \sin 2t, \ y(t) = -B \cos 2t + A \sin 2t;$
$x(t) = \cos 2t, y(t) = \sin 2t$

14. $x(t) = A \cos 10t + B \sin 10t, \ y(t) = B \cos 10t - A \sin 10t;$
$x(t) = 3 \cos 10t + 4 \sin 10t, y(t) = 4 \cos 10t - 3 \sin 10t$

15. $x(t) = A \cos 2t + B \sin 2t, \ y(t) = 4B \cos 2t - 4A \sin 2t$

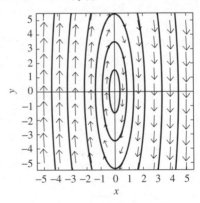

16. $x(t) = A \cos 4t + B \sin 4t, \ y(t) = \frac{1}{2} B \cos 4t - \frac{1}{2} A \sin 4t$

17. $x(t) = Ae^{-3t} + Be^{2t}$, $y(t) = -3Ae^{-3t} + 2Be^{2t}$; $x(t) = e^{2t}$,
$y(t) = 2e^{2t}$

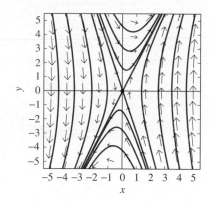

18. $x(t) = Ae^{-2t} + Be^{-5t}$, $y(t) = 2Ae^{-2t} + 5Be^{-5t}$; $A = \frac{17}{3}$ and
$B = -\frac{11}{3}$ in the particular solution.

19. $x(t) = -e^{-2t}\sin 3t$, $y(t) = e^{-2t}(3\cos 3t + 2\sin 3t)$

20. $x(t) = (A + Bt)e^{3t}$, $y(t) = (3A + B + 3Bt)e^{3t}$

27. $2(I_1' - I_2') + 50I_1 = 100\sin 60t$, $2(I_2' - I_1') + 25I_2 = 0$

28. $I_1' = -20(I_1 - I_2)$, $I_2' = 40(I_1 - I_2)$

Section 4.2

1. $x(t) = a_1 e^{-t} + a_2 e^{2t}$, $y(t) = a_2 e^{2t}$

2. $x(t) = (c_1 + c_2 t)e^{-t}$, $y(t) = (c_1 - \frac{1}{2}c_2 + c_2 t)e^{-t}$

3. $x(t) = \frac{4}{5}(e^{3t} - e^{-2t})$, $y(t) = \frac{2}{5}(6e^{3t} - e^{-2t})$

4. $x(t) = \frac{1}{2}(3e^{2t} - e^{-2t})$, $y(t) = \frac{1}{2}(3e^{2t} - 5e^{-2t})$

5. $x(t) = e^{-t}(a_1 \cos 2t + a_2 \sin 2t)$,
$y(t) = -\frac{1}{2}e^{-t}[(a_1 + a_2)\cos 2t + (a_2 - a_1)\sin 2t]$

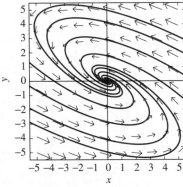

6. $x(t) = e^{-2t}(3\cos 3t + 9\sin 3t)$, $y(t) = e^{-2t}(2\cos 3t - 4\sin 3t)$

7. $x(t) = a_1 e^{2t} + a_2 e^{3t} - \frac{1}{3}t + \frac{1}{18}$, $y(t) = -2a_1 e^{2t} - a_2 e^{3t} - \frac{2}{3}t - \frac{5}{9}$

8. $x(t) = c_1 e^{t} + c_2 e^{3t} + e^{2t}$, $y(t) = -c_1 e^{t} + c_2 e^{3t}$

9. $x(t) = 3a_1 e^{t} + a_2 e^{-t} - \frac{1}{5}(7\cos 2t + 4\sin 2t)$,
$y(t) = a_1 e^{t} + a_2 e^{-t} - \frac{1}{5}(2\cos 2t + 4\sin 2t)$

10. $x(t) = e^{t}$, $y(t) = -e^{t}$

11. $x(t) = a_1 \cos 3t + a_2 \sin 3t - \frac{11}{20}e^{t} - \frac{1}{4}e^{-t}$,
$y(t) = \frac{1}{3}[(a_1 - a_2)\cos 3t + (a_1 + a_2)\sin 3t] + \frac{1}{10}e^{t}$

12. $x(t) = c_1 e^{2t} + c_2 e^{-2t} + c_3 e^{3t} + c_4 e^{-3t}$,
$y(t) = -c_1 e^{2t} - c_2 e^{-2t} + \frac{3}{2}c_3 e^{3t} + \frac{3}{2}c_4 e^{-3t}$

13. $x(t) = a_1 \cos 2t + a_2 \sin 2t + b_1 \cos 3t + b_2 \sin 3t$,
$y(t) = \frac{1}{2}(a_1 \cos 2t + a_2 \sin 2t) - 2(b_1 \cos 3t + b_2 \sin 3t)$

14. $x(t) = c_1 \cos 2t + c_2 \sin 2t + \frac{1}{3}\sin t$,
$y(t) = c_1 \cos 2t + c_2 \sin 2t + c_3 \cos 2t\sqrt{2} +$
$c_4 \sin 2t\sqrt{2} + \frac{4}{21}\sin t$

15. $x(t) = a_1 \cos t + a_2 \sin t + b_1 \cos 2t + b_2 \sin 2t$,
$y(t) = a_2 \cos t - a_1 \sin t + b_2 \cos 2t - b_1 \sin 2t$

17. $x(t) = a_1 \cos t + a_2 \sin t + b_1 e^{2t} + b_2 e^{-2t}$,
$y(t) = 3a_2 \cos t - 3a_1 \sin t + b_1 e^{2t} - b_2 e^{-2t}$

18. $x(t) = \frac{1}{6}(4c_1 e^{3t} - 3c_2 e^{-4t})$, $y(t) = c_1 e^{3t} + c_2 e^{-4t}$,
$z(t) = \frac{1}{6}(-4c_1 e^{3t} + 3c_2 e^{-4t})$

19. $x(t) = a_1 + a_2 e^{4t} + a_3 e^{8t}$, $y(t) = 2a_1 - 2a_3 e^{8t}$,
$z(t) = 2a_1 - 2a_2 e^{4t} + 2a_3 e^{8t}$

20. $x(t) = a_1 e^{2t} + a_2 e^{-t} + \frac{2}{3}te^{-t}$, $y(t) = a_1 e^{2t} + b_2 e^{-t} - \frac{1}{3}te^{-t}$,
$z(t) = a_1 e^{2t} - \left(a_2 + b_2 + \frac{1}{3}\right)e^{-t} - \frac{1}{3}te^{-t}$

23. Infinitely many solutions

24. No solution

25. Infinitely many solutions

26. Two arbitrary constants

27. No arbitrary constants

28. No solution

29. Four arbitrary constants

31. $I_1(t) = 2 + e^{-5t}\left[-2\cos\left(10t/\sqrt{6}\right) + 4\sqrt{6}\sin\left(10t/\sqrt{6}\right)\right]$,
$I_2(t) = \left(20/\sqrt{6}\right)e^{-5t}\sin\left(10t/\sqrt{6}\right)$

32. $I_1(t) = \frac{1}{1321}(120e^{-25t/3} - 120\cos 60t + 1778\sin 60t)$,
$I_2(t) = \frac{1}{1321}(-240e^{-25t/3} + 240\cos 60t + 1728\sin 60t)$

33. $I_1(t) = \frac{2}{3}(2 + e^{-60t})$, $I_2(t) = \frac{4}{3}(1 - e^{-60t})$

37. (a) $x(t) = a_1 \cos 5t + a_2 \sin 5t + b_1 \cos 5t\sqrt{3} + b_2 \sin 5t\sqrt{3}$,
$y(t) = 2a_1 \cos 5t + 2a_2 \sin 5t - 2b_1 \cos 5t\sqrt{3} - 2b_2 \sin 5t\sqrt{3}$;
(b) In the natural mode with frequency $\omega_1 = 5$, the masses
move in the same direction, whereas in the natural mode with
frequency $\omega_2 = 5\sqrt{3}$ they move in opposite directions. In each
case the amplitude of the motion of m_2 is twice that of m_1.

39. $x(t) = a_1 \cos t + a_2 \sin t + b_1 \cos 2t + b_2 \sin 2t$,
$y(t) = 2a_1 \cos t + 2a_2 \sin t - b_1 \cos 2t - b_2 \sin 2t$.
In the natural mode with frequency $\omega_1 = 1$ the masses move in
the same direction, with the amplitude of motion of the second
mass twice that of the first mass. In the natural mode with
frequency $\omega_2 = 2$ they move in opposite directions with the

same amplitude of motion.

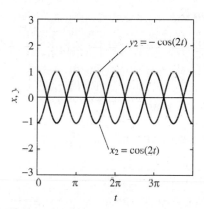

40. $x(t) = a_1 \cos 5t + a_2 \sin 5t + b_1 \cos 10t + b_2 \sin 10t,$
$y(t) = 2a_1 \cos 5t + 2a_2 \sin 5t - b_1 \cos 10t - b_2 \sin 10t.$

41. $x(t) = a_1 \cos t + a_2 \sin t + b_1 \cos 3t + b_2 \sin 3t,$
$y(t) = a_1 \cos t + a_2 \sin t - b_1 \cos 3t - b_2 \sin 3t.$
In the natural mode with frequency $\omega_1 = 1$ the masses move in
the same direction, while in the natural mode with frequency
$\omega_2 = 3$ they move in opposite directions. In each case the
amplitudes of motion of the two masses are equal.

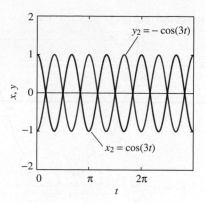

42. $x(t) = a_1 \cos t + a_2 \sin t + b_1 \cos 2t + b_2 \sin 2t,$
$y(t) = a_1 \cos t + a_2 \sin t - \frac{1}{2}b_1 \cos 2t - \frac{1}{2}b_2 \sin 2t.$

43. $x(t) = a_1 \cos t + a_2 \sin t + b_1 \cos t\sqrt{5} + b_2 \sin t\sqrt{5},$
$y(t) = a_1 \cos t + a_2 \sin t - b_1 \cos t\sqrt{5} - b_2 \sin t\sqrt{5}.$
In the natural mode with frequency $\omega_1 = 1$ the masses move in
the same direction, while in the natural mode with frequency
$\omega_2 = \sqrt{5}$ they move in opposite directions. In each case the
amplitudes of motion of the two masses are equal.

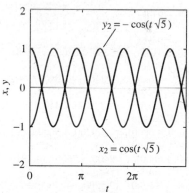

44. $x(t) = a_1 \cos t\sqrt{2} + a_2 \sin t\sqrt{2} + b_1 \cos 2t + b_2 \sin 2t,$
$y(t) = a_1 \cos t\sqrt{2} + a_2 \sin t\sqrt{2} - b_1 \cos 2t - b_2 \sin 2t.$

45. $x(t) = a_1 \cos t\sqrt{2} + a_2 \sin t\sqrt{2} + b_1 \cos t\sqrt{8} + b_2 \sin t\sqrt{8},$
$y(t) = a_1 \cos t\sqrt{2} + a_2 \sin t\sqrt{2} - \frac{1}{2}b_1 \cos t\sqrt{8} - \frac{1}{2}b_2 \sin t\sqrt{8}.$
In the natural mode with frequency $\omega_1 = \sqrt{2}$ the two masses
move in the same direction with equal amplitudes of oscillation.
In the natural mode with frequency $\omega_2 = \sqrt{8} = 2\sqrt{2}$ the two
masses move in opposite directions with the amplitude of m_2

being half that of m_1.

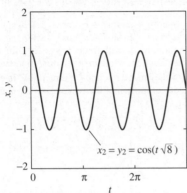

46. $x(t) = a_1 \cos 2t + a_2 \sin 2t + b_1 \cos 4t + b_2 \sin 4t$,
$y(t) = a_1 \cos 2t + a_2 \sin 2t - b_1 \cos 4t - b_2 \sin 4t$.

Section 4.3

The format for the first eight answers is this: $(x(t), y(t))$ at $t = 0.2$ by the Euler method, by the improved Euler method, by the Runge-Kutta method, and finally the actual values.

1. $(0.8800, 2.5000)$, $(0.9600, 2.6000)$, $(1.0027, 2.6401)$, $(1.0034, 2.6408)$

2. $(0.8100, -0.8100)$, $(0.8200, -0.8200)$, $(0.8187, -0.8187)$, $(0.8187, -0.8187)$

3. $(2.8100, 2.3100)$, $(3.2200, 2.6200)$, $(3.6481, 2.9407)$, $(3.6775, 2.9628)$

4. $(3.3100, -1.6200)$, $(3.8200, -2.0400)$, $(4.2274, -2.4060)$, $(4.2427, -2.4205)$

5. $(-0.5200, 2.9200)$, $(-0.8400, 2.4400)$, $(-0.5712, 2.4485)$, $(-0.5793, 2.4488)$

6. $(-1.7600, 4.6800)$, $(-1.9200, 4.5600)$, $(-1.9029, 4.4995)$, $(-1.9025, 4.4999)$

7. $(3.1200, 1.6800)$, $(3.2400, 1.7600)$, $(3.2816, 1.7899)$, $(3.2820, 1.7902)$

8. $(2.1600, -0.6300)$, $(2.5200, -0.4600)$, $(2.5320, -0.3867)$, $(2.5270, -0.3889)$

9. At $t = 1$ we obtain $(x, y) = (3.99261, 6.21770)$ $(h = 0.1)$ and $(3.99234, 6.21768)$ $(h = 0.05)$; the actual value is $(3.99232, 6.21768)$.

10. At $t = 1$ we obtain $(x, y) = (1.31498, 1.02537)$ $(h = 0.1)$ and $(1.31501, 1.02538)$ $(h = 0.05)$; the actual value is $(1.31501, 1.02538)$.

11. At $t = 1$ we obtain $(x, y) = (-0.05832, 0.56664)$ $(h = 0.1)$ and $(-0.05832, 0.56665)$ $(h = 0.05)$; the actual value is $(-0.05832, 0.56665)$.

12. We solved $x' = y$, $y' = -x + \sin t$, $x(0) = y(0) = 0$. With $h = 0.1$ and also with $h = 0.05$ we obtain the actual value $x(1.0) \approx 0.15058$.

13. Runge-Kutta, $h = 0.1$: about 1050 ft in about 7.7 s

14. Runge-Kutta, $h = 0.1$: about 1044 ft in about 7.8 s

15. Runge-Kutta, $h = 1.0$: about 83.83 mi in about 168 s

16. At $40°$: 5.0 s, 352.9 ft; at $45°$: 5.4 s, 347.2 ft; at $50°$: 5.8 s, 334.2 ft (all values approximate)

17. At $39.0°$ the range is about 352.7 ft. At $39.5°$ it is 352.8; at $40°$, 352.9; at $40.5°$, 352.6; at $41.0°$, 352.1.

18. Just under $57.5°$

19. Approximately 253 ft/s

20. Maximum height: about 1005 ft, attained in about 5.6 s; range: about 1880 ft; time aloft: about 11.6 s

21. Runge-Kutta with $h = 0.1$ yields these results:
(a) 21400 ft, 46 s, 518 ft/s; (b) 8970 ft, 17.5 s; (c) 368 ft/s (at $t \approx 23$).

Chapter 5

Section 5.1

1. (a) $\begin{bmatrix} 13 & -18 \\ 23 & 17 \end{bmatrix}$; (b) $\begin{bmatrix} 0 & -1 \\ 2 & 19 \end{bmatrix}$;

(c) $\begin{bmatrix} -9 & -11 \\ 47 & -9 \end{bmatrix}$; (d) $\begin{bmatrix} -10 & -37 \\ 14 & -8 \end{bmatrix}$

2. $(\mathbf{AB})\mathbf{C} = \mathbf{A}(\mathbf{BC}) = \begin{bmatrix} -33 & -7 \\ -27 & 103 \end{bmatrix}$;

$\mathbf{A}(\mathbf{B} + \mathbf{C}) = \mathbf{AB} + \mathbf{AC} = \begin{bmatrix} -18 & -4 \\ 68 & -8 \end{bmatrix}$

3. $\mathbf{AB} = \begin{bmatrix} -1 & 8 \\ 46 & -1 \end{bmatrix}$; $\mathbf{BA} = \begin{bmatrix} 11 & -12 & 14 \\ -14 & 0 & 7 \\ 0 & 8 & -13 \end{bmatrix}$

4. $\mathbf{Ay} = \begin{bmatrix} 2t^2 - \cos t \\ 3t^2 - 4\sin t + 5\cos t \end{bmatrix}$, $\mathbf{Bx} = \begin{bmatrix} 2t + 3e^{-t} \\ -14t \\ 6t - 2e^{-t} \end{bmatrix}$

5. (a) $\begin{bmatrix} 21 & 2 & 1 \\ 4 & 44 & 9 \\ -27 & 34 & 45 \end{bmatrix}$; (b) $\begin{bmatrix} 9 & 21 & -13 \\ -5 & -8 & 24 \\ -25 & -19 & 26 \end{bmatrix}$;

(c) $\begin{bmatrix} 0 & -6 & 1 \\ 10 & 31 & -15 \\ 16 & 58 & -23 \end{bmatrix}$; (d) $\begin{bmatrix} -10 & -8 & 5 \\ 18 & 12 & -10 \\ 11 & 22 & 6 \end{bmatrix}$;

(e) $\begin{bmatrix} 3 - t & 2 & -1 \\ 0 & 4 - t & 3 \\ -5 & 2 & 7 - t \end{bmatrix}$

7. $\det(\mathbf{A}) = \det(\mathbf{B}) = 0$ **8.** $\det(\mathbf{AB}) = \det(\mathbf{BA}) = 144$

9. $(\mathbf{AB})' = \begin{bmatrix} 1 - 8t + 18t^2 & 1 + 2t - 12t^2 + 32t^3 \\ 3 + 3t^2 - 4t^3 & 8t + 3t^2 + 4t^3 \end{bmatrix}$

11. $\mathbf{x} = \begin{bmatrix} x \\ y \end{bmatrix}$, $\mathbf{P}(t) = \begin{bmatrix} 0 & -3 \\ 3 & 0 \end{bmatrix}$, $\mathbf{f}(t) = \begin{bmatrix} 0 \\ 0 \end{bmatrix}$

13. $\mathbf{x} = \begin{bmatrix} x \\ y \end{bmatrix}$, $\mathbf{P}(t) = \begin{bmatrix} 2 & 4 \\ 5 & -1 \end{bmatrix}$, $\mathbf{f}(t) = \begin{bmatrix} 3e^t \\ -t^2 \end{bmatrix}$

15. $\mathbf{x} = \begin{bmatrix} x \\ y \\ z \end{bmatrix}$, $\mathbf{P}(t) = \begin{bmatrix} 0 & 1 & 1 \\ 1 & 0 & 1 \\ 1 & 1 & 0 \end{bmatrix}$, $\mathbf{f}(t) = \begin{bmatrix} 0 \\ 0 \\ 0 \end{bmatrix}$

17. $\mathbf{x} = \begin{bmatrix} x \\ y \\ z \end{bmatrix}$, $\mathbf{P}(t) = \begin{bmatrix} 3 & -4 & 1 \\ 1 & 0 & -3 \\ 0 & 6 & -7 \end{bmatrix}$, $\mathbf{f}(t) = \begin{bmatrix} t \\ t^2 \\ t^3 \end{bmatrix}$

19. $\mathbf{x} = \begin{bmatrix} x_1 \\ x_2 \\ x_3 \\ x_4 \end{bmatrix}, \mathbf{P}(t) = \begin{bmatrix} 0 & 1 & 0 & 0 \\ 0 & 0 & 2 & 0 \\ 0 & 0 & 0 & 3 \\ 4 & 0 & 0 & 0 \end{bmatrix}, \mathbf{f}(t) = \begin{bmatrix} 0 \\ 0 \\ 0 \\ 0 \end{bmatrix}$

21. $W(t) = e^{3t};\ \mathbf{x}(t) = \begin{bmatrix} 2c_1 e^t + c_2 e^{2t} \\ -3c_1 e^t - c_2 e^{2t} \end{bmatrix}$

22. $W(t) = -5e^{3t};\ \mathbf{x}(t) = \begin{bmatrix} c_1 e^{3t} + 2c_2 e^{-2t} \\ 3c_1 e^{3t} + c_2 e^{-2t} \end{bmatrix}$

23. $W(t) \equiv 4;\ \mathbf{x}(t) = \begin{bmatrix} c_1 e^{2t} + c_2 e^{-2t} \\ c_1 e^{2t} + 5c_2 e^{-2t} \end{bmatrix}$

24. $W(t) = e^{5t};\ \mathbf{x}(t) = \begin{bmatrix} c_1 e^{3t} + c_2 e^{2t} \\ -c_1 e^{3t} - 2c_2 e^{2t} \end{bmatrix}$

25. $W(t) = 7e^{-3t};\ \mathbf{x}(t) = \begin{bmatrix} 3c_1 e^{2t} + c_2 e^{-5t} \\ 2c_1 e^{2t} + 3c_2 e^{-5t} \end{bmatrix}$

26. $W(t) = 16e^{9t};\ \mathbf{x}(t) = \begin{bmatrix} 2c_1 e^t - 2c_2 e^{3t} + 2c_3 e^{5t} \\ 2c_1 e^t - 2c_3 e^{5t} \\ c_1 e^t + c_2 e^{3t} + c_3 e^{5t} \end{bmatrix}$

27. $W(t) \equiv 3;\ \mathbf{x}(t) = \begin{bmatrix} c_1 e^{2t} + c_2 e^{-t} \\ c_1 e^{2t} + c_3 e^{-t} \\ c_1 e^{2t} - (c_2 + c_3) e^{-t} \end{bmatrix}$

28. $W(t) = -84e^{7t};\ \mathbf{x}(t) = \begin{bmatrix} c_1 + 2c_2 e^{3t} - c_3 e^{4t} \\ 6c_1 + 3c_2 e^{3t} + 2c_3 e^{4t} \\ -13c_1 - 2c_2 e^{3t} + c_3 e^{4t} \end{bmatrix}$

29. $W(t) = e^{2t};\ \mathbf{x}(t) = \begin{bmatrix} 3c_1 e^{-2t} + c_2 e^t + c_3 e^{3t} \\ -2c_1 e^{-2t} - c_2 e^t - c_3 e^{3t} \\ 2c_1 e^{-2t} + c_2 e^t \end{bmatrix}$

30. $W(t) = 1;\ \mathbf{x}(t) = \begin{bmatrix} c_1 e^{-t} + c_4 e^t \\ c_3 e^t \\ c_2 e^{-t} + 3c_4 e^t \\ c_1 e^{-t} - 2c_3 e^t \end{bmatrix}$

31. $\mathbf{x} = 2\mathbf{x}_1 - \mathbf{x}_2$

32. $\mathbf{x} = 7\mathbf{x}_1 - 2\mathbf{x}_2$

33. $\mathbf{x} = 15\mathbf{x}_1 - 4\mathbf{x}_2$

34. $\mathbf{x} = \frac{8}{7}(3\mathbf{x}_1 - 2\mathbf{x}_2)$

35. $\mathbf{x} = \mathbf{x}_1 + 2\mathbf{x}_2 + \mathbf{x}_3$

36. $\mathbf{x} = 7\mathbf{x}_1 + 3\mathbf{x}_2 + 5\mathbf{x}_3$

37. $\mathbf{x} = 3\mathbf{x}_1 - 3\mathbf{x}_2 - 5\mathbf{x}_3$

38. $\mathbf{x} = -2\mathbf{x}_1 + 15\mathbf{x}_2 - 4\mathbf{x}_3$

39. $\mathbf{x} = 3\mathbf{x}_1 + 7\mathbf{x}_2 + \mathbf{x}_3 - 2\mathbf{x}_4$

40. $\mathbf{x} = 13\mathbf{x}_1 + 41\mathbf{x}_2 + 3\mathbf{x}_3 - 12\mathbf{x}_4$

41. (a) $\mathbf{x}_2 = t\mathbf{x}_1$, so neither is a constant multiple of the other.
(b) $W(\mathbf{x}_1, \mathbf{x}_2) \equiv 0$, whereas Theorem 2 would imply that $W \neq 0$ if \mathbf{x}_1 and \mathbf{x}_2 were independent solutions of a system of the indicated form.

Section 5.2

1. $x_1(t) = c_1 e^{-t} + c_2 e^{3t},\ x_2(t) = -c_1 e^{-t} + c_2 e^{3t}$

2. $x_1(t) = c_1 e^{-t} + 3c_2 e^{4t},\ x_2(t) = -c_1 e^{-t} + 2c_2 e^{4t}$

3. General solution $x_1(t) = c_1 e^{-t} + 4c_2 e^{6t},\ x_2(t) = -c_1 e^{-t} + 3c_2 e^{6t}$
Particular solution $x_1(t) = \frac{1}{7}(-e^{-t} + 8e^{6t})$,
$x_2(t) = \frac{1}{7}(e^{-t} + 6e^{6t})$.

4. $x_1(t) = c_1 e^{-2t} + c_2 e^{5t},\ x_2(t) = -6c_1 e^{-2t} + c_2 e^{5t}$

5. $x_1(t) = c_1 e^{-t} + 7c_2 e^{5t},\ x_2(t) = c_1 e^{-t} + c_2 e^{5t}$

6. General solution $x_1(t) = 5c_1 e^{3t} + c_2 e^{4t},\ x_2(t) = -6c_1 e^{3t} - c_2 e^{4t}$
Particular solution $x_1(t) = -5e^{3t} + 6e^{4t},\ x_2(t) = 6e^{3t} - 6e^{4t}$.

7. $x_1(t) = c_1 e^t + 2c_2 e^{-9t}$, $x_2(t) = c_1 e^t - 3c_2 e^{-9t}$

8. $x_1(t) = 5c_1 \cos 2t + 5c_2 \sin 2t$,
$x_2(t) = (c_1 - 2c_2) \cos 2t + (2c_1 + c_2) \sin 2t$

9. General solution $x_1(t) = 5c_1 \cos 4t + 5c_2 \sin 4t$,
$x_2(t) = c_1(2 \cos 4t + 4 \sin 4t) + c_2(2 \sin 4t - 4 \cos 4t)$.
Particular solution $x_1(t) = 2 \cos 4t - \frac{11}{4} \sin 4t$,
$x_2(t) = 3 \cos 4t + \frac{1}{2} \sin 4t$

10. $x_1(t) = -2c_1 \cos 3t - 2c_2 \sin 3t$,
$x_2(t) = (3c_1 + 3c_2) \cos 3t + (3c_2 - 3c_1) \sin 3t$

11. General solution $x_1(t) = e^t(c_1 \cos 2t - c_2 \sin 2t)$,
$x_2(t) = e^t(c_1 \sin 2t + c_2 \cos 2t)$
Particular solution $x_1(t) = -4e^t \sin 2t$, $x_2(t) = 4e^t \cos 2t$

12. $x_1(t) = e^{2t}(-5c_1 \cos 2t - 5c_2 \sin 2t)$,
$x_2(t) = e^{2t}[(c_1 + 2c_2) \cos 2t + (-2c_1 + c_2) \sin 2t]$

13. $x_1(t) = 3e^{2t}(c_1 \cos 3t - c_2 \sin 3t)$,
$x_2(t) = e^{2t}[(c_1 + c_2) \cos 3t + (c_1 - c_2) \sin 3t]$

14. $x_1(t) = e^{3t}(c_1 \cos 4t + c_2 \sin 4t)$,
$x_2(t) = e^{3t}(c_1 \sin 4t - c_2 \cos 4t)$

15. $x_1(t) = 5e^{5t}(c_1 \cos 4t - c_2 \sin 4t)$,
$x_2(t) = e^{5t}[(2c_1 + 4c_2) \cos 4t + (4c_1 - 2c_2) \sin 4t]$

16. $x_1(t) = c_1 e^{-10t} + 2c_2 e^{-100t}$, $x_2(t) = 2c_1 e^{-10t} - 5c_2 e^{-100t}$

17. $x_1(t) = c_1 e^{9t} + c_2 e^{6t} + c_3$,
$x_2(t) = c_1 e^{9t} - 2c_2 e^{6t}$,
$x_3(t) = c_1 e^{9t} + c_2 e^{6t} - c_3$

18. $x_1(t) = c_1 e^{9t} + 4c_3$,
$x_2(t) = 2c_1 e^{9t} + c_2 e^{6t} - c_3$,
$x_3(t) = 2c_1 e^{9t} - c_2 e^{6t} - c_3$

19. $x_1(t) = c_1 e^{6t} + c_2 e^{3t} + c_3 e^{3t}$,
$x_2(t) = c_1 e^{6t} - 2c_2 e^{3t}$,
$x_3(t) = c_1 e^{6t} + c_2 e^{3t} - c_3 e^{3t}$

20. $x_1(t) = c_1 e^{9t} + c_2 e^{6t} + c_3 e^{2t}$,
$x_2(t) = c_1 e^{9t} - 2c_2 e^{6t}$,
$x_3(t) = c_1 e^{9t} + c_2 e^{6t} - c_3 e^{2t}$

21. $x_1(t) = 6c_1 + 3c_2 e^t + 2c_3 e^{-t}$,
$x_2(t) = 2c_1 + c_2 e^t + c_3 e^{-t}$,
$x_3(t) = 5c_1 + 2c_2 e^t + 2c_3 e^{-t}$

22. $x_1(t) = c_2 e^t + c_3 e^{3t}$,
$x_2(t) = c_1 e^{-2t} - c_2 e^t - c_3 e^{3t}$,
$x_3(t) = -c_1 e^{-2t} + c_3 e^{3t}$

23. $x_1(t) = c_1 e^{2t} + c_3 e^{3t}$,
$x_2(t) = -c_1 e^{2t} + c_2 e^{-2t} - c_3 e^{3t}$,
$x_3(t) = -c_2 e^{-2t} + c_3 e^{3t}$

24. $x_1(t) = c_1 e^t + c_2(2 \cos 2t - \sin 2t) + c_3(\cos 2t + 2 \sin 2t)$
$x_2(t) = -c_1 e^t - c_2(3 \cos 2t + \sin 2t) + c_3(\cos 2t - 3 \sin 2t)$
$x_3(t) = c_2(3 \cos 2t + \sin 2t) + c_3(3 \sin 2t - \cos 2t)$

25. $x_1(t) = c_1 + e^{2t}[(c_2 + c_3) \cos 3t + (-c_2 + c_3) \sin 3t]$,
$x_2(t) = -c_1 + 2e^{2t}(-c_2 \cos 3t - c_3 \sin 3t)$,
$x_3(t) = 2e^{2t}(c_2 \cos 3t + c_3 \sin 3t)$

26. $x_1(t) = 4e^{3t} - e^{-t}(4 \cos t - \sin t)$,
$x_2(t) = 9e^{3t} - e^{-t}(9 \cos t + 2 \sin t)$,
$x_3(t) = 17e^{-t} \cos t$

27. $x_1(t) = 15e^{-0.2t}$, $x_2(t) = 15(e^{-0.2t} - e^{-0.4t})$.
The maximum amount ever in tank 2 is $x_2(5 \ln 2) = 3.75$ lb.

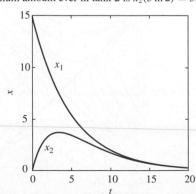

28. $x_1(t) = 15e^{-0.4t}$, $x_2(t) = 40(-e^{-0.4t} + e^{-0.25t})$.
The maximum amount ever in tank 2 is about 6.85 lb.

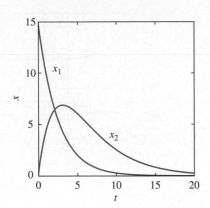

29. $x_1(t) = 10 + 5e^{-0.6t}$, $x_2(t) = 5 - 5e^{-0.6t}$

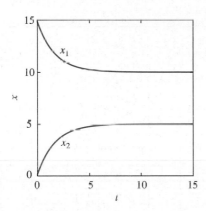

30. $x_1(t) = 5c_1 + c_2e^{-0.65t}$, $x_2(t) = 8c_1 - c_2e^{-0.65t}$.

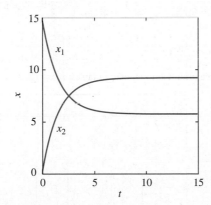

31. $x_1(t) = 27e^{-t}$,
$x_2(t) = 27e^{-t} - 27e^{-2t}$,
$x_3(t) = 27e^{-t} - 54e^{-2t} + 27e^{-3t}$.
The maximum amount of salt ever in tank 3 is $x_3(\ln 3) = 4$

pounds.

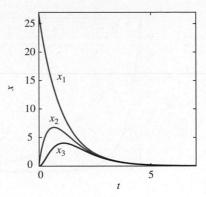

32. $x_1(t) = 45e^{-3t}$,
$x_2(t) = -135e^{-3t} + 135e^{-2t}$,
$x_3(t) = 135e^{-3t} - 270e^{-2t} + 135e^{-t}$.
The maximum amount of salt ever in tank 3 is $x_3(\ln 3) = 20$ pounds.

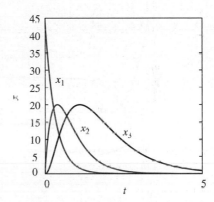

33. $x_1(t) = 45e^{-4t}$,
$x_2(t) = 90e^{-4t} - 90e^{-6t}$,
$x_3(t) = -270e^{-4t} + 135e^{-6t} + 135e^{-2t}$.
The maximum amount of salt ever in tank 3 is $x_3\left(\frac{1}{2}\ln 3\right) = 20$ pounds.

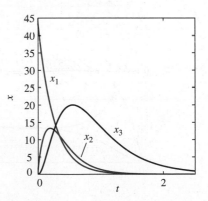

34. $x_1(t) = 40e^{-3t}$,
$x_2(t) = 60e^{-3t} - 60e^{-5t}$,
$x_3(t) = -150e^{-3t} + 75e^{-5t} + 75e^{-t}$.
The maximum amount of salt ever in tank 3 is

$x_3(\frac{1}{2}\ln 5) \approx 21.4663$ pounds.

and 15 lb.

35. $x_1(t) = 10 - \frac{1}{7}\left(55e^{-18t} - 216e^{-11t}\right),$
$x_2(t) = 3 - \frac{1}{7}\left(165e^{-18t} - 144e^{-11t}\right),$
$x_3(t) = 20 + \frac{1}{7}\left(220e^{-18t} - 360e^{-11t}\right).$
The limiting amounts of salt in tanks 1, 2, and 3 are 10 lb, 3 lb, and 20 lb.

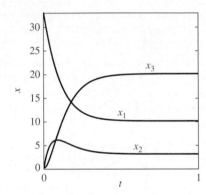

36. $x_1(t) = 4 + e^{-3t/5}[14\cos(3t/10) - 2\sin(3t/10)],$
$x_2(t) = 10 - e^{-3t/5}[10\cos(3t/10) - 10\sin(3t/10)],$
$x_3(t) = 4 - e^{-3t/5}[4\cos(3t/10) + 8\sin(3t/10)].$
The limiting amounts of salt in tanks 1, 2, and 3 are 4 lb, 10 lb, and 4 lb.

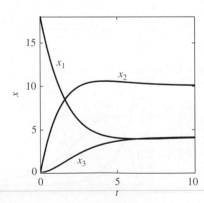

37. $x_1(t) = 30 + e^{-3t}[25\cos(t\sqrt{2}) + 10\sqrt{2}\sin(t\sqrt{2})],$
$x_2(t) = 10 - e^{-3t}[10\cos(t\sqrt{2}) - \frac{25}{2}\sqrt{2}\sin(t\sqrt{2})],$
$x_3(t) = 15 - e^{-3t}[15\cos(t\sqrt{2}) + \frac{45}{2}\sqrt{2}\sin(t\sqrt{2})].$
The limiting amounts of salt in tanks 1, 2, and 3 are 30 lb, 10 lb,

38. $x_1(t) = c_1e^t,$
$x_2(t) = -2c_1e^t + c_2e^{2t},$
$x_3(t) = 3c_1e^t - 3c_2e^{2t} + c_3e^{3t},$
$x_4(t) = -4c_1e^t + 6c_2e^{2t} - 4c_3e^{3t} + c_4e^{4t}$

39. $x_1(t) = 3c_1e^t + c_4e^{-2t},$
$x_2(t) = -2c_1e^t + c_3e^{2t} - c_4e^{-2t},$
$x_3(t) = 4c_1e^t + c_2e^{-t},$
$x_4(t) = c_1e^t$

40. $x_1(t) = c_1e^{2t},$
$x_2(t) = -3c_1e^{2t} + 3c_2e^{-2t} - c_4e^{-5t},$
$x_3(t) = c_3e^{5t},$
$x_4(t) = -c_2e^{-2t} - 3c_3e^{5t}$

41. $x_1(t) = 2e^{10t} + e^{15t} = x_4(t), x_2(t) = -e^{10t} + 2e^{15t} = x_3(t)$

42. $\mathbf{x}(t) = c_1\begin{bmatrix} 3 \\ -1 \\ 2 \end{bmatrix} + c_2\begin{bmatrix} 1 \\ 1 \\ 1 \end{bmatrix}e^{2t} + c_3\begin{bmatrix} 2 \\ -3 \\ 1 \end{bmatrix}e^{5t}$

43. $\mathbf{x}(t) = c_1\begin{bmatrix} 3 \\ -1 \\ 5 \end{bmatrix}e^{-2t} + c_2\begin{bmatrix} 1 \\ 1 \\ 1 \end{bmatrix}e^{4t} + c_3\begin{bmatrix} 1 \\ -1 \\ 3 \end{bmatrix}e^{8t}$

44. $\mathbf{x}(t) = c_1\begin{bmatrix} 3 \\ -2 \\ 2 \end{bmatrix}e^{-3t} + c_2\begin{bmatrix} 7 \\ 1 \\ 5 \end{bmatrix}e^{6t} + c_3\begin{bmatrix} 5 \\ -3 \\ 3 \end{bmatrix}e^{12t}$

45. $\mathbf{x}(t) =$
$c_1\begin{bmatrix} 1 \\ 1 \\ 1 \\ -1 \end{bmatrix}e^{-3t} + c_2\begin{bmatrix} 1 \\ 2 \\ -1 \\ 1 \end{bmatrix} + c_3\begin{bmatrix} 2 \\ 1 \\ 1 \\ 1 \end{bmatrix}e^{3t} + c_4\begin{bmatrix} 1 \\ -1 \\ 2 \\ -1 \end{bmatrix}e^{6t}$

46. $\mathbf{x}(t) =$
$c_1\begin{bmatrix} 3 \\ 2 \\ -1 \\ 1 \end{bmatrix}e^{-4t} + c_2\begin{bmatrix} 1 \\ 2 \\ 2 \\ -1 \end{bmatrix}e^{2t} + c_3\begin{bmatrix} 1 \\ 1 \\ -1 \\ 1 \end{bmatrix}e^{4t} + c_4\begin{bmatrix} 3 \\ -2 \\ 3 \\ -3 \end{bmatrix}e^{8t}$

47. $\mathbf{x}(t) =$
$c_1\begin{bmatrix} 2 \\ 2 \\ 1 \\ -1 \end{bmatrix}e^{-3t} + c_2\begin{bmatrix} 1 \\ 2 \\ -1 \\ 1 \end{bmatrix}e^{3t} + c_3\begin{bmatrix} 2 \\ 1 \\ 1 \\ 1 \end{bmatrix}e^{6t} + c_4\begin{bmatrix} 1 \\ -1 \\ 2 \\ -1 \end{bmatrix}e^{9t}$

48. $\mathbf{x}(t) =$
$c_1\begin{bmatrix} 1 \\ 2 \\ -1 \\ 2 \end{bmatrix}e^{16t} + c_2\begin{bmatrix} 2 \\ 5 \\ 1 \\ -1 \end{bmatrix}e^{32t} + c_3\begin{bmatrix} 3 \\ -1 \\ 1 \\ 2 \end{bmatrix}e^{48t} + c_4\begin{bmatrix} 1 \\ 1 \\ 2 \\ -3 \end{bmatrix}e^{64t}$

49. $\mathbf{x}(t) = c_1\begin{bmatrix} 1 \\ 0 \\ 3 \\ 1 \\ 1 \end{bmatrix}e^{-3t} + c_2\begin{bmatrix} 0 \\ 3 \\ 0 \\ -1 \\ 1 \end{bmatrix} + c_3\begin{bmatrix} 1 \\ 7 \\ 1 \\ 1 \\ 1 \end{bmatrix}e^{3t} +$

$$c_4 \begin{bmatrix} 0 \\ 1 \\ 0 \\ 1 \\ 1 \end{bmatrix} e^{6t} + c_5 \begin{bmatrix} 2 \\ 0 \\ 5 \\ 2 \\ 1 \end{bmatrix} e^{9t}$$

50. $\mathbf{x}(t) = c_1 \begin{bmatrix} 0 \\ 1 \\ 1 \\ 1 \\ 0 \\ 1 \end{bmatrix} e^{-7t} + c_2 \begin{bmatrix} 1 \\ 0 \\ 0 \\ 0 \\ 1 \\ 1 \end{bmatrix} e^{-4t} + c_3 \begin{bmatrix} 0 \\ 1 \\ 0 \\ 1 \\ 0 \\ 1 \end{bmatrix} e^{3t} +$

$c_4 \begin{bmatrix} 0 \\ 0 \\ 1 \\ 0 \\ 1 \\ 0 \end{bmatrix} e^{5t} + c_5 \begin{bmatrix} 1 \\ 1 \\ 0 \\ 0 \\ 0 \\ 1 \end{bmatrix} e^{9t} + c_6 \begin{bmatrix} 0 \\ 0 \\ 1 \\ -1 \\ -1 \\ 0 \end{bmatrix} e^{11t}$

Section 5.3

1. The natural frequencies are $\omega_0 = 0$ and $\omega_1 = 2$. In the degenerate natural mode with "frequency" $\omega_0 = 0$ the two masses move linearly with $x_1(t) = x_2(t) = a_0 + b_0 t$. At frequency $\omega_1 = 2$ they oscillate in opposite directions with equal amplitudes.

2. The natural frequencies are $\omega_1 = 1$ and $\omega_2 = 3$. In the natural mode with frequency ω_1, the two masses m_1 and m_2 move in the same direction with equal amplitudes of oscillation. At frequency ω_2 they move in opposite directions with equal amplitudes.

3. The natural frequencies are $\omega_1 = 1$ and $\omega_2 = 2$. In the natural mode with frequency ω_1, the two masses m_1 and m_2 move in the same direction with equal amplitudes of oscillation. In the natural mode with frequency ω_2 they move in opposite directions with the amplitude of oscillation of m_1 twice that of m_2.

4. The natural frequencies are $\omega_1 = 1$ and $\omega_2 = \sqrt{5}$. In the natural mode with frequency ω_1, the two masses m_1 and m_2 move in the same direction with equal amplitudes of oscillation. At frequency ω_2 they move in opposite directions with equal amplitudes.

5. The natural frequencies are $\omega_1 = \sqrt{2}$ and $\omega_2 = 2$. In the natural mode with frequency ω_1, the two masses m_1 and m_2 move in the same direction with equal amplitudes of oscillation. At frequency ω_2 they move in opposite directions with equal amplitudes.

6. The natural frequencies are $\omega_1 = \sqrt{2}$ and $\omega_2 = \sqrt{8}$. In the natural mode with frequency ω_1, the two masses m_1 and m_2 move in the same direction with equal amplitudes of oscillation. In the natural mode with frequency ω_2 they move in opposite directions with the amplitude of oscillation of m_1 twice that of m_2.

7. The natural frequencies are $\omega_1 = 2$ and $\omega_2 = 4$. In the natural mode with frequency ω_1, the two masses m_1 and m_2 move in the same direction with equal amplitudes of oscillation. At frequency ω_2 they move in opposite directions with equal amplitudes.

8. $x_1(t) = 2\cos t + 3\cos 3t - 5\cos 5t$,
$x_2(t) = 2\cos t - 3\cos 3t + \cos 5t$.
We have a superposition of three oscillations, in which the two masses move (1) in the same direction with frequency $\omega_1 = 1$ and equal amplitudes; (2) in opposite directions with frequency $\omega_2 = 3$ and equal amplitudes; (3) in opposite directions with frequency $\omega_3 = 5$ and with the amplitude of motion of m_1 being 5 times that of m_2.

9. $x_1(t) = 5\cos t - 8\cos 2t + 3\cos 3t$,
$x_2(t) = 5\cos t + 4\cos 2t - 9\cos 3t$.

We have a superposition of three oscillations, in which the two masses move (1) in the same direction with frequency $\omega_1 = 1$ and equal amplitudes; (2) in opposite directions with frequency $\omega_2 = 2$ and with the amplitude of motion of m_1 being twice that of m_2; (3) in opposite directions with frequency $\omega_3 = 3$ and with the amplitude of motion of m_2 being 3 times that of m_1.

10. $x_1(t) = \cos 2t - 15\cos 4t + 14\cos t$,
$x_2(t) = \cos 2t + 15\cos 4t + 16\cos t$.
We have a superposition of three oscillations, in which the two masses move (1) in the same direction with frequency $\omega_1 = 1$ and with the amplitude of motion of m_2 being $8/7$ times that of m_1; (2) in the same direction with frequency $\omega_2 = 2$ and equal amplitudes; (3) in opposite directions with frequency $\omega_3 = 4$ and equal amplitudes.

11. (a) The natural frequencies are $\omega_1 = 6$ and $\omega_2 = 8$. In mode 1 the two masses oscillate in the same direction with frequency $\omega_1 = 6$ and with the amplitude of motion of m_1 being twice that of m_2. In mode 2 the two masses oscillate in opposite directions with frequency $\omega_2 = 8$ and with the amplitude of motion of m_2 being 3 times that of m_1.
(b) $x(t) = 2\sin 6t + 19\cos 7t$, $y(t) = \sin 6t + 3\cos 7t$
We have a superposition of (only two) oscillations, in which the two masses move (1) in the same direction with frequency $\omega_1 = 6$ and with the amplitude of motion of m_1 being twice that of m_2; (2) in the same direction with frequency $\omega_3 = 7$ and with the amplitude of motion of m_1 being $19/3$ times that of m_2.

12. The system's three natural modes of oscillation have (1) natural frequency $\omega_1 = \sqrt{2}$ with amplitude ratios $1:0:-1$; (2) natural frequency $\omega_2 = \sqrt{2 + \sqrt{2}}$ with amplitude ratios $1: \sqrt{2}:1$; (3) natural frequency $\omega_3 = \sqrt{2 - \sqrt{2}}$ with amplitude ratios $1:\sqrt{2}:1$.

13. The system's three natural modes of oscillation have (1) natural frequency $\omega_1 = 2$ with amplitude ratios $1:0:-1$; (2) natural frequency $\omega_2 = \sqrt{4 + 2\sqrt{2}}$ with amplitude ratios $1:-\sqrt{2}:1$; (3) natural frequency $\omega_3 = \sqrt{4 - 2\sqrt{2}}$ with amplitude ratios $1:\sqrt{2}:1$.

15. $x_1(t) = \frac{2}{3}\cos 5t - 2\cos 5\sqrt{3}\,t + \frac{4}{3}\cos 10t$,
$x_2(t) = \frac{4}{3}\cos 5t + 4\cos 5\sqrt{3}\,t + \frac{16}{3}\cos 10t$.
We have a superposition of two oscillations with the natural frequencies $\omega_1 = 5$ and $\omega_2 = 5\sqrt{3}$ and a forced oscillation with frequency $\omega = 10$. In each of the two natural oscillations the amplitude of motion of m_2 is twice that of m_1, while in the forced oscillation the amplitude of motion of m_2 is four times that of m_1.

21. (a) $\omega_1 \approx 1.0293$ Hz; $\omega_2 \approx 1.7971$ Hz.
(b) $v_1 \approx 28$ mi/h; $v_2 \approx 49$ mi/h

23. $\omega_1 = 2\sqrt{10}$, $v_1 \approx 40.26$ (ft/s (about 27 mi/h),
$\omega_2 = 5\sqrt{5}$, $v_2 \approx 71.18$ ft/s (about 49 mi/h)

24. $\omega_1 \approx 6.1311$, $v_1 \approx 39.03$ ft/s (about 27 mi/h)
$\omega_2 \approx 10.3155$, $v_2 \approx 65.67$ ft/s (about 45 mi/h)

25. $\omega_1 \approx 5.0424$, $v_1 \approx 32.10$ ft/s (about 22 mi/h),
$\omega_2 \approx 9.9158$, $v_2 \approx 63.13$ ft/s (about 43 mi/h)

Section 5.4

1. Repeated eigenvalue $\lambda = -3$, eigenvector $\mathbf{v} = \begin{bmatrix} 1 & -1 \end{bmatrix}^T$;
$x_1(t) = (c_1 + c_2 + c_2 t)e^{-3t}$, $x_2(t) = (-c_1 - c_2 t)e^{-3t}$

2. Repeated eigenvalue $\lambda = 2$, single eigenvector $\mathbf{v} = \begin{bmatrix} 1 & 1 \end{bmatrix}^T$;
$x_1(t) = (c_1 + c_2 + c_2t)e^{2t}$, $x_2(t) = (c_1 + c_2t)e^{2t}$

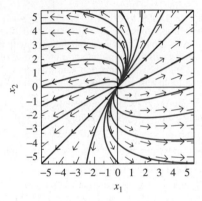

3. Repeated eigenvalue $\lambda = 3$, eigenvector $\mathbf{v} = \begin{bmatrix} -2 & 2 \end{bmatrix}^T$;
$x_1(t) = (-2c_1 + c_2 - 2c_2t)e^{3t}$, $x_2(t) = (2c_1 + 2c_2t)e^{3t}$

4. Repeated eigenvalue $\lambda = 4$, single eigenvector $\mathbf{v} = \begin{bmatrix} -1 & 1 \end{bmatrix}^T$;
$x_1(t) = (-c_1 + c_2 - c_2t)e^{4t}$, $x_2(t) = (c_1 + c_2t)e^{4t}$

5. Repeated eigenvalue $\lambda = 5$, eigenvector $\mathbf{v} = \begin{bmatrix} 2 & -4 \end{bmatrix}^T$;
$x_1(t) = (2c_1 + c_2 + 2c_2t)e^{5t}$, $x_2(t) = (-4c_1 - 4c_2t)e^{5t}$

6. Repeated eigenvalue $\lambda = 5$, single eigenvector $\mathbf{v} = \begin{bmatrix} -4 & 4 \end{bmatrix}^T$;
$x_1(t) = (-4c_1 + c_2 - 4c_2t)e^{5t}$, $x_2(t) = (4c_1 + 4c_2t)e^{5t}$

7. Eigenvalues $\lambda = 2, 2, 9$ with three linearly independent eigenvectors; $x_1(t) = c_1e^{2t} + c_2e^{2t}$, $x_2(t) = c_1e^{2t} + c_3e^{9t}$, $x_3(t) = c_2e^{2t}$

8. Eigenvalues $\lambda = 7, 13, 13$ with three linearly independent eigenvectors; $x_1(t) = 2c_1e^{7t} - c_3e^{13t}$, $x_2(t) = -3c_1e^{7t} + c_3e^{13t}$, $x_3(t) = c_1e^{7t} + c_2e^{13t}$

9. Eigenvalues $\lambda = 5, 5, 9$ with three linearly independent eigenvectors; $x_1(t) = c_1e^{5t} + 7c_2e^{5t} + 3c_3e^{9t}$, $x_2(t) = 2c_1e^{5t}$, $x_3(t) = 2c_2e^{5t} + c_3e^{9t}$

10. Eigenvalues $\lambda = 3, 3, 7$ with three linearly independent eigenvectors; $x_1(t) = 5c_1e^{3t} - 3c_2e^{3t} + 2c_3e^{7t}$, $x_2(t) = 2c_1e^{3t} + c_3e^{7t}$, $x_3(t) = c_2e^{3t}$

11. Triple eigenvalue $\lambda = -1$ of defect 2;
$x_1(t) = (-2c_2 + c_3 - 2c_3t)e^{-t}$,
$x_2(t) = (c_1 - c_2 + c_2t - c_3t + \frac{1}{2}c_3t^2)e^{-t}$,
$x_3(t) = (c_2 + c_3t)e^{-t}$

12. Triple eigenvalue $\lambda = -1$ of defect 2;
$x_1(t) = e^{-t}(c_1 + c_3 + c_2t + \frac{1}{2}c_3t^2)$
$x_2(t) = e^{-t}(c_1 + c_2t + \frac{1}{2}c_3t^2)$,
$x_3(t) = e^{-t}(c_2 + c_3t)$

13. Triple eigenvalue $\lambda = -1$ of defect 2;
$x_1(t) = (c_1 + c_2t + \frac{1}{2}c_3t^2)e^{-t}$,
$x_2(t) = (2c_2 + c_3 + 2c_3t)e^{-t}$,
$x_3(t) = (c_2 + c_2t)e^{-t}$

14. Triple eigenvalue $\lambda = -1$ of defect 2;
$x_1(t) = e^{-t}(5c_1 + c_2 + c_3 + 5c_2t + c_3t + \frac{5}{2}c_3t^2)$,
$x_2(t) = e^{-t}(-25c_1 - 5c_2 - 25c_2t - 5c_3t - \frac{25}{2}c_3t^2)$,
$x_3(t) = e^{-t}(-5c_1 + 4c_2 - 5c_2t + 4c_3t - \frac{5}{2}c_3t^2)$

15. Triple eigenvalue $\lambda = 1$ of defect 1;
$x_1(t) = (3c_1 + c_3 - 3c_3t)e^t$,
$x_2(t) = (-c_1 + c_3t)e^t$, $x_3(t) = (c_2 + c_3t)e^t$

16. Triple eigenvalue $\lambda = 1$ of defect 1;
$x_1(t) = e^t(3c_1 + 3c_2 + c_3)$
$x_2(t) = e^t(-2c_1 - 2c_3t)$,
$x_3(t) = e^t(-2c_2 + 2c_3t)$

17. Triple eigenvalue $\lambda = 1$ of defect 1;
$x_1(t) = (2c_1 + c_2)e^t$, $x_2(t) = (-3c_2 + c_3 + 6c_3t)e^t$,
$x_3(t) = -9(c_1 + c_3t)e^t$

18. Triple eigenvalue $\lambda = 1$ of defect 1;
$x_1(t) = e^t(-c_1 - 2c_2 + c_3)$,
$x_2(t) = e^t(c_2 + c_3t)$,
$x_3(t) = e^t(c_1 - 2c_3t)$

19. Double eigenvalues $\lambda = -1$ and $\lambda = 1$, each with defect 1;
$x_1(t) = c_1e^{-t} + c_4e^t$,
$x_2(t) = c_3e^t$,
$x_3(t) = c_2e^{-t} + 3c_4e^t$,
$x_4(t) = c_1e^{-t} - 2c_3e^t$

20. Eigenvalue $\lambda = 2$ with multiplicity 4 and defect 3;
$x_1(t) = (c_1 + c_3 + c_2 t + c_4 t + \frac{1}{2} c_3 t^2 + \frac{1}{6} c_4 t^3) e^{2t}$,
$x_2(t) = (c_2 + c_3 t + \frac{1}{2} c_4 t^2) e^{2t}$,
$x_3(t) = (c_3 + c_4 t) e^{2t}$, $x_4(t) = c_4 e^{2t}$

21. Eigenvalue $\lambda = 1$ with multiplicity 4 and defect 2;
$x_1(t) = (-2c_2 + c_3 - 2c_3 t) e^t$, $x_2(t) = (c_2 + c_3 t) e^t$,
$x_3(t) = (c_2 + c_4 + c_3 t) e^t$, $x_4(t) = (c_1 + c_2 t + \frac{1}{2} c_3 t^2) e^t$

22. Eigenvalue $\lambda = 1$ with multiplicity 4 and defect 2;
$x_1(t) = (c_1 + 3c_2 + c_4 + c_2 t + 3c_3 t + \frac{1}{2} c_3 t^2) e^t$,
$x_2(t) = -(2c_2 - c_3 + 2c_3 t) e^t$, $x_3(t) = (c_2 + c_3 t) e^t$,
$x_4(t) = -(2c_1 + 6c_2 + 2c_2 t + 6c_3 t + c_3 t^2) e^t$

23. $\mathbf{x}(t) = c_1 \mathbf{v}_1 e^{-t} + (c_2 \mathbf{v}_2 + c_3 \mathbf{v}_3) e^{3t}$ with $\mathbf{v}_1 = \begin{bmatrix} 1 & -1 & 2 \end{bmatrix}^T$,
$\mathbf{v}_2 = \begin{bmatrix} 4 & 0 & 9 \end{bmatrix}^T$, $\mathbf{v}_3 = \begin{bmatrix} 0 & 2 & 1 \end{bmatrix}^T$

24. $\mathbf{x}(t) = c_1 \mathbf{v}_1 e^{-t} + (c_2 \mathbf{v}_2 + c_3 \mathbf{v}_3) e^{3t}$ with $\mathbf{v}_1 = \begin{bmatrix} 5 & 3 & -3 \end{bmatrix}^T$,
$\mathbf{v}_2 = \begin{bmatrix} 4 & 0 & -1 \end{bmatrix}^T$, $\mathbf{v}_3 = \begin{bmatrix} 2 & -1 & 0 \end{bmatrix}^T$

25. $\mathbf{x}(t) = \left[c_1 \mathbf{v}_1 + c_2 (\mathbf{v}_1 t + \mathbf{v}_2) + c_3 \left(\frac{1}{2} \mathbf{v}_1 t^2 + \mathbf{v}_2 t + \mathbf{v}_3 \right) \right] e^{2t}$ with
$\mathbf{v}_1 = \begin{bmatrix} -1 & 0 & -1 \end{bmatrix}^T$, $\mathbf{v}_2 = \begin{bmatrix} -4 & -1 & 0 \end{bmatrix}^T$, and
$\mathbf{v}_3 = \begin{bmatrix} 1 & 0 & 0 \end{bmatrix}^T$

26. $\mathbf{x}(t) = \left[c_1 \mathbf{v}_1 + c_2 (\mathbf{v}_1 t + \mathbf{v}_2) + c_3 \left(\frac{1}{2} \mathbf{v}_1 t^2 + \mathbf{v}_2 t + \mathbf{v}_3 \right) \right] e^{3t}$ with
$\mathbf{v}_1 = \begin{bmatrix} 0 & 2 & 2 \end{bmatrix}^T$, $\mathbf{v}_2 = \begin{bmatrix} 2 & 1 & -3 \end{bmatrix}^T$, and
$\mathbf{v}_3 = \begin{bmatrix} 1 & 0 & 0 \end{bmatrix}^T$

27. $\mathbf{x}(t) = [c_1 \mathbf{v}_1 + c_2 (\mathbf{v}_1 t + \mathbf{v}_2) + c_3 \mathbf{v}_3] e^{2t}$ with
$\mathbf{v}_1 = \begin{bmatrix} -5 & 3 & 8 \end{bmatrix}^T$, $\mathbf{v}_2 = \begin{bmatrix} 1 & 0 & 0 \end{bmatrix}^T$, $\mathbf{v}_3 = \begin{bmatrix} 1 & 1 & 0 \end{bmatrix}^T$

28. $\mathbf{x}(t) = \left[c_1 \mathbf{v}_1 + c_2 (\mathbf{v}_1 t + \mathbf{v}_2) + c_3 \left(\frac{1}{2} \mathbf{v}_1 t^2 + \mathbf{v}_2 t + \mathbf{v}_3 \right) \right] e^{2t}$ with
$\mathbf{v}_1 = \begin{bmatrix} 119 & -289 & 0 \end{bmatrix}^T$, $\mathbf{v}_2 = \begin{bmatrix} -17 & 34 & 17 \end{bmatrix}^T$, and
$\mathbf{v}_3 = \begin{bmatrix} 1 & 0 & 0 \end{bmatrix}^T$

29. $\mathbf{x}(t) = [c_1 \mathbf{v}_1 + c_2 (\mathbf{v}_1 t + \mathbf{v}_2)] e^{-t} + [c_3 \mathbf{v}_3 + c_4 (\mathbf{v}_3 t + \mathbf{v}_4)] e^{2t}$ with
$\mathbf{v}_1 = \begin{bmatrix} 1 & -3 & -1 & -2 \end{bmatrix}^T$, $\mathbf{v}_2 = \begin{bmatrix} 0 & 1 & 0 & 0 \end{bmatrix}^T$,
$\mathbf{v}_3 = \begin{bmatrix} 0 & -1 & 1 & 0 \end{bmatrix}^T$, $\mathbf{v}_4 = \begin{bmatrix} 0 & 0 & 2 & 1 \end{bmatrix}^T$

30. $\mathbf{x}(t) = [c_1 \mathbf{v}_1 + c_2 (\mathbf{v}_1 t + \mathbf{v}_2)] e^{-t} + [c_3 \mathbf{v}_3 + c_4 (\mathbf{v}_3 t + \mathbf{v}_4)] e^{2t}$, with
$\mathbf{v}_1 = \begin{bmatrix} 0 & 1 & -1 & -3 \end{bmatrix}^T$, $\mathbf{v}_2 = \begin{bmatrix} 0 & 0 & 1 & 2 \end{bmatrix}^T$,
$\mathbf{v}_3 = \begin{bmatrix} -1 & 0 & 0 & 0 \end{bmatrix}^T$, $\mathbf{v}_4 = \begin{bmatrix} 0 & 0 & 3 & 5 \end{bmatrix}^T$

31. $\mathbf{x}(t) = \left[c_1 \mathbf{v}_1 + c_2 (\mathbf{v}_1 t + \mathbf{v}_2) \right.$
$\left. + c_3 \left(\frac{1}{2} \mathbf{v}_1 t^2 + \mathbf{v}_2 t + \mathbf{v}_3 \right) + c_4 \mathbf{v}_4 \right] e^t$ with
$\mathbf{v}_1 = \begin{bmatrix} 42 & 7 & -21 & -42 \end{bmatrix}^T$,
$\mathbf{v}_2 = \begin{bmatrix} 34 & 22 & -10 & -27 \end{bmatrix}^T$, $\mathbf{v}_3 = \begin{bmatrix} -1 & 0 & 0 & 0 \end{bmatrix}^T$,
$\mathbf{v}_4 = \begin{bmatrix} 0 & 1 & 3 & 0 \end{bmatrix}^T$

32. $\mathbf{x}(t) = (c_1 \mathbf{v}_1 + c_2 \mathbf{v}_2) e^t + (c_3 \mathbf{v}_3 + c_4 \mathbf{v}_4 + c_5 \mathbf{v}_5) e^{3t}$ with
$\mathbf{v}_1 = \begin{bmatrix} 8 & 0 & -3 & 1 & 0 \end{bmatrix}^T$, $\mathbf{v}_2 = \begin{bmatrix} 1 & 0 & 0 & 0 & 3 \end{bmatrix}^T$,
$\mathbf{v}_3 = \begin{bmatrix} 3 & -2 & 1 & 0 & 0 \end{bmatrix}^T$,
$\mathbf{v}_4 = \begin{bmatrix} 2 & -2 & 0 & -3 & 0 \end{bmatrix}^T$, $\mathbf{v}_5 = \begin{bmatrix} 1 & -1 & 0 & 0 & 3 \end{bmatrix}^T$

33. $\mathbf{x}_1(t) = \begin{bmatrix} \cos 4t & \sin 4t & 0 & 0 \end{bmatrix}^T e^{3t}$,
$\mathbf{x}_2(t) = \begin{bmatrix} -\sin 4t & \cos 4t & 0 & 0 \end{bmatrix}^T e^{3t}$,
$\mathbf{x}_3(t) = \begin{bmatrix} t \cos 4t & t \sin 4t & \cos 4t & \sin 4t \end{bmatrix}^T e^{3t}$,
$\mathbf{x}_4(t) = \begin{bmatrix} -t \sin 4t & t \cos 4t & -\sin 4t & \cos 4t \end{bmatrix}^T e^{3t}$

34. $\mathbf{x}_1(t) = \begin{bmatrix} \sin 3t \\ 3\cos 3t - 3\sin 3t \\ 0 \\ \sin 3t \end{bmatrix} e^{2t}$,

$\mathbf{x}_2(t) = \begin{bmatrix} -\cos 3t \\ 3\sin 3t + 3\cos 3t \\ 0 \\ -\cos 3t \end{bmatrix} e^{2t}$,

$\mathbf{x}_3(t) = \begin{bmatrix} 3\cos 3t + t\sin 3t \\ (3t - 10)\cos 3t - (3t + 9)\sin 3t \\ \sin 3t \\ t\sin 3t \end{bmatrix} e^{2t}$,

$$\mathbf{x}_4(t) = \begin{bmatrix} -t\cos 3t + 3\sin 3t \\ (3t + 9)\cos 3t + (3t - 10)\sin 3t \\ -\cos 3t \\ -t\cos 3t \end{bmatrix} e^{2t}$$

35. $x_1(t) = x_2(t) = v_0(1 - e^{-t})$; $\lim\limits_{t \to \infty} x_1(t) = \lim\limits_{t \to \infty} x_2(t) = v_0$

36. $x_1(t) = v_0(2 - 2e^{-t} - te^{-t})$,
$x_2(t) = v_0(2 - 2e^{-t} - te^{-t} - \frac{1}{2}t^2 e^{-t})$;
$\lim\limits_{t \to \infty} x_1(t) = \lim\limits_{t \to \infty} x_2(t) = 2v_0$

Section 5.5

1. $\mathbf{\Phi}(t) = \begin{bmatrix} e^t & e^{3t} \\ -e^t & e^{3t} \end{bmatrix}$, $\mathbf{x}(t) = \frac{1}{2} \begin{bmatrix} 5e^t + e^{3t} \\ -5e^t + e^{3t} \end{bmatrix}$

2. $\mathbf{\Phi}(t) = \begin{bmatrix} 1 & e^{4t} \\ 2 & -2e^{4t} \end{bmatrix}$, $\mathbf{x}(t) = \frac{1}{4} \begin{bmatrix} 3 + 5e^{4t} \\ 6 - 10e^{4t} \end{bmatrix}$

3. $\mathbf{\Phi}(t) = \begin{bmatrix} 5\cos 4t & -5\sin 4t \\ 2\cos 4t + 4\sin 4t & 4\cos 4t - 2\sin 4t \end{bmatrix}$,
$\mathbf{x}(t) = \frac{1}{4} \begin{bmatrix} -5\sin 4t \\ 4\cos 4t - 2\sin 4t \end{bmatrix}$

4. $\mathbf{\Phi}(t) = e^{2t} \begin{bmatrix} 1 & 1 + t \\ 1 & t \end{bmatrix}$, $\mathbf{x}(t) = e^{2t} \begin{bmatrix} 1 + t \\ t \end{bmatrix}$

5. $\mathbf{\Phi}(t) = \begin{bmatrix} 2\cos 3t & -2\sin 3t \\ -3\cos 3t + 3\sin 3t & 3\cos 3t + 3\sin 3t \end{bmatrix}$,
$\mathbf{x}(t) = \frac{1}{3} \begin{bmatrix} 3\cos 3t - \sin 3t \\ -3\cos 3t + 6\sin 3t \end{bmatrix}$

6. $\mathbf{\Phi}(t) = e^{5t} \begin{bmatrix} \cos 4t - 2\sin 4t & 2\cos 4t + 2\sin 4t \\ 2\cos 4t & 2\sin 4t \end{bmatrix}$,
$\mathbf{x}(t) = 2e^{5t} \begin{bmatrix} \cos 4t + \sin 4t \\ \sin 4t \end{bmatrix}$

7. $\mathbf{\Phi}(t) = \begin{bmatrix} 6 & 3e^t & 2e^{-t} \\ 2 & e^t & e^{-t} \\ 5 & 2e^t & 2e^{-t} \end{bmatrix}$, $\mathbf{x}(t) = \begin{bmatrix} -12 + 12e^t + 2e^{-t} \\ -4 + 4e^t + e^{-t} \\ -10 + 8e^t + 2e^{-t} \end{bmatrix}$

8. $\mathbf{\Phi}(t) = \begin{bmatrix} 0 & e^t & e^{3t} \\ e^{-t} & -e^t & -e^{3t} \\ -e^{-2t} & 0 & e^{3t} \end{bmatrix}$, $\mathbf{x}(t) = \begin{bmatrix} e^t \\ -e^t + e^{-t} \\ -e^{-2t} \end{bmatrix}$

9. $e^{\mathbf{A}t} = \begin{bmatrix} 2e^{3t} - e^t & -2e^{3t} + 2e^t \\ e^{3t} - e^t & -e^{3t} + 2e^t \end{bmatrix}$

10. $e^{\mathbf{A}t} = \begin{bmatrix} -2 + 3e^{2t} & 3 - 3e^{2t} \\ -2 + 2e^{2t} & 3 - 2e^{2t} \end{bmatrix}$

11. $e^{\mathbf{A}t} = \begin{bmatrix} 3e^{3t} - 2e^{2t} & 3e^{3t} + 3e^{2t} \\ 2e^{3t} - 2e^{2t} & -2e^{3t} + 3e^{2t} \end{bmatrix}$

12. $e^{\mathbf{A}t} = \begin{bmatrix} -3e^t + 4e^{2t} & 4e^t - 4e^{2t} \\ -3e^t + 3e^{2t} & 4e^t - 3e^{2t} \end{bmatrix}$

13. $e^{\mathbf{A}t} = \begin{bmatrix} 4e^{3t} - 3e^t & -4e^{3t} + 4e^t \\ 3e^{3t} - 3e^t & -3e^{3t} + 4e^t \end{bmatrix}$

14. $e^{\mathbf{A}t} = \begin{bmatrix} -8e^t + 9e^{2t} & 6e^t - 6e^{2t} \\ -12e^t + 12e^{2t} & 9e^t - 8e^{2t} \end{bmatrix}$

15. $e^{\mathbf{A}t} = \begin{bmatrix} 5e^{2t} - 4e^t & -10e^{2t} + 10e^t \\ 2e^{2t} - 2e^t & -4e^{2t} + 5e^t \end{bmatrix}$

16. $e^{\mathbf{A}t} = \begin{bmatrix} -9e^t + 10e^{2t} & 15e^t - 15e^{2t} \\ -6e^t + 6e^{2t} & 10e^t - 9e^{2t} \end{bmatrix}$

17. $e^{\mathbf{A}t} = \frac{1}{2} \begin{bmatrix} e^{4t} + e^{2t} & e^{4t} - e^{2t} \\ e^{4t} - e^{2t} & e^{4t} + e^{2t} \end{bmatrix}$

18. $e^{\mathbf{A}t} = \frac{1}{2} \begin{bmatrix} e^{2t} + e^{6t} & -e^{2t} + e^{6t} \\ -e^{2t} + e^{6t} & e^{2t} + e^{6t} \end{bmatrix}$

19. $e^{\mathbf{A}t} = \frac{1}{5} \begin{bmatrix} 4e^{10t} + e^{5t} & 2e^{10t} - 2e^{5t} \\ 2e^{10t} - 2e^{5t} & e^{10t} + 4e^{5t} \end{bmatrix}$

20. $e^{\mathbf{A}t} = \dfrac{1}{5}\begin{bmatrix} e^{5t}+4e^{15t} & -2e^{5t}+2e^{15t} \\ -2e^{5t}+2e^{15t} & 4e^{5t}+e^{15t} \end{bmatrix}$

21. $e^{\mathbf{A}t} = \begin{bmatrix} 1+t & -t \\ t & 1-t \end{bmatrix}$

22. $e^{\mathbf{A}t} = \begin{bmatrix} 1+6t & 4t \\ -9t & 1-6t \end{bmatrix}$

23. $e^{\mathbf{A}t} = \begin{bmatrix} 1+t & -t & -t-t^2 \\ t & 1-t & t-t^2 \\ 0 & 0 & 1 \end{bmatrix}$

24. $e^{\mathbf{A}t} = \begin{bmatrix} 1+3t & 0 & -3t \\ 5t+18t^2 & 1 & 7t-18t^2 \\ 3t & 0 & 1-3t \end{bmatrix}$

25. $e^{\mathbf{A}t} = \begin{bmatrix} e^{2t} & 5te^{2t} \\ 0 & e^{2t} \end{bmatrix}$, $\mathbf{x}(t) = e^{\mathbf{A}t}\begin{bmatrix} 4 \\ 7 \end{bmatrix}$

26. $e^{\mathbf{A}t} = \begin{bmatrix} e^{7t} & 0 \\ 11te^{7t} & e^{7t} \end{bmatrix}$, $\mathbf{x}(t) = e^{\mathbf{A}t}\begin{bmatrix} 5 \\ -10 \end{bmatrix}$

27. $e^{\mathbf{A}t} = \begin{bmatrix} e^t & 2te^t & (3t+2t^2)e^t \\ 0 & e^t & 2te^t \\ 0 & 0 & e^t \end{bmatrix}$, $\mathbf{x}(t) = e^{\mathbf{A}t}\begin{bmatrix} 4 \\ 5 \\ 6 \end{bmatrix}$

28. $e^{\mathbf{A}t} = \begin{bmatrix} e^{5t} & 0 & 0 \\ 10te^{5t} & e^{5t} & 0 \\ (20t+150t^2)e^{5t} & 30te^{5t} & e^{5t} \end{bmatrix}$, $\mathbf{x}(t) = e^{\mathbf{A}t}\begin{bmatrix} 40 \\ 50 \\ 60 \end{bmatrix}$

29. $e^{\mathbf{A}t} = \begin{bmatrix} 1 & 2t & 3t+6t^2 & 4t+6t^2+4t^3 \\ 0 & 1 & 6t & 3t+6t^2 \\ 0 & 0 & 1 & 2t \\ 0 & 0 & 0 & 1 \end{bmatrix}e^t$,

$\mathbf{x}(t) = e^{\mathbf{A}t}\begin{bmatrix} 1 \\ 1 \\ 1 \\ 1 \end{bmatrix}$

30. $e^{\mathbf{A}t} = e^{3t}\begin{bmatrix} 1 & 0 & 0 & 0 \\ 6t & 1 & 0 & 0 \\ 9t+18t^2 & 6t & 1 & 0 \\ 12t+54t^2+36t^3 & 9t+18t^2 & 6t & 1 \end{bmatrix}$,

$\mathbf{x}(t) = e^{\mathbf{A}t}\begin{bmatrix} 1 \\ 1 \\ 1 \\ 1 \end{bmatrix}$

33. $\mathbf{x}(t) = \begin{bmatrix} c_1\cosh t + c_2\sinh t \\ c_1\sinh t + c_2\cosh t \end{bmatrix}$ **35.** $e^{\mathbf{A}t} = \begin{bmatrix} e^{3t} & 4te^{3t} \\ 0 & e^{3t} \end{bmatrix}$

36. $e^{\mathbf{A}t} = e^t\begin{bmatrix} 1 & 2t & 3t+4t^2 \\ 0 & 1 & 4t \\ 0 & 0 & 1 \end{bmatrix}$

37. $e^{\mathbf{A}t} = \begin{bmatrix} e^{2t} & 3e^{2t}-3e^t & 13e^{2t}-(13+9t)e^t \\ 0 & e^t & 3te^t \\ 0 & 0 & e^t \end{bmatrix}$

38. $e^{\mathbf{A}t} = \begin{bmatrix} e^{5t} & 4e^{10t}-4e^{5t} & 16e^{10t}-(16+50t)e^{5t} \\ 0 & e^{10t} & 4e^{10t}-4e^{5t} \\ 0 & 0 & e^{5t} \end{bmatrix}$

39. $e^{\mathbf{A}t} =$
$\begin{bmatrix} e^t & 3te^t & 12e^{2t}-(12+9t)e^t & (51+18t)e^t-(51-36t)e^{2t} \\ 0 & e^t & 3e^{2t}-3e^t & 6e^t-(6-9t)e^{2t} \\ 0 & 0 & e^{2t} & 4e^{3t}-4e^{2t} \\ 0 & 0 & 0 & e^{2t} \end{bmatrix}$

40. $e^{\mathbf{A}t} =$
$\begin{bmatrix} e^{2t} & 4te^{2t} & (4t+8t^2)e^{2t} & 100e^{3t}-(100+96t+32t^2)e^{2t} \\ 0 & e^{2t} & 4te^{2t} & 20e^{3t}-(20+16t)e^{2t} \\ 0 & 0 & e^{2t} & 4e^{3t}-4e^{2t} \\ 0 & 0 & 0 & e^{3t} \end{bmatrix}$

Section 5.6

1. $x(t) = \frac{7}{3},\ y(t) = -\frac{8}{3}$

2. $x(t) = \frac{1}{8}(1+12t),\ y(t) = -\frac{1}{4}(5+4t)$

3. $x(t) = \frac{1}{756}(864e^{-t}+4e^{6t}-868+840t-504t^2)$,
$y(t) = \frac{1}{756}(-864e^{-t}+3e^{6t}+861-882t+378t^2)$

4. $x(t) = \frac{1}{84}(99e^{5t}-8e^{-2t}-7e^t),\ y(t) = \frac{1}{84}(99e^{5t}+48e^{-2t}-63e^t)$

5. $x(t) = \frac{1}{3}(-12-e^{-t}-7te^{-t}),\ y(t) = \frac{1}{3}(-6-7te^{-t})$

6. $x(t) = -\frac{1}{256}(91+16t)e^t,\ y(t) = \frac{1}{32}(25+16t)e^t$

7. $x(t) = \frac{1}{410}(369e^t+166e^{-9t}-125\cos t-105\sin t)$,
$y(t) = \frac{1}{410}(369e^t-249e^{-9t}-120\cos t-150\sin t)$

8. $x(t) = \frac{1}{3}(17\cos t+2\sin t),\ y(t) = \frac{1}{3}(3\cos t+5\sin t)$

9. $x(t) = \frac{1}{4}(\sin 2t+2t\cos 2t+t\sin 2t),\ y(t) = \frac{1}{4}t\sin 2t$

10. $x(t) = \frac{1}{13}e^t(4\cos t-6\sin t),\ y(t) = \frac{1}{13}e^t(3\cos t+2\sin t)$

11. $x(t) = \frac{1}{2}(1-4t+e^{4t}),\ y(t) = \frac{1}{4}(-5+4t+e^{4t})$

12. $x(t) = t^2,\ y(t) = -t^2$

13. $x(t) = \frac{1}{2}(1+5t)e^t,\ y(t) = -\frac{5}{2}te^t$

14. $x(t) = \frac{1}{8}(-2+4t-e^{4t}+2te^{4t}),\ y(t) = \frac{1}{2}t(-2+e^{4t})$

15. (a) $x_1(t) = 200(1-e^{-t/10}),\ x_2(t) = 400(1+e^{-t/10}-2e^{-t/20})$
(b) $x_1(t) \to 200$ and $x_2(t) \to 400$ as $t \to +\infty$
(c) Tank 1: about 6 min 56 s; tank 2: about 24 min 34 s

16. (a) $x_1(t) = 600(1-e^{-t/20}),\ x_2(t) = 300(1+e^{-t/10}-2e^{-t/20})$
(b) $x_1(t) \to 600$ and $x_2(t) \to 300$ as $t \to \infty$
(c) Tank 1: about 8 min 7 sec; tank 2: about 17 min 13 sec

17. $x_1(t) = 102-95e^{-t}-7e^{5t},\ x_2(t) = 96-95e^{-t}-e^{5t}$

18. $x_1(t) = 68-110t-75e^{-t}+7e^{5t},\ x_2(t) = 74-80t-75e^{-t}+e^{5t}$

19. $x_1(t) = -70-60t+16e^{-3t}+54e^{2t}$,
$x_2(t) = 5-60t-32e^{-3t}+27e^{2t}$

20. $x_1(t) = 3e^{2t}+60te^{2t}-3e^{-3t},\ x_2(t) = -6e^{2t}+30te^{2t}+6e^{-3t}$

21. $x_1(t) = -e^{-t}-14e^{2t}+15e^{3t},\ x_2(t) = -5e^{-t}-10e^{2t}+15e^{3t}$

22. $x_1(t) = -10e^{-t}-7te^{-t}+10e^{3t}-5te^{3t}$,
$x_2(t) = -15e^{-t}-35te^{-t}+15e^{3t}-5te^{3t}$

23. $x_1(t) = 3+11t+8t^2,\ x_2(t) = 5+17t+24t^2$

24. $x_1(t) = 2+t+\ln t,\ x_2(t) = 5+3t-\dfrac{1}{t}+3\ln t$

25. $x_1(t) = -1+8t+\cos t-8\sin t$,
$x_2(t) = -2+4t+2\cos t-3\sin t$

26. $x_1(t) = 3\cos t-32\sin t+17t\cos t+4t\sin t$,
$x_2(t) = 5\cos t-13\sin t+6t\cos t+5t\sin t$

27. $x_1(t) = 8t^3+6t^4,\ x_2(t) = 3t^2-2t^3+3t^4$

28. $x_1(t) = -7+14t-6t^2+4t^2\ln t$,
$x_2(t) = -7+9t-3t^2+\ln t-2t\ln t+2t^2\ln t$

29. $x_1(t) = t\cos t-(\ln\cos t)(\sin t)$,
$x_2(t) = t\sin t+(\ln\cos t)(\cos t)$

30. $x_1(t) = \frac{1}{2}t^2\cos 2t,\ x_2(t) = \frac{1}{2}t^2\sin 2t$

31. $x_1(t) = (9t^2+4t^3)e^t,\ x_2(t) = 6t^2e^t,\ x_3(t) = 6te^t$

32. $x_1(t) = (44+18t)e^t+(-44+26t)e^{2t}$,
$x_2(t) = 6e^t+(-6+6t)e^{2t},\ x_3(t) = 2te^{2t}$

33. $x_1(t) = 15t^2+60t^3+95t^4+12t^5,\ x_2(t) = 15t^2+55t^3+15t^4$,
$x_3(t) = 15t^2+20t^3,\ x_4(t) = 15t^2$

34. $x_1(t) = 4t^3+(4+16t+8t^2)e^{2t}$,
$x_2(t) = 3t^2+(2+4t)e^{2t}$,
$x_3(t) = (2+4t+2t^2)e^{2t},\ x_4(t) = (1+t)e^{2t}$

Chapter 6

Section 6.1

1. 6.1.13 **2.** 6.1.15 **3.** 6.1.18 **4.** 6.1.12
5. 6.1.11 **6.** 6.1.17 **7.** 6.1.14 **8.** 6.1.16

9. Equilibrium solutions $x(t) \equiv 0, \pm 2$. The critical point $(0, 0)$ in the phase plane looks like a center, whereas the points $(\pm 2, 0)$ look like saddle points.

10. Equilibrium solution $x(t) \equiv 0$. The critical point $(0, 0)$ in the phase plane looks like a spiral sink.

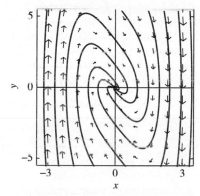

11. Equilibrium solutions $x(t) \equiv \ldots, -2\pi, -\pi, 0, \pi, 2\pi, \ldots$. The phase portrait shown in the solutions manual suggests that the critical point $(n\pi, 0)$ in the phase plane is a spiral sink if n is even, but is a saddle point if n is odd.

12. Equilibrium solution $x(t) \equiv 0$. The critical point $(0, 0)$ in the phase plane looks like a spiral source, with the solution curves emanating from this source spiraling outward toward a closed curve trajectory.

13. Solution $x(t) = x_0 e^{-2t}$, $y(t) = y_0 e^{-2t}$. The origin is a stable proper node similar to the one illustrated in Fig. 6.1.4.

14. Solution $x(t) = x_0 e^{2t}$, $y(t) = y_0 e^{-2t}$. The origin is an **unstable saddle point.**

15. Solution $x(t) = x_0 e^{-2t}$, $y(t) = y_0 e^{-t}$. The origin is a **stable node.**

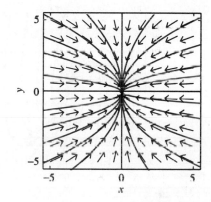

16. Solution $x(t) = x_0 e^{t}$, $y(t) = y_0 e^{3t}$. The origin is an **unstable improper node.**

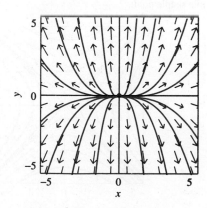

17. Solution $x(t) = A \cos t + B \sin t$, $y(t) = B \cos t - A \sin t$. The

origin is a stable center.

18. Solution $x(t) = A \cos 2t + B \sin 2t$,
$y(t) = -2B \cos 2t + 2A \sin 2t$. The origin is a stable center.

19. Solution $x(t) = A \cos 2t + B \sin 2t$, $y(t) = B \cos 2t - A \sin 2t$.
The origin is a stable center.

20. Solution $x(t) = e^{-2t}(A \cos t + B \sin t)$,
$y(t) = e^{-2t}[(-2A + B) \cos t - (A + 2B) \sin t]$. The origin is a
stable spiral point.

23. The origin and the circles $x^2 + y^2 = C > 0$; the origin is a stable
center.

24. The origin and the hyperbolas $y^2 - x^2 = C$; the origin is an
unstable saddle point.

25. The origin and the ellipses $x^2 + 4y^2 = C > 0$; the origin is a

stable center.

26. The origin and the ovals of the form $x^4 + y^4 = C > 0$; the origin
is a stable center.

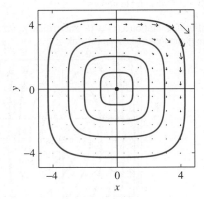

Section 6.2

1. Asymptotically stable node

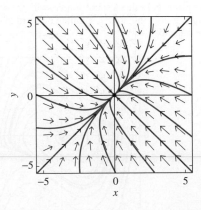

2. Unstable improper node

3. Unstable saddle point

4. Unstable saddle point

5. Asymptotically stable node

6. Unstable node

7. Unstable spiral point

8. Asymptotically stable spiral point

9. Stable, but not asymptotically stable, center

10. Stable, but not asymptotically stable, center

11. Asymptotically stable node: $(2, 1)$

12. Unstable improper node: $(2, -3)$

13. Unstable saddle point: $(2, 2)$

14. Unstable saddle point: $(3, 4)$

15. Asymptotically stable spiral point: (1, 1)

16. Unstable spiral point: (3, 2)

17. Stable center: $\left(\frac{5}{2}, -\frac{1}{2}\right)$

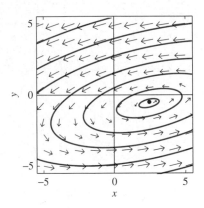

18. Stable, but not asymptotically stable, center: (−2, −1)

19. (0, 0) is a stable node. Also, there is a saddle point at (0.67, 0.40).

20. (0, 0) is an unstable node. Also, there is a saddle point at

(−1, −1) and a spiral sink at (−2.30, −1.70).

21. (0, 0) is an unstable saddle point. Also, there is a spiral sink at (−0.51, −2.12).

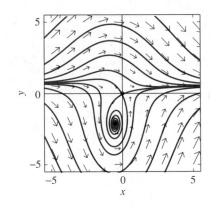

22. (0, 0) is an unstable saddle point. Also, there are nodal sinks at (±0.82, ±5.06) and nodal sources at (±3.65, ∓0.59).

23. (0, 0) is a spiral sink. Also, there is a saddle point at

$(-1.08, -0.68)$.

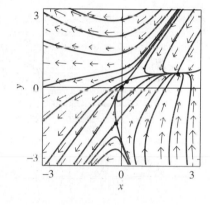

24. $(0, 0)$ is an spiral source. No other critical points are visible.

26. Theorem 2 implies only that $(0, 0)$ is an unstable source. The phase portrait for $-3 \leq x, y \leq 3$ shows also saddle points at $(0.20, 0.25)$ and $(-0.23, -1.50)$, as well as a nodal sink at $(2.36, 0.58)$.

25. Theorem 2 implies only that $(0, 0)$ is a stable sink—either a node or a spiral point. The phase portrait for $-5 \leq x, y \leq 5$ shows also a saddle point at $(0.74, -3.28)$ and spiral sink at $(2.47, -0.46)$. The origin looks like nodal sink in a second phase portrait for $-0.2 \leq x, y \leq 0.2$, which also reveals a second saddle point at $(0.12, 0.07)$.

27. Theorem 2 implies only that $(0, 0)$ is a center or a spiral point, but does not establish its stability. The phase portrait for $2 \leq x, y \leq 2$ shows also saddle points at $(-0.25, -0.51)$ and $(-1.56, 1.64)$, plus a nodal sink at $(-1.07, -1.20)$. The origin looks like a likely center in a second phase portrait for $-0.6 \leq x, y \leq 0.6$.

28. Theorem 2 implies only that $(0, 0)$ is a center or a spiral point, but does not establish its stability (though in the phase portrait it looks like a likely center). The phase portrait for $-0.25 \leq x \leq 0.25$, $-1 \leq y \leq 1$ also shows saddle points at $(0.13, 0.63)$ and $(-0.12, -0.47)$.

29. There is a saddle point at $(0, 0)$. The other critical point $(1, 1)$ is indeterminate, but looks like a center in the phase portrait.

30. There is a saddle point at $(1, 1)$ and a spiral sink at $(-1, 1)$.

31. There is a saddle point at $(1, 1)$ and a spiral sink at $(-1, -1)$.

32. There is a saddle point at $(2, 1)$ and a spiral sink at $(-2, -1)$.

37. Note that the differential equation is homogeneous.

Section 6.3

1. Linearization at $(0, 0)$: $x' = 200x$, $y' = -150y$; phase plane portrait:

Linearization at $(75, 50)$: $u' = -300v$, $v' = 100u$; phase plane portrait:

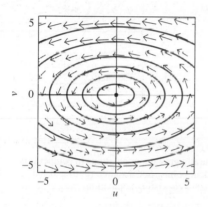

5. The characteristic equation is $\lambda^2 + 45\lambda + 126 = 0$

7. The characteristic equation is $(-24 - \lambda)^2 - 2 \cdot (18)^2 = 0$. Phase plane portrait.

Phase plane portrait for the nonlinear system in Problems 4–7:

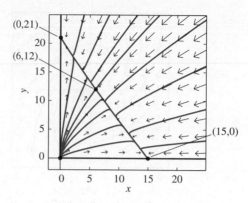

9. The characteristic equation is $\lambda^2 + 58\lambda - 120 = 0$.

10. The characteristic equation is $(\lambda + 36)(\lambda + 18) - 576 = 0$. Phase plane portrait:

Phase plane portrait for the nonlinear system in Problems 8–10:

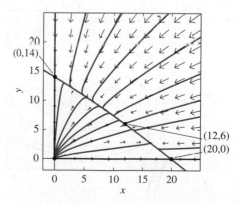

12. The characteristic equation is $\lambda^2 + 2\lambda - 15 = 0$.

13. The characteristic equation is $\lambda^2 + 2\lambda + 6 = 0$. Phase plane

portrait:

portrait:

15. The characteristic equation is $\lambda^2 + 2\lambda - 24 = 0$.

17. The characteristic equation is $\lambda^2 - 4\lambda + 6 = 0$. Phase plane portrait:

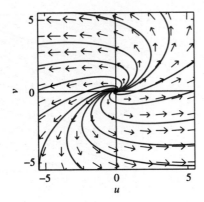

19. The characteristic equation is $\lambda^2 + 10 = 0$. Phase plane portrait:

21. The characteristic equation is $\lambda^2 - \lambda - 6 = 0$.

22. The characteristic equation is $\lambda^2 - 5\lambda + 10 = 0$. Phase plane

24. The characteristic equation is $\lambda^2 + 5\lambda - 14 = 0$.

25. The characteristic equation is $\lambda^2 + 5\lambda + 10 = 0$. Phase plane portrait:

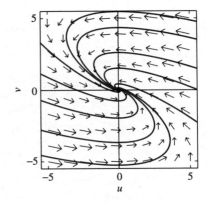

26. Naturally growing populations in competition
Critical points: nodal source $(0, 0)$ and saddle point $(3, 2)$
Nonzero coexisting populations $x(t) \equiv 3$, $y(t) \equiv 2$

27. Naturally declining populations in cooperation
Critical points: nodal sink $(0, 0)$ and saddle point $(3, 2)$
Nonzero coexisting populations $x(t) \equiv 3$, $y(t) \equiv 2$

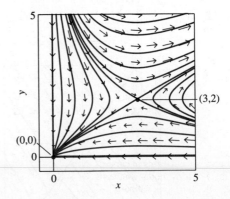

28. Naturally declining predator, naturally growing prey population
Critical points: saddle point $(0, 0)$ and apparent stable center $(4, 8)$

Nonzero coexisting populations $x(t) \equiv 4$, $y(t) \equiv 8$

29. Logistic and naturally growing populations in competition
Critical points: nodal source $(0, 0)$, nodal sink $(3, 0)$, and saddle point $(2, 2)$
Nonzero coexisting populations $x(t) \equiv 2$, $y(t) \equiv 2$

30. Logistic and naturally declining populations in cooperation
Critical points: saddle point $(0, 0)$, nodal sink $(3, 0)$, and saddle point $(5, 4)$
Nonzero coexisting populations $x(t) \equiv 5$, $y(t) \equiv 4$

31. Logistic prey, naturally declining predator population
Critical points: saddle points $(0, 0)$ and $(3, 0)$, spiral sink $(2, 4)$
Nonzero coexisting populations $x(t) \equiv 2$, $y(t) \equiv 4$

32. Logistic populations in cooperation
Critical points: nodal source $(0, 0)$, saddle points $(10, 0)$ and $(0, 20)$, nodal sink $(30, 60)$
Nonzero coexisting populations $x(t) \equiv 30$, $y(t) \equiv 60$

33. Logistic prey and predator populations
Critical points: nodal source $(0, 0)$, saddle points $(30, 0)$ and $(0, 20)$, nodal sink $(4, 22)$

Nonzero coexisting populations $x(t) \equiv 4$, $y(t) \equiv 22$

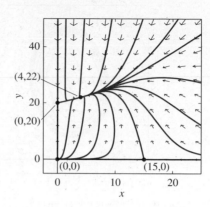

34. Logistic prey and predator populations
Critical points: nodal source $(0, 0)$, saddle points $(15, 0)$ and $(0, 5)$, spiral sink $(10, 10)$
Nonzero coexisting populations $x(t) \equiv 10$, $y(t) \equiv 10$

Section 6.4

1. Eigenvalues: -2, 3; stable node

2. Eigenvalues: $1, 3$; unstable node

3. Eigenvalues: $-3, 5$; unstable saddle point

4. Eigenvalues: $-1 \pm 2i$; stable spiral point

5. Critical points: $(0, n\pi)$ where n is an integer; an unstable saddle point if n is even, a stable spiral point if n is odd

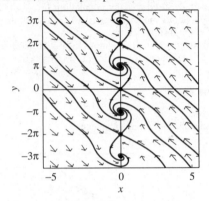

6. Critical points: $(n, 0)$ where n is an integer; an unstable saddle point if n is even, a stable spiral point if n is odd

7. Critical points: $(n\pi, n\pi)$ where n is an integer; an unstable saddle point if n is even, a stable spiral point if n is odd

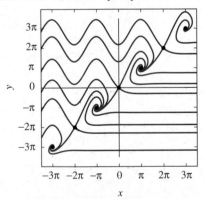

8. Critical points: $(n\pi, 0)$ where n is an integer; an unstable node if n is even, an unstable saddle point if n is odd

9. If n is odd then $(n\pi, 0)$ is an unstable saddle point.

10. If n is odd then $(n\pi, 0)$ is a stable node.

11. $(n\pi, 0)$ is a stable spiral point.

12. Unstable saddle points at $(2, 0)$ and $(-2, 0)$, a stable center at $(0, 0)$

13. Unstable saddle points at $(2, 0)$ and $(-2, 0)$, a stable spiral point at $(0, 0)$

14. Stable centers at $(2, 0)$ and $(-2, 0)$, an unstable saddle point at $(0, 0)$

15. A stable center at $(0, 0)$ and an unstable saddle point at $(4, 0)$

16. Stable centers at $(2, 0)$, $(0, 0)$, and $(-2, 0)$, unstable saddle points at $(1, 0)$ and $(-1, 0)$

17. $(0, 0)$ is a spiral sink.

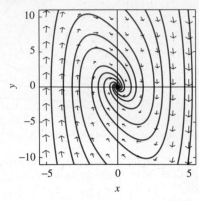

18. $(0, 0)$ is a spiral sink; the points $(\pm 2, 0)$ are saddle points.

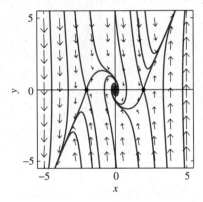

19. $(0, 0)$ is a spiral sink.

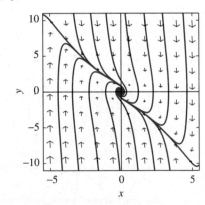

20. $(n\pi, 0)$ is a spiral sink if n is even, a saddle point if n is odd.

Chapter 7

Section 7.1

1. $1/s^2$, $s > 0$ **2.** $2/s^3$, $s > 0$

3. $e/(s - 3)$, $s > 3$ **4.** $s/(s^2 + 1)$, $s > 0$

5. $1/(s^2 - 1)$, $s > 1$

6. $\frac{1}{2}[1/s - s/(s^2 + 4)]$, $s > 0$

7. $(1 - e^{-s})/s$, $s > 0$

8. $(e^{-s} - e^{-2s})/s$, $s > 0$

9. $(1 - e^{-s} - se^{-s})/s^2$, $s > 0$

10. $(s - 1 + e^{-s})/s^2$, $s > 0$

11. $\frac{1}{2}\sqrt{\pi}s^{-3/2} + 3s^{-2}$, $s > 0$

12. $(45\pi - 192s^{3/2})/(8s^{7/2})$, $s > 0$

13. $s^{-2} - 2(s - 3)^{-1}$, $s > 3$

14. $3\sqrt{\pi}/(4s^{5/2}) + 1/(s + 10)$, $s > 0$

15. $s^{-1} + s(s^2 - 25)^{-1}$, $s > 5$ **16.** $(s + 2)/(s^2 + 4)$, $s > 0$

17. $\cos^2 2t = \frac{1}{2}(1 + \cos 4t)$; $\frac{1}{2}\left[s^{-1} + s/(s^2 + 16)\right]$, $s > 0$

18. $3/(s^2 + 36)$, $s > 0$

19. $s^{-1} + 3s^{-2} + 6s^{-3} + 6s^{-4}$, $s > 0$

20. $1/(s - 1)^2$, $s > 1$ **21.** $(s^2 - 4)/(s^2 + 4)^2$, $s > 0$

22. $\frac{1}{2}\left[s/(s^2 - 36) - s^{-1}\right]$ **23.** $\frac{1}{2}t^3$

24. $2\sqrt{t/\pi}$ **25.** $1 - \frac{8}{3}t^{3/2}\pi^{-1/2}$

26. e^{-5t} **27.** $3e^{4t}$

28. $3\cos 2t + \frac{1}{2}\sin 2t$ **29.** $\frac{5}{3}\sin 3t - 3\cos 3t$

30. $-\cosh 2t - \frac{9}{2}\sinh 2t$ **31.** $\frac{3}{5}\sinh 5t - 10\cosh 5t$

32. $2u(t - 3)$

37. $f(t) = 1 - u(t - a)$. Your figure should indicate that the graph of f contains the point $(a, 0)$, but not the point $(a, 1)$.

38. $f(t) = u(t - a) - u(t - b)$. Your figure should indicate that the graph of f contains the points $(a, 1)$ and $(b, 0)$, but not the points $(a, 0)$ and $(b, 1)$.

39. Figure 7.2.8 shows the graph of the unit staircase function.

Section 7.2

1. $x(t) = 5\cos 2t$ **2.** $x(t) = 3\cos 3t + \frac{4}{3}\sin 3t$

3. $x(t) = \frac{2}{3}(e^{2t} - e^{-t})$ **4.** $x(t) = \frac{1}{2}(7e^{-4t} - 3e^{-5t})$

5. $x(t) = \frac{1}{3}(2\sin t - \sin 2t)$ **6.** $x(t) = \frac{1}{3}(\cos t - \cos 2t)$

7. $x(t) = \frac{1}{8}(9\cos t - \cos 3t)$ **8.** $x(t) = \frac{1}{9}(1 - \cos 3t)$

9. $x(t) = \frac{1}{6}(2 - 3e^{-t} + e^{-3t})$

10. $x(t) = \frac{1}{4}(2t - 3 + 12e^{-t} - 9e^{-2t})$

11. $x(t) = 1$, $y(t) = -2$

12. $x(t) = \frac{2}{9}(e^{2t} - e^{-t} - 3te^{-t})$, $y(t) = \frac{1}{9}(e^{2t} - e^{-t} + 6te^{-t})$

13. $x(t) = \left(2/\sqrt{3}\right)\sinh\left(t/\sqrt{3}\right)$,

$y(t) = \cosh\left(t/\sqrt{3}\right) + \left(1/\sqrt{3}\right)\sinh\left(t/\sqrt{3}\right)$

14. $x(t) = \frac{1}{4}(2t - 3\sin 2t)$, $y(t) = -\frac{1}{8}(2t + 3\sin 2t)$

15. $x(t) = \frac{1}{3}\left(2 + e^{-3t/2}\left[\cos(rt/2) + r\sin(rt/2)\right]\right)$,

$y(t) = \frac{1}{21}\left(28 - 9e^t + 2e^{-3t/2}\left[\cos(rt/2) + 4r\sin(rt/2)\right]\right)$ where $r = \sqrt{3}$

16. $x(t) = \cos t + \sin t$, $y(t) = e^t - \cos t$, $z(t) = -2\sin t$

17. $f(t) = \frac{1}{3}(e^{3t} - 1)$

18. $f(t) = \frac{3}{5}(1 - e^{-5t})$

19. $f(t) = \frac{1}{4}(1 - \cos 2t) = \frac{1}{2}\sin^2 t$

20. $f(t) = \frac{1}{9}(6\sin 3t - \cos 3t + 1)$

21. $f(t) = t - \sin t$

22. $f(t) = \frac{1}{9}(-1 + \cosh 3t)$

23. $f(t) = -t + \sinh t$

24. $f(t) = \frac{1}{2}(e^{-2t} - 2e^{-t} + 1)$

Section 7.3

1. $24/(s - \pi)^5$ **2.** $\frac{3}{4}\sqrt{\pi}(s + 4)^{-5/2}$

3. $3\pi/[(s + 2)^2 + 9\pi^2]$ **4.** $\sqrt{2}(2s + 5)/(4s^2 + 4s + 17)$

5. $\frac{3}{2}e^{2t}$ **6.** $(t - t^2)e^{-t}$

7. te^{-2t} **8.** $e^{-2t}\cos t$

9. $e^{3t}\left(3\cos 4t + \frac{7}{2}\sin 4t\right)$ **10.** $\frac{1}{36}e^{2t/3}\left(8\cos\frac{4}{3}t - 5\sin\frac{4}{3}t\right)$

11. $\frac{1}{2}\sinh 2t$ **12.** $2 + 3e^{3t}$

13. $3e^{-2t} - 5e^{-5t}$ **14.** $2 + e^{2t} - 3e^{-t}$

15. $\frac{1}{25}(e^{5t} - 1 - 5t)$ **16.** $\frac{1}{125}\left[e^{2t}(5t - 2) + e^{-3t}(5t + 2)\right]$

17. $\frac{1}{16}(\sinh 2t - \sin 2t)$ **18.** $e^{4t}\left(1 + 12t + 24t^2 + \frac{32}{3}t^3\right)$

19. $\frac{1}{3}(2\cos 2t + 2\sin 2t - 2\cos t - \sin t)$

20. $\frac{1}{32}\left[e^{2t}(2t - 1) + e^{-2t}(2t + 1)\right]$

21. $\frac{1}{2}e^{-t}(5\sin t - 3t\cos t - 2t\sin t)$

22. $\frac{1}{64}e^{t/2}\left[(4t + 8)\cos t + (4 - 3t)\sin t\right]$

27. $\frac{1}{4}e^{-3t}(8\cos 4t + 9\sin 4t)$

28. $\frac{1}{4}(1 - 2e^{2t} + e^{4t})$ **29.** $\frac{1}{8}(-6t + 3\sinh 2t)$

30. $\frac{1}{10}\left[2e^{-t} - e^{-2t}(2\cos 2t + \sin 2t)\right]$

31. $\frac{1}{15}(6e^{2t} - 5 - e^{-3t})$ **32.** $\frac{1}{2}(\cosh t + \cos t)$

33. $x(t) = r(\cosh rt \sin rt - \sinh rt \cos rt)$ where $r = 1/\sqrt{2}$

34. $\frac{1}{2}\sin 2t + \frac{1}{3}\sin 3t$ **35.** $\frac{1}{16}(\sin 2t - 2t\cos 2t)$

36. $\frac{1}{50}\left[2e^{2t} + (10t - 2)\cos t - (5t + 14)\sin t\right]$

37. $\frac{1}{50}\left[(5t - 1)e^{-t} + e^{-2t}(\cos 3t + 32\sin 3t)\right]$

38. $\frac{1}{510}e^{-3t}(489\cos 3t + 307\sin 3t) + \frac{1}{170}(7\cos 2t + 6\sin 2t)$

39.

Section 7.4

1. $\frac{1}{2}t^2$ **2.** $(e^{at} - at - 1)/a^2$

3. $\frac{1}{2}(\sin t - t\cos t)$ **4.** $2(t - \sin t)$

5. te^{at} **6.** $(e^{at} - e^{bt})/(a - b)$

7. $\frac{1}{3}(e^{3t} - 1)$ **8.** $\frac{1}{4}(1 - \cos 2t)$

9. $\frac{1}{54}(\sin 3t - 3t\cos 3t)$ **10.** $(kt - \sin kt)/k^3$

11. $\frac{1}{4}(\sin 2t + 2t\cos 2t)$ **12.** $\frac{1}{5}[1 - e^{-2t}(\cos t + 2\sin t)]$

13. $\frac{1}{10}(3e^{3t} - 3\cos t + \sin t)$ **14.** $\frac{1}{3}(\cos t - \cos 2t)$

15. $6s/(s^2 + 9)^2$, $s > 0$ **16.** $(2s^3 - 24s)/(s^2 + 4)^3$, $s > 0$

17. $(s^2 - 4s - 5)/(s^2 - 4s + 13)^2$, $s > 0$

18. $\dfrac{2(3s^2 + 6s + 7)}{(s + 1)^2(s^2 + 2s + 5)^2}$, $s > 0$

19. $\frac{1}{2}\pi - \arctan s = \arctan(1/s)$, $s > 0$

20. $\frac{1}{2}\ln(s^2 + 4) - \ln s$, $s > 0$

21. $\ln s - \ln(s - 3)$, $s > 3$

22. $\ln(s + 1) - \ln(s - 1)$, $s > 1$

23. $-(2\sinh 2t)/t$ **24.** $2(\cos 2t - \cos t)/t$

25. $(e^{-2t} + e^{3t} - 2\cos t)/t$ **26.** $(e^{-2t}\sin 3t)/t$

27. $2(1 - \cos t)/t$ **28.** $\frac{1}{8}(t\sin t - t^2\cos t)$

29. $(s + 1)X'(s) + 4X(s) = 0$; $x(t) = Ct^3e^{-t}$, $C \neq 0$

30. $X(s) = A/(s + 3)^3$; $x(t) = Ct^2e^{-3t}$, $C \neq 0$

31. $(s - 2)X'(s) + 3X(s) = 0$; $x(t) = Ct^2e^{2t}$, $C \neq 0$

32. $(s^2 + 2s)X'(s) + (4s + 4)X(s) = 0;$
$x(t) = C(1 - t - e^{-2t} - te^{-2t}), C \neq 0$

33. $(s^2 + 1)X'(s) + 4sX(s) = 0; x(t) = C(\sin t - t \cos t), C \neq 0$

34. $x(t) = Ce^{-2t}(\sin 3t - 3t \cos 3t), C \neq 0$

Section 7.5

1. $f(t) = u(t - 3) \cdot (t - 3)$

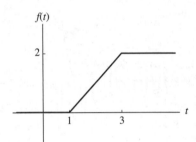

2. $f(t) = (t - 1)u(t - 1) - (t - 3)u(t - 3)$

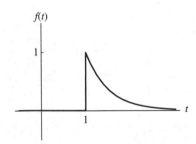

3. $f(t) = u(t - 1) \cdot e^{-2(t-1)}$

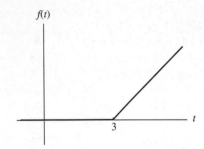

4. $f(t) = e^{t-1}u(t - 1) - e^2 e^{t-2}u(t - 2)$

5. $f(t) = u(t - \pi) \cdot \sin(t - \pi) = -u(t - \pi) \sin t$

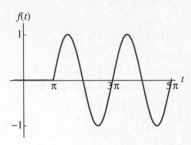

6. $f(t) = u(t - 1) \cdot \cos \pi(t - 1) = -u(t - 1) \cos \pi t$

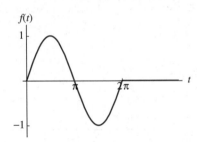

7. $f(t) = \sin t - u(t - 2\pi) \sin(t - 2\pi) = [1 - u(t - 2\pi)] \sin t$

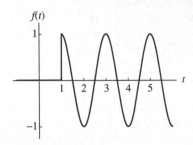

8. $f(t) = \cos \pi t - u(t - 2) \cos \pi(t - 2) = [1 - u(t - 2)] \cos \pi t$

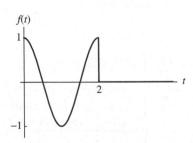

9. $f(t) = \cos \pi t + u(t - 3) \cos \pi(t - 3) = [1 - u(t - 3)] \cos \pi t$

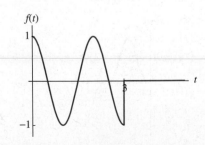

10. $f(t) = 2u(t - \pi)\cos 2(t - \pi) - 2u(t - 2\pi)\cos 2(t - 2\pi)$
$\quad = 2[u(t - \pi) - u(t - 2\pi)]\cos 2t$

11. $f(t) = 2[1 - u_3(t)]; F(s) = 2(1 - e^{-3s})/s$
12. $F(s) = (e^{-s} - e^{-4s})/s$
13. $F(s) = (1 - e^{-2\pi s})/(s^2 + 1)$
14. $F(s) = s(1 - e^{-2s})/(s^2 + \pi^2)$
15. $F(s) = (1 + e^{-3\pi s})/(s^2 + 1)$
16. $F(s) = 2(e^{-\pi s} - e^{-2\pi s})/(s^2 + 4)$
17. $F(s) = \pi(e^{-2s} + e^{-3s})/(s^2 + \pi^2)$
18. $F(s) = 2\pi(e^{-3s} + e^{-5s})/(4s^2 + \pi^2)$
19. $F(s) = e^{-s}(s^{-1} + s^{-2})$
20. $F(s) = (1 - e^{-s})/s^2$
21. $F(s) = (1 - 2e^{-s} + e^{-?s})/s^?$
28. $F(s) = (1 - e^{-as} - ase^{-as})/[s^2(1 - e^{-2as})]$
31. $x(t) = \frac{1}{2}[1 - u(t - \pi)]\sin^2 t$

32. $x(t) = g(t) - u(t - 2)g(t - 2)$ where
$\quad g(t) = \frac{1}{12}(3 - 4e^{-t} + e^{-4t}).$

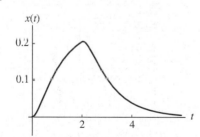

33. $x(t) = \frac{1}{8}[1 - u(t - 2\pi)]\left(\sin t - \frac{1}{3}\sin 3t\right)$

34. $x(t) = g(t) - u(t - 1)[g(t - 1) + h(t - 1)]$ where
$\quad g(t) = t - \sin t$ and $h(t) = 1 - \cos t.$

35. $x(t) = \frac{1}{4}\{-1 + t + (t + 1)e^{-2t} +$
$\quad u(t - 2)\left[1 - t + (3t - 5)e^{-2(t-2)}\right]\}$

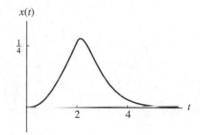

36. $i(t) = e^{-10t} - u(t - 1)e^{-10(t-1)}$
37. $i(t) = [1 - u(t - 2\pi)]\sin 100t$
38. $i(t) = \frac{10}{99}[1 - (u - \pi)](\cos 10t - \cos 100t)$
39. $i(t) = \frac{1}{50}\left[(1 - e^{-50t})^2 - u(t - 1)(1 + 98e^{-50(t-1)} - 99e^{-100(t-1)})\right]$
40. $i(t) = \frac{1}{50}\left[(1 - e^{-50t} - 50te^{-50t}) - u(t - 1)(1 - e^{-50(t-1)} + 2450te^{-50(t-1)})\right]$
41. $x(t) = 2|\sin t|\sin t$

42. $x(t) = g(t) + 2\sum_{n=1}^{\infty}(-1)^n u(t - n\pi)g(t - n\pi)$ where
$\quad g(t) = 1 - \frac{1}{3}e^{-t}(3\cos 3t + \sin 3t)$

Section 7.6

1. $x(t) = \frac{1}{2}\sin 2t$

2. $x(t) = \frac{1}{2}[1 + u(t - \pi)]\sin 2t$

3. $x(t) = \frac{1}{4}[1 - e^{-2t} - 2te^{-2t}] + u(t - 2)(t - 2)e^{-2(t-2)}$

4. $x(t) = -2 + t + 2e^{-t} + 3te^{-t}$

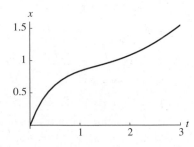

5. $x(t) = 2u(t - \pi)e^{-(t-\pi)}\sin(t - \pi)$

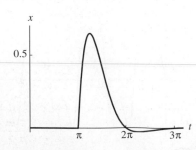

6. $x(t) = \frac{1}{6}(t - 2u(t - 3\pi))\sin 3t$

7. $x(t) = \left[2 - e^{2\pi}u(t - \pi) + e^{4\pi}u(t - 2\pi)\right]e^{-2t}\sin t$

8. $x(t) = (2 + 5t)e^{-t} - u(t - 2)(t - 2)e^{-(t-2)}$

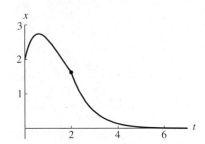

9. $x(t) = \int_0^t \frac{1}{2}(\sin 2\tau)f(t - \tau)\,d\tau$

10. $x(t) = \int_0^t \tau e^{-3\tau}f(t - \tau)\,d\tau$

11. $x(t) = \int_0^t (e^{-3\tau}\sinh \tau)f(t - \tau)\,d\tau$

12. $x(t) = \int_0^t (e^{-2\tau}\sin 2\tau)f(t - \tau)\,d\tau$

13. (a) $mx_\epsilon(t) = p[t^2 - u_\epsilon(t)(t - \epsilon)^2]/(2\epsilon)$;
 (b) If $t > \epsilon$, then $mx_\epsilon(t) = p(2\epsilon t - \epsilon^2)/(2\epsilon)$, and hence $mx_\epsilon(t) \to pt$ as $\epsilon \to 0$;
 (c) $mv = (mx)' = (pt)' = 0$.

15. The transform of each of the two given initial value problems is $(ms^2 + k)X(s) = mv_0 = p_0$.

17. (b) $i(t) = e^{-100(t-1)}u_1(t) - e^{-100(t-2)}u_2(t)$.
 If $t > 2$, then $i(t) = -(e^{100} - 1)e^{100(1-t)} < 0$.

18. $i(t) = [1 - u(t - \pi)]\sin 10t$

19. $i(t) = \sum_{n=0}^{\infty} u(t - n\pi/10)\sin 10t$

20. $i(t) = \sum_{n=0}^{\infty} (-1)^n u(t - n\pi/5)\sin 10t$

21. $i(t) = \sum_{n=0}^{\infty} u(t - n\pi/10)e^{3n\pi}e^{-30t}\sin t$

22. $x(t) = \sum_{n=0}^{\infty} u(t - 2n\pi)\sin t$

Chapter 8

Section 8.1

1. $y(x) = c_0\left(1 + x + \frac{x^2}{2} + \frac{x^3}{3!} + \cdots\right) = c_0 e^x; \rho = +\infty$

2. $y(x) = c_0\left(1 + \frac{4x}{1!} + \frac{4^2 x^2}{2!} + \frac{4^3 x^3}{3!} + \frac{4^4 x^4}{4!} + \cdots\right) = c_0 e^{4x};$
$\rho = \infty$

3. $y(x) = c_0\left(1 - \frac{3x}{2} + \frac{(3x)^2}{2!2^2} - \frac{(3x)^3}{3!2^3} + \frac{(3x)^4}{4!2^4} - \cdots\right) =$
$c_0 e^{-3x/2}; \rho = +\infty$

4. $y(x) = c_0\left(1 - \frac{x^2}{1!} + \frac{x^4}{2!} - \frac{x^6}{3!} + \cdots\right) = c_0 e^{-x^2}; \rho = \infty$

5. $y(x) = c_0\left(1 + \frac{x^3}{3} + \frac{x^6}{2!3^2} + \frac{x^9}{3!3^3} + \cdots\right) = c_0 \exp\left(\frac{1}{3}x^3\right);$
$\rho = +\infty$

6. $y(x) = c_0\left(1 + \frac{x}{2} + \frac{x^2}{4} + \frac{x^3}{8} + \frac{x^4}{16} + \cdots\right) = \frac{2c_0}{2-x}; \rho = 2$

7. $y(x) = c_0(1 + 2x + 4x^2 + 8x^3 + \cdots) = \frac{c_0}{1 - 2x}; \rho = \frac{1}{2}$

8. $y(x) = c_0\left(1 + \frac{x}{2} - \frac{x^2}{8} + \frac{x^3}{16} - \frac{5x^4}{128} + \cdots\right) = c_0\sqrt{1+x};$
$\rho = 1$

9. $y(x) = c_0(1 + 2x + 3x^2 + 4x^3 + \cdots) = \frac{c_0}{(1-x)^2}; \rho = 1$

10. $y(x) = c_0\left(1 - \frac{3x}{2} + \frac{3x^2}{8} + \frac{x^3}{16} + \frac{3x^4}{128} + \cdots\right) = c_0(1-x)^{3/2};$
$\rho = 1$

11. $y(x) = c_0\left(1 + \frac{x^2}{2!} + \frac{x^4}{4!} + \frac{x^6}{6!} + \cdots\right) +$
$c_1\left(x + \frac{x^3}{3!} + \frac{x^5}{5!} + \frac{x^7}{7!} + \cdots\right)$
$= c_0 \cosh x + c_1 \sinh x; \rho = +\infty$

12. $y(x) = c_0\left(1 + \frac{(2x)^2}{2!} + \frac{(2x)^4}{4!} + \frac{(2x)^6}{6!} + \cdots\right) +$
$\frac{c_1}{2}\left((2x) + \frac{(2x)^3}{3!} + \frac{(2x)^5}{5!} + \frac{(2x)^7}{7!} + \cdots\right) =$
$c_0 \cosh 2x + \frac{c_1}{2}\sinh 2x; \rho = \infty$

13. $y(x) = c_0\left(1 - \frac{(3x)^2}{2!} + \frac{(3x)^4}{4!} - \frac{(3x)^6}{6!} + \cdots\right) +$
$\frac{c_1}{3}\left(3x - \frac{(3x)^3}{3!} + \frac{(3x)^5}{5!} - \frac{(3x)^7}{7!} + \cdots\right)$
$= c_0 \cos 3x + \frac{1}{3}c_1 \sin 3x; \rho = +\infty$

14. $y(x) = x + c_0\left(1 - \frac{x^2}{2!} + \frac{x^4}{4!} - \frac{x^6}{6!} + \cdots\right)$
$+ (c_1 - 1)\left(x - \frac{x^3}{3!} + \frac{x^5}{5!} - \frac{x^7}{7!} + \cdots\right)$
$= x + c_0 \cos x + (c_1 - 1)\sin x; \rho = \infty$

15. $(n+1)c_n = 0$ for all $n \geq 0$, so $c_n = 0$ for all $n \geq 0$.

16. $2nc_n = c_n$ for all $n \geq 0$, so $c_n = 0$ for all $n \geq 0$.

17. $c_0 = c_1 = 0$ and $c_{n+1} = -nc_n$ for $n \geq 1$, thus $c_n = 0$ for all $n \geq 0$.

18. $c_n = 0$ for all $n \geq 0$

19. $(n+1)(n+2)c_{n+2} = -4c_n;$
$y(x) = \frac{3}{2}\left[(2x) - \frac{(2x)^3}{3!} + \frac{(2x)^5}{5!} - \frac{(2x)^7}{7!} + \cdots\right] = \frac{3}{2}\sin 2x$

20. $(n+1)(n+2)c_{n+2} - 4c_n;$
$y(x) = 2\left[1 + \frac{(2x)^2}{2!} + \frac{(2x)^4}{4!} + \frac{(2x)^6}{6!} + \cdots\right] = 2\cosh 2x$

21. $n(n+1)c_{n+1} = 2nc_n - c_{n-1};$
$y(x) = x + x^2 + \frac{x^3}{2!} + \frac{x^4}{3!} + \frac{x^5}{4!} + \cdots = xe^x$

22. $n(n+1)c_{n+1} = -nc_n + 2c_{n-1}; y = e^{-2x}$

23. As $c_0 = c_1 = 0$ and $(n^2 - n + 1)c_n + (n - 1)c_{n-1} = 0$ for $n \geq 2$, $c_n = 0$ for all $n \geq 0$

Section 8.2

1. $c_{n+2} = c_n; y(x) = c_0 \sum_{n=0}^{\infty} x^{2n} + c_1 \sum_{n=0}^{\infty} x^{2n+1} = \frac{c_0 + c_1 x}{1 - x^2}; \rho = 1$

2. $c_{n+2} = -\frac{1}{2}c_n; \rho = 2;$
$y(x) = c_0 \sum_{n=0}^{\infty} \frac{(-1)^n x^{2n}}{2^n} + c_1 \sum_{n=0}^{\infty} \frac{(-1)^n x^{2n+1}}{2^n}$

3. $(n+2)c_{n+2} = -c_n;$
$y(x) = c_0 \sum_{n=0}^{\infty} \frac{(-1)^n x^{2n}}{n!2^n} + c_1 \sum_{n=0}^{\infty} \frac{(-1)^n x^{2n+1}}{(2n+1)!!}; \rho = +\infty$

4. $(n+2)c_{n+2} = -(n+4)c_n; \rho - 1;$
$y(x) = c_0 \sum_{n=0}^{\infty} (-1)^n (n+1)x^{2n} + \frac{1}{3}c_1 \sum_{n=0}^{\infty} (-1)^n (2n+3)x^{2n+1}$

5. $3(n+2)c_{n+2} = nc_n; \rho = \sqrt{3}; y(x) = c_0 + c_1 \sum_{n=0}^{\infty} \frac{x^{2n+1}}{(2n+1)3^n}$

6. $(n+1)(n+2)c_{n+2} = (n-3)(n-4)c_n; \rho = \infty;$
$y(x) = c_0(1 + 6x^2 + x^4) + c_1(x + x^3)$

7. $3(n+1)(n+2)c_{n+2} = (n-4)^2 c_n;$
$y(x) = c_0\left(1 - \frac{8x^2}{3} + \frac{8x^4}{27}\right) +$
$c_1\left(x - \frac{x^3}{2} + \frac{x^5}{120} + 9\sum_{n=3}^{\infty} \frac{(-1)^n (2n-5)!! x^{2n+1}}{(2n+1)!3^n}\right)$

8. $2(n+1)(n+2)c_{n+2} = (n-4)(n+4)c_n; y(x) = c_0(1 - 4x^2 +$
$2x^4) + c_1\left(x - \frac{5x^3}{4} + \frac{7x^5}{32} + \sum_{n=3}^{\infty} \frac{(2n-5)!!(2n+3)!! x^{2n+1}}{(2n+1)!2^n}\right)$

9. $(n+1)(n+2)c_{n+2} = (n+3)(n+4)c_n;\ \rho = 1;$

$$y(x) = c_0 \sum_{n=0}^{\infty}(n+1)(2n+1)x^{2n} + \frac{c_1}{3}\sum_{n=0}^{\infty}(n+1)(2n+3)x^{2n+1}$$

10. $3(n+1)(n+2)c_{n+2} = -(n-4)c_n;$

$$y(x) = c_0\left(1 + \frac{2x^2}{3} + \frac{x^4}{27}\right) +$$

$$c_1\left(x + \frac{x^3}{6} + \frac{x^5}{360} + 3\sum_{n=3}^{\infty}\frac{(-1)^n(2n-5)!!x^{2n+1}}{(2n+1)!3^n}\right)$$

11. $5(n+1)(n+2)c_{n+2} = 2(n-5)c_n;$

$$y(x) = c_1\left(x - \frac{4x^3}{15} + \frac{4x^5}{375}\right) +$$

$$c_0\left(1 - x^2 + \frac{x^4}{10} + \frac{x^6}{750} + 15\sum_{n=4}^{\infty}\frac{(2n-7)!!\,2^n x^{2n}}{(2n)!5^n}\right)$$

12. $c_2 = 0;\ (n+2)c_{n+3} = c_n;$

$$y(x) = c_0\left(1 + \sum_{n=1}^{\infty}\frac{x^{3n}}{2\cdot 5\cdots(3n-1)}\right) + c_1\sum_{n=0}^{\infty}\frac{x^{3n+1}}{n!3^n}$$

13. $c_2 = 0;\ (n+3)c_{n+3} = -c_n;$

$$y(x) = c_0\sum_{n=0}^{\infty}\frac{(-1)^n x^{3n}}{n!3^n} + c_1\sum_{n=0}^{\infty}\frac{(-1)^n x^{3n+1}}{1\cdot 4\cdots(3n+1)}$$

14. $c_2 = 0;\ (n+2)(n+3)c_{n+3} = -c_n;$

$$y(x) = c_0\left(1 + \sum_{n=1}^{\infty}\frac{(-1)^n x^{3n}}{3^n\cdot n!\cdot 2\cdot 5\cdots(3n-1)}\right) +$$

$$c_1\sum_{n=0}^{\infty}\frac{(-1)^n x^{3n+1}}{3^n\cdot n!\cdot 1\cdot 4\cdots(3n+1)}$$

15. $c_2 = c_3 = 0;\ (n+3)(n+4)c_{n+4} = -c_n;$

$$y(x) = c_0\left(1 + \sum_{n=1}^{\infty}\frac{(-1)^n x^{4n}}{4^n\cdot n!\cdot 3\cdot 7\cdots(4n-1)}\right) +$$

$$c_1\sum_{n=0}^{\infty}\frac{(-1)^n x^{4n+1}}{4^n\cdot n!\cdot 5\cdot 9\cdots(4n+1)}$$

16. $y(x) = x$

17. $y(x) = 1 + x^2$

18. $y(x) = 2\sum_{n=0}^{\infty}\frac{(-1)^n(x-1)^{2n}}{n!\,2^n};$ converges for all x

19. $y(x) = \frac{1}{3}\sum_{n=0}^{\infty}(2n+3)(x-1)^{2n+1};$ converges if $0 < x < 2$

20. $y(x) = 2 - 6(x-3)^2;$ converges for all x

21. $y(x) = 1 + 4(x+2)^2;$ converges for all x

22. $y(x) = 2x + 6$

23. $2c_2 + c_0 = 0;\ (n+1)(n+2)c_{n+2} + c_n + c_{n-1} = 0$ for $n \geq 1;$

$$y_1(x) = 1 - \frac{x^2}{2} - \frac{x^3}{6} + \cdots;\ y_2(x) = x - \frac{x^3}{6} - \frac{x^4}{12} + \cdots$$

24. $y_1(x) = 1 + \frac{x^3}{3} + \frac{x^5}{5} + \frac{x^6}{45} + \cdots;$

$$y_2(x) = x + \frac{x^3}{3} + \frac{x^4}{6} + \frac{x^5}{5} + \cdots$$

25. $c_2 = c_3 = 0,\ (n+3)(n+4)c_{n+4} + (n+1)c_{n+1} + c_n = 0$ for

$n \geq 0;\ y_1(x) = 1 - \frac{x^4}{12} + \frac{x^7}{126} + \cdots;$

$$y_2(x) = x - \frac{x^4}{12} - \frac{x^5}{20} + \cdots$$

26. $y(x) = c_0\left(1 - \frac{x^6}{30} + \frac{x^9}{72} + \cdots\right) + c_1\left(x - \frac{x^7}{42} + \frac{x^{10}}{90} + \cdots\right)$

27. $y(x) = 1 - x - \frac{x^2}{2} + \frac{x^3}{3} - \frac{x^4}{24} + \frac{x^5}{30} + \frac{29x^6}{720} + \frac{13x^7}{630} - \frac{143x^8}{40320} + \cdots;$

$y(0.5) \approx 0.4156$

28. $y(x) = c_0\left(1 - \frac{x^2}{2} + \frac{x^3}{6} + \cdots\right) + c_1\left(x - \frac{x^3}{6} + \frac{x^4}{12} + \cdots\right)$

29. $y_1(x) = 1 - \frac{1}{2}x^2 + \frac{1}{720}x^6 + \cdots;$

$$y_2(x) = x - \frac{1}{6}x^3 - \frac{1}{60}x^5 + \cdots$$

30. $y(x) = c_0\left(1 - \frac{x^2}{2} + \frac{x^3}{6} + \cdots\right) + c_1\left(x - \frac{x^2}{2} + \frac{x^4}{18} + \cdots\right)$

33. The following figure shows the interlaced zeros of the 4th and 5th Hermite polynomials.

34. The figure below results when we use $n = 40$ terms in each summation. But with $n = 50$ we get the same picture as Fig. 8.2.3 in the text.

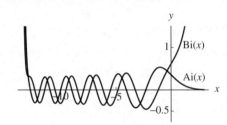

Section 8.3

1. Ordinary point

2. Ordinary point

3. Irregular singular point

4. Irregular singular point

5. Regular singular point; $r_1 = 0,\ r_2 = -1$

6. Regular singular point; $r_1 = 1,\ r_2 = -2$

7. Regular singular point; $r = -3, -3$

8. Regular singular point; $r = \frac{1}{2}, -3$

9. Regular singular point $x = 1$

10. Regular singular point $x = 1$

11. Regular singular points $x = 1, -1$

12. Irregular singular point $x = 2$

13. Regular singular points $x = 2, -2$

14. Irregular singular points $x = 3, -3$

15. Regular singular point $x = 2$

16. Irregular singular point $x = 0$, regular singular point $x = 1$

17. $y_1(x) = \cos\sqrt{x},\ y_2(x) = \sin\sqrt{x}$

18. $y_1(x) = \sum_{n=0}^{\infty}\frac{x^n}{n!(2n+1)!!},\ y_2(x) = x^{-1/2}\sum_{n=0}^{\infty}\frac{x^n}{n!\,(2n-1)!!}$

19. $y_1(x) = x^{3/2}\left(1 + 3\sum_{n=1}^{\infty} \dfrac{x^n}{n!\,(2n+3)!!}\right),$

$y_2(x) = 1 - x - \sum_{n=2}^{\infty} \dfrac{x^n}{n!\,(2n-3)!!}$

20. $y_1(x) = x^{1/3}\sum_{n=0}^{\infty} \dfrac{(-1)^n 2^n x^n}{n!\cdot 4\cdot 7\cdots(3n+1)},$

$y_2(x) = \sum_{n=0}^{\infty} \dfrac{(-1)^n 2^n x^n}{n!\cdot 2\cdot 5\cdots(3n-1)}$

21. $y_1(x) = x\left(1 + \sum_{n=1}^{\infty} \dfrac{x^{2n}}{n!\cdot 7\cdot 11\cdots(4n+3)}\right),$

$y_2(x) = x^{-1/2}\left(1 + \sum_{n=1}^{\infty} \dfrac{x^{2n}}{n!\cdot 1\cdot 5\cdots(4n-3)}\right)$

22. $y_1(x) = x^{3/2}\left(1 + \sum_{n=1}^{\infty} \dfrac{(-1)^n x^{2n}}{n!\cdot 9\cdot 13\cdots(4n+5)}\right),$

$y_2(x) = x^{-1}\left(1 + \sum_{n=1}^{\infty} \dfrac{(-1)^{n-1} x^{2n}}{n!\cdot 3\cdot 7\cdots(4n-1)}\right)$

23. $y_1(x) = x^{1/2}\left(1 + \sum_{n=1}^{\infty} \dfrac{x^{2n}}{2^n\cdot n!\cdot 19\cdot 31\cdots(12n+7)}\right),$

$y_2(x) = x^{-2/3}\left(1 + \sum_{n=1}^{\infty} \dfrac{x^{2n}}{2^n\cdot n!\cdot 5\cdot 17\cdots(12n-7)}\right)$

24. $y_1(x) = x^{1/3}\left(1 + \sum_{n=1}^{\infty} \dfrac{(-1)^n x^{2n}}{2^n\cdot n!\cdot 7\cdot 13\cdots(6n+1)}\right),$

$y_2(x) = 1 + \sum_{n=1}^{\infty} \dfrac{(-1)^n x^{2n}}{2^n\cdot n!\cdot 5\cdot 11\cdots(6n-1)}$

25. $y_1(x) = x^{1/2}\sum_{n=0}^{\infty} \dfrac{(-1)^n x^n}{n!\cdot 2^n} = x^{1/2}e^{-x/2},$

$y_2(x) = 1 + \sum_{n=1}^{\infty} \dfrac{(-1)^n x^n}{(2n-1)!!}$

26. $y_1(x) = x^{1/2}\sum_{n=0}^{\infty} \dfrac{x^{2n}}{n!\cdot 2^n} = x^{1/2}\exp\left(\tfrac{1}{2}x^2\right),$

$y_2(x) = 1 + \sum_{n=1}^{\infty} \dfrac{2^n x^{2n}}{3\cdot 7\cdots(4n-1)}$

27. $y_1(x) = \dfrac{1}{x}\cos 3x, \; y_2(x) = \dfrac{1}{x}\sin 3x$

28. $y_1(x) = \dfrac{1}{x}\cosh 2x, \; y_2(x) = \dfrac{1}{x}\sinh 2x$

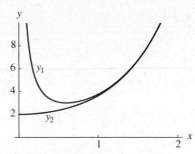

29. $y_1(x) = \dfrac{1}{x}\cos\dfrac{x}{2}, \; y_2(x) = \dfrac{1}{x}\sin\dfrac{x}{2}$

30. $y_1(x) = \cos x^2, \; y_2(x) = \sin x^2$

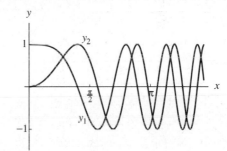

31. $y_1(x) = x^{1/2}\cosh x, \; y_2(x) = x^{1/2}\sinh x$

32. $y_1(x) = x + \dfrac{x^2}{5},$

$y_2(x) = x^{-1/2}\left(1 - \dfrac{5x}{2} - \dfrac{15x^2}{8} - \dfrac{5x^3}{48} + \cdots\right)$

33. $y_1(x) = x^{-1}\left(1 + 10x + 5x^2 + \dfrac{10x^3}{9} + \cdots\right),$

$y_2(x) = x^{1/2}\left(1 + \dfrac{11x}{20} - \dfrac{11x^2}{224} + \dfrac{671x^3}{24192} + \cdots\right)$

34. $y_1(x) = x\left(1 - \dfrac{x^2}{42} + \dfrac{x^4}{1320} + \cdots\right),$

$$y_2(x) = x^{-1/2}\left(1 - \frac{7x^2}{24} + \frac{19x^4}{3200} + \cdots\right)$$

Section 8.4

1. $y_1(x) = \dfrac{1+x}{x^2}$, $y_2(x) = 1 + 2\displaystyle\sum_{n=1}^{\infty}\frac{x^n}{(n+2)!}$

2. $y_1(x) = \dfrac{1}{x^2}\left(1 + x + \dfrac{x^2}{2} + \dfrac{x^3}{6}\right)$, $y_2(x) = 1 + 24\displaystyle\sum_{n=1}^{\infty}\frac{x^n}{(n+4)!}$

3. $y_1(x) = \dfrac{1}{x^4}\left(1 - 3x + \dfrac{9x^2}{2} - \dfrac{9x^3}{2}\right)$,

$y_2(x) = 1 + 24\displaystyle\sum_{n=1}^{\infty}\frac{(-1)^n 3^n x^n}{(n+4)!}$

4. $y_1(x) = \dfrac{1}{x^5}\left(1 - \dfrac{3x}{5} + \dfrac{9x^2}{50} - \dfrac{9x^3}{250} + \dfrac{27x^4}{5000}\right)$,

$y_2(x) = 1 + 120\displaystyle\sum_{n=1}^{\infty}\frac{(-1)^n 3^n x^n}{(n5)!\cdot 5^n}$

5. $y_1(x) = 1 + \dfrac{3x}{4} + \dfrac{x^2}{4} + \dfrac{x^3}{24}$,

$y_2(x) = x^5\left(1 + 120\displaystyle\sum_{n=1}^{\infty}\frac{(n+1)x^n}{(n+5)!}\right)$

6. $y_1(x) = x^4\left(1 + \dfrac{8}{5}\displaystyle\sum_{n=1}^{\infty}\frac{(2n+5)!!x^n}{n!\,(n+4)!2^n}\right)$

7. $y_1(x) = \dfrac{1}{x^2}(2 - 6x + 9x^2)$, $y_2(x) = \displaystyle\sum_{n=1}^{\infty}\frac{(-1)^{n-1}3^n x^n}{(n+2)!}$

8. $y_1(x) = 3 + 2x + x^2$, $y_2(x) = \dfrac{x^4}{(1-x)^2}$

9. $y_1(x) = 1 + \dfrac{x^2}{2^2} + \dfrac{x^4}{2^2\cdot 4^2} + \dfrac{x^6}{2^2\cdot 4^2\cdot 6^2} + \cdots$,

$y_2(x) = y_1(x)\cdot\left(\ln x - \dfrac{x^2}{4} + \dfrac{5x^4}{128} - \dfrac{23x^6}{3456} + \cdots\right)$

10. $y_1(x) = x\left(1 - \dfrac{x^2}{2^2} + \dfrac{x^4}{2^2\cdot 4^2} - \dfrac{x^6}{2^2\cdot 4^2\cdot 6^2} + \cdots\right)$,

$y_2(x) = y_1(x)\cdot\left(\ln x + \dfrac{x^2}{4} + \dfrac{5x^4}{128} + \dfrac{23x^6}{3456} + \cdots\right)$

11. $y_1(x) = x^2\left(1 - 2x + \dfrac{3x^2}{2} - \dfrac{2x^3}{3} + \cdots\right)$,

$y_2(x) = y_1(x)\cdot\left(\ln x + 3x + \dfrac{11x^2}{4} + \dfrac{49x^3}{18} + \cdots\right)$

12. $y_1(x) = x^2\left(1 - \dfrac{x}{2} + \dfrac{3x^2}{20} - \dfrac{x^3}{30} + \dfrac{x^4}{168} - \cdots\right)$,

$y_2(x) = y_1(x)\cdot\left(-\dfrac{1}{3x^3} + \dfrac{1}{20x} + \dfrac{x}{700} + \cdots\right)$

13. $y_1(x) = x^3\left(1 - 2x + 2x^2 - \dfrac{4x^3}{3} + \cdots\right)$,

$y_2(x) = y_1(x)\cdot\left(2\ln x - \dfrac{1}{2x^2} - \dfrac{2}{x} + \dfrac{4x}{3} + \cdots\right)$

14. $y_1(x) = x^2\left(1 - \dfrac{2x}{5} + \dfrac{x^2}{10} - \dfrac{2x^3}{105} + \dfrac{x^4}{336} - \cdots\right)$,

$y_2(x) = y_1(x)\cdot\left(-\dfrac{1}{4x^4} + \dfrac{1}{15x^3} + \dfrac{1}{100x^2} - \dfrac{13}{1750x} + \cdots\right)$;

$y_2(x)$ contains no logarithmic term.

16. $y_1(x) = x^{3/2}\left(1 + \displaystyle\sum_{n=1}^{\infty}\frac{(-1)^n x^{2n}}{2^n\cdot n!\cdot 5\cdot 7\cdots(2n+3)}\right)$,

$y_2(x) = x^{-3/2}\left(1 + \displaystyle\sum_{n=1}^{\infty}\frac{(-1)^n x^{2n}}{2^n\cdot n!\cdot(-1)\cdot 1\cdot 3\cdots(2n-3)}\right)$

Section 8.5

2.

4.

5. $J_4(x) = \dfrac{1}{x^2}(x^2 - 24)J_0(x) + \dfrac{8}{x^3}(6 - x^2)J_1(x)$

12. The following figure corroborates the calculated value $y(0) = 3$.

13. $x^2 J_1(x) + x J_1(x) - \displaystyle\int J_0(x)\,dx + C$

14. $(x^3 - 4x)J_1(x) + 2x^2 J_0(x) + C$

15. $(x^4 - 9x^2)J_1(x) + (3x^3 - 9x)J_0(x) + 9\displaystyle\int J_0(x)\,dx + C$

16. $-xJ_1(x) + \displaystyle\int J_0(x)\,dx + C$

17. $2xJ_1(x) - x^2 J_0(x) + C$

18. $3x^2 J_1(x) + (3x - x^3)J_0(x) - 3\displaystyle\int J_0(x)\,dx + C$

19. $(4x^3 - 16x)J_1(x) + (8x^2 - x^4)J_0(x) + C$

20. $-2J_1(x) + \displaystyle\int J_0(x)\,dx + C$

21. $J_0(x) - \dfrac{4}{x}J_1(x) + C$

26.

Section 8.6

1. $y(x) = x\left[c_1 J_0(x) + C_2 Y_0(x)\right]$

2. $y(x) = \dfrac{1}{x}\left[c_1 J_1(x) + c_2 Y_1(x)\right]$

3. $y(x) = x\left[c_1 J_{1/2}(3x^2) + c_2 J_{-1/2}(3x^2)\right]$

4. $y(x) = x^3\left[c_1 J_2(2x^{1/2}) + c_2 Y_2(2x^{1/2})\right]$

5. $y(x) = x^{-1/3}\left[c_1 J_{1/3}\left(\tfrac{1}{3}x^{3/2}\right) + c_2 J_{-1/3}\left(\tfrac{1}{3}x^{3/2}\right)\right]$

6. $y(x) = x^{-1/4}\left[c_1 J_0(2x^{3/2}) + c_2 Y_0(2x^{3/2})\right]$

7. $y(x) = x^{-1}\left[c_1 J_0(x) + c_2 Y_0(x)\right]$

8. $y(x) = x^2\left[c_1 J_1(4x^{1/2}) + c_2 Y_1(4x^{1/2})\right]$

9. $y(x) = x^{1/2}\left[c_1 J_{1/2}(2x^{3/2}) + c_2 J_{-1/2}(2x^{3/2})\right]$

10. $y(x) = x^{-1/4} \left[c_1 J_{3/2} \left(\frac{2}{5} x^{5/2} \right) + c_2 J_{-3/2} \left(\frac{2}{5} x^{5/2} \right) \right]$

11. $y(x) = x^{1/2} \left[c_1 J_{1/6} \left(\frac{1}{3} x^3 \right) + c_2 J_{-1/6} \left(\frac{1}{3} x^3 \right) \right]$

12. $y(x) = x^{1/2} \left[c_1 J_{1/5} \left(\frac{4}{5} x^{5/2} \right) + c_2 J_{-1/5} \left(\frac{4}{5} x^{5/2} \right) \right]$

Chapter 9

Section 9.1

1. $\frac{2}{3} \pi$

2. 1

3. $\frac{4}{3} \pi$

4. 6

5. π

6. $\frac{1}{2}$

7. Not periodic

8. Not periodic

9. π

10. $\frac{1}{3} \pi$

11. $a_0 = 2;\ a_n = b_n = 0$ for $n \geq 1$

12. $-\dfrac{12}{\pi} \left(\sin t + \dfrac{1}{3} \sin 3t + \dfrac{1}{5} \sin 5t + \cdots \right)$

13. $\frac{1}{2} + \frac{2}{\pi}\left(\sin t + \frac{1}{3}\sin 3t + \frac{1}{5}\sin 5t + \cdots\right)$

14. $\frac{1}{2} - \frac{10}{\pi}\left(\sin t + \frac{1}{3}\sin 3t + \frac{1}{5}\sin 5t + \cdots\right)$

15. $2\left(\sin t - \frac{1}{2}\sin 2t + \frac{1}{3}\sin 3t - \cdots\right)$

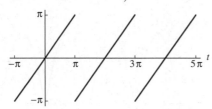

16. $\pi - 2\left(\sin t + \frac{1}{2}\sin 2t + \frac{1}{3}\sin 3t + \cdots\right)$

17. $\frac{\pi}{2} - \frac{4}{\pi}\left(\cos t + \frac{1}{9}\cos 3t + \frac{1}{25}\cos 5t + \cdots\right)$

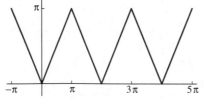

18. $\frac{\pi}{2} + \frac{4}{\pi}\left(\cos t + \frac{1}{9}\cos 3t + \frac{1}{25}\cos 5t + \cdots\right)$

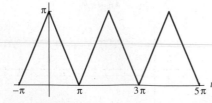

19. $a_0 = \frac{\pi}{4}$; $a_n = \frac{1 - (-1)^n}{n^2\pi}$ for $n \geq 1$; $b_n = -\frac{1}{n}$ for $n \geq 1$.

20. $\frac{1}{2} + \frac{2}{\pi}\left(\cos t - \frac{1}{3}\cos 3t + \frac{1}{5}\cos 5t - \cdots\right)$

21. $\frac{1}{3}\pi^2 - 4\left(\cos t - \frac{1}{4}\cos 2t + \frac{1}{9}\cos 3t - \cdots\right)$

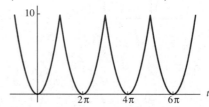

22. $a_0 = \frac{8}{3}\pi^2$; $a_n = \frac{4}{n}$ for $n \geq 1$; $b_n = -\frac{4\pi}{n}$ for $n \geq 1$.

23. $a_0 = \frac{1}{3}\pi^2$; $a_n = \frac{2 \cdot (-1)^n}{n^2}$ for $n \geq 1$;

$b_n = 2 \cdot \frac{(-1)^n - 1}{\pi n^3} - \pi \cdot \frac{(-1)^n}{n}$ for $n \geq 1$

24. $b_n = 0$ for $n \geq 1$; $a_n = 0$ if n is odd;

$a_n = -\frac{4}{\pi(n^2 - 1)}$ for $n = 0, 2, 4, \ldots$.

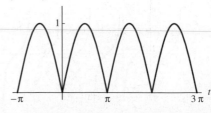

25. $a_0 = 1$, $a_4 = \dfrac{1}{2}$, and all other coefficients are zero.

26. $\dfrac{1}{2} \sin t + \dfrac{1}{\pi} -$
$\dfrac{2}{\pi} \left(\dfrac{1}{1 \cdot 3} \cos 2t + \dfrac{1}{3 \cdot 5} \cos 4t + \dfrac{1}{5 \cdot 7} \cos 6t + \cdots \right)$

Section 9.2

1. $\dfrac{8}{\pi} \left(\sin \dfrac{\pi t}{3} + \dfrac{1}{3} \sin \dfrac{3\pi t}{3} + \dfrac{1}{5} \sin \dfrac{5\pi t}{3} + \cdots \right)$

2. $\dfrac{1}{2} + \dfrac{2}{\pi} \left(\sin \dfrac{\pi t}{5} + \dfrac{1}{3} \sin \dfrac{3\pi t}{5} + \dfrac{1}{5} \sin \dfrac{5\pi t}{5} + \cdots \right)$

3. $\dfrac{1}{2} - \dfrac{6}{\pi} \left(\sin \dfrac{t}{2} + \dfrac{1}{3} \sin \dfrac{3t}{2} + \dfrac{1}{5} \sin \dfrac{5t}{2} + \cdots \right)$

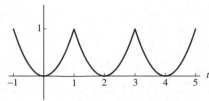

4. $\dfrac{4}{\pi} \left(\sin \dfrac{\pi t}{2} - \dfrac{1}{2} \sin \dfrac{2\pi t}{2} + \dfrac{1}{3} \sin \dfrac{3\pi t}{2} - \cdots \right)$

5. $4 \left(\sin \dfrac{t}{2} - \dfrac{1}{2} \sin \dfrac{2t}{2} + \dfrac{1}{3} \sin \dfrac{3t}{2} - \cdots \right)$

6. $\dfrac{3}{2} - \dfrac{3}{\pi} \left(\sin \dfrac{2\pi t}{3} + \dfrac{1}{2} \sin \dfrac{4\pi t}{3} + \dfrac{1}{3} \sin \dfrac{6\pi t}{3} + \cdots \right)$

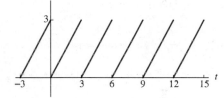

7. $\dfrac{1}{2} - \dfrac{4}{\pi^2} \left(\cos \pi t + \dfrac{1}{9} \cos 3\pi t + \dfrac{1}{25} \cos 5\pi t + \cdots \right)$

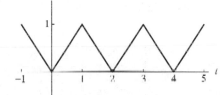

8. $\dfrac{1}{3} - \dfrac{\sqrt{3}}{\pi} \left(\cos \dfrac{2\pi t}{3} - \dfrac{1}{2} \cos \dfrac{4\pi t}{3} \right.$
$\left. + \dfrac{1}{4} \cos \dfrac{8\pi t}{3} - \dfrac{1}{5} \cos \dfrac{10\pi t}{3} + \cdots \right)$

9. $\dfrac{1}{3} - \dfrac{4}{\pi^2} \left(\cos \pi t - \dfrac{1}{4} \cos 2\pi t + \dfrac{1}{9} \cos 3\pi t - \cdots \right)$

10. $\dfrac{2}{3} - \dfrac{8}{\pi^2} \left(\cos \dfrac{\pi t}{2} - \dfrac{1}{4} \cos \dfrac{2\pi t}{2} + \dfrac{1}{9} \cos \dfrac{3\pi t}{2} - \cdots \right) +$

$$\frac{4}{\pi}\left(\sin\frac{\pi t}{2}-\frac{1}{2}\sin\frac{2\pi t}{2}+\frac{1}{3}\sin\frac{3\pi t}{2}-\cdots\right)-$$
$$\frac{16}{\pi^3}\left(\sin\frac{\pi t}{2}+\frac{1}{27}\sin\frac{3\pi t}{2}+\frac{1}{125}\sin\frac{5\pi t}{2}+\cdots\right)$$

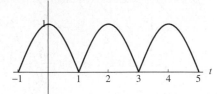

11. $a_0=\dfrac{4}{\pi}$; $a_n=\dfrac{4\cdot(-1)^{n+1}}{\pi(4n^2-1)}$; $b_n=0$ for all $n\geqq1$.

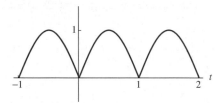

12. $L=\dfrac{1}{2}$; $a_0=\dfrac{4}{\pi}$; $a_n=-\dfrac{4}{\pi(4n^2-1)}$ for $n\geqq1$;
$b_n=0$ for all $n\geqq1$.

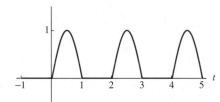

13. $a_0=\dfrac{2}{\pi}$; $a_1=0$; $a_n=-\dfrac{1+(-1)^n}{\pi(n^2-1)}$ for $n\geqq2$;
$b_1=\frac{1}{2}$; $b_n=0$ if $n\geqq2$.

14. $\dfrac{1}{2}\sin t+\dfrac{4}{\pi}\left(\dfrac{1}{3}\cos\dfrac{t}{2}-\dfrac{1}{5}\cos\dfrac{3t}{2}-\dfrac{1}{21}\cos\dfrac{5t}{2}-\cdots\right)$

15.

16.

17.

18.

19.

20.

21.

24.

25.

Section 9.3

1. Cosine series: $f(t) = 1$

Sine series:
$$f(t) = \frac{4}{\pi}\left(\sin t + \frac{1}{3}\sin 3t + \frac{1}{5}\sin 5t + \frac{1}{7}\sin 7t + \cdots\right)$$

2. Cosine series: $f(t) =$
$$\frac{1}{2} + \frac{4}{\pi^2}\left(\cos \pi t + \frac{\cos 3\pi t}{3^2} + \frac{\cos 5\pi t}{5^2} + \frac{\cos 7\pi t}{7^2} + \cdots\right)$$

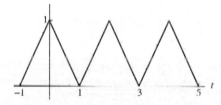

Sine series:
$$f(t) = \frac{2}{\pi}\left(\sin \pi t + \frac{\sin 2\pi t}{2} + \frac{\sin 3\pi t}{3} + \frac{\sin 4\pi t}{4} + \cdots\right)$$

3. Cosine series: $f(t) =$
$$\frac{8}{\pi^2}\left(\cos \frac{\pi t}{2} + \frac{1}{3^2}\cos \frac{3\pi t}{2} + \frac{1}{5^2}\cos \frac{5\pi t}{2} + \frac{1}{7^2}\cos \frac{7\pi t}{2} + \cdots\right)$$

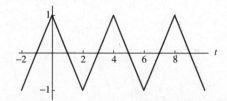

Sine series:
$$f(t) = \frac{4}{\pi}\left(\frac{\sin \pi t}{2} + \frac{\sin 2\pi t}{4} + \frac{\sin 3\pi t}{6} + \frac{\sin 4\pi t}{8} + \cdots\right)$$

4. Cosine series: $f(t) =$
$$1 - \frac{16}{\pi^2}\left(\frac{\cos \pi t}{2^2} + \frac{\cos 3\pi t}{6^2} + \frac{\cos 5\pi t}{10^2} + \frac{\cos 7\pi t}{14^2} + \cdots\right)$$

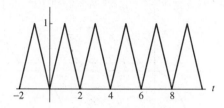

Sine series: $f(t) =$
$$\frac{8}{\pi^2}\left(\sin \frac{\pi t}{2} - \frac{1}{3^2}\sin \frac{3\pi t}{2} + \frac{1}{5^2}\sin \frac{5\pi t}{2} - \frac{1}{7^2}\sin \frac{7\pi t}{2} + \cdots\right)$$

5. Cosine series: $f(t) = \frac{1}{3} - \frac{2\sqrt{3}}{\pi}\left(\frac{1}{2}\cos \frac{2\pi t}{3} - \frac{1}{4}\cos \frac{4\pi t}{3} + \frac{1}{8}\cos \frac{8\pi t}{3} - \frac{1}{10}\cos \frac{10\pi t}{3} + \cdots\right)$

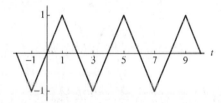

Sine series:
$$f(t) = \frac{2}{\pi}\left(\sin \frac{\pi t}{3} - \frac{2}{3}\sin \frac{3\pi t}{3} + \frac{1}{5}\sin \frac{5\pi t}{3} + \frac{1}{7}\sin \frac{7\pi t}{3} - \frac{2}{9}\sin \frac{9\pi t}{3} + \frac{1}{11}\sin \frac{11\pi t}{3} + \cdots\right)$$

6. Cosine series: $f(t) =$
$$\frac{\pi^2}{3} - 4\left(\cos t - \frac{1}{2^2}\cos 2t + \frac{1}{3^2}\cos 3t + \frac{1}{4^2}\cos 4t + \cdots\right)$$

Sine series:
$$f(t) = 2\pi\left(\sin t - \frac{1}{2}\sin 2t + \frac{1}{3}\sin 3t - \frac{1}{4}\sin 4t + \cdots\right) - \frac{8}{\pi}\left(\sin t + \frac{1}{3^2}\sin 3t + \frac{1}{5^3}\sin 5t + \frac{1}{7^3}\sin 7t + \cdots\right)$$

7. Cosine series:
$$f(t) = \frac{\pi^2}{6} - 4\left(\frac{\cos 2t}{2^2} + \frac{\cos 4t}{4^2} + \frac{\cos 6t}{6^2} + \frac{\cos 8t}{8^2} + \cdots\right)$$

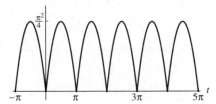

Sine series: $f(t) = \dfrac{8}{\pi}\left(\sin t + \dfrac{\sin 3t}{3^3} + \dfrac{\sin 5t}{5^3} + \dfrac{\sin 7t}{7^3} + \cdots\right)$

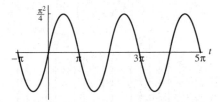

8. Cosine series: $f(t) =$
$$\frac{1}{6} - \frac{4}{\pi^2}\left(\frac{\cos 2\pi t}{2^2} + \frac{\cos 4\pi t}{4^2} + \frac{\cos 6\pi t}{6^2} + \frac{\cos 8\pi t}{8^2} + \cdots\right)$$
Sine series:
$$f(t) = \frac{8}{\pi^3}\left(\sin \pi t + \frac{\sin 3\pi t}{3^3} + \frac{\sin 5\pi t}{5^3} + \frac{\sin 7\pi t}{7^3} + \cdots\right)$$
The graphs of these two series look precisely the same as those in Problem 7, except with a scale of -1 to 5 (rather than $-\pi$ to 5π) on the t-axis, and with arches of height 1/4 (rather than $\pi^2/4$).

9. Cosine series:
$$f(t) = \frac{2}{\pi} - \frac{4}{\pi^2}\left(\frac{\cos 2t}{3} + \frac{\cos 4t}{15} + \frac{\cos 6t}{35} + \frac{\cos 8t}{63} + \cdots\right)$$
Sine series: $f(t) = \sin t$
The graphs of these two series look essentially the same as those in Problem 7, except with sine arches rather than quadratic arches.

10. Cosine series:
$$f(t) = \frac{1}{\pi} - \frac{4}{\pi}\left(-\frac{1}{3}\cos \frac{t}{2} + \frac{1}{5}\cos \frac{3t}{2} + \frac{2}{12}\cos \frac{4t}{2} + \frac{1}{21}\cos \frac{5t}{2} + \frac{1}{45}\cos \frac{7t}{2} + \frac{2}{60}\cos \frac{8t}{2} + \cdots\right)$$

Sine series:
$$f(t) = \frac{1}{2}\sin t - \frac{4}{\pi}\left(\frac{1}{3}\sin \frac{t}{2} + \frac{1}{5}\sin \frac{3t}{2} - \frac{1}{21}\sin \frac{5t}{2} + \frac{1}{45}\sin \frac{7t}{2} - \frac{1}{77}\sin \frac{9t}{2} + \cdots\right)$$

11. $x(t) = \dfrac{4}{\pi}\displaystyle\sum_{n\text{ odd}}\frac{\sin nt}{n(2-n^2)} = \dfrac{4}{\pi}\left(\sin t - \dfrac{\sin 3t}{21} - \dfrac{\sin 5t}{115} - \dfrac{\sin 7t}{329} - \cdots\right)$

12. $x(t) = -\dfrac{4}{\pi}\left[\dfrac{\sin t}{5} + \dfrac{\sin 3t}{39} + \dfrac{\sin 5t}{145} + \dfrac{\sin 7t}{371} + \cdots\right]$

13. $x(t) = \dfrac{2}{\pi}\displaystyle\sum_{n=1}^{\infty}\frac{(-1)^n \sin n\pi t}{n(n^2\pi^2 - 1)}$

14. $x(t) = \dfrac{16}{\pi}\displaystyle\sum_{n=1}^{\infty}\frac{(-1)^n \sin(n\pi t/2)}{n(n^2\pi^2 - 8)}$

15. $x(t) = \dfrac{\pi}{4} + \dfrac{4}{\pi}\displaystyle\sum_{n\text{ odd}}\frac{\cos nt}{n^2(n^2 - 2)} = \dfrac{\pi}{4} + \dfrac{4}{\pi}\left(-\cos t + \dfrac{\cos 3t}{63} + \dfrac{\cos 5t}{575} + \dfrac{\cos 7t}{2303} + \cdots\right)$

17. *Suggestion*: Substitute $u = -t$ in the left-hand integral.

Section 9.4

1. $x_{\text{sp}}(t) = \dfrac{12}{\pi}\displaystyle\sum_{n\text{ odd}}\frac{\sin nt}{n(5-n^2)} = \dfrac{12}{\pi}\left(\dfrac{\sin t}{4} - \dfrac{\sin 3t}{12} - \dfrac{\sin 5t}{100} - \dfrac{\sin 7t}{308} - \cdots\right).$

2. $x_{sp}(t) = \dfrac{48}{\pi} \displaystyle\sum_{n \text{ odd}} \dfrac{(-1)^{(n-1)/2}}{n(40 - \pi^2 n^2)} \cos \dfrac{n\pi t}{2}.$

3. $x_{sp}(t) = 4 \displaystyle\sum_{n=1}^{\infty} \dfrac{(-1)^{n-1} \sin nt}{n(3 - n^2)} =$

$4 \left(\dfrac{\sin t}{2} + \dfrac{\sin 2t}{2} - \dfrac{\sin 3t}{18} + \dfrac{\sin 4t}{52} - \cdots \right).$

4. $x_{sp} = \dfrac{1}{2} - \dfrac{64}{\pi^2} \displaystyle\sum_{n \text{ odd}} \dfrac{\cos n\pi t/2}{\pi^2 n^2 (16 - \pi^2 n^2)}$

5. $x_{sp}(t) = \dfrac{8}{\pi^3} \displaystyle\sum_{n \text{ odd}} \dfrac{\sin n\pi t}{n^3 (10 - n^2 \pi^2)}$

6. $x_{sp} = \dfrac{4}{\pi} - \dfrac{4}{\pi} \displaystyle\sum_{n \text{ even}} \dfrac{\cos nt}{(n^2 - 1)(n^2 - 2)}$

$= \dfrac{4}{\pi} \left(1 - \dfrac{\cos 2t}{6} - \dfrac{\cos 4t}{210} - \dfrac{\cos 6t}{1190} - \cdots \right).$

7. The Fourier series $F(t) = \dfrac{4}{\pi} \displaystyle\sum_{n \text{ odd}} \dfrac{\sin nt}{n}$ contains a $\sin 3t$ term, so resonance does occur.

8. No resonance occurs.

9. The Fourier series $F(t) = \dfrac{4}{\pi} \displaystyle\sum_{n \text{ odd}} \dfrac{\sin nt}{n}$ contains no $\sin 2t$ term, so no resonance occurs.

10. Resonance does occur.

11. The Fourier series $F(t) = \dfrac{\pi}{2} - \dfrac{4}{\pi} \displaystyle\sum_{n \text{ odd}} \dfrac{\cos nt}{n^2}$ contains no $\cos 4t$ term, so no resonance occurs.

12. Resonance does occur.

13. $x_{sp}(t) \approx (1.2725) \sin(t - 0.0333) + (0.2542) \sin(3t - 3.0817) + (0.0364) \sin(5t - 3.1178) + \cdots$

14. $x_{sp}(t) \approx (0.2500) \sin(t - 0.0063) - (0.2000) \sin(2t - 0.0200) + (4.444) \sin(3t - 1.5708) - (0.0714) \sin(4t - 3.1130) + \cdots$

15. $x_{sp}(t) \approx (0.08150) \sin(\pi t - 1.44692) + (0.00004) \sin(3\pi t - 3.10176) + \cdots$

16. $x_{sp}(t) \approx 0.5000 + 1.0577 \cos(\pi t/2 - 0.0103) - 0.0099 \cos(3\pi t/2 - 3.1390) - 0.0011 \cos(5\pi t/2 - 3.1402) \cdots$

17. (a) $x_{sp}(t) \approx (0.5687) \sin(\pi t - 0.0562) + (0.4721) \sin(3\pi t - 0.3891) + (0.01396) \sin(5\pi t - 2.7899) + (0.0318) \sin(7\pi t - 2.9874) + \cdots$

18. $x_{sp}(t) \approx 0.0531 \sin(t - 0.0004) - 0.0088 \sin(3t - 0.0019) + 1.0186 \sin(5t - 1.5708) - 0.0011 \sin(7t - 3.1387) + \cdots$

Section 9.5

1. $u(x, t) = 4e^{-12t} \sin 2x$

2. $u(x, t) \equiv 7$

3. $u(x, y) = 5 \exp(-2\pi^2 t) \sin \pi x - \frac{1}{5} \exp(-18\pi^2 t) \sin 3\pi x$

4. $u(x, t) = 2e^{-4t} \sin 2x + 2e^{-36t} \sin 6x$

5. $u(x, t) = 4 \exp\left(-\dfrac{8\pi^2 t}{9}\right) \cos\left(\dfrac{2\pi x}{3}\right) - 2 \exp\left(-\dfrac{32\pi^2 t}{9}\right) \cos\left(\dfrac{4\pi x}{3}\right)$

6. $u(x, t) = \exp(-2\pi^2 t) \sin 2\pi x + \frac{1}{2} \exp(-8\pi^2 t) \sin 4\pi x$

7. $u(x, t) = \dfrac{1}{2} + \dfrac{1}{2} \exp\left(-\dfrac{16\pi^2 t}{3}\right) \cos 4\pi x$

8. $u(x, t) = 5 \exp(-4\pi^2 t) \cos 2\pi x + 5 \exp(-16\pi^2 t) \cos 4\pi x$

9. $u(x, t) = \dfrac{100}{\pi} \displaystyle\sum_{n \text{ odd}} \dfrac{1}{n} \exp\left(-\dfrac{n^2 \pi^2 t}{250}\right) \sin \dfrac{n\pi x}{5}$

10. $u(x, t) = \dfrac{80}{\pi} \displaystyle\sum_{n=1}^{\infty} \dfrac{(-1)^{n+1}}{n} \exp\left(-\dfrac{n^2 \pi^2 t}{500}\right) \sin \dfrac{n\pi x}{10}$

11. $u(x, t) = 20 - \dfrac{160}{\pi^2} \displaystyle\sum_{n \text{ odd}} \dfrac{1}{n^2} \exp\left(-\dfrac{n^2 \pi^2 t}{500}\right) \cos \dfrac{n\pi x}{10}$

12. $u(x, t) = \dfrac{80000}{\pi^3} \displaystyle\sum_{n \text{ odd}} \dfrac{1}{n^3} \exp\left(-\dfrac{n^2 \pi^2 t}{10000}\right) \sin \dfrac{n\pi x}{100}$

13. (a) $u(x, t) = \dfrac{400}{\pi} \displaystyle\sum_{n \text{ odd}} \dfrac{1}{n} \exp\left(-\dfrac{n^2 \pi^2 kt}{1600}\right) \sin \dfrac{n\pi x}{40}$,

(b) with $k = 1.15$, $u(20, 300) \approx 15.16°C$;

(c) about 19 h 16 min

14. (b) $u(10, 60) \approx 25.15°C$; (c) 6 min 54.3 s

16. (b) 15 h

Section 9.6

1. $y(x, t) = \frac{1}{10} \cos 4t \sin 2x$

2. $y(x, t) = \frac{1}{10} \cos \pi t \sin \pi x - \frac{1}{20} \cos 3\pi t \sin 3\pi x$

3. $y(x, t) = \dfrac{1}{10} \left(\cos \dfrac{t}{2} + 2 \sin \dfrac{t}{2} \right) \sin x$

4. $y(x, y) = \frac{1}{10} \cos \pi t \sin 2\pi x$

5. $y(x, t) = \dfrac{1}{4} \cos 5\pi t \sin \pi x + \dfrac{1}{\pi} \sin 10\pi t \sin 2\pi x$

6. $y(x, t) = \displaystyle\sum_{n \text{ odd}} \dfrac{8}{\pi n^3} \cos 10nt \sin nx$

7. $y(x, t) = \displaystyle\sum_{n=1}^{\infty} \dfrac{(-1)^{n+1}}{5\pi^2 n^2} \sin 10n\pi t \sin n\pi x$

8. $y(x, t) = \cos 2t \sin x + \dfrac{4}{\pi} \displaystyle\sum_{n \text{ odd}} \dfrac{\sin 2nt \sin nx}{n}$.

9. $y(x, t) = \dfrac{4}{\pi^4} \displaystyle\sum_{n \text{ odd}} \dfrac{\sin 2n\pi t \sin n\pi x}{n^4}$.

10. $y(x, t) = \dfrac{8}{5\pi} \displaystyle\sum_{n \text{ odd}} \dfrac{(5n \cos 5nt + \sin 5nt)}{n^2(4 - n^2)}$.

11. Fundamental frequency: 256 Hz; velocity: 1024 ft/s

Section 9.7

1. $u(x, y) = \displaystyle\sum_{n=1}^{\infty} c_n \sinh \dfrac{n\pi x}{b} \sin \dfrac{n\pi y}{b}$ where

$$c_n = \dfrac{2}{b \sinh(n\pi a/b)} \int_0^b g(y) \sin \dfrac{n\pi y}{b} \, dy$$

2. $u(x, y) = \displaystyle\sum_{n=1}^{\infty} c_n \sinh \dfrac{n\pi(a - x)}{b} \sin \dfrac{n\pi y}{b}$ where

$$c_n = \dfrac{2}{b \sinh(n\pi a/b)} \int_0^b g(y) \sin \dfrac{n\pi y}{b} \, dy$$

3. $u(x, y) = \displaystyle\sum_{n=1}^{\infty} c_n \sin \dfrac{n\pi x}{a} \sinh \dfrac{n\pi y}{a}$ where

$$c_n = \dfrac{2}{a \sinh(n\pi b/a)} \int_0^a f(x) \sin \dfrac{n\pi x}{a} \, dx$$

4. $u(x, y) = b_0 y + \displaystyle\sum_{n=1}^{\infty} b_n \cos \dfrac{n\pi x}{a} \sinh \dfrac{n\pi y}{a}$ where $b_0 = a_0/2b$, $b_n = a_n/(\sinh n\pi b/a)$ and $\{a_n\}_0^\infty$ are the Fourier cosine coefficients of $f(x)$ on the interval $0 \leq x \leq a$.

5. $u(x, y) = \dfrac{a_0}{2a}(a - x) + \displaystyle\sum_{n=1}^{\infty} a_n \dfrac{\sinh(n\pi[a - x]/b)}{\sinh(n\pi a/b)} \cos \dfrac{n\pi y}{b}$,

where $a_n = \dfrac{2}{b} \displaystyle\int_0^b g(y) \cos \dfrac{n\pi y}{b} \, dy$

6. $u(x, y) = A_0 + \displaystyle\sum_{n=1}^{\infty} A_n \cos \dfrac{n\pi x}{a} \cosh \dfrac{n\pi y}{a}$ where $A_0 = a_0/2$, $A_n = a_n/(\cosh n\pi b/a)$ and $\{a_n\}_0^\infty$ are the Fourier cosine coefficients of $f(x)$ on the interval $0 \leq x \leq a$.

7. $u(x, y) = \displaystyle\sum_{n=1}^{\infty} b_n \exp\left(-\dfrac{n\pi y}{a}\right) \sin \dfrac{n\pi x}{a}$

where $b_n = \dfrac{2}{a} \displaystyle\int_0^a f(x) \sin \dfrac{n\pi x}{a} \, dx$

8. $u(x, y) = \dfrac{a_0}{2} + \displaystyle\sum_{n=1}^{\infty} a_n e^{-n\pi y/a} \cos \dfrac{n\pi x}{a}$ where $\{a_n\}_0^\infty$ are the Fourier cosine coefficients of $f(x)$ on the interval $0 \leq x \leq a$.

9. $u(0, 5) \approx 41.53$; $u(5, 5) = 50$; $u(10, 5) \approx 58.47$

11. $u(x, y) = \displaystyle\sum_{n \text{ odd}} c_n \sinh \dfrac{n\pi(a - x)}{2b} \cos \dfrac{n\pi y}{2b}$ where

$$c_n = \dfrac{2}{b \sinh(n\pi a/(2b))} \int_0^b g(y) \cos \dfrac{n\pi y}{2b} \, dy$$

13. $c_n = \dfrac{2}{\pi a^n} \displaystyle\int_0^\pi f(\theta) \sin n\theta \, d\theta$

14. $c_n = \dfrac{2}{\pi a^n} \displaystyle\int_0^\pi f(\theta) \cos n\theta \, d\theta$

15. $c_n = \dfrac{2}{\pi a^{n/2}} \displaystyle\int_0^\pi f(\theta) \sin \dfrac{n\theta}{2} \, d\theta$

Chapter 10

Section 10.1

7. $1 = 2hL \displaystyle\sum_{n=1}^{\infty} \dfrac{1 - \cos \beta_n}{\beta_n(hL + \cos^2 \beta_n)} \sin \dfrac{\beta_n}{L}$, $0 < x < L$

8. $1 = \displaystyle\sum_{n=1}^{\infty} \dfrac{4 \sin \beta_n}{2\beta_n + \sin 2\beta_n} \cos \dfrac{\beta_n x}{L}$, $0 < x < L$

9. $x = 2h(1 + h) \displaystyle\sum_{n=1}^{\infty} \dfrac{\sin \beta_n \sin \beta_n x}{\beta_n^2(h + \cos^2 \beta_n)}$, $0 < x < 1$

10. $x = \displaystyle\sum_{n=1}^{\infty} \dfrac{4(\beta_n \sin \beta_n + \cos \beta_n - 1)}{\beta_n(2\beta_n + \sin 2\beta_n)} \cos \beta_n x$, $0 < x < 1$

Section 10.2

2. $u(x, y) = \displaystyle\sum_{n=1}^{\infty} c_n \sin \dfrac{\beta_n x}{L} \sinh \dfrac{\beta_n(L - y)}{L}$, where $\{\beta_n\}_1^\infty$ are the positive roots of the equation $\tan x = -x/hL$, and

$$c_n = \dfrac{4\beta_n}{L(\sinh \beta_n)(2\beta_n - \sin 2\beta_n)} \int_0^L f(x) \cos \dfrac{\beta_n x}{L} \, dx.$$

3. $u(x, y) = \displaystyle\sum_{n=1}^{\infty} c_n \sinh \dfrac{\beta_n(L - x)}{L} \cos \dfrac{\beta_n y}{L}$, where $\{\beta_n\}_1^\infty$ are the positive roots of the equation $\tan x = x/hL$, and

$$c_n = \dfrac{2h}{(\sinh \beta_n)(hL + \sin^2 \beta_n)} \int_0^L g(y) \cos \dfrac{\beta_n y}{L} \, dy.$$

4. $u(x, y) = \displaystyle\sum_{n=1}^{\infty} c_n \sin \dfrac{\beta_n x}{L} \exp\left(-\dfrac{\beta_n y}{L}\right)$, where $\{\beta_n\}_1^\infty$ are the positive roots of the equation $\tan x = -x/hL$, and

$$c_n = \dfrac{4\beta_n}{L(2\beta_n - \sin 2\beta_n)} \int_0^L f(x) \cos \dfrac{\beta_n x}{L} \, dx.$$

5. $u(x, t) = \displaystyle\sum_{n=1}^{\infty} c_n \exp\left(-\dfrac{\beta_n^2 kt}{L^2}\right) X_n(x)$, where

$X_n(x) = \beta_n \cos \dfrac{\beta_n x}{L} + hL \sin \dfrac{\beta_n x}{L}$, $\{\beta_n\}_1^\infty$ are the positive roots of the equation $\tan x = -x/hL$, and

$$c_n = \left(\int_0^L f(x) X_n(x) \, dx\right) \Big/ \int_0^L X_n(x)^2 \, dx.$$

6. $u(x, t) = \displaystyle\sum_{n=1}^{\infty} c_n \exp\left(-\dfrac{\beta_n^2 kt}{L^2}\right) X_n(x)$, where

$X_n(x) = \beta_n \cos \dfrac{\beta_n x}{L} + hL \sin \dfrac{\beta_n x}{L}$, $\{\beta_n\}_1^\infty$ are the positive roots of the equation $\tan x = 2hLx/(x^2 - h^2 L^2)$, and

$$c_n = \left(\int_0^L f(x) X_n(x) \, dx\right) \Big/ \int_0^L X_n(x)^2 \, dx.$$

7. $u(1, 1) \approx 30.8755 + 0.4737 + 0.0074 + 0.0002 + 0.0000 + \cdots$
$\approx 31.4°$

10. (a) $a \approx 5.08 \times 10^5$ cm/sec ≈ 11364 mph.
(b) $a \approx 1.50 \times 10^5$ cm/sec ≈ 3355 mph.

15. $\displaystyle\int_0^L \sin \dfrac{\beta_m x}{L} \sin \dfrac{\beta_n x}{L} \, dx = -\dfrac{LM \cos \beta_m \cos \beta_n}{m\beta_m \beta_n} \neq 0$

20. (b) $\omega_1 \approx 1578$ rad/sec ≈ 251 cycles/sec. Thus we hear middle C (approximately).

Section 10.3

In Problems 1–6, ω_n denotes the nth natural frequency of vibration.

1. $\omega_n = \dfrac{n\pi}{L}\sqrt{\dfrac{E}{\delta}}$

2. $\omega_n = \dfrac{n\pi}{L}\sqrt{\dfrac{E}{\delta}}$

3. $\omega_n = \dfrac{(2n-1)\pi}{2L}\sqrt{\dfrac{E}{\delta}}$

4. $\omega_n = \dfrac{\beta_n}{L}\sqrt{\dfrac{E}{\delta}}$, where β_n is the nth positive root of the equation
$\tan x = \dfrac{M/m}{x}$.

5. $\omega_n = \dfrac{\beta_n}{L}\sqrt{\dfrac{E}{\delta}}$, where β_n is the nth positive root of the equation
$AE\beta \tan \beta = kL$.

6. $\omega_n = \dfrac{\beta_n}{L}\sqrt{\dfrac{E}{\delta}}$, where β_n is the nth positive root of the equation
$(m_0 m_1 z^2 - M^2)\sin z = M(m_0 + m_1)z\cos z$.

15. 1.03 Hz

Section 10.4

3. (a) $u(r,t) = \displaystyle\sum_{n=1}^{\infty} c_n J_0\left(\dfrac{\gamma_n r}{c}\right)\sin\dfrac{\gamma_n at}{c}$ where
$c_n = \dfrac{2aP_0}{\pi c\rho a^2 \gamma_n J_1(\gamma_n)^2}\cdot\dfrac{J_1(\gamma_n\epsilon/c)}{\gamma_n\epsilon/c}$.
(b) How do you know that $J_1(x)/x \to 1/2$ as $x \to 0$?

4. (a) $u(r,t) = \displaystyle\sum_{n=1}^{\infty} c_n \exp\left(\dfrac{\gamma_n^2 kt}{c^2}\right)J_0\left(\dfrac{\gamma_n r}{c}\right)$ where
$c_n = \dfrac{2q_0}{s\pi c^2 J_1(\gamma_n)^2}\cdot\dfrac{J_1(\gamma_n\epsilon/c)}{\gamma_n\epsilon/c}$.

11. $u(r,t) = \dfrac{F_0}{\omega^2 J_0(\omega b/a)}\left[J_0\left(\dfrac{\omega r}{a}\right) - J_0\left(\dfrac{\omega b}{a}\right)\right]\sin\omega t$, where b is the radius of the circular membrane.

Appendix

1. $y_0 = 3$, $y_1 = 3 + 3x$, $y_2 = 3 + 3x + \frac{3}{2}x^2$,
$y_3 = 3 + 3x + \frac{3}{2}x^2 + \frac{1}{2}x^3$,
$y_4 = 3 + 3x + \frac{3}{2}x^2 + \frac{1}{2}x^3 + \frac{1}{8}x^4$; $y(x) = 3e^x$

3. $y_0 = 1$, $y_1 = 1 - x^2$, $y_2 = 1 - x^2 + \frac{1}{2}x^4$,
$y_3 = 1 - x^2 + \frac{1}{2}x^4 - \frac{1}{6}x^6$,
$y_4 = 1 - x^2 + \frac{1}{2}x^4 - \frac{1}{6}x^6 + \frac{1}{24}x^8$; $y(x) = \exp\left(-x^2\right)$

5. $y_0 = 0$, $y_1 = 2x$, $y_2 = 2x + 2x^2$,
$y_3 = 2x + 2x^2 + \frac{4}{3}x^3$,
$y_4 = 2x + 2x^2 + \frac{4}{3}x^3 + \frac{2}{3}x^4$; $y(x) = e^{2x} - 1$

7. $y_0 = 0$, $y_1 = x^2$, $y_2 = x^2 + \frac{1}{2}x^4$, $y_3 = x^2 + \frac{1}{2}x^4 + \frac{1}{6}x^6$,
$y_4 = x^2 + \frac{1}{2}x^4 + \frac{1}{6}x^6 + \frac{1}{24}x^8$; $y(x) = \exp(x^2) - 1$

9. $y_0 = 1$, $y_1 = (1+x) + \frac{1}{2}x^2$, $y_2 = (1+x+x^2) + \frac{1}{6}x^3$,
$y_3 = \left(1 + x + x^2 + \frac{1}{3}x^3\right) + \frac{1}{24}x^4$;
$y(x) = 2e^x - 1 - x = 1 + x + x^2 + \frac{1}{3}x^3 + \cdots$

11. $y_0 = 1$, $y_1 = 1 + x$, $y_2 = (1 + x + x^2) + \frac{1}{3}x^3$,
$y_3 = (1 + x + x^2 + x^3) + \frac{2}{3}x^4 + \frac{1}{3}x^5 + \frac{1}{9}x^6 + \frac{1}{63}x^7$;
$y(x) = \dfrac{1}{1-x} = 1 + x + x^2 + x^3 + x^4 + x^5 + \cdots$

12. $y_0 = 1$, $y_1 = 1 + \frac{1}{2}x$, $y_2 = 1 + \frac{1}{2}x + \frac{3}{8}x^3 + \frac{1}{8}x^3 + \frac{1}{64}x^4$,
$y_3 = 1 + \frac{1}{2}x + \frac{3}{8}x^2 + \frac{5}{16}x^3 + \frac{13}{64}x^4 + \cdots$; $y(x) = (1-x)^{-1/2}$

13. $\begin{bmatrix} x_0 \\ y_0 \end{bmatrix} = \begin{bmatrix} 1 \\ -1 \end{bmatrix}$, $\begin{bmatrix} x_1 \\ y_1 \end{bmatrix} = \begin{bmatrix} 1 + 3t \\ -1 + 3t \end{bmatrix}$,
$\begin{bmatrix} x_2 \\ y_2 \end{bmatrix} = \begin{bmatrix} 1 + 3t + \frac{3}{2}t^2 \\ -1 + 5t - \frac{1}{2}t^2 \end{bmatrix}$,
$\begin{bmatrix} x_3 \\ y_3 \end{bmatrix} = \begin{bmatrix} 1 + 3t + \frac{1}{2}t^2 + \frac{1}{3}t^3 \\ -1 + 5t + \frac{1}{2}t^2 + \frac{5}{6}t^3 \end{bmatrix}$

14. $\mathbf{x}(t) = \begin{bmatrix} e^t + te^t \\ e^t \end{bmatrix}$

16. $y_3(1) \approx 0.350185$

INDEX

Boldface page numbers indicate where terms are defined.